GEOTECHNICAL SPE(

GEOSUPPORT 2004

DRILLED SHAFTS, MICROPILING, DEEP MIXING, REMEDIAL METHODS, AND SPECIALTY FOUNDATION SYSTEMS

PROCEEDINGS OF SESSIONS OF THE GEOSUPPORT CONFERENCE:
INNOVATION AND COOPERATION IN THE GEO-INDUSTRY

January 29-31, 2004
Orlando, Florida

SPONSORED BY
International Association of Foundation Drilling (ADSC)
The Geo-Institute of the American Society of Civil Engineers

EDITED BY
John P. Turner
Paul W. Mayne

Published by the American Society of Civil Engineers

Cataloging-in-Publication Data on file with the Library of Congress.

American Society of Civil Engineers
1801 Alexander Bell Drive
Reston, Virginia, 20191-4400
www.pubs.asce.org

ISBN 0-7844-0713-4
Manufactured in the United States of America.

Geotechnical Special Publications

1 *Terzaghi Lectures*
2 *Geotechnical Aspects of Stiff and Hard Clays*
3 *Landslide Dams: Processes, Risk, and Mitigation*
6 *Use of In Situ Tests in Geotechnical Engineering*
7 *Timber Bulkheads*
8 *Foundations for Transmission Line Towers*
9 *Foundations & Excavations in Decomposed Rock of the Piedmont Province*
11 *Dynamic Response of Pile Foundations–Experiment, Analysis and Observation*
14 *Geotechnical Aspects of Karst Terrains*
15 *Measured Performance Shallow Foundations*
16 *Special Topics in Foundations*
17 *Soil Properties Evaluation from Centrifugal Models*
18 *Geosynthetics for Soil Improvement*
19 *Mine Induced Subsidence: Effects on Engineered Structures*
21 *Hydraulic Fill Structures*
22 *Foundation Engineering*
23 *Predicted and Observed Axial Behavior of Piles*
24 *Resilient Moduli of Soils: Laboratory Conditions*
25 *Design and Performance of Earth Retaining Structures*
26 *Waste Containment Systems: Construction, Regulation, and Performance*
27 *Geotechnical Engineering Congress*
28 *Detection of and Construction at the Soil/Rock Interface*
29 *Recent Advances in Instrumentation, Data Acquisition and Testing in Soil Dynamics*
32 *Embankment of Dams–James L. Sherard Contributions*
33 *Excavation and Support for the Urban Infrastructure*
34 *Piles Under Dynamic Loads*
35 *Geotechnical Practice in Dam Rehabilitation*
37 *Advances in Site Characterization: Data Acquisition, Data Management and Data Interpretation*
39 *Unsaturated Soils*
40 *Vertical and Horizontal Deformations of Foundations and Embankments*
41 *Predicted and Measured Behavior of Five Spread Footings on Sand*
42 *Serviceability of Earth Retaining Structures*
43 *Fracture Mechanics Applied to Geotechnical Engineering*
44 *Ground Failures Under Seismic Conditions*
45 *In Situ Deep Soil Improvement*
46 *Geoenvironment 2000*
47 *Geo-Environmental Issues Facing the Americas*
48 *Soil Suction Applications in Geotechnical Engineering*
49 *Soil Improvement for Earthquake Hazard Mitigation*
50 *Foundation Upgrading and Repair for Infrastructure Improvement*
51 *Performance of Deep Foundations Under Seismic Loading*
52 *Landslides Under Static and Dynamic Conditions–Analysis, Monitoring, and Mitigation*
53 *Landfill Closures–Environmental Protection and Land Recovery*
54 *Earthquake Design and Performance of Solid Waste Landfills*
55 *Earthquake-Induced Movements and Seismic Remediation of Existing Foundations and Abutments*
56 *Static and Dynamic Properties of Gravelly Soils*
57 *Verification of Geotechnical Grouting*
58 *Uncertainty in the Geologic Environment*
59 *Engineered Contaminated Soils and Interaction of Soil Geomembranes*
60 *Analysis and Design of Retaining Structures Against Earthquakes*
61 *Measuring and Modeling Time Dependent Soil Behavior*
62 *Case Histories of Geophysics Applied to Civil Engineering and Public Policy*
63 *Design with Residual Materials: Geotechnical and Construction Considerations*
64 *Observation and Modeling in Numerical Analysis and Model Tests in Dynamic Soil-Structure Interaction Problems*
65 *Dredging and Management of Dredged Material*
66 *Grouting: Compaction, Remediation and Testing*
67 *Spatial Analysis in Soil Dynamics and Earthquake Engineering*
68 *Unsaturated Soil Engineering Practice*
69 *Ground Improvement, Ground Reinforcement, Ground Treatment: Developments 1987-1997*

70 *Seismic Analysis and Design for Soil-Pile-Structure Interactions*
71 *In Situ Remediation of the Geoenvironment*
72 *Degradation of Natural Building Stone*
73 *Innovative Design and Construction for Foundations and Substructures Subject to Freezing and Frost*
74 *Guidelines of Engineering Practice for Braced and Tied-Back Excavations*
75 *Geotechnical Earthquake Engineering and Soil Dynamics III*
76 *Geosynthetics in Foundation Reinforcement and Erosion Control Systems*
77 *Stability of Natural Slopes in the Coastal Plain*
78 *Filtration and Drainage in Geotechnical/Geoenvironmental Engineering*
79 *Recycled Materials in Geotechnical Applications*
80 *Grouts and Grouting: A Potpourri of Projects*
81 *Soil Improvement for Big Digs*
82 *Risk-Based Corrective Action and Brownfields Restorations*
83 *Design and Construction of Earth Retaining Systems*
84 *Effects of Construction on Structures*
85 *Application of Geotechnical Principles in Pavement Engineering*
86 *Big Digs Around the World*
87 *Jacked Tunnel Design and Construction*
88 *Analysis, Design, Construction, and Testing of Deep Foundations*
89 *Recent Advances in the Characterization of Transportation Geo-Materials*
90 *Geo-Engineering for Underground Facilities*
91 *Special Geotechnical Testing: Central Artery/Tunnel Project in Boston, Massachusetts*
92 *Behavioral Characteristics of Residual Soils*
93 *National Geotechnical Experimentation Sites*
94 *Performance Confirmation of Constructed Geotechnical Facilities*
95 *Soil-Cement and Other Construction Practices in Geotechnical Engineering*
96 *Numerical Methods in Geotechnical Engineering: Recent Developments*
97 *Innovations and Applications in Geotechnical Site Characterization*
98 *Pavement Subgrade, Unbound Materials, and Nondestructive Testing*
99 *Advances in Unsaturated Geotechnics*
100 *New Technological and Design Developments in Deep Foundations*
101 *Slope Stability 2000*
102 *Trends in Rock Mechanics*
103 *Advances in Transportation and Geoenvironmental Systems Using Geosynthetics*
104 *Advances in Grouting and Ground Modification*
105 *Environmental Geotechnics*
106 *Geotechnical Measurements: Lab & Field*
107 *Soil Dynamics and Liquefaction 2000*
108 *Use of Geophysical Methods in Construction*
109 *Educational Issues in Geotechnical Engineering*
110 *Computer Simulation of Earthquake Effects*
111 *Judgment and Innovation: The Heritage and Future of the Geotechnical Engineering Profession*
112 *Soft Ground Technology*
113 *Foundations and Ground Improvement*
114 *Soils Magic*
115 *Expansive Clay Soils and Vegetative Influence on Shallow Foundations*
116 *Deep Foundations 2002: An International Perspective on Theory, Design, Construction, and Performance*
117 *Discrete Element Methods: Numerical Modeling of Discontinua*
118 *A History of Progress: Selected U.S. Papers in Geotechnical Engineering*
119 *Soil Behavior and Soft Ground Construction*
120 *Grouting and Ground Treatment*
121 *Probabilistic Site Characterization at the National Geotechnical Experimentation Sites*
122 *Sinkholes and the Engineering and Environmental Impacts of Karst*
123 *Recent Advances in Materials Characterization and Modeling of Pavement Systems*
124 *Geosupport 2004: Drilled Shafts, Micropiling, Deep Mixing, Remedial Methods, and Specialty Foundation Systems*

Foreword

The design and construction of drilled foundations and ground support systems are important facets of the geo-industry that interrelate geotechnical specialty contractors, engineering consultants, drilling construction firms, and testing agencies. Geo-Support 2004 was organized to bring these groups together in a spirit of cooperation and mutual interest in promoting innovation in design and construction. A special tribute is made to the late Professor Michael W. O'Neill (1940-2003) who was one of the few individuals who truly spanned the bridge between the foundation construction industry and geotechnical engineering. Mike was sought equally for his expertise in the analysis and performance of drilled shafts and piling by both contractors and consultants alike. He represented the best of ADSC and the ASCE Geo-Institute and his work was cited as often in *Foundation Drilling* as in the wide selection of ASCE GSPs and the *Journal of Geotechnical & Geoenvironmental Engineering*.

This collection of 63 technical papers offers a new look at the analysis and performance of foundation technologies, including topics related to drilled shafts, micropiles, underground construction, repair, deep mixing, grouting, soil nailing, and hybrid foundations. A number of papers address interesting case histories involving the construction of new foundation systems or remediation works for replacement of existing infrastructure. New methods of ground modification and innovative techniques for drilling and anchor support are also discussed in the proceedings. Situations include projects in both soil and rock, in varied geologic settings, and under diverse working conditions.

The papers were prepared by individuals from research, practice, and the construction industry, thus offering diversity in their viewpoints and purposes. All of the papers underwent a peer-review process similar to the ASCE Journals. In many cases, edits and/or revisions were required prior to acceptance for publication. In producing this Geotechnical Special Publication, the editors would like to acknowledge the assistance of Donna Dickert at ASCE and Carol Maddox at GT.

The Editors

John P. Turner, PhD, P.E.
Professor
Geotechnical Engineering
Dept. of Civil & Architectural Engineering
University of Wyoming
Box 3295
Laramie, WY 82071-3295
Email: turner@uwyo.edu
Phone : 307.766.4265
Fax: 307.766.2221

Paul W. Mayne, PhD, P.E.
Professor
Geosystems Engineering Program
Civil & Environmental Engineering
Georgia Institute of Technology
790 Atlantic Drive, Room 241
Atlanta, GA 30332-0355
Email: paul.mayne@ce.gatech.edu
Phone: 404-894-6226
Fax: 404-894-2281

Acknowledgments

The editors would like to thank the following individuals who reviewed one or more papers for this geotechnical special publication:

Brian Anderson
Tom Anderson
Tom Armour
Clyde Baker
Timoth Bedenis
Maral Bedian
Jonathan Bennett
Richard Berry
John Bickford
Tom Bird
Dan Brown
Donald Bruce
Mary Ellen Bruce
Paul Bullock
George Burke
Allen Cadden
Dan Cadenhead
Jim Cahill
Raymond Castelli
Billy Camp
K. Ronald Chapman
Marcelo Chuaqui
Jose Clemente
Len Cobb
Hubert Deaton
Eric Drooff
Elizabeth Dwyre
Tuncer Edil
Conrad Felice
Rudolph Frizzi
Luis Garcia
Sarah Gassman
Christian Girsang
Jesus Gomez
Jie Han
Youssef Hashash
Bill Heckman
Mohamad Hussein
Magued Iskander
Zia Islam
Kerop Janoyan
Lawrence Johnsen
Byung-chul Kim
Linggang Kong
Dabra Laefer
Guoming Lin
San-Shyan Lin
Ulf Lindblom
Jim Long
Phillip Lowery
Alan Lutenegger
Alan Macnab
Bill Maher
Raymond Mankbadi
Antonio Marinucci
Justice Maswoswe
Paul Mayne
Oscar McConnell
Ross McGillivray
Jim Melcher
Matthew Meyer
Gerald Miller
Ally Mohammad
Robert Mokwa
Charles Ng
Philip Ooi
Jorj Osterberg
Ali Porbaha
Annand Puppala
Tong Qiu
Eric Reuther
Frederick Rhyner
Thomas Richards
Kwang Ro
Tye Savage
Vern Schaefer
Michael Sharp
Dawn Shuttle
Richard Sisson
J. Andrew Steele
Bryan Sweeney
Todd Swoboda
Frank Townsend
John Turner
Ed Ulrich
Mike Waldren
Joe Waxse
Gary Weinstein
David White
Ragui Wilson-Fahmy
Kord Wissmann
Tom Witherspoon
Limin Zhang
Gang Zuo

In Tribute to

Michael W. O'Neill

(1940-2003)

Michael W. O'Neill was a world class contributor to the technology of deep foundations and was at the height of his creative abilities at his untimely death. Expressions honoring him have come from around the world and the geotechnical community will miss his presence now and into the future.

It was my great good fortune to be his mentor during his graduate years when he set forth on his many years of productive work. During that time he was smart, energetic, pleasant, patient, well-liked, and an inspiration among his fellow students. Mike's capacity for work was unexcelled; his dissertation of 750 pages in two volumes remains a standard of careful and diligent effort.

Mike was a leader during his years on the campus at the University of Houston. He served a term as Chairman of the Department of Civil Engineering and was named the Hugh Roy & Lillie Cranz Cullen Distinguished Professor of Civil Engineering. His research, accomplished with graduate students from many corners of the globe, led to numerous awards: M.S. Kapp Award, ASCE, 2003; Distinguished Service Award, Deep Foundations Institute, 2002; The 34th Terzaghi Lecture, ASCE, 1998, entitled "Side Resistance in Piles and Drilled Shafts" (published in the Journal of Geotechnical & Geoenvironmental Engineering, Vol. 127, No. 1, 2001); Professional Service Award, International Association of Foundation Drilling, 1990; W.L. Huber Research Prize, ASCE, 1986; and State-of-the-Art in Civil Engineering Award, ASCE, 1984. He was slated to receive the award of GeoHero from the Geo-Institute of ASCE in 2004, and had been nominated for membership in the National Academy of Engineering.

The proposed citation in his nomination to NAE read: "For contributions to the design of deep foundations through an understanding of the impact of construction procedures on foundation performance." When his sudden death was being discussed with one of his research sponsors, the company president in great distress said: "But who will we go to now for fundamental research on the behavior of drilled shafts under axial load?" Many of us can ask the same question: who will we go to for a pleasant, quiet, thoughtful discussion of vexing problems in foundation engineering? Who, indeed, can replace Mike?

Lymon C. Reese
Emeritus Professor of Civil Engineering

University of Texas-Austin

Contents

Special Invited Papers

Drilled Shafts I: Design and Construction Case Histories 1

Micropiles I: Micropiles in Anchored Geo-Support

Auger-Cast-in-Place and Drilled Displacement Piles

Drilled Shafts II: Design and Construction Case Histories 2

Drilled Shafts III: Earth Retention Using Drilled Shafts

Deep Underground Construction

Drilled Shafts V: Field Performance

Micropiles II: Case Histories

Geotechnical Engineering Education

Indexes

DRILLED AND DRIVEN FOUNDATION BEHAVIOR IN A CALCAREOUS CLAY

W. M. Camp, III, P.E., Member, Geo-Institute[1]

ABSTRACT: Most major structures in the Charleston, SC area of the U.S. are supported on deep foundations bearing in the Cooper Marl. The fine grained Cooper Marl, which typically classifies as highly plastic clay, has a calcium carbonate content of 60 to 80 percent. Unlike some other calcareous soils, the Cooper Marl is generally an effective foundation material for both driven and drilled foundations. In general, the performance of driven and drilled foundations is comparable but the marl's cementation does cause some differences. The relatively open structure collapses upon shearing which causes large excess pore pressures. As a result, the capacity of a driven pile is highly time dependent and increases as the excess pore pressures dissipate. Vibratory installation, which generates a large number of shear reversals, is particularly damaging to the soil structure and negatively impacts the performance of foundations installed in this manner. Drilled foundations often use a vibratory installed steel casing as a construction aid and performance data consistently indicate that the cased zone of the shaft within marl, even when the casing was extracted during concrete placement, experiences a reduction in side resistance as compared to an uncased excavation.

INTRODUCTION

Charleston, South Carolina, lies in the Coastal Plain of the Southeastern United States. Locally, the greater Charleston area is appropriately known as the "Low Country". The elevation of the "high-ground" is typically less than 5 m and saltwater marshes, tidal creeks, and rivers are abundant. As to be expected for such a topography, the shallow subsurface conditions (i.e., to depths of 10 to 15 m) frequently consist of very soft clays and loose sands. This is particularly true of much of today's undeveloped land since the best land was developed first. As an additional complication, the region, which experienced a major earthquake in 1886, is considered one of the most seismically active in the eastern U.S.

The population of the Low Country has grown considerably in the last few decades and this growth has produced a corresponding increase in commercial, industrial, and

[1] Senior Engineer, S&ME, Inc., 620 Wando Park Blvd., Mount Pleasant, SC 29464

infrastructure construction projects. As a result of the difficult geotechnical conditions, deep foundations are commonly required in the region and the vast majority of these deep foundations are supported in a common bearing stratum, the Cooper Marl. The site characterization and foundation testing performed in conjunction with the numerous construction projects over the last few decades have produced data that are valuable in understanding the behavior of this very interesting formation. Within this paper data from various projects are combined to better understand deep foundation performance in the area. Specifically, differences in the behavior of drilled foundations as compared to driven foundations are examined.

GEOLOGY

The Coastal Plain in South Carolina consists of a wedge of late Cretaceous and younger sediments that thickens to the southeast, or in a direction normal to the present day shoreline (Heron, 1962). The marine sediments within this wedge primarily consist of non-indurated siliciclastic materials and carbonates with varying quantities of terrigenous matter (Horton and Zullo, 1991). Within the Charleston area, the first formation within this wedge with a thickness, uniformity, and strength capable of providing foundation support is the Cooper Marl. The Cooper Marl nomenclature was created in 1848 (Tuomey, 1848) and later became the Cooper Formation (Ward et al., 1979). In 1984, the deposit was elevated to group rank and it is now formally known as the Cooper Group, which consists of three formations: the Oligocene Ashley Formation, the Late Eocene Parkers Ferry Formation and the Eocene Harleyville Formation (Weems & Lemon, 1984).

In 1848 when Tuomey named the Cooper Marl, the term "marl" was used to describe any deposit containing economically important quantities of "carbonate of lime" (for use as fertilizer) that was soft enough to be excavated with the ordinarily available equipment of the day (Heron, 1962). Today, according to the *Glossary of Geology*, the term marl is used to refer to a material that contains significant quantities of both carbonates and clay. Although meeting the original definition, the Cooper Marl clay content is too small to meet today's definition of marl. Nevertheless, the geotechnical and construction community generally continues to refer to all members of the Cooper Group as the Cooper Marl. Within the depths of interest, the engineering differences between the formations are relatively minor and the use of the older and now more generic term is an acceptable convention.

The geologic literature generally classifies the Cooper Marl as a calcarenite or calcilutite and lists its constituents as the skeletal remains of microscopic sea organisms (alternatively – microfossils or foraminifers), quartz sand, phosphate and clay minerals (Weems and Lemon, 1989; Weems and Lemon, 1996; Ward et al., 1979; Gohn et al., 1977). The abundance of microfossils is readily apparent in the scanning electron micrograph of a Cooper Marl sample shown in Figure 1.

FIG. 1. SEM Photograph of a Cooper Marl Sample

GEOTECHNICAL CHARACTERIZATION

The available Cooper Marl data, with regard to both site characterization and foundation performance, have primarily been generated through consultants working on various industrial and government projects (relevant reports listed in References). Consequently, for any given project, with a few exceptions, the data are relatively limited. Some of the deficiencies can be reduced, however, by combining the data from multiple projects and sites. The creation of such a database must be based on the premise that the properties and characteristics of the Cooper Marl are relatively uniform across the Charleston area. For decades, the local geotechnical state-of-the-practice has been to treat the Cooper Marl as a relatively uniform stratum and experience has generally justified this approach. And in fact, a comparative review of the data from across the area, some of which are presented herein, does support the premise of a reasonable areal uniformity.

For decades, subsurface conditions in the area were explored by standard penetration testing (SPT) with split spoon sampling supplemented with undisturbed samples and laboratory testing. The SPT N-value of the Cooper Marl is usually in the range of 10 to 20 and the samples are visually classified as stiff to very stiff, olive, sandy clay or sandy silt. During the last 10 years, the area has begun to use cone penetration testing (CPT) on a regular basis and a representative sounding is shown in Figure 2.

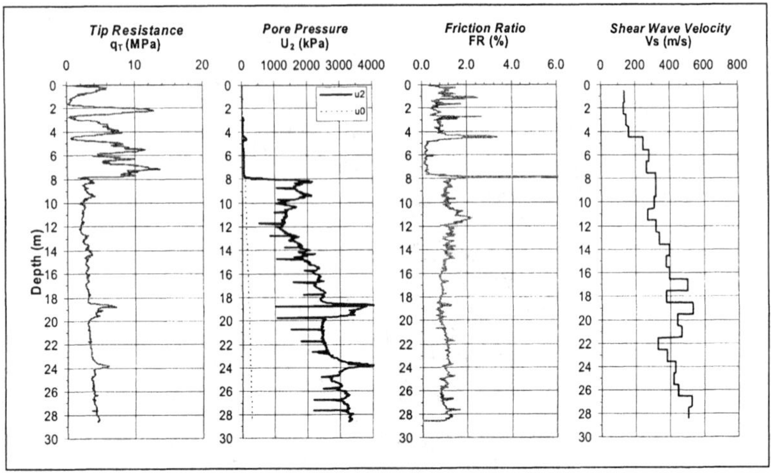

FIG. 2. Representative CPT Sounding with Top of Cooper Marl at 8 m.

The most notable feature of the CPT sounding is the large increase in the u_2 pore pressure, which identifies the top of the Cooper Marl. This is a consistent characteristic of the Cooper Marl and is seen in all cone data from the Cooper Marl. The tip resistance within the marl is relatively uniform, although thin stiffer "lenses" are occasionally encountered as seen at depths of 19 m and 24 m in Figure 2. It should be noted that there is a corresponding pore pressure increase at each of these depths, which is also a typical characteristic of CPT data in the Cooper Marl.

A substantial number of deep (i.e., ± 30 m) CPT soundings are available from projects throughout the Charleston area. The Cooper Marl data (i.e., data from formations above the Cooper Marl are not included) from five of these sites are plotted versus elevation in Figure 3. For sites with multiple soundings, the data were first averaged together to produce a representative sounding for the site. Even though the data were collected across an approximately 400 km^2 area, the variations between the soundings are relatively small, which indicates that the Cooper Marl may be considered a relatively uniform deposit.

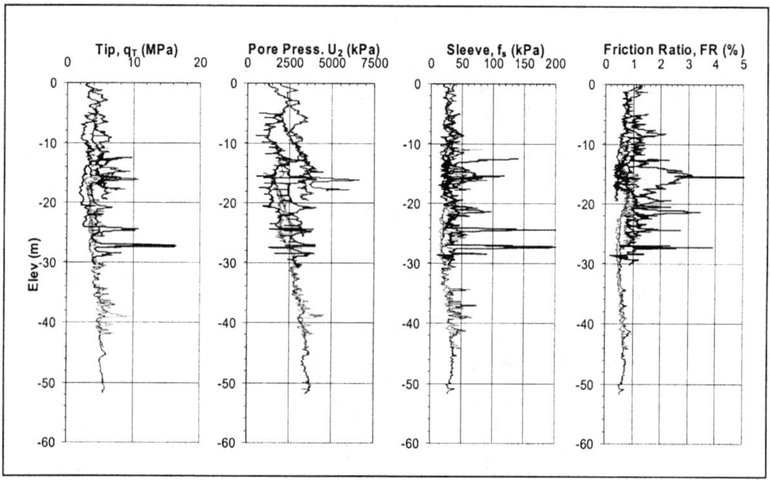

FIG. 3. Cooper Marl CPT Data from 5 Different Sites

The large pore pressure response of the CPT within the Cooper Marl is due to cementation, which also helps explain many other aspects of the Cooper Marl characteristics. The data shown in Figures 4 through 6 support this conclusion. A primary constituent (60 to 80 percent) of the Cooper Marl is calcium carbonate (Fig. 4) which is a common cementing agent. Additionally, the maximum preconsolidation stress (Fig. 5), as measured by oedometer tests, is substantial (resulting in overconsolidation ratios of 3 to 6) and difficult to explain by conventional consolidation. The area has not been subjected to glaciation since the deposition of the Cooper Marl nor is there any evidence of desiccation. And to completely account for the level of preconsolidation, more than 50 m of soil would have had to have been eroded from the Charleston area during the Miocene and/or Pliocene Epochs. The geological evidence does not support such an erosional event(s). Further, the preconsolidation stress, albeit scattered, is essentially constant with depth, which is consistent with an apparent overconsolidation due to cementation. The shear wave velocity (Fig. 6) is also roughly constant with depth, or stress independent, as is typical of cemented soils (Rinaldi & Santamarina, 2003). Finally, the fabric of the Cooper Marl is relatively open (void ratio of 1 to 2, moisture content of 40 to 60 percent) yet its shear wave velocity is in the range of 500 m/s and its undrained shear strength (from triaxial CIUC testing) is typically around 200 kPa (and is also generally constant with depth).

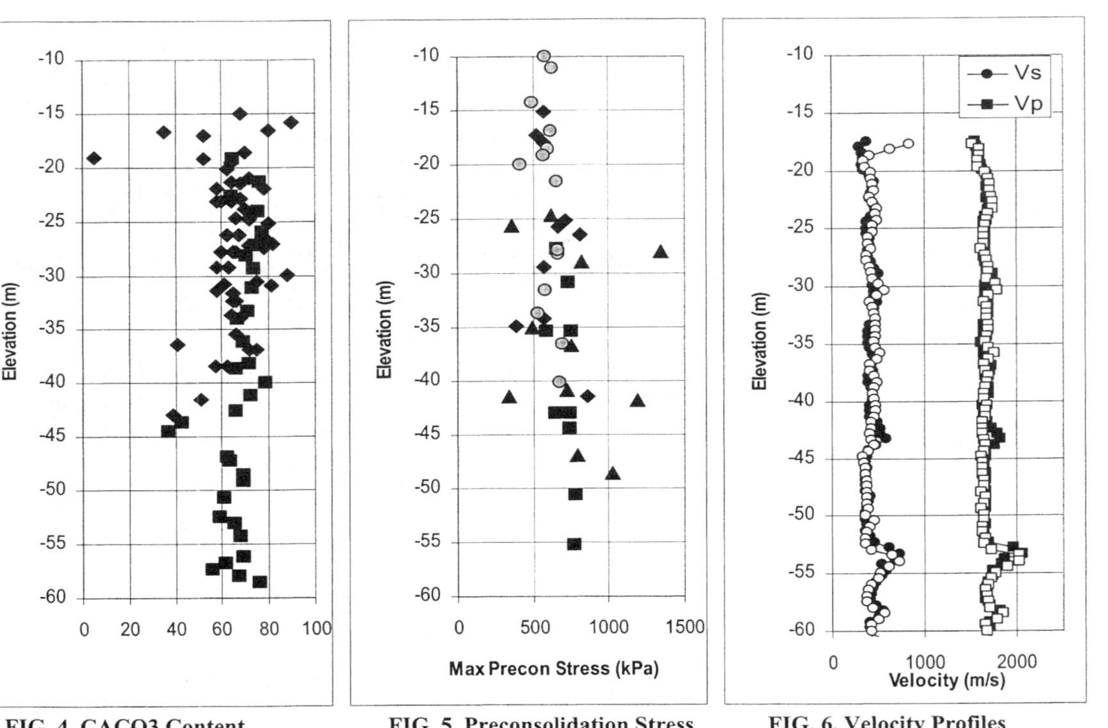

FIG. 4. CACO3 Content

FIG. 5. Preconsolidation Stress

FIG. 6. Velocity Profiles

During cone penetration testing (or pile driving or any other shearing), the soil skeleton is stressed until the cementation bonds are broken. Once broken, the skeleton collapses and forms new particle contacts. However, the sudden collapse of the cementation bonds in the saturated soil results in an increase in pore pressure, hence the large pore pressure measurements during cone penetration testing.

Another notable aspect of the Cooper Marl is its clay-like behavior. Grain size data are summarized in Figure 7. The fines content is usually in the range of 60 to 90 percent and the clay size fraction is generally in the range of 10 to 30 percent. As shown in Figure 4, the calcium carbonate content is generally in the range of 60 to 80 percent and therefore, a significant portion of the fines content and clay size fraction must be carbonate material rather than clay minerals. This is surprising considering the highly plastic nature of the Cooper Marl which is illustrated by the data from multiple projects plotted Figure 8. Although there is scatter within the data, there are no apparent trends with respect to project location. In other words, all projects tend to have a similar amount of scatter. In general, the material is highly plastic with liquid limits often greater than 100 and PI's greater than 60.

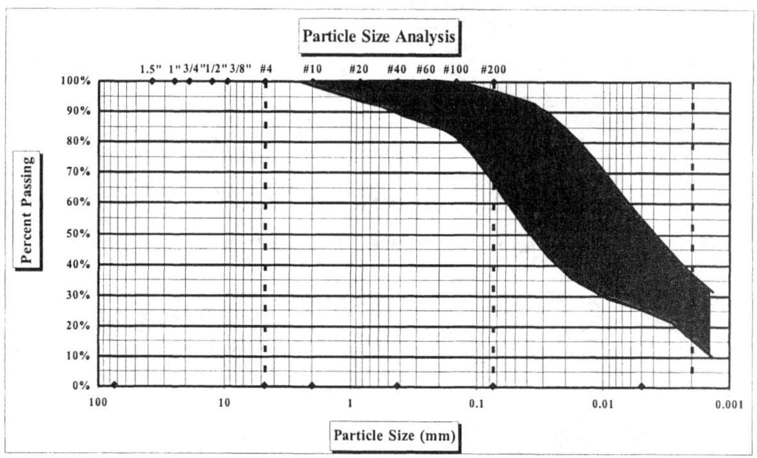

Fig. 7. Summary of Cooper Marl Grain Size Distribution Data

The seemingly contradictory properties (high plasticity and low clay content) can be explained in two ways. First, X-ray diffraction analyses confirm that the Cooper Marl contains a relatively small amount of phyllosilicates; 4 to 9 percent by weight. However, smectite is one of the primary clay minerals found in the Cooper Marl accounting for approximately 20 to 60 percent of the total phyllosilicate content. The highly plastic properties of smectite (synonymous with montmorillonite) are well known. Additionally, in SEM photographs of the Cooper Marl, many of the microfossils appear to be hollow. The interstitial water trapped in these hollow particles will be included in laboratory moisture measurements but it will not necessarily be free to influence the soil's behavior.

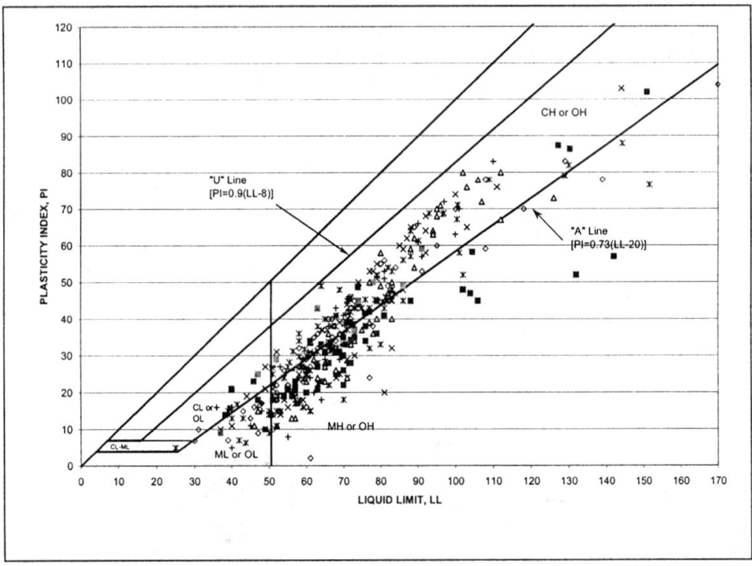

Fig. 8. Cooper Marl Plasticity Data

Finally, the Cooper Marl has a surprisingly high effective stress friction angle, 44 degrees as measured in triaxial CIUC tests. This is a result of the rough, irregular shapes of the microfossils and other particles, as pictured in the SEM micrograph shown in Figure 1.

LOCAL DEEP FOUNDATION PRACTICES AND OBSERVATIONS

An overview of the local deep foundation practice, in terms of both driven piles and drilled shafts, as well as common construction observations is an informative start to a comparison of drilled and driven foundations in the Cooper Marl.

Driven Foundations

The most common driven pile foundation used in the Charleston area is a square prestressed concrete pile. Sizes range from 254 mm to 610 mm. Steel H-sections and open-end pipe piles are also used but less frequently. Pile driving into the Cooper Marl is generally very easy. Relatively small hammers are capable of driving piles to depths that ultimately yield a high capacity even though the final driving resistance may be quite small (often less than 60 blows per meter). As seen during cone penetration testing, excess pore pressures are large at the time of pile installation. Pore pressure measurements made within the Cooper Marl around driven piles have confirmed that pore pressures are suddenly and significantly elevated (Camp et al., 1992) during pile installation. As a result of the reduction in the effective stress, the pile capacity at the time of installation is very small but increases as the pore pressures dissipate. As shown in Figure 9, the capacity gain with time is substantial and the beneficial impact must be considered for an economical foundation design. Conversely, the low capacity available soon after pile installation must be considered when the piles are loaded shortly after installation (e.g., top-down construction, temporary support of construction equipment).

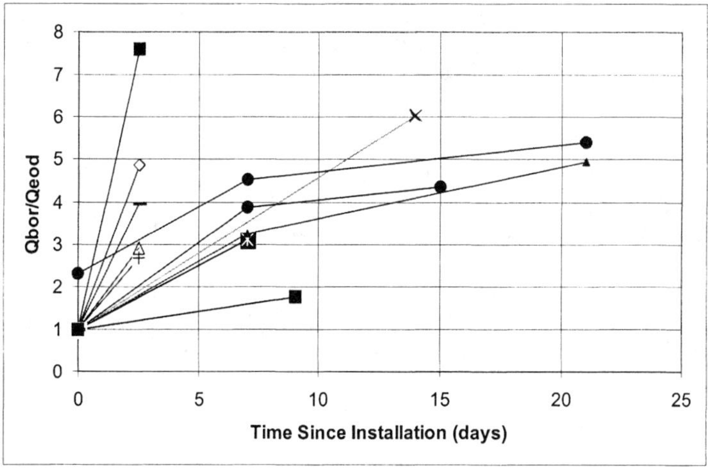

Fig. 9. Ratio of Capacity at the Beginning of Restrike to the Capacity at the End-of-Drive for Piles Driven into the Cooper Marl (Camp & Parmar, 1999)

Other, more detrimental effects of the driving-induced pore pressures are sometimes seen as well. In particular, two "problem" scenarios occasionally develop. During pile installation, pile "bounce" may occur. The term "bounce" in this situation refers to an extremely high quake condition in which the pile moves down under a given hammer blow a significant distance (e.g., ± 30 mm) and then rebounds an equal amount during the hammer upstroke. The pile is essentially acting like a large elastic spring. Although proper hammer selection can help avoid the problem, the typical solution is to move on to the next pile and drive it until it also refuses due to bounce. This process is continued until eventually, the initial pile is driven once again. Usually, after 1+ hours, the pile can be driven some additional depth before bounce will once again occur. This problem is only seen with displacement piles and in particular with the larger displacement piles.

Secondly, with open-end pipe piles and cylinder piles, a "head" of very wet soil or water may develop within the pile. In other words, the level of the soil and/or water within the pipe pile or cylinder pile may exceed the level of the existing ground surface. An extreme example of this phenomenon was observed during the installation of a 914 mm diameter concrete cylinder pile (127 mm wall thickness). The water rose within the interior of the pile to a height of more than 9 m (Soil & Material Engineers, 1984). The water was prevented from going higher by the hammer helmet and the resulting pressure became so high that vent holes had to cut near the pile top to relieve the water pressure.

In the bounce scenario, large pore pressures are generated but cannot dissipate and in the pipe pile case, the large pore pressures are dissipated by flowing up through the interior of the pile.

A representative load-deflection curve for a statically top-loaded driven pile is shown in Figure 10. The shape of the curve is typical for types of driven piles bearing in the Cooper Marl. The pile initially exhibits a relatively stiff response, then softens considerably and eventually plunges.

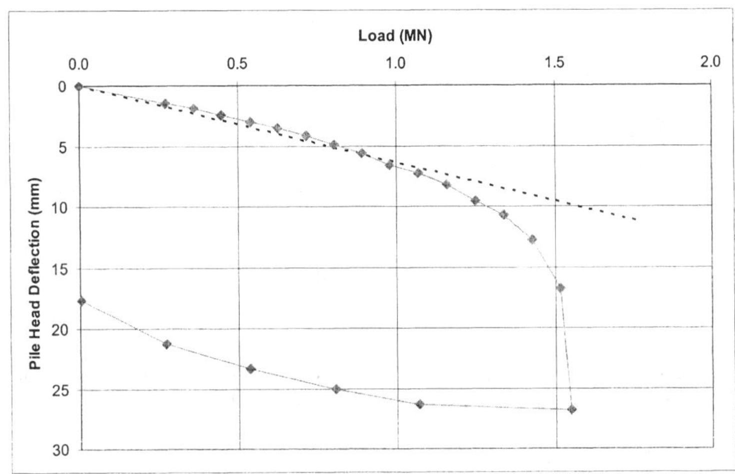

FIG. 10. Representative Load-Deflection Response of an Axially-Loaded, 305 mm square PSC Driven Pile

Drilled Foundations

Drilled shafts are the most commonly used drilled foundation in the Charleston area. Auger cast-in-place (ACIP) piles have occasionally been used on commercial projects but they typically bear within the soils above the Cooper Marl or at very shallow depths within the Cooper Marl and relevant data on ACIP piles are therefore limited. Most drilled shafts in the area are constructed without the use of drilling fluids. Typically, casing is driven into the top 1 to 3 m of the Cooper Marl. This eliminates any potential shaft wall stability problems within the very weak overburden soils and effectively forms a seal within the marl. Over typical foundation depths, the Cooper Marl is easily excavated with conventional augers or digging buckets. The sidewalls remain stable and bells have been successfully constructed. Water, polymer slurries, and mineral slurries have been used as drilling fluids as a precaution for very deep shafts and for commercial projects where the use of casing is undesirable. As presented in Camp et al. (2002), any influence of the type of drilling fluid (or lack thereof) on the shaft capacity is negligible.

The Osterberg-Cell™ load test method has been used on the majority of the tested shafts and many have used two levels of O-Cells. Representative load-deflection curves from a multi-level O-Cell test are shown in Figure 11.

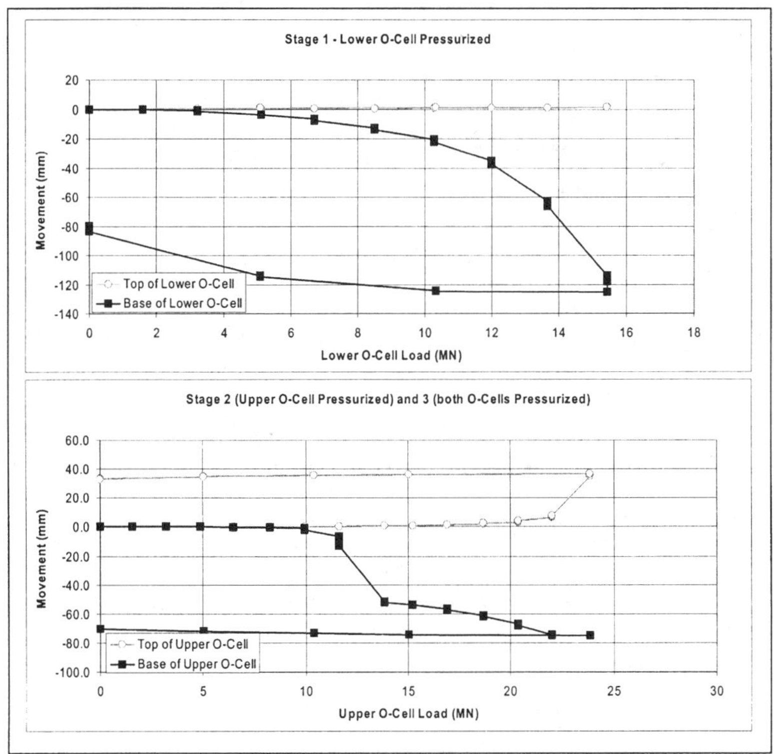

FIG. 11. Representative Load-Deflection Curves from a 2.4 m diam. Drilled Shaft loaded with 2 Levels of Osterberg Cells™.

DEEP FOUNDATION PERFORMANCE

The axial performance of driven and drilled foundations bearing within the Cooper Marl has been evaluated on many projects by load testing instrumented piles. The majority of the available results are from O-Cell tests but the driven pile data and two drilled shaft data sets are from conventional top-down load tests. From these load tests, average unit side shear and unit end values have been back-calculated for various foundation segment lengths. The available data have been reviewed and summarized in Figures 12 and 13 as plots of unit end resistance versus the vertical effective stress at the pile tip elevation and unit side shear versus the vertical effective stress at the elevation of the midpoint of the relevant side shear segment.

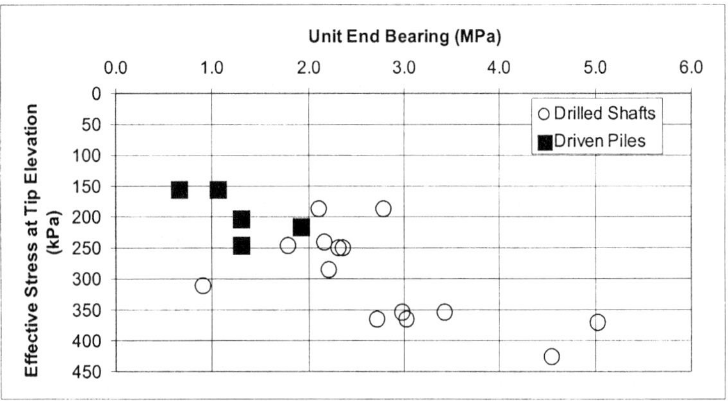

Fig. 12. Unit End Resistance (mobilized at a normalized tip displacement of 2%) vs Effective Stress for Drilled and Driven Piles Bearing within the Cooper Marl.

There is obvious trend between the unit end resistance and the effective stress at the pile tip elevation. This is true for both the driven pile and the drilled shaft data. Since the undrained shear strength is relatively constant with depth, the end bearing results indicate that the unit end resistance can be estimated using an effective stress approach:

$$q_{eb} = N_q \sigma_{vo}' \tag{1}$$

where the bearing capacity factor, N_q, is approximately 10.

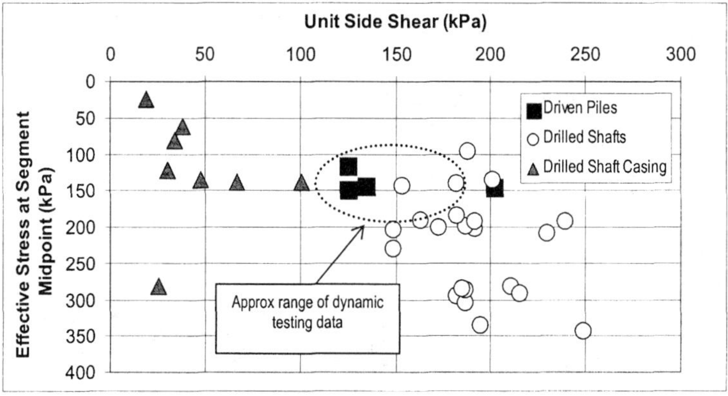

Fig. 13. Unit Side Shear Resistance vs Segment Midpoint Effective Stress for Drilled and Driven Piles Bearing within the Cooper Marl.

As shown in Figure 13, the unit side shear for driven and drilled piles appears to increase only slightly at higher effective stress, although there is large scatter within the data. The driven pile data plotted in Figure 13 are from instrumented static load tests and as a result, are relatively limited. However, unit side shear data have been estimated from the hundreds of dynamic tests performed on driven piles in the Charleston area and the resulting values, whether from CAPWAP analyses or conventional static back-calculation, fall within the range of about 120 to 180 kPa. Therefore, it appears that the driven pile unit side shear values are generally within the range of the lower bound of the drilled shaft unit side shear values.

The third data set shown in Figure 13 is for drilled shaft casing. Of course a drilled shaft casing is simply a driven open-end pipe pile but on some shafts, the casing was removed after concrete placement. The drilled shaft casing data plotted in Figure 13 includes the unit side shear calculated for the segment of the shaft that was below the top of the Cooper Marl but within the zone of the casing, regardless of whether the casing was permanent or temporary. Additionally, unlike the other driven pile data, the drilled shaft casings were always installed with a vibratory hammer, which is also the reason why the unit side shear values within the cased zone are lower.

Canivan and Camp (2002) presented data from two small dynamic pile testing programs specifically designed to evaluate the differences in capacity of open-end pipe piles driven with impact hammers versus identical piles driven with vibratory hammers. Both programs were performed in the Charleston area and the bearing stratum was the Cooper Marl. As shown in Figure 14, there is a marked difference between the capacities of the piles installed with the impact hammer as compared to those installed with a vibratory hammer. At any given time, the piles installed with the impact hammer had capacities that were at least 30% and, in some cases, more than 200% higher than the companion piles installed with a vibratory hammer.

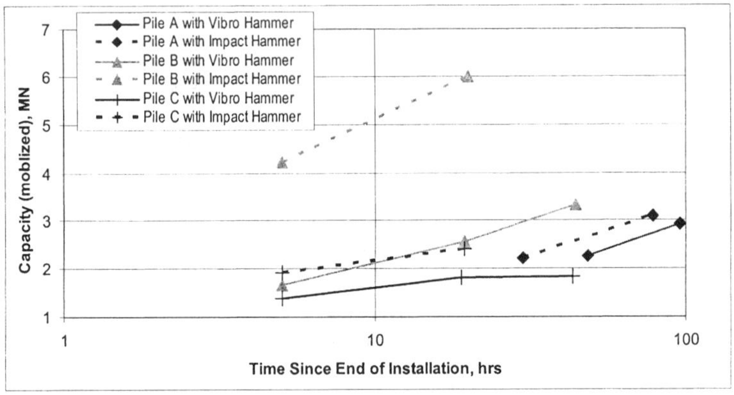

Fig. 14. Comparison of Pile Capacity vs Time for Identical Open-End Pipe Piles (diameters of 610 mm or 762 mm) Driven with Vibratory or Impact Hammers into the Cooper Marl (Canivan and Camp, 2002).

The difference in capacity between the two installation methods was attributed to differences in the level of disturbance or remolding. Specifically, under the vibratory hammer, a pile will experience a much greater number of load-unload cycles and the degree of shear reversal during the unload cycles would be much larger as compared to conditions under an impact hammer loading. For a cemented material like the Cooper Marl, the more severe loading induced by the vibratory hammer would likely be even more detrimental (i.e., larger pore pressures, a greater degree of destructuring).

The multi-level Osterberg-Cell testing affords an opportunity to evaluate the consequences of cyclic loading along the shaft interface and several tests have been performed on shafts bearing within the Cooper Marl to evaluate the effects of cyclic loading with and without shear stress reversal. Specifically, three shafts constructed for the Cooper River Bridge Load-Test program were subjected to supplemental load sequences after the completion of the specified testing. Details of the load test program can be found in Camp, Brown and Mayne, 2002; Camp, Mayne, and Brown, 2002; and Brown and Camp, 2002. The supplemental testing consisted of pushing the "socket" (i.e., the portion of the shaft located between the upper and lower level of O-cells) up and down between the two levels of O-cells.

For one shaft, the load sequence consisted of three load-unload cycles following the completion of the Stage 3 loading (i.e., the final stage in a conventional two-level O-cell test in which both levels are pressurized resulting in an upward failure of the portion of the shaft above the uppermost O-cell). The direction of shear along the socket interface for all three load cycles was therefore downward, which was the same as the direction at the end of the Stage 3 loading. At the completion of the third load-unload cycle, the shaft was loaded from the bottom and the socket was subjected to three more load-unload cycles, but in the opposite direction.

For the other two shafts, at the completion of the Stage 3 loading, the shaft was loaded from the bottom, subjecting the socket to a load in the opposite direction of the Stage 3 load. At the completion of the bottom loading, the shaft was again loaded from the top (i.e., from the upper level of O-cells), thereby subjecting the socket to another shear reversal. This process was repeated two more times.

Additionally, at the start of the construction of the new Cooper River Bridge, the designers (Parsons Brinckerhoff Quade and Douglas) had two more test shafts constructed and loaded. Both shafts were 1.5 m in diameter and were subjected to multi-level O-cell testing (Loadtest 2002a and Loadtest 2002b). Both of these shafts were subjected to a second cycle of loading at each O-cell level. In particular, after mobilizing the ultimate side shear on the segment of the shafts above the upper level of O-cells, the segment was unloaded and loaded again. As with one of the Cooper River Bridge Load-Test program shafts, this resulted in a second load cycle in the same direction as the first load cycle.

The data from these five test shafts are summarized in Figure 15. For each cycle, the load was increased until the shaft continued moving under a constant load or in some cases, a decreasing load. This loading, normalized to the peak loading obtained during the first load cycle, is plotted in Figure 15. The solid lines connect points that were not subjected to shear stress reversals and the dashed lines connect those that were.

The cyclic loading with full shear reversal is obviously much more detrimental to the shaft capacity than cyclic loading without the shear reversal. These data support the vibratory hammer/impact hammer hypothesis and help explain why the unit side shear within the cased zone is so much lower. And the fact that the side shear is still reduced even when the casing is removed indicates that the reduction is not attributable to an interface mechanism.

CONCLUSIONS

The available in situ and laboratory testing data indicate that the Cooper Marl is cemented. As a result of the cementation, the deposit has a relatively open structure, as indicated by a void ratio of 1 to 2, but also has a high small strain stiffness, a relatively large undrained shear strength, and is capable of providing significant support for foundations. Upon shearing, the structure collapses, which produces very large excess pore pressures. The pore pressure generation makes driven pile capacity highly time dependent and sometimes creates installation difficulties. In general, driven and drilled foundation performance is comparable although the unit side shear resistances tend to be slightly higher in the drilled piles. As evidenced by differences in the performance of impact driven and vibratory driven piles, shear stress reversal appears to have a significant detrimental effect on the unit side shear within the Cooper Marl. Cyclic testing of drilled shafts with Osterberg-cells confirms that the unit side shear is greatly reduced under conditions of shear stress reversal. Even without shear stress reversal under working loads, the consideration of the reduction is important since it appears within the cased zones of drilled shafts as a result of the vibratory installation (and sometimes removal) of the steel casing.

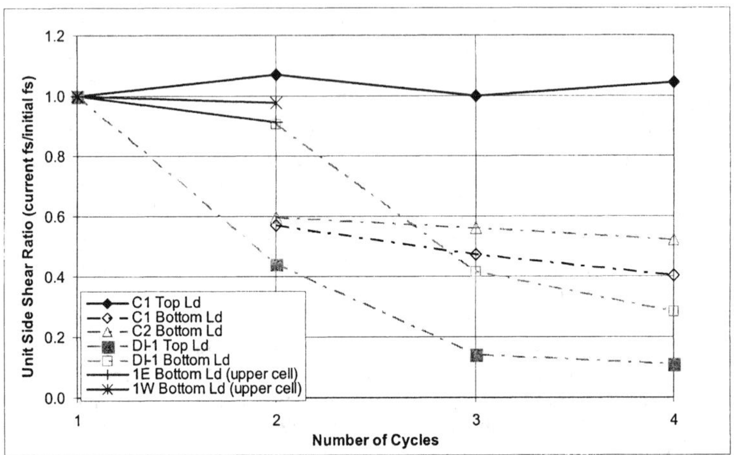

FIG. 15. Summary of Cyclic Loading with (Dashed Lines) and without (Solid Lines) Shear Stress Reversal

ACKNOWLEDGEMENTS

The research presented in this paper has been supported by the ADSC and S&ME, Inc. and their contributions are greatly appreciated. The Case Atlantic Company kindly provided access to project sites and test data. Loadtest, Inc. collected data beyond their contracted scope of work and provided data reduction assistance. The author also wishes to acknowledge the assistance of Dr. Paul Mayne and Dr. Dan Brown.

REFERENCES

Brown, D.A. and Camp, W. M. (2002) "Lateral Load Test Program for the Cooper River Bridge, Charleston, SC", *Proceedings of the International Deep Foundations Congress 2002*, Orlando, FL, ASCE GSP 116, Vol 1, 95-109.

Camp, W. M.; Brown, D.A and Mayne, P. W. (2002) "Construction Method Effects on Axial Drilled Shaft Performance", *Proceedings of the International Deep Foundations Congress 2002*, Orlando, FL, ASCE GSP 116, Vol 1, 195-208.

Camp, W. M.; Mayne, P. W. and Brown, D.A. (2002) "Drilled shaft axial design values: predicted versus measured response in a calcareous clay", *Proceedings of the International Deep Foundations Congress 2002*, Orlando, FL, ASCE GSP 116, Vol 2, 1518-1532.

Camp, William and Parmar, Harpal (1999). "Characterization of Pile Capacity with Time in the Cooper Marl: A Study of the Applicability of a Past Approach to Predict Long-Term Pile Capacity" *Transportation Research Record,* Journal of the Transportation Research Board, No. 1663, Washington, DC, 16-24.

Camp, William M., Wright, William B., and Hussein, Mohamad (1993). "The Effect of Overburden on Pile Capacity in a Calcareous Marl", Proceedings of the Deep Foundations Institute (DFI) 18th Annual Member's Conference held in Pittsburgh, Pennsylvania.

Canivan, Gregory and Camp, William. "Three Case Histories Comparing Impact and Vibratory Driven Pile Resistances In Marl," Deep Foundations Institute 27th Annual Conference – The Time Factor in Design and Construction of Deep Foundations, San Diego, CA. 51-65.

Gohn, G. S., Higgins, B. B., Smith, C. C., and Owens, J. P. (1977). "Lithostratigraphy of the Deep Corehole (Clubhouse Crossroads Corehole 1) Near Charleston, South Carolina" *Studies Related to the Charleston, South Carolina, Earthquake of 1886 – A Preliminary Report*, Rankin, D. W. editor, U. S. Geological Survey Professional Paper 1028-E, p 59-70.

Heron, S. D., Jr., Limestone Resources of the Coastal Plain of South Carolina", (1962), Bulletin No. 28, Division of Geology, State Development Board, Columbia, SC 128 pp.

Horton, J. Wright, Jr. and Zullo, Victor A., (1991), "An Introduction to the Geology of the Carolinas", *The Geology of the Carolinas: Carolina Geological Society 50th Anniversary Volume*, edited by J. Wright Horton, Jr. and Victor A. Zullo. pp 1-10, University of Tennessee Press, Knoxville, TN

Law Engineering, Inc. (1991) "Report of Test Shaft Program – Isle of Palms Connector", Report submitted to The LPA Group, Inc., Columbia, SC, January.

Loadtest, Inc., (2002a), "Report on Drilled Shaft Load Testing (Osterberg Method) – Test Shaft 1W – Cooper River Bridge", Report submitted to Case Atlantic Company, Clearwater, FL, March.

Loadtest, Inc., (2002b), "Report on Drilled Shaft Load Testing (Osterberg Method) – Test Shaft 1E – Cooper River Bridge", Report submitted to Case Atlantic Company, Clearwater, FL, March.

Parsons Brinckerhoff Quade & Douglas (1993), "Report on Preliminary Geotechnical Investigation for U.S. 17 Bridges over Cooper River, Charleston, South Carolina", report submitted to the South Carolina DOT, October.

Parsons Brinckerhoff Quade & Douglas (1999), "Supplemental Geotechnical Investigation Data Report for U.S. 17 Cooper River Bridges Charleston, South Carolina", report submitted to the South Carolina DOT, March.

Rinaldi, Victor A., and Santamarina, J. Carlos, (2003), "Cemented Soils: Behavior and Conceptual Framework", in press.

S&ME, Inc. (2000a), "Phase II Geotechnical Data Summary Report – Cooper River Bridge Replacement Project, Charleston, South Carolina", Report submitted to Sverdrup Civil, Inc., Charleston, SC

S&ME, Inc. (2000b), "Final Report of Geotechnical Exploration – Mount Pleasant Interchange – Cooper River Bridge Project, Charleston, South Carolina", Report submitted to Sverdrup Civil, Inc., New York, NY

S&ME, Inc., (1993), "Report of Preliminary Geotechnical Exploration – Mark Clark Expressway (I-526) - U.S. Highway 17 to Folly Road – Charleston, South Carolina", report submitted to Kimley-Horn Associates, Raleigh, NC.

S&ME, Inc., (1999) "Report of Final Geotechnical Exploration – Maybank Highway Replacement Bridge over the Stono River, Charleston, South Carolina", report submitted to HDR Associates, November.

S&ME, Inc., (2001) "Report of Geotechnical Exploration – Bridge Foundations and Embankments: Ashley Phosphate/I-26 Interchange Improvements", Report submitted to Earth Tech, Inc., Raleigh, NC, August.

S&ME, Inc. (2003) "Report of Geotechnical Exploration – Bridge Foundations: Aviation Avenue & Remount Road/I-26 Interchange Improvements", Report submitted to Arcadis G&M, Raleigh, NC, October.

Soil & Material Engineers, (1984) "Report on Test Pile Program – North Charleston Test Site, I-526 Bridges", report submitted to HNTB, Atlanta, GA.

Tuomey, Michael, (1848), "Report on the geology of the South Carolina", South Carolina Geological and Agricultural Survey Report, No. 1, 293 pp.

Ward, L. W., Blackwelder, B. W., Gohn, G. S., and Poore, R. Z., (1979), "Stratigraphic Revision of Eocene, Oligocene and Lower Miocene Formations of South Carolina", *Geologic Notes*, Vol 23, No. 1., South Carolina Geological Survey, Columbia, SC, pp 2 – 31

Weems, R. E., and Lemon, E. M., Jr., (1984). "Geologic map of the Stallsville quadrangle, Dorchester and Charleston Counties, South Carolina", U. S. Geological Survey, Miscellaneous Investigations Series, Map GQ-1581, 1:24,000.

Weems, R. E., and Lemon, E. M., Jr., (1989). "Geology of the Bethera, Cordesville, Huger and Kitteredge Quadrangles, Berkley County, South Carolina", U. S. Geological Survey, Miscellaneous Investigations Series, Map I-1854, 1:24,000.

Weems, R. E., and Lemon, E. M., Jr., (1996). "Geology of the Clubhouse Crossroads and Osborn Quadrangles, Charleston and Dorchester Counties, South Carolina", U. S. Geological Survey, Miscellaneous Investigations Series, Map I-2491, 1:24,000.

ZEN AND THE ART OF DRILLED SHAFT CONSTRUCTION: THE PURSUIT OF QUALITY

Dan A. Brown, P.E., M. ASCE

ABSTRACT: With apologies to author Robert Pirsig (Zen and the Art of Motorcycle Maintenance, 1974), this paper pursues the concept of quality with respect to construction of drilled shaft foundations. With the evolution of more sophisticated techniques for integrity testing and load testing, it is possible now to better observe the end result of our construction activities and make judgments about the effectiveness of techniques and materials at achieving quality. This paper describes some aspects of construction techniques and materials that can lead to defects or less than optimal performance for drilled shaft foundations. Examples are cited of some of the more common problems encountered in drilled shaft construction in order that lessons can be learned from these problems. The case is made for designers and contractors to emphasize constructability in designs, workability in construction materials, and individual responsibility toward quality on the jobsite.

INTRODUCTION

In the book "Zen and the Art of Motorcycle Maintenance", author Robert Pirsig describes a personal spiritual journey (and a motorcycle trip) in a quest for "quality". In the construction of drilled shaft foundations, developments in integrity testing and load testing have afforded the industry an improved opportunity to assess the "quality" of the end product of our labors in terms of the integrity and load carrying capacity of the foundation. While this improved "vision" has often lead to disputes over the definition of "quality" that was specified in the contract documents and that is generally provided by current practice, most engineers and constructors can recognize those attributes of quality that are desirable in a drilled foundation:

- The foundation should consist of a relatively uniform mass of sound concrete,
- The concrete should have good bond and load transfer to the bearing formation,
- The reinforcement should be in the intended position and should be bonded to the concrete.

This paper will address some aspects of design and construction which influence the tendency to achieve quality in drilled foundations. A brief description of some selected case histories is included in order that lessons may be learned from these experiences. The general theme which emerges from these case histories is that designers and contractors need to consider and emphasize constructability in designs, workability in construction materials, and individual responsibility toward quality on the jobsite.

KEY ELEMENTS FOR QUALITY IN DRILLED SHAFTS

The writer's experiences suggest that the majority of construction problems which compromise the quality of drilled shafts come from a failure to adequately consider one or more of the following categories:

- Workability of concrete for the duration of the pour
- Compatibility of congested rebar and concrete
- Control of the stability of the hole during excavation and concrete placement, especially with the use of casing
- Drilling fluid which avoids contamination of the bond between the concrete and bearing material or excessive suspended sediment

To this list can be added a broader category, which is human attentiveness to any or all of the above. Inattentiveness can be the result not only of carelessness in workmanship, but also to contractual arrangements that do not encourage attentiveness and to inadequate resources devoted to inspection and quality control.

WORKABILITY OF CONCRETE

Workability can be defined as the ability of the concrete to readily flow through the tremie, the rebar cage, and to all places within the hole where it needs to go. With drilled shaft construction, this must be achieved without the need for external sources of energy such as a vibrator. Most commonly, slump is the measured property associated with concrete workability, although high slump is not always sufficient to ensure workability. Aggregate shape and gradation is important to avoid a tendency to segregate in high slump mixes. Experienced workers often describe a desirable mix as having a "creamy" paste rather than "boney" texture, in reference to the tendency of the mix to flow without segregation. When concrete has inadequate workability, several problems can ensue:

1. During tremie placement of concrete, there is a tendency for debris to become entrapped within the concrete and thus produce flaws in the structural integrity of the foundation. This can occur as the oldest concrete in the shaft is riding on top of the rising column of concrete, and as this old concrete becomes stiff then the fresh concrete can tend to "burp through" and trap the debris and/or contaminated concrete on top. Loss of workability can also lead to plugging in the tremie itself, which may cause the contractor to breach the tremie in order to get flow going again. The result of the breach would also be to trap debris and/or contaminated concrete as some concrete would tend to flow through water and lose cement.
2. Even with placement of concrete into a cased hole without the use of a tremie, there is a need for concrete workability to be maintained from start to finish. When the casing is removed, the concrete must have adequate workability to flow through the rebar cage, displace the water that may be present outside the casing, and produce lateral stress against the soil or rock so as to provide a good bond within the bearing stratum. If the concrete workability has been

lost by the time the casing is pulled, it may be very difficult to remove the casing. The concrete could tend to arch within the casing and be lifted with the casing, thus forming a neck. Even if the casing is recovered without necking the concrete, a column of stiff concrete that has been "slip-formed" into an oversized hole is not likely to provide good bond to the bearing formation. The presence of a heavy rebar cage can complicate the problem, as the lateral concrete flow after removal of the casing will be restricted by the cage.

3. A concrete mix which lacks sufficient cohesiveness in the paste and which has a tendency to segregate will pose a workability problem. Segregation in the rising column of concrete during placement can lead to large quantities of bleed water and paste at the top of the concrete column. These materials can become trapped as pockets within the concrete, particularly if there is a loss of workability. Segregation and bleeding can lead to problems during transport if concrete must be delivered to a remote location and cannot be continuously mixed during the entire time. During tremie placement or pumping, any delay in continuous delivery of the concrete to the tremie can result in segregation within the concrete column inside the tremie. Segregation inside the tremie can lead to plugging of the tremie and inclusions of non-uniform concrete within the shaft.

The FHWA guidelines (O'Neill and Reese, 1999) for drilled shaft concrete suggest that a slump of around 200 mm (8 inches) should be used for tremie placed concrete. Many state DOT's use specifications which routinely call for a slump of at least 100 mm (4 inches) to be maintained for a period of 4 hours after batching. It is the opinion of the writer that a 100 mm (4 inch) slump is probably not adequate for most conditions. If concrete with 200 mm slump is being placed into concrete which now has 100 mm slump, there will be two dissimilar fluids interacting within the hole with potentially undesirable consequences.

Rather, it is suggested that the concrete mix be designed to have a very high workability (slump loss of no more than 50 mm, or 2 inches) for the duration of the period required for placement, whatever that period may be. These days, it is quite possible to use admixtures to retard concrete for many hours. The concrete mix design should have workability and the time required for the construction sequence as a primary component of the mix design process.

Aggregate type and gradation is important for drilled shaft concrete workability, and must be considered in ways differing from normal concrete. Rounded gravel aggregates are much preferred over crushed stone aggregates due to the enhanced workability from using gravel and thus reducing water needed to achieve a given slump. The aggregate gradation may need to be adjusted to achieve workability without segregation. There is currently no test used for evaluation of segregation, although standard tests are under development specifically for use with self compacting concrete. Evaluation of the tendency for segregation is presently done as a part of the "art" of development of a workable mix, based upon visual evaluation of the cohesiveness of the paste. Bleeding would also tend to occur with segregation, and vice versa. In order to control segregation and bleeding, it is beneficial that a drilled shaft concrete mix be adjusted to include

Figure 1 Surface flaws in concrete (left), after removal for repair (right)

smaller maximum aggregate size and a greater proportion of fines, especially cementitious fines such as fly ash or slag.

Example 1 Surface Flaws in Concrete

Observation

Single drilled shafts were used to support individual columns for a bridge over a lake in the southwest. The shafts utilized casing extending through the lake and the relatively thin alluvial soil overlying rock. The upper portion of the shafts were formed using a removable casing so that no permanent steel casing would be visible within the zone of water fluctuation. The shafts were drilled using water only as a drilling fluid and the contractor appeared to do an excellent job of cleaning the hole. The rebar cage was not particularly restrictive to concrete flow, with relatively wide openings of at least 200 mm between longitudinal and transverse rebars.

The project was a 45 minute drive from the concrete plant. Upon arrival at the jobsite, the concrete trucks were placed on a barge and ferried to the foundation location. Concrete was placed using a tremie, with concrete delivered from the truck to the tremie by using a bucket. The tremie was maintained at least 2 m below the surface of the concrete in the hole at all times. Each shaft required 5 to 6 concrete trucks to complete the pour and the concrete placement took approximately 4 to 6 hours from start to finish.

Upon removal of the forms, the inspectors noted the presence of pockets of weak concrete which could be chipped away quite easily with a hammer. These pockets appeared to be a weak, cemented grout-like material that had no aggregates within it. Photographs of this material, along with a typical pattern after the weak material had been chipped away, are provided on Figure 1.

Explanation

At first it was suspected that either bleed water was contributing to this problem, or the contractor's removal forms were somehow not sealing and water was mixing with concrete during placement. However, further investigation revealed that the concrete mix did not maintain sufficient workability for the duration of the pour during the hot summer months. As tremie placement of concrete continued, the tremie would be lifted from the bottom of the shaft but always maintaining the base of the tremie at least 2 m below the surface of the concrete. As the initial charge of concrete (now riding on top, above the tremie) started to lose workability, the freshly place concrete entering the shaft through the tremie below the surface of this now-stiff concrete tended to erupt through this old, stiff concrete like a volcano rather than lifting the entire surface upwards. As this fresh, fluid concrete vents through the surface of older, less fluid concrete, the latent cement/water mixture on the surface of the rising concrete plug flowed to the lower surface outside the cage and became trapped below the fresh concrete. This flow pattern was visibly revealed in one of the shafts when the contractor pumped off the water above the concrete after the concrete was well up into the casing, and the placement continued using the tremie. The inspector reported the "volcano" of fresh concrete with the lateral displacement and subsequent trapping of latent water/concrete mixture that was present on top of the old concrete.

Implications for Performance

The concrete outside the rebar cage serves primarily to transfer load to the soil and as cover to protect the rebar from corrosion. In this case, the concern is for the long term durability of this cover. Discussions with a bridge inspection diver with the Texas DOT indicates that at least one bridge (Lake Houston) built in 1988 using similar construction techniques is now suffering spalling of concrete from the surface of the shafts.

Lessons Learned

It is critical that the concrete mix have sufficient retarder that the concrete maintain its workability for the duration of the pour. In this case, that was at least 6 hours from the time of batching because of the slow delivery of concrete to the foundation location. Also, the slump life of the concrete mix varies with the temperature, and increased retarder dosage is required during hot weather.

Example 2 Poor Bond in Rock Socket

Observation

A drilled shaft was installed through about 12 m of soil and socketed approximately 3 m into an underlying rock formation. In order to allow downhole visual inspection of the bottom of the shaft, the contractor was required to case the hole for the full length. Once satisfied with the inspection, a load test shaft was constructed using an Osterberg cell placed at the base of the socket. After placement of the O-cell and rebar, concrete was placed within the rock socket and the casing subsequently removed. The O-cell was found to mobilize only less than 0.5 MN of side shear resistance, a small fraction of the amount which had been expected.

Another test shaft was constructed, only this time a wet hole method was used with tremie placement of concrete and without casing into the rock. In this case the O-cell test indicated over 10 MN of side resistance in the socket.

Explanation

The amount of time required after concrete placement to extract the casing allowed the concrete workability to diminish to the point that the shaft was almost like a "slip-formed column" within the rock socket. Because of the lack of lateral pressure between the concrete and the rock, the side shearing resistance of the socket was very low. There may have been some additional detrimental effects of using the casing, such as trapping of debris behind the casing which contaminated the bond between concrete and rock.

Lessons Learned

Even though the casing provides a "dry hole" the need for concrete workability for the entire duration of the construction process remains. And although a shaft is constructed without any observable structural defects in the concrete, it may not be a "quality" shaft.

COMPATIBILITY OF CONGESTED REBAR AND CONCRETE

In recent years, it seems that contractors have become more well equipped to construct very large diameter drilled shafts and so engineers have become more prone to design and specify very large diameter drilled shafts. Large shafts have some compelling advantages for structures such as highway bridges, where large lateral and overturning forces are produced by design conditions for seismic, vessel impact, wind, etc. And a single large diameter shaft can have a smaller footprint than a pile footing, an advantage when working on congested sites or nearby existing structures. However, with the use of large diameter shafts designed for large bending moments, the rebar cages can become quite dense. Added to the rebar is the frequent addition into the cage a number of access tubes for integrity testing.

Problems can arise from restrictive rebar cages in the following ways:

1. If the lateral flow of concrete is significantly impeded, then there is an increased likelihood that debris will become trapped in the annular space outside the cage. This trapping of debris can result from the fact that the rising column of concrete inside the cage tends to be at a higher elevation than the concrete outside the cage, so there would be a natural tendency for any accumulated sediment on top of the concrete to slough off toward the side. Even a small accumulation outside the cage can be detrimental to the bond in the bearing formation.
2. Even with a clean slurry, the concrete can be impeded to such a degree that voids form outside the cage or the lateral stress at the concrete/rock/soil interface is diminished.

The FHWA guidelines (O'Neill and Reese, 1999) recommend that the clear space between bars be at least 5 times the size of the maximum aggregate. The writer has seen this guideline routinely violated in practice. In particular where seismic loads are important, there is a tendency for designers to use spiral confinement with a 90 mm pitch (3.5 inches), leaving only about 75 mm (3 inches) or less clear between spirals. The FHWA guidelines would suggest a mix design using a pea-gravel size aggregate for this case. Some state DOT's are using such a mix with success. Workability of the concrete is enhanced in such severe cases if the aggregate is specified to be a rounded gravel rather than a crushed stone. It would also be prudent for designers to consider the implications of the use of such tight spiral reinforcement, and consider if the needed confinement of the interior concrete can be provided in a way which is more easily constructed.

It should also be noted that it is not sufficient for an agency to ALLOW the use of a pea gravel mix, and then place the burden entirely upon the contractor. Because a pea gravel concrete mix is more expensive on a materials basis, the result of such practice is that the winning bid on the job goes to the contractor who uses the least expensive mix allowed by the project specifications rather than the one which is needed. Subsequent problems can lead to poor quality, disputes about who is responsible, and claims.

Figure 2 Surface of the Shaft with #57 Stone

Example 3 Observations on the Flow of Tremie-Placed Concrete

Observations

During the winter of 2002-2003, several drilled shafts were cast at the Auburn University National Geotechnical Experimentation Site (NGES) at Spring Villa, Alabama for the purpose of closely observing concrete flow during tremie placement. Concrete consistent with the Alabama DOT standard mix was used, with a #57 crushed aggregate (19 mm (¾ inch) maximum size) and approximately 200 mm (8 inch) slump was placed using a tremie within a 1.1 m diameter hole and the process filmed using a downhole camera. The hole was dry so that the concrete could be observed, but the placement was conducted as if in a wet hole environment. Rebar included longitudinal bars with about 200 mm (8 inches) clear between bars and hoops with approximately 125 mm (5 inches) clear between hoops.

The concrete was observed to flow up alongside the tremie pipe and produce a rolling action of the top of the rising concrete column, with the concrete surface rolling from the center of the vent near the tremie radially towards the perimeter of the shaft. There was typically about ½ m difference in head between the concrete within the rebar cage and the concrete in the annular space outside the cage. The concrete could be seen to "cascade" over the hoop steel to fill the annular space outside the cage.

Another shaft was constructed similarly and using a similar slump, but with a gravel aggregate of 12 mm (½ inch) maximum size. There was less noticeable "rolling" of the concrete from the center of the shaft towards the perimeter and the difference in head between the interior and the annular space was significantly less.

Subsequently, the shafts were exhumed and the surface of the concrete examined (see Figure 2). The surface texture of the shaft with the pea gravel aggregate appeared less permeable. There was no trapped debris, since both shafts were constructed using tremie placement in the dry.

Figure 3 Defect in Shaft with Double Rebar Cages

Lesson Learned

Concrete does not flow upward above the tremie as a nice homogeneous mass. Rather, the surface is a rolling, boiling three dimensional thing which could easily entrap any debris which resides on top and which can build up a head differential between the interior and the exterior of the rebar cage.

Example 4 Extreme Rebar Cage

Observations

A bridge on the west coast was designed to utilize a single 4 m diameter shaft to support an oval shaped column. In order to mate the rebar cage in the shaft to the shape of the column, a double rebar cage was utilize in which two cages overlap in a sort of "fat figure 8" configuration. As a result, concrete was forced to flow through not one, but two rebar cages, both of which were had extremely congested cages due primarily to the closely spaced transverse steel. The shafts were socketed through an overlying sandy alluvial soil into sandstone. The concrete mix had an aggregate which used crushed stone with a 25 mm (1 inch) maximum size. The contractor attempted to case through the overlying sand with a temporary casing; due to aesthetic reasons, permanent casing to the ground surface was not permitted.

Integrity testing indicated the presence of voids around the perimeter of several shafts. Access shafts were constructed to examine the problem and pockets of sand were discovered within the zone where the two cages overlap, as illustrated on Figure 3. Coring was performed from the top of the shaft and showed good concrete of consistent compressive strength within the center of the shaft but very erratic compressive strengths around the perimeter.

Lesson Learned

A simple cage with considerations of constructability is needed in order to make drilled shafts less vulnerable to defects and to promote quality construction. A permanent casing socketed into the underlying rock could also have reduced the likelihood of defects in this case because the hole could have been kept dry, although the difficulty with the rebar and concrete would still be present. The concrete mix in the case of severely congested rebar should incorporate good flow characteristics with small rounded gravel aggregates.

Example 5 Concrete Placement in a Large Shaft with Tight Spiral Reinforcement

Observation

During construction of the test shafts for a new bridge in South Carolina, careful observations were made of the concrete behavior during tremie placement (Camp et al, 2002). The shafts were 2.4 m diameter (8 feet) and up to 50 m deep (160 feet), with a rebar cage which included a heavy longitudinal reinforcement tied with a spiral transverse rebar at a 90 mm (3.5 inch) pitch. The concrete used a small crushed stone maximum aggregate less than 12 mm size (½ inch) and 200 mm slump (8 inch). In spite of the small aggregate and high slump, measurements indicated as much as a 1.4 m difference in head between the concrete levels on the interior and exterior of the cage.

Lesson Learned

If any adverse performance of the test shafts resulted from the tight spiral reinforcement in the cage, it was masked by other factors. Nevertheless, the observations indicate that the potential exists for debris to be trapped on the outside of the cage under these circumstances, and designers and contractors were alerted to the need to take extra precaution to maintain a clean slurry during construction of the production shafts.

CONTROL THE STABILITY OF THE HOLE

The successful installation of a drilled shaft is predicated on the ability of the contractor to maintain a stable hole in order that the foundation can be cast-in-place with quality materials and workmanship. However, it is not sufficient to only gain stability of the hole prior to concrete placement; the hole must be stable from the start throughout the completion of the shaft. Quality foundation construction requires that the stability of the hole be maintained at all times in order to preserve the integrity of the bearing formation and to avoid defects resulting from voids and irregularities in the overlying strata.

In wet hole construction, it is essential that a positive pressure be maintained against the sides of the hole at all times. Groundwater should not be allowed to seep from surrounding soil into the hole, as cave-ins and sidewall sloughing can occur. Even if sloughing does not occur, the surrounding soil can become loosened and lateral stresses can be reduced around the shaft and around nearby structures. Ground subsidence is sometimes observed with augered cast-in-place piles (although not the subject of this paper) in water-bearing sands. An unstable and loosened sidewall could result in sloughing during concrete placement which would result in defects within the concrete.

With temporary casing used to provide stability of the hole, it is often attempted to complete the shaft excavation and place concrete in the dry. However, seepage from the bearing formation below the casing can result in softening of the bearing materials and the accumulation of debris on the base of the hole. In shales or fractured rock materials, there may not be a large quantity of seepage observed, but the unbalanced fluid pressures and the stress relief provided by the hole may be sufficient to produce softening. If seepage is observed, it is often preferable to complete the excavation in the wet to avoid uncontrolled seepage into the hole.

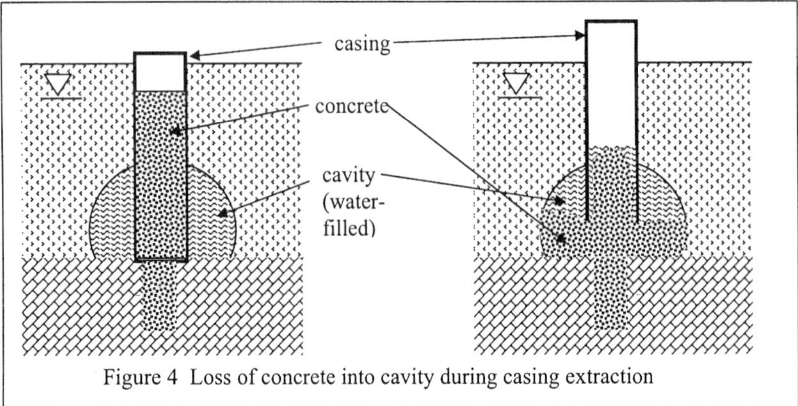

Figure 4 Loss of concrete into cavity during casing extraction

Where temporary casing is sealed into the top of a relatively impervious formation, it is important that the seal be successful so as to avoid seepage into the hole around the base of the casing. Such seepage could result in a large cavity forming around the outside of the casing, and large cavities can result in large concrete overruns and lead to potential defects in the concrete. As illustrated in Figure 4, a large loss of concrete volume into a hole around the casing could result in the head of concrete within the casing becoming less than the head of water on the outside of the casing. If this condition occurs, there will be flow of water into the casing, potentially displacing or mixing with the concrete.

Example 6 Cavity Around Casing

Observation

A drilled shaft was installed at a site in Florida by vibro-driving a casing through the overlying water-bearing sand into the top of limerock. The shaft excavation was completed within the limerock and concrete placed to near the top of the casing. The casing was pulled upwards approximately 0.6 m (2 feet) and the concrete level within the casing dropped to near the bottom of the casing. Although the contractor attempted to place additional concrete with a pump, the hole eventually was abandoned. No significant inflow of sand had been noted during the excavation into the limerock.

Explanation

Although the contractor had used care to avoid causing a cavity around the casing during excavation, the limerock in this area had natural cavities formed by solution activity. These cavities tend to be found most often near the top of the formation. The contractor used good practice when withdrawing the casing only by a small amount so that additional concrete could be added to maintain the head, but the volume lost was so large that the head of concrete dropped below that of the surrounding groundwater and the structural integrity of the shaft was compromised.

Lesson Learned

Because this cavity was a natural occurrence, the design of the shaft needs to accommodate the possibility of large cavities as a construction consideration. This condition would be more effectively handled with the use of a permanent casing.

A CLEAN SLURRY HELPS AVOID CONTAMINATION

The use of drilling slurry can be a very effective means to maintain control over the stability of the hole. However, it is important that the slurry properties be controlled in order to avoid potential contamination of the bond between the shaft and surrounding soil or rock, or within the concrete due to excessive suspended sediment.

Several recent comparative tests have demonstrated that polymer slurries can significantly outperform bentonite in terms of side shearing resistance, as illustrated by the data on Figure 5 (Brown et al, 2002). This trend appears most notably in granular soils where there is some fluid loss into the surrounding formation and a bentonite filter cake is likely to form (Brown, 2002; Majano et al, 1994; Meyers, 1996). Where slurry is used in a relatively cohesive formation and there is little opportunity for fluid loss, differences in unit side shear have been insignificant (Camp et al, 2002). Although current design procedures do not generally delineate values used for design on the basis of construction technique, it is clear that the construction procedure used can have a major effect on performance. Where bentonite slurry is used within the bearing formation, the quality of the bond is enhanced by minimizing the exposure time.

The quality of the shaft also depends on the control of the amount of suspended sediment in the slurry. Years ago, contractors were routinely permitted to have as much as 10% sand within bentonite slurries. In recent years, most state DOT's have adopted a 4% criterion for suspended sand within the slurry. However, the key concern is not so much the amount of sand in suspension, but the amount of sediment which can settle out during concrete placement. As drilled shafts have become larger diameter and deeper, the time required to place concrete has increased and thus the time opportunity for sediment to settle out of suspension has increased. Because of considerations described previously of the rolling surface and the potential for differential head across the top of the rising concrete column, and sediment which occurs during concrete placement may be subject to become included within the concrete. As shafts become larger and deeper, the allowable sand content within bentonite slurry will likely need to be reduced.

With polymer slurries, there is a tendency for a false sense of security about suspended sediment, since sand particles tend to settle out quickly. However, silt sized particles (smaller than the #200 sieve) and fine sands have been observed to remain in suspension for extended periods and result in contamination atop the concrete column. There have been numerous cases of projects with polymer slurry where removal of silts and fine sands from the slurry has been very difficult. On one recent project near the Atlantic coast, a contractor constructing shafts using a polymer slurry was forced to overpour the shafts by over 2 m (6 feet) in order to remove concrete contaminated with silt. On another project at a site in Florida with very fine silty sands, the base of several shafts were inspected using a downhole camera to evaluate the condition of the base of the shaft excavation prior to concrete placement. The camera revealed the bottom 0.6 m (2 feet) of the slurry to be heavily laden with silt in an appearance which resembled that of a gel. These and many other anecdotal experiences with polymer slurries in silty fine sands suggest that engineers and contractors should exercise caution in the use of these materials.

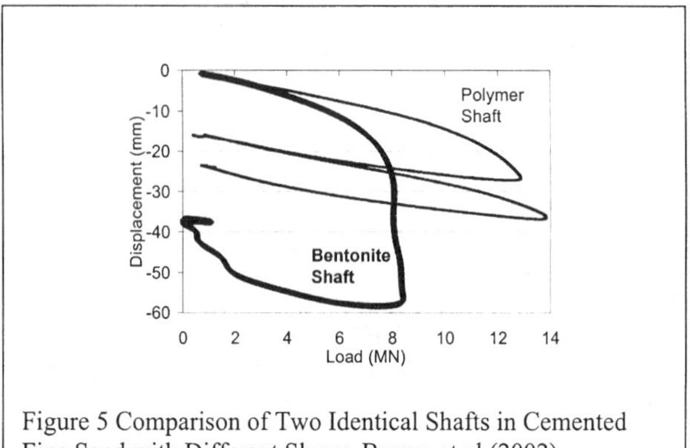

Figure 5 Comparison of Two Identical Shafts in Cemented Fine Sand with Different Slurry, Brown et al (2002)

Example 7 You Broke My Shaft!

Observation

A small diameter drilled shaft was constructed for a load test at a site with sand and clay overlying limerock. Temporary casing was seated into the rock, and the shaft drilled into the stable limerock formation in the wet. The limerock was stable, but dewatering the excavation was not attempted due to the potential for seepage. Since the rock was stable and the overburden was cased, the slurry was not given much attention; water was used, although some mixing with the native materials occurred. The shaft was completed and the casing was pulled. No integrity testing was performed.

When the load testing was performed, the shaft failed suddenly at approximately 150% of the design load. Strain gauges within the shaft suggested that very little load was reaching the lower portions of the limestone socket. Upon excavating the upper 2 m of the shaft (6 feet), the shaft was observed to have suffered structural failure at a load corresponding to less than 14 Mpa (2000 psi) over the theoretical cross sectional area of the shaft (see Figure 5). The shaft was cut off cleanly at about 2 m below grade, a new top formed, and the load test was successfully carried out to over 2 times the design load without achieving failure (either geotechnical or structural).

Lesson Learned

The shaft was constructed with a false sense of security regarding concrete placement because of the casing and the stable limerock formation, and no testing was performed on the slurry or the completed shaft. Even though only plain water was used during drilling, the properties of the resulting slurry are important because of the potential effect on the concrete during placement. It appears that the concrete integrity was compromised within the upper 2 m of the shaft, resulting in structural failure when loaded to high stresses. The lesson learned from this is example is that any wet pour has the potential to impact concrete integrity and the drilling fluid must be clean and the concrete must have good workability.

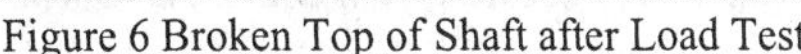

Figure 6 Broken Top of Shaft after Load Test

CONTRACTUAL AND ORGANIZATIONAL FACTORS

In order to achieve quality construction, the work must be organized so that all parties involved have incentive to achieve quality. This aspect of quality is perhaps the most difficult to achieve. If only the structural integrity of the concrete between the crosshole sonic logging tubes is tested and observed, the contractor's workers often have little concern for other aspects of quality, such as preservation of the integrity of the bearing stratum. Engineers may sometimes focus so much on the optimal arrangement of reinforcement for bending forces that constructability issues are overlooked.

Quality construction of drilled shafts requires that:

- The design engineer should be knowledgeable regarding constructability issues and would produce a design for which ease of construction is a key element,
- The general contractor should appreciate the need for a qualified sub and provide the resources and support needed to ensure that this critical part of the work is performed without interruptions,
- The drilling subcontractor should be conscientious and genuinely interested in producing a quality product, and would have well trained workers who are properly equipped for the job,
- The inspector should be well trained and knowledgeable regarding drilled shaft construction, the critical aspects of the design, and the geologic conditions at the site,
- The project should include provisions for measuring quality using the latest techniques for inspection, non-destructive testing, load testing, and test installations where necessary.

The entire project team must work together to achieve quality, with each party accepting individual responsibility for their own role in the process. Responsible team members expect and demand quality from other members of the team while at the same time working cooperatively to resolve difficulties. It is too often the case that the various parties adopt and adversarial position early in the project in order to position themselves for the anticipated battles (claims, damages, disputes over problems) with other team members. These expectations for disputes seem to be self-fulfilling.

There is no magic solution to resolve problems in these areas. Quality construction requires preparation and persistent attention to details. Just as an army succeeds because of extensive training for its mission, designers, inspectors, and constructors can succeed by preparing themselves with extensive training. And a final key ingredient is that quality be measured in a timely fashion so that adjustments to construction techniques can be adopted before small problems become major ones.

Example 8 I Didn't Do It, Nobody Saw Me Do It, You Can't Prove a Thing

Observation

A bridge project in the Midwest was constructed using a group of drilled shafts installed into a rock formation through water and overlying soil using permanent casing. The general contractor hired a drilled shaft subcontractor to provide services only consisting of drilling the holes. In order to plan their work efficiently, the general contractor had the sub drill the entire bent (with up to 12 shafts), after which the concrete would be placed and the bent completed. Although the bridge was over water, the holes were expected to be dry due to the permanent casing.

Fortunately for the owner, there was at least post-construction integrity testing performed using crosshole sonic logging. These logs indicated that several shafts had zones of very poor quality concrete. Coring indicated that there was weak concrete in some areas and some zones had washed aggregate in places where sound concrete should be present. The shafts required extensive grouting and underpinning with micropiles.

Explanation and Lessons Learned

Because the holes had been expected to be dry, the concrete was placed using free-fall into the hole. Some of the holes were open for weeks between drilling and completion. In fact, the rock had some small fractures that resulted in seepage into the holes. In some holes, concrete was dropped through water resulting in defective zones of concrete within these shafts.

The divided responsibility for completion of the drilled shaft foundations is a very undesirable arrangement. The work of constructing a drilled shaft cannot be easily subdivided into drilling the hole and placing the concrete as if they were two independent operations; once the hole is drilled to the required depth, it is important that the shaft be completed in a timely fashion with a single point of responsibility for making this happen.

Although the inspection failed to prevent concrete placement through water, it was fortunate that the designers had included crosshole testing, which allowed the defects to be discovered. Many state DOT's require integrity testing only where tremie placement of concrete is used. The effects on the axial capacity from the seepage and extended period of open holes is not determined, but the conditions at this site are not conducive to quality in this respect.

CONCLUSIONS

Like a motorcycle trip across America, the pursuit of quality in drilled shaft construction is a long and arduous journey, accomplished in small incremental steps.

More powerful drilling equipment and better techniques have lead designers to utilize drilled shaft foundations in larger diameters and to greater depths than ever before. The challenges for quality in construction have increased. However, improvements in technology with respect to integrity testing and load testing have made it possible to measure quality in ways that were never before possible. Improvements in construction materials, including better concrete and more sophisticated slurry products, have made it possible to construct quality drilled shafts in more difficult conditions. With attention to details on these items, improved quality in drilled shafts is being realized.

But perhaps the most influential component in the process and the most difficult to control is the human element. All of the examples cited in this paper in which quality was compromised could have been avoided or corrected by engineers, constructors, and inspectors who were knowledgeable of their craft and attentive in their work. The challenge remains to put into place systems of training (for all members of the team) and jobsite control to encourage and emphasize quality in construction.

ACKNOWLEDGEMENTS

The author would like to acknowledge the contributions of Mike Muchard of Applied Foundation Testing and Jack Hayes of Loadtest, Inc. The author is indebted to Dr. Anton Schindler of Auburn University for his insight into concrete materials and many stimulating discussions on the subject. The author would especially like to acknowledge the many years of mentoring and wise council of the late Mike O'Neill, to whom the author will be forever indebted.

REFERENCES

Brown, D. A. (2002). "The Effect of Construction on Axial Capacity of Drilled Foundations in Piedmont Soils," *J. of Geotechnical and Geoenvironmental Engineering*, 128(12), pp967-973.

Brown, D., Muchard, M., and Khouri, B. (2002). The Effect of Drilling Fluid on Axial Capacity, Cape Fear River, NC. *Proceedings* Deep Foundation Institute Annual Conference, San Diego, CA, Oct.

Camp, W.M., Brown, D.A., and Mayne, P.W. (2002). "Construction Method Effects on Axial Drilled Shaft Performance" Geotechnical Special Publication No. 116, ASCE, pp. 193-208.

Majano, R.E., O'Neill, M.W., and Hassan, K.M. (1994). "Perimeter Load Transfer in Model Drilled Shafts Formed Under Slurry," *Journal of Geotechnical Engineering,* ASCE, Vol. 120, No. 12, pp. 2136-2154.

Meyers, B. (1996). "A Comparison of Two Shafts: Between Polymer and Bentonite Slurry Construction and Between Conventional and Osterberg Cell Load Testing," Paper Presented at the Southwest Regional FHWA Geotechnical Conf., Little Rock, AR, April.

O'Neill, M.W. and L.C. Reese (1999) *Drilled Shafts: Construction Procedures and Design Methods*, Technical Manual Prepared for Federal Highway Administration, 758 p.

Pirsig, R. M. (1974) *Zen and the art of motorcycle maintenance: an inquiry into values*, Morrow, New York, 412 p.

ON THE AXIAL BEHAVIOR OF DRILLED FOUNDATIONS

Fred H. Kulhawy[1], Fellow ASCE

ABSTRACT: Four representative types of drilled foundations (drilled shafts, augered cast-in-place piles, pressure-injected footings, and micropiles) are examined from the standpoint of axial capacity evaluation, generalized behavior, and normalized load-displacement response. As shown, the capacities can be computed rationally, taking into account the particular characteristics of each foundation type, and the results agree well with the interpreted failure load from field load tests. The interpreted failure load also is linked consistently and simply to the elastic limit and the conventional slope tangent methods for interpreting compression and uplift load test results. The displacements also exhibit consistent patterns of behavior in compression and uplift, at both the interpreted failure load and elastic limit. These results will be useful for better understanding axial foundation behavior over a range of drilled foundation types.

INTRODUCTION

Since the mid-1970s, our research group at Cornell has been working on the general behavior of foundation systems and the mechanics of soil (and rock) - foundation interaction. We began with drilled shafts, but then expanded into other types of drilled foundations, spread foundations, ground anchors, and driven piles. At first, we addressed terrestrial installations for the electric utility industry through EPRI, but then the focus was broadened for other sponsors, and marine applications were included as well. A recent overview of our EPRI work is given by Kulhawy et al. (2002). As with all long-term and broadly-based research efforts, ideas and concepts get refined, expanded, and hopefully integrated as newer data become available and as newer technologies evolve.

In this keynote lecture, I want to discuss four types of drilled foundation systems

[1] Professor, School of Civil and Environmental Engineering, Cornell University, Hollister Hall, Ithaca, New York 14853-3501; fhk1@cornell.edu

- drilled shafts, augered cast-in-place (ACIP) or augercast piles, pressure-injected footings (PIF), and micropiles - all under axial loading and installed exclusively in soil. These systems illustrate the wide range of drilled foundations employed and also represent an array of construction technologies that, on the surface, may suggest the need for widely different methods for analysis and design. However, we have been able to examine some of the available data on these systems and have been able to determine some consistent and interesting trends in their behavior. Many of these details we have described in numerous publications. Herein, I will give a concise summary of some of the key issues involved, focusing on axial capacity evaluation, generalized behavior, and normalized load-displacement response. These issues will be discussed first in general terms, and then they will be tailored to specific foundation types. Drilled shafts will be addressed first, because they have been evaluated most comprehensively, and then the discussion will be generalized to include the other drilled foundation types.

AXIAL CAPACITY

The axial capacity of a drilled foundation is a function of the foundation weight (W), side resistance (Q_s), and tip resistance (Q_t). By force equilibrium, the compression capacity (Q_c) is given by:

$$Q_c = Q_{sc} + Q_{tc} - W \quad (1)$$

while the uplift capacity (Q_u) is given by:

$$Q_u = Q_{su} + Q_{tu} + W \quad (2)$$

The side and tip resistances are discussed below, with full details given elsewhere (e.g., Kulhawy et al. 1983; Kulhawy 1991). As in good engineering practice, all geotechnical properties should be evaluated properly over the full foundation depth and beneath the tip.

For the weight, W is the effective weight for drained loading, given by the total weight above the water table and the buoyant weight below the water table. For undrained loading, W is the total weight, above or below the water table.

Side Resistance

For drained loading, the side resistance is given by the effective stress or β method, expressed as the frictional strength over a cylindrical shaft surface as follows:

$$Q_{sc} = Q_{su} = \pi\, B\, (K/K_o)_e \int_0^D \{\sigma_v'(z)\, K_o(z) \tan [\phi'(z) \cdot \delta/\phi']\}\, dz \quad (3)$$

in which B = shaft diameter, D = shaft depth, $(K/K_o)_e$ = stress factor that represents the change in the horizontal effective stress as a function of the construction method, σ_v' = vertical effective stress, K_o = in-situ coefficient of horizontal soil stress ($= \sigma_h'/\sigma_v'$), ϕ' = effective stress friction angle for the soil, δ = effective stress friction angle for the soil-shaft interface, δ/ϕ' = interface roughness factor, and z = depth. For convenience in some analyses, the terms in Eq. 3 can be grouped as follows:

$$Q_{sc} = Q_{su} = \pi B \int_0^D \beta(z)\, \sigma_v'(z)\, dz = \pi B \int_0^D f(z)\, dz \tag{4}$$

in which β is a coefficient ($= K \tan \delta$) and f = unit side resistance. In Eq. 3, the interface friction angle (δ) normally is related to the effective stress friction angle of the soil (ϕ'), as the interface roughness factor (δ/ϕ'), which is ≤ 1.

For undrained loading, the side resistance is given by the total stress or α method, expressed as the undrained strength over a cylindrical shaft surface as follows:

$$Q_{sc} = Q_{su} = \pi B \alpha (K/K_o)_t \int_0^D s_u(z)\, dz \tag{5}$$

in which B = shaft diameter, D = shaft depth, α = empirical adhesion factor, $(K/K_o)_t$ = factor that represents the change in the horizontal total stress as a function of the construction method, s_u = undrained shear strength, and z = depth. The adhesion factor is obtained from an α-s_u correlation, in which α is selected based on the mean s_u over the depth of the shaft.

To standardize the α-s_u relationship, the CIUC (consolidated-isotropically, undrained triaxial compression) test was selected because it is the most common, good quality, reference test. The interrelationships among the different s_u values from various test types is given elsewhere (e.g., Kulhawy & Mayne 1990; Chen & Kulhawy 1993, 1994).

Tip Resistance in Compression

The tip resistance in compression is provided by the bearing capacity of the soil beneath the tip, as given by:

$$Q_{tc} = q_{ult} A_{tip} = q_{ult} \pi B^2 / 4 \tag{6}$$

in which q_{ult} = ultimate bearing capacity and A_{tip} = shaft tip area. The general solution for q_{ult} (e.g., Vesic 1975) is the Terzaghi-Buisman equation given below:

$$q_{ult} = c\,N_c + 0.5\,B\,\gamma\,N_\gamma + q\,N_q \qquad (7)$$

in which c = soil cohesion, γ = soil unit weight, q = vertical stress at the shaft tip (= γ D), and N_c, N_γ, N_q = bearing capacity factors.

Further studies have extended Eq. 7 to actual field conditions, and modifiers (ζ) that include foundation shape (s), depth (d), and rigidity (r) have been introduced by a number of authors (e.g., Hansen 1970; Vesic 1975; Kulhawy et al. 1983). With these modifiers, the general form of the bearing capacity equation becomes:

$$q_{ult} = cN_c\zeta_{cs}\,\zeta_{cd}\,\zeta_{cr} + 0.5\,B\,\gamma\,N_\gamma\,\zeta_{\gamma s}\,\zeta_{\gamma d}\,\zeta_{\gamma r} + q\,N_q\,\zeta_{qs}\,\zeta_{qd}\,\zeta_{qr} \qquad (8)$$

The detailed values of N_c, N_γ, and N_q are given elsewhere (e.g., Kulhawy, 1991).

For drained compression loading, with c = 0, Eq. 8 becomes:

$$q_{ult} = 0.5\,B\,\gamma\,N_\gamma\,\zeta_{\gamma s}\,\zeta_{\gamma d}\,\zeta_{\gamma r} + q\,N_q\,\zeta_{qs}\,\zeta_{qd}\,\zeta_{qr} \qquad (9)$$

The soil parameters to use represent the average values over a depth B below the shaft tip. Also, in accordance with proper modeling of the stress systems along the bearing capacity failure surface, ϕ' is evaluated as the average from triaxial compression (TC), direct simple shear (DSS), and triaxial extension (TE) tests (e.g., Ladd et al. 1977). However, this correction is minor compared to the TC value and conservative (on the order of 5%), and so it is normally neglected.

For undrained compression loading, Eq. 8 becomes:

$$q_{ult} = 6.17\,s_u\,\zeta_{cd}\,\zeta_{cr} + q \qquad (10)$$

The soil parameters to use represent the average values over a depth B below the shaft tip. Furthermore, in accordance with proper modeling of the stress systems along the bearing capacity failure surface, s_u is evaluated as the average value from TC, DSS, and TE tests (e.g., Ladd et al. 1977). This average s_u is on the order of 0.55 to 0.70 s_u(CIUC) for typical clays.

Tip Resistance in Uplift

The tip resistance in uplift is very different from that in compression because it can develop from tension and suction, as given by:

$$Q_{tu} = s_t\,A_{tip} \text{ or } s_s\,A_{tip} \qquad (11)$$

in which s_t = tip tension and s_s = tip suction. The value of s_t is the minimum tensile strength of the soil or the concrete. Since s_t is low and normal construction practices potentially lead to a thin, low strength soil zone at the tip, it is prudent to disregard tip tension under normal circumstances.

Tip suction can develop in saturated finer-grained soils in undrained loading, but it is zero in drained loading. For the undrained case, the tip suction can be approximated as follows (Stas & Kulhawy 1984):

$$s_s = (W / A_{tip}) - u_i = (W / A_{tip}) - \gamma_w (D - h) \tag{12}$$

in which h = depth to water table. Suction dissipates promptly with time, and therefore it should be considered only for transient live-loading cases.

GENERALIZED LOAD-DISPLACEMENT BEHAVIOR

The load-displacement response of drilled foundations in axial compression and uplift can exhibit any one of the three shapes shown in Fig. 1. The peak of A and the asymptote of B define clearly the maximum resistance of the foundation. However, if the response resembles C, the maximum resistance is not defined clearly. Drilled foundations most often resemble curve C.

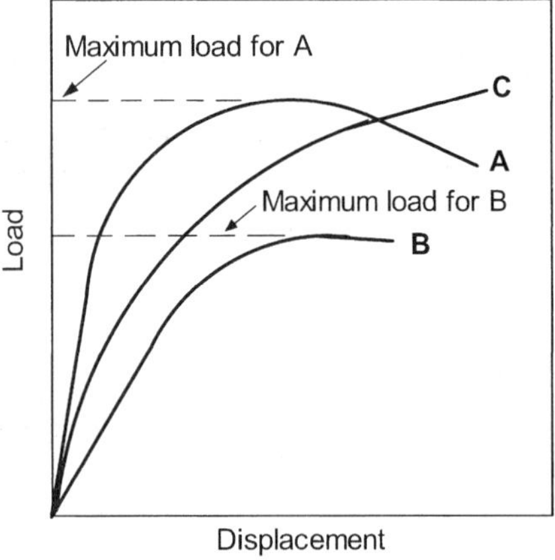

FIG. 1. Typical Load-Displacement Curves for Drilled Foundations

In the late 1980s, a set of consistent criteria were proposed for interpreting the results of load tests conducted on drilled shafts under axial loading (Hirany & Kulhawy 1988, 1989; Kulhawy & Hirany 1989). The term "ultimate capacity" was avoided and discouraged because it lacks a universally accepted definition. Instead, the term "interpreted failure load" was used to emphasize that any selected failure load really is an interpreted value. These criteria have been used widely (correctly and incorrectly) in the interim. Our recent paper (Hirany &

Kulhawy 2002) re-examined the interpretation criteria, provided additional guidelines and clarifications, and recommended detailed procedures for applying these criteria in a consistent manner. Since the initial proposal, these criteria also have been applied to other types of drilled foundations, including augercast piles, pressure-injected footings, and micropiles, as described later. It appears that, in principle, these interpretation criteria are appropriate across the full range of drilled foundations. These general interpretation criteria are described below.

Drilled foundation compression and uplift load-displacement curves normally can be simplified into three distinct regions: initial linear, transition, and final linear, as shown in Fig. 2. Point L_1 corresponds to the load (Q_{L1}) and butt displacement (ρ_{L1}) at the end of the initial linear region, while point L_2 corresponds to the load (Q_{L2}) and butt displacement (ρ_{L2}) at the initiation of the final linear region. The failure loads interpreted by most existing methods lie either in the transition region, the final linear region, or sometimes beyond the final linear region (extrapolated failure loads). For drilled shafts, Hirany and Kulhawy (1989) showed that, for six other common interpretation methods, the ratio of Q_{L2} to the interpreted failure load by the other methods varied, on average, from 0.93 to 1.14.

During a load test, creep displacements generally become significant at load levels beyond the transition region, where it is difficult to maintain a constant load with a hydraulic jack. Consequently, the displacements measured in this region generally would not be representative of the true behavior because of fluctuating applied load. Any interpreted failure loads in this region therefore would be

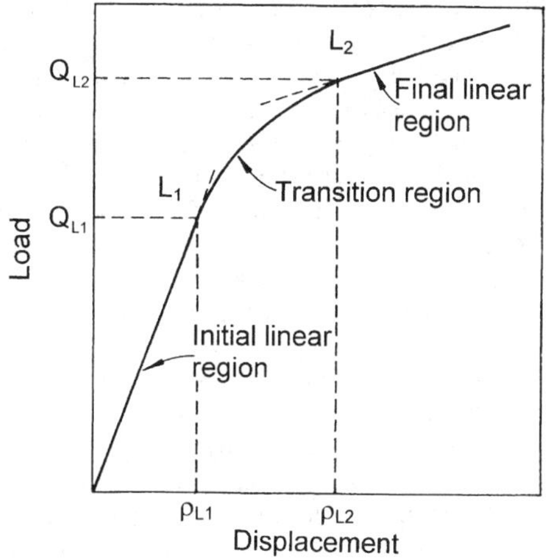

FIG. 2. Regions of Load-Displacement Curve

subject to error, which will depend upon the geotechnical conditions (soil type, creep characteristics, etc.) and the care exercised during the conduct of the test. Because of these problems, it would be desirable to use an interpretation method that gives a "failure" load close to the upper limit of the transition region of the load-displacement curve.

In Fig. 2, points L_1 and L_2 essentially designate the "elastic limit" and the "failure threshold", respectively, with failure being defined qualitatively as the load beyond which a small increase in load produces a significant increase in displacement. These L_1 and L_2 points are evaluated by graphical interpretation of the load-displacement curves. With good-quality field data, L_1 normally can be interpreted in a straightforward manner. However, interpretation of L_2 sometimes is difficult because of the problems cited above and the need for high-quality data at relatively large displacements.

Because of this difficulty with some data, it is useful to consider the basic interpretation concept proposed by Davisson (1972) for driven steel piles in compression. The Davisson concept was to plot through the origin the pile elastic compression line, PD/AE, in which P = load, D = depth, A = area, and E = Young's modulus. A second line would be drawn parallel to the elastic line, at an offset of 0.15 inches (4 mm) + B (in inches or mm)/120. The intersection of this offset line with the load-displacement curve defines the "Davisson load", Q_D. This load is higher than Q_{L1} and lower than Q_{L2}, and it always occurs well within the transition region in Fig. 2, so it is a low "failure" value. However, this is a useful predictive method and can be applied to nearly all load test data. By contrast, the graphical L_2 interpretation is a back-plotting method that requires relatively high-quality data at larger displacements. Therefore, it would be useful to correlate Q_D and Q_{L2} so that Q_D could be used to estimate the "interpreted failure load" when the large displacement data are poor or non-existent. For drilled shafts, Hirany and Kulhawy (1989) found that Q_{L2}/Q_D averaged 1.14 over a range of soil and loading conditions. However, subsequent studies at Cornell have shown that the elastic line is not generally appropriate for drilled foundations because it is too steep at lower D/B ratios. Instead, the initial slope line should be plotted tangent to the data, with the same offset then applied to this line. This method defines the "compression slope tangent load", Q_{STC}, which is used in place of Q_D. For the Hirany and Kulhawy database, Q_{STC} and Q_D are similar.

For uplift loading, a similar slope tangent concept was used (Kulhawy et al. 1983; O'Rourke & Kulhawy 1985). For $D/B > 10$ or so, this slope tangent line is basically equal to the elastic line. However, for $D/B < 10$, the elastic line is steeper. This slope tangent line is offset by 0.15 inches (4 mm), and the intersection of the offset line with the load-displacement curve defines the "uplift slope tangent load", Q_{STU}. Again, this load is higher than Q_{L1} and lower than Q_{L2}, and it always occurs well within the transition region in Fig. 2, so it is a low "failure" value. However, this method also is a useful predictive method and can be applied to nearly all load test data. Therefore, it would be useful to correlate

Q_{STU} and Q_{L2} so that Q_{STU} could be used to estimate the "interpreted failure load" when the large displacement data are poor or non-existent. For drilled shafts, Kulhawy and Hirany (1989) found that Q_{L2}/Q_{STU} averaged 1.13 over a range of soil and loading conditions.

Fig. 3 illustrates these loads for a PIF in uplift. As can be seen, the initial slope line is plotted tangent to the data and L_1 is defined. Then the final slope line is plotted and L_2 is defined. Finally, the intersection of the offset line with the curve determines the "uplift slope tangent load", Q_{STU}.

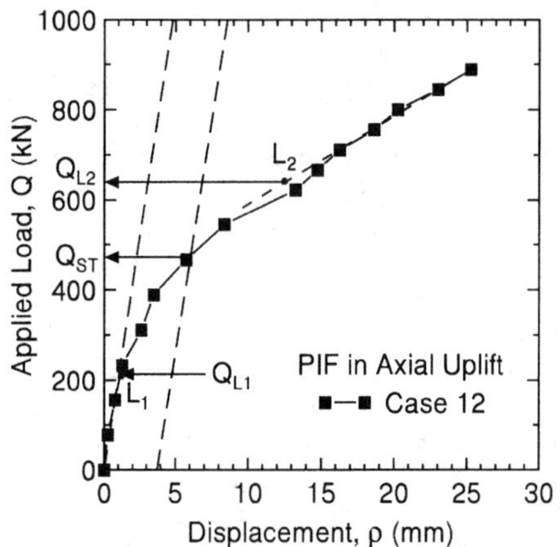

FIG. 3. Illustration of Interpreted Failure Loads (Chen & Kulhawy 2002)

DRILLED SHAFTS

Drilled shafts are constructed by augering a cylindrical shaft into the ground to the design depth, with or without wall support. After inserting the reinforcing steel cage, as needed, the shaft is concreted. The overall behavior of drilled shafts was our initial focus in deep foundation research. Subsequently the data and evaluations are most comprehensive for this type of drilled foundation, which will be the "baseline" against which the other foundations will be evaluated. For reference, 234 tests were evaluated for capacity and 188 of these were evaluated for displacements, under both drained and undrained loading, in uplift and compression. For the uplift tests, B varied from 0.20 to 1.62 m, with a mean of 0.64 m, and D/B varied from 2 to 42, with a mean of about 9; for the compression tests, B varied from 0.18 to 2.00 m, with a mean of 0.81 m, and D/B varied from 4 to 64, with a mean of about 18.

Capacity Evaluation

The capacity of drilled shafts was evaluated in some detail by Chen and Kulhawy (1994, 2002) and Kulhawy and Chen (2003), using a substantial load test data base. Following the basic equilibrium equations (Eq. 1 and 2), there was general agreement with Eqs. 3 through 12, but there were some refinements warranted for specific terms.

For side resistance in drained loading, $\delta/\phi' = 1$ because it is a cast-in-place soil-concrete interface. Assessment of the stress factor $(K/K_o)_e$ has evolved with time. Initial examination of the available load test data (Kulhawy et al. 1983) showed that the stress factor normally varied between 2/3 and 1 and is a function of the construction method and its influence on the in-situ stress. For dry construction, minimal sidewall disturbance, and prompt concreting, the soil disturbance is minimized and $(K/K_o)_e$ is about 1. For slurry or wet-hole construction, using proper construction techniques, the soil disturbance also is minimized, and therefore $(K/K_o)_e$ still is close to 1. However, when slurry or wet-hole procedures are not applied properly, the soil stress may relax significantly, and therefore $(K/K_o)_e$ could be reduced to about 2/3. Casing construction under water may be intermediate between these ranges. Subsequent Chen-Kulhawy examinations showed that, for properly constructed shafts by whatever means, the stress factor is essentially 1. Only when poor construction practices are employed, or unusual ground conditions are encountered, should the stress factor be less than 1.

For side resistance in undrained loading, α factors have been calibrated specifically for drilled shafts, and therefore $(K/K_o)_t = 1$. Early relationships were proposed by Stas and Kulhawy (1984) and Kulhawy and Stas-Jackson (1989), before it was possible to normalize the α and s_u data to standardized reference values. Subsequent studies have developed these standardized α-s_u relationships calibrated to the CIUC triaxial test, as given below (Chen & Kulhawy 1994; Kulhawy & Chen 2003 for Eq. 13 a,b, and Kulhawy & Phoon 1993 for Eq. 13c):

arithmetic $$\alpha_{CIUC} = 0.31 + 0.17 / [s_u(CIUC)/p_a] \quad (13a)$$

logarithmic $$\alpha_{CIUC} = 0.52 - 0.51 \log [s_u(CIUC)/p_a] \quad (13b)$$

exponential $$\alpha_{CIUC} = 0.5 / [s_u(CIUC)/p_a]^{0.5} \quad (13c)$$

in which p_a = atmospheric stress in the desired units. For these three relationships, $r^2 \approx 0.7$. For $0.25 < s_u/p_a < 2.5$ or so, all three equations give very similar results. For $s_u/p_a > 2.5$, Eq. 13c may be the most appropriate, followed by Eq. 13a.

For tip resistance, the procedure is outlined with Eqs. 6 through 12, using the geotechnical properties based on the initial in-situ conditions.

Capacities calculated with these procedures agreed well, on average, with the interpreted values from field tests, with means varying by 5% or less.

Load-Displacement Evaluation

Fig. 4 shows the average interpreted load-displacement curves for drilled shafts in axial compression, based largely on the Hirany and Kulhawy (1988) and Chen and Kulhawy (1994) data bases. This figure shows that $Q_{L2}/Q_{L1} \approx 2$, $\rho_{L2} \approx 4\%$ B, and $\rho_{L1} \approx 0.4\%$ B. [Re-examination of the original data indicates an average $\rho_{L2} \approx$ 25-30 mm.] In addition, $Q_{L2}/Q_{STC} = 1.14$. Further examination of the figure shows that, at L_1, about 89% of Q_{L1} is supported in side resistance, with only 11% in tip resistance, for both undrained loading of cohesive soils and drained loading of cohesionless soils. At L_2 and larger displacements, the side resistance was about 76% of Q_{L2} and is essentially given by Eqs. 3 through 5. For undrained loading, the remaining 24% of Q_{L2} is equal to the computed tip capacity, given by Eq. 10. However, for drained loading, only 29% of the computed tip capacity is mobilized at L_2. The tip must be displaced about 10% of the diameter B before the full tip capacity is mobilized. This large displacement has important design ramifications, because most designs will be acceptable only at much smaller displacements, and therefore a rather large factor of safety must be employed on tip resistance in drained loading to control displacements. Further discussion is given by Chen and Kulhawy (1994, 2002).

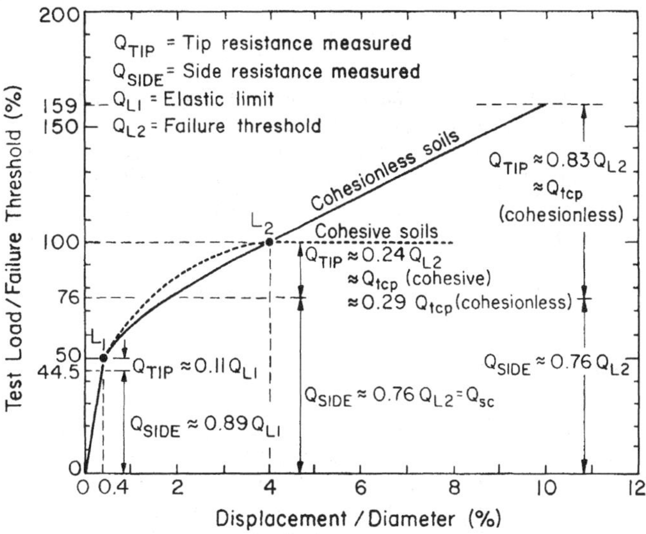

FIG. 4. Average Interpreted Load-Displacement Curves for Drilled Shafts in Axial Compression (Chen & Kulhawy 2002)

For uplift loading, the tip resistance is small or negligent, so the load is supported by the shaft weight and side resistance, which is equal to Q_{STU}. The side resistance in uplift and compression is the same. In addition, $Q_{L2}/Q_{L1} \approx 2.4$, $Q_{L2}/Q_{STU} = 1.13$, and ρ_{L2} occurs at about 0.5 inch (13 mm) displacement. Initially, ρ_{L1} was not evaluated because it was so small and of little design consequence. Table 1 summarizes these compression and uplift results across the first row, designated for References "c".

Subsequent studies by Cushing (2001) and Cushing and Kulhawy (2001, 2002) amplified these results. As shown in Table 1a, rows 2 through 4, there is some difference between the undrained and drained load ratios, separately and grouped together. However, one can still see that $Q_{L2}/Q_{L1} \approx 2$ to 2.5, with undrained on the low side and drained on the high side. Rows 2 and 3 in Table 1b give expanded data for ρ_{L1}. Evaluation of these data indicates that undrained and drained displacements are similar, ρ_{L2} in compression $\approx$ 4% B, ρ_{L2} in uplift $\approx$ 13 mm (0.5 inch), and $\rho_{L1} \approx 0.1\ \rho_{L2}$ in both compression and uplift.

It should be noted that the values in Table 1 are mean values developed from substantial populations covering a range of soil types and shaft geometries. The coefficient of variation (COV) on these means is in the range of 15 to 40% for loads and 40 to 80% for displacements. Although the COVs on displacements are relatively large, the absolute values of the displacements are relatively small.

AUGERED CAST-IN-PLACE PILES

Augered cast-in-place (ACIP) piles are constructed by drilling a hollow stem continuous flight auger into the ground to the design depth. Then, as the auger is withdrawn steadily, sand-cement grout or concrete is pumped down the hollow stem and reinforcing may be installed, as needed. Design of these foundations is based largely on empirical correlations, and no rigorous analytical models exist. The data also are limited. Chen (1998) evaluated 56 tests for capacity and displacements, under drained compression only. For these tests, B varied from 0.36 to 0.61 m, with a mean of 0.44 m, and D/B varied from 5 to 68, with a mean of about 24.

Capacity Evaluation

ACIP piles are somewhat similar to drilled shafts. The shape is cylindrical, B is generally smaller, and D/B is somewhat larger. During construction, the wall is always supported, and grouting traditionally is done at approximately gravity stresses. With these similarities, the capacity evaluation should follow the drilled shaft methodology. Following the basic equilibrium equation (Eq. 1), there was general agreement with Eq. 3, but there were some refinements warranted for specific terms.

For side resistance in drained loading, $\delta/\phi' = 1$ because it is a cast-in-place soil-

Table 1. Mean Normalized Loads and Displacements

a. Mean Normalized Loads

Fndn Type & Ref[a]	Mode[b]	Compression Tests		Uplift Tests	
		Q_{L2}/Q_{L1}	Q_{L2}/Q_{STC}	Q_{L2}/Q_{L1}	Q_{L2}/Q_{STU}
DS[c]	U+D	2.06	1.14[h]	2.44	1.13
DS[d]	U	1.89	-	1.79	-
DS[d]	D	2.75	-	2.27	-
DS[d]	U+D	2.43	-	2.10	-
ACIP[e]	D	~2.1	1.22	-	-
PIF[f]	D	~2.1-2.5	1.18	~2.1	1.20
MP[g]	U+D	~2.1	1.18	-	-

b. Mean Normalized Displacements

Fndn Type & Ref[a]	Mode[b]	Compression Tests				Uplift Tests			
		ρ_{L1}		ρ_{L2}		ρ_{L1}		ρ_{L2}	
		mm	% B	mm	% B	mm	% B	mm	% B
DS[c]	U+D	-	~0.4	~25-30	~4.0	-	-	~13	-
DS[d]	U	3.6	0.6	-	-	1.4	0.2	-	-
DS[d]	D	4.9	0.6	-	-	1.5	0.2	-	-
ACIP[e]	D	2.8	0.7	14.0	3.4	-	-	-	-
PIFuc[f]	D	~3	~0.6	~22	~4.0	1.4	0.3	14.0	2.6
PIFc[f]	D	~4	~1.0	~26	~6.3	-	-	-	-
MP[g]	U	~3	~2	17.0	10.5	-	-	-	-
MP[g]	D	~3	~2	20.0	12.0	-	-	-	-

a - DS - drilled shaft; ACIP - augered cast-in-place pile; PIF - pressure-injected footing (uc - uncased, c - cased); MP - micropile
b - U - undrained; D - drained
c - Hirany & Kulhawy (1988, 1989); Kulhawy & Hirany (1989)
d - Cushing (2001); Cushing & Kulhawy (2001, 2002)
e - Chen (1998)
f - Chen (1998); Chen & Kulhawy (2001, 2002, 2003)
g - Kulhawy & Jeon (1999); Jeon & Kulhawy (2001)
h - value for Q_{L2}/Q_D (Q_D and Q_{STC} are similar in this database)

concrete interface. As with drilled shafts, the stress factor $(K/K_o)_e$ should be about 1. Only when poor construction practices are employed, or unusual ground conditions are encountered, should the stress factor be less than 1. If the grout is pressurized during auger withdrawal, $(K/K_o)_e$ could exceed 1, but there are no data on this issue at present.

Because of the relatively small diameter B and large D/B, the tip resistance commonly is ignored. However, for smaller D/B, the tip resistance could be significant.

Following these procedures, it was found that the mean calculated Q_{sc} / field interpreted $Q_{L2} = 0.92$.

Load-Displacement Evaluation

Within the general framework of Fig. 4, the Chen studies found that $Q_{L2}/Q_{L1} \approx 2.1$, $\rho_{L2} = 3.4\%$ B = 14.0 mm, and $\rho_{L1} = 0.7\%$ B = 2.8 mm. In addition, $Q_{L2}/Q_{STC} = 1.22$. Table 1 summarizes these results in the ACIP rows. These values are approximately equal to those for drilled shafts and, at most, suggest a slightly stiffer load-displacement response for ACIP piles. However, it should be noted that, for the slope tangent interpretation, the elastic slope is greater than the initial slope for D/B < 20 to 25. For D/B > 20 to 25, these slopes are essentially the same. For drilled shafts in uplift, the transition occurred at about D/B = 10.

PRESSURE-INJECTED FOOTINGS

Pressure-injected footings (PIFs) are constructed by augering a straight shaft, with or without casing, creating an expanded base (generally by mechanical compaction), and concreting the base and shaft. As with the ACIP piles, the design of these foundations is based largely on empirical correlations, and no rigorous analytical models exist. Comprehensive data also are limited. Chen (1998) and Chen and Kulhawy (2001, 2002, 2003) evaluated 25 tests for capacity and displacements, under drained uplift loading. For these tests, B_s varied from 0.41 to 0.56 m, with a mean of 0.53 m, and D_b/B_s varied from 8 to 42, with a mean of about 16. B_s is the shaft diameter and D_b is the depth to the center of the expanded base, assuming a sphere. From the available data, it appeared that B_b (expanded base diameter) $\approx 2\ B_s$.

In addition, Chen (1998) evaluated 215 tests under drained compression loading. These tests only had limited information, so a detailed capacity evaluation was not possible. However, they were useful for general behavior and inference. For the 99 uncased tests, B_s varied from 0.41 to 0.64 m, with a mean of 0.55 m, and D_b/B_s varied from 4 to 21, with a mean of about 12; for the 116 cased tests, B_s varied from 0.27 to 0.53 m, with a mean of 0.41 m, and D_b/B_s varied from 7 to 64, with a mean of about 22.

Capacity Evaluation

PIFs have fewer similarities to drilled shafts than ACIP piles. The upper shaft is cylindrical, B is generally smaller, D/B is somewhat larger, and the shaft is concreted as a drilled shaft. However, the base construction is very different. Still, the capacity evaluation should follow broadly the general drilled shaft methodology. Following the basic equilibrium equation (Eq. 1), there was general agreement with Eqs. 3, 6, and 9, but there were appropriate modifications warranted to address the construction and geometry differences.

For side resistance in drained loading, $\delta/\phi' = 1$ for an uncased shaft because it is a cast-in-place soil-concrete interface. For cased shafts, $\delta/\phi' \approx 0.8$ for corrugated steel and ≈ 0.6 for smooth steel. Then the different characteristics of PIFs need to be addressed.

The Chen-Kulhawy studies identified three important characteristics of PIFs: compaction/strength effect, shape effect, and stress effect. First, the base construction alters the soil density and subsequently the strength for capacity evaluation. A "compacted ring" model was developed that could estimate the density increase around the shaft, which then is used to estimate the strength increase. Second, the base effect is important for shallower PIFs and can be addressed by assuming that the shaft and base are modeled by an equivalent shaft of uniform diameter, $B_e = B_s + (B_b - B_s)/3$, which would be applicable for $D_b/B_s < 15$. For deeper PIFs, the base effect is minimal, and therefore $B_e = B_s$. Third, the stress effect also is important for shallower PIFs and can be addressed by a stress factor $(K/K_o)_e = 1 + (15 - D_b/B_s)/4$ for $D_b/B_s < 15$. Alternatively, a conservative $(K/K_o)_e \approx 1.5$ can be selected. For deeper PIFs, the stress effect is minimal, and therefore $(K/K_o)_e = 1$.

Following these procedures, it was found that the mean calculated Q_u / field interpreted $Q_{L2} = 0.90$.

For the compression tests, the methodology developed from the uplift tests was applied directly. The side resistance in uplift and compression was considered to be equal. The tip resistance in compression considers the strength increase from compaction and uses B_b.

Unfortunately, reasonable comparisons of computed and measured values can not be made for these tests because there was no instrumentation to evaluate side and tip resistances separately. The results seem to follow the behavior described previously for drilled shafts, with the side resistance being a very small percentage of the capacity, and the measured capacity (Q_{L2}) being much less than the predicted, at the small L_2 displacements. These issues can only be resolved with large displacement data and instrumented tests that differentiate the side and tip resistances.

Load-Displacement Evaluation

Within the general framework of Fig. 4, it was found for the uplift tests that $Q_{L2}/Q_{L1} \approx 2.1$, $\rho_{L2} = 2.6\%$ B = 14.0 mm, and $\rho_{L1} = 0.3\%$ B = 1.4 mm. In addition, $Q_{L2}/Q_{STU} = 1.20$. For the compression tests, the results were more approximate and showed Q_{L2}/Q_{L1} on the order of 2.1 - 2.5, $\rho_{L2} \approx 4.0\%$ B = 22 mm (uncased) and ≈ 6.3% B = 26 mm (cased), and ρ_{L1} ≈ 0.6% B = 3 mm (uncased) and ≈ 1.0% B = 4 mm (cased). In addition, Q_{L2}/Q_{STC} ≈ 1.18. Table 1 summarizes these results in the PIF rows. These values are comparable to those for drilled shafts. For uplift with $D_b/B_s < 15$, the PIF is a bit stiffer, but they are similar at greater depths. For compression, the behavior seems similar. However, it should be noted that, for the slope tangent interpretation in compression, the elastic slope is greater than the initial slope for D/B < 20 to 25. For D/B > 20 to 25, the elastic and initial slopes are essentially the same. For the slope tangent interpretation in uplift, the transition is at D/B =10, which is similar to the drilled shafts.

MICROPILES

Micropiles are constructed by drilling a relatively small-diameter hole with casing to the design depth. After installing a central reinforcing bar, the casing is withdrawn while the hole is pressure-grouted. Design of these foundations is based largely on empirical correlations, and no rigorous analytical models exist. The data also are limited. Jeon and Kulhawy (2001) evaluated 21 tests for capacity and displacements, under both drained and undrained compression loading. For these tests, B varied from 0.14 to 0.18 m, with a mean of 0.16 m. For the 13 drained tests, D/B varied from 44 to 139, with a mean of about 77. For the 8 undrained tests, D/B varied from 60 to 200, with a mean of about 107.

Capacity Evaluation

Micropiles also are somewhat similar to drilled shafts. The shape is cylindrical, B is much smaller, D/B is much larger, pressure grouting is used, and there may be axial flexibility because of the large D/B ratios. With these characteristics, the capacity evaluation generally should follow the drilled shaft methodology. Following the basic equilibrium equations (Eq. 1 and 2), there was general agreement with Eq. 3 through 5, but there were some refinements warranted for specific terms.

For side resistance in drained loading, $\delta/\phi' = 1$ because it is a cast-in-place soil-concrete interface. Because of pressure grouting during installation, the stress factor $(K/K_o)_e$ increases and is in the range of 1.5 to 2, with values up to 6, for D/B < 100 or so. At greater depths, $(K/K_o)_e \approx 1$. For side resistance in undrained loading, the α in Eq. 13 can be used. However, because of the grouting during installation, the stress factor $(K/K_o)_t$ increases and is in the range of 1.25 to 1.5, with values up to 2.5, for D/B < 100 or so. At greater depths, $(K/K_o)_t \approx 1$.

Because of the relatively small diameter B and large D/B, the tip resistance commonly is ignored.

With the limited data, and the stress factors being developed from these data, a comparison of calculated and measured capacities is not appropriate.

Load-Displacement Evaluation

Within the general framework of Fig. 4, the Jeon-Kulhawy studies found that $Q_{L2}/Q_{L1} \approx 2.1$, ρ_{L2} = 10-12% B = 17-20 mm, and $\rho_{L1} \approx$ 2% B = 3 mm. In addition, Q_{L2}/Q_{STC} = 1.18. Table 1 summarizes these results in the MP rows. These load ratios and elastic displacements are similar to those for drilled shafts, but the L_2 displacements are larger as a % B, possibly suggesting more axial flexibility. However, the absolute values of ρ_{L2} are comparable. Also, the undrained case is slightly stiffer. Furthermore, it should be noted that, for the slope tangent interpretation, the elastic and initial slopes are similar.

SUMMARY

Four representative types of drilled foundations (drilled shafts, augered cast-in-place piles, pressure-injected footings, and micropiles) exhibit many similarities from the standpoint of axial capacity evaluation, generalized behavior, and normalized load-displacement response. The capacities can be computed rationally, taking into account the particular characteristics of each foundation type, and the results agree well with the interpreted failure load Q_{L2} from field load tests.

For general load behavior, on average, $Q_{L2}/Q_{L1} \approx$ 2 to 2.5, and both Q_{L2}/Q_{STC} and $Q_{L2}/Q_{STU} \approx$ 1.13 to 1.22. The latter result will be particularly useful for interpreting load tests conducted to limited displacements.

For general displacement behavior, in compression, $\rho_{L2} \approx$ 4% B, except for micropiles at 10-12% B, and $\rho_{L1} \approx$ 0.1 to 0.2 ρ_{L2}. However, the absolute values of ρ_{L2} are comparable. In uplift, $\rho_{L2} \approx$ 13 mm (0.5 inch) and $\rho_{L1} \approx$ 0.1 ρ_{L2}. The absolute value of ρ_{L1} is always small, typically < 5 mm.

These patterns suggest the following conclusion for drilled foundations in soil, on average. If these foundations are designed for load levels $\leq Q_{L1}$, then the factor of safety will be $\geq Q_{L2}/Q_{L1}$ ($\approx$ 2 to 2.5), the displacements will be $\leq \rho_{L1}$, and the load-displacement response will be essentially linear.

ACKNOWLEDGMENTS

I thank J-R Chen, AG Cushing, A Hirany, and JP Turner for their detailed review comments that helped to improve this final version.

REFERENCES

Chen, J-R (1998). "Case History Evaluation of Axial Behavior of Augered-Cast-In-Place Piles & Pressure-Injected Footings", *MS Thesis*, Cornell University, Ithaca, NY.

Chen, J-R & Kulhawy, FH (2001). "Compaction Effects Induced in Cohesionless

Soil by Installation of Pressure-Injected Footings", *Foundations & Ground Improvement (GSP 113),* Ed. TL Brandon, ASCE, Reston, 230-244.

Chen, J-R & Kulhawy, FH (2002). "Axial Uplift Behavior of Pressure-Injected Footings in Cohesionless Soil", *Deep Foundations 2002 (GSP 116)*, Ed. MW O'Neill & FC Townsend, ASCE, Reston, 1275-1289.

Chen, J-R & Kulhawy, FH (2003). "Significance of Construction Effects on Uplift Behavior of Drilled Foundations", *Proc., 12th Asian Regional Conf. Soil Mech. & Geotech. Eng.,* Singapore, in press.

Chen, Y-J & Kulhawy, FH (1993). "Undrained Strength Interrelationships Among CIUC, UU & UC Tests", *J. Geotech. Eng.*, ASCE, 119(11), 1732-1750.

Chen, Y-J & Kulhawy, FH (1994). "Case History Evaluation of Behavior of Drilled Shafts Under Axial & Lateral Loading", *Rpt TR-104601*, EPRI, Palo Alto.

Chen, Y-J & Kulhawy, FH (2002). "Evaluation of Drained Axial Capacity of Drilled Shafts", *Deep Foundations 2002 (GSP 116)*, Ed. MW O'Neill & FC Townsend, ASCE, Reston, 1200-1214.

Cushing, AG (2001). "Small-Strain Elastic Deformation of Drilled Shafts in Axial Uplift & Compression Loading", *MS Thesis*, Cornell University, Ithaca, NY.

Cushing, AG & Kulhawy, FH (2001). "Undrained Elastic Behavior of Drilled Shaft Foundations in Cohesive Soils", *Proc., 15th Intl. Conf. Soil Mech. & Geotech. Engrg.(2)*, Istanbul, 873-876.

Cushing, AG & Kulhawy, FH (2002). "Drained Elastic Behavior of Drilled Shafts in Cohesionless Soils", *Deep Foundations 2002 (GSP 116)*, Ed. MW O'Neill & FC Townsend, ASCE, Reston, 22-36.

Davisson, MT (1972). "High Capacity Piles", *Proc. Lecture Series on Innovations in Foundation Construction*, ASCE Illinois Section, Chicago, 52 p.

Hansen, JB (1970). "Revised & Extended Formula for Bearing Capacity", *Bulletin 28*, Danish Geotech. Inst., Copenhagen, 5-11.

Hirany, A & Kulhawy, FH (1988). "Conduct & Interpretation of Load Tests on Drilled Shaft Foundations: Detailed Guidelines", *Rpt EL-5915(1)*, EPRI, Palo Alto, CA.

Hirany, A & Kulhawy, FH (1989). "Interpretation of Load Tests on Drilled Shafts - Pt 1: Axial Compression", *Fndn. Engrg.: Current Principles & Practices (GSP 22)*, Ed. FH Kulhawy, ASCE, New York, 1132-1149.

Hirany, A & Kulhawy, FH (2002). "On the Interpretation of Drilled Foundation Load Test Results", *Deep Foundations 2002 (GSP 116)*, Ed. MW O'Neill & FC Townsend, ASCE, Reston, 1018-1028.

Jeon, SS & Kulhawy, FH (2001). "Evaluation of Axial Compression Behavior of Micropiles", *Fndns. & Ground Improvement (GSP 113),* Ed TL Brandon, ASCE, Reston, 460-471.

Kulhawy, FH (1991). "Drilled Shaft Foundations", Chap. 14 in *Fndn. Eng. Handbook (2/E)*, Ed. HY Fang, Van Nostrand Reinhold, New York, 537-552.

Kulhawy, FH & Chen, Y-J (2003). "Evaluation of Undrained Side & Tip

Resistances for Drilled Shafts", *Proc., 12th Pan-Am Conf. Soil Mech. & Geotech. Eng.,* Cambridge, MA, in press.

Kulhawy, FH & Hirany, A (1989). "Interpretation of Load Tests on Drilled Shafts - Pt 2: Axial Uplift", *Fndn. Engrg.: Current Principles & Practices (GSP 22),* Ed. FH Kulhawy, ASCE, New York, 1150-1159.

Kulhawy, FH & Jeon, SS (1999). "Some Observations on Axial Compression Behavior of Micropiles", *Proc., 2nd Intl. Workshop on Micropiles*, Ube (Japan), 89-92.

Kulhawy, FH & Mayne, PW (1990). "Manual on Estimating Soil Properties for Foundation Design", *Rpt EL-6800*, EPRI, Palo Alto.

Kulhawy, FH & Phoon, KK (1993). "Drilled Shaft Side Resistance In Clay Soil to Rock", *Design & Performance of Deep Foundations: Piles & Piers in Soil & Soft Rock (GSP 38),* Ed. PP Nelson, TD Smith & EC Clukey, ASCE, New York, 172-183.

Kulhawy, FH & Stas-Jackson, C (1989). "Some Observations on Undrained Side Resistance of Drilled Shafts", *Fndn. Eng.: Current Principles & Practices (GSP 22)*, Ed. FH Kulhawy, ASCE, New York, 1011-1025.

Kulhawy, FH, Trautmann, CH, Beech, JF, O'Rourke, TD, McGuire, W, Wood, WA & Capano, C (1983). "Transmission Line Structure Foundations for Uplift - Compression Loading", *Rpt EL-2870*, EPRI, Palo Alto.

Kulhawy, FH, Trautmann, CH & Hirany, A (2002). "Overview of Some EPRI Research for Transmission Line Structure Foundations", *Electrical Transmission in the New Age*, Ed. DE Jackman, ASCE, Reston, 282-291.

Ladd, CC, Foote, R, Ishihara, K, Schlosser, F & Poulos, HG (1977). "Stress-Deformation & Strength Characteristics", *Proc., 9th Intl. Conf. Soil Mech. & Fndn. Eng. (2)*, Tokyo, 421-494.

O'Rourke, TD & Kulhawy, FH (1985). "Observations on Load Tests for Drilled Shafts", *Drilled Piers & Caisssons II*, Ed CN Baker, ASCE, New York, 113-128.

Stas, CV & Kulhawy, FH (1984). "Critical Evaluation of Design Methods for Foundations Under Axial Uplift & Compression Loading", *Rpt EL-3771*, EPRI, Palo Alto.

Vesic, AS (1975). "Bearing Capacity of Shallow Foundations", Chap. 3 in *Fndn. Eng. Handbook*, Ed. H Winterkorn & HY Fang, Van Nostrand Reinhold, New York, 121-147.

"VALUE ENGINEERING…?" DURING CONSTRUCTION

Maral Papazian Bedian[1], P.E., Member, ASCE

ABSTRACT: Construction contracts in the USA frequently contain a clause on "Value Engineering" — an interesting and curious clause — allowing contractor-initiated design changes. Misleading is the interpretation of value engineering to imply cost savings shared with the owner, and its implementation, just before or during actual construction, is problematic. It is not surprising that such a clause would simply be ignored because it involves changes in design, often major changes in very short time; and change is feared, and moreover, vehemently resisted by all parties, owner, designer, and contractor. The problem may lie in the divergence and separation of the two entities, designer/engineer and builder/contractor, assigned to the "one" engineering project. Their priorities and incentives are very different. The engineer spends years, even decades, in design and prepares contract documents often without a deep-seated understanding of construction methods, including geotechnical construction. Even worse, given the tremendous computational advancements, the designer submits plans that are inexcusably exaggerated code-based design with excessive safety factors. The contractor on the other hand, often builds, without appreciation of design principles or regard for design engineers. Owner budget and schedule constraints (not commensurate with his demands) and the ever-increasing litigious climate have exacerbated the situation. Adverse and hostile relationship between the various groups is often the norm resulting in extended disputes and claims, not to mention the excessive costs these entail. Redesign to apply a new technology or optimization of an inferior design just before construction becomes unthinkable. Major design changes were nevertheless successfully implemented in record speed on several very large projects in the metropolitan New York area. Four case histories spanning from 1998 to 2002 are presented.

[1] Chief Geotechnical Engineer, Perini Corp. Civil Construction, Peekskill, New York
mbedian@perini.com

INTRODUCTION

Four case histories are presented in this paper where major design changes were implemented during construction. They include two Design/Build projects: **Case History 1)** the redesign of large diameter drilled shafts for the Hudson-Bergen Light Rail Transit System (a $1.1 billion total value and $343 million for the Initial Operating Segment for this case); **Case History 2)** the elimination of deep caissons in favor of spread footings for a new $90 million MTA bus depot in Manhattan; and, two conventional Design/Bid/Build projects: **Case History 3)** maintaining in lieu of removal of a 100-year old abutment of the $72 million Queens Boulevard Bridge Replacement; and **Case History 4)** the complete redesign of major retaining walls and actual use for the first time outside Asia, of the "Giken" tubular pressed-in pipe piles as very high cantilever retaining walls, for the $150 million expansion of the Long Island Expressway.

Before describing the four case histories, it is noted that projects of such large magnitude involve armies of people with different backgrounds. For many, their basis of experience in design and/or construction is largely to follow codes and specifications and paper tracking is the occupation for many more. In fact, most of the parties involved in large projects do not feel the need nor have the incentive to consider cost-effective solutions. Often, they are not even aware that changes are necessary or possible. In their attempt to be "safe", most do not advocate or employ the very advances that are presented and discussed in journal articles and conferences. Moreover, in a litigious society such as in the USA, liability concerns impede innovation, much less implementing design changes once construction has begun. Owner budget constraints — not commensurate with his demands — and fast-track schedules — often unrealistic — have exacerbated the situation.

It is perhaps helpful to mention that the "valuable engineering" design changes for the cases described herein, are the personal account of a geotechnical engineer "defecting" to the construction side where the possibility for adequate change could be detected, combined with the least welcome attitude of a female passion and insistence. The process of change was unconventional, painful, and even comical.

CASE HISTORY 1 – Hudson-Bergen Light Rail Transit System, New Jersey: Redesign of Newport Viaduct Drilled Shaft Foundations

The Project

The Hudson-Bergen Light Rail Transit System (HBLRTS) is a 33 Km Design-Build-Operate-Maintain (DBOM) light rail project in northern New Jersey. The Initial 16 Km of the project extends through Jersey City and includes a 0.8 Km long elevated Viaduct, a multiple-span bridge carrying the light rail in a north-south direction just west of the Hudson River. Near the mid-length, the bridge crosses over the entrance to Holland Tunnel leading to Manhattan, New York.

For the multi-span Newport viaduct, deep foundations consisting of drilled shafts were selected with loads ranging from 4,450 KN to 16,300 KN (500 to 1,800 tons) from each pier. (In general, the design/build team followed the original bid reference documents including project design criteria and mandatory codes and specifications.) The initial design required the drilled shafts to extend through a thick zone of "completely weathered rock" and then be socketed into "sound" sandstone. The assigned design parameters were generally consistent with the bid reference documents, a unit end bearing resistance for rock of 0.8 MPa (8 tsf), increased to 1.2 MPa (12 tsf), after minimum 3 m penetration into "sound" bedrock. For "sound" rock, an allowable unit shaft resistance of 275 KPa (2.9 tsf) was specified.

The subsurface conditions along the bridge alignment consist of an upper 6 m thick granular fill over 3 m soft marine clay, underlain by 12 m of medium dense to very dense glacial deposits comprised of alternating layers of silty sands with gravel to clayey silts. Weathered sandstone, with an average thickness of about 7 m and described in the test boring logs as "completely weathered rock", extends below the glacial deposits with SPT-N values of 100 blows for only 25 mm to 150 mm penetration. Sandstone bedrock exists below a depth of about 28 m. The groundwater table at this site is shallow and within 3 m of the ground surface (Figure 1.1).

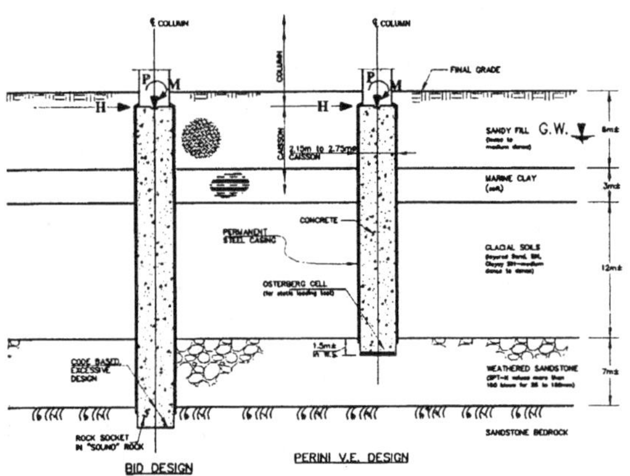

FIG 1.1. Typical Subsurface Section with Drilled Shafts

The Change

The initial design for the drilled shaft foundations supporting the viaduct was questioned by this author, given the highly variable nature of the sandstone rock and the implications of searching for "sound" sandstone at great depths. Encouraged by recent studies of intermediate geomaterials (O'Neill et al., 1996), it was deemed unnecessarily conservative to bypass the weathered sandstone in search of deeper "sound" rock. Also, a report (Baker, 1988) on foundations for an adjacent 30-story office tower supported on 1.2 m diameter drilled shafts in the same "completely weathered rock" zone showed resistances that are much higher than the code-based values. The report contained two conventional head-down static loading tests that measured unit toe resistance ranging from 7.8 MPa (80 tsf) @ 35 mm movement, to 10.4 MPa (108 tsf) @ 15 mm movement. The higher unit toe resistance was measured in the upper parts of the weathered layer having SPT N-indices over 100 for 150 mm penetration. From the same loading tests, a unit shaft resistance of 2.2 MPa (23 tsf) was deduced for the same weathered material. With this and other design information at hand, the code-based design was rejected and higher design values were implemented subject to verification in full-scale static loading tests. The intent of the testing was to demonstrate that the drilled shafts can be supported on top of or just within the "Weathered Sandstone", believed to provide adequate shaft and toe resistances to support the bridge column loads. More important, the testing was also intended to evaluate the impact of the proposed construction procedures on the axial capacity of the drilled shafts. Because of the large design loads, the Osterberg-cell (O-cell) test method was selected.

It is perhaps not a surprise that the above change was strongly resisted by all parties involved, including the designers, whose "safe" code design in "sound" rock was challenged. What was not anticipated, at least by the author, was the drilled-shaft subcontractor's reluctance to accept "reduced drilling quantities". The first loading test with the O-cell was not successful because of premature mobilization of shaft resistance caused by excessive drilling disturbances. However, after making all the necessary adjustments/refinement in the construction procedures, the second loading test, which immediately followed, achieved the intended "valuable" engineering change, as described below.

Testing Program

Principles of the O-cell method can be found elsewhere (Osterberg 1994, Schmertmann 1997, Fellenius 2001). In summary, the testing is conducted from bottom up with the use of a hydraulic jack—i.e. the Osterberg Cell, placed near the base of the shaft to be tested. As the O-cell expands, it pushes the shaft upward and the base downward. Unlike the classic head-down static loading test, the O-cell allows the separate measurements of load-movement behavior of the shaft and the base. The upward load movement is governed by the shear resistance characteristics of the soil or rock along the shaft, whereas the downward load movement is governed by the compressibility of the soil or rock below the shaft toe (Fellenius et al., 1999).

For this project, the O-cell loading test was conducted on one of the central 2.15 m diameter drilled shafts, which was prepared as a production caisson. The subsurface profile at the test location, as determined from a nearby test boring and as observed from the drill cuttings, consisted of 4.6 m thick granular fill, over 5.5 m marine clay, underlain by 10.8 m glacial deposits of silty sands with gravel.

What had been described in a nearby boring as "completely weathered rock" at about 21 m depth below the existing grade, was recovered as a 1 m diameter solid sandstone core (Photo 1.1), immediately underlain by soil-like, completely weathered rock. The test shaft was advanced about 3 m into this weathered sandstone. The total length of the test shaft was about 24 m below the ground surface.

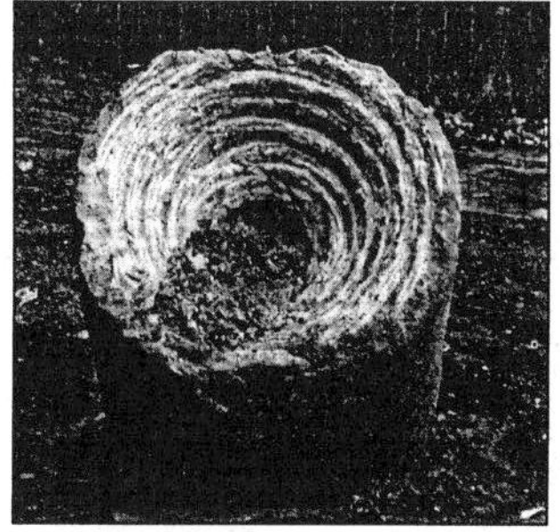

PHOTO 1.1 – Recovered 1m dia. Core in the "Completely Weathered Rock" zone

The test shaft (as well as all the remaining 44 production shafts with diameters ranging from 2.15 m to 2.75 m) was constructed using the slurry method to maintain a stable hole. Drilling began with soil augers to 1.5 m depth before introducing the polymer slurry, followed by further drilling and installing a slightly oversized 10 m long temporary casing to support the upper fill and marine clay. Both soil and rock augers were used to advance the shaft below the temporary casing and through the glacial soils. The slurry level inside the shaft was maintained just below the existing ground surface, approximately 1.5 m above the outside groundwater level.

Once weathered rock was reached, a permanent steel casing (Photo 1.2) with welded teeth was inserted into the shaft and twisted for about 0.3 m into the weathered sandstone. Drilling below the permanent casing was then continued at slightly reduced 2 m diameter using rock augers, sometimes assisted with core barrels to core the harder rock. For the test shaft, the length of the socket extending into the weathered sandstone was 2.5 m. The sides of the socket were scraped to remove softened material. At the end of drilling, the bottom of the socket was cleared of cuttings and accumulated sediments were removed using a clean-out bucket. It was estimated that approximately 25 mm of sediments remained at the base of the test shaft before beginning the O-Cell static loading test.

PHOTO 1.2 (left) **-Permanent Steel Casing**

Following drilling and clean-out of the test shaft, three 533 mm (21 inch) diameter sacrificial O-cells (welded and contained between two circular steel plates (Photo 1.3), and mounted on a steel frame), were lowered to the bottom of the test shaft.

The three O-cells were capable of applying a total combined load of 34,200 KN (3,845 tons). The O-cell assembly was complete with instrumentation including sister bar vibrating wire strain gages at three levels along the shaft, to measure the shaft resistance (side shear load transfer) in the various layers.

PHOTO 1.3–Three O-Cells welded & contained between two steel plates

Once the O-cell assembly was positioned inside the test shaft, concrete placement by pumping from the bottom up proceeded, until the level reached to within 2.2 m below the final head level. (The upper 2.2 m part of the shaft was later reinforced and concreted monolithic with the bridge pier. It is noted that the design optimization had also eliminated the use of full-length steel reinforcing cages; instead, the drilled shafts consisted of 16 mm thick permanent steel casing filled with 34.5 MPa concrete and reinforcement limited to the upper 2.2 m part of the shaft.) Following placement of the concrete, the annulus between the outer temporary casing and the inner permanent casing in the top 10 m of the shaft was grouted via a tremie pipe and the temporary casing was gradually pulled as grout filled the voids that were created during drilling.

The O-cell static loading test was performed by Loadtest Inc., on December 21, 1998, five days after placement of the shaft concrete and its attainment of the necessary compressive strength. The quick loading test procedure was followed and readings of all gages were obtained at 1, 2 and 4 minutes for each of the 14 loading increments. The three O-cells were pressurized to a total test load of 20,600 KN (2,315 tons) (half upward and half downward). The loading was halted when it was determined that the side shear in the overburden soils surrounding the permanent casing was approaching its full resistance mobilization.

Results

The results of the O-cell loading test on the full production drilled shaft are presented on Figure 1.2 which shows the Load vs. Movement curves, separately, for each of the upward and downward movement directions.

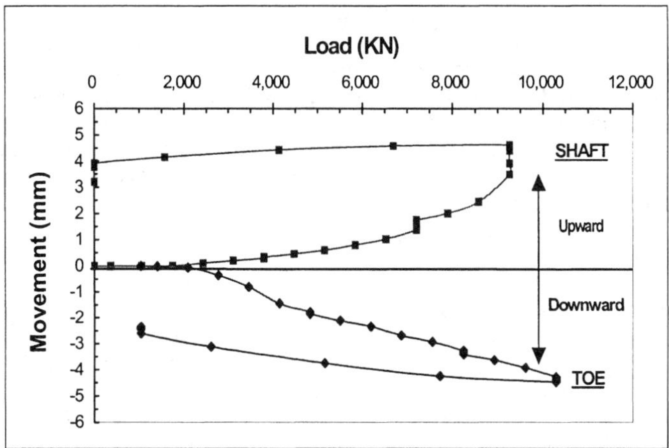

Figure 1.2 –Load-Movement Curves from O-Cell Test Results

Table 1 following summarizes the results of the Side Shear Transfer data for the various subsurface layers along the test shaft from the strain gage instrumentation.

TABLE 1 – Side Shear Transfer from O-Cell Strain Gage Data

Load Transfer Zone	Unit Side Shear Resistance
Grouted Zone outside Permanent Casing (upper ~ 9.3 m of overburden soils)	30 KPa (0.31 tsf)
Non-Grouted Zone outside Permanent Casing (~11.6 m glacial soils; drilled under slurry)	6 KPa (0.07 tsf)
Upper ~ 1.7 m Socket in Weathered Sandstone (includes 0.3m casing embedded into W.S.)	333 KPa (3.48 tsf)
Lower ~ 1.1 m Socket in Weathered Sandstone	302 KPa (3.16 tsf)

Based on the results of the O-cell static loading test, the total upward movement of the shaft was 4.6 mm at a maximum upward net load (gross load minus the buoyant weight of the shaft) of 9,300 KN (1,040 tons). The total downward movement of the shaft base at the maximum downward gross load (net load plus buoyant weight of shaft) of 10,300 KN (1160 tons) was 4.5 mm. This movement is only 0.2% of the shaft socket base diameter of 2 meter. Because of such small base movement and no apparent creep, it can be concluded that the ultimate end-bearing resistance of the shaft founded in the "completely weathered sandstone" was never reached.

The interpreted average unit bearing resistance in the weathered sandstone at the measured nominal 4.5 mm movement was 3.35 MPa (35 tsf). While not approaching its anticipated ultimate capacity, this deduced unit bearing resistance in the "completely weathered rock" was much higher than the bid design value, of 0.8 MPa to 1.2 MPa (8 to 12 tsf), specified for "sound" rock.

It is further noted that the unit shaft resistances measured during the loading with the O-cells, were lower than measured for the previously described conventional head-down loading tests at the nearby office tower. This could be due to that, in a conventional head-down test the instrumentation will not measure the load locked-in the pile (residual load) before the start of the test, and when residual load is neglected, the shaft resistance is overestimated. The use of polymer slurry and running the test only five days after drilling and concreting could also have resulted in a lower range of shaft resistance values.

Conclusion

Despite the strong resistance to change by the parties involved, the main objective of demonstrating that the "Weathered Sandstone" should not be bypassed in search of "Sound" rock was achieved. Newport Viaduct is now supported on large diameter drilled shafts with only nominal 1.5 m penetration or "sockets" in the "completely weathered rock". The unnecessary search for "sound" sandstone at great depths was eliminated, together with unavoidable disputes and delays. In addition to considerable time and cost savings, the redesign contributed to some valuable insight and experience in design and construction of large diameter drilled shafts in northern New Jersey.

CASE HISTORY 2. – 100th Street Bus Depot, New York City: Elimination of Deep Caissons in Favor of Spread Footings

The Project

The 100th Street Bus Depot is a new five-story bus terminal for the New York City Metropolitan Transit Authority occupying one city block in Manhattan between 99th and 100th Streets, and between Lexington and Park Avenues. It replaces a former two-story garage, built in the 1890's and initially used as a trolley-car barn. The new structure is a steel-framed building with concrete floors and includes a partial basement near its middle. The easternmost column line with design loads of up to 10 MN (1,125 tons) is within 1.5 m of the underlying multi-tube Lexington Avenue subway system. To support the new columns along the subway line, the project bid plans had called for 1 m diameter and 17 m deep caissons. Each caisson unit would have required 50 mm separation from the surrounding rock in the upper 14 m length, thus transferring the column loads to below the base of the existing tunnels, via 3 m long sockets in bedrock (Figure 2). (This follows a routine requirement by MTA to prevent stress from being imposed on the roof and walls of their tunnels.)

The bid documents contained extensive test boring information, which revealed that the project site was underlain by massive Manhattan mica Schist bedrock within just 3 m below street level. In fact, the Lexington Avenue two-level and multi-tube subway had been tunneled through this rock circa 1910, leaving an about 6 m thick solid rock roof over the upper-level subway (Figure 2). The NX rock core recovery values were near 100 % with an average RQD of 75 %.

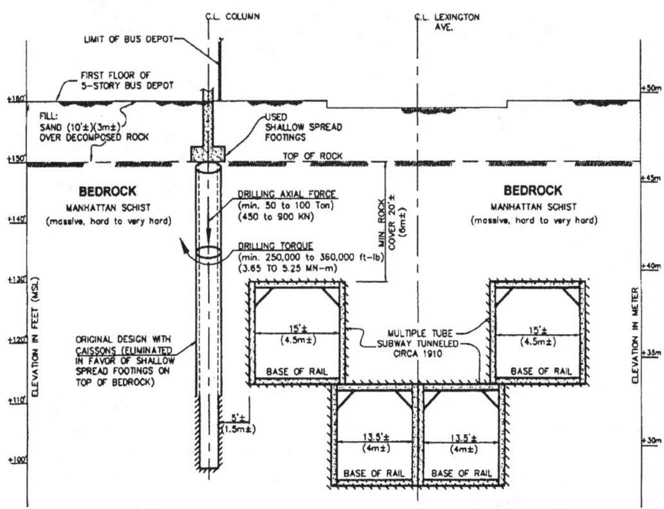

FIG. 2 – Typical Subsurface Section at Bus Depot

The Change

The Bus Depot project was let as a design/build contract with a very aggressive fast-track schedule. Both the pre-bid preliminary design and the subsequent final design by the Contractor's own hired design engineers relied heavily on very restrictive codes, including unnecessarily conservative approaches to seismic design. Any suggestion for value engineering change was undesirable and was strongly resisted. Yet, objecting to the blind reliance on codes in foundation design, the author questioned the real need for the deep caissons to support the columns along the subway. The unsupported statement that no stresses from the new structure can be imposed on the roof and walls of the adjacent tunnel was not sufficient reason for a deep foundation design.

Unconvinced that deep shafts were necessary, the author rushed to MTA's warehouse where many rows of the project rock core boxes were neatly stacked. Careful inspection and some simple testing of the rock cores with the blows of a geologic hammer gave support to the reservations about the design. Later, as expected, the schist bedrock at this site was found to be hard to very hard with unconfined compressive strength values estimated to range from at least 70 MPa to over 100 MPa (10,000 psi to 15,000 psi). The tensile strength of the rock was assessed to be minimum 5 MPa (750 psi).

It may be necessary to mention here that prior to the bid, more than 40 deep test borings including extensive coring of the rock had been carried out. However, not a single test was conducted on the cores to determine the strength of the rock. It is discouraging to realize that in this and in many other projects, test borings are performed to simply satisfy code requirements. Then, the boring logs become just part of the bid package for the contractor to review and, in essence, to become responsible and liable for the subsurface conditions encountered. It is further noted that contract specifications including this project, often demand requirements such as "no damage", "no movement" or "no vibration", thereby shifting all liability of underground work onto the contractor. It is not surprising that such shifting of liability has resulted in unwarranted contingency for the contractor and is a source of claims and disputes. The need for careful evaluation of all aspects of a project, from inception to construction completion, including the very often-neglected structure/foundation interaction could perhaps alleviate some pain during construction, where the real test for any design begins.

Results

Convinced that deep foundations were not warranted for the 100th Street Bus Depot because good hard rock was so shallow, the fight for using shallow footings instead, began. At the pre-construction meeting, the author rolled a sample of the drilled rock core onto the conference table to show the more than 25 attendees what actually lay beneath the surface. An unrelenting argument was presented, dramatizing the nominal stresses on the roof and walls of the tunnel induced by shallow spread footings supported only on top of bedrock, as opposed to the substantial stresses caused due to drilling of the specified deep shafts so very near the tunnel (Figure 2). To overcome the tensile strength and actually cut the hard rock by drilling or coring, large axial forces and a substantial torque from the drilling machine would be required, it was argued. These drilling induced stresses immediately adjacent to the tunnel would be at least ten times larger than the maximum stresses induced from a shallow spread footing, under extreme loading condition, including the code-based unrealistically large seismic loads. It was, therefore, clear that the strict limitations imposed by the bid regarding impact on the adjacent tunnel were not realistic, nor were they consistent.

Conclusion

Eventually, all deep caissons were eliminated and shallow spread footings founded on top of the mica schist bedrock were used to support the columns of the new structure.

CASE HISTORY 3 – Reconstruction of Queens Boulevard Bridge, New York: Saving the100-year old South Abutment

The Project

The reconstruction of the 100-year old Queens Boulevard Bridge (QBB) for the New York City Department of Transportation was let as a conventional design-bid-build contract with a very aggressive schedule. The QBB is a major bridge crossing Sunnyside Railroad Yard where an extensive network of railway tracks is in constant use. Immediately above the QBB runs the New York City Transit Authority "elevated subway" line, leading to the underground subway system in Manhattan. The QBB itself is situated in one of the most congested parts of Queens, NY. This, combined with an active railway system 7 m below the bridge, and an active train system less than 9 m above the bridge, can make any construction activity in the middle a very difficult task. These physical constraints are further complicated because different agencies own and operate the various infrastructures.

The contract bid plans had specified the complete removal and replacement of the existing bridge superstructure and the complete removal and replacement of only one of the two abutments, the South Abutment (a 30 m long concrete gravity structure with shallow spread footing, Figure 3). The north abutment and the substructures of all the 18 intermediate piers (supported on timber piles) were to be maintained. According to the contract, the existing south abutment was to be replaced with a new reinforced concrete structure, supported on deep concrete-filled steel pipe piles. For removal and replacement of the abutment, the bid plans had specified an excavation support system comprised of soldier piles / lagging, with tiebacks. This excavation support system was in very close proximity to the deep pile foundations of the existing overhead "elevated subway" bents, built circa 1915.

The subsurface conditions at the 100-year old railroad yard within which all the foundations of QBB are founded, are comprised of an upper 8 m thick silty sand fill, overlying a 1.5 m thin layer of organic clayey silt with peat, over 15 m thick dense glacial deposits of silty sands and gravel, over schist bedrock (Figure 3).

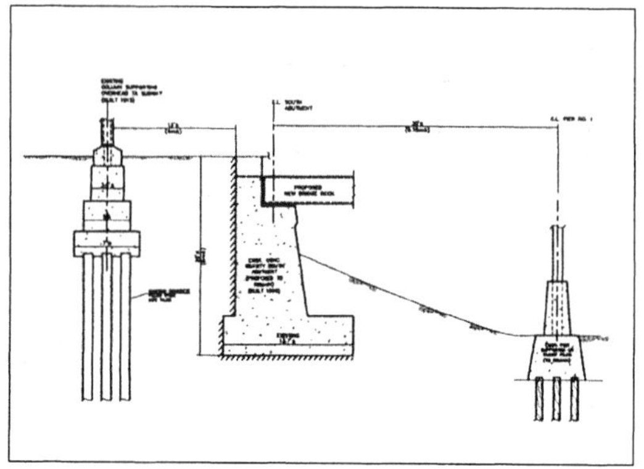

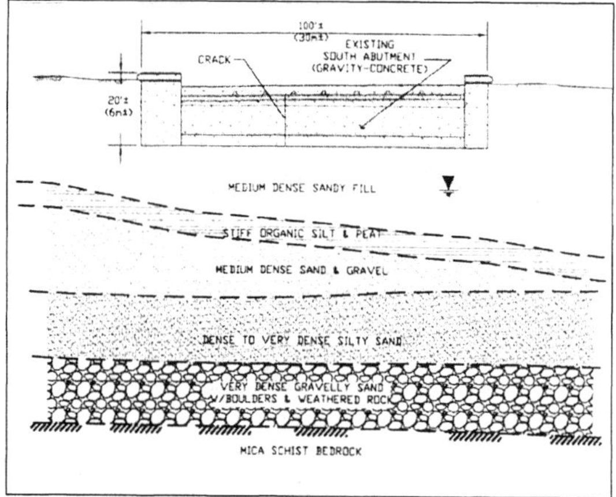

FIG. 3 – Typical Section and Elevation of South Abutment

The Change

Upon award of the construction contract, finding the reason behind the complete removal and replacement of only one of the twenty supporting elements of the new lighter superstructure of the QBB began. The complexity of the site and the very real potential for conflicts between such closely situated structures administered by so many different agencies provided ample reason to question the bid design. Indeed, except for one obvious crack, which was easily repairable, there did not appear to be much justification for the complete removal of the south abutment. Value engineering to maintain this abutment was initiated but it received an unbelievably fierce resistance. Unfortunately, this is an only too-common reaction whenever spread footings in lieu of deep foundations are proposed. Very often, deep foundations are selected without proper evaluation of constructability or impact due to construction.

In many situations, shallow footings perform as well, if not better than deep foundations, yet they are consistently ignored. More recently, the excuse of "seismic consideration" is immediately presented without real understanding or evaluation of dynamic behavior.

Despite the obstacles, however, saving the historic structure became a personal mission for the author. Extensive settlement, and static, dynamic, and liquefaction analyses were performed in support of maintaining it.

Results

The record drawings of the existing South Abutment contained a valuable note indicating that the massive concrete gravity structure had experienced about 75 mm settlement during construction from May 1909 to October 1910. This was not unusual, since the abutment is underlain by a sandy fill layer, which probably was in a loose state some 100 years ago. (It is likely that the subsequent use of timber piles to support the remaining piers and the north abutment was a result of the observed settlement.) Yet, since the south abutment and the bridge had remained functional and in successful service for over 90 years, any further settlement must have been small. The probable cause for the 10 mm wide vertical crack in the abutment is differential settlement resulting from consolidation of the thin organic layer below the fill. However, after nearly 100 years, the structure had reached a complete state of equilibrium. Furthermore, because the new loads from the replacement bridge superstructure were lighter than the existing to be removed, no further settlements were to be anticipated. It is noted that the net bearing stress on the foundation soils imposed by the existing or the new bridge do not exceed a nominal 100 KPa (1 tsf).

The seismic analysis of the abutment proved to be very contentious. Efforts were not spared (by the original design team) to impede the consistent conclusions that the existing abutment had adequate dynamic resistance. Even with unrealistic peak ground acceleration, and exaggerated seismic loads (much higher than that used for the other elements of the same bridge), the dynamic performance of the abutment was still more than adequate. The "drama" of soil liquefaction and the suggestion that 1.2 m settlements would occur due to liquefaction, was even more unrealistic. Response analysis revealed that such settlements, if any, would not exceed 25 mm under a "fictitious" Magnitude 6.5 Earthquake in Queens, New York. Ironically, the "elevated subway" bents above the subject abutment cannot withstand an earthquake magnitude of 5.0 or 5.5, typically assumed for this site. Moreover, the original bid design involving abutment removal had easily overlooked potential movements inherent with deep excavation and tieback installation, not to mention the vibrations from driving new piles, so near the vulnerable existing structures.

Conclusion

Needless to say, the existing South Abutment of Queens Boulevard Bridge was saved.

CASE HISTORY 4 – Expansion of the Long Island Expressway, New York: Redesign of Retaining Walls with "Giken" Tubular Pressed-in Steel Pipe Piles

The Project

The expansion of the ever-congested Long Island Expressway (LIE) over Cross Island Parkway (CIP) just outside Manhattan, New York included numerous retaining walls and several deep cuts for bridge expansions. The bid plans for this conventional design/bid/build project contained complete designs for both permanent retaining walls and temporary excavation support systems. The main specified support method consisted of very deep, 1 m diameter drilled-in soldier beams/lagging, complemented with tiebacks or bracing. Unusually heavy, 1,100 kg/m (740 lb/ft), and 25 m long H-beams with welded-on cover plates were specified. To install such beams, deep drilling in granular soils below the water table would have been required.

Soon after start of construction in the fall of 2000, serious supply problems for the specified heavy beams ensued. It became necessary to explore alternative systems for three major structures: a 78 m long permanent retaining wall SP-1 at maximum 8.5 m height; an excavation system to support a 11 m deep cut for construction of the new westerly abutment of the LIE bridge over CIP; and an excavation system to support a 9 m deep cut for the reconstruction of the Marathon Parkway Bridge abutments over LIE.

The subsurface conditions at this site are characterized as terminal moraine glacial deposits consisting of medium dense, becoming very dense with depth, sands and gravel with varying proportions of silt, numerous cobbles, and occasional boulders. Groundwater is present just below the proposed excavation levels.

The Change

For this case, a value engineering change proposal was welcomed by the owner, the New York State Department of Transportation (NYSDOT). As usual, the fast track schedule of this major project and producing a new design during construction became a challenge, requiring the full and amicable cooperation of NYSDOT. Interestingly, the beginning of redesign and the value engineering process for the LIE retaining walls coincided with the Deep Foundations Institute's Eighth International Conference in New York on October 5 to 7, 2000. Agonizing over the strict limitations imposed by NYSDOT not only on soil parameters, but on the methodology of design based on codes, the author felt compelled to openly express her frustration with some attendees at that conference, which apparently included the Giken "Press-in Method" representative. Thus, Giken responded to a request to provide "*not just video animation or colorful brochures, but good technical literature*". Read and absorbed during that weekend, a presentation of the concept of the innovative technology was made to NYSDOT. In less than one week following the conference, the complete redesign of one of the deep excavation support systems in full cantilever was submitted as value engineering change proposal. The installation of the permanent wall and the first phase of the two temporary support systems were completed in early July 2001. Phase 2 was completed in the spring of 2002 making the Giken "Press-in Method" a first outside the Far East.

Design and Installation
The Giken "Press-in Method", also known as the "Silent Piler" (because it is virtually noiseless and vibration-free) is described elsewhere (White et al. 2000; ENR 2001; Bearss et al. 2002). For this project, a constant 914 mm outside diameter open-ended steel pipe piles were pressed almost contiguously into the ground with a powerful 260 metric ton capacity hydraulic push piler. The nominal gaps of 180 mm between adjacent piles were closed with pre-welded Pipe-Tee (P-T) interlocks in the case of the permanent wall, and with flat bars for the temporary support, extending a nominal 1.5 m below the excavation level. To aid in penetrating into the very dense granular soils, two built-in water jets with maximum 7,000 KN/m^2 (1,000 psi) pressure per jet were used. The pile installation was guided by laser beam resulting in remarkable small alignment deviation of less than 3 mm (1/8"). Photos 4.1 and 4.2 show the Giken piler during installation of the LIE retaining wall SP-1.

PHOTO 4.1 – Begin of Installation

PHOTO 4.2 – Close up of Giken Piler with Pile

Results

The 0.914 m outside diameter steel pipe piles possess significant bending resistance to lateral loads and as such, support in full cantilever of the deep cuts, was achieved — the most desirable excavation support system for a contractor. Deflection based methods for the analysis of piles under lateral loads (Reese et al. 1974; API 1993; Reese et al. 2000; Reese and Van Impe 2001), were employed and a parametric analysis was performed with varying loading and subsurface conditions. The total length and wall thickness of each pile was selected based on maximum allowable deflection at the pile head and allowable bending stresses in the pile itself. Consistently, the results of the analysis indicated that a depth of embedment for each pile about equal to the cantilever height would provide an adequate performance. In general, thin-wall pipe piles offered a better flexible performance (Bedian 2002).

Upon final excavation, measured movements at the pile heads were remarkably close to the predicted values (based on estimated p-y curves for the granular soils at this site.) Measured movements at the permanent wall were less than 25 mm. For the new west abutment excavation where large movements (150 mm to 180 mm) were allowed, the maximum measured pile top movements were as follows (Figure 4.1): along the centerline of LIE where 25 mm thick wall tubular piles were pressed-in, the maximum pile-head movement was 125 mm. Along the back of the new abutment where 17.5 mm thick wall piles were pressed-in, the average measured pile head movement was 161 mm. These total lateral pile head movements are inclusive of about 25 mm deflection experienced by each tubular pile immediately following the driving of the new abutment pile foundations (324 mm diameter steel pipe piles) at the bottom of the cut. The additional movement was probably due to the pile driving vibration-induced temporary loosening/liquefaction of the submerged sands below the excavation, upon removal of the 11 m overburden soils.

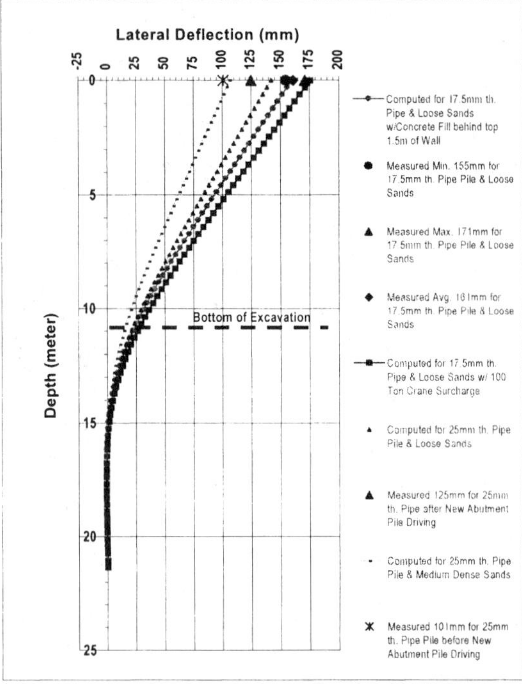

FIG. 4.1 – Pile Lateral Deflection Diagram

Figures 4.2 and 4.3 below present the computed pile moment and mobilized soil reaction diagrams, respectively.

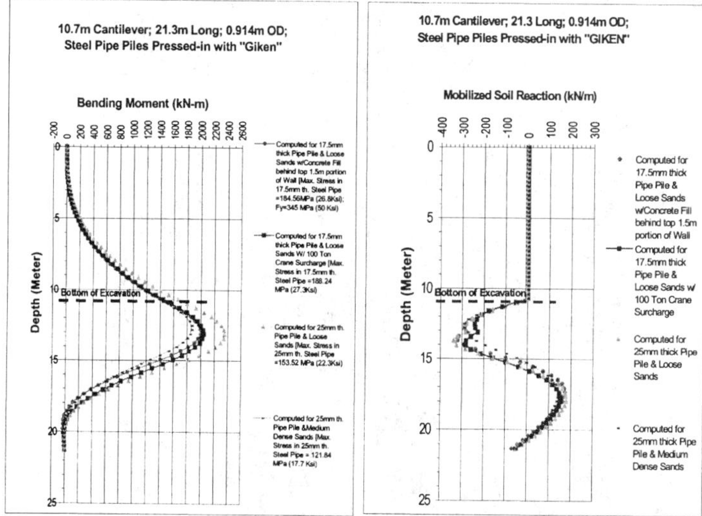

FIG. 4.2 – Pile Moment Diagram **FIG. 4.3 – Mobilized Soil Reaction**

Photos 4.3 and 4.4 show impressive views of the completed retaining wall in full cantilever at 10.7 m (35 foot) height, with average pile embedment of 10.7 m.

Photo 4.3 – Overall View of the Retaining Walls upon Final Excavation

Photo 4.4 – 10.7m Full-Cantilever Support with "Giken" Pressed-in Tubular Steel Piles

(The wall is comprised of 914 mm diameter, 21.5 m total length Tubular Piles, embedded average 10.7 m below the excavation level; note the 324 mm diameter pipe piles for the new abutment, driven and concreted following excavation/removal of the overburden soils)

Conclusion

Sound new technology and advanced geotechnical engineering design (considering soil-pile interaction), made this case history a true value engineering change.

REFERENCES:

Case History 1:

Baker Jr., C. N. (1988). "Test caisson design analysis, Newport Office Tower, Jersey City, New Jersey." (un*published report* by STS Consultants, Chicago, Illinois).

Fellenius, B. H., Altaee, A., Kulesza, R., and Hayes, J. A. (1999). "O-Cell testing and FE Analysis of 28 m deep barrette in Manila, Philippines." *Journal of Geotechnical & Geoenvironmental Engineering,* ASCE, 125(7), 566-575.

Fellenius, B. H., (2001). "The O-Cell – an innovative engineering tool." *Geotechnical News Magazine,* Vol. 19, No. 6, 55-58.

O'Neill, M. W., Townsend, F. C., Hassan, K. M., Buller, A., and Chan, P. S. (1996). "Load transfer for drilled shafts in intermediate geomaterials." *USDOT Publication* No. FHWA-RD-95-172, McLean, Virginia.

Osterberg, J. O. (1994). "Recent advances in load testing driven piles & drilled shafts using the Osterberg load cell method." *Geotechnical Lecture Series,* ASCE, 1994, Illinois, USA.

Schmertmann, J. H. and Hayes, J. A. (1997). "The Osterberg cell and bored pile testing – A Symbiosis." Presented at the *Third International Geotechnical Engineering Conference*, Cairo, Egypt.

Case History 4:

American Petroleum Institute. (1993). "Recommended practice for planning, designing and constructing fixed offshore platforms – working stress design." *API, (RP 2A- WSD),* 20th Edition, Washington, D.C.

Engineering News Record. (2001). "Japanese system quietly breaks ground on highway job" *ENR*, July 9, 2001, 16; also *ENR* July 30, 2001, "Viewpoint", 63.

Bearss, G.R.J., Bedian, M. P., Carter, M.W.T.,and Takahiko, I., (2002). "Urban legends: two case studies that redefine "impossible"". *Proceedings, Deep Foundations Institute, Ninth International Conference,* Nice, France. 351-361.

Bedian, M. P. (2002). "Value engineering in United States of America". *Proceedings, Deep Foundations Institute, Ninth International Conference on Piling and Deep Foundations,* Nice, France. 429-443.

Reese, L. C., Cox, W. R., and Koop, F. D. (1974). "Analysis of laterally loaded piles in sand". *Proceedings, 6th Offshore Technology Conference*, Houston, Texas, Vol. II, OTC 2080.

Reese, L. C., Wang, S.-T., Isenhower, W.M., Arréllaga, J.A., and Hendrix, J. (2000). "A Program for the analysis of piles and drilled shafts under lateral loads," *Computer Program LPILE Plus, Version 4.0,Ensoft Inc.,* Austin, Texas.

Reese, L. C., Van Impe, W. F. (2001). "Single piles and pile groups under lateral loading," A.A. Balkema, Rotterdam. The Netherlands.

White, D. J., Sidhu, H.K., Finlay, T.C.R., Bolton, M.D. and Nagayama, T. (2000). "Press-in piling: the influence of plugging on driveability" *Proceedings, Deep Foundations Institute, Eighth International Conference,* New York, USA. 299-310.

DRILLED SHAFT CONSTRUCTION WITH BLASTING

Frederick C. Rhyner, P.E., Member, ASCE[1]

ABSTRACT

Construction of a transmission line for a new power plant presented the opportunity to utilize a relatively new construction technique involving blasting for drilled shafts. Drilled shafts are often used as foundations for transmission line pole foundations, where they are commonly constructed by auger and rock coring techniques. However, rock coring, especially for large diameter shafts in hard metamorphic and igneous rock, is slow and expensive. This paper describes how a contractor utilized blasting techniques to assist the construction of drilled shafts for transmission line pole foundations. Controlled blasting techniques generally improved the production rate of drilled shaft excavation in hard schist and quartzite rock, without adverse impacts. However, the resulting rock sockets tended to be irregular in shape and a few had significant amounts of overbreak. Recommendations are provided for improved controlled blasting techniques for drilled shafts.

INTRODUCTION

Con Edison Development, Inc. constructed a new 525 MW, combined cycle, natural gas power plant in Newington, New Hampshire in 2000 and 2001. Figure 1 shows that the plant includes a 1.25 mile long transmission line connecting to the existing power grid. Preliminary engineering studies determined that steel poles, spaced at 360 feet to 900 feet spans would provide the most effective support system for the 345 kV transmission line. The designers adjusted pole locations and the alignment several times during design to accommodate easements and landowner preferences. Final design settled on twelve steel poles with heights ranging from approximately 100 feet to 150 feet. Nine of the poles were at located at angle points on the alignment, consequently generating high overturning moments.

1. Vice President, Haley & Aldrich, Inc., 340 Granite Street, Manchester, NH, 03102, FCR@HaleyAldrich.com

Drilled shafts socketed into bedrock were an obvious economical foundation for proposed pole locations. Drilled shafts were the only suitable choice for the two poles that were situated within 10 feet of a high-pressure gas transmission main and the four poles that were adjacent to a railroad. Other potential issues included a nearby sanitary sewer and industrial buildings.

The owner's geotechnical engineer made a detailed subsurface investigation consisting of a boring at each pole location and laboratory testing, because bedrock depths varied significantly. They also prepared a preliminary foundation design for each drilled shaft to use as a basis for bidding during selection of an Engineer-Procure-Construct (EPC) contractor.

During bidding, the low bid contractor proposed an alternate to construct the drilled shafts using blasting prior to conventional rock coring tools. A limited review during evaluation of bids, consisting of discussion with certain drilled shaft contractors, researchers and other utility companies, revealed that the use of blasting to assist the construction of drilled shafts has been limited to few transmission pole foundations, mostly in the southeastern United States. We were not aware of instances where blasting for drilled shafts has been utilized in other infrastructure projects or building construction. Nevertheless, the potential savings were substantial and the risks appeared manageable. The owner decided to accept the low bidder's proposal, but with new design criteria for controlled blasting and full-time inspection by the geotechnical engineer.

This paper describes the subsurface conditions, the design methods used for drilled shafts subjected to large overturning moments, the controlled blasting methods actually used, the conventional drilled shaft methods used and the results in terms of rock socket production, shape and overall acceptability. Recommendations are provided for improved controlled blasting techniques for drilled shafts.

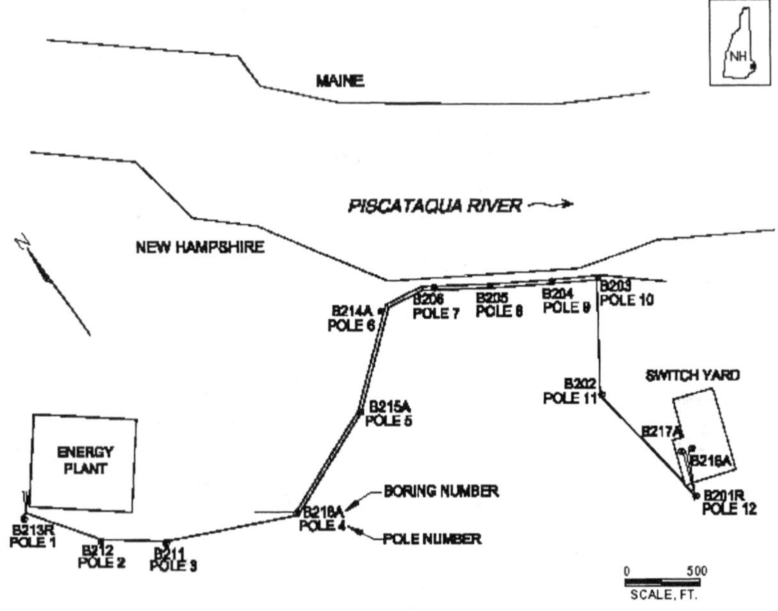

FIG. 1. Location Plan

SUBSURFACE CONDITIONS AND LABORATORY TESTING

Geologically, the site is in the seacoast lowland section of the New England physiographic province. Overburden includes fill of various sort and swamp deposits, which are underlain by Pleistocene marine silt and clay. Glacial till and bedrock underlie the marine deposits.

Bedrock consists of folded and faulted metamorphic rocks of the Paleozoic era that have been intruded by igneous rocks. Bedrock is part of the Eliot formation, which is most likely of Silurian age. Rock types were gray schist and fine-grained quartzite. No outcrops were observed at the project site.

Figure 2 shows typical subsurface conditions as interpreted from borings. Twelve borings were made; one at each proposed pole foundation. Boring depths ranged from 19 feet to 61 feet. Borings included split spoon sampling in overburden soils and rock core in bedrock with an NX size double-tube core barrel. Overburden soils were generally medium-dense, fine to coarse silty sands, or stiff clayey silts. Numerous cobbles and boulders were present. Weathered bedrock was encountered at depths ranging from 4 feet to 26.5 feet and varied in thickness from one foot to ten feet. Bedrock was encountered at depths ranging from 8 feet to 31 feet. It was predominantly hard, fractured and weathered, gray quartzite interbedded with schist. Average bedrock core recovery for all core runs was 80% and average RQD was 39%.

At the nearby plant site, the subsurface investigation for design of turbine foundations included downhole seismic tests to determine the dynamic shear modulus of the bedrock.

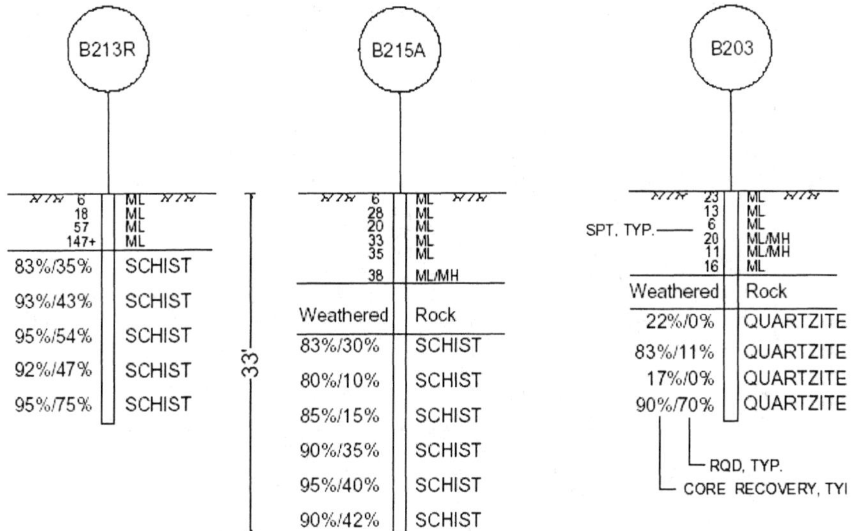

FIG. 2. Selected borings illustrate range of subsurface conditions.

A modest laboratory test program was made to aid in design of pole foundations and to provide prospective bidders with sufficient information on which to base their bids. Laboratory tests of soil samples were limited to a few index and classification tests. Rock core samples were selected generally from within the proposed rock socket depths at each of the 12 transmission pole locations. Most of the samples selected were of gray aphanitic quartzite but two samples of schist were also tested.

Table 1 provides a summary of laboratory test results on rock core samples. Rock unit weights varied from 149 pounds per cubic foot (pcf) to 172 pcf and averaged 164 pcf. Unconfined compressive strengths for the 16 tests averaged 10,200 psi, with a range from 1,110 psi to 25,200 psi. Only two samples of schist were tested, yielding unconfined compressive strengths of 4,720 psi and 7,680 psi. The modulus of elasticity reported in the table represents vertical strain at 50% of ultimate strength. It ranged from 2,980,000 psi to 7,010,000 psi with an average of 5,030,000 psi. The low strain, dynamic shear modulus measured in borehole geophysical tests at the plant site was 125,000 psi for weathered bedrock and 3,450,000 for bedrock. Poisson's ratio was determined in seven of the unconfined compressive strength tests, by including lateral strain gages on the samples. The values of Poisson's ratio ranged from 0.01 to 0.16 for the core samples tested, with an average of 0.08.

Table 1. Summary of Laboratory Tests on Rock Core Samples

Boring No.	Sample No.	Depth (ft)	Rock Type	Core Run Recovery (%)	Core Run RQD (%)	Unit Weight (pcf)	Unconfined Comp. Strength (10^3 psi)	Tan. Modulus (10^6 psi)	Poisson's Ratio
(1)	(2)	(3)	(4)	(5)	(6)	(7)	(8)	(9)	(10)
B201R	C1	18 to 23	Quartzite	95	15	160	6.12	4.82	0.09
B202	C1	9 to 14	Quartzite	83	20	172	8.96	---	---
B202	C2	14 to 19	Quartzite	95	25	152	1.11	---	---
B203	C2	21 to 25.5	Quartzite	83	11	149	6.47	2.98	0.08
B204	C2	20 to 25	Quartzite	80	45	156	6.30	---	---
B205	C3	20.5 to 25.5	Quartzite	89	80	167	6.29	2.98	0.01
B211	C1	15 to 20	Quartzite	100	50	169	25.20	---	---
B211	C2(a)	20 to 25	Quartzite	97	70	170	17.04	6.54	0.16
B211	C2(b)	20 to 25	Quartzite	97	70	169	14.62	6.03	0.04
B212	C3	18 to 23	Quartzite	90	47	170	14.91	---	---
B213R	C2	13 to 18	Schist	93	43	169	7.68	4.83	0.14
B214A	C1	31 to 36	Quartzite	98	75	168	6.66	7.01	0.06
B214A	C2	36 to 41	Quartzite	99	70	169	7.38	---	---
B215A	C1	19 to 24	Schist	83	30	163	4.72	---	---
B216A	C1	13 to 18	Quartzite	92	50	168	12.47	---	---
B217A	C2	19 to 24	Metaquartzite	20	0	159	17.09	---	---

In summary, the quartzite and schist samples were hard and brittle in lab tests but the bedrock stratum includes weathered and fractured seams as evidenced by the moderate core recoveries and low RQD values.

DESIGN OF DRILLED SHAFTS FOR LARGE OVERTURNING MOMENTS

Transmission pole foundations are subject to high overturning moments, moderate lateral shear and light vertical loads. For example, on this project, the overturning moments ranged from 22,945 ft.-kips at an angle pole to 3,923 ft.-kips at a tangent pole. Standardized guides for foundation design are available for situations where poles are founded in soil (e.g. Teng 1969). The Broms method of analysis is appropriate in such cases when drilled shafts in soil are short and stiff. However where poles are founded in rock, design depends on an analysis that can model lateral load transfer through a short, stiff pier into equally stiff bedrock, that may be overlain by soil.

The owner's geotechnical engineer prepared a preliminary foundation design for each drilled shaft to use as a basis for bidding during selection of an Engineer-Procure-Construct (EPC) contractor. Drilled shafts for these preliminary designs were analyzed using the p-y curve method (e.g. O'Neill and Reese 1999). This common method is based on analysis of laterally loaded piles as adapted to rock sockets. Four current software packages based on the p-y curve method are: COM624P, LPILE, GROUP and FBPIER. These contain guidance for analysis of rock sockets, but only for so-called "weak" rock, such as shale or Florida limestone as based on Reese (1997). The weak rock analysis relies on an assessment of the rock mass strength as inferred from core recoveries, RQD values and unconfined compressive strength tests on rock core samples. The recommended procedure is to model the rock stratum as a very hard clay in terms of strength and deformation parameters, then iterate and adjust the parameters until obtaining reasonable results.

It is difficult to provide proven design criteria for laterally loaded rock sockets. Only a small number of tests have been published in the literature, probably because the cost of performing such tests is rather large and few problems have occurred with this type of foundation (e.g. Reese (1997), Schaffer, Greene and Green (2000) and Zhang, Ernst and Einstein (2000)). A lateral load test was considered for this project (in view of the use of blasting), but the cost of the test was prohibitive compared to the small number of pole foundations. For this project, the owner's geotechnical engineer used a cohesive strength of 2,000 psi for the bedrock stratum and a strain at 50% of failure of 0.1% for modeling purposes with COM624. A different drilled shaft design was necessary for each pole because subsurface conditions and rock quality varied to a significant degree, and the loads on each pole were different.

Design criteria included an allowable lateral deflection at the ground surface of 1 inch and an allowable rotational deflection at the ground surface of 1 degree. These values are considerably higher than would be used on other types of projects, such as transportation structures or buildings. However, they were reasonable as long as the deflections were recoverable (in the elastic range) and were well below the ultimate capacity of the foundation. Furthermore, the diameters of the drilled shafts were partly controlled by the base diameter of the steel poles. Shaft depths were determined to provide an adequate factor of safety against overturning and to limit deflections to the design criteria. But, the shafts barely qualify as "deep foundations" because the ratio of depth to diameter is only about 2:1 to 3:1. They are more correctly analyzed as rigid piers than laterally loaded piles.

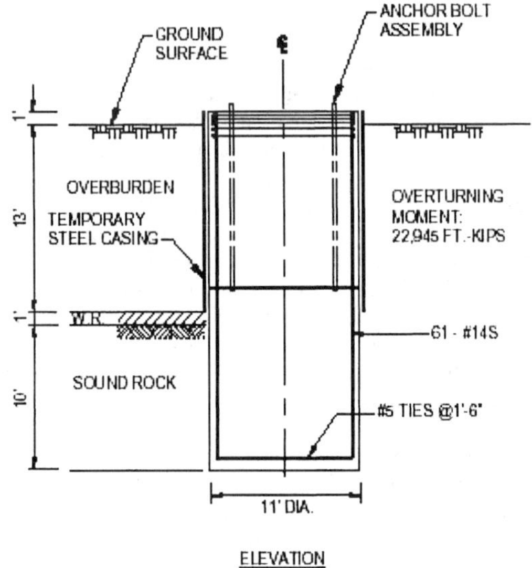

FIG. 3. Typical drilled shaft foundation design for transmission line pole.

Figure 3 shows a typical design for a drilled shaft pole foundation on this project. The EPC designer made the final design of steel poles and foundations. Drilled shaft sizes ranged from 8 feet diameter by 17 feet deep, to 11 feet diameter by 30 feet deep. Longitudinal reinforcing consisted of varying numbers of #14S bars and shear reinforcement was provided by #5 bar ties at 1'-6" spacing. This reinforcement provided an ample clear space between bars for concrete placement.

The design engineer used an Electrical Power Research Institute (EPRI) software package, CUFAD, (Compression and Uplift Foundation Analysis and Design), which is based on a finite element analysis of drilled shafts, as opposed to the p-y method. The EPRI method is based on research by Carter and Kulhawy (1992) using drilled shafts of different widths, depths and stiffnesses (EI values) in rock of varying shear modulus. A parameter called the modified shear modulus is used to represent the effect of varying Poisson's ratio. Main parameters for use in the EPRI method are the shear modulus and Poisson's ratio of the rock mass. Dynamic shear modulus of the subsoils and bedrock was measured at the power plant site in downhole seismic tests. Results were correlated approximately with stratigraphy, resulting in values of G, dynamic shear modulus, for weathered rock of 125,000 psi and for bedrock of 3,450,000 psi. For design, Poisson's ratio was 0.25 for weathered rock and 0.20 for bedrock.

The p-y method and the EPRI finite element method yielded designs that were similar.

BIDDING AND CONSTRUCTION PHASES

The utility company constructed the transmission line via an Engineer-Procure-Construct contract, common to the power industry. Prospective bidders were supplied with preliminary designs, design specifications and geotechnical report. Although prohibited in the design specifications, the low bidder proposed using blasting to construct the drilled shaft foundations and a savings of approximately $500,000, roughly 30 percent of the foundation cost. The utility owner and geotechnical engineer discussed the merits of the cost savings and potential risks.

A limited review, consisting of discussion with certain drilled shaft contractors, researchers and other utility companies, revealed that the use of blasting to assist the construction of drilled shafts has been limited to few transmission pole foundations, mostly in the southeastern United States. Staff were not aware of instances where blasting for drilled shafts has been utilized in other infrastructure projects or building construction.

Potential problems related to blasting on this project included proximity to a high-pressure gas line (10 feet), railroad (12 feet), sanitary sewer (40 feet) and a building (75 feet). On the other hand, the concept appeared manageable through the use of controlled blasting techniques. Other favorable factors for the use the blasting on this project: the drilled shaft specialty contractor had an excellent reputation and references from successful blasting on earlier projects, and the transmission pole foundations were in an industrial area where there were no residences.

The owner decided to accept the low bidder's proposal, but with new design criteria for controlled blasting and full-time inspection by the geotechnical engineer. Design criteria for blasting were developed to minimize the risk of diminishing the lateral load capacity of the rock sockets. We considered performing a lateral load test on a production-drilled shaft to determine the effect of blasting, but the cost was prohibitive compared to the small size of this transmission line project.

Controlled Blast Design

Blasting to assist with rock socket excavation is more similar to blasting for a vertical tunnel shaft, than to, say, excavating rock slopes. In a rock socket, there is no room for blast energy to dislodge rock particles horizontally, as there when blasting a rock slope. When blasting a rock socket, the major amount of blast energy is directed upward towards the free face which is the ground surface. Another concern is that blast gasses will penetrate existing fractures and seams in bedrock surrounding the rock socket, weakening its structure. Finally, there is essentially no rock burden to absorb the blast energy.

To overcome these problems, we developed criteria for controlled blasting consisting of central blast holes and perimeter cushion blast holes. The concept was to detonate the central holes first, allowing some rock to dislodge upwards. Subsequently when the perimeter cushion holes were detonated, rock dislodged radially toward the center of the rock socket. Central blast holes were planned to be loaded heavier than the cushion holes, to minimize the risk of gasses fracturing the rock forming the socket. Thus, this plan created the greatest blast relief, while minimizing ground vibrations and lessening the risk of damage to adjacent structures.

The drilled shaft specialty contractor retained a local, blasting contractor to perform the blasting. The Contractor submitted a detailed blasting plan for review and approval prior to construction.

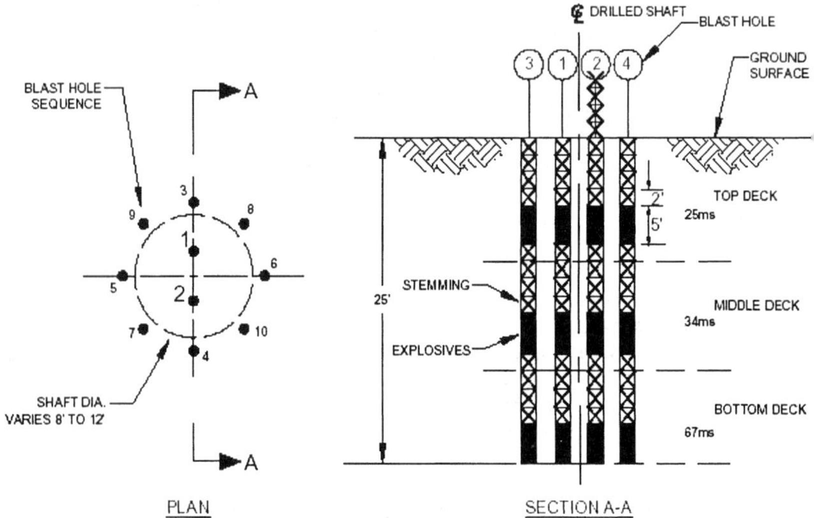

FIG. 4. Typical controlled blast design used by contractor.

Figure 4 shows the blast design for Pole No. 7, which is typical of blasts on this project. Blast design consisted of two central holes and eight equally-spaced perimeter holes. Perimeter holes were spaced approximately 6 to 18 inches outside the design diameter of the drilled shaft to assure that there would be no rock protrusions into the shaft. The circumferential spacing between perimeter blast holes was approximately 4.3 feet. Each blast hole was 3 inches diameter and approximately 25 feet deep, to the proposed bottom of rock socket. Each hole was loaded with the same quantity of explosives, 37 pounds, spaced in three vertical groups even through the overburden soils. Figure 4 shows that the detonation sequence was the first two central blast holes followed by perimeter holes in diametric opposition to each other. Within each blast hole, the detonation sequence was in three equal decks: beginning with the top deck, followed by the middle deck and finally the bottom deck. Thirty delays (2.5 milliseconds each) were used with a maximum of three delays per hole and 12.43 pounds of explosive per delay. For this drilled shaft, the charge amounted to 1.48 pounds of explosives per foot of blast hole. The powder factor was approximately 4.21 pounds per cubic yard.

Drilled Shaft Construction

The contractor developed a construction sequence to maximize productivity, since completion of the transmission line was on the critical path for the entire power plant project. Initially, the blasting subcontractor conducted a pre-blast survey of structures within 250 feet of each blast site. Drilling and blasting was performed from the existing ground surface at each pole location using a track-mounted air compression drill rig. Blasting was performed through soil overburden and bedrock. Rubber tire mats were used to limit fly rock. Peak particle velocities were measured with a seismograph at the nearest structure. No damage or reports of damage occurred. Drilling and blasting was performed several days or more in advance of the drilled shaft excavation.

After drilling and blasting, the drilled shaft specialty contractor used a Watson 3000 truck-mounted drill rig and conventional drilling techniques to excavate the drilled shaft. Overburden soils were excavated with large diameter augers, rock was excavated with a combination of rock augers and rock core barrels with carbide-tipped teeth. All drilled shafts used temporary casing; sometimes two or three different diameters of temporary casing were telescoped to the top of bedrock. Shaft excavations were cleaned with a muck bucket. The excavation and cleaning methods produced acceptable results and no workers entered the shaft excavations. Submersible pumps controlled the small amount of water infiltration. The contractor used the dry hole method to concrete the shafts. A ring-shaped steel template held the anchor bolts in position for the pour. Temporary casing was withdrawn as the shafts were concreted.

Drilled shafts were constructed under the full time observation of the geotechnical engineer. For the first two shafts, the geotechnical engineer performed a down-hole inspection of the excavated shafts to determine the condition, quality and soundness of the rock socket as a result of the blasting. After observing satisfactory results on the two shafts, and because the other shafts were relatively shallow and visible from the ground surface, the remaining shafts were constructed without downhole entry by workers or inspectors.

Results

Figure 5, 6, 7 and 8 show examples of rock sockets during excavation. Generally, the rock sockets were somewhat oversized because the perimeter blast holes were made 6 to 18 inches beyond the design circumference of the drilled shaft. The sidewall of the rock socket tended to be irregular, with rock fractured along the wavy foliation planes. The sockets were not truly cylindrical in shape as normally occurs with core barrel excavations. Nevertheless, the rock forming the socket was sound and intact. There were no loose zones or remnant loose blocks in the rock socket as a result of blasting. Most of the rock excavated from the shafts had been fractured into

FIG. 5. Fractured rock in spoils pile shows results of blasting.

FIG. 6. Drilled shaft excavation after blasting and spoil removal with auger.

FIG. 7. Irregularly shaped rock socket with overbreak.

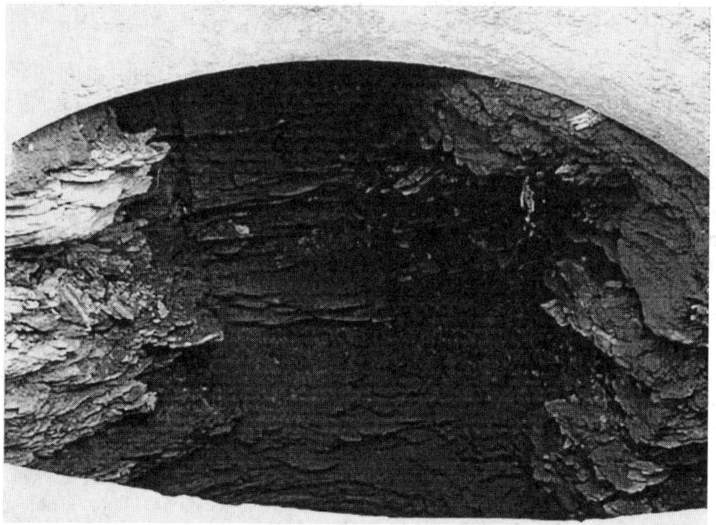

FIG. 8. Rock excavation before reaming with core barrel.

TABLE 2. Summary of Productivity Rates and Concrete Volumes

Drilled Shaft No.	Design Dimensions			Productivity Rates		Concrete Volumes		
	Total Length (ft.)	Dia. (ft.)	Rock Socket Length (ft.)	Excavation (hours)	Total (hours)	Design (cy)	Actual (cy)	Percent Difference (%)
(1)	(2)	(3)	(4)	(5)	(6)	(7)	(8)	(9)
1	19.8	9	12	7.0	11.0	51	63	24
2	19.4	8	7.4	7.0	13.5	42	57	36
3	19	9	4	6.0	11.0	53	59	11
4	23	10.5	10	28.5	34.5	82	106	29
5	23	9	8	-	-	62	73	17
6	34	10	5	-	19.0	117	166	42
7	32.5	10	10.5	16.0	22.0	108	140	30
8	18.7	8.5	4.5	2.0	7.0	47	47	0
9	20	9	7.5	-		54	54	0
10	33	11	14	30.0	40.0	129	149	16
11	21	10	10	8.0	14.0	68	80	18
12	33.5	11	16	16.0	34.0	130	150	16

small pieces, less than about 8 inches across. Nevertheless, excavation of the rock at some locations was fairly difficult because it appeared that sections of rock remained intact after blasting. In these locations, rock augers proved ineffective and core barrels were necessary to excavate the rock.

Table 2 summarizes the productivity rates of excavation for the drilled shafts. Because all shafts were constructed with blasting, it is not possible to compare a measured improvement in productivity as a result of blasting. Nevertheless, excavation times in Table 2 show that many drilled shafts of about 20 feet deep with 4 foot to 12 foot long rock sockets were excavated within one day. Considering that the rock sockets were 9 foot and 10 foot diameters, these are excellent production rates. Also, the drilled shaft contractor believes that the blasting did significantly improve productivity when compared to other projects.

Table 2 also presents the estimated volumes of concrete actually used compared to the volumes based on designed shaft dimensions. On average for the 12 shafts, the actual concrete take was 19 percent greater than volume based on shaft design dimensions. The largest take was at Shaft No. 6 where the actual take was 42 percent greater than design volume. The increase in takes resulted from moderate amounts of overbreak in a few of the rock sockets (for example see Fig. 7) and from the use of oversized telescoping casing in the overburden.

More cylindrical rock sockets with fewer overbreaks can probably be formed with slightly different controlled blasting techniques. Three suggestions: 1) provide closer spacing of perimeter blast holes, say 16 inches to 24 inches, 2) use more lightly loaded perimeter holes than the central holes, and 3) design the blast sequence so that the entire top deck is detonated first, followed by the middle deck and lastly the bottom deck.

CONCLUSIONS

1. Blasting was used effectively to improve the productivity of constructing drilled shafts in hard quartzite and schist bedrock.
2. Controlled blasting techniques created a relief face at the ground surface, and subsequently at two lower decks thereby minimizing both ground vibrations and damage to the supporting bedrock.
3. Blast design consisted of central holes detonated first, followed by cushion holes around the perimeter of the shaft. Conventional rock augers and core barrels subsequently excavated the rock sockets.
4. Rock sockets formed by blasting techniques on this project tended to be irregularly shaped, and generally oversized, with some overbreak, but otherwise showed no adverse impacts from blasting.
5. More cylindrical rock sockets with less overbreak can probably be formed with slightly different controlled blasting techniques: 1) provide closer spacing of perimeter blast holes, 2) use lighter-loaded perimeter holes than the central holes, and 3) design the blast sequence so that the entire top deck is detonated first, followed by the middle deck and lastly the bottom deck.

REFERENCES

Carter, J. P. and Kulhawy, F. H. (1992). "Analysis of Laterally Loaded Shafts in Rock." *Journal of Geotechnical Engineering*, ASCE, Vol. 118, No. 6, p. 839 – 855.

O'Neill, M. W., and Reese, L. C. (1999). *Drilled Shafts: Construction Procedures and Design Methods*, FHWA, Pub. No. FHWA-IF-99-025.

Reese, L. C. (1997). "Analysis of Laterally Loaded Piles in Weak Rock." *Journal of Geotechnical and GeoEnvironmental Engineering*, ASCE, Vol. 123, No. 11, p. 1010 – 1017.

Schaffer, A., Greene, B. H., and Green, D. (2000). "Testing the Load." *Civil Engineering,* ASCE, p. A2 – A9.

Teng and Associates. (1969). *Tapered Steel Poles Foundation Design*, United States Steel.

Zhang, L., Ernst, H., and Einstein, H. H. (2000). "Nonlinear Analysis of Laterally Loaded Rock-Socketed Shafts." *Journal of Geotechnical and GeoEnvironmental Engineering*, ASCE, Vol. 126, No.11, p. 955 – 968.

DRILLED SHAFT CONSTRUCTION FOR THE BANDRA-WORLI SEA LINK PROJECT

Conrad W. Felice[1], P.E. Member, ASCE and Henry Brenniman[2]

ABSTRACT: The Bandra Worli Sea Link Project (BWSL) includes a 500 m long cable stayed bridge with two 250 m cable supported main spans on either side of a single central tower with a height of 158 m. Foundations for the project consist of 676 drilled shafts, 72 with a diameter of 2 m and 604 with a diameter of 1.5 m. Working loads on the shafts range from 3.3 MN to 25 MN. The shafts were constructed using a reverse circulation drilling unit and ranged from 4 m to 34 m in length. Rock encountered at the site included highly weathered, fractured and oxidized volcanic material with RQD's of less than 25 percent and unconfined compressive strengths of 1 MPa to massive intact material with unconfined compressive strengths in excess of 100 MPa.

INTRODUCTION

Located in Mumbai, India the Bandra Worli Sea Link Project (BWSL) includes a 500 m long cable stayed bridge with two 250 m cable supported main spans on either side of a single central tower with a height of 158 m. The cable stay structure is to be built by the balanced cantilever method. The approach bridge from the north is 850 m long and a 2939 m long approach

---

[1] Managing Principal, C. Felice & Company, LLC, 11411 NE 124th Street, Suite 275, Kirkland, WA 98034; PH 425-820-0800; Email: cfelice@cfelice.com

[2] Senior Geologist, AMEC Earth & Environmental, Inc., 11335 NE 122nd Way, Suite 100, Kirkland, WA 98034, PH 425-820-4669; Email: henry.brenniman@amec.com

bridge extends from the main span to the south. The approach spans consist of precast segmental box girders and will be constructed span-by-span using an overhead gantry. The project is designed to carry eight lanes of traffic on two carriageways each with a 4 lane capacity. The project, currently in construction is under the direction of the Maharashtra State Road and Development Corporation which was appointed responsibility for the project by the Government of Maharashtra.

Foundations for the project consist of 2 m and 1.5 m diameter drilled shafts. Two meter diameter drilled shafts were designed for the approaches and the main tower of the cable stay bridge. The approach structures to the north and south of the cable stay bridge are founded on 1.5 m diameter drilled shafts. Overall, the project consists of 676 drilled shafts, 72 with a diameter of 2.0 m and 604 with a diameter of 1.5 m. Working loads on the shafts range from 3.3 MN to 25 MN. Each bridge section, except at the cable-stayed portion, is supported on piers typically spaced at 50 m.

This paper describes the construction procedures and techniques employed to install the 2.0 m and 1.5 m diameter drilled shafts for the main tower of the cable stay bridge and the approach structures, respectively. The discussion will include a brief overview of the site conditions and subsurface exploration program. Subsequently, the design of the drilled shafts is presented followed by the procedures and equipment used in their installation. Finally, conclusion and constructability issues encountered during the installation of the drilled shafts are described.

SITE CONDITIONS

The project is located along the western flank of the Deccan Trap Province within the central west coastal region of India. Extensive work by Sukheswala and Poldervaart (1958), Subbarao et al. (1988, 1994) and Sethna (1999) in Mumbai and surrounding areas has revealed that a series of volcanic eruptions and ensuing lava flows covered the project area after the rifting of the main plateau basalts (Cretaceous age ~ 65 million years ago, Ma) while the subcontinent was moving towards the northeast. These lava flows were primarily deposited under subaqueous conditions giving rise to hyaloclastites, spilitic pillow flows and tholeiitic basalt flows, and are believed to represent a much younger phase in the eruptive sequence (62 Ma) of the Deccan Trap basalts. The rocks of Bombay, Salsette and Trombay Islands are the weathered remnants of these lava flows. Successive sheet flow lava accumulations of the Deccan Trap basalts are found to range in thickness from 5 m to 30 m. Periods of quiescence between sequential volcanic events allowed chemical weathering to occur and the deposition of some sedimentary deposits of tuffaceous and/or fossiliferous materials. These interbedded sediments within the Deccan Trap basalts have been designated as intertrappean beds.

The project site geology falls within the Malabar Hill and Nana Chowk lava flows of the Bombay Island Formation. Rock types consist of basalts, volcanic tuffs and breccias and some intertrappean deposits. The fresh basalts, volcanic tuffs, breccias and intertrappean deposits are overlain by completely weathered rock and residual soils. The strength of these rocks ranged from extremely weak to extremely strong at conditions ranging from highly weathered and fractured to fresh, massive and intact.

These weathered rock beds are overlain by transported soils, calcareous sandstone and by thin beds of coarse-grained conglomerate. The tops of these strata are overlain by a marine soil layer ranging in thickness from 0 m to 9 m in thickness and consisting of dark brown clayey silt with some fine sand overlying weathered dark brown basaltic boulders embedded in clayey silt. Water depth ranges from 3 m to 6 m below mean sea level.

SUBSURFACE EXPLORATION

An extensive subsurface exploration and drilling program was undertaken to define the subsurface stratigraphy, identify the rock types, and obtain material properties for design. The program consisted of 191 borings at selected locations along the alignment and at each pier location. Selected rock samples at various depths from the exploratory borings were tested in the laboratory to further classify the material and to estimate relevant engineering properties. Figure 1 shows the range of unconfined compressive strengths for the rock types encountered at the north side of the main pylon and illustrates the extent of the variability that was encountered.

Coring operations were conducted from a jack-up platform equipped with two PVT Limited Drill Max hydraulic drill rigs. The boring program consisted of using NW drill rods and a 3.1 m long, triple tube core barrel with a face discharging diamond core bit. A 55 mm diameter rock core was recovered. Drilling fluids consisted of seawater circulated to the core bit face to wash the cuttings away. No other drilling muds were used. Core runs varied up to 3 m in length depending on drilling conditions.

DRILLED SHAFT DESIGN

Drilled shafts with diameters of 2 m and 1.5 m were selected as the foundation elements for this project because of their high load-carrying

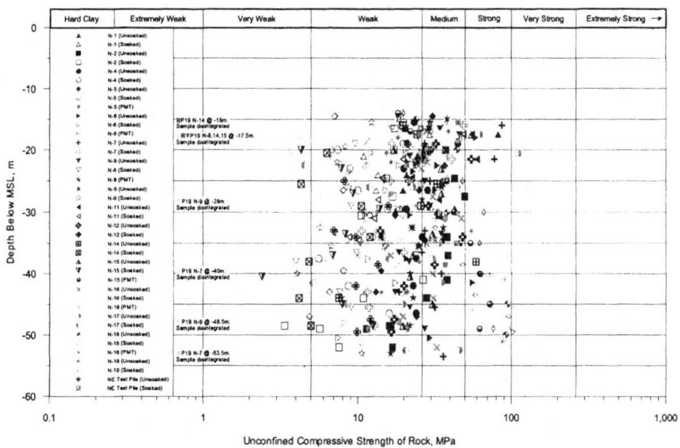

FIG. 1. Range of rock strength on north side of the main pylon

capacity. Two-meter diameter drilled shafts were selected for the approaches and the main tower pylon of the cable stay bridge. The approach structures to the north and south of the cable stay bridge are to be founded on 1.5 m diameter drilled shafts. Overall, the project consists of 676 drilled shafts, 72 with a diameter of 2.0 m and 604 with a diameter of 1.5 m. Working loads on the shafts range from 3.3 MN to 25 MN.

The approach piers are supported by four 1.5 m diameter drilled shafts placed within a pile cap and spaced with a center-to center spacing of not less than 3 times the shaft diameter (i.e., 4.5 m). Figure 2 shows a typical pile cap with dimensions and shaft layout. A similar configuration and spacing was used for the approach piers to the cable stay bridge, where 2 m shafts were placed within a pile cap. The working loads for the approach piers range from 3.3 MN to 14.9 MN.

Figure 3 is a plan and elevation view of the foundation to support the main pylon for the cable stay bridge. As shown, there are twenty (20), 2 m diameter drilled shafts supporting each (north and south) section of the foundation. The working load per shaft is 25 MN.

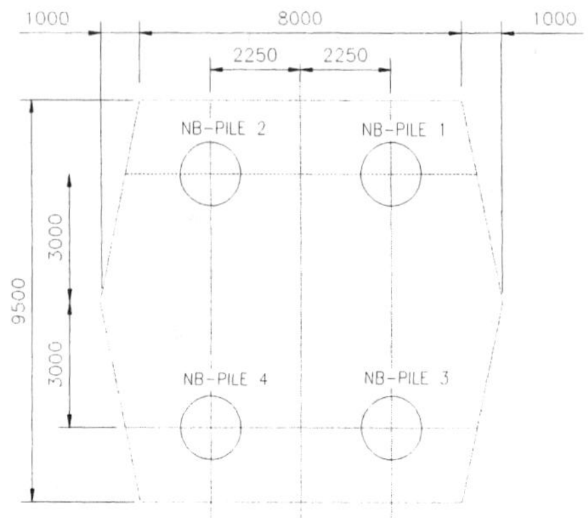

FIG.2. Typical pile cap layout

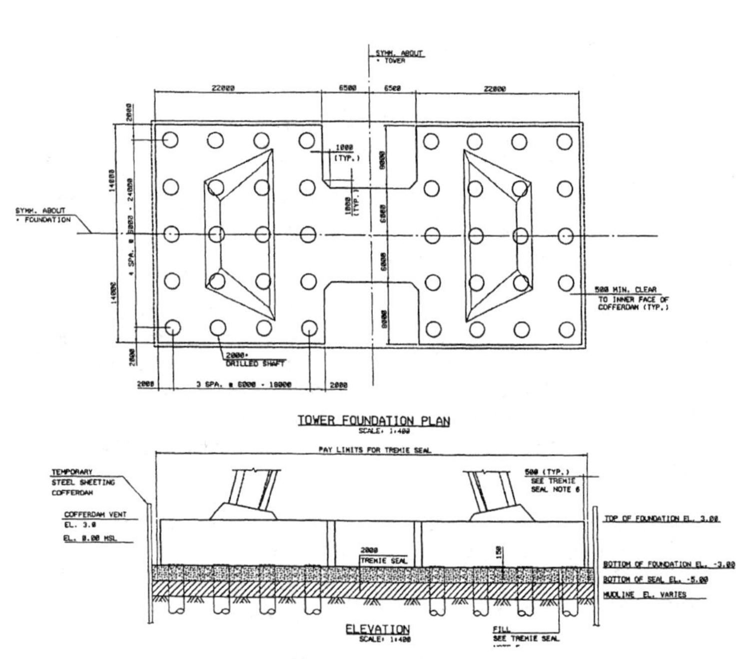

FIG. 3. Pylon foundation layout.

The drilled shafts were designed following the methods proposed by the Canadian Geotechnical Society (1992) and by O'Neill et al. (1996) to compute the capacity and settlement response of a drilled shaft founded in rock. The Canadian Geotechnical Society method was followed when the rock was determined to be sound, with closed and favorably spaced and oriented discontinuities. The direct load-settlement method proposed by O'Neill et al. was used for all other rock types. Based on the extensive subsurface explorations and accompanying load test program, a factor of safety of 2.0 was used in the design calculations for establishing the production tip elevations.

The unit side shear values measured in the load test program that were used for design ranged from 850 kPa for to 1420 kPa. Due to the highly variable geology, the design calculations were performed on a pier by pier basis and the unit side shear values were checked to ensure that they did not exceed the load test results where rock conditions were similar.

The final design of the drilled shafts was based on the assumption that the axial compression load would be carried primarily by the side shear developed between the rock socket and the reinforced concrete shaft. This assumption was based on the allowable settlement criterion of 10 mm and that movement in excess of 10 mm would be required to fully mobilize the available side resistance before additional capacity would be available to resist the load in end bearing. The rock socket lengths for the approach piers ranged from 4 m to 34 m and 9 m to 16 m at the pylon.

The concrete mix yielded an unconfined compressive strength of the 50 MPa and the slump was set at 200 mm in accordance with the project specifications. Based on the use of the reverse circulation drilling procedure for shaft construction, a smooth sidewall surface was assumed in the borehole after drilling, along with a conservative estimate that joints in the rock would be open rather than closed. In addition, the contribution to side shear was neglected over the cased portion of the shaft, as well as the bottom 1 m of the shaft.

DRILLED SHAFT CONSTRUCTION

Casing installation

Prior to drilling each shaft, a permanent casing was vibrated into place using the 60 ton Maxi Traction vibratory hammer (see Figure 4). The average installation time was approximately 1 hour after the casing and vibratory hammer were delivered to the jack-up platform and positioned for driving.

Typical penetrations in below the top of the rock were generally on the order of 1m to 3m, with the occasional refusal on rock fragments in the soil layer.

FIG. 4. Casing installation

Drilling equipment and operation

Drilling of the shafts was accomplished with a Bu-Ma BM R200 pile top drilling unit using the reverse circulation drilling (RCD) procedure. The BM R200 is designed to drill both 1.5 m and 2.0 m diameter shafts with nominal dimensions of 1.485 m and 1.95 m, respectively.

Set-up procedures for the BM R200 consisted of placing the base of the drilling unit on top of the casing using the crane positioned on the jack-up platform and hydraulically clamping the unit to the outer casing wall. The top of the drilling unit is tiltable above the working platform. The BM R200 is pivoted back to allow the bottom-hole assembly to be placed inside the casing (see Figure 5). The bottom-hole assembly consisted of the drill bit, stabilizer, and drill string. The crane on the jack-up platform was used to lower the bottom-hole assembly into the casing and drill pipe is added until the drill bit is resting on the seabed floor. Once the bottom-hole assembly is in place, the working platform is returned to the vertical position and locked in place. Final preparations for drilling include connecting air and water lines from the external hydraulic power pack and compressor to the BM R200.

FIG. 5. RDC assembly

Air pressure is applied around the perimeter of the drill bit assembly forcing water (and the cuttings during drilling) up through an uptake port and the drill pipe. This is connected to a discharge hose at the surface, thus completing the reverse circulation process. Once the circulation process is active, power is delivered to the drill bit engaging the array of tungsten carbide rolling button cutters at the face of the bit (see Figure 6). The BM R200 has seven rolling button cutters: two central cutters and five side cutters on the 1.5 m diameter unit. Controlled thrust delivered to the drill bit by the operator initiates the drilling operation, advancing the boring at the full diameter of the shaft. Under the force of the applied thrust and the rolling action of the button cutters, the rock is crushed and the cuttings airlifted

FIG. 6. Roller drill bit

through the uptake tube of the drill bit assembly and the drill string and discharged at the surface. Cuttings for examination can be collected by screening at the discharge outlet. This operation continues, adding sections of drill pipe (drill pipe sections are 3 m in length) until the shaft is advanced to the specified founding level. At the specified founding level, the bottom-hole assembly is raised approximately 1.5 m to 2.0 m and the base of the shaft is flushed with water under pressure to yield a clean rock surface free of accumulated debris from the drilling operation.

Prior to drilling, a clamshell grab bucket was used to remove the soft marine sediments from within the casing. Removal of the surface soils required up to 2 hours and 30 minutes. Set-up time for the RCD rig prior to drilling was approximately 1 hour and 30 minutes. This included mounting the BM R200 on the casing, connecting the power and hydraulic systems and positioning the drill pipe and bottom-hole assembly in the casing. After the drill string was in position, water was circulated within the casing for approximately 1 hour to remove any remaining loose soils. Drilling rates ranged from 1m to 3m per hour depending on the rock type. At the completion of drilling, the boring was circulated with water for approximately 30 minutes to flush and clean the hole prior to the installation of the reinforcement cage.

Concreting

Table 1 lists the constituents and proportions for the concrete mix used for the drilled shafts. The design unconfined compressive strength was 50 MPa with a slump of 200 mm. Concreting was batched and mixed on shore and barged to each pier location. Transit times ranged up to 2 hrs. Neglecting

TABLE 1. Concrete mix

Concrete Mix M50 proportions (per m^3)	
Constitutents	**Design mix**
Cement 53 grade	370 kg
Microsilica	37 kg
Aggregate 20 mm	617 kg
Aggregate 10 mm	192 kg
Crushed Sand	517 kg
Natural Sand	517 kg
Admixture (Fosroc superplasticiser SP500)	8.954 litres
Water	168 liters

transit time from shore, the average time for placing approximately 42 m^3 of concrete was 2 hours and 30 minutes.

The concrete was tremied into the shaft through a 250 mm tremie pipe. The tremie pipe was gradually withdrawn as the level of concrete rose while maintaining a head of concrete above the bottom of the pipe. During concrete placement, the tremie pipe tip was maintained below the top of the fresh concrete. Cube samples of the concrete (150 mm x 150 mm x 150 mm) were taken to obtain the unconfined compressive strength of the concrete as a function of age. Samples were tested at 7 days and 28 days.

Integrity testing

Integrity testing of the concrete was performed using the cross hole sonic logging method. During fabrication of the reinforcing cage, steel tubes with an inside diameter of 40 mm were tied to the inside perimeter. Four tubes were used for the 1.5 m shafts and 6 tubes were used for the 2 m shafts. CSL testing was performed using equipment manufactured by Testconsult Limited in the UK.

CONCLUSIONS

Challenging aspects of the project included the highly variably geology. Rock encountered included highly weathered, fractured and oxidized volcanic material with RQD's of less than 25 percent and unconfined compressive strengths of 1 MPa to massive intact material with unconfined compressive strengths in excess of 100 MPa. In addition, during the early phases of construction a significant number of issues and concerns were raised that required adjustment and changes to construction practices for the installation of production shafts. These observation are summarized below.

1. Inspection records must be accurate and complete and include reference points for determining tip elevations of the casing and drilled shafts. Also, they should have a tally sheet form for determining accurate casing and drill stem lengths.
2. Improvement in the installation of the surface casing can be accomplished by using clamshell to remove disturbed soils down to bedrock. The RCD drilling unit is not designed to drill through silts and clays and needless time is wasted trying to drill through soft materials.
3. A discharge system should be employed to collect representative samples for comparison against the recovered borehole cores. A log of penetration rates must be maintained to correlate with in situ rock conditions.
4. Each shaft must be properly flushed and sounded to avoid soft bottom effects. The tremie tube can be placed at the bottom of the drilled shaft and used to sound the bottom of the shaft and flush/airlift to enhance spoil removal.
5. In order to accurately determine and compare theoretical and actual concrete volumes, a caliper log of the shaft should be performed to confirm diameter of the borehole. Concurrently, a directional survey of the hole should be conducted to determine if the shaft meets plans and specifications for plumbness.
6. Care is needed in the handling of reinforcement cages, especially with regard to the methods used for picking up and transporting the cages. Reinforcement cages should be kept as straight as possible to avoid bending the reinforcing steel and CSL tubes. Multiple pick points and the use of a spreader bar should be considered.
7. Concrete must be tremied into the hole in a timely manner. Proper embedment (e.g., 1.5 m) into the concrete column must be maintained at all times while extracting the tremie tube. Prior to grouting, the tip of the tremie tube shall be sealed to prevent the influx of water or a plug/mouse shall be placed in front of the grout to drive the water out of the tremie tube prior to grout placement. Grout mix should be such that it will flow and not plug in the tremie tube or hopper.
8. CSL tubes should be filled with water prior to or concurrent with the concreting of the drilled shafts. The water is necessary during CSL testing to provide coupling between the CSL probes and the steel tubes. The water also acts as a temperature buffer while the concrete is curing. Without water in the tubes, the heat of hydration in the concrete causes the steel CSL tubes to expand. When the concrete has cured and cooled

down, the tubes return to their original size. This could lead to a condition called "debonding" in which the CSL signal cannot be passed across the gap between the CSL tubes and the concrete.

ACKNOWLEDGMENTS

The authors would like to thank the authorities of the Maharashtra State Road and Development Corporation (MSDRC) for granting permission to publish this information. The opinions expressed in the paper are entirely those of the authors and may not necessarily constitute a position of MSRDC.

REFERENCES

Canadian Geotechnical Society, (1992). *Canadian Foundation Engineering Manual, 3rd Edition*, Technical Committee on Foundations, Richmond, British Columbia,

O'Neill, M.W., Townsend, F. C., Hassan, K. M., and Chan, P. S. (1996). *Load transfer for drilled shafts in intermediate geomaterials*, FHWA-RD-95-172, Federal Highway Administration, Washington, DC.

Sethna, S. F. (1999). "Geology of Mumbai and Surrounding Areas and its Position in the Deccan Volcanic Stratigraphy, India", *Journal of the Geological Society of India*, Vol. 53, March, pp. 359-365.

Subbarao, K. V. and Hooper P. R. (1988). "Reconnaissance map of the Deccan Basalt Group in the Western Ghats, India," *Mem. Geol. Soc.*, India. No. 10.

Subbarao, K. V., Chandrasekharam, D., Navaneetha-Krishnan, P., and Hooper, P. R. (1994). "Stratigraphy and structure of part of the Central Deccan Basalt Province: Eruptive Models," *Volcanis: Radhakrishna Volume*, K. V. Subbarao, editor, Wiley Eastern Ltd., pp. 321-332.

Sukheswala, R. N., and Poldervaart, A. (1958). "Deccan basalts of Bombay area, India," *Bulletin Geological Society of America*, Vol. 69, pp. 1475-1494.

CAPACITY OF DRILLED SHAFTS FOR THE PROPOSED SUSQUEHANNA RIVER BRIDGE

Benjamin B. Gordon[1], Associate Member ASCE, Joan L. Hawk[2], P.G., and Oscar T. McConnell, Jr.[3], P.E., Fellow ASCE

ABSTRACT: In an effort to reduce the required rock socket lengths for a new bridge over the Susquehanna River near Harrisburg, Pennsylvania, full scale Osterberg Cell load tests were performed on two rock sockets in strata of the Gettysburg Formation. Within each socket, two different zones of interest were identified. One of the objectives of the testing was to compare the side resistance obtained from the load tests with that obtained using empirical methods found in AASHTO and Pennsylvania Department of Transportation publications. Based on the results of this testing, ultimate side resistance exceeded empirical calculations in three of the four zones (977, 757, and 1,006 kPa (20, 16, and 21 ksf) observed versus 407, 304, and 359 kPa (8, 6, and 7 ksf) predicted). For one zone, side resistance was approximately 40% of predicted values (192 kPa (4 ksf) observed versus 497 kPa (10 ksf) predicted), apparently due to construction difficulties.

INTRODUCTION

The foundation investigation associated with the design of replacement structures for the existing Pennsylvania Turnpike's Susquehanna River Bridge project was initiated in 2001. The existing Susquehanna River Bridge is owned by the Pennsylvania Turnpike Commission and at 1,380 meters (4,526 feet) in length it is the longest bridge on the Pennsylvania Turnpike. The existing two-lane bridge, constructed in 1949, consists of a 46-span steel girder structure founded on spread footings. The bridge has been in use for more than 50 years and although it is in reasonably good condition, it is no longer capable of handling projected increased traffic flow on the Turnpike system.

[1] Geotechnical Engineer-in-Training, L. Robert Kimball & Associates, Inc., 615 West Highland Avenue, P.O. Box 1000, Ebensburg, PA 15931, gordob@lrkimball.com
[2] Senior Geologist, L. Robert Kimball & Associates, Inc., 615 West Highland Avenue, P.O. Box 1000, Ebensburg, PA 15931
[3] Senior Technical Leader, L. Robert Kimball & Associates, Inc., 615 West Highland Avenue, P.O. Box 1000, Ebensburg, PA 15931

The existing bridge will be replaced with two new structures on a new alignment, north of the existing alignment. The new structures will be precast concrete segmental bridges across the Susquehanna River. Each structure will consist of 40 spans carrying three travel lanes. Each of the new bridge structures will carry three, four meter (12-foot) travel lanes with a minimum two meter (6-foot) left shoulder and a four meter (12-foot) right shoulder. The total deck width of each structure will be 17 meters (57 feet) out to out. The new precast segmental bridge structures will be the first vehicular bridge construction project in the Commonwealth of Pennsylvania using precast segmental design and construction methods.

Spread footings were originally proposed as the foundations for the new structures. When it became evident that the potential area of disturbance associated with construction of new spread footings in the river and on Calver Island (areas of high archaeological sensitivity), the design team considered the use of cast-in-place drilled shafts to reduce surface area disturbance. The preliminary assessment of available bearing capacity associated with drilled shafts, using Pennsylvania Department of Transportation design methods, resulted in an uneconomical design caused by very deep shaft embedment lengths. Since the drilled shaft foundation option was the preferred foundation system by designers, it was concluded that the best way to assess the viability of a drilled shaft foundation system was to perform in-situ load tests.

GEOLOGIC CONDITIONS

The Susquehanna River Bridge is located within the Newark-Gettysburg Basin. The Triassic Gettysburg Formation is the uppermost formation filling the former Newark-Gettysburg Basin, a narrow, arcuate rift basin that stretches across southeastern Pennsylvania. Ancient river systems draining the arid landscape dumped large amounts of poorly-sorted, sub-angular sediments into the basin. The Gettysburg Formation strikes approximately N62°E and dips approximately 36° to the northwest. The proposed new structures lie at an approximate 19° angle to strike. As a result, borings penetrate increasingly younger portions of the Gettysburg Formation proceeding across the proposed bridge alignment (west to east) and older strata from the westbound lanes to the eastbound lanes. In addition, the dip resulted in the sampling of an approximate 520-meter (1,700-foot) section of the Gettysburg Formation along the proposed alignment.

The Gettysburg Formation at the project site is a heterogeneous unit composed of reddish brown sandstone, siltstone, and shale beds. The results of the subsurface investigation indicate that sandstone is the dominant lithology, but the percentage of the different lithologies varies from east to west along the profiles, and from the westbound to eastbound side of the proposed alignment. Individual beds seldom exceed two meters (5 feet) in thickness, and at many locations the sandstone beds contain thin conglomeratic layers and slickensided shale layers and laminations. There is very little matrix, and the cementing agent is predominantly calcium carbonate ($CaCO_3$).

Diabase sheets have intruded the Gettysburg Formation in the vicinity of the west abutment and have thermally altered (metamorphosed) the strata in a zone extending to the approximate middle of the west channel area. Diabase encountered in the west abutment area appears to be colluvium that has fallen from the steep valley walls, rather than in-place. Evidence of thermal metamorphism of the siltstones and sandstones was found in the abutment borings and a portion of some borings in the west channel. No diabase or evidence of thermal metamorphism was found in borings in the vicinity of Calver Island, the east channel or the east shore exploration areas.

Bedrock is generally softer and weathered to greater depths on the east shore than within the other areas of the project. This is likely due to the overlying mantle of alluvium, which facilitates chemical weathering along micro- and macro-scale discontinuities within the bedrock. These discontinuities provide avenues for free oxygen and water to enter, react with, and alter the minerals of the constituent rock particles or the matrix. Extremely weathered bedrock (saprolite), although a relatively small portion of the bedrock sampled, comprises an increasingly larger percentage within the eastern half of the project area.

Rock strength in the project area appears to be a function of lithology and thermal metamorphism. A comparison of unconfined compressive strength data from samples taken at the west abutment with those taken in other areas of the project areas shows a trend of decreasing unconfined compressive strength, in lithologically similar strata, away from the abutment and the diabase intrusion. In general, fine-grained sandstones and siltstones have medium-high to high unconfined compressive strengths, with moderate to medium-high unconfined compressive strengths predominating away from the intrusion. Sandstones containing siltstone or shale layers are more variable in unconfined compressive strength, ranging from low to medium-high. Weathered sandstones containing weathered shale or siltstone layers are consistently within the low to very low unconfined compressive strength range. Unconfined compressive strengths across the site range from approximately 172 MPa (25,000 psi) near the west abutment to approximately 5 MPa (700 psi) near the east abutment, with a mean value of approximately 34 MPa (5,000 psi).

LOAD TEST PROGRAM

Using the data obtained from the drilling program, two test shaft locations were selected. Test shaft locations were selected so that one test shaft, Test Shaft S-2, could be constructed through poor rock strata thought to be representative of the poorest rock conditions at the site, and the other test shaft, Test Shaft S-6, through rock strata representing the average rock conditions. The assessment of rock conditions was based on lithology, recovery and RQD, and unconfined compressive strength. Other factors affecting the location of the test shafts included minimizing existing utility interference, staying clear of the proposed bridge foundations, accessibility for the drilling equipment, permission from property owners, and the availability of space for the lay-down areas.

The two locations selected were both on the east shore area of the project site, approximately 180 meters (600 feet) from each other. Prior to the construction of the test shafts, three test borings were drilled in a triangular pattern at each test shaft location. The borings were oriented such that one boring was located at the proposed test shaft location, one along the strike of the rock strata, and one along the dip. The borings were field logged by the project geologist and rock samples returned to the laboratory for unconfined compressive strength testing. This testing was used to help delineate the zones of varying strength within rock strata, and also provide the basis for comparison with empirical correlations of shaft skin friction.

The test shaft configurations utilized a singe Osterberg Cell, capable of exerting a force of 16,000 kN (3,600 kips) over a 150-millimeter (6-inch) displacement. The socket lengths were selected so that the single Osterberg Cell would cause the sections above and below the cell to fail at approximately the same load. The test shaft diameter and reinforcement were based on production shaft specifications.

Test Shaft S-2

Test Shaft S-2 was excavated to a depth of 13.31 meters (43.67 feet). The nominal 1,829-millimeter (72-inch) diameter excavation was cased eight meters (25 feet) through soil overburden and into the underlying rock strata. However, the casing did not completely prevent groundwater infiltration, which was observed to flow into the excavation from around the bottom of the casing. Pumps were used to lower the water level to allow for camera inspection of the sides of the rock socket, and then the water level was allowed to return to its static level of approximately four meters (13 feet) below ground surface.

The Osterberg Cell testing arrangement used for this shaft consisted of a single 660-millimeter (26-inch) diameter cell located at a depth of approximately 10.95 meters (35.92 feet). The cell effectively separated the shaft into two sections, a lower 2.13-meter (7.00-foot) long section through medium-grained sandstone having an average unconfined compressive strength of 11 MPa (1,657 psi), and an upper 2.41-meter (7.92-foot) long section through siltstone having an average unconfined compressive strength of 20 MPa (2,904 psi).

Since the primary objective of this testing was to measure skin friction values, compressible end-bearing material was added to the end of the load cell frame to minimize the effects of end-bearing on the test. Rebar cages similar to those proposed for construction shafts were welded to the top and bottom plates of the Osterberg cell, and the whole assembly welded to a steel channel carrying frame, which was used to support the three linear vibrating wire displacement transducers (LVWDTs) and four embedded compression telltales (ECTs) used to monitor movement of the shaft.

The carrying frame was lowered into the excavation, and 28 MPa (4,000 psi) concrete with a 203-millimeter (8-inch) slump was tremie poured via a pump through a 254-millimeter (10-inch) pipe from the top of the compressible end-bearing material to a depth of nine meters (28 feet). The concrete was allowed to cure for at least seven days prior to the load test.

Test Shaft S-6

Test Shaft S-6 was excavated to a depth of 16.98 meters (55.70 feet). The nominal 1829-millimeter (72-inch) diameter excavation was cased eight meters (25 feet) through soil overburden and into the underlying rock strata. However, the casing did not completely prevent groundwater infiltration, which was observed to flow into the excavation from around the bottom of the casing. Several attempts were made to dewater the excavation, but the rate of infiltration was greater than the pump's capacity. The water level quickly returned to its static level of approximately three meters (10 feet).

The Osterberg Cell testing arrangement used for this shaft consisted of a single 660-millimeter (26-inch) diameter cell located at a depth of approximately 16.50 meters (54.15 feet). A rebar cage similar to that proposed for the construction shafts was welded to the top plate of the Osterberg cell, and the whole assembly welded to a steel channel carrying frame, which was used to support the three LVWDTs and two ECTs used to monitor movement of the shaft.

The carrying frame was lowered into the excavation, and 28 MPa (4,000 psi) concrete with a 203-millimeter (8-inch) slump was tremie poured via a pump through a 254-millimeter (10-inch) pipe from the base of the test shaft to a depth of 14.32 meters (47.00 feet). This resulted in an approximate 2.18-meter (7.15-foot) shaft section through fine- to medium-grained sandstone with an average unconfined compressive strength of 33 MPa (4,743 psi). Difficulties were encountered during concrete placement, which resulted in the entire carrying frame moving upwards approximately 0.6 meters (2 feet) while the concrete pump was operating. After the concrete was finally placed, the carrying frame was observed to be approximately 150 millimeters (6 inches) higher than originally planned. The concrete was allowed to cure for at least seven days prior to the load test.

During load testing of this shaft, excessive lateral movement and upward translation was observed, thought to be due to problems resulting from the concrete placement. The decision was made to add a second shaft section on top of the first, and perform additional load testing.

After removing approximately 1.2 meters (4 feet) of silt and sand from the top of the first section with an airlift, measurements to the top of concrete in the first section revealed approximately a 300-millimeter (one-foot) variance across the top of the test shaft. Numerous pieces of limestone gravel, some with thin coatings of Portland cement, were also observed flowing through the air lift system. Two additional traditional telltale casings were placed on top of the existing shaft section and an approximate 2.35-meter (7.70-foot) shaft section, through fine- to medium-grained sandstone having an average unconfined compressive strength of 17 MPa (2,409 psi), was then tremie poured via a pump and allowed to cure at least seven days before testing.

LOAD TEST RESULTS

The Osterberg Cell load testing was performed by LOADTEST, Inc., on July 1, 2002 for Test Shaft S-2, July 8, 2002, for the bottom section of Test Shaft S-6, and on July 24, 2002, for the upper section of Test Shaft S-6. Dial gauges were attached to a 12-meter (40-foot) reference beam supported by steel dunnage and used to monitor top of shaft movement. The instrumentation was connected through a data logger and data recorded automatically every 30 seconds.

The Osterberg Cell was pressurized in predetermined loading increments and the pressure maintained for four minutes at each increment. After the maximum skin friction was achieved, or the Osterberg Cell capacity maximized, the cell was unloaded in five decrements and the test concluded. A summary of the load test parameters is presented in Table 1.

Table 1. Load Test Parameters

Shaft	S-2 (Poor Condition)	S-6 (Average Condition)
Depth to shaft tip	13.31 m (43.67 ft)	16.98 m (55.70 ft)
Depth to water	3.96 m (13.00 ft)	3.05 m (10.00 ft)
Depth to Osterberg Cell	10.95 m (35.92 ft)	16.50 m (54.15 ft)
Depth to top of upper shaft section	8.53 m (28.00 ft)	11.98 m (39.30 ft)
Upper shaft section length	2.41 m (7.92 ft)	2.35 m (7.70 ft)
Upper shaft rock strata	Siltstone	Sandstone
Average upper shaft unconfined compressive strength	20 MPa (2,904 psi)	17 MPa (2,409 psi)
Depth to top of lower shaft section	10.95 m (35.92 ft)	14.33 m (47.00 ft)
Lower shaft section length	2.13 m (7.00 ft)	2.18 m (7.15 ft)
Lower shaft rock strata	Sandstone	Sandstone
Average lower shaft unconfined compressive strength	11 MPa (1,657 psi)	33 MPa (4,743 psi)

Test Shaft S-2

The maximum load in Test Shaft S-2 was 13,629 kN (3,064 kips), with resulting displacements of 19.56 millimeters (0.77 inches) in the shaft section above the Osterberg Cell, and 62.99 millimeters (2.48 inches) in the shaft section below the cell. However, some load was taken up in end bearing. The maximum load taken in skin friction for the lower shaft section was estimated at approximately 8,896 kN (2,000 kips). The resulting maximum unit skin friction values were 976.8 kPa (20.4 ksf) and 756.5 kPa (15.8 ksf) in the upper and lower shaft sections, respectively. A tilt of approximately 2.54 millimeters (0.10 inches) was calculated at the bottom Osterberg Cell plate. Load-Displacement and Unit Skin Friction Curves are presented in Figures 1 and 2, respectively.

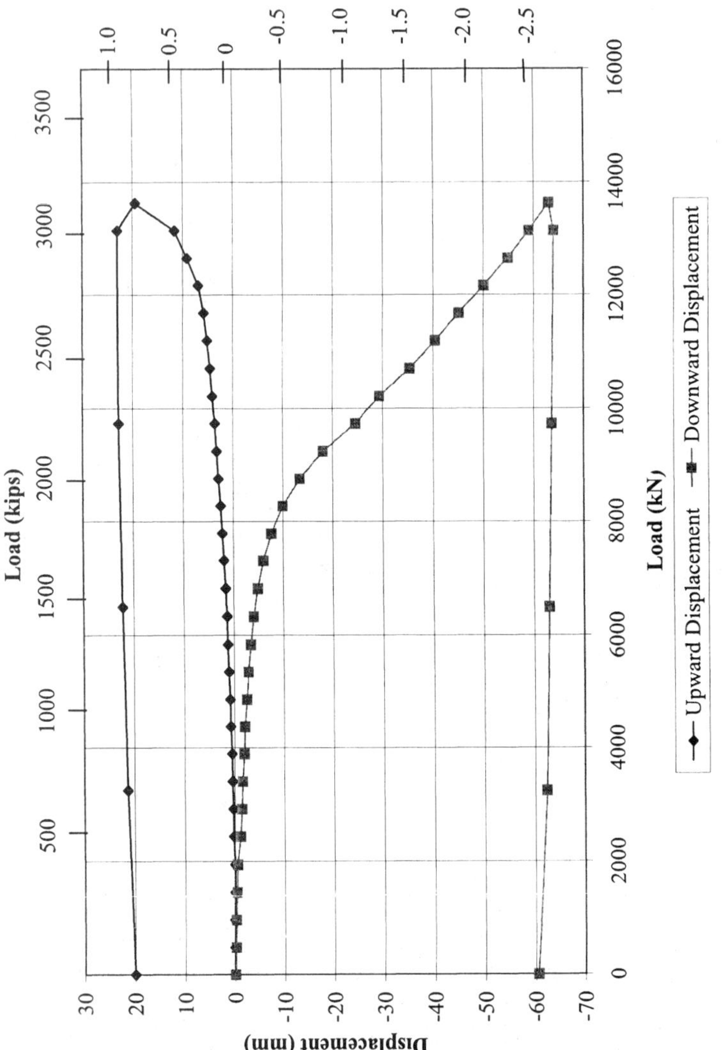

FIG. 1 Shaft S-2 Load-Displacement Curves

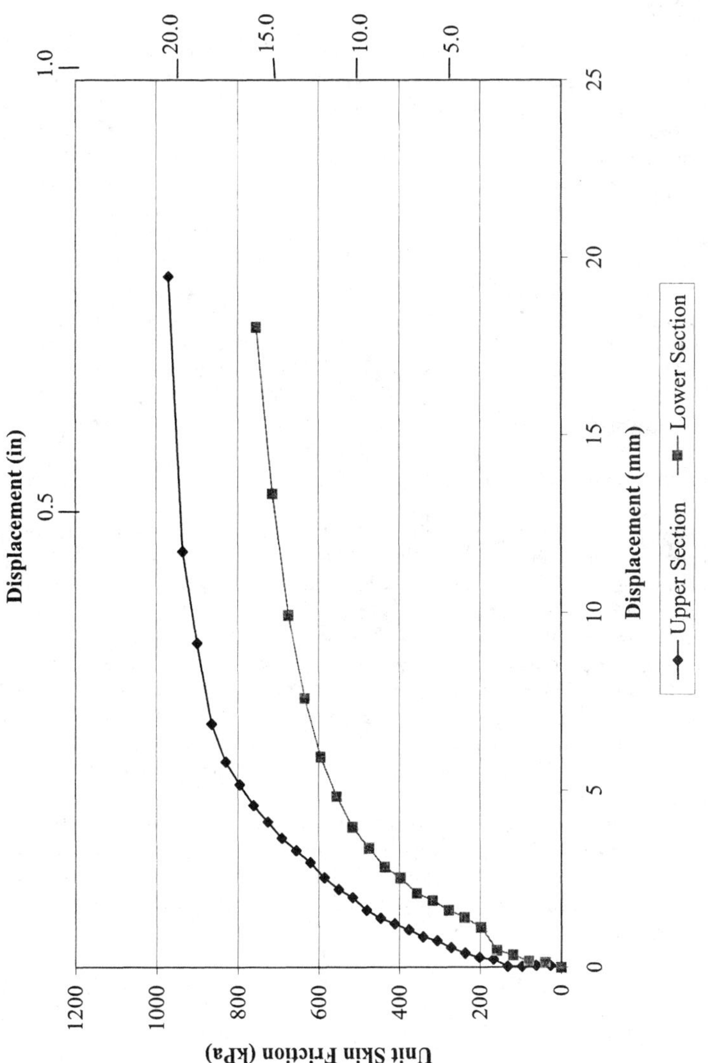

FIG. 2 Shaft S-2 Skin Friction Curves

Test Shaft S-6

The maximum load applied to the bottom section of Test Shaft S-6 was 2,268 kN (510 kips), with resulting displacements of 27.18 millimeters (1.07 inches) above the Osterberg Cell, and 5.33 millimeters (0.21 inches) below the cell. Testing was halted at this point due to excessive upward and lateral displacement. The resulting maximum unit skin friction value was 210.7 kPa (4.4 ksf). The bottom plate of the Osterberg Cell was calculated to tilt approximately 26.16 millimeters (1.03 inches). These results are attributed to the difficulties encountered during concrete placement.

A second test section was cast above the first, and additional load testing performed. The maximum load was 16,044 kN (3607 kips), with net displacements of 13.46 millimeters (0.53 inches) above the Osterberg Cell, and 23.62 millimeters (0.93 inches) below the cell. Testing was terminated once the maximum capacity of the Osterberg Cell was achieved. The resulting maximum unit skin friction values were 1,005.5 kPa (21.0 ksf) and 191.5 kPa (4.0 ksf) in the upper and lower shaft sections, respectively. The end bearing pressure was 5,903.6 kPa (123.3 ksf). Load-Displacement and Unit Skin Friction Curves are presented in Figures 3 and 4.

Measured Design Parameters

Although the original objective of the load testing program was to determine design parameters for both poor and average rock conditions, the test results for the average condition were deemed unreliable due to the excessive deflections observed and poor correlation with empirical design parameters. Therefore, design parameters for the entire structure were based on test results from the poor rock strata.

Since the unconfined compressive strengths from the upper sections of Test Shafts S-2 and S-6 were similar, and the maximum unit skin friction values were 976.8 kPa (20.4 ksf) and 1005.5 kPa (21.0 ksf), respectively, these test results were averaged together. Test results from the lower section of Test Shaft S-2 were used for the design of foundation elements at the east abutment and Pier 39, which are in areas that generally exhibit the poorest rock conditions.

The design parameters were based on net unit skin friction values at a displacement of 10.2 millimeters (0.4 inches). This displacement was selected because it represents the approximate point at which load begins to transfer from skin friction to end bearing. A factor of safety of two was then applied. A summary of measured design parameters is presented in Table 2.

Table 2. Summary of Measured Design Parameters

Shaft Section	S-2 Upper	S-2 Lower	S-6 Upper	S-6 Lower
Ultimate Unit Skin Friction	976.8 kPa (20.4 ksf)	756.5 kPa (15.8 ksf)	1005.5 kPa (21.0 ksf)	191.5 kPa (4.0 ksf)
Unit Skin Friction at 10 mm (0.4 inches)	909.7 kPa (19.0 ksf)	670.3 kPa (14.0 ksf)	861.8 kPa (18.0 ksf)	119.7 kPa (2.5 ksf)
Allowable Unit Skin Friction	442.89 kPa (9.25 ksf)	335.2 kPa (7.0 ksf)	442.89 kPa (9.25 ksf)	59.85 kPa (1.25 ksf)

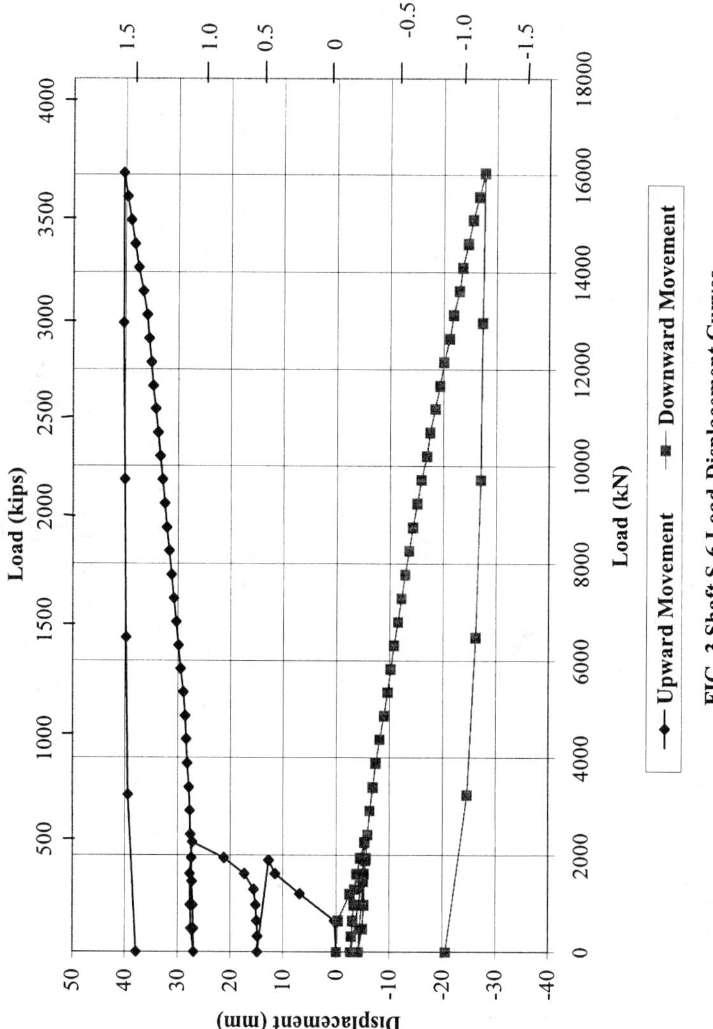

FIG. 3 Shaft S-6 Load-Displacement Curves

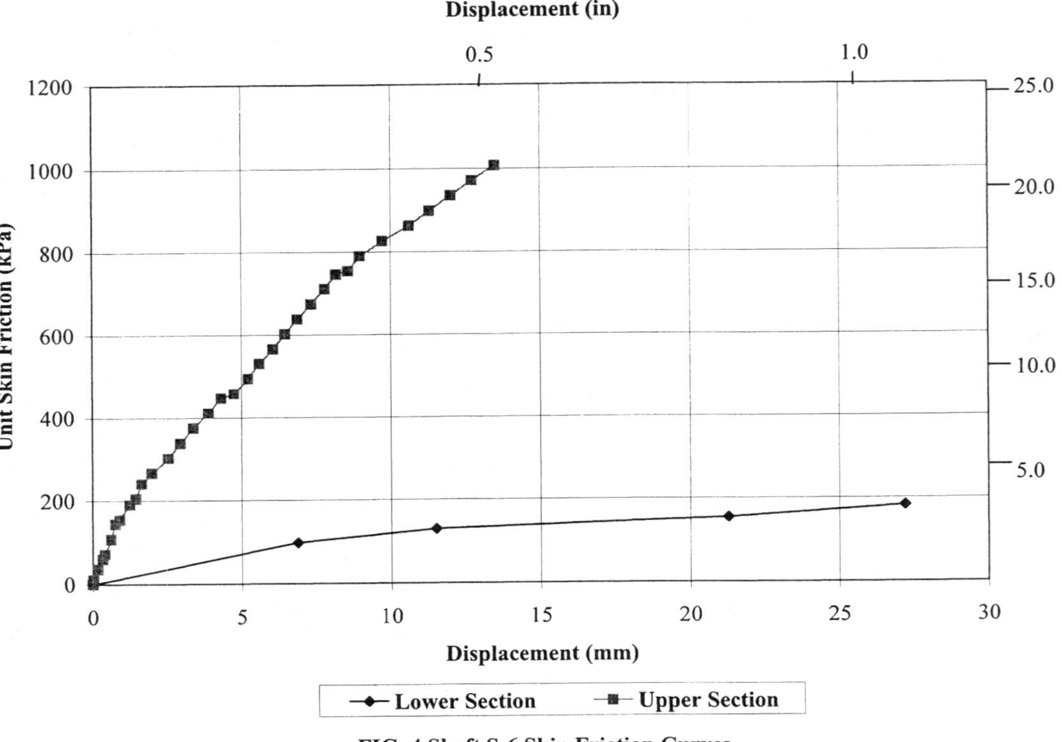

FIG. 4 Shaft S-6 Skin Friction Curves

Empirical Design Parameters

This project was designed in accordance with Pennsylvania Department of Transportation Design Manual Part 4, 1993 edition, which references AASHTO 1993 Interim Bridge Design Specifications. The ultimate skin friction for shafts socketed into rock (Q_{sr}) is presented in Equation 1, where B_r and D_r are the shaft diameter and length, respectively.

$$Q_{sr} = \pi B_r D_r (0.144 q_{sr}) \qquad (1)$$

Values for q_{sr} are from Figure 4.6.5.3.1A of Design Manual Part 4 and are based on the weaker of the unconfined compressive strength of the rock mass or concrete. The unconfined compressive strength of the rock mass is given by:

$$C_m = \alpha_E C_o \qquad (2)$$

C_o is the uniaxial compressive strength of intact rock, and α_E is expressed as:

$$\alpha_E = 0.0231(RQD) - 1.32 \geq 0.15 \qquad (3)$$

Since the RQD of the rock at the shaft locations does not exceed 60%, α_E is taken as 0.15. These ultimate values of skin friction are further reduced by a factor of safety of 2.5. A summary of results using this empirical design procedure is presented in Table 3 using English units, as the 0.144 factor presented in Equation 1 converts the rock socket skin friction resistance from Figure 4.6.5.3.1A from psi to ksf. The allowable skin friction is based on a factor of safety of 2.5, in accordance with Design Manual requirements.

Table 3. Summary of Empirical Design Parameters

Shaft Section	S-2 Upper	S-2 Lower	S-6 Upper	S-6 Lower
C_o	2904 psi	1657 psi	2409 psi	4743 psi
C_m	436 psi	249 psi	361 psi	711 psi
q_{sr} (low)	52 psi	39 psi	46 psi	65 psi
q_{sr} (high)	65 psi	47 psi	58 psi	81 psi
q_{sr} (average)	59 psi	44 psi	52 psi	72 psi
Average ultimate skin friction q_{sr}	406.99 kPa (8.50 ksf)	303.56 kPa (6.34 ksf)	358.62 kPa (7.49 ksf)	496.52 kPa (10.37 ksf)
Allowable skin friction q_{sr}	162.79 kPa (3.40 ksf)	121.62 kPa (2.54 ksf)	143.64 kPa (3.00 ksf)	198.22 kPa (4.14 ksf)

CONCLUSIONS

The load test program conducted for this project is projected to result in significant savings to the client. Ultimate net side friction values used in design are more than double the design values obtained using empirical methods. Since site specific load testing was performed, a lower factor of safety (2.0 versus 2.5) was permitted. Using this reduced factor of safety, in conjunction with the load test data, resulted in allowable side friction values 270 to 300 percent greater than the allowable values from empirical calculations.

Although it is expected that similar increases in side friction will occur in rock strata of higher unconfined compressive strength, the results from the poor rock strata were used across the project site. Using these results, the calculated shaft lengths were within a meter of the four meter (12 foot) limiting length based on design manual requirements. The design manual requires a minimum embedment of two diameters in strata where the bedding planes are not perpendicular to the shaft axis, so additional load testing in rock strata with higher unconfined compressive strength was not warranted. It is estimated that the reduction in production shaft lengths will result in over a million dollars of savings to the client.

Another benefit of conducting the load tests during the design phase was the opportunity to observe construction methods. The test shafts were designed to be very similar to the proposed production shafts. By using the proposed production shaft specifications for the test shafts, specification sections requiring additional clarification were readily identified as a result of the bidders' questions.

Also, it appears construction technique greatly affects the load test results. In the lower section of Test Shaft S-6, the ultimate skin friction was approximately 2.5 times lower than the skin friction calculated from empirical methods. We believe this is directly attributable to the difficulties encountered when placing the concrete in this shaft, and is likely the result of excessive groundwater infiltration. Since this problem was encountered prior to issuing the construction bid documents, additional specifications were prepared to address infiltration, shaft cleanliness, and shaft integrity testing. The end result is a better design with relevant construction requirements that addresses the needs of the project while reducing costs to the client.

ACKNOWLEGMENTS

The load tests were funded by the Pennsylvania Turnpike Commission. Shaft design was performed by Figg Bridge Engineers of Tallahassee, Florida and Media, Pennsylvania. Construction of the test shafts was completed by Brayman Construction Corporation of Saxonburg, Pennsylvania. LOADTEST Inc. of Gainesville, Florida and Baltimore, Maryland, monitored the load tests.

REFERENCES

American Association of State Highway Transportation Officials. (1993). *Interim Specifications – Bridges*, AASHTO, Washington, DC, 81-82.

Attewell P.B. and Farmer, I.W. (1976). *Principles of Engineering Geology*, Chapman and Hall, London, 1045 pp.

Bell, F.G. (1992). *Engineering properties of soils and Rocks,* 3rd ed., Butterworth-Heinemann Ltd., Oxford, pp. 169-189.

Bissell, M.H. (1921). *The Triassic Area of the New Cumberland Quadrangle, Pennsylvania*, A Dissertation presented to the Faculty of the Graduate School of Yale University, in Candidacy for the Degree of Doctor of Philosophy, 153 pp.

Commonwealth of Pennsylvania Department of Transportation. (1993). *Publication No. 15: Design Manual Part 4*, PDT, Harrisburg, B.4-69 – B.4-76.

Glaeser, J.D. (1966). *Provenance, dispersal, and depositional environments of Triassic sediments in the Newark-Gettysburg Basin*, PA. Geol. Survey, 4th ser., General Geology Report 43.

Loadtest, Inc. (2002). *Report on Drilled Shaft Load Testing (Osterberg Method), Test Shaft S-2- PA Turnpike Over Susquehanna River, Dauphin County, PA (LT-8856-1).*

Loadtest, Inc. (2002). *Report on Drilled Shaft Load Testing (Osterberg Method), Test Shaft S-6- PA Turnpike Over Susquehanna River, Dauphin County, PA (LT-8856-2).*

Loadtest, Inc. (2002). *Report on Drilled Shaft Load Testing (Osterberg Method), Test Shaft S-6- PA Turnpike Over Susquehanna River, Dauphin County, PA (LT-8856-2B) Stage 2.*

Maclachlan, D.B. (1999). *Mesozoic, Part VI Geologic History*, Chapter 34, in, the Geology of Pennsylvania, C.H. Schultz, ed., Pennsylvania. Geol. Survey, 4th ser., Special Publication 1, pp. 435-449.

Root, S.I. and MacLachlan, D.B. (1999). *Gettysburg-Newark Lowland, Part III Structural Geology and Tectonics*, Chapter 22, in, the Geology of Pennsylvania, C.H. Schultz, ed., Pennsylvania. Geol. Survey, 4th ser., Special Publication 1, pp. 299-305.

APPLICATIONS OF A SIMPLIFIED DYNAMIC LOAD TESTING METHOD FOR CAST-IN-PLACE PILES

Mohamad Hussein[1], P.E., M. ASCE; Brent Robinson[2],P.E., A.M. ASCE, and Garland Likins[3], P.E., M. ASCE

ABSTRACT

Dynamic loading tests take advantage of the high impact load that a relatively small mass can generate by falling from a preselected height. For driven piles, where the readily available driving hammer itself is the loading apparatus, this test method is particularly convenient. The uniform geometry and material properties of driven piles makes pile load and motion measurements and their interpretation relatively easy. For drilled shafts, however, the testing effort is somewhat more involved because a ram with a weight equal to 1 to 2% of the test load must be available and dropped from heights of 1 to 3 m by using a crane. Additionally, potentially irregular shape and non-uniform material properties of cast-in-place pile tops make accurate force measurements more challenging. This paper further describes how the dynamic testing procedure has been simplified and made more accurate by the APPLE™ (Advanced Pile Proof Loader/Evaluator) method, which combines the force measuring device with the loading system. In effect, the pile top force is calculated from the product of measured deceleration and mass of the drop weight; motion is also measured at the pile top. The paper describes two tests, conducted on different field sites, and presents correlations that further verify the soundness of this approach.

[1] Vice-President, GRL Engineers, Inc., 8000 South Orange Avenue, Suite 108, Orlando, Florida 32809 mhgrlfl@pile.com

[2] Project Engineer, GRL Engineers, Inc., 4535 Renaissance Parkway, Cleveland, Ohio, 44128 brent@pile.com

[3] President, Pile Dynamics, Inc., 4535 Renaissance Parkway, Cleveland, Ohio 44128 garland@pile.com

INTRODUCTION

Cast-in-place deep foundations (known as drilled shafts, bored piles, caissons, piers, auger-cast piles, pressure-injected displacement piles, etc.) are produced by forming holes in the ground and filling them with fluid concrete or grout. In most cases, reinforcing steel is used. The common types are typically 5 to 30 m in length and are constructed with diameters of 250 to 1500 mm, although much larger shafts are also in use. A variety of installation equipment, construction techniques, design and analysis methods are available today (Deep Foundation Institute, 1990; O'Neill and Reese, 1999). Properly designed and constructed cast-in-place shafts provide deep foundation alternatives in a wide variety of subsurface conditions (soils, intermediate geomaterials, and rock) to support a wide range of structures (e.g., bridges, buildings, etc.).

Since deep foundations are often employed where subsurface conditions are difficult, it is the very nature of these conditions that sometimes prompts questions regarding the adequacy and reliability of cast-in-place shafts. Structural strength and geotechnical capacity are functions of grout/concrete material quality and method of placement, shaft structural integrity, subsurface conditions, pile-soil/rock interaction characteristics, design and construction practices, and workmanship. In addition to uncertainties inherent in the design and construction of cast-in-place shafts, the method of installation also has a pronounced effect on their performance under load (Camp et al., 2002). When a single shaft is used to replace a group of smaller deep foundation units to resist large axial and lateral forces, issues regarding foundation adequacy and reliability become more conspicuous due to reduced redundancy in the foundation.

Quality control methods during installation range from simple visual observations to automated instrumentation monitoring systems (Brettmann, 2000). Methods for assessment of the structural integrity of constructed shafts include top surface dynamic low-strain, and down-hole sonic-logging methods (ASTM D5882-00: "Standard Test Method for Low Strain Integrity Testing of Piles, and ASTM D6760-02: “Standard Test Method for Integrity Testing of Concrete Deep Foundations by Ultrasonic Crosshole Tests”). Concern over the behavior of structurally defective cast-in-place shafts under applied loads has motivated research in recent years (Anwar, 1996; Hassan and O'Neill, 1998; Iskander et al., 2001; Sarhan et al., 2002).

Load bearing capacity of cast-in-place shafts can be evaluated using conventional static loading (Crowther, 1988), or dynamic load test methods (ASTM D4945-00: Standard Test Method for High-Strain Dynamic Testing of Piles). Quick execution, simple application, random testing ability, accuracy of results, and low cost make dynamic load testing a preferred method for evaluation and quality assurance of cast-in-place shafts worldwide. The first "production" use of high-strain dynamic testing of cast-in-place shafts was made in 1975 on a project in Charleston, West Virginia (Goble et al., 1993). Extensive experiences from many countries show that high-strain dynamic testing of cast-in-place shafts has economically produced reliable and accurate results (Seidel and Rausche, 1984; Seitz, 1988; Lee et al., 1991; Mukaddam and Iskandarani, 1996; Niyama et al., 2000).

DYNAMIC LOAD TESTING

High-strain dynamic loading tests take advantage of the high impact force that a relatively small mass can generate after falling from a prescribed drop height. Testing is traditionally based on the dynamic measurements of pile strain and motion under hammer impact loads. For driven piles, this test method is particularly convenient since the pile driving hammer itself is the loading apparatus, and testing is done during the normal pile installation process (or during restrike). The uniform geometry and material properties of driven piles make the measurements and real-time data analysis relatively easy for comprehensive evaluations of the hammer-pile-soil system performance during construction (Hussein and Likins, 1995). Reusable strain transducers and accelerometers are attached near the pile top. Field testing is performed with a Pile Driving Analyzer® (PDA) system using real-time data processing and analysis according to the Case Method (Goble et al., 1975). Further data analysis is made with CAPWAP® (CAse Pile Wave Analysis Program) which combines the advantages of field measurements and rigorous numerical "signal matching" analysis to model the soil response (Rausche, 1970; Rausche et al., 1994).

Testing of cast-in-place shafts requires that a ram with a weight of 1 to 2% of the pile capacity be available on site. The weight is handled and dropped using a crane capable of providing drop heights of heights of 1 to 3 meters. Ram weight, drop height, and pile top plywood cushion thickness can be rationally selected by using wave equation analysis for a desired ultimate test load to ensure shaft structural safety and sufficient set under impacts for a successful dynamic test (Hussein et al., 1996).

Typically, each test consists of two or three hammer impacts, applied after the concrete (or grout) has acquired sufficient strength. Instrumentation and analysis methods are similar to those used for dynamic testing of driven piles. The permanent set for each blow is independently measured. In this regard, testing is similar to that of a driven pile during restrike some time following initial installation.

However, potentially irregular shape and non-uniform material properties near the top of cast-in-place shafts make accurate force measurement more challenging. This paper describes how the dynamic testing procedure has been simplified and made more accurate by a recently developed method that eliminates the dynamic measurement of strain (or force) at the shaft top, and reduces the need for a shaft build-up (or excavation below the shaft top). Also presented are discussions on special features in the CAPWAP program that are particularly effective in analyzing test records from cast-in-place shafts. Data from actual cases will also be delineated and discussed.

APPLE - ADVANCED PILE PROOF LOADER EVALUATOR

The APPLE is modification to the dynamic test method that combines the force measuring device with the loading system (Robinson et al., 2002). It improves and simplifies the field performance of traditional high-strain dynamic loading tests of cast-in-place shafts. Obtaining the pile top force by measuring the deceleration of the ram reduces the time and cost of testing by significantly simplifying the pile top preparation required prior to dynamic load testing.

The easily transported and assembled APPLE system consists of a guide frame, modular ram, and a free release mechanism. Figure 1 shows a photograph of it on a site and Figure 2 presents a schematic. The modular ram can be configured to weigh between 45 and 265 kN with drop heights of up to 3 meters (with possible increments of 100 mm), which can test a shaft for up to 26.5 MN static capacity. The apparatus can be easily moved on site for rapid testing using a relatively small crane. The guide frame assures uniform ram impacts and serves to independently support the ram weight prior to the free fall impact; in this way, the crane is not subjected to a sudden load release. Depending on site access conditions, several piles can be tested in one day.

FIG. 1. APPLE On-Site

In preparation for testing, the shaft top must be level and perpendicular to the pile's axis. The frame is then positioned axially and concentrically over the test shaft and evenly supported on the ground. The drop weight is then raised to the height chosen based on wave equation analysis and pinned into position on the frame such that the ram's weight is removed from the crane and transferred through the frame to the ground. Once all personnel are clear of the shaft, a steel wire loop is hydraulically severed causing the ram to fall freely. The ram impacts the top of the shaft, which is cushioned with a few plywood sheets, and the shaft's response is recorded by the PDA. If the base of the ram has an area less than 80% of that of the shaft top, a steel striker plate is typically used to distribute the impact load uniformly at the shaft top.

Instrumentation consists of up to six accelerometers which are attached in pairs to the ram, striker plate (if present), and shaft top; a PDA unit; and a means of

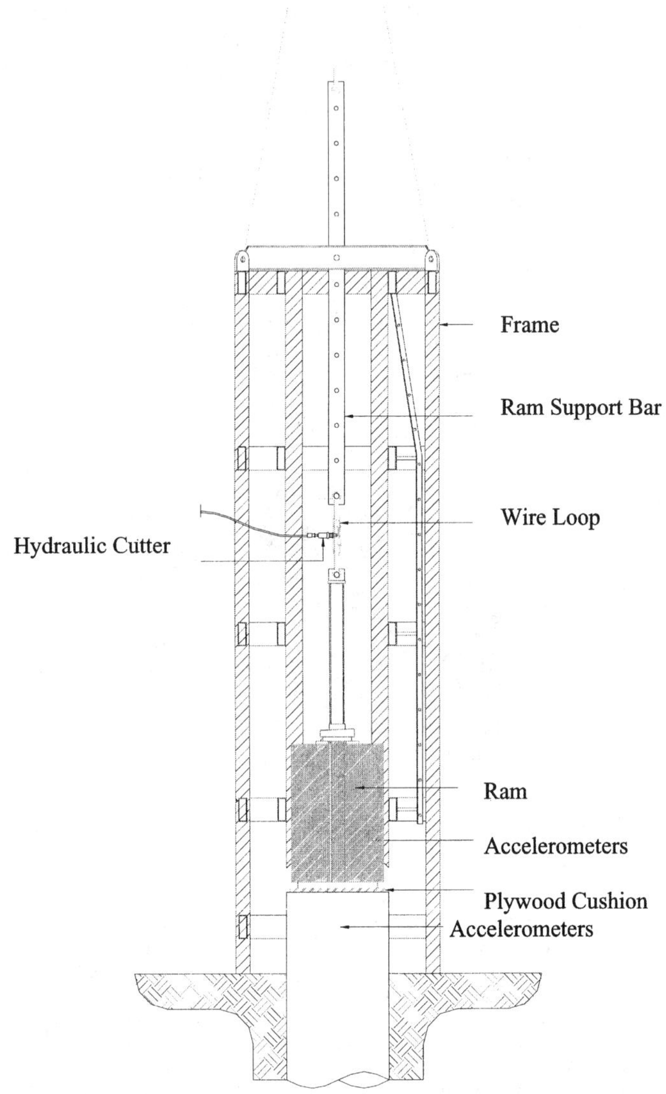

FIG. 2: APPLE Load System Diagram

independently measuring permanent pile set (usually a surveyor's instrument) after each impact. Testing typically consists of obtaining dynamic measurements for two to four impacts per shaft. In most cases, it is desirable to start with a relatively low drop height and incrementally increase the drop height with successive drops.

Utilizing Newtonian Principles, the force (F) acting on top of the shaft can be calculated as the product of the mass of the ram (m) and its measured deceleration (a), **F = ma**. If a steel striker plate is present between the ram and the shaft top, the force is reduced by the plate's inertia, or the product of the plate mass and its measured acceleration. In many cases, however, it has been observed that the striker plate is not needed, further simplifying interpretation of the records. Figure 3 shows that force records independently obtained from ram accelerometer and pile top strain gage measurements under a hammer impact are essentially the same. However, due to cushion deformation effects, ram and pile motions are different from each other. To obtain the required velocity record for dynamic analysis, it is necessary to attach two accelerometers on the shaft as close to the top as possible to minimize phase shift with the ram force record. This location also minimizes the correction for the inertia of the concrete above the location of the accelerometer. The final derived force and velocity records are used in the PDA for Case Method evaluation of stresses and hammer energy, and CAPWAP program for later capacity evaluation.

The CAPWAP computer analysis program employs a sophisticated soil model to analyze high-strain test measurements in a system identification process employing signal matching techniques (Rausche, 1970). It combines the benefits of field measurements with wave equation type simulation in an interactive dynamic environment. Analysis results include the determination of static load bearing capacity, soil resistance forces and their distribution along the shaft length in skin friction and end bearing, and a simulated static load test result showing shaft top load-movement relationship. Analysis results may also produce a simulated uplift load-movement relationship.

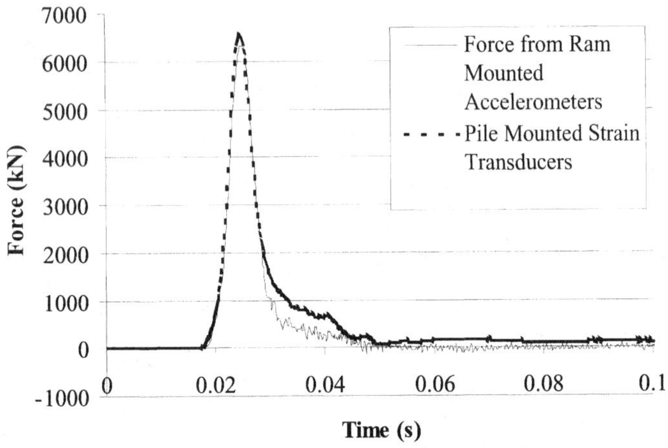

FIG. 3. Force Measurement Comparison

CAPWAP incorporates advanced modeling features and special analysis options that make it particularly suitable include: Smith-viscous damping which allows for a more realistic representation of the dynamic portion of soil resistance, radiation damping model to account for motion of surrounding soil, and the multiple blow analysis option for the combined analysis of test results under several hammer impacts (which is particularly useful for cases of long, mostly friction shafts). The combination of the APPLE testing system and CAPWAP data analysis procedure provides a simple, economical, and accurate high-strain dynamic loading test for cast-in-place shafts of varying diameters and lengths.

CASE HISTORIES

Case 1

A 560 mm diameter auger-cast production pile was tested with a 67 kN APPLE system. The pile extended 26.5 m below the ground. The grout had a nominal compressive strength of 34.5 MPa. The ultimate capacity required for this pile was 3560 kN, a value most likely determined by static analyses from the soil borings. To minimize the amount of excavation around the shaft needed to perform the dynamic load test, a 1 m long, 610 mm diameter steel shell was filled with high early strength concrete and attached to the existing pile. For this pile, ram measurements were used to obtain the force. Another pile on site had been dynamically tested using strain measurements near the pile top to obtain the force and was subsequently calibrated with a static load test.

The pile was installed in soils consisting of approximately 7.6 m of fill material which was underlain by layers of organic clays and silts to depths of 12 m. Medium to stiff clays or silts extended to depths of 26 m, followed by very dense silts at 26.5 m where the pile was founded.

Three blows from successively greater drop heights were applied to this pile, ranging from 0.9 to 1.25 m. The permanent set under each blow was very low, on the order of 1 mm or less. From these three blows, the first blow was selected for CAPWAP analysis. The CAPWAP results are shown in Figure 4.

CAPWAP analyses showed a mobilized capacity of 4350 kN and that approximately 70% of the resistance mobilized was gained along the shaft. This mobilized capacity shows the static analysis methods used to design this pile were likely conservative. To best fit the CAPWAP computed curve with the measured curve, the pile model was changed from "uniform" to "enlarged" over the upper 8 m. This increase in pile impedance (the product of concrete mass density, cross sectional area and pile material wave speed) was justified by the porous fills to depths of 7.6 m and by the pile's total grout volume observed on the site during installation. Increasing the pile impedance beyond the nominal, planned impedance is not unusual for auger-cast piles, where the pressurized grout can infiltrate granular soils and result in a larger than planned cross section.

Case 2

Dynamic load testing was performed with a 67 kN APPLE system on a 914 mm diameter drilled shaft. The shaft was 10.7 m long and constructed using concrete with a compressive strength of 27.6 MPa. The desired ultimate capacity was 2500 kN. This shaft was excavated for transducer attachment to the pile; additional concrete was not poured to extend the pile top for dynamic testing.

The shaft was installed through a 2.5 m thick layer of fine to coarse sand followed by a 7.2 m thick layer of silts and clayey silts. A 1.5 m thick layer of clay described as very hard followed. It was reported that the shaft was installed 5.1 m into dense materials.

Three impacts were applied to this shaft during the dynamic testing. The drop heights ranged from 0.6 to 1.1 m and resulted in sets of approximately 3 mm each. Data obtained under the highest drop height was selected for CAPWAP analysis. The CAPWAP results are shown in Figure 5. The CAPWAP calculated mobilized static resistance for this shaft was 2350 kN, with 1625 kN acting along the shaft. To obtain this match, radiation damping along the shaft and Smith-viscous damping at the toe were used. The high clay content of the soil and the apparent enlarged cross-section of the drilled shaft justified the use of these models. A slight increase in pile impedance was also made near the pile top to improve the overall quality of the signal matching.

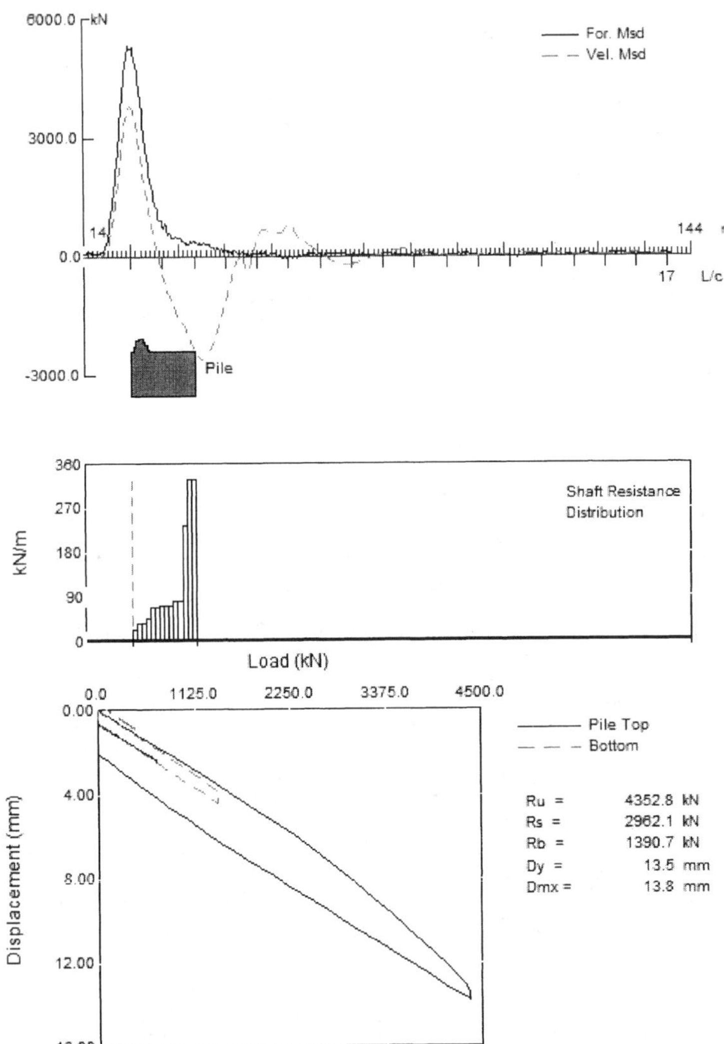

FIG. 4. CAPWAP Results, 560 mm Auger-Cast Pile

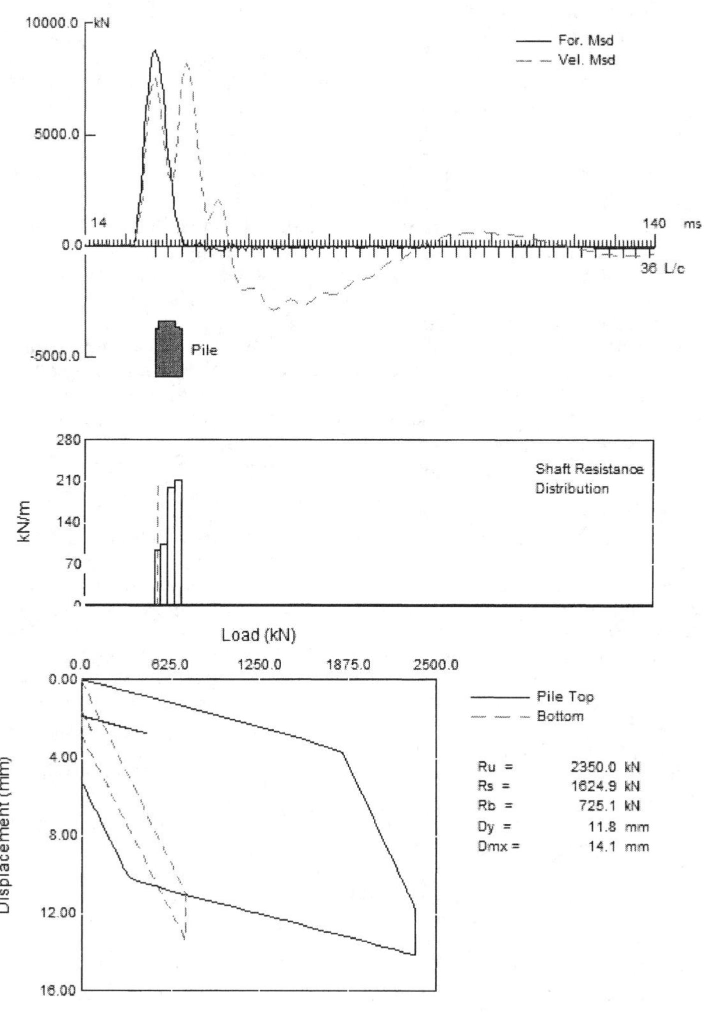

FIG 5. CAPWAP Results, 910 mm Drilled Shaft

SUMMARY

High-strain dynamic testing of cast-in-place shafts is a well established method in contemporary foundation engineering practice around the world for both design and construction control. Advantages over other types of load testing include low cost, convenience and speed of testing, assessment of structural integrity in addition to bearing capacity and resistance distribution, and the ability to randomly test shafts after installation. The analytical methods have been proven to be reliable and accurate through their long track record of scientific study and practical application spanning over a quarter of a century.

The APPLE - Advanced Pile Proof Loader Evaluator - improves and simplifies the high-strain dynamic testing process. It combines the pile force measuring device with the loading system, eliminating the installation of strain transducers on the shaft. Utilizing Newton's principles, the force acting on top of the shaft is calculated as the product of the ram mass and measured deceleration. Pile top velocity is also measured with an accelerometer. The loading apparatus is easily transported and assembled and consists of a guide frame, modular ram, and a free release mechanism. Instrumentation consists of accelerometers affixed to the ram, striker plate (if present), and shaft just below its top, a PDA unit, and a means of independently measuring permanent pile set (usually with a surveyor's instrument) after each impact. The combination of the APPLE system and CAPWAP data analysis procedure provides an economical, accurate, and simple high-strain dynamic loading test for cast-in-place shafts.

REFERENCES

Anwar, N. (1996). "Strength Evaluation of a Defective Reinforced Concrete Bored Pile", ADSC's *Foundation Drilling*, Dallas, Texas, 26(5), 13-18.

Brettmann, T. (2000). "Old and New Quality Control Options for Augered Cast-in-Place Piles", *ASCE Geo-Institute Geotechnical Special Publication No.100*, N. Dennis, R. Castelli, and M. O'Neill editors, 458-470.

Camp, W., Brown, D., and Mayne, P. (2002). "Construction Method Effects on Axial Drilled Shaft Performance", *ASCE Geo-Institute Geotechnical Special Publication No. 116*, M. W. O'Neill and F. C. Townsend editors, 193-208.

Crowther, C. L. (1988). Load Testing of Deep Foundations, John Wiley & Sons publishers, New York, NY.

Deep Foundations Institute (1990). "Augered Cast-in-Place Piles Manual", DFI Augered Cast-in-Place Pile Committee, Sparta, New Jersey.

Goble, G. G., Likins, G., and Rausche, F. (1975). "Bearing Capacity of Piles from Dynamic Measurements - Final Report", Case Western Reserve University, Cleveland, Ohio, ODOT Report No. OHIO-DOT-05-75.

Goble, G.G., Rausche, F., and Likins, G. (1993). "Dynamic Capacity Testing of Drilled Shafts", Foundation Drilling, ADSC Foundation Drilling, September/October, Dallas, Texas.

Hassan, K. M. and O'Neill, M. W. (1998). "Structural Resistance Factors for Drilled Shafts with Minor Defects - Final Report, Phase I", Department of Civil Engineering, University of Houston.

Hussein, M. and Likins, G. (1995). "Dynamic Testing of Pile Foundations During Construction", *Proc. of the ASCE Structures Congress XIII*, Boston, MA, 1349-1364.

Hussein, M., Likins, G., and Rausche, F. (1996). "Selection of a Hammer for High-Strain Dynamic Testing of Cast-in-Place Shafts", *Proc. of the 5th Int. Conference on the Application of Stress-Wave Theory to Piles*, F. Townsend et al., editors, Orlando, Florida, 759-772.

Lee, S.I., Chow, Y.K., Somehsa, P., Kog, Y.C., Chan, S.F., and Lee P. C. S. (1991). "Dynamic Testing of Large Diameter Bored Piles", Proc. of Piletalk Int. '91 Conference, Kuala Lumpur, Malaysia, 63-68.

Mukaddam, M. and Iskandarani, W. (1996). "High Strain Dynamic Testing of Cast-in-Situ Piles in the UAE", *Proc. of the Fifth Int. Conference on the Application of Stress-Wave Theory to Piles*, Orlando, Florida, F. Townsend et al. editors, 805-822.

Niyama, S., de Campos, G., Navajas, S., Paraiso, S., Costa, C., and Barbosa, G. (2000). "Dynamic Load Test of Cast in Place Piles Using a Free Fall Hammer", *Proc. of the Sixth Int. Conference on the Application of Stress-Wave Theory to Piles*, Sao Paulo, Brazil, 429-434.

O'Neill, M. W., and Reese, L. C. (1999). Drilled Shafts: Construction Procedures and Design Methods - Technical Manual, U.S. Federal Highway Administration, Washington, D.C., Report No. FHWA-IF-99-025, 2 volumes.

Rausche, F. (1970). "Soil Response from Dynamic Analysis and Measurements on Piles", Ph.D. Dissertation, Division of Solid Mechanics, Structures, and Mechanical Design, Case Institute of Technology (Case Western Reserve University), Cleveland, Ohio.

Rausche, F., Hussein, M., Likins, G., and Thendean, G. (1994). "Static Pile Load-Movement from Dynamic Measurements", *ASCE Geotechnical Special Publication No. 40*, 291-302.

Robinson, B., Rausche, F., Likins, G. and Ealy, C. (2002). "Dynamic Load Testing of Drilled Shafts at National Geotechnical Experimentation Sites", *ASCE Geo-Institute Geotechnical Special Publication No. 116*, M. W. O'Neill and F. C. Townsend editors, 851-867.

Sarhan, H., O'Neill, and Hassan, K. (2002). "Flexural Performance of Drilled Shafts with Minor Flaws in Stiff Clay", *ASCE Journal of Geotechnical and Geoenvironmental Engineering*, volume 128, no. 12, 974-985.

Seidel, J. and Rausche, F. (1984). "Correlation of Static and Dynamic Pile Tests on Large Diameter Drilled Shafts", *Proc. of the Second Int. Conference on the Application of Stress Wave Theory to Piles*, Stockholm, Sweden, 313-318.

Seitz, J. (1988). "Dynamic Testing of Bearing Capacity of Bored Piles in the FRG - A State-of-the-Art", *Proc. of the Third Int. Conference on the Application of Stress-Wave Theory to Piles*, Ottawa, Canada, B. Fellenius editor, 499-512.

RESULTS OF LATERAL LOAD TESTS ON MICROPILES

James Long[1], Member, Geo-Institute, Massimo Maniaci[2], Glen Menezes[2], Rory Ball[2]

ABSTRACT: Results of ten lateral load tests conducted on micropiles are presented. The micropiles were 15.2 m long and reinforced with a centered high-strength threaded bar along the entire length. Bending stiffness and capacity in the upper 9 m of the micropile were increased by including a 244 mm-OD casing with a 13.8 mm thick wall. Lateral load behavior measured during the load tests were compared with behavior predicted using conventional p-y methods originally developed for laterally loaded piles. Predicted and measured displacements usually agreed within 10 percent.

Most of the lateral load tests were terminated when lateral displacements exceeded the displacement capacity of the jacks; however, two of the ten micropiles experienced excessive displacement at relatively small lateral load. Excavating beside these two micropiles revealed that excessive rotation occurred at the threaded connection in the casing about 2 meters below the ground surface. The threaded connections in these two micropiles experienced excessive rotation at one-fourth of the first-yield moment as determined from structural tests on an intact casing.

INTRODUCTION

Ten lateral load tests were conducted on 244 mm (9.625-inch) diameter micropiles approximately 15.2 meters (50 ft) in length. The micropiles were installed for a seismic retrofit program of critical bridges in Illinois. The tests were located along Interstate 57 about four miles north of the Illinois-Missouri state line. Purposes for the load test program were to investigate the lateral-load behavior of micropiles, to compare the measured lateral behavior with behavior predicted using "state of practice" methods, and to determine the structural behavior of the composite section for the micropile. This paper summarizes the analyses, site investigation, field testing, and results of the lateral load test program. Relevant measurements of lateral load

[1]P.E., Associate Professor of Civil Engineering, University of Illinois at Urbana/Champaign, 205 North Mathews, Urbana, Illinois, 61801, *jhlong@uiuc.edu*
[2]Graduate Research Assistant, University of Illinois at Urbana/Champaign.

versus displacement are compared with predictions using the computer program LPILE. Results of the lateral load test program emphasize the importance of the strength and stiffness of the threaded connection joining micropile sections.

LOCATION AND SOIL PROFILE

The lateral load test program was conducted at an Illinois D.O.T. construction project site near Cairo, Illinois [Section (02-1HB) in Alexander County]. The construction project consisted of a seismic retrofit of a section of Interstate Highway 57 [structures S.N. 002-0001 (NB) and S.N. 002-0002 (SB)]. The test site was located on the southeast corner of Pier No. 17 of the Northbound (NB) lane.

Soil borings had been performed prior to the installation of the test piles, both as part of the original and revised sections of Interstate Highway 57. A total of 29 borings were drilled along the alignment of the north and south-bound lanes of the interstate when the structure was first built (about 1968). Seven additional borings were drilled along the centerline of the interstate as part of the retrofit (1999). Boring depths ranged from 12.6 to 20.3 m (41.5 - 66.5 ft) for the borings advanced in 1968 and 20.3 to 24.8 m (66.5 - 81.5 ft) for the borings conducted in 1999.

The subsurface investigation program consisted of soil sampling, visual classification, strength tests, and Standard Penetration Tests (SPT). Water contents and unconfined compressive strengths, q_u, were obtained for the fine-grained soils. A summary soil profile is given in Fig. 1 showing results of boring logs. Descriptions for the soil profile in 1968 and in 1999 consistently identify a medium-strength clay overlying sand. However, the 1999 soil boring yields higher unconfined clay strengths than reported in the 1968 borings. Strength results from the 1968 borings were used since there were more borings in close proximity to the load tests than for the 1999 boring program. Compressive strengths for the 1968 borings vary from about 67 kPa (0.7 tsf) near the ground surface to about 105 kPa (1.1 tsf) at about 3.3 m (11 ft) below the ground surface, and then decrease in strength to about 38 kPa (0.4 tsf) at the bottom of the clay layer. The standard penetration resistance values in the sand increase with depth and range from 8 to 35 blows per foot.

The lateral load test program required an excavated pit area; therefore, the ground was removed to Elevation 93 m (305 ft). Strengths and thicknesses of soil layers used in the LPILE analyses are discussed later in the section on analysis.

CONSTRUCTION, GEOMETRY, AND MATERIAL PROPERTIES

Micropiles in the test section were originally tested in axial compression, axial tension, or served as reaction piling. Piles 1, 2, 3, 4, 8, 9, 10, and 11 were used as reaction piling in an axial load test program. Piles 5, 6, and 7 were tested in axial compression and/or tension. The lateral load test program was conducted after axial load tests were completed. Fig. 2 is a top view of the test area with the pile positions, numbering, and dimensions.

Micropiles at the test site were constructed in two stages - an upper section with thick-walled casing, and a lower, uncased section with a centered high-strength bar only. The high-strength bar extended the full length of the micropile.

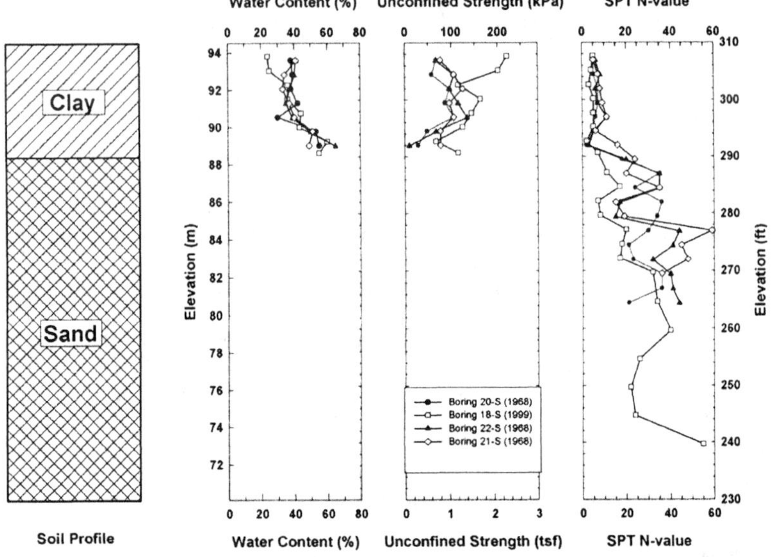

FIG. 1. Soil Profile at Micropile Site

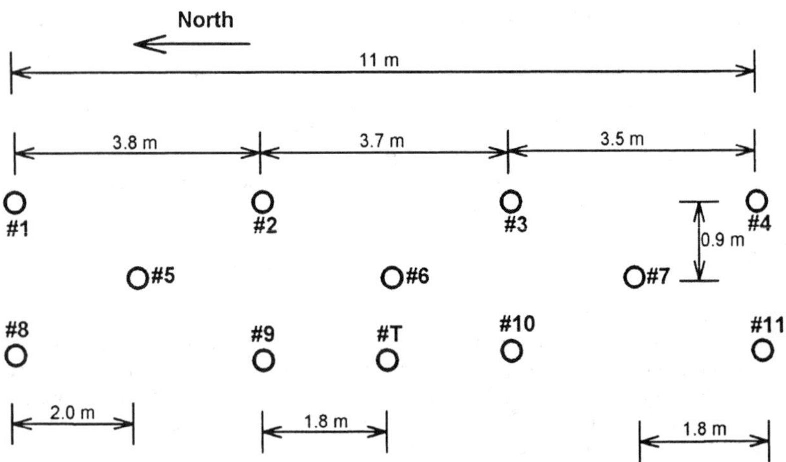

FIG. 2. Plan View Illustrating Micropile Positions

While drilling the upper section with water, a casing with a diameter of 244 mm (9.625 inches) followed closely behind the drill bit to minimize ground loss. Drilling with water and casing continued until a depth of about 9 m (30 ft). The casing was 13.8 mm (0.545 in) thick to provide bending stiffness and capacity to the upper section of the micropile where bending moments from seismic loads would be greatest. The micropile casing was fabricated with high strength steel (yield strength, f_y = 1.02 GPa {147.7 ksi}) and lengths of 3.05 m (10 ft). The top and bottom end of each 3.05 m (10 ft) section was threaded to allow joining casing sections together.

The uncased portion of the micropile extended below the bottom of the casing. The drillhole was advanced using polymer slurry to stabilize the borehole wall. The length of uncased section was typically about 6 m (20 ft). After completion of drilling, the borehole was tremie filled with 27.6 MPa (4000 psi) grout and allowed to set. The lower section was later post-grouted using multi-port grout tube allowing the micropile to develop high bond strengths for developing axial capacity.

Center reinforcement consisted of a 34.9 mm diameter (#11 – 1.375 in) threaded bar for the test piles 1 through 4, and 8 through 11. Micropiles 5, 6, and 7 used a 63.5 mm diameter (#20 - 2.5 in) threaded bar. Threaded bar lengths extended the full length of the micropile.

One additional micropile (pile no. T) was constructed later at the site to provide additional information for moment versus depth for the lateral load test program. The upper 6 m (30 ft) section of the micropile was installed similarly to the existing micropiles; however, no lower section was constructed. A reinforcing cage, instrumented with strain gages, was lowered down the casing and the micropile was tremie grouted. The instrumented reinforcing cage consisted of four- 12.7 mm (1/2 in) diameter reinforcing bars 6m (20 ft) in length tied to a spiral cage that resulted in a 12.7 mm (0.5 in) clearance between the outside of the reinforcing bars and the inside of the casing. The bars were oriented at 0, 90, 180, and 270 degrees with respect to the direction of the applied lateral load. Strain gages were mounted on two opposing sides of the cage (0 and 180 degrees) with a separation distance of 190 mm (7.5 in). The first pair of opposing strain gages was mounted at distances of 305 mm (1 ft) from the top of the cage. Subsequent pairs of gages were positioned every 914 mm (3 ft). A total of 7 pairs of strain gages were installed.

LATERAL LOAD TEST PROGRAM

The pile layout and geometry at the test site are given in this section. The testing system, load frames and platforms, hydraulics, instrumentation, data acquisition, and testing procedure are described.

Pile Layout and Test Pit Geometry

A total of 12 micropiles were installed at the test site (Fig. 2). The test piles were installed along the southeast corner of pier 17 of the northbound lane of the interstate. Strain gauges were installed in one test pile (pile no. T) for determining bending moment along the length of the micropile.

FIG. 3. Photograph of Lateral Load Test in Progress

The area around the test piles was excavated to a depth of 1.8 m (6 ft) below the ground surface prior to testing to allow access to the top of the piles. A 50mm (2 in) layer of crushed stone was placed along the bottom of the excavation to provide suitable mobility for personnel conducting the lateral load tests. However, the crushed stone was removed from the immediate area around the micropile before testing.

Testing System

The major components used to conduct the lateral load tests were as follows: 1) load frames and platforms, 2) hydraulic jacks, hoses, pressure gauges, and hand pumps, 3) instrumentation (load cells, dial gauges), and 4) data acquisition/signal conditioning system. Each component is discussed below.

Load frames and platforms. The loading system was designed to be simple and lightweight to address the issues of tight working conditions and the lack of available heavy equipment at the site. The setup for the lateral load test is illustrated in Fig. 3. The loading system consisted of two identical yokes, one on each pile, connected by two 25.4 mm (1.0 in) diameter high strength threaded rods. The rods were secured on each side with nuts and washers. The loading system was designed to house a hydraulic jack, that when extended, would pull the two piles together. The jack's piston rested against an aluminum cap that fitted into a pin connector allowing the connector to apply load to the micropile without applying moment.

Loads were applied to each pile by a 130 kN (30 kips) hydraulic jack with a maximum piston travel of 150 mm (6 in). The jack was connected to a hand-operated pump equipped with a pressure gauge that read to an accuracy of 690 kPa (100 psi).

Instrumentation. Loads were monitored with two, center-hole load cells that were calibrated prior to testing. Deflections of each pile were measured with two dial gauges, one mounted above the other along the length of each pile head. For the instrumented pile (pile T), seven strain gauge stations were used to measure bending moments along the length of the pile. As mentioned previously. the first pair of strain gages were mounted at a distance of 305 mm (1 ft) from the top of the cage, and subsequent pairs of gages were positioned every 915 mm (3 ft) along the length of the micropile.

Testing Procedure and Program

The lateral loads were applied to the test piles by pulling two piles together. Loads were controlled with a hand-operated pump. The test procedure was to extend the first jack in equal load increments and then repeat the procedure for the second jack. At each load application, the flow valve at the hand pump was locked for approximately five minutes prior to the next pressure increment. The loads and deflections were recorded immediately after the application of the interval and at the end of the interval. After the final set of readings was recorded for a given load increment, the pressure was increased to the next pressure level and the process repeated. This process continued until the maximum travel of the first jack was reached. The flow valve to the first jack was closed, and loading increments were continued using the second jack. Upon reaching the maximum travel for the second jack, the pile was unloaded by opening the flow valves until no further load was recorded on the pressure gauges. A summary of the lateral load test program is given in Table 1.

TABLE 1. Testing Sequence for Lateral Load Tests on Micropiles

Date (1)	Test No. (2)	Pile No. (3)
19 June, 2002	1	3, 4
20 June, 2002	2	1, 2a
20 June, 2002	3	10, 11
21 June, 2002	4	2b, T
21 June, 2002	5	5, 6

STRUCTURAL TEST ON A MICROPILE SECTION

A test section of micropile was constructed on April 26, 2002 in-site by welding a steel plate to one end of a 3.05 m (10 ft) steel casing. A steel bar having cross-sectional area of 820 mm^2 (1.27 in^2) was centered in the casing and grout was poured into the casing. The dimensions of the steel casing were: external diameter = 244 mm (9.625 in), internal diameter = 231 mm (9.08 in), and length = 3.05 m (120 in). The yield strength of the steel casing was 1.01 GPa (147 ksi) and the ultimate strength of the steel rebar was 1.03 (150 ksi). The ultimate strength of grout was 27.6 MPa (4000 psi). The grouted micropile section was transported to the University of Illinois where the micropile was instrumented and tested structurally.

Values for flexural rigidity estimated from theory and back-calculated from deflection and strain measured during the structural bending tests are summarized in Table 2. The computer program LPILE contains a structural model that includes the contribution of both steel and grout for predicting bending stiffness. The agreement between computed and measured is good and both give values of bending stiffness of approximately $14.3x10^3$ (5.0×10^6 in^2-kips). A value of $14.3x10^3$ ($5.0x10^6$ in^2-kips) was used for the modeling the micropile in LPILE.

TABLE 2. Predicted and Measured Bending Stiffness for Micropile Sections

Average Flexural Rigidity m^2-kN (in^2-kips) (1)	Method used for determination of flexural rigidity (2)
$14.3x10^3$ (5.00×10^6)	Calculated by LPILE
$14.1x10^3$ (4.92×10^6)	Determined from bending strains
$15.1x10^3$ (5.28×10^6)	Determined from displacement
$13.4x10^3$ (4.67×10^6)	Theoretical value for steel casing only

LATERAL LOAD PREDICTION

The computer program LPILE was used to predict the lateral load behavior for the micropiles at the site. Predictions require estimates for soil properties, pile properties, load geometry and load.

Soil strength was estimated from the soil exploration programs managed by IDOT. Four borings (20-s, 21-s, 22-s, and 18-s) were used to estimate the soil profile and its strength versus depth. More emphasis was placed on the boring drilled closest to the test site area (borings 20-s and 22-s). The soil profile for the lateral load tests uses a ground surface elevation of 93 m (305 ft). This elevation corresponds to the ground surface in the test pit. Clay soils extend 4 m (13 ft) below the test pit elevation. The shear strength at the ground surface is taken as 34 kPa (700 psf) increasing to 53 kPa (1100 psf) at a depth of 2.4 m (8 ft). The shear strength then decreases to 19 kPa (400 psf) at a depth of 4 m (13 ft). Dense sands are below the clay with standard penetration values varying from 28 to 36.

A flexural rigidity (EI) of $14.3x10^3$ m^2-kN ($5.0x10^6$ in^2-kips) was determined from structural bending tests on a 3.05 m (10 ft) section of micropiling as described earlier in this paper. This value represents the stiffness provided by the upper section of micropiling. The lower section of micropiling was not tested to determine its stiffness, and in fact, the bending stiffness of the lower section has negligible effect on the lateral load behavior. Accordingly, the bending stiffness used for modeling the lower section was one-tenth the bending stiffness of the upper section.

The high bending stiffness of the micropile was used to a depth of 7 m (23 ft) below the ground surface and then the reduced bending stiffness for the uncased section of micropile was used for the remaining 6.1 m (20 ft) for all piles tested, except pile T. Pile T was 9.1 m (30 ft) long with about 1.5 m (5 ft) cantilevered above the ground level in the test pit. Pile T was constructed without the 6.1 m (20 ft) uncased section.

RESULTS OF LATERAL LOAD TESTS

The lateral load tests were conducted by pulling two piles together. The distances between piling were great enough to ensure minimal interaction between the pairs of piles. Specific results (load, displacement, rotation, etc.) for each lateral load test were recorded during the load tests, however, primary attention is drawn to the load versus displacement results and moment versus load results.

Comparison of Predicted and Measured Lateral Behavior

Plots of lateral load versus displacement and lateral load versus slope are shown for all the lateral load tests in Fig. 4. In most cases, the displacement measured at any given load was in reasonably good agreement with the displacement predicted using LPILE. Some piles, notably piles 2 and 3, experienced significantly greater deflections than predicted and greater deflections than the other piles. Pile T was instrumented for moment along the length of piling. Results of moment versus depth are shown in Fig. 5. A short summary for each lateral load test is given in Table 3.

Piles 2 and 3 exhibited greater deflections for the same level of load than the other micropiles. After completion of the lateral loading test program, the soil around piles 2 and 3 was excavated to a depth of 2.4 m (8 ft) and inspected. The threaded connection joining the casing had ruptured - slipping between the female and male threads at the connection. The excessive displacement exhibited by these two piles can be explained by the presence of these structurally weak locations in the micropiles.

Bending moments imposed by the load test program were compared with the bending moments imposed in the structural load test on the micropile. Based on the field lateral load tests, the rate of lateral deflection became excessive at 71 kN (16 kips) for pile 2 and at 102 kN (23 kips) for pile 3. These loads correspond to predicted bending moments of 124 m-kN (1100 in-kips) for pile 2 and 141 m-kN (1250 in-kips) for pile 3 at the threaded connection (1.5 m {5 ft} below the ground surface). For comparison, structural tests on the 3.05 m (10 ft) micropile section resulted in first yield occurring at 565 m-kN (5000 in-kips). Accordingly, these sections failed at about 20 -25 percent of the ultimate moment capacity of the section.

The cause of the damage at the threaded connections in piles 2 and 3 is unknown. Evidence of weak threaded connections was not observed in the other original micropile tested, even with loading some piles to 124.5 kN (28 kips). However, the strain readings for the instrumented pile provided evidence the structural section at the threaded connection was less stiff than the continuous casing (Fig. 4). There were no obvious structural defects in the casing. Accordingly, some explanations for the ruptured connection are:

1) It is possible that the threaded connections were structurally sound, but were ruptured during the process of excavating and leveling the test pit area. The bucket of the excavating equipment may have caught on the micropile, subjecting the micropile to excessive load and causing structural damage to the micropile section. This possibility seems unlikely because the load required to damage the micropile connection would be great enough that the operator of the excavation equipment would have noticed before imposing a damaging lateral load to the pile. Furthermore, no evidence (scratches or scrapes) was observed along the length of the pile.

2) It is possible there was a defect in the structural steel at these threaded connections that resulted in weak structural properties. In this case, either the process of excavation or the lateral load tests could have ruptured the section. It seems unlikely that a manufactured material (the steel casing) would have such a high incidence of structural defects to affect 20 percent of the piles tested.

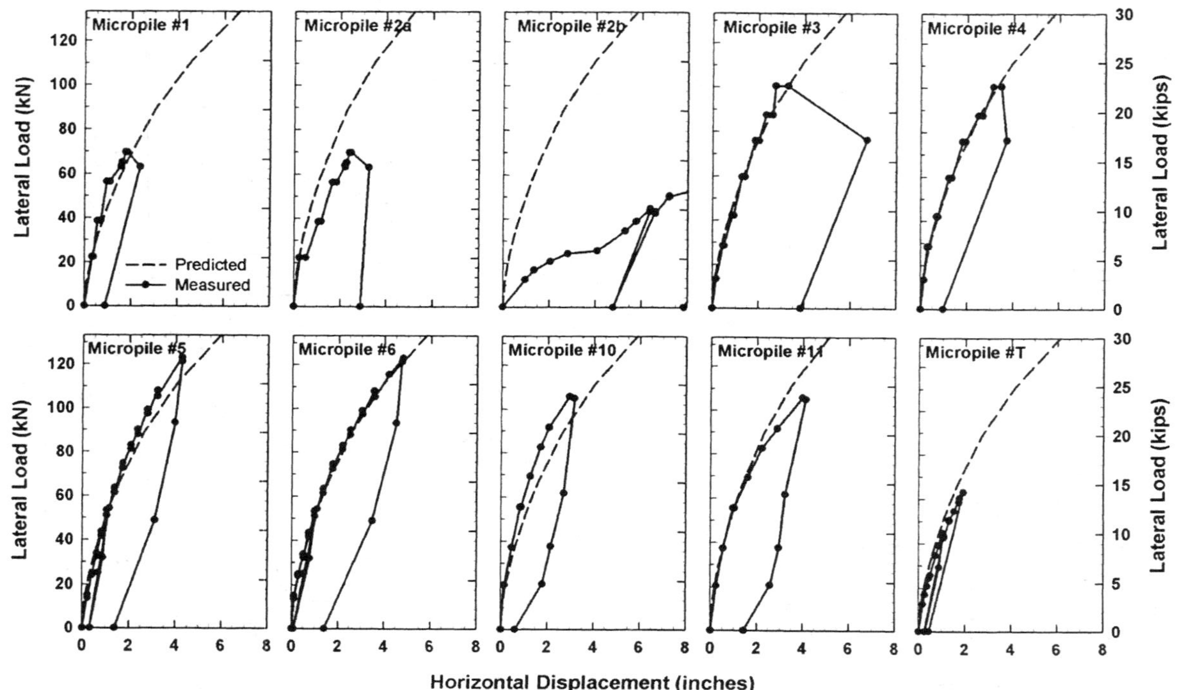

FIG. 4. Load versus displacement

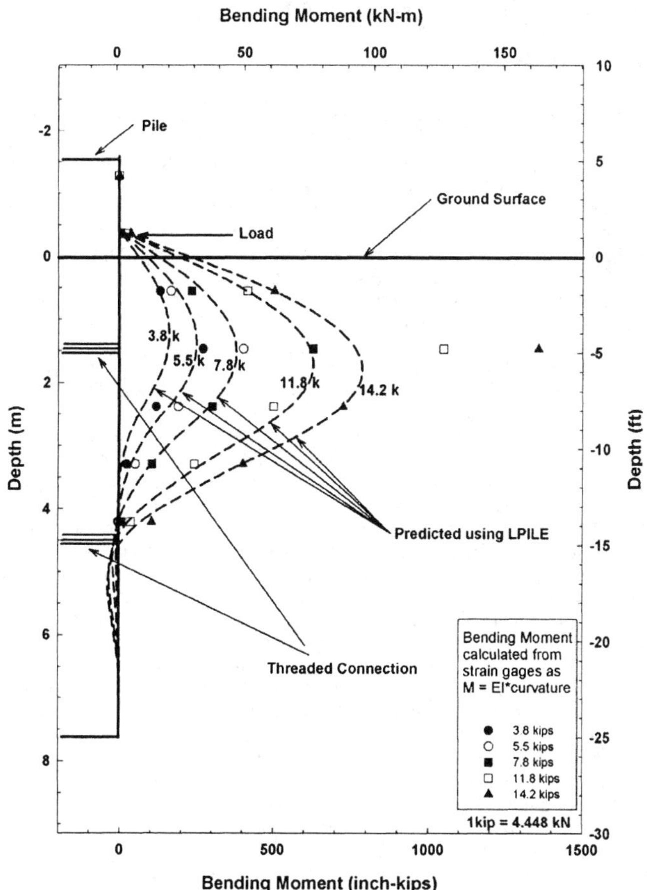

FIG. 5. Bending Moment versus Depth Plot for Instrumented Micropile

3) It is possible that the threaded connections were cross-threaded during installation resulting in an inherently weak connection. It is possible the design details of the threaded connection result in a weak connection in bending, or result in an inherent susceptibility to cross threading in the field.

Although any of these scenarios are possible, it is believed the reduction in bending capacity is a characteristic of the flush-tapered threaded connections used for these micropiles. There are different types of threaded connections available for casing and it is prudent to determine their strength.

TABLE 3. Comments on Each Lateral Load Test

Pile (1)	Comments on Load Test (2)
1	Load test conducted between micropile 1 and 2. Lateral Load applied at 380 mm (15 in) above groundline and displacement measured at 743 mm (29.25 in) above groundline. Loading was stopped because of excess jack travel. Maximum lateral load occurred about 69.8 kN (15.7 kips), then lateral load dropped to 63.2 kN (14.2 kips).
2a	Load test conducted between pile 2 and 1. Lateral Load applied and lateral displacement plotted at 267 mm (10.5 in) above groundline. Loading was stopped because of excess jack travel. Maximum lateral load occurred about 69.8 kN (15.7 kips), then loss of lateral load to 63.2 kN (14.2 kips). Greater than predicted displacements indicate structural weakness along the micropile length.
2b	Load test conducted between micropile 2 and T. This was the second load test on micropile 2. Micropile 2 was loaded in the opposite direction, so the previous loading was not expected to affect the lateral response significantly. Lateral load applied and lateral displacement plotted at 356 mm (14 in) above groundline. Loading was stopped because of excess jack travel. Jacks were readjusted, and the test was continued. Large displacements for these loads indicate structural weakness along the micropile
3	Load test conducted between micropile 3 and 4. Lateral load applied and lateral displacement plotted at 343 mm (13.5 in) above groundline. Loading was stopped because of excess jack travel. Maximum lateral load occurred about 101 kN (22.6 kips), then reduced to 76.5 kN (17.2 kips) at a much greater displacement. The behavior indicates the development of a structural weakness along the micropile length.
4	Load test conducted between micropile 4 and 3. Lateral load applied and lateral displacement plotted at 356 mm (14 in) above groundline. Loading was stopped because of excess jack travel. Maximum lateral load occurred at about 101 kN (22.6 kips), then dropped to 76.5 kN (17.2 kips) because micropile 3 developed a structural weakness.
5	Load test conducted between micropile 5 and 6. Lateral load applied and lateral displacement plotted at 381 mm (15 in) above groundline. Loading was stopped because of excess jack travel. Maximum lateral load occurred about 123 kN (27.7 kips). There was no evidence of structural weakness in the micropile.
6	Load test conducted between micropile 6 and 5. Lateral load applied and lateral displacement plotted at 356 mm (14 in) above groundline. Loading was stopped because of excess jack travel. Maximum lateral load occurred about 123 kN (27.7 kips). There was no evidence of structural weakness in the micropile.
10	Load test conducted between micropile 10 and 11. Lateral load applied and lateral displacement plotted at 381 mm (15 in) above groundline. Loading was stopped because of excess jack travel. Maximum lateral load occurred about 106 kN (23.8 kips). There was no evidence of structural weakness in the micropile.
11	Load test conducted between micropile 11 and 10. Lateral load applied and lateral displacement plotted at 254 mm (10 in) above groundline. Loading was stopped because of excess jack travel. Maximum lateral load occurred about 106 kN (23.8 kips). There may be structural weakness developing in the micropile starting at 53 kN (12 kips).
T	Load test conducted between micropile T and 2. This test pile was instrumented with strain gages attached to a reinforcement cage inside the micropile section. Lateral load applied and lateral displacement plotted at 406 mm (16 in) above groundline. Maximum lateral load occurred about 53 kN (12 kips). Loading was stopped because of excess jack travel, primarily because micropile 2 was structurally weak. Values of moment were estimated for the micropile by multiplying the curvature, as determined from the strain gages, by the bending stiffness (EI). Fig. 5 shows the moment in the micropile increases significantly at the first threaded connection. This is most likely due to a reduced bending stiffness in this zone, and may be indicative of structural weakness.

SUMMARY AND CONCLUSIONS

Lateral load tests were conducted on 244 mm (9.625 in) OD, 13.8 mm (0.545 in) wall thickness micropiles on June 19-21, 2002. Structural tests were conducted on June 24, 25. Except for piles 2 and 3, there was good, overall agreement between lateral load behavior among most of the micropiles. The micropiles were found to displace within about ±10% of displacements predicted using a computer program for lateral loaded piles (LPILE).

The magnitude of flexural rigidity (EI) was dominated by the steel cross section. EI estimated from structural tests agreed reasonably well with EI determined from LPILE. The EI value was predicted and determined to be approximately 14.3×10^3 m^2-kN (5×10^6 in^2-kips). EI of the steel casing was calculated to be 14.3×10^3 m^2-kN (4.7×10^6 in^2-kips), therefore 94% of the EI was contributed by the steel casing.

Most of the lateral load tests were terminated because the travel of the jack was exceeded. However, two of the piles failed structurally, at the threaded connection. Possible sources for structural failure of the threaded connections in the two piles are: 1) excessive loads imposed by excavation equipment during the excavation of the test pit, or 2) due to a defect of the structural section, or 3) due to cross-threading, incomplete threading, or design details of the threaded connection. While there is incomplete information to conclude with certainty which mechanism is responsible for the weakened section, the flush-tapered threaded connection was the source. Additional evidence of a weak connection was exhibited in the instrumented micropile as demonstrated by the strain gages located at the threaded connection. However, in spite of the weak connection, the load displacement behavior of the instrumented micropile exhibited reasonable agreement with predicted behavior when the connections remained intact. Structural distress was not exhibited by any of the other micropiles.

ACKNOWLEDGMENTS

Several people and agencies were responsible for supporting the lateral load tests. Notably Comer Phillips, Bill Kramer, and Chris Hahin, and Joe Lenzini of the Illinois Department of Transportation, Mark Boushele and Al DiMillio of the Federal Highway Administration, Mark Goudschaal and Dave Weatherby of Schnabel Foundation Company. Additional contributions were made by the Mid-America Earthquake Center at the University of Illinois through the Earthquake Engineering Research Centers Program of the National Science Foundation under NSF Award No. EEC-9701785. Any opinions, findings and conclusions or recommendations expressed in this material are those of the authors and do not necessarily reflect those who have supported this effort.

EFFECT OF MICROPILES ON SEISMIC SHEAR STRAIN

Kevin J. McManus[1], Guillaume Charton[2], and John P. Turner[3], Member ASCE

ABSTRACT: The use of inclined micropiles as reinforcement to prevent soil liquefaction in level ground has been investigated experimentally. Deposits of loose (D_r = 0.2 to 0.4), dry sand were prepared inside a large (2.0 m deep by 1.8 m long by 0.8 m wide) laminated box and subjected to shaking of different intensities on a one-dimensional shake table. For low intensity shaking (up to 0.12 g) the cyclic shear strains were modest (up to 0.11 percent) and there was a modest settlement (0.31 percent). For higher intensity shaking, (0.16 g) there was a significant transformation in response with much greater cyclic shear strain (0.65 percent) and settlement (3.1 percent).

Other deposits were reinforced by use of Titan 26-14 self-drilling micropiles installed at 30 degrees inclination. Reinforcement by one inclined micropile was found to have little effect on response to shaking but installation of two diagonally opposed, inclined micropiles was found to reduce cyclic shear strain by half and settlement to one fifth that of similar unreinforced deposits.

INTRODUCTION

Soil liquefaction is a significant hazard in earthquake prone regions. The recognized extent of the hazard is growing rapidly in size as regional studies continue to identify large areas of liquefiable soils. There is a growing problem of knowing how to treat sites where small, low cost structures including dwellings are planned. Large projects can more easily absorb the costs of traditional ground improvement techniques such as deep dynamic compaction, stone columns, and vibro-compaction and large structures can economically be founded on piles. But these techniques are seldom found to be economical for smaller projects and they are not applicable to retrofitting numerous existing affected structures.

[1]Senior Lecturer and [2]Graduate Student, Dept. of Civil Engineering, Univ. of Canterbury, Christchurch, New Zealand; k.mcmanus@civil.canterbury.ac.nz
[3]Professor, Dept. of Civil & Arch. Engineering, Univ. of Wyoming, Laramie, WY 82071; turner@uwyo.edu

Traditional ground improvement techniques are highly invasive, require large-size equipment, generate considerable amounts of noise and vibration, make a big mess, and need a large site to operate in. They are unsuited to small or congested sites or where there are near neighbors. By contrast micropiles can be installed with lightweight equipment, quietly, and in confined spaces, even inside of existing buildings.

Horizontal micropiles (usually called "soil nails") have become widely accepted as a means of reinforcing slopes against sliding failures both from static gravity induced forces and earthquakes. This study has investigated the possibility of adapting micropiles to stabilize level ground during earthquakes by installing them as diagonal reinforcement.

Traditional installation techniques for micropiles and soil nails involve drilling, insertion of steel reinforcing, followed by grouting, and are not suited for loose granular soil below the water table without the use of temporary casing. However, self-drilling micropiles (e.g. Ischebeck Titan micropiles) are now available which are ideally suited to installation in loose, liquefiable sands without use of temporary casing.

There is growing understanding that cyclic shear strains rather than cyclic shear stresses determine the onset of soil liquefaction. Soil liquefaction is a result of the tendency of loose sands to densify with shaking, the resulting effort of the soil to expel the excess pore water causing a temporary increase in pore water pressure and loss of effective confining stress. A number of researchers (Silver and Seed, 1971, Youd, 1972) have shown experimentally that the densification of dry sands is controlled by cyclic shear strains and not shear stresses. Further, the existence of a threshold cyclic shear strain has been found below which soil densification does not occur.

Therefore, if the cyclic shear strain in the soil can be kept below this threshold value, then pore pressure should not be generated and liquefaction should not occur. Dobry and Ladd (1980) have found that for different sands, prepared by different methods, and tested at different effective confining pressures the threshold cyclic shear strain for significant pore pressure generation is approximately 0.1 percent.

Dobry et. al. (1982) have proposed a method for estimating the cyclic shear strain amplitude at a point in the ground as:

$$\gamma_{cyc} = 0.65 \frac{a_{\max}}{g} \frac{\sigma_v r_d}{G(\gamma_{cyc})} \tag{1}$$

in which a_{max} = estimated peak ground acceleration at the site and r_d = an empirical reduction factor. The soil shear modulus G is highly non-linear and is a function of the cyclic shear strain. $G_{\max}$, the small strain modulus may be found for the soil profile by use of seismic CPT profiling, for instance. Modulus reduction curves as a function of cyclic shear strain are available (e.g. Vucetic and Dobry, 1991).

From Equation (1), if the shear modulus, G, for the soil mass is enhanced sufficiently by diagonal reinforcement, and the cyclic shear strain maintained below the threshold value of 0.1 percent, then there should be minimal generation of pore water pressure and no liquefaction. Further, by reducing cyclic shear strain the reinforcement should also act to maintain the soil's own initial stiffness which otherwise tends to degrade rapidly.

This study has examined experimentally the effectiveness of inclined micropiles as reinforcement to reduce soil cyclic shear strains during shaking. Full-size prototype micropiles (Titan 26-14) were installed in loose sand in a large (2 m deep x 1.8 m long, x 0.8 m wide) laminated shear box then subjected to different levels of shaking. Results were compared to similar soil deposits without micropile reinforcement.

TESTING PROGRAM

Laminated Sand Tank

Prototype inclined micropiles were tested in a large laminar sand tank on a one degree of freedom shaking table at the University of Canterbury. The purpose of the laminar tank was to simulate free-field shaking response by allowing the soil to deform in simple shear with minimal boundary effects from the tank (e.g. Hushmand et. al., 1988, Iai, 1991, Whitman et. al. 1981). The tank design used for this study follows from that of Hushmand et al., and is shown in Figure 1. The tank has internal dimensions of 1.8 m long by 0.8 m wide by 2.0 m deep.

Laminates were made from 100 mm by 50 mm cold-formed steel channel that was laid on its flat and welded into rectangular frames. Teflon strips of 150 mm long by 10 mm wide by 1 mm thick were glued to both sides of the laminates at six locations to minimize friction. Tests showed that the strips produced a coefficient of friction of 0.07, indicating that at normal stresses equivalent to those at the base of the tank, the load required to shear the laminates was only 2 percent of the load required to shear the soil mass.

Soil was contained within the tank by a flexible membrane liner. Latex rubber sheets, each 1 mm thick, were draped over the inside of the laminates and glued to the top laminate.

The laminates were supported by a steel frame that was constructed from 50 mm by 50 mm rectangular hollow section (RHS), with 10 mm diameter rods acting as cross-bracing members. The frame restricted the laminates to move only in the direction of the shaking table and also supported the stack of laminates when the tank was empty. Both the top-cap and the side members of the supporting frame were coated with Teflon strips to reduce friction during shaking.

Instrumentation

Five potentiometers (Showa type 50LP300) were placed in contact with tank laminates at various heights above the tank base (1.07 m, 1.34 m, 1.55 m, and 1.97 m) and fixed rigidly to the shaking table in order to measure relative lateral displacements of the tank laminates during shaking. An accelerometer (Kyowa AS-5GA) was fixed to one of the tank laminates near to the soil surface (1.77 m above tank base) in order to measure soil acceleration and a similar accelerometer was fixed directly to the shaking table.

Settlement of the soil surface was measured by fixing a vertically oriented potentiometer to the tank support frame and making contact with an aluminium plate resting on the sand surface.

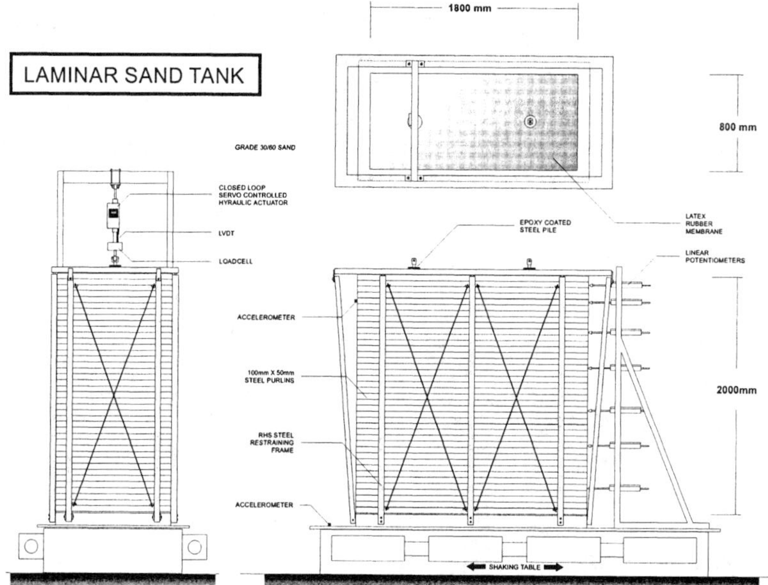

FIG. 1. Laminar sand tank.

Soil Deposits

The soil used was an industrial grade 30/60 silica sand supplied by Commercial Minerals Ltd, Auckland, New Zealand, with properties given in Table 1. This soil was selected because it is suitable for air pluviation and can be re-used without degradation.

Soil deposits were prepared by air pluviation. Sand flowed from a hopper through a gate, was collected in a suspended funnel, then flowed down a 95 mm diameter flexible hose, discharging through a wire mesh diffuser into the laminated tank. The diffuser was made from a 100 mm diameter by 300 mm long section of plastic tube that was packed with wire mesh. By discharging sand from the diffuser directly onto the surface of the deposit, a low initial relative density (six deposits, Dr = 0.17 – 0.26) was achieved. The sand densified somewhat during each episode of shaking enabling some tests to be performed in higher densities (as high as $D_r = 0.4$).

Two cone penetrometer tests (CPT) were performed in one of the deposits. The penetration was found to be consistent throughout the deposit with q_c ranging from 1 MPa to 1.4 MPa.

Shaking Table

Characteristics of the University of Canterbury shaking table are given in Table2. . The table is driven by a closed-loop, servo-controlled hydraulic actuator with an MTS Teststar 2 system controller. Each test was performed under displacement control, with the cyclic table displacements generated by entering the required amplitude, frequency and number of cycles into the controller.

Table 1. Soil Properties

Property	Symbol	Value
Density of Solid Particles	ρ_s	2.65 t/m^3
10% finer	D_{10}	0.30 mm
60% finer	D_{60}	0.45 mm
minimum voids ratio	e_{min}	0.53
maximum voids ratio	e_{max}	0.83
Steady State Friction Angle	ϕ_{ss}	33°

RESULTS

Soil response to shaking

Four deposits were constructed without micropile reinforcement and subjected to shaking to verify behaviour of the laminar tank and to determine baseline soil response. Soil deposits were subjected to individual "earthquakes" consisting of 26 cycles of 1 Hz sine wave shaking at three amplitudes: +/- 20 mm, +/- 30 mm, and +/- 40 mm corresponding to accelerations of 0.08 g, 0.12 g, and 0.16 g.

During shaking, the displacement measurements showed that the soil mass deformed in a linear, simple shear mode from the tank base to a height of 1.6 m, then deformed in a non-linear, irregular mode from 1.6 m to the surface at 2.0 m. The deformation of the surface soil seems to have been affected by surface waves of complex shape that were observed during shaking. Typical displacement measurements are shown in Figure 2.

Cyclic shear strain during shaking is shown in Figure 3 for the three amplitudes of shaking. The peak displacement of each transducer for each cycle of shaking was captured and divided by the height above the tank base then averaged over all of the transducers to give an average peak shear strain for each cycle. Peak strain decreased during each "earthquake" as the initially loose soil densified. For the lower amplitude shaking (0.08 g and 0.12 g) the cyclic shear strains were modest (0.08 percent and 0.11 percent for cycle 13) but for the higher amplitude shaking (0.16 g) the cyclic shear strain was much greater (0.65 percent for cycle 13).

Significant settlements occurred at the surface of each soil deposit during shaking as the initially loose sand densified. The amount of settlement varied significantly depending on the amplitude of shaking, as shown in Figure 4. For the shaking at 0.08 g and 0.12 g the settlement was modest and similar(0.31 percent and 0.35 percent), but, for the higher level shaking at 0.16 g the settlement was much greater (3.1 percent).

Table 2. Characteristics of the University of Canterbury Shaking Table

Property	Value
Plan Dimensions	4.0 m x 2.0 m
Maximum Allowable Load	200 kN
Maximum Horizontal Force	200 kN
Maximum acceleration with a mass of 5 tonne	2.7 g
Maximum Velocity	1.0 m/s
Maximum Displacement	0.30 m

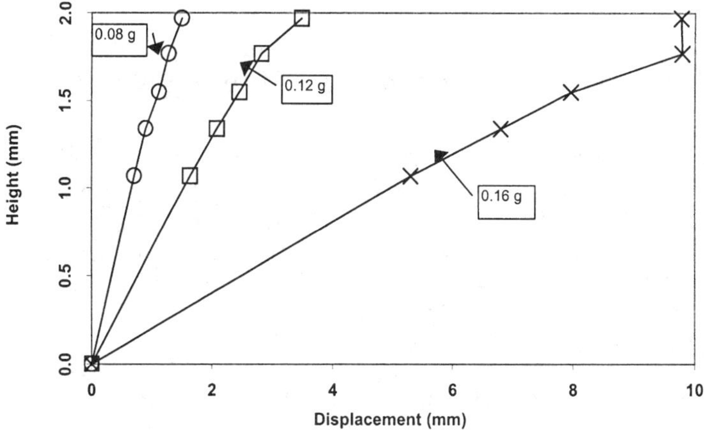

FIG. 2. Displacement profiles for unreinforced soil deposits.

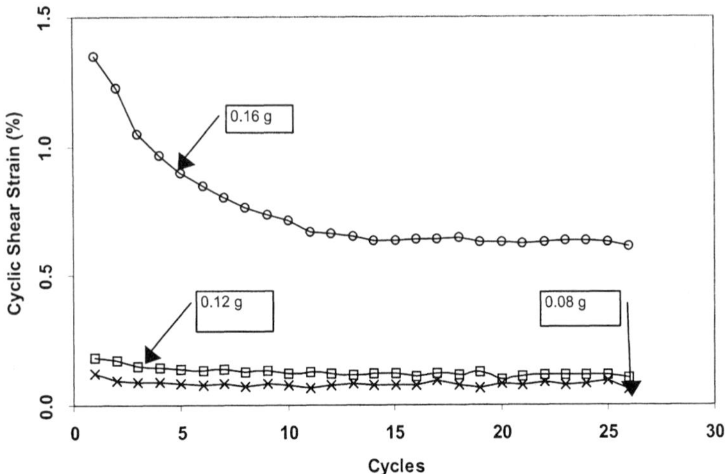

FIG. 3. Average peak cyclic shear strain for unreinforced soil deposits (D_r = 0.2)

Clearly, a significant transformation in response occurred between the shaking at 0.12 g and the shaking at 0.16 g, with settlement jumping from 0.35 percent to 3.1 percent. The shaking at 0.12 g caused a cyclic shear strain of 0.11 percent which is very close to the threshold value for liquefaction of 0.1 percent suggested by Dobry and Ladd (1980). Increasing the shaking intensity further to 0.16 g may have triggered some "collapse" of the soil fabric with a large reduction in shear stiffness and increase in settlement. This "collapse" may be equivalent to liquefaction occurring in a saturated sand deposit.

Micropile Installation

Two soil deposits were reinforced with diagonal micropiles. One deposit was reinforced with a single micropile and was reinforced with two diagonally opposed micropiles, as shown in Figure 5. Titan 26-14 self-drilling micropiles supplied by Ischebeck (NZ) Ltd were used as reinforcement. Titan micropiles consist of high-strength hollow steel threaded bars installed by a self-drilling process with a sacrificial drill bit. Grout is injected during drilling at low pressure to mix with the surrounding soil and provide bonding and corrosion protection.

For this study, the micropiles were installed into loose sand at shallow (2 m) depth and so the bars were installed simply by pushing with hydraulic rams (a cone penetrometer pushing rig), as shown in Figure 6. An oversize cone-shaped drill head was fixed to the pile tip to create an annular space that was progressively filled with grout during pushing. A photograph of the completed installation with two diagonally opposed micropiles is shown in Figure 7.

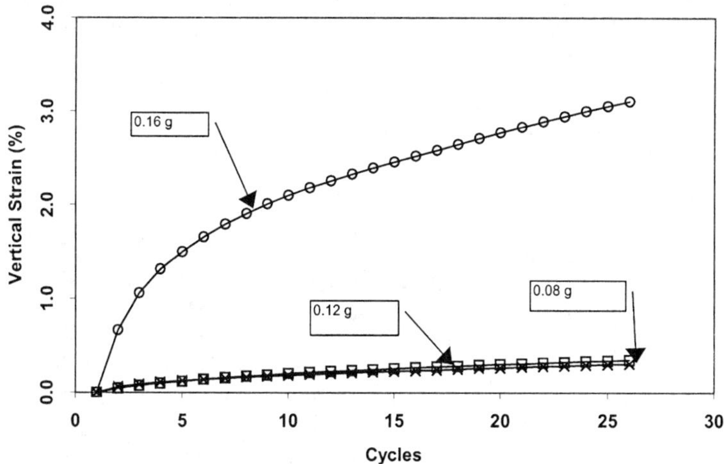

FIG. 4. Settlement of unreinforced soil deposits ($D_r = 0.2$)

The grout mix was 50:50 by weight of ordinary Portland cement:water, with 3 percent bentonite by weight of cement added to stabilize the grout and reduce water loss to the dry sand. The unconfined compressive strength of the cured grout was 9 MPa at 7 days and 11 MPa at 28 days. Micropiles were exhumed from the soil after each test and were found to be highly uniform in cross-section with a diameter of 100 mm +/- 15 mm.

The two reinforced soil deposits (one micropile and two micropiles) were subjected to the same levels of shaking as unreinforced deposits of similar density ($D_r = 0.4$). The response of the reinforced deposits is compared to equivalent unreinforced deposits in Figures 8 and 9, showing average peak cyclic shear strain and settlement during shaking. Response of the soil deposit with one inclined micropile was quite similar to the unreinforced deposit. The average peak cyclic shear strains were similarly high (0.54 percent for the unreinforced deposit and 0.64 percent for the reinforced deposit after 13 cycles) and the total settlements were also quite similar (1.2 percent and 1.1 percent). The main notable difference in response was that the deposit with one micropile initially had a larger response in terms of both cyclic shear strain and settlement than the unreinforced deposit, with the response steadily declining during the test.

Response of reinforced soil deposits

The response of the soil deposit with two diagonally opposed micropiles was reduced to about half of the cyclic shear strain of the unreinforced deposit (0.24 percent after 13 cycles) and about one fifth of the settlement (0.24 percent). This level of cyclic shear strain is somewhat above the threshold of 0.1 percent for liquefaction suggested by Dobry and Ladd (1980), but the settlement was reduced substantially suggesting that liquefaction in an equivalent saturated deposit might have been prevented.

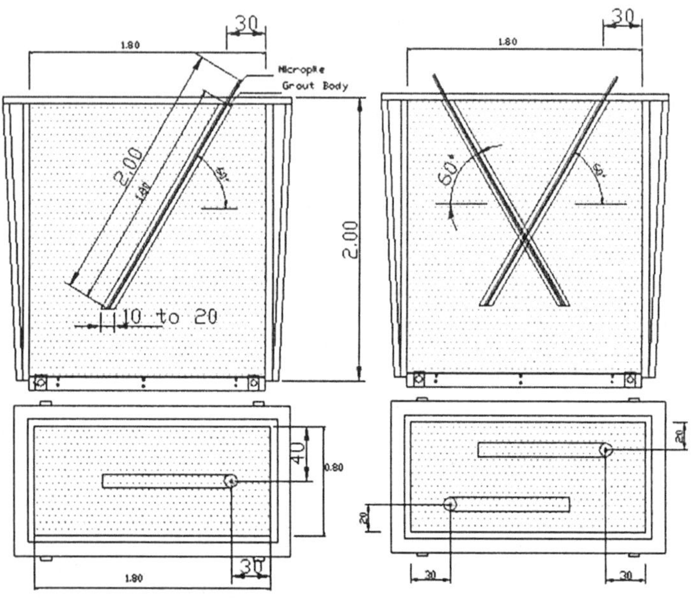

FIG. 5. Layout of micropile reinforcement in laminar tank

Nevertheless, the increase in shear stiffness of the deposit provided by the reinforcement was low considering the steel cross-section introduced. From the measured cyclic shear strains shown in Figure 8, the equivalent shear modulus, *G*, at mid-depth of the deposit (1 m) may be calculated as 380 KN/m^2 for the unreinforced deposit and 1010 KN/m^2 for the deposit with two micropiles. The equivalent shear modulus provided by the steel reinforcement, if the steel cross-section were fully mobilised, is 52,000 KN/m^2. Obviously, the capacity of the micropiles is hardly mobilised suggesting that reinforcement of much lower strength and stiffness may provide similar benefit at lower cost.

CONCLUSIONS

A large-size laminar tank performed well on the shaking table with linear simple shear being generated in deposits of loose sand in all but the upper 0.2 m. Shaking of loose, unreinforced sand deposits with accelerations of 0.08 g and 0.12 g caused peak cyclic shear strains of up to 0.11 percent and settlements of up to 0.35 percent. Shaking at higher level (0.16 g) caused a transformation in response with large cyclic shear strains (0.65 percent) and large settlements (3.1 percent). Installation of Titan

FIG. 6. Installation of inclined micropiles

FIG. 7. Completed installation of two inclined micropiles

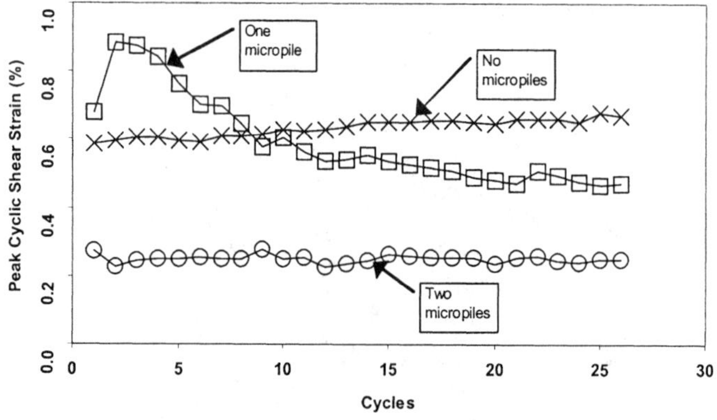

FIG. 8. Average peak cyclic shear strain: Acceleration = 0.16 g, D_r = 0.4

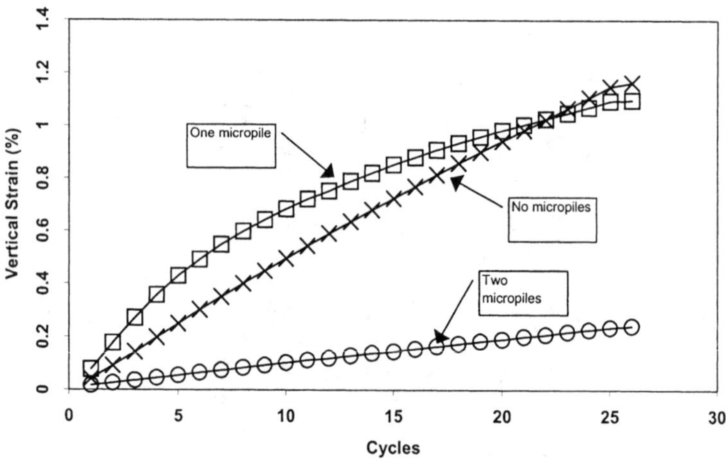

FIG. 9. Settlement during shaking: Acceleration = 0.16 g, D_r = 0.4

self-drilling micropiles at inclinations of 30 degrees was achieved readily by direct-push with simultaneous grout injection at low pressure. Reinforcement of sand deposits with a single inclined micropile had little effect on response to shaking. Reinforcement with two, diagonally opposed micropiles had a significant effect, reducing cyclic shear strain by half and settlement to one fifth that of a similar unreinforced deposit. It is probable that the two micropiles would have prevented liquefaction of a saturated soil deposit in this case (D_r = 0.4, 0.16 g). Reinforcement efficiency was low, with relatively little of the potential increase in stiffness from the steel cross-section utilised. Future research should investigate use of lighter reinforcement elements which may provide similar benefits at greater economy.

ACKNOWLEDGMENTS

This project received funding from the New Zealand Earthquake Commission Research Foundation. Ischebeck(NZ) Ltd. provided the micropiles and John Yonge provided advice on installation. Alistair Chambers designed and developed the laminar tank. John Maley assisted with construction and setup of the apparatus and operated the shake table. Richard Pascoe operated the CPT rig and installed the micropiles.

REFERENCES

Dobry, R. and Ladd, R.S. (1980). Discussion to "Soil liquefaction and cyclic mobility evaluation for level ground during earthquakes, : by H.B. Seed and "Liquefaction potential: science versus practice," by R.B. Peck, *J. of the Geotechnical Engineering Division,* ASCE, Vol. 106, No. GT6, pp. 720-724.

Dobry, R., Ladd, R.S., Yokel, F.Y., Chung, R.M., and Powell, D. (1982). "Prediction of pore water pressure buildup and liquefaction of sands during earthquakes by cyclic strain method," *NBS Building Science Series 138,* National Bureau of Standards, Gaithersburg, Maryland, 150 p.

Hushmand, B., Scott R.F. & Crouse C.B. (1988). "Centrifuge liquefaction tests in a laminar box", *Geotechnique,* Vol. 38, No 2, pp. 253-262.

Iai, S. (1991). "A Strain Space Multiple Mechanism Model For Cyclic Behavior of Sand and its Application," *Research Note No.43*, Earthquake Engineering Research Group, Port and Harbour Research Institute, Ministry of Transport, Japan. May 1991.

Silver, N.L. and Seed, H.B. (1971). "Volume changes in sands during cyclic loading," *J. of the Soil Mechanics and Foundations Division,* ASCE, Vol. 97, No. SM9, pp. 1171-1182.

Vucetic, M. and Dobry, R. (1991). "Effect of soil plasticity on cyclic response," *J. of Geotechnical Engineering,*ASCE, Vol. 117, No. 1, pp. 89-107.

Whitman, R.V., Lambe, P.C. & Kutter, B.L. (1981). "Initial Results from a Stacked-Ring Apparatus for Simulation of a Soil Profile". *Proceedings, International Conference on Recent Advances in Geotechnical Earthquake Engineering and Soil Dynamics*, St Louis, Mo., April 26 – May 3, 1981.

Youd, T.L. (1972). "Compaction of sands by repeated shear straining," *J. of the Soil Mechanics and Foundations Division,* ASCE, Vol. 98, No. SM7, pp. 709-725.

LOW ENERGY COMPACTED CONCRETE GROUT MICROPILES

Alan J. Lutenegger[1]

ABSTRACT: Results from a series of small diameter vertical drilled and compacted grout micropiles that were installed and load tested to failure in axial tension are presented. Tests were performed in the clay crust at the National Geotechnical Experimentation Site at the University of Massachusetts. Micropiles with a diameter of 76 mm and length of 3.05 m were tested. The micropiles were constructed by incrementally applying dynamic compaction energy from the SPT hammer of a conventional drill rig to the concrete grout as the hole was filled. The results of the tension tests showed that in comparison with a micropile constructed using gravity free-fall placement of the concrete, the pullout capacity of the micropiles was significantly increased when compaction energy was used. This increase is primarily due to the increase in lateral stress produced by the compaction.

INTRODUCTION

Small diameter drilled and grouted micropiles and anchors have a number of applications in geotechnical construction for resisting tensile loads applied to various constructed facilities. There is a general lack of test results for micropiles used as anchors that have been load tested to failure so that engineers can verify design procedures. Small diameter grouted micropiles are often constructed using low pressure or high pressure grouting, and in some cases the unit side resistance has been related to the grouting pressure. For micropiles constructed in soils that maintain an open drilled hole, it was hypothesized that a simple alternative to using pressure grouting could be to apply dynamic compaction energy to compact the concrete grout. A test program was performed to determine the axial tension capacity and load-displacement behavior of nine grouted micropiles installed in a stiff clay using different compaction energy. After testing, four of the micropiles were exhumed and the as-built dimensions were obtained to determine the increase in diameter resulting from the compaction. The paper describes the soil conditions at the site, provides a description of the micropile construction and load testing procedures, and presents results of the load tests. The tests clearly show that the tension capacity is related to the compaction energy and that the increase in capacity is primarily the result of an increase in lateral stress produced by the compaction.

---

[1] Prof. and Head, Dept. of Civil Engrg., Univ. of Massachusetts, 224 Marston Hall, Amherst, Ma. 01003; email: lutenegg@ecs.umass.edu

BACKGROUND

Expanded base piles (sometimes referred to as Pedestal Piles, Pressure Injected Footings, Compacted Concrete Piles, or Franki Piles) have been used for about the past 60 years. The installation procedure consists of advancing a drive tube, making an expanded base and then forming the shaft. A dense plug of zero-slump concrete is typically placed at the bottom of the tube and then the drive tube is withdrawn slightly and the concrete is driven out the end to form the expanded base.

In most cases, the shaft is made by adding small amounts of concrete and sequentially driving the concrete out the end of the drive tube. This produces a compacted concrete shaft which gives good contact between the soil and the shaft. Typical drop hammer weights are on the order of 22.3 to 31.2 kN and drop heights may be as much as 10 m. The design and behavior of Franki piles has been described in detail (e.g., Meyerhof 1964; Nordlund 1982; McAnally and Douglas 1984; Neely 1990).

In the present study, the objective was to investigate the use of low-energy to compact the shaft of a small diameter drilled and grouted foundation element to enhance the tension capacity over that obtained from concrete placement by simple gravity free-fall. This construction procedure would be an alternative to the use of pressure injection of the concrete grout and could be achieved with readily available drilling equipment. Because of the small size, the element is considered a micropile or microanchor. No attempt was made to construct an enlarged base.

FIELD TESTING

Test Site

Tests were performed at the National Geotechnical Experimentation Site (NGES) at the University of Massachusetts-Amherst (Area A). The stratigraphy at the site consists of about 1m of mixed fine-grained fill overlying a thick deposit of late Pleistocene lacustrine varved clay. This deposit is locally known as Connecticut Valley Varved Clay (CVVC) and is composed of alternating layers of silt and clay as a result of lacustrine deposition into glacial Lake Hitchcock. Below the fill, the CVVC has a well developed overconsolidated crust that was formed as a result of surface erosion, desiccation, ground water fluctuations and other physical and chemical processes. Geotechnical characteristics of the upper 3 m at the site have been well documented and previously described in considerable detail (e.g., Lutenegger 1995; 2000; Lutenegger and Miller 1994). The site has previously been used to study the tension capacity of grouted anchors (Lutenegger and Miller 1994; Lutenegger and Miller 1998) and pipe piles (Miller and Lutenegger 1997).

Installation-Series 1

Five micropiles were installed for the initial series of tests. A 76 mm open sided auger was used to drill each of the micropile holes. A different level of compaction energy was used for each micropile. All micropiles were installed vertically. A control micropile was installed using zero compaction energy by placing the concrete using gravity free-fall. Each of the next four micropiles was installed using a different level of compaction energy. After drilling, a No. 6 reinforcing bar with a 76 mm diameter 12 mm thick plate attached to the base was inserted into the hole. A threaded rod section was welded to the top of the bar to allow attachment of the load frame.

After placing the bar, an initial increment of 0.6 m of concrete was placed using gravity free-fall. Compaction energy was applied to the concrete column using conventional Standard Penetration Test equipment consisting of a rope and cathead and a 63.6 kg SPT hammer. The hammer was raised a distance of 0.76 m and was then allowed to free-fall. The downhole head of the hammer had a diameter of about 75 mm with an internal clearance hole of 25 mm. This procedure was repeated for each 0.6 m increment until the anchor was filled to the ground surface. After the last increment of compaction, the top surface of the anchor was topped off with concrete. This construction sequence is illustrated in Figure 1.

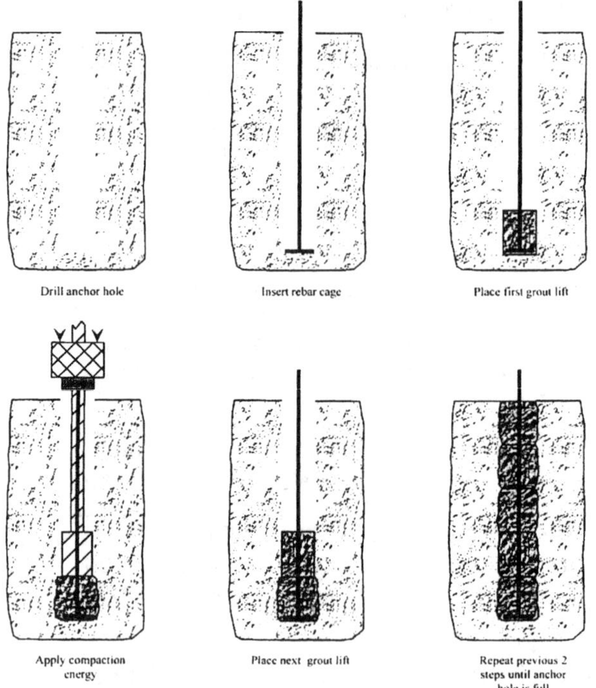

FIG. 1. Sequence of Constructing Compacted Grout Micropiles

A summary of the compaction energy used on each of the anchors is given in Table 1. All of the anchors were constructed on the same day. The concrete was mixed on site using a portable concrete mixer and using commercially available bags of premixed materials (Sacrete) having a maximum coarse aggregate size of about 9.5mm. Extra Portland Cement (Type I) was added to the dry mix and then water was added to give a slump between 15 and 20 mm. The mix showed a 28 day compressive strength of 26.9 MPa. The compaction energies given in Table 1 are theoretical values assuming no loss of energy in the system (i.e., 100% efficiency). It is likely that during the compaction, some friction losses occurred in the hammer system even though extra care was taken to align the hammer and obtain the full hammer drop height with each hammer blow. Additionally, there was likely some energy loss as a result of the compactor head dragging on the sides of the hole or along the reinforcing bar.

Table 1. Energy Used for Series-1 Micropiles

Test No.	No. of Hammer Drops per Lift	Energy per Lift (kJ)	Total Energy (kJ)
CGA1	0	0	0
CGA2	2	949	4745
CGA3	5	2372	11863
CGA4	10	4745	23727
CGA5	20	9491	47543

Installation-Series 2

A second series of four micropiles was installed with the same dimensions and procedure described for Series 1, with the exception that constant compaction energy of 11863 kJ was used. In this series of tests, the concrete grout mix was varied to produce a different concrete slump. The water/cement ratio was varied to give target slumps of 25.4, 50.8, 101.6, and 152.4mm. The slumps measured in the field for each of the mixes is given in Table 2. Compressive strength for each mix is also given.

Table 2. Summary of Concrete Grout Slumps Used for Series-2

Test No.	Total Energy (kJ)	Target Slump (mm)	Measured Slump (mm)	Compressive Strength (MPa)
CGAS1	11863	25.4	25.4	29.1
CGAS2	11863	50.8	63.5	25.5
CGAS3	11863	101.6	114.3	25.1
CGAS4	11863	152.4	165.1	22.9

Load Testing

Pullout tests were performed on each micropile in the period between 30 and

34 days following installation. Tests were performed using the procedures described in ASTM D3689. Load was applied by a single acting, hollow ram 250kN hydraulic jack that was placed in the center of two a reaction beams centered over the micropile and resting on wood cribbing. Load was transferred from the jack to the micropile through a threaded extension rod that was positioned through the hydraulic cylinder. Deformation measurements were made using two digital dial indicators attached to an independent reference beam and placed equidistance on opposite sides of the micropile. Loads were applied incrementally in the range of approximately 5 to 10% of the estimated ultimate capacity. Each load increment was maintained for 20 min. The load was measured using a Geokon load cell placed on top of the hydraulic jack and was read using an electronic digital indicator.

RESULTS

Series - 1

The load vs. displacement curves for each of the load tests in Series-1 are shown in Figure 2. Tests were conducted until a minimum displacement of approximately 35 mm had occurred at which time the load increment could not be

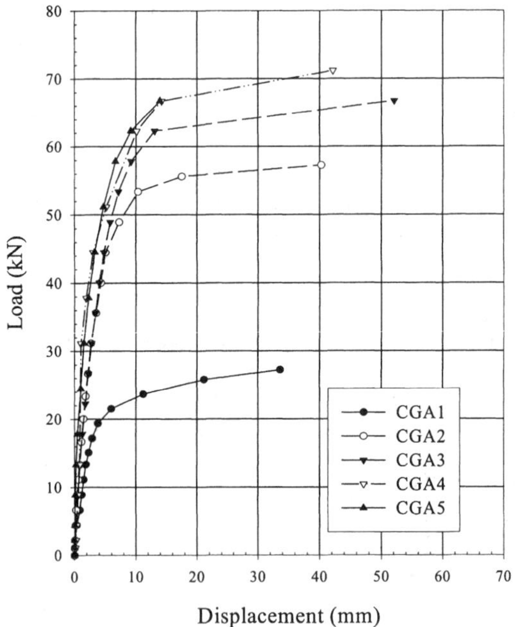

FIG. 2. Load Test Results – Series-1.

maintained. It can be seen from the load curves that a relatively sharp failure occurred in each test. Based on observations of over 40 tests performed at this site, this is typical behavior of micropiles and anchors in this clay. Although there are a number of methods for interpreting the ultimate capacity from the results of a load test, the ultimate capacity was obtained from the curves using a simple tangent intersection method which defines the ultimate capacity at the intersection of tangents drawn to the initial and final portions of the load curve. A summary of the interpreted ultimate capacities is given in Table 3.

Table 3. Summary of Test Results

Test No.	Ultimate Capacity (kN)	Displacement at Failure (mm)	Unit Side Resistance (kPa)
Series-1			
CGA1	24.0	12.0	32.5
CGA2	54.5	14.0	74.3
CGA3	61.0	12.0	83.2
CGA4	65.0	12.0	88.6
CGA5	66.0	13.5	90.0
Series-2			
CGAS1	58.0	16.5	79.1
CGAS2	74.5	10.0	101.7
CGAS3*	79.5	10.0	108.5
CGAS4	71.0	8.0	96.9

*estimated

The test results shown in Figure 2 and summarized in Table 3 indicate that there is a rapid increase in ultimate capacity with only a small amount of compaction energy. Thereafter, the rate of increase drops off. This is further illustrated in Figure 3.

Series-2

Figure 4 gives the load displacement curves for the four micropiles tested in Series-2. Failure was achieved in three of the tests, and in one of the tests a failure of the reinforcing bar just above the ground surface occurred. However, as in the Series-1, sufficient movement of the micropiles occurred to allow interpretation of the failure load. Interpreted ultimate capacities are given in Table 3. The results indicated that the use of concrete grout with a slump greater that about 50 mm produced higher capacities; this may be related to better bonding between the soil and grout for higher slump mixes.

The ultimate capacity of a micropile in tension can be given as:

$$Q_{pullout} = Q_{shaft} + W \qquad (1)$$

where:
Q_{shaft} = total side resistance in tension
W = weight of the micropile

This expression ignores any contribution to pullout capacity from the tip that can only occur if some suction is present. The average unit side resistance is obtained by dividing Q_{shaft} by the total surface area of the pile as:

$$f_s = Q_{shaft} /(\pi DL) \qquad (2)$$

where:
D = diameter
L = length

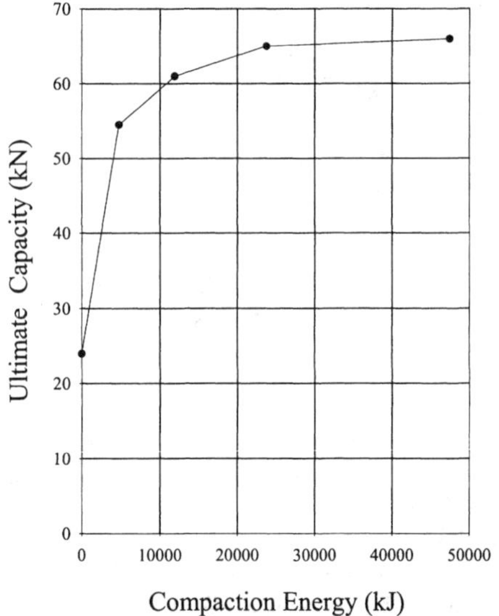

FIG. 3. Increase in Ultimate Capacity with Energy – Series 1.

Values of unit side resistance back calculated from the interpreted capacity are also given in Table 3. In order to calculate the average unit skin friction an estimate of the anchor diameter is needed. The values given in Table 3 were determined using the initial diameter of the borehole.

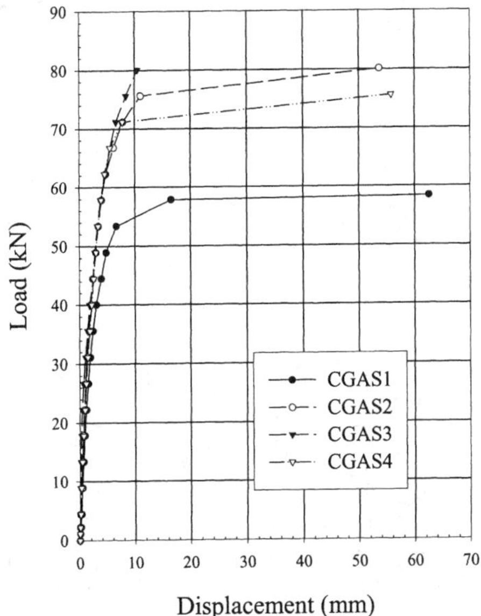

FIG. 4. Load Test Results – Series-2

EXHUMED ANCHORS

Following completion of the tension tests, the micropiles from Series-1 were pulled from the ground so that the shape of each anchor could be observed and measurements of the anchor diameter could be obtained. One of the micropiles could not be removed. A summary of the measured diameters is shown in Figure 5. It can be seen that the compaction produced an increase in pile diameter as expected. It was observed that there was a significant amount of soil adhered to the sides of each of the micropiles. The diameter of each micropile was first measured with the soil still attached. A pressure washer was then used to remove the soil and measurements were then taken of only the concrete shaft. In all cases, it was observed that the shafts had a very rough and irregular surface texture.

The results shown in Figure 5 indicate that the compaction only increased the diameter of the grout column by a relatively small amount. By comparison, the diameter of the shaft with the soil adhered increased continuously for each energy level. If it is assumed that this diameter represents the location of the failure surface, then it can be concluded that the compaction also produced and increase in lateral stress immediately adjacent to the shaft. This then would drive the failure surface outward, away from the interface between the grout and the soil.

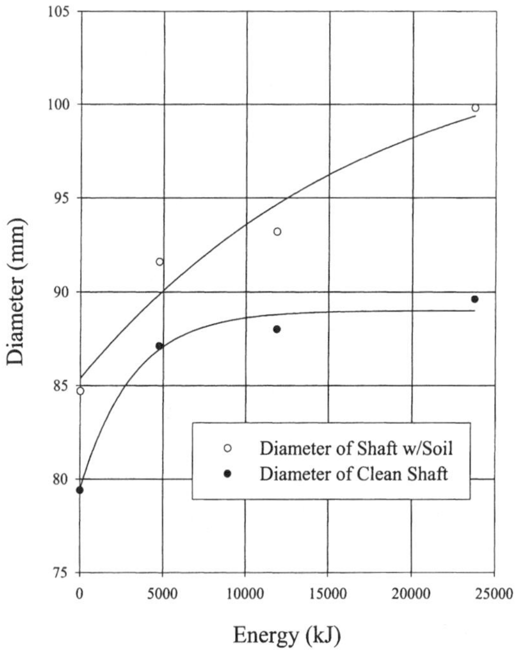

FIG. 5. Measured Diameters of Micropiles – Series-1

INTERPRETATION OF RESULTS AND DESIGN

From an effective stress approach, the available skin friction of a grouted anchor is a function of the lateral stress acting between the pile and the surrounding ground. This lateral stress is related to the construction procedure used to install the pile. For concrete placed using gravity free-fall procedures, it should be expected that the lateral stress would be close to the in situ at-rest condition, provided that no concrete shrinkage occurs. For a pressure-grouted anchor, the lateral stress may be related in some way to the grouting pressure. For a compacted micropile, the lateral stress should be greater than the at-rest condition but lower than the limiting condition of passive earth pressure as a result of lateral displacement of the concrete during compaction. Similar arguments have been made in analyzing compaction grouting (Schmertmann and Henry 1992) and compacted granular piers (Lawton et al. 1994). The difficult question that must be addressed is where in relation to at-rest and limiting passive stress is the operational lateral stress resulting from compaction and acting at the time of loading. The operational lateral stress is likely to be related to the soil conditions (stiffness) at the time of construction; the compaction energy used; the properties of the concrete (slump); and time. Time is included as a parameter to acknowledge the possibility of any relaxation in stress after construction and during the design life of the anchor. One estimate of the upper bound limiting value of lateral stress may be taken as the limit pressure created by the expansion of a cylindrical cavity in the soil such as obtained using the limit pressure, P_L, from pressuremeter tests.

As previously noted, compaction of the concrete produced some increase in the pile diameter over the original diameter of the bored hole as grout and soil is displaced laterally. Using the results of measurements obtained on the four exhumed anchors previously described, it appears that in this soil, increasing the compaction energy had only a minor effect on anchor diameter. Using the measured diameters for both the clean shaft and soil adhered shaft the back calculated average side resistance may be obtained and is given in Table 4. These results indicate that even after adjustment for the observed dimensions there is still a significant increase in unit side resistance that can not be attributed to an increase in pile diameter and which therefore must be attributed to an increase in lateral stress. On average, the results given in Table 4 suggest an average increase in lateral stress by about a factor of 2.2 over the initial lateral stress; i.e., from a noncompacted pile. This appears to be the primary source of increase in capacity.

Table 4. Back Calculated Side Resistance for Different Conditions – Series-1

	Unit Side Resistance (kPa)		
Test No.	Initial Hole Diameter	Clean Grout Shaft	Soil Adhered Shaft
CGA1	32.5	31.1	29.2
CGA2	74.3	64.8	61.6
CGA3	83.2	71.9	67.8
CGA4	88.6	75.2	67.5

An alternative design approach, which would not require measuring or estimating the limit pressure might be to use an “Equivalent or Effective Diameter” of the pile and the original at-rest lateral stress. This approach simply considers that the compaction of the concrete has produced a larger surface area resulting from lateral enlargement of the pile. This is not a true increase in diameter, but represents an “Equivalent Diameter” which would be needed to produce the observed ultimate capacity if all other variables operating on a noncompacted (i.e., free-fall) micropile were constant. For example, an increase in capacity of a compacted micropile by a factor of 2 over a noncompacted micropile would result from an increase in diameter by a factor of 2. This approach simply applies the same unit side resistance calculated for a noncompacted pile to a compacted pile.

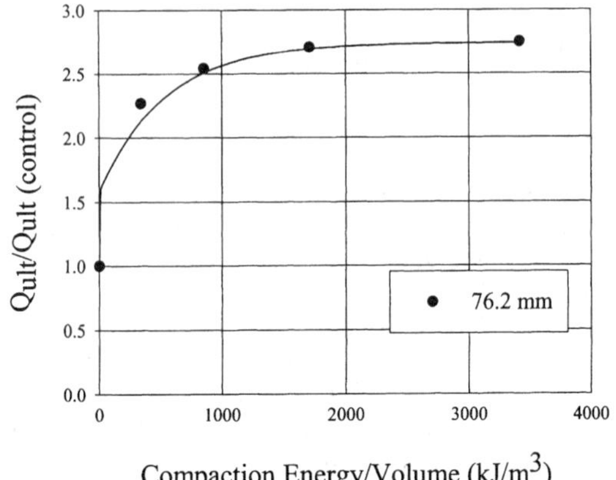

FIG. 6. Normalized Capacity of Individual Micropiles as Related to Energy/Volume – Series-1

Figure 6 shows the increase in ultimate capacity as a function of the energy per unit volume of the micropile. Expressed in this way the results show how much gain in capacity can be achieved relative to an equivalent grouting "pressure". Again, since this increase is much larger than the increase in diameter shown in Figure 5, this illustrates that the increase in capacity is the result of an increase in lateral stress. It is clear that more testing and analysis is needed using different size piles different energy levels, and in different soils, however, it is appropriate to consider potential design approaches based on the test results presented in this paper.

CONCLUSIONS

The results of field tests conducted on a series of drilled and grouted micropiles constructed using low energy compaction in clay showed that the axial tension capacity can be significantly enhanced using a modest amount of compaction energy. The results indicate that the compaction produces an increase in both the pile diameter and lateral stress between the soil and the pile but suggest that the increase in lateral stress is the most important factor accounting for the observed increase in capacity. This increase in capacity means that fewer micropiles would be required or that smaller micropiles could be used, both of which result in a substantial cost savings. The results indicated that this type of construction, which is simple to apply in the field, is a viable technique and can greatly enhance the use of small diameter micropiles in many design situations.

ACKNOWLEDGMENTS

Test results presented in this paper were conducted at the NGES UMass-Amherst. The continuing research support provided by Michael Adams and Al DiMillio of the Federal Highway Administration is gratefully acknowledged. The author also wishes to acknowledge the assistance of Scott Smith, Patrick Drury, Mike Mitchell, and Shawn Kelley during field construction and testing.

REFERENCES

Lawton, E.C., Fox, N.S. and Handy, R.L. (1994). "Control of settlement and uplift of structures using short aggregate piers". *In-Situ Deep Soil Improvement*, ASCE, 121-132.

Lutenegger, A. J. (1995). "Geotechnical behavior of overconsolidated clay crusts". *Transportation Research Record No. 1479*, 61-74.

Lutenegger, A.J. (2000). "National geotechnical experimentation Site - University of Massachusetts". *National Geotechnical Experimentation Sites*, ASCE, 102-129.

Lutenegger, A.J. and Miller, G.A. (1994). "Uplift capacity of small diameter drilled shafts from in situ tests". *J. Geotech. Engrg.*, ASCE, 120(8), 1362-1380.

Lutenegger, A.J. and Miller, G.A. (1998). "Tension tests on drilled micropiles in a stiff clay". *Proc. 4th Int. Conf. on Case Histories in Geotech. Engrg*, 901-906.

McAnally, P.A. and Douglas, D.J (1984). "The design of Franki piles in clays". *Proc. 4th Australia-New Zealand Conf. on Geomechanics*, 2, 402-407.

Miller, G.A. and Lutenegger, A.J. (1997). "Influence of pile plugging on skin friction in overconsolidated clay". *J. Geotech. and Geoenvir. Engrg.*, ASCE, 123(6), 525-533.

Neely, W.J. (1989). "Bearing capacity of expanded-base piles with compacted concrete shafts". *J. Geotech. Engrg.*, ASCE, 116(9), 1309-1324.

Nordlund, R.L. (1982). "Dynamic formula for pressure injected footings". *J. Geotech. Div.*, ASCE, 108(3), 419-437.

Schmertmann, J.H. and Henry, J.F. (1992). "A design theory for compaction grouting". *Grouting, Soil Improvement and Geosynthetics*, ASCE, 1, 215-228.

LATERAL LOADS ON PIN PILES (MICROPILES)

Thomas D. Richards, Jr., P.E., M. ASCE[1], Mark J. Rothbauer, P.E., M. ASCE[2]

ABSTRACT

Due to the small diameters, Pin Piles, generically known as micropiles, are often considered to have little to no lateral capacity. This paper focuses on lateral load performance and design of Pin Piles. The intent is to demonstrate that micropiles and micropile groups can be designed to support lateral loads and provide options and considerations for lateral load design. The results of lateral load tests are compared to the predictions using various available methods. Design considerations of combined stresses, options for increasing lateral resistance, and analysis for battered piles are then discussed.

INTRODUCTION

Pin Piles are small-diameter, typically 127 to 305 mm, high-capacity drilled and grouted piles. They are ideal for foundations on sites where it is not practical or economical to install more traditional deep foundation elements such as driven piles or drilled shafts. These site conditions include subsurface obstructions or difficult ground, limited overhead clearance, vibration or noise sensitivity, settlement sensitivity, limited plan access, and the need to install elements in close proximity to or through existing footings, columns, walls, or other structures. Additional background information can be found in References by Pearlman, Bruce, and Tarquino.

Due to their diverse and flexible capabilities, the use of Pin Piles has dramatically increased in recent years. The FHWA Micropile Design and Construction Guidelines, Publication # FHWA-SA-97-070 (Armour et al., 2000) has contributed to increased use and shows the FHWA's interest in the use of micropiles. The manual

---

[1] Chief Engineer, Nicholson Construction Company, 12 McClane St., Cuddy, PA 15031 email : trichards@nicholson-rodio.com

[2] Design Engineer, Nicholson Construction Company, 12 McClane St., Cuddy, PA 15031 email : mrothbauer@nicholson-rodio.com

documents the design and installation methods of micropiles. Lateral loads and bending are briefly presented in this Manual.

Due to the increased use of Pin Piles, combined with the increased consideration of seismic loads, lateral loads on piles or pile groups often govern the design, requiring a higher pile section modulus. Lateral loads on foundations can be resisted by battering the piles, considering partial passive resistance of footing, and/or designing the piles for lateral loads. Even when battered piles are used, bending can occur in the piles especially for non-sustained and variable loads such as wind and seismic loads.

This paper focuses on lateral load performance and design of Pin Piles and presents:

- results of lateral load tests including load and deflection
- comparison of lateral tests results to predictions using LPILE, NAVFAC, and Characteristic Load Method (CLM)
- combined stresses
- options for increasing lateral resistance
- analysis for battered piles

The intent is to demonstrate that micropiles and micropile groups can be designed to support lateral loads

LATERAL LOADS TESTS

Lateral load test were performed on eight projects mostly during the last few years. All of the micropiles consisted of steel casing(s) filled with grout. The steel casing was industry standard mill secondary casing with minimum yield strength of 552 MPa and typically had flush threaded joints at 1.52 or 3.05 meter centers. For double cased piles, the threaded joints were typically staggered by approximately 0.6 meters as a minimum. The grout was colloidally mixed neat cement and water with a water to cement ratio of 0.45 and typical unconfined compressive strength of cubes and cylinders of 34.5 MPa. All of the piles had length to diameter ratios greater than 20.

The drill methods are summarized in Table 1. Detailed descriptions of the drill methods can be found in Bruce (2003). In most cases even when duplex drilling was used, the initial 3 to 6 meters of drilling still experienced some external flush.

A typical lateral load test is shown in Figures 1 and 2. All of the tests were performed in pits excavated to the bottom of footing elevation, so that the piles pushed against the same soil as they would in the final structure and to at least partially have the effect of the top of pile being below the ground surface resulting in a passive surcharge. The depth of the excavations is "dpit" in Table 1. These tests were essentially free (pinned) head tests. The collar around the pipe had a weld bead at the line of load application and the bolts were typically loosened after an alignment load was applied. The axis of applied load was typically 150 to 300 mm above the ground surface, as shown in Table 1 as "zP". In each test, two piles were loaded simultaneously. At sites where more than one test was performed, piles 1 and 2 were a pair and piles 3 and 4 were a pair.

FIG. 1. Lateral Load Test on Micropile Overview

FIG 2. Lateral Load Test Setup Detailed View

PILE	PILE PROPERTIES			SOIL PROPERTIES					ASSIGNED SOIL PARAMETERS						TEST	
	D Mm	EI kN mm^2	DRILL METHOD	TYPE	N Min	N max	N typ.	Dw M	Su KPa	φ deg	γ kN/m³	γ avg kN/m³	f kN/m³	kh kPa	zP cm	dpit cm
A1	244	1.914E+10	Rotary Duplex with water	Sandy Lean Clay	12	25	19.0	6.7	129	0	19.6	19.6		4525	18	122
A2	244	1.914E+10			12	25	19.0	6.7	129	0	19.6	19.6		4525	24	122
C1	244	1.914E+10	Rotary Duplex with water	Sandy clay or silty clay	8	15	13.3	8.7	86	0	18.9	18.9		3016	24	137
C2	244	1.929E+10			8	15	13.3	8.7	86	0	18.9	18.9		3016	21	134
MR1	244	1.914E+10	Rotary Duplex with water	Flyash	4	4	4.0	3.0	0.0	25	14.1	13.8	1923		30	131
MR2	244	1.927E+10			4	4	4.0	3.0	0.0	25	14.1	13.8	1923		30	131
Z1	244	2.056E+10	Rotary Duplex with water	Silty sand with gravel	41	61	50.3	13.3	0.0	35	19.6	19.6	15043		30	107
Z2	244	2.058E+10			41	61	50.3	13.3	0.0	35	19.6	19.6	15043		27	107
G1	244	1.929E+10	Rotary Eccentric Percussive Duplex with Air	silt & sand to 2.4 m, then dense sand with silt & gravel	3	57	13.4	0.0	0.0	30	19.6	9.8	8014		18	130
G2	244	1.929E+10			3	57	13.4	0.0	0.0	30	19.6	9.8	4701		15	130
MC1	197	7.662E+09	Rotary Duplex with water	Fill – Silty Clay with sand	5	12	9.3	21.5	100	0	19.6	19.6		4525	52	134
MC2	197	7.662E+09			5	12	9.3	21.5	100	0	19.6	19.6		4525	55	137
MC3	197	7.662E+09			5	12	9.3	21.5	100	0	19.6	19.6		4525	40	143
MC4	197	7.662E+09			5	12	9.3	21.5	100	0	19.6	19.6		4525	27	131
B1	254	4.718E+09	Single Tube = Ext Flush	Fill – silty sand to silty sandy gravel	3	16	8.0	2.4	0.0	30	18.9	16.9	2645		15	122
B2	254	4.718E+09			3	16	8.0	2.4	0.0	30	18.9	16.9	2645		15	122
O1	381	4.348E+10	Open Hole with Air	Stiff silty clay/ clayey silt with chert fragments	12	44	24.5	15.2	96	0	17.3	17.3		3352	23	76
O2	381	5.051E+10			12	44	24.5	15.2	96	0	17.3	17.3		3352	23	76
O3	381	4.348E+10			12	44	24.5	15.2	96	0	17.3	17.3		3352	23	76
O4	381	5.051E+10			12	44	24.5	15.2	96	0	17.3	17.3		3352	23	76

TABLE 1. Summary of Test Pile Data

The load was typically applied by a hydraulic jack with a hand pump. An electric resistance load cell was used to control the load since the jack pressures were at the low end of the pressure gauges and calibration. Two dial gages were placed at fixed elevations, typically near the top of the pile and at the point of the applied load, to measure the pile deflection and rotation. All deflections reported were calculated at the ground surface by extrapolating from the two dial gauge readings. For any tests where deflections were recorded starting at an alignment load, the deflection from zero load to alignment load was added by curve fitting to the initial load increments.

COMPARISON OF LATERAL TESTS RESULTS TO PREDICTIONS

Subsurface Conditions

Generally, the sites had limited subsurface investigation and testing, but represented typical data available on projects. The soil type, uncorrected N values, ground water level, and assigned soil parameters are shown in Table 1 and are from project borings in the upper 4.6 meters below pile subgrade. For granular soils, these parameters were based on typical correlations between N, total unit weight (g) and angle of shearing resistance (ϕ). For cohesive soils, the undrained shear strength (su) was based on average pocket penetrometer readings if available or correlations between N and su. No attempt was made to force a fit of pile analyses to the actual load test data. g'avg is the weighted average effective unit weight in the upper three meters and is used only for CLM analyses.

Transformed Section

For consistency and to eliminate a source of difference, the composite pile stiffness (EI) was determined using the LPILE program. The EI used for all analyses is summarized in Table 1. The result was typically near the average of the uncracked transformed section and the steel only section. All analysis neglected the reduced EI over discrete lengths at the threaded joints of the drilled pipe. The only method that would be able to consider this is LPILE by using variable EI along the pile length. The effect of this unconservative assumption is discussed in the "Comparison of Results" section below.

LPILE Method

LPILE uses the p-y method for lateral soil resistance and is a Windows based and slightly improved version of COM624. A complete description of LPILE is beyond the scope of this paper and is available in Reese et.al. (2000) and Reese (1984). LPILE Plus 4.0 was used.

Each analysis was modeled with the top of pile below ground surface thus considering the passive surcharge. The moment due to the load being applied above the ground surface was also input. The k values for LPILE analyses were those presented in the LPILE User's Manual. For clay sites, the "stiff clay without free water" p-y curves were used.

This method also allows the effect of the watertable to be considered at the appropriate depth.

NAVFAC Method

The "NAVFAC" method is from NAVFAC (1986) and based on Reese and Matlock (1956). This method uses linear elastic coefficient of subgrade reaction and assumes "*that the lateral load does not exceed about 1/3 of the ultimate lateral load capacity.*" For granular soil and normally to slightly overconsolidated cohesive soils, NAVFAC states "*the coefficient of subgrade reaction, Kh, increases linearly with depth in accordance with:*

$$Kh = \frac{f * z}{D} \tag{1}$$

where: *Kh = coefficient of lateral subgrade reaction [F/L^3]*
f = coefficient of variation of lateral subgrade reaction [F/ L^3]
z = depth [L]
D = width/diameter of loaded area [L]"

For overconsolidated cohesive soils, NAVFAC states "*for heavily overconsolidated hard cohesive soils, the coefficient of lateral subgrade reaction can be assumed to be constant with depth. The methods presented in Chapter 4 can be used for the analysis; Kh, varies between 35c and 70c (units of force/length^3) where c is the undrained shear strength.*" NAVFAC Chapter 4 presents traditional elastic modulus of subgrade reaction equations. The "free end, concentrated load" case was used. The units of 35c appear to be force/length^2. Therefore, the modulus of subgrade reaction used was Kb = 35su/b where b = pile diameter.

This method estimates the moment diagram versus depth and does not consider the effect of passive surcharge. This method does not easily deal with the applied moment from the applied load being above the ground surface and this was not considered.

Characteristic Load Method (CLM)

This method is available as a spreadsheet from the Virginia Tech, Center for Geotechnical Practice and Research. Per Clarke and Duncan (2001), "*The characteristic load method (CLM) of analysis of laterally loaded piles (Duncan et al.,1994) was developed by performing nonlinear p-y analyses for a wide range of free-head and fixed-head piles and drilled shafts in clay and sand. The results of the analyses were used to develop nonlinear relationships between dimensionless measures of load and deflection. These relationships were found to be capable of representing the nonlinear behavior of single piles and drilled shafts quite accurately, producing essentially the same values of deflection and maximum moment as p-y analysis computer programs like COM624 and Lpile Plus 3.0. The principal limitation of the CLM method is that it is applicable only to uniform soil conditions.*" They also reported that "*only the soil within eight pile diameters of the ground surface is important for lateral load conditions, because almost all lateral load is transferred from the piles to the soil within eight diameters below the ground surface.*" Based on the authors' experience, LPILE analysis of micropiles with moderate lateral loads typically shows significant bending moments to depths of 10 to 20 pile diameters.

The parameters for this method are soil strength parameters (ϕ or su), unit weight (for sand only), and pile structural properties including diameter, elastic modulus, moment of inertia, and cracked to uncracked moment of inertia. As stated above, the EI used was from LPILE section analyses. When the water table was within 3 meters of pile subgrade, the weighted average effective unit weight (g'avg) was used as suggested in the CLM Manual, Clarke and Duncan (2001) and as shown in Table 1.

The deflections were determined both with the applied moment from the point of load application above the ground surface. This method does not provide rotations or bending moments versus depth.

Comparison of Results

The calculated deflections for the various methods and measured deflections are shown in Figures 3 through 12. Generally, the measured deflections were typically significantly less than predicted by CLM or NAVFAC. The LPILE analysis tended to provide the best fit. However, the measured deflections often exceeded the LPILE predictions, due primarily to the "passive surcharge" considered in LPILE. By comparing LPILE to CLM curves, the impact of this surcharge is significant even on clay sites. The pits did not provide a pure surcharge and were typically often 0.6 meters beyond the edge of the pile. The underestimated predictions with LPILE were also due to the fairly high undrained shear strengths, especially at site MC

The exceptions to the above fit were the MC piles (Figure 8) where measured significantly exceeded calculations by LPILE and were near NAVFAC & CLM predictions. This is caused by fairly high undrained shear strength used when compared to the blow count; the soil being a clayey fill, therefore pocket penetrometer readings may represent "chunks" versus the mass; perched water near the bottom of the pit; and the limits of the pit excavation was approximately 2 meters beyond the piles. LPILE analysis without the surcharge would be similar to CLM.

The tests and analyses as well as other literature show that the lateral load performance is very sensitive to the soil type and shear strength in the upper 2 to 5 meters of the pile. Therefore, this zone should be well sampled and characterized in subsurface investigations including laboratory testing for projects expecting deep foundations.

Since the measured deflections were close typically close to predicted, ignoring the reduction in EI of the threads in predicting deflections appears appropriate. The performance is judged to be dominated more by the soil strength.

COMBINED STRESSES

The combined stresses on a pile due to axial load and bending moment must be checked in design. The simple method to determine the combined stresses is:

$$\frac{P}{Pall} + \frac{M}{Mall} \leq 1.0 \tag{2}$$

Where: P = applied axial load
$Pall$ = allowable axial structural load of pile
M = bending moment from analysis

Mall = allowable bending strength of the pile

The allowable bending moment must consider the threaded joint section of the pile. An approximation for the section modulus of the flush joint thread length is 50% of the section modulus of the solid pipe.

Often designers allow higher bending stresses than axial stresses. This is not clear in various Codes and may not be appropriate since the bending moments are known with less accuracy. However, the bending moments seem to be conservatively predicted since measured pile deflections are typically less than predicted. Also, the load cases (wind, seismic, etc.) that cause most of the lateral load are short term and infrequent in nature. This should be considered in the allowable overstresses for load combinations listed in Building Codes or in the load factor combinations for LRFD.

A secondary moment occurs due to P-delta effects. That is the axial load is now applied on a deflected and curved pile. This can be directly considered and evaluated using LPILE. For the CLM method, this can be input as a moment at the pile top. For the NAVFAC method, superposition may be used with the P-delta as an applied moment. Considering P-delta as an applied moment is considered to be conservative by the authors.

Group effects also influence the lateral deflection and bending moments of the individual piles in the group. Leading piles (along the pile group side in the direction of movement) behave stiffer and thus support more of the lateral load and have higher bending moment. Further information on group effects is provided by Reese et al (2002) and Clarke and Duncan (2001). Group effects can be considered in the GROUP 6 program or the CLM method.

OPTIONS FOR INCREASING LATERAL RESISTANCE OF PILES OR PILE GROUPS

Lateral capacity of an individual micropile or a micropile group can be increased by installing an oversized casing in the top portion of the pile where moments are high, constructing a larger pile diameter at the top (bending moment decreases with increased diameter and passive resistance), embedding the pile cap deeper, or creating a "fixed" connection. Although pure fixity between the pile and pile cap with zero rotation is unrealistic. Lateral capacity of a pile group can also be increased by battering piles or making the group larger, i.e. increasing the pile spacing to decrease the group reduction effects.

ANALYSIS FOR BATTERED PILES

A simple graphical procedure for estimating the compressive and tensile forces in micropile groups containing not more than three rows of micropiles is described in Tomlinson (1987) and Teng (1962).

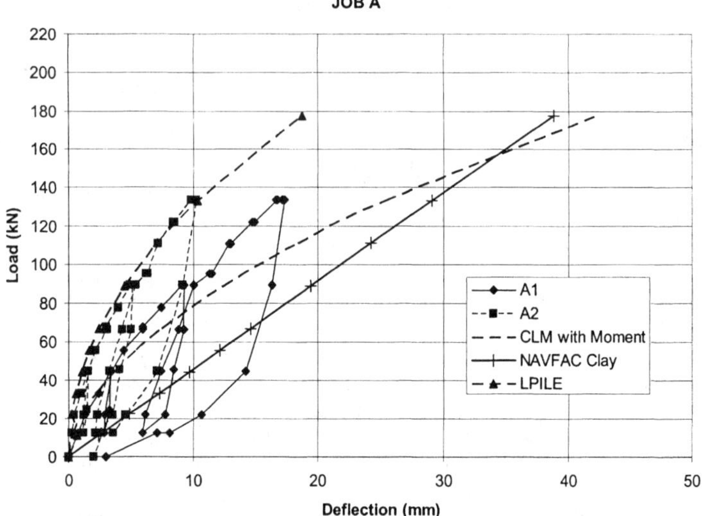
JOB A
Load (kN)
Deflection (mm)
A1
A2
CLM with Moment
NAVFAC Clay
LPILE

FIG. 3.

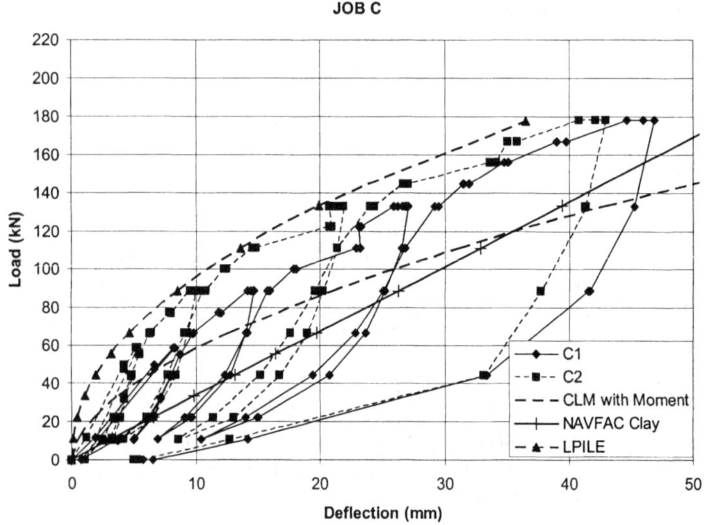
JOB C
Load (kN)
Deflection (mm)
C1
C2
CLM with Moment
NAVFAC Clay
LPILE

FIG. 4.

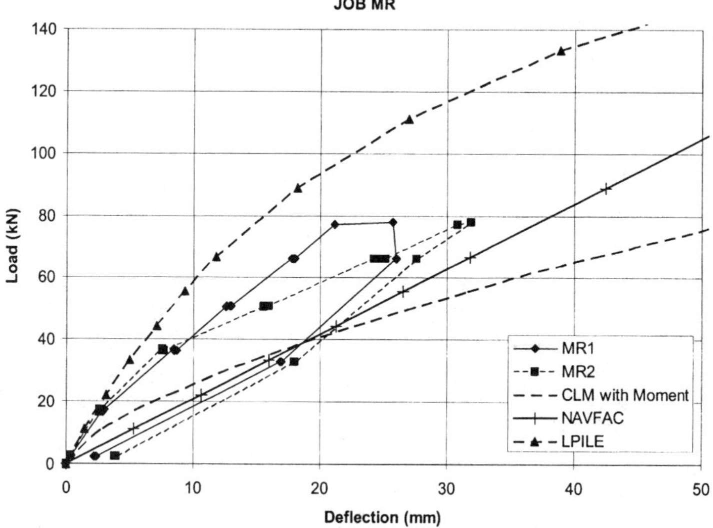
JOB MR
Load (kN)
Deflection (mm)
MR1
MR2
CLM with Moment
NAVFAC
LPILE

FIG. 5.

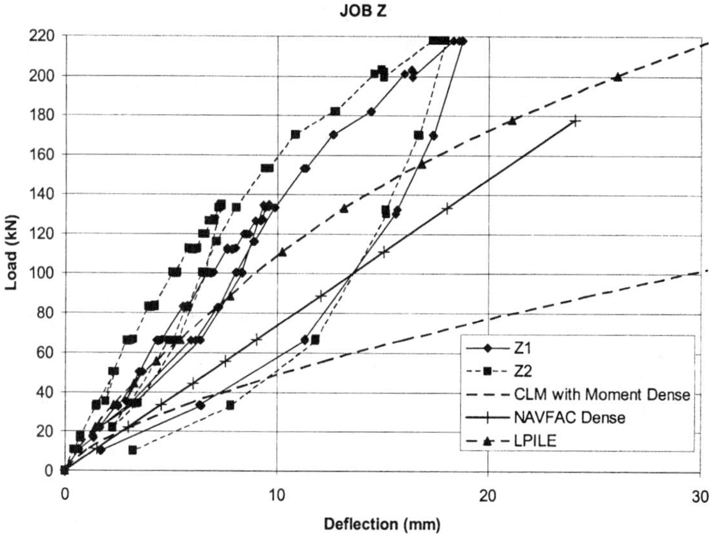
JOB Z
Load (kN)
Deflection (mm)
Z1
Z2
CLM with Moment Dense
NAVFAC Dense
LPILE

FIG. 6.

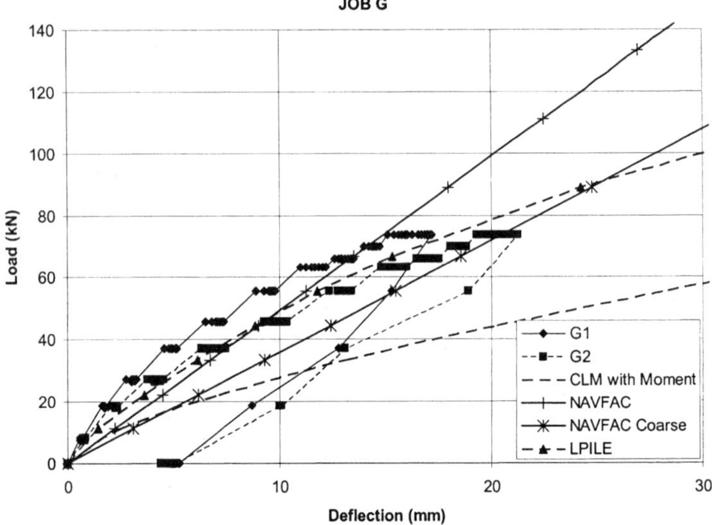
JOB G
Load (kN)
Deflection (mm)
0
20
40
60
80
100
120
140
0
10
20
30
G1
G2
CLM with Moment
NAVFAC
NAVFAC Coarse
LPILE

FIG. 7.

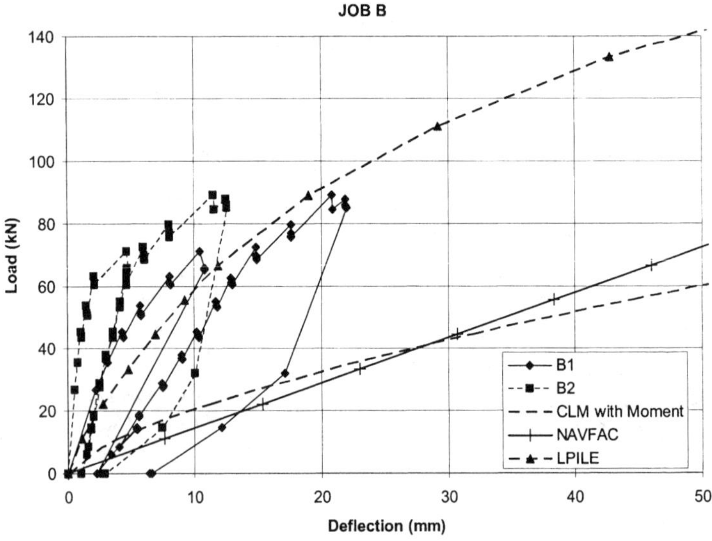
JOB B
Load (kN)
Deflection (mm)
0
20
40
60
80
100
120
140
0
10
20
30
40
50
B1
B2
CLM with Moment
NAVFAC
LPILE

FIG. 8.

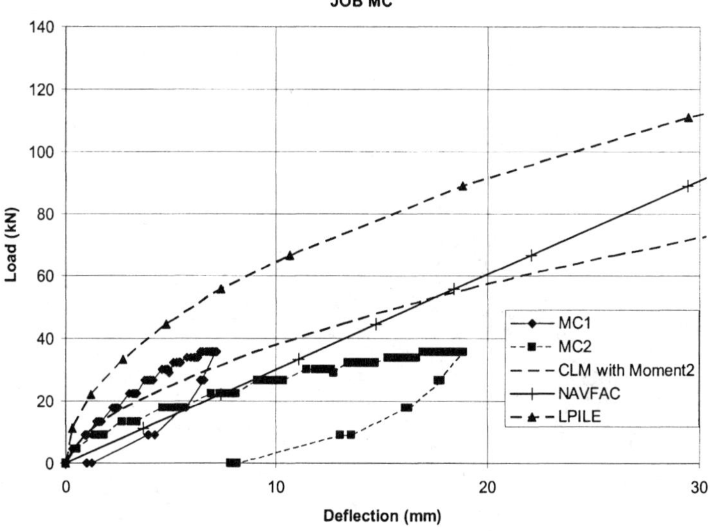
JOB MC
Load (kN)
140
120
100
80
60
40
20
0
0
10
20
30
Deflection (mm)
MC1
MC2
CLM with Moment2
NAVFAC
LPILE

FIG. 9.

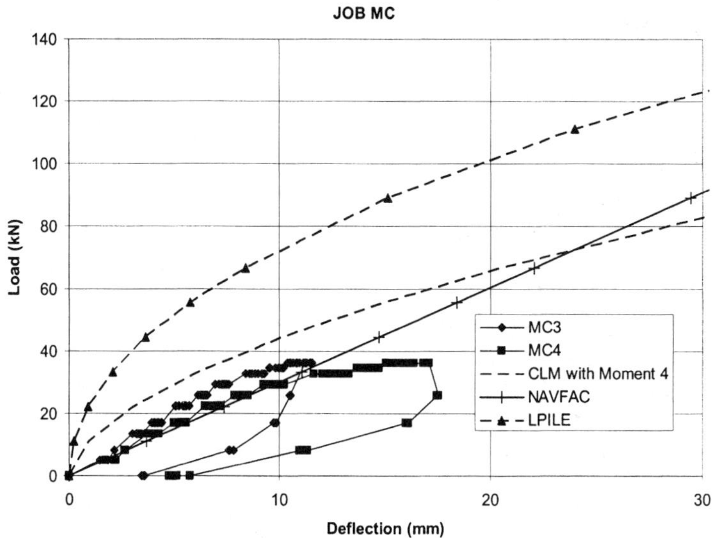
JOB MC
Load (kN)
140
120
100
80
60
40
20
0
0
10
20
30
Deflection (mm)
MC3
MC4
CLM with Moment 4
NAVFAC
LPILE

FIG. 10.

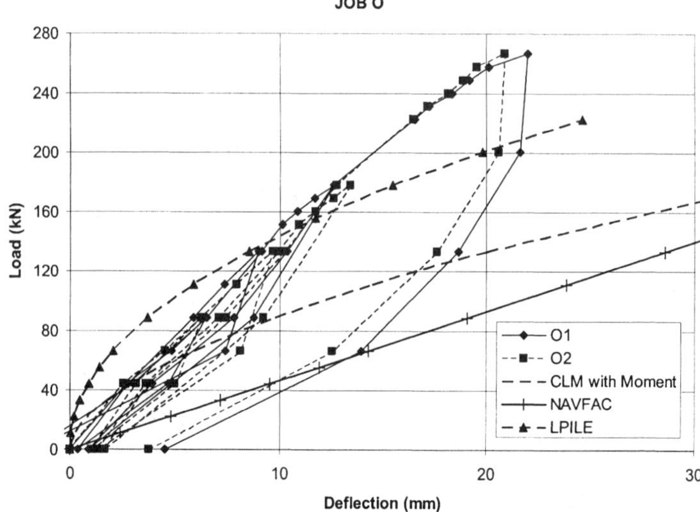
JOB O
Load (kN)
Deflection (mm)
280
240
200
160
120
80
40
0
0
10
20
30
O1
O2
CLM with Moment
NAVFAC
LPILE

FIG. 11.

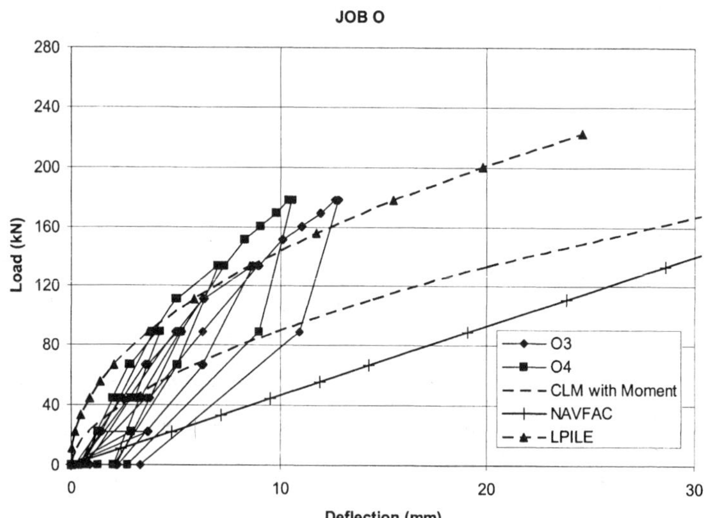
JOB O
Load (kN)
Deflection (mm)
280
240
200
160
120
80
40
0
0
10
20
30
O3
O4
CLM with Moment
NAVFAC
LPILE

FIG. 12

For analysis of three-dimensional pile groups that considers nonlinear soil response and micropile-soil-micropile interaction, the GROUP 6 program can be used. An interesting outcome from working with the GROUP program is the realization that even battered pile groups have bending moments in the piles. Consider a battered pile group with vertical load only applied to the group. The piles must deflect to support the load and in doing so, the pile cap moves vertically. The soil under the pile resists this downward movement and creates bending moments in the pile. This can also be analyzed by resolving a vertical load on a battered pile into an axial and a lateral load on the pile, but with the deflection constrained to a vertical plane. These same effects occur when there are lateral, axial and overturning moments applied to the pile group. Battered piles can substantially reduce and balance, but not eliminate, the bending moments in the piles.

CONCLUSIONS

Micropile foundations can be and have been designed to carry substantial lateral loads. The loads can be resisted by the lateral load resistance of the micropile and/or by battering the piles. In either case, the micropiles must be designed for the resulting combined stresses often resulting in the need to include casing near the top of the pile for bending strength.

Lateral tests on micropiles have generally shown less deflection than predicted due to typical conservatism in assigned soil parameters or neglecting "passive surcharge" due to the top of the pile being below ground surface. The elastic solutions generally greatly overestimate deflection.

The tests and analyses as well as other literature show that the lateral load performance is very sensitive to the soil type and shear strength in the upper 2 to 5 meters of the pile. Therefore, this zone should be well sampled and characterized in subsurface investigations including laboratory testing for projects expecting deep foundations.

ACKNOWLEDGMENTS

The authors thank

- Nicholson Construction Company for the opportunity to perform, analyze and report these tests
- Owners of the projects that involved lateral tests:
  American Electric Power; Louisville Gas and Electric; NYSDOT; Take One, LLC &York Hunter Construction Services, Inc.; U.S. Department of Energy (DOE), Oakridge National Laboratory
- Ensoft and Dr Wang for their help in teaching and troubleshooting LPILE and GROUP
- David Tarasovic, SAI Consulting Engineers, Inc. for assistance performing the LPILE analysis
- Hayward Baker for joint venture partnering at Oakridge
- Virginia Tech, Center for Geotechnical Practice and Research for access to CLM

REFERENCES

Armour, T.A., Groneck, P., Keeley J., and Sharma, S. (2000). "Micropile Design and Construction Guidelines Implementation Manual." Priority Technologies Program Project, *FHWA-SA-97-070*, FHWA Office of Technology Applications, FHWA Federal Lands Highway Division, 2000, 376 p.

Brettmann, T. and Duncan, J. M. (1996). "Computer Application of CLM Lateral Load Analysis to Piles and Drilled Shafts." *ASCE Journal of Geotechnical Engineering,* Vol. 122, No. 6, 496-500

Bruce, D.A., and Juran, I. (1996). "Drilled and Grouted Micropiles: State-of-Practice Review", Four Volumes, *FHWA-RD-96-019*, Nicholson Construction Company under contract with FHWA Office of Engineering R & D.

Bruce, D.A. (2003) "The Basics of Drilling for Specialty Geotechnical Construction Processes." Grouting and Ground Treatment, *Geotechnical Special Publication No 120*, 752-771

Bruce, D. A., Pearlman, S. L., and Clark, J. H. (1990). "Foundation Rehabilitation of the Pocomoke River Bridge." *Proceedings of the 7th Annual International Bridge Conference*, June 1990, 208 - 216, Paper#IBC-90-42.

Clarke, J.A., and Duncan, J.M. (2001). "Revision of the CLM Spreadsheet for Lateral Load Analyses of Deep Foundations" *Report of Research*, Virginia Tech Center for Geotechnical Practice and Research, Virginia Polytechnic Institute and State University, August 2001

Duncan, J. M., Evans, Jr., L. T. and Ooi, P. S. K. (1994). "Lateral Load Analysis of Single Piles and Drilled Shafts," *ASCE Journal of Geotechnical Engineering*, Vol. 120, No. 5, 1018-1033.

Hrennikoff, A. (1950). "Analysis of Pile Foundations with Batter Piles." *Transactions of the ASCE,* Vol. 115, 351-376.

Kulicki, J. M. (1988). "Aspects of the Construction of the I-78 Bridge Across the Delaware River", *Proceedings of the 5th Annual International Bridge Conference*, June 1988, 289-295, Paper #IBC-88-54.

Mokwa, R. L. and Duncan, J. M. (2001), "Laterally Loaded Pile Group Effects and p-yMultipliers," *ASCE Geotechnical Special Publication No 113*, 728-742.

NAVFAC (1986) *Foundations & Earth Structures, Design Manual 7.02*, Revalidated by Change 1 September 1986, Naval Facilities Engineering Command

O'Neill, M. W., Ghazzaly, O.I., and Ha, H. B. (1977) "Analysis of Three-Dimensional Pile Groups with Non-Linear Soil Response and Pile-Soil-Pile Interaction." *Proceedings, Ninth Annual Offshore Technology Conference*, 245-256.

O'Neill, M. W., and Tsai, C. N. (1984). "An Investigation of Soil Nonlinearity and Pile-Soil-Pile Interaction in Pile Group Analysis." *Research Report No. UHUC 84-9*, Department of Civil Engineering, Univ. of Houston. Prepared for U.S. Army Engineer Waterways Experiment Station, Vicksburg, Mississippi

Ooi, P.S.K, and Duncan, J. M. (1994). "Lateral Load Analysis of Groups of Piles and Drilled Shafts." *Journal of Geotechnical Engineering*, Vol. 120, No. 6, 1034-1050.

Pearlman, S.L. (2000), "Pin Piles for Structural Underpinning", *Proceedings of the 25th Annual Deep Foundations Institute Meeting*, Deep Foundations Institute, October 2000

Pearlman, S.L., Richards, T.D., Wise, J.D., and Vodde, W.L. (1997). "Pin Piles for Bridge Foundations: A Five Year Update", *Proceedings of the 14th Annual International Bridge Conference*, June 1997, 472-480, Paper #IBC-97-53

Pearlman, S.L., Wolosick, J.R., and Groneck, P.R. (1993). "Pin Piles for Seismic Rehabilitation of Bridges", *Proceedings of the 10th Annual International Bridge Conference*, June 1993, 307 - 312, Paper#IBC-93-53.

Pearlman, S.L., and Wolosick, J.R. (1992). "Pin Piles for Bridge Foundations", *Proceedings of the 9th Annual International Bridge Conference*, June 1992, 247 - 254, Paper#IBC-92-40.

Reese, L. C., Wang, S. T., Arrellaga, J. A., and Hendrix, J. (2002), "Group6.0: A Program for the Analysis of a Group of Piles Subjected to Axial and Lateral Loading," *Ensoft, Inc.*, P. O. Box 180348, Austin, Texas, 78718.

Reese, L. C. , Wang, S. T., Isenhower W.M., Arrellaga, J. A., and Hendrix, J (2000), "LPILEPlus Version 4.0: A Program for the Analysis of Piles and Drilled Shafts under Lateral Loading," *Ensoft, Inc*, P. O. Box 180348, Austin, Texas, 78718.

Reese, L. C. (1984) "Handbook on Design of Piles and Drilled Shafts Under Lateral Load. ",*FHWA-IP-84/11*, FHWA, U.S. Department of Transportation, Washington, D.C.

Reese, L.C. and Matlock, H. (1956). "Non-Dimensional Solutions for Laterally Loaded Piles with Soil Modulus Assumed Proportional to Depth." *Proceedings, Eighth Texas Conference on Soil Mechanics and Foundation Engineering*, Austin, Texas, ASCE, 1956.

Saul, W. E. (1968) "Static and Dynamic Analysis of Pile Foundations." *Journal of the Structural Division*, ASCE, Vol. 94, No. ST5, 1968, pp. 1077-1100.

Tarquinio, F.S. and Pearlman, S.L. (1999). "Pin Piles for Building Foundations", *Proceedings of the 7th Annual Great Lakes Geotechnical and Geoenvironmental Conference*, May 1999

Teng, W.C. (1962) *Foundation Design.* Prentice Hall, Inc. 466 pp.

Tomlinson, M. J. (1987) *Pile Design and Construction Practice.* Viewpoint Publication, 1987, 415 pp.

ROCK SOCKETED MICROPILES

Nasser Massoudi[1], P.E., Member, ASCE

ABSTRACT: Micropiles are small diameter piles often used for support of foundations in difficult conditions. They are one of the more efficient deep foundation systems available in the industry. They can tolerate very large loads for their size, especially when socketed in rock. This paper presents three case histories on micropiles; all piles were socketed into weathered rock or bedrock. The micropiles were installed as foundations for a power plant, a retail shopping mall, and an industrial building. The case histories present information on the subsurface conditions, foundation selection, and micropile construction in schist, gneiss, and karstic limestone. The case histories also present information on pile drilling issues particularly that, from a production perspective, the success of any micropile application is a function of the tools and equipment that are used to drill the piles in an efficient, reasonable time. Pile load test data along with other pertinent information is also presented and discussed. The results illustrate that micropiles can be successfully utilized for virtually all kinds of ground conditions and that they are highly flexible to modification at anytime during construction in response to changing ground conditions.

INTRODUCTION

Micropiles are small diameter, drilled-in piles that were originated in Italy in the 1950s. They are typically in the 5- to 10-inch diameter range. Their load-carrying capability is essentially through side-friction (bonding) with the surrounding materials, with very little, if any, end-bearing contribution. They can be installed to great depths, on the order of 200 feet. Despite their small size, these piles can support large foundation loads, on the order of 200 tons, and sometimes even larger,

[1] Principal Geotechnical Engineer, URS Corporation, 200 Orchard Ridge Drive, Gaithersburg, MD 20878, nasser_massoudi@urscorp.com

owing to the utilization of high-strength steel, relatively high steel-cement grout ratio, and grout pressurization.

Many factors govern the selection of appropriate foundations for any project. Yet, constructability is the single most important factor that influences the selection of a foundation type over another. And, more often than not, when all other alternatives are exhausted, it is this factor that leads to the selection of micropiles as appropriate foundations for a project. It is when other foundation systems cannot be relied upon to readily penetrate obstructions, adapt to drastically varying geologic conditions, access restricted spaces, or mobilize within very short notices that micropiles are called upon. It has only been a short time, probably 20 years or so, that micropiles have been in more common use; during this period they have found many applications in the foundation industry owing primarily to the development of smaller yet more powerful equipment for their installation. Increases in urban construction, retrofitting, and emergency repairs have also played significant roles in the development of micropiles. Examples of such construction are given by Bruce (2002), Tarquinio and Pearlman (1999, 2002), Pearlman et al. (1997), and Munfakh and Soliman (1987).

This paper presents three case histories, illustrating the application of micropiles in various rock environments, for different projects, and with varying loading and construction requirements.

POWER PLANT: MICROPILES IN SCHIST

A power plant was constructed in Bethesda, Maryland. The plant was designed to be housed in a new building with structural steel framing. The equipment included turbine generator, heat recovery systems, compressors, 150-foot high stack, and a host of other mechanical units. Foundation pressures were estimated to be about 4,000 psf. The facility was constructed at the location of an existing building upon demolition. The existing building was supported on driven piles; construction of the power plant required utilizing some of the existing piles to the extent possible.

Geologically, the area is within the eastern Piedmont physiographic province, underlain by the metamorphic rocks of the Whissihickon Formation, consisting of schist and gneiss. Results of a subsurface investigation by test borings including Standard Penetration Tests (SPT) and supplemented by information taken in 1950, indicated the presence of random fill, underlain by weathered rock, and bedrock. The fill varied in thickness from about 20 to 25 feet, consisted of loose to medium dense low plasticity micaceous silt (ML) with varying sand contents, and included debris such as concrete, wood, wire, and cement grout from previous construction. The grout had been installed a few years earlier using jet grouting in support of a remedial construction, resulting in unconfined compressive strengths for soilcrete of about 3,000 psi. Weathered rock (residual soils and friable decomposed rock) was present below the fill, extending to a depth from 60 to 75 feet, and consisted of dense to very dense low plasticity micaceous sand (SM) with silt and gravel. SPT values exceeding 100 blows per foot were common with greater depth, indicating increase in density and a lower degree of decomposition with depth in these materials. Other indices for these soils found in laboratory tests included water content ranging from 10 to 20%, liquid limits of 30 to 40, plasticity index of 7 to 10, and fines content of 30 to 33%. The actual depth to intact rock was not known at the site; however, refusal to SPT sampling, defined as 100 blows per 2 inches or less of penetration, was recorded in some borings at a depth of about 60 to 75 feet, indicating variability in depth to bedrock. Groundwater level at the site was at depths from about 25 to 45 feet, probably influenced by on-going dewatering for nearby construction. A typical subsurface profile of the site is shown in Fig. 1.

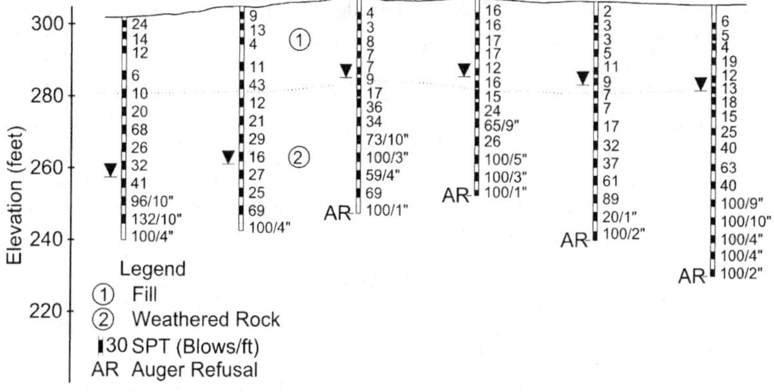

FIG. 1. Subsurface Profile at the Power Plant Site

From a foundations perspective, the most significant feature at the site was the extent and variability of the fill material. The thick, uncontrolled fill and its variability presented unfavorable conditions for using shallow foundations. In addition, the new foundations needed to be constructed immediately adjacent to existing buildings, with potential to cause damage to existing foundations in the process of any excavations that would be necessary, not to mention that an on-going deep excavation neighboring the power plant allegedly had resulted in loss of ground, adversely impacting foundations of an adjacent building. Some of the new foundations also needed to be constructed within areas restricted in access. Therefore, deep foundations appeared feasible. Various pile options were considered, however, they were mostly rejected due to associated excessive vibration, noise, large size construction equipment, restricted access, uncertainty in penetrating the very dense/hard zones, and overcoming obstructions in the fill, especially in the 3,000 psi soilcrete. Micropiles were selected for the project given the associated reduced noise and vibration, equipment maneuverability in restricted foundation areas, and ready penetration capability in virtually all types of materials present at the site, particularly obstructions. Micropiles were designed for a compression and tension design load of 50 and 10 tons, respectively. They were designed to derive their capacity from the weathered and/or intact rock.

The micropiles consisted of 7-inch diameter, 0.5-inch thick wall, Grade 80 steel pipe casings. The casings were flush-joint and threaded; the lead casing was fitted with cutting shoes at the tip. Duplex drilling with nominal 6-inch diameter tri-cone roller bit was used to advance the casings and remove the cuttings using water. Each pile was fitted with a No. 14, Grade 75 central steel bar, extending the full length of the pile. Each casing was initially seated into the weathered rock and then a minimum 20 feet of bond length was drilled for the pile. The bond diameter was slightly less than 6 inches. The piles were tremied with 5,000 psi grout; the grout was pressurized to 75 to 100 psi. About 100 micropiles were installed, including test piles. Different drilling rigs were used for the pile installation. In areas with restricted access, a remote-controlled miniature rig, slightly larger than a household refrigerator, was utilized.

Three test piles were constructed, two of which were load tested (quick test). Results of one of the load tests are presented in Fig. 2. The test pile was initially loaded in tension to 20 tons, twice the design tensile load and then unloaded (not shown in Fig. 2). Less than 0.05 inch of pile head movement was observed at the maximum tension test load. The pile was then loaded in compression to a maximum test load of 140 tons, as shown in Fig. 2.

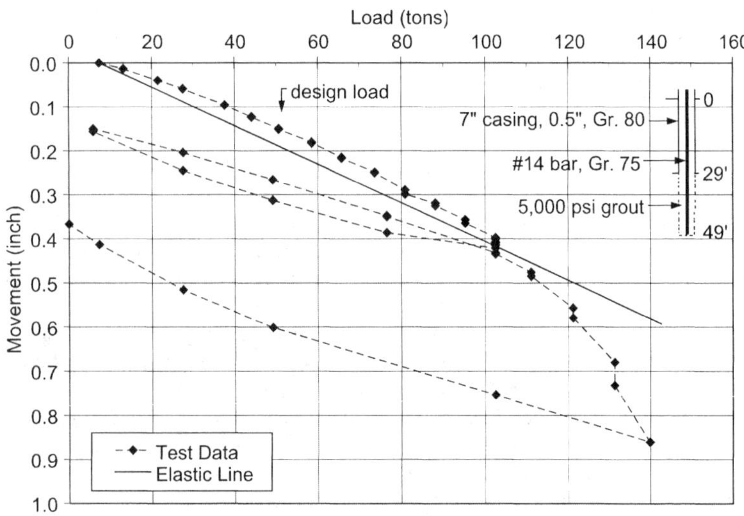

FIG. 2. Pile Load Test Results for Power Plant Project

The load test results indicate a safe working load of over 50 tons, the design compression load of the pile. The data also indicates that the pile primarily behaved elastically during a substantial portion of the initial loading. Up to a load of about 100 tons (twice the design load), the pile exhibited an essentially linear response. This pile met the project needs with more than the required safety factor. The bond length for this test pile was 20 feet; the tremie pipe was grouted in with the test pile due to construction logistics. It is noted that a similar pile, but with 15 feet of bond length, was also load tested which failed at a load of about 90 tons.

Back-calculations based on the test loads suggested an ultimate bond stress of 4 tsf, or an equivalent friction angle of 35 degrees, assuming that all loads were resisted by the pile socket with no end-bearing contribution. These values are consistent with the adopted engineering parameters for weathered rock at the site based on correlations with SPT values (FHWA,1996). A key factor in the achievement of the desired bond stress was believed to be due to the 75 to 100 psi pressure that was applied to the grout.

An important aspect of any micropile installation is observation of the rate of drilling. Information on drilling rate early in the job serves several purposes. One purpose would be that it assists with the selection of appropriate drilling tools. Information collected on rate of drilling for this project is presented in Fig. 3, for Duplex drilling. The two drill rig types that were utilized were a Klemm KR806-3 with motor type BF6M 1013 and an electric Klemm KR704-1E.

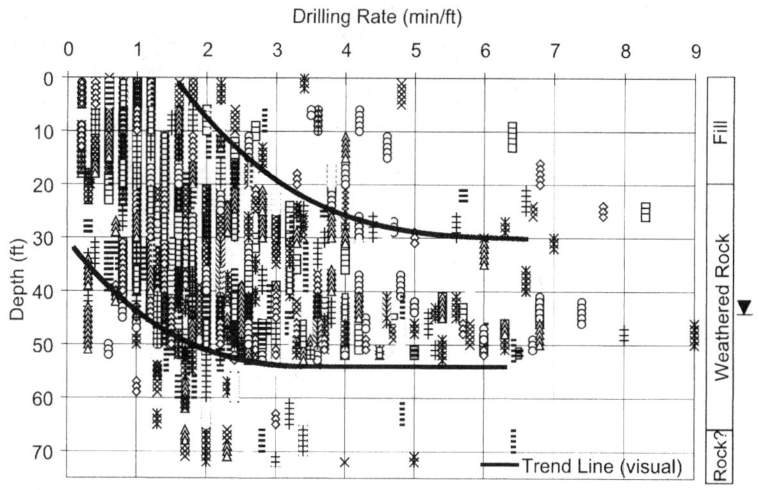

FIG. 3. Drilling Rate for Power Plant Project

The drilling data suggested that on average it took about 110 minutes to drill a 50-foot long pile. In comparison, it took about 10 minutes to grout the same pile. In other words, over 90% of the time required to complete a pile was spent on its drilling and less than 10% on grouting it. Given that often a much greater proportion of the time is invested into drilling the pile, this information is of interest to the pile contractor to seek improvements in drilling time, to enhance production. The contractor uses this information to review drilling tool options and to select the most appropriate tools for the subsurface condition. Equally, the engineer is interested in the same information for qualifying ground condition and assessing its adequacy. Although not as common or as important, owners may use the information to verify or revise construction schedule or progress. It is important to note that the relatively little time devoted to grouting the pile should not render this operation insignificant. Good quality grouting, including pressurization, is essential to forming quality piles. Given that grout pressurization in known to improve bond quality and load-carrying capability, it is considered good construction practice to pressurize the grout in all micropiles particularly in light of the relatively little time that is required to accomplish this task.

SHOPPING MALL: MICROPILES IN GNEISS

Renovations at a shopping mall in Springfield (a suburb of Philadelphia), Pennsylvania necessitated designing new foundations for the facility. The original mall was supported on caisson (drilled shaft) foundations. The modification consisted of erecting a new 2-story structure, including the construction of a new plaza. The column loads for the new facility were about 150 kips.

Geologically, the site is located within the eastern Piedmont physiographic province, underlain by the metamorphic rocks of the Whissihickon Formation, consisting of gneiss. Subsurface information at the site indicated, in descending order, fill, natural alluvial soils, and weathered to intact rock. The subsurface information included borings and records of caisson construction for the original building; caissons records were particularly helpful in characterizing the bedrock condition. A typical subsurface profile of the site is shown in Fig. 4.

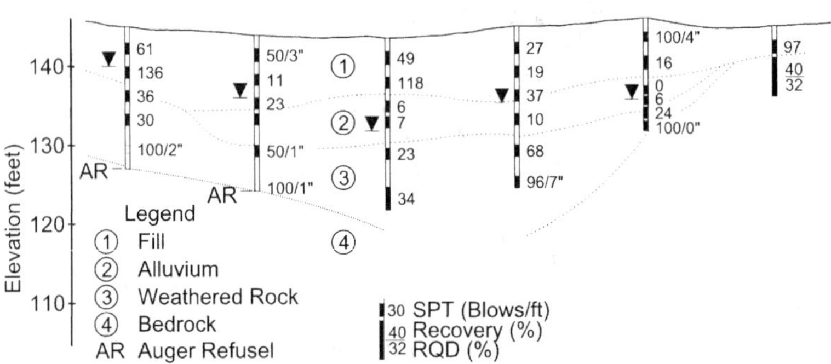

FIG. 4. Subsurface Profile at the Shopping Mall Site

As shown in Fig. 4, the fill material varied in thickness from about 7 to 10 feet, consisted of dense silty sand (SM), and included rock fill. The alluvium was primarily 3 to 7 feet thick, consisted of soft to firm sandy silt (ML). The decomposed rock, derived from weathering of gneiss, was from 5 to more than 10 feet thick, and consisted of dense to very dense, friable silty sand (SM) and sandy silt (ML). The bedrock was hard gneiss, with occasional seams, and with Rock Quality Designation

(RQD) values in the range of about 30 to 90%. Sometimes, highly weathered zones of rock were present below essentially intact rock. Groundwater was at about 5 to 12 feet below the ground surface.

The uncontrolled fill at the site and its variability presented unfavorable conditions for shallow foundations. Therefore, deep foundation was selected for support of the structure. Given need for maneuverability in restricted access areas and penetration capability in the random fill, micropiles were selected for foundations. Micropiles were designed for a compression load of 25 tons, deriving their capacity from bond stress between grout and rock. The piles were designed for an allowable bond stress of 3 tsf.

The micropiles consisted of 5.5-inch diameter, threaded, flush joint, 0.244-inch thick wall, and A-36 steel pipe casings. Each pile was fitted with a No. 9, Grade 60 central steel bar, extending the full length of the pile. The bond diameter was about 4.5 inches. Casings were initially set into the top of bedrock and then a minimum 8 feet of bond length was drilled for each pile. Pile lengths of about 40 to 60 feet were installed, including 8 to 20 feet of rock socket. When the rock quality appeared poor, based on observation of the drilling, the bond zone was lengthened. These resulted in adjustments of the bond length in the field, sometimes by as much as 20 feet. The piles were tremied with 4,500 psi grout.

Due to the relatively small design load, pile load tests were not performed. More than 30 micropiles were installed. Drilling time was also monitored for this project, as shown in Fig. 5.

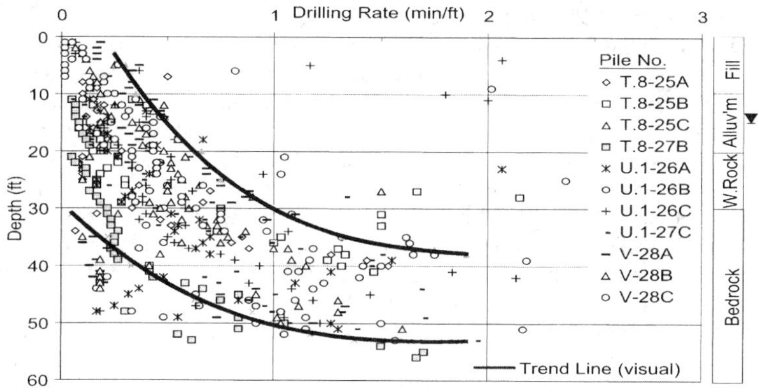

FIG. 5. Drilling Rate for the Shopping Mall Project

An Odex air percussion hammer was used to advance the casings and remove the cuttings. The equipment consisted of a Davey-Kent DK 90RG drill rig and an Ingersoll Rand 375 air compressor. The collected information, as shown in Fig. 5, suggests that on average it took about 30 minutes to drill a 40-foot long pile. In comparison, it took about 15 minutes to grout the same pile. In other words, about

70% of the time required to complete a pile was spent on its drilling and the remaining 30% on grouting it. As stated earlier, drilling information early in the job serves several important functions. The function it served for this project was assisting with the evaluation of the bond zone quality. This information was used to establish qualitative correlations between adopted pile design parameters and the required ground conditions to meet the design needs, as subsequently described.

The drilling rate information in the bond zone, as shown in Fig. 5, is also indicative of the rock quality in that zone, with the rates varying from a fraction of 1 min/ft to more than 2 min/ft. More important than the selection of suitable drilling tools, such observations are critical for validating design assumptions and for selecting appropriate quality control/quality assurance measures. In this case, the drilling rates in the bond zone were considered inconsistent with the adopted design parameters, probably caused by seams, highly weathered zone(s), and/or discontinuities within the rock mass. The collected drilling rate information was used to modify the design bond length on a case-by-case basis for each pile. Based on observed drilling rates and correlation with results of the rock cores, an empirical relationship was established between drilling rate and the "value" of every foot of bond zone drilled, i.e., drilling rates of 25 sec/ft or higher were considered equal to one foot of acceptable bond length, drilling rates of less than 12 sec/ft were considered as zero bond length (ignored), and drilling rates between 12 and 25 sec/ft were counted as one-third of a foot of acceptable bond length. As stated earlier, this condition resulted in increased bond length for some piles. An example was Pile No. T.8-27B, shown in Fig. 5, which encountered drilling rates from 11 to 16 sec/ft between depths of 30 to 40 feet, resulting in increased pile length.

Difficult drilling was encountered in a number of piles, caused by the presence of obstructions in the fill. Also, during drilling the socket for some piles, geyser-like conditions were observed, water ejecting from the adjacent already-drilled piles. The condition was attributed to fractures in the bedrock, providing conduits among adjacent piles. A similar, yet less pronounced, feature was observed during grouting operations, where grout placed in one pile resulted in pushing water out of adjacent pile(s). This "leakiness" in piles resulted in large grout take in some piles. In a few piles, the sockets simply could not hold grout, in which case grouting was terminated, and the piles were redrilled and regrouted the next day. Additionally, the quality of socket in some piles was judged objectionable due to the leakiness, resulting in redrilling these piles. The grout factor (ratio of actual grout volume to the theoretical volume of pile) in most piles was estimated as about 1.3. In the leaky piles, however, the grout factor exceeded 5.0.

INDUSTRIAL BUILDING: MICROPILES IN KARSTIC LIMESTONE

Renovations and the addition of new buildings at an existing manufacturing facility in Lancaster, Pennsylvania, required the construction of new foundations. The addition was a one-story building, with column loads on the order of 500 kips.

Geologically, the site is underlain by the Ledger and Conestoga Formations; the Ledger Formation is coarse, crystalline dolomite and the Conestoga Formation is limestone with shale partings and has open joints. Test borings indicated the presence of random fill underlain by weathered rock and bedrock. The fill varied in thickness from about 10 to 15 feet, and consisted of soft to stiff clay (CL) and loose to medium dense low plasticity sand and gravel (SM), including debris such as bricks, coal, and slag. Weathered rock (residual soils and decomposed rock) present below the fill, extended to depths of about 15 to 30 feet, and consisted of very soft to stiff silty clay (CL) with sand and gravel. The cored rock was moderately hard to hard, weathered to fresh karstic limestone with fractures. RQD values ranged from 0 to 100%, indicative of very poor to excellent quality rock, with rock quality generally improving with depth. Groundwater level at the site was recorded at a depth of about 10 feet. A typical subsurface profile of the site is shown in Fig. 6.

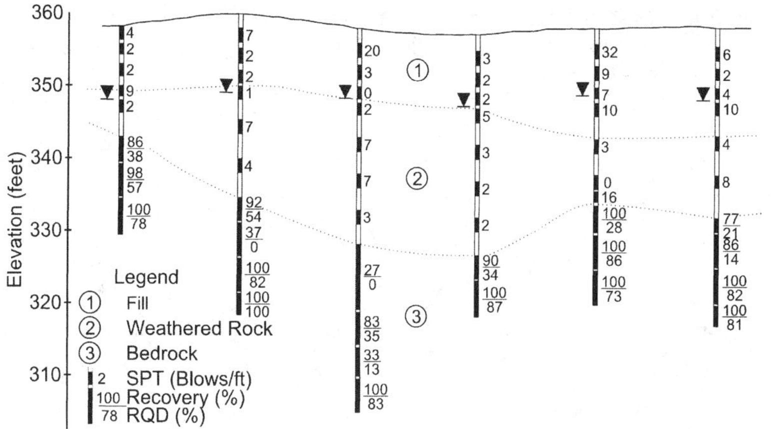

FIG. 6. Subsurface Profile at the Industrial Building Site

The presence of random, uncontrolled fill, as well as occasional soft residual soils rendered shallow foundations unsuitable. The bedrock was known to be pinnacled with solution channels, therefore susceptible to developing sinkholes. Caissons were considered problematic should extensive pinnacles be present in the rock, requiring significant rock excavation. Driven piles were dismissed given questionable capacity in solutioned rocks and proximity to existing buildings. Micropiles were, therefore, considered for their greater flexibility in seeking suitable bearing within the limestone. Micropiles were designed for a compression load of 62.5 tons, deriving their capacity from bonding with bedrock. Based on available information, average pile lengths of about 35 feet were expected, including 8 feet of rock socket.

The micropiles consisted of 7-inch diameter, 0.362-inch thick wall, A-36 steel pipe casings. The casings were threaded, flush joint, and the lead casings were fitted with carbide teeth to facilitate drilling. Duplex rotary drilling with nominal 6-inch diameter tri-cone roller bit was used to advance the casing. Compressed air was used to remove the cuttings. Each pile was fitted with a No. 14, Grade 60 central steel bar, extending from the pile tip to about 8 feet into the casing. The casings were initially seated into the rock and a minimum 8-foot long socket was subsequently drilled for each pile, using a nominal 6-inch diameter down-hole hammer. The rate of drilling in the bond zone was about 30 to 35 sec/ft. The piles were tremied with 5,000 psi grout. The grout was not pressurized. About 50 micropiles were installed, including a test pile.

One test pile was installed. The test pile, the first pile at the site, exceeded anticipated pile lengths. Although variability is commonplace, particularly in karstic environments, significant increases in pile lengths, especially in the first pile, are distinctly striking. A casing length of about 68 feet was installed to reach the top of rock for the test pile, compared to the assumed average 27 feet. Also, the test pile socket length was about 12 feet, compared to the designed 8 feet. The overall test pile length was about 80 feet compared to the average design length of 35 feet, more than twice that surmised from the available data. The test pile was load tested; the results are presented in Fig. 7.

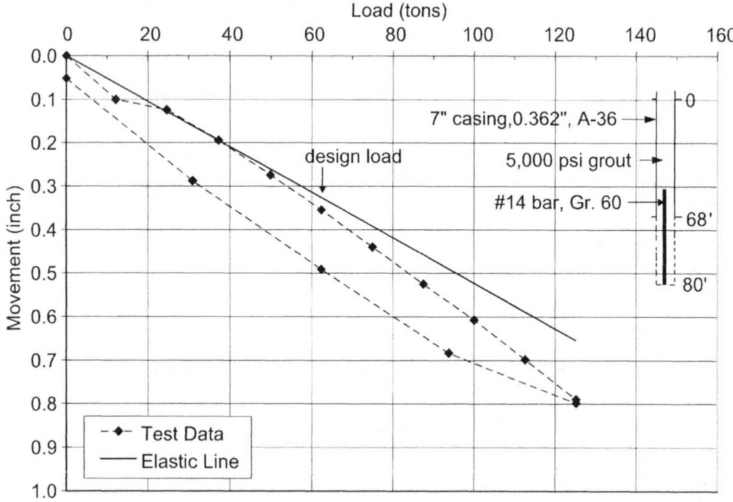

FIG. 7. Pile Load Test Results for Industrial Building Project

The load test results, shown in Fig. 7, indicated less than 0.75 inch of pile head movement at the maximum test load of 125 tons. The results indicated a safe working load of over 62.5 tons, the design compression load. Based on the maximum test load, and an assumed load distribution entirely along the pile socket, the ultimate bond stress was calculated as 7 tsf. A review of the rock recovery and RQD values from Fig. 6 indicates that the bedrock is capable of offering such side friction, particularly after the improvement in rock quality that is realized following grouting.

Large grout takes were observed during the construction of the piles. This is not uncommon in carbonate rocks that are known to contain cavities. To minimize grout loss, upon observation of significant grout injection, grouting was performed intermittently, i.e., the grouting operation would stop to allow the grout to "gel" and then resume after a few minutes. This process was repeated until the pile was completely grouted. At times, it was noticed that grout level in piles dropped, in which case grout was added to the top of the pile to make up for the loss. Further examples of micropile construction in karst geology are given by Tarquinio and Pearlman (2001).

CONCLUSIONS

Micropiles have developed to the extent that they are nowadays considered acceptable foundations for practically any situation, particularly when other foundation types fail to meet or are found marginal for the needs. Small drilling tools, highly mobile equipment, and high strength steel render these piles fitting to difficult grounds, restricted access, low settlement tolerance, emergency repairs, retrofitting, and low noise and vibration situations. They are now within the tool kit of every foundation practitioner. This is believed to be partly due to the advancement of the micro drilling technology and, to some degree, the advocacy and utilization that these piles enjoy in noticeable quantities by transportation authorities.

Micropiles are superior load bearing elements for their size. To ensure their success, superior quality processes during their construction are needed. There are many factors that control the successful completion of this pile and, therefore, they are more dependent on competent, continuous observation and monitoring. Processes considered to have the greatest potential to render these piles unacceptable, therefore requiring close scrutiny, include

- Variations in ground conditions compared to those assumed in initial design: This requires the observation and monitoring of each pile, including continuous evaluation of the drilling tool's suitability for the ground, seating the casing at proper depth and in proper materials, attention to visible and audible changes in drilling the bond zone, and examination of the drill fluid return.
- Condition of the pile upon completion of drilling and prior to grouting: This requires measurement and verification of the pile hole to ensure adequate construction of bond length, removal of the cuttings, and stability of the pile hole until such time that the pile is grouted.
- Grouting operation and rebar installation: Grouting is a very brief yet significant step in forming the pile. It requires observation of the grout preparation and use of admixtures, grout sampling for testing, tremie operation, grout pressurization, and the final grout level in the pile. Unless there are obvious reasons not to, it is highly advantageous to pressurize the grout for better quality (higher capacity) piles. Reinforcement installation, in addition to meeting the design needs, is also a critical

indicator of the quality and continuity in the pile hole.

- Protection of piles: Constructed piles should be protected, so that they are not damaged, until such times that they receive the pile cap.

Little should be left to fate in the design of these piles; possible scenarios that may be encountered during construction should be evaluated during design. While the more prevalent ground conditions should be assumed for the pile design, possible variations that could be encountered during construction should be cited in construction documents so that they are looked after and flexibility in meeting these variations included for.

REFERENCES

Bruce, D.A. (2002). "Micropiles: Think Small for Big Loads." *Geo-Strata*, ASCE, July 2002, 15-18.

FHWA (1996). "Drilled and Grouted Micropiles: State-of-Practice Review." FHWA-RD-96-016 through FHWA-RD-96-19.

Munfakh, G.A. and Soliman, N.N. (1987). "Back on Track at Coney Island." *Civil Engineering Magazine*, ASCE, December 1987.

Pearlman, S.L., Richards, T.D., Wise, J.D., and Vodde, W.F. (1997). "Pin Piles for Bridge Foundations: A Five Year Update." *Proceedings 14th Annual International Bridge Conference*, 472-480.

Tarquinio, F. (2002). "Mastering Marginal Soils." *Structural Engineer Magazine*, September 2002.

Tarquinio, F.S. and Pearlman, S.L. (1999). "Pin Piles for Building Foundations." *Proceedings 7th Annual Great Lakes Geotechnical and Geoenvironmental Conference.*

Tarquinio, F.S. and Pearlman, S.L. (2001). "Pin Piles in Karst Topography." *Proceedings 8th Multidisciplinary Conference on Sinkholes and the Engineering and Environmental Impacts of Karst, Louisville, Kentucky, AA Balkema, Rotterdam, Netherlands, 177-182.*

SOURCES OF UNCERTAINTY IN LATERAL RESISTANCE OF SLENDER REINFORCEMENT USED FOR SLOPE STABILIZATION

Jorge R. Parra[1], P.E., Member, ASCE, Eng-Chew Ang[2], Erik Loehr[3], Member, ASCE

ABSTRACT: Significant uncertainty remains in predicting the stabilizing forces that can be developed in slender reinforcing members used for slope stabilization. This uncertainty arises from the high number of potential failure modes as well as from uncertainties present in predicting the resistance contributed by the individual failure modes themselves. The aim of this research has been to identify the most significant sources of uncertainty in the design of slender reinforcement through parametric analyses so that the uncertainty can be addressed by modifications to current design and construction practice, and future research. Results of the parametric analyses showed that the degree of improvement in the factor of safety of a reinforced slope depends on a combination of the design parameters evaluated. The parameters with greater influence on the factor of safety for a given soil condition were slope inclination, method for predicting limiting soil resistance, and capacity of reinforcement used. While the uncertainty in limiting resistance for weak reinforcing members is greatly reduced due to the controlling effect of the structural failure mode, significant uncertainty still exists in predicting the limiting resistance on strong reinforcing members

INTRODUCTION

The current state of practice in designing slender reinforcement like micropiles, soil nails, or steel piles for slope stabilization involves a number of uncertainties that often times are overcome by assuming very conservative design postures. Aside from the ever-present uncertainties in the soil strength parameters and pore pressure

[1] Doctoral Candidate, University of Missouri-Columbia, Department of Civil and Environmental Engineering, EBE 2509, Columbia, MO, 65203, e-mail: jrpb66@mizzou.edu
[2] Doctoral Candidate, University of Missouri-Columbia, Department of Civil and Environmental Engineering, EBE 2509, Columbia, MO, 65203, e-mail: engchew@hotmail.com
[3] Assistant Professor, University of Missouri-Columbia, Department of Civil and Environmental Engineering, EBE 2509, Columbia, MO, 65203, e-mail: eloehr@missouri.edu

conditions, there are other significant sources of uncertainty that stem from predicting the resistance that each reinforcing member contributes to the stability of the slope. These uncertainties arise first from the different potential failure modes that may develop, as well as from uncertainties in predicting the limit condition of each individual failure mode.

The objective of this paper is to identify and quantify the most significant sources of uncertainty found in design of slender reinforcement for slope stabilization. To accomplish this, a parametric study was performed to investigate the effect of varying different design parameters on the overall factor of safety of a slope and identify those parameters that have more impact on the factor of safety. The analyses were performed for purely frictional soil profiles (cohesion intercept $c'=0$), and for fully drained cohesive soil profiles (c' and angle of internal friction $\phi'>0$). The parameters investigated in the study were slope inclination, soil conditions, capacity of reinforcement, reinforcement spacing, and the method used to predict the limiting soil resistance.

ANALYSIS METHOD

A design methodology for predicting the resistance that can be provided by slender reinforcement for slope stabilization has been developed (Loehr et al., 2000). The method uses a limit state design approach wherein a series of potential failure mechanisms are considered in developing the overall distribution of lateral resistance along the reinforcing member. The three limit states considered include:

1) Failure of soil around or between reinforcing members
2) Failure of the reinforcing member either in bending or shear
3) Failure of soil due to insufficient anchorage

Figure 1 illustrates typical limiting resistance distributions for each limit state for a 2.4-m (8-ft) long member installed in cohesionless soil with $\phi'=20.4$ degrees. A description of each limit state is given below.

Failure Mode 1- Failure of Soil Around or Between Reinforcing Members.

Estimation of this limit state requires knowledge of the limiting soil pressure, which is the maximum pressure on the soil before it fails in a plastic mode around or between reinforcing members. Limiting resistance at a particular depth of sliding is computed from the distribution of limiting soil pressure along the reinforcing member by integrating the area under the limiting soil pressure curve. Different methods for predicting limiting soil pressure have been developed for purely cohesive soils, cohesionless soils, and soils that exhibit friction and cohesion, (Ito and Matsui, 1975; De Beer and Carpentier, 1975; Reese et al., 1974; Broms, 1964; Fleming, et al., 1992, Poulos, 1996; Hansen (1961). All methods rely on different assumptions and are generally a function of the soil unit weight γ, angle of internal friction ϕ, cohesion intercept c, center to center spacing in the longitudinal (strike) direction, and depth. A comparison of limiting soil pressure predicted from different methods for 2.4-m (8-ft) long reinforcement installed at a longitudinal spacing of 0.9-m (3-ft) in a purely

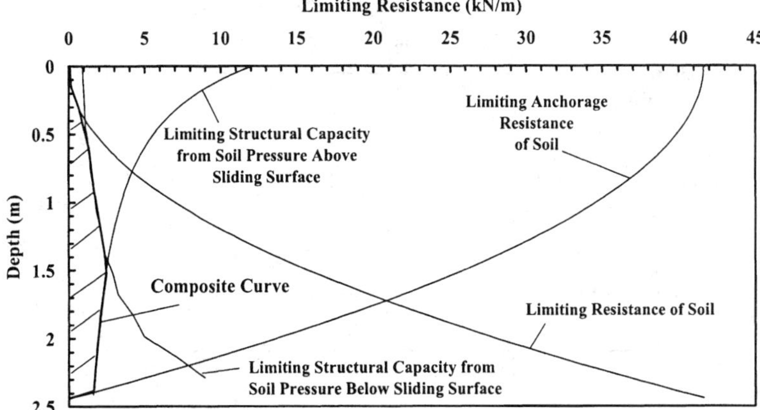

FIG. 1. Example of Limiting Resistance Curves for Weak Reinforcing Members at Longitudinal Spacing of 1-m, Based on Broms' Method for $c = 0$ Soil, $\phi' = 20.4$ degrees, and γ=17.3 kN/m^3, and Composite Limiting Resistance Curve

frictional soil with an angle of internal friction of 23 degrees is shown in Figure 2.

For a fixed set of soil parameters, the distribution of limiting pressure varies linearly with depth, and the values of limiting soil pressure from different methods may differ by an order of magnitude. In the case presented, methods by DeBeer and Carpentier and by Ito and Matsui give similar values and serve as a lower bound of limiting pressure. Reese and Fleming's methods provided intermediate values, while Broms and Poulos' values serve as an upper bound.

Failure Mode 2-Failure of the Soil Due to Insufficient Anchorage Length

The resistance developed in the reinforcing member due to load applied by the soil underlying the sliding surface is computed in a similar fashion to the first soil failure mode except that the limiting soil pressures are integrated over the length of the reinforcing member below the sliding surface. The limit resistance curve computed in this manner is shown in Figure 1.

Failure Mode 3-Failure of the Reinforcing Member

Structural failure of the reinforcing member may result from bending or shear stresses imposed by the retained soil and the moving mass above the sliding surface. The approach used to account for the potential of the reinforcing member to fail structurally in bending is to consider a factored distribution of lateral soil pressure, given as a multiple of the limiting soil pressure. The factored pressure is determined by computing a factor α that equates the moment capacity of the

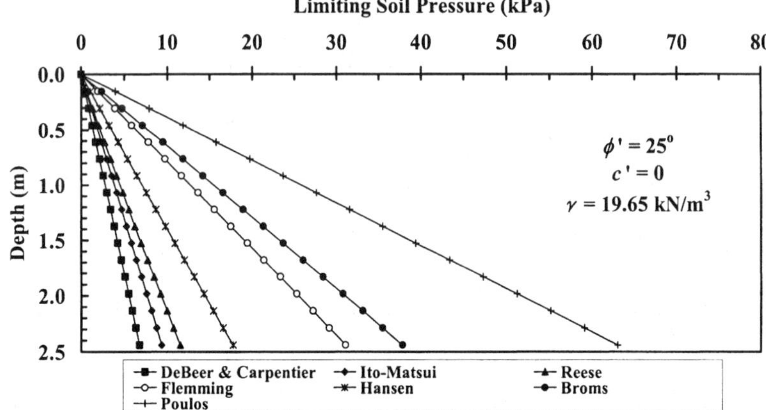

FIG. 2 Comparison of Methods for Predicting Limiting Pressure in Homogeneous, Purely Frictional Soil ($c' = 0$, $u = 0$).

reinforcing member to the maximum moment generated in the member due to lateral loading of the soil. The maximum moment due to lateral loading from soil was computed by assuming the soil response to lateral loads (p-y curve) to be elastic, and assuming a pile of infinite length. The resisting force is then computed by integrating the factors pressure distribution for different depths to produce the curve shown in Figure 1.

From the three failure modes discussed above, a composite limiting resistance curve that corresponds to the most critical component at each depth is established by taking the limiting resistance for the mode with the least resistance at each depth. The thick line on Figure 1 comprises the composite limiting resistance curve. The composite curve shows that the limiting soil resistance controls the part of the curve close to the ground surface, while the member capacity controls over the middle portion of the member. The anchorage capacity controls the lower portion of the curve. The depths at which different failure modes control varies with soil conditions, reinforcement type, and reinforcement spacing.

STABILITY CALCULATIONS

The general approach in analyzing the stability of an unreinforced slope and calculating factor of safety is based on limit equilibrium. A similar approach is adopted for slopes reinforced with slender members. A force due to a reinforcing member defined by the limiting resistance at the point where the sliding surface intersects the reinforcing member, is added to the other forces that resist sliding. Therefore, inclusion of reinforcement through the shear zone produces a positive effect on the overall stability by increasing the factor of safety as compared to an unreinforced slope.

PARAMETRIC ANALYSES

Parametric analyses were performed for a number of different slope configurations. Two different profiles were considered: a homogeneous profile, and a two-layer profile consisting of a 1.2-m (4-ft) thick layer underlain by an infinitely deep layer of higher strength. The slope inclinations analyzed were 1.5 (Horizontal): 1(Vertical), 2.5(H):1(V), and 3.5(H):1(V), all with a fixed slope height of 6.8-m (22.4-ft). A constant unit weight of 17.3 kN/m^3 (110 lb/ft^3) was assumed for all soils, and all pore pressures were assumed to be zero. The soil strength conditions selected for each profile consisted of purely frictional soil (c = 0), and a c-ϕ soil ($c \neq 0$, $\phi \neq 0$).

The theories proposed by Ito and Matsui, and by Broms were selected for the parametric study for predicting limiting soil pressure because they are representative of the lower and upper bounds, respectively. Figure 3 shows plots of the composite limiting resistance curves of a 2.4-m (8-ft) long for both weak and strong reinforcement based on these methods. This figure shows that the composite limiting resistance for weak reinforcement is considerably reduced by the structural capacity of the member. It also shows that while there is only a minimum difference in limiting resistance for weak reinforcement as predicted from different methods, there is significant difference for strong reinforcement.

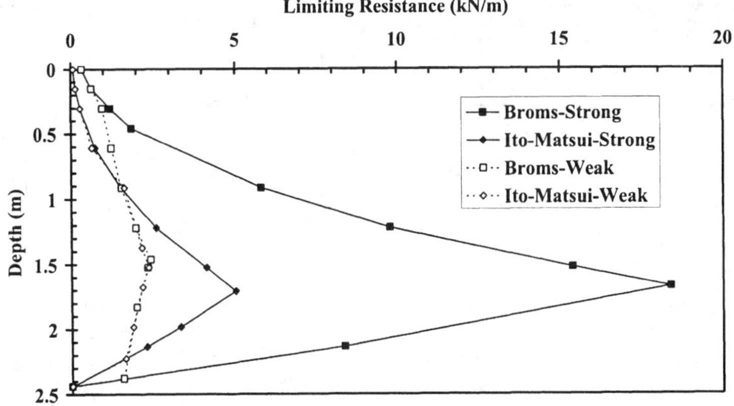

FIG. 3. Composite Limiting Resistance Curves Developed for Homogeneous, c-ϕ Soil in a 2.5(H):1(V) Slope, c'=1.2 kPa, ϕ' = 17.5 degrees, S_l= 0.91-m.

The first step for analyzing slopes with schemes of reinforcement was back analyzing each slope geometry without reinforcement to get the soil parameters that would result in a factor of safety of unity. All analyses performed for the homogeneous c - ϕ material included a small cohesion intercept c' = 1.2 kPa (25 psf). All analyses for the two-layer case with c = 0 and c - ϕ included the same underlying layer consisting of a dense cohesionless soil with ϕ' of 40 degrees. Table 1 shows the results of the back analyses for $c = 0$ and c-ϕ conditions, respectively. A greater back analyzed angle of internal friction was obtained for the steeper slopes.

The soil strength parameters from the back analyses were then used to develop the limiting resistance curves for each reinforcing member based on the two methods. Reinforcement size was 0.1-m x 0.1-m x 2.4-m long (4-in x 4-in x 8-ft long). Stability analyses were performed for each of the slopes assuming different schemes of reinforcing members represented by the reinforcement spacing in the longitudinal (strike) and transverse (dip) spacing, S_l and S_t, respectively. The longitudinal spacing was varied at 0.3-m (1-ft), 0.9-m (3-ft), and 1.83-m (6-ft). For each longitudinal spacing selected, the transverse spacing was varied at 0.3-m, 0.9-m, 1.83-m, and 3.66-m. The same limiting resistance was assigned to all reinforcement with a given longitudinal spacing S_l regardless of the transverse spacing. Axial loads were ignored in the analyses.

TABLE 1. Soil Parameters Obtained from Back Analyses for c-ϕ and $c = 0$ Soils in Homogeneous and Two-Layer Profiles

	c - ϕ Soil					c = 0 Soil				
	Homogeneous		Two-Layers			Homogeneous		Two-Layers		
Slope Inclination (1)	c' (kPa) (2)	ϕ'(°) (3)	Depth of Layers (m) (4)	c' (kPa) (5)	ϕ'(°) (6)	c' (kPa) (7)	ϕ'(°) (8)	Depth of Layers (m) (9)	c' (kPa) (10)	ϕ'(°) (11)
1.5(H):1(V)	1.2	27.5	0 - 1.22	1.2	26.5	0	31.9	0 - 1.22	0	31.9
			>1.22	0	40			>1.22	0	40
2.5(H):1(V)	1.2	17.5	0 - 1.22	1.2	16.3	0	20.4	0 - 1.22	0	20.4
			>1.22	0	40			>1.22	0	40
3.5(H):1(V)	1.2	12.8	0 - 1.22	1.2	11.5	0	15.2	0 - 1.22	0	15.2
			>1.22	0	40			>1.22	0	40

HOMOGENEOUS PROFILES, $c = 0$ AND $c - \phi$ SOILS

The results of stability analyses performed for homogeneous profiles in cohesionless $c = 0$ and $c - \phi$ soils were similar. Figure 4 shows the results obtained for cohesionless soil in the form of factors of safety versus transverse spacing for different slopes at a longitudinal spacing of 0.9-m (3-ft). In general, the factor of safety increased in all cases by reducing the longitudinal and transverse spacing, with greater improvement at a transverse spacing of less than 0.9-m (3-ft). Factors of safety also increased with the slope angle. Two reasons produce the result. One reason is that the back-analyzed angles of internal friction were significantly greater for steeper slopes; consequently, the limiting resistance and sliding resistance derived from these parameters will also be greater. However, factors of safety are also greater for steep slopes because moving the critical surface produces larger increases in the factor of safety for steeper slopes as evidenced by the maximum factor of safety noted in Figure 4.

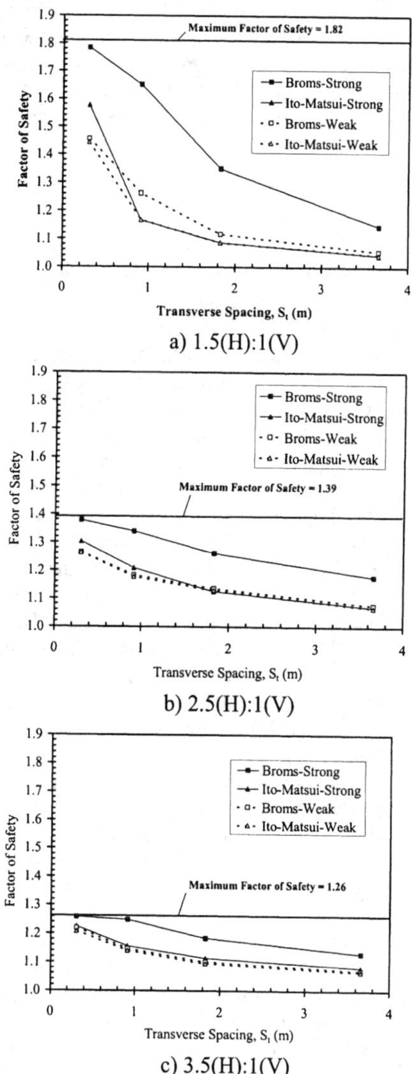

FIG. 4. Factor of Safety versus Transverse Spacing for Longitudinal Spacing S_l = 0.91-m, Homogeneous c = 0 Soil. a) 1.5(H): 1(V) Slope, b) 2.5(H): 1(V) Slope, c) 3.5(H): 1(V) Slope.

It is also interesting to note that factors of safety determined for weak reinforcing members were only slightly affected by the method used to predict the limiting soil pressure. Conversely, factors of safety determined for strong reinforcing members varied as much as 50% for the two methods used to predict the limiting soil pressure.

This observation suggests that while higher factors of safety can be obtained using strong reinforcing members, these factors of safety are highly dependent on the method used to predict the limiting soil pressure.

The largest improvement in the factor of safety for the reinforced slopes as compared to the unreinforced slopes is indicated on each plot by a thick horizontal line. This maximum factor of safety is obtained from closely spaced reinforcing members that force the critical sliding surface deeper and beyond the reinforced zone of soil. The maximum factors of safety obtained were 1.26 for the 3.5(H):1(V) slope, 1.39 for the 2.5(H):1(V), and 1.82 for the 1.5(H):1(V) slope.

Figure 5 illustrates the critical surfaces for a 2.5(H):1(V) slope at a longitudinal spacing of 3-ft. In general, critical sliding surfaces for 2.5(H):1(V) and flatter slopes with transverse spacing of 6-ft or less tended to pass through or very close to the toe, intersecting the reinforcement in the upper third and lower third portions of the slope (Figures 5a and 5b). For steeper slopes, or slopes with wider transverse spacing, and applying lower limiting resistance distributions, the critical surface generally tend to be relatively shallow, passing through or near the toe of the slope, and intersecting all the reinforcement in the reinforced zone (Figures 5c and 5d). The deepest surfaces were always obtained for the closest spacing independent of the slope inclination.

The location of the critical surfaces had an important influence on the trends observed in the factors of safety. The critical surface dictates what limit state for the reinforcing member governs the stability of the slope. The example presented in Figure 3 shows the remarkable difference in limiting resistance at different depths from different methods and for different types of reinforcement. The limiting resistance curves predicted for weak reinforcement are very similar because the structural capacity controls the resistance. This explains why the factors of safety for weak reinforcement are very similar. On the other hand, the composite limiting resistance curves developed for strong reinforcement yielded greater differences throughout the entire reinforcement length. This explains why the factor of safety varies over a wide range when the critical surface intersects the reinforcing members.

TWO-LAYER PROFILES

Figures 6 shows typical trends of variation in factor of safety with transverse spacing at different slopes, for a longitudinal spacing S_l = 0.91-m (3-ft). These results again show that the method for predicting the limiting soil pressure has a significant effect on the computed factors of safety for strong reinforcing members but very little effect for weak strong members. However, factors of safety for schemes of closely spaced weak and strong reinforcing members from two-layer profiles were generally higher than those obtained from homogeneous profiles for the same scheme and slope

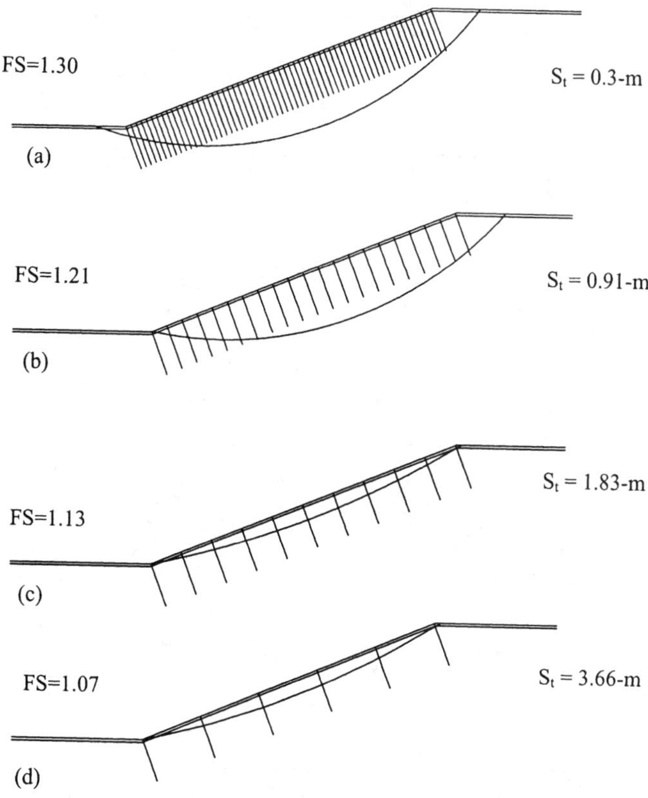

FIG. 5. Critical Surfaces for 2.5(H): 1(V) Slope with Weak Reinforcement At Longitudinal Spacing S_l = 1.83-m, Homogeneous $c = 0$ Soil, Ito and Matsui Method

inclination. This is a result of the fact that, because of the stronger lower layer, factors of safety for deep sliding surfaces are quite high. The critical sliding surfaces therefore tend to pass through the reinforcing members, which produces a higher factor of safety and places more importance on the method used for predicting limiting pressure. Evidence of this is shown in Figure 7, which shows the location of the critical sliding surfaces. For all but the most closely spaced reinforcement schemes, the critical sliding surface is forced to pass through the reinforcement because of the strong lower layer.

It is also interesting to note from Figures 4 and 6 that the computed factors of safety reach a threshold value for closely spaced reinforcing members. This value

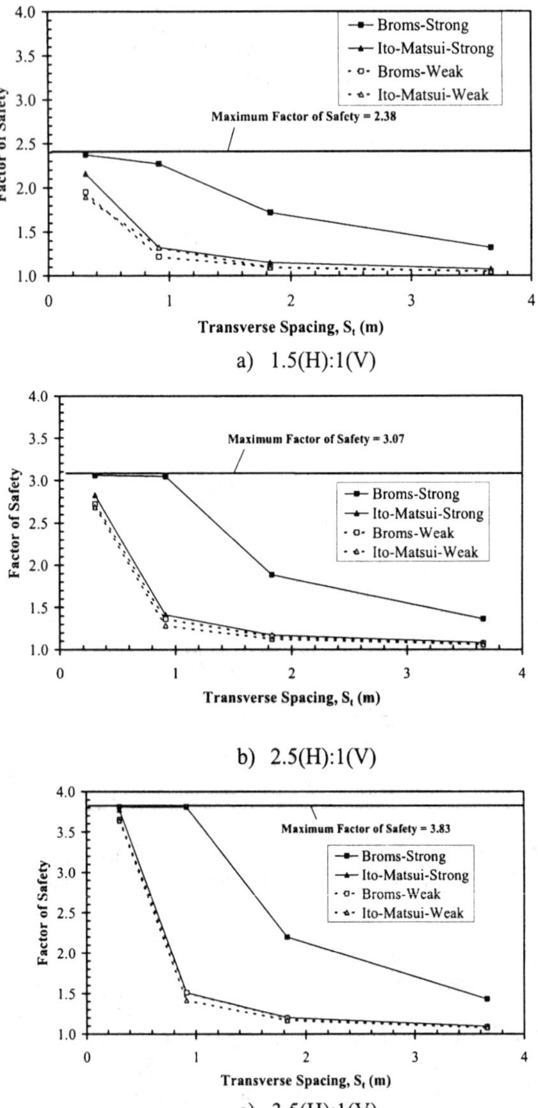

FIG. 6. Factor of Safety with Variation of Transverse Spacing for Two-Layered, *c*-ϕ Soil, S_l = 0.91-m. a) 1.5(H):1(V) Slope, b) 2.5(H):1(V) Slope, c) 3.5(H):1(V) Slope

corresponds to a "fully reinforced" condition wherein the critical sliding surface is forced to pass outside of the reinforced zone. In these cases, the factor of safety is fully controlled by the properties of the soil and the geometry of the slope and thus, these factors of safety are not affected by the limit resistance of reinforcing members. In this light, it is interesting to note that factors of safety for the fully reinforced condition tend to increase with decreasing slope inclination for the two-layer profile, in contrast to what was observed for the homogeneous profile.

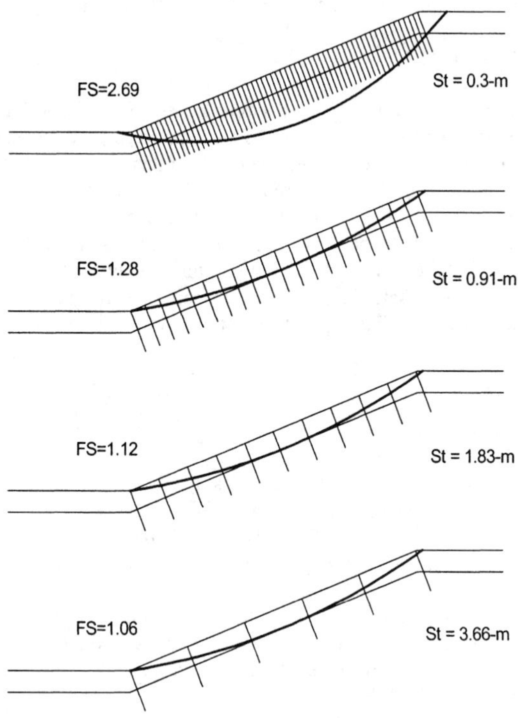

FIG. 7. Critical Sliding Surfaces in 2.5(H):1(V) Slope for Two-Layered, c-ϕ Soil with Weak Reinforcement at Longitudinal Spacing S_l= 0.91-m, Using Ito and Matsui's Method

CONCLUSIONS

The results of the slope stability analyses performed for different soil conditions and reinforcement schemes showed that the factor of safety for homogeneous and two-layer slopes generally increases with closer reinforcement spacing. Applying different methods for predicting limiting resistance does not have a significant impact on the factor of safety that can be obtained from weak reinforcement schemes because the structural capacity tends to controls the resistance provided by the reinforcing member throughout most of the reinforcement length. However, the method for predicting the limiting soil resistance has a large influence on factors of safety when strong reinforcement is used. The largest increases were observed when the reinforcement caused the critical sliding surface to move beyond the reinforced zone. The factor of safety for homogeneous slopes increased with steeper slope inclination. An opposite effect occurred in two-layer profiles where the factor of safety increased with decreasing slope inclination due to effect of the location of the critical surface either along the upper layer, or within the underlying stronger layer.

ACKNOWLEDGMENTS

The work described in this paper is supported by the Missouri Department of Transportation (Project No. RI98-007b). This support is gratefully acknowledged. The authors are appreciative of the help of numerous individuals who have contributed significantly to the work described including (among others) Mr. Thomas Fennessey and Mr. Bill Billings of the Missouri Department of Transportation. The help of the undergraduate student Michelle Stover in the slope stability analyses is also appreciated. The opinions, findings and conclusions expressed in this publication are not necessarily those of the Department of Transportation, Federal Highway Administration. This paper does not constitute a standard, specification or regulation.

REFERENCES

Broms, B. B. (1964)."Lateral resistance of piles in cohesionless soil". *Journal of the Soil Mechanics and Foundations Division*, ASCE, (SM 3), 123-156.

De Beer E., and Carpentier R. (1975). "Discussion on methods to estimate lateral force action on stabilizing piles by Ito, T. and Matsui T". *Soils and Foundations*, 17(1), 68-82.

Fleming, W. G. K., Weltman, A.J., Randolph, M.F., and Elson, W.K. (1992). *Piling Engineering*, Surrey University Press, Glasgow and London, UK.

Hansen B. (1961). "The ultimate resistance of rigid piles against transversal forces". *Geoteknisk Institut,* Copenhagen, 5-9.

Ito, T., and Matsui, T. (1975). "Methods to estimate lateral force acting on stabilizing piles". *Soils and Foundations*, 15 (4), 43-59.

Loehr, J.E., Bowders, J.J., Owen, J.J., Sommers, L., Liew, W. (2000). "Stabilization of slopes using recycled plastic pins. *Transportation Research Record: Journal of the Transportation Research Board*, TRB,1714, 1-8.

Poulos H. (1996). "Behaviour of piles subjected to lateral soil movements". *National Technical Information Service*, US Dept. of Commerce, (Publication PB97137228), 1-7.

Reese, L. C., Cox, W.R.,and Koop, F.D. (1974). "Analysis of laterally loaded piles in sand". *Proceedings, 6th Offshore Technology Conference*, 2, Houston, TX, 473-483.

MULTI-METHOD STRENGTH CHARACTERIZATION FOR SOFT CRETACEOUS ROCKS IN TEXAS

Emin Cavusoglu,[1] Moon S. Nam,[2] Michael W. O'Neill[3] and Mark McClelland[4]

ABSTRACT: Modern methods for designing drilled shafts, ACIP piles and similar foundations in soft rock require knowledge of the compressive strength and modulus of the rock. However, jointing at many sites prohibits the recovery of samples of sufficient length and integrity to test rock cores in either unconfined or triaxial compression. Since rational design procedures usually require values of compressive strength, surrogate methods must be employed to estimate the compressive strength of the rock. The surrogate methods considered here are the splitting tension test, the point load index test and the TxDOT dynamic penetrometer test (in which a 76-mm-diameter solid steel cone is driven into rock at the bottom of a borehole in much the same way as a split spoon is driven during the performance of a standard penetration test in cohesionless soil).

Soft rock formations typical of those for which such substitutions might be used are the upper Cretaceous formations of North Central Texas, including the Eagle Ford (clay shale) and Austin (limestone) formations. Correlations of the results of the tests listed above with compression strengths of soft rock cores from these formations are provided in the paper. The correlations are formation-dependent, most likely through the degree of cementation present in the geomaterial. The strongest and apparently most reliable correlation was between compressive strength and the TxDOT cone penetration test, although separate correlations were observed in limestone and in clay shale. A clear correlation, but with considerable variance, was found between point load index strength and compressive strength in clay shale. No clear correlations between either point load index or splitting tension strength and compressive strength appeared in the limestone.

[1] Graduate Research Assistant, Dept. of Civil and Env. Eng., University of Houston, Houston, Texas 77204-4003

[2] Graduate Research Assistant, Dept. of Civil and Env. Eng., University of Houston, Houston, Texas 77204-4003

[3] Cullen Dist. Prof., Dept. of Civil and Env. Eng., University of Houston, Houston, Texas 77204-4003

[4] Geotechnical Branch Manager, Bridge Division, Texas DOT, Austin, Texas 78701

INTRODUCTION

The design of drilled foundations, footings and rafts in soft rock requires quantification of the strength properties of the rock in order to be rational. For example, Seidel and Collingwood (2001), Rowe and Armitage (1987), Kulhawy and Phoon (1993), Hassan and O'Neill (1997), and O'Neill and Reese (1999), among others, suggest design methods to establish side resistance in drilled foundations that utilize the compressive strength of soft rocks and hard soils, sometimes in conjunction with Young's modulus E and RQD of the cores (to convert core compression strength and E to rock mass values). When rocks are massive, cores can be recovered and tested in the laboratory to ascertain compressive strength. However, at many locations rock cores are recovered in fragments if the soft rock has a strong closed joint structure or contains open joints or joints with gouge. The core fragments are therefore often too short for the performance of compression tests. When such conditions occur, the foundation engineer is faced with the question of how to quantify geomaterial strength, which can conceivably be accomplished by alternative laboratory testing on short core samples, or by performing *in situ* tests.

This paper explores the use of point load index tests (ASTM D-5731) and splitting tension tests (ASTM D-3967) on relatively short core segments to quantify the compressive strength of soft rock by direct comparison of the results of such alternative tests with compression tests in two soft rock formations in Texas. It also considers a simple *in situ* test, the TxDOT cone penetrometer test, which is a dynamic penetration test that is used routinely for foundation design by the Texas Department of Transportation. Use of these tests in principle allows the strength of the rock to be quantified without the recovery of cores of sufficient length to perform compression tests.

TEST SITE LOCATIONS

Five test sites were established. They were all located in the vicinity of Dallas, Texas. The locations of the sites and their names are shown in Fig. 1. At four sites, denoted Denton Tap, Belt Line Road, Hampton Road and Lone Star Park, cores were recoverd from the Eagle Ford formation, which is found near the ground surface. At a fifth site, the Rowlett Creek Site, cores were recovered from the Austin formation, which is found immediately below the surface. TxDOT cone penetrometer tests and compression tests on the cores were conducted at all five sites. Point load index and splitting tension tests were performed on samples from all sites except Lone Star Park.

Both the Eagle Ford and Austin formations are marine deposits of upper Cretaceous age, with the Austin lying unconformably on top of the Eagle Ford. Geotechnically, the two formations differ primarily by their relative contents of calcium carbonate and clay, with the Austin being very rich in the former and the Eagle Ford being very rich in the latter but with thin layers in which carbonates produce true cementation of the soft rock. Otherwise, the Eagle Ford behaves not as a true rock but as a very hard clay without true cementation. The Austin geomaterials are considered true rocks with cementation. Horizontal bedding is evident in both formations, but it is more obvious in the Eagle Ford in the form of closely spaced laminations. In places, the Eagle Ford contains layers of soft bentonite or friable sandstone. Friable sandstone was present at the Denton Tap Site below the zone that was studied in detail, but no bentonite seams or layers were encountered at any of the test sites. The Eagle Ford materials are temed "clay shale," while the Austin materials are weak limestones that are sometimes referred to as "chalk." A typical section of the Eagle Ford is shown in Fig. 2.

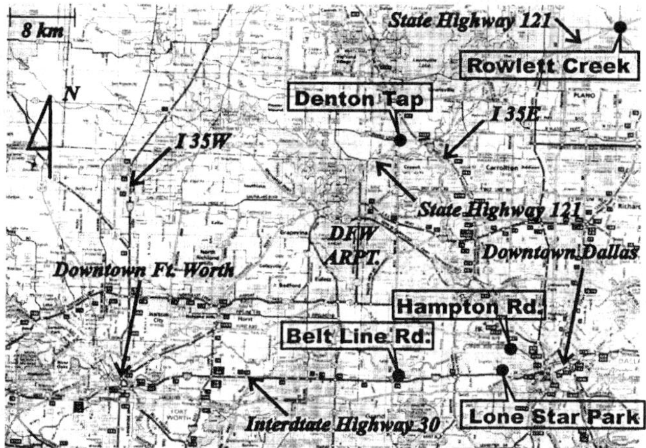

Fig. 1. Regional Map Showing Test Site Locations

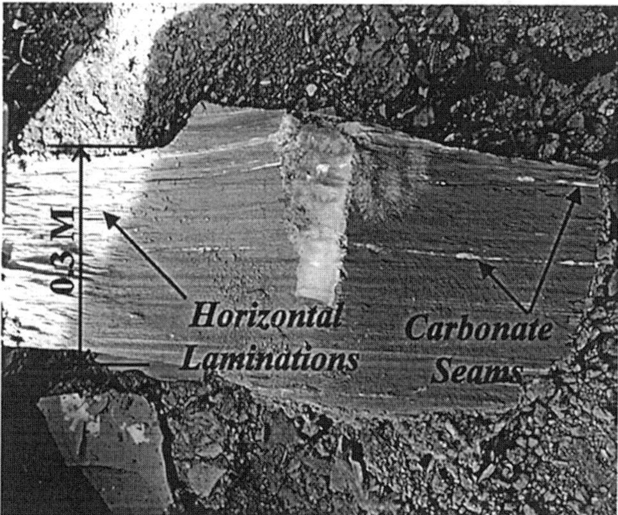

Fig. 2. Section of Clay Shale from Denton Tap Site

SAMPLING AND TESTING

Two continuous cores were recovered at each of the test sites. A third core hole was made in which TxDOT cone penetration tests were conducted. The three sample holes were in an equilateral triangular pattern 8 m on a side.

Cores were recovered with a double-walled, NWD4 (52.3 mm ID) core barrel using a light bentonitic drilling fluid. The samples were cut into short lengths (approximately 200 mm), sealed and then heavily taped to prevent them from expanding and fracturing along bedding planes before testing. The cores were nominally 50 mm in diameter. Rock quality designation (RQD) values are summarized in Table 1. The high RQDs suggest that jointing likley had little effect on the results of the TxDOT cone penetrometer tests. In the laboratory, the core segments were trimmed and subjected to (1) compression testing [UU triaxial testing for samples of clay shale; unconfined testing for samples of limestone], (2) splitting tension testing and (3) point load index testing. Splitting tension test samples were 38 mm long, while point load index test samples were 25 mm long. These can be considered representative of "short" samples recovered from cores in highly jointed rock. Compression test samples were 100 mm long. Total confining pressures in the UU triaxial compression tests were equal to the estimated total overburden pressure at the depth from which the sample was recovered.

TxDOT cone penetrometer tests are peformed as illustrated in the schematic in Fig. 3. The penetrometer is a steel cone. It is driven with approximately the same energy per blow as is used in the standard penetration test. The solid cone is used in preference to the split spoon because of its robustness. However, rock samples cannot be recovered and examined. After seating, the penetrometer is struck 100 blows, and the penetration is recorded. One such reading was made every 0.76 m of depth for this study.

Table 1. RQD Data

Site	Depth Range (m)	RQD (%)
Denton Tap (clay shale)	6.1 – 8.3	72.2
	8.3 – 11.0	84.2
Hampton Road (clay shale)	7.6 – 10.7	94.2
	11.1 – 14.2	100
Rowlett Creek (limestone)	3.1 – 5.3	85.4
	5.8 – 7.3	96.5

TEST RESULTS

The best estimates of the variation of soft rock strength with depth at the test sites are considered to be the profiles of the compressive strength, $(\sigma_1 - \sigma_3)_f$ (peak) for purposes of this paper. These are plotted for the sites studied in Figs. 4 – 8. The mean compressive strengths range from about 1500 kPa at Hampton Road and Lone Star Park to 2000 kPa at Denton Tap and Belt Line Road (clay shale) to about 11,000 kPa at Rowlett Creek (limestone). The strength generally decreases with depth at the clay shale sites because the surficial rock contains some calcareous materials, probably

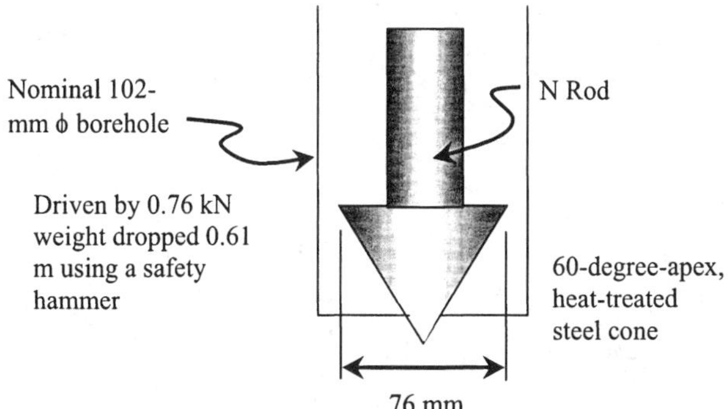

Fig. 3. The TxDOT Cone Penetrometer

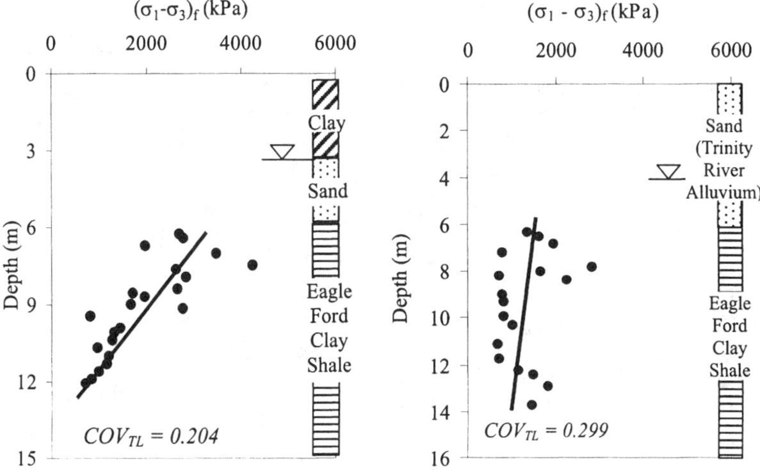

Fig. 4. $(\sigma_1-\sigma_3)_f$ vs. Depth, Belt Line Rd.

Fig. 5. $(\sigma_1-\sigma_3)_f$ vs. Depth, Hampton Rd.

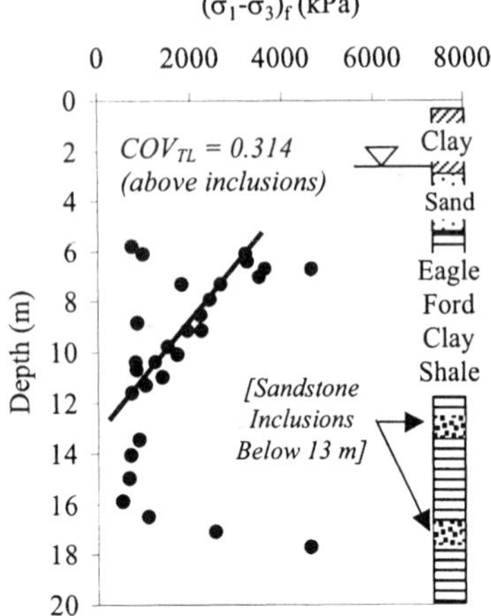

Fig. 6. $(\sigma_1\text{-}\sigma_3)_f$ vs. Depth, Denton Tap

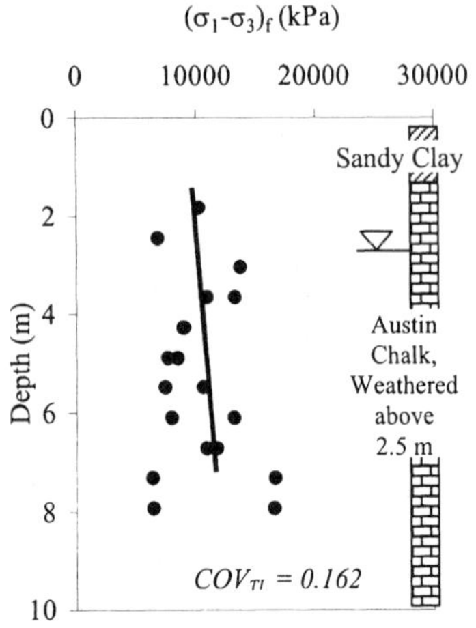

Fig. 7. $(\sigma_1\text{-}\sigma_3)_f$ vs. Depth, Rowlett Creek

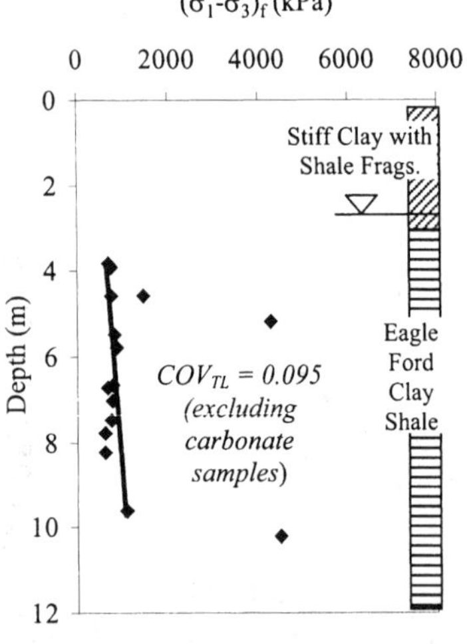

Fig. 8. $(\sigma_1 - \sigma_3)_f$ vs. Depth, Lone Star Park

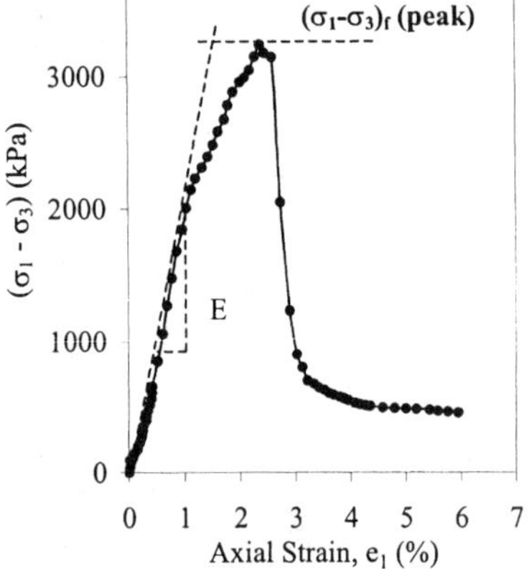

Fig. 9. Typical Stress-Strain Curve

due to desiccation in the geologic past. The exception is the Lone Star Park Site, where the upper calcareous surface of the clay shale was removed by human mining early in the 20th Century.

At the Denton Tap and Lone Star Park sites, some carbonate seams were visible in the cores, which yielded much higher compressive strengths than cores without visible carbonate cementation. In the lower part of the cores at Denton Tap, friable sandstone was observed in layers and seams. This material influenced the compressive strength in some samples below a depth of 13 m. That rock is not considered in the correlations that follow.

Values of COV_{TL} (coefficient of variation based on offset values from the linear trend line defined by fitting the data with a least squares function) are shown in Figs. 4 - 8. This method of expressing variability, proposed by Phoon and Kulhawy (1996), indicates that the variance in compressive stength in the soft rock studied is approximately equivalent to that obsered in apparently uniform overconsoliodated clays in Texas.

A typical stress-strain curve for the clay shale (Denton Tap, depth of 10.4 m) is shown in Fig. 9. The characteristic shape of these curves revealed an essentially linear initial portion, followed by the intitiation of fracturing, followed by a secondary slope, less steep than the initial slope, to peak failure. Failure was generally along well-defind shear planes, after which most of the peak strength was almost immediately lost. All of the compressive strength values shown in Figs. 4 – 8 are peak values. Typical Young's moduli, defined as the slopes of the initial portions of the stress-strain curves (Fig. 9), are shown in ratio to compressive stength $[\sigma_1 - \sigma_3)_f]$ in Table 2. Such information is needed, in addition to compressive strength information, in some models for predicting side shear capacity in drilled shafts [e. g., Seidel and Collingwood, 2001; Hassan and O'Neill, 1997]. Note that the Young's modulus / compressive strength ratios are much higher for the limestone than for the clay shale, most likely because the cementation in the limestone prevented expansion of cores and consequent microcraking after sampling. In all cases the COV of that ratio was consistently high because of the combination of two variables, each of which exhibited relatively high COVs.

Table 2. Mean $E/(\sigma_1 - \sigma_3)_f$ Values from UU Triaxial Compression for Selected Sites

Site	No. of Tests	Mean $(\sigma_1 - \sigma_3)_f$ (MPa)	Mean E (MPa)	Mean $\frac{E}{(\sigma_1 - \sigma_3)_f}$	COV of $\frac{E}{(\sigma_1 - \sigma_3)_f}$
Belt Line (clay shale)	12	1.977	224	124	0.515
Denton Tap (clay shale)	17	1.850	141	77	0.497
Hampton Rd. (clay shale)	8	1.553	171	120	0.407
Rowlett Creek (limestone) (unconfined)	11	11.412	5006	447	0.443

CORRELATIONS

The main objective of this paper is to determine whether correlations can be established between compressive strength and (1) dynamic field TxDOT cone penetrometer tests, and (2) laboratory index tests on short cores. If such correlations exist, then data from these simpler tests can be transformed into values of $(\sigma_1 - \sigma_3)_f$ and, approximately through the information in Table 2, to E, both of which could then be used in foundation design models for the upper Cretaceous formations that were studied.

Figure 10 shows graphically and in equation form the correlations obtained between the TxDOT cone penetration resistance (PR) (mm / 100 blows) and $(\sigma_1 - \sigma_3)_f$. One additional site denoted as SH 45 was added on Fig. 10. It locates in the vicinity of Austin, Texas and consits of about 2 m clay overburden overlying a deposite of limestone that is about 3.5 m and which become a deposite of clay shale. The data of TxDOT cone penetration resistance (PR) (mm / 100 blows) and $(\sigma_1 - \sigma_3)_f$ were obtained from TxDOT.

Considering the obvious scatter in the data, a reasonable design relationship for the uncemented clay shale may be below the linear least-squares lines shown in Fig. 10, with exact values to be used being coupled to the reliability of the overall design process.

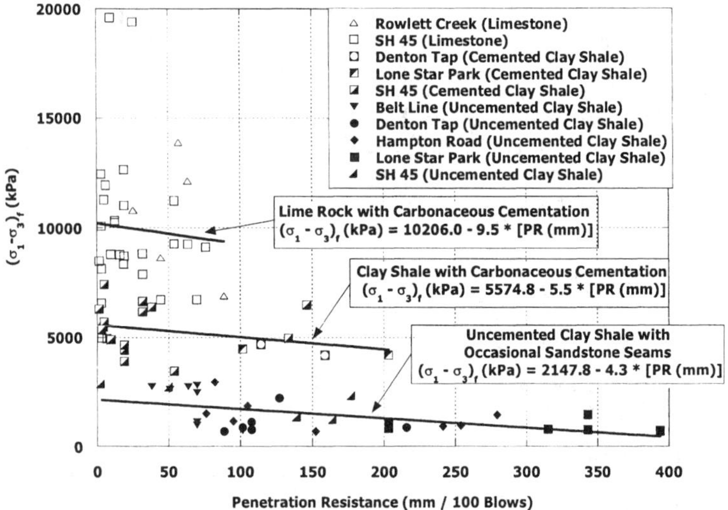

Fig. 10. Correlations between TxDOT Cone Penetration Resistance and Compression Strength of Soft Rock Cores

The correlations of the point load index strength σ_{pl} and compression strength $(\sigma_1 - \sigma_3)_f$ are shown in Fig. 11, in which $\sigma_{pl} = P / D_e^2$, where P is the value of the point load at failure and D_e is the diameter of the core, per ASTM D-5731. The correlations are clearly different for the clay shale and limestone samples, but in the limestone it is virtually impossible to distinguish $(\sigma_1 - \sigma_3)_f$ based on the point load index strength. The linear least-squares relationship for the uncemented clay shale samples shows a positive correlation but with significant scatter. Therefore, the point load index correlation equations shown in Fig. 11 should probably not be used for short samples in limestone and should be used with considerable caution in uncemented clay shale.

Figure 12 contains correlations between σ_t, the splitting tension strength, and $(\sigma_1 - \sigma_3)_f$. σ_t is computed from the relationship $\sigma_t = 2P/ (\pi\ D\ L)$, in which P is the value of the applied diametric load at failure, D is sample diameter and L is sample length, per ASTMD-3967. A weak positive correlation was observed in both the limestone and the clay shale; however, the scatter was significant in both geomaterials.

The primary problem with both the point load index strength and the splitting tension strength is that they are related to the tensile strength of the rock under test. In the geomaterials considered here, the tensile strength was obviously quite variable, which makes the use of these tests problematical. In order to compare the behavior of these geologic materials to a controlled, manufactured, cemented material, several point load and splitting tension tests were performed on samples of grout made from portland cement, ASTM C-33 sand and water, with the supposition that such samples would be more uniform than the geomaterial samples. The correlations between $(\sigma_1 - \sigma_3)_f$, σ_{pl} and σ_t in the grout samples are shown in Fig. 13. Although the

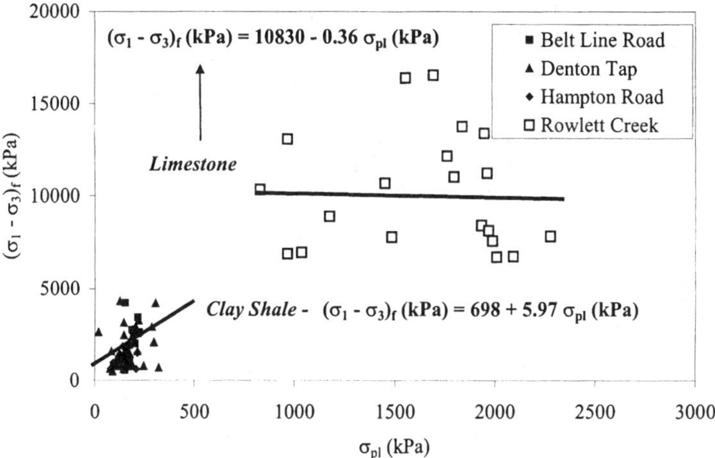

Fig. 11. Correlations between Point Load Index (σ_{pl}) and Compression Strength of Soft Rock Cores

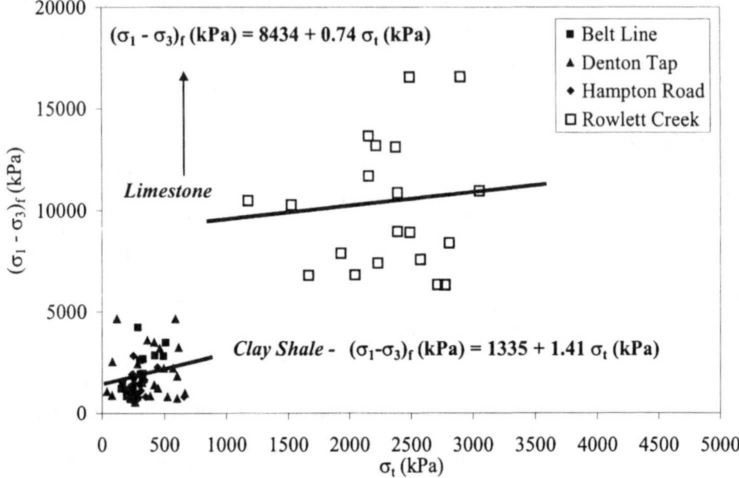

Fig. 12. Correlations between Splitting Tension Strength (σ_t) and Compressive Strength of Soft Rock Cores

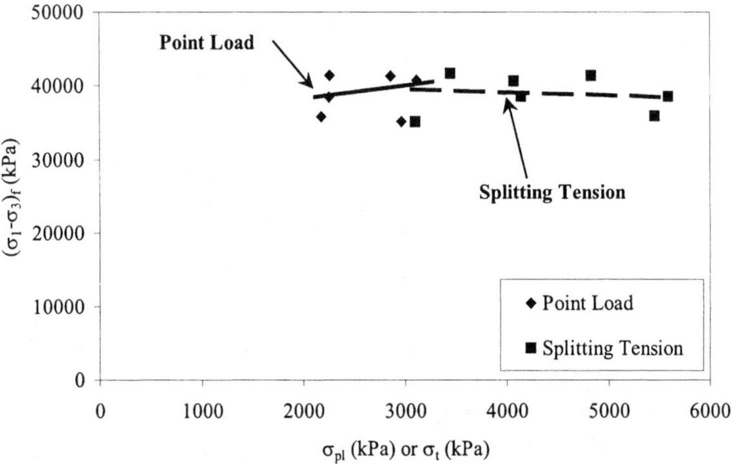

Fig. 13. Correlations between Point Load Index, Splitting Tension Strength, and Compressive Strength of Portland Cement Grout

scatter is smaller than for the limestone samples, the correlation is extremely weak, suggesting that the low degree of correlation in the upper Cretaceous limestone considered here may apply to other cemented soft rock formations of similar geologic history as well.

CONCLUSIONS

Analysis of compression test data from five sites in the upper Cretaceous soft rock formations of North Texas indicated that the variability of compressive stength from cores was of about the same magnitude as would be expected of generally uniform clay soils in the region. The ratios of E to $(\sigma_1 - \sigma_3)_f$ (Table 2) were found to be dependent on the type of rock (clay shale or limestone), most likely due to the degree of cementation in the rock. These ratios exhibited relatively high coefficients of variation.

Concerning the possibility of using simple surrogate tests for compression tests on cores whose lengths are twice their diameters, when only short, fractured cores can be recovered, the TxDOT cone penetrometer test exhibited relatively strong correlations. These correlations, however, were different in the limestone (and cemented clay shale) than in the uncemented clay shale, which requires that the formation and its general cementation characteristics be identified before applying a correlation formula such as shown in Fig. 10. Care should also be exercised in such application because of the scatter that was observed.

There was also a relatively strong correlation between point load index strength and compressive strength in the uncemented clay shale (Fig. 11). However, the variance in the data was large, which requires that such a correlation be applied with considerable judgment. Correlations between any type of tension test (point load or splitting tension) were very weak in the limestone. A separate series of tests on manufactured cementitious specimens (grout, intended to simulate cemented rock) showed almost no correlation between the tension index strength and compression stength (Fig. 13), which suggests that a high degree of success with correlating σ_{pl} and σ_t with compressive stength in similar limestone formations may also be problematical.

ACKNOWLEDGMENTS

We will forever be grateful to Dr. Michael W. O'Neill (1940 - 2003) for his numerous contributions to our research project. Also, we feel honored to have worked with him as his graduate students and project committee. May his soul rest in peace.

Most of the sampling and testing work reported here was funded by the Texas Department of Transportation as a part of a larger research project to improve the design of drilled foundations in soft rock. Cores were recovered and TxDOT cone tests were performed by HBC, Inc., of Dallas. All laboratory tests were performed in the University of Houston geotechnical and materials laboratories.

REFERENCES

Hassan, K. M., and O'Neill, M. W. (1997). "Side Load Transfer Mechanisms in Drilled Shafts in Soft Argillaceous Rock," *Journal of Geotechnical and Geoenvironmental Engineering*, Vol. 123, No. 2, ASCE, pp.145 – 152.

Kulhawy, F. H., and Phoon, K. K (1993). "Drilled Shaft Side Resistance in Clay Soil to Rock," *Geotechnical Special Publication No. 38, Design and Performance of Deep Foundations,* Ed. by P. P. Nelson, T. D. Smith and E. C. Clukey, ASCE, October, pp. 172 - 183.

O'Neill, M. W., and Reese, L. C. (1999). "Drilled Shafts: Construction Procedures and Design Methods," *Publication No. FHWA-IF-99-025*, Office of Infrastructure, U. S. Department of Transportation, Federal Highway Administration, Washington, D. C., August, 537 p. plus appendices

Phoon, K. K., and Kulhawy, F. H. (1996). "On Quantifying Inherent Soil Variability," *Geotechnical Special Publication No. 58, Uncertainty in the Geologic Environment – From Theory to Practice*, Ed. by C. D. Shackelford, P. P. Nelson, and M. J. S. Roth, ASCE, Vol. 1, pp. 326 – 340.

Rowe, P. K., and Armitage, H. H. (1987). "A Design Method for Drilled Piers in Weak Rock," *Canadian Geotechnical Journal,* Vol. 24, pp. 126 – 142.

Seidel, J. P., and Collingwood B. (2001). "A New Socket Roughness Factor for Prediction of Rock Socket Shaft Resistance," *Canadian Geotechnical Journal,* Vol. 38, February, pp.138 – 153.

DESIGN, MONITORING, AND INTEGRITY TESTING OF DRILLED SOIL DISPLACEMENT PILES (DSDP) FOR A GAS-FIRED POWER PLANT

Guoming Lin, Ph.D, P.E.[1], Member, ASCE, Edward L. Hajduk, P.E.[2], Member, ASCE, and Willie NeSmith[3], P.E., Member, ASCE

ABSTRACT: Increasingly, Cast-In-Place (CIP) pile foundations are being used in place of traditional driven pile foundations to support settlement sensitive structures. The advantages of CIP pile foundations include quick installation, flexibility to adjust pile length for varying subsurface conditions, and reduced noise and vibration. There are inherent uncertainties in all CIP systems and thus detailed testing and production monitoring programs may be appropriate to verify the integrity of each individual CIP pile. This paper presents the design considerations, load testing program, production monitoring, and integrity testing of Drilled Soil Displacement Piles (DSDP) installed for a gas-fired power plant in Bartow, Florida. Efficient planning of the load testing and integrity testing programs resulted in the foundation construction being completed ahead of schedule while verifying the piles met the design requirements.

INTRODUCTION

Cast-In-Place (CIP) pile foundations are advantageous to conventional driven piles in terms the speed of construction, ease of adjusting pile length during construction and potential cost savings. CIP pile foundations are particularly attractive for sites where noise and vibration are of concern. However, the installation process of CIP piles through soft/loose soils can give rise to uncertainties with respect to pile integrity, such as necking. To address these uncertainties, more stringent and detailed quality control (QC) monitoring and testing programs may be appropriate.

A variety of CIP pile systems have been developed over the past decades. These pile systems differ in the method of installation, equipment used, and/or the materials

---

[1] Principal and Senior Geotechnical Engineer, WPC, 5205 Waters Avenue, Savannah, GA 31404

[2] Geotechnical Engineer, WPC, 1017 Chuck Dawley Blvd, Mt Pleasant, SC 29464

[3] Chief Geotechnical Engineer, Berkel & Company, 1503 Milner Crescent, Birmingham, AL 35205

and methods used in casting. Augered Cast-In-Place (ACIP) piles are the most commonly used CIP pile system in the Southeast US. Conventional ACIP piles form the borehole via a continuous augering system that brings the auger cuttings (spoils) to the ground surface. The piles are formed by pumping grout through the center of the auger as the auger is removed. Drilled Soil Displacement Piles (DSDP) are relatively new to the United States, but are rapidly coming into broader use. DSDP systems contain an element in the lower portion of the tooling that causes full lateral displacement of the materials penetrated, in contrast to bringing the soils that occupy the pile area to the ground surface. This displacement process densifies responsive soils in the area of the pile, thus increasing the pile capacity from the added shaft resistance. DSDP is particularly attractive for sites with potentially contaminated soils as disposal of soil cuttings is typically not required. The process used at this site is a pressure-grouted displacement system (NeSmith, 2002).

This paper focuses on the pile integrity aspect of DSDP in the design, testing and monitoring phases of a power plant in Bartow, Florida. Specific quality control procedures were developed in the project specifications and implemented in the monitoring program. These special procedures included the use of Singlehole Sonic Logging (SSL) testing to verify the integrity of both the test piles and production piles.

PROJECT INFORMATION

The project consists of the addition of a new 530 MW gas-fired combined cycle electric generation unit (Power Block 2) to the existing Florida Power Hines Energy Complex in Bartow, Florida. The new power block includes two combined cycle combustion turbine generators (CTG), two heat recovery steam generators (SRSG), one steam turbine generator (STG), auxiliary boilers, cooling towers, water treatment equipment, tanks, transformers, underground sumps and utilities.

The site is part of a reclaimed phosphate mine with 6.0 to 9.1 meters (20 to 30 ft) of cast overburden soils. The bottom of the original mine pit was uneven, varying in elevation by approximately 4.6 m (15 ft). The overburden soils are predominantly silty fine sands and clayey fine sands without phosphoric clays deposited in the power block areas. The site was initially filled with loose overburden soils and later prepared for power plant development in 1992. The site preparation in 1992 consisted of compaction of the upper overburden fill, placement of 1.5 to 2.4 m (5 to 8 ft) of engineered fill, and preloading the site with 3.0 to 4.6 m (10 to 15 ft) of rolling surcharge. The site was graded to the finished grade elevation during this site work.

SUBSURFACE CONDITIONS

The site was explored using a combination of soil borings, cone penetration test (CPT) soundings and flat blade dilatometer (DMT) soundings. The history of the site as a reclaimed phosphate mine resulted in unique and challenging subsurface conditions. Overburden fill was encountered from 7.6 to 10.7 m (25 to 35 ft) below ground surface with an average thickness of approximately 9.8 m (32 ft). The overburden fills are predominantly silty fine sands (SM) with slight clay contents at intermitted depth intervals. The undisturbed soils below the original mine pits are classified as clayey sandy silts or silty clayey sands with limestone fragments. The upper 30.5 m (100 ft) of soils can be roughly divided in terms of lithology and densities into the following four strata presented in Table 1.

TABLE 1. Site Lithology Summary

Depth (m, BGS) (1)	Soil Type (2)	Average SPT N (bpf) (3)	Average CPT q_t (MPa) (4)	Average DMT M (MPa) (5)
0 to 5.2	Compacted overburden soils or controlled fill, silty fine sands	16 to 35	12.3	148.4
5.2 to 9.8	Uncompacted cast overburden soils, silty fine sands (SM)	5 to 8	4.4	20.1
9.8 to 18.1	Undisturbed sandy clayey silt	21 to 39	14.0	45.4
18.1 to 30.5	Clayey limestone	33 to 50/25mm	N/A	N/A

The upper 5.2 m (17 ft) of soils are comprised of compacted fill or cast overburden compacted by the previous construction activities. However, the soils between depths of 5.2 to 9.8 m (17 to 32 ft) are generally in a condition of loose relative density. Apparently, the soils in this depth interval are uncompacted cast overburden. The soils below a depth of 9.8 m (32 ft) to approximately 18.1 m (62 ft) are undisturbed stiff to very stiff sandy clayey silts. This layer is underlain by clayey limestone to the bottom of the borings, which terminated at 30.5 m (100 ft) below existing grades.

FOUNDATION DESIGN CONSIDERATIONS

All major equipment areas (i.e. the CTG, STG and HRSG) have a mat foundation of at least 12.2 m (40 ft) in width and 30.5 m (100 ft) in length. These foundation sizes will allow the contact pressure to transmit below the cast overburden fill layer. A majority of the loads will be incurred during the construction stage, with the exception of the water tanks and the clarifiers.

Settlements were the controlling factor in the foundation evaluation. Settlements were calculated by summation of increased stress divided by the modulus of compression derived empirically from the SPT N values, CPT tip resistance and DMT data. Two different approaches were used in calculating the distribution of stress within the soil. The first approach, conservative in nature, assumed a uniform distribution of the load-induced stresses into the soils overlying the limestone. A weighted average modulus based on thickness was used in the settlement calculation. Settlements at different stages of construction were evaluated using this approach. The second approach used the Boussinesq equation to calculate stress distribution along the soil profile to the top of limestone. Settlements calculated in this method offer an insight into the distribution of the settlement between different layers. Table 2 presents the settlements calculated with a uniform distribution of soil stress.

TABLE 2. Estimated Settlements for Major Equipment Foundations

Area	Size of Mat (m)			Soil Pressure (kPa)		Settlement (mm)		
	L (1)	W (2)	H (3)	Constr.[1] (4)	Oper.[2] (5)	Constr.[1] (6)	Oper.[2] (7)	Total (8)
STG	33.5	13.7	2.4	97.7	19.2	42.9	8.4	51.3
HRSG	45.7	12.2	1.2	94.3	4.8	41.4	<6.4	<47.8
CTG	30.5	15.2	1.8	60.3	1.9	26.4	<6.4	<32.8

NOTES:
1. Construction Conditions
2. Operation Conditions

Based on the settlement analyses, pile foundations were recommended for the support of the major equipment while shallow foundations were recommended to support the tanks and other auxiliary structures. CIP piles were recommended over driven piles to ease concerns regarding noise and vibration effects to the existing power plant. Both conventional auger cast piles and drilled soil displacement piles (DSDP) were evaluated. The pile capacities were estimated based on CPT data, using the LCPC method (Bustamante and Gianeselli, 1982), a method developed specifically for the system used (NeSmith, 2002) and other empirical relationships between CPT penetration resistance and pile capacity, such as Schmertmann's method. The design engineer selected 406 mm (16 inch) diameter DSDP piles for a 0.89 MN (100 ton) axial compression capacity.

Due to the site conditions, concerns regarding quality control and constructability of the piles were raised. The soil stratigraphy featuring loose fill placed over the original uneven mine pits made it difficult to determine a fixed pile length or tip elevations. In addition, the upper 6.1 m (20 ft) of medium dense to dense soils gave rise to concerns about the DSDP equipment ability to drill through this layer. To address these concerns, it was recommended that all piles penetrate through the overburden fills and be embedded into the underlying undisturbed soils. A minimum tip elevation of the lesser of 38.1 m (125 ft) MSL or 1 m (3 ft) below the "refusal" depth was established. The 1 m (3 ft) penetration into the "refusal" layer criteria was intended to ensure the pile tip was embedded within the competent soils underlying the uncompacted overcast fill. The "refusal depth" was defined with the following criteria.

1. Required use of a drill with a rated torque of at least 244 kn-m (180,000 ft-lb).
2. Drive unit required to exert 267-356 kN (60,000-80,000 lb) downward force (crowd) on the tool.

3. Refusal defined as a torque of (150,000 ft-lb) or a hydraulic pressure of 1725 kPa (250 psi).

To address potential difficulties in drilling through the upper 6.1 m (20 ft) of dense soils, predrilling provisions were added in the project specifications. Predrilling with an auger smaller than the pile diameter was permitted in the upper 6.1 m (20 ft). However, based on communications with experienced contractors, it was understood that the DSDP system would not be cost-competitive to conventional auger cast piles if predrilling were to be required.

TEST PILE PROGRAM

The test pile program consisted of the following: conducting test probes at five locations across the site, installation of two test piles and eight reaction piles, and axial compressive static load testing of the two test piles. In addition, integrity testing of the two test piles was conducted using SSL testing.

Test Probes

The test probing was performed on February 12, 2002 and the test piles were installed on February 13, 2002. The purpose of the test probing was to test the capability of the equipment and drilling conditions. The piling contractor used a Bauer BG 25 installation platform for this project. This equipment develops 25 meter-tons (180,000 ft-lbs) of torque and 356 kN (80,000 lbs) of crowd.

The drilling process was closely observed by contractors, project geotechnical engineers and the owner's representatives. During each probing, attention was directed to the motor torque reading, ease and difficulty of drilling as indicated by the rate of drilling and the type of material recovered at the end of the auger bit. Table 3 summarizes the observed drilling conditions for the pile test program.

Following the completion of the field drilling, the torque readings and the rate of drilling were printed and reconciled with field notes. The following features were noted from the observed probing conditions:

1. In general, the rig was capable of drilling through the upper dense layer, although the drilling rate was considerably slower in this layer than in the underlying layers. The thickness of the upper crust varies between 3.0 to 6.1 m (10 and 20 ft). Due to the densification effects from the closely spaced piles to be installed during production, drilling conditions were anticipated to become significantly more difficult.
2. The torque pressure gage readings were relatively high, with the majority of the pressure gage readings exceeding 250 bars. However, torque readings were observed to reduce below 250 bars at depths 9.1 to 11.3 m (30 to 37 ft) within two probes.
3. Four probes, P-2 through P-5, encountered noticeable hard drilling at depths of approximately 10.7 to 11.3 m (33 to 37 ft), as evidenced by torque reading of greater than 250 bars and slower drilling rate. However, probe P-1

encountered uniform drilling resistance below the upper crust without noticeable difficulty above 16.2 m (53 ft).

4. Practical refusal (no advance in auger penetration in 1 to 2 minutes) was encountered at a depth of 12.8 m (42 ft) within probing P-2.
5. The drilling conditions encountered are relatively consistent with the soil density and consistency conditions reported in the Geotechnical Report.

TABLE 3. Summary of Observed Probe Drilling Conditions

Probe No. (1)	Install Time (min) (2)	Probe Depth (m) (3)	Torque Pressure Gage Reading (Bar) (4)	Drilling Rate (5)	Material at end of Auger (6)
P-1	8	16.2	250 to 300 over the entire depth	Slow in upper 3.4m Uniform below 3.4m	Clayey fine sand, v. dense
P-2	7	13.0	250 to 280 at upper 6.1m <250 at 11.3m	Slow in upper 6.1m Increased below 6.1m Practical refusal @ 13 m	
P-3	15	11.6	> 250 over the entire depth	11 minutes in upper 5.5m Inc. rate below 5.5m Vibrating @ 4.3m during auger extraction.	Very dense sand
P-4	6	11.3	> 250 over the entire depth	Uniform rate in upper 10.4m Substantially slowed down at 10.4 – 11.3m	
P-5	8	12.5	<200 @ 10.1-11.6m 250 to 300 @ 11.6 – 12.5m	Noticeably harder drilling at 11.6 – 12.5m	

Test Pile Selection and Installation

Based on the observed drilling conditions, P-1 and P-4 locations were selected for the installation of test piles. The test pile at location P-1, designated as TP-1, was located at the northwest part of the power block. This location represents a drilling condition without noticeable drilling difficulty between 9.1 to 16.2 m (30 and 53 ft). The test pile at location P-4, designated as TP-2, was located at the southeast part of the power block, which represents a "typical" or "average" drilling condition (i.e. encountering hard drilling below 10.4 m (34 ft)).

Each test pile setup consisted of one test pile and four reaction piles. The reaction piles (designated as T-R) were spaced 4.9 m (16 ft) apart in the long direction and 1.8 m (6 ft) apart in the short direction. Each reaction pile was installed with a length of 11.9 m (39 ft), and reinforced with one full length #8 grade 1034 MPa (150 ksi) center bar sleeved with a PVC pipe over the upper portion of the pile. Both test piles (i.e. TP-1 and TP-2) were 11.3 m (37 ft) long, reinforced with a center bar tied with a 38 mm (1.5 inch) diameter Schedule 40 PVC pipe along the bar for the Singlehole Sonic Logging (SSL) tests.

During drilling and grouting of the reaction piles and test piles, attention was directed to the drilling conditions, as indicated by torque pressure reading and drilling rate, as well as volume of grout placed in each pile. Grout factor is defined as the actual volume of the grout divided by the theoretic volume of the pile. Table 4 presents a summary of the observed conditions.

TABLE 4. Summary of Observed Test Pile Installation Conditions

Pile (1)	Torque Pressure (bar) (2)	Drilling Rate (3)	Length (m) (4)	Total Strokes (5)	Grout Factor (6)
T2-R1	N/A	Hard drilling at 10.1 m	11.9	117	1.31
T2-R2	N/A	Hard drilling at 10.1 m	11.9	110	1.23
T2-R3	N/A	Hard drilling at 10.1 m	11.9	110	1.23
T2-R4	N/A	Hard drilling at 10.1 m	11.9	108	1.21
TP-2	<200 bars in upper 8.8m >250 below 10.1m	Hard drilling at 11.1 m	11.3	104	1.22
T1-R1	>250 bars over entire depth	N/A	11.9	103	1.15
T1-R2	280 bars to 11.0m <200 bars in 11.0-11.9m	Easy drilling at 11.0 m	11.9	104	1.22
T1-R3	280 bars to 11.6m 200 bars at 11.6-11.9m	Easy drilling at 11.6 m	11.9	105	1.18
T1-R4	280 bars to 11.6m 150 bars at 11.6-11.9m	Easy drill at 11.0 m	11.9	104	1.22
TP-1	>250 bars over entire depth	Uniform drilling rate	11.3	103	1.21

Both test piles were installed after the reaction piles. During installation of both test piles, heaving and flow of grout was noticed from the previously installed reaction piles 1.8 m (6 ft) away. No heave was observed in the reaction piles located 3.0 to 4.9 m (10 or 16 ft) away. This observation had important implication in pile installation sequence and spacing between newly installed piles; the 1.8 m (6 ft) spacing was inadequate in controlling group effects from the adjacent piles. Therefore, a 3.0 m (10 ft) minimum spacing was recommended for production piles to avoid interference between adjacent piles.

As stated previously, the location of TP-1 was selected as no noticeably harder layer was encountered during probe drilling at this location. At this location, it was noted that all four reaction piles encountered relatively easier drilling (torque pressure reading below 250 bars and drilling rate increased) below 11.0 m (36 ft). These drilling conditions suggested a relatively weaker stratum below 11.0 m (36 ft). However, these conditions were less apparent when drilling test pile TP-1, which was drilled after the four reaction piles. It appears soils in this region at the test pile location had been densified from the reaction pile installation. TP-1 was terminated at 11.3 m (37 ft) long to evaluate the pile capacity when the pile tip is not embedded in a very dense layer. The drilling conditions of TP-2 and reaction piles were consistent with the previously noted soil conditions. No group interference effects were noted during TP-2 pile installation. The SSL testing results for TP-1 and TP-2 did not indicate any unusual pile conditions.

Axial Compressive Load Testing

Axial compressive load testing was performed on the two test piles in accordance with the methods described by ASTM D1143 "Standard Test Method for Piles under Static Axial Compressive Load". Loading was performed using the cyclic loading method described in Section 5.2 of ASTM D1143. Figure 1 presents the graphical results from the axial compression load test on TP-1 while Figure 2 shows TP-2 static load test results.

The axial capacities for TP-1 and TP-2, as determined by the Davisson's criteria, were 1.95 MN (219 tons) and 2.24 MN (252 tons), respectively. Based on a required axial capacity of 0.89 MN (100 ton), the test piles had factors of safety (FS) of 2.19 for TP-1 and 2.24 for TP-2. Total settlement at the failure loads were 14 mm (0.55 inches) and 15.2 mm (0.6 inches) for TP-1 and TP-2, respectively.

PRODUCTION PILE MONITORING

Production pile monitoring of the DSDP's consisted the following:

- A full-time pile inspector, who documented the installation of each DSDP on individual logs and verified that they met the project specifications. In general, the pile inspectors followed the guidelines provided in the *Inspector's Guide to Augered Cast-In-Place Piles* (DFI, 1994).
- Grout samples for compressive strength testing were taken at selected volume intervals.
- Singlehole Sonic Logging (SSL) integrity testing of 7% of the production piles selected at random and those selected by the project geotechnical engineer based on the pile inspector's data. Initially, pulse echo testing (PET) was initially specified as quality control testing method but was changed to Singlehole Sonic Logging (SSL) at the request of the DSDP contractor. The DSDP contractor indicated that SSL testing would be more effective in evaluating pile integrity given the soil conditions at the site.

The project engineer reviewed the individual DSDP logs within 2 days of pile installation to verify that the pile met the project specifications. DSDP's that did not meet the project specifications and/or had unusual installation circumstances were "flagged" for SSL testing. Of the approximately 900 production piles chosen, only 60 were selected for SSL testing.

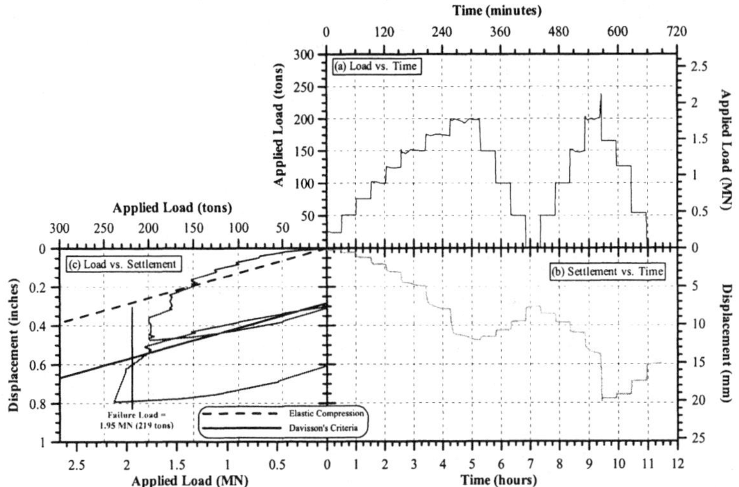

FIGURE 1. TP-1 Axial Compressive Load Test Results

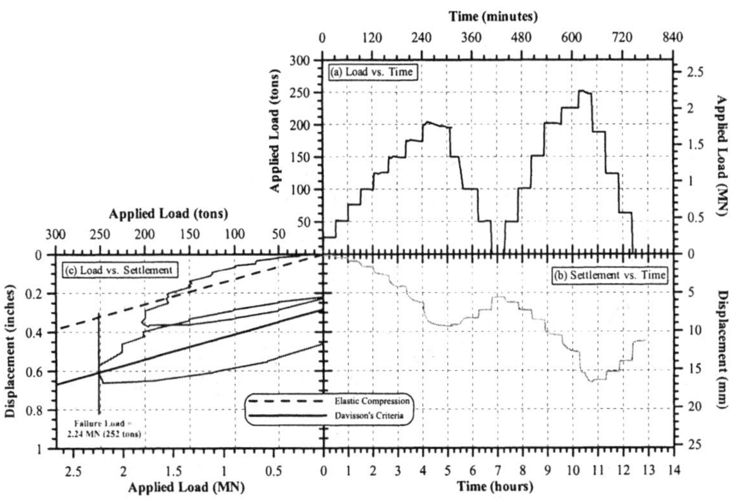

FIGURE 2. TP-2 Axial Compressive Load Test Results

Singlehole Sonic Logging (SSL) Testing

Singlehole Sonic Logging (SSL) testing, also known as Single-Hole Ultrasonic Testing (SHUT), is performed by placing an ultrasonic transmitter and receiver down an access tube placed within the deep foundation member. The transmitter and receiver are lowered to the bottom of the access tubes and then slowly pulled to the surface. At selected depth intervals, an ultrasonic pulse is sent from the transmitter to receiver. A depth wheel at the top of the shaft records the distance traveled. Figure 3 presents a typical SSL testing setup for this project. A detailed explanation of SSL testing is provided by Chernauskas and Paikowsky (1999) and Amir (2002).

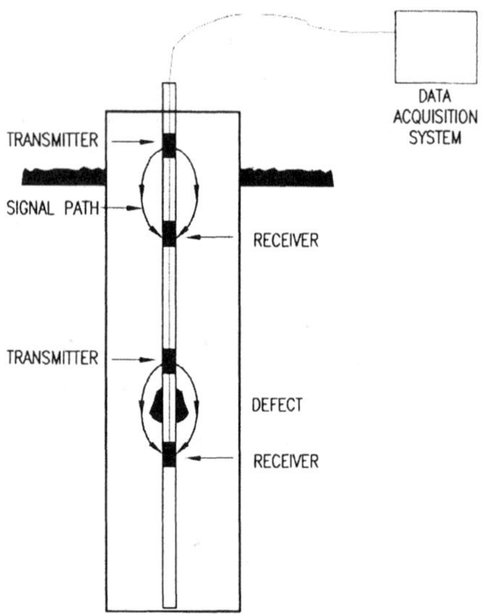

FIGURE 3. Typical SSL Testing Setup (after Chernauskas and Paikowsky, 1999).

SSL testing was conducted using the Pile Integrity Sonic Analyzer (PISA), manufactured by Piletest.com. The PISA is a lightweight, portable, pen touch computer that operates in a Windows based environment (Chernauskas and Paikowsky, 2000). To facilitate SSL testing, 3.2 cm (1¼ inch) nominal diameter Polyvinyl Chloride (PVC) tubing was attached to the DYWIDAG steel reinforcement. This combination PVC/DYWIDAG rod was then placed in the center of the production DSDP's shortly after installation and the access tube was then filled with water shortly after installation. PVC access tubing was selected as the access tube material based cost considerations and recent research that has shown steel tubing is not conducive to SSL testing (Amir, 2002).

During SSL testing of the test piles, the separation distance of the transmitter and receiver was varied to establish an optimal distance for the testing. Past research has shown that increased gage separation leads to enhanced anomaly definition. However, the increased gage separation also results in inaccurate vertical anomaly definition (Amir, 2002). Based on the results of our initial DSDP testing on the test piles and consultation with the North American equipment distributor of the PISA, a gage separation distance of 46 cm (18 inches) was used in the SSL testing.

Anomalies in SSL testing are defined as areas that experience an increase in the First Arrival Time (FAT) and/or a reduction in relative energy of the ultrasonic signal. Anomaly definition guidelines for this project were set by the SSL testing engineer in consultation with the geotechnical engineer. These guidelines were based on FAT increases and are summarized in Table 5.

TABLE 5. SSL Anomaly Definition Guidelines.

FAT Increase (%)	Remarks
0–10%	Not significant
10-30%	Requires additional investigation by either SSL re-test or pulse echo testing (PET)
> 30%	Significant anomaly – requires inspection coring or replacement

In accordance with the project specifications, SSL testing was conducted within a 3 to 10 day window from DSDP installation. SSL testing was performed on a total of 60 of the 882 production DSDP's installed for the project. This amount is approximately 6.8%. Of the 60 piles tested, only 1 pile exhibited an anomaly with a FAT increase greater than 10%. This is an anomaly rate of 1.7%, which was deemed acceptable for the project. The contractor selected to replace the DSDP with the anomaly to keep pace with the accelerated construction schedule rather than further evaluate the anomaly with additional testing.

PROJECT TIMELINE

The test pile program was conducted within 10 days. Production piles were installed within 32 days from the start of production. The total time for the project, from start of the testing program to the end of production DSDP installation took 6 weeks, which was considered extremely quick for pile foundation installation for a combined cycle power generating unit.

CONCLUSIONS

The design, monitoring, and integrity testing of Drilled Soil Displacement Piles (DSDP) for a Gas-Powered Power Plant has been described in detail. The integrated process showed that DSDP piles can be used effectively to carry loads through loose and soft soils to underlying bearing stratum. An efficient test pile program was used to determine the installation conditions and group effects across the site as well as ascertain the axial compressive capacity of the piles in typical soil conditions. It was also demonstrated that installations parameters (torque and tool penetration rate) are reliable indicators of subsurface conditions, and, in combination with a good site characterization, can be used as a guide for setting pile toe levels. An extensive monitoring program, consisting of detailed pile installation inspection and Singlehole Sonic Logging (SSL) testing verified that the piles were installed to the project specifications. The integrated design, monitoring, and integrity testing of the DSDP allowed the pile test program and production pile installation completed within six weeks.

ACKNOWLEDGMENTS

The authors would like to give thanks to the following organizations that provided support for this paper: Gemma Power Systems, Progress Energy, Parsons Energy & Chemical Group, Berkel Contracting Company, and WPC. In particular, the authors wish to acknowledge the following people for their support of the project and contributions to this paper: Mr. William Carter and Mr. Alan Smithe of Gemma Power Systems; Mr. Wade Anundson and Mr. Yogesh Shan of Parsons Energy & Chemical Group; and Mr. William Wright and Mr. Donovan Ledford of WPC. In addition, the authors would like to express gratitude to the following people for their assistance: Mr. Matt Christie and Mr. Mark Knussmann for data collection; Mrs. Amy Lockhart for administrative support; and Mr. Les Chernauskas for technical evaluation.

REFERENCES

Amir, J., (2002), "Single-tube Ultrasonic Testing of Pile Integrity", Proceeding of Deep Foundations 2002, Orlando, FL, ASCE GSP No. 116, pp 836-850.

ASTM D1143-81 (1994) *"Standard Test Method for Piles under Static Axial Compressive Load"*

Bustamante, M., and L. Gianeeselli (1982) "Pile Bearing Capacity Predictions by Means of Static Penetrometer CPT", Proceedings of the 2nd European Symposium on Penetration Testing, ESOPT-II, Amsterdam, Vol. 2, pp. 493-500.

Chernauskas, L.R. and S.G. Paikowsky (1999) "Deep Foundations Integrity Testing: Techniques and Case Histories", Civil Engineering Practice Spring/Summer 1999, vol. 14, no. 1, Boston Society of Civil Engineers Section/ASCE, pp 39-56.

Chernauskas, L.R. and S.G. Paikowsky (2000) "Defect Detection and Examination of Large Drilled Shafts Using a New Cross-Hole Sonic Logging System", Proceedings of Performance Confirmation of Constructed Geotechnical Facilities, ASCE GSP No. 94, pp.124-147.

Deep Foundations Institute (1994) *Inspector's Guide to Augered Cast-In-Place Piles*, Deep Foundations Institute, Englewood Cliffs, New Jersey.

NeSmith, W.M. (2002) "Static Capacity Analysis of Augered, Pressure-Injected Displacement Piles", Proceeding of Deep Foundations 2002, Orlando, FL, ASCE GSP No. 116, pp 1174-1186.

RESIDUAL LOAD DEVELOPMENT IN ACIP PILES IN A BRIDGE FOUNDATION

Min-Gu Kim[1], Member, ASCE, Emin Cavusoglu[2], Michael W. O'Neill[3], Fellow, ASCE, Tim Roberts[4], Member, ASCE, and Stanley Yin[5], Member, ASCE

ABSTRACT: It has traditionally been assumed that when cast-*in-situ* piles, such as drilled shafts or ACIP piles, are constructed, the residual stresses after curing are zero. When conducting load tests on instrumented cast-*in-situ* piles, therefore, the readings in the strain gauges at the beginning of the load test are assumed to represent an unstressed condition. The implementation of ACIP piles in a bridge foundation in the Texas DOT system recently has allowed for the measurement of residual loads in several ACIP piles during curing. These loads were found to be relatively small, but not insignificant, and appear to have been related to site-specific soil profile characteristics.

INTRODUCTION

The existence of post-driving residual loads in driven piles has long been known [e. g., Briaud and Tucker, 1984]. The "capture" of stress waves, particularly in flexible piles, produces these residual loads. Fellenius (2002) and Falconio and Mandolini (2003) have recently suggested that residual loads may also develop in drilled shafts and that part of the residual or "locked-in" load in a deep foundation is caused by post-construction volume change of an ambient soil and other time-dependent phenomena. Fellenius also stressed that the residual loads must be considered in the analysis of load test data to avoid erroneous conclusions, for example, regarding ratios of developed unit side resistance to the shear strength of the soil at various points along the pile.

---

[1] Research Assistant, Dept. of Civ. and Env. Eng., University of Houston, Houston, Texas 77204-4003
[2] Research Assistant, Dept. of Civ. And Env. Eng. University of Houston, Houston, Texas 77204-4003
[3] Cullen Dist. Prof., Dept. of Civ. and Env. Eng., University of Houston, Houston, Texas 77204-4003
[4] Project Manager, Fugro South, Inc., 6100 Hillcroft, Houston, Texas, 77036
[5] District Materials Engineer, Houston District Laboratory, Texas Department of Transportation, 7721 Washington Avenue, Houston, Texas 77007

In the past, it has generally been assumed that residual loads do not develop in cast-*in-situ* piles such as drilled shafts and ACIP piles, since there are no stress waves to capture and since reconsolidation of the overconsolidated soil should be minimal. Therefore, when conducting a load test on an instrumented cast-*in-situ* pile, the zero readings in the strain gauges at the beginning of the load test have always been assumed to represent an unstressed condition, and load transfer patterns have been derived from the test data based on that assumption. The implementation of ACIP piles in the Krenek Road bridge foundation in the Texas DOT system recently has allowed for the measurement of residual loads in several ACIP piles to test the validity of this assumption.

SITE CONDITIONS

The bridge site consists of a mixed soil profile of generally stiff clays and medium dense sands. The site is in the Beaumont formation, a Pleistocene-aged deposit on the Texas Coastal plain, where ACIP piles are well suited (O'Neill et al., 2002). Fig. 1 summarizes the subsurface profile along with the sister bar instrumentation system for a test pile and typical central bent piles that support the bridge. All piles had a nominal diameter of 0.46 m (18 inches). All instrumentation levels had two diametrically opposite sister bars, except for the top levels, which had four sister bars at 90 degrees.

Subsurface data are summarized in Table 1. The surface layer is a stiff, sandy clay to a depth of 2.1 m below the ground surface. Stiff plastic clay layers were observed from 2.1 m to 12.8 m. A 1.5-m-thick layer of waterbearing, medium dense clayey sand was found from 12.8 m to 14.3 m. A layer of stiff clay was present from 14.3 m to 15.8 m. Below 15.8 m is a stratum of dense, fine and relatively clean waterbearing sand. The piles in question were designed to penetrate a short distance into this sand stratum. The piezometric level in this sand stratum was about 1.5 m below the natural ground surface.

Table 1. Soil Data for the ACIP Pile Bridge Site

Interval (m)	Soil	$s_{u\,avg}$ (kPa)	N_{avg}	w (%)	w_L	w_P	I_P
0-2.1	Stiff Sandy Clay	57.6	-	21	42	14	24
2.1-3.7	Stiff Clay	110.1	-	21	49	16	33
3.7-6.7	Stiff Clay	87.2	-	27	67	26	41
6.7-12.8	Stiff Clay	108.4	-	23	56	19	37
12.8-14.3	Med. Dense Clayey Sand	-	-	20	-	-	-
14.3-15.8	Stiff Clay	84.2	-	31	75	29	46
15.8-23.5	Dense Sand	-	47	18	-	-	-

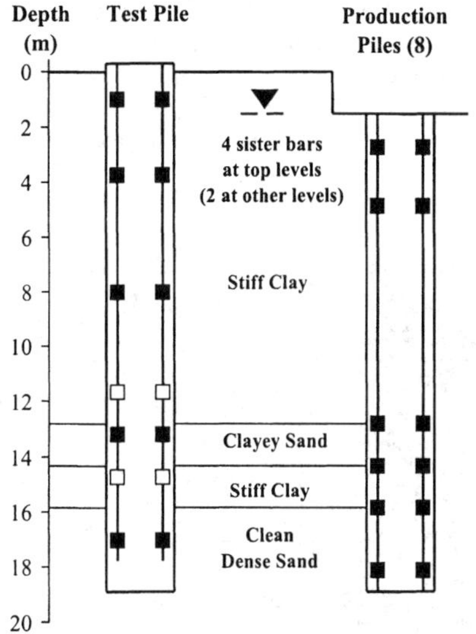

Fig. 1. Geotechnical and Instrument Profile

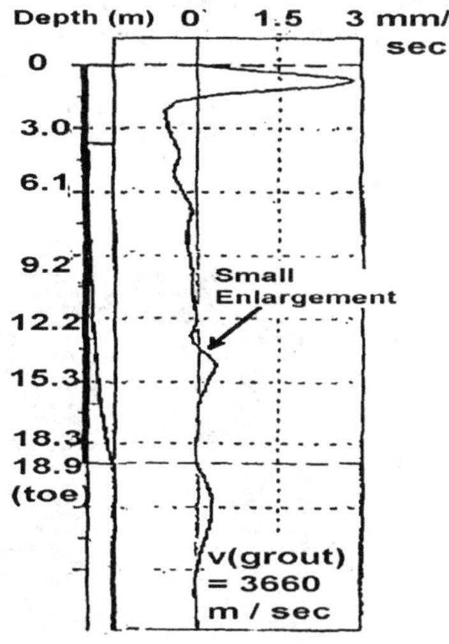

Fig. 2. Pulse-Echo Test – Test Pile

PILE INSTALLATION AND MONITORING

The central bent, or production, piles consisted of two groups of four piles each, in square grid arrays with an approximately 3-diameter center-to-center spacing, under adjacent column footings, denoted FA and FB, respectively. They, and the test pile, which were all 0.46 m in nominal diameter, were installed by rotating a continuous-flight, hollow stem auger into the ground until the required depth was achieved and then pumping grout through the hollow stem under pressure as the auger was withdrawn, a process that took about 15 to 18 minutes for each pile installed. After the auger was fully withdrawn an instrumented full-depth reinforcing cage was inserted into the grout column by pushing. The cages were all 18 - 19 m long and consisted of 6 #6 grade 60 steel longitudinal deformed bars with a # 4 deformed bar spiral at a 152-mm pitch.

For purposes of understanding residual loads, it was important that the structural integrity of the piles be verified. Following installation, the test pile was subjected to a pulse-echo test, with the result that no necks or voids could be detected in the top 15 m (the depth of resolution of the pulse-echo test). See Fig. 2. Furthermore, during installation, incremental grout ratios (IGR's) were monitored for all piles with a prototype commercial ACIP pile installation recorder using a magnetic flow meter. The IGR records are displayed in Fig. 3, which shows that the IGR's generally exceeded 1.15, the design agency standard, except at a few locations, most notably the IGR at 9.1 m depth for the test pile, where it was less than 1.0. Since IGR's were recorded for every 0.61-m increment of auger tip depth, and since a relatively large-stroke pump was used, an IGR slightly less than 1.0 was not deemed unsatisfactory, since the IGR's in the increments just below and above 9.1 m exceeded 1.0 considerably. Examination of Fig. 2 confirms that no significant neck, if any, existed at that or any other level. Note that an IGR > 1 does not necessarily imply that the borehole expanded when grouted, as much of the grout was observed to be pumped up the auger flights.

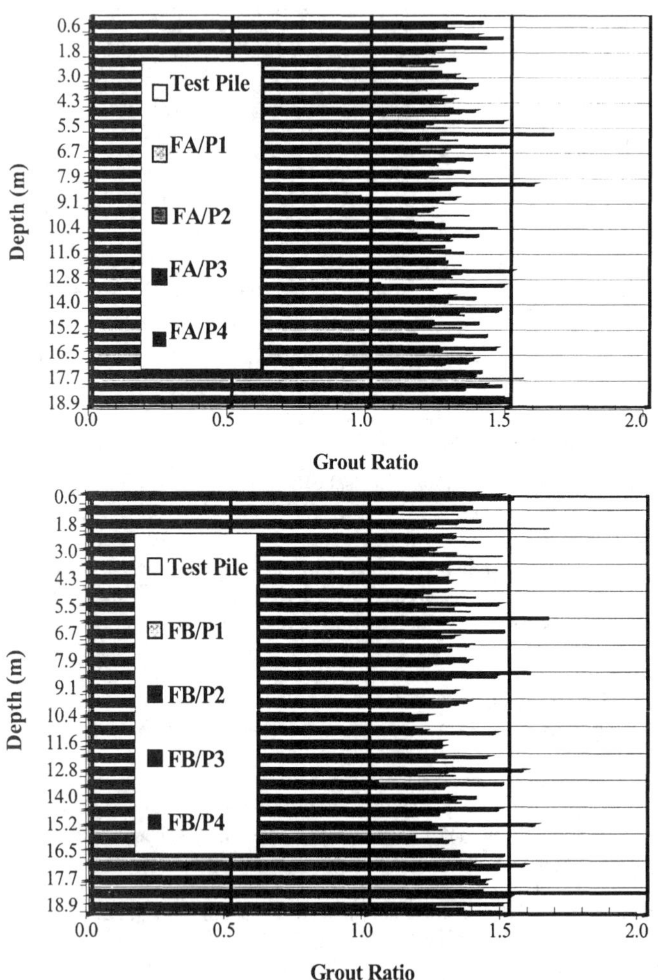

Fig. 3. Incremental Grout Ratios for Piles in Footings FA, FB and Test Pile

INSTRUMENTATION

Calibrated, vibrating wire strain gages ("sister bars") were used to instrument the ACIP piles. Those sensors have an advantage over more conventional electrical resistance or semi-conductor types in that the sensor output is a frequency rather than a resistance or voltage. The frequency output is stable over long periods (months or years) because it is unaffected by voltage drops brought about by corrosion of terminal contacts, moisture intrusion into the sensor or the cable, or temperature effects on the cable (Hayes and Simmonds, 2002). The test pile and eight central bent piles in FA and FB (among others not considered here) were instrumented as indicated in Fig. 1, by tying the sister bars securely to the rebar cage. The test pile protruded about 0.3 m above the ground surface. The top of each production pile was 1.5 m below the ground surface to accommodate placement of the column footings (pile caps). The toes of all piles are located at 18.9 m below natural grade.

The sister bars were all oriented parallel to the longitudinal steel. The instrumented cages were pushed into the grout within five minutes after the piles were grouted. Fig. 1 shows instrument locations. For the test pile, 16 strain gages were placed at 7 levels along the pile depth. Among the eight central bent piles, half of them (denoted FA/P1, FA/P3, FB/P1, and FB/P3) were instrumented with 14 sister bars at 6 levels as shown in Fig. 1. One of them (denoted FA/P4) was instrumented with 10 sister bars at 4 levels and the rest of them (denoted FA/P2, FB/P2, and FB/P4) with 8 bars at 3 levels. During data reduction, a few sister bar readings (< 10 %) were spurious and were eliminated for the residual load analysis (e. g., those shown by white squares in Fig. 1).

GROUT MIX DESIGN

The constituents of the grout mix may affect strain development in ACIP piles during curing. The mix design for the grout used in this project is shown in Table 2. The fluidifier is a proprietary product but serves the same purpose as a high-range water reducer in a concrete mix. The field efflux value (19-mm-ϕ orifice) for this mix was 8 – 18 seconds per 950 mL.

Table 2. Grout Mix Design

Constituents	Amount by Weight (%)
Portland Cement (Type II)	20.3
Fine Aggregate (ASTM C-33)	61.4
Water	12.1
Fly Ash (Class F)	6.1
Additive (Fludifier)	0.05

LOAD TEST

A conventional top-down axial compression load test was carried out on the test pile when it was 18 days old by jacking it against beams anchored by four reaction piles using a pneumatically powered jacking system. The jacking system contained a hydraulic jack and an electronic load cell. Four dial gauges attached to the top of the test pile and suspended from reference beams were used to monitor the settlement of the pile. The test pile was loaded in 222-kN increments to failure. Each load increment was held for 20 minutes, during which the vibrating wire sister bars were read. The test pile was unloaded in 512-kN decrements until the applied load reached zero. The load test results will be summarized later.

GROUT STRENGTH AND MODULUS TESTS

In order to deduce residual loads from sister bar readings, strains were converted to stresses using appropriate moduli for the grout and steel. 76 mm ϕ X 152 mm cylindrical grout specimens were acquired at the site and tested in compression in the laboratory at various ages of the grout. The testing machine was operated at a constant displacement rate of 0.03 mm/min. Sets of three compression tests were conducted at various times after pile installation [4 days, 7 days, 14 days, 18 days (for the test pile), and 28 days (for the central bent piles)], with two additional tests at 14 days for the test pile and eight additional tests at 28 days for the central bent piles.

The objective of the grout tests was to obtain a relation between the low-strain (ε_l = 0.0001) secant modulus of the grout and time. This was accomplished by placing bonded foil strain gauges on two opposite sides of reference samples (five for the test pile and eleven for the production piles, respectively) collected from the field during pile installation. These samples were tested on the day of the load test (test pile) or capping the production piles, after curing and storing the samples in humid room until the time of testing. The samples were loaded in compression in a small compression-testing machine while simultaneously making strain readings from the bonded strain gauges on the specimens. Ratios of initial, small-strain secant modulus (termed ISM) to compressive strength (f'_{cr}) were computed from these tests, where f'_{cr} refers to the compressive strength of the reference grout samples (on the day of the load test or day of capping the production piles).

The sets of grout samples tested at intermediate times (4, 7, 14 days after casting) were not strain-gauged but instead were subjected to compression tests to failure. ISM's corresponding to these intermediate times were then determined for the test pile and for the production piles (separately) by multiplying the ratio of measured intermediate compressive strength, f'_{ci}, to the compressive strength of the reference specimen, f'_{cr}, times the mean ISM of the instrumented reference specimen, on the assumption that ISM/f'_c is approximately constant over time in a given grout batch. Separate ISM/f'_{cr} values were employed for the test and production piles because the grout was batched at different times and by different suppliers.

Table 3 shows mean values of ISM and f'_{cr} for the test pile at the time of the load test (18 days after pile installation). The mean value of ISM/f'_{cr} at 18 days was 598. Table 4 summarizes the values of f'_{ci} measured at various intermediate times after casting and values of ISM computed from those values and the measured values of f'_{cr}

and ISM for the reference specimens. The computed values of ISM for the grout were converted to ISM values for the pile by considering the relative moduli of the longitudinal steel and the grout and the relative nominal cross-sectional areas.

Table 3. ISM / f'_{cr} for Test Pile Grout at Age 18 Days (5 Samples)

Value	f'_{cr} (MPa)	ISM (GPa)	ISM/f'_{cr}
Mean	36.0	21.5	598
St. Dev.	1.5	0.9	35
C.O.V.	0.04	0.04	0.06

Table 4. Summary of f'_{ci} for Grout, ISM of Grout, and ISM of Pile, at Various Times after Casting Test Pile

Time after Casting (days)	f'_{ci} (MPa)	Computed ISM of Grout (GPa)	Computed ISM of Pile (GPa)
0	0.0	0.0	0.0
4	28.7	17.2	19.1
7	30.4	18.2	20.1
14	32.6	19.5	21.4
18 (Load Test)	36.0	21.5	23.4

The ISM's of the grout and production piles were determined in a similar manner. Table 5 shows the f'_{cr} and ISM values measured for these piles. The average value of ISM/f'_{cr} was 683 at 28 days after pile installation. The residual loads were evaluated in the production piles 28 days after casting. Table 6 shows information for the production piles that is similar to that in Table 4 for the test pile.

PROCEDURE FOR CONVERTING STRAIN TO LOAD

Vibrating wire strain gauges were read at various times after casting the piles. Initial strain readings, used as zero readings, were made when the grout was fluid with the sister bars attached to the rebar cage immediately after the cage had been thrust into the grout column.

Table 5. ISM / f'_{cr} of Production Pile Grout at Age 28 days (11 Samples)

Value	f'_{cr} (MPa)	ISM (GPa)	ISM/f'_{cr}
Mean	38.6	26.4	683
St. Dev	3.4	2.5	92
C.O.V.	0.09	0.09	0.13

Small strains were measured throughout the curing process; therefore, the stress-strain curve for the grout was assumed to be linear (expressed using the ISM) but time dependent due to the curing. The pile ISM values from either Table 4 (test pile) or Table 6 (production piles) were plotted against time, as illustrated in Fig. 4. Axial stress, σ_a, in the pile at each level of instrumentation was then computed incrementally in time from Eq. 1.

$$\sigma_a = \sum_{t=1}^{t=t_{max}} \left(\varepsilon_t - \varepsilon_{t-1}\right) \left[\frac{\frac{(ISM_{t-1} + ISM_t)}{2} A_g + E_s A_s}{A_g + A_s} \right], \tag{1}$$

in which ε_t and ε_{t-1} are the measured strains (average of all gauges at a level), considered positive in extension, at discrete times t and t-1, respectively. The ISM's are the ISM's of the grout at the appropriate times after casting (t). A_g and A_s are the nominal areas of grout and steel in the cross-section, respectively (164173 mm^2 and 1703 mm^2, respectively) and E_s is the Young's modulus of the longitudinal steel (200,000 MPa).

Stresses were summed from time increment to time increment because the ISM changes with time. Within any time increment, the ISM was taken to be constant, as ISM_τ, shown in Fig. 4.

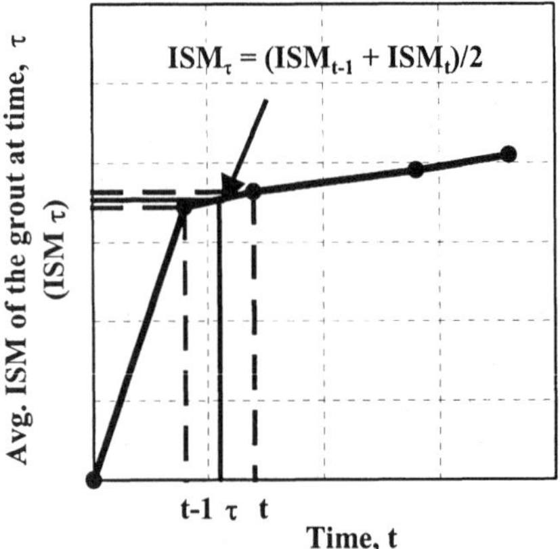

Fig. 4. ISM vs. Time

The resulting value of σ_a was a composite value, considering both steel and grout, and was therefore multiplied by the nominal gross cross-sectional area of each pile (0.164 m^2 or 254 in.^{2}) to obtain the residual load in the pile at time t.

In Fig. 5, the load vs. depth relationship at compressional failure (1913 kN) in the load test is also shown based on (1) pre-test zeros and (2) zeros taken when the instruments were first placed in the grout (true zeros). Residual loads were generally tensile along the lengths of all nine piles considered. During the load test increments of load in the test pile were computed from the measured increments of strain, nominal cross-sectional area and the entire laboratory stress-strain curve (considering stress-related nonlinearity) measured at 18 days after casting.

Table 6. Summary of f'_{ci} of Grout, ISM of Grout, and ISM of Pile at Various Times after Casting Production Piles

Time after Casting (days)	f'_{ci} (MPa)	Computed ISM of Grout (GPa)	Computed ISM of Pile (GPa)
0	0.0	0.0	0.0
4	28.0	19.1	21.0
7	29.6	20.2	22.0
14	32.5	22.2	24.0
28	38.6	26.4	28.2

RESIDUAL LOADS AND LOAD TRANSFER ANALYSIS

Distributions of residual loads along the test and production piles calculated using procedures described in the previous section are plotted in Figs. 5 and 6.

The pattern of residual load in the test pile (Fig. 5) suggests in general that the clay soils at the site were exerting upward-directed shear stresses on the upper sections of the pile (generally above 8 m depth) when it was in the unloaded condition and that below that depth the side shear stresses were generally directed downwards, with some local variations.

In the production piles (Fig. 6) the side shear stresses in the unloaded condition also acted upward near the ground surface and in the lower clay stratum, while throughout most of the upper clay stratum and in the sand strata, the side shear loads were directed downwards (per an anchor). One would expect the opposite pattern of developed side shear during curing if the ACIP piles were expanding (downward directed shear stresses near the head and upward directed shear stresses near the toe of the pile to restrain movement). It is speculated that the grouting process enhanced minor fractures in the overconsolidated clay that allowed free water to intrude and cause portions of the clay layers to swell.

The source of free water to produce such a phenomenon is suggested in Fig. 5. Very little load was transferred from the pile to the soil within the bottom 2 m (dense, clean sand, Fig. 1). This suggests that during construction the density of this waterbearing sand decreased dramatically, possibly because of a rapid depressurization of the sand layer as the auger broke through the overlying clay. This would have caused a flow of sand upward through the auger, reducing the density of the dense sand layer and producing a corresponding upward flow of groundwater prior to the time grouting through the hollow stem of the auger began. This speculation is strengthened by the fact that a column of free water, about 3 m high, was observed rising on top of the grout column for all nine piles as grouting proceeded.

It is noted that the largest residual tensile force in any pile was only about 325 kN, giving a maximum tensile stress of approximately 1980 kPa (287 psi). It is further noted that temperature effects in the vibrating wire gauges and their circuitry account for a maximum of only about 50 kPa (7 psi) of the measured residual stress. Even considering the pseudo stress indicated by temperature changes, 1980 kPa (287 psi) was well below the mean tensile strength of the grout, which was measured in splitting tension on seven grout samples to be 4380 kPa (635 psi) at an age of 18 days [with a minimum value of 3120 kPa (452 psi)]. Therefore, the residual strain readings should represent values from continuous, not cracked, structural elements, making the assumptions regarding computation and application of ISM for the piles valid (which may not have been true had the piles been cracked). It is also noted indirectly from Fig. 2 that the shape of the test pile appeared to be relatively uniform except for a small enlargement at a depth of around 13.7 m. This corresponds to the depth of the upper sand layer (Fig. 1), which may have been enlarged by the construction process. This enlargement, or minor "bell," may have enhanced the anchoring capacity of the soil at that level, which is clearly indicated in the reduction in tensile load in the production piles at the level of that sand stratum in Fig. 6. It is apparent, therefore, that the patterns of residual loads in the test and production piles are controlled to some extent by site conditions.

Load-depth curves and load-settlement data from the load test have been transformed into relationships of unit side resistance (f) versus local settlement (w) (so-called "f-w" or "t-z" curves). Figures 7 and 8 show f-w curves for two representative depths, using instrument zeros prior to the load test and at the time of installation in the fluid grout ("true"). As with flexible driven piles, the effects of the residual loads are evident in the t-z curves. Near the head of the pile (Fig. 7), the true values of f are higher than those for the pre-test zero values at all values of w, whereas near the toe of the pile (Fig. 8), the opposite is true. Mean undrained shear strength values for the clay at the two levels at which f-w curves are shown are also plotted in Figs. 7 and 8 to provide a general indication of the error involved in neglecting residual loads in an ACIP pile during a load test.

A further indication of the effect of errors in f-w curves on pile design is indicated in Fig. 9, in which is shown the measured load-settlement curve for the test pile and the load-settlement curve synthesized by using APILE (Ensoft, 1997). APILE allows for input of arbitrary values of f_{max} along the pile (values obtained from true and pre-test zeros in this case) but not the shape of the f-w curve or initial strains in the pile, so that it models loading of residually stressed piles only approximately. With this constraint, it is observed that use of unit side resistances from true zeros predicts the Davisson failure load marginally better than use of side resistances from pre-test zeros. In both APILE analyses in Fig. 9, the unit toe load-settlement curve that was input was very soft (maximum unit toe resistance of 200 kPa at 40 mm of toe settlement). This reflected the loss-of-density condition in the lower sand layer described earlier.

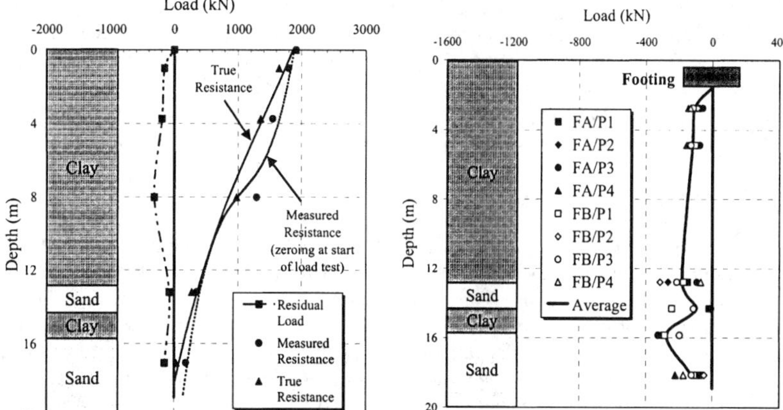

Fig. 5. Residual Load Distribution Prior to Load Test on Test Pile (Age 18 Days) and Load Distribution at Compression Failure

Fig. 6. Mean Residual Load Distribution along Eight Production Piles at Age 28 Days

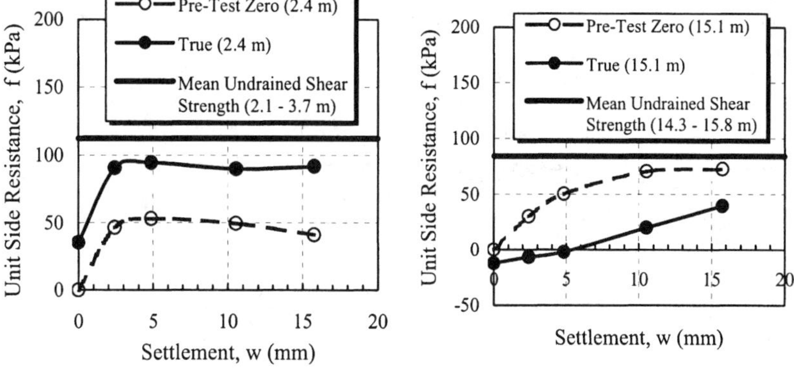

Fig. 7. f vs. w at Depth of 2.4 m

Fig. 8. f vs. w at Depth of 15.1 m

CONCLUSIONS

Residual loads were shown to exist in ACIP piles. These residual loads were computed by using vibrating wire strain gauges to read strains along the piles at several times, both during and after casting, developing an approximation of time-dependent grout modulus, and computing stress and load increments from the incremental strains as indicated in Eq. 1. Residual loads in the specific ACIP piles considered here (installed in typical Texas Gulf Coast layered clay and sand strata) varied along the pile length and were found to be generally tensile, most likely because of expansion of the clay soils, although the specific patterns appear to be influenced by site stratigraphy. Those residual load values could be considered as minimum values since the cross-sectional areas of the piles could have been larger than nominal cross-sectional areas that were used to calculate the residual loads.

The load distribution pattern along the compression test pile at failure, whether computed using pre-test or true zeros (Fig. 5), showed very low load transfer at the pile toe despite the toe being located in dense sand. This effect was likely due to a rapid pore water pressure reduction in the sand resulting from the drilling process, followed by an upward flow of sand and groundwater through the auger. Had dense sand remained below the pile toe, the pattern of residual loads may have been different than that observed. This conclusion suggests that future ACIP piles (of the non-displacement type considered here) constructed in Texas coastal soils with stratigraphies similar to those at the site described here should terminate in clay strata overlying sand strata, rather than in the waterbearing sand strata, if the sand strata function as confined aquifers.

Although residual axial stresses in the piles were generally tensile throughout the curing process, computed stresses did not exceed the splitting tensile strength of the grout, which has positive implications regarding durability of the pile material and the validity of using embedment strain gauges such as sister bars to measure load along the piles.

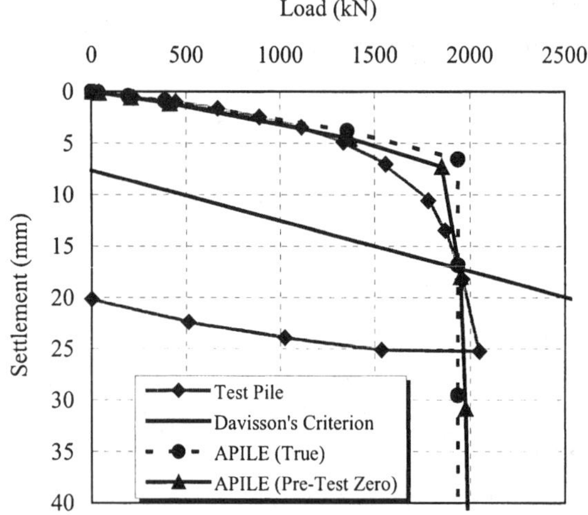

Fig 9. Comparison of Measured Load-Settlement Curve with Load-Settlement Curves Synthesized Using APILE

ACKNOWLEDGMENT

We would like to express our sincere gratitude to one of co-authors of this paper, Dr. Michael W. O'Neill (1940 - 2003) who had served as our advisor, mentor, and principal investigator in the Texas Department of Transportation implementation project on ACIP piles. We are honored to have the opportunity to have worked closely with him in this project as his graduate students, contractor, and project director and to co-author this paper with him. We will all miss his guidance, friendship and leadership. May his warm soul rest in peace.

REFERENCE

Briaud, J-L. and Tucker, L. (1984). "Piles in Sand: A Method Including Residual Stresses," *Journal of Geotechnical Engineering*, Vol. 110, No. 11, ASCE, pp. 1666-1680.

Ensoft (1997). *User's Guide for APILE*, Ensoft, Inc., Austin, Texas.

Falconio, G., and Mandolini, A. (2003). "Influence of Residual Stresses for Non-Displacement Cast In Situ Piles," *In Proceedings*, Deep Foundations on Bored and Auger Piles, Balkema, Rotterdam, pp. 145-152.

Fellenius, B. H. (2002). "Determining the True Distributions of Load in Instrumented Piles," Deep Foundations Congress 2002, *Geotechnical Special Technical Publication No. 116*, Volume 2, ASCE, Orlando, FL, pp. 1455-1470.

Hayes, J. and Simmonds, T. (2002). "Interpreting Strain Measurements from Load Tests in Bored Piles," *In Proceedings*, The Ninth International Conference on Piling and Deep Foundations, Deep Foundations Institute, June, Nice, France, pp. 663-669.

O'Neill, M.W., Ata, A., Vipulanandan, C. and Yin, S. (2002). "Axial Performance of ACIP Piles in Texas Coastal Soils," Deep Foundations Congress 2002, *Geotechnical Special Technical Publication No. 116*, Volume 2, ASCE, Orlando, FL, pp. 1290-1304.

AUGER PRESSURE GROUTED DISPLACEMENT PILES: AN ACCEPTABLE DEEP FOUNDATION ALTERNATIVE

Osama El-Fiky, P.E.[1], Michael Majchrzak, P.E., G.E.[2]

ABSTRACT: There are many types of deep foundations currently being used in the San Francisco Bay Area, one of the most common being driven precast, prestressed concrete (PC/PS) piles. PC/PS piles are frequently selected for use due to of ease of installation, availability, and relatively low cost. Disadvantages of using PC/PS piles are noise, vibrations, front-end delays due to indicator pile programs, and reduced flexibility. An acceptable alternative to driven PC/PS piles was selected for construction of foundations for the heavy housing and administrative buildings at the San Francisco County Jail No. 3 replacement project in San Bruno, California. The pile type selected, the Auger Pressure Grouted Displacement (APGD) pile, was chosen to reduce noise, vibrations, installation time, and cost, and to allow for greater flexibility of installed pile lengths. This paper presents a case history of the use of APGD piles that were substituted for driven PC/PS concrete piles for the subject project. Standard and quick loading pile load tests were performed to confirm adherence with the design criteria originally used for the driven PC/PS concrete piles. The APGD piles were found to increase shear strength around the piles similarly to driven displacement piles. In addition, this paper presents the methodology used in monitoring APGD pile installation.

PROJECT DESCRIPTION

The San Francisco County Jail facility is located at One Moreland Drive in San Bruno, California, as shown in the Site Vicinity Map – Figure 1. The facility covers an approximate 150-acre parcel of land owned by the City and County of San

---

[1] Project Manager , Kleinfelder, Inc. 7133 Koll Center Parkway, Suite 100, Pleasanton, CA 94566

[2] Central Bay Area Regional Manager, Kleinfelder, Inc., 7133 Koll Center Parkway, Suite 100, Pleasanton, CA 94566, mmajchrzak@kleinfelder.com

Francisco in San Mateo County. The area is bounded by the San Francisco State Fish and Game Refuge, the Golden Gate National Recreation Area, Skyline College, and residences. Several jail structures currently exist at the facility including the existing Jail No. 3 housing complex, Jail No. 7, and the Learning Center. These three structures are located within a relatively level area, and are surrounded by moderately steep hills on the south, west, north, and northeast.

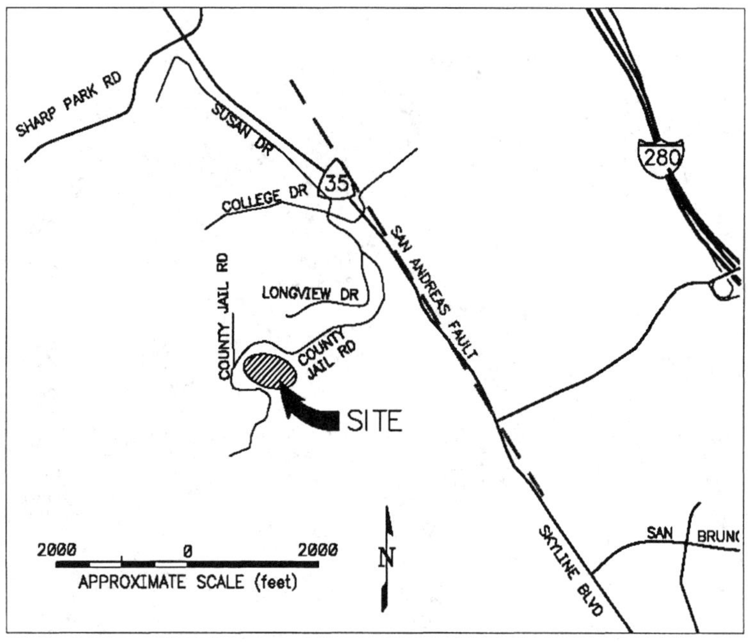

FIG. 1 - Site Vicinity Map

The project consists of constructing the following:

- Six new roadway alignments
- New Staff and Public Parking Lots
- A four-story Jail No. 3 Housing Complex
- A two-story Administration Building
- A bridge connecting the new Jail No. 3 Housing Complex to the Administration Building
- A one-story Central Plant

- A 70-foot diameter, 800,000-gallon water storage tank
- Eight retaining walls

The layout of the three buildings is shown on the site plan – Figure 2.

This project is a design-build project. The general contractor is AMEC, the architect is Kaplan, McLaughlin, Diaz, the structural engineers are The Crosby Group and SOHA Engineering, and the civil engineer is Telamon Engineering Consultant.

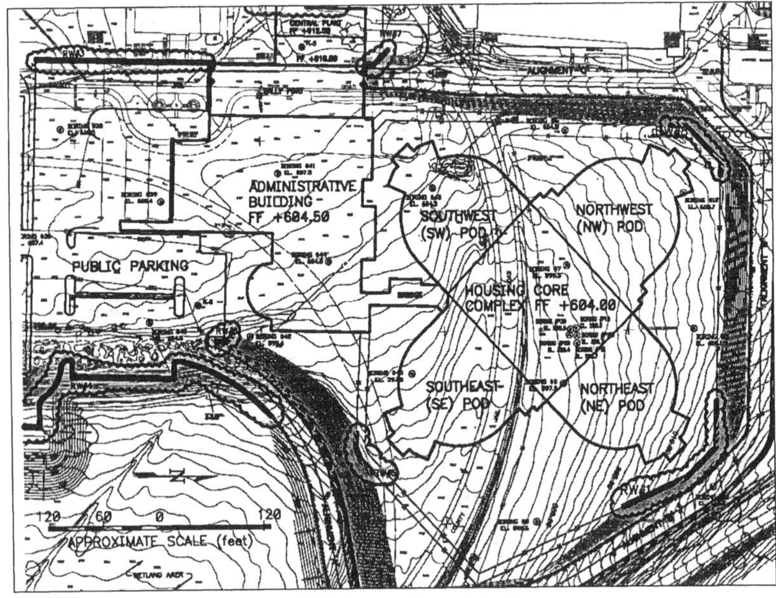

FIG. 2 – Site Plan

GEOLOGIC SETTING AND SUBSURFACE CONDITIONS

The San Francisco County Jail site is located along the western foothills of the Coast Ranges Geomorphic Province in the County of San Mateo, California. The site is located approximately 3 miles east of the Pacific Ocean on the San Francisco Peninsula.

Geologic maps covering the site area have been prepared by Schlocker (1970), Brabb and Pampeyan (1972 and 1983), LaJoie and others (1974), Wagner and others (1991), Ellen and Wentworth (1995) and Wentworth and others (1998). These maps generally agree that the hills that surround the jail site are comprised of Franciscan

Complex greenstone (altered fine-grained volcanic rock) of Cretaceous geologic age, while the low lying areas of the site are underlain by Quaternary surficial deposits such as alluvium and slope wash materials.

Wentworth and others (1985) described the surficial deposits present in the low lying areas in the site vicinity as slope debris and ravine fill. In addition, the jail facility has undergone extensive grading with fills reported to be up to about 30 feet thick. Small to moderate-size, old landslide deposits with subdued topographic expression were observed along the hillside areas of the site. A wetlands area is located at the southeast quadrant of the facility.

The site is located approximately 0.31 miles west of the active San Andreas fault, about 5.9 miles northeast of the San Gregorio fault, and about 18.6 miles southwest of the active Hayward fault. No known active faults are present within the boundaries of the site. Based on the information provided in Hart and Bryant (1997), the site is not within an Alquist-Priolo Earthquake Fault Zone. The Working Group on California Earthquake Probabilities (1999) estimated probabilities of large earthquakes (magnitude 6.7 or greater) in the San Francisco Bay Region for segments of the San Andreas, Hayward, and Calaveras faults. The study estimated that there is a 70 percent probability of a large earthquake in the San Francisco Bay Region in the next 30 years (from 2000). With respect to individual fault segments considered in the study, estimates are provided of 21 percent for the San Francisco Peninsula segment of the San Andreas fault, 32 percent for the Hayward fault, and 18 percent for the Calaveras fault. A major seismic event on these faults could cause significant ground shaking at the site.

The Housing complex and Administration Buildings are underlain by up to 17 feet of fill consisting of very loose to loose organic silty sand. The fill was found to overlie about 12 to 35 feet of interbedded layers of medium stiff to very stiff lean clay, sandy lean clay, and loose to medium dense clayey sand. These soils were found to be underlain by an approximately 3- to 6-foot thick layer of very soft to soft organic silt. The organic silts were found to be underlain by highly weathered and fractured Greenstone/Franciscan bedrock occurring at depths below the existing ground surface of about 32 to 62 feet at the Housing Building site, and about 32 to 47 feet at the Administration Building site. Groundwater was observed during drilling at depths of approximately 4 to 16 feet below the ground surface.

Unconfined compressive tests performed on the fill material indicated average shear strengths of 840 pounds per square foot (psf), while unconfined compressive strengths of the native soils were approximately 2,000 psf. A direct shear test on the native soil resulted in a cohesion of 860 psf with a friction angle of 26 degrees.

INITIAL FOUNDATION RECOMMENDATIONS

The fill soils at the site were judged incapable of directly supporting the building loads. Because of the soft consistency of the fill soils, and due to the presence of the soft zone of clayey soils overlying the Greenstone/Franciscan bedrock, it was decided to support the buildings on deep foundations extending into the weathered bedrock. A driven pile system was favored over a drilled pier system due to the shallow depth of groundwater, the presence of soils with moderate caving potential, and the unknown composition of the fill from an environmental standpoint. Pre-cast, pre-stressed 14-inch square concrete driven piles were originally recommended to support all of the compressional building loads and a portion of the lateral loads for the Administration Building and Housing Complex. To provide additional lateral resistance during a seismic event, short 24-inch diameter, 10-foot deep piers (lugs) were also to be recommended. The pile layout for the Housing Complex is shown on Figure 3.

The piles were designed to derive support both through friction and end bearing in the alluvium and the underlying bedrock. Based on the results of our analyses, the piles needed to be socketed a minimum of 3 feet into bedrock, except where significant amounts of new fill were to be placed, which would lead to a potential downdrag condition, and would therefore necessitate deeper penetration into the bedrock. The piles were designed for an ultimate skin friction of 425 pounds per square foot (psf) for the upper 10 feet of the existing fill, 1,000 psf for the native sandy gravelly clay material above the bedrock, and 3,000 psf for the weathered rock. In addition, the piles were designed for an allowable end bearing value of 75,000 to 85,000 psf depending on the condition of the bedrock which depends on the depth of the embedment. A factor of safety of 2.0 was used for the skin friction portion and 3.0 for the end-bearing.

For easy of future comparison with the pile load tests, we estimated an equivalent skin friction using a converted end bearing along with the skin friction of the material above the bedrock. We used pile lengths of 41 and 49 feet for estimating the equivalent skin friction. The equivalent skin friction for the piles was estimated to be 1,000 psf allowable or 2,000 psf ultimate. An average ultimate skin friction value for the piles used for the design was about 1000 psf. In areas where new fill was placed, the skin friction in the new fill was ignored. These values reflect higher unit resistances as compared to those used for design of conventional drilled pier foundations due to the beneficial effect of displacement of the soils surrounding the piles. For drilled piers, lateral pressure factors typically ranging from 0.5 to 0.7 are used in skin friction evaluations. For driven piles, this factor may be increased to reflect an at-rest or even a passive state. The factors associated with this condition can be between 1.45 and 1.8 (Bowles [1996]). Generally, these factors apply to granular soils and not necessarily applicable to cohesive soils.

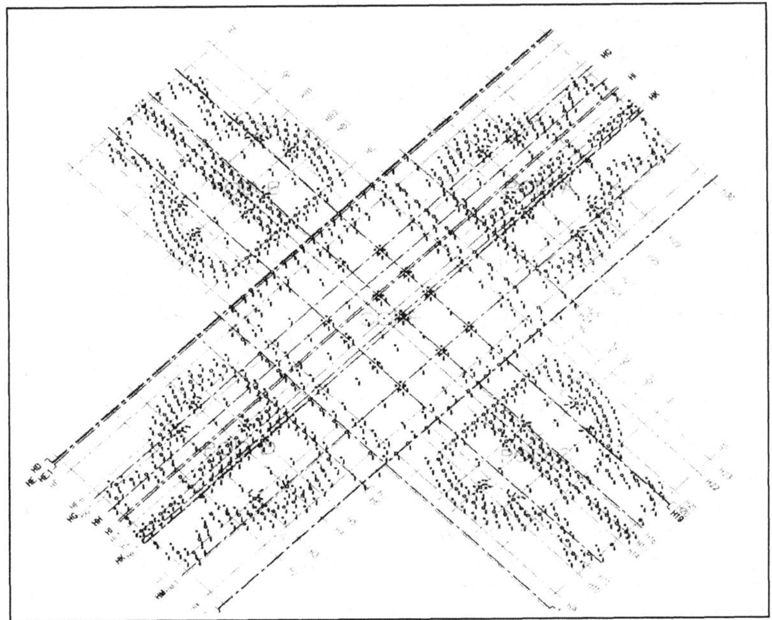

FIG. 3 – Pile Locations at Housing Complex

The 14-inch driven piles were designed for allowable downward capacities of 200 kips, with a one-third increase for the seismic loads. For areas in the Housing Complex that would experience downdrag, the allowable capacities were reduced up to 60 kips to account for the downdrag loads.

The depth to bedrock varies at the sites of the planned structures from 32 to 60 feet. Because of this variation, to confirm the estimated depth of the bedrock, an indicator pile program of 20 piles was recommended.

AUGER PRESSURE GROUTED DISPLACEMENT (APGD) PILES

Because of the potential for complaints from the project neighbors during installation of the piles, an accelerated project schedule, and cost concerns, an alternative foundation system consisting of Auger Pressure Grouted Displacement (APGD) piles was considered to replace the precast, prestressed 14-inch square driven piles. This system was suggested by Berkel & Company Contractors, with the contention that the skin friction values would be equal to or greater than those used in our design of the driven piles. The APGD piles were to be 18 inches in diameter and reinforced by six #6 vertical rebars in the upper 24 feet, with a #8 rebar extending the entire depth of the pile.

APGD piles are constructed by drilling a displacement auger into the subsurface soils to the desired depth and pumping grout under pressure while withdrawing the auger. The displacement auger includes a 16-inch diameter flight auger tip that is about 10 feet long. At about 4 feet above the auger tip, an approximately 2 ½-foot long, 18-inch diameter steel tube section placed over the auger acts as a displacement element. The displacement auger is fixed to a 12-inch drill stem.

The drill rig used for APGD pile installation for this project was a Bauer drill rig with a capacity of approximately 335 bar of torque. During drilling, the soil is displaced laterally. Thus creating an at-rest or passive condition. Because this installation process does not allow for observing the type of material encountered, the torque of the drill-rod is recorded throughout the process. The torque is initially calibrated by locating the first piles adjacent to test borings drilled during the field investigation and comparing the torque at various depths with the soils encountered in the borings. Although this system of monitoring seems unsophisticated, it worked well for this project due to the significant increase in torque going from the native soil to the bedrock (a minimum of 3 feet in the bedrock was required). During construction of the pile, the total volume of concrete used is compared to the theoretical volume to insure that caving has not occurred. The torque, drilling depth, drilling rate, and grout pressure were continuously electronically monitored at the drill rig cab. A sample print out of the electronically acquired installation monitoring data is shown in Figure 4. As indicated in this printout, the slope of the depth curve shows the drilling rate. The steeper the slope, the harder the material is to drill through. From the torque curve, an estimate can be made of the average maximum torque the machine is delivering as the displacement auger is drilled through hard material such as bedrock. A torque value of 300 bars was found to be indicative of a very hard material such as weathered bedrock (Greenstone/Franciscan formation) for this project.

TEST PILE INSTALLATION

Two test piles were installed to measure the capacity of the APGD pile in the field. One test pile was installed within the footprints for both the housing and administration buildings. The test piles at the housing and administration buildings extended to about 49 and 41 feet below finish subgrade elevation, respectively. The piles were embedded into the bedrock approximately 8 and 3 feet for the housing and administration buildings, respectively. The embedment in the bedrock was based on the amount of fill that specific area would receive. For areas that would receive less than or equal to 3 feet of fill, pile embedment into bedrock was 3 feet and where the new fill thickness was greater than 3 feet pile embedment into bedrock was 8 feet.

As part of the test pile set-up, four reaction piles were installed around each test pile, which extended to about the same depth as the test pile. The reaction piles spanned 20 feet apart in one direction and 10 feet in the other direction forming a rectangular shape with plan dimensions as shown on Photograph 1. The reaction piles did not have the same reinforcement as the test pile but just a # 10 rebar that extended to the bottom of the reaction pile. Torque used to achieve embedment into the bedrock at each test pile location was on the order of about 280 to 300 bars of torque. The volume of grout used to fill each test pile was about 120 and 135 cubic feet. The pressure of the grout at the bottom of the pile was about 44 pounds per square inches (psi) and near the top of pile was on the order of about 5 psi. Grout cubes were sampled at each test pile location for compressive strength measurements before the pile load test was performed.

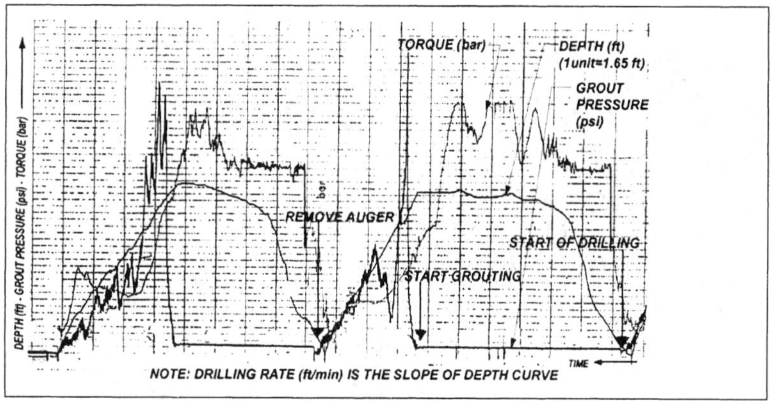

FIG. 4 – Drill Rig Printout

LOAD TESTS - SETUP

Pile load tests were performed on test piles constructed at each of the building sites. The purpose of the load testing was to confirm that the APGD pile installation method resulted in equivalent or greater unit resistances than used in our design of the driven piles. The setup of one of the pile load tests is shown in Photograph 1, below. Both the standard loading and the quick load test procedures were used for the first test pile (located at the Housing Building site), while only a quick load test procedure was used on the test pile at the Administration Building site. Compression pile load tests were conducted in general conformance with ASTM D1143 using both the standard loading procedure and the quick load test method. Each compression load increment for the standard test method was maintained until the rate of pile displacement was about 0.01 inches per hour based on a time period of at least 15 minutes. The hold periods of load increments for the quick load test method were about 5 minutes. Loading increments of about 50 to 55 kips were used on all tests. Nominal load decrements were about 25 percent of the maximum load attained during the subject loading cycle.

As shown in Photograph 2, a load cell was used to measure applied compression loads during each test. Load and displacement measurements for each test pile were obtained using a data acquisition system connected to an on-site laptop computer. Pile load test data was processed and the results viewed in real time on a laptop display using a program developed by GEODAQ, Inc. Primary and secondary displacement readings were obtained during all pile load tests. Primary readings were made using four DC powered displacement transducers (DCDT) mounted close to the pile top, and backup measurements were made by reading a scale through a level.

Photograph 1 – Load Test Set-Up

Photograph 2 – Load Cell

LOAD TESTS – RESULTS

A maximum load of 450 kips was applied to the test pile at the Housing Building location. Axial displacement at 200 kips (approximate design load) varied from 0.065 to 0.071 inches. The maximum axial displacement was about 0.21 inches at the end of the 450 kips hold period of 12 hours. After unloading the pile to zero load, the axial displacement was about 0.03 inches, as shown graphically in Figure 5.

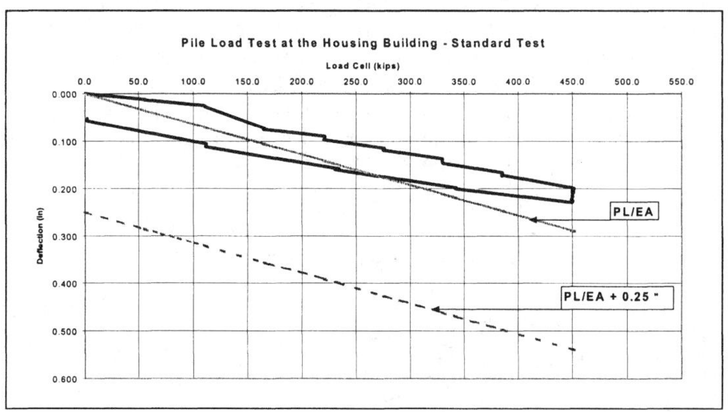

FIG. 5 – Pile Load Test at the Housing Building – Standard Test

A maximum load of 520 kips was applied to the same pile during a subsequent quick load test. Axial displacement at the design load of 200 kips varied from 0.071 to 0.073 inches during the quick load test. The maximum axial displacement was about 0.21 inches at the maximum test load of 520 kips. After unloading the pile to zero load, the axial displacement was less than 0.01 inches as shown graphically in Figure 6. The load-displacement plots for the standard test method and the quick load test method were nearly identical, which indicates that secondary loading effects such as grain crushing, consolidation, and soil creep were minimal. The load test results indicate that the ultimate load capacity for the test pile is significantly greater than the required ultimate capacity of 450 kips. Results from both tests indicate that the pile-soil system remains essentially elastic within the range of applied compression load.

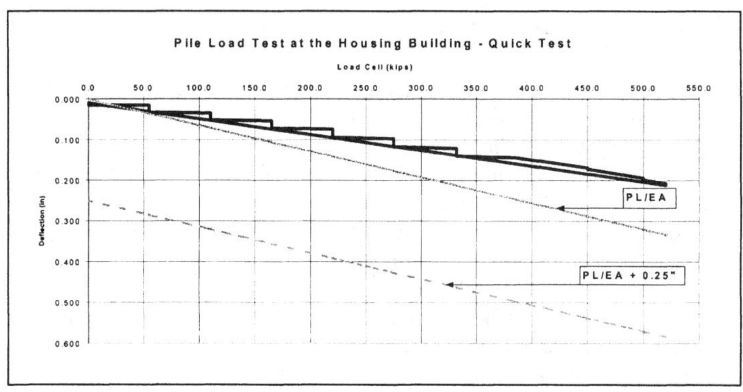

FIG. 6 – Pile Load Test at the Housing Building – Quick Test

The second test pile, located in the Administration Building area, was tested to a maximum load of about 490 kips. Axial displacement at 200 kips (approximate design load) was about 0.07 inches. The maximum axial displacement was about 0.34 inches at 490 kips, the maximum test load. After unloading the pile to zero, the axial displacement was about 0.10 inches as shown in Figure 7. The load test results indicate that the ultimate load capacity for the test pile is significantly greater than the required ultimate capacity of 400 kips. Results of this test indicate that the pile-soil system was slightly beyond the elastic range of behavior at the maximum test load.

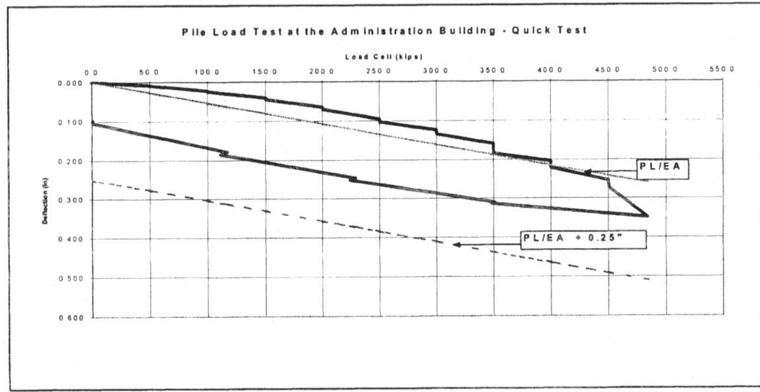

FIG. 7 – Pile Load Test at the Administration Building – Quick Test

Both test pile load test results were interpreted based on Tomlinson (1969) using a value of ¼ inch settlement plus the elastic compression of pile (PL/EA) curve to develop a failure line as drawn in Figures 5, 6, and 7.

LOAD TESTS – CONCLUSIONS

The pile load test results indicate that the APGD piles tested exceed the required pile capacities used in our analysis. The load test at the Administration Building shows that it took about twice the design load to fully mobilize resistance over the entire length of the pile, as indicated by the intersection of the PL/EA curve with the load-displacement curve. These results also indicate that the piles have achieved twice the design capacity utilizing skin friction only without mobilizing end-bearing resistance. The standard load test at the Housing Building site shows that at twice the design load, the full length of the pile had not been mobilized, and, therefore, end bearing had not been engaged. When a quick load test was repeated on the same pile at the Housing Building site (after completing the standard load test), the pile deflection was similar to the standard test. The interpreted ultimate skin friction values from the load tests were in the range (averaged over the length of the pile) of 1,800 to 2,400 psf. An average ultimate skin friction value of about 2,000 psf was used for our analyses.

PRODUCTION PILE INSTALLATION

The piles were installed between September 20, 2001 and March 6, 2002 using a Bauer hydraulic drill rig. During installation, torque measurements were continuously made and recorded versus depth of penetration, and the drilling rate was also monitored. As shown in Photograph 3, monitoring was performed by both the drill rig operator and our field representative. A torque of about 280 to 300 bars with a drill rate of about one to three feet per minute was established as an indication of penetration into bedrock. As soon as the required embedment was achieved, the hole was filled through the stem of the auger with grout using a grout pump while the auger was being withdrawn. The grout pressure was recorded, and the grout volume calculated. After filling the hole with grout, the steel cage and center bar were inserted into the hole. After the auger was pulled out of the hole, the material retained on the auger tip was observed as a secondary means of confirmation that bedrock material was penetrated. Photograph 4 shows the tip of the auger. Pile production averaged about 20 to 25 piles per day for a total of 800 APGD piles at the housing complex and 160 piles at the administration building. The 24-inch diameter short piers (lugs) for additional lateral load resistance were also installed by Berkel & Company by the auger pressure grouted (APG) piling method. Pile lengths ranged from 25 to 68 feet in the housing complex and between 25 to 48 feet in the administration building. In most cases, these lengths were required to reach the recommended embedment into the bedrock, which in most cases were sufficient to support the design loads. Embedment into the bedrock was needed for producing consistent settlement characteristics.

To avoid design changes in the original pile layout, the number of APGD were kept the same as the PC/PS piles.

Photograph 3 – APDG Pile Installation

Photograph 4 – Displacement Auger

CONCLUSIONS

Auger Pressure Grouted Displacement (APGD) piles installed and tested for this project demonstrated unit capacities equal to or greater than those that would be expected for a driven PC/PS concrete pile foundation. In addition, the selection of the APGD piling method allowed for a quick start, reduced noise and vibration, and was less costly relative to driven PC/PS concrete piles. The pile load tests indicate that the shear strengths of the soils at the site were improved during the drilling process which resulted in higher soil resistances than indicated by our laboratory tests. It is essential that when the APGD piling method is used, an installation monitoring procedure be established similar to that followed for this project.

REFERENCES

Bowles, Joseph E. (1996), Foundation Analysis and Design, Fifth edition.

Brabb, E.E. and Pampeyan, E.H., 1972, Preliminary Geologic Map of San Mateo County, California, U.S. Geological Survey Miscellaneous Field Studies Map MF-328.

Brabb, E.E. and Pampeyan, E.H., 1983, Geologic Map of san Mateo County, California, U.S. Geological Survey Miscellaneous Investigations Series Map I-1257-A.

California Department of Conservation Division of Mines and Geology (CDMG, 1982), Special Studies Zones Maps, San Francisco South Quadrangle, scale 1:24,000.

Ellen, S. D., and Wentworth, C. M., (1995), Hillside Materials of the Northeastern Part of the san Francisco Bay Region, USGS, Professional Paper 1357, Plate 2.

Hart, E.W. and Bryant, W.A. (1997), Fault-Rupture Hazard Zones in California: California Division of Mines and Geology, Special Publication 42, 1997 revised edition.

LaJoie, K.R., Helley, E.J., Nichols, D.R., and Burke, D.B., 1974, Geologic Map of Unconsolidated and Moderately Consolidated Deposits of San Mateo County, California, U.S. Geological Survey Basic data Contribution 68, Miscellaneous Field Studies MF-575.

Schlocker, J. (1970), Generalized Geologic Map of the San Francisco Bay Region, California: United States Geological Survey Open File Report.

Tomlinson, M.J., Foundation Design and Construction, 2nd edition, Sir Isaac Pitman and Son, London England, 1969.

Wagner, D. L., Bortugno, E, J. and Mc Junkin, R. D. (1990), Geologic Map of the San Francisco-San Jose Quadrangle: California Division of Mines and Geology Regional Geologic Map Series, scale 1:250,000.

Wentworth, C.M., Ellen, S., Frizell, V.A. and Schlocker, J., 1985, Map of Hillside Materials and Descriptions of their Engineering Character, San Mateo County, California: USGS Map I-1257D, scale 1:62,500.

9-Foot (2.75m) Diameter Drilled Shafts at Cranston Viaduct: Design, Load Testing, and Construction

Aly M. Mohammad, Ph.D., P.E.[1], Member ASCE/Geo-Institute, and Karen C. Armfield[2], Member ASCE/Geo-Institute

ABSTRACT: Analytical and empirical design methods are available to determine side shear and tip resistances of large diameter drilled shafts in cohesionless soils. The replacement of Cranston Viaduct in Rhode Island provided a good opportunity to test these design methods and compare it with a full-scale load test in the field. The foundation support for the replacement Viaduct consists of nine-foot (2.75m) diameter drilled shafts. As required by the Federal Highway Administration design guidelines, nine-foot (2.75m) diameter trial shaft was drilled at the site to establish the proper construction procedures and to verify design assumptions. The construction of the trial shaft and the production shafts for the southbound of the viaduct has already been completed. The paper will present a description of the unique design and construction methods adopted for the foundations. In addition, the paper will present the results of the Osterberg load test performed on the nine-foot diameter trial shaft and will compare side shear resistance and end bearing values obtained from the test to the values computed during design.

The design equations for drilled shafts in cohesionless soils, adopted by the Federal Highway Administration and AASHTO, proved to provide a good design tool in this project and compared very well with actual field measurements.

INTRODUCTION

The existing Cranston Viaduct, located in the City of Cranston, Rhode Island, was originally built in 1959 and carries Rhode Island Route 10 (a four-lane divided highway) over Amtrak Railroad Right of Way. The viaduct is comprised of two

[1] Geotechnical Department Manager, DMJM+HARRIS, 605 Third Avenue, NY, NY 10158
EM:aly.mohammad@dmjmharris.com
[2] Geotechnical Engineer, DMJM+HARRIS, 605 Third Avenue, NY, NY 10158
EM:Karen.armfield@dmjmharris.com

parallel span structures with a total span length of 486 feet (148.2m) along the centerline. The substructure consists of reinforced concrete stub abutments founded on piles and reinforced concrete multi-column piers supported on spread footings. A new Freight Dedicated Track will be constructed parallel to and west of the existing two Amtrak mainline tracks. The existing viaduct will be replaced due to its deteriorated condition as well as to accommodate the proposed three-track alignment. DMJM+HARRIS, Inc. provided engineering services to Rhode Island Department of Transportation for the replacement of Cranston Viaduct. Work included the replacement of the existing steel multi-span structure in its entirety with a three-span, pre-cast, pre-stressed New England Bulb Tee superstructure supported on reinforced concrete abutments and piers founded on drilled shafts. Three-foot diameter (0.92m), 50 feet (15.25m) long shafts were recommended for the abutments foundation. Monoshafts were recommended as a foundation support for the viaduct piers to accommodate the limited available footage area. Due to the large axial and lateral loads, nine-foot diameter(2.75m), 130 feet (39.6m) long shafts were recommended for the pier foundations.

The design of the drilled shafts follows the guidelines outlined in the latest Federal Highway Administration (FHWA) publication and in AASHTO design manual. Both side shear resistance and end bearing in the sand stratum are utilized in the analysis to develop the axial load. Factors of safety are computed for two loading conditions: the static case (summation of dead load and live load) and the seismic case (summation of dead load and earthquake load).

As required by the FHWA design guidelines, a nine-foot (2.75m) diameter trial shaft was drilled at the site to establish the proper construction procedures. The close proximity to Amtrak mainline tracks provided a unique challenge in the construction of the drilled shaft. Permanent steel casing was installed for a specified depth to protect the tracks. In addition, a 3300-ton (29,357 kN) Osterberg load test was performed on the trial shaft to verify design assumptions and confirm the shaft capacity. The results of the test were utilized to establish load transfer characteristics within the sand and to develop foundation design parameters to be used for the production shafts.

It is believed that the results and finding presented in this paper will benefit both the design and construction industries by providing a comparison between geotechnical design assumptions and actual field measurements. Figures 1 and 2 show the construction of the southbound of the Viaduct.

Figure 1:
SB Viaduct Pier 2.
Looking Southeast at Pier 2

Figure 2: *(left)*
SB Viaduct North Abutment. Looking South from North Approach behind North Abutment

Figure 3: *(right)*
9-foot (2.75m) diameter auger used to drill shafts.

SUBSURFACE CONDITIONS

A brief description of the soil strata encountered during the subsurface investigation program is summarized below.

Fill - This layer consists of miscellaneous materials including Bricks, Sand, Gravel, Wood, and Cobbles. The thickness of this layer varies from nine feet (2.75m) to twenty-one feet (6.41m). The fill soils are generally unsatisfactory as bearing foundations because of variations in their composition. *Note: Below the fill layer a concrete slab was encountered during the drilling of two borings, with a thickness of 3 feet (.92m) at one location and 4 feet (1.22m) at the other location.*

Brown to Gray Fine to Medium SAND, Trace of Fine Gravel, and Little Silt - The Sand stratum was encountered below the fill layer and extended to the bottom of the borings located at a depth of 140 feet (42.7m) below existing ground surface. The top thirty feet (9.15m) of this stratum consists of brown to gray, fine to medium, SAND with little Silt and trace fine to medium Gravel. The average Standard Penetration Test N-values recorded during the investigation ranged from 22 to 35 blows per foot (0.305m). Design values for angle of internal friction varied between 32 and 35 degrees.

Brown, Compact Fine SAND with Little Silt- The bottom fifty feet (15.25m) of the Sand stratum consists of brown compact fine SAND with little Silt. The average Standard Penetration Test N-values recorded during the investigation ranged from 27 to 55 blows per foot (0.305m). Design values for angle of internal friction varied between 33 and 36 degrees.

GROUNDWATER CONDITION

The groundwater table was recorded at a depth varying from 15 to 20 feet (4.6 to 6.1m) during the subsurface investigation. The presence of the groundwater would necessitate the selection of the appropriate drilling equipment and type of slurry to be used to drill through the saturated sand.

DRILLED SHAFT CONSTRUCTION

Case Foundation Company began excavating the dedicated test shaft on January 29, 2002 and performed the final cleanout and concreting on February 1, 2002. The ground surface at the shaft location is at elevation +74.0ft(22.55m). A 126"(3.2m) diameter temporary casing was installed for a depth of 8 ft(2.44m) to protect the ground around the hole from caving. A 114"(2.89m) diameter permanent casing was installed to a depth of 30ft (9.15m). Polymer slurry was introduced into the hole and a 108"(2.75m) diameter shaft was drilled to an elevation –59.5ft(-18.13m) bringing the total shaft length to 133.5ft(40.72m). The shaft was drilled with an auger and the base was cleaned with a cleanout bucket. After cleaning the base, the reinforced cage with attached O-cell assembly was inserted into the excavation. The section of the cage below the O-cell assembly consisted of 9 steel CSL (cross-hole sonic logging) pipes with two hoops for attaching instrumentation. The reinforcing cage was subsequently lowered further before being temporarily supported by beams and chains at the top of the shaft. Concrete was then delivered by pumping into a 10"(0.25m) O.D. tremie pipe into the base of the shaft until the top of the concrete reached an elevation of 70.70ft (21.54m).

OSTERBERG LOAD TEST SET-UP

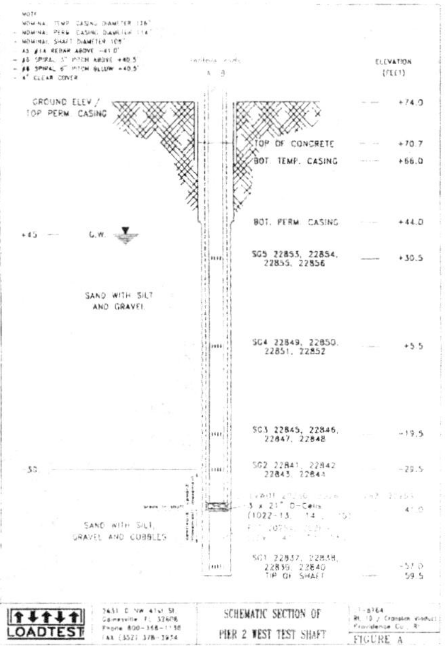

Figure 4: Schematic Diagram of Load Test Instrumentation Set-up of Test Shaft

Shaft Instrumentation: Test shaft instrumentation and assembly was carried out under the direction of Mr. John F. Usab and Mr. William G. Ryan of LOADTEST, Inc. on January 29 through February 1, 2002. The loading assembly consisted of three 21-inch (0.53m) O-cells™ located 18.5 feet (5.64m) above the tip of shaft.

A point was determined along the length of the shaft where if the O-cell expanded vertically, applying equal forces both in the upward and downward directions, the resisting force of the side shear resistance would equal the resisting force of the end bearing. In the test shaft, where the ground elevation was 74.0ft (22.56m) and the tip elevation was -59.5ft (-18.1m), this point of equilibrium was estimated to be elevation -41.0ft (-12.5m). Consequently, as shown in Figure 4 above this was where three 21" (0.53m) O-cells were installed, 18.5ft (5.64m) above the tip of the shaft. The instrumentation set-up was as described in the excerpt from the Contractor's report:

> O-cell™ testing instrumentation included four Linear Vibrating Wire Displacement Transducers (LVWDT) – (Geokon Model 4450 series) positioned between the lower and upper plates of the O-cell™ assembly to measure expansion. Two telltale casings were attached to the reinforcing cage, diametrically opposed, extending between the top of the O-cell™ assembly to the top of concrete and used to measure upper shaft compression. Compression of the pile below the O-cell™ assembly was measured by one section of Embedded Compression Telltales (ECTs), consisting of ¼ inch telltale rods in ½ inch steel casing, with an LVWDT attached.
>
> Strain gages were used to assess the side shear load transfer of the shaft above and below the Osterberg Cell assembly. One level of four sister bar vibrating wire strain gages (Geokon Model 4911 Series) was installed, 90 degrees apart, in the shaft below the base of the O-cell™ assembly and four levels of four were installed in the shaft above it.
>
> Two lengths of PVC pipe were also installed, extending from the top of the shaft to the top of the bottom plate, to vent the break in the shaft formed by the expansion of the O-cells™. In addition, steel CSL pipes extended through the fracture plane (fitted with a PVC section) and also served as vent pipes. All pipes were full of water prior to the start of the test.

The load was applied in increments using the Quick Load Test Method for Individual Piles (ASTM D1143 Standard Test Method for Piles Under Static Axial Load), holding each successive load increment constant for eight minutes by manually adjusting the O-cells™ pressure. The data logger automatically recorded the instrument readings every 30 seconds, but herein is reported only 1, 2, 4, and 8-minute readings during each increment of maintained load. Figures 4 through 8 describe the load test set-up.

Figure 5:
Test Set-up and Data Logging

Figure 6:
Test Site in Relation to Existing Viaduct

Figure 7:
Load Test Set-up

DRILLED SHAFTS DESIGN

The axial capacity of the drilled shafts was computed utilizing the American Association of State Highway and Transportation Officials (AASHTO) Guidelines and FHWA Design Manual for drilled shafts in cohesionless soils.

$$Q_{ult} = Q_s + Q_T - W \quad \ldots\ldots\ldots\ldots\ldots(1)$$

where,

Q_{ult} is the ultimate axial load capacity
Q_s is the ultimate side shear resistance
Q_T is the ultimate tip resistance
W is the weight of shaft

Figure 8:
Instrumentation measures Shaft Movement Relative to the Stationary Beam.

The allowable axial load (Q_{all}) was determined as:

$$Q_{all} = Q_{ult} / FS \quad \ldots\ldots\ldots\ldots\ldots(2)$$

where,

FS is the factor of safety

Side Shear Resistance

The ultimate side shear resistance of axially loaded drilled shafts in cohesionless soils was determined using the following:

$$Q_s = \pi B \sum_{i=1}^{N} \gamma'_i \, z_i \beta_i \Delta z_i \qquad \ldots\ldots(3)$$

where,

B is the shaft diameter in feet (ft)
γ' is the effective soil unit weight in ith interval, kips per cubic foot (kcf)
z_i is the depth to midpoint of ith interval in feet (ft)
β_i is the load transfer factor in the ith interval
Δz_i is the ith increment of shaft length in feet (ft)

The Load transfer factor, β_i can be calculated using:

$$\beta_i = 1.5 - 0.135\sqrt{z_i} \;, (1.2 > \beta_i > 0.25) \qquad \ldots\ldots(4)$$

Tip Resistance

The ultimate tip resistance was computed using the following equation:

$$Q_T = q_T A_t \qquad \ldots\ldots(5)$$

where,

q_T is the ultimate unit tip resistance for shafts
A_t is the area of shaft tip

The value of q_T can be determined from the results of standard penetration tests using uncorrected blow count readings as recommended by AASHTO.

Recommended values of q_T* for Estimation of Drilled Shaft Tip Resistance in Cohesionless Soil after Reese O'Neill(1988)

Standard Penetration Resistance (N) (Blows/foot) (uncorrected)	Value of q_T (ksf)
0 to 75	1.20 N
Above 75	90

* Ultimate value or value at settlement of 5 percent of base diameter.

OSTERBERG LOAD TEST RESULTS

The results of the Osterberg load test are presented as follows:

Figure 9 – Osterberg Cell Load Movement Curves

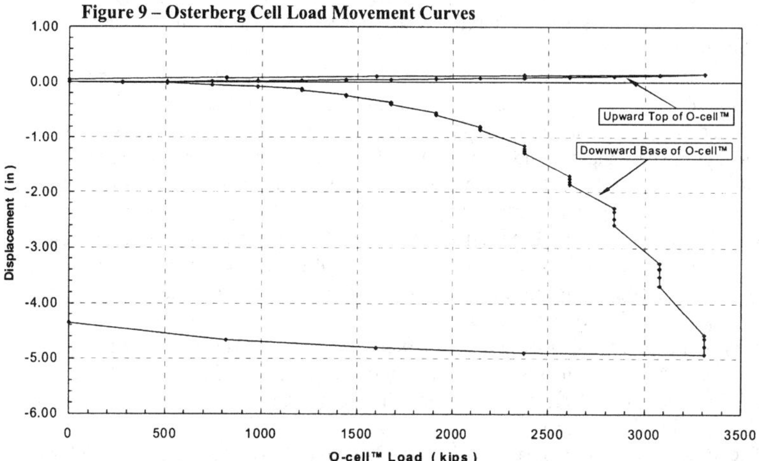

Figure 9 shows both the upward and downward movement under different loads. The maximum load applied in each direction was 3308 kips (14,714kN). Maximum upward movement was 0.15" (3.811mm) while the maximum downward movement reached 4.93" (125mm).

Figure 10 – Strain Gage Distribution Curves

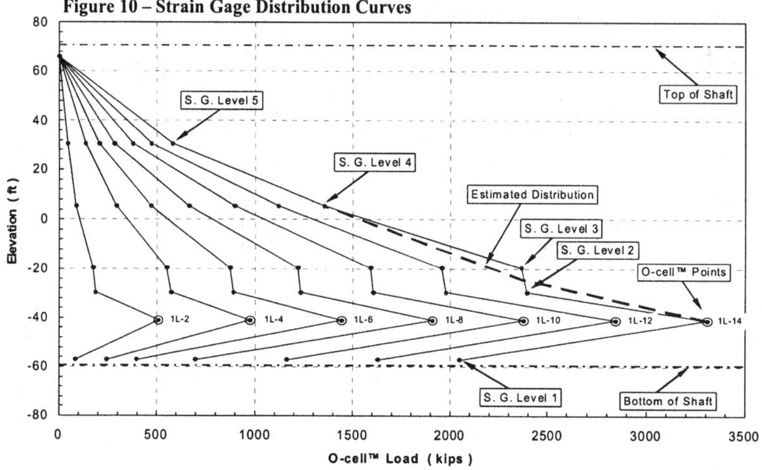

Figure 10 illustrates the side shear resistance distribution along the shaft length.

COMPARISON BETWEEN LOAD TEST RESULTS AND DESIGN VALUES

The Osterberg Load Test provides the opportunity to not only confirm the design assumptions used to calculate the capacity of the shaft, but to determine the distribution of side shear resistance dissipation over the length of the shaft and to isolate the end-bearing capacity.

In this paper the foundation of the southbound of the Viaduct will be discussed. At the southbound, there are 4 foundation piers: Pier 1-East, Pier 1-West, Pier 2-East, and Pier 2-West. Each pier is supported on a single 9-foot (2.75m) diameter shaft.

Each pier was reviewed and compared to the load test results. The graphs in Figures 11 through 18 illustrate the design unit side shear resistance of each shaft versus the mobilized unit side shear resistance under 0.1" (2.54mm) of movement and the extrapolated data of side shear resistance under 0.5" (12.71mm) of movement and the cumulative design side shear resistance versus the cumulative friction mobilized under 0.1" (2.54mm) of movement and the extrapolated data of side shear resistance under 0.5" (12.71mm) of movement.

Measurements recorded from the Osterberg load test indicates that the ultimate unit side shear resistance in sand increased with depth from 1.5 to 1.8 ksf (71.82 to 86.18kPa) at 0.5" (12.71mm) movement. These numbers are almost identical to the values used in the design.

The downward force measured in the Osterberg Load Test was resisted by a combination of side shear resistance and tip resistance of the shaft. The measured ultimate unit tip resistance was 12 ksf (574.6kPa) for 0.5" (12.71mm) movement. During the design, the computed ultimate tip resistance was reduced to control settlement. This reduction is recommended by the FHWA and AASHTO design guidelines. By doing so, the computed design value of unit tip resistance was very close to actual measurement recorded from load test in the field.

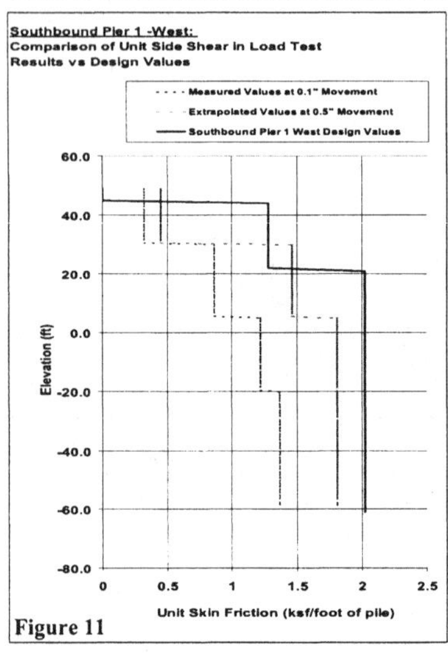

Figure 11

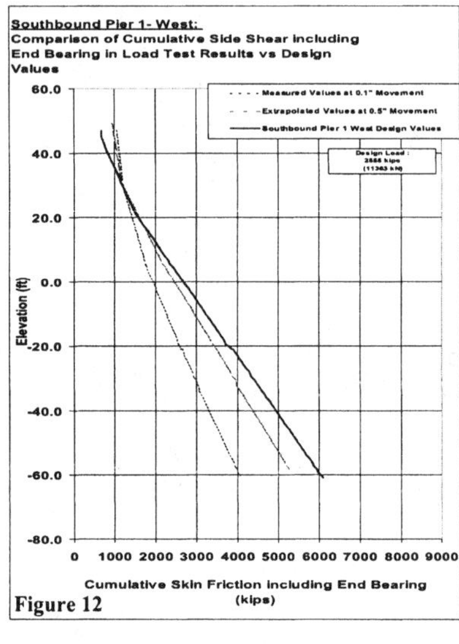

Figure 12

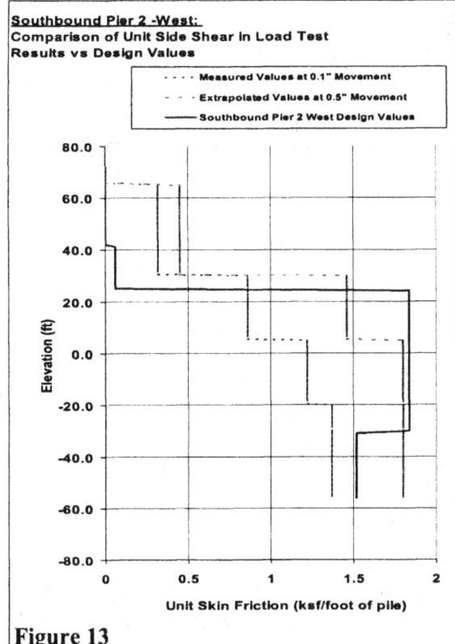

Figure 13

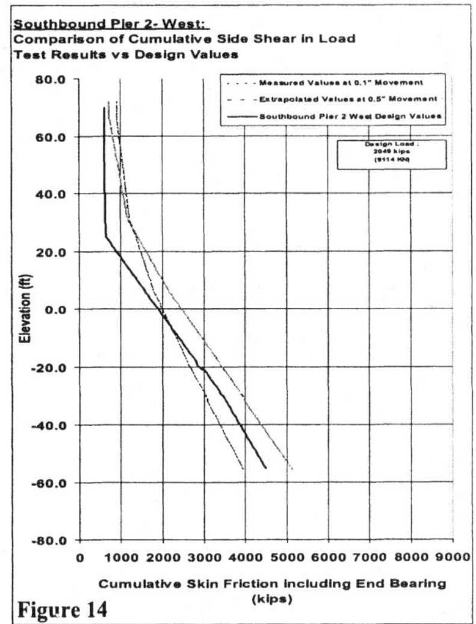

Figure 14

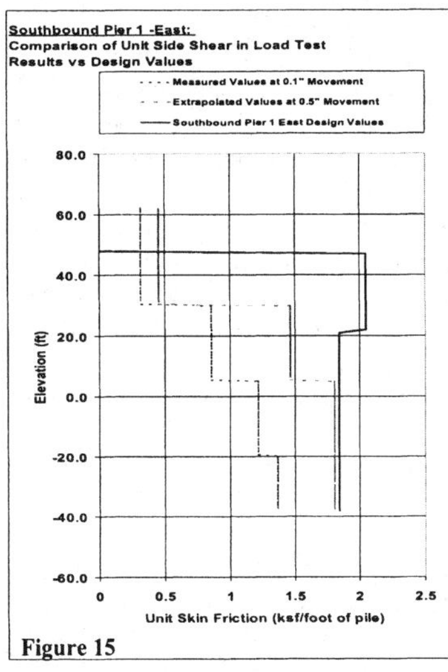

Figure 15

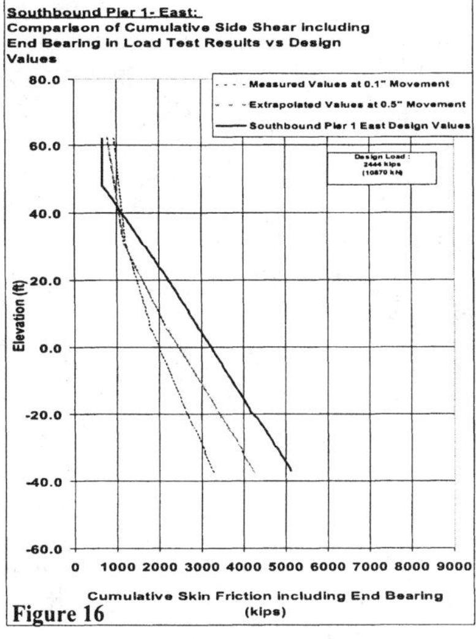

Figure 16

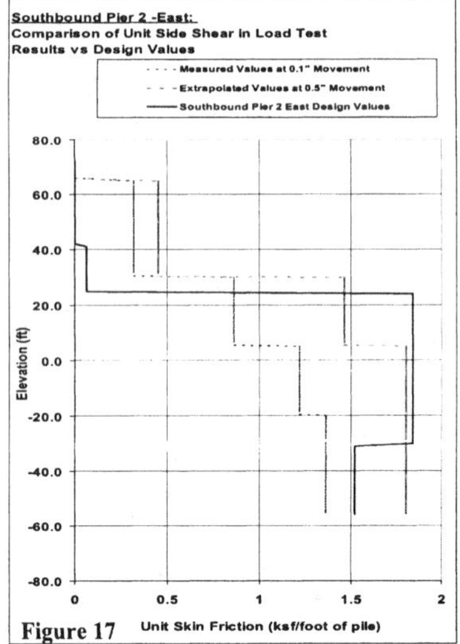

Figure 17

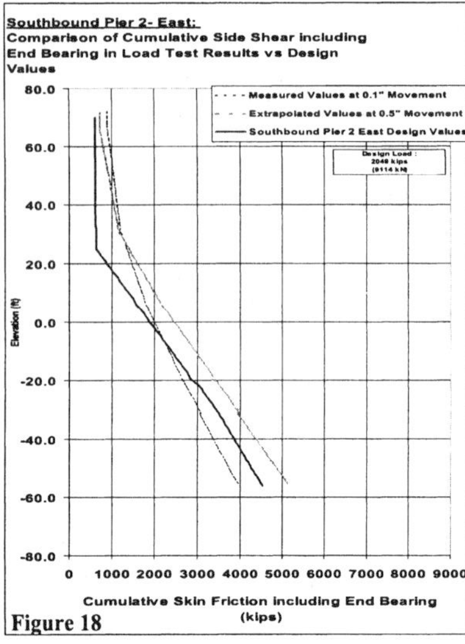

Figure 18

CONCLUSIONS

Analytical and empirical design methods are available to determine side shear and tip resistances of drilled shafts in cohesionless soils. These methods are dependent on: soil unit weight and density, depth of groundwater table, shaft length, and method of construction. The Engineer has to verify the design assumptions in the field and ensure that the shaft length is adequate to carry the design load.

The Osterberg load test provided a unique opportunity to separate the tip resistance from side shear resistance along the shaft length. Based on the test results, values of unit side shear and tip resistance in sand measured in the load test were very close to the design values in this project. Therefore, the Engineer was able to confirm the shaft length.

The design equations for drilled shafts in cohesionless soils, adopted by the Federal Highway Administration and AASHTO, proved to provide a good design tool in this project and compared very well with actual field measurements.

ACKNOWLEDGMENTS

The Rhode Island Department of Transportation owns the project described in this paper. The vision and commitment provided by the Department made it possible to develop this unique project from a concept to reality. The authors would like to express their gratitude for the support and encouragement provided by Mr. Kazem Farhoumand, P.E. Deputy Chief Engineer, Ms. Georgette Chahine, P.E. Principal Engineer, and Mr. Andranik Tahmassian, Principal Engineer.

Mr. Peter Osborn of the Federal Highway Administration provided valuable and constructive ideas during the design review process. The implementation of the Federal Highway Administration Design Guidelines and their State-of-the-Art construction specifications were important for the success of this project.

We would also like to sincerely thank Robert Simpson and John (Rick) Usab of LOADTEST, INC. for their outstanding efforts in performing the Osterberg load test at the site.

Firooz Panah and Ray Palmer, lead bridge engineers with DMJM+HARRIS were very instrumental in providing the structural loads during the design of this challenging project.

REFERENCES

FEDERAL HIGHWAY ADMINISTRATION (FHWA) MANUAL, 1999. Drilled Shafts: Construction Procedures and Design Methods. Volume II, Chapter 14. Prepared by Michael W. O'Neill and Lymon C. Reese.

AMERICAN ASSOCIATION OF STATE HIGHWAY AND TRANSPORTATION OFFICALS (AASHTO), 1996. Standard Specifications for Highway Bridges, Sixteenth Edition. Section 4: Foundations.

LOADTEST, INC. of Gainesville, Florida. Data Report on Drilled Shaft Load Testing (Osterberg Method), Pier 2 West Test Shaft – Rote 10 / Cranston Viaduct, Providence Rhode Island, February 2002.

OSTERBERG, J. O., 1999. What has been learned about drilled shafts from the Osterberg load test. Deep Foundations Institute, Annual Meeting.

OSTERBERG, J. O., 1998. The Osterberg load test method for drilled shafts, and driven piles- the first ten years. Proceedings of the Seventh International Conference on Piling and Deep Foundations, Vienna, Austria. Published by the Deep Foundations Institute.

SCHMERTMANN, J. H., HAYES, J. A., and MOLNIT, T., 1998. O–Cell testing case histories demonstrate the importance of bored piles (drilled shafts) construction techniques. Proceedings of the Fourth International Conference on Case Histories in Geotechnical Engineering. University of Missouri-Rolla, Rolla, Missouri.

SCHMERTMANN, J. H., HAYES, J. A., 1997. The Osterberg cell and bored pile testing, a symbiosis. Proceedings of the Third International Geotechnical Engineering Conference, Cairo, Egypt.

A DEEP FOUNDATION SURPRISE, ENGINEERED RESPONSE AND FOUNDATION PERFORMANCE

Clyde N. Baker, Jr.,[1] P.E, S.E., Tony A. Kiefer[2], P.E., William H. Walton[3], P.E., S.E. Charles E. Anderson[4], S.E.

ABSTRACT: The paper describes the foundation design and construction history of a 19-story residential high rise on deep foundations on the south side of downtown Chicago, including: 1) a value engineering foundation design innovation; 2) the changes in anticipated conditions actually encountered during construction; 3) the additional subsurface exploration and special testing and new engineering analyses performed to better predict ultimate foundation performance; and 4) the foundation design modification required to make the predicted foundation performance acceptable. The paper also presents the actual measured foundation performances, compares prediction with performance and offers possible explanations where there are significant differences.

BACKGROUND

A 19-story residential high rise (5 levels of parking and 14 levels of condominiums) located on Chicago's near south side was originally designed supported on conventional steel encased caisson foundations socketed 1 foot into sound dolomite at the Chicago Code allowable design bearing of 100 tons per square foot (tsf). STS Consultants, Ltd. (STS) was retained to determine whether an allowable bearing pressure of approximately 50 tsf could be confirmed at the interface of the underlying dolomite bedrock as an alternative design proposed by one of the bidding foundation contractors. The concept of straight shaft caissons constructed under polymer slurry (without full length permanent steel casing) sitting on the surface of the underlying bedrock rather than cased and socketed into the bedrock was a value engineering

[1]Senior Principal Engineer, STS Consultants, Ltd., 750 Corporate Woods Parkway, Vernon Hills, IL 60061, baker@stsltd.com
[2]Senior Project Engineer
[3]Principal Engineer
[4]Principal Engineer, C.E. Anderson and Associates, 175 N. Franklin, Chicago, IL 60606

alternative that required confirmation. STS performed supplementary explorations with pressuremeter testing and confirmed 50 tsf at the bedrock interface and also confirmed a value of 47 tons per square foot (tsf) sitting on boulder till just above the bedrock. At a meeting with the City, it was agreed that the design and construction intent would be to sit on the bedrock surface even though the actual bearing pressure on most of the caissons was less than the 47 tsf test value for sitting on boulder till above the bedrock. The caisson contractor felt confident that using his largest and most powerful equipment available locally that drilling to refusal with his earth auger would reach the surface of bedrock.

The first caisson was inspected personally by the lead writer to see that the procedures followed did result in reaching the rock interface. The first two caissons drilled were J-5 and J-3 and they both encountered auger refusal in the elevation range of -66 to -67 Chicago City Datum (CCD). while the closest boring (B-7) indicated the surface of weathered rock at -69 CCD. It was clear that drilling to refusal with the earth auger was not going to be sufficient to reach the bedrock surface. The contractor agreed to expand their procedures, and after reaching refusal with the earth auger, used a rock auger with special carbide rock bits to see if they could penetrate deeper. By drilling an additional 15 to 20 minutes with the rock auger, they were able to penetrate to the weathered rock surface at approximate elevation -69 CCD. Based on this experience, the contractor agreed to follow up their earth auger refusal by drilling with their rock bit for an additional 20 minutes at all production caissons, and this was the procedure that was used for the remainder of the project. The procedure appeared to be successful in getting the caisson bottoms at or very close to the bedrock interface where the caissons were close to borings (bedrock interface of -68 to -72 CCD) with the exception of the central building core area where there were no pre-construction test borings with which to correlate.

Caisson excavation bottom cleaning was achieved with a cleanout bucket with bottom cleanliness checked with a 10-pound steel bar attached to a cable with taped depth marks. The cleanliness was "felt" by lifting the bar several inches and letting it drop on the bottom. If it felt sticky, more cleanup was necessary. However, for most of the caissons, the best that could be achieved with the cleanout bucket was "slight" stickiness.

INITIAL POST CONSTRUCTION TESTING PROGRAM

Project specifications called for sonic logging of all caissons and coring of two representative core caissons to confirm caissons were on rock. Four sonic logging tubes were attached equal distance around the full length rebar cages used in the caissons. The results of the sonic logging tests performed tube to tube around the caisson and across the caisson confirmed the continuity and integrity of the tremie-poured concrete. Since the caissons were constructed under polymer slurry, sonic logging was performed to confirm that no soil cave-ins occurred during concreting. This was further confirmed by the plots made of concrete volume placed versus depth.

Caissons D.2-3.2 and E.8-3.8 in the highly loaded core area were selected for coring and testing because they were two of the shallowest caissons. One hundred percent (100%) core recovery was achieved in the concrete which appeared sound with no evidence of contamination. Beneath the caissons, very dense clayey sandy silt with rock layers (could also be boulders) was noted with Standard Penetration Test (SPT) values in excess of 100 blows per foot. Because of the rock fragments or boulders, pressuremeter testing was difficult with only one successful test completed in the 3-foot thick zone underlying the caissons which indicated a creep pressure (P_f) of 39 tsf at caisson location E.8-3.8. A pressuremeter test performed 4 feet below the caisson bottom at E.8-3.8 at the rock interface was loaded to 39 tsf without approaching the creep pressure and with zero movement indicating very high capacity and probable bedrock.

Based on the pressuremeter test results and the SPT tests, the bearing material could be approved for 40 tsf but not for the design value of 50 tsf. Since it was anticipated that the caissons would be resting at the rock interface rather than 3 to 4 feet above the rock interface, it was decided to expand the testing program as described in the following sections. Figure 1 depicts the original boring location diagram and the as-built caisson locations and bottom elevations. Figure 2 shows the geologic profile

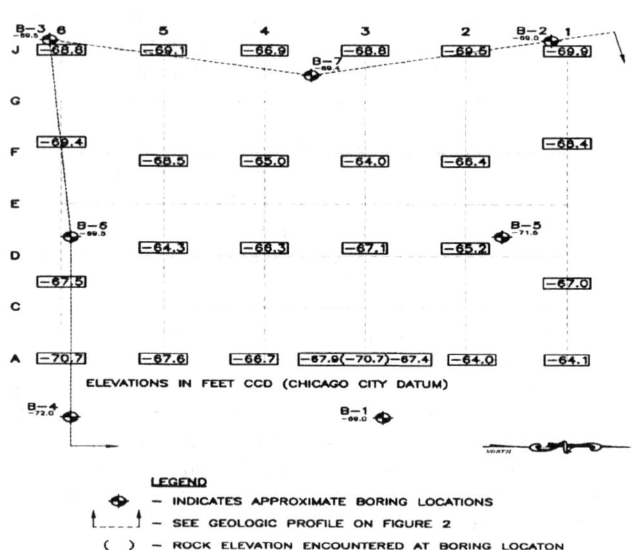

FIG. 1. Soil Boring Location Plan

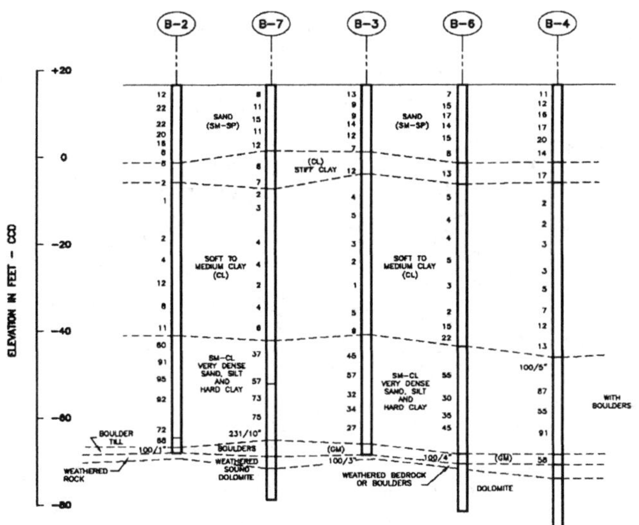

FIG. 2. Geologic Profile

ADDITIONAL POST CONSTRUCTION TESTING PROGRAM

In addition to the initial two caissons cored, four additional caissons were cored (J-2, D.2-4.8, D.2-2.2 and E.8-3.1) and seven additional borings were performed alongside of caissons to speed up the pressuremeter testing process below the caisson bottoms. The pressuremeter data is summarized and presented on Figures 3 and 4.

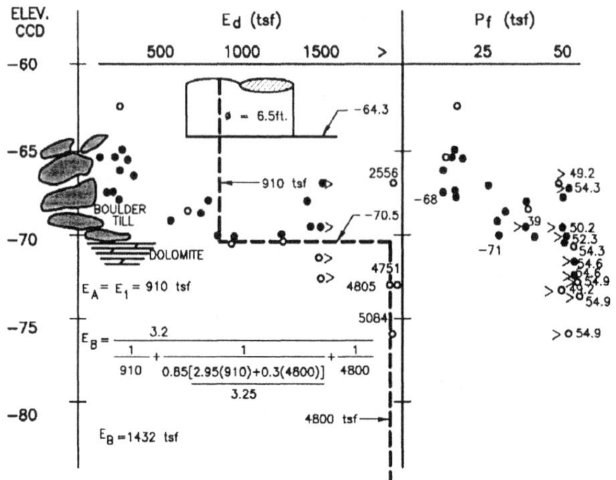

LEGEND

- • POST-CONSTRUCTION PMT
- o PRE-CONSTRUCTION PMT

E_d – DEFORMATION MODULUS

P_f – CREEP PRESSURE

NOTES:
BEARING −64.3 ft. CCD
P = 37.8 tsf
α = 0.34

$$\delta = \frac{1.33\ (37.8)\ (3.25)^{0.34}}{3\ (1432)} + \frac{0.34\ (37.8)\ (3.25)}{4.5\ (910)} = 0.017' + 0.010' = 0.028' = 0.332''$$

k = 1659 x 2/(0.136" + 0.332") = 7,090 k/in.
USE k = 7,100 k/in.

FIG. 3. Representative Spring Development Based on a Two-Layer Soil/Rock Model

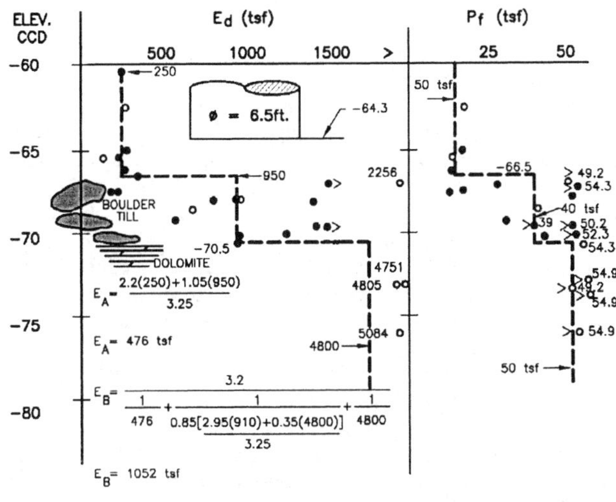

LEGEND

- ● POST-CONSTRUCTION PMT
- o PRE-CONSTRUCTION PMT

E_d DEFORMATION MODULUS

P_f CREEP PRESSURE

NOTES:

BEARING −64.3 ft. CCD

P = 32.3 tsf (AT TIP)

α = 0.34

$$\delta = \frac{1.33(32.3)(3.25)^{0.34}}{3(1052)} + \frac{0.34(32.3)(3.25)}{4.5(476)}$$

$$\delta = 0.020 + 0.017 = 0.037' = 0.444''$$

k= 1659x2/(0.136+0.444)=5726 k/in. USE 5,700 k/in.

FIG. 4. Representative Spring Development Based on a Three-Layer Soil/Rock Model

DEVELOPMENT OF CAISSON SPRING MODEL FOR USE IN STRUCTURAL ANALYSIS

Interpretation of Data

When first looking at the data, it is obvious that there is a very large range in both pressuremeter data and SPT data ranging from 30 to more than 300 blows per foot and pressuremeter creep pressure (P_f) ranging from as low as 15 tsf to beyond the capacity of the equipment (approximately 55 tsf). It is difficult to see how the low test values could be representative of material dense enough to cause refusal for a powerful caisson drill rig. The answer may lie with the influence of boulders in the till above rock. While boulders are only occasionally mentioned on the boring logs, they can be much more pervasive when considering 4.5 to 7-foot diameter auger rig equipment. To illustrate this point, consider that a 2-foot diameter boulder on 6-foot centers across the site would mean that approximately 9% of the area would encounter boulders or approximately one out of eleven 3-inch diameter borings. On the other hand, 100% of caissons with diameters in the 4.5 to 7 foot range would encounter boulders. Considering the low compressibility of boulders relative to adjacent silt, it is easy to see how it is possible to have lower blow count silt adjacent to a boulder compared to silt away from the boulder. The silt away from the boulder would feel the full weight of past overlying glaciers whereas, the silt immediately adjacent to the boulder would be protected from the weight of the glacier as the glacier load was transferred through boulder to boulder contact bypassing the more compressible silt.

With this geologic history in mind and considering the large difference between small scale tests in a 3 inch borehole compared to large diameter caissons, it is reasonable to look at average test data and properties rather than any one specific test.

Possible Multiple Layer Soil-Rock Foundation Model

Since all of the caissons reached refusal in the elevation range of -64 to -71 CCD, one approach would be to average all of the test data in the entire zone from -64 to –71 CCD into a two layer model of soil over rock. Another possible approach is to break up the soil deposit into two layers since there appears to be a break in the data about -66.5 CCD. This three layer approach is most critical for the shallowest caissons supported above elevation -66.5. To see what difference it makes, calculations were performed to estimate spring values using both of these soil models as shown in Figures 3 and 4. To check the worst case scenario, two cases were included where additional soil compression was added in to soften the springs. This is the approach recommended by Menard when using his modulus averaging procedure where there is a soft layer between two hard layers.

In addition to the soil model utilizing average test properties in the different layers, it was necessary to estimate how much load was carried in the caisson shaft in side friction since any such load will reduce the load at the tip and the amount of settlement predicted at the tip based on that load. Initially, analyses were made based on maximum estimated values extrapolated from other nearby projects where side friction measurements were available. These maximum side friction values were divided by a factor of safety of 2 in the calculation for the load assumed acting at the shaft base in order to be conservative and allow for some long term creep effects. The spring value was then determined by predicting the settlement of the caisson shaft following the Menard pressuremeter rules that have been successfully used by STS in the Chicago area for the past 30 years. A representative basic calculation is shown on the same graph that shows a plot of the test data and caisson foundation (Figures 3 and 4).

Grade Beam Springs

In addition to the springs under the actual caissons, there will also be resistance and therefore additional support springs as the grade beams try to settle with the caissons. Calculations were made estimating this effect but the influence was less than 10% of the total load and was not considered significant.

DYNAMIC LOAD TESTING TO CALIBRATE SPRINGS AND SIDE FRICTION DEVELOPMENT

Because of the importance of the friction value in determining realistic springs and also in determining what the real factors of safety are with regard to bearing capacity, it was decided to perform several representative full scale dynamic load tests on existing caissons. This procedure had been used successfully on 22-inch diameter auger cast piles at a project approximately three blocks from the site with a similar soil profile. In a meeting with City officials to discuss the planned testing program and use of the dynamic load test results, it was agreed that a factor of safety of 3 would be applied to dead load considerations and a factor of safety of 2 to total load including dead and live load considerations.

The dynamic load tests were performed by Goble Rausche Likins & Associates, Inc. (GRL) the developer of the pile driving analyzer and the CAPWAP computer program. Their equipment and technology was used in interpreting the dynamic results and constructing a simulated static load test.

Interpretation of Dynamic Load Test Results

The test results which are summarized on Figure 9 can be used to calculate a static spring by taking the calculated static capacity developed during the test and dividing it by the total deflection. These calculated springs range from 7,500 kips per inch for the 4.5 foot diameter A-6 shaft to an average of approximately 12,900 kips per inch for the 6.5 foot diameter shafts at caissons F.2-6 and C.5-6. These values plot very well on the spring charts developed using the two soil models developed for the project and not including the additional Menard soft clay layer compression thereby confirming the reasonableness of the assumptions utilized (Figures 5 and 6). Leaving off the additional Menard soft layer compression is reasonable given the dynamic short term loading nature of the test.

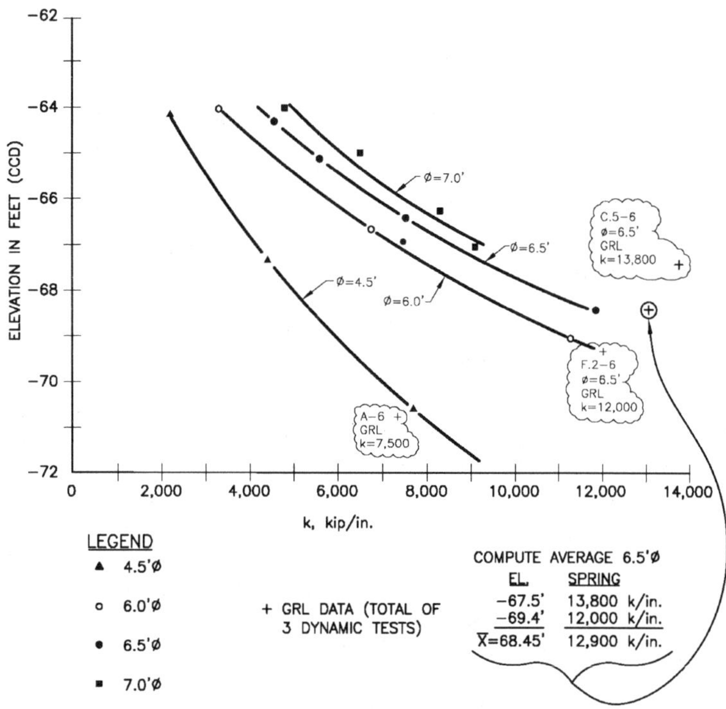

FIG. 5. Load Test Calibration With Model 250/950/4800

USING FS=2 ON SIDE FRICTION FROM GRL TESTS FROM 1845 S. MICHIGAN AND PRAIRIE HOUSE (AVERAGE OF 10 TESTS).
USING 50 tsf AT TOP OF CAISSON. AVERAGE 66 FT. LENGTH. WE ADOPTED $E_d=910/E_m$ MODEL. WITH COMPRESSION OF BOULDER TILL METHOD, $\alpha=0.34$, $E_m=4,800$ tsf.

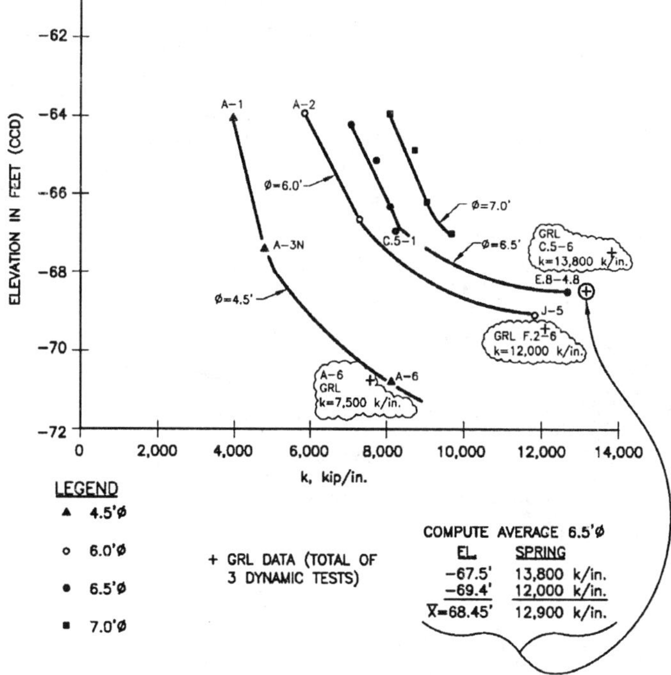

FIG. 6 . Load Test Calibration With Model 910/4800

Development of Plot of Friction vs. Deflection

The dynamic load test results cannot be used to predict the ultimate capacity for the test shafts since the forces used were not sufficient to significantly move the tips. These test results can also not be used to compute the ultimate shaft friction that might be developed since the entire shaft was not moved sufficiently to develop the maximum side friction. However, the results can be utilized to generate a load deflection plot when correlated with the dynamic load test results from the nearby 22-inch diameter auger cast piles project where it was possible to move the piles sufficiently to develop maximum side friction in some cases. A plot of friction versus deflection utilizing the available data is shown in Figure 7.

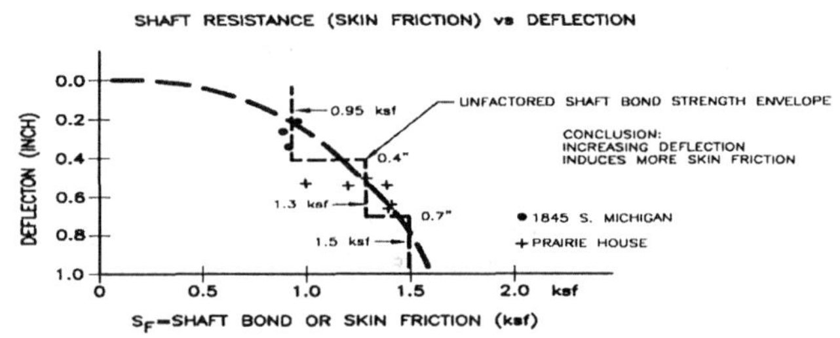

CAISSON NO.	DYNAMIC FORCE (k) GRL 11/19/01	STATIC REACTION (k)/DEFLECTON ANDERSON (11/14/01)
A–6	2,500^k(1000^{k}SHAFT, 1450^{k}END)	>1,649^k/0.19" OK
C.5–6	2,900^k(1450^{k}SHAFT, 1450^{k}END)	>1,978^k/0.19" OK
F.2–6	3,000^k(1365^{k}SHAFT, 1645^{k}END)	>2,484^k/0.22" OK

	CAISSON	FS TEST	δ VERTICAL
	A–6	1.52	0.33" > 0.19"
SUMMARY:	C.5–6	1.47	0.21" > 0.19"
	F.2–6	1.21	0.25" > 0.22"

DEFORMATIONS DURING DYNAMIC TESTING MATCH PREDICTIONS

FIG. 7. Summary of Dynamic Load Test

PREDICTED PERFORMANCE

Analysis of Critical Caissons

The predicted deflection at each of the caissons utilizing the recommended springs was developed by the structural engineer using their building structure model for all of the various soil models. The average predicted settlement for the different cases is shown on Figure 8. A representative result from the cases studied is shown in Figure 9.

		AV. PREDICTED SETTLEMENT (FT)
CASE 1	3 LAYER MODEL, Ed OF 250/950/4800 TSF, MAXIMUM FRICTION FROM NEARBY DYNAMIC LOAD TEST ON AUGER CAST PILES AND F.S.=2, NO GRADE BEAM SPRINGS	
CASE 2 OLD	1 LAYER 910/4800+ ADDITIONAL CONSOLIDATION ADDED FOR CASE OF SOFT LAYER BETWEEN 2 HARDER LAYERS BEYOND THAT OBTAINED USING THE MENARD EFFECTIVE AVERAGING TECHNIQUE. (REF. 1 P-37). SAME FRICTION AND NO GRADE BEAM SPRINGS.	0.032
CASE 3	SAME AS 2 EXCEPT 3 LAYER SYSTEM USED OF 250/950/4800 TSF.	0.039
CASE 4	SAME AS 2 EXCEPT BASED ON REDUCED FRICTION USING SITE CORRELATED DYNAMIC LOAD TEST DATA AND NO ADDED CONSOLIDATION.	
CASE 5	SAME AS 3 EXCEPT USING REDUCED FRICTION VALUES FROM DYNAMIC LOAD TESTS AND NO ADDED CONSOLIDATION.	
CASE 6	SAME AS 5 BUT 26K/FT/IN SPRING INCLUDED UNDER GRADE BEAMS.	0.024
CASE 10	CASE 6 WITHOUT SOIL SPRING BELOW GRADE BEAM.	0.023
CASE 11	OLD CASE 2 BUT WITH SOIL SPRINGS BELOW GRADE BEAMS.	0.029
CASE 2 NEW	CASE 2 BUT MODIFIED FOR DOUBLE CAISSON AND GRADE BEAM AT GRID A-3.	0.031

FIG. 8. Foundation Springs for Different Soil Models

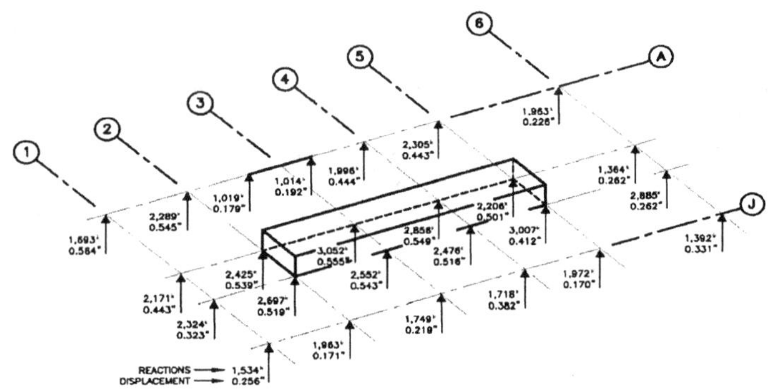

AVERAGE OF 16 POINTS WHERE SETTLEMENT WAS MEASURED = 0.031 ft. = 0.369 inches

FIG. 9. Case 2 New Settlement Prediction

Remedial Measures Where Predicted Differential Settlements Are Excessive

The structural engineer had set 0.4 inch as a maximum tolerable differential settlement between adjacent caissons. This value was exceeded only between A-2 and A-3 where worst case predicted deflection was 0.7 inch at caisson A-2 and only 0.2 inch at caisson A-3 where an unrelated caisson construction problem required two adjacent caissons be installed creating a relative hard spot. This was resolved at caisson A-2 where predicted settlement was reduced nearly in half by installing three rock socketed micropiles. The micropiles consisted of 7.6-inch diameter steel casing which was grouted with a centered #18 rebar. The micropiles were incorporated into an enlarged caisson cap.

Comparison of Observed Settlement versus Predicted Settlement

Settlement of the building during construction was measured at 16 points around the perimeter and core and the settlement data with time shown in Table 1. A plot of average building settlement versus floors constructed is shown on Figure 10. Since settlement is only recorded to 0.01 feet (1/8 inch), averaging the settlement at each date should provide a more reliable picture of settlement versus time. From the data it appears there are three distinct settlement trends; up to the 9th floor total settlement was only 1/4 inch which probably represents the initial load carried in friction; between the 9th and 12th floors there is a sharp increase in the amount of settlement occurring; and from the 12th floor to completion at the 19th floor there is a sharp reduction in the amount of additional settlement per floor load.

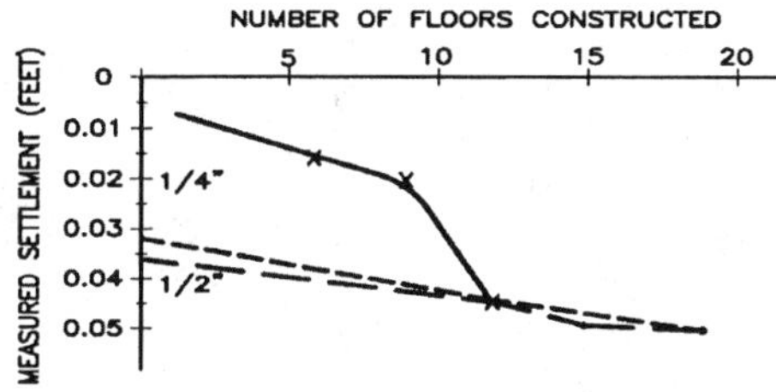

BASED ON AVERAGE MEASURED SETTLEMENT AND THE SHAPE OF THE FLOORS vs SETTLEMENT CURVE. IT APPEARS THAT THE INITIAL 9 FLOORS ARE CARRIED IN SIDE FRICTION (INITIAL 1/4" DEFLECTION). FROM 9–12 FLOORS, SIDE FRICTION IS EXCEEDED AND LOAD GOES TO BOTTOM AND THE THIN LAYER OF SOFTER MATERIAL ON BOTTOM THAT WAS NOT REMOVED BY ONE EYE BUCKET SQUEEZES AND COMPRESSES. FROM FLOORS 12 TO 19, THE LOAD IS TRANSFERRED FULLY TO THE UNDISTURBED BEARING MATERIAL. IF THIS EXPLANATION IS CORRECT, AND THE SOFTENED BOTTOM MATERIAL WERE COMPLETELY REMOVED, AVERAGE ACTUAL SETTLEMENT WOULD BE REDUCED BY 0.021 FT, YIELDING A TOTAL SETTLEMENT OF 0.031 FT, AGREEING BEST WITH THE CASE 2 PREDICTION BUT ONLY SLIGHTLY MORE THAN THE CASE 11 PREDICTION.

FIG. 10 Average Building Settlement Versus Number of Floors Constructed

Table 1 - Settlement Versus Time and Number of Floors Constructed

Pt.#	Elevation	6 Floors 4-24-02	9 Floors 5-23-02	12 Floors 6-11-02	15 Floors 6-25-02	19 Floors 7-25-02	19 Floors 9-05-02
S1	501.00	0.03	0.03	0.05	0.05	0.05	0.05
S2	501.00	0.04	0.04	0.06	0.06	0.06	0.06
S3	501.00	0.03	0.04	0.06	0.07	0.07	0.07
S4	501.00	0.03	0.04	0.06	0.06	0.06	0.06
S5	501.00	0.03	0.04	0.06	0.06	0.06	0.06
S6	501.00	0.03	0.04	0.05	0.05	0.05	0.05
S7	511.00	0.02	0.02	0.06	0.07	0.07	0.07
S8	511.00	0.00	0.00	0.05	0.05	0.05	0.05
S9	511.00	0.00	0.00	0.03	0.04	0.04	0.04
S10	511.00	0.00	0.01	0.04	0.04	0.04	0.04
S11	509.00	0.00	0.01	0.04	0.05	0.05	0.05
S12	507.00	0.01	0.01	0.04	0.05	0.05	0.05
S13	505.00	0.01	0.01	0.04	0.05	0.05	0.05
S14	505.00	n/a	n/a	n/a	n/a	n/a	n/a
S15	505.00	0.02	0.02	0.05	0.06	0.06	0.06
S16	505.00	0.02	0.02	0.05	0.06	0.06	0.06
Average (ft.)		0.016	0.021	0.046	0.051	0.052	0.052
Average (in.)		0.19	0.25	0.55	0.61	0.62	0.62

From this data it appears likely that there is a thin layer of softened material at the base of the shafts that could not be removed by the one eye cleanout bucket and that this material was squeezed and compressed after the side friction capacity of the caisson shaft was exceeded. By the time 12 floors of building dead load was in place, the layer was sufficiently compressed so that additional load was transferred to the undisturbed bearing material below (the material on which the caisson drilling rig met auger refusal). Assuming this settlement model is correct, if the compression in the thin softened spoil was not present (had been adequately removed by the cleanout procedures) average measured settlement of 5/8 inch (0.052 feet) would be reduced to 3/8 inch (0.031 feet) which is close to the Case 2 predicted settlement based on a 2-layer model with an average pressuremeter modulus value with additional consolidation added in, but no provision of grade beam springs (Figure 9). Inclusion of grade beam springs reduces the predicted settlement to slightly below the measured settlement less the estimated compression in the disturbed material at the base of the shafts.

CONCLUSIONS

Based on analysis of all the data collected, the following conclusions can be made:

1. The majority of the caissons reached refusal on a dense boulder till and did not reach the bedrock interface as anticipated in the design.
2. The density and stiffness of the boulder till deposit and the multitude of boulders in the deposit immediately overlying bedrock was sufficient to offer drilling refusal to large diameter augers even though powered with the largest and most powerful locally available caisson drilling rigs.
3. Using calculated springs based on average pressuremeter test properties in the zone between final caisson bottom and weathered rock surface and Menard settlement calculation procedures, and utilizing confirmed friction values with appropriate factors of safety results in predicted settlement values in line with average measured settlement provided estimated compression in the thin layer of disturbed material at the base of the shafts is subtracted from the total measured value.
4. The specified post construction test procedures were essential in determining that construction was not in accordance with design.
5. The criteria of drilling to earth auger refusal plus 20 minutes more with a rock auger, while not sufficient to assure bearing on rock, was adequate to assure bearing material was sufficient to carry the design loads in conjunction with the available side friction on this project.

RECOMMENDATIONS

1. More boring data to measure the distance to top of bedrock is recommended for design based on caissons sitting on rock so that more correlation of caisson construction to a site specific rock contour map is practical. This would help assure that the possibility of shafts stopped short on boulders or rock "floaters" could be avoided.
2. Air lift cleanout procedures, rather than bucket cleanout procedures are recommended if the potential for a thin layer of softened material (1 inch plus or minus that cannot be removed by a cleanout bucket) is to be avoided.

REFERENCES

Baker, C.N., "Measured vs. Predicted Long Term Load Distribution in Drilled Shafts", Proceedings of the 6th Annual Great Lakes Geotechnical and Environmental Conference, 1998.

Menard, L., "The Pressuremeter - Use and Interpretation", Sols Soils, 1975.

Baker, C.N. and Walton, W.H., "Caisson Confirmation Testing and Analysis for Residential High Rise Project at 1845 S. Michigan Avenue, Chicago, IL - STS Project 31721-B, November, 2001.

FOUNDATION SEISMIC RETROFIT OF BOEING FIELD CONTROL TOWER

Dominic M. Parmantier[1], Member, ASCE, Tom A. Armour[2], P.E., Member, ASCE, Bill J. Perkins[3], L.E.G., and Jim A. Sexton[4], P.E., Member, ASCE

ABSTRACT: The existing control tower at King County International Airport in Seattle, Washington, was severely damaged during the Nisqually Earthquake, February 28, 2001. As a result of the damage, the control tower was immediately closed and the existing timetable for a seismic retrofit of the control tower was expedited. The existing tower and associated support structures were founded on timber piles of unknown lengths (probably extending to depths between 25 and 35 feet); subsurface explorations indicated liquefiable soils to a depth of approximately 35 feet. The foundation retrofit included compaction grouting to densify the loose, liquefiable sands beneath the existing structures and the installation of drilled shafts adjacent to the tower to support the new structural steel bracing which was added to increase the resistance of the tower to overturning during the design earthquake. With the air traffic control operations being performed in a temporary facility, the urgency of performing the retrofit led the Federal Aviation Administration to negotiate a contract directly with the general contractor. This paper addresses the method of contract delivery, compaction grouting methods and results for liquefaction mitigation, and the construction of drilled shafts adjacent to an existing structure.

INTRODUCTION

The King County Boeing Field International Airport air traffic control tower in Seattle, Washington was constructed in the early 1960's. From its dedication in 1928, Boeing Field served as the primary airfield in the Seattle area for commercial aircraft

[1] Project Manger, Condon-Johnson & Associates, Inc., 651 Strander Blvd., Suite 110, Tukwila, WA 98188, dparmantier@condon-johnson.com
[2] President and CEO, Donald B. Murphy Contractors, Inc., 1220 S. 356th, Federal Way, WA 98063
[3] Project Manager, Shannon & Wilson, Inc., 400 North 34th Street, Suite 100, Seattle, WA 98103
[4] Project Manager, Donald B. Murphy Contractors, Inc., 1220 S. 356th, Federal Way, WA 98063

until the opening of Sea-Tac International Airport in the late 1940s. Since that time, Boeing Field has remained open and currently serves as a hub for several freight carriers, home to Boeing's only paint facility for commercial aircraft constructed in the Puget Sound region, and home to many charter services and private planes.

The FFA operates the air traffic control tower around the clock at Boeing Field. FAA facilities at the site include the air traffic control tower and a single story support building attached to the base of the tower. The control tower and the support building were constructed using reinforced concrete and CMU block construction techniques. Both buildings were supported by driven timber piles although there are no records indicating the actual length of these piles.

As part of a planned FAA upgrade of the facility, a comprehensive seismic evaluation of the facility was performed. In addition to reviewing geotechnical information from the original construction, addition borings and CPT's were performed to assist in the geotechnical analysis. Structural analysis involved developing accurate as-builts and evaluating the response and loading of the structure using the appropriate ground motion spectrum developed using the design event in combination with the site specific soil profile. This evaluation indicated that the site was susceptible to liquefaction and that the control tower exhibited an unacceptable factor of safety against overturning and/or structural failure of the tower in the design seismic event.

NISQUALLY EARTHQUAKE DAMAGE TO CONTROL TOWER

During the evaluation and design phase of the project, a 6.3 magnitude earthquake occurred approximately 70 miles SW of the King County Airport on February 27, 2001. The Nisqually Earthquake produced a peak ground acceleration of 0.15g at the airport.(need reference for this information) Earthquake induced liquefaction produced sand boils and voids beneath portions of the main runway. The ground motion also caused extensive damage to the glass in the cab of the control tower which was immediately taken out of service. Emergency remedial actions included the installation of a temporary control tower and the repair of the main runway using compaction grouting to densify the runway foundation soils in those areas where excessive sand blows and/or pavement displacement indicated the occurrence of liquefaction.

Following several weeks of emergency repairs to the runway, the airport resumed full operations utilizing the temporary air control tower. Subsequently, the pace of the seismic evaluation already underway was quickened so as to expedite repairs to the control tower and return the air traffic controllers to the permanent control tower.

SEISMIC EVALUATION AND RETROFIT DESIGN

Subsurface Conditions

The King County Airport is located in the Duwamish River valley. Historically the river channel meandered across the valley prior to flood control and river bank hardening. Subsequent to those efforts, previously unusable valley property was filled and converted to industrial uses. As a consequence, the site is generally underlain by fill consisting of approximately 15 ft of loose silty-sand which is underlain by approximately 65 to 70 ft of alluvial deposits consisting of loose to dense interbedded layers of clean sand, sitly sand, and sand. Based on the blow counts and CPT tip resistances measured during the site investigation, it was determined that the cleaner sands located from a depth of 15 to 35 ft were susceptible to liquefaction during the design seismic event.

The existing control tower was founded on driven timber piles. Each corner of the tower was supported by a pile cap and a cluster of sixteen piles. Although driving resistances from the original installation were available, there were no accurate records of the actual pile lengths. Based on construction practices and similar structures in the area with better records, it was believed that the timber piles were 25-ft to 35-ft in length. Without the actual pile length records, it was impossible to determine whether the piles were achieving their end bearing in the liquefiable sands or the non-liquefiable stiff silt below.

Even if the existing timber piles were founded below the depth of liquefiable soils, lateral loads in excess of the lateral restraint provided by the timber piles would be generated during the design event. The predicted lateral loads were the result of lateral spread in the liquefiable sands and seismic loading from the structure itself.

Evaluation Results

The results of the seismic analysis can be summarized below:

1) Liquefiable soils from 15 to 35 ft below ground surface.
2) Buildings founded on timber piles which were thought to be 25 to 30 ft in length. Therefore, existing timber piles were obtaining all bearing capacity in liquefiable sands.
3) Even with mitigation of the liquefaction potential in the soils, the existing foundation does not provide sufficient lateral capacity for the tower structure.
4) The existing tower structure has an unacceptable risk of collapse during the design seismic event.

Seismic Retrofit Design

After identifying the risks posed to the facility by the design seismic event, a remediation plan was conceived which consisted of three main goals:

1) Mitigate the potential for liquefaction of the soils between 15 and 35 ft underneath and around the tower and associated support building.
2) Reinforce the structure of the tower to prevent collapse in an earthquake.
3) Increase the lateral capacity of existing foundation.

Mitigation of Liquefaction Potential

In clean sands with less than 10% fines content, liquefaction potential is typically mitigated by densifying the sand with either vibro-compaction methods or compaction grouting methods. Liquefaction potential can also be mitigated by soil stabilization using permeation grouting or soil mixing although these methods are less common for reasons of economics. At the Boeing Field site, the need to densify under the existing timber piles and around the immediate perimeter of the structures dictated that compaction grouting be utilized. The installation of the injection pipes can be performed using small, rotary drilling rigs. The injection pipes can be installed on a batter to actually treat the soils beneath the perimeter of a structure. In addition, the operations building remained occupied and in service during construction so the ground improvement methods could not disturb the building occupants.

Since the site was confined, densification was desired under the tower and the building perimeter, and the operations building attached to the tower was to be occupied during construction, the use of compaction grouting methods to densify the liquefiable sands was selected. The small size of the drilling equipment used to install the casing would limit noise impact on the occupants as would the fact that no percussion or vibratory equipment was required.

By densifying the loose sands, the potential for loss of vertical capacity in the exsiting timber piles would be mitigated. Additionally, the densification would provided greater lateral capacity for the new drilled shafts and prevent the potential of liquefaction and lateral spread loading of the drilled shafts which were founded below the liquefiable soils.

Increasing Capacity of Existing Foundation

While densifying the loose sands below and adjacent to the existing timber piles would improve the seismic response of the site and increase the lateral capacity of the timber piles, the structural response of the tower during the design event would produce loads vertical and horizontal loads which were in excess of both the capacity of the timber piles as well as the capacity of the connection between the structure and the piles. Due to the size of the loads which the tower was capable of producing in the design seismic event, the use of several drilled shafts located outside the existing tower pile cap offered the most reasonable means of providing additional required vertical and lateral capacity to the existing foundation system.

The final design incorporated a total of eight new drilled shafts. Each drilled shaft was 4 ft in diameter with a total depth of 45 ft. A group of four drilled shafts was located on the west and the east side of the tower. Each set of four drilled shafts was connected by a common pile cap which served as the foundation for the new structural steel bracing columns. The layout of the proposed drilled shafts included a drilled shaft adjacent to each corner of the control tower. The clear space between these new shafts and the closest timber pile at each corner was only 3-ft.

Reinforcing Existing Structure

The large vertical and horizontal forces which were generated by the elevated cab of the control tower had to be transferred from the control tower to the new drilled shaft foundations. The existing structure of the control tower itself was insufficient to transfer the lateral loading from the cab of the tower to the foundation. With the decision to utilize drilled shafts to provide additional lateral capacity to the foundation, it was possible to design two pairs of inclined steel columns which would connect to the existing tower just below the cab and carry the horizontal and vertical loading to the new drilled shafts. Due to the proximity of the adjacent support building and fire station, the new drilled shafts could only be installed on the east and west side of the tower. As a result, each pair of inclined steel columns splayed out to provide restraint in both the east-west and north-south directions. The earthquake loads from the control tower were transferred to the new steel columns by means of reinforced concrete collar constructed around the base of the cab.

CONTRACT DELIVERY

Prior to the Nisqually earthquake on February 28, 2001, plans for a seismic retrofit for the Boeing Field Air Traffic Control Tower were in the evaluation and design phase. In addition to evaluating the structural adequacy of the control tower and the existing subsurface soil conditions, the designers had discussions with DBM Contractors, Inc., a local drilling and specialty steel erection contractor, regarding constructability and schedule issues associated with the design options under consideration at the time. When the earthquake struck, causing sufficient structural damage to the control tower to render it uninhabitable, the timetable for completion of the seismic retrofit was accelerated significantly.

The first priority was to return air traffic control capabilities to Boeing Field. Fortunately, a scheme for the erection of a temporary control tower had previously been discussed by the designers and DBM as part of a constructability review. The FAA issued an emergency contract to DBM to erect the temporary control tower on a time and material reimbursable contract. Work began on the temporary tower installation on March 5 and was ready for use by the time the remedial compaction grouting work on the main runway was complete.

The design effort for the permanent control tower retrofit was now required to proceed on a priority basis. As design work progressed over the next several months, DBM provided technical, scheduling and budgeting support. In August 2001, the FAA issued a single-source Request for Offers to DBM. DBM tendered an offer based on the 95% design documents. FAA awarded the contract on September 4 based on the 95% design, and gave a notice to proceed date of September 10. Final design documents were completed on September 7. As the construction phase of the project began to unfold, the events of September 11, 2001 resulted in the need for the FAA to reassess the security requirements at its facilities nationwide. Construction activity was suspended during this time and a revised notice to proceed was issued on October 11. Mobilization of the site began on October 15 and revised pricing for the changes between the 95% and final design documents was completed on October 17.

CONSTRUCTION

After negotiation of the contract, the project was started following a short delay as additional security measures were implemented following the events of September 11, 2001. While the preliminary work of locating utilities and submittal preparation was being performed, a discussion on the sequence of the foundation construction arose after a review of the proposed construction sequence and methods.

The contractor had submitted a work sequence calling for the installation of the drilled shafts before the performing the compaction grouting. The drilled shafts were to be constructed using a surface casing and polymer drilling fluids. The drilled shafts were to be installed before the compaction grouting to achieve the greatest permanent densification from the grouting. If the drilled shafts were installed after the compaction grouting, some loosening of the previously densified soils was expected. However, the owner's representatives became concerned about the potential loss of vertical capacity in the existing piles during installation of the four drilled shafts located directly adjacent to the existing pile groups.

During subsequent discussions, several methods of mitigating the perceived risk associated with drilling next to the existing piles were assessed along with their associated costs. The use of cased hole drilling methods for the shafts was rejected due to cost and increased risk to the quality of the completed shaft. Pretreating the soils from 25 to 35 ft at each of the shaft locations was not selected due to cost and increased schedule duration. Ultimately, it was decided to perform the compaction grouting prior to the drilled shaft installation. By performing the compaction grout first, the loose soils would be densified prior to drilling the shafts which would reduce the potential for over break during shaft construction. The compaction grouting beneath the pile tips would increase the vertical capacity of the existing piles so those piles which are outside of the influence of the shaft construction can carry additional load safely should excessive over break occur. In addition, the compaction grout layout was altered and several locations added to create a partial ring of grout columns between the drilled shafts and the existing piles.

Compaction Grouting

Equipment & Methods

Prior to beginning the work a series of monitoring points were established on the perimeter of the control tower and the support building and baseline elevations recorded. During the performance of the compaction grouting, the elevations of the monitoring points within the treatment zone were measured and recorded using conventional survey equipment at least once per shift. In addition to this daily survey of the monitoring points, a rotating laser system was used at all times during grout injection to provide real time monitoring of the ground surface and adjacent structures within 50 ft of the injection point.

Since the choice of compaction grouting for densification of the subsurface soils was a result of the ability to treat under the perimeter of the structures with battered holes, work on a tight site, and minimize disturbance of the occupied building, the grouting subcontractor utilized a small rotary drilling system mounted on and powered by a skid steer unit. This drilling unit installed the flush joint casing using rotary drilling methods with water flush. Once the casing was installed, a low-slump grout consisting of dirty sand having about 25% fines mixed with 3% to 6% cement was injected in an up-stage manner. The grout was mixed on-site using a continuous mixing system which consists of two conveyors which feed cement and sand from separate hoppers into an inclined mixing auger which discharges directly in the holding hopper on the pump.

Since the compaction grouting was governed by a performance specification which only established the treatment depths and the level of improvement required, the actual pattern and spacing of compaction grout holes was selected by the grouting subcontractor. The 8 ft triangular pattern, with extensive field adjustments and battered installations for existing structures, used for the compaction grouting is shown in Figure 1. The casing was retracted and the grout injected in 1 ft stages. The original specifications also indicated that grout was to be injected in each stage until one of the following criteria was achieved:

1) Header pressure exceeds 700 psi
2) Volume of grout injected at a given stage exceeds 25% of the ground being treated by that stage.
3) Total building movement exceeds 0.25 in.

Construction Issues Related to Compaction Grouting

The compaction grouting was started around the control tower since this offered the only area which could be tested due to the proximity of the remainder of the holes to the perimeter of the buildings. During the initial phase of grouting, one corner of the tower experienced a cumulative upward movement in excess of the 0.25 in total allowed by the contract. After the project team met to discuss this issue, it was decided that the tower could tolerate several inches of vertical movement as long as the relative movement between the corners did not exceed the threshold value of 0.25 in. In order to accommodate this requirement and prevent differential movement in excess of 0.25 in, a pattern of hole injection which alternated between the sides was implemented. Although this worked well on the tower which had a small enough footprint to permit treatment of all of the soil under the tower, the support building experienced some internal cosmetic damage as the compaction grouting around the perimeter moved the outside building wall up to 0.75 in while the interior walls and columns experienced no heave since they were outside the zone of influence of the compaction grout points.

A review of the grouting records also indicates that most building movement occurred when grout was being injected between depths of 25 to 30 ft. This depth corresponds to the likely tip elevation of the timber pile on which the tower and the support building were constructed. It seems that the grouting actually lifted structure and the piles as a unit from the pile tips. While this may raise some questions about the stresses induced in the piling by the grouting, it also allays the concerns about the grouting process actually separating the pile cap from the timber piles.

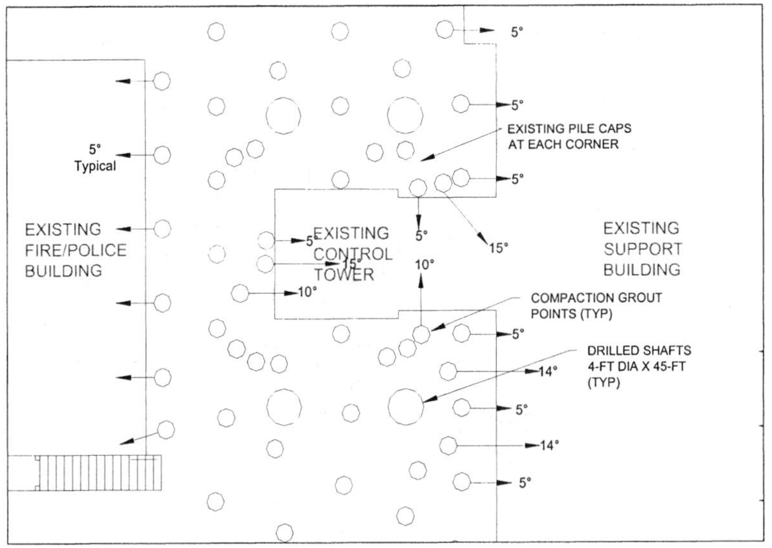

FIG. 1. COMPACTION GROUT POINT LAYOUT

Results of Compaction Grouting

Since the purpose of the compaction grouting was to densify the loose, saturated sands to prevent liquefaction, post-treatment CPT's located between the treatment points were used to verify that a minimum level of densification had been achieved. The specifications required that the post-treatment soils have an average, corrected clean sand equivalent Normalized Cone Penetrometer Resistance $(q_{c1n})_{cs}$ exceeding 125 tsf. Each measured CPT tip resistance was averaged with the resistances measured in the 2 ft above and the 2 ft. There were no performance requirements stipulated for soils with permeabilities less than 1 x 10^{-2} cm/sec. This permeability cutoff served as an indicator of a fine grained soil which is commonly thought not to be a risk of liquefaction and which can only be minimally densified by compaction grouting efforts.

A graph of the raw values for the pre-treatment CPT-01 is shown in Figure 2. The graph of the friction ratio indicates that the material from 15 to 25 ft has a friction ratio less than 0.75% which indicates a clean sand which is very amenable to improvement by compaction grouting. The material from 25 to 30 feet has several spikes greater than one in the friction ratio which indicate a layering of the clean sand with silty material, and the material from 30 to 35 feet has a friction ratio of just under 1% which indicates a silty sand. The silty material has a lower permeability

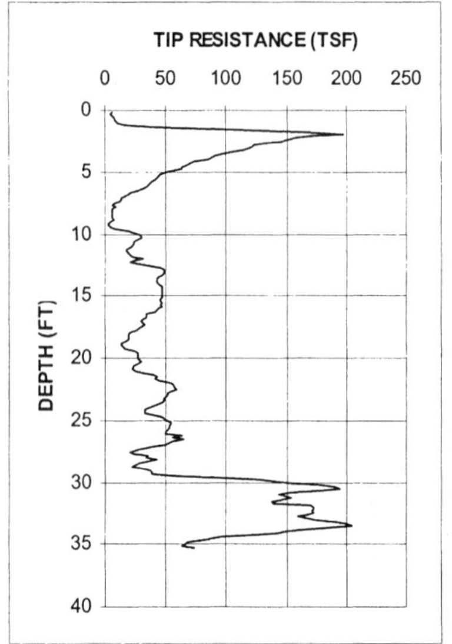

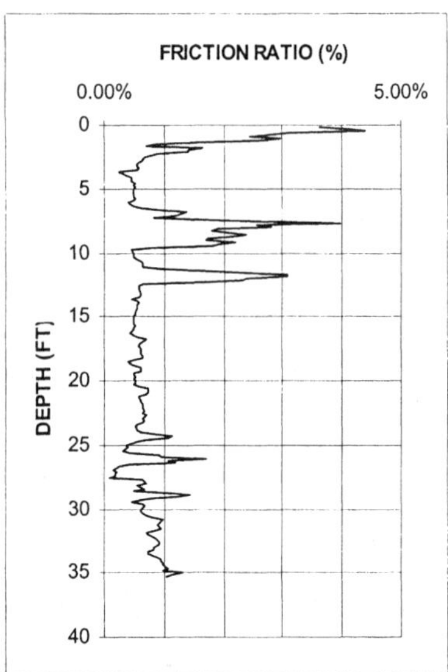

FIG. 2. CPT-01 PRE-TREATMENT RESULTS

which inhibits the effects of densification by reducing the rate at which excess pore pressures dissipate during the injection of the grout bulb.

The corrected tip resistance for the pre-treatment CPT-01 and the post-treatment CPT-04 performed on the east side of the tower are shown in Figure 3 along with the target value of the post-treatment CPT. The post treatment CPTs used to measure the performance of the compaction grouting indicated generally acceptable levels of improvement as compared to the required performance level. There are zones in the treatment zone between 25 and 35 ft where the required densification was not achieved. In fact, the pre and post treatment CPTs show little difference at 28 ft. This zone of no improvement corresponds well with the fine grained soil layers indicated by the higher friction ratios recorded in the pre-treatment CPT. Although the CPT's showed improvement which generally met the performance requirement, the presence of some zones of material which did not achieve the required performance led to some remedial points being added in the critical area of the control tower.

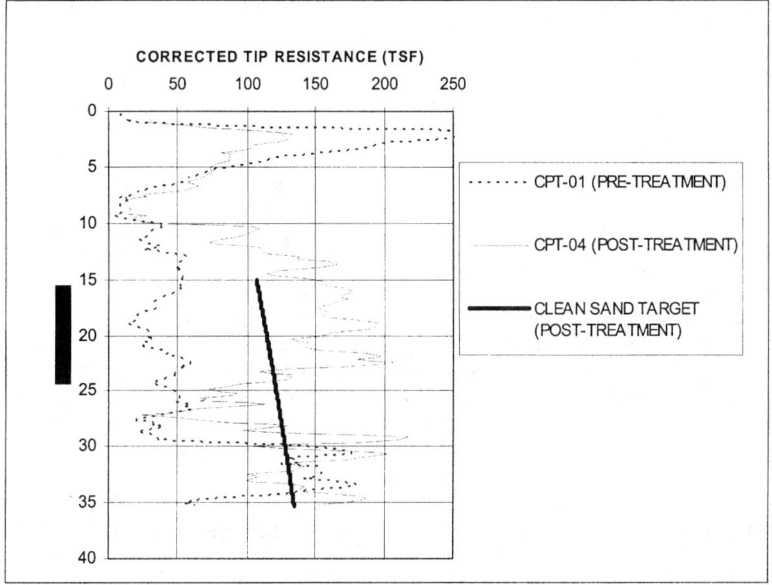

FIG. 3 . COMPARISON OF PRE-TREATMENT AND POST-TREATMENT CPTS

Drilled Shafts

Equipment & Methods

The primary considerations in the selection of the drilling methods and equipment for drilled shaft construction were the anticipated presence of groundwater between 12 and 15 feet below existing grade, and the limited headroom between existing grade and the soffit of the control tower for the four shafts immediately adjacent to the existing structure.

The presence of groundwater dictated the use of either temporary casing or drilling slurry to provide shaft sidewall stability during drilling and concrete placement operations. The use of casing, however, proved impractical for a host of reasons. First, the headroom distance from existing grade to the underside of the cantilevered portion of the tower structure was approximately 46 ft. The planned depth of the shafts was 45 feet. The minimal headroom therefore precluded the use of full-depth temporary casing for the four shafts nearest to the existing structure. The use of telescoping casing was also investigated, but was discounted due to the proximity of the existing timber piles as well as the complications associated with casing removal during concrete placement. Ultimately, it was decided to use polymer slurry for shaft sidewall stability, combined with additional compaction grouting points adjacent to the existing structure to address concerns regarding the induced lateral soil pressure near the bottom of the existing timber piles.

The shafts were drilled using an IMT AF-10 hydraulic drill rig. Approximately two feet of soil was excavated at the shaft locations adjacent to the existing structure to allow enough head room for the boom of the rig to clear the tower overhang. A temporary casing six feet in diameter was installed down to a depth of approximately 12 feet to support the soil around the top of the shaft. The top of the shaft concrete was held down to approximately eight feet below grade to allow for subsequent splicing of the shaft rebar with that required for the pile caps

Upon completion of drilling, rebar cages were placed in the slurry-supported holes. The specifications required Crosshole Sonic Logging (CSL) tests to be performed after concrete placement, so 1 ½" schedule 40 steel pipes were tied inside of the vertical reinforcing bars for subsequent testing. Concrete was placed using a tremie pipe and concrete pump truck. The polymer slurry was pumped out of the shaft and into a holding tank as concrete was placed. CSL test data was collected and analyzed by an independent testing firm. Each shaft was accepted based upon the recommendation of the independent firm without the need for any remedial work.

CONCLUSION

A foundation retrofit system, which combined ground improvement by means of compaction grouting with eight 4 ft diameter drilled shafts, provided a cost effective means of mitigating liquefaction potential and providing increased lateral load capacity to the existing foundation. The foundation contractor was able to negotiate a fixed price contract with a government agency based on past performance with the agency and a proven team of subcontractors. The project was performed on time and on budget, and it resulted in an air traffic control tower that is more likely to survive and maintain operations following the next earthquake in the Pacific Northwest.

DRILLED SHAFT VALUE ENGINEERING DELIVERS SUCCESS TO WAHOO, NEBRASKA BRIDGE

By Joseph A. Waxse[1], P.E., Member, ASCE; Jorj Osterberg[2], Ph.D., P.E., Hon. Mem., ASCE; Omar Qudus[3], P.E.

ABSTRACT: This case history illustrates how a successful "Value Engineering" redesign with slurry-drilled shafts and an O-Cell™ test saved considerable costs and expedited early completion of a large bridge project for the Nebraska Department of Roads (NDOR). The paper first describes the original project and its geologic setting. It then relates the inception, development and verification of the redesign concept and its implementation via NDOR's "Value Engineering Proposal" (VEP) standard specification provisions. The comparative advantages of drilled shafts versus driven piles relative to the site geology, some benefits of using polymer drilling slurry and the practicality of foundation verification with an O-Cell™ test and cross-hole sonic logging (CSL) are discussed. The lack of redundancy in drilled shafts and the increased importance of having an experienced contractor using appropriate equipment and procedures under an effective quality control program are also emphasized. The paper concludes with a discussion of the overall impact of the VEP redesign on the project cost and schedule, as well as on NDOR's subsequent bridge design and construction practices.

FIG. 1 HWY 77 "Wahoo South" Bridge

[1]Principal, Terracon, Inc., P.O Box 2506, Omaha, NE 68130
[2]Professor Emeritus, Northwestern University
[3]Asst. Materials and Research Division Engineer, Nebraska Department of Roads, P.O Box 94759, Lincoln, NE 68509

INTRODUCTION

As has often been the case in Nebraska, the city of Wahoo's growth has been concentrated along the state highway and has caused traffic congestion. One of the NDOR's ongoing priorities has been to build bypasses around such cities to improve highway safety and capacity. Wahoo's Hwy 77-Bypass route required an 1800-foot long bridge to span a valley containing two creeks and a railroad line. The NDOR performed a geotechnical site exploration and designed this bridge in-house. Plans for both pre-stressed concrete and continuous steel girder design alternatives were advertised for competitive bid in February 2001. An Omaha-based construction company won the bid to be the General Contractor for the concrete design alternative.

ORIGINAL BRIDGE DESIGN

The 85-foot wide bridge consisted of 12 spans ranging from 118 to 162 feet long. Each of the 11 bridge piers had 6 columns. Each column was supported on a group of 9 to 11 pre-stressed pre-cast concrete driven piles. The 12-inch square piles were to be driven to a design compressive axial load capacity of 60 tons. Pile Dynamic Analysis was to be performed on one test pile per pier during pile installation. The bottoms of the pile caps were designed, for scour protection, to extend to depths of about 15 to 25 feet below the ground surface. Figure 2 shows a pier foundation pile plan and the typical pier cross-section from the original plans (dimensioned in mm, divide by 328 for feet).

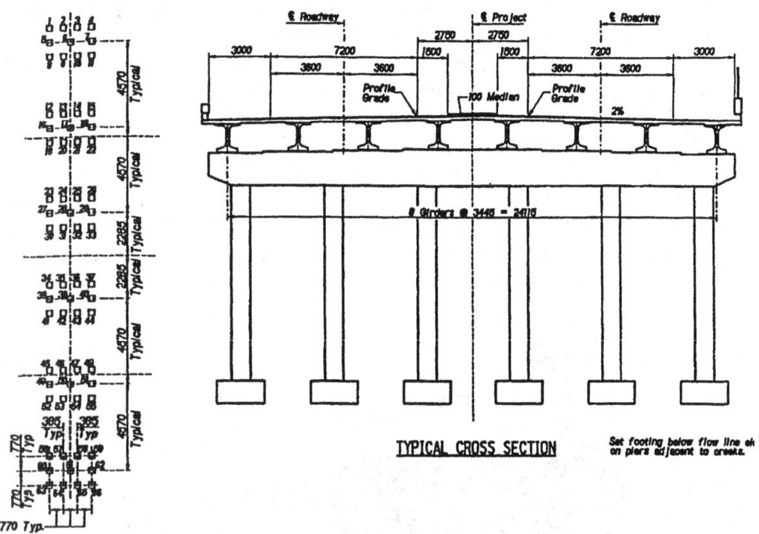

FIG. 2. Original Design - Pier Pile Plan and Typical Pier Cross-section

SITE GEOLOGY

The valley that the bridge spans is a tributary of a prominent regional geographic feature cited by Souders, et. al.(1971) as the Todd Valley. During Pleistocene post-glacial periods, large stream flows scoured this channel about 60 to 90 feet deep and into Dakota shale and sandstone bedrock. The subsequent alluvial deposits in the bridge area are composed predominantly of silts and clays in approximately the upper 20 to 30 feet. These recent cohesive soils are underlain with alluvium of Wisconsin age, and older. These are predominantly granular deposits that generally consist of fine to coarse sand with gravel. Locally, and at variable depth, the sands contain clayey zones and some relatively coarse scour-bottom ("thalweg") deposits of larger gravel with occasional "nested" cobbles and boulders. The upper surface of the Dakota shale and sandstone is typically highly to moderately weathered. The groundwater table is typically about 10 to15 feet below the ground surface.

VEP INCEPTION & DESIGN DEVELOPMENT

Due to the low strength of the upper cohesive soils and shallow groundwater conditions, excavations for pile driving and pile cap construction would have required installation of sheet pile shoring and dewatering systems for each of the 66 column locations. The deepened pile caps next to the creek channels would have required relatively deep and internally braced sheet pile shoring and large capacity dewatering wells. Test pits that the General Contractor (GC) excavated at the site also heightened their concerns due to the levels of soil instability and groundwater seepage that they observed.

The GC solicited review of the situation and input on potential value alternatives from their geotechnical consultant and a local deep foundation contractor. They both independently recommended the consideration of slurry-drilled shafts as a potential means of eliminating the piles and pile caps and the associated shoring and dewatering systems. The geotechnical consultant also warned of the risk of potential driving interference and/or damage of concrete piles due to the noted presence of boulders above the design pile tip elevation.

The GC obtained preliminary cost estimates for drilled shafts from the foundation contractor, a proposal for structural redesign of the sub-structure from a structural engineering firm, and a proposal for associated services from their geotechnical consultant. They then approached the NDOR about the alternative and its potential time and cost savings. The NDOR initially expressed some reservations about the use of slurry-drilled shafts, since they had not used them on their previous projects. They were, however, aware of their successful use in other locales, and they granted permission for the GC to develop a detailed Value Engineering Proposal (VEP) for their consideration. The GC then authorized their geotechnical and structural consultants to proceed with the foundation and sub-structure redesign.

The geotechnical consultant first performed supplemental geotechnical subsurface exploration at the site, including soil borings and electronic cone penetration test (CPT) soundings. They then analyzed the new information and the original NDOR borings and generated recommended slurry-drilled shaft design parameters in general accordance with the procedures cited by O'Neill and Reese (1999) in the FHWA report "Drilled Shafts: Construction Methods and Design Procedures". These parameters included average allowable compressive unit skin friction values of 300 psf in the upper clays and 900 psf in the deeper granular alluvium. An allowable end bearing value of 25 ksf was recommended for shaft tips, which were required to penetrate to a depth at least equal to the shaft diameter into the sandstone or shale bedrock, or granular alluvium with a Standard Penetration Test (SPT) N value greater than 25. Based on the resulting load capacities achievable with shafts of reasonable diameter and length, The structural engineer determined that it was viable to reduce the number of columns per bridge pier from 6 down to 4. Drilled shafts with a 5.5-foot diameter were designed for up to a 1230 kip vertical axial compressive capacity. Figure 3 shows a typical redesigned drilled shaft foundation plan and pier cross-section (dimensioned in mm, divide by 328 for feet).

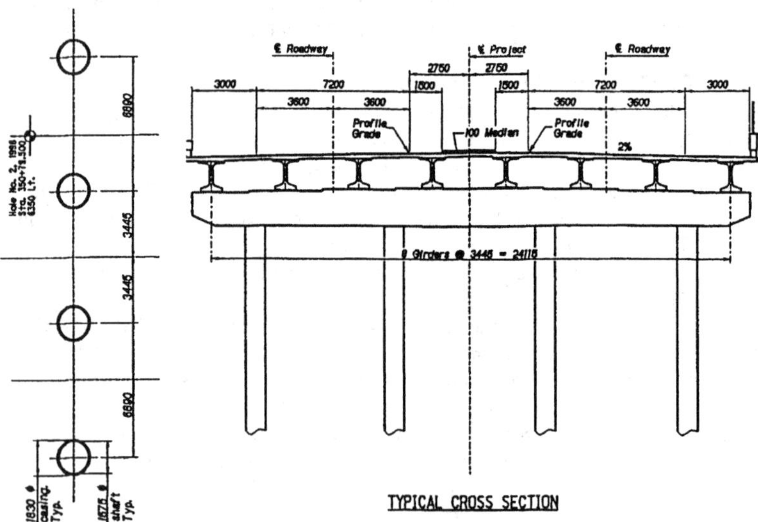

FIG. 3. Redesign - Drilled Shaft Plan and Typical Pier Cross-section

The geotechnical consultant developed specifications for slurry-drilled shaft construction. The use of polymer slurry, rather than conventional bentonite slurry, was specified for two reasons. The primary reason was that prior experience with polymer slurry indicated that silts and fine sands could be expected to settle out of suspension more rapidly than typically occurs with bentonite slurry. This effect was desired in order to reduce potential problems and delays associated with achieving effective final cleaning of the bottoms of the completed drilled shaft excavations. The second reason for specifying polymer slurry was to avoid a perceived risk of incurring a decrease in mobilized shaft skin friction. Due to the relatively high hydraulic conductivity of the sands, there was a concern that a slurry "cake" could form on the sides of the excavation. This low-strength layer of bentonite could possibly have been trapped between the poured concrete and the soil. The potential for this effect was mitigated with polymer slurry since the long-chain molecules that produce the seepage-limiting viscosity are broken down to an innocuous state when exposed to the high alkalinity of the fluid concrete.

To demonstrate that adequate design capacity would be achieved, the geotechnical consultant recommended (and the NDOR specified) that an Osterberg cell (O-Cell™) load test be performed on the first drilled shaft installed for the bridge. Cross-hole sonic logging was also specified for at least the first six production shafts for verification of suitable installation procedures. The O-Cell™ test and cross-hole sonic logging are described in the "Verification and Implementation" section below.

Upon completion of the redesign, the GC developed final costs for the new sub-structure and the installation, monitoring and testing of the drilled shaft foundations. They then presented the VEP, representing a $385,000 cost savings and 6-month early completion, to the NDOR. It was approved.

VEP DESIGN VERIFICATION AND IMPLEMENTATION

O-cell Test Background- Dr. Jorj Osterberg invented the Osterberg Cell (O-cell) in the 1980s to provide a practical method of performing load tests to determine the ultimate capacity of large-diameter, high-capacity drilled shafts. Osterberg (1989) describes the O-cell as a "jack-like hydraulic loading device" that is cast within the shaft. The O-cell is instrumented with redundant remote electronic (vibrating-wire displacement transducers) and mechanical (metal rod and pipe "telltale") systems that measure jack expansion during the test. Electronic strain gauges can also be positioned at various levels in the shaft to measure representative local shaft stress and evaluate the general proportion of total shaft capacity that is developed in the defined zones during the test.

The O-cell is ideally located in the shaft near the "balance point" between the ultimate load capacities of the upper and lower shaft segments. The mutual opposition and axial loading reaction of the shaft segments thereby essentially doubles the effective capacity of the loading device (compared to a top-down loading device) and eliminates the need for an aboveground reaction system. Historically, such aboveground systems have consisted of multiple reaction shafts and massive load transfer structures or enormous kentledge (dead weight) systems. These systems were typically so expensive and time-consuming (and sometimes dangerous) to build that their use was practically inhibited for all but the largest of projects. Although the alternative O-cell test method is relatively new, its effectiveness and reliability has been well documented by Osterberg, (1998).

O-cell Test Performance The O-cell test was specified for the Wahoo project to demonstrate the performance of slurry-drilled shafts to the NDOR, who had not previously used them, and to reduce the conservatism that would have otherwise been appropriately applied to the unit load capacity design parameters. The resulting savings in the cost of the 44 drilled shafts more than paid for the load test. The test was performed on an actual bridge foundation shaft by grouting the O-cell full of "neat" cement after conclusion of the test. Figure 4 is a schematic representation of the Wahoo O-cell test configuration.

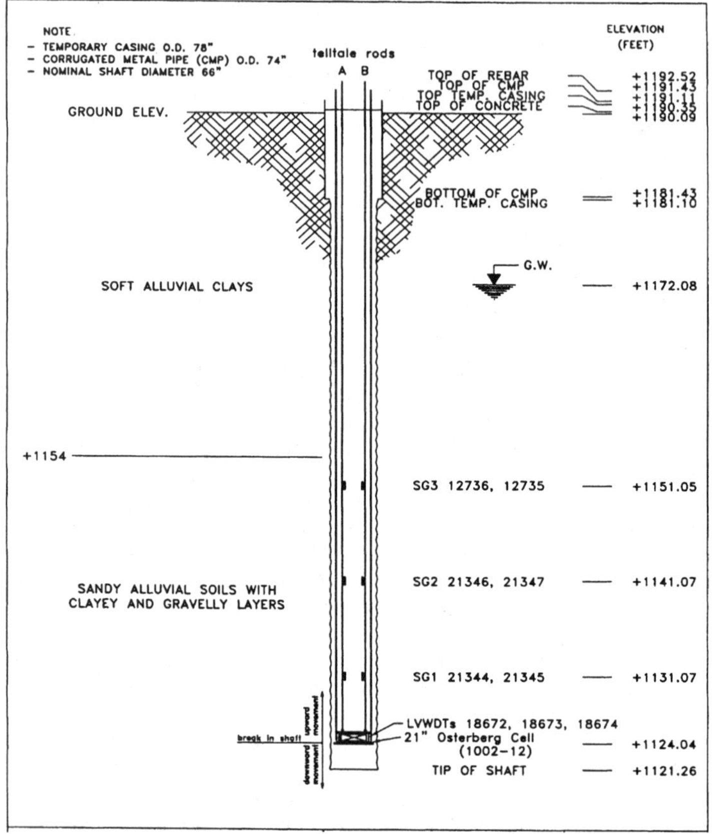

FIG. 4. Schematic Section of Wahoo O-cell Test Configuration

The test shaft was excavated using vinyl acrylate co-polymer slurry. The geotechnical consultant monitored the drilling and slurry management operations. The drilling operation did not experience any notable problems with obstructions, slurry loss or sidewall instability, even when nested cobbles and up to 12-inch diameter boulders were encountered well above the design tip depth. A slurry sample taken from the bottom of the excavation less than fifteen minutes after final auger removal showed negligible sand content.

The geotechnical consultant probed the shaft bottom with a heavy, cone-tipped sounding weight before and after two trips were made with the clean-out or "spin" bucket. Initial contact of the probe with bottom sediments was easily detectable and the corresponding depth was measured. The probe was then "jigged" to penetrate loose material until the tip encountered sound penetration refusal. This method was used to confirm that final clean-out achieved less than the specified maximum 1 inch thickness of loose sediment on the shaft bottom. Approximately 2 ½ feet of concrete was then tremmied into the bottom of the shaft excavation. The instrumented O-cell and rebar cage was then installed and concrete with a minimum slump of 7 inches was tremmied into place to complete the test shaft. Loadtest, Inc. supervised installation the O-cell test equipment and later performed the load test after the shaft concrete had reached specified strength. The test was performed using the ASTM Quick test Method D1143 (ASTM 1993). Figure 5 shows the resulting O-cell test load movement curves and Figure 6 shows an equivalent top-down load deflection curve.

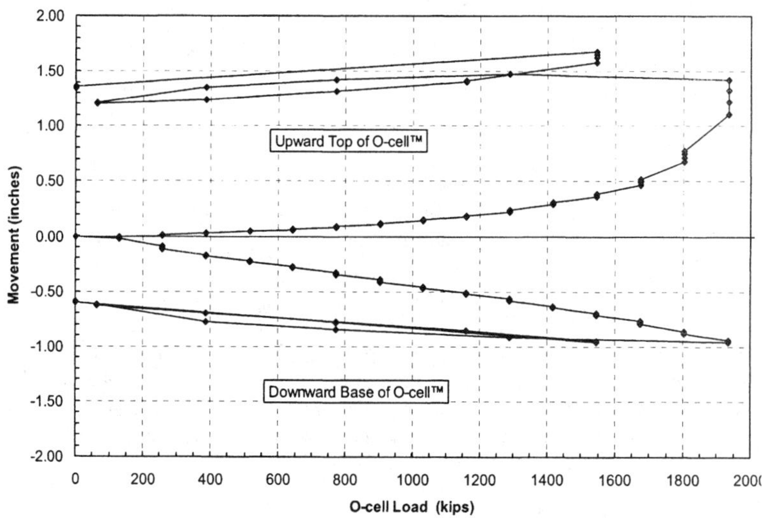

FIG. 5. O-cell Test Load Movement Curves

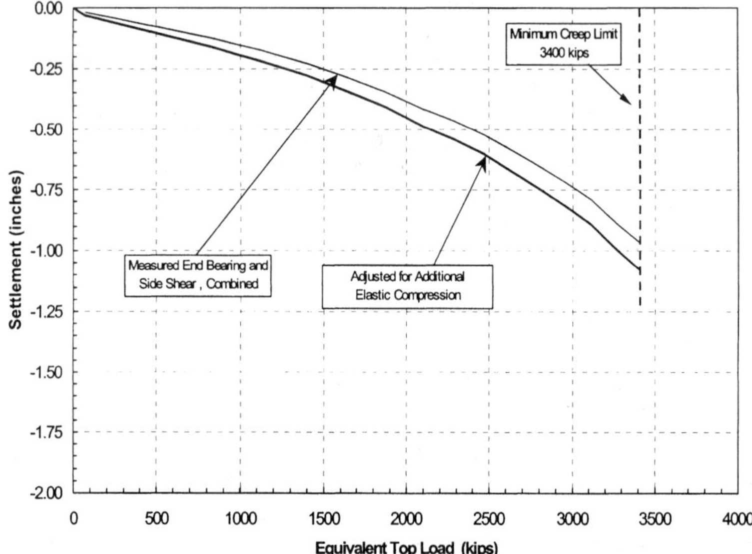

FIG. 6. Equivalent Top Load/Movement Curves

The results of the O-cell test confirmed that use of the design unit skin friction values would produce an appropriate factor of safety. The end bearing was tested to over 80 ksf without indication of approaching ultimate capacity. However, due to the anticipated variation in bedrock condition across the bridge site, the recommended design end-bearing value was not increased above 25 ksf.

Cross-hole Sonic Logging Cross-hole sonic logging is a non-destructive remote-sensing method of evaluating the as-built concrete uniformity and strength of completed shafts. The system used on this project was an Olson Instruments, Inc. CSL-1/Freedom NDT PC™ logging system. To accommodate the testing, six, 2-inch diameter PVC pipes were tied vertically at 60 degree spacing to the full-height of the inside of the rebar cage. The pipes were filled with water and capped prior to concrete placement. After the concrete reached adequate initial strength (4 to 7 days), the two probes that emit and receive sonic signals were lowered down a pair of the pipes. The devices were then raised in tandem and the time that sound took to travel between the pipes was recorded in the data logger and divided by the known pipe spacing to determine the sonic velocity of the concrete. This procedure was repeated between all possible pairs of the six pipes.

Since there is a direct correlation between the concrete's compressive strength and its sonic velocity, anomalous zones of low velocity indicate corresponding low strength. This is commonly caused by contamination of the concrete with slurry or spoils and can indicate trapping of material due to improper cleaning, improper concrete placement methods or intrusion due to sidewall caving. Cross-hole sonic testing was performed on the first 6 production shafts and confirmed that the equipment, materials and construction methods being used were producing uniform shafts free of significant anomalies.

Quality Assurance For the first several weeks of the project, the geotechnical consultant provided full-time monitoring of drilled shaft installation and concurrent familiarization of the assigned NDOR inspectors with the general procedures and principals involved in the process.

Although the redesign of Wahoo South for drilled shafts provided multiple benefits, the lack of redundancy in a typical drilled shaft system greatly increased the seriousness of any significant potential shaft defect going undetected and uncorrected. If one driven pile in a ten-pile group were to fail, the results would typically not be catastrophic. Conversely, if one of the drilled shafts supporting the Wahoo Bridge experienced excessive settlement, the foundation performance would be compromised. It is the authors' opinion that undertaking a slurry-drilled shaft project is particularly ill-advised without confident assurance that a well-qualified and experienced foundation contractor will use the proper equipment, materials and methods. Those responsible for quality control and quality assurance must also be competent, conscientious and uncompromising toward their critical responsibilities on this type of project.

Conclusions The value engineering redesign was quite successful, with the foundation installation actually finishing ten months ahead of schedule and the Contractor and the NDOR sharing $378,000 in savings. With as many boulders as were encountered in the shaft excavations, the redesign also avoided a significant risk of the project having incurred added costs or delays due to problems with premature refusal and/or damage of driven concrete piles. The NDOR also derived the broader, long-term value of developing acceptance of slurry-drilled shafts as a practical foundation option for their subsequent projects. The NDOR used slurry-drilled shafts on three bridge projects that they designed the year following completion of the Wahoo South project. The NDOR also has sponsored a drilled shaft short course, the National Highway Institute Course No. 132014A, for their staff and interested local consultants and contractors. This project illustrates how the VEP provisions of the NDOR "Standard Specifications for Highway and Bridge Construction" promote innovation and provide a means of delivering true "Value Engineering" to Nebraska's transportation projects.

APPENDIX I. REFERENCES

O'Neill, M. W. and Reese, L. C. (1999). "Drilled shafts: construction procedures and design methods." *Report No. FHWA-IF-99-025,* ADSC, Dallas.

Osterberg, J. O., (1989). "New device for load testing driven piles and drilled shafts separates friction and end bearing". *Proceedings of International Conference on Pilings and Deep Foundations,* London, A.A.Balkema

Osterberg, J. O., (1998). "The Osterberg load test method for drilled shafts and driven pile – the first ten years". *Proceedings of 7th Int. Conference on Deep Foundations"*. Vienna, Austria. Deep Foundations Institute.

Souders, V. L., Elder, J. A. and Dreeszen, V. H. (1971). "Guidebook to selected pleistocene paleosols in eastern Nebraska." *The University of Nebraska Conservation and Survey Division, Lincoln Nebraska Geological Survey,* 15

APPENDIX II. CONVERSION TO SI UNITS

Feet (ft) x 0.305 = meter (m)
Pounds per square foot (PSF) x 47.85 = Pascal (Pa)
1000 pounds (1 kip) x 4.45 = kiloNewtons (kN)

SECANT PILES SUPPORT ACCESS SHAFTS FOR TUNNEL CROSSING IN DIFFICULT GEOLOGIC CONDITIONS

Thomas C. Anderson, M.ASCE 1

ABSTRACT: A new 84 inch water pipeline project for Tampa Bay Water included a tunnel section under the Alafia River in Riverview, FL. Access shafts were required on each side of the river for the pressure-balanced tunnel boring machine. Since the 70 ft. deep access shafts extended roughly 25 ft. into the confined Floridan Aquifer (Tampa Limestone) under an artesian head, control of groundwater flows was a serious concern.

Two circular cofferdams (one 40 ft. and one 24 ft. ID) were designed and built for these access shafts. Alternating, overlapping hard/soft secant piles (36 inch diameter) were utilized to act as a lateral cutoff wall. Circular concrete compression ring walers were constructed to support the secant piles at five (5) levels. In order to maintain bottom stability and to control bottom water inflows, a pressure relief well system was installed at each cofferdam.

The smaller cofferdam was instrumented with an inclinometer in one of the secant piles to monitor lateral deflections. In addition, vibrating wire strain gages were installed in the concrete ring walers to measure compressive strains and the resulting stresses during the excavation process.

---

1 Area Manager, Schnabel Foundation Company, Celebration, FL.

INTRODUCTION

The Hillsborough County South Central Intertie-Contract 2 for Tampa Bay Water required the installation of a new 84 inch water pipeline in a tunnel section under the Alafia River in Riverview, FL. The tunnel invert was situated at a depth of 70 ft. at El. -59.0 below the existing grade at El. +11.0 (Note: all elevations are in NGVD datum). A pressure-balanced tunnel boring machine (TBM) with a 102 inch diameter cutting head was planned to be used for the tunnel section to install a 96 inch OD steel casing into which the 84 inch pipeline would be installed. In order to create access for the TBM, a 40 ft. long launching shaft was required on the south side of the river while a 24 ft. long receiving shaft was necessary on the north side of the river.

Area Geology

The soil stratigraphy in Hillsborough County generally consists of alternating layers of sand and clay of Holocene to Miocene age overlaying limestone layers of Miocene to Oligocene age. The surficial sands are generally clean, fine to very fine quartz sand. The clay content of this deposit increases with depth. The surficial sands are underlain by a sandy clay layer called the Hawthorn formation. Evidence indicates that this clay may be the weathered residue of the underlying limestone.

Beneath the near-surface sand and clay layers, the Tampa and Suwanee Limestone formation comprise the Floridan Aquifer, the principal source of water in the area. This is a confined aquifer by virtue of the overlying Hawthorn formation. The Tampa Limestone of Miocene age is a white to gray, fossiliferous limestone, while the Suwanee Limestone is a fossiliferous, yellow to white, fine-grained limestone of Oligocene age. The limestone is characterized by solution cavities and fissures. The highly porous nature of the limestone has frequently resulted in sinkhole formation.

Soil Stratigraphy

From existing grade at approximately El. +11.0, a variable thickness (10 to 15 ft.) veneer of near-surface sandy soils overlay relatively thick beds of predominately stiff to hard and variably cemented clays with occasional thin stringers of limestone. The clay unit commonly includes lenses or seams of clayey sands and sands. The thickness of this upper clayey unit is typically in excess of 20 ft. and extended to El. -22 to -34.

Variations above and below this range certainly can occur as a result of erosion and deposition occurring during the Miocene epoch.

A soft to relatively hard limestone formation occurred below the upper predominantly clayey unit and extended to completion of the borings (100 ft.). Relatively poor core recovery and low RQD values were obtained within the limestone formation. This is due to the variable lithification or variable degrees of cementation which occur within the limestone unit and the fact that the limestone is commonly riddled with thin, clayey seams that tend to washout during the coring process. Further, zones of very high circulation fluid losses were observed in the limestone formation, along with thin cherty seams.

Groundwater Conditions

The groundwater level in the borings varied from El. +5 to +7 at the time the borings were performed during an extended period of minimal rainfall. The test borings within the Alafia River from El. 0 showed an artesian head as high as El. +5 to +7. Piezometric levels could certainly occur at higher elevations during a wet season as a result of recharge of the upper Floridan Aquifer.

Design

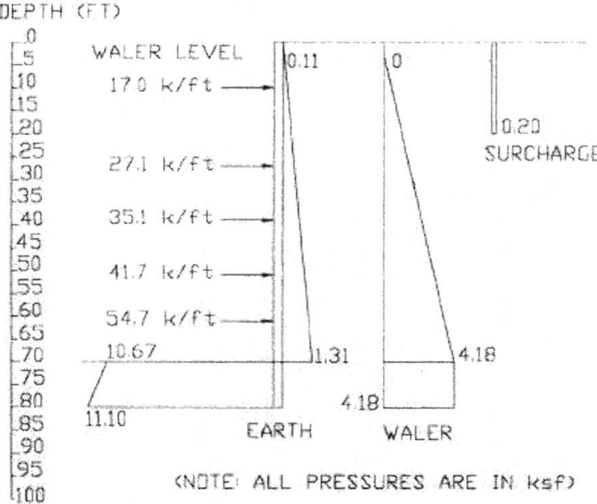

FIG. 1. Design pressure diagrams

In order to create unobstructed access shafts without internal bracing and facilitate excavation and tunnel construction, circular cofferdams (40 ft. ID for the launching shaft and 24 ft. ID for the receiving shaft) were selected with compression ring walers for the internal supports. The retention system had to be able to penetrate the soft to hard limestone below an approximate depth of 45 ft. and had to provide a cutoff for the lateral groundwater flow. For these purposes, an overlapping hard-soft secant pile wall was utilized.

The secant pile wall was designed for active earth pressures corresponding to a total unit weight of soil or rock of 125 pcf (submerged unit weight of 65 pcf) and an effective friction angle of 32.5 degrees (see Figure 1). In addition, full hydrostatic water pressures were considered corresponding to a groundwater table at El. +8. Further, equipment surcharge pressures were added to the earth and water pressures. A cohesion of 8000 psf was assumed for the rock below subgrade.

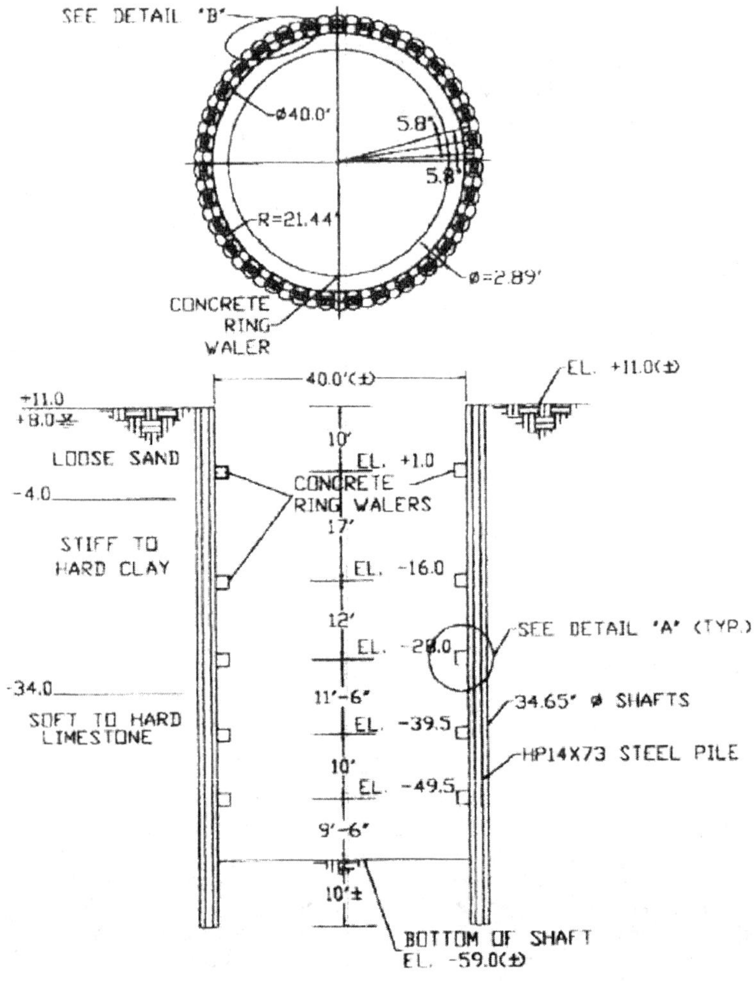

FIG. 2. Plan and elevation view of launching shaft

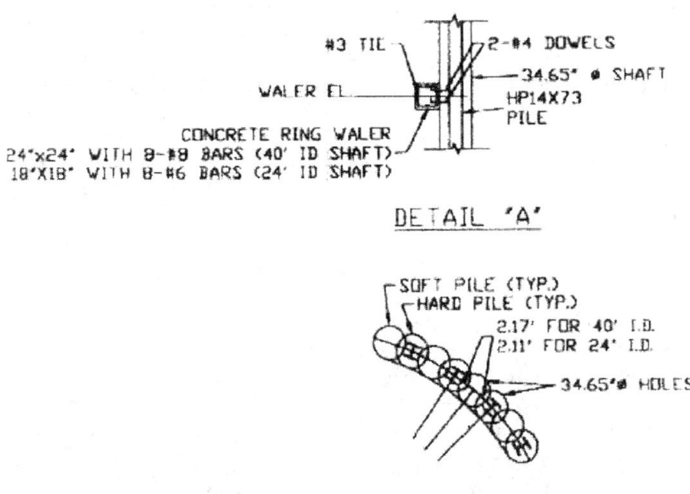

FIG. 3. Details of secant piles and concrete ring walers

The resulting design consisted of 80 ft. long soldier beams (HP14x73 GR50) at approximately 4.25 ft. on center in alternating secant pile holes with five (5) levels of supports at El. +1.0, -16.0, -28.0, -39.5 and -49.5. The bottom support was located to provide access for the TBM. Figure 2 shows a plan view for the launching shaft with a corresponding elevation view. Figure 3 provides the details of the secant piles and the concrete ring walers.

Bottom stability and controlling the upward water flows from the bottom of the shafts were also other critical features of the shaft design. Initially, a bottom grout plug was considered to be the best solution. Before starting construction of the access shafts, a boring and simple pump test were done in the center of each shaft to a depth of 100 ft. to further investigate the nature of the rock below the shaft bottom since the original contract borings were terminated just below the proposed tunnel invert. This additional investigation indicated that the rock below the bottom of the shafts consisted of soft to hard, cream colored, clayey limestone with seams of calcareous clay. The pump tests showed a fairly consistent specific capacity of 0.5 gpm/ft. of drawdown, which is relatively low for the limestone in this area. The limestone had voids and fissures which had the ability to carry water. The void structure of the limestone apparently is filled with silts and clays and decomposing limestone. As a result, it was decided that pregrouting the rock below the bottom of the shafts may not be completely effective. There was absolutely no way to guarantee the complete effectiveness of the bottom grout plug of rock, since there could easily be ungrouted crevices in the rock subject to the full hydrostatic head at El. +8 which could cause a "blow up" of the bottom of the shaft. Therefore, it was decided to utilize pressure relief wells around the outside of the shafts to reduce the hydrostatic head in the limestone to maintain bottom stability and control the water inflows from the bottom. Properly filtered and screened wells were essential for this purpose, since the voids in the limestone can be "cleaned out" or scoured due to open pumping, causing the specific capacity and the volume of water to be pumped to drastically increase.

Construction

For the secant pile walls, adjacent shafts were designed to overlap 8 to 9 inches, which required a very tight drilling tolerance equal to or less than 1.0 percent of the depth of the hole. This requirement was to insure that the adjacent secant holes overlapped all the way to bottom of the access shaft and, thus, there would be no "windows" which could lead to heavy, concentrated water

FIG. 4. Guide template for receiving shaft

inflows. To accomplish this requirement, an elaborate template was built for the overlapping secant shafts to greatly reduce the tolerance in the top location of each drill hole (see Figure 4). Further, a Bauer BG30 drill rig was used to drill the secant holes since this rig could drill-in a 34.65 inch OD heavy wall casing for the full depth of the drill holes in coupled 20 ft. long sections. Rock augers or coring tools were used to remove the material inside the casing. A large hydraulic oscillator was utilized to pull the casing as the concrete was tremied by pumping through a tremie pipe extending to the bottom of the hole. The soft

Figure 5 - BG 30 Rig Drilling Hole and Oscillator Removing Casing in Another Hole

Fig. 6. Exposed Upper Secant Wall at Launching Shaft

secant shafts were drilled first and filled with 1500 psi concrete. Next, the hard secant shafts were drilled, the H-pile installed and then the hole backfilled with 3000 psi concrete. Figure 5 shows the BG30 rig drilling a hole while the oscillator removed the casing from another hole.

As the excavation inside the access cofferdams was carried down to the bottom of each of the compression ring walers, the rebar cage was placed and the walers were shotcreted using a 4000 psi concrete. Figures 6 and 7 show an exposed portion of the upper secant wall and a view looking to the bottom of the shaft with the TBM in position, respectively, at the launching shaft.

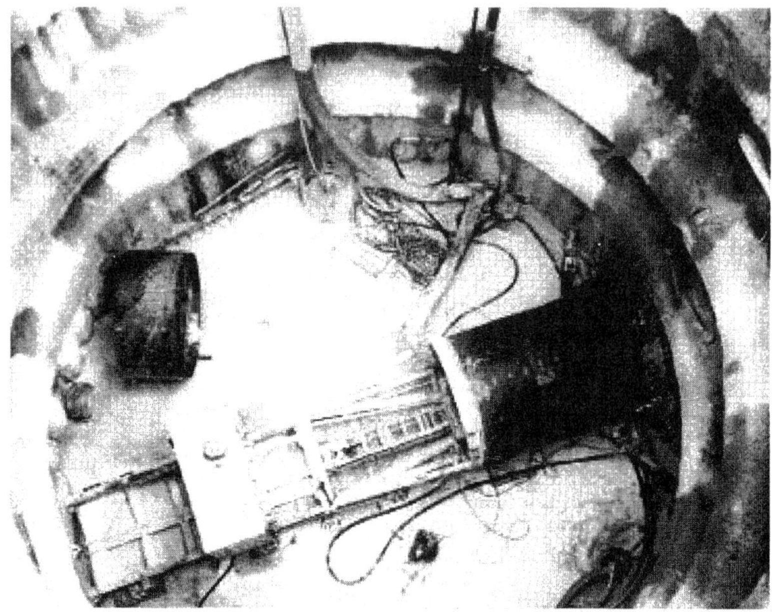

FIG. 7. View looking to Bottom of Launching Shaft with TBM in Position

For the pressure relief well system at both cofferdams, the first step was to run a full scale pump test using two wells that would function as part of the production well system. The pump tests indicated that the limestone aquifer has the characteristics of a confined aquifer with some indication of leakage. Drawdowns are directly proportional to pumping volume with this condition. The resulting relief well system for the south launching shaft consisted of eight 4 inch diameter wells around the outside of the shaft at a radius of 37 ft., along with two 6 inch diameter interior wells for sumps. All of the wells had a tip at El. -77 and were sealed in the clay formation just above the top of the limestone. The total volume pumped for all of the wells was on the order of 800 gpm, with the maximum and minimum volumes for any well being about 300 gpm and 50 gpm, respectively. At this site, a number of well installations did not contact voids capable of passing extraordinary volumes of water. After extended pumping, the measured equilibrium water levels in the wells varied from El. -30 to -70. It should be noted that during the well installation, the artesian head in the limestone had risen to roughly El. +12.

At the north receiving shaft, the relief well system consisted of six 4 inch diameter wells around the outside of the shaft at a radius of 25 ft., along with one 6 inch diameter sump well on the inside of the shaft. The total pumped volume for the well system was roughly 325 gpm. At this shaft, the measured equilibrium water levels in the wells varied from El. -67 to -74 after extended pumping.

Instrumentation and Monitoring

The north receiving shaft was instrumented with one inclinometer tube attached to a H-pile and a pair of vibrating wire embedment strain gauges attached to the rebar in each of the five waler levels. The maximum inward lateral displacement measured by the inclinometer was 0.05 inches, less than 0.01 percent of the excavation depth.

The compression forces calculated from the strains measured by the embedment strain gauges were only on the order of thirteen (13) percent of the original design loads. These small loads are believed to be due to the following factors:

a) Lateral pressures in the hard clay and soft limestone are much lower than assumed based upon active pressures for a friction angle of 32.5 degrees.

b) Earth pressures on a circular shaft are smaller than for the two dimension wall assumption.

c) Pressure relief wells lowered the hydrostatic head in the rock to well below the design level of El. +8.

d) Significant compression loads are taken by the hard/soft secant piles due to ring compression.

Summary and Conclusions

Two circular cofferdams for a tunnel below the Alafia River were successfully designed and constructed using hard/soft secant piles supported by compression ring walers. Only very minor seepage was observed on the exposed secant walls and no "windows" due to secant piles not overlapping were found.

A pressure relief well system successfully lowered the hydrostatic head in the limestone formation so that bottom stability was maintained and water infows from the bottom were controllable. Standard sump pit and pump operation kept the bottom of the shafts essentially dry. Observations of the rock within the tunnel excavation confirmed the fact that a grout plug would not be effective and that the pressure relief well system was the best solution.

Measured lateral displacements for the receiving shaft were extremely small. The compression loads calculated from the measured strains in the compression ring walers were also very small.

Acknowledgments

Owner: Tampa Bay Water
Program Manager: Hill/MK JV
Design Consultant: CDM
Geotechnical Consultant: Driggers Engineering Service Inc.
General Contractor: Kenko Inc.
Access Shaft D/B Contractor: Schnabel Foundation Company
Secant Shaft Drilling Contractor: Coastal Caisson Corp.
Pressure Relief Wells: Moretrench Environmental Services, Inc.

References

Miller, James A., "Groundwater Atlas of the United States - Alabama, Florida, Georgia, South Carolina", US. Geologic Survey, HA730-G, 1990.

Driggers Engineering Services Inc., "Test Boring Investigation for Alafia River Crossing of Tampa Bay Pipeline", March 2000.

Department of the Navy, Naval Facilities engineering Command, Design Manual DM-7.2, May, 1982.

DESIGN OF CANTILEVER SOLDIER PILE RETAINING WALLS IN STIFF CLAYS AND CLAYSTONES

Richard C. Sisson[1], P.E. Member, ASCE, Clint J. Harris[2], P.E., Member, ASCE and Robert L. Mokwa[3], P.E., Member, ASCE

ABSTRACT: The T-REX Project is a $1.67 billion project in the Denver Metropolitan area that consists of approximately 19 miles of a new Light Rail Transit (LRT) line, and improvements to approximately 17 miles of two interstate highways. Several miles of new retaining walls were required to provide for widened roadways and new light rail track. Because the right-of-way was restricted, tied-back walls were often infeasible, and so cantilever caisson walls with heights of over 30 feet and diameters up to 4 feet were employed. The caisson walls represented a considerable expense to the design-build team, and significant studies were performed to optimize the design.

Foundation conditions along most of the alignment consisted primarily of stiff clays and relatively weak claystone, often with moderate to high swell potential. A study of design methods available for walls in these materials indicates that uncertainties exist in selecting the appropriate long-term loading, soil strength, and even the method of analysis for design of long-term soldier pile walls. This paper describes the evaluation of the uncertainty and selection of a design approach for the project.

INTRODUCTION

The T-REX Project is a $1.67 billion project in the Denver Metropolitan area to replace outdated, aging highways in a heavily congested, growing business corridor with modern, efficient highways and a light rail transit system. The corridor is one of the most heavily used sections of roadway in the state, with average daily traffic (ADT) of more than 230,000 vehicles. The project consists of approximately 19 miles of a new Light Rail Transit (LRT) line, including 13 new stations and improvements to an existing station, and improvement to approximately 17 miles of two interstate highways. Highway improvements will result in one to two additional through lanes in each direction, and reconfiguration of

---

[1] Lead Geotechnical Engineer, Canadian Natural, Ltd. Calgary, AB, Canada, rick.sisson@cnrl.com (formerly Senior Geotechnical Engineer, Terracon Inc.)
[2] Geotechnical Project Manager, Terracon Inc., Wheatridge, CO
[3] Assistant Professor, Montana State University, Bozeman, MT

two major interchanges. Because of the critical nature of the infrastructure, a design-build approach was selected to provide speed, economy, and innovation in construction.

Extensive developments have occurred along this urban corridor over the past half-century such that space for expansion related to this project was very expensive. Several miles of new retaining walls were required to provide for widened roadways and light rail track. Where possible, mechanically stabilized earth (MSE) was employed as an economical solution. However, right-of-way restrictions required use of cast-in-place concrete cantilever walls supported on spread footings in some instances, and drilled shaft soldier pile retaining walls in others. Because the right-of-way was restricted, tie-backs for the soldier pile retaining walls were often infeasible.

Cantilever soldier pile retaining walls, which consist of a retaining wall constructed with vertical piles that are not tied back, were often utilized for the T-Rex project. The soldier piles were constructed using drilled shafts, although steel piles are also frequently used. This type of wall is often called a tangent wall when the center to center spacing of the piles or drilled shafts is one diameter. As the walls on this project had heights sometimes exceeding 30 feet and diameters up to 4 feet, they represented a considerable expense to the design-build team, and significant studies were performed to optimize the design.

GEOLOGIC SETTING

Surficial soils along the southern portion of the alignment typically consist of Pleistocene loess, composed of fine eolian silt and clay that is commonly mixed with sand. The deposit may be up to 20 feet thick, but more generally is less than 10 feet thick. Soils at the northern end of the project are frequently granular and include Quatenary eolian sands as well as alluvial deposits associated with the South Platte River and its tributaries.

Bedrock in the vicinity consists of the Denver Formation, a Paleocene to Upper Cretaceous sedimentary deposit that includes both sandstone and claystone facies. The deposit is typically a very weak rock, with RQD often zero and unconfined compressive strength below 20,000 psf. A more competent facies known locally as the "Denver Blue" is also encountered.

With the exception of the granular deposits at the north end of the project, most of the geologic materials at or near the ground surface along the project alignment contain varying amounts of clay, frequently with potential for swelling and shrinkage as a result of moisture content changes. Numerous buildings and other man-made structures in the area have been severely damaged due to swelling clays.

METHODS FOR ANALYZING LATERALLY LOADED PILES

Several widely used methods exist that may be used to calculate the lateral capacity of cantilever soldier pile retaining walls. In order to select a method for the T-REX design project, the theoretical basis for several methods was reviewed, and a comparative analysis was performed.

Limit Equilibrium Analyses

Teng (1962) describes a limit equilibrium method for analyzing sheet piles with uniform foundation conditions that is often applied to soldier pile design. This "conventional" method satisfies both force and moment equilibrium. However, the method is not readily extended to the analyses of non-uniform foundation conditions. Teng presents a "simplified" method that uses a less accurate passive pressure distribution that allows application to a layered foundation system. Teng's limit equilibrium methods assume plane-strain conditions. Thus, they may be most applicable to tangent soldier pile walls. The Teng method could potentially be applied to widely spaced soldier piles by using the inverse of group load factors described below, but this approach is unproven.

Sabatini et al. (1999) describes the Wang-Reese method for evaluating ultimate passive resistance of soldier beams embedded in cohesionless and cohesive soils. The method explicitly includes pile spacing. Based on force-equilibrium, the method includes three potential failure mechanisms:

1. A wedge failure in front of the pile.
2. An overlapping wedge failure for deep or closely-spaced shafts.
3. Plastic flow around the shaft.

Calculations performed using limit equilibrium methods yield a pile embedment depth where lateral loads are equal to soil resistance (i.e., a safety factor of 1). Classically, this embedment depth is increased by 20% to 40% to provide a margin of safety. Alternatively, the calculations may be performed using either reduced shear strength or an increased load. Often, a value of 1.5 is used to reduce the shear strength or increase the load. Limit equilibrium methods offer only very approximate methods to calculate pile-head deflection.

Brom's Method

Broms (1964) presents a method to calculate ultimate lateral resistance of piles assuming that failure takes place when two plastic hinges form in the pile, or when lateral resistance of the supporting soil is exceeded along the total length of the pile.

The failure mode depends on the pile length, the stiffness of the pile section, and the load deformation characteristics of the soil. The failure mode for soldier pile walls typically falls into the classification of short piles, where pile lateral capacity is a function of the lateral resistance provided by the soil. Broms' method assumes no resistance to a depth of 1.5D, and then $9.5c_uD$ below that depth, where c_u is the undrained soil strength, and D is the pile diameter.

Broms' method considers the fundamental behavior of laterally loaded piles, and can be applied with simple charts. The method also provides a means of calculating pile-head deflections. However, the method is limited to uniform soil deposits, and the approach does not account for nonlinear soil stress-strain behavior. With the availability of modern computers, Brom's method is gradually being replaced by more rigorous analytic and numeric approaches.

P-Y Analyses

The p-y method evolved from research sponsored by the petroleum industry in the 1950's and 1960's. The advent of the desktop computer has led to widespread application of the method. P-Y analyses predict soil resistance along a pile as a function of pile deflection by treating the soil resistance as a series of functions based on discrete elements and non-linear Winkler springs. Semi-empirical formulations have been developed for establishing properties of the resistance elements based on the results of full-scale instrumented loading

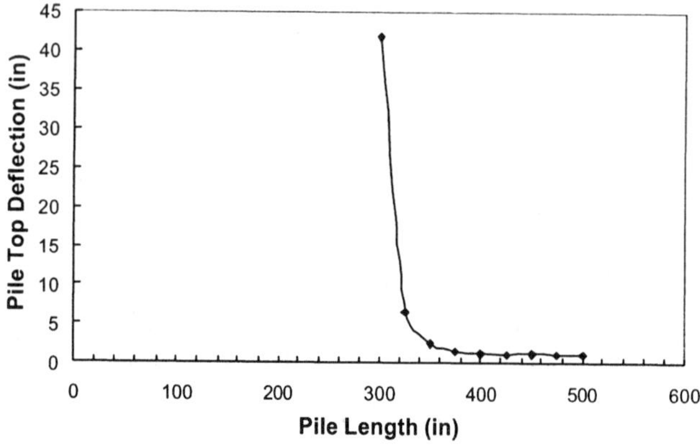

FIG. 1. Estimate of pile critical length based on p-y analyses.

tests. A significant advantage of the method is that it allows consideration of non-linear soil stress-deformation response. Reese and Wang (1977) describe implementation of the method in detail.

The p-y method does not explicitly yield embedment depth. Instead, pile head deflection is plotted against pile length, as shown in Figure 1. The curve typically indicates a critical length where deflection begins to increase rapidly. Generally, a pile is selected with a length somewhat greater than the critical length.

Finite Element and Finite Difference Solutions

Finite element and finite difference methods provide approaches that allow for full three-dimensional, non-linear analyses of soil-pile interaction. Reese and Wang (1997) note that difficulties lie in selection of an appropriate constitutive model, methods for dealing with tensile stresses, modeling layered soils, separation between pile and soil during repeated loading, and changes in soil characteristics associated with various types of loading. While these methods provide powerful research tools, the complexity limits application for routine design.

PILE SPACING EFFECTS

Both theory and experiment indicate that the capacity of a group of closely spaced piles is less than would be calculated based on the sum of the capacities of the same number of widely spaced piles. Of the available theoretical methods, only the Wang-Reese method directly includes spacing in calculation of pile capacity. Capacity is reduced when the assumed plastic failure zones for adjacent piles overlaps. Typical group reduction factors calculated using the Wang-Reese method are shown on Figure 2. Because the Teng limit equilibrium methods are based on plane strain analyses, they directly apply to tangent pile walls. No documented method is known for applying the Teng approach to widely spaced piles. Group effects in p-y analyses are indirectly considered by multiplying the soil resistance by a p-multiplier, a factor that is one for widely spaced piles and decrease with pile spacing. This factor is determined from experimental data as described below.

Field tests to measure group effects on multi-row pile groups indicate that capacity reduction due to group effects is greater for trailing piles than for leading piles as a result of the shadowing effect (Mokwa and Duncan 2001). Mokwa and Duncan (2001) analyzed the results of four separate experimental studies on single rows of piles oriented normal to the direction of load (side-by-side arrangement). As shown in Figure 2, their study indicates that piles oriented in a side-by-side arrangement at a spacing greater than or equal to 3D are not affected by adjacent piles. However, when the spacing is less than 3D, there

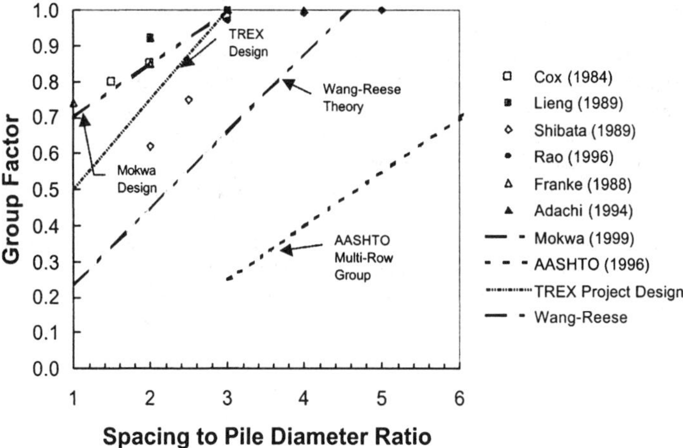

FIG. 2. Group efficiency for a single row of piles oriented perpendicular to the direction of load. (Modified from Mokwa 1999)

is a reduction in pile capacity as demonstrated by Franke's (1988) model tests, and as supported by elasticity-based methods (Poulos 1971).

For project design, p-y analyses were performed using a p-multiplier of 0.5 for a pile center-to-center spacing of 1.0, and a p-multiplier of 1.0 for spacing of 3.0 or greater. Because the data is limited and contains scatter, other relationships might also be selected. Mokwa (1999) recommends somewhat higher values than used for T-REX design, as shown in Figure 2. Both the theoretical results from the Wang-Reese analyses and field data indicate that use of average values for a multi-row group such as those in AASHTO (1996) is likely over-conservative for soldier pile walls, as leading piles in a group are likely to have a larger capacity than average.

COMPARISON OF ANALYSIS METHODS

A comparison of the various methods of lateral pile design was performed in order to select a technique for design of soldier pile walls on the T-REX project. The results of a simple parametric comparison are shown in Figure 3. The embedment ratio is defined as the length of pile below the base of the wall to the wall height. This ratio was calculated using several analytical techniques to determine a safety factor of 1.0 against lateral instability. The results show that the required embedment is greater for p-y analyses than for limit equilibrium methods. For example, for the Teng (1962) simplified limit equilibrium method, an embedment to wall height ratio of 0.2 is required for piles embedded in a uniform foundation with a cohesive strength of 6000 psf. The embedment ratio calculated using Wang-Reese is below 0.1. The ratio calculated by p-y analyses with a p-y multiplier of 0.5 is 0.4.

Poulos and Davis (1980) compared results of limit equilibrium analyses and those obtained from Brom's theory, and note that the calculated resistance from the latter method is larger. The larger calculated resistance by Brom's method is attributed to the higher ultimate resistance ($9c_u$ vs $2\ c_u$) used by Broms. Poulos and Davis suggest that plasticity (limit equilibrium) solutions are unduly conservative for typical pile sizes, but may be relevant for shallowly embedded sheet piling or groups of piles closely spaced in a single long row.

Because the various methods do not yield the same design embedment depth, the fundamentals of the methods were examined to establish the basis for the discrepancy. The primary reasons for the differences appear to be the assumed variation of ultimate soil resistance with depth, and the mobilization of resistance with pile deformation.

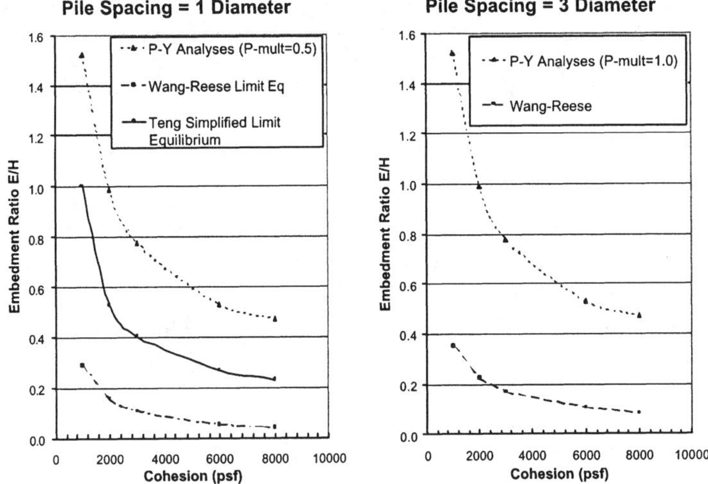

FIG. 3. Variation of Embedment Length with Cohesion

Ultimate Soil Resistance

As noted by Poulos and Davis (1980), one difference between the various methods for calculating lateral pile capacity is the variations of ultimate soil resistance with depth. The variation calculated using methods appropriate for tangent piles is shown in Figure 4A, and for piles spaced at three diameters is shown on Figure 4B. The variation shown for the p-y method is generally similar to the distribution measured in tests. The distribution calculated using the Wang-Reese method is similar to p-y analyses. The distributions assumed by Broms and Teng represent simplifications adopted for convenience of calculation.

Mobilized Soil Resistance

The failure mode for soldier piles typically involves rotation of the pile around a point below the base of the wall. Near the rotation point, soil deformations may not be large enough to fully mobilize the ultimate passive resistance. Of the available methods excluding finite element analyses, only p-y analyses consider variation in soil resistance with pile deformation. Teng's conventional method considers a transition region around the rotation point, but the extent of the transition region is established by equilibrium considerations, not by soil deformation. Teng's simplified method assumes full mobilization of ultimate soil resistance, as does the Wang- Reese method and the Brom's method.

Of the methods evaluated, p-y analyses, Teng's conventional method, Brom's method, and p-y analyses satisfy both force and moment equilibrium. The Teng simplified method does not satisfy both equilibrium equations simultaneously. The Wang-Reese method satisfies only force equilibrium, and assumes full passive resistance is mobilized over the full depth of the embedded pile. This is likely an important reason that the latter method indicates the lowest embedment ratio of any method, substantially less than calculated by the p-y method.

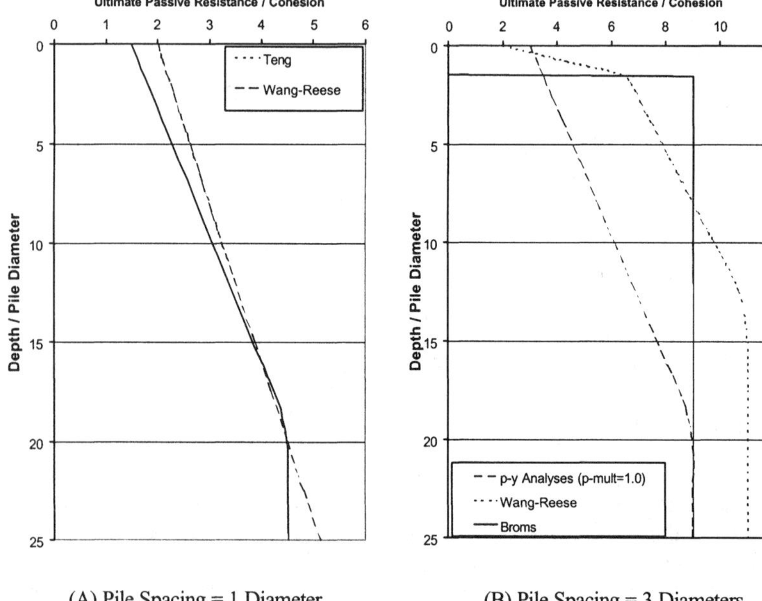

(A) Pile Spacing = 1 Diameter (B) Pile Spacing = 3 Diameters

FIG. 4. Variation of Ultimate Passive Resistance with Depth

Figure 5 shows a comparison of mobilized resistance versus depth for a soldier pile wall in a uniform cohesive foundation. The results show that the mobilized soil resistance calculated using p-y analyses for pile head displacement of about 6% of the wall height is less than that calculated by limit equilibrium. Consequently, longer embedment depths are required for stability based on the p-y approach. This is because maximum soil resistance is only mobilized over a limited length of the pile for typical design loads. The p-y approach inherently accounts for the partial mobilization of soil resistance. It may be possible to develop the resistance as predicted by limit equilibrium analyses, but the wall displacement would be much larger than could be tolerated.

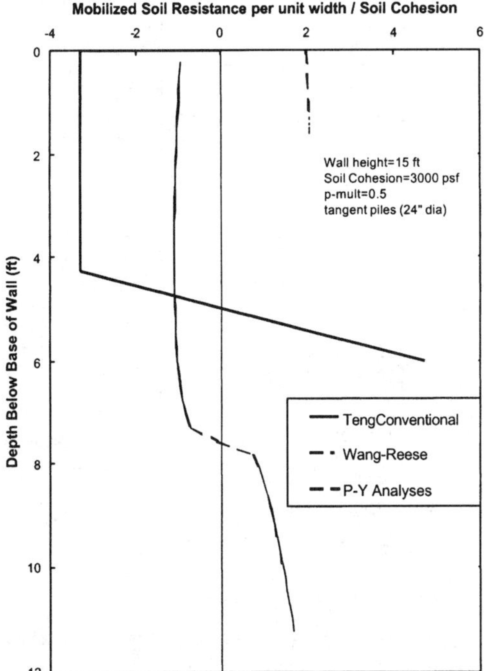

FIG. 5. Variation of Ultimate Passive Resistance with Depth

DESIGN APPROACH

Analysis Method

The p-y method of analysis was selected for design because it provides a rational method for determining pile embedment length with a reasonable factor of safety. Figure 6 shows the variation of top-of-wall deflection as a function of pile length. For the case analyzed, p-y analyses indicate that very large deflections occur at an embedment to height ratio of 0.56. At an embedment to height ratio of 0.87, the analyses indicate that further increases in embedment do not significantly decrease deflection. This difference represents about a 50 percent increase in embedment ratio required to move from near instability to acceptable deflections. The embedment selected for design should be based on deflection control, and also for other uncertainties such as variation in soil properties and stratigraphy, and construction tolerances. Thus, while a 30% increase in embedment length would likely reduce deflections to an acceptable value, it would not provide any margin of safety for uncertainty in design parameters and construction tolerances. Limit equilibrium methods

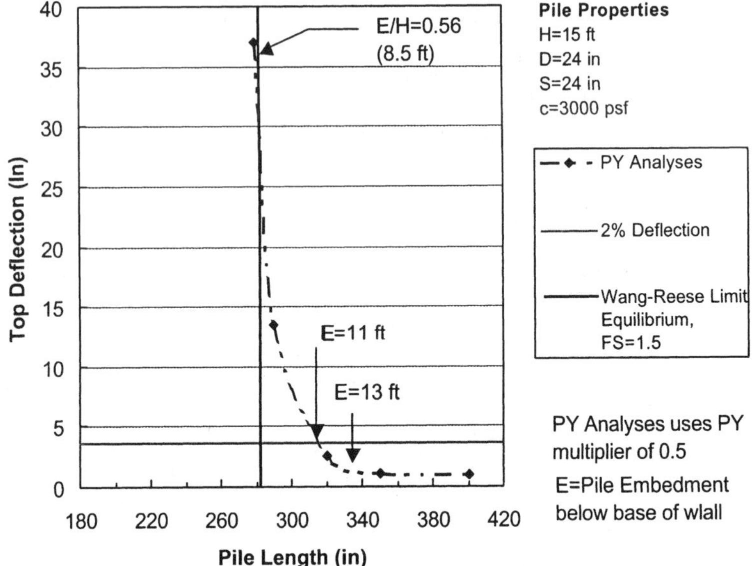

FIG. 6. P-Y Design of Soldier Pile

provide only very approximate methods for separating safety margin for deflection control and the safety margin for uncertainty in soil properties and stratigraphy. P-y analyses were adopted for the T-Rex walls because this method provides a more rational basis than limit equilibrium for evaluating a proposed wall design. The basis for selecting soil parameters is described below. Embedment ratios were then selected at the point where the deflection versus pile length curve becomes essentially horizontal.

Soil Strength

The stress-deformation behavior of stiff, fissured clay and clay shale is complex and involves considerable uncertainty. Unfortunately, performance data for laterally loaded walls is insufficient to establish appropriate methods for selecting strength parameters. While short-term, undrained strength of stiff clays and clay shales is typically high, the following factors can affect the long-term strengths.

1. Strength reduction may occur as a result of progressive failure during strain under loading.
2. A loss of cohesive strength may occur due to swelling resulting from moisture content changes.
3. Negative pore water pressures may occur as a result of excavations or desiccation. Subsequent decreases in negative pore water pressure will result in strength loss.

4. Strength loss over time may occur due to weathering.

Current practice for analyses of slopes in stiff fissured clay is to design using strength between fully softened and residual (Stark and Eid, 1997), but it is not clear that this approach is appropriate for laterally loaded piles. Bjerrum (1966) describes the progressive failure mechanism in detail, particularly in regard to slope failures. One important requirement for progressive failure is that boundary conditions allow large lateral strains to develop without development of additional lateral restraint –i.e., the failure zone is not confined. While the failure zone is probably relatively unconfined above the base of a retaining wall, contained plastic flow likely exists around piles with significant embedment below the base of the wall. Under the conditions of contained plastic flow, the lateral load may be transferred through a highly strained zone, which is approaching the ultimate strength, to a zone where strains are sufficiently small and peak strength is applicable. Progressive failure seems unlikely under these conditions, and as a result, the use of fully softened strength (or less) would be overly conservative. The use of fully softened strength results in pile embedment depths that are much larger than customary, providing some support to this hypothesis.

Although progressive failure may not lead to widespread strength reduction in the clay shale below the base of the wall, other mechanisms such as a decrease in negative pore pressure, swelling, or weathering may reduce the strength to some extent. However, quantifying the reduction is not possible with conventional design approaches. On the basis of the above considerations, the following judgment-based approach was used to establish strength parameters for calculating active earth pressures and lateral pile resistance for soldier pile walls:

1. Lateral earth pressures were calculated based on a fully softened friction angle, with an added component that depended on the measured swell potential of the soil as described below.
2. Pile resistance below the base of the wall was calculated using the undrained strength established assuming a weathered zone thickness of at least 3 ft, and a lower bound estimate below the weathered zone.

For design purposes, undrained shear strength estimates were determined considering results of both penetration tests and unconfined shear strength tests. Penetration resistance was considered an upper bound indication of strength, because of the effects of thin, hard siltstone and sandstone layers within the clay shale. Unconfined shear strength tests were considered a lower bound, because of sample disturbance effects. Blow counts for a 2.8-inch sampler with brass liners ranged between 50 and 100, while measured unconfined shear strengths (i.e., half the unconfined compressive strength) typically ranged between 2000 and 8000 psf. Equivalent SPT blow counts are estimated to be about one-half to two-thirds of those for the 2.8-inch sampler. Based on relationships between blow count and undrained strength published in NAVFAC (1982), this indicates undrained shear strengths ranging from about 3500 psf to 10,000 psf or more. Typically, a shear strength of 2000 to 3000 psf was assigned to the weathered zone, and 6000 psf was used for relatively unweathered material.

Swell-Induced Lateral Earth Pressures

The potential was recognized for developing additional pressure and wall movement due to swell induced by moisture content increase. In accordance with routine engineering practice, swell potential was identified by measuring vertical deformation in response to saturation in an oedometer. While this test provides a reasonable index of swell potential, the boundary conditions do not replicate those for soldier pile walls where lateral movement is only partially restrained. Therefore, lateral pressures caused by swell were estimated based on judgement and experience using the design swell pressures shown in Table 1.

TABLE No. 1 Design Swell Pressures

Measured Swell at in-situ vertical stress	Clayey Loess	Claystone Bedrock
<1%	45 pcf	45 pcf
1% to 3%	65 pcf	75 pcf
>3%	75 pcf	90 pcf

CONCLUSIONS AND RECOMMENDATIONS

The p-y method of analysis offers an approach for design of cantilever soldier pile walls that is useful for comparing effects of variations in factors such as stratigraphy, wall height, wall loads, and wall displacement. However, uncertainties that exist in establishing both long-term lateral earth pressures and soil strength are such that careful consideration must be given to prior experience under similar conditions. For hard clays and relatively weak claystone, pile embedment to wall height ratios on the order of 0.7 to 1.2 are within the range of precedence. Although lower embedment ratios may be calculated depending on the choice of analytical method and soil strength parameters, considerable uncertainty exists in implementing such designs.

The p-y approach is preferable over limit equilibrium methods for establishing pile lengths, because the p-y method:

- is supported by published field tests,
- directly considers displacements,
- is mechanistically superior, and
- provides a semi-rational treatment of layered stratigraphy.

Further research is required to address several issues, including:

- measurement of actual swell pressures developed on unrestrained retaining walls,
- selection of long term strength parameters for cost-effective design of laterally loaded piles in stiff clays, and
- treatment of group effects for loading of single row drilled shafts.

ACKNOWLEDGMENT

William Attwooll, P.E., M.ASCE. contributed significant advice and experience during the preparation of this paper.

APPENDIX I. CONVERSION TO SI UNITS

Inch (in) x 25.4=millimeter (mm)
Feet (ft) x 0.305=meter (m)
Pounds per square foot (psf) x 0.0479=kilopascal (kPa)
Pounds per cubic foot (pcf) x 0.157 = kilonewton per cubic meter (kN/m^3)

REFERENCES

American Association of State Highway and Transportation Officials (AASHTO). (1996). *Standard Specifications for highway bridges,* 16th ED., AASHTO, Washington D.C.

Bjerrum, Laurits, (1967). "Progressive Failure of Overconsolidated Plastic Clay and Clay". *ASCE Journal of the Soil Mechanics and Foundations Division,* 93 (SM5), 3-49

Broms, B.B. (1964) "Lateral Resistance of Piles in Cohesive Soils", *ASCE Journal of the Soil Mechanics and Foundations Division.* 90 (SM2), 27-63

Franke, E. (1988). "Group action between vertical piles under horizontal loads". W.F. Van Impe, ed., A.A. Balkema, Rotterdam, The Netherlands, 83- 93.

NAVFAC DM-7.1, (1982)."Soil Mechanics". *Department of the Navy Naval Facilities Engineering Command.* 7.1-88.

Mokwa, R.L. (1999). "Investigation of the resistance of pile caps to lateral loading". Ph.D. Dissertation, Dept. of Civil and Environmental Engineering, Virginia Tech, Blacksburg, VA, 383 pages.

Mokwa, R.L. and Duncan, J.M. (2001). "Laterally loaded pile group effects and p-y multipliers." *ASCE Geotechnical Special Publication (GSP No. 113),* Foundations and Ground Improvement, 728-742.

Poulos, H.G. and Davis, E.H. (1980) "Pile Foundation Analysis and Design". John Wiley and Sons

Poulos, H. G. (1971). "Behavior of laterally loaded piles: II – group piles". *ASCE Journal of the Soil Mechanics and Foundations Division.* 97 (SM5), 733-751.

Reese, L.C. and Wang, S.T. (1997) "Computer Program LPILE Plus, A Program for the Analysis of Piles and Drilled Shafts under Lateral Loads", Ensoft, Inc.

Sabatini, P.J., Pass, D.G., and Bachus, R.C. (1999) "*Geotechnical Engineering Circular No. 4*, Ground Anchors and Anchored Systems". FHWA-IF-99-015

Stark, T.D. and Eid, H.T, (1997). "Slope Stability Analyses in Stiff Fissured Clays ". *ASCE Journal Geotechnical and Geoenvironmental Engingeering* 123 (4), 335

Teng, W.C. (1962). *Foundation Design*. Prentice Hall, Inc.

DESIGN, CONSTRUCTION, AND PERFORMANCE OF AN ANCHORED TANGENT PILE WALL FOR EXCAVATION SUPPORT

J.P. Turner[1], Member, ASCE, J.A. Steele[2], Member, ASCE, W.F. Maher[3], M.R. Zortman[4], Member, ASCE, J.R. Carpenter[5], Member, ASCE

ABSTRACT: Excavation for a building in Washington, D.C. required support systems to limit the deformation of several adjacent structures. One adjacent structure would have been difficult to underpin conventionally because column footings are located approximately 13 feet behind the face of the excavation. A geo-support system consisting of tangent piles and two rows of tieback anchors was presented as an alternative to direct underpinning. In order to predict potential ground movements adjacent to the 27-foot deep excavation, tangent piles were analyzed as beam-columns subjected to lateral earth pressures, footing surcharge loads, and anchor loads. Deformations of the adjacent building were monitored during construction for comparison with the predictions. In addition, the drilled shafts of the tangent pile wall were used as a permanent foundation for the new building columns. This paper describes the design, analysis, construction and performance of the anchored tangent pile wall. This case history demonstrates the effectiveness of using this configuration of drilled shafts to limit ground movements in deep excavations close to existing structures.

INTRODUCTION AND BACKGROUND

This paper presents a design/construction case history in which an anchored tangent pile wall was used to support a portion of an urban excavation. The purpose of the excavation was to allow construction of a 14-story residential building located at 1499 Massachusetts Avenue in downtown Washington, D.C., including three stories of underground parking. The tangent pile wall was used to solve the difficult problem of limiting ground movements in the portion of the excavation adjacent to an existing 9-story steel-frame building. An aerial view of the site is shown in Fig. 1.

[1] Professor, Dept. of Civil & Arch. Engrg., Univ. of Wyoming, Laramie, WY 82071
[2] Vice President, Steele Foundations, Inc., 3299 K St., N.W., Suite 601, Washington, D.C. 2007
[3] Vice President, McKinney Drilling Co., P.O. Box C, Delmont, PA 15626
[4] Project Manager and [5]Chief Engineer, Engineering Consulting Services, Ltd., 14026 Thunderbolt Place, Suite 100, Chantilly, VA 20151

The adjacent building consists of a 2-story wing supported on strip footings located right at the edge of the excavation and a 9-story structure supported on individual column footings that are approximately 13 ft back from the edge of the excavation. The possibility of settlement beneath the column footings and associated damage made it necessary to either underpin the footings or to construct an excavation support system that would limit ground movements to acceptable levels.

FIG. 1. View of construction site, 1499 Massachusetts Avenue

SITE CONDITIONS

Regional Geology

The site is located in the Coastal Plain Physiographic Province of the District of Columbia. This area is typically underlain by variable thicknesses of fine to coarse grained sediments associated with step terrace deposits of the ancestral Potomac River. Older sediments associated with the drowned ancestral Potomac system, typically identified as Potomac Series sands and clays, are often exposed at the lower elevations of this general area. The near surface soils in this area have also been disturbed by previous construction, clearing and/or fill placement activities.

Soil Conditions

Soil borings for this site indicate stratification consistent with the regional geologic weathering profile to a depth of 85 feet from existing grades where the borings were terminated. In general, the soils encountered increased in density with depth.

Existing fill materials were present in the upper 3 to 19 ft. Below the fill materials, laminations of silts and clays with sands were encountered to depths of approximately 8 to 24 feet below existing grades. These soils were underlain by fine to coarse gravels and sands to depths of approximately 60 feet, followed by a clay stratum with fine laminations of sand and silt. This clay layer was the bearing stratum for the drilled shafts used to form the tangent pile wall. Fig. 2 shows the soil profile from a boring located along the alignment of the tangent pile wall and is typical of the conditions encountered.

Within the clay stratum, drilled shafts were designed to bear on natural soils that have a minimum Standard Penetration Test (SPT) N-value of 50 blows per foot. Recommendations for this material were provided for shafts to be sized for a maximum allowable bearing pressure of 35 kips per square foot (ksf). With a 35 ksf bearing pressure, total settlements were estimated to be on the order of 1 inch. Differential settlement between any two adjacent shafts was expected to be minor, although differential settlement along the length of the tangent pile wall was expected to approach ½ inch. Minimum penetration values were confirmed in every other drilled shaft, utilizing split spoon sampling at the time of construction.

Groundwater Observations

Groundwater ranged in elevation from as low as approximately EL. 42 feet to as high as approximately EL. 66 feet. Groundwater was a challenge during installation of drilled shafts for the tangent pile wall. The well-defined layers of gravel and sand described above were mostly water-bearing. The District of Columbia requires metering of water that leaves construction sites. During construction, metered groundwater flows as high as 70 gpm were observed. Groundwater flow calculations based on large well theory produced a steady state flow rate of about 30 gpm. Historically, in the District. initial flows are on the order of 2.5 to 3 times the steady state flow rate. For this site the current flow rate is about 25 gpm, which validates through field monitoring these historical observations.

DESIGN

The solution proposed by the excavation support subcontractor for controlling ground movements adjacent to the 9-story building consisted of an anchored tangent pile wall. As shown in Fig. 3, tangent piles consisted of 30-inch diameter drilled shafts on 32-inch centers. Fig. 4 shows an elevation view of a typical section. Two rows of tieback anchors were used to resist lateral earth pressures, the top row at a depth of 6 ft and the bottom row at a depth of 16 ft. Anchors were spaced on 8-ft centers, except in the top row where one additional anchor was located directly in front of each column footing, reducing the horizontal spacing to 4 ft in these locations (Fig. 3). Four column footings were supported in this manner. The proposed wall is 62 ft in length and required installation of 41 tangent piles and 30 tieback anchors. Maximum excavation depth was 27 ft.

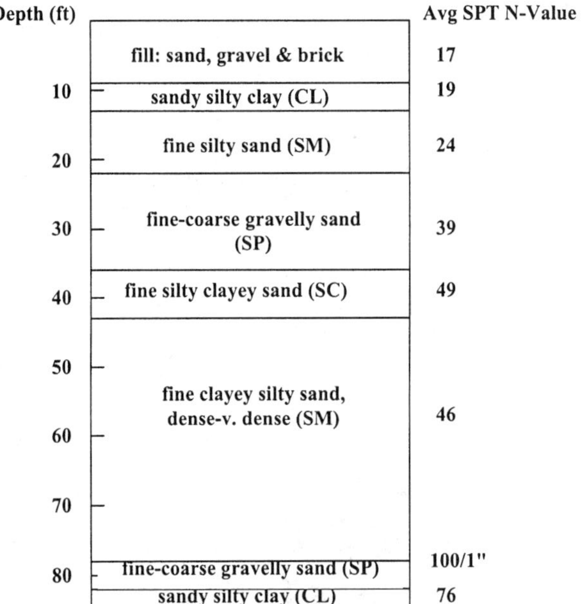

FIG. 2. Subsurface conditions adjacent to tangent pile wall (elev. 96.6 ft).

The principal design objective was to control ground movements to a level that would prevent damage to the existing structure. The proposed design was analyzed to predict lateral deflections during various stages of excavation. Structural capacities of the tangent piles and tieback anchors were analyzed and the design was optimized to obtain a cost-effective system that would provide the necessary capacity and stiffness.

Design Lateral Earth Pressures

Three components of lateral earth pressure were considered in this case: (1) apparent earth pressures associated with the retained soil, (2) earth pressures caused by surcharge of the adjacent wall footing, and (3) surcharge of the column footings. The apparent earth pressure (AEP) envelope used successfully in the Washington D.C. area for shoring design in granular soils consists of a trapezoidal distribution with an ordinate (maximum lateral earth pressure in psf) of 25H, where H = depth of excavation (ft). For the full excavation depth of 27 ft, this gives a maximum lateral earth pressure of 675 psf. Lateral earth pressure caused by the wall footing was analyzed on the basis of elastic theory (Scot 1963) for the stress on a wall due to a vertically loaded strip of infinite length oriented parallel to the wall. The bearing stress was estimated to be 2,218 psf on the 3-ft wide wall footing. Lateral earth pressure caused by column footings was also approximated as an infinitely long strip load under a uniform bearing stress of 4,500 psf. Although the column footings are not infinite strips, the approximate analysis was considered a conservative

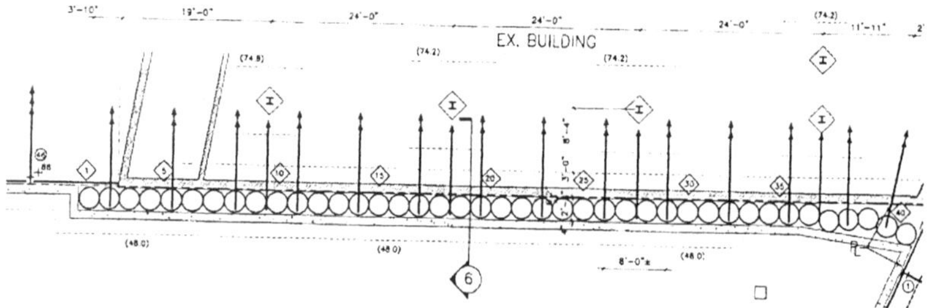

FIG. 3. Plan view of proposed anchored tangent pile wall.

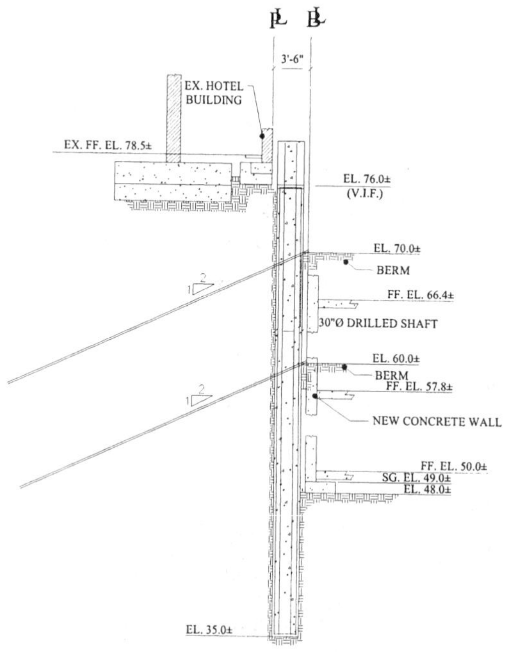

FIG. 4. Typical section, anchored tangent pile wall.

representation of lateral earth pressures. Superposition of the three components of lateral earth pressure results in the design pressure envelope shown in Fig. 5. This pressure envelope was used to design the tieback anchors and for predicting wall deflections for the final excavation depth. Horizontal components of the design anchor loads determined by taking the tributary areas beneath the design earth pressure diagram are: T_{H1} = 103 kips and T_{H2} = 130 kips.

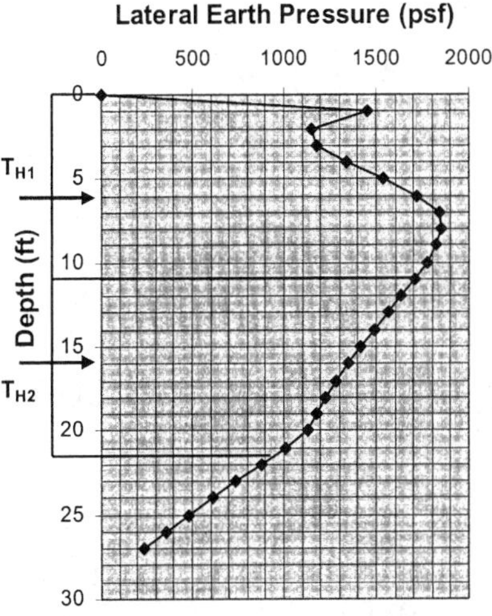

FIG. 5. Design lateral earth pressure diagram resulting from superposition of apparent earth pressure diagram and surcharge loads due to wall footing and column footings.

Analysis

The anchored tangent pile wall was analyzed as a beam-column subjected to lateral loading consisting of: (*i*) design lateral earth pressures as discussed above, (*ii*) tieback anchor loads, and (*iii*) passive earth pressure below the subgrade elevation. Loads are illustrated in Fig. 6. The computer program PYWALL (Ensoft Inc., 1999) was used to perform these analyses. Required input data includes: geometry of the retaining wall, structural properties of the wall element, restraint conditions applied to the wall, concentrated or distributed forces acting on the wall (e.g., anchor loads), subsurface geometry, soil properties including unit weight, strength parameters, and modulus, and soil resistance (*p-y*) curves for the subgrade soil which can be entered by the user or generated by the program on the basis of soil strength and stiffness properties. The computer program accounts for soil-structure interaction by modeling the soldier pile (drilled shafts) as a generalized beam-column on a nonlinear foundation, in which the nonlinear soil reaction is modeled by *p-y* curves. The governing differential equation of the bending beam is solved numerically using a finite difference scheme. Output consists of wall deflection, shear force, bending moment, and strut forces.

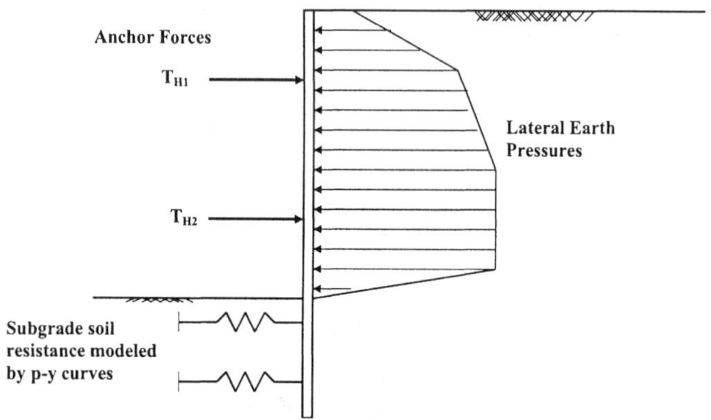

FIG. 6. Model for analysis of wall deflections.

Input parameters are summarized as follows:

Wall Stiffness (EI): The program STIFF1 (Ensoft Inc., 1999) was used to determine the nonlinear moment/EI relationship for a 30-inch diameter shaft with six No. 10 bars. For values of moment less than 1500 in-kips, EI was conservatively assigned a value of .15E+09 K-IN2, or 1.5E11 lb-in^2.

Anchor Restraint: Anchors were modeled using p-y curves at the nodes corresponding to the anchor locations. It was assumed that each anchor load is distributed to 2 tangent piles.

Soil Properties: The soil was modeled as a single layer of dense sand with $\phi = 38$ degrees, $\gamma = 125$ pcf, modulus of subgrade reaction k = 175 lb/in^3. The p-y curves for the soil below the depth of the excavation are then generated by the program.

Lateral Earth Pressures: For each depth that was analyzed, the total lateral earth pressure was determined by summing the apparent earth pressure diagram and the lateral earth pressures caused by the two surcharge loads (wall footing and column footing). For PYWALL the earth pressures in lb/ft^2 are converted to lb/in^2 then multiplied by the width of wall pressure carried by a single tangent pile (32 inches) and input in units of lb/in. Table 1 summarizes maximum deflection and moment for five stages of construction.

In the initial proposed design, it was assumed that the top row of anchors would be located at a depth of 8 ft. Analysis using PYWALL yielded a predicted deflection of 1.2 inches, which was considered risky with respect to potential damage to the adjacent building. This assessment of risk was based on the fact that the adjacent building was relatively new and that a significant portion of the design lateral earth pressure was a result of loads transmitted from the adjacent building. Clough and O'Rourke (1990) suggest that horizontal movements in well-designed insitu walls in

TABLE 1. Results of PYWALL Analysis, Anchored Tangent Pile Wall.

Loading Case	Max Deflection (in)	Max Moment (in-k)
6 ft cantilever	0.59	1,280
6 ft, top row installed	0.17	732
18 ft, top row installed	0.16	737
18 ft, both rows installed	0.16	737
27 ft (full depth)	0.18	1010

all values are for a single drilled shaft tangent pile

"stiff clays, residual soils, and sands" tend to average about 0.2% of wall height when the soil mass behaves approximately as an elastic material. This corresponds to 0.65 inches of lateral movement for the 27-ft deep wall of this case. The analysis (Table 1) showed that a small reduction in the cantilever depth (8 ft to 6 ft) reduced the predicted deflection significantly, to approximately 0.6 inches, which is within 0.2% of wall height. This case illustrates the benefit of conducting this type of analysis during design. A small change in design suggested by the analysis provides improved control of ground movements, an issue that is of paramount importance to all parties involved in the project.

A limitation of this type of analysis can also be observed in the results given in Table 1. The numerical analysis predicts that lateral deflections will actually decrease when the top row of anchors is installed. This occurs because of the increase in force applied by the anchors, the assumed elastic behavior of the system, and the two-dimensional model of the anchored wall. Experience shows that wall deflections are difficult or impossible to reverse once they occur, regardless of anchor load. Deflections may be reversed locally in the vicinity of the anchor, but maximum deflections, which typically occur near the top of the wall, will not be reversed significantly, if at all. However, the analysis is a valuable tool in that it predicts: (1) that the cantilever condition that exists prior to installation of the top row of anchors is a critical consideration for wall deflection and (2) that no significant increase in deflection will occur once the top row of anchors is installed. Subsequent measurements of wall deflections made during construction support this prediction.

CONSTRUCTION

Tangent piles consisted of 30-inch diameter drilled shafts. Initially, it was proposed to construct the shafts using the slurry method of construction. However, it quickly became evident that slurry construction was not feasible, primarily because of cobble size material that was not identified in the borings and larger than expected groundwater inflow. Casing was installed as an alternative to the use of slurry and this approach worked well for temporary excavation support. A 25-ft long top casing was installed through the soil. Upon reaching the layer identified as dense silty sand (Fig. 4) it was observed that this layer in fact consisted mostly of decomposed rock, mostly cobble sized material. The casing was sealed into this layer and drilling continued using a 28-in diameter rock auger and, in some cases, a core barrel, as shown in Fig. 7. This method allowed all shafts to be poured in the dry.

Shafts were drilled using a track-mounted Williams LDH. The wall was constructed by drilling alternate shafts then back stepping to drill the intervening shafts. Intervening shafts were drilled while the concrete in the initial shafts was green, usually one day later. Despite the unanticipated subsurface conditions, no claims were filed pertaining to construction of the tangent pile wall.

Tieback anchors were installed by drilling 4-inch diameter holes with the Interoc AN 109 B anchor drill rig. A combination of rotary drilling, top hammer, and a button bit was used to install 4-inch diameter casing. This method proved highly effective, particularly in the cobble layers. Each anchor consisted of four 0.5-inch diameter 7-wire strands, which were installed through the cased hole and grouted as the casing was removed. All anchors were proof tested to 1.2 times the design load and locked off at the design load. Fig. 8 shows the completed anchored tangent pile wall.

Following construction of the below-grade portions of the building, all anchors were cut. The drilled shafts were incorporated into the permanent foundation system of the building by supporting column loads along the perimeter wall adjacent to the 9-story structure. Design column loads ranged from approximately 250 to 800 kips and each column is supported by at least five drilled shafts.

DEFORMATION MONITORING

A monitoring program was conducted to monitor lateral and vertical deformations of the 9-story building, which was contiguous to, and supported by, the anchored tangent pile wall. The primary means of monitoring consisted of surveying a series of prisms (reflectors) attached to the 9-story building monitored from a fixed reference station.

FIG. 7. Drilling with core barrel. Note the cobble-size material in foreground. Inset shows core barrel and rock auger.

FIG. 8. Completed anchored tangent pile wall.

Utilizing a total surveying station, the horizontal and vertical deflection of each of the monitoring points was measured. The location of the reflectors at each reading were measured and compared with the initial baseline reading, which was obtained prior to construction on the site. Monitoring points were surveyed from April 20, 2001 through December 1, 2001. Horizontal deflections perpendicular to the face of each of the structures are shown in Fig. 9. Vertical deflection of each of the monitoring points are shown in Fig. 10.

Minor scatter in the surveying data is discernable, which is a function of the accuracy of the surveying equipment, meteorological conditions at the time of reading, measurement techniques, and operator consistency. The magnitude of the individual readings is not as important as the trends of movement or settlement. It is the trends of movement or settlement with time that are meaningful.

Due to the proximity of the tangent pile wall to the existing 9-story building, any movement from the tangent pile wall would be directly reflected by the 9-story building. During the monitoring program of the 9-story building, only limited minor lateral or vertical movement was noted. The maximum lateral and vertical deflections noted during the monitoring program was 0.3 inches laterally with no measurable settlement.

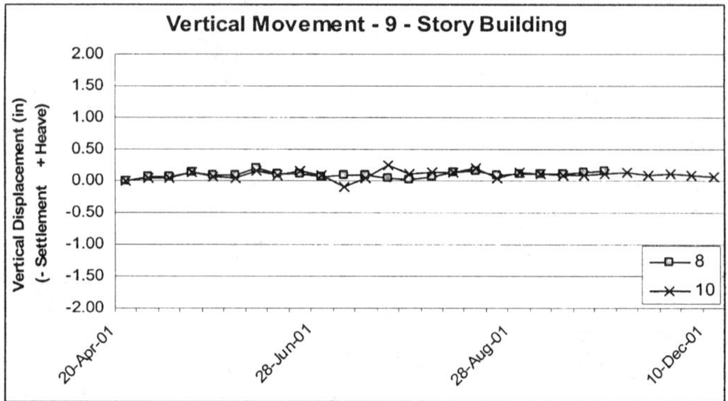

FIG. 9. Vertical displacement versus time, adjacent 9-story building.

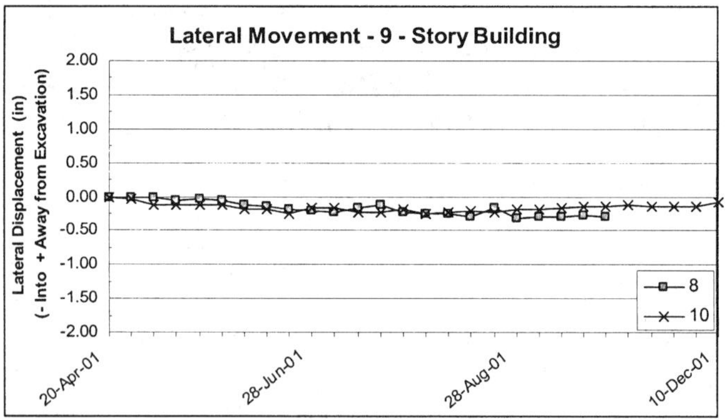

FIG. 10. Horizontal displacement versus time, adjacent 9-story building.

SUMMARY AND CONCLUSIONS

A case study is presented in which an anchored tangent pile wall was designed, built, and monitored, for the purpose of providing excavation support adjacent to an existing 9-story building. The primary factor involved in selecting this type of system was the location of large column footings that were set back from the face of the excavation approximately 13 ft., making conventional underpinning not feasible. To limit deformations of the adjacent building, a relatively stiff shoring system was required. An anchored tangent pile wall was selected for this reason.

Design of the wall required analysis of soil-structure interaction to predict ground movements. The program PYWALL was used to conduct deformation analysis and to analyze bending moment in the tangent piles. The usefulness and some of the limitations of this type of analysis are demonstrated. Results of the analysis showed that the cantilever condition prior to anchor installation was critical with regard to lateral deformations and the depth to the top row of anchors was decreased from 8 ft to 6 ft on the basis of these predictions. One limitation of the two-dimensional elastic analysis is that predictions of wall deformation in response to anchor installation are not consistent with field observations. Once lateral deformation has occurred, it is not likely to be reversed in response to anchor installation and stressing.

Construction of the drilled shafts was challenging due to the presence of cobbles that were not identified in the borings and because of large groundwater inflows. Both of these conditions were addressed by changing the construction method from slurry method to casing. This case clearly demonstrates the importance of adequate site investigation, particularly as related to cobble and boulder size material and groundwater conditions. It also shows how a competent drilled shaft contractor can handle unexpected conditions and still provide a quality product.

Results of deformation monitoring of the adjacent 9-story structure showed that vertical and horizontal deformations were well within acceptable levels and no damage occurred. It is concluded that the anchored tangent pile wall provided a highly effective excavation support system and that the design criteria of limiting ground movements was achieved. The ability to incorporate the tangent piles into the permanent foundation system for the new structure increased the cost-effectiveness of the support system.

REFERENCES

Clough, G.W., and O'Rourke, T.D. (1990). "Construction Induced Movements of Insitu Walls", *Design and Performance of Earth Retaining Structures (GSP No. 25)*, Ed. P.C. Lambe and L.A. Hansen, ASCE, New York, 439-470.

ENSOFT, Inc. (1999). Computer Program PYWALL, Version 1, Austin, TX.

Scott, R.F. (1963). *Principles of Soil Mechanics.* Addison-Wesley Publishing Cmpany, Inc., Reading, MA. 550 p.

DESIGN OF DRILLED SHAFT CUTOFF WALLS FOR SLOPE STABILITY IN MARGINAL SOILS

Philip S. Lowery[1], P.E., Member, Geo-Institute of ASCE

ABSTRACT: The use of vertical or near-vertical buried structural members to stabilize slope failures is an established and effective approach to landslide correction and can be economical in urban areas constrained by property limits, access restrictions, and high land costs. Traditional designs for such slope stabilization methods have relied on assumed active and passive pressures mobilized by the failing soil mass. Increasingly, sites are being developed in soils of marginal slope stability where some form of stability enhancement is required to achieve a minimum safety factor against deep-seated failure. In such pre-failure situations, a different set of design criteria is required because the ground movements necessary to mobilize active and passive pressures might exceed structural tolerances and even result in strains that might initiate soil failure.

A practical method is presented for the design of drilled shaft cutoff walls used to enhance slope stability in marginally stable soils. The method specifically addresses the pre-failure condition and applies limiting deflection criteria to structural members to preclude soil failure. The use of commercially available slope stability and soil-structure interaction programs is adopted. Emphasis is placed on practical application by the experienced practitioner.

INTRODUCTION

The cutoff wall, typically consisting of reinforced concrete drilled shafts, is a common and proven method to resist a failing soil mass in an active landslide. The

[1]Manager, Geotechnical Services, Facility Engineering Associates, P.C.
11001 Lee Highway, Suite D, Fairfax, Virginia 22030-5018
lowery@feapc.com

approach is particularly useful in locations where existing upslope and downslope development limit the potential use of more straightforward methods such as re-grading and toe buttressing.

Design of such systems typically begins with an analysis of active and passive lateral earth pressures generated by the failure geometry and applies these pressures to an embedded cantilever wall comparable to a soldier pile wall used in excavation support. The driving or destabilizing force due to the failing soil mass is analogous to the active earth pressure wedge in classical Rankine-Coulomb limit equilibrium retaining wall analysis. The resisting pressure below the active failure plane on the downslope side of the cutoff wall is equivalent to the Rankine passive resistance. Soil-arching between the discrete structural members is assumed for load transfer between the soil mass and the cutoff wall.

The requirements of the traditional stabilization design using cutoff walls are:

(i) to find an adequate penetration of the cutoff wall below the observed failure surface that provides an acceptable global factor of safety for slope stability, and

(ii) to select a structural section and configuration that satisfy requirements for bending, shear, fixity, and deflection under the soil loading.

In the case of a marginally stable soil slope that has not experienced shear failure but which nonetheless demonstrates an unacceptably low factor of safety, a similar need for some form of stabilization is apparent. For a cutoff wall in marginally stable ground, the fundamental design requirements are the same as for the stabilization of an active landslide: adequate penetration to improve global stability and selection of an appropriate wall configuration to satisfy internal structural criteria. Consideration of the soil mechanics of the two situations, however, suggests that in the pre-failure, marginally stable condition the traditional design approach is inapplicable in at least one important respect that undermines the assumption that an improved factor of safety has been provided.

FAILURE AND PRE-FAILURE CONDITIONS

Retaining Wall Analogy

For comparison of a soil slope in the failure and pre-failure conditions, it is useful first to consider the analogy of a retaining wall undergoing deflection to mobilize active pressure and passive resistance. Published data suggest considerable variation in the wall deflection necessary to fully mobilize active and passive states. Values from commonly used sources are shown in Table 1.

The range of published values for both conditions is striking (not to mention the scarcity of published sources on the subject) and indicates the degree of uncertainty present in any evaluation adopting limit equilibrium analysis. More instructive is the consistently and significantly higher deflection required to mobilize passive resistance. Displacement at least two to four times that required for active conditions is necessary to mobilize passive resistance, up to about 4 percent of the wall height with an extreme of 10 percent in the cited references.

TABLE 1. Typical Displacement[a] to Mobilize Active and Passive Conditions

Soil Type (1)	Active (2)	Passive (3)
Granular	0.0005H[b] to 0.004H	0.002H to 0.1H
Cohesive	0.004H to 0.05H	0.02H to 0.04H
[a] Displacement measured at top of wall. [b] H is the exposed height of wall.		
Sources: NAVFAC, Wu, Bowles (active only), ASCE/Corps of Engineers.		

The study of lateral earth pressure under limit equilibrium conditions has naturally concentrated on its application to retaining walls. Under typical conditions for conventional walls, however, designers generally neglect the contribution of the 'passive wedge' to sliding resistance for the valid practical reason that the soil mass in front of the toe of a retaining wall could be excavated at some point during the service life of the wall. For retaining wall applications, therefore, it follows that the existence of a fully mobilized passive condition is generally unnecessary to satisfy structural requirements. Design procedures incorporate structural load factors and geotechnical safety factors such that any portion of a retaining wall that does rely on some form of passive resistance (for example, a shear key to resist sliding) does not need to approach the limit equilibrium passive condition. These procedures, either through happy accident or engineering insight, avoid having to reconcile the inconsistency of a fully mobilized active condition at a degree of displacement that would be insufficient to mobilize full passive resistance.

Active Slope Failure Condition Stabilized by a Cutoff Wall

Consider the case of an active slope failure in clay for which a subgrade cutoff wall is proposed as a means of stabilization (Fig. 1). Relative displacement at the shear zone between the failing soil and the underlying stable material is, by definition, at least equal to the displacement required to mobilize the soil's drained shear strength. Through subsequent additional displacement the shear zone will reach the ultimate condition, i.e., the soil's residual drained shear strength.

(General shear failure in the ultimate condition is defined as increasing strain under constant load. The narrower, analogous definition of increasing shear displacement under constant load is applicable for an active slope failure with a clearly defined slip surface. It is noted, however, that the entire soil mass above the failure plane is not necessarily in that state of volumetric strain required to satisfy the plastic failure criterion. A useful comparison can be made between the drained direct shear test, which creates a localized shear zone within the sample, and the drained triaxial shear test which, in principle, induces shear throughout the sample.)

Once the residual drained shear strength along the critical failure surface has been fully mobilized, displacement may be expected to continue until, depending on the failure mode, failure geometry, and the slope configuration, the failing soil reaches a point of restored equilibrium through relocation of mass (what might be called the 'metastable slump' condition). Under the condition of mobilized residual shear strength, it could be argued that the displacements are large enough to validate an active/passive analysis of mobilized earth pressures when a cantilevered structural member is embedded into stable soil below the failure surface.

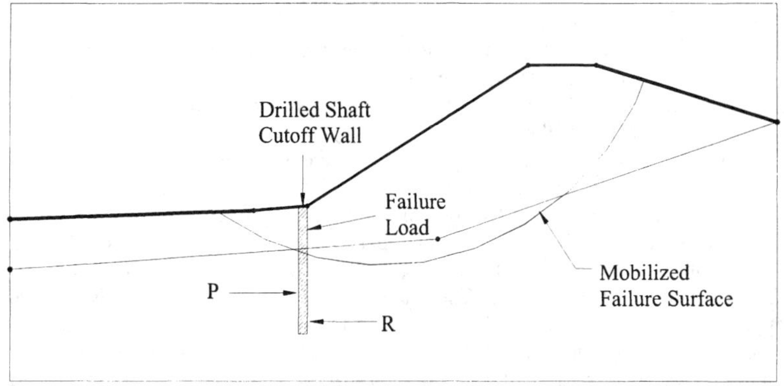

FIG. 1. Conceptual Slope Stabilization Using Active/Passive Analysis (after Abramson et al, 1996)

For a cutoff wall installed close to the toe of the slope, the equivalent 'active' pressure zone is defined by the point at which the mobilized failure surface passes through the line of the cutoff wall. The destabilizing forces due to the shear failure might or might not be similar in magnitude to a theoretical Rankine-Coulomb active wedge acting on the cantilevered portion of the cutoff wall above the failure surface. The mode and geometry of the given slope failure would determine whether this assumption is valid.

More important to the verification of the soil mechanics applied to such an analysis, however, is that the relatively large displacements along the failure surface (which can be physically observed through field instrumentation) may be expected to generate passive resistance downslope of the cutoff wall and below the observed failure surface. Note that failed material downslope of the cutoff wall but above the failure surface would be assigned no resistance in the analysis.

The principal assumption of the design method is that through load transfer from the failing soil mass to the cutoff wall the resulting horizontal deflection of the cutoff wall initiates passive resistance sufficient to balance the destabilizing force. Since any design would incorporate a reasonable factor of safety, it is unlikely that the mobilized passive resistance would ever reach the theoretical maximum because the displacement of the cutoff wall would have to be so large. Conversely, if the deflection of the cutoff wall *were* to become so large, it would then be difficult to make the case that the slope stabilization effort had been successful in practical terms.

Any stabilization of an active, ongoing slope failure begins to function when the actual or 'working' factor of safety has reached unity (1.0). It is therefore incorrect, for the purposes of analysis, to assume actual mobilization of the full theoretical passive resistance from which an applied safety factor provides the required margin of error. Rather, it is more realistic to acknowledge that:

(i) stresses in the passive zone will have developed *just enough* to balance the destabilizing forces,
(ii) the actual stresses represent an indeterminate condition between the theoretical 'at-rest' and 'passive' conditions, and therefore
(iii) the factor of safety of the stabilization design (within the assigned parameters of the model used for analysis) cannot be computed.

The inherent assumption in the use of mobilized passive resistance below the failure plane is that deflection of the cutoff wall is required post-installation and therefore that further slope movement is permissible following construction of the cutoff wall. This assumption is valid only when additional ground movement can be tolerated. Presumably this is not always the case, particularly when upslope structures are threatened by the slope failure. The widespread use of the active/passive design approach in practice, however, would suggest that applied factors of safety are conservative enough to account for the discrepancy, whether or not the designer knowingly makes allowances for the additional displacement. In the case of a cutoff wall consisting of discrete structural members, for example, the conservatism would result in closer shaft spacing and/or heavier sections.

The concept of a theoretical maximum passive resistance nonetheless provides the necessary upper limit to allow the designer to quantify the degree of conservatism necessary for a given design, and from which the 'working' resistance is estimated through application of reasonable factors of safety.

Marginally Stable Pre-Failure Condition

The slope stability problem that is the subject of the design method presented in this paper may be described as: a slope that is not currently experiencing an active failure but for which slope stability analysis indicates potential critical failure surfaces with unacceptably low factors of safety. The definition of 'unacceptably low' will vary from case to case, but is generally dependent upon the engineer's degree of confidence in the subsurface characterization, the method of slope stability analysis, the relative risk associated with current and proposed use, and the available (usually local or regional) knowledge of a given soil's behavior under similar conditions. The following situations may be considered to represent marginally stable conditions in the pre-failure state:

(i) An existing slope that is determined through exploration and analysis to contain significant residual-strength soils and which exhibits a critical factor of safety of 1.0 or slightly greater. Ancient landslides and slopes containing fissured, slickensided clays would fall into this category.
(ii) A proposed slope reconfiguration for which analysis indicates an unacceptable critical factor of safety. This may include a steepened slope, a cut to be supported by a retaining wall, or the introduction of building loads or other surcharges that adversely affect the global stability of an existing slope.

It is possible, of course, to have the two conditions occur simultaneously. In highly developed areas (for example, the urban and suburban centers of the east and west coasts of the U.S.), it has become a noticeable trend that developers are returning to sites that previously were rejected due to the presence of 'problem' soils. Increasing land costs and the high marketability of suburban and urban development now make such difficult sites economically viable. The trend towards high-density development with preservation of adjacent open spaces has also contributed to the increasingly dramatic changes to grades that site designers are proposing, which result in large retaining walls supporting cuts and fills to create level building lots. Developments of this nature frequently require some form of ground improvement or stabilization. Small, irregular building lots in conjunction with significant grade reconfiguration create conditions appropriate to the use of cutoff walls to ensure slope stability.

The fundamental difference in design philosophy between stabilization of an existing slope failure with a cutoff wall and the introduction of a cutoff wall to enhance a marginally stable slope is simply that the assumption of passive resistance developed downslope of the cutoff wall below the design failure surface is invalid if the slope is to be prevented from failing in the first place.

Implicit in the above statement is that the slope and cutoff wall deflections necessary to develop the active/passive state are unacceptably large. In fact, the criterion of 'failure' for the stabilized marginal slope should be redefined in terms of *tolerable deflection* rather than the shear failure criterion established by traditional soil mechanics. The designer is therefore faced with an apparent paradox:

(i) The cutoff wall should be designed to resist the maximum theoretical loading anticipated in the event of a slope failure.

(ii) Performance criteria for the deflection of the cutoff wall dictate that the failure condition cannot be assumed for the purposes of providing resistance to failure.

The problem now becomes a question of quantifying the available soil resistance supporting the cutoff wall in terms of predicted deflection. A relatively straightforward approach is to apply the principle of soil-structure interaction, which models the non-linear response of soil under load and which has been adapted and refined specifically to predict the behavior of laterally loaded piles and shafts.

The concept of non-linear p-y (load-deflection) curves for specific soil types is well documented elsewhere. Computer software is readily available to perform the lengthy, iterative finite-difference calculations necessary to find a solution to the fourth order differential equation describing shaft behavior under lateral load. In essence, the soil-structure interaction analysis allows specific structural shapes with known initial loading conditions to be applied to soil in a known initial equilibrium condition. The response of the shaft depends on the soil resistance, which in turn is a function of the applied lateral load.

The important application to the pre-failure, marginal slope condition is in the ability to directly relate the theoretical failure load to the predicted response beginning from initial equilibrium conditions rather than a state of fully (or even partially) mobilized passive pressure. The model is more realistic than a factored theoretical maximum passive pressure because it permits direct evaluation of the induced horizontal soil displacement. The designer is therefore able to establish and design for a maximum shaft deflection criterion, thereby avoiding the risk of inadvertently permitting excessive displacement that, ironically, could actually result in initiation of shear failure (or propagation of an upslope shear failure downslope of the cutoff wall).

CUTOFF WALL DESIGN

A simplified soil slope is used for clarity in the following design steps. (A thorough slope stability analysis that considers all potentially critical soil conditions and load configurations--including a determination of the feasibility of other methods of stabilization--is a prerequisite for design of the cutoff wall.) Figure 2 shows a marginally stable slope consisting of a clayey sand underlain by fissured, slickensided clay exhibiting residual drained shear strength characteristics. (Stability analysis for the example was performed using the STABL program with the STEDWin editing and graphics package.)

Slope Stability Analysis

The slope stability analysis is a two-step process:

(i) Determine the design critical failure surface from which cutoff wall loading will be developed.

(ii) Determine the minimum penetration below the design critical failure surface necessary to provide an acceptable global factor of safety.

Determine the design critical failure surface through conventional limit equilibrium analysis. Search routines should be set up to find the deepest failure surface that meets the definition of 'marginally stable' appropriate to the conditions under investigation. For the configuration in Figure 2, the presence of residual-strength soils suggests a factor of safety of near 1.0 is realistic. The goal is to find the highest potential failure load that may be expected to act on the cutoff wall in the event shear failure were to mobilize. Shallow slip surfaces, therefore, are not of interest in the current discussion.

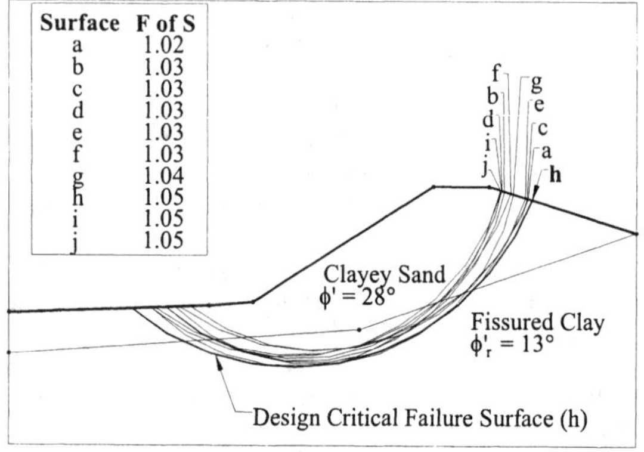

FIG. 2. Determination of Design Critical Failure Surface

Next, determine the depth of cutoff wall necessary to achieve the desired global factor of safety (Figure 3). Introduce a boundary line in the analysis to represent the depth of the cutoff wall. Trial failure surfaces should deflect below the boundary line. A logical cutoff wall location for the section under consideration would be at the toe of the slope. For grade reconfiguration problems involving retaining walls to support cuts, the logical location would be coincident with the retaining wall alignment. Several trial lengths may be required to reach the desired global safety factor (greater than 1.30 for the example).

Note that the minimum penetration relates only to the global limit equilibrium for slope stability. Structural requirements for drilled shaft fixity and deflection may dictate that additional penetration is necessary.

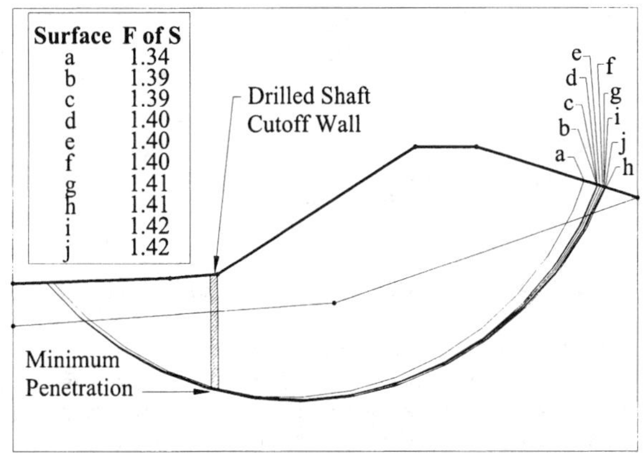

FIG. 3. Minimum Penetration of Cutoff Wall for Global Stability

Theoretical Failure Load

The soil model adopted for the design procedure calls for the application of the horizontal load resulting from a mobilized shear failure along the design critical failure surface identified in the slope stability analysis. The load is applied to that portion of the cutoff wall extending above the elevation of the design critical failure surface. A direct determination of the failure load at the line of the cutoff wall is available from an analysis of the design critical failure surface using an appropriate method of slices (e.g., Spencer's). Most slope stability programs incorporate an option to analyze a specific failure surface (easily imported from a previous general search), which includes a detailed output providing inter-slice forces. It is a straightforward matter to examine the output and find the corresponding side force for the slice at the line of the cutoff wall and the height of force application on the slice (Figure 4a). (The analysis geometry is typically such that the cutoff wall alignment coincides with a slice boundary--at the toe of the slope, for example, or at the base of a retaining wall. User-specified input, however, allows this coincidence to be pre-arranged for unusual cases, for instance mid-slope cutoff walls or irregular ground surface profile.)

A suitable check on the reasonableness of the failure load can be made by comparison with an approximate Rankine-Coulomb active wedge load (Figure 4b). The assumption that the soil within the active Rankine wedge is in a fully plastic state produces an upper bound solution for the lateral thrust, i.e., an over-estimate of the load. The comparison can also be used in situations involving retaining walls above the subgrade cutoff wall, where the retaining wall and backfill can be assumed to be a surcharge acting on a horizontal surface defined by the top of the cutoff wall.

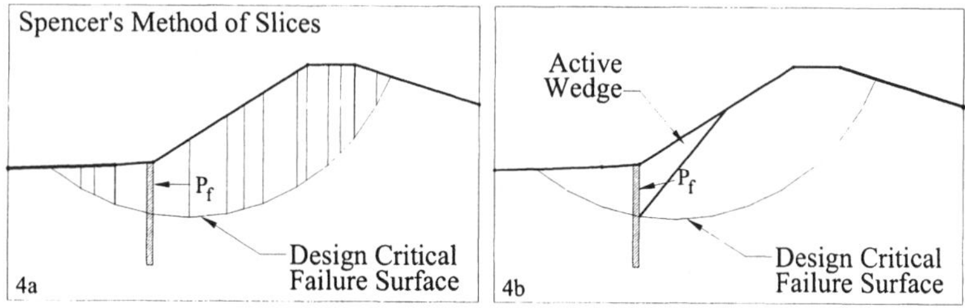

FIG. 4. Comparison of Predicted Failure Loads

It is worth noting that the calculation of the theoretical failure load is in one sense a search for a reasonable order of magnitude of load. The various limit equilibrium methods of stability analysis are known to exhibit considerable variability, largely due to the initial assumptions regarding indeterminacy in the moment and force equilibrium equations. The comparison with calculated loads using earth pressure theory allows a quick approximation to a number that will, after all, be increased by an appropriate factor to determine the design load.

Selection of this load factor is, once again, a question of engineering judgment as well as design performance criteria. Adopting load factors from ACI strength design methods for dead and live loads is suggested as a way to quantify the apparent degree of risk and as a qualitative measure of structural tolerance. A dead load factor of 1.4 is suggested for general use; a live load factor of 1.7 would be appropriate for situations where upslope development within the potential failure zone is threatened.

The final step in computing the design failure load is to calculate a distributed load on the drilled shaft wall for input into COM624P or an equivalent soil-structure interaction program. This may be most easily accomplished by assuming a triangular distribution from the top of the shaft to the design critical failure surface. (Alternative distributions might be applicable when surcharge conditions apply.) The maximum distributed failure load on the cutoff wall (per unit length of wall),

$$p_f = \frac{2\,(LF)\,P_f\,s}{D_c} \tag{1}$$

where:

D_c = depth to design critical failure surface, measured from the top of the shaft
P_f = failure load (by Spencer's, or other method)
LF = Applied load factor (e.g. 1.4 for dead load)
s = drilled shaft spacing, center-to-center

In such a back-calculation, there is no need to account for changes in soil stratigraphy or the presence of groundwater because these values would be incorporated into the initial slope stability analysis. The calculation of p_f in Eq. 1 is strictly to permit a realistic input of distributed load for the soil-structure interaction computation. Use of the resultant lateral force, P_f, applied at the line of thrust would be an over-simplification of the failure model adopted for the design method and would not accurately represent the loading mechanism.

Note that in the case of a surcharge acting at the top of the cutoff wall, an initial lateral pressure would be generated that must be added to the load distribution described in Eq. 1.

Trial Shaft Configuration

The selection of shaft size and spacing for analysis is essentially a process of trial and error with the initial restriction of minimum shaft depth determined for global slope stability. A maximum shaft spacing, s, of 2.5 to 3 times the shaft diameter may be adopted (AASHTO; Abramson et al) to ensure that soil-arching effects can be assumed. Local experience is invaluable in assessing the applicability of this rule of thumb. Construction cost savings are realized with increased shaft spacing for a given shaft diameter. Ultimately, however, the selection of a limiting deflection criterion and the structural performance of the drilled shaft in bending and shear may govern the selected configuration.

Structural Properties of Drilled Shafts

The Federal Highway Administration's design publications for drilled shafts provide a thorough description of the application of ACI strength design methods (ACI 318) to the design of reinforced concrete drilled shafts under lateral loading. For the purposes of this paper, it is relevant to note the following structural requirements for input into the COM624P program: moment of inertia, concrete compressive strength, steel reinforcement yield strength, concrete and steel elastic moduli, and layout of longitudinal reinforcement steel (number of bars and bar size). Evaluation of the program output also requires calculation of the shear modulus of the trial shaft section, which may also be determined using ACI methods. The COM624P program includes the PMEIX subroutine, which calculates the moment capacity of the trial section based on user input.

Selection of Limiting Deflection Criterion

As discussed earlier, the deflection of the drilled shaft cutoff wall (measured at the shaft head) is the critical parameter for the design of cutoff walls in the marginal, pre-failure condition as presented in this paper. In the absence of a controlling deflection criterion, the designer is unable to provide a rational justification for selection of a given cutoff wall configuration.

The limiting deflection criterion may be set based on the performance requirements of the enhanced slope and the known stress-strain properties of the marginal soil within the slope stratigraphy. For example, a marginal slope that is not expected to support habitable structures may reasonably be permitted deflection up to 25 mm (1 inch) or more without serious consequences. On the other hand, a slope that is to be re-graded and loaded with a significant retaining wall that in turn supports buildings or infrastructure might require a limiting deflection of less than 12.5 mm (0.5 inches).

Soil-Structure Interaction

A single, free-headed (unrestrained) drilled shaft subject to the design loading is analyzed for deflection, bending, and shear. The input parameters and operational procedures for COM624P or a commercially available equivalent (e.g., LPILE) are described in detail in the user manuals and will not be repeated here.

Soil properties are entered for the stratigraphy downslope of the cutoff wall. While the active/passive design approach quite properly neglects any resistance above the failure surface, the model of the pre-failure condition allows resistance in this zone because there has been no initial failure. Conceptually there will be no failure, based on resistance provided in part by the intact soils above the design critical failure surface. The writer acknowledges, however, that in several practical applications of the design method, the resistance attributed to this zone was minimized by using conservative input parameters applicable to soft clays. This decision was made primarily due to lack of confidence in the relevance of the programmed p-y curves and the absence of available load-deflection information for the local soils. It is therefore an opportunity to introduce a degree of conservatism appropriate to the structural tolerance of the slope under consideration and the reliability of the available soils information.

It is useful to note that COM624P permits the user to override the programmed p-y curves and to specify known values. The designer's preference should be to use as much reliable and realistic site-specific and soil-specific information as possible.

Iteration to Final Selection

The iterative process to analyze combinations of shaft-diameter/shaft-spacing/shaft-length is governed by the computed deflection at the shaft head but must also satisfy requirements for bending and shear in the drilled shaft. Inspection of the computed deflected shape and bending moment distribution also allows the designer to check for shaft fixity.

Iteration through a matrix of shaft-diameters/shaft-spacings provides sufficient data to conduct a thorough economic comparison of workable cutoff wall configurations. In general, the larger the shaft diameter the greater the cost despite the increased spacing; this effect is readily apparent from an analysis of concrete volume per unit length of cutoff wall.

CONCLUSIONS

The design method presented for drilled shaft cutoff walls to enhance the stability of marginally stable soil slopes adopts a soil model that reflects actual initial pre-failure conditions and analyzes the likely response of that model under ultimate loading.

The initially counter-intuitive proposition of designing for failure without being able to adopt the failure state for a portion of the analysis demonstrates the importance of developing a rational soil model. To be considered a 'successful' design, the cutoff wall introduced to enhance global stability of a marginal slope should not deflect beyond a magnitude dictated by performance criteria, which can be very strict. Without permissible displacement of the cutoff wall to mobilize resistance to design loads, the soil model is clearly very different from the limit-state model applicable to active failure conditions.

ACKNOWLEDGMENTS

The writer would like to acknowledge the helpful comments of Paul Swanson, Tom Larson, Emad Saadeh, and Ed O'Malley during development of the design method presented in this paper.

REFERENCES

AASHTO. *Standard Specifications for Highway Bridges*, 16th Edition.

Abramson, L.W., Lee, T.S., Sharma, S., Boyce, G.M. (1996). *Slope Stability and Stabilization Methods*, John Wiley and Sons, Inc., 492-495, 502-507

American Concrete Institute, ACI 318-71, Strength Design Method

ASCE/US Army Corps of Engineers (1996). *Design of Sheet Pile Walls*, ASCE Press, 21-23

Bowles, J.E. (1988). "Chapter 11, Lateral Earth Pressure." *Foundation Analysis and Design*, 4th Ed., McGraw-Hill, 474-476

Craig, R.F. (1987). "Chapter 6, Lateral Earth Pressure." *Soil Mechanics*, 4th Ed., Van Nostrand Reinhold UK, 175-191

Federal Highway Adminstration (1986). COM624P *User's Manual FHWA-SA-91-048.* USDOT.

Federal Highway Adminstration (1988). *Drilled Shafts Student Workbook FHWA HI-88-042*, National Highway Institute (USDOT), 303-352.

Lambe, T.W. and Whitman, R.V. (1969). *Soil Mechanics*, John Wiley & Sons, 328-330

NAVFAC. (1986). *Design Manual 7.2, Foundations and Earth Structures*, Department of the Navy, 7.2-60

Wu, T.H. (1975). "Chapter 12, Retaining Walls." *Foundation Engineering Handbook*, Winterkorn, H.F. and Fang, H-Y (Eds.), Van Nostrand Reinhold, 404-406

STABILIZATION OF RETAINING WALL ALONG TAILRACE CHANNEL OF HODENPYL DAM

Jesús Gómez[1], Ph.D., P.E.; Mary Ellen Bruce[2], P.E.; Donald Bruce[3], Ph.D., C.Eng.; Donald Basinger[4], P.E.; Allen Cadden[5], P.E.

ABSTRACT

A cantilever wall located along the tailrace channel of the Hodenpyl Hydroelectric Dam was undergoing significant lateral displacement. An unsuccessful repair attempt was carried out in 1996 using conventional strand tiebacks. Remediation of the movements was accomplished in 2002 by installing additional tiebacks through the wall. The tieback system selected for this project was Single Bore Multiple Anchors (SBMA), bonded into high-plasticity clays behind the potential slip surface. The SBMA system provided the desired tieback capacities in the clay with a reduced potential for long-term creep. This was the first application of post-grouted SBMA technology in the United States. The limited monitoring data collected after remediation suggests that the SBMA system has performed remarkably well.

To expedite remediation of the wall, the design had to be completed in phases and adjusted according to the monitoring data obtained until the start of construction. This paper presents a description of the investigation and analyses performed for design of the remediation. It also describes briefly the SBMA system and the results obtained as interpreted from the wall monitoring performed up to this date. A companion paper by Bruce et al. (2004) describes in more detail the SBMA design and construction details for this project.

[1] Member ASCE, Associate, Schnabel Engineering, Inc., 510 East Gay Street, West Chester, PA 19380; jgomez@schnabel-eng.com

[2] Member ASCE, President Geotechnica, s.a., Inc., P.O Box 178, Venetia, PA 15367; mebruce@cobweb.net

[3] Member ASCE, President Geosystems, L.P., P.O Box 237, Venetia, PA 15367; dabruce@geosystemsbruce.com

[4] Member ASCE, President, Schnabel Engineering, Inc., 405-A Parkway Drive, Greensboro, NC 27401; dbasinger@schnabel-eng.com

[5] Member ASCE, Principal, Schnabel Engineering, Inc., 510 East Gay Street, West Chester, PA 19380; acadden@schnabel-eng.com

INTRODUCTION

The Hodenpyl Hydroelectric Dam is located on the Manistee River in Michigan. The facility, constructed between 1923 and 1925, consists of an earthen embankment running roughly in the north-south direction. The powerhouse is located on the southern end of the embankment and discharges the water to a tailrace channel excavated in natural soils. Unlike most dams, the tailrace channel is nearly parallel to the core of the dam. Cantilever retaining walls of varying height were originally constructed on the east and west sides of the tailrace channel. The east wall is approximately 250 feet long and 30 to 50 feet high (see Figure 1).

FIG. 1. View of the tailrace channel and east retaining wall. Note the existing tiebacks installed in 1996. The sheetpiles were also added in 1996 as a facing and do not penetrate below the footing

The east wall was first repaired in 1996. The wall had undergone visible lateral displacements. A repair attempt was carried out using conventional strand tiebacks. Subsequent monitoring of the wall revealed that lateral displacement of the wall continued to develop, and that additional stabilization measures were required.

Design of the additional stabilization was performed in stages. Initially, a geotechnical investigation was performed to assess the potential causes for movement while collection of movement data continued. Inclinometers and piezometers were installed to determine the geometry of the sliding mass, and variation of the groundwater table behind the wall. Several potential mechanisms that could cause the observed movements were identified. It was concluded that a global slide encompassing the existing wall and tiebacks was the most likely cause for the movements.

Based on the results from these analyses, a preliminary remediation design was completed that was used as part of the bid package. These analyses also aided in establishing additional measures to enhance the existing monitoring program and to verify the preliminary design assumptions. Construction drawings were furnished to the bidders along with a set of well-defined performance specifications. The bidders were required to provide a final design to match or exceed the established performance requirements.

Pre-selection and final selection of contractors took place concurrently with monitoring of the wall and redesign of the remediation. After the contract was awarded, final adjustments to the remediation design were introduced based on interpretation of the latest results from the extended monitoring program. It was understood, however, that movements of the wall might continue and that additional remediation measures might be necessary in the future. As designed, the stabilization was expected to improve the present conditions of the wall while additional monitoring information was collected.

This approach presented several advantages. It reduced costs, as design was completed for the most likely slide mechanism, while some risk related to unknown conditions was accepted and managed. It shortened the total time required for procurement and construction, as procurement was started before final design was completed. Finally, it involved the owner and contractor in the decision process while making them aware of the importance of wall monitoring and testing of the proposed solution.

Because the slip surface was relatively deep, stabilization of the wall required tiebacks that developed relatively large capacities within the high-plasticity clay predominant at the site. The solution adopted for this project consisted of the use of post grouted Single Bore Multiple Anchors (SBMA), bonded behind the potential slip surface. This tieback system was selected due to the need to obtain the desired capacities in the high-plasticity clay with reduced potential for long-term creep.

The proposed stabilization was successful and reduced significantly the rate of movement of the wall in the months following tieback installation. This paper describes the approach followed during the geotechnical study and design of the stabilization. The paper also includes an evaluation of the performance of the SBMA system based on monitoring and inclinometer data before and after stabilization. Details regarding the design, installation, and testing of the SBMA system are given in a companion paper by Bruce et al. (2004).

SOIL CHARACTERISTICS

Figure 2 shows a cross section of the east tailrace wall. The section depicts the soil strata identified from the site exploration. The soils behind the wall consisted of a layer of poorly graded sand, underlain by an interval of overconsolidated clay. The clay typically contained a 10- to 30-percent fraction of fine to coarse sand and traces of gravel-size particles. Sporadic lenses and/or seams of sand were also detected at some locations within the clay mass. The results of the laboratory testing revealed that the plasticity of the clay materials increased with depth. An approximate and somewhat arbitrary boundary between low and high plasticity clay is represented in the figure.

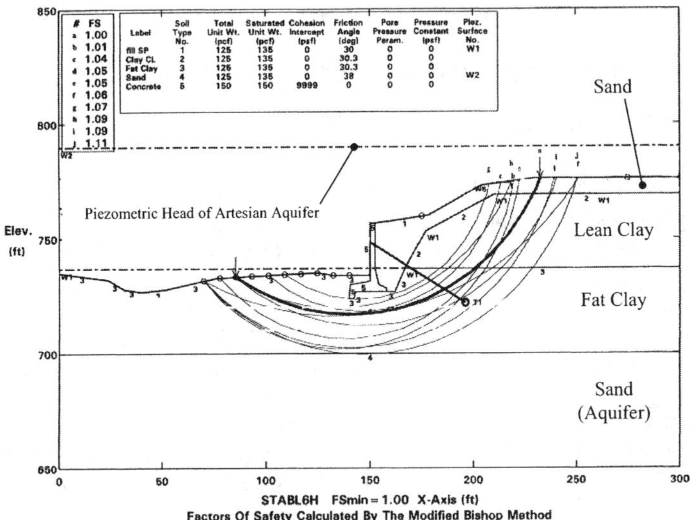

FIG. 2. Typical Section through East Tailrace Wall looking Downstream. Preliminary Stability Analyses Results are shown.

The clay stratum was underlain by a confined aquifer consisting of fine to coarse sands. The hydrostatic head of the aquifer was not known accurately. However, based on readings from nearby piezometers, an elevation of 790 ft (237 m) for the piezometric surface was estimated conservatively.

A number of tests were performed on disturbed and undisturbed specimens of the clay. In particular, ring shear tests on remoulded specimens were performed to determine the residual friction angle of the clay (ASTM D6467-99). Table 1 summarizes the interpreted property values of the clay, which were used for the preliminary stability analyses described in the next section. It is interesting to note the relatively large value of residual friction angle obtained for the clay. This is likely due to the sand content of the tested specimen. It must also be noted that the strength parameters obtained for lean and fat clay specimens were similar.

The properties of the sand strata used for the preliminary stability analyses were estimated based on correlations with the Standard Penetration Test (SPT) blowcount (Mc Gregor and Duncan 1998). They are also listed in Table 1.

ASSESSING THE MECHANISM OF WALL MOVEMENT

Several preliminary analyses were performed to ascertain the potential cause or causes for the observed movements. These analyses considered several potential mechanisms of movement of the wall. Progressive sliding and/or overturning of the wall were analyzed. Potential failure or creep of the tiebacks installed during the first remediation attempt was a concern, given the characteristics of the clay at the site. Finally, global failure along a slip surface passing below the bottom of the wall footing was also considered a potential failure mechanism.

TABLE 1. Soil Properties used for Preliminary Stability Analyses

Material (1)	Properties (2)
Stratum 1 Surface Sand Layer and Wall Backfill (SP)	Total Unit Weight, γ_{tot} = 125 pcf Saturated Unit Weight, γ_{sat} = 135 pcf Effective Friction Angle, ϕ' = 30°
Stratum 2 Clay (CL)	Total Unit Weight, γ_{tot} = 125 pcf Saturated Unit Weight, γ_{sat} = 135 pcf Residual Friction Angle, ϕ'_{res} = 30° to 33°
Stratum 3 Fat Clay (CH)	Total Unit Weight, γ_{tot} = 125 pcf Saturated Unit Weight, γ_{sat} = 135 pcf Residual Friction Angle, ϕ'_{res} = 30° to 33°
Stratum 4 Sand (SP) (Aquifer)	Total Unit Weight, γ_{tot} = 125 pcf Saturated Unit Weight, γ_{sat} = 135 pcf Effective Friction Angle, ϕ' = 38°

The results of these preliminary analyses suggested that the most likely cause of the observed movements was a global failure mechanism. To analyze this mechanism, the clay was assigned an effective friction angle corresponding to the residual condition. The intention of these analyses was to model the long-term behavior of the clay assuming that large displacements had already occurred along the slip surface. Although it was presumed that the magnitude of movement along the slip surface had reached several inches at the time, there was no supporting data on cumulative movement. However, the assumption of a fully developed residual condition along the slip surface was considered reasonable given the magnitude of wall movements observed and the response of the remoulded specimens during ring shear testing.

The analyses were performed in terms of effective stresses. To model pore pressures within the clay, it was assumed that a linear piezometric gradient existed across the clay interval. This assumption, although not entirely accurate, was found to provide reasonable results for this case.

Circular slip surfaces were analyzed in terms of effective stresses using the Bishop's modified procedure in the program PCSTABL. The analyses suggested that the critical slip surface passed below the wall footing and intercepted the bond zone of the existing tiebacks, as illustrated in Figure 2. A factor of safety of one against circular failure was obtained for a residual friction angle of 30 to 31 degrees, which were the lowest expected values according to the laboratory test results.

During this preliminary phase, only limited horizontal displacement data of the top of the wall were available. The results of these preliminary analyses were used to develop an improved monitoring program to confirm the assumed mechanism of wall displacement. The analyses were also the basis for development of preliminary design and testing requirements for use by contractors during the bid process.

Two inclinometers were installed behind the wall that extended beyond the potential slip surface. The survey program in place at that time was extended to include vertical and horizontal movements of the base of the wall and the soil slope above the wall. For final design, more refined analytical techniques were used that took into account the data from the instrumentation and the extended survey.

INSTRUMENTATION DATA

Approximately six months after the preliminary analyses were completed, the initial analytical assumptions were revised to account for additional data provided by the inclinometers and the extended survey.

The inclinometer readings suggested that sliding was taking place along a surface passing approximately 10 ft below the bottom of the wall footing. In addition, it appears that the sliding mass rotated as a 'rigid body' about a point located directly above the wall. The results of the inclinometer readings were consistent with the surveying data. Inclinometer and surveying data along other sections of the wall suggested that most of the movements were occurring toward the downstream end of the wall, and that the thickness of the materials involved in the movement varied along the wall. Therefore, it was concluded that the movements were taking place along a spoon- or wedge-shaped surface.

BACK ANALYSES OF WALL STABILITY

The stability of the wall was re-analyzed considering the surveying data and the location of the slip surface revealed by the inclinometers, as illustrated in Figure 3. These analyses were performed for final design of the remediation solution. The procedure followed consisted of establishing slip surfaces and back calculating the strength parameters of the clay to obtain a factor of safety of one. The shape and location of the slip surfaces was estimated based, primarily, on the inclinometer data.

Two mechanisms of sliding were analyzed. A circular slip surface using Bishop's modified method was first considered as illustrated in Figure 3. In the figure it can be observed that the circular slip surface daylights behind the crest of the slope. This condition was not confirmed by field observations, which indicated that most of the surface movement appeared to occur on the slope itself. Therefore, composite or block slide mechanisms were also analyzed using the modified Janbu method, where the extent of the block behind the wall was limited to match field observations. The composite slide mechanism is not represented in the figure. Further details on stability analysis techniques can be found in Duncan (1996).

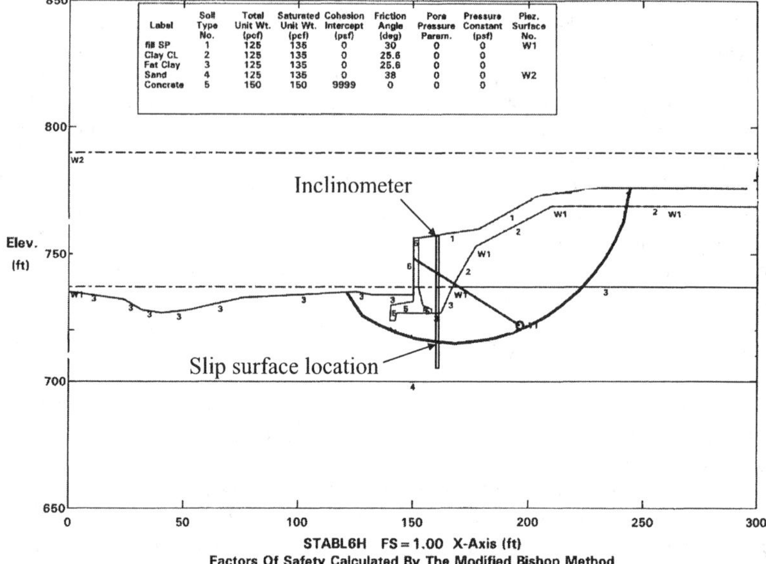

Label	Soil Type No.	Total Unit Wt. (pcf)	Saturated Unit Wt. (pcf)	Cohesion Intercept (psf)	Friction Angle (deg)	Pore Pressure Param.	Pressure Constant (psf)	Piez. Surface No.
fill SP	1	125	135	0	30	0	0	W1
Clay CL	2	125	135	0	25.6	0	0	
Fat Clay	3	125	135	0	25.6	0	0	
Sand	4	125	135	0	38	0	0	W2
Concrete	5	150	150	9999	0	0	0	

FIG. 3. Results of Back Analysis of the East Tailrace Wall.

Although the existing tiebacks were included in the analyses, their contribution to stability was negligible as the sliding mass typically encompassed them entirely.

As indicated previously, the friction angle of the clay was adjusted until a factor of safety of one was obtained. Table 2 contains the soil strength parameters estimated from the back analyses. It is seen that the estimated friction angle value for the clay is lower than those presented in Table 1. It is possible that the sand fraction of the specimens subject to shear testing may have been higher than the average sand fraction within the clay interval; thus, the specimens may not have been representative of the stratum. It is also possible that the slip surface developed through a 'weak' layer or through isolated 'weak' spots within the clay. Finally, it is possible that some of the sand seams within the clay may have been connected to the underlying artesian aquifer; therefore, the pore pressures close to the base of the wall would be significantly larger than estimated in the analyses. This last possibility was not supported by field observations during exploration or tieback installation.

REMEDIATION

Initially, the preferred alternative for remediation consisted of installing a row of caissons along the toe of the wall and in contact with the wall footing. The main advantage of this solution was avoiding the installation of tiebacks bonded into clay, which required consideration of the potential for creep and consequent load relaxation. In addition, local contractors were more experienced and better equipped

TABLE 2. Soil Properties estimated from Back Analyses

Material (1)	Properties (2)
Stratum 1 Surface Sand Layer and Wall Backfill (SP)	Total Unit Weight, γ_{tot} = 125 pcf Saturated Unit Weight, γ_{sat} = 135 pcf Effective Friction Angle, ϕ' = 30°
Stratum 2 Clay (CL)	Total Unit Weight, γ_{tot} = 125 pcf Saturated Unit Weight, γ_{sat} = 135 pcf Residual Friction Angle, ϕ'_{res} = 25° to 27°
Stratum 3 Fat Clay (CH)	Total Unit Weight, γ_{tot} = 125 pcf Saturated Unit Weight, γ_{sat} = 135 pcf Residual Friction Angle, ϕ'_{res} = 25° to 27°
Stratum 4 Sand (SP) (Aquifer)	Total Unit Weight, γ_{tot} = 125 pcf Saturated Unit Weight, γ_{sat} = 135 pcf Effective Friction Angle, ϕ' = 38°

in drilled shaft installation techniques. However, this alternative was discarded for two reasons. Firstly, the thickness of the materials involved in the movement was, as established from the inclinometer data, greater than previously estimated. Therefore, the design of the caissons would need to be revised to account for shear and moment magnitudes larger than initially estimated. Secondly, the costs of the caisson remediation were relatively high. Consequently, remediation using tiebacks was considered a viable alternative, provided that an adequate tieback system was used and validated through careful field testing.

Stabilization using tiebacks presented several challenges. The tiebacks should develop their capacity within the high plasticity clays below the wall foundation. Creep was a concern given the relatively large capacities needed. In addition, the tiebacks should not penetrate the underlying confined aquifer. The length and orientation of the tiebacks would then be controlled by the presence of the aquifer below, the existing tiebacks above, and the location of the slip surface.

Figure 4 illustrates the results of one of the stability analyses performed to establish the remediation requirements. Analyses were performed using the same slip surfaces and procedures used for the back analyses described in the previous section. For clarity, only the circular slip surface is represented in the figure. The analyses were performed using the corresponding soil properties listed in Table 2.

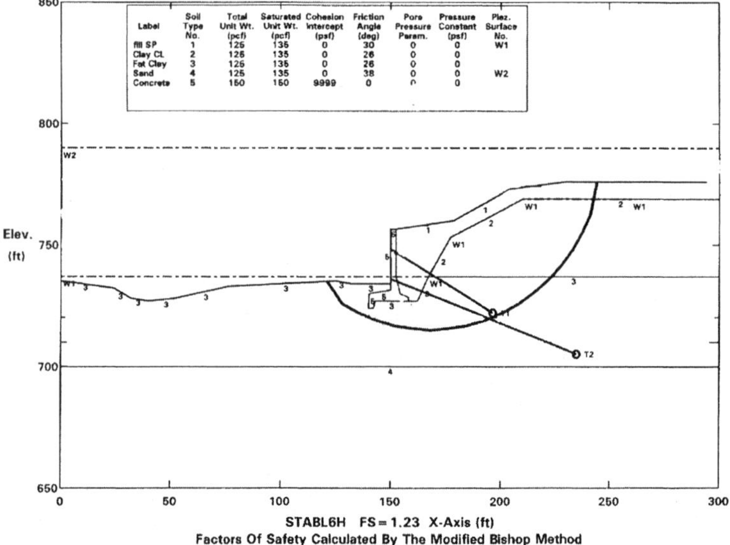

FIG. 4. Stability Analysis of the East Tailrace Wall for Remediation Design.

After consultation with the Owner, the target factor of safety was established at 1.2 to 1.25. Such a factor of safety could be achieved using one row of tiebacks with reasonably attainable capacity requirements. At that time, it was important to provide additional and immediate support for the wall rather than designing a final remediation solution, which would have required the acquisition and analysis of more monitoring data for a longer period, and installation of additional instrumentation. A well-defined monitoring/action plan after construction was established in order to assess the effectiveness of the remediation. The plan included threshold displacement values for performing further analyses and development of additional remedial measures if necessary.

The tiebacks forces required for the desired factor of safety were thus determined for the different slip mechanisms and types of analyses. The tiebacks would be installed at an inclination of 20 degrees from horizontal. The free length was established at approximately 15 m (50 ft). A maximum bond length of 12 m (40 ft) was required to avoid penetration into the aquifer. The tiebacks would have a minimum capacity of 600 kN (135 kip), and would be installed at a center-to-center spacing of 1.8 m (6 ft).

A set of remediation plans and performance specifications was prepared as part of the bid package. The specifications called for tiebacks with the length and capacity requirements determined from the stability analyses. In addition, the tiebacks should be retensionable in case creep-induced relaxation of the tieback load occurred over time. The contract specification required the contractor to be responsible for the design of an adequate tieback system that met these requirements.

SINGLE BORE MULTIPLE ANCHOR (SBMA) SYSTEM

In conventional tiebacks, the bonded portion is a single unit where all the tieback strands are encased together within the grout. Upon tensioning of the tieback, the resulting load transfer distribution along each of the strands is dependent on the magnitude of the load, the response of the grout-soil and strand-grout interfaces, and the stiffness of the strand. Conventional tiebacks have been used extensively and successfully for many years. However, the load transfer distribution along the bond zone is often non-uniform and the efficiency of the conventional tieback is reduced (see Barley and Windsor 2000). Furthermore, the length of the bond zone in traditional tiebacks is often limited to a 3 to 10 m (10 to 33 ft) range, as greater bond zone lengths do not increase significantly the tieback capacity.

The SBMA technology was developed to reduce these limitations. In this system, the bond length comprises a set of shorter and staggered individual unit anchors, each of which is fitted with one or more strands and is tensioned using an independent jack (see Figure 5). A more detailed description of the SBMA system is presented in a companion paper by Bruce et al. (2004).

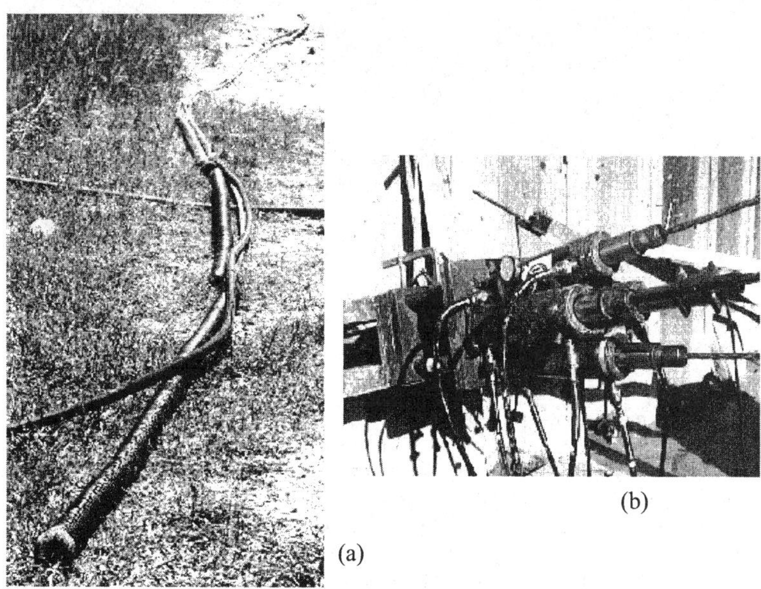

FIG. 5. (a) String of SBMA units almost ready for installation; (b) Performance testing of a production SBMA. Note the multiple jacks for stressing of each strand.

This system was particularly attractive for the Hodenpyl project. By using a bond length similar to that of a conventional anchor, the required individual tieback capacity could be achieved maintaining a relatively low bond stress along the grout-soil interface of the SBMA elements, thus reducing the potential for long-term creep of the tiebacks.

The selected contractor proposed the use of post-grouted SBMA elements consisting of four unit anchors with bond lengths approximately 10 feet long. This was the first application of post-grouted tiebacks using the SBMA technology in the United States.

A sacrificial test anchor was installed and tested before the start of construction to verify the capacity and response of the proposed SBMA design (see Figure 6). Due to space restrictions, the sacrificial anchor was located on the embankment above the wall and was installed in a vertical position. To prevent drilling into the aquifer, a shorter bond length consisting of three 10-ft bond units was used for the sacrificial anchor. Two strands were used for each of the upper two units. This modification with respect to the production design was introduced for research purposes in order to attempt to induce bond failure of the upper bond units during a secondary test phase.

(a) (b)

FIG. 6. (a) View of the setup for sacrificial anchor testing; (b) Detail of the stressing arrangement. Note the two additional strands not engaged during the first portion of the test.

In addition, two performance tests (PTI 1996) were carried out on production anchors. Details of procedures and results from the sacrificial and production anchor tests are presented in the companion paper by Bruce et al. (2004). The test results indicated that the required capacity of the tiebacks could be achieved and that creep would not be an issue. The tiebacks were installed from within a cofferdam constructed at the toe of the wall (see Figure 7). A barge was used to provide access for materials and equipment. The tiebacks were all installed in a period of approximately 15 working days without disruption to the plant operation. A waler beam was used to connect the anchorage to the facing of the wall (see Figure 8). All the anchorage elements of the tiebacks were encased within a grease-filled protective cover, which was large enough to house the excess strand length for eventual lift-off testing and retensioning, as shown in Figure 9.

FIG. 7. View of the site during installation of the SBMA elements. a sheet piling cofferdam provided a dry working area below the tailrace water level. Note service barge on the left.

FIG. 8. View of the wall at near completion of the installation of the SBMA elements near the bottom of the wall. The strands are to be saw-cut and covered. The old tieback installation is visible near top of photo.

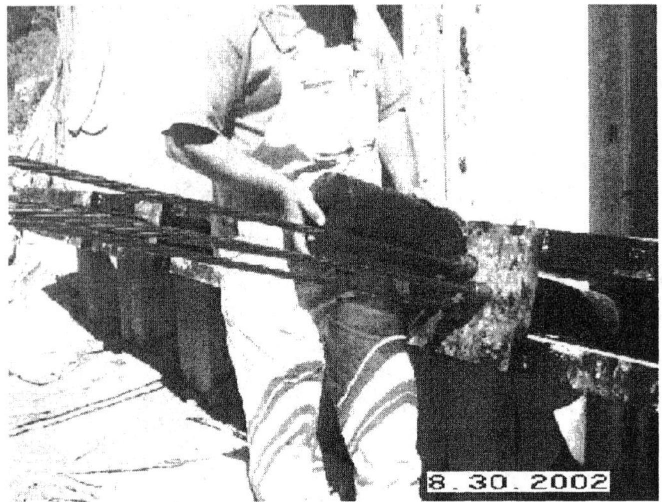

FIG. 9. Detail showing the anchorage elements. The strands are about to be cut and covered with the grease-filled housing.

PERFORMANCE OF THE REMEDIATION

Figure 10 summarizes the survey data at selected locations along the wall. It is seen that displacements of the wall were steadily increasing before remediation. After a temporary and expected disturbance during installation of the tiebacks, the rate of movement decreased significantly.

The inclinometer data shows a similar trend. However, at the time this paper was written, some displacement at the slip surface continued to occur at a reduced rate. It is believed that this movement is due to redistribution of stresses within the clay mass.

At the time of preparation of this paper, the period elapsed since installation of the SBMA system was relatively short, especially considering the long-term nature of the movements. However, the preliminary conclusion is that the remediation was successful. Further monitoring data over the next two years will provide indication as to the long-term performance of the tiebacks and/or the accuracy of the model used to analyze the mechanism of movement of the wall.

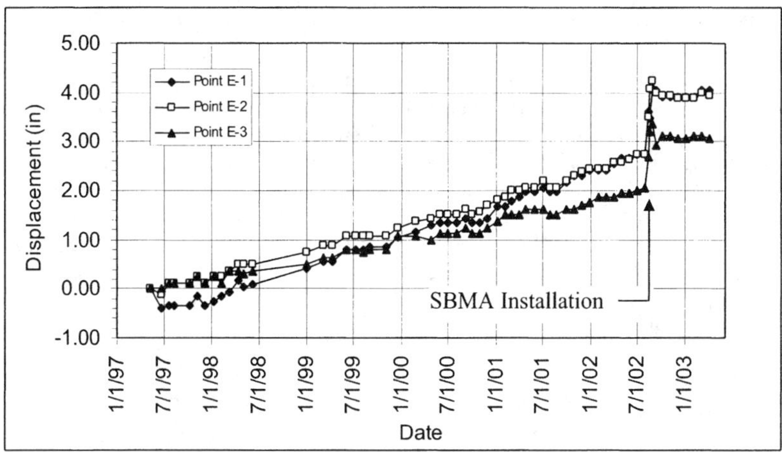

FIG. 10. Relative Displacement of the East Tailrace Wall (Values measured along the Cap of the Downstream End of the Wall).

CONCLUSIONS

Remediation of the East Tailrace Wall was successful on several levels. The philosophy followed for design allowed to reduce costs significantly and permitted an accelerated procurement and construction schedule. The factor of safety against a global slide was defined in concordance with the owner of the dam, and was intended to provide immediate improvement of the stability of the wall. In addition, a well-defined monitoring/action program was established to track the performance of the remediation over time.

The design called for restressable tiebacks, which would allow load adjustments if they were subject to relaxation due to creep or unanticipated movement patterns. Although it is entirely possible that no further remediation will be needed, the monitoring program would allow detection of additional movements that could be cause for concern. If additional and immediate remediation is required, the tiebacks could be restressed depending on the causes for the additional movement, while other stabilization measures are implemented as needed.

The SBMA system selected for this project performed remarkably well during the tests. Additionally, in spite of its sophistication, the installation of the system was simple and few issues developed during construction.

Finally, the project was completed as a team effort, where all the parties involved collaborated in accomplishing precise goals established during the design.

ACKNOWLEDGEMENTS

We would like to thank the owner of this project, Consumers Energy Company, and especially to Mr. Stuart Johnson for kindly allowing the publication of the information contained in this paper. Additional thanks must also go to Gerace Construction Company for the cooperation during the construction phase. Tony Barley (SBMA, LLC) provided important insights during load testing interpretation. Carlos Englert and Russ Preuss developed much of the analysis work summarized in this paper. We want to thank Schnabel Engineering for providing significant resources during the preparation of this paper. Finally, we would like to thank Robert Traylor (Traylor, LLC) for his support and expertise during testing and installation of the SBMA system.

REFERENCES

Barley, A.D., Windsor, C.R. (2000). "*Recent advances in ground anchor and ground reinforcement technology with reference to the development of the art*," GeoEng 2000, International Conference on Geotechnical and Geological Engineering, Melbourne, November 12-19, pp, 1084-1094.

Bruce, M.E., Traylor, R.P., Barley, A.D., Bruce, D.A., Gómez, J.E. (2004). "*Use of post grouted Single Bore Multiple Anchors at Hodenpyl Dam, Michigan*," GeoSupport 2004, Florida.

Duncan, J.M. (1996). "*Soil slope stability analysis*," Transportation Research Board, Special Report 247, National Academy Press, Washington, D.C.

Mc Gregor, J.A., and Duncan, J.M. (1998). "*Performance and use of the Standard Penetration Test in Geotechnical Engineering practice*," Center for Geotechnical Practice and Research, Department of Civil Engineering, Virginia Polytechnic Institute and State University, Blacksburg.

Post-Tensioning Institute (1996). "*Recommendations for prestressed rock and soil anchors*," Third Edition, June, 1996.

Post GroutedSingle Bore Multiple Anchors at Hodenpyl Dam, Michigan

M.E. Bruce[1], R.P. Traylor[2], A.D. Barley[3], D.A. Bruce[4], and J. Gomez[5]

ABSTRACT

Despite the installation of a row of anchors in 1996, the right downstream retaining wall of Hodenpyl Dam, MI, continued to move inward and downstream. The remediation design called for installation of additional anchors. However, there were concerns about creep-induced relaxation of the anchor load over time given the existence of high-plasticity clays at the site. The solution was the installation of post grouted multi-anchors along the base of the retaining wall (Single Bore Multiple Anchors – SBMAs).

SBMAs utilize several "unit" anchors within the same borehole, each with its own short efficient bond length positioned at staggered intervals along the bond zone. This staggered arrangement allows each unit load component to be transferred to the soil in a controlled manner over a discrete length of the borehole, thereby producing a very efficient load transfer mechanism. A larger factor of safety against creep is therefore attainable using SBMAs as compared to that provided by conventional tendons.

One sacrificial anchor and 13 production anchors were installed. An innovative testing setup and program were developed to allow extended creep testing of the sacrificial anchor followed by load testing to 2.8 times the design load without anchor failure. This paper describes the design and construction of the SBMAs, the load testing setup, and results of extended creep load testing.

1. INTRODUCTION

Hodenpyl Dam, owned and operated by Consumers Electric Company of Cadillac, MI, is located in Wexford and Manistee Counties, MI, and is one of a series of dams

[1] President, Geotechnica, s.a., Inc., P.O. Box 178, Venetia, PA 15367, mebruce@cobweb.net.
[2] President, Traylor LLC, 17204 Hunter Green Rd, Upperco, MD 21155, geobtraylor@msn.com
[3] Director, SBMA Single Bore Multiple Anchor, United Kingdom, tony.barley@sbmasystems.com and sbmallc@cobweb.net
[4] President, Geosystems, L.P. and Vice President, SBMA LLC, P.O. Box 237, Venetia, PA 15367, dabruce@geosystemsbruce.com
[5] Associate, Schnabel Engineering, Inc., 510 East Gay Street, West Chester, Pa. 19380, jgomez@schnabel-eng.com.

along the Manistee River. Investigation and analysis of the wall movement was conducted by Schnabel Engineering, Inc. (West Chester, PA) and Applied Engineering and Geosciences, Inc. (Greensboro, NC). The details of these studies (including descriptions of soil properties and design methodologies) are presented in a companion paper by Gómez et al. (2004). A total of 13 restressable, regroutable anchors were specified to improve the stability of the wall. The philosophy behind this design was to provide an immediate improvement of the wall stability while controlling construction costs. A detailed monitoring/action plan was also specified which would allow to determine whether subsequent treatments were necessary.

Figure 1. Downstream face of Hodenpyl Dam, showing right downstream retaining wall on left.

2. ANCHOR DESIGN

2.1 System Design Criteria

The specified working load for each tieback was 135 kips. Anchor geometry requirements dictated a minimum free length of 50 feet; a minimum bond length of 40 feet; a minimum drill hole diameter of 5 inches; and an anchor inclination of 20° (or less). The specified elevation of anchor entry was El. 734.2 feet. Bar tendons were recommended originally since these elements were initially assumed to be easier to regrout and restress. Class I corrosion protection (PTI, 1996) was required for the permanent anchors.

The expected anchor bond zone materials (between approximately El. 718 and 693 ft) consisted of a clay stratum generally stiff to very stiff, moist. Geotechnical properties of these materials are described in detail in the companion paper (Gomez et al., 2004). The clay stratum was underlain by a sand aquifer.

2.2 Single Bore Multiple Anchor (SBMA) System

2.2.1 SBMA Concept

Although the anchor requirements presented above may have been satisfied by using a conventional multi-strand tendon with a 40-foot bond length, Single Bore Multiple Anchors (SBMAs) were proposed by the contractor as an alternate to conventional tendons to provide a more efficient and uniform mode load transfer and therefore enhanced performance (i.e., reduced potential for creep).

A typical SBMA (Figure 2) consists of a multiple of unit anchors (single or double strand) with varying lengths installed in a borehole (4- to 8-inch diameter) such that their respective load transfer lengths are located at predetermined positions within the total fixed length.

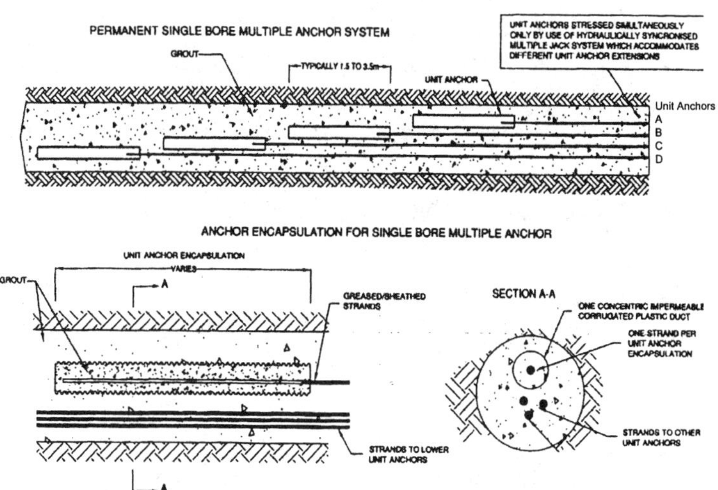

Figure 2. Elevations and cross section of a typical 4-unit SBMA.

It is fully acknowledged by researchers who have investigated grout/ground load transfer that the distribution of stress along the fixed anchor is non-uniform due to general incompatibility between elastic moduli of the anchor tendon, anchor grout, and the ground (Littlejohn and Bruce, 1977; Barley, 2000). In the majority of conventional anchors, after initial loading, the bond stress is concentrated at the proximal end of the fixed anchor, while the distal end of the fixed length remains unstressed. As load is increased, the ultimate bond stress at the proximal end of the fixed length along either (or both) the steel/grout interface or the grout/ground interface is exceeded. At that time, the bond stress reduces to a residual value at that location, and movement occurs: the capacity in that section of the anchor decreases, and the load is transferred distally. As load on the tendon is further increased, the bond stress concentration zone progresses farther along the fixed anchor. Just prior to ultimate pull-out, the load is concentrated at the distal end of the fixed length. Figure 3 depicts this load transfer phenomenon, referred to as "progressive debonding." The area under the bond stress distribution line is representative of the ultimate load in the anchor. It can be seen that the load does not increase uniformly with increasing length.

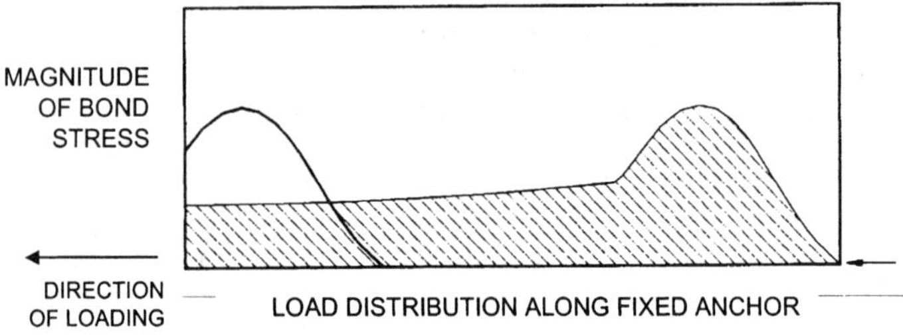

Figure 3. Load distribution and progressive debonding in conventional anchors.

Data collected and analyzed over a 10-year period were used to develop a simple mathematical expression to reduce τ_{ult} by accounting for the occurrence of progressive debonding (Barley, 1995):

$$\tau_{ult} \propto f_{eff} \text{ x } L$$

where, f_{eff} = efficiency factor, which itself is a function of L
L = fixed length (in meters)

Efficiency factors were back-analyzed from data collected from eight projects where anchors of different fixed lengths were tested to failure in clays, silty clays, and sandy clays, boulder clay and glacial till. The analysis of the data and comparison of curves from other researchers were described by Barley and Windsor (2000). These efficiency factors were plotted versus fixed length, and the best fit curve is represented by the following equation:

$$f_{eff} = 1.6\, L^{-0.57}$$

For a conventional tendon with a 40-foot (13-m) length, the efficiency factor is

$$f_{eff} = 1.6 \text{ x } (13\text{ m})^{-0.57}$$
$$= 0.37$$

For an SBMA with four fixed anchor lengths of 3 m (total fixed length = 40 feet), the efficiency factor for each 10-foot unit anchor is:

$$f_{eff} = 1.6 \times (3\ m)^{-0.57}$$
$$= 0.86$$

i.e., an SBMA with 4 unit anchors will be 2.3 times more efficient than a conventional anchor with a single 40-foot fixed length (0.86 / 0.37 = 2.3).

SBMAs were developed to transfer load to the grout over a series of short lengths at staggered intervals along the borehole, and to carry the same load on each tendon simultaneously – thereby reducing or eliminating the occurrence of progressive debonding and substantially increasing the efficiency of the overall anchor. A comparison of load distribution along an SBMA and a conventional anchor is depicted in Figure 4.

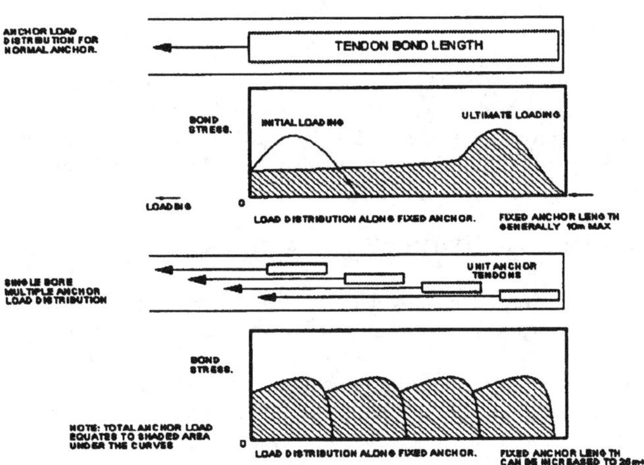

Figure 4. Comparison of load distribution along a conventional anchor and an SBMA (Barley, 2000).

2.2.2 SBMA Design

As shown in Figure 2, the encapsulation of each unit anchor consists of a 10-foot-long, 2-inch-diameter corrugated PVC duct grouted prior to delivery. The spacing of the encapsulations in the borehole defines the fixed length for each unit anchor; i.e., the fixed length = the encapsulation length plus the distance between adjacent encapsulations. A computer program developed by Single Bore Multiple Anchor Ltd. was used to design fixed lengths of unit anchors in a range of soil conditions. This program relates bond stress at the grout/soil interface to either N-values, undrained shear strength (c_u), or undrained shear strength plus an enhancement of bond stress due to post grouting. The latter program was used for the Hodenpyl anchor design. Table 1 lists the unit anchor lengths and SBMA geometry generated by this program.

Table 1. Summary of unit anchor lengths and SBMA geometry.

Unit Anchor	Free Length (ft)	Fixed Length* (ft)	Total Tendon Length (ft)**
A	55	11.5	71.5
B	66.5	10	81.5
C	76.5	10	91.5
D	86.5	10	101.5

* Total tendon length = 5-foot tail + free length + fixed length
Overall SBMA = 101.5 feet long.

Design parameters for the SBMAs for this project are summarized in Table 2. In addition to the design of the physical components of the anchors, a very important part of the design process was the installation and testing of a sacrificial anchor as discussed in Section 3, the performance of which provided site-specific information to confirm the design assumptions of the production anchors.

3. SACRIFICIAL ANCHOR INSTALLATION AND TESTING

A sacrificial anchor was required by the Owner and his Engineer. This anchor was installed using identical construction methods and materials and bonded in the same soils as those proposed during production with the following exceptions a) the sacrificial anchor was installed vertically, as opposed to being inclined at an angle of 20° below horizontal; b) three 10-foot-long unit encapsulations were installed as opposed to four (to avoid penetrating the underlying sand aquifer); and c) the upper two unit anchors contained two strands in each encapsulation to allow load testing to very high grout/ground bond stress (described in Section 3.3).

3.1 Anchor Installation

The borehole was installed through a steel-reinforced concrete pad. Rotary drilling with end-of-casing water flush was performed using a diesel/hydraulic rig. A 7-inch o.d. N80 steel casing (0.45-inch thick wall) was advanced to the full depth of hole. The complete tendon comprising the three unit anchors [Top (A), Middle (B), Bottom (C)], the Primary grouting pipe, and the post grouting pipe (tube à manchette) was assembled in the field and installed through the casing. The drill casing was then withdrawn as the borehole was grouted via the primary grout pipe. The 1½-inch tube à manchette had sleeves at 3-foot intervals along the length of the four bond lengths. The grout was delivered to each sleeve through a double-packer placed at the corresponding sleeve location. Two post-grouting events (primary and secondary) were performed for each SBMA anchor. Within 24 hours of initial grouting, water was applied to each sleeve at pressures of 1000 to 2000 psi to fracture the initial casing grout. Neat cement grout was then injected at a target volume and pressure of 2 ft^3 per sleeve and 50 psi, respectively. After the refusal criteria for each sleeve were obtained, the double packer was advanced to the next sleeve in the post-grouting sequence. The secondary sleeve grouting followed the primary sleeve grouting by 24 hours.

Table 2. Summary of SBMA design parameters for Hodenpyl Dam.

Grout/Ground Bond

- Factor of Safety on grout/ground bond = 3
- Post grouting enhancement factor (based on two phases of post grouting) = 2
- Ultimate grout/ground bond strength (τ_{ult}) = 3 ksf x 2 (post grouting) = 6 ksf
- Primary Grout
  - Type I/II cement; Master Builders XR100 admixture, high speed, high shear mixer
  - Water/cement ratio (by weight of cement) = 0.45
  - Unconfined compressive strength (28 days) > 4000 psi
- Post Grout
  - Type I/II cement, high speed, high shear mixer
  - Water/cement ratio (by weight of cement) < 0.55
  - Target volume = 15 gallons/sleeve; maximum pressure = 800 psi
- Encapsulation Grout (SBMA)
  - Type II cement
  - Water/cement ratio (by weight of cement) = 0.475

Grout/Steel Bond (within encapsulation)

- Verified by recent in-house testing performed by tendon supplier
- Maximum test load was reached over a bond length of 10 feet without pullout.

Tendon

- 4-strand anchor tendons
- Strand within the 10-foot fixed length encapsulation contained three evenly spaced nodes (localized areas of untwisted strand that provide increased mechanical interlock)
- Guaranteed Ultimate Tensile Strength (GUTS) = 58.6 kips; therefore, maximum test load = 80% GUTS = 46.8 kips.

Corrosion Protection

- Tendons and top anchorage (bearing plate, anchor head, wedge plate, and wedges) meet Class I (PTI, 1996)
- Encapsulation = 2-inch i.d. 10-foot long corrugated duct

3.2 Load Test Set Up

The jack arrangement for a three-unit SBMA includes three hydraulic rams that are synchronized by coupling to the same hydraulic powerpack, so that the same load is applied simultaneously to each unit anchor. The jacking arrangement is shown in Figure 5. A primary gauge and a reference gauge were calibrated with one of the jacks. The ram extensions were recorded using a stiff steel rule, and during creep testing by using a vernier caliper. Measurements were corrected for reaction pad movement measured by dial gauges mounted on an independent reference beam.

Figure 5. Sacrificial test anchor stressing jack.

3.3 Load Test Sequence

In addition to the Performance and Extended Creep Testing as detailed in the project specifications, load testing to high bond capacity was attempted on the sacrificial anchor. The stressing sequence for these tests is detailed in Figure 6.

The "Performance and Extended Creep Tests" included a Performance Test performed in general accordance with PTI (1996) to a maximum load of 1.33 x Design Load (≡ 44.9 kips), with Extended Creep Testing involving additional load holding periods at each load maxima (i.e., to 300 minutes). To obtain still longer-term creep measurement, the load-hold period for the sacrificial anchor was extended overnight (total load-hold time 810 minutes) at maximum load.

The Ultimate Load Test was performed in two parts. In Part 1, the previously unstressed strands (A2 and B2) were subjected to a cyclic loading/unloading sequence identical to that imposed upon the stressed strands (A1 and B1) (although without the load hold periods) to 1.33 x Design Load. This extra cycling was performed to impart similar stress histories into the strands so that, when these strands were subjected to further stressing, the pairs of A and B strands would each exhibit similar behavior. In Part 2, the maximum test load of 2 x 80% x 58.6 kips = 93.8 kips/unit anchor was applied. This test load therefore equaled 2.78 x Design Load (93.8 kips / 33.75 kips = 2.78).

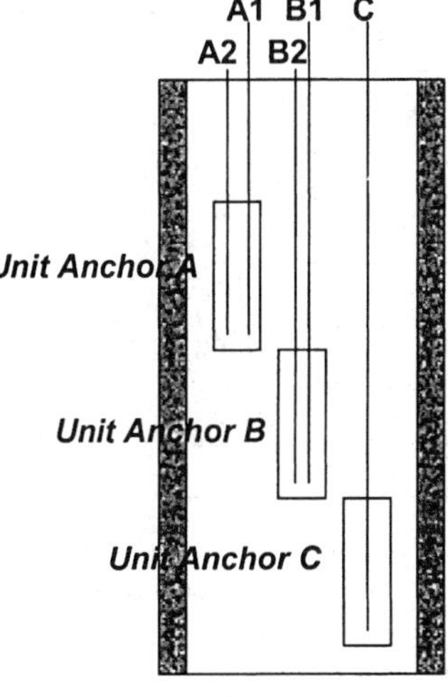

Performance and Extended Creep Testing
Strands A1, B1 and C stressed as detailed in specifications for Performance and Extended Creep tests to 1.33 x Design Load. The second strands of Unit Anchors A and B remain unloaded throughout this testing).

Ultimate Load Testing
Upon completion of specified Performance and Extended Creep testing, the jacks were removed and re-installed to grip all four strands in Unit Anchors A and B (i.e., A1, A2, B1, and B2). A second load test was performed on these units, and the maximum test load was 2 x 80% x 58.6 kips/unit anchor (GUTS) = 93.8 kips/unit anchor. Test load applied to Unit Anchors A and B (and therefore the stress applied to their corresponding grout/soil interfaces) is doubled.

Figure 6. Stressing sequence for sacrificial anchor.

3.4 Performance and Extended Creep Test Results

Part 1

- The behavior of each unit anchor was extremely linear and repeatable, i.e., the plots of successive load cycles exhibit similar slopes and shapes. Figure 7 shows typical total movement versus load for the C Strand (A and B strands were similar but at different slope gradients due to different elastic lengths).
- Elastic and permanent movements for each unit anchor are shown in Figure 8. The plots of elastic movements are extremely linear and therefore indicate virtually no progressive debonding into the unit fixed anchor. As expected, the slopes of the curves decrease with increasing free stressing length, since at a given load, greater movement will be generated by a strand with a longer free stressing length.
- The permanent movements recorded at the maximum test load ranged from 0.2 to 0.35 inch. The readings were occasionally erratic due to the low Alignment Load. At the higher loads, the data are consistent and logical.

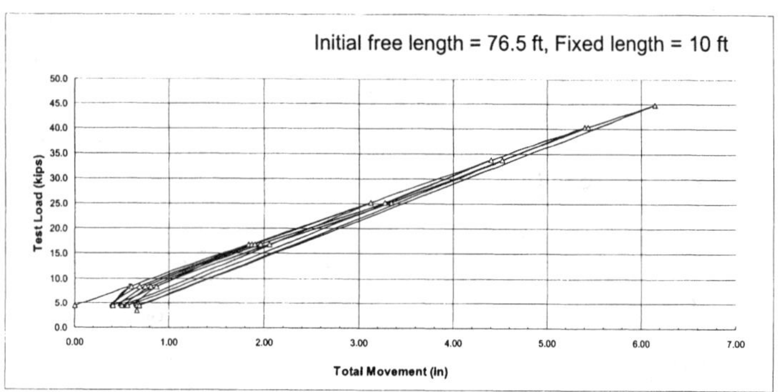

Figure 7. Total movement vs. load for Unit Anchor C during the Performance and Extended Creep Test.

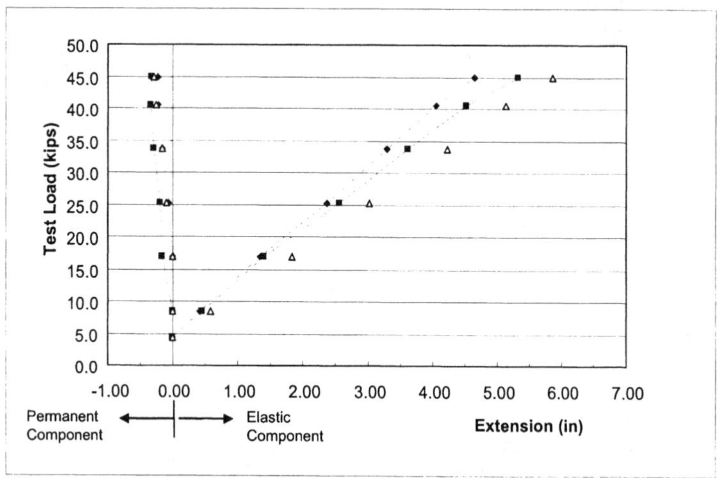

◆ Strand A1 (Initial free stressing length = 60.5 feet); ■ Strand B1 (Initial free stressing length = 70.5 feet); Δ Strand C (Initial free stressing length = 80.5 feet)

Figure 8. Elastic and permanent movements of Unit Anchors A, B, and C during the Performance Test.

- No creep was measured during the load-hold periods of the specified testing. During the additional optional load hold period, creep measurements were extremely low: 0.020 inch, 0.020 inch, and 0.082 inch, for the A1, B1, and C strands, respectively.

Part 2

- Figure 9 shows permanent and elastic movements to a maximum test load of 2.78 x Design Load on the upper two unit anchors. The plots from this loading were extremely linear and repeatable.
- Real permanent movements ranged from approximately 0.002 to 0.178 inch and were again somewhat erratic (similarly to those measured during the Performance and Extended Creep Test).
- There was no indication of impending failure at the maximum permissible test load, equivalent to 80% GUTS of the enhanced tendon capacity (i.e., 93.8 kips); therefore, no ultimate load was established.

With respect to the analysis of elastic movements, permanent movements (manipulated to compensate for the low Alignment Load effects), and creep, each unit SBMA proved entirely successful and compliant with the acceptance criteria.

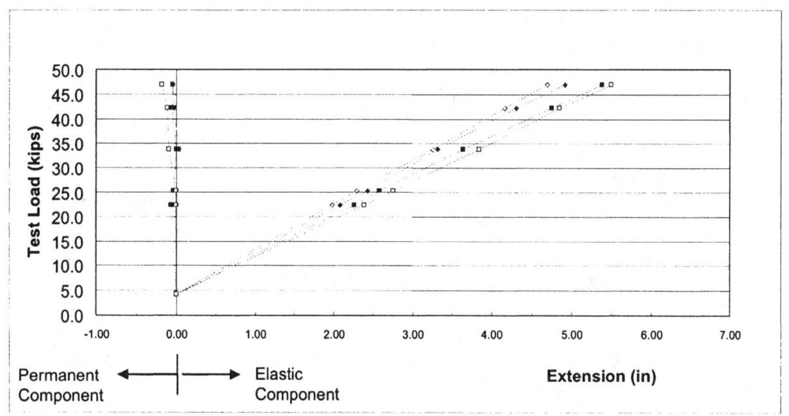

◆ Strand A1 (Initial free stressing length = 60.5 feet); ■ Strand B1 (Initial free stressing length = 70.5 feet)

Figure 9. Elastic and permanent movements of strands of Unit Anchors A and B during the Ultimate Test. (Note: Each unit anchor has two strands; therefore the maximum ptest load on each unit anchor = 2 strands x 80% x 58.6 kips (GUTS) = 93.8 kips).

4. PRODUCTION ANCHOR INSTALLATION AND TESTING

4.1 Drilling and Tendon Installation

The 13 production anchors were installed between July 29 and August 23, 2002. Testing and lock off were conducted August 29 and 30, 2002. Anchors were installed sequentially from downstream to upstream starting with even numbered anchors (Anchors 2 through 12), followed immediately by odd numbered anchors (Anchors 1 through 13).

Anchor holes were drilled using rotary duplex with water flush and 7-inch casing with a 1-inch overcut from the casing shoe teeth, resulting in an 8- to 8½-inch hole diameter. Casing was advanced to the target tip elevation for each anchor.

4.2 Load Testing

Anchors 2 and 6 were subjected to Extended Creep Testing. As noted during sacrificial anchor testing, the permanent movement for each unit anchor appears to be artificially exaggerated due to low alignment loads and friction within the jack chair assembly. Remaining anchors were subjected to proof testing. Each anchor performed elastically at or above the required 80% of theoretical extension of the free length. Anchor 2 was the only anchor to show debonding, with magnitudes of 4.5 and 1.8 feet in Unit Anchors A and B, respectively. All other anchors had theoretical debonding values less than 0. Permanent movements at maximum test load were less than 1 inch for all anchors. No measurable creep at the maximum test load was recorded for any anchor.

4.3 Lock Off and Final Assembly

All unit anchors were locked off at 34 kips (136 k anchor load) using a monostrand jack. The strands were then trimmed to approximately 8 inches beyond the wedges. The trumpets were grouted, and the steel caps were installed as designed and filled with grease.

5. CONCLUSIONS AND FINAL REMARKS

Single Bore Multiple Anchors (SBMAs) incorporating a post-grouting program was used to satisfy the specified load requirements of the remedial anchors founded in a cohesive stratum. No load loss due to creep was encountered within a normal time period at 1.33 x design load. PTI and specified load carrying and movement acceptance criteria were met. This excellent anchor performance is considered due to the beneficial effects of a very concentrated post-grouting program, and the constructional and operational features of each unit SBMA. Analysis of the load-movement data confirmed no discernible debonding at any structural interface; very small and gradually increasing permanent movements with increasing load; and therefore, absolutely no evidence of impending failure at Test Load (i.e., at an average grout/ground bond of 135 kips/30 feet = 4.5 kips/ft).

ACKNOWLEDGMENTS

The authors thank Consumers Electric Company for permitting the publication of this paper, and for their cooperation during design and construction. Schnabel Engineering and Applied Engineering and Geosciences provided valuable input to this project. The general contractor was Gerace Construction Company of Midland, MI, and the anchor design was performed by Geotechnica, s.a., Inc. in association with SBMA LLC, both of Venetia, PA. Field direction and oversight was provided by Traylor, LLC of Upperco, MD. The testing apparatus and SBMA components were manufactured and supplied by the system's licensee, Lang Tendons, Inc. of Toughkenamon, PA.

REFERENCES

Barley, A.D. (1995). Anchors in Theory and Practice. International Symposium on "Anchors in Theory and Practice," Salzburg, Austria, October.

Barley, A.D. (2000). "Trial soil nails for tunnel face support in London Clay and the detected influence of tendon stiffness and bond length on load transfer." Proceedings of the Third International Geosystems, London, UK, June.

Barley, A.D., and C.R. Windsor. (2000). "Recent advances in ground anchor and ground reinforcement technology with reference to the development of the art." GeoEng2000, International Conference on Geotechnical and Geological Engineering, Melbourne, November 19-12, pp. 1048-1094.

Littlejohn, G.S. and Bruce, D.A. (1977). "Rock Anchors - State of the Art." Foundation Publications, Essex, England, 50 pp. (Previously published in Ground Engineering in 5 parts, 1975-1976.)

Post Tensioning Institute (PTI) (1996). "Recommendations for prestressed rock and soil anchors." Post Tensioning Manual. Fourth Edition. Phoenix, Arizona. 41 p.

WAX PROTECTION FOR GROUND ANCHORS

Heinz Nierlich[1]

ABSTRACT: Special Applications Of Permanent Anchors And Micro Piles. These products are used mainly for foundations and to secure construction, but also can be used for a variety of additional geotechnical applications. The following are three noteworthy case histories.In tension piles for harbor construction with a service load of 270 Kips and a length of 86 feet, the firm was able to improve the load transfer into the soil by increasing the diameter of the pile in the upper part by high-pressure injection up to a depth of 40 feet. Through the use of the post-tensioning permanent anchor system with a service load of 142 Kip and a length of 86 feet, the company was able to provide protection of the permanent anchors against aggressive concrete-attacking groundwater. In underpinning historical buildings in Berlin with micro piles with a service load of 194 Kip and a length of 105 feet, the firm was able to fabricate and install piles within cramped cellars, apply special constructive measures for connecting the piles and provide for corrosion protection within the aggressive organic soil. All details of the construction are described. As a basis for reliable, controlled performance, details regarding the quality control program are also discussed.

1.0 INTRODUCTION

This paper intends to explain why grease cannot guarantee a long-term protection against corrosion for ground anchors, why wax is a better solution and what the benefits are in changing from grease to wax protection. I will show the difference in their chemical compositions and how this affects their suitability for a long-term protection.

To make ground anchors "permanent", they need an absolutely reliable corrosion protection system. The best protection system is still cement grout within a strong plastic sheath or tubing, assembled and grouted at the tendon supplier's factory. The grout embeds the high tensile steel in an alkaline media, which passivates the steel surface and prevents corrosion from taking place. The Double Corrosion Protected bar anchor, pre-grouted within a corrugated PVC tube is an example for this, but also short

[1]Dipl.-Ing., Manager of Foreign Sales, DWYIDAG Systems International, Inc. USA 15 Industrial Road Fairfield, NJ 07004-3017, heinz.nierlich@dsiamerica.com

strand anchors can be supplied this way. This solution, however, is not practical for long and heavy strand tendons. When trying to protect both, the bond and the free length of strand tendons solely with grout after their installation (2-stage grouting), the spacers around the tendon may not guarantee adequate grout cover over the strand and the free length would be permanently bonded to the surrounding ground. Further, pressure grouting of a soil anchor with a common corrugated tube will almost inevitably collapse that tube. A more efficient method is to protect the individual strands in the free stressing length of an anchor tendon with a permanently plastic corrosion protective compound inside an extruded PE sheath, which also will result in a permanently unbonded free length. Easy melting substances like wax (petrolatum) or grease (metal soaps) are used for this purpose, since, because of their low viscosity and low internal friction, they can easily penetrate the interstices between the individual strand wires.

Petrolatum, for years, has been used predominantly as a corrosion protection for buried steel pipelines. The experience gained here has been adopted for ground anchors in Europe and other regions in the world, since the corrosion protection problems for both cases are closely related. Both systems are buried in the ground and their corrosion protection system has to be absolutely reliable. Both are within a moist environment that, if of a highly alkaline nature, can be very aggressive to corrosion protective compounds. This comparison allows the conclusion, that corrosion protective compounds which have performed successfully for pipelines would also be a good choice for protecting ground anchors.

2.0 COMPOUNDS FOR CORROSION PROTECTION

The compounds most widely used for protecting strand tendons are metallic soaps (grease) and petrolatums (wax). Both compounds have low friction, are easily melted, have a low viscosity for easy pumping, are resistant to heat, have excellent wetting capability in regards to the steel surface, evidence permanent plasticity and are insensitive to freezing conditions. The following, however, will take a look at their long-term performance in protecting the post-tensioning steel against corrosion and show the differences.

2.1 Metal Soaps

Petroleum greases are metal soaps. They consist of a mixture of petroleum oils and a thickener. Under a microscope the matrix looks like a sponge, which holds the oil portion of the grease. Soap thickeners are formed by reacting a metallic hydroxide (lithium, calcium or lead, for example) and a fatty acid. All metal soaps have a molecular structure consisting of a hydrophobic or water repellant carbohydrate tail and a hydrophilic or water absorbing carboxylate end group. During a short time exposure to moisture, the hydrophobic effect of the carbohydrate tail dominates by far. However, at a prolonged contact, also the hydrophilic nature of the carboxylate group shows, as the soap slowly but steadily absorbs water. As the soap structure starts breaking down, the oil with its additives is released. After a water-in-oil emulsion begins to form, the system will eventually completely deteriorate and the protective coating will disappear. Adding hydrophobic oils, corrosion protective additives or petrolatum to the metal soaps or forming the soaps from heavy metals improves their water resistance. If grease is heated over its melting point and it breaks down, it will not reconstitute itself.

2.2 Petroleum Wax

Petroleum waxes are a byproduct at refining heavy crude oil. They are solid at room temperature, totally hydrophobic and cannot be emulsified in water. There are three basic categories of petroleum waxes, Paraffin (crystalline), Microcrystalline (small crystals) and Petrolatum (amorphous). Corrosion preventives are made from microcrystalline wax, which provides the high melting point characteristics and petrolatum. Waxes also need the addition of some oil, not for keeping it from absorbing water, but as a softener and lubricant. Waxes can be melted and cooled at normal temperatures, far below the cracking temperature of 760 degrees F, without being destroyed.

3.0 STABILITY UNDER HEAT AND DE-OILING

Corrosion protective compounds for filling voids underneath the wedge plate have to be able to resist heat from being exposed to the sun without loosing their qualities. Loss of oil will result in loss of volume and the vacated space can absorb moisture. Loss of oil also will cause the compound to become porous and crack under temperature changes. The corrosion preventive additives are lost together with the oil and the compound remains are now more sensitive to moisture and especially to alkaline liquids. The tendency for the oil to separate from the soap matrix increases with increasing temperatures. This phenomenon has to be considered also in conjunction with capillary forces, since the compounds can be in contact with absorbent concrete, grout or rock surfaces. Tests, where various corrosion protective compounds were placed on cement cubes, did show the de-oiling, but were hard to evaluate quantitatively. A better and faster evaluation of the de-oiling properties is to place an open metal cylinder, 30mm high and 20mm in diameter, on filter paper and measure the diameter of the oil ring that has formed on the filter paper around the cylinder after a certain time.

This method of using filter paper, of course, does not reflect field conditions, but allows a fast and repeatable laboratory evaluation of the de-oiling properties of a corrosion protective compound. Tests done that way show that metal soaps loose much oil already at room temperature and that the oil ring increases in diameter at a moderate way at higher temperatures. In contrary, the oil in waxes begins to separate slowly only at a temperature above 35∞ C (94∞ F) with the rate increasing rapidly with higher temperatures. The real difference in elevated temperature performance is that grease looses oil at room temperatures, while waxes only start to become fluid at higher temperatures and have a satisfactory performance up to 40∞ C (102∞ F).

4.0 LONG TERM BEHAVIOR IN WATER OR MOISTURE

As a law of nature, two different substances with mobile molecules have the tendency to diffuse into each other to a certain degree, even if they are not mixable with each other. The faster diffusion partner enters the slower one, but the slower one determines the rate of diffusion. In the case of plastic compounds in a moist environment, water is by far the faster diffusion partner, while the viscosity of the compound determines the diffusion rate. With increasing viscosity, the speed and depth of water penetration decreases at an exponential rate. Softer compounds, such as grease, therefore absorb more water and faster at a far higher degree than waxy substances. With their low water resistance and their hydrophilic molecule groups, metal soaps start to swell and to emulsify until a total collapse of the system occurs. An evaluation of the water resistance of a new type of grease will need at least a one-year test in a laboratory, while tests with waxes with low oil content, over years, have not shown an appreciable amount of water absorption. For pipelines no longtime behavior tests are required for waxes due to their proven case history.

The questionable performance of grease has also been confirmed by the many failures of greased mono-strand tendons in concrete slabs. Especially earlier installations, where the strand had been protected with just paper wrapped around the greased strand, demonstrate that the surrounding concrete had absorbed the oils in the grease, leaving the strand without protection.

5.0 LONG TERM BEHAVIOR IN ALKALINE ENVIRONMENT

There can be two sources for high alkaline medias in the ground. Highly alkaline water from fresh concrete is one of them, but also stray currents or DC currents, easily attracted by ground anchors, can create a highly alkaline environment in oxygen rich soils. Stray currents can be caused by water flowing perpendicular to the magnetic direction of an iron ore deposit, electric air currents shortly before a thunderstorm or geomagnetic disturbances in tact with magnetic storms in the ionosphere. The fact, that for buried steel pipelines always the highest possible alkaline resistance for corrosion protective compounds is specified, shows how important this issue is being considered in that industry.

The strong aggressivity of alkaline environments towards corrosion protective compounds results in swelling of the compound, saponification and flocculation. Tests show that swelling in the hydrophilic metal soaps, when placed in an alkaline liquid with a pH factor of 13, is three times as much as when placed in distilled water. The alkaline liquid penetrates and emulsifies the metal soap (grease) fairly quick, while moisture absorption in the hydrophobic petrolatum (wax) remains small, even after years, and remains limited to the outer layer. Saponification, the chemical decomposition of an ester into soap and alcohol, is an important characteristic for judging the long-term performance of corrosion proof compounds. Longtime experience with corrosion protective systems for pipelines has shown a good correlation between the saponification value for a compound and its long-term behavior. Substances with a high saponification value swell over time in cold alkaline environment with a pH factor of 13 and substances with a low saponification value are only resistant to alkaline liquids, if they do not de-oil on filter paper or if they are not a soap. Flocculation of the soap in a test may indicate a good performance of a grease, since it prevents swelling, but emulsified and flocculated grease looses its corrosion protective properties.

6.0 CORROSION PROTECTION IN ELECTROLYTIC SOLUTIONS

Tests have been performed to confirm the validity of the theoretical aspects presented here. Clean bar sections were coated with different types of metal soap (grease) and petrolatum (wax) and then suspended in a test liquid representing aggressive soil until corrosion started to show. The results demonstrate that for both compound types the oilier one resisted corrosion better. This also means that, when oil is lost through emulsification, the ability to protect is diminished. Petrolatums performed better than metal soaps and within the metal soaps the duration of the corrosion resistance decreased with increasing saponification values.

7.0 WAX FOR STRAND ANCHORS

Besides the superior long-term corrosion protection of petrolatum based wax, also the practical aspects for the use of waxed strand anchors shall be discussed here.

7.1 Application

For a complete coverage of all surfaces of the seven-wire strand with wax, the six outer wires are separated from the center wire and all wires are coated with the hot wax. After closing the strand again, more wax is applied and a PE sheath is extruded over the waxed strand. This process ensures a complete filling of the interstices of the strand and of the annular space between the strand and the PE sheath. Depending on the type of wax and the fabrication method used, the strand may also be pushed into a PE tube filled with hot wax to achieve a void free system.

7.2 Resistance to Hydrostatic Pressure

For strand anchors, especially where the anchorage is located below the water table, ground water that can infiltrate the anchor tendon will tend to travel up the strand. Wax will resist hydrostatic intrusion much better than the soft grease and so will the extruded waxed strand compared to strand stuffed into a wax filled PE tube. A high resistance to hydrostatic pressure is of particular importance for the area just underneath the anchorage, which is the portion of an anchor tendon most susceptible to corrosion. If water is allowed to collect there, corrosion can occur in this oxygen rich environment.

7.3 Friction

Wax is normally a more viscous compound than grease. For that reason, higher friction values have been observed during stressing of a wax protected anchor tendon compared to grease filled extruded strand. Tests have shown that a three times higher pull-out force is required for wax extruded strand embedded in grout than for grease extruded strand. Extrapolated to a 100 ft free length, this results in a friction loss of about 1% for greased strand compared to 3% for waxed strand. Actual values will depend on the magnitude of the free stressing length, the viscosity of the compound and the tightness of the PE sheath over the strand. However, in proof loading the anchor tendons to 80% of their ultimate strength and then locking it off to 70%, this friction is overcome and is not more a factor for a successful stressing of the anchor tendon.

7.4 COSTS

Compared to grease, waxed strand anchors will be somewhat more expensive. Depending on the type of project, this cost increase is estimated with about 1, 2 or 3%

of the total project cost, but the longevity of the anchor tendon will improve un-proportionately to this cost increase.

8.0 CONCLUSIONS

There is always the possibility of water having access to the high strength steel of the anchor tendon. When there is a way, water will find it. The first and most important aspect here is workmanship in the assembly, installation, stressing and grouting of the anchor, followed by the quality of the materials used. Workmanship in the field is difficult to control and, once the anchor tendon disappears into the drill hole, any damage to the plastic sheath or tube by the insertion and the subsequent grouting and stressing cannot be observed any more. Equally difficult is it to verify a perfect corrosion protection underneath the anchorage. Therefore the longevity of the protective materials becomes a very important factor. Metal soaps (grease) have shown to become unstable when in contact with moisture, while petrolatum (wax) is very resistant to moisture and water (hydrophobic), even at points where the outer protective layers (plastic sheath) have been damaged. Therefore the existing standards for corrosion protection need to undergo a critical review and upgrading, considering the long-term behavior of their components, in particular the corrosion protective compounds.

Compared with a required service life of more than 100 years in many cases, tests can only be performed for a fraction of this time. A good understanding of the behavior of the compound during the test is required in order to make meaningful and reliable extrapolations. Anchors as well as buried pipelines are objects of value. Since they cannot be inspected, a blind trust has to be placed on their corrosion protection system. The suitability of corrosion protective compounds depends in essence on their resistance to de-oiling, absorption of water, emulsification and saponification. The better behavior of petrolatum observed in the tests coincides with the practical experience gained from buried steel pipelines.

REPAIR OF FAILING MSE RAILROAD BRIDGE ABUTMENT

Tom A. Armour[1], P.E., Member Geo-Institute, John Bickford[2], P.E., Member Geo-Institute and Tom Pfister[3], P.E., Member, ASCE

ABSTRACT: In 1978, Idaho Transportation Department (ITD) designed and constructed the first U.S. "true" MSE bridge abutment over the Union Pacific Railroad (UPRR) tracks in Soda Springs. This project was one of the six original FHWA Demonstration Projects introducing reinforced earth technology to the U.S. transportation industry.

In the summer of 2002, approximately six of the precast concrete facing panels "popped out" in a localized area of one of the MSE earth retaining walls supporting the bridge approaches. It was determined that the galvanized steel soil reinforcing strips had corroded at the panel connection to a point where the lateral earth pressures exceeded the connection's remaining capacity. Concerned, ITD and UPRR officials immediately prepared the contract documents for the repair of approximately 3,700 m^2 of MSE walls and bridge abutments.

The scope of work in the ITD contract documents included installing horizontal drilled and grouted crosstie ground anchors in the approach walls, anchored soldier piles at the abutments and a reinforced shotcrete facing structurally attached to the existing MSE facing panels. The crosstie anchors connected the "back-to-back" MSE approach walls located approximately 25 m from panel to panel. Challenges included traffic control and rail traffic sequencing, winter construction, monitoring the condition of the existing structures, drilling ground anchors (from panel to panel) horizontally through a compacted fill with thousands of steel soil reinforcing elements and installing the soldier pile and rock anchor elements from the existing bridge deck.

---

[1] President, DBM Contractors, Inc., Federal Way, Washington

[2] Engineering Manger, DBM Contractors, Inc. Federal Way, Washington

[3] Bridge Design Engineer, Idaho Transportation Department, Boise, Idaho

INTRODUCTION

This paper presents a case history of the failure and rehabilitation of a 25-year-old Mechanically Stabilized Earth (MSE) bridge abutment and approach retaining walls in Soda Springs Idaho. The paper will provide the background of the original wall design and construction, the failure mode for the MSE walls, the structural and geotechnical design for the rehabilitation of the reinforced soil earth retention structures and the retrofit construction methods and results.

PROJECT BACKGROUND

Idaho State Highway SH-34 crosses the main line of the Union Pacific Railroad in the small town of Soda Springs located in south eastern Idaho. The railroad, with over forty trains a day divided the town in half. The at-grade crossing prevented and delayed traffic and emergency response vehicles from the other side of the tracks. To improve the safety of the existing at grade railroad crossing and alleviate the traffic bottleneck an overpass was designed and constructed by ITD in 1978 (Figure 1).

FIG. 1 Aerial view of Idaho State Highway 34 bridge abutments and approach walls

Soda Springs is located in the northern part of the Great Basin with an elevation of 1740 m above sea level. The precipitation is less than 250 mm per year and heaviest in the cooler months coming mostly as snow. The soils in the region are mostly classified as sodic soils, characterized by high alkalinity and low permeability.

The bridge abutment spread footings were designed to bear directly on the MSE Walls, a new design concept at the time. The project was one of the six original FHWA demonstration projects introducing reinforced earth technology to the U.S. transportation industry. The MSE wall manufacturer designed the bridge abutments and approach earth retention walls for the ITD to place in the contract documents. The "horseshoe" shaped retaining walls vary in height from 2.5 m to 12 m at the abutment and consisted of over 3,700 m^2 of pre-cast concrete facing panels.

The sizes of metallic soil reinforcing strips used in the MSE approach walls were 60 mm and 80 mm wide by 3 mm thick, and the sizes of reinforcing strips used in the bridge abutment face were 80 mm, 90 mm, and 100 mm wide by 3 mm thick. Soil reinforcement lengths varied from 5 m to 13.5 m. The soil reinforcement consisted of smooth steel strips cut and hot dipped galvanized from steel coils in accordance with ASTM A446 Grade C/ Coating Designation G210 (ASTM 1985). The hot dip galvanized coating is equivalent to approximately 0.09 mm of galvanization on each side of the steel reinforcing strip. MSE wall facing elements consisted of 1.5 m x 1.5 m x 220 mm thick cruciform shaped pre-cast concrete panels and were manufactured off site.

The backfill used to construct the MSE approach fills and bridge abutments was a slag waste product purchased from Monsanto Corporation's nearby phosphate plant. At the time of construction the only requirement for the MSE backfill material was cohesionless material with a specified gradation.

WALL FAILURE MODE

On July 9, 2002 a section of the southwest approach wall located approximately 45 m from the bridge abutment failed when six of the precast concrete MSE wall facing panels "popped out". The wall at this location was approximately 8 m tall. The failure occurred in the first row of panels above the existing ground line (Figure 2). The collapse created an unstable condition for the wall facing panels immediately above the first row. Two days after the panels in the first row failed two additional panels in the second row failed from loss of vertical support. The panels in the rows above the second row formed an arch-shaped support mechanism that supported the remaining portion of wall in this area. Approximately 15 m^3 of reinforced soil backfill sloughed out through the void in the wall. Backfill above (and to the sides) of the wall opening remained in place.

After investigation ITD engineers determined that the failure was caused by the corrosion of the metallic soil reinforcing strips attached to the lower concrete facing panels. Each panel in this area had four 60 mm wide by 7.3 m long soil reinforcing strips. The hot dipped zinc galvanizing had been completely consumed. In some of the soil reinforcement the corrosion had completely "eaten" through the steel (Figure 3). Any remaining strips that had not been completely consumed from the corrosion process failed in tension at the connection with the galvanized tie strip embedded in the panel. The soil reinforcing strips above the second row of panels all remained attached to their respective panels. These reinforcing strips exhibited severe corrosion on the top surface of the strips only with the bottom surface still galvanized.

FIG. 2. MSE wall after initial failure on July 12, 2002

To best determine the reason for the accelerated corrosion rates of the metallic soil reinforcements, the backfill material in the area of the panel failure was sampled and tested to evaluate the electrochemical properties. The results of these tests as well as the 2002 AASHTO LRFD Bridge Construction Specifications and current ITD MSE wall backfill electrochemical requirements are given in Table 1.

TABLE 1. MSE wall backfill electrochemical properties, measured vs. specified

	Test Sample	AASHTO LRFD Bridge Construction Specifications (AASHTO 1998)	Current ITD Specifications (ITD 1999)
Resistivity (ohm-cm)	820	> 3000	> 3000
Soluble Chlorides (ppm)	690	< 100	< 100
Sulfates (ppm)	1770	< 200	< 200
PH	9.9	5 to 10	4.5 to 9.5

FIG. 3. Corroded metallic soil reinforcements

According to *Corrosion/Degradation of Soil Reinforcements for Mechanically Stabilized Earth Walls and Reinforced Soil Slopes* (Elias 1996)) resistivities in this range are rated as corrosive to moderately corrosive. Chlorides and sulfates in this range are equivalent to a resistivity of less than 700 ohm-cm, which would be classified as a very corrosive environment for metallic reinforcing reinforcing strips. Soils with high alkalinity and/or pH are associated with significant corrosion rates.

The apparent lack of construction quality control of the backfill during construction led to corrosion and premature failure of the soil reinforcement system. Therefore, the original design life of 75 years was greatly reduced to only 25 years.

STRUCTURE REHABILITATION DESIGN

The first concern was to stabilize the wall in the area of the failure to prevent additional wall panels from "popping out" and reinforced soil backfill in the unsupported area from unraveling. An earthen berm was placed against the wall in the area of the "pop out" to stabilize the local failure area. The remainder of the wall was monitored daily for deviations or changes in alignment or any other indication of pending failure. This temporary fill, or the "Wart" as it became known, was completed on July 13, 2002. This allowed for the continued use of the two center lanes of the four-lane highway by cars and light trucks until the walls were retrofitted.

A meeting was held to determine a retrofit design scheme and a schedule for the rehabilitation project. A number of major design considerations were identified:

- The design life of any retrofit scheme must produce a design with a life of at least 100 years as per the current AASHTO (1998) and FHWA (Elias 1996) recommendations.
- Structurally, any settlement of the abutment of greater than 13 mm was considered unacceptable; therefore, the walls around the abutment had to be designed to prevent deflections of the wall of greater than 13 mm from their current position.
- The existing horizontal clearance from the face of the abutment wall to centerline of the railroad track was 6.5 m. U.P.R.R. currently has a minimum horizontal clearance requirement of 6.8 m. Any revision in the horizontal clearance had to be minimized and approved by U.P.R.R. The railroad also restricted any construction equipment from within 6 m of the track when a train was passing.
- A major concern expressed by the local community was that the bridge remain open for emergency vehicles and school buses during construction.
- Design and construction of any retrofit to the MSE bridge abutments and approach walls had to be completed prior to the end of November 2002, because it was possible that the harsh winter weather conditions in Soda Springs could shut down the project. This requirement only allowed 20 weeks to complete the design and construction.

The solution to the design challenges for the bridge abutments was to design an anchored soldier pile wall (Figure 4). The wall was designed using horizontal loads based on an active pressure of 5.5 kN/m^3. The design included a vertical surcharge

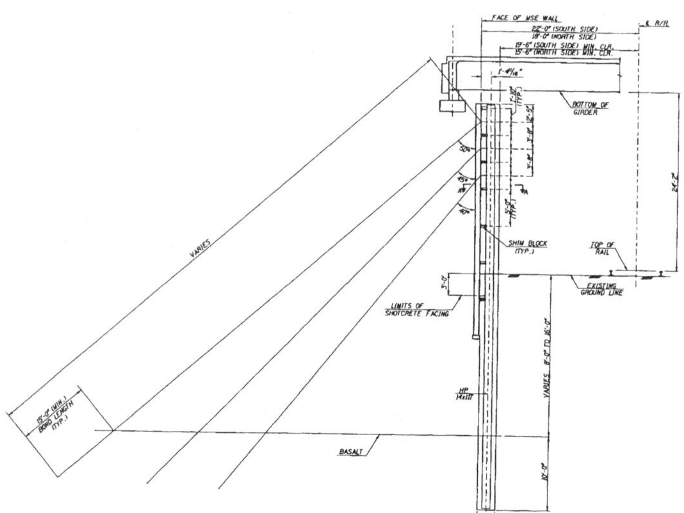

FIG. 4. Retrofit scheme at MSE bridge abutments

due to the abutment dead and live loads of 190 kPa applied over the bottom of the abutment spread footing.

The retrofit design was accomplished using a finite element program with the MSE wall modeled as plate elements and the rock anchors as tension members. The deflections and loads generated from the finite element analysis were compared to the expected pile movements from the L-Pile program. In order to balance the loads and the deflections in the piles, three rock anchors were located at the top of each pile with a lock off load of 534 kN. The maximum spacing of the soldier piles was 3 m. The pile spacing was adjusted so that the piles were located between the AASHTO Type III pre-stressed bridge girders. This allowed access holes to be cut out in the bridge deck and permitted the drill rig for the soldier pile and the rock anchors to drill from above. This eliminated any overhead clearance restrictions and problems associated with U.P.R.R. sequencing requirements for construction equipment on the railroad right-of-way.

The first attempt at a design solution for the MSE approach walls was to use a soil nailed retrofit scheme. This solution was quickly eliminated because it was believed that the open gradation of the wall backfill material would not permit the soil nails to be completely encased by the grout. It was determined that ground anchors (crossties) placed completely through the back-to-back MSE walls would be more efficient and easier to install (Figure 5). The horizontal crossties were designed on a grid with a maximum spacing of 3 m horizontally throughout the taller areas of the wall. The crossties used four 15 mm diameter 1860 MPa pre-stressing strands. Prior to horizontal drilling, cores were drilled through the existing precast concrete panels. The crossties were stressed to a test load of 133 kN with a design lock off load of 67 kN.

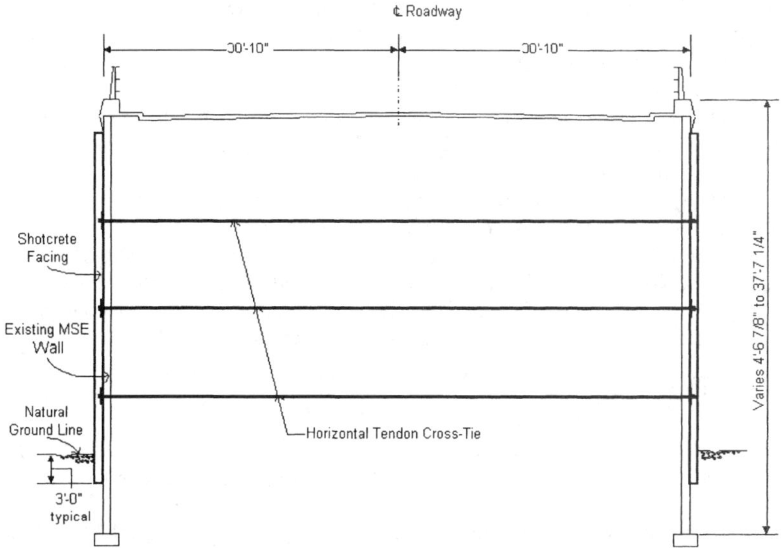

FIG. 5 Retrofit scheme at MSE approach walls

A Class 1 corrosion protection system was utilized for all of the rock anchors and crossties. The first level of corrosion protection for the rock anchors and crossties was to coat each strand with a corrosion inhibitor and then encapsulating each strand in a plastic sheath. The four-sheathed strands were then encased in a grout matrix inside a PVC sheathing. The grout used consisted of a Portland cement, water and expansion admixture, and was required to have a minimum 28 day cube strength of 190 kPa. The grout was pumped into the crossties after the crossties were installed into the wall and stressed.

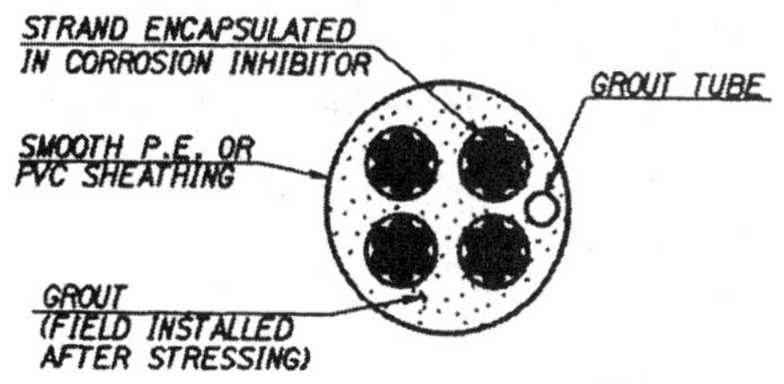

FIG. 6. Typical section of Class 1 crosstie anchor

To complete the structure rehabilitation a 200 mm to 300 mm thick reinforced shotcrete curtain wall was designed to span between the crossties and rock anchors. The wall extended from one meter below the existing ground line to the bottom of the original pre-cast concrete MSE wall coping.

In addition, a staged construction plan that required the general contractor to use only two lanes of the roadway and bridge at any given time was developed and included in the contract drawings. The bidding contractors were given the option of developing their own plan provided that vehicular traffic across the bridge was maintained throughout the project.

Forty-nine days after the wall failure bid documents were completed and the structure rehabilitation plans were advertised for bids. Bids for the project were opened on September 9, 2002. The successful low bidder was DBM Contractors Inc. of Federal Way, Washington, with a total bid price of $2,455,954.60.

CONSTRUCTION

Risk factors analyzed by the general contractor during the bid process for this project included 1.) installation of the crossties, soldier piles and rock anchors, 2.) monitoring the existing condition of the MSE wall systems for potential movements, 3.) schedule and weather constraints, and 4.) traffic control and U.P.R.R. rail car sequencing.

Traffic control was a major concern during construction. This bridge provided the City's only above grade crossing of the U.P.R.R. tracks. Emergency vehicles and school bus schedules would be severely impacted by the train traffic (40 trains per day) if the bridge was closed during construction. The City of Soda Springs therefore specified that a minimum of two lanes of traffic be maintained throughout the construction of the bridge rehabilitation project.

DBM was required to work very closely with U.P.R.R. to ensure a safe construction site. Construction sequencing played a major role in scheduling construction activities around the busy rail traffic that ran through the middle of the project site about once every 30 to 40 minutes.

Another issue on the contractor's mind was the financial requirements if the project was not completed on time. ITD specified construction to be substantially complete by the first week of December due to potential weather conditions and the upcoming holiday season. Liquidated damages of $2500/day would apply for every calendar day the project was extended beyond the specified completion date.

The ITD designed retrofit scheme for the two MSE bridge abutments consisted of 18 each 0.8 m diameter soldier piles socketed in rock, 54 each steeply inclined rock anchors (three each 534 kN anchor per soldier pile) and reinforced shotcrete pilasters at each soldier pile.

Installation of these elements could not be accomplished (cost effectively) from existing grade due to overhead constraints of the bridge and U.P.R.R. rail traffic restrictions. Therefore, 1.2 m x 1.8 m cutouts were saw cut in the bridge deck. Pile and rock anchor installation could then begin through the deck cutouts. Cutout locations were coordinated with the existing bridge girders, soldier pile location and anchor inclination.

Existing ground conditions at the bridge location consisted of 3 m to 5 m of loose to medium dense silty sand over basalt bedrock. Each soldier pile had a 3 m rock socket. DBM used an IMT AF-180 drill and conventional soil and rock augers/core barrels to install the soldier piles from the bridge deck (Figure 7). Steel casings were installed through the overburden material and seated into the basalt.

FIG. 7. Soldier pile installation from the bridge deck

Rock anchors with a contractor designed 7 m bond length in the basalt bedrock were installed with a Klemm 806 drill rig atop the bridge deck after installation and proper curing of the soldier piles. The rock anchors were installed by advancing 2 m sections of 133 mm diameter casings. Inside the casing was a 90 mm diameter inner drill rod with a down-the-hole hammer at the tip. The drill casing was advanced and seated in the basalt bedrock. The remainder of the anchor was drilled open hole with the drill rod and hammer.

The first three rock anchors installed at each abutment were performance tested to a maximum load of 150% design capacity. The remaining anchors were proof tested to 150% of design capacity. Upon completion of the anchors and soldier piles, the deck cutouts were formed, poured and cured for re-establishment of traffic on the bridge.

To retrofit the back-to-back MSE approach walls, DBM had to develop an installation method to accurately drill and grout the 200 horizontal crosstie anchors. Anchor locations were on approximate 2 m centers horizontally and varied in elevation up to 8 m from existing ground. The approach walls are located approximately 25 m face to face and required drilling the crossties from one side of the approach fill through the concrete MSE facing panel on the other side.

DBM used a track mounted Interoc 109B anchor drill to install crosstie anchors located less than 4 m above existing ground level (Figure 8). For crossties located greater than 4 m above existing grade DBM modified a standard Caterpillar 325 excavator by removing the digging bucket and replacing it with a drill mast unit. With a reach of nearly 6.5 m, this allowed DBM to install the upper horizontal crossties from existing grade and from the top of the roadway. To ensure drilling accuracy through the compacted slag backfill and "sea" of metallic soil reinforcing strips DBM utilized standard rotary methods with 2 m sections of 152 mm drill casing and standard 114 mm roller bits attached to 90 mm diameter drill rods.

FIG. 8 Installation of crosstie anchors.

At the completion of drilling each crosstie hole, 75 mm diameter PVC smooth plastic sheaths were installed and externally grouted with neat cement. The four-strand Class 1 greased and sheathed crosstie anchors were then installed, attached to bearing plates on each face and grouted with neat cement (Figure 9). Upon curing, anchors were load tested to 133 kN and locked off at the design load of 67 kN.

FIG. 9. Crosstie anchor head

Monitoring of the failing MSE walls condition for excessive movements was maintained throughout the duration of the project. Traffic was not allowed in the outside lanes (nearest the MSE wall facings) until the successful completion of all of the crosstie anchors. Crossties were installed from the bottom row upward. This process allowed the added loads from the drill rig (placing the crossties from the upper roadway) to be adequately supported. Once the crosstie anchors were completed, traffic was switched from the middle two lanes to the two outside lanes and soldier piles and rock anchors were completed in these areas.

After completion of the crosstie and rock anchors DBM installed approximately 3,700 m^2 of 200 mm – 300 mm thick wood float finished reinforced shotcrete facing. 1 m x 0.5 m rectangular pilasters were also installed by the shotcrete method at each soldier pile location. With Soda Springs located in the foothills of the Cascade Mountain Range at an elevation of approximately 1740 m, cold weather protection for the shotcrete operations had to be considered. DBM designed a scaffolding system that was protected from wind, snow and the cold climates with heavy plastic sheeting and heating system that was maintained at a constant temperature of 18°-19°C. Figure 10 illustrates the shotcrete operations and required weathered protection.

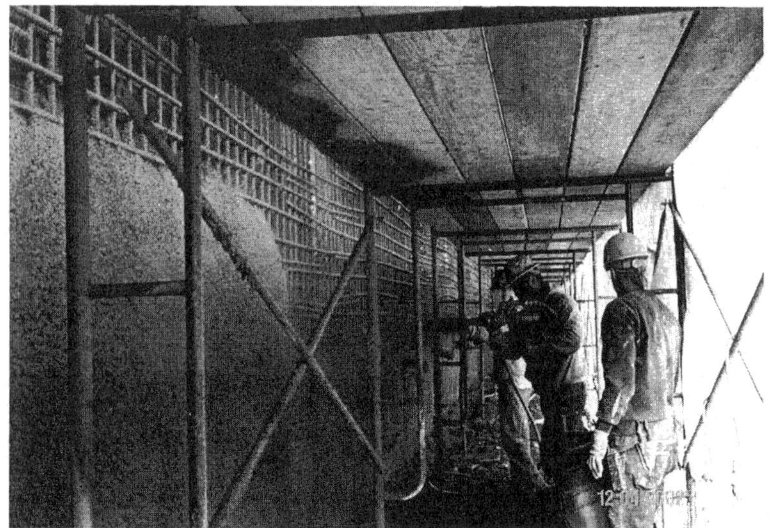

(a) Shotcrete placement

(b) Weather protection

FIG. 10. Shotcrete operations

PERFORMANCE AND CONCLUSIONS

The three major challenges during construction were the drilling accuracy of the horizontal crossties, drilling of the basalt bedrock, and preventing the MSE panels on the opposite side of the approach fill from “popping out” when the drill hit them. The first issue was controlled by the general contractor’s selected drilling method and speed of the drilling. Drilling deviations from design location varied up to one meter with an average deviation of approximately 300 mm. In some cases (very few) where deviations were greater than 600 mm the reinforcing steel in the shotcrete wall was revised to accommodate a greater spacing than was designed for. Placing a front-end loader bucket against the MSE panel where the drill rods exited was a simple solution to the issue of panels “popping out”.

This unique rehabilitation project was completed on time and under the ITD budget. Both ITD and DBM worked effectively together in identifying and dealing with all of the project challenges. This working relationship helped create a project that met or exceeded FHWA’s, City of Soda Springs, ITD’s and DBM’s expectations. Currently the retrofitted bridge abutments and approach earth retention structures are performing satisfactorily (Figure 11).

FIG. 11. Finished bridge abutment and approach walls

A major design issue that was not addressed in this project was the elimination of surface water that is migrating from the roadway surface down through the pavement into the approach fills. Another project has been scheduled for the Summer of 2005 to remove the existing asphalt pavement and install an impervious membrane to intercept any flows that may contain deicing chemicals. At that time more samples of the approach fill material and metallic soil reinforcements will be taken and additional testing will be performed.

In conclusion, the reinforced soil backfill specifications for MSE walls have been improved greatly since the late 1970's. However, the corrosive nature of the fill material should be closely monitored on all future projects of this type to prevent the rapid deterioration of the metallic soil reinforcements. Other MSE walls with metallic soil reinforcements constructed in the same time frame as the project should be examined to determine if they have similar corrosion problems.

REFERENCES

AASHTO, "*LRFD Bridge Design Specifications*," Second Edition, American Association of State Highway and Transportation Officials, Washington, DC, 1998.

AASHTO, "*LRFD Bridge Construction Specifications*," First Edition, American Association of State Highway and Transportation Officials, Washington, DC, 1998.

ASTM, "*Steel Sheet, Zinc-Coated (Galvanized) By The Hot-Dip Process, Structural (Physical) Quality* (ASTM A44/A446M-85)", Volume 01.06, American Society for Testing and Materials, West Conshohocken, PA, 1985.

ASTM, "*General Requirements for Steel Sheet, Zinc-Coated (Galvanized) By The Hot-Dip Process* (ASTM A525-86)", Volume 01.06, American Society for Testing and Materials, West Conshohocken, PA, 1986.

Elias, V., "*Corrosion/Degradation of Soil Reinforcements for Mechanically Stabilized Earth Walls and Reinforced Soil Slopes*," FHWA-SA-96-072.

Idaho Transportation Department, "*Standard Specifications for Highway Construction*", Idaho Transportation Department, Boise, ID, 1999.

INFLUENCE OF GROUND WATER TABLE FLUCTUATION ON LATERAL LOAD BEHAVIOR OF RIGID DRILLED SHAFTS IN STIFF CLAY

Alan J. Lutenegger[1] and Amy E. Dearth[2]

ABSTRACT: A series of static one-way lateral load tests was performed on prototype-scale drilled shafts installed in a stiff surficial clay crust to investigate the effect of ground water on lateral load behavior. Tests were conducted under different ground water conditions at the National Geotechnical Experimentation Site at the University of Massachusetts, Amherst. The load-displacement results of the tests are presented herein. The results of the tests show that the load-displacement behavior was significantly influenced by the ground water regime at the time of testing. The ultimate lateral load capacity, defined as the load at a displacement of 10% of the shaft diameter, and the initial stiffness increased as the water table dropped. A drop in the ground water table from 0.4 m to 2.3 m below ground surface increased the ultimate lateral load from 30 kN to 95 kN. The paper presents a summary of the testing and illustrates the influence that ground water can have on field tests.

INTRODUCTION

The behavior of near surface foundations may be influenced by fluctuations in the ground water table as the effective stress regime changes due to seasonal ground water fluctuations. This influence is likely to be more pronounced in a fine-grained soil than in a granular soil because of the potential for large soil matric suction to develop during the dry season. Many structures require special design considerations to take into account lateral loads that may be imposed over the life of the structure. In fact, for some structures, lateral loading may be the most important mode of loading that the structure will undergo over its service life. In the past 25 years, results of a number of interesting studies have been presented regarding the lateral

[1] Professor and Head, Department of Civil and Environmental Engineering, University of Massachusetts, 224 Marston Hall, Amherst, MA 01002; lutenegg@ecs.umass.edu

[2] Geotechnical Engineer, McPhail Associates, Inc., Cambridge, Ma. 02139; adearth@mcphailgeo.com

load behavior of drilled shafts in clays however, none of these studies have addressed the influence of ground water conditions on the response.

At most real sites, the ground water table fluctuates as a result of natural fluctuations in the ground water regime. These fluctuations are brought about by seasonal changes in precipitation and other effects related to the geology and topography of an individual site. Surface water and ground water entering or leaving a site may have a dramatic effect on short term and long term water levels and piezometric pressures. Other influences, such as local base level or artificial drawdown, as from pumping, may also influence ground water fluctuations. In order to evaluate the influence of ground water conditions on the lateral load behavior of drilled shafts, a series of load tests were performed at different times of the year, when the ground water level was different.

Ten drilled shafts were constructed on the same day using identical procedures. All of the shafts had a diameter of 0.30 m and a length of 1.5 m and were installed in the stiff clay crust at the National Geotechnical Experimentation Site (NGES) at the University of Massachusetts. The length was chosen so that the full length of the shafts would be within the zone of anticipated ground water fluctuation, based on past measurements at the site. The length to diameter ratio of 5 used in these tests means that the shafts most likely behave as rigid elements. Lateral load tests were performed by applying incremental loads between pairs of shafts at the ground line. The paper presents a summary of the site conditions, a description of the shaft construction and testing procedures and the load-displacement curves of the shafts. The results illustrate dramatically that for real soils and real site conditions, the response of these foundations are significantly influenced by ground water conditions. This influence should be considered in design and is considered particularly important in situations where load test results are being compared with design predictions, such as in prediction events.

INVESTIGATION

The testing consisted of a conventional site investigation to determine soil characteristics within the zone of embedment of the drilled shafts, construction of the drilled shafts, and lateral load testing. The details of each of these components of the investigation are described herein.

Test Site and Soil Conditions

Tests were performed at the National Geotechnical Experimentation Site (NGES) at the University of Massachusetts. The stratigraphy and soil properties at the site have been well documented (Lutenegger 2000). Tests were performed in Area B where the stratigraphy at the location of the drilled shaft tests consists of approximately 1.5 m of stiff silty-clay fill overlying a thick deposit of late Pleistocene lacustrine varved clay. This deposit is locally known as Connecticut Valley Varved Clay (CVVC) and is composed of alternating layers of silt and clay as a result of deposition into glacial Lake Hitchcock. The individual varves of the CVVC are on the order of 2 to 8 mm in thickness and generally occur in a horizontal position. At the location of the test shafts, the fill consists of CVVC placed about 30 years ago after excavations at the Town of Amherst Wastewater Treatment Plant, adjacent to the site. Below the fill, the CVVC has a well-developed stiff overconsolidated crust that was formed as a result of surface erosion, desiccation, ground water fluctuations and other physical and chemical processes. The NGES site has been previously used to study the tension capacity of vertical drilled and grouted anchors (Lutenegger and Miller 1994) and for other field studies involving prototype scale foundation behavior (e.g., Lutenegger and Miller 1993; Miller and Lutenegger 1997; Lutenegger and Dearth 2002). Figure 1 gives a summary of site characteristics.

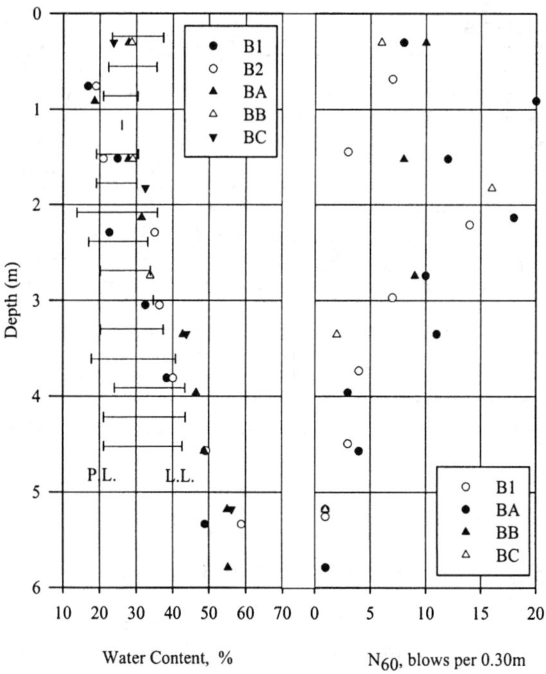

FIG. 1. NGES (Area B) Site Characteristics

Ground Water Table Fluctuations

As previously noted, one of the characteristics of a natural site is that the ground water regime may show annual fluctuations depending on the topography and the site geology. This is largely the result of seasonal fluctuations in precipitation at the site and surface runoff characteristics. Field tests such as foundation loading tests or even in situ tests may be affected by such fluctuations if tests are performed at different times under differing ground water conditions. During a drying regime, large soil matric suctions develop which can dramatically change the effective stress conditions and soil strength and stiffness characteristics. Fluctuations in the ground water table at the NGES have been described previously in some detail (e.g., Lutenegger 1995;

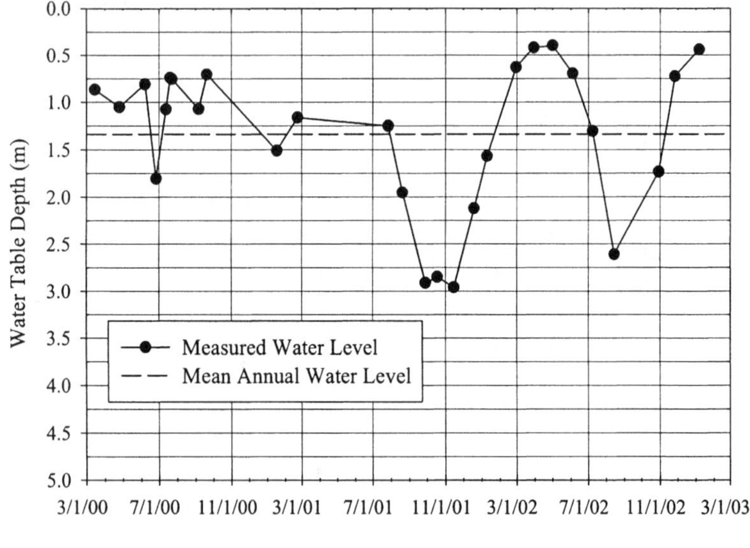

FIG. 2. Typical Ground Water Table Fluctuations at NGES

DeGroot and Lutenegger 2002). To illustrate these variations, Figure 2 shows fluctuations in the depth (below ground surface) to the water table at the site over a period of three years. These fluctuations are significant and may influence the results of any type of field testing performed at the site. For the current work, the depth to the ground water table at the time of each test was measured in an open standpipe piezometer installed to a depth of 3 m and located adjacent to the test shafts. Results are given in Table 1.

Construction of Drilled Shafts.

All shafts were constructed using helical flight augers operated from an all-terrain drill rig. After drilling to the desired depth, concrete was placed in each of the holes using gravity free-fall. Six No. 6 steel reinforcing rods were placed in each of the shafts. All shafts were constructed on the same day. The drill holes were left open for approximately 1 hour prior to placement of concrete. The concrete was mixed to give a slump between 100 and 150 mm and showed a 28-day compressive strength of 200 MPa. Characteristics of the shafts are given in Table 1. Pairs of shafts having the same diameter were constructed to provide redundancy in the testing and to act as self reaction for the load testing. The shafts were located 3 m apart. One of the shafts was defective and was not included in the results presented here.

Lateral Load Testing Procedures

Lateral load tests were conducted on pairs of shafts over a period of about 10 months. Tests were performed using the general procedures described in ASTM D3966-90 (1995) *Standard Test Method for Piles Under Lateral Loads*. Load was applied by a single 250kN hydraulic jack placed inline between the two shafts. Load from the jack was applied at the ground line using a simple steel frame. Deformation measurements were made using two digital dial indicators attached to independent reference beams placed on the backside of each of the shafts. Loads were applied incrementally in the range of approximately 5 to 10% of the estimated ultimate capacity. Each load increment was maintained for 16 min. The tests were conducted until a minimum displacement on the order of 10% of the shaft diameter had occurred. The load was measured using a Geokon load cell placed between the hydraulic jack and the reaction beam. Deflection was read using an electronic digital indicator. No cracks were observed in any of the shafts during or following the tests.

Table 1. Summary of Drilled Shafts Tested.

Shaft	Diameter (m)	Length (m)	Date Installed	Date Tested	Depth to GW Table (m)
DS-9	0.30	1.52	9/19/00	4/10/01	0.4
DS-10	0.30	1.52	9/19/00	4/10/01	0.4
DS-3	0.30	1.52	9/19/00	10/22/00	0.8
DS-4	0.30	1.52	9/19/00	10/22/00	0.8
DS-15	0.30	1.52	9/19/00	10/26/00	0.8
DS-11	0.30	1.52	9/19/00	8/7/01	1.5
DS-12	0.30	1.52	9/19/00	8/7/01	1.5
DS-17	0.30	1.52	9/19/00	8/30/01	2.3
DS-18	0.30	1.52	9/19/00	8/30/01	2.3

RESULTS

Figure 3 presents the load-displacement curves for the drilled shafts tested in the current work. Individual tests and the average of tests on each test date are given. Ground water conditions at the time of testing fluctuated between 0.37 to 2.29 m (1.2 to 7.5 ft) below the ground surface, dropping below the base of the shafts. The load-displacement curves show the typical response of laterally loaded deep foundations and indicate that the depth to the ground water table has a significant influence on the response. The results also show that the repeatability of results for each pair of shafts is very good. Table 2 summarizes the load on each shaft at different lateral displacements.

The load curves shown in Figure 3 do not indicate an obvious failure load as is usually the case with deep foundations under lateral load. Since the ultimate lateral load capacity of a drilled shaft may be defined in a number of ways, the results given in Table 2 may be used to make different comparisons of the behavior. For the purposes of providing a simple comparison of the behavior, the ultimate lateral load capacity was defined as the load which produced a lateral displacement, s, of 10% of the shaft diameter, D. The results presented in Table 2 show that there is a dramatic influence on the lateral load response depending the ground water table position. The load at any absolute or relative displacement of the shaft increases as the depth to the ground water table increases.

Table 2. Summary of Drilled Shaft Capacities.

Shaft	Q @ s/D = 2.5% (kN)	Q @ s/D = 5% (kN)	Q @ s/D = 10% (kN)	Q @ s = 25.4 mm (kN)	Q @ R = 1° (kN)
DS-9	16	22	32	29	30
DS-10	14	20	29	26	27
DS-3	21	33	52	46	48
DS-4	23	32	46	43	43
DS-15	21	30	42	42	39
DS-11	20	34	55	48	51
DS-12	26	45	73	64	69
DS-17	33	56	100	78	80
DS-18	35	57	93	83	85

Figure 4 shows the effect of the depth to the ground water table on the interpreted ultimate lateral load capacity. The trend line shows that within the zone of water table measurements, the ultimate lateral load increased linearly with increased depth to the ground water table as the water table dropped below the base of the shafts. It would not be expected that there would be an additional linear increase if the water table were to drop further. The use of the water table measurement is only a practical and simple means to provide an indication that some change in water conditions has occurred. As the depth to ground water increases and soil suction increases, there is a positive change in vertical effective stress, resulting in greater shaft capacity as seen in the increase in shaft capacity with increased depth to ground water. This relationship would be valid for any definition of failure using the data from Table 2.

On three of the test dates, water content samples were obtained throughout the soil profile using a small hand augur. These results are shown in Figure 5. It can be seen that the use of water content alone is not sufficient to provide a consistent measure of changes in moisture regime. Wetting versus drying fronts may produce similar water contents but different soil suction. The use of reliable field soil suction measurements would be the more desirable to determine the moisture regime. Simple use of water content is not sufficient to give the actual indication of the in situ conditions of a complex problem. Time is not considered a factor influencing the test results in this case since some shafts which gave lower capacity were tested after some shafts that had higher capacity.

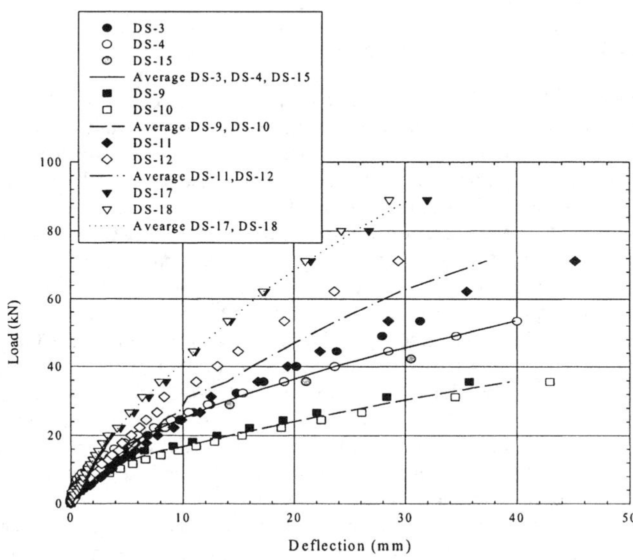

FIG. 3. Load-Displacement Curves

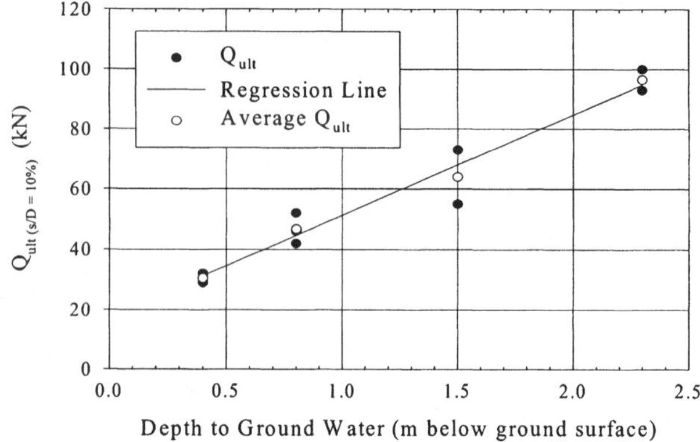

FIG. 4. Increase in Ultimate Capacity with Increase in Depth to Water Table

SUMMARY

The results presented in Figures 3 and 4 demonstrate the effect that moisture regime can have on the behavior of foundation elements located in a zone of fluctuating ground water. Under ideal conditions of soil saturation, in this case with a water table located at the ground surface, it would be expected that the lateral load capacity would be a minimum. This is illustrated by the trend line shown in Figure 4. As the ground water table drops under a drying front, the upper soil becomes unsaturated and matric soil suction develops. For design, it would be simple to assume the worst conditions and expect that the soil remains saturated throughout the life of the structure. However, consider the implications that a site investigation and design are performed at a time when extreme drying has occurred and at some time thereafter, the ground water table rises reducing the matric suction. This could produce additional unexpected displacements from the loss of soil strength and stiffness.

Additionally, it would be expected that the results of any in situ tests, such as Pressuremeter or Dilatometer tests used to predict the lateral load behavior must be performed under similar moisture conditions as the foundations in order to make reliable estimates of performance. It is suggested that some variations between predicted and observed response of foundations may be related to ground water and thus soil moisture conditions that have not been previously considered in any detail.

Even though the results presented herein are for relatively small diameter and short shafts, it can be expected that similar results, but perhaps showing different magnitude of influence, would be obtained for larger production scale shafts. The lengths of shafts tested in the current work were kept intentionally short in order to keep the anticipated range in ground water fluctuation within the full shaft length. At this site, it is not expected that natural ground water fluctuations would be much more that those shown in Figure 2. Additional tests are planned using larger diameter and longer shafts but the testing schedule will be controlled by ground water location.

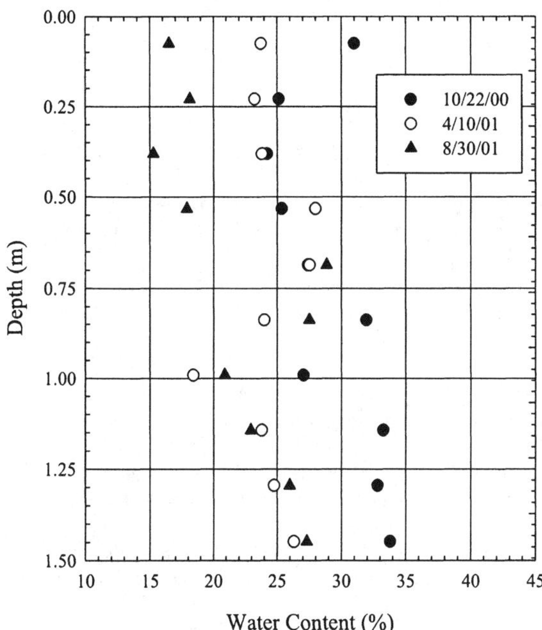

FIG. 5. Variation in Water Content on Different Test Dates

CONCLUSIONS

A series of lateral load tests was performed on drilled shafts in a stiff clay. The results indicate that the lateral capacity of shafts can be significantly affected by the moisture regime at the time of testing. The results presented herein were obtained in a stiff clay which is expected to show may dramatic effects than at a site composed of more granular materials, such as sand. This is clearly related to the ability of the soil at an individual site to develop high matric suction. A drop in the water table in this case resulted in an increase in ultimate lateral capacity nearly three times that of the capacity under high ground water conditions. The results presented herein are intended only to draw attention to this issue and demonstrate that at sites where ground water conditions can change the moisture condition should be taken into account. A more directed and detailed test program is required to further document changes in matric suction and moisture conditions over longer period of time.

REFERENCES

Ahams, J.I. and Radhakrishna (1973). "The lateral capacity of deep augered footings." *Proc. 8th Int. Con. on Soil Mech. and Found. Engrg.*, 2.1, 1-8.

Bhushan, K., Haley, S.C., and Fong, P.T. (1979). "Lateral load tests on drilled piers in stiff clays." *J. Geotech. Engrg. Division*, ASCE, 105 (GT8), 969-985.

DeGroot, D. J. and Lutenegger, A.J. (2002). "Geology and engineering properties of Connecticut Valley Varved Clay." *Proc. Int. Workshop on Characterization and Engineering Properties of Natural Soil.*

Lutenegger, A.J. (1995). "Geotechnical behavior of overconsolidated clay crusts". *Transportation Research Record No. 1479*, 61-74.

Lutenegger, A.J. (2000). "National geotechnical experimentation site - University of Massachusetts." *National Geotechnical Experimentation Sites*, ASCE, 102-129.

Lutenegger, A.J. and Adams, M.T. (1998). "Bearing capacity of footings on compacted sand." *Proc. 4th Int. Conf. on Case Histories in Geotech. Engrg.*, 1216-1224.

Lutenegger, A.J. and Adams, M.T. (1999). "Tension tests on bored piles in residual soil." *Behavioral Characteristics of Residual Soils*, ASCE, 43-53.

Lutenegger A.J. and Miller, G.A. (1993). "Behavior of laterally loaded drilled shafts in stiff soil." *Proc. 3rd Int. Conf. on Case Histories in Geotech. Engrg*, 1, 147-152.

Lutenegger, A.J. and Miller, G.A. (1994). "Uplift capacity of small diameter drilled shafts from in situ tests. *J. Geotech. Engrg.*, ASCE, 120 (8), 1362-1380.

Lutenegger, A.J. and Dearth, A. E. (2001). "Scale effects of laterally loaded drilled shafts in clay." *Foundations and Ground Improvement*, ASCE, 554-564.

Miller, G.A. and Lutenegger, A.J. (1997). "Influence of pile plugging on skin friction in overconsolidated clay." *J. Geotech. and Geoenv. Engrg.*, ASCE, 123 (6), 525-533.

PILOT HOLES FOR DRILLED PIER FOUNDATIONS IN KARST CONDITIONS

James G. Bierschwale[1], P.E., Member, ASCE, and
Todd E. Swoboda[2], P.E., Member, ASCE

ABSTRACT: Development on land underlain by limestone formations that exhibit karst conditions can present unique engineering and construction challenges. These challenges often include concerns regarding questionable subgrade for foundation support and economical methods for foundation construction.

Over the past decade, many structures have been constructed in the Central Texas area over limestone formations exhibiting karst features such as voids and solution cavities, as well as soil seams and layers. The most common geologic formation to exhibit such properties in Central Texas is the Edwards Group limestone. Significant foundation design challenges may be encountered for structures constructed over such formations, particularly larger structures with significant column loads. Many such structures often require a pilot hole program for evaluation of the bearing stratum below each foundation unit. Through multiple projects in the Central Texas area, the authors have compiled experience with three major categories of pilot hole programs, including 1) the use of geotechnical drilling rigs, 2) air percussion drilling equipment, and 3) drilling pilot holes during construction with pier drilling equipment. This paper discusses the benefits and drawbacks of utilizing each of these three different pilot hole methods, and presents several case histories.

INTRODUCTION

Since the early 1980's, significant development has occurred in areas of Central Texas that are in the Edwards Geologic Group of lower Cretaceous Age. These areas

1 Principal, HBC/Terracon, 5307 Industrial Oaks Boulevard, Suite 160, Austin, Texas 78749, (512) 442-1122, jgbierschwale@terracon.com.

2 Senior Geotechnical Engineer, HBC/Terracon, 11555 Clay Road, Suite 100, Houston, Texas 77043, (713) 690-8989, teswoboda@terracon.com.

in Central Texas that are experiencing an increase in development include the northwest, west and southwest portions of Austin, as well as the north and northwest portions of San Antonio. The Edwards Group is common in these areas. The Edwards Group is comprised primarily of limestones and dolomitic limestones that frequently contain karst features which are caused by dissolution of the limestone resulting from years of subsurface groundwater flow. Karst (or solution) features include solution cavities and channels, sinkholes, and caverns. Many of these features have since been filled with soil (often reddish brown clay mixed with limestone fragments) or secondary crystallization of limestone.

The subsurface variations that can result from the solution activity discussed above can significantly complicate foundation design for buildings to be constructed in this geologic environment. Obviously, karst features within the limestone can significantly impair the bearing capacity and settlement characteristics of the rock mass. In some situations, nondestructive geophysical methods can be used to identify and isolate solution features that would affect foundation performance. However, it is common to discover subsurface variations in this formation that are too extensive to be reliably characterized by geophysical means. The presence of karst conditions is further complicated by situations in which structures with significant concentrated loads require large capacity foundation units (typically drilled piers or spread footings) for support. In such situations, it has become conventional to drill small-diameter pilot holes at the individual foundation locations to evaluate the subsurface conditions at each foundation unit and allow for design modifications in situations where karst features or other anomalies impact the bearing stratum. This paper focuses on the benefits and drawbacks of three different methods of performing pilot holes for drilled pier foundations that are commonly used in the Central Texas area.

PREVIOUS LITERATURE

Several papers were identified that address the use of pilot holes to identify subsurface variations for foundation design purposes (Weber, 1988; Fischer, et. al., 2000; Sowers, 1996; and Fischer, et. al., 1988). These authors have documented karst conditions in limestone in locations ranging from Ontario to Alabama (along the valleys of the Appalachian range) as well as karstic limestones in the western United States. These authors report the performance of standard geotechnical borings as providing the most valuable information when evaluating limestone containing karst features. Air percussion drilling methods were often discussed as being used as a means to provide supplemental information. As stated by Weber (1988), "The most accurate and useful information about the subsurface will be derived from the conventional, sampled and core-drilled test borings. ...where extensive exploration is required, air track drilling can be used to augment conventional exploration in order to save time and money."

Other methods of evaluating limestone containing karst features such as geophysical methods and geohydrological investigations were also mentioned. However, as stated by Fischer, et. al., (2000) "The basic geophysical tools such as reflection and refraction surveys, resistivity, and ground penetrating radar are of marginal use. Thus, while geophysical tools may be helpful in an auxiliary role, ...they are rarely as useful as conventional investigation tools unless one is looking for large, relatively shallow cavities (i.e. near-surface caves)." Fischer goes on to state "One must also recognize that the variations that can occur over short distances at most karst sites will usually preclude the development of an all-encompassing geologic model."

These authors discuss the necessity to evaluate each site containing karst conditions individually to customize additional studies at the site to provide useful information for benefit of the foundation design. These additional studies often consist of drilling pilot holes. Sowers (1996) states "In badly solutioned or deeply fractured limestone formations, it is essential that the foundation rock be examined for signs of weakness and explored below the bottom of each foundation for defects. These include cavities below the rock surface that could allow the rock to crush or break under the future foundation load or clay-filled seams that would allow the foundation to subside as the clay consolidates or extrudes outward under the concentrated foundation load."

PILOT HOLE PROCEDURES COMMONLY USED

The authors are experienced with three methods of drilling pilot holes prior to drilled pier installation. Typically, these procedures are employed after construction of the building pad and immediately prior to pier installation. Often, both pilot holes and foundation installation take place concurrently. These methods are listed below.

- Drilling pilot holes with geotechnical drilling rigs.
- Drilling pilot holes with air percussion drilling equipment.
- Utilizing the pier drilling equipment to drill pilot holes at the bottom of the pier excavations during construction.

This paper is based upon monitoring of pilot hole programs by the authors for 23 projects performed in Central Texas. Of these projects pilot holes were completed using air percussion rigs for eleven projects, pier drilling rigs for ten projects, and geotechnical drilling rigs for two projects.

For each of these procedures, the philosophy behind the pilot hole procedure is generally the same. The pilot hole is normally extended to a depth of 1.5 to 2.0 pier diameters below the proposed bearing elevation of the pier (typically with a minimum penetration of 1.5 meters below the bearing surface). When clay layers, voids, etc. are observed, the pilot hole is extended until the subsurface conditions observed appear conducive to providing adequate capacity for the proposed foundation unit. The extent/thickness of observed voids and clay layers, along with the quality of the rock, are generally considered on a pier by pier basis when evaluating the capacity of the foundation unit and the need to extend the pier beyond its original design embedment to achieve the desired capacity.

The following sections describe each of the above procedures and outline advantages and disadvantages that the authors have observed through experiences with the these methods.

GEOTECHNICAL DRILLING RIGS

This procedure basically involves the use of the same geotechnical drilling/sampling rigs that are utilized to perform geotechnical investigations. During drilling of the pilot hole, the rig will typically collect continuous samples of the limestone strata through the use of rock coring procedures. This provides the same quality of data that is typically obtained during a standard geotechnical field program. Listed below are the advantages and disadvantages that the authors have experienced with this method of pilot hole drilling.

Advantages

- *More accurate characterization of subsurface conditions* – The sampling program allows the geotechnical engineer to evaluate recovery and rock quality designation (RQD), view the solution features in the rock cores, and perform laboratory testing if desired.
- *Pier depths are known ahead of time* – Since the geotechnical engineer can evaluate the pier depths after pilot hole drilling, the Contractor will know the actual depth of the pier prior to beginning each foundation unit. Therefore, the reinforcing steel length, concrete volume, etc. will be known ahead of time and a more accurate estimate of the schedule and cost of the foundation installation can be developed.
- *Less depth adjustments during pier installation* – Although some of the piers may need to be deepened due to solution features observed in the pier excavation which were not encountered in the pilot hole, this is infrequent when geotechnical sampling techniques are utilized.
- *Pier capacity can be interpreted more accurately* – Since the recovery and RQD will typically be measured and the engineer given an opportunity to observe the rock cores and perform additional strength tests, the pier capacity at each location can be evaluated more accurately.

Disadvantages

- *Time Consuming* – Geotechnical drilling rig production rates will in some cases be slower than the actual pier drilling. In situations where a short lead time is given before initiation of pier construction, the pilot hole program can cause a delay in the foundation installation.
- *Expensive* – This method is significantly more expensive than the other two methods discussed.
- *Site Access* – Although the pier rig will have to set up over the same location as the geotechnical drilling rig, clear access to some of the foundation locations is at times not possible at the time the pilot hole drilling is in progress.
- *Wet rotary drilling saturates the building pad* – Drilling fluid will typically escape from the pit/borehole during the coring process, which can saturate the building pad to the point where access becomes a problem. This is less of a concern when crushed limestone base material is used at the surface of the pad (which is common in Central Texas). Air-rotary coring alleviates this problem but is often difficult in Edwards limestone because the clay layers easily clog the air ports during drilling.

- *Availability of Drilling Equipment* – Due to the sometimes limited availability of quality geotechnical drilling rigs in the area, it is at times difficult to place one or more rigs on a project for what could be up to several weeks to complete a pilot hole program.

AIR PERCUSSION DRILLING EQUIPMENT

An air percussion drilling rig is typically a compressor-driven piece of equipment that is mounted on tracks. The rig typically drills a 2-inch diameter hole utilizing rotary-percussion drilling. Compressed air is used to force cuttings out of the drill hole. No samples can be collected except for the dust/cuttings which are blown out of the borehole. The penetration rate of the air percussion drilling is monitored to detect voids, clay seams, etc. A rapid drop of the drill bit typically indicates a void whereas variations in the penetration rate and the manner in which cuttings rise to the surface often indicates the presence of weathered rock zones or clay layers.

Advantages

- *Drilling time* – Air percussion drilling rates are generally significantly faster than geotechnical drilling rigs. Thus, the pilot hole program can be completed in a shorter time frame.
- *Pier depths are known ahead of time* – As with the geotechnical drilling equipment, the pier depths will typically be known ahead of time, although often with less accuracy.
- *Cost* – The cost of air percussion drilling is usually significantly less on a per foot basis when compared to geotechnical drilling equipment.
- *Maneuverability* – The small, compact nature of this equipment, compounded with the ability to turn sharp on tracks, makes it very maneuverable.
- *No water* – The use of compressed air for drilling negates the need to use water as drilling fluid, thus eliminating the possibility of saturating the building pad.

Disadvantages

- *Characterization of the formation* – Since no samples can be taken, characterization of the formation can be difficult, particularly in situations where the driller is inexperienced in identifying the conditions based upon the drilling rates, cutting consistency, etc. Generally, the most effective subsurface characterization under these conditions would be to differentiate between competent rock, poor or weathered rock, clay layers, and open voids. However, sometimes the distinction between the above general categories can be difficult.
- *Adjustments in pier depths more common* – Due to the fact that the subsurface conditions cannot be identified as accurately, pier excavations with conditions that vary from the pilot hole data are more common when an air percussion rig is utilized versus when the pilot holes are drilled with a geotechnical rig. In these situations, if the pier needs to be deepened, a pilot hole is typically

drilled through the bottom of the pier with the pier drilling rig to confirm subsurface conditions below the bearing elevation.

- *Noise and dust* – Air percussion drilling is very noisy and creates a large quantity of dust. This can be problematic at locations which are sensitive to such conditions.
- *Lead Time* – Although not as much of a problem as with geotechnical drilling rigs, scheduling can be a problem.
- *Experienced driller required* – Personnel that operate air percussion equipment are often focused on producing drilling footage with little attention paid to identifying subsurface conditions. Therefore, they may not be experienced in characterizing the rock conditions during drilling. It is important to use experienced drillers who are capable of characterizing the strata.
- *More conservative* – Since the characterization of the subsurface conditions is less accurate than with geotechnical rigs or direct observation of the pier sidewalls, the resulting pier design will often be more conservative (thus more costly) than for the other methods.
- *Clay layers* – The existence of frequent clay layers often causes the air ports to clog during drilling. When this occurs, the bit has to be brought up to the surface and cleaned, which impedes drilling production.

PIER DRILLING RIG

This method of pilot hole drilling occurs during construction of the pier. The subsurface conditions at each pier location are documented by observations of the pier sidewalls and drilling penetration rate during drilling. Once the desired penetration into rock has been achieved, a small diameter pilot hole is augered in the center of the pier excavation to the prescribed depth. If solution features are encountered, the pilot hole is deepened until subsurface conditions are acceptable for proper end bearing of the pier, then the drilled pier is extended accordingly.

Advantages

- *Cost Effective* – Since the pilot hole only has to extend below the bottom of the pier hole, the actual amount of pilot hole footage is substantially reduced.
- *No lead time* – It is not necessary to drill the pilot holes in advance using this procedure, thus the foundation installation may commence once pad preparation and foundation layout has been completed.
- *Accurate characterization of subsurface conditions* – Since the rock condition can be observed along the sidewalls of the piers and the hardness of the rock monitored through penetration rates and ease of drilling, the rock condition at each pier location can be interpreted easily.

Disadvantages

- *Pier depth not known in advance* – This can create several complications. When many piers need to be extended significantly on a given day, it creates problems for the contractor in terms of tying steel cages, ordering concrete, production rates, etc. Another drawback is that the total pier footage of the foundation system is not known in advance, thus the cost of the foundation system is unknown until the foundation system has been installed. In situations where the degree of pier extension due to solution features is underestimated, this can lead to significant cost overruns and schedule delays.

- *Loss of efficiency during pier excavation* – Once a pier has been excavated to its planned depth, the auger must be removed and replaced with a small diameter auger to drill the pilot hole. The larger diameter auger must then be replaced prior to drilling the next pier. An alternative to this procedure would be to employ multiple drilling rigs at a site, one of which drills pilot holes exclusively.
- *Pilot hole diameter* – The diameter of the auger utilized to drill the pilot hole should be significantly smaller than the actual pier diameter. Typically, the cross-sectional area of the auger should be less than 10 percent of the cross-sectional area of the pier. This can present difficulties with smaller pier diameters (e.g. 18-inch) because the kelly bar is often at least 6 inches in diameter and requires a minimum 6 to 8-inch diameter bit to drill.
- *Questions need quick resolution* – In cases where unusual circumstances are encountered that require engineering interpretation, a very quick response is required by the geotechnical engineer to evaluate the conditions and recommend adjustments in the pier design to allow for completion of the foundation unit.

CASE HISTORY 1: PILOT HOLES DRILLED WITH PIER DRILLING RIG

A geotechnical study for a nine-story building in north Austin revealed the presence of a major fault across the building area. In the eastern portion of the site, the bearing stratum was limestone of the Austin Group of Upper Cretaceous Age, which was generally competent with weathered seams and layers in the upper portions. The western portion of the site was characterized by limestone of the Edwards Group, which contained numerous clay seams and layers as well as other signs of extensive karst activity. These features became more prominent closer to the fault zone. Also, varying amounts of limestone boulders, seams, and fragments were encountered to significant depths within the "gouge zone" of the fault.

The geotechnical engineer recommended a pilot hole program to insure placement of the piers into competent limestone. The Contractor for the project decided to use a pier drilling rig to drill pilot holes. Due to the significant variation in the subsurface conditions, a graduate geotechnical engineer was dispatched full-time to the site during pier and pilot hole drilling for quick interpretation of the subsurface conditions. The drilling contractor drilled the piers until sufficient limestone was observed for the required embedment and then began the pilot holes. The pier drilling was slowed considerably due to the hardness of the Edwards limestone and the variability of the subsurface conditions. Productivity was also slowed each time the rig changed from a large-diameter pier auger to a small-diameter pilot hole auger.

The majority of the piers founded in the Edwards limestone had to be extended significantly beyond the recommended minimum pier embedments due to multiple zones of clay layers and fractured rock. Piers bearing in the Edwards limestone were typically extended 4 to 13 meters deeper than the required embedments. A few piers within the gouge zone had to be extended to depths of up to 19 meters. Pier drilling was much more productive in the Austin Group limestone, with fewer pier extensions.

Although the geotechnical report went to great lengths to characterize the poor quality of much of the Edwards limestone and the probability that many of the piers would need to be extended significantly (particularly in the fault zone), the Contractor's base bid included only pier depths equal to the minimum design rock penetration. As a result, the Contractor's schedule was impacted due to the increased scope of foundation installation. In addition, the cost of the foundation installation was significantly higher than the Contractor's original estimate.

CASE HISTORY 2: PILOT HOLES DRILLED WITH AIR PERCUSSION DRILLING EQUIPMENT

During a geotechnical study at an 8-acre site in northwest Austin for three multi-story tilt-wall office buildings, numerous clay seams and layers, a significant amount of solution activity, and a few voids were observed in the borings. The frequency of karst features was observed to be significantly greater in the northern portion of the site. Due to time constraints associated with the project, it was decided to begin the pilot hole program utilizing air-percussion drilling equipment at the northernmost building as soon as the building pad was completed and the pier locations staked.

The Contractor provided the equipment and an operator. A senior geotechnical technician was at the site full-time while the pilot holes were drilled. The drilling equipment operator was responsible for interpreting the subsurface conditions being penetrated, while our senior technician monitored the drilling operations and logged the pilot holes based on the descriptions provided by the driller. With a requirement for the pilot holes to extend at least 1.5 meters and 1.5 pier diameters below the bottom of the pier, the pilot holes subsequently ranged in depth from about 4 to 10 meters.

An estimated pier depth was developed for each pier location once the pilot hole program was completed. For the building at the northern end of the site (where the karst features were more pronounced), approximately 25 percent of the piers had to be extended beyond the depths estimated from the pilot hole program due to karst features observed during pier drilling. This indicated that the driller had difficulty interpreting some of the subsurface strata being penetrated. To extend the pilot holes deeper during construction, a small diameter auger was attached to the pier drilling rig, and the pilot hole was extended through the bottom of the excavated pier. If additional karst features were observed in the pilot hole, then the appropriate sized diameter auger was reattached and the pier depth extended accordingly.

For the building located in the central portion of the site, approximately 10 percent of the piers had to be extended beyond the depths estimated from the pilot hole program. At the building in the southern portion of the site (where the least amount of karst features were observed) approximately 4 percent of the piers had to be extended. Based upon these observations, it can be concluded that as the rock quality decreases, interpretation of the subsurface conditions becomes more difficult when utilizing air-percussion drilling equipment.

CASE HISTORY 3: PILOT HOLES DRILLED WITH GEOTECHNICAL DRILLING RIGS

This project involved the construction of several three to four story office buildings and a parking garage on a site in west Austin. The geotechnical study revealed Edwards limestone with significant clay layers and numerous other karst features. A pilot hole program was recommended in the geotechnical study. It was decided to use geotechnical drilling rigs to perform the pilot holes.

The pilot hole drilling began about one week before pier drilling commenced. After each pilot hole was drilled, the information was relayed to the geotechnical engineer in charge of the pilot hole program. A spreadsheet was developed that included the design pier embedments, along with the depth to rock and extensions of the piers due to karst features. This was updated daily and provided to the Contractor. During pier drilling, conditions reported in the pilot holes were checked by a field technician; any variations were discussed with the geotechnical engineer. Only a few pier locations resulted where the pier depth estimated from the pilot hole program was modified.

While this project was a success from an engineering standpoint, it was very difficult in terms of field coordination. The Contractor was behind schedule due to poor weather and added several pier drilling rigs to increase production. As a result, the pilot hole drilling operations had difficulty keeping ahead of the foundation installation. The Contractor could have been delayed due to this situation, which could have impacted the overall construction schedule.

CASE HISTORY 4: PILOT HOLES DRILLED IN GLEN ROSE LIMESTONE

This project involved several four to five-story buildings and a multi-level parking garage in west Austin constructed on a site underlain by limestone of the Glen Rose Group. Although it is unusual to perform pilot hole programs in this geologic formation, the geotechnical study revealed the existence of multiple clay layers within the limestone. The clay layers represented a situation where decreased end bearing capacity would result where the piers were founded within or just above such a layer. An attempt was made to correlate the locations of the clay layers between borings, however, no pattern could be developed. As a result, the geotechnical report considered an option for a drilled pier foundation system with lowered end bearing pressure taking into account the existence of the clay layers. Another option was presented with a significantly higher end bearing provided pilot holes were drilled to check for clay layers under each pier. (Spread footings were not feasible due to the significant cut/fill conditions required to achieve final grade). It was determined that the pilot hole program could significantly reduce foundation installation costs.

The Contractor decided to use pier drilling rigs to auger a small diameter pilot hole at each location. During pier drilling, the technician recorded clay layers that were encountered in the limestone both along the pier sidewalls and within the pilot hole. For clay layers observed within the pier sidewalls, the design rock penetration was adjusted to account for their existence. In situations where clay layers were observed in the pilot hole, the geotechnical engineer was contacted for recommendations for pier extension. The pier installation proceeded smoothly and the foundation system was completed with only a small increase in cost due to extensions of the piers.

CONCLUSIONS

The authors have presented a general overview of pilot hole procedures based upon experience in the Central Texas area for subsurface conditions comprised of karstic limestone and other formations where characterizing subsurface conditions for foundation design purposes is difficult. Three different methods of drilling pilot holes have been presented, along with advantages and disadvantages of each method that the authors have experienced. The use of pilot holes is an effective means of evaluating subsurface conditions at individual foundation locations in situations where other means of site evaluation cannot adequately characterize subsurface variations. When choosing a method of drilling pilot holes, the advantages and disadvantages of each method should be carefully considered in developing a pilot hole program that is well suited to the site conditions, the Owner's expectations, and schedule considerations of the Contractor. Construction documents and cost estimates should consider the probability of foundation modifications due to karst features which will be identified in the pilot hole program.

REFERENCES

Barnes, V.E. (1983). *Geologic Atlas of Texas: San Antonio Sheet,* Bureau of Economic Geology, The University of Texas at Austin, Austin, Texas.

Fischer, J.A., Fischer, J.J., and Ottoson, R.S. (2000). "Karst Investigations – Folded or Flat, Young or Old." *Proc. Specialty Conference on Geotechnical Site Characterization,* ASCE, GSP No. 97, 229-244.

Fischer, J.A., Greene, R.W., Ottoson, R.S., and Graham, T.C. (1988). "Planning and Design Considerations in Karst Terrain." *Environmental Geology and Water Sciences.* 12(2), 123-128.

Garner, L.E. and Young, K.P. (1976). *Environmental Geology of the Austin Area: An Aid to Urban Planning,* Bureau of Economic Geology, The University of Texas at Austin, Austin, Texas.

Sowers, G.F. (1996). *Building on Sinkholes: Design and Construction of Foundations in Karst Terrain,* ASCE Press, New York, 167-174.

Weber, L.C. (1988). "Exploring Karst Terrain with Air Track Drilling," *Proc. Specialty Conference on Geotechnical Aspects of Karst Terrains: Exploration, Foundation Design and Performance, and Remedial Measures,* ASCE, GSP No. 14. 68-73.

LATERAL DRILLED SHAFT RESPONSE FROM DILATOMETER TESTS

Paul W. Mayne[1], PhD, P.E., M. ASCE

ABSTRACT: The procedures used for a successful Class "A" prediction are documented for the lateral load-displacement response of four drilled shafts (d = 0.914 m and L = 11 m) constructed in Piedmont residual sandy silts at the National Geotechnical Experimentation Site (NGES) near Opelika, Alabama. The approach was based on the results of flat plate dilatometer tests (DMT) to evaluate the lateral soil resistance ($p_u = p_0$) for determining the static capacity (H_u) and equivalent soil modulus ($E_s = E_D$) to represent stiffness within the context of elastic continuum solutions. A more generalized Class "C" approach is discussed using an approximate closed-form expression and data obtained from seismic dilatometer tests (SDMT), since the shear wave velocity provides a fundamental initial stiffness of the ground in terms of the small-strain shear modulus (G_{max}).

INTRODUCTION

A series of drilled shafts were constructed as part of an ongoing research series on deep foundations for the Alabama DOT and Auburn University at the NGES south of Opelika, Alabama. Different installation methods of shaft construction included the dry method with temporary casing, bentonite slurry, and polymeric slurry (Brown, 2002). The completed shafts were 0.914-m in diameter and 11-m in length, embedded in residual fine sandy silts of the Piedmont geology. An extensive site characterization program has been underway at the NGES with a variety of both laboratory and in-situ field tests (Brown & Vinson, 1998; Mayne et al., 2000). The laboratory series includes index data (grain size, water contents, plasticity) on driven and tube samples and higher order testing on thin-walled tube samples comprised of triaxial, direct shear, consolidation, permeability, and resonant column. The field test

[1] Professor, Geosystems Engineering, School of Civil & Environmental Engineering, Georgia Institute of Technology, 790 Atlantic Drive, Atlanta, GA 30332-0355; Email: paul.mayne@ce.gatech.edu

program has included soil borings with standard penetration testing, electrical cone, single- and dual-element piezocone penetration, prebored pressuremeter, flat dilatometer, shear wave measurements (crosshole, spectral analysis of surface wave, downhole tests), surface resistivity surveys, and field permeameters. Several specialized devices have also been used at the site, including borehole shear, full-displacement pressuremeter, electrical conductivity soundings, impulse shear testing, dielectric soundings, SPTs with torque measurements, and seismic flat dilatometer.

This paper documents the Class "A" predictions submitted to the NGES site manager (Professor Dan Brown) and the procedures used to evaluate the lateral load-deflection curves from capacity calculations and stiffnesses obtained by flat dilatometer tests (DMT). With an elastic solution by Poulos & Davis (1980), good agreement is obtained in comparison with the full-scale tests. A modified procedure is discussed that employs a closed-form solution (Randolph, 1981) and the initial stiffness derived from shear wave profiles. Nonlinear soil stiffness is handled approximately using a hyperbolic expression (Fahey & Carter, 1993).

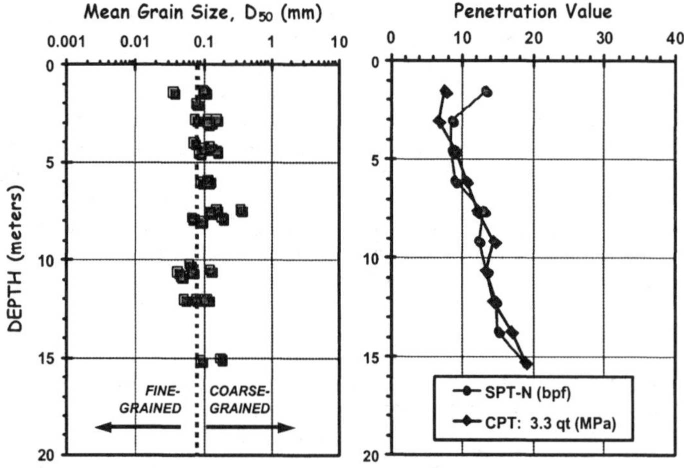

FIG. 1. Profiles of Mean Grain Size and Penetration Values at Opelika NGES.

SITE AND SOIL CONDITIONS

The 130-hectare national geotechnical experimentation site (NGES) at Opelika is located at the south end of the Piedmont geologic province (Mayne, et al., 2000). The region is underlain by natural residuum derived from the inplace weathering of metamorphic rocks (gneiss and schist) and igneous rocks (granite). Overburden soils (residuum) are comprised of fine sandy silts (ML) to silty fine sands (SM), with trace to some mica content, that transition with depth back into saprolite and partially-weathered rock (Sowers & Richardson, 1983). At the specific test locations discussed herein, the depth to bedrock exceeds 32 meters and the groundwater table was approximately 2.5 to 3 meters below grade.

The latest set of index tests (n = 47) shows an average mean grain size D_{50} = 0.086 mm and mean percent fines content of PF = 48.2 % in the upper 15 meters, although variations are evident with depth as depicted in Figure 1a. The liquid limit averages about 46% and plasticity index about 8%, although many samples are nonplastic. Water contents generally vary from 22 to 36 % within the upper 15 m. The mean profile of SPT-N-value derived from six borings is compared with the mean profile of cone tip stress (q_t) averaged from 22 CPTs at the site in Figure 1b. The observed ratio of q_t/N = 3 in Piedmont soils (where q_t is in bars) agrees well with empirical correlations based on mean grain size (e.g., Kulhawy & Mayne, 1991). As the onsite value of D_{50} is so close to the demarcation between fine-grained and coarse-grained soils as specified by a US No. 200 sieve size (i.e., 0.075 mm), the Piedmont residuum is more suitably categorized as a dual symbol in the Unified Soil Classification System; thus ML-SM. As such, the soils respond in either undrained or drained behavior, depending upon the applied rate of loading. For in-situ penetration testing and quick load tests, undrained conditions can prevail, whereas under normal (slow) rates of construction and earthwork operations, drained behavior is prevalent.

The conventional DMT involves pushing a steel blade into the ground and inflating a flexible membrane to obtain two pressure readings (denoted "A" and "B")

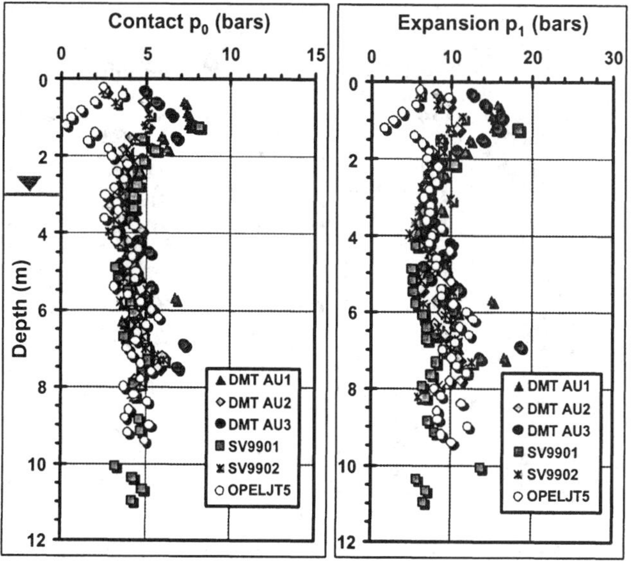

FIG. 2. Standard Readings from DMTs in Piedmont Silts at Opelika.

at regular 0.2-m or 0.3-m intervals with depth (Marchetti, 1980). The DMT data are interpreted to provide information on the lateral resistance characteristics of the ground, as well as the stiffness at working loads (Mayne, et al. 1999). The corrected A and B readings taken from six DMT soundings at Opelika are presented in Figure 2. The contact pressure is denoted as p_0 and the expansion pressure is p_1 (Schmertmann, 1986). A shallow crust is evident in the upper 2 to 3 meters underlain by fairly uniform residual soils. Statistical analysis (n = 177) gives mean values of $p_0 = 4.3 \pm 1.2$ bars and $p_1 = 8.9 \pm 3.1$ bars, essentially constant with depth.

The processing of the DMT readings provides indices in the interpretation of soil type, strength, and stiffness, where:

$$I_D = (p_1 - p_0)/(p_0 - u_0) = \text{material index} \quad (1)$$

$$E_D = 34.7\,(p_1 - p_0) = \text{dilatometer modulus} \quad (2)$$

where u_0 = hydrostatic porewater pressure. The material index is determines soil type: clays ($I_D < 0.6$), silts ($0.6 < I_D < 1.8$), and sands ($I_D > 1.8$). The dilatometer modulus is derived from elasticity solutions and provides a reasonable equivalent elastic modulus (E_s) for foundations situated in Piedmont soils at working load levels (e.g., Mayne & Frost, 1988; Mayne, et al. 1999). Profiles of I_D and E_D at Opelika are shown in Figure 3. Statistical evaluations give $I_D = 1.14 \pm 0.48$ and $E_D = 160 \pm 75$ bars. The soil type is seen to be reasonably characterized by the material index.

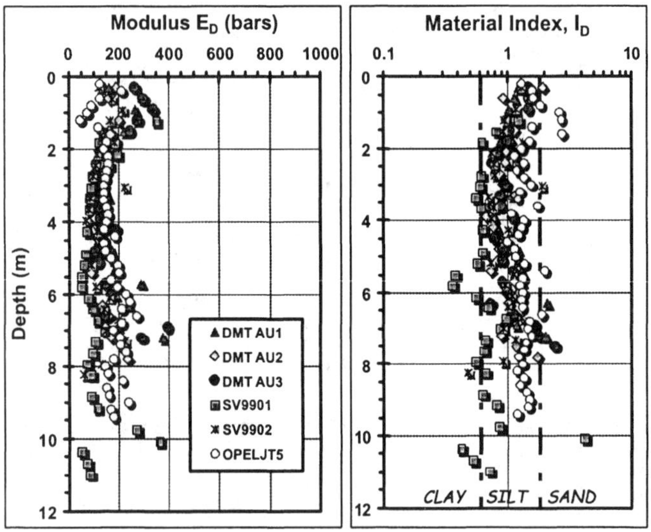

FIG. 3. Profiles of Dilatometer Modulus and Material Index at Opelika NGES.

LATERAL PILE ANALYSIS

The analysis of drilled shafts and piles under lateral loading can be handled using either subgrade reaction models or elastic continuum solutions. The author prefers the latter as site-specific data can be collected to ascertain the soil modulus for analysis, either by laboratory or field test methods. An approximate approach to providing nonlinear lateral load-deflection representations is given by (Mayne & Kulhawy, 1991):

$$H = \frac{1}{\frac{1}{\delta \cdot K_i} + \frac{1}{H_u}} \tag{3}$$

where H = lateral force, H_u = lateral capacity, δ = lateral deflection, $K_i = E_s \cdot L/I^*$ = initial pile-soil stiffness, E_s = equivalent soil modulus, L = pile length, and I^* = displacement influence factor obtained from boundary element or finite element solutions. For a pure horizontal loading (eccentricity e = 0), the solution for $I^* = I_{\rho H}$ from Poulos (1971) is shown in Figure 4 and depends upon both the slenderness ratio (L/d) and relative pile-soil stiffness (K_R), where:

$$K_R = \frac{E_P \cdot I_P}{E_s \cdot L^4} \tag{4}$$

where E_P = pile modulus and $I_P = (\pi d^4/64)$ = pile moment of inertia. An additional term must be included for moment loading or lateral eccentricity is to be considered.

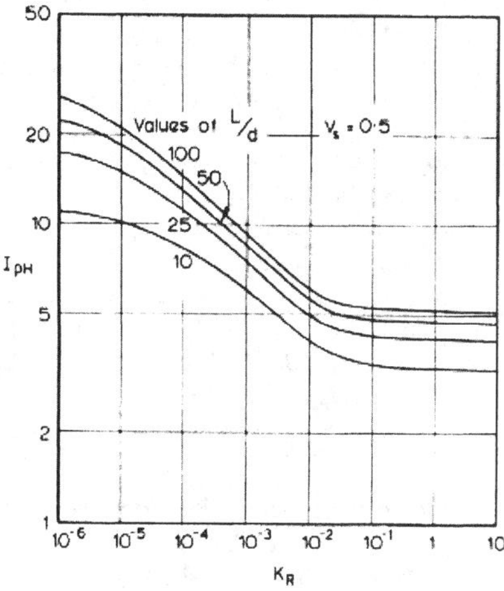

FIG. 4. Displacement Influence Factor for Free-Headed Lateral Pile Embedded in a Homogeneous Elastic Medium (Poulos, 1971).

The lateral capacity (H_u) for a free-headed condition can be obtained from a static analysis of a rigid pile (Poulos & Davis, 1980). For a uniform distribution of lateral soil resistance along the shaft (p_u = constant with depth), the capacity is:

$$H_u = p_u \cdot d \cdot L \cdot \left[\sqrt{(1+2e/L)^2 + 1} - (1+2e/L) \right] \quad (5)$$

For the case of no eccentricity of loading (e = 0), this reduces to:

$$H_u = 0.414 \cdot p_u \cdot d \cdot L \quad (6)$$

CLASS "A" PREDICTION

As a first approximation, the p_0 value from the DMT can be considered a measure of the limit pressure (P_L) related to bearing capacity via cavity expansion theory. In fact, pressuremeter and dilatometer data on soft to firm clays (Lutenegger, 1988; Kalteziotis, et al., 1991) show considerable agreement in a direct relationship between these readings, such that p_0 (DMT) = P_L (PMT). On this basis, the author adopted the lateral soil resistance as a constant $p_u = p_0$ = 4.3 bars (430 kPa) at the Opelika site. For the given geometry (d = 0.914 m; L = 11.0 m, and e = 0.3 m), the calculated capacity is $H_u = 0.399\ p_u \cdot d \cdot L$ = 1723 kN.

Adopting the equivalent soil modulus $E_s = E_D$ for the Piedmont soils and a value of E_P = 35 GPa for lightly-reinforced concrete (I_P = 0.037 m^4), the calculated pile flexibility factor is K_R = 0.0056. This is an intermediate stiffness on the Poulos (1971) chart in Figure 4, as rigid piles exhibit $K_R > 0.10$ and flexible piles are defined for $K_R < 10^{-6}$. For the case at hand, the influence factor $I_{\rho H}$ = 4.57, giving an initial stiffness $K_i = E_s L / I_{\rho H}$ = 38.5 MN/m. The stiffness K_i and capacity H_u are input into (3) to obtain load H for each deflection δ, thus giving the Class "A" predictions shown in Figure 5. The measured responses for four shafts that were subjected to lateral loading are seen to be in good compliance with the elastic continuum solution.

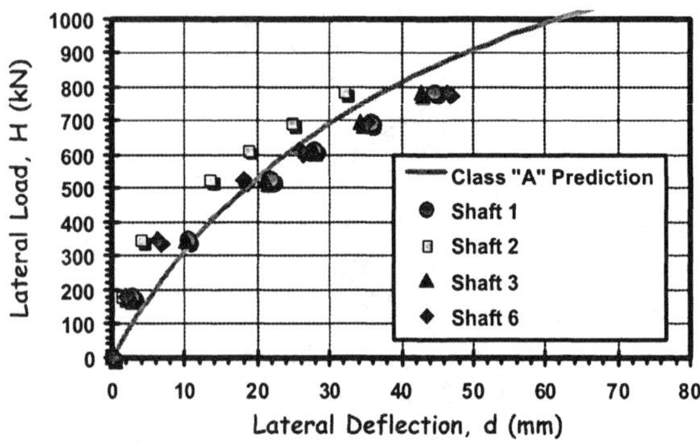

FIG. 5. Class "A" Prediction of Lateral Load Response from Opelika DMTs

Even though the Class "A" prediction was sufficient for this case study, a more generalized approach is desirable in order to account for nonhomogeneity of stiffness and strength in the ground, as well as to accommodate field data obtained from other than dilatometer tests. Therefore, a critique of the shortcomings in the aforementioned solution is warranted. First, reliance on a chart (i.e., Figure 4) was necessary in determining the relevant displacement influence factor ($I_{\rho H}$) for the elastic solution. This is inconvenient for routine design in geotechnical practice, particularly if additional loads due to moments and eccentricity of applied forces must be considered, since additional charts are needed. While alternate methods are available for obtaining displacement factors (e.g., Poulos & Hull, 1989), it is of interest to utilize the approximate closed-form elastic solution of Randolph (1981), as flexible to rigid piles can be considered and to the variation of soil modulus with depth can be homogeneous (constant) or linearly-increasing with depth (Gibson soil).

In the prior simplified approach, the profile of lateral soil resistance (p_u) was assumed constant with depth. In more rigorous solutions, however, p_u is reduced near the ground surface because of the boundary condition at z = 0 (Mayne, et al. 1992). This reduction has been observed experimentally as well (e.g., Stevens and Audibert, 1979). The limit plasticity solution developed by Randolph & Houlsby (1984) is adopted henceforth as it accounts for pile roughness and reduced near-surface p_u. In addition, the undrained strength mode corresponding to passive loading conditions is employed as it best realizes strength anisotropy for this situation.

So that the method can be generalized for use in other soil types and geologic origins, the fundamental stiffness is the initial tangent shear modulus (G_{max}) since this represents the beginning of all stress-strain-strength curves in geotechnique and is applicable to all deformation problems in the ground (Burland, 1989). The small-strain shear modulus is obtained from:

$$G_{max} = \rho_T (V_s)^2 \tag{7}$$

where ρ_T = total mass density of the soil and V_s = shear wave velocity.

CLOSED-FORM ELASTIC SOLUTION

A generalized closed-form solution for lateral (H) and moment (M) loading has been presented by Randolph (1981) based on finite-element analyses. In this approach, the ground stiffness is represented by a modified shear modulus (G*), such that:

$$G^* = G\,(1+3\nu/4) = \text{modified shear modulus} \tag{8}$$

Since the lateral pile behavior is controlled by shallow soil conditions and the pile-soil relative flexibility, a characteristic ground stiffness (G_c) is defined as the value of the modified shear modulus at a depth equal to half the critical pile length (L_c), or $G_c = G^*_{Lc/2}$, as depicted in Figure 6. The critical pile length represents the active zone over which the pile responses to the lateral loading and depends upon the elastic pile modulus (E_P) such that:

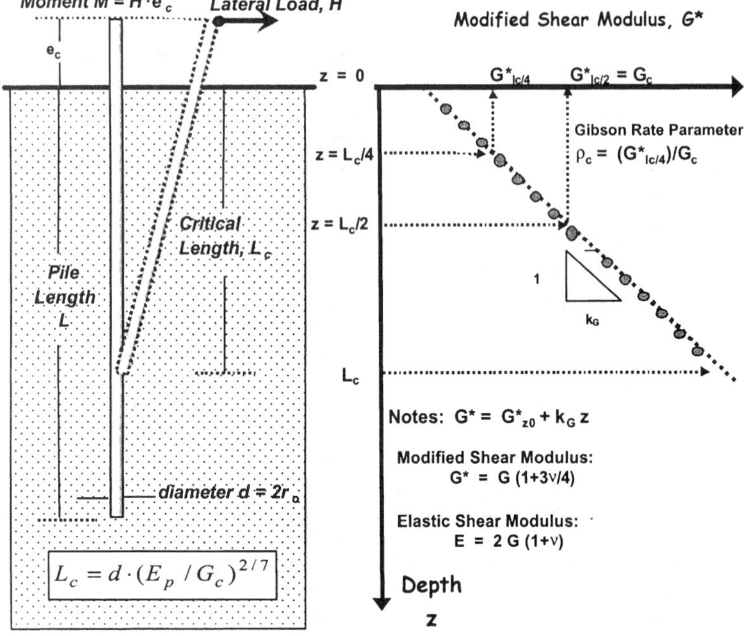

FIG. 6. Defined Parameters for Lateral Pile Solution by Randolph (1981)

$$L_c = d\,(E_P/G_c)^{2/7} = \text{critical pile length} \tag{9}$$

The degree of homogeneity of the ground stiffness is handled by the ratio of the modified shear modulus at one-quarter the critical length to that at the characteristic shear modulus (Figure 6), thus defining a parameter:

$$\rho_c = G^*_{Lc/4}/G_c \tag{10}$$

With all parameters now defined, the lateral deflection ($u = \delta$) and rotation (θ) of the pile at the ground surface may be conveniently expressed:

$$u = \delta = \frac{(E_P/G_c)^{(1/7)}}{\rho_c G_c}\left[0.27\cdot H\cdot\left({}^{l_c}\!/_2\right)^{-1} + 0.3\cdot M\cdot\left({}^{l_c}\!/_2\right)^{-2}\right] \tag{11}$$

$$\theta = \frac{(E_P/G_c)^{(1/7)}}{\rho_c G_c}\left[0.3\cdot H\cdot\left({}^{l_c}\!/_2\right)^{-2} + 0.8\cdot(\rho_c)^{0.5}\cdot M\cdot\left({}^{l_c}\!/_2\right)^{-3}\right] \tag{12}$$

Lateral soil resistance in clays varies with depth and soil-pile roughness. A wedge-plasticity solution ($N_P = p_u/s_u$) that accounts for both depth effects and adhesion has been developed by Randolph & Houlsby (1984). Experimental studies have verified the applicability of this solution and documented that strength anisotropy effects are significant, thus best handled by use of an undrained shear strength corresponding to triaxial extension (Mayne, et. al. 1994). The value of s_{uTE} can be measured by laboratory triaxial tests on high-quality samples, or alternatively estimated from constitutive soil models (e.g., Mayne, et al., 1992), or else empirical correlations (Jamiolkowski, et al. 1985) that depend upon the profile of overconsolidation ratio (OCR). The calculation of lateral capacity by limit equilibrium analyses for varying soil resistance with depth is detailed elsewhere (Poulos & Davis, 1980) and can be easily implemented on a spreadsheet; Mayne & Kulhawy, 1991).

In order to model nonlinear stress-strain-strength behavior of soils starting from the fundamental stiffness (G_{max}), a modified hyperbola has been successfully used (Fahey & Carter, 1993):

$$G = G_{max}\,[1 - f(FS)^{g}] \tag{13}$$

where G = shear modulus at the particular load level, and f and g are fitting parameters. In the case of static monotonic loading of uncemented and insensitive geomaterials ($f = 1$), the exponent g may take on assumed values of 0.3, 0.2, and 0.1, respectfully, for compression, simple shear, and extension modes of loading.

OPELIKA CLASS "C" FITTING

The residual soils at Opelika NGES show evidence of an apparent preconsolidation stress that can be explained on the basis of groundwater fluctuations at the site (Mayne & Brown, 2003). A recent drought confirmed a lowering of the ambient groundwater from about 2.5 m to 10 m during the period from 1998 to 2002. Thus, a postulated 20-m drop in groundwater matches well with available lab and field evidence of apparent overconsolidation ratios ($AOCR = \sigma_p'/\sigma_{vo}'$) that decrease from about 7 at the ground surface to 2.5 at the level of the shaft bases at z = 11 m. The normalized undrained strength ratio for triaxial extension can be evaluated from the effective stress friction angle ($\phi' = 35.5^{\circ}$) and AOCR (Kulhawy & Mayne, 1990):

$$s_u/\sigma_{vo}'_{TE} = 0.17\,OCR^{0.85} \tag{14}$$

In terms of the DMT, the apparent preconsolidation stress (σ_p') in the Piedmont can be directly evaluated from an expression derived from a database on clay soils (Kulhawy & Mayne, 1990):

$$\sigma_p' = 0.509\,(p_0 - u_0) \tag{15}$$

Results from three sets of seismic flat dilatometer tests (SDMT) have been obtained at Opelika and presented in Figure 7 (Martin & Mayne, 1998). The developed profile of undrained shear strength in extension mode using (14) and (15) is also shown.

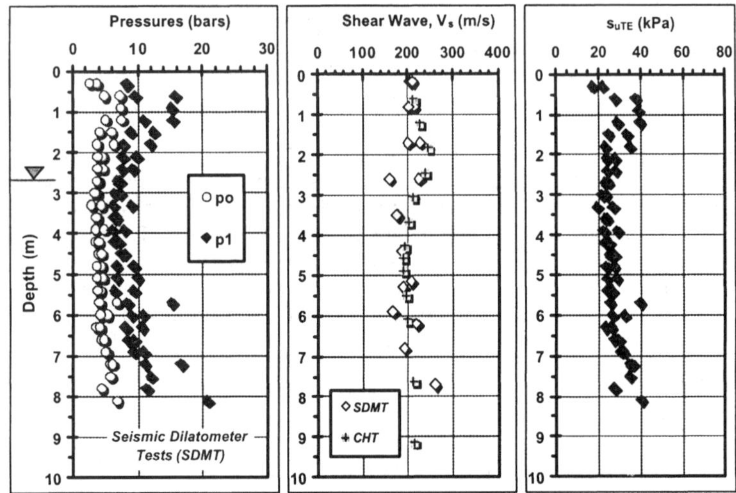

FIG. 7. Results from Seismic Flat Dilatometer Tests at Opelika NGES.

The derived AOCRs from the DMT interpretations agree well with the calculated profile from capillarity conditions (Mayne & Brown, 2003). An overall representative s_{uTE} = 34 kPa is obtained by this procedure. Using the wedge-plasticity p_u profile with depth, an ultimate lateral capacity of H_u = 1015 kN is obtained from a spreadsheet calculation, as partially for the 1000-rows of data shown in Figure 8.

LATERAL PILE CAPACITY
Opelika Shafts - Feb. 1999

Groundwater z_w =	3	m
Max GWT (σ_p') =	20	m
e = eccentricity =	0.3	m
L = Length =	9.82	m
d = Diameter =	0.914	m
Unit Weight γ_T =	17.5	kN/m³

PROFILE FOR p_u/s_u (Randolph & Houlsby, Geot. 1984)
Statics Solution Described by Poulos & Davis (1980)

e/d = 0.031
L/d = 11
s_u/σ_{vo}' = 0.17 NC (TE)
Λ = 0.85
AOCR = apparent OCR
p_u = lateral soil resistance

Lateral Capacity, H_u (kN)		
$H_u = \Sigma F_{right} - \Sigma F_{left}$ =	1015.72	(kN)
Depth Rotation z_r =	7.16	(m)
Normalized z_r/L =	0.73	(-)

Horiz. ΣF =	2128.3	1112.6	Seek
	Right (kN)	*Left* (kN)	@MIN 7.16

Depth z/d	Depth z (m)	Norm. p_u/s_u	Δ_{Mom} Moment	top ΣM	bottom ΣM	σ_{vo}' kN/m²	σ_p' kN/m²	s_u kN/m²	AOCR	p_u kN/m²	Σ Forces Right	Σ Forces Left	Find z_r
Means=	4.91	10.77	1130	355902	775417	67.23	234.00	32.87	3.81	361.08			
0.011	0.010	3.530	27.030	27	1130189	29.51	196.28	25.11	6.65	88.65	0.7957	0.0000	9.82
0.021	0.020	3.560	27.720	55	1130162	29.58	196.35	25.13	6.64	89.47	0.8030	0.0000	9.82
0.032	0.029	3.591	28.418	83	1130134	29.66	196.43	25.15	6.62	90.29	0.8104	0.0000	9.82
0.043	0.039	3.621	29.125	112	1130105	29.73	196.50	25.16	6.61	91.12	0.8178	0.0000	9.82
0.054	0.049	3.651	29.841	142	1130076	29.81	196.58	25.18	6.59	91.94	0.8252	0.0000	9.82
0.064	0.059	3.681	30.564	173	1130046	29.88	196.65	25.20	6.58	92.77	0.8327	0.0000	9.82
0.075	0.069	3.712	31.297	204	1130016	29.96	196.73	25.22	6.57	93.60	0.8401	0.0000	9.82
0.086	0.079	3.742	32.037	236	1129985	30.03	196.80	25.24	6.55	94.43	0.8475	0.0000	9.82
0.097	0.088	3.772	32.787	269	1129953	30.11	196.88	25.25	6.54	95.26	0.8550	0.0000	9.82
0.107	0.098	3.802	33.544	302	1129920	30.19	196.96	25.27	6.52	96.09	0.8624	0.0000	9.82

FIG. 8. Partial Spreadsheet Calculations for H_u by Wedge-Plasticity Solution.

LATERAL PILE DEFLECTION ANALYSIS

Elastic Continuum Solution (after Randolph, 1981, Geotechnique 31, No. 2)

d, Pile Diameter (m) =	0.912	I_p = Moment of Inertia (m^4) =	0.0340
L, Pile Length (m) =	11	E_pI_p = pile stiffness (kN-m^2) =	1018757
Resistance, p_u (kPa) =	382	Modified Shear Modulus G* (kPa)=	80265
Eccentricity, e (m) =	0.3	Equivalent pile radius, r_o (m) =	0.456
E_p = pile mod (kPa) =	3E+07	Length to Diameter Ratio, L/d =	12.06
Mid-depth G (kPa) =	69795.9	Slenderness Ratio = L/r_o =	24.12
Poisson's ratio, ν =	0.2	Critical Slenderness Ratio, $(l/r_o)_c$ =	10.87
$(RHO)_c$ Parameter =	1	Initial Critical Length, L_c (m) =	4.95
γ_T =Gamma (kN/m^3) =	17.1	Flex. Factor $K_R = E_pI_p/(E_sL^4)$ =	0.0004
Shear Wave V_s (m/s) =	200	Modulus Reduction Exponent, g =	0.1
G_{max} (kPa) = $\rho_t V_s^2$ =	69796	*Notes:* $(Rho)_c$ = Ratio $G^*_{Lc/4}$ over $G^*_{Lc/2}$	
H_u (kN) = Capacity =	1015	Mid-depth G_c = value of G* at $L_c/2$	

u (mm)	K =H/u	H/H_u	H (kN)	G_c (kPa)	L_c	$(L/r_o)_c$	G^*/G^*_{max}
0.00	278.00	0	0.0	80265	4.95	10.87	1.00
0.19	109.47	0.02	20.3	25986	6.84	15.00	0.32
0.56	90.87	0.05	50.8	20778	7.29	15.99	0.26
1.35	75.01	0.1	101.5	16508	7.78	17.07	0.21
3.55	57.19	0.2	203.0	11932	8.54	18.73	0.15
6.68	45.60	0.3	304.5	9105	9.23	20.24	0.11
11.06	36.70	0.4	406.0	7028	9.94	21.79	0.09
17.32	29.30	0.5	507.5	5375	10.73	23.52	0.07
21.49	25.98	0.55	558.3	4658	11.18	24.51	0.06
26.66	22.84	0.6	609.0	3997	11.67	25.60	0.05
33.23	19.85	0.65	659.8	3384	12.24	26.85	0.04
41.83	16.99	0.7	710.5	2812	12.91	28.31	0.04
53.57	14.21	0.75	761.3	2276	13.71	30.07	0.03

FIG. 9. Calculated Lateral Load-Deflections By Elastic Continuum Solution.

The shear wave data from the SDMTs are utilized to obtain the initial shear modulus profile, per equations (7) and (8). The derived stiffnesses give a homogeneous profile with G_{max} = 70 MPa constant with depth. Equivalent secant shear moduli (G) are reduced from G_{max} accordingly with increasing load level per the modified hyperbola given by (13) and adopting H/H_u = 1/FS. The individual lateral load–deflection calculations per the Randolph closed-form procedure are shown in the spreadsheet solution given in Figure 9 for adopted exponent g = 0.1. As seen by comparison with the measured full-scale lateral load tests at Opelika presented graphically in Figure 10, good agreement is achieved when $0.1 < g < 0.2$ at this site.

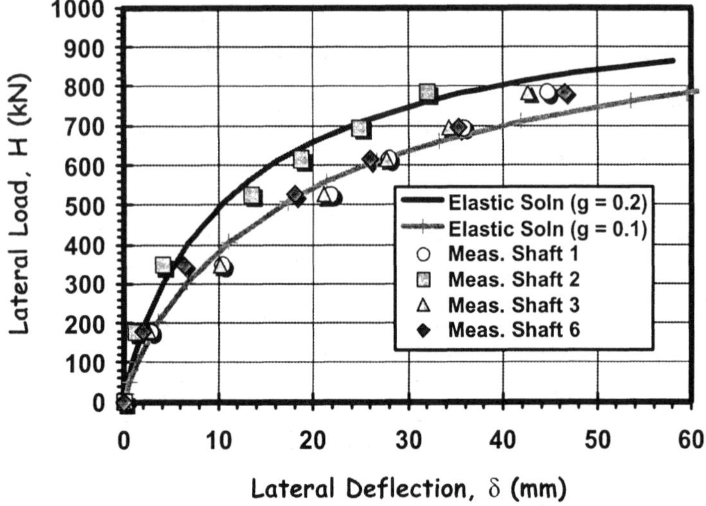

FIG. 10. Class "C" Fitting of Continuum Solution to Lateral Load Response

CONCLUSIONS

Results of a Class "A" prediction of the lateral response of drilled shaft foundations using flat dilatometer data at the Opelika national test site are presented in comparison with measured full-scale load tests. Procedures to extend the approach to a generalized closed-form solution are presented to allow nonhomogeneity in the strength and modulus profiles of the ground, as well as include the initial shear modulus which is a fundamental measurement of soil stiffness. A Class "C" fitting of the data is presented for the case study.

ACKNOWLEDGMENTS

The author appreciates the support of the National Science Foundation, Mid-America Earthquake Center, and Georgia Department of Transportation. The conclusions, recommendations, and results provided in this report are not necessarily endorsed by NSF, MAE, or GDOT. Additional gratitude is given to Dr. Dan Brown for providing the lateral load test data used in this exercise.

REFERENCES

Burland, J.B. (1989). "Small is beautiful: The stiffness of soils at small strains". *Canadian Geotechnical Journal* 26 (4), 499-516.

Brown, D.A. (2002). "Effect of construction on axial capacity of drilled foundations in Piedmont soils". *Journal of Geotechnical & Geoenvironmental Engineering* 128 (12), 967-973.

Brown, D.A. and Vinson, J. (1998). "Comparison of strength and stiffness parameters for a Piedmont residual soil". *Geotechnical Site Characterization*, Vol. 2, Balkema, Rotterdam, 1229-1234.

Fahey, M. and Carter, J.P. (1993). "A finite element study of the pressuremeter in sand using a nonlinear elastic plastic model". *Canadian Geotechnical Journal* 30 (2), 348-362.

Jamiolkowski, M., Ladd, C.C., Germaine, J.T. and Lancellotta, R. (1985). "New developments in field and lab testing of soils". *Proceedings, 11th International Conference on Soil Mechanics & Foundation Engrg* (1), San Francisco, 57-153.

Kalteziotis, N.A., Pachakis, M.D., and Zervogiannis, H.S. (1991). "Applications of the flat dilatometer test in cohesive soils". *Proceedings, 10th European Conf. on Soil Mechanics & Foundation Engineering* (1), Balkema, Rotterdam, 125-128.

Kulhawy, F.H. and Mayne, P.W. (1990). Manual on Estimating Soil Properties for Foundation Design, *Report EL-6800*, Electric Power Research Institute, Palo Alto, 306 pages.

Kulhawy, F.H. and Mayne, P.W. (1991). "Relative density, SPT, and CPT interrelationships". *Calibration Chamber Testing*, (Proceedings, ISOCCT, Clarkson University), Elsevier, New York, 197-211.

Lutenegger, A.J. (1988). "Current status of the Marchetti flat dilatometer test". *Penetration Testing 1988* (1), Balkema, Rotterdam, 137-155.

Marchetti, S. (1980). "In-situ tests by flat dilatometer". *Journal of Geotechnical Engineering* 106 (GT3), 299-321.

Martin, G.K. and Mayne, P.W. (1998). "Seismic flat dilatometer tests in Piedmont residual soil. *Geotechnical Site Characterization*, Vol. 2, Balkema, Rotterdam, 837-843.

Mayne, P.W. and Frost, D.D. (1988). "Dilatometer experience in Washington, DC and vicinity". *Transportation Research Record* 1169, National Academy Press, Washington, D.C., 16-23.

Mayne, P.W. and Kulhawy, F.H. (1991). "Load-displacement behavior of laterally-loaded drilled shafts in clay". *Piling and Deep Foundations*, Vol. 1, (Proceedings, 4th International Conf. DFI, Stresa), Balkema, Rotterdam, 409-413.

Mayne, P.W., Kulhawy, F.H. and Trautmann, C.H. (1992). Experimental Study of undrained lateral and moment behavior of drilled shafts during static and cyclic loading. *Report EPRI TR-100221*, Electric Power Research Institute and Cornell University, Ithaca, New York.

Mayne, P.W., Kulhawy, F.H. and Trautmann, C.H. (1994). "Nonlinear undrained lateral response of rigid drilled shafts using continuum theory". *Vertical & Horizontal Deformations of Foundations & Embankments*, Vol. 1, ASCE, Reston, Virginia, 663-676.

Mayne, P.W., Martin, G.K. and Schneider, J.A. (1999). "Flat dilatometer modulus applied to drilled shaft foundations in Piedmont residuum". *Behavioral Characteristics of Residual Soils* (GSP 92), ASCE, Reston/Virginia, 101-112.

Mayne, P.W., Brown, D.A., Vinson, J., Schneider, J.A. and Finke, K.A. (2000). "Site characterization of Piedmont residual soils at the national experimentation site, Opelika, Alabama". *National Geotechnical Experimentation Sites*, GSP 93, ASCE, Reston/Virginia, 160-185.

Mayne, P.W. and Brown, D.A. (2003). "Site characterization of Piedmont residuum of North America". *Characterization and Engineering Properties of Natural Soils*, Vol. 2, Swets & Zeitlinger, Lisse, Netherlands, 1323-1339.

Poulos, H.G. (1971). "Behavior of laterally-loaded piles: single piles". *Journal of the Soil Mechanics & Foundations Division* (ASCE), Vol. 97 (SM5), 711-731.

Poulos, H.G. and Davis, E.H. (1980). *Pile Foundation Analysis and Design*, Wiley & Sons, New York, 397 pages.

Poulos, H.G. and Hull, T.S. (1989). "Role of analytical geomechanics in foundation engineering". *Foundation Engineering: Current Principles & Practices*, Vol. 2, GSP 22, ASCE, Reston/Virginia, 1578-1606.

Randolph, M.F. (1981). "The response of flexible piles to lateral loading". *Geotechnique* 31 (2), 247-259.

Randolph, M.F. and Houlsby, G.T. (1984). "The limiting pressure on a circular pile loaded laterally in cohesive soil". *Geotechnique* 34 (4), 613-623.

Schmertmann, J.H. (1986). "Suggested method for performing the flat dilatometer test". ASTM *Geotechnical Testing Journal* 9 (2), 93-101.

Sowers, G.F. and Richardson, T.L. (1983). "Residual soils of the Piedmont and Blue Ridge". *Transportation Research Record* 919, National Academy Press, Washington, D.C., 10-16.

Stevens, J.B. and Audibert, J.M.E. (1979). "Re-examination of p-y curve formulations". *Proceedings, 11th Offshore Technology Conference*, Vol. 1, Houston, 397-403.

AXIAL SHAFT RESPONSE FROM SEISMIC PIEZOCONE TESTS

Paul W. Mayne[1], PhD, P.E., M.ASCE, and Guillermo Zavala[2], Student M., ASCE

ABSTRACT: The axial-load-displacement-capacity response of drilled shaft foundations can be analyzed efficiently using the results of seismic piezocone tests (SCPTu) which provide four independent readings in the ground: cone tip stress (q_t), sleeve friction (f_s), penetration porewater pressure (u_b), and downhole shear wave velocity (V_s). The penetration data are used in a direct CPT method to assess side resistance and end-bearing capacity components, while the shear wave provides the initial stiffness for an elastic continuum analysis of vertical displacements and axial load transfer distributions. Results from instrumented drilled shaft load tests conducted at the national geotechnical experimentation sites at College Station, Texas and Opelika, Alabama are utilized to illustrate the methodology.

INTRODUCTION

The rational and economical design of drilled shaft foundations requires an evaluation of the axial capacity, performance, and load transfer prior to the installation of production shafts at the project site. For this purpose, various analytical and numerical procedures have been developed to address these calculations (O'Neill & Reese, 1999). Of growing interest is the utilization of cone penetration test (CPT) data, since continuous stratigraphic profiling and multiple readings are taken in a single sounding. The CPT data may either be used in a *direct* method or *indirect* method to assess the unit side resistance (f_p) and unit end bearing (q_b) for the deep foundations (e.g., Robertson, et al. 1988). In the indirect (or rational) CPT methods,

[1]Professor, Geosystems Engineering Program, School of Civil & Environmental Engineering, Georgia Institute of Technology, 790 Atlantic Drive, Atlanta, GA 30332-0355; Email: paul.mayne@ce.gatech.edu
[2]Research Assistant, Geosystems Engineering Program, School of Civil & Environmental Engineering, Georgia Institute of Technology, 790 Atlantic Drive, Atlanta, GA 30332-0355; Email: guillermo.zavala@ce.gatech.edu

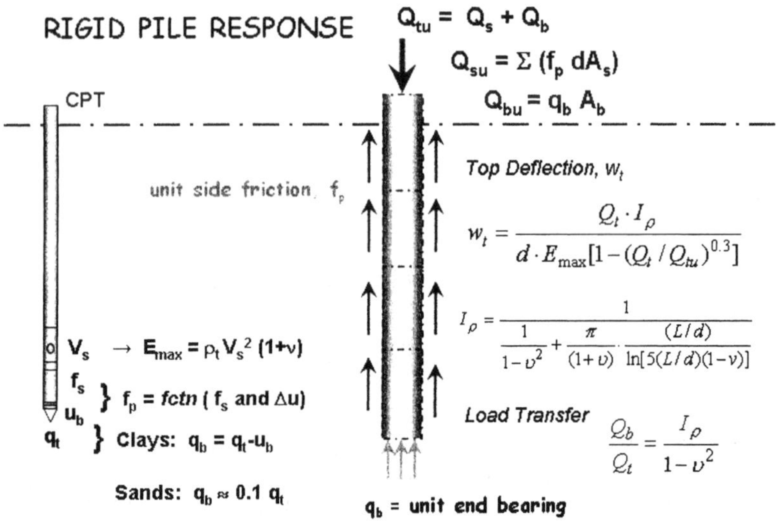

FIG. 1. Concept of Direct CPT Methodology for Axial Shaft Response.

CPT data are first used to evaluate soil engineering parameters (i.e., K_0, ϕ', OCR, s_u, E', etc) which in turn are input into static equilibrium and limit equilibrium solutions to determine shaft and base capacities (Poulos, 1989). This approach is preferred from a fundamental and rational procedure to axial pile analysis, but requires careful considerations in the proper interpretation of parameters from CPT data, particularly in geomaterials of differing geologic origins (Lunne, et al., 1997; Mayne 2001).

Direct CPT methods rely on empirical expressions to scale the penetrometer readings (up or down) for application to the full-scale pile foundations. In the earliest direct methods, reliance was placed solely on the measured cone tip resistance (q_c) to obtain both f_p and q_b (e.g., Bustamante & Gianeselli, 1982). As the electronics became more reliable, the measured q_c and f_s were used to estimate q_b and f_p (e.g., Schmertmann, 1978). With the advent of the piezocone test, the importance and significance of the necessary porewater pressure correction to q_c to obtain the total cone tip stress (q_t) became recognized (Lunne, et al. 1997), especially for clays and silts. These findings led to re-examination of prior correlations in term of q_t readings (Almeida, et al. 1996). As the piezocone test (CPTu) provides three continuous measurements with depth (q_t, f_s, u_b), the latest relationships (Eslami & Fellenius, 1997; Takesue, et al., 1998) utilize all three readings in an attempt to better evaluate axial capacity. A recent case study involving an axial drilled shaft for the Georgia DOT showed the Takesue method was useful in obtaining axial side shear in Piedmont residual soils (Mayne & Schneider, 2001).

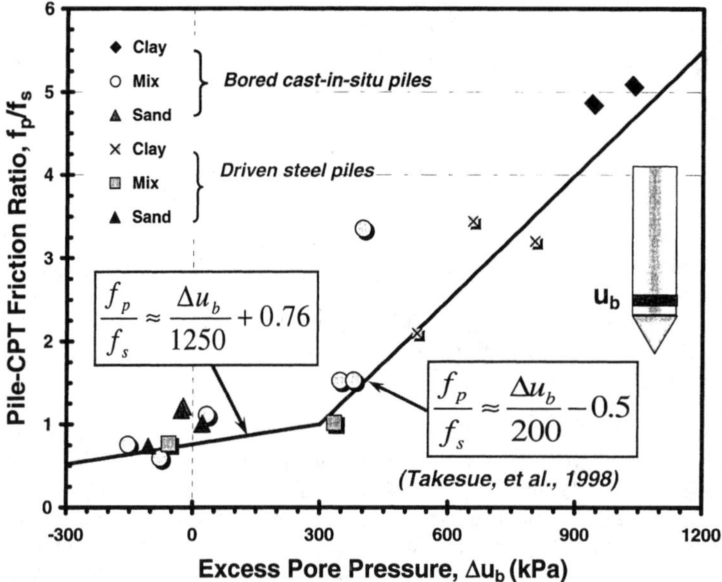

FIG. 2. Direct CPT Method for Unit Pile Side Friction from f_s and Δu Readings. (after Takesue, Sasao, and Matsumoto, 1998).

While direct CPT methods address the calculations of capacity, in fact, the entire axial load-displacement-capacity curves are desired by the geotechnical designer. The seismic piezocone test (SCPTu) provides additional measurements of shear wave velocity at 1-m intervals, thus a fundamental stiffness of the ground is obtained. It is therefore of interest to develop and calibrate a direct CPT method with the capabilities of generating the complete axial response of drilled shafts including load-displacement, capacity, and axial load transfer. This can be done within the context of elastic continuum solutions, as depicted by Figure 1 and detailed in this paper.

AXIAL CAPACITY BY CPT

The total axial capacity (Q_{tu}) of a deep foundation is comprised of a shaft component (Q_{su}) and base component (Q_{bu}). In the method of Takesue, et al. (1998), the unit side resistance (f_p) of the pile is estimated from the measured CPT f_s that is scaled up or down depending upon the measured excess porewater pressures during penetration (Δu). The data used to derive the correlation were obtained from both bored and driven pile foundations in clays, sands, and mixed ground conditions. From Figure 2, the scaling factors are:

For $\Delta u < 300$: $$\frac{f_p}{f_s} = \frac{\Delta u + 950}{1250} \quad (1a)$$

For $300 < \Delta u < 1250$: $$\frac{f_p}{f_s} = \frac{\Delta u - 100}{200} \quad (1b)$$

The summation of the unit side resistance along the perimetric area of the sides of the shaft give the total shaft capacity (Q_{su}).

In clays, the unit end bearing resistance can be obtained directly from the effective cone tip resistance (Eslami & Fellenius, 1997):

Clays: $$q_b = q_t - u_b \quad (2)$$

Alternatively, Powell et al. (2001) recommend unit end-bearing in terms of net cone resistance, with q_b on the order of one-half to one-third (q_t-σ_{vo}).

In sands, the full mobilization of bearing capacity beneath the base will not be realized because of the very large displacements required. Therefore only a fraction of the available resistance will be available. For example, recent analyses by Lee & Salgado (1999) show that the unit end bearing available depends upon the base tip movement, with approximately 10 percent available at a base deflection equal to 5% of the pile diameter (s/d = 0.05) and about 12 to 15 percent available for a movement equal to 10% of the pile diameter (s/d = 0.10). Experimental data confirm these values (e.g., Fioravante, et al. 1995). For practical use, the authors recommend:

Sands: $$q_b = 0.1\, q_t \quad (3)$$

The unit end bearing is applied to the end area to obtain the base capacity, Q_{bu}.

AXIAL DISPLACEMENTS AND LOAD TRANSFER

Elastic continuum theory provides a convenient means for representing the load-displacement response of pile foundations under axial loading (Poulos & Davis, 1980; Poulos, 1989). An approximate closed-form solution has been developed that can account for piles floating in homogeneous soil, end-bearing piles, and Gibson-type soils, as well as pile compressibility effects and belled pier situations (Randolph & Wroth, 1978, 1979; Fleming, et al. 1992). For a pile of diameter d and length L residing within an elastic medium, the top displacement (w_t) is given by:

$$w_t = \frac{Q_t \cdot I_\rho}{d \cdot E_s} \quad (4)$$

where Q_t = applied axial load and I_ρ = displacement influence factor. For the simple case of a rigid pile embedded in a homogeneous soil:

$$I_\rho = \frac{1}{\dfrac{1}{1-\upsilon^2} + \dfrac{\pi}{(1+\upsilon)} \cdot \dfrac{(L/d)}{\ln[5 \cdot (L/d) \cdot (1-\upsilon)]}} \tag{5}$$

and the percentage of load transferred to the base is given by:

$$\frac{Q_b}{Q_t} = \frac{I_\rho}{1-\upsilon^2} \tag{6}$$

For more generalized cases involving end-bearing type shafts, Gibson soil profiles with E_s increasing with depth, and pile compressibility effects, the more complex solutions for displacement factor I_ρ and load transfer (Q_b/Q_t) are given elsewhere (Randolph & Wroth, 1978, 1979; Fleming, et al. 1992; O'Neill & Reese, 1999; Mayne & Schneider, 2001).

SOIL STIFFNESS

The stiffness of soils is highly nonlinear at all levels of loading. The most fundamental stiffness is that measured at small strains (Burland, 1989), as it represents the beginning of all stress-strain curves at the initial state. This small-strain shear modulus is given by:

$$G_{max} = \rho_t V_s^2 \tag{7}$$

where ρ_t = total mass density of the soil and V_s = shear wave velocity. The mass density can be estimated from an empirical relationship (Burns & Mayne, 1996):

$$\rho_t \text{ (g/cc)} = 0.277 + 0.648 \log (V_s) \tag{8}$$

An equivalent elastic modulus can be obtained from:

$$E_{max} = 2\, G_{max}\, (1+\nu) \tag{9}$$

Since the value of E_{max} is very stiff and applicable only at very small strains, a modulus reduction formulation must be adopted to reflect higher strain levels at foundation loading levels. One convenient means for this is the modified hyperbola (Fahey & Carter, 1993) which is of the simple form:

$$E/E_{max} = 1 - f\,(Q/Q_u)^g \tag{10}$$

where f and g are empirical fitting parameters. Based on a review of static monotonic laboratory and field loading data, initial assumed values of $f = 1$ and $g = 0.3$ can be used for unstructured and uncemented soils (Fahey, 1998; Mayne, 2001). If stiffnesses at intermediate level strains can be measured by pressuremeter or dilatometer tests, the possibility to obtain site-specific f and g values further exists. Combining (10) with (4) allows for an approximate nonlinear load-displacement-capacity representation of the form:

$$w_t = \frac{Q_t \cdot I_\rho}{d \cdot E_{\max} \cdot [1-(Q_t/Q_{tu})^g} \tag{11}$$

AXIAL SHAFT AT COLLEGE STATION, TEXAS

Results of an axial load test on an instrumented drilled shaft at the national geotechnical experimentation site near Texas A&M University in College Station, Texas can be used to verify this method. The site has been used for foundation load testing and has been extensively subjected to site characterization (Briaud & Gibbens, 1994). The soil profile consists of 12.5 m of silty to clean to clayey sands overlying hard clay shale.

A representative seismic piezocone profile is shown in Figure 3 for the site. The data were obtained from recent series of seismic CPTs conducted by Tumay & Bynoe (1998). The penetration porewater pressure profile was obtained from prior field testing at the site (Briaud & Gibbens, 1994) using a Fugro Type 1 penetrometer with

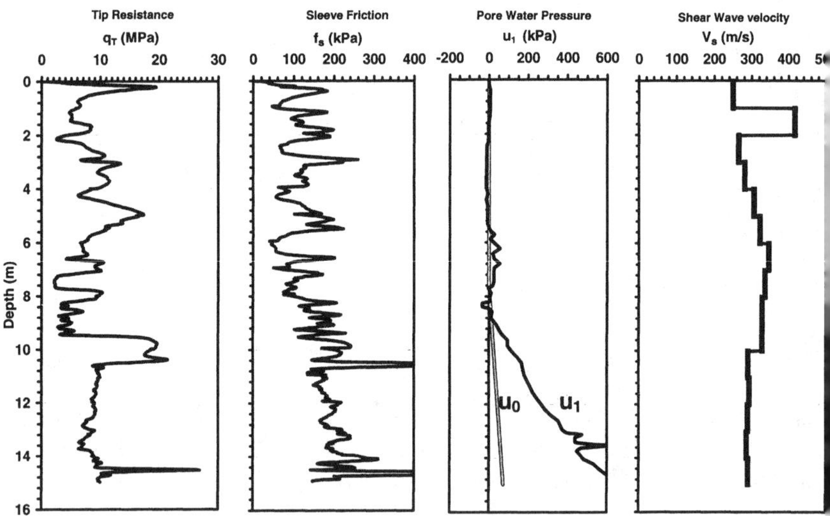

FIG. 3. Seismic Piezocone Results in Sands at NGES, College Station, Texas. (after Tumay & Bynoe, 1998; Briaud & Gibbens, 1994).

midface element (u_1) that showed essentially measured u = 0 above the water table and hydrostatic conditions beneath the phreatic surface. For clean sands, the midface and shoulder elements are of similar magnitude and both follow the trend of the hydrostatic line (Bruzzi & Battaglio, 1987).

A 0.914-m diameter drilled shaft was constructed with a length of L = 10.4 m at the site (Briaud, et al. 2000). Using the Takesue et al. (1998) expressions on a line-by-line basis with the CPT data, the derived f_p/f_s ratio is 0.76 and mean f_p = 115 kPa. For a total surface area A_s = 29.9 m^2, a shaft capacity of Q_{su} = 3434 kN is obtained. The mean measured tip stress beneath the base is q_t = 9.5 MPa. Taking an allowable 10 percent from the Lee & Salgado (1999) analysis gives an unit end bearing of q_b = 950 kPa, which when applied over the end area of A_b = 0.66 m^2 gives a base capacity of Q_{bu} = 622 kN. The total axial compression capacity is then Q_{tu} = 4056 kN. Utilizing the shear wave velocity profile and an assumed (homogeneous) constant modulus profile with depth gives an initial E_{max} = 387 MPa.

The predicted load-displacement-capacity relationship is shown in Figure 4 with evidently good agreement with the measured results from the instrumented load tests. The proportions of side and base shear appear to be well described using the simplified elastic analysis. The approximate method of accounting for nonlinearity is also apparent. The measured and predicted axial load distributions along the length of the drilled shaft are presented in Figure 5. In this case, the loads at the bottom depth (z = L = 10.4 m) were obtained by extrapolation from the lowest level of strain gages at z = 9.0 m depth.

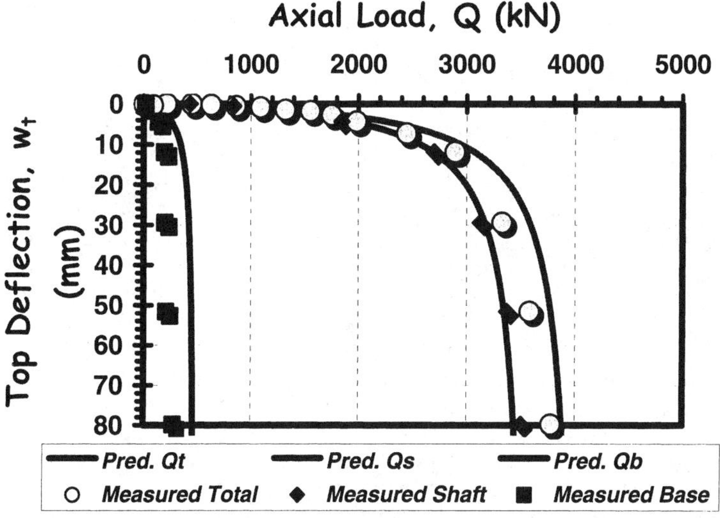

FIG. 4. Measured and Predicted Axial Load Response at Sand NGES, Texas.

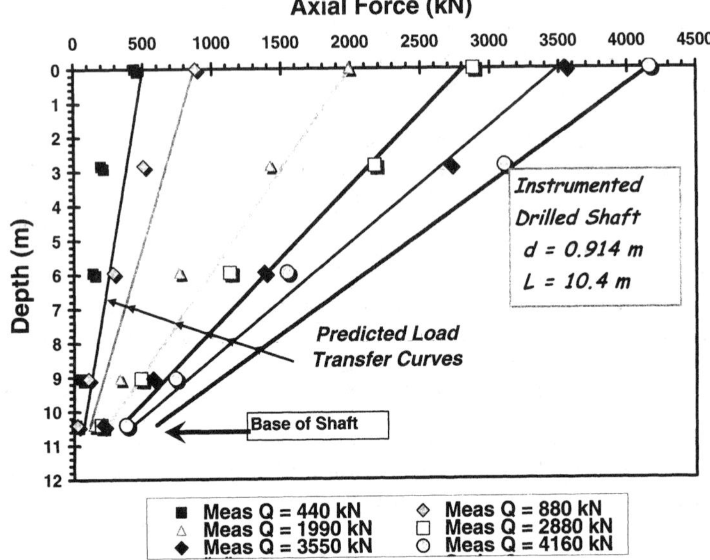

FIG. 5. Measured and Predicted Axial Load Transfer at College Station, TX.

AXIAL SHAFT AT OPELIKA, ALABAMA

The 150-hectare national geotechnical experimentation site near Opelika, Alabama is underlain by residual soils derived from the weathering of gneiss, schist, and granites. The soils are a mixture of very fine sandy silts to silty fine sands with trace mica and grade to saprolite with depth. Detailed laboratory and field testing programs have been conducted at the site (Brown & Vinson, 1998; Mayne, et al. 2000; Mayne & Brown, 2003). A representative seismic piezocone test at the site is presented in Figure 6. Penetration porewater pressure response in these residual soils is unusual in that the readings are negative below the water table (Finke, et al. 2001). Upon the halt at each new rod at 1-m intervals, the porewater pressures fully dissipate rapidly to hydrostatic conditions in only about 1 minute. The porous filter element remains fully saturated, as when the push resumes, the readings again return to negative during penetration, yet cavitation does not occur.

A series of ten drilled shafts were constructed at the site for a load test program (Brown, 2002). These were constructed using a variety of installation methods, including dry (cased) method, bentonite slurry, and polymer (dry and liquid) slurries. The shafts were built with 0.914-m diameters and embedded lengths of 11.0 meters.

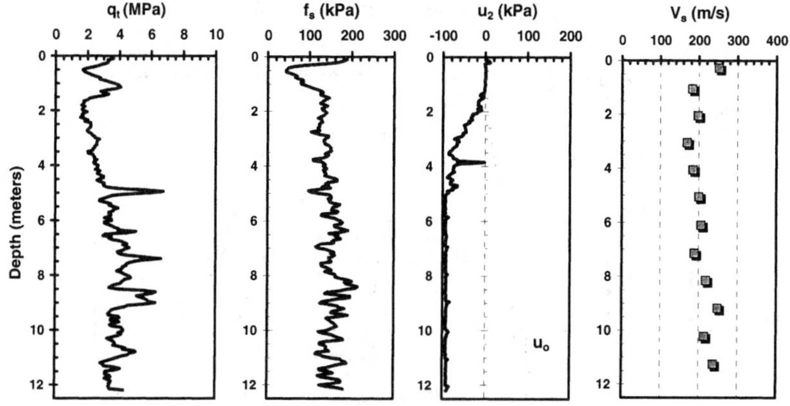

FIG. 6. Seismic Piezocone Tests in Residuum at Opelika NGES, Alabama.

With the CPT data and aforementioned methodology, the calculations of unit side resistance give f_p = 96 kPa and available unit end bearing of q_b = 338 kPa. For the geometry given, the shaft capacity is Q_{su} = 3032 kN and base capacity Q_{bu} = 222 kN, providing a total calculated value of Q_{tu} = 3254 kN. The shear wave profile was interpreted to provide an initial E_{max} = 190 MPa, assumed constant with depth.

Measured and predicted load-displacement responses for one Opelika shaft installed by liquid polymer method are shown in Figure 7. These data are from the manually-recorded load test file, not the automated Megadac system (Brown, 2002). Once again, the comparison indicates good agreement in the overall deflection behavior and relative proportions of side and base components. The approximate nonlinear curves are quite reasonable considering their rather simplistic form.

CONCLUSIONS

The results of seismic cone tests can be used to assess the entire load-deflection, capacity, and axial load transfer of drilled shafts within an elastic continuum framework. A direct method facilitates the calculations of side and base capacities and the initial stiffness is obtained from the profile of shear wave velocity. A modified hyperbola provides an expedient means to reduce the equivalent elastic modulus with increasing load levels. The method is applied to shaft load test results obtained at the national test sites at College Station, Texas and Opelika, Alabama.

ACKNOWLEDGMENTS

The author appreciates the support of the National Science Foundation, Mid-America Earthquake Center, and Georgia Department of Transportation. Any of the

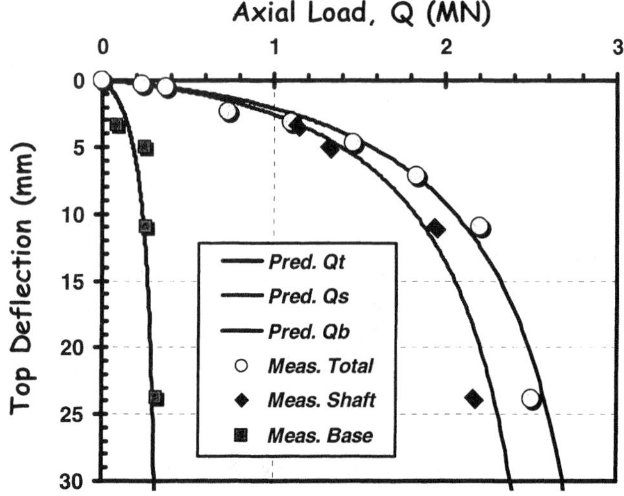

FIG. 7. Measured and Predicted Axial Response of Opelika Shaft, Alabama.

conclusions, recommendations, and results provided in this report are not necessarily endorsed by NSF, MAE, or GDOT. Thanks are also given to Dr. Mehmet T. Tumay for providing CPT data and to Dr. Dan Brown for load test files.

REFERENCES

Almeida, M.S.S., Danziger, F.A.B. and Lunne, T. (1996). "Use of the piezocone test to predict the axial capacity of driven and jacked piles in clay". *Canadian Geotechnical Journal* 33 (1), 23-41.

Briaud, J-L., Ballouz, M. and Nasr, G. (2000). "Static capacity prediction by dynamic methods for three bored piles". ASCE *Journal of Geotechnical & Geo-environmental Engineering* 126 (7), 640-649.

Briaud, J-L. and Gibbens, R.M., editors (1994). *Predicted and Measured Behavior of Five Spread Footings on Sand*, GSP 41, ASCE, Reston, Virginia, 255 pages.

Brown, D.A. (2002). "Effect of construction on axial capacity of drilled foundations in Piedmont soils". *Journal of Geotechnical & Geoenvironmental Engineering* 128 (12), 967-973.

Brown, D.A. and Vinson, J. (1998). "Comparison of strength and stiffness parameters for a Piedmont residual soil". *Geotechnical Site Characterization*, Vol. 2, Balkema, Rotterdam, 1229-1234.

Bruzzi, D. and Battaglio, M. (1987). "Pore pressure measurements during cone penetration tests". *ISMES Report No. 229*, Bergamo, Italy, 265 pages.

Bustamante, M. and Gianeselli, L. (1982). "Pile bearing capacity predictions by means of static penetrometer CPT". *Penetration Testing*, Vol. 2 (ESOPT-2, Amsterdam), Balkema, Rotterdam, 493-500.

Burland, J.B. (1989). "Small is beautiful: The stiffness of soils at small strains." *Canadian Geotechnical Journal* 26 (4), 499-516.

Burns, S.E. and Mayne, P.W. (1996). "Small- and high-strain measurements of in-situ soil properties using the seismic cone penetrometer". *Transportation Research Record* 1548, National Academy Press, Washington, D.C., 81-88.

Campanella, R.G., Robertson, P.K., and Gillespie, D. (1986). "Seismic cone penetration test". *Use of In-Situ Tests in Geotechnical Engineering*, GSP 6, ASCE, Reston, Virginia, 116-130.

Eslami, A. and Fellenius, B.H. (1997). "Pile capacity by direct CPT and CPTu methods applied to 102 case histories." *Canadian Geotechnical Journal* 34 (6), 886-904.

Fahey, M. and Carter, J.P. (1993). "A finite element study of the pressuremeter in sand using a nonlinear elastic plastic model". *Canadian Geotechnical Journal* 30 (2), 348-362.

Fahey, M. (1998). "Deformation and in-situ stress measurement". *Geotechnical Site Characterization*, Vol. 1 (Proc. ISC, Atlanta), Balkema, Rotterdam, 49-68.

Fleming, W.G.K., Weltman, A.J., Randolph, M.F., and Elson, W.K. (1985). *Piling Engineering*, Surrey University Press, Wiley & Sons, New York, 380 p.

Finke, K.A., Mayne, P.W., and Klopp, R.A. (2001). "Piezocone penetration testing in Atlantic Piedmont residuum". *Journal of Geotechnical & Geoenvironmental Engineering* 127 (1), 48-54.

Fioravante, V., Ghionna, V.N., Jamiolkowski, M. and Pedroni, S. (1995). "Load carrying capacity of large diameter bored piles in sand and gravel". *Proceedings, 10th Asian Regional Conference on Soil Mechanics & Foundation Engineering,* Beijing, China.

Lee, J.H. and Salgado, R. (1999). "Determination of pile base resistance in sands." *Journal of Geotechnical & Geoenvironmental Engineering* 125 (8), 673-683.

Lunne, T., Robertson, P.K. and Powell, J.J.M. (1997). *Cone Penetration Testing in Geotechnical Practice*, Blackie Academic/Routledge, New York, 306 pages.

Mayne, P.W., Brown, D.A., Vinson, J., Schneider, J.A. and Finke, K.A. (2000). "Site characterization of Piedmont residual soils at the national experimentation site, Opelika, Alabama". *National Geotechnical Experimentation Sites*, GSP 93, ASCE, Reston/Virginia, 160-185.

Mayne, P.W. (2001). "Stress-strain-strength-flow parameters from enhanced in-situ tests". *Proceedings, International Conference on In-Situ Measurement of Soil Properties and Case Histories*, Bali, Indonesia, 27-48.

Mayne, P.W. and Schneider, J.A. (2001). "Evaluating axial drilled shaft response by seismic cone". *Foundations and Ground Improvement,* GSP No. 113, ASCE, Reston, Virginia, 655-669.

Mayne, P.W. and Brown, D.A. (2003). "Site characterization of Piedmont residuum of North America". *Characterization and Engineering Properties of Natural Soils*, Vol. 2, Swets & Zeitlinger, Lisse, Netherlands, 1323-1339.

O'Neill, M.W. and Reese, L.C. (1999). *Drilled Shafts: Construction Procedures & Design Methods*, Volumes I & II, Publication No. FHWA-IF-99-025, U.S. Dept. of Transportation, published by ADSC, Dallas, 758 p.

Poulos, H.G. and Davis, E.H. (1980). *Pile Foundation Analysis and Design*, Wiley & Sons, New York, 397 pages.

Poulos, H.G. (1989). "Pile behavior: theory and applications." 29th Rankine Lecture. *Geotechnique* 39 (3), 363-416.

Powell, J.J.M., Lunne, T. and Frank, R. (2001). "Semi-empirical design for axial pile capacity in clays." *Proceedings, 15th International Conference on Soil Mechanics and Geotechnical Engineering*, Vol. 2, Istanbul, Balkema/Rotterdam, 991-994.

Randolph, M.F. and Wroth, C.P. (1978). "Analysis of deformation of vertically loaded piles." *Journal of the Geotechnical Engineering Division*, ASCE, Vol. 104 (GT12), 1465-1488.

Randolph, M.F. and Wroth, C.P. (1979). "A simple approach to pile design and the evaluation of pile tests." *Behavior of Deep Foundations*, STP 670, ASTM, West Conshohocken/PA, 484-499.

Robertson, P.K., Campanella, R.G., Davies, M.P. and Sy, A. (1988). "Axial capacity of driven piles in deltaic soils using CPT." *Penetration Testing 1988* (2), Balkema, Rotterdam, 919-928.

Schmertmann, J.H. (1978). "Guidelines for cone penetration test: performance and design". *Report FHWA-TS-78-209*, Federal Highway Administration, Washington, D.C., 146 pages.

Takesue, K., Sasao, H., and Matsumoto, T. (1998). "Correlation between ultimate pile skin friction and CPT data." *Geotechnical Site Characterization* (2), Balkema, Rotterdam, 1177-1182.

Tumay, M.T. and Bynoe, Y. (1998). "In-situ testing at the national geotechnical experimentation sites (Phase 2)". *Contract Report DTFH61-97-00161* to Federal Highway Administration, Louisiana Transportation Research Center, Baton Rouge.

TEXAS CONE PENETROMETER-PRESSUREMETER CORRELATIONS FOR SOFT ROCK

Gerald A. Miller[1], Member, Geo-Institute, and James M. Smith[2], Member, ASCE

ABSTRACT: Results from Texas Cone Penetration Tests are employed for assessing the bearing capacity of drilled shafts in relatively soft sedimentary rocks. The test involves driving a solid, nominally 76-mm diameter cone tip into the rock at the bottom of a borehole and recording the number of blows associated with the measured penetration. While the test is common in Texas and Oklahoma, it has many shortcomings due to the dynamic nature of the test and the small penetrations that can occur. Alternatively, the Pressuremeter Test can be employed in test holes within the shale to assess the stress-strain behavior and bearing capacity. While not without limitations, the Pressuremeter Test seems inherently better for assessing bearing capacity of shale, mainly because it involves carefully controlled quasi-static loading of a substantial portion of the formation to failure, resulting in well defined stress-strain behavior. This paper presents results of Texas Cone Penetration and Pressuremeter Tests conducted primarily in Permian age shale and sandstone at nine sites in central Oklahoma. Correlations between results of the two tests and comparisons of predicted end-bearing capacities for drilled shafts are presented and discussed.

INTRODUCTION

A large amount of near-surface geologic units in Oklahoma are comprised of soft rocks made up of shale, sandstone, and mudstone. Because of their proximity to the ground surface, they are typically used to support heavy foundation loads. Due to the difficulty obtaining intact cores for laboratory testing, prediction of engineering properties is generally achieved using in situ test results. Typical in situ tests include the Standard Penetration Test (SPT), Texas Cone Penetration Test (TCPT), and the Pressuremeter Test (PMT). Nevels and Laguros (1993) developed correlations between the SPT N-values and properties derived from the PMT for clays and shales from the

[1] Associate Professor, University of Oklahoma, School of Civil Engineering and Environmental Science, 202 West Boyd St., Rm. 334, Norman, OK 73019, gamiller@ou.edu
[2] District Manager, Professional Service Industries, Inc., 801 SE 59th St., Oklahoma City, OK 73129.

Hennessey Unit in Oklahoma. This paper presents similar correlations developed between TCPT and PMT results obtained at nine sites in Oklahoma. The TCPT method is commonly employed in Oklahoma to determine load carrying capacity of the foundation materials.

Correlations of the type presented in this paper are useful for three major reasons: 1) In conducting PMTs in soft rock formations it is common to reach expansion pressures near the upper limit of the testing equipment. Thus, via correlations, TCPT or SPT test results can be used to estimate the maximum pressures expected during pressuremeter testing. These estimates can then be used to anticipate whether yielding of the rock formation will likely occur during the PMTs and also to anticipate problems associated with testing at high pressures. It is common to rupture membranes during pressuremeter testing at high pressures in these formations. 2) Correlations can be used to estimate pressuremeter-derived parameters, such as the pressuremeter modulus, which cannot be readily estimated from the penetration test results. The estimated pressuremeter modulus can, for example, be used for preliminary estimates of foundation settlement. 3) Assuming the pressuremeter can more accurately characterize the actual stress-strain behavior, the scatter of data in the correlations can serve as a qualitative means of assessing the uncertainty in parameters derived from penetration test results.

TEST SITES AND GEOLOGY

Seven of the nine test sites in this study were located in Oklahoma and Cleveland Counties in and around Oklahoma City and Norman, Oklahoma. At six of these seven sites, tests were conducted in the shales of the Hennessey Unit. This Permian-age unit consists primarily of red platy to blocky clay shales and mudstone. The red clay shale of the Hennessey unit is characterized by numerous bands or streaks of white or light green color ranging from a few centimeters to 1.2 meters in thickness (Hartronft et al. 1967). The remaining site, out of these seven, involved testing in sandstone of the Permian-age Garber Unit. The sandstone is characterized as massive, typically cross-bedded and lenticular (Hartronft et al. 1967).

The remaining two test sites, out of the nine, were outside of the Oklahoma City area; one in Lawton in Comanche County, and one in Tulsa, in Tulsa County. In Lawton, tests were conducted in shale that is part of the Permian-age Addington Unit. The unit consists primarily of soft red-brown sandstone, shale, and mudstone conglomerate (Hartronft et al. 1969). In Tulsa, tests were conducted in Pennsylvanian-age shales of the Nellie Bly Unit. The unit consists of shales and sandy shales containing some sandstone and siltstone (Hartronft et al. 1965).

FIELD INVESTIGATION

Test borings were made with continuous flight hollow-stem augers. Preliminary test holes were performed with TCPTs taken at 1.5-meter intervals in the shale strata. The TCPT is performed by attaching a 76-mm diameter solid penetrometer cone with a 60° apex angle to NW-size drill rod and lowering it to the bottom of the borehole. The anvil and hammer is then attached to the top of the drill stem. The penetrometer cone is driven for 12 blows or 152 mm, whichever comes first, to seat it in the material. The drill rod is then marked at 152-mm increments to prepare for the test. The cone is then driven 305 mm into relatively soft material or for 100 blows into relatively hard materials. In hard materials, the cone is driven with the resulting penetration recorded for the first and second of 50 blows. TCPT correlation graphs based upon research and past experience are included in the Texas Department of Transportation (TexDot) Geotechnical Manual (2000). These graphs show the relationship between bearing capacity based on laboratory shear strengths and the measured dynamic driving resistance.

The procedure developed by the Texas Department of Transportation (TexDot Manual of Testing Procedures, Test Method Tex-132) consists of driving the cone with a 77-kilogram hammer dropped a regulated 600 mm. However, since most drilling rigs are equipped with standard 63.5-kilogram hammers, the local industry standard test procedure consists of dropping the 63.5-kilogram hammer a regulated 760 mm in accordance with SPT procedures in ASTM D1586 (ASTM 2002). Thus, local practice theoretically provides 2.7% more energy per blow, which should result in slightly more conservative estimates of foundation capacity.

Secondary test holes were then drilled within an approximately 1.5-meter radius for pressuremeter tests. After reaching the top of rock using a hollow stem auger, the test cavities for the pressuremeter tests were made with 75-mm diameter continuous flight solid stem augers. Typically, the center of the pressuremeter probe was placed at the elevation of adjacent TCPTs. Pressuremeter tests were conducted using an NX-size monocell pressuremeter having a deflated length to diameter ratio of eight for the expanding portion of the probe. The membrane was enclosed in a metallic sheath. The test was conducted in accordance with ASTM Standard D 4719, Method A (ASTM 2002), which involves increasing the probe pressure incrementally and recording the volume change during a 60-second interval. Raw data is then corrected for membrane resistance, device compressibility, and hydrostatic fluid pressure to produce a corrected pressure-volume curve, such as shown in Fig. 1. The limit pressure and pressuremeter modulus were interpreted from these curves using the procedures described in Standard D 4719 (ASTM). For calculating the pressuremeter modulus a Poisson's ratio of 0.33 was used.

CORRELATION OF TCPT AND PMT RESULTS

In Fig. 2 the TCPT N-values are plotted against corresponding limit pressures obtained at each test location. Fifty comparisons representing the nine sites are presented. While there is considerable scatter in the test data, a general trend of exponentially decreasing limit pressure with increasing N-value is observed with a coefficient of determination, r^2, of 0.59. It is important to note that upper portion of the trend results from tests in the sandstone and the scatter in the shale, representing the lower N-values is substantial. This same data is presented in Fig. 3 on a Log-Log plot along with 95% prediction intervals. Assuming the limit pressures represent the load bearing capacity of the formation, the prediction intervals indicate that predictions using

N-values based on the trend line can vary substantially. Generally, the prediction intervals show that limit pressures can be over or under predicted by a factor of roughly two.

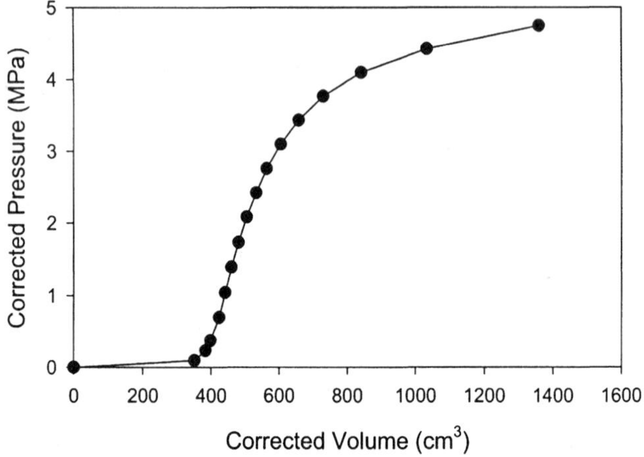

FIG. 1 Typical Pressuremeter Curve for Shale

In Fig. 4, the correlation of N-value and pressuremeter modulus shows that a well-defined trend exists. However, estimates of pressuremeter modulus based on N-values using the best-fit line can at best be expected to be correct within one order of magnitude.

DISCUSSION OF RESULTS

Correlations presented show that reasonable trends between TCPT N-values and pressuremeter parameters exist; however, consideration of the data scatter should be included in the use of these correlations. Sources of scatter may be attributed to aspects of each testing method as well as natural variations in geomaterials.

While borings for TCPTs and PMTs were in close proximity at each test site, natural variations in the geomaterials at corresponding test depths probably contributed to the scatter. Since the pressuremeter effectively loads the formation over a vertical zone of approximately 610 mm and the Texas cone penetrates vertically from as little as 7 to as much as 250 mm, depending on formation hardness, it is expected that vertical variations at the test locations would effect each test differently. These differences are likely compounded by variations in stress and material anisotropy in the formation since the mode of shearing is different for each test.

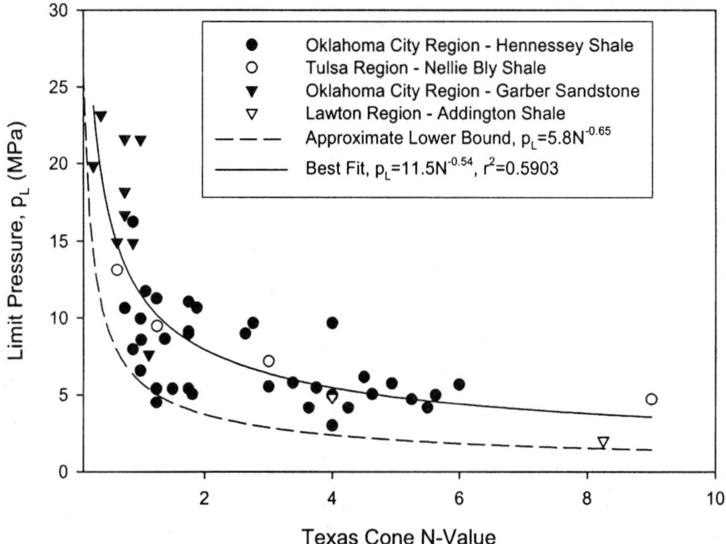

FIG. 2 Texas Cone Penetrometer N-Value versus Limit Pressure

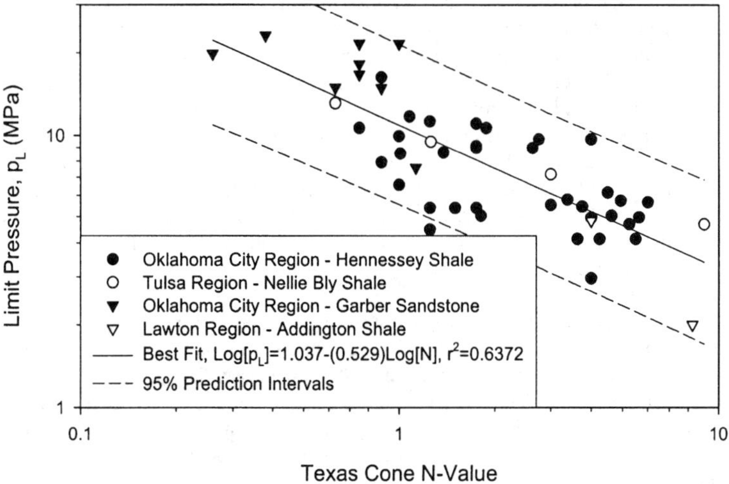

FIG. 3 N-Value versus Limit Pressure on Log-Log Plot

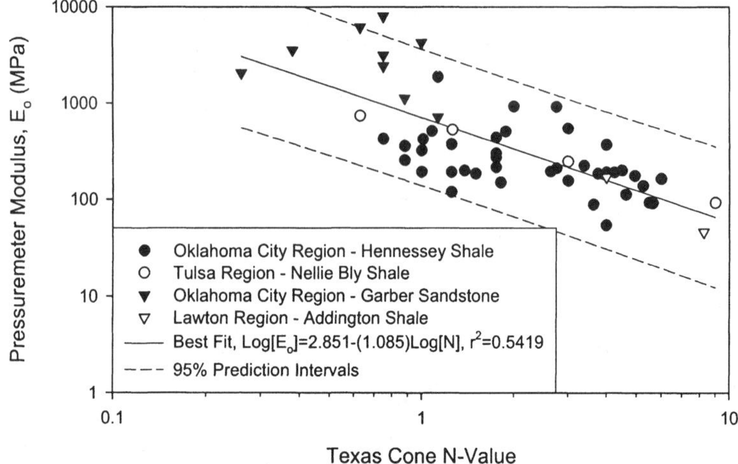

FIG. 4 N-Value versus Pressuremeter Modulus on Log-Log Plot

Regarding the TCPT test method, scatter can be attributed to factors similar to most dynamic penetration testing methods. Others have witnessed scatter of similar magnitude in SPT-PMT correlations for residual soils and soft rocks (Martin 1977, Nevels and Laguros 1993). Martin (1977) partly attributed the scatter to the inherent variability in the SPT procedure. Important among the factors influencing N-values are variations in the efficiency of energy transfer from the ground surface to the penetrometer. This efficiency depends on the efficiency of the hammer, length and inclination of drill rods, integrity of the drill rod connections, borehole size, and extent of soil friction along the drill rods, among others. Another factor of significant importance is the amount of wear on the cone tip. In addition, if the cone tip bounces up and down on a hard formation it is possible that rock fragments breaking ahead of the tip and along the drill rod will affect the final resting position of the cone. These factors may be especially important where measured penetrations are small and subject to greater inaccuracy of measurement.

The pressuremeter, while considered a much more precise method of ascertaining formation properties, is not without problems. Foremost among these is the degree of disturbance that occurs during formation of the test cavity. This disturbance may be greater in more brittle geomaterials where significant fracturing along the borehole cavity can occur during drilling. An advantage of the pressuremeter is that the degree of disturbance can to some extent be ascertained by the quality of the pressuremeter expansion curve.

COMPARISON OF END-BEARING CAPACITY PREDICTIONS

The end-bearing capacity of a drilled shaft can be predicted using the net limit pressure as given in Eq. 1 (Briaud 1989).

$$q_u = kp_L^* - \sigma_{vo} \qquad [1]$$

where: q_u = ultimate unit end-bearing capacity,
k = bearing capacity factor,
p_L^*= net limit pressure = $p_L - \sigma_{ho}$,
p_L= average limit pressure within a short distance above and below the shaft base,
σ_{ho}= at-rest total horizontal earth pressure estimated from the PMT curve, and
σ_{vo}= total vertical overburden pressure at the base of the shaft.

For weathered rocks, recommended k-values vary between 1.1 and 1.8. Because of the lack of full-scale load test data in the local area a k-value of 1.2 is was employed. Thus, predicted end-bearing is typically close to the limit pressure, p_L.

Allowable unit end-bearing based on the TCPT correlations with end-bearing is given by Eq. 2 (TexDot 2000).

$$q_a = Cf(\phi)/N \qquad [2]$$

where: q_a=allowable unit end-bearing capacity=q_u/F.S.,
q_u=ultimate unit end-bearing capacity,
F.S.=factor of safety=2,
C=constant=70,
$f(\phi)$=0.876, and
N=penetration in inches per 100 blows.

Thus, the ultimate unit end-bearing capacity based on the TCPT correlation is given by Eq. 3.

$$q_u = 122.6/N \qquad [3]$$

Using Eqs.1 and 3, the ultimate end-bearing capacity was computed for corresponding PMT and TCPT data presented in Fig. 2. The comparison of ultimate end-bearing capacity is shown in Fig. 5. Looking at the position of data points relative to the line of equality, this graph shows that 70% of the predictions based on the PMT results were greater than corresponding TCPT predictions, while 30% were below. It is observed that 92% of the data points (46 out of 50) fall in the band delineated by the dashed lines in Fig. 5. The upper line is about 7 Mpa above the line of equality, while the lower line is about 4 Mpa below the line of equality. This indicates that the degree to which the TCPT method under predicts the end-bearing determined by the PMT is about half the degree of over prediction.

SUMMARY

A comparison of N-values determined by the Texas Cone Penetration Test was made with Limit Pressures and Moduli determined from the Pressuremeter Test in soft shale and sandstone. The correlations presented are reasonable given the numerous factors that may influence the test methods. However, the scatter observed in these correlations is significant and reveals the uncertainty associated with estimating geotechnical properties. It is argued that a great deal of the uncertainty observed in these correlations can probably be attributed to the inherent variability in the Texas Cone Penetration Test; however, other sources of variation most certainly contribute to the scatter as well. Comparisons of ultimate end-bearing capacities show that for 70% of the data set the end-bearing predicted using the pressuremeter was greater than that predicted using the Texas Cone.

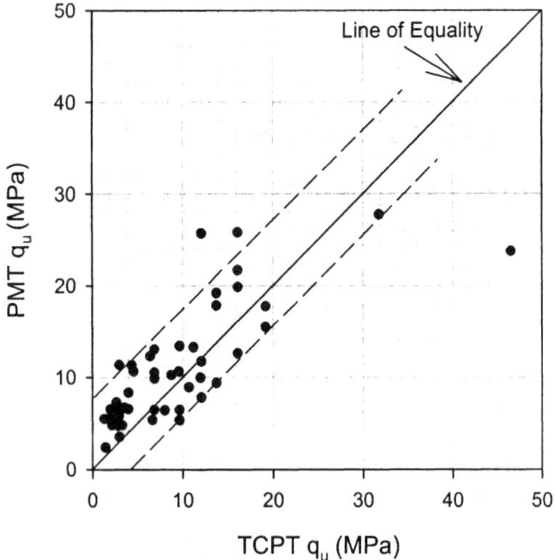

FIG. 5 Comparison of Ultimate End-Bearing Capacity from Pressuremeter and Texas Cone Penetrometer

REFERENCES

ASTM (2002), *Standards on Disc, Section Four Construction*, ASTM International, Vol. 4.08.

Briaud, J.L. (1989), *The Pressuremeter Test for Highway Applications*, FHWA Report No. FHWA-IP-89-008.

Hartronft, B.C., Barber, I.E., Buie, L.D., Hayes, C.J. and McCasland, W. (1965), *Engineering Classification of Geologic Materials – Division Eight*, Research and Development Division, Oklahoma Highway Department in cooperation with the U.S. Bureau of Public Roads.

Hartronft, B.C., Hayes, C.J. and McCasland, W. (1967), *Engineering Classification of Geologic Materials – Division Four*, Research and Development Division, Oklahoma Highway Department in cooperation with the U.S. Bureau of Public Roads.

Hartronft, B.C., Smith, M.D., Hayes, C.J. and McCasland, W. (1969), *Engineering Classification of Geologic Materials – Division Seven*, Research and Development Division, Oklahoma Highway Department in cooperation with the U.S. Bureau of Public Roads.

Martin, R. E. (1977), "Estimating Foundation Settlements in Residual Soils," *Journal of the Geotechnical Engineering Division*, ASCE, Vol. (103), No. GT3, pp. 197-212.

Nevels, J. and Laguros, J.G. (1993), "Correlation of Engineering Properties of the Hennessey Formation Clays and Shales," *Geotechnical Engineering of Hard Soils-Soft Rocks*, Anagnostopoulos et al. (eds), Balkema, Rotterdam.

Texas Department of Transportation (2000), *Geotechnical Manual.*

Texas Department of Transportation (2002), *Manual of Testing Procedures*.

GEOTECHNICAL ENGINEERS, WAKE UP -- THE SOIL EXPLORATION PROCESS NEEDS DRASTIC CHANGE

Jorj O. Osterberg[1], Honorary Member, ASCE

ABSTRACT: The usual process for obtaining subsurface soil and rock information is to obtain bids for soil borings on a cost per foot basis and awarding the contract to the lowest bidder. Thus, there is no incentive to do good quality work and there is every incentive to somehow get a hole down as quickly as possible without regard to obtaining the best soil, rock, water levels and collateral information. As a result, perched water tables and important soil layers and other important information are often missed. In general the equipment, techniques, and procedures for obtaining better quality samples and other subsurface information has long been available, but there is no incentive on the part of the drilling contractor to make use of them. Many new techniques for geophysical prospecting have been developed but the geotechnical engineer seldom uses them except for large or special projects. The geotechnical engineer often fails to use his knowledge of geology in laying out an exploratory program and interpreting the borings and soil test results.

INTRODUCTION: This paper is concerned with the poor quality of soil exploration, the lack of initiative on the part of the geotechnical engineer to work for improvement in the quality and the failure to use the many new developments in geophysical equipment available. Here, concern is with the ordinary every day project involving or should involve a geotechnical engineer and not with the large and special projects. In the entire process of site characterization, drilling supervision, evaluation of samples, laboratory testing, and inspection during construction, there often is no overall responsibility or accountability. To illustrate, let's go through the whole process on a typical job.

The architect or the design engineer or the structural engineer lays out a boring plan and asks for bids from various drilling companies. The work is let to the lowest bidder based on the lowest price per foot of drilling. On a small job, the geotechnical engineer is often not engaged or consulted during this process and in some cases not engaged until the borings have been completed. Most of the time the driller isn't

[1] Professor of Civil Engineering Emeritus, Northwestern University, Evanston, Illinois

even told what type of structure is to go on the site. It could be a skyscraper or an outhouse – it makes no difference to the driller. He has a standard protocol to follow no matter what. He is told to make the borings to a certain depth or rock, whichever occurs first. He classifies the samples and keeps a field log. He usually is asked to record the water level in the borehole after completion. Most of the time there is no field inspector to monitor the boring operations, and if there is, he is usually an engineer or technician at the bottom of the totem pole who is just out of school and has had little or no previous contact with soil borings and has been given little or no training. From the field logs, the final boring logs are prepared and sent to the client. In many cases no one even examines and reclassifies the samples before preparing the final boring logs. After this, no one looks at the samples. On many contracts, the foundation contractor is told that the samples are available for his inspection but in the vast majority of cases no one asks to examine the samples. This is the worst scenario of the process, but it is not uncommon. What are the consequences?

THE DRILLING CONTRACTOR AND THE DRILLER

Since the contract for drilling is based on the price per foot, why should the drilling contractor be concerned with obtaining the best possible samples and using superior devices for obtaining the samples unless this saves time and money. The driller will do almost anything to speed up the work because he gets rewarded for it, either directly or indirectly. Why should the drilling contractor spend time and money to train the driller since this doesn't give him more work or more profit? Here are some of the consequences experienced.

Case 1. For a new toll road, a contract was let to a drilling company for borings for a number of bridges. When the contractor for the foundations for one of the bridges started work, he found that the ground conditions encountered were nothing like that portrayed by the borings. On investigation it was determined that the borings were never made and that the driller had fabricated the field logs. The writer was called by the design engineers for the project and asked his opinion and recommendation of what to do. When asked "Where was the drilling inspector?", I was told that there wasn't any inspector because the toll road authorities wouldn't pay for it. "I guess you are stuck," I said. "The drilling company is small and if they get sued they will just declare bankruptcy, and probably you would be next to be sued." Even though I asked to be informed about the outcome, I never heard from any of the parties involved.

Case 2. The writer was engaged to recommend foundations for a manufacturing building with a basement, located near the Illinois River north of Peoria. The borings had already been made and I had the opportunity of looking at the samples. The boring logs all showed the water table at or very near the same level as the nearby Illinois River and the basement level was above the water table. I recommended spread footings which were incorporated in the design drawings. When excavating, water flowed freely into the hole from above the water table. I questioned the driller and learned that water did flow into the borehole but when he cased the hole, water came to the level he reported. He didn't report what he observed. The boring logs indicated a stiff clay layer about several feet thick just below the foundation level indicating a perched water table. The water problem was solved by placing coarse sand in holes drilled at intervals in the foundation area to allow the perched water to flow freely into the water table below and no further water problems occurred during construction and after the building was completed. If the driller had been told to report water coming in the hole while drilling, the perched water table would have been discovered.

Case 3. I was in charge of a soils investigation for a new cement plant in Milwaukee in a river valley and only a short distance from Lake Michigan. I visited the site periodically and had the driller contact me by phone almost daily. During one of my visits, the driller told me he had an experience he couldn't explain. He noticed that the water level in each borehole was about 10 feet below the surface in each boring which was about the same level at the nearby river. When the driller reached 30 to 40 feet depth in each boring, the water level would drop to about 20 feet below the river level. Immediately, I realized there was a perched water table. But why would the water table right next to the river and only a short distance from the lake be consistently lower than the lake level in every boring? After some searching, I learned that the breweries in the Milwaukee area pumped water from wells which penetrated an aquifer just above and into the underlying limestone. This explained why in the immediate area there were a lot of old buildings only one story in height showing cracks caused by differential settlement. Also an old bridge only a few blocks away had settled over a foot but did not exhibit much distress since the settlement was quite uniform. This is an example how an alert and intelligent driller was the key to discovering a perched water table. The particular driller did not even graduate from high school, yet he was observant, curious, and showed some intelligence. Such a driller is rare.

The above cases indicate how important it is that the driller is trained to observe occurrences during drilling and to record them and to alert the inspector (if there is one), the person in charge of the borings, or his immediate superior. Also, it is important to have a trained and experienced boring inspector present, or at a minimum, a geotechnical engineer visit the site periodically during boring operations. Again, why should the drilling contractor train the driller to do anything more than just make the borings the way he has in the past when there are no incentives for doing so?

THE GEOTECHNICAL ENGINEER

Now, let's look at the geotechnical engineer, if there is one on a project. Many projects don't have a geotechnical engineer. His duties are frequently performed by the structural engineer, who may or may not know much about geotechnical engineering. Boring plans often are laid out, executed and interpreted without thinking that the boring information may be erroneous, or may not reveal the true subsurface conditions. Often, the geotechnical engineer depends only on the boring logs and does not use any redundancy in the exploratory planning, and very importantly, does not have a good knowledge of geology.

Case 4. When I worked for the Chicago District of the Corps of Engineers, I was in charge of the Soils and Foundations Division. One of my projects was the foundation investigation for a structure in central Wisconsin. I instructed the driller to phone me daily and to not start another boring until he phoned me. When he phoned about a boring he had just finished, he said he hit granite and had two feet of rock core to prove it. I knew that the underlying rock was limestone and said it must be a granite boulder. I ordered him to go back to the hole and drill deeper. He drilled a total of six feet through the granite and then reached sand. After continuing drilling, limestone was reached. A 6-foot boulder in the glacial deposit is not rare. A knowledge of geology and glaciation was essential in this situation.

Case 5. On another occasion, I was in charge of a preliminary subsurface investigation for a proposed earth dam. The dam was to be across a river valley in glaciated country. Since the valley had been scoured by glaciers and filled with glacial drift, I suspected that there might be a buried pre-glacial river channel. Since I only had the Corps of Engineers drilling crew at my disposal and no other resources, I persuaded the State of Illinois to make a geophysical survey. The five borings I laid out did not indicate any channel in the underlying rock. The geophysical survey indicated a river channel in the rock not under the present river but off on the side of the present valley. A boring at this location confirmed the buried channel and showed it was filled with sand and gravel. In this case, knowledge of geology and use of more than one method of exploration made it possible to find an important feature which could possibly have caused large under seepage and possibly endanger the safety of the dam. The dam however, was never built.

Case 6. Foundations for a new City Hall were designed based only on the boring logs and the standard penetration tests indicated on the logs. No water content or other laboratory tests were made. Figure 1 shows a typical boring log. A sufficient number of borings were made which indicated that conditions over the site were quite uniform. It was decided that the structure should rest on drive piles. A contract was let for mandrel driven corrugated shell piles to be driven into the very hard glacial till at a depth of approximately 45 feet. The piles could only be driven 14 to 16 feet into the "stiff gray very silty clay" before they met refusal. After two weeks of frustration at not being able to drive the piles to the required depths, the writer was consulted.

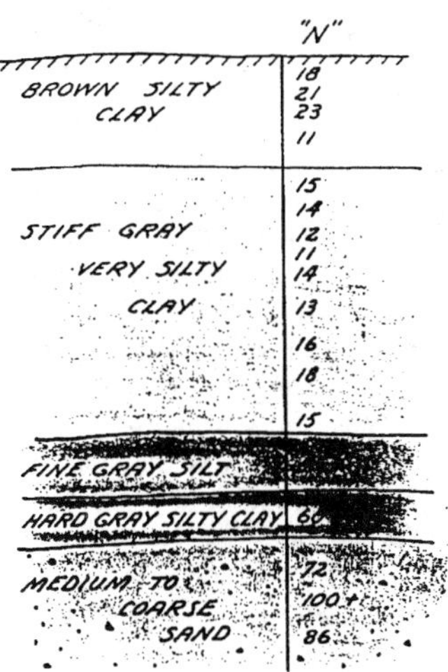

Fig. 1 Boring Log Used for Design

Undisturbed tube samples were taken in the "very dense silty clay" from which water content and consolidation tests were made. From the results (Figure 2), it was apparent that the "stiff gray very silty clay" was in fact a very dense silt with water contents of 12-15%, unconfined compression test results of 4.7 to 7.0 tons/sq.ft. and preconsolidation stresses of about 6 tons/sq.ft.

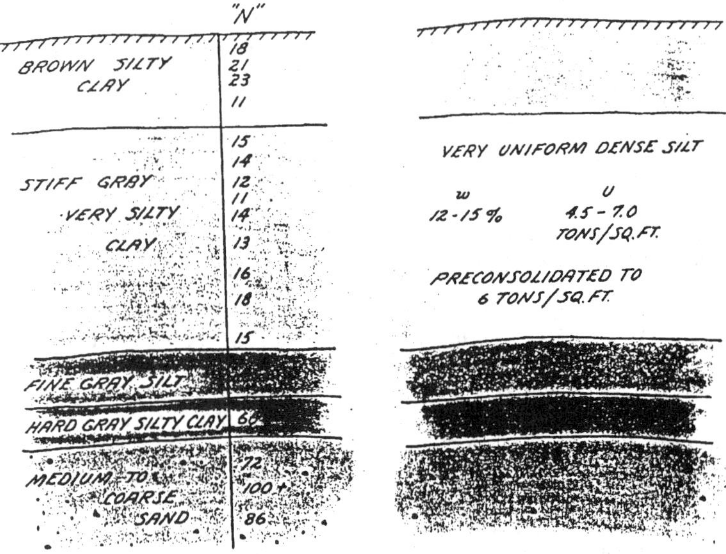

Fig. 2 Soil Profile before (left) and after Tests were made (right)

It was common knowledge that the area had been covered with at least 2000 feet of ice during the last ice age, and thus the reason for the high preconsolidation stresses. It is obvious that the structure need not rest on piles and could be placed on spread footings. I recommended using 6,000 lbs./sq.ft. for footing design. The nearby old city hall, a heavy stone faced building five or six stories high constructed in the 19th century rested on the same formation and showed no signs of distress due to differential settlement. My recommendations were accepted by the architect, structural engineer, foundation contractor and the superintendent of public works for the city, The only dissenter was the city engineer who was the one who originally decided that the structure be placed on piles. He threatened to resign his job if the building was not placed on piles. After considerable argument, which managed to get coverage in the local newspaper, everyone yielded to his wishes (except the writer) and the structure was placed on piles, but not without further difficulty. At each pile location, it was necessary to auger a hole through the dense silt, drop the pile into the pre-drilled hole, and drive the pile the last few feet into the glacial till. The contractor was paid all the additional cost of pre drilling. If placed on spread footings the saving would have been in the millions of dollars.

Why was the foundation investigation insufficient? The soil was misclassified. Apparently, no one with a proper background classified the soil. If only water contents on the dense silt samples had been determined, it would have been found that the soil was very dense. If a grain-size distribution test had been made, it would be found that the soil was silt and not clay. If consolidation tests had been made, it would have indicated large over-consolidation. If a rudimentary knowledge of glacial geology had been applied, the large over-consolidation could have been explained. If cone penetration tests had been made to check the standard blow counts, it would have indicated that the silt was very dense and inconsistent with the 10-15 blow counts obtained. Why were the blow counts low? One would expect that for such a dense soil, the blow counts would have been in the 50-60 range. The reason can only be speculated. It is quite possible that the water in the drill hole actual penetrated into the silt several feet so that when sampled, the sampler penetrated a softened soil yielding a lower blow count. This case illustrates the need for redundancy in testing, the lack of qualified inspector, lack of geological knowledge, and lack of general geotechnical knowledge.

Case 7. I was asked by an owner to review the borings and make recommendations for the foundations for a one story industrial building with light loads. The column loads were not very large but because of underlying soft soils, using spread footings for the columns would lead to unacceptable settlements. I recommended that the building columns be placed on 18-inch drilled shafts extending to limestone bedrock. The borings showed that the rock surface sloped over the site and there was a dense sand layer a few feet thick at a depth of about 15 feet under which was a organic silty clay to the top of bedrock. I had no further contact with the project and had nothing to do with the construction of the drilled shafts. After completion of the shafts, serious settlements occurred under many columns before the building was completed. I learned that there was no inspection of the drilling and concreting of the shafts and it was left entirely to the driller. After examining the boring logs and the drilled shaft depth records, it was obvious that most of the shafts did not reach rock. The shafts rested on top of the dense sand and the large settlement was due to the underlying organic soil. The first 10 shafts reached rock where the sloping rock surface was at the shallow depths. As the driller continued, he encountered the dense sand layer he assumed was rock. He did not look at the boring logs to see that the rock surface became deeper as he progressed with the shaft installations. It is obvious that the soft organic soil under the dense sand layer was the cause of the settlement. To remedy the situation, I recommended that a six-inch hole be drilled vertically through the center of each defective shaft and to continue drilling to make a 2-foot socket in the limestone. A 4-inch heavy wall pipe was inserted in the hole and hammered to make sure it reached to the bottom of the 2-foot socket. It was then grouted in place. I insisted that a competent inspector be present when this work was done.

It is obvious that the driller was incompetent and most likely didn't even look at the borings. The structural engineer (there was no geotechnical engineer) did not have an inspector present during the drilling. Perhaps I might have been partly to blame because I did not recommend in my report that a competent inspector be present during the drilling of the shafts.

Case 8. Figure 3 shows the results of the failure of a braced excavation in Texas. It was made to construct a permanent entrance shaft into rock and later as a permanent entrance shaft to a large sewer tunnel. The soldier beam and lagging wall was to be 120 ft. diameter and extending to the claystone 34 feet deep. Only one boring was made at the center of the shaft (Figure 4). No other borings were made within 300 feet. The plan was to extend the shaft beginning at the top of the claystone and deep enough to start tunnel construction at 76 foot depth. Since the one boring did not indicate any free water or water bearing sand, the soldier beam and lagging method was deemed the most economical. This system requires continuous support around the entire periphery for stability. If there is loss of soil from behind, there is danger that the whole system could collapse, and that is exactly what happened.

Fig. 3 – Failure of Circular Wall

When the contractor had excavated almost to the claystone, and the last section of wooden lagging was being installed, water started to flow profusely into the excavation, thus washing the sand from under the bottom of the lagging into the excavation. The soil from behind the lagging fell in the void created below and the support behind the wall was lost. The ring beam walers twisted resulting in the entire wall collapsing. The cost of the failed wall was $2,500,000. After the debris was removed, the excavation was finished by open cut.

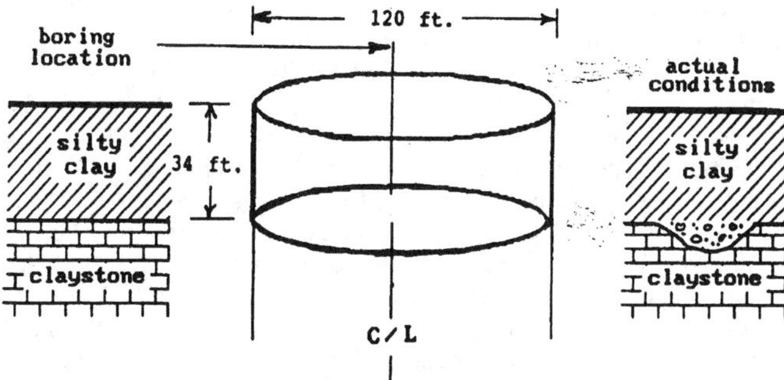

Fig. 4 - Profile According to Soil Boring (Left) and Conditions as Found (Right)

Clearly, this was an exploration failure. One boring for such a large and important structure was hardly enough to indicate the true soil conditions around the periphery. However, three or four more borings around the perimeter might not have discovered the buried streambed. It was later learned that a geological report made for the tunneling phase of the project had indicated that there were buried stream beds in the general area, but this information was not relayed to the contractor. This is another case of where important information was not passed on to the proper parties. You might call the failure a "people failure".

THE PRESENT SITUATION

The following are my observations on the present day status of subsurface exploration. My comments refer to the majority of projects which are generally small to mid size and do not apply to the larger and special projects where ground characterization utilizes the many exploratory tools available.

1. Soil boring methods, soil and rock sampling, and the quality of work has not changed much in the last 50 years. This is in spite of the availability of more and better tools, and newer and better techniques. This is due mostly to the almost universal process of awarding exploration contracts to the lowest bidder. Also, the geotechnical engineer doesn't seem to demand anything better.
2. Much of the geotechnical engineering involved in soil and subsurface characterization is being done by non-geotechnical engineers who mostly are not qualified.
3. Geotechnical engineers often do only what they are asked to do and do not assert themselves on matters and decisions which are clearly geotechnical.
4. Drillers often lack proper training and are not informed what is expected of them and what they should look for. They do not write down important

information on their daily log sheets, information which could possibly be important and useful to the geotechnical engineer.

5. Soil and rock samples are frequently classified by unqualified personnel. Soil samples and classifications are often not checked or inspected by the geotechnical engineer.
6. Geotechnical engineers are often unaware of how the laboratory tests are made and seldom visit the testing laboratory to observe the testing.
7. The importance of redundancy in subsurface exploration is underestimated.
8. "People problems" are underestimated and often ignored. Examples of "people problems" are: general lack of communication between field and office; lack of design personnel in giving field personnel important information necessary to carry out their job functions; lack of field personnel in informing those in charge those field observations which may be important.
9. Geotechnical engineers generally lack sufficient knowledge of geology.

WHAT NEEDS TO BE DONE

1. Geotechnical engineers need to assert themselves on those matters which concern geotechnics. If they are to be held liable for geotechnical matters, then they should be involved in all geotechnical matters.
2. Every effort should be made to abandon the policy of awarding drilling contracts to the lowest bidder on a per foot basis. The geotechnical engineer should ask the owner or the person with purchasing authority: if he were ill, would he ask for bids for his x-rays and lab tests. The owner is buying information and should want the best information available with cost being secondary Pre-qualified drilling companies should submit proposals for a specific project and awarded to the best overall proposal with price being a consideration. This type of contract award will give an incentive for the contractor to offer innovation, quality work, alternative equipment, new techniques, and better qualified drillers. Owners should be persuaded that they should buy the best and necessary subsurface information needed to provide a practical, safe and economical design. Flexibility should be allowed and provided for in the contract to allow for changes that may be required as the exploratory program proceeds and yields conditions which indicate that these changes are needed.
3. Geotechnical engineers need to use more redundancy in subsurface exploration for verification and to possibly discover unanticipated conditions.
4. When borings are made, a competent and qualified inspector should be present or at a minimum, a geotechnical engineer or other qualified person should visit at intervals to asses the quality of work and the possible need to make changes in the exploratory program as conditions indicate.
5. During construction, those items for which the geotechnical engineer is

responsible should be inspected at intervals or if necessary, should have continuous inspection to assure that the intentions of the designer are carried out.

6. To obtain a master's degree in geotechnical engineering, a course in general geology and engineering geology should be required. These courses can be obtained at the undergraduate or the graduate level.
7. Either a course in soil exploration or sufficient time for soil exploration in a foundation engineering course should be required for a master's degree in geotechnical engineering. At least one field visit to observe drilling operations should be required.

MODULUS LOAD TEST RESULTS FOR *RAMMED AGGREGATE PIERS™* IN GRANULAR SOILS

Craig S. Shields[1], P.E. Member, ASCE, Brendan T. FitzPatrick[2], Associate Member, ASCE, Kord J. Wissmann[3], P.E. Member, ASCE

ABSTRACT: In the past five years, over 25 structures in the states of California, Oregon, and Washington, have been supported by *Rammed Aggregate Piers™* constructed in granular soils. The piers are installed by drilling 60 to 90 cm diameter holes and ramming thin lifts of highway base course stone within the drilled cavities. The elements are used to support conventional shallow footings. The system is unique and innovative because it incorporates features associated with the design and construction of shallow and deep foundation systems. Accordingly, the design procedures include concepts derived from conventional shallow foundation design, historical stone column soil improvement system design, and cast-in-drilled-hole concrete shaft design. Unlike design values for drilled deep foundation systems, which are well-documented in the literature, parameter values for rammed aggregate piers are established from the results of modulus tests conducted at each project site.

This paper presents results of 19 rammed aggregate pier modulus tests performed at sites underlain by granular soils. Test results are correlated to matrix soil characteristics and length of the piers. This paper is of particular significance because it presents a database of in-situ modulus values used in the design of a cost-effective and increasingly popular drilled foundation system.

---

[1]Principal Engineer, Treadwell & Rollo, Inc., 510 14th Street, Third Floor, Oakland, CA 94612, csshields@treadwellrollo.com
[2]Associate Project Engineer, Geopier Foundation Company, Inc., 200 Country Club Drive, Suite D-1, Blacksburg, VA 24060, bfitzpatrick@geopiers.com
[3]President, Geopier Foundation Company, Inc., 200 Country Club Drive, Suite D-1, Blacksburg, VA 24060, kwissmann@geopiers.com

INTRODUCTION

The use of *Rammed Aggregate Piers*™ as a cost-effective foundation support option has gained widespread acceptance over the past five years in the United States. The aggregate piers provide settlement control for the support of conventional shallow spread foundations. The design methodology for the rammed aggregate pier system combines design aspects used for shallow spread footings, historical stone columns, and cast-in-drilled-hole (CIDH) concrete shafts. Figure 1 shows a photograph of the Justice Center Parking Garage, Hillsboro, Oregon, which is supported on rammed aggregate piers.

FIG. 1: Structure supported on rammed aggregate piers in granular soils

The design methodology and project-specific performance of the rammed aggregate pier system are well documented in the literature (Lawton et al. 1994, Lawton and Fox 1994, Wissmann et al. 2000). However, the literature contains no detailed study of governing mechanisms or design parameter values to be used in granular soils. This paper presents a study of the rammed aggregate pier modulus test results for 19 piers installed in granular soils. The data is used to provide a database of pier performance categorized by matrix soil characteristics and pier length.

RAMMED AGGREGATE PIER CONSTRUCTION

The construction sequence of the proprietary *Geopier*® rammed aggregate pier system is shown in Figure 2. The aggregate piers are installed by drilling 60 cm (24 inch) to 90 cm (36 inch) diameter holes to depths ranging between 2 m and 8 m (7 feet and 26 feet) below working grade elevations, placing controlled lifts of well-graded highway base course aggregate within the cavities, and compacting the aggregate using a specially designed high-energy beveled impact tamper. The first lift consists of open-graded stone that is rammed into the soil to form a bottom bulb below the excavated shaft. The piers are completed by placing additional 30 cm (12 inch) thick lifts of aggregate over the bottom bulb and densifying the aggregate with the beveled tamper. In granular soils, casing is often required to maintain the stability of the sidewalls of the cavity during construction. A steel casing that is slightly larger than the drill tool is either vibrated or lowered into the cavity during drilling. The casing is withdrawn incrementally during aggregate tamping.

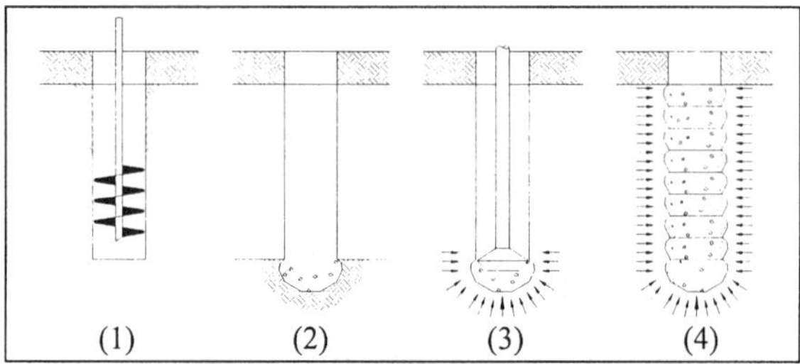

FIG. 2: Rammed Aggregate Pier Construction Process

During densification, the beveled shape of the tamper forces the stone laterally into the sidewall of the drilled cavity. This action increases the lateral stress in the matrix soil, thus providing additional stiffening and increased normal stress perpendicular to the perimeter shearing surface. The development of the increased horizontal pressure and the undulated shape resulting from construction provide an efficient mechanism for the development of shaft resistance along the perimeter of the pier. The installation of the very stiff aggregate piers, which exhibit high internal angles of friction (White et al. 2002, Fox and Cowell 1998), increases the composite shear resistance beneath the foundation, providing an increase in the allowable bearing pressure of the reinforced zone to values ranging from 240 kPa (5,000 psf) to 480 kPa (10,000 psf).

DESIGN METHODS

Rammed aggregate piers are installed in groups beneath conventional shallow foundations to increase the allowable bearing pressure for design and control foundation settlement. The design methodology uses a two-layer settlement approach that consists of evaluating the settlement of both the aggregate pier-reinforced soil (upper zone) and the unreinforced matrix soil (lower zone) below the aggregate pier-reinforced zone. The upper zone design methodology (see Eqs. 1 and 2) are described by Lawton and Fox (1994), Lawton et al. (1994), Lawton and Merry (2000), Wissmann et al. (2000), and Minks et al. (2001).

The settlement in the upper zone is computed using Eq. 1, where q_g is the stress applied to the aggregate piers and k_g is the stiffness modulus of the aggregate pier elements:

$$s_{uz} = \frac{q_g}{k_g} \tag{1}$$

The stress on top of the aggregate piers (q_g) depends on the average bearing pressure of the rigid footing (q), the area coverage of the aggregate piers (R_a), and the stiffness ratio between the aggregate piers and the matrix soil (R_s):

$$q_g = q\left(\frac{R_s}{R_s R_a - R_a + 1}\right) \tag{2}$$

The stiffness ratio, R_s, is the ratio of the stiffness modulus of the rammed aggregate piers (k_g) to the stiffness modulus of the matrix soil (k_m). The stiffness modulus of the aggregate piers is therefore an important parameter value because it plays a role in determining the top-of-pier stress (Eq. 2) and the upper zone settlement (Eq. 1). The stiffness modulus is typically established at each project site with a modulus test.

MODULUS TESTING

Test procedures

The modulus test set-up (Figure 3) is similar to a pile load test configuration and the test is performed in general accordance with ASTM D-1143. Rammed aggregate pier elements outfitted with uplift assemblages or steel anchors are installed to serve as reactions for the test beam. During the installation of the compression test pier, a steel telltale is positioned on top of the bottom bulb with sleeved telltale rods extending to the surface. This allows for deflection measurements to be made at both the top and the bottom of the pier. Following telltale installation, the test pier is constructed in the same manner as production piers. A concrete cap is constructed at the top of the completed pier to provide a platform for the hydraulic jack.

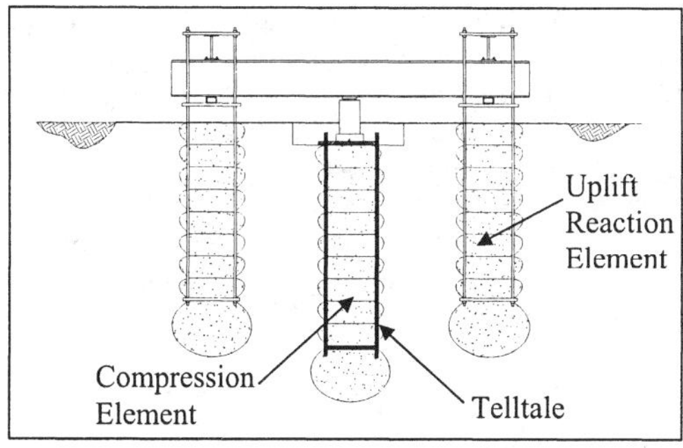

FIG. 3: Modulus Test Setup

The modulus test is performed by applying loads of up to 150 percent of the design stress to the top of the installed aggregate pier. During the application of loads, the deflections of the concrete cap at the top of the pier are measured. The deflections of the bottom of the pier are measured by monitoring the movement of the telltale rods at the surface. The rammed aggregate pier system is unique in the use of the telltale rods to measure deflections at depths within the piers. Plots of the stress versus deflection are constructed from the modulus test results to evaluate the stiffness of the modulus and deformation behavior of the aggregate pier.

Modulus test interpretation

The modulus test affords the opportunity to not only evaluate the stiffness modulus of the pier, but also to identify the governing behavior of the aggregate pier. The relationship between stress and deflection of the aggregate pier is typically characterized by a bi-linear response (Figures 4a and 4b). The stress level at the intersection of the two legs of the bi-linear stress-deflection curve is referred to as the inflection stress (σ_i). At stress levels less than the inflection stress the aggregate pier is characterized by elastic deformation. At stress levels greater than the inflection stress, the pier experiences non-recoverable plastic deformation. For foundation support, the top-of-pier stress for production elements is limited to values less than the inflection stress. Stiffness modulus (k_g) is defined as the ratio of the applied stress (σ) to deflection, and is expressed in units of F/L^3.

The respective movements of the top of the pier and the telltales at levels exceeding the inflection stress provide an indication of the governing deformation mechanism. The two types of modulus test responses are illustrated in the modulus test curves shown in Figures 4a and 4b. For an aggregate pier that undergoes plastic deformation with very little movement of the telltales, as shown in Figure 4a, the post-inflection stress deformation behavior results from radial bulging of the element into the matrix soil. The propensity for bulging is related to the in-situ matrix soil horizontal stress and the shear strength of the matrix soil. In some cases, bulging is attributed to the presence of an interbedded soft layer along the shaft. Bulging behavior is common in soft cohesive soils where resistance to bulging is low or in granular soils with very long shaft lengths where the shaft capacity exceeds the bulging resistance of the stiff or dense matrix soil. For an aggregate pier that undergoes plastic deformation with movement experienced by the telltale, as shown in Figure 4b, the post-inflection stress deformation behavior results from the development of tip stresses at the bottom of the pier. This type of behavior occurs when the applied load exceeds the frictional resistance along the perimeter of the shaft, which is common for short aggregate piers in granular soils where the bulging resistance exceeds the shaft capacity of the short pier.

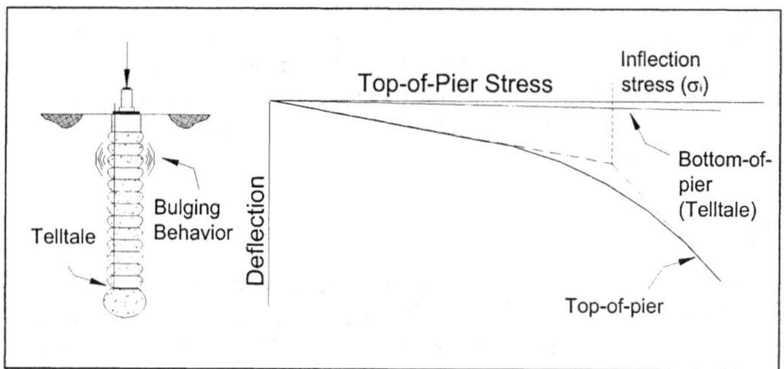

FIG. 4a: Modulus Test Results for Bulging Behavior

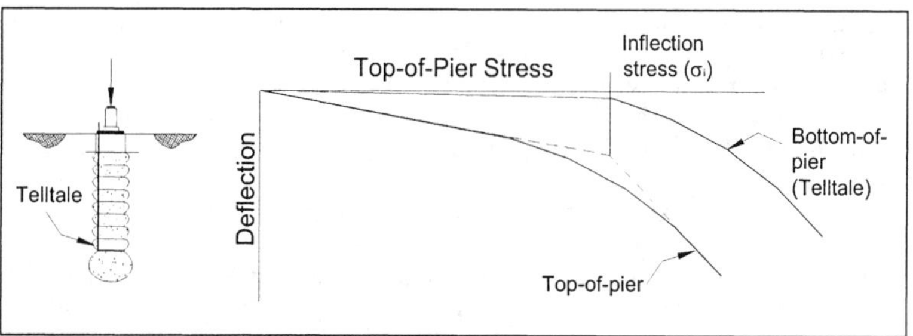

FIG. 4b: Modulus Test Results for Tip-Stress Behavior

Inflection Stress for Piers Characterized by Tip Stresses at High Stress

The estimated inflection stress for piers characterized by the development of tip stresses may be calculated by evaluating the ratio of the total shaft resistance along the perimeter of the pier (Q_s) to the cross-sectional area of the pier (A_g):

$$\sigma_i = \frac{Q_s}{A_g} \quad . \tag{3}$$

The shaft capacity of the pier is calculated as the product of the unit shaft friction (f_s) and the area of the assumed cylindrical shearing surface along the sides of the pier:

$$Q_s = f_s d\pi H_s \quad , \tag{4}$$

where d is the pier diameter and H_s is the shaft length. The unit frictional resistance (f_s) is calculated as the product of the average effective horizontal stress (σ'_h) and the tangent of the matrix soil friction angle (tan ϕ'_m). Past research has shown that the post-construction lateral pressure is increased to a value approximately equal to the Rankine passive earth pressure until a maximum lateral pressure of 120 kPa (2,500 psf) is reached (Handy 2001). As a result of the high lateral stress, significantly greater frictional resistance is developed along the aggregate shaft compared to other ground improvement systems.

CASE HISTORIES

A total of 19 modulus test results were obtained from West Coast project sites in granular soils (Table 1). As shown in Table 1, the majority of the soils are classified as naturally occurring silty sand (SM) or silty sand fill. Some of the sites are characterized by silty sand / sandy silt (SM/ML) soils. Five sites are characterized by poorly graded sand with silt (SP-SM) soils. Average SPT $(N_1)_{60}$-values recorded prior to pier installation within the upper zone of soil requiring aggregate pier reinforcement ranged from 4 blows per foot to greater than 40 blows per foot. The angles of internal friction of the matrix soils (Column 5 of Table 1) for each site are based on correlations with SPT $(N)_{60}$-values (Terzaghi et al. 1996) or project-specific laboratory shear tests when available.

TEST RESULTS AND INTERPRETATION

Inflection Stress for Different Deformation Mechanisms

As shown in Table No. 1, an inflection point was reached in less than half of the modulus tests. For the tests exhibiting an inflection point at high stress levels, two tests showed bulging behavior, while three tests showed the development of tip stresses. The deformation behavior of the remaining two tests exhibiting an inflection point was not apparent because telltale data was not available.

For tests that showed the development of tip stresses, measured inflection point stresses were compared to the calculated inflection stresses using the procedures shown in Eqs. 3 and 4. Figure 5 presents a comparison of the estimated inflection stress and the observed inflection stress. The diagonal line on the figure indicates a 1:1 correlation between the estimated and the observed inflection stresses. The three shaded points represent tests that exhibited inflection points and showed the development of tip stresses. The open-square points represent tests where no inflection point was exhibited because the maximum applied stresses were less than the estimated inflection stress. The open-triangle points represent tests where no inflection point was exhibited even though the applied stress exceeded the estimated inflection stress. The comparison shows reasonable agreement between the estimated and the observed inflection stresses for tests that exhibit tip stresses. In many instances, the estimated inflection stress exceeded the maximum applied stress during testing and no inflection point was observed (open-square points). The lack of an observed inflection stress is indicative of a testing stress level that was not sufficiently high to exceed the shaft resistance or cause plastic deformation of the

TABLE 1: Summary of Modulus Test Results

Proj. No. (1)	Location (2)	ASTM Soil Classification (3)	Average $(N_1)_{60}$[a] (blows/0.3m) (4)	Estimated Matrix Soil Friction Angle (degrees) (5)	Aggregate Pier Diameter (mm) (4)	Shaft Length (m) (5)	Break in curve (Y/N) (6)	Deformation Behavior[c] (TS, B, U) (7)	Inflection Stress[d] (kPa) (8)	Stiffness Modulus (MN/m^3) (9)	Aggregate Pier Elastic Modulus (MPa) (10)
1	Tacoma, WA	Fill (SM)	16	33	760	4.1	Y	B	1124	151	621
2	Tacoma, WA	SM	46	41	610	1.8	N	-	1524	417	762
3	Bellevue, WA	Fill (SM)	22	35	610	2.9	N	-	1524	150	434
4	Manchester,WA	Fill (SM)/SM	6	30	760	2.4	N	-	938	60	147
5	Seattle, WA	Fill (SM)	27	37	610	2.4	N	-	1524	297	723
6	Kent, WA	ML/SM	4	29	760	4.0	N	-	487	73	290
7	Seattle, WA	Fill (SM)	9	31	760	4.4	Y	U	973	96	423
8	Vancouver, WA	Fill (SM/SP)	46	41	760	2.9	Y	TS	1274	257	745
9	Seattle, WA	Fill (SM/ML)	10	32	760	2.4	Y	B	919	82	201
10	Hillsboro, OR	SM/ML	12	32	840	2.7	Y	TS	763	98	268
11	Hillsboro, OR	SM/ML	12	32	840	1.8	Y	TS	551	72	131
12	Roseville, CA	SM/SC	18	34	610	2.1	Y	U	1032	239	510
13	Sherwood, OR	SM/SP-SM	15	33	760	2.7	N	-	957	140	383
14	San Diego, CA	Fill (SC)	50/0.15m	42	760	5.2	N	-	1340	403	2087
15	West Covina,CA	Fill (SC)	30	37	760	2.4	N	-	969	98	239
16	Palm Desert, CA	SP/SM	33[b]	40	760	2.4	N	-	1122	260	634
17	Salem, OR	Fill (ML/SM)	17	34	760	2.7	N	-	1178	157	430
18	Anaheim, CA	SM/SP	15	34	760	2.4	N	-	1178	238	580
19	Anaheim, CA	SM/SP	15	34	760	2.4	N	-	1178	211	514

[a] SPT $(N_1)_{60}$-value not corrected for fines content.
[b] Average SPT $(N_1)_{60}$-value obtained from correlations with CPT tip resistances (Robertson and Campanella 1989).
[c] TS: Tip stress, B: Bulging, U: Unavailable for lack of telltale data.
[d] Maximum recorded stress when inflection point not reached.

pier. An inflection point would have eventually been observed in all cases if a higher stress were applied during the testing. Additional data also indicate that applied stress levels exceeded the estimated inflection stress without reaching the inflection stress (open-triangle points). The inflection stress calculations for these tests were conservative, yielding inflection stress estimates that were exceeded in the field tests.

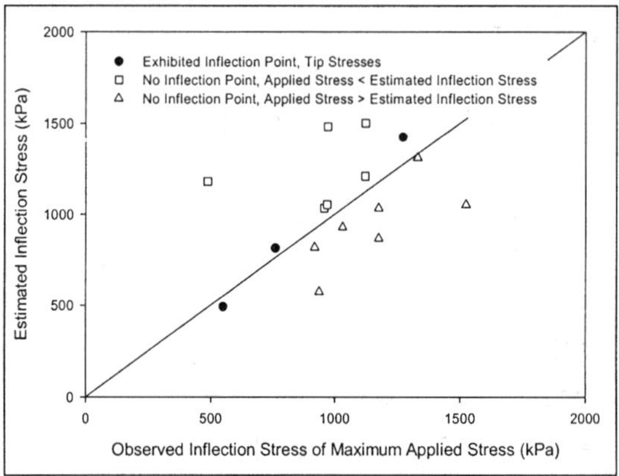

FIG. 5: Comparison of Estimated Inflection Stress and Observed Inflection Stress or Maximum Applied Stress

Figure 6 shows the relationship between the inflection stress and matrix soil friction angle in order to estimate the propensity for bulging. The two shaded points represent tests that exhibited inflection points and showed bulging. The open-square points represent tests that exhibited inflection points and showed the development of tip stresses. The open-triangle points represent tests where no inflection point was exhibited because the maximum applied stresses were not sufficiently large to induce bulging. The line that intersects the two points that represent tests exhibiting bulging deformations provides an empirical trend that relates propensity for bulging at different friction angles.

Figure 7 provides a tool that can be used to distinguish between piers that have the propensity for the development of tip stresses and piers that have the propensity for bulging. The near-vertical lines present contours of *predicted* inflection point stresses corresponding to the development of tip stresses as calculated using Eqs. 3 and 4. The single bold diagonal line represents the *empirical* relationship that describes the bulging propensity for different values of matrix soil friction angle from Figure 6. For a given matrix soil friction angle and shaft length, the propensity for tip stresses exists for a point located to the left of the bulging line, while a point located to the right of the bulging line is expected to bulge prior to exhibiting tip stresses. The plot suggests that the propensity for bulging increases with increasing shaft length and decreasing matrix soil friction angle.

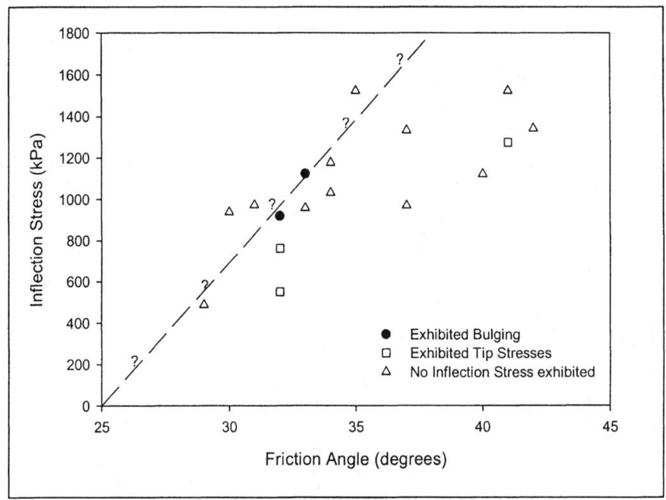

FIG. 6: Bulging Propensity in Granular Soils

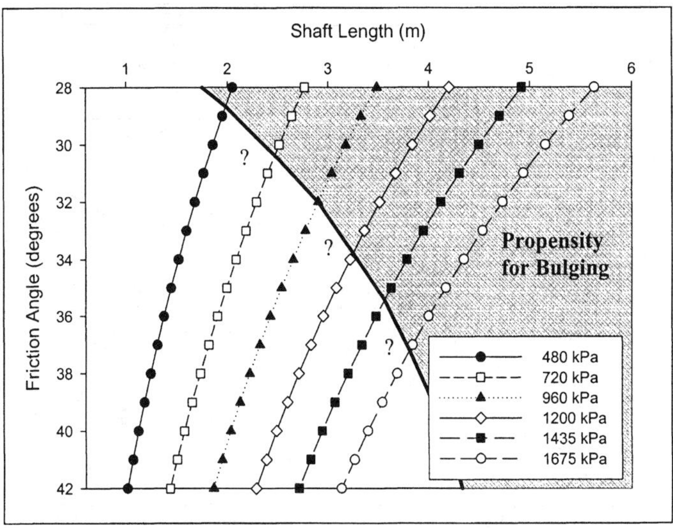

FIG. 7: Inflection Stress Prediction Chart

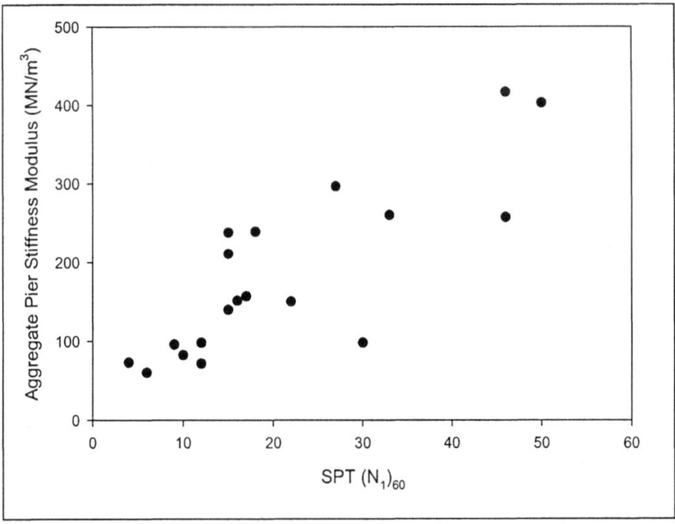

FIG. 8: Aggregate Pier Stiffness Modulus versus SPT $(N_1)_{60}$

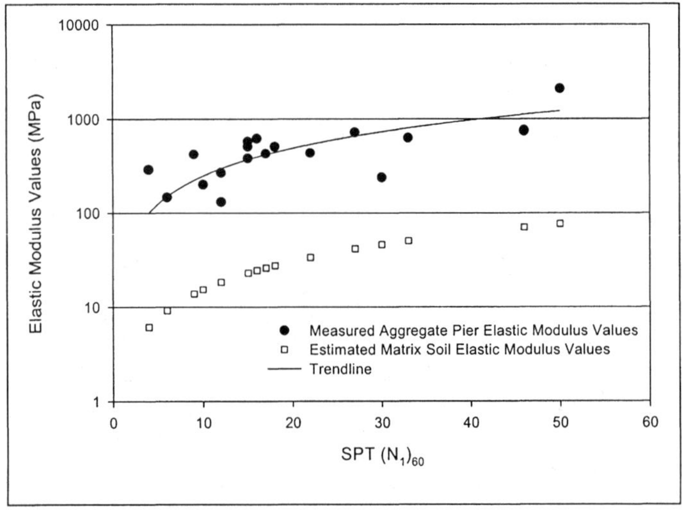

FIG. 9: Elastic Modulus Values versus SPT $(N_1)_{60}$

Aggregate Pier Stiffness

The plot shown in Figure 8 describes the relationship between the aggregate pier stiffness modulus and the measured SPT $(N_1)_{60}$-value of the unreinforced matrix soil prior to pier installation. The data clearly show that the pier stiffness modulus increases with the matrix soil $(N_1)_{60}$-value.

Figure 9 provides a comparison of the *measured* elastic modulus values of the rammed aggregate piers with the *estimated* matrix soil elastic modulus values. The elastic modulus value of the aggregate pier is conservatively estimated as the product of the stiffness modulus and the shaft length. The matrix soil elastic modulus value is calculated based on correlations between SPT N-values, CPT tip resistances, and elastic modulus values as reported by Robertson and Campanella (1989). The ratio between the measured aggregate pier elastic modulus values and the estimated matrix soil elastic modulus values ranges from 5 to 45. The largest ratios occur at sites exhibiting the lowest matrix soil SPT N-values.

SUMMARY AND CONCLUSIONS

The use of rammed aggregate pier elements to provide cost-effective support of shallow foundations has gained wide-spread acceptance over the past few years. Modulus tests are used to evaluate the inflection stress, measure the stiffness modulus, and identify the deformation behavior of rammed aggregate piers installed in granular soils. This paper presents a summary of 19 modulus test results for rammed aggregate pier elements in granular soils from project sites on the West Coast of the United States. The test results and interpretation indicate the following:

1. Aggregate piers subjected to high stresses deform either by developing tip stresses or by bulging radially.

2. A comparison of the estimated inflection stress and the observed inflection stress is shown in Figure 5. The data indicate the inflection stress may be estimated using Eqs. 3 and 4. This approach to estimate the inflection stress consistently predicts the development of tip stresses at high applied loads as indicated by the strong correlation between the estimated and observed inflection stresses.

3. Theoretical approaches and site-specific modulus test results confirm the inflection stress values for piers not subject to bulging increase with increasing shaft length and with increasing matrix soil SPT $(N_1)_{60}$-values.

4. An empirical trend that describes the bulging propensity for aggregate piers is shown in Figure 6. The trend suggests that the propensity for bulging decreased with increasing matrix soil friction angle for a given shaft length.

5. The stiffness modulus of rammed aggregate piers is related in large degree to the matrix soil SPT $(N_1)_{60}$-value.

6. Measured elastic modulus values for rammed aggregate piers are on the order of 5 to 45 times greater than estimated elastic modulus values for matrix soils. The installation of the significantly stiffer aggregate piers provides a greater degree of settlement control and predictability.

ACKNOWLEDGEMENTS

Modulus test data and soil information used for this research was made available by Geopier Foundation Company, Inc., Geopier Foundation Company – Northwest and Geopier Foundation Company – West.

REFERENCES

Fox, N.S., and Cowell, M.J. (1998). Geopier Foundation and Soil Reinforcement Manual. Geopier Foundation Company, Inc., Scottsdale, AZ.

Handy, R.L. (2001). "Does Lateral Stress Really Influence Settlement?" ASCE Journal of Geotechnical and GeoEnvironmental Engineering. Vol. 127 No. 7, July.

Lawton, E. C., and N. S. Fox (1994). "Settlement of structures supported on marginal or inadequate soils stiffened with short aggregate piers." *Geotechnical Specialty Publication No. 40: Vertical and Horizontal Deformations of Foundations and Embankments*, ASCE, 2, 962-974.

Lawton, E. C., N. S. Fox, and R. L. Handy (1994). "Control of settlement and uplift of structures using short aggregate piers." *In-Situ Deep Soil Improvement,* Proc. ASCE National Convention, Atlanta, Georgia, 121-132.

Lawton, E.C. and Merry, S.M. (2000). "Performance of Geopier® Supported Foundations During Simulated Seismic Tests on Northbound Interstate 15 Bridge Over South Temple, Salt Lake City." Final Report No. UUCVEEN 00-03. University of Utah. December.

Minks A.G., Wissmann, K.J., Caskey, J.M., and Pando, M.A. (2001). "Distribution of Stresses and Settlements Below Floor Slabs Supported by Rammed Aggregate Piers." Proceedings, 54th Canadian Geotechnical Conference, Calgary, Alberta. Sept. 16-19.

Robertson, P.K. and Campanella, R.G. (1989). "Guidelines for Geotechnical Design Using the Cone Penetrometer Test and CPT with Pore Pressure Measurement." Hogentogler and Com., Gaithersburg, MD 20877, Fourth Edition, pp. 193.

Terzaghi, K., Peck, R.B., Mesri, G. (1996). Soil Mechanics in Engineering Practice. John Wiley & Sons, Inc., New York, NY. p.151.

White, D.J., Suleiman, M.T., Pham, H.T., and Bigelow, J. (2002). "Shear Strength Envelopes for Aggregate used in Geopier® Foundation Construction." Final Report. Iowa State University. September.

Wissmann, K.J., N.S. Fox, and J.P. Martin (2000). "Rammed Aggregate Piers Defeat 75-foot Long Driven Piles." *Performance Confirmation of Constructed Geotechnical Facilities.* ASCE Geotechnical Special Publication No. 94. April 9-12. Amherst Massachusetts.

GROUND IMPROVEMENT UTILIZING VIBRO-CONCRETE COLUMNS

Raymond Mankbadi[1], P.E., Member, ASCE, Geo-Institute; Jack Mansfield[2], P.E., Member, ASCE, Geo-Institute; Ragui Wilson-Fahmy[3], P.E., Ph.D., Member, ASCE, Geo-Institute; Sherif Hanna[4], Associate Member, ASCE; and Vedrana Krstic[5], Ph.D., Associate Member, ASCE.

ABSTRACT: The use of Vibro-Concrete Columns (VCC) to improve soft ground is growing in popularity in various parts of the world due to its cost effectiveness and proven performance. A case history involving the first use of VCC by a US Department of Transportation to support an approach embankment is presented. The project involves the installation of VCC supporting 10 m high back-to-back mechanically stabilized earth walls underlain by a geotextile-reinforced sand platform. The design concept of the embankment supporting system is outlined. Static load tests results performed on two VCC are also presented and the acceptance criteria are discussed. The performance of the embankment supporting system is assessed based on monitoring data obtained from various instruments installed during construction including settlement platforms, probe extensometers, inclinometers, piezometers and strain gages attached to the sand platform geotextile reinforcing elements. Finally, recommendations are presented regarding the design aspects of the VCC and the geotextile-reinforced sand platform.

---

[1] Department Head, Geotechnical Department, Parsons Brinckerhoff, Inc., 506 Carnegie Center Blvd., Princeton, NJ 08540, mankbadi@pbworld.com

[2] Manager, Geotechnical Unit, New Jersey Department of Transportation, 1035 Parkway Avenue, Trenton, NJ 08625

[3] Lead Geotechnical Engineer, Parsons Brinckerhoff, Inc., 506 Carnegie Center Blvd.,Princeton, NJ 08540

[4] Geotechnical Engineer, Parsons Brinckerhoff, Inc., 506 Carnegie Center Blvd.,Princeton, NJ 08540

[5] Senior Geotechnical Engineer, Parsons Brinckerhoff, Inc., 506 Carnegie Center Blvd.,Princeton, NJ 08540

INTRODUCTION

Ground Improvement using Vibro-Concrete Columns (VCC) represents a cost effective solution for supporting highway approach embankments (Mitchell et al. 1998). The performance of embankments supported on soft soils using VCC has been extensively investigated during the past decade (Hussin 1994; Guido et al. 1987). Overall soil and column responses during installation have been measured in details (Munfakh, 2000). However, a number of difficulties related to highway applications, including performance prediction methods, the lack of well documented case histories, and standardized procedures for quality control, still need to be addressed. This paper expands upon previous literature by providing additional case history details supplemented by observations obtained from VCC load tests and instrumentation data.

The case history involves the construction of an approach embankment of a new bridge. The embankment is underlain by soft organic soil. The bridge replaces an eighty years old existing single leaf movable wooden bridge in carrying US Route 9 over Nacote Creek. The new bridge is a fixed high level concrete bridge with five 30.5 meter spans supported on four piers. The project is located in US Route 9 Section 15D, Galloway Township, New Jersey, one mile south of the Garden State Parkway. It should be noted that the National Partnership for Highway Quality (NPHQ) has awarded its 2003 National Achievement Award to this project.

The new alignment of Route 9 was originally planned to coincide with the existing roadway alignment. The existing roadway approach embankments are underlain by granular fill material overlying alternating layers of granular and cohesive soils. It is not known whether the granular fill resulted from soil replacement or soil displacement of the organic material known to exist in this area. It was required to shift the alignment along the south approach due to the existence of a historical building, active marina, wetlands, and parklands. The alignment shift resulted in supporting the embankment partially on fill material (old alignment) and partially on marshlands as shown in Figure 1.

The marshlands in the area consist of 3 to 6 meters of organic silt and peat sediments. Accordingly, ground improvement was necessary for the south approach to accommodate the above situation. Different ground improvement techniques were evaluated and it was concluded that the Vibro-Concrete Column technique was the most appropriate solution.

The south approach embankment at the south abutment location consists of 10 meters high by 16 meters wide back-to-back mechanically stabilized earth walls (MSE). The walls are supported on one meter thick geotextile-reinforced sand platform. The fill load is transmitted to the VCC through the sand platform. A schematic of the supporting platform is shown in Figure 2. It should be noted that geogrid could be used in lieu of geotextile.

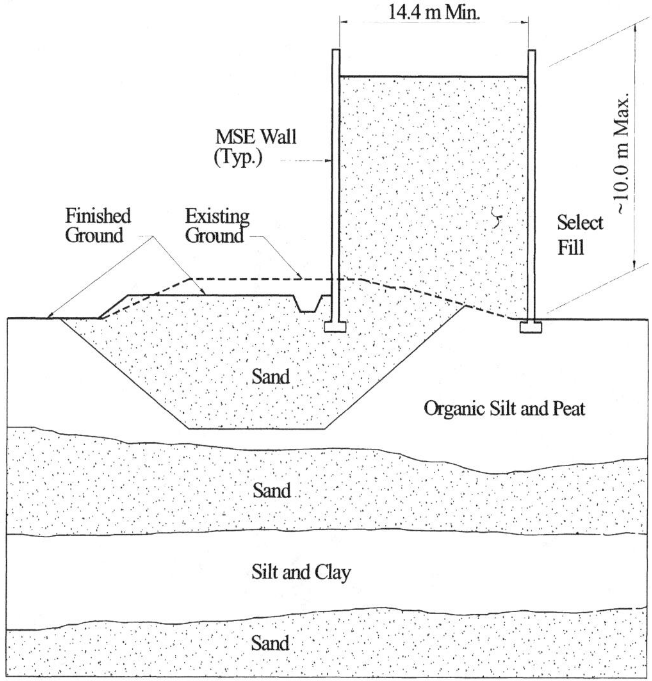

FIG. 1. Schematic of Site Conditions

This paper presents the design concept of the VCC solution. The results of two load tests conducted to evaluate the degree of success of the VCC installation technique are also presented. Prediction of the design capacity of the VCC during the design phase is assessed in view of the ultimate capacity determined from the load tests. Finally, performance of the VCC-sand platform system is evaluated based on extensive instrumentation monitoring data of ground deformation and geotextile strains.

SUBSURFACE CONDITIONS

The site is located within the Atlantic City Plain Geological Province (Widmer 1964). The soils in the Nacote Creek area can be generally categorized as Pleistocene lowland alluvial deposits, overlying Tertiary marine sediments. In the majority of the site, the alluvial deposits are covered by recent marsh soils.

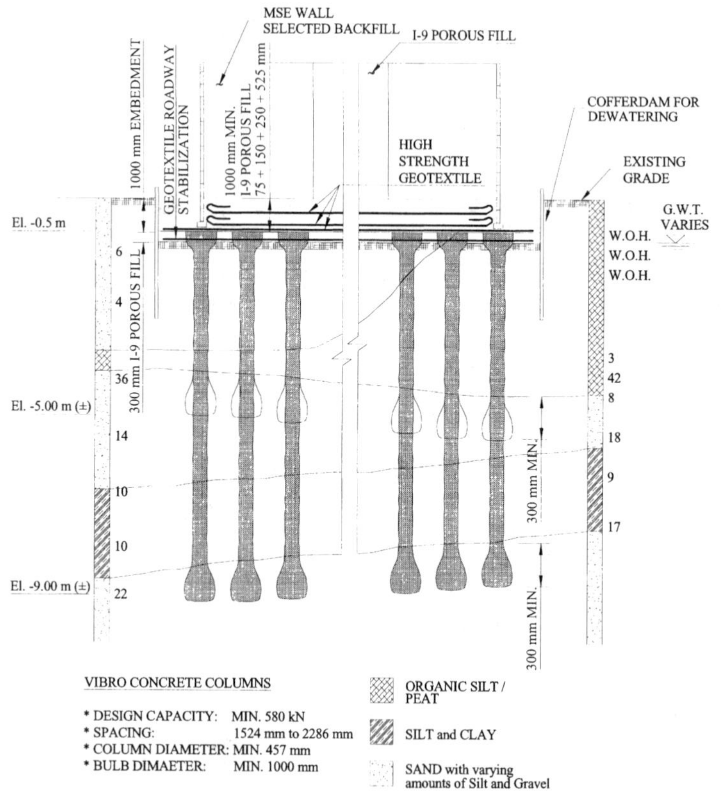

FIG. 2. Typical Cross Section

The tidal marsh deposits in the Atlantic County area consist of compressible organic silt and peat extending to a depth of 3 to 9 meters. Underlying Pleistocene alluvial deposits are mapped as the Cape May formation. The formation is composed of layers of yellow-brown to gray-white quartz sand and gravel, occurring in well defined alternating or intermixed layers. Silts and Clays are encountered in lenses and pockets. The Cohansey Sand formation underlies the alluvial deposits. This formation is a Tertiary marine deposit consisting primarily of medium to fine silty sand, white to yellow in color, with localized gray to brown clay layers and lenses. Typical soil profile encountered along the east side of the south approach is given in Figure 3.

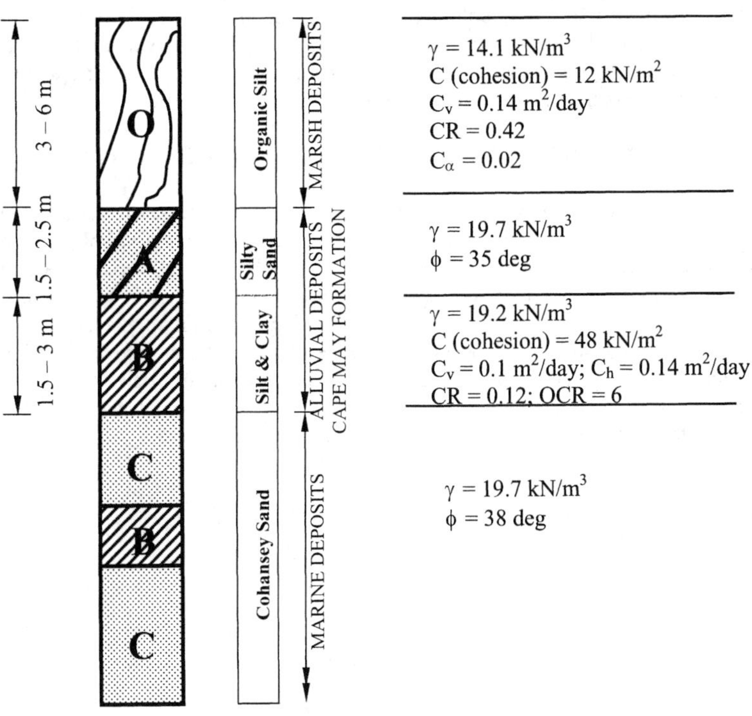

FIG. 3. Typical Soil Profile and Design Parameters

Based on the results of the subsurface exploration program and laboratory testing program, design soil parameters were determined as shown in Figure 3. Based on such design parameters, settlement and stability analyses were performed at several locations along the alignment of the south approach. Analysis results showed that the MSE wall constructed on virgin ground may experience differential settlement of up to 1.2 meters. These differential settlements are considered excessive for MSE walls; accordingly, ground improvement and treatment were required.

EVALUATION OF GROUND IMPROVEMENT OPTIONS

Several ground improvement techniques were explored for supporting the south approach embankment. Soil replacement with competent granular material was excluded due to environmental restrictions. Preloading embankments were eliminated due to excessive construction easement requirements. Ground improvement using stone columns would result in intolerable differential settlement. Soil improvement using soil mixing was considered. However, laboratory test results showed that the maximum compressive stress of the soil-cement-sand mixtures were not adequate for embankment support.

VCC DESIGN CONCEPT AND THEORETICAL CONSIDERATIONS

The main concept consists of bypassing the organic material that is not suitable for supporting the embankment using the Vibro-Concrete Column technique. The Vibro-Concrete Columns are constructed using a probe called "Vibroflot" which penetrates the cohesive materials under its self-weight aided, if needed, with a hydraulic thrust. This is accompanied by vibration of the probe to facilitate its inclusion into the soil. After penetration of the organic material into the suitable bearing stratum, concrete is introduced into the ground from the tip of the vibrator, and a continuous shaft is constructed to the ground level. Schematic of the construction procedure is given in Figure 4.

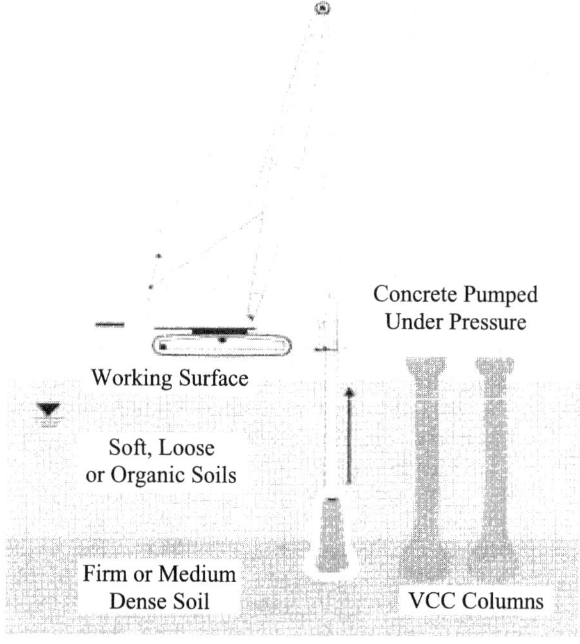

FIG. 4. Schematic of the Construction Procedure (Courtesy of Hayward Baker, Inc.)

The main advantage of the system is that it combines advantages of in-situ ground improvement vibro system with the load carrying characteristics of piles. The VCC can be installed with a relatively large diameter bottom bulb in order to increase its end bearing capacity. An enlarged head can also be formed with the purpose of reducing the exposed area of the weak soil below the embankment fill. This would lead to a better load transfer of the weight of the fill to the VCC through an overlying geotextile reinforced sand platform.

Because the method of installation of the VCC involves densification of the sandy soil during the formation of the bottom mushroom, the end bearing capacity can be closer to that of a driven pile than that of a drilled shaft. A value of 73 for the bearing capacity factor, Nq, was assumed in the calculations based on the Nq value given in DM-7.2 (NAVFAC, 1986) for a soil friction angle of 35 degrees. It involves an arbitrary reduction of 15% from the value of 86 given in DM-7.2. The resulting ultimate capacity was 1100 kN (250 kips), neglecting skin friction, for a 4.5 meter long column having its tip in sandy stratum A. Spacing between the VCC was chosen such that the maximum load on the column does not exceed 580 kN (130 kips) which corresponds to a factor of safety of 1.9 against bearing capacity failure.

This factor of safety was considered adequate taking into account the fact that installation of relatively closely spaced VCC would result in substantial improvement of the soil in between the columns. Accordingly, it can be postulated that the VCC and the surrounding soil will act, to some extent, as one body. This renders the situation to a settlement problem rather than a bearing capacity problem.

It was decided to increase the length of the columns to 8.5 meters within a 60 meters long stretch behind the south abutment where the fill is highest. By doing so, the columns will have their tips in sandy stratum C thereby minimizing potential consolidation settlement of lower stratum B deposits. With this configuration of short and long columns, it was estimated that the total settlement of the MSE back-to-back walls along the south approach would be in the range of 20 to 50 millimeters with the elastic settlement ranging between 12 and 20 millimeters and the consolidation settlement ranging between 8 and 30 millimeters. A maximum differential settlement of 25 millimeters was expected at the location of the change in column length, and was anticipated to be accommodated by the flexible structure of the MSE wall. The settlement was expected to occur over a period of 15 to 30 days after constructing the full height of the fill.

The geotextile-reinforced sand platform overlying the VCC was designed assuming that the geotextiles will deform sufficiently to promote arching between the Vibro-Concrete Columns. As a result, the soil weight will be transferred to the relatively unyielding Vibro-Concrete Columns. The load transmitted by the geotextiles to the columns will induce tensile stresses which should be considered in determining the required mechanical properties of the geotextiles. Design methods are available to determine the vertical pressure carried by the geotextile and the resulting geotextile tensile stresses. One such method which is commonly adopted is recommended in the current British Standard BS 8006, Code of Practice for strengthened/reinforced soils and other fills to design geosynthetics over piles.

In addition to transmitting the fill load to the Vibro-Concrete Columns, the geotextiles should be sufficiently strong to resist the frictional force at their interface with the overlying MSE wall which is induced by the lateral earth pressure acting at the back of the wall. It is not desirable to transmit a large portion of the frictional force to the Vibro-Concrete Columns considering the fact that their top portion is embedded in weak organic soil and, hence they have relatively low lateral capacity.

VCC LOAD TESTS

Accurate design and prediction methods for both ultimate column resistance and load-deformation characteristics are limited. The main unknown parameters are related to the installation procedures, which have significant effect on altering and/or changing the initial in-situ properties for the cohesive and cohesionless soils encountered. Therefore, a load testing program was essential for this particular site. A subsequent paper will discuss in details the quality control for VCC installation and the difficulties encountered during construction.

The goals of the load testing program were to:

1. Establish an installation procedure, based on columns performance;
2. Verify that the VCC is capable of sustaining the design axial load.

Two sacrificial demonstration test columns were installed before constructing production columns. One static load test was conducted on a short column (4.5m long), and one on a long column (8.5m long). The short column top and bottom elevations were El.-0.5 m and El.-5.0 m respectively. The long column top and bottom elevations were El. -0.5 m and El. -9.0 m respectively. Locations of test columns were on the wetland side of the south approach, where relatively thick soft organic silt and peat layers exist.

The load tests were performed according to ASTM D 1143-81. Each test load increment was held until column movement reached 0.25 mm/hr, but not more than two hours per increment up to the design load. The definition of failure for the short column corresponded to the applied load that resulted in excessive settlement without increase in load or a sudden decrease in jack pressure for the same settlement.

Figures 5 and 6 show the load–settlement curves for the short and the long columns respectively. As depicted from the figures, the ultimate capacities predicted using the Davisson (1972) and Kyfor et al. (1992) criteria are as presented in Table 1.

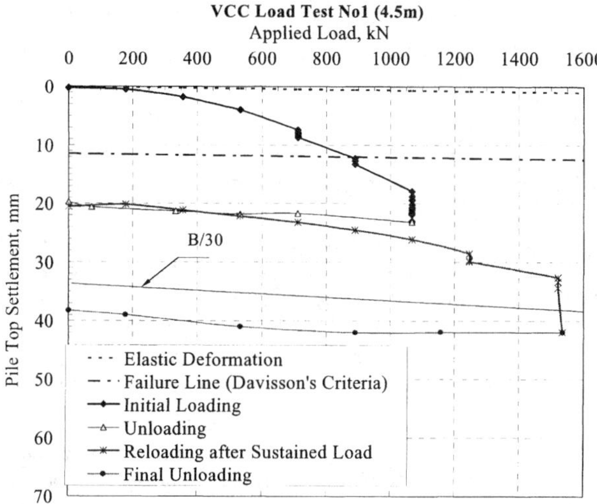

FIG. 5. Load-Settlement Curve for Short Concrete Column

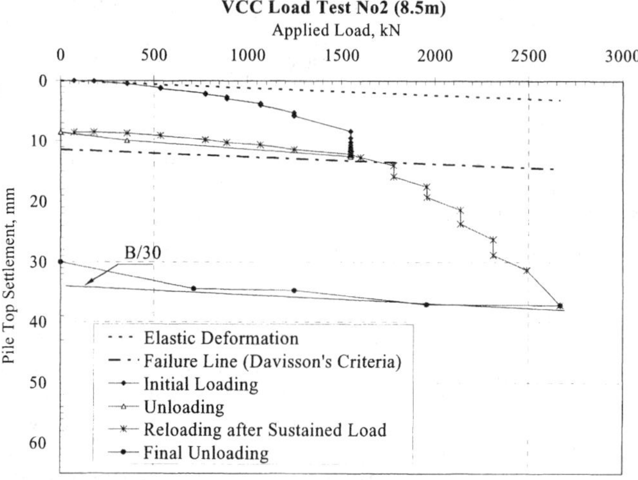

FIG. 6. Load-Settlement Curve for Long Concrete Column

TABLE 1. Predicted VCC Ultimate Capacities

Column Type	Ultimate Capacity (kN)		
	Davisson Criterion	Kyfor et al.	Actual Failure Load
(1)	(2)	(3)	(4)
Short	900	N/A	1530
Long	1800	2600	N/A

Note that Davisson criterion is applicable to driven piles of diameters less than 0.6 meters and the criterion by Kyfor et al. (1992) is applicable to larger diameter driven piles. Since the estimated bottom bulb of the VCC is about one meter in diameter, applying the Davisson criterion would be overly conservative resulting in a factor of safety against end bearing capacity of 1.55 for the short column and 3.1 for the long column. Considering the fact that much higher factors of safety would be obtained using the actual failure load or the criterion by Kyfor et al. (1992), it was concluded that the VCCs are adequate to carry the applied loads.

In addition, as previously mentioned, it is logical to assume that installation of relatively closely spaced VCC would result in substantial improvement of the soil constrained with the columns and hence it can be postulated that the VCC and the surrounding soil will act to some extent as one body under the embankment fill. This renders the situation to a settlement problem rather than a bearing capacity problem.

INSTRUMENTATION AND PERFORMANCE EVALUATION

Settlement platforms, probe extensometers, inclinometers, piezometers, and strain gages attached to the geotextiles were used to monitor the VCC system performance.

Most instruments were located on the east (wetland) side of the south approach embankment where the thickest organic layer exists. Instrumentation monitoring was performed during and after construction of the approach embankment.

It is beyond the scope of this paper to discuss in details the instrumentation monitoring data. However, it may be noted that the maximum settlement of the embankment was 40 mm. This value is close to the maximum predicted settlement of 50 mm. The maximum lateral deformation of the ground between the VCC located on the wetland side determined from the inclinometer data was 13 mm. No predictions were made for the lateral deformation. However, it is felt that this is a relatively small value indicating stability of the system. The maximum strain in the geotextile was 1%. It occurred in the transverse direction where the embankment fill is highest. A one percent strain corresponds to a force of about 75 kN/m in a wide width tension test conducted on the geotextile used in this project. Considering the fact that the tension test is carried out at a much faster rate than that corresponding to the actual construction rate, it can be postulated that the actual load at 1% strain would be no greater than 75 kN/m, which is lower than the design value of 140 kN/m. It is known that the slower the tension load is applied, the flatter the load-extension curve would be.

Based on the above mentioned instrumentation data and all other data obtained during and after construction of the embankment fill, it can be concluded that the VCC solution was adequate for supporting the embankment fill.

CONCLUSIONS

The following conclusions can be made regarding the Vibro-Concrete Column concept:

1. The Vibro-Concrete Column solution is an effective and viable solution where approach embankments are to be constructed over soft ground within a limited right of way.
2. Vibro-Concrete Columns can be designed in a similar fashion to driven piles relying on their end bearing capacity. This is confirmed by load tests conducted on two demonstration shafts.
3. It is logical to assume that installation of relatively close spaced VCC would result in substantial improvement of the soil in between the columns and hence it can be postulated that the VCC and the surrounding soil would act to some extent as one body under the embankment fill. This renders the situation to a settlement problem rather than a bearing capacity problem.
4. In order to account for soil densification around and below the columns, it is recommended to perform the load test on a column within an array of already installed columns.
5. Instrumentation monitoring data indicated that the Vibro-Concrete Column solution in conjunction with a geotextile reinforced sand platform can be successful in supporting high embankments.

ACKNOWLEDGMENT

The paper is based on the design and construction work of Parsons Brinckerhoff, Inc., New Jersey Department of Transportation, Hayward Baker, Inc., and JH REID General Contractor. We are grateful to Tony Guerrieri, Bruce Riegel, Kuangyu Yang and Kim Sharp of NJDOT and Joseph Cavey of Hayward Baker for their support and dedication. We are also grateful to Nassef Soliman and Marty Maloney from Parsons Brinckerhoff for their valuable comments and suggestions.

REFERENCES

ASTM D 1143 – 81 (1997). "Standard Test Method for Piles Under Static Axial Compressive Load", American Society for Testing and Materials (ASTM), Annual Book of ASTM Standards, Section 4, Volume 04.08 Soil And Rock (I), 95-105.

Barksdale, R. D., and Bachus, R. C. (1983). "Design and Construction of Stone Columns", *FHWA Report No. RD-83/027.*

British Standard BS 8006 (1995). "Code of practice for strengthened/reinforced soils and other fills", BSI, London.

Davisson, M. T. (1972). "High Capacity Piles", *Proc. Soil Mechanics Lecture Series on Innovations in Foundation Construction*, ASCE, Illinois Section, Chicago, 81-112.

Guido, V. A., and Sweeney, M. A. (1987). "Plate Loading Tests on Geogrid and reinforced Earth Slabs", Proc. of Geosynthetics, Vol. 1.

Hussin, J. D. (1994). "Ground Modification with Vibro Concrete Columns", ASCE South Florida Section Meeting, Ft. Lauderdale, FL.

Kyfor, Z. G., Schnore, A. S., Carlo, T. A., and Bailey, P. F. (1992). "Static Testing of Deep Foundations", *FHWA Report No. SA-91-042*, U.S. Department of Transportation, Federal Highway Administration, Office of Technology Applications, Washington, D. C., 174.

Mitchell, J., Cooke,H., and Schaeffer, J. (1998) "Design Considerations in Ground Improvement for Seismic Risk Mitigation" Geotechnical Special Publications No. 75, *Geotechnical Earthquake Engineering and Soil Dynamics III.*

Munfakh, G. (2000) "Ground Improvement: Design and Construction", Journal of Ground Improvement, No-2, Vol. 4.

NAVFAC Design Manual DM-7.2 (1986). "Foundations and Earth Structures".

Widmer, K. (1964). "The Geology and Geography of New Jersey", D. Van Nostrand Co., Inc.

MODELING OF THE SEISMIC RESPONSE OF THE AGGREGATE PIER FOUNDATION SYSTEM

Christian H. Girsang[1], Marte S. Gutierrez[2], Member, ASCE, and Kord J. Wissmann, P.E.[3]

ABSTRACT: There are a lot of ground improvement techniques that have been developed. One of them is the aggregate pier foundation system. The response of an aggregate pier foundation system during seismic loading was investigated. Comprehensive numerical modeling using FLAC were performed. The research was divided into three parts: 1) ground acceleration, 2) excess pore water pressure ratio, and 3) shear stress distribution in the soil matrix generated during seismic loading. Two earthquake time histories scaled to different maximum acceleration (pga) were used in the numerical modeling: the 1989 Loma Prieta earthquake (pga = 0.45g) and the 1988 Saguenay earthquake (pga = 0.05g). The results of the simulation showed that: 1) the aggregate pier amplifies the peak horizontal acceleration on the ground surface (a_{max}), 2) the aggregate pier reduces the liquefaction potential up to depth where it is installed, 3) pore pressures are generally lower for soils reinforced with aggregate pier than unreinforced soils, except when the applied shear stresses exceed the cyclic shear resistance of the aggregate materials, and 4) the maximum shear stresses in soil are much smaller for reinforced soils than unreinforced soils.

INTRODUCTION

One of the most dramatic cause of damage of structures during earthquakes is the development of liquefaction in saturated cohesionless deposits. Because of the damages that are caused by liquefaction, specialized construction procedures have

[1] Head of Soil Mechanics Laboratory, Pelita Harapan University, Department of Civil Engineering, UPH Tower, Lippo Karawaci, Tangerang 15811, Banten, Indonesia, c_girsang@yahoo.com

[2] Associate Professor, Virginia Polytechnic Institute and State University, The Charles E. Via, Jr., Department of Civil and Environmental Engineering, 200 Patton Hall, Blacksburg, VA 24061-0105

[3] President, Geopier Foundation Company, Inc., 200 Country Club Drive, Suite D-1, Blacksburg, VA 24060

been developed to reduce them. Various techniques have been developed to densify the liquefiable soil and to provide drainage path to accelerate pore pressure dissipation during seismic loading. Aggregate pier foundation system is one of the examples of ground improvement techniques that have been used to reinforce matrix soils and increase the resistance to liquefaction.

The use of aggregate pier foundation system has been gaining wide acceptance in the past decades. Its capabilities to increase the bearing pressure of weak soil, to reduce settlement, and to provide uplift capacity have been studied extensively (e.g., Lawton and Fox 1994; Fox and Edil 2000). Because of the success of aggregate pier foundation system in improving the static response of foundations, there has been an increasing interest to use aggregate pier to improve the seismic response of ground. The behavior of aggregate piers under seismic loading has been a subject of recent research and is summarized by Lawton and Merry (2000).

The typical construction process of aggregate pier foundation system can be divided into four main stages:

1. A cylindrical or rectangular prismatic (linear) cavity in the soil matrix is created by augering or trenching,
2. Aggregate (clean stone) is placed at the bottom of cavity,
3. A bottom bulb is constructed by ramming the aggregate with tamper, which has 45° beveled foot, and
4. The shaft is constructed with undulating layers in thin lifts (30 cm or less) consisting of well-graded or open-graded aggregate, typically stone as used for highway base course material.

NUMERICAL MODELING

Comprehensive numerical modeling using FLAC (FLAC 2000), a finite difference computer code, were performed. For this research, the matrix soil was modeled as an elasto-plastic material with a Mohr-Coulomb failure criterion. The Mohr-Coulomb criterion incorporates dilation at failure but not densification during cyclic loading at stress below failure. Modification of the Mohr-Coulomb model was done to simulate volumetric strains in drained cyclic loading or pore pressures in undrained cyclic loading. The changes in volumetric strains or pore pressures were modeled using the Finn model (Martin et al. 1975). The aggregate pier was represented by elements in the FLAC grids that were assigned a Mohr Coulomb response without including the Finn model modification.

In dynamic analyses, the silent boundaries and free field boundaries (FLAC 2000) were applied so that the outward waves propagating from inside the model may be properly absorbed by the side boundaries.

Figures 1 and 2 show the grid generation used in the numerical model. The darker color indicates the aggregate pier elements and the lighter color shows the soil elements. The models simulate reinforced "cells" of soil that are 2.5 m wide with two different heights: 4.5 m for Figure 1 and 8 m for Figure 2.

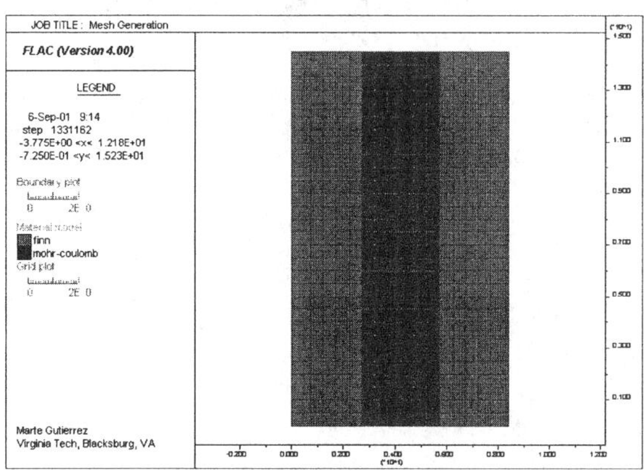

FIG. 1. Grid generation 1

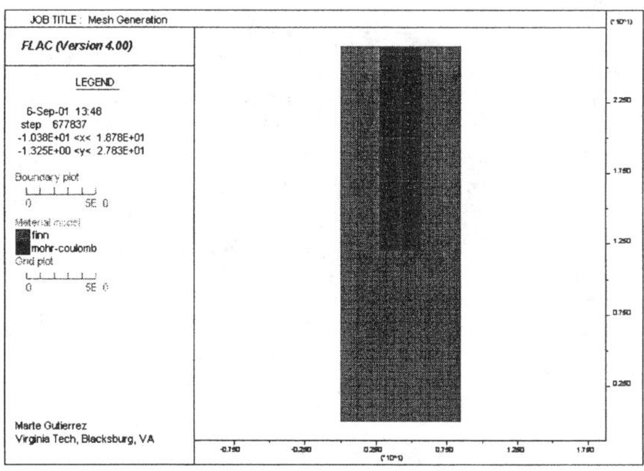

FIG. 2. Grid generation 2

The following strong motions from two earthquakes were used in the numerical modeling:

1. The 17 October 1989 Loma Prieta earthquake as recorded at Corralitos station, California with magnitude (Mw) of 7.1 and peak ground acceleration of 0.64g scaled down to 0.45g.
2. The 25 November 1988 Saguenay earthquake as recorded at Chicoutimi-Nord, Quebec with magnitude (Mw) of 6.0 and peak ground acceleration of 0.05g.

Both earthquake records were filtered and baseline corrected following the procedures previously explained using the software Bandpass (Olgun 2001) written in MATLAB version 5.3 (1999).

For the Loma Prieta earthquake records, the model was shaken for 16 seconds. For the Saguenay earthquake, the model was shaken for 13 seconds.

The models that were analyzed using FLAC version 4.0 (FLAC 2000) are shown on Figure 3. The models incorporate the following soil conditions: loose silty sand with 20% fines content (Figures 3a, 3b, 3e, and 3f), silty sand over soft clay (Figures 3c and 3g), and silt (Figures 3d and 3h). Hence, the determination of the soil parameter values is different for each model.

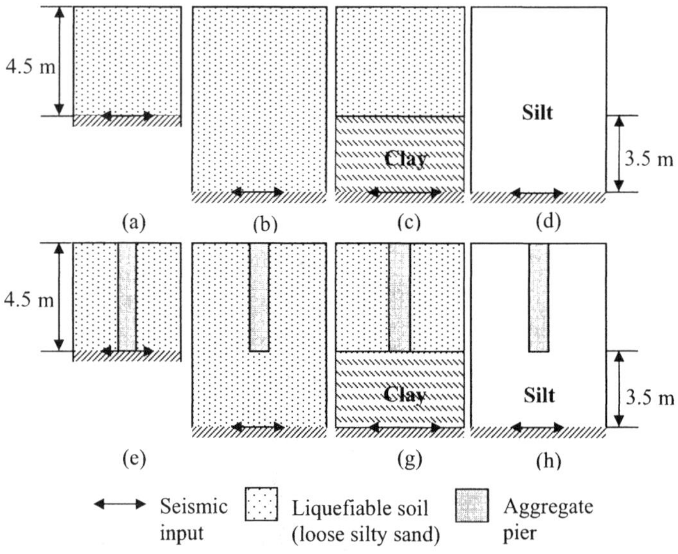

FIG. 3. Models used in FLAC analyses

Table 1 shows the summary of parameters and values that were used in FLAC analyses.

TABLE 1. Summary of parameter values used in FLAC analyses

Parameters (1)	Loose silty sand (2)	Soft clay (3)	Silt (4)	Aggregate pier: Silty sand[a] (5)	Aggregate pier: Silt[b] (6)	Water (7)
Saturated unit weight, γ_{sat} (kN/m^3)	19	17	18	23	23	10
Dry unit weight, γ_d (kN/m^3)	15	12	13	21	21	N/A
Young's modulus, E (kPa)	15080	5030	10055	44700	29800	N/A
Bulk modulus, K (kPa)	15080	9790	12510	24000	16000	10^{-7}
Shear modulus, G (kPa)	5655	1780	3680	18790	12525	N/A
Cohesion, c (kPa)	0	10	3	0	0	N/A
Friction angle, ϕ (°)	30	17	25	50	50	N/A
Dilation angle, ψ (°)	0	0	0	0	0	N/A
SPT $(N_1)_{60}$ (blows/30 cm)	10	4	8	N/A	N/A	N/A
Void ratio, e	0.7	1.1	0.9	0.25	0.25	N/A
Porosity, n	0.412	0.524	0.474	0.2	0.2	N/A
Permeability, k (cm/sec)	10^{-4}	10^{-7}	10^{-6}	10^{-1}	10^{-1}	N/A

[a] Aggregate pier with silty sand as soil matrix
[b] Aggregate pier with silt as soil matrix

INTERPRETATION OF RESULTS OF NUMERICAL ANALYSES

There were total of 20 cases analyzed using FLAC. The Loma Prieta and Saguenay earthquake records were used to analyze 9 cases each as shown on Table 2 and Figure 3. Cases with C indicate cases using Loma Prieta earthquake and cases with S indicate cases using Saguenay earthquake. The last two cases were analyzed with emphasis on the shear stress in soil matrix using the Loma Prieta earthquake records.

TABLE 2. Analyses using FLAC

Case no. (1)	Figures used in FLAC (2)	Case no. (3)	Figures used in FLAC (4)
1C	Figure 3(a)	1S	Figure 3(a)
2C	Figure 3(e)	2S	Figure 3(e)
3C	Figure 3(b)	3S	Figure 3(b)
4C	Figure 3(f)	4S	Figure 3(f)
5C	Figure 3(c)	5S	Figure 3(c)
6C	Figure 3(g)	6S	Figure 3(g)
7C	Figure 3(d)	7S	Figure 3(d)
8C	Figure 3(h)	8S	Figure 3(h)
9C	Figure 3(f) with drainage allowed	9S	Figure 3(f) with drainage allowed
1C2	Figure 3(a)	2C2	Figure 3(e)

The following sections discuss the results of the numerical analyses.

Ground Acceleration

The installation of aggregate pier amplifies the ground motions (a_{max}) when subjected to seismic loading. This is because the aggregate piers stiffen the system. Table 3 summarizes the values of a_{max}. Note that the peak acceleration on rock outcrops is designated as pga and the peak acceleration at the soil surface is designated as a_{max}. Table 3 shows that the input ground acceleration (pga) is de-amplified for cases with Loma Prieta earthquake and is amplified for cases with Saguenay earthquake except for cases with silty sand layer underlain by soft clay, namely Cases 5S and 6S.

TABLE 3. Values of ground acceleration from FLAC analyses

Case (1)	pga (2)	a_{max} (3)	Note (4)	Case (5)	pga (6)	a_{max} (7)	Note (8)
1C	0.45g	0.163g	D	1S	0.05g	0.086g	A
2C	0.45g	0.381g	D	2S	0.05g	0.100g	A
3C	0.45g	0.155g	D	3S	0.05g	0.059g	A
4C	0.45g	0.380g	D	4S	0.05g	0.098g	A
5C	0.45g	0.134g	D	5S	0.05g	0.0384g	D
6C	0.45g	0.419g	D	6S	0.05g	0.040g	D
7C	0.45g	0.269g	D	7S	0.05g	0.0562g	A
8C	0.45g	0.416g	D	8S	0.05g	0.0616g	A
9C	0.45g	0.433g	D	9S	0.05g	0.0993g	A

(Note: D = Deamplification; A = Amplification)

It is apparent that the use of aggregate pier amplifies the ground acceleration (a_{max}) by a factor ranging from 1.55 to 3.13 for the Loma Prieta earthquake and 1.04 to 1.67 for the Saguenay earthquake. The average value for both earthquake ranges from 1.4 to 2.3. This phenomenon is not surprising and has been shown by other researchers, for example by Liu and Dobry (1997). Liu and Dobry (1997) studied a model of footing and showed that the amplification ratios of the footing resting on compacted sand zone within a liquefiable soil mass increase as the depth of the compacted zone increases.

Excess Pore Water Pressure Ratio

The second parameter that will be discussed in this section is the excess pore water pressure ratio (r_u). The excess pore water pressure ratio (r_u) is defined as the ratio between excess pore water pressure and the initial effective vertical stress. The r_u values discussed here are the peak values.

The peak values of r_u were collected and compared between cases with and without aggregate piers to observe the degree of improvement that occurs by installing aggregate piers as shown on Figure 4. A best-fit curve was plotted as indicated by the solid line.

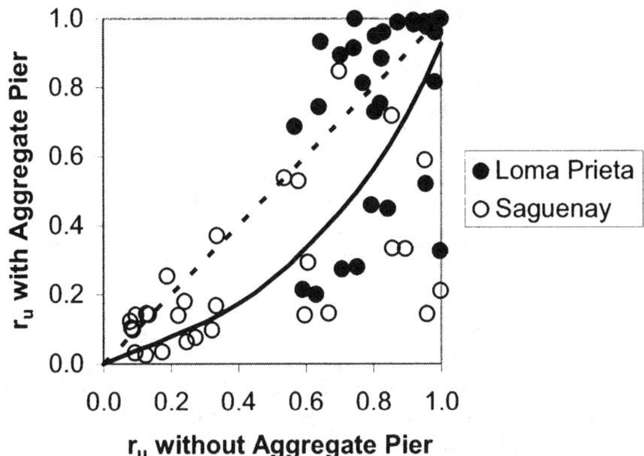

FIG. 4. Comparison of peak values of r_u between cases with and without aggregate pier for Loma Prieta and Saguenay earthquakes

From Figure 4 it is apparent that for Loma Prieta earthquake most of the data points actually lay above the 1:1 line indicated by the dashed line. It means that most r_u values increase due to the installation of aggregate pier. It appears that for cases where the input acceleration (pga) overcomes the shear strength of the reinforced soil, no reduction in r_u value is obtained.

Figure 4 also shows that for Saguenay earthquake most of the data points lay beneath the 1:1 line (the dashed line). It means that for Saguenay earthquake most r_u values decrease due to the installation of aggregate piers.

Since Figure 4 shows that most data points lie beneath the 1:1 line, it can be generally concluded that improvement occurs due to the installation of aggregate pier. Note that improvement is defined as a decrease in the values of r_u as a result of the installation of aggregate pier.

Shear Stress in Soil Matrix

In the Simplified Procedure proposed by Seed and Idriss (1971), the maximum shear stress in soil matrix (τ_{max}) can be estimated by multiplying the stress reduction coefficient (r_d), total overburden pressure (σ_0), and the peak horizontal acceleration at the ground surface (a_{max}) as shown in (1).

$$\tau_{max} = r_d * \sigma_0 * \frac{a_{max}}{g} \quad (1)$$

Baez and Martin (1993) and Goughnour and Pestana (1998) introduced a new parameter, which they defined as the shear stress reduction factor (K_G), as shown on eqs. 2 and 3, respectively.

$$K_G = \frac{\tau_s}{\tau} = \frac{1}{1+R_a(R_s-1)} \quad (2)$$

$$K_G = \frac{\tau_s}{\tau} = \frac{1+R_a(n-1)}{1+R_a(R_s-1)} \quad (3)$$

where:

τ_s = shear stress in the soil matrix

τ = the input shear stress

n = the vertical stress ratio which is the ratio of the effective overburden pressure within the stone column to the effective overburden pressure within the soil matrix. The value of n varies between 4 and 10 based on model tests and 2 to >10 based on field measurement.

The ratio between the area of reinforcing element and the total plan area can be written as $R_a = A_r/A$ and the ratio between the shear modulus of the reinforcing element and the shear modulus of the soil can be written as $R_s = G_r/G_s$.

The shear stress in soil matrix (τ_s) can be estimated by multiplying K_G with the average shear stress (τ_{ave}) calculated using the following equation:

$$\tau_{ave} = 0.65 * \sigma_0 * \frac{a_{max}}{g} * r_d \quad (4)$$

Eq. 4 can be written in terms of maximum shear stress (τ_{max}), which is shown by (1). Hence, (2) can also be written in terms of maximum shear stress (τ_{max}).

$$K_G = \frac{\tau_s}{\sigma_0 \frac{a_{max}}{g} r_d} \quad (5)$$

By applying values of τ_{max} calculated using FLAC into (5), the values of K_G can be calculated. This procedure was applied to the case with aggregate pier (Case 2C2) with value of a_{max} of 0.37g. Figure 5 shows plot of average K_G values versus depth. It can be seen that the average value of K_G is 0.17 throughout all depths.

The values of K_G calculated using (5) were based on shear stress calculated using the Simplified Procedure (Seed and Idriss 1971), which does not take into account the effects of reinforcing elements. To overcome this problem a modification to the shear stress reduction factor (K_G) is introduced, the shear stress reduction factor, which takes into account the reinforcement factor (K_{GR}). The value of K_{GR} is nothing more than the ratio of the maximum shear stress for cases with aggregate pier to cases without aggregate pier obtained from FLAC analyses.

$$K_{GR} = \frac{(\tau_{\max})_{with\ aggregate\ pier}}{(\tau_{\max})_{without\ aggregate\ pier}} \quad (6)$$

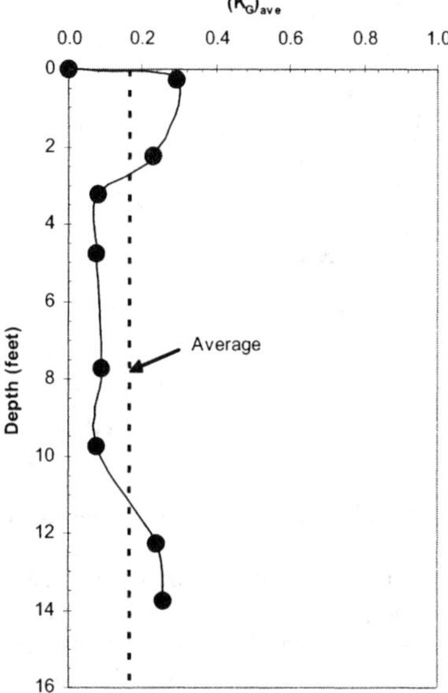

FIG. 5. Plot of values of $(K_G)_{ave}$ versus depths for Case 2C2

To make the discussion easier, the case analyzed was given a new name. Case C2 represents the ratio of Case 2C2 and Case 1C2 in (6).

Figure 6 shows plot of average K_{GR} values versus depth. It is apparent that the K_{GR} values in the aggregate pier are much larger than those in the soil matrix. This shows that the aggregate pier carries more shear stresses than the soil matrix under seismic loading. From Figure 6, it can be seen that the average values of K_{GR} are approximately 0.6 and 3.6 for the soil matrix and the aggregate pier, respectively.

It can be concluded that the average value of K_{GR} for matrix soil (0.6) is much larger than the value of K_G of 0.17 for Case 2C2. It is apparent that the value of K_G calculated using (5) gives smaller value than the value of K_{GR} calculated using (6). The use of K_{GR} value is more preferable since it depicts the "real" reinforcing effects of aggregate pier as shown in (6).

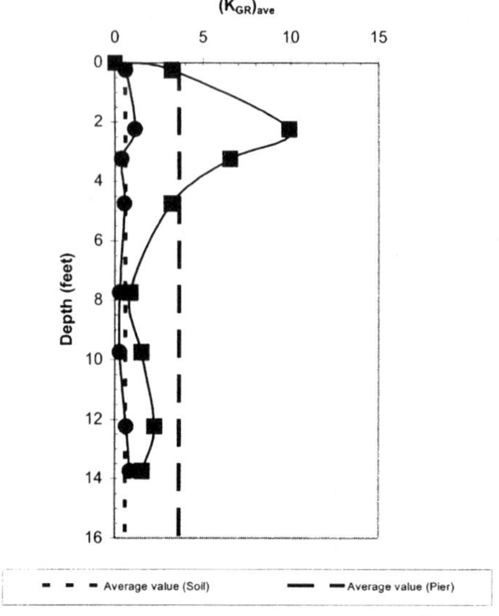

FIG. 6. Plot of values of $(K_{GR})_{ave}$ versus depths for Case C2

A general procedure should be developed to estimate the value of K_{GR}. For this purpose, the procedures proposed by Baez and Martin (1993) and Goughnour and Pestana (1998) were reviewed. Using the procedure proposed by Baez and Martin (1993), a value of K_G of 0.6 is calculated using (2). If the procedure proposed by Goughnour and Pestana (1998) is used, a value of K_G of 0.8 is obtained from (3).

It is observed that the value of K_G calculated using the shear reinforcement approach (2) gives the same value as the value of K_{GR} calculated using (6). Therefore, (2) can be used in designing the reinforcing effects of aggregate pier foundation system during seismic loading.

Figure 7 presents (2) in form of a chart. It is apparent that at any given R_a, the reinforcement factor (K_{GR}) decreases with increasing shear modulus ratio (R_s). It can also be seen that at any given R_s, the reinforcement factor (K_{GR}) decreases with increasing area replacement ratio (R_a).

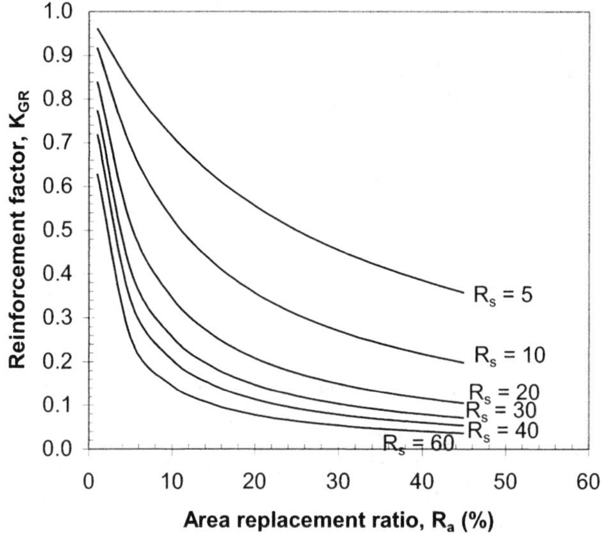

FIG. 7. The reinforcement factor, K_{GR} (after Baez and Martin 1993)

CONCLUSIONS

- The use of aggregate pier generally amplifies the ground acceleration (a_{max}). The input acceleration time history (pga) is amplified for cases with Loma Prieta earthquake. The values of pga are amplified for cases with Saguenay earthquake, except for cases with the presence of soft clay underlying the silty sand layer.
- The effects of installation aggregate pier are much more significant in cases with Saguenay time history. It can be concluded that generally improvement occurs due to the installation of aggregate pier.
- The values of K_G can be estimated by applying values of τ_{max} calculated using FLAC. It can be concluded that the K_G values in the aggregate pier are larger than those in the soil matrix. This shows that the installation of aggregate pier is effective in that the aggregate pier carries more shear stresses than the soil matrix under seismic loading.
- A modification to the shear stress reduction factor (K_G) was introduced that is the shear stress reduction factor, which takes into account the reinforcement factor (K_{GR}). The K_{GR} values in the aggregate pier are much larger than those in the soil matrix. This again shows that the aggregate pier carries more shear stresses than the soil matrix under seismic loading.
- The value of K_G calculated based on shear stresses from FLAC normalized by shear stresses from the Simplified Procedure is smaller than the value of K_{GR}.

The use of K_{GR} value is more preferable since it depicts the "real" reinforcing effects of aggregate pier.

- The equation using the shear reinforcement approach (Eq. 2) agrees well with the value of K_{GR} calculated using FLAC. The equation proposed by Goughnour and Pestana (1998) overestimates the value of K_{GR}.

REFERENCES

Baez, J. I. and Martin, G. R. (1993). "Advances in the design of vibro systems for the improvement of liquefaction resistance", *Proc. 7th Annual Symp. of Ground Improvement*, 1-16.

FLAC User's Manual – version 4.0. (2000). Itasca Consulting Group, Inc., Minneapolis, Minnesota.

Fox, N. S. and Edil, T. B. (2000). "Case histories of rammed aggregate pier soil reinforcement construction over peat and highly organic soils", *Proc. Soft Ground Technology*, 146-157.

Goughnour, R. R. and Pestana, J. M. (1998). "Mechanical behavior of stone columns under seismic loading", *Proc. 2nd Int. Conf. on Ground Improvement Techniques*, 157-162.

Lawton, E. C. and Fox, N. S. (1994). "Settlement of structures supported on marginal or inadequate soils stiffened with short aggregate piers", *Vertical and Horizontal Deformations of Foundations and Embankments*, Yeung, A. T. and Felio, G. Y., eds., ASCE, GSP No. 40, 2, 962-974.

Lawton, E. C. and Merry, S. M. (2000). "Performance of geopier-supported foundations during simulated seismic tests on Northbound Interstate 15 bridge over South Temple, Salt Lake City, Utah", Final Report, *Rep. No. UUCVEEN 00-03*, Univ. of Utah, Salt Lake City, Utah, 73pp.

Liu, L. and Dobry, R. (1997). "Seismic response of shallow foundation on liquefiable sand", *J. Geotech. and Geoenv. Engrg.*, ASCE, 123(6), 557-567.

Martin, G. R., Finn, W. D. L., and Seed, H. B. (1975). "Fundamentals of liquefaction under cyclic loading", *J. Geotech. Engrg. Div.*, ASCE, 101(GT5), 423-438.

Olgun, C. G. (2001). "Bandpass – Software to process earthquake time histories", Earthquake Center of Southeastern of United States (ECSUS), Virginia Tech, Blacksburg, Virginia.

Seed, H. B. and Idriss, I. M. (1971). "Simplified procedure for evaluating soil liquefaction potential", *J. Soil Mech. & Found. Div.*, ASCE, 97(SM9), 1249-1273.

COMPRESSION AND UPLIFT OF RAMMED AGGREGATE PIERS IN CLAY

Christopher Lillis[1], Alan J. Lutenegger[2] and Michael Adams[3]

ABSTRACT: Full scale compression and uplift load tests were conducted on 3 Rammed Aggregate Piers in the upper surficial clay fill at the NGES located at the University of Massachusetts – Amherst. The tests were conducted to evaluate the performance of this intermediate foundation system in a fine-grained soil. The field investigation also included measurements of vertical and lateral soil deformation to evaluate the active zone of soil resistance. The results show viable uplift and bearing capacity application for the foundation system in fine grained soils.

INTRODUCTION

In recent years, Rammed Aggregate Piers or Geopier® elements have been used as an intermediate foundation system between conventional shallow spread footings and deep driven or drilled foundations. In places where marginal soils are present near the surface, these foundation elements may provide an economic alternative for foundation support. This type of ground improvement is considered an "intermediate" foundation and has been previously described (e.g., Lawton and Fox 1994; Lawton et. el. 1994; Blackburn and Fassell 1998; Wissmann et al. 2001). Geopier elements have been used as a ground improvement technique in marginal soil conditions to reduce settlements and allow the use of shallow foundations as well as provide uplift resistance against wind and other lateral loads.

An investigation was performed to evaluate the load-displacement behavior of Rammed Aggregate Piers in clay at the NGES. Overall, a total of six uplift and one compression test were conducted. This paper discusses the results of two of the uplift

[1]Graduate Research Assistant, Department of Civil and Environmental Eng., University of Massachusetts, 28 Marston Hall, Amherst, MA 01003, c.lillis@lycos.com
[2]Professor and Head, Department of Civil and Environmental Eng., University of Massachusetts, 224 Marston Hall, Amherst, MA 01003
[3]Research Geologist, Turner Fairbanks Highway Research Center, FHWA, 3600 Georgetown Pike, McLean, VA 22203

tests and the one compression test.

RAMMED AGGREGATE PIERS

A Rammed Aggregate Pier or Geopier is conceptually similar to a compacted column of gravel (e.g., Klabena and Mica 1998: Kumar and Ranjan 1999) however this is a proprietary foundation system that uses special equipment for installation. Piers are constructed by drilling an open hole, placing controlled lifts of aggregate stone within the open hole, and compacting the aggregate with a specially designed high energy beveled impact tamper. The first lift of aggregate consists of clean stone and is rammed into the soil to form a bottom bulb below the borehole while imparting lateral and vertical soil stresses (Lawton et al. 1994; White et al. 2000). Additional 0.3 m (1.0 ft.) lifts are then placed and densified with the tamper. The high energy impact ramming action necessary for proper construction pre-stresses and pre-strains the surrounding soil matrix, increasing skin friction and density while reducing settlement (White et al. 2000). This construction sequence is illustrated in Figure 1. According to Wissmann et al. (2001) Rammed Aggregate Piers are typically designed to cover approximately 30 to 40 % of the gross area of the overlying footing element.

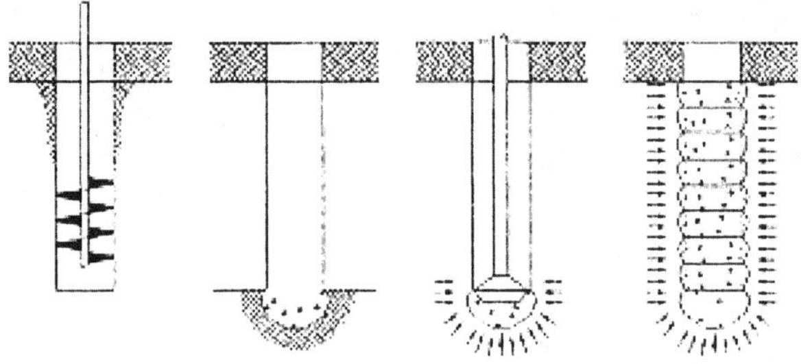

FIG. 1. Geopier Construction Sequence

INVESTIGATION

Site Characteristics

Tests were performed at the National Geotechnical Experimentation Site (NGES) located at the University of Massachusetts – Amherst. The site is situated in a thick deposit of the Connecticut Valley Varved Clay (CVVC). Geotechnical characteristics of the site have been extensively documented (Lutenegger 2000). The site currently consists of two test areas; Area A and Area B. Tests described in this paper were performed at Area B which represents an extension of the original NGES.

Figure 2 shows the soil characteristics and SPT N_{60} values obtained at the site located near the Geopier elements. The upper 2 m (6.6 ft.) consists of a stiff compacted clay fill placed approximately 30 years ago from clay excavated at the adjacent Town of Amherst Wastewater Treatment Plant. Beneath the clay fill the CVVC extends to a depth of about 25 m (82 ft.) and is followed by a deposit of glacial sand and gravel. The upper 3 m (10 ft.) of the CVVC is overconsolidated as a result of desiccation, freezing, etc. The CVVC below a depth of about 8 m (26.2 ft.) is very soft and near normally consolidated. Results of electric Cone Penetrometer Test (CPT) profiles conducted adjacent to the test area are shown in Figure 3. It can be seen that the ground conditions are very similar at all three locations. The upper fill and the clay crust are clearly seen in the CPT profiles. Typically, the water table shows a seasonal fluctuation between 0.5 m (1.6 ft.) and 2.5 m (8.2 ft.) below grade.

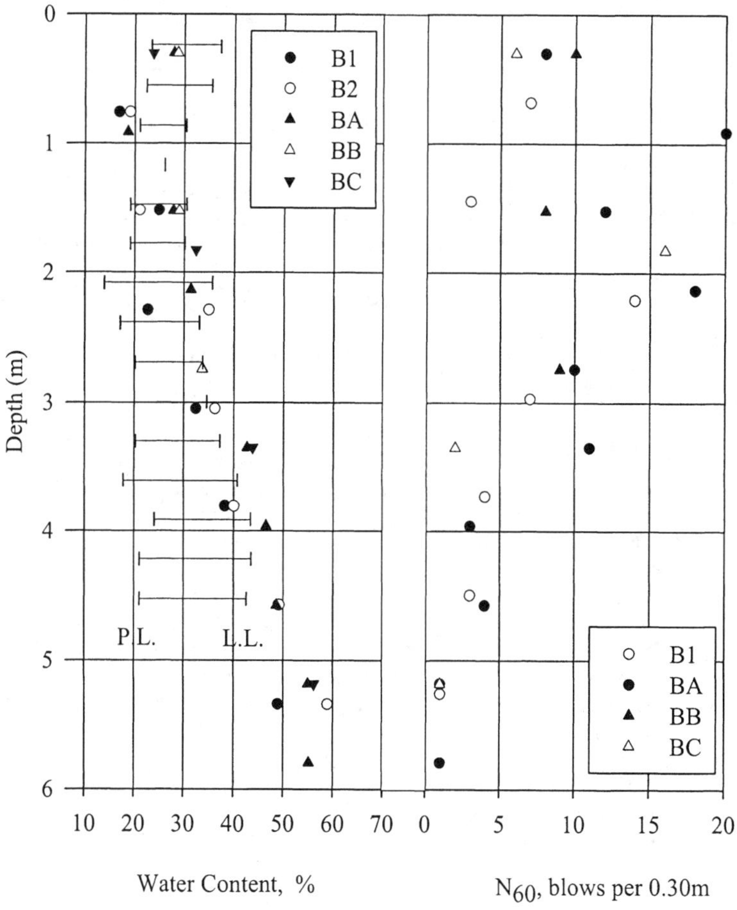

FIG. 2. Soil Characteristics and SPT N_{60} Values, NGES Area B

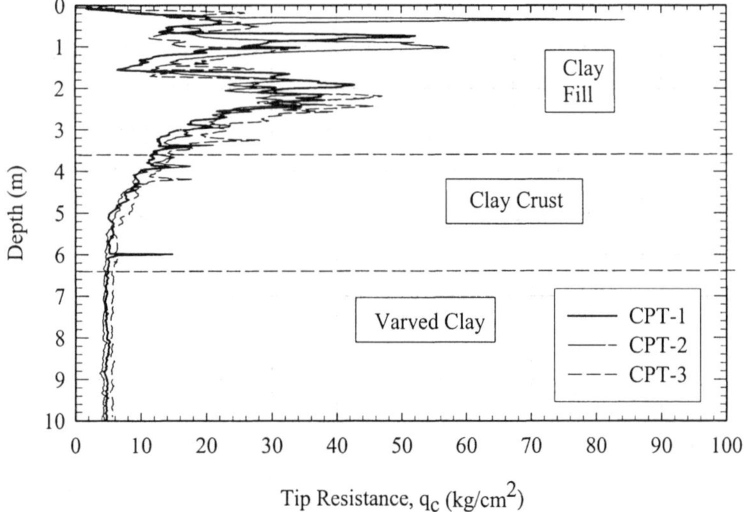

FIG. 3. CPT Profiles Conducted at NGES, Area B

Geopier Construction

Each Geopier for this project was 0.61 m (2 ft.) in diameter by 3.0 m (10 ft.) long. For this project, the Geopier foundation company designed and installed the Geopier elements using the procedure shown in Figure 1. The auger and tamper were operated from a skid steer uniloader; however, in most cases Geopier elements are installed with larger equipment such as a trackhoe. Installation required two laborers, one-day and about 6 metric tons of gravel for all seven Geopier elements.

Construction specific to the Geopier elements to be tested for uplift included the placement of a 0.61 m (2 ft.) long section of 20.3 cm (8 in.) deep channel at the base of each of the piers. Two pieces of threaded rod were then fastened to the channel and extended above ground, serving as the load transfer mechanism during uplift.

Instrumentation

In order to evaluate the active zone of soil resistance, inclinometer casings were installed adjacent to two of the Geopier elements and tell tales were installed.

Uplift Testing

A steel plate or top plate was placed directly on top of the aggregate piers and deflection of the plate was measured during uplift using digital dial gauges attached to an external reference frame. The deflection of the threaded tension rods extending to the base of the pier was also measured during uplift using digital dial gauges attached to an external reference frame.

Inclinometer casing was installed approximately 0.3 m (1.0 ft) away from the edge of one Geopier tested in uplift to a depth of 4.5 m (14.8 ft.). Lateral deflection was monitored over the course of the test using a Slope Indicator Digitilt Sensor and recorded with a Digitilt DataMate.

Compression Testing

For the compression test, a steel plate or top plate was placed directly on top of the aggregate pier to provide a base for applying the load and settlement was measured during loading using a digital dial gauge attached to an external reference frame. Tell tales were installed during construction of the Geopier tested in compression at depths of 1.5 m (5 ft.) and 3.0 m (10 ft.). Tell tale deflection was measured using digital dial gauges attached to an external reference frame.

Inclinometer casing was installed approximately 0.15 m (0.5 ft) and 0.3 m (1.0 ft) away from the edge of the Geopier tested in compression to a depth of 4.5 m (14.8 ft.). Lateral deflection was monitored over the course of the test using a Slope Indicator Digitilt Sensor and recorded with a Digitilt DataMate.

Load Testing

A reaction frame was constructed consisting of a 7.6 m (24.9 ft.) steel I beam supported by 0.15 m x 0.15 m (6 in. x 6 in.) wood cribbing and a 0.91m (3.0 ft.) long section of 0.28 m (0.91 ft.) deep channel. Wood cribbing was placed directly on the surrounding soil at a distance of approximately 3.0 m (10 ft.) from the edge of the Geopier to avoid any influence during testing. Load was applied using a hydraulic jack and pump. Applied load was measured using a load cell with a 1334 kN (300 kips) capacity and a P3500 digital readout. Load was applied initially in increments of approximately 4.5 kN (1 kip) and later in increments of approximately 9 kN (2 kips). Each load increment was held for a period of 15 minutes. The load test setup for a Geopier tested in uplift is shown in Figure 4.

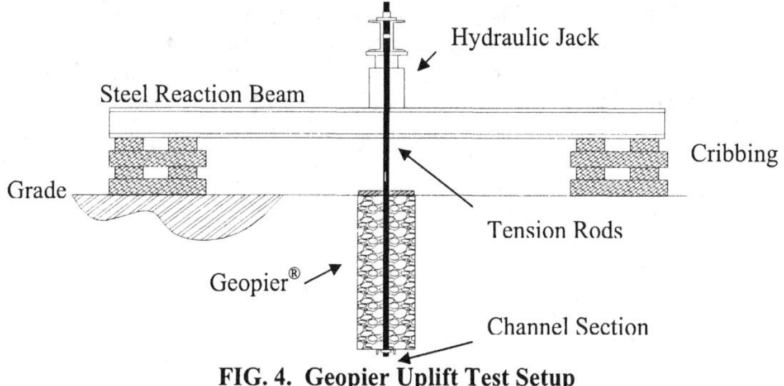

FIG. 4. Geopier Uplift Test Setup

The Geopier tested in compression used four surrounding piers constructed for uplift testing and three I beams placed on cribbing for a reaction system. In turn, while testing the center Geopier in compression, the four surrounding piers were placed in uplift. All five piers were monitored for rod deflection, top plate deflection and applied load as noted in the above mentioned procedures for uplift testing. The load test setup for the Geopier tested in compression is shown in Figure 5.

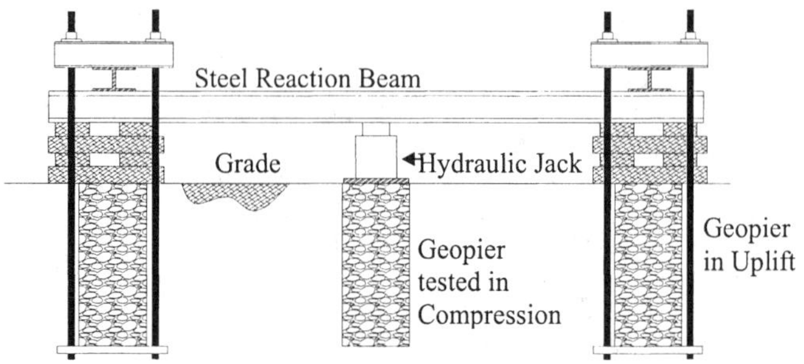

FIG. 5. Geopier Compression Test Setup

RESULTS

Uplift

The deflection of both the tension rods and the top plates of the Geopier elements tested in uplift are shown in Figures 6 and 7. Uplift Test #1 was loaded to a maximum of approximately 130 kN (30 kips) with resulting deflections of approximately 10mm and 85 mm (3.3 in.) for the top plate and tension rods, respectively. Uplift Test #2 was loaded to a maximum of approximately 130 kN (30 kips) with resulting deflections of approximately 30 mm (1.2 in.) and 150 mm (5.90 in.) for the top plate and tension rods, respectively. These results indicate that very little movement of the top of the pier occurred during uplift. This suggests either compression of the Geopier itself (which is unlikely) or failure by lateral bulging.

Lateral deflection at an applied load of 111 kN (25 kips) for Uplift Test #2 is shown in Figure 8. A maximum deflection of 3 mm (0.12 in.) occurred at the base of the Geopier at a depth of 3 m (10 ft.). The majority of the deflection occurred below a depth of 1.5 m (5 ft.) or L/2 with less than 1 mm (0.04 in.) of lateral deflection occurring above this point. These results are consistent with deflection results shown in Figures 7 and 8 and indicate that the pier bulged laterally in the lower meter.

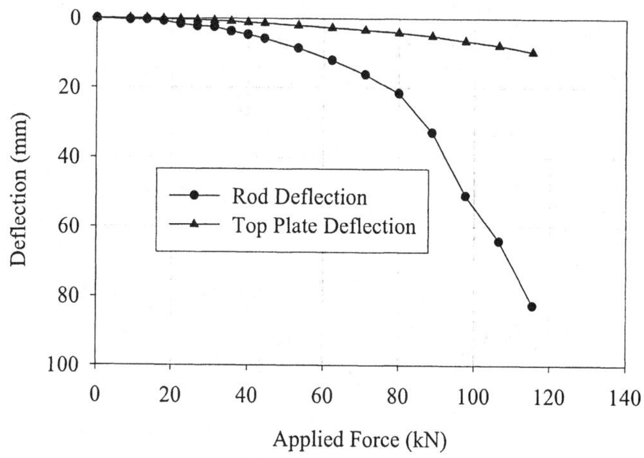

FIG. 6. Load Settlement Curve Uplift Test #1

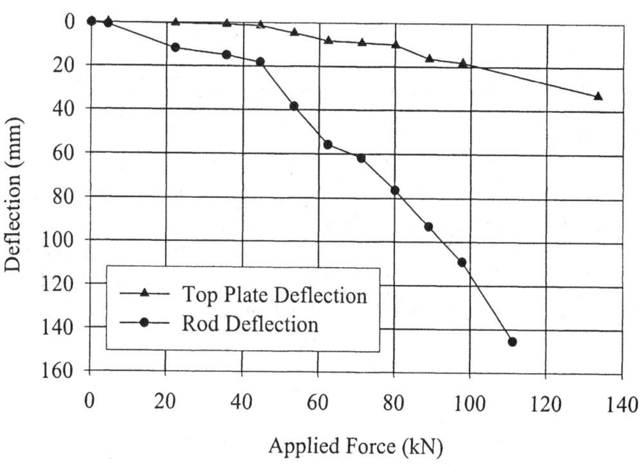

FIG. 7. Load Settlement Curve Uplift Test #2

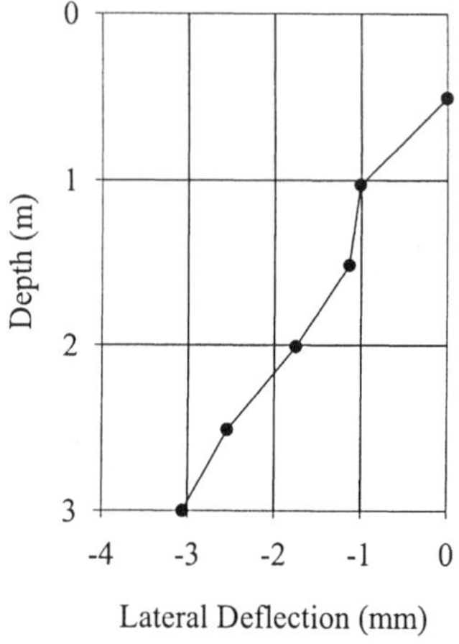

FIG. 8. Lateral Deflection Uplift Test #2

Compression

The settlement curve for the Geopier tested in compression is shown in Figure 9. A maximum settlement of approximately 90 mm (3.5 in.) occurred at an applied load of 311 kN (70 kips).

Soil settlement, as measured by the tell tales, is shown in Figure 10. A maximum settlement of approximately 28 mm (1.1 in.) was recorded at a depth of 1.5 m (5 ft.) or L/2. A maximum settlement of approximately 7 mm (0.28 in.) was recorded at a depth of 3.0 m (10 ft.) or L. Very little settlement of the base or the mid height occurred throughout the loading with essentially no movement until a load of approximately 100 kN (22 kips). This indicates that the pier was bulging laterally in the upper meter.

Lateral deflection at the maximum applied load of 311 kN (70 kips) for the Geopier tested in compression is shown in Figure 11. The S-Inclinometer and N-Inclinometer were placed approximately 0.15 m (0.5 ft.) and 0.3 m (1.0 ft.) away from the edge of the Geopier, respectively. Maximum deflections of 16 mm (0.63 in.) and 22 mm (0.87 in.) occurred at the top of the Geopier for the N-Inclinometer and S-Inclinometer, respectively, with the majority of the lateral deflection occurring in the upper meter (3.3 ft.). This indicates that significant lateral bulging occurred in the upper meter (3.3 ft.).

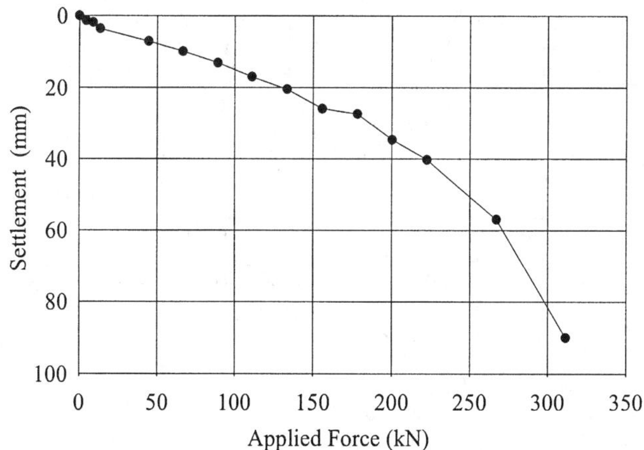

FIG. 9. Load Settlement Curve Compression Test

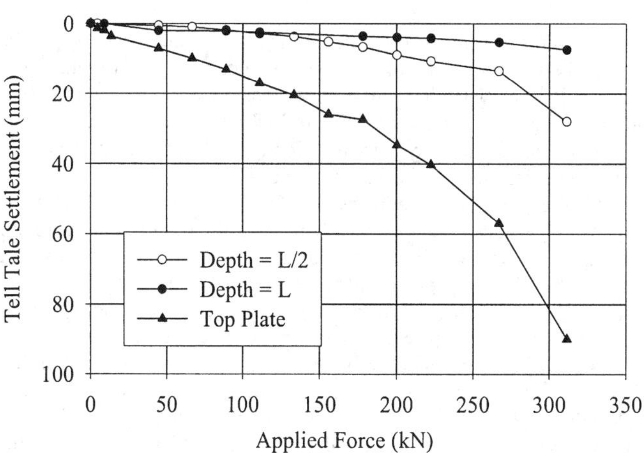

FIG. 10. Tell Tale Settlement Compression Test

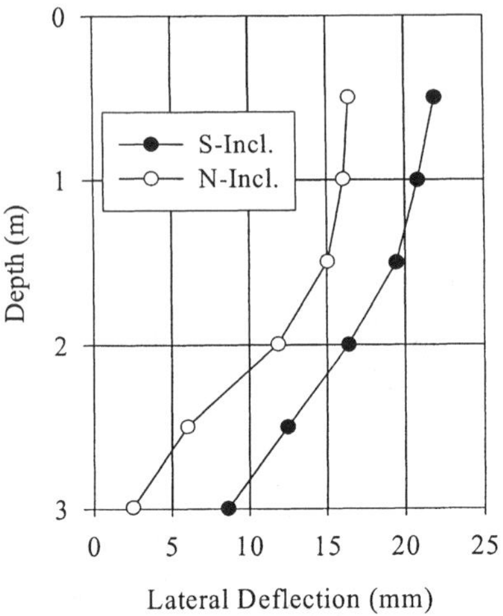

FIG. 11. Lateral Deflection Compression Test

SUMMARY AND CONCLUSIONS

Failure of the Geopier elements in uplift, defined here as the inflection point of the straight line portions of the load/settlement curve, occurred at approximately 80 kN (18 kips). At failure, very little deflection was seen at the top plates with corresponding large deflections in the tension rods. This indicates that for the Geopier elements tested in uplift in this clay, very little load is transferred from the base to the top plate, which remains outside of the active zone of influence.

Failure of the Geopier tested in compression (again defined at the inflection point of the straight line portions of the load/settlement curve) occurred at approximately 266 kN (60 kips). At failure, the tell tale located at a depth of 1.5m (5ft) settled approximately 33% of the top plate settlement as opposed to 10% of the top plate settlement from the tell tale located at the base (3.0 m (10 ft.)) of the Geopier. This indicates that for the Geopier placed in compression, very little load is transferred from the top plate to the base, which remains outside of the active zone of influence.

In both uplift and compression the majority of the lateral deflection occurred within L/2 of the applied load and show failure by lateral bulging. The level of influence with respect to lateral deflection decreases from 0.15m (0.5ft) to 0.3m (1ft) as measured by the N-Inclinometer and S-inclinometer during compression testing. It can also be noted that the lateral active zone of influence extends beyond 0.3m (1ft).

The Geopier elements performed well in both uplift and compression and appear to be very viable for clays. Possible applications include foundation support and/or as a form of ground improvement.

TABLE 1. Geopier Load Test Summary Table

Mode	Test No.	Load at Inflection Point (kN)
Uplift	#1	80
	#2	80
Compression	#1	266

ACKNOWLEDGMENTS

The authors would like to thank the Geopier Foundation Co. of Scottsdale, AZ. and Peterson Contractors, Inc. of Reinbeck, IA for installing the Geopier elements. Dr. Kord Wissman was responsible for arranging the construction of Geopier elements and his efforts and input throughout are greatly appreciated.

REFERENCES

Blackburn, T.W. and Farrell, T.M. (1998). "Shopping center saved by short aggregate piers." *Proc. of the 4th International Conference on Case Histories in Geotechnical Engineering*, 988-991.

Klabena, P. and Mica, L. (1998). "Driven compacted-in-place sand-gravel piers." *Proc. of the 7th International Conference and Exhibition on Piling and Deep Foundations*, 5.23.1-5.23.4.

Kumar, P. and Ranjan, G. (1999). "Granular pile system for uplifting loads – A case study." *Proc. of the International Conference on Offshore and Nearshore Geotechnical Engineering*. 427-432.

Lawton, E.C. and Fox, N.S. (1994). "Settlement of structures supported on marginal or inadequate soils stiffened with short aggregate piers." *Vertical and Horizontal Deformations of Foundations and Embankments*, ASCE, Vol. 2, 962-974.

Lawton, E.C., Fox, N.S. and Handy, R.L. (1994). "Control of settlement and uplift of structures using short aggregate piers." *In Situ Deep Soil Improvement*, ASCE, 121- 132.

Lutenegger, A.J. (2000). "National Geotechnical Experimentation Site - University of Massachusetts." *National Geotechnical Experimentation Sites,* ASCE, 102-129.

White, D.J., Lawton, E.C. and Pitt, J.M. (2000). "Lateral earth pressure induced by Rammed Aggregate Piers." *Proc. of the 53rd Canadian Geotechnical Conference*, Vol. 2, 871-876.

Wissmann, K.J., Moser, K. and Pando, M.A. (2001). "Reducing settlement risks in residual Piedmont soils using Rammed Aggregate Pier elements." *Foundations and Ground Improvement*, ASCE, 943-957.

Wissmann, K.J., Fox, N.S. and Martin, J.P. (2000). "Rammed Aggregate Piers defeat 75-foot long driven piles." *Performance Confirmation of Constructed Facilities*, ASCE, 198-210.

LESSONS LEARNED FROM A STONE COLUMN TEST PROGRAM IN GLACIAL DEPOSITS

Barry S. Chen[1], P.E., Member, Geo-Institute and Michael J. Bailey[2], P.E., Member, Geo-Institute

ABSTRACT

A stone column test program was conducted for ground improvement below an earth embankment project in western Washington. Stone columns were initially designed to mitigate soil liquefaction, improve shear strength, and reduce subgrade settlements to provide a stable foundation for the embankment. The test program includes four test areas of stone columns installed with varying control parameters such as diameter, spacing, and area replacement ratio. Each test area consists of 32 to 41 vibro-replacement stone columns. Stone columns were installed through the post-glacial, loose to medium dense, silty sand and soft to medium stiff silt and embedded into glacially overridden, dense, silty sand and stiff to hard, sandy silt. Standard Penetration Tests (SPT) and Cone Penetration Tests (CPT) were used as verification tests in each of the four test areas before and after the installation of stone columns. Results indicated densification of the matrix in soil zones containing less than approximately 15 percent fines and little to no improvement in soil zones containing over approximately 15 percent fines. In some deeper zones where native soils were over-consolidated by glacial actions, disturbance in the stiff to hard silt actually caused a decrease in penetration resistance.

INTRODUCTION

For decades, vibro-replacement stone columns have been used as an effective ground improvement technique for increasing bearing capacity, reducing ground settlement, and mitigating soil liquefaction of foundation soils. These design goals are typically achieved by providing (1) densification of the soil in-between columns, (2) increase in shear strength and stiffness of the soil-stone matrix, and (3) dissipation of excess pore water pressure in the soil through the more permeable stone column.

---

[1] Principal, Hart Crowser, Inc., 1910 Fairview Avenue East, Seattle, Washington 98102-3699, barry.chen@hartcrowser.com

[2] Senior Principal, Hart Crowser, Inc., 1910 Fairview Avenue East, Seattle, Washington 98102-3699, mike.bailey@hartcrowser.com

Densification of soils can be easily verified using penetration tests such as Standard Penetration Test (SPT) and Cone Penetration Test (CPT) for construction quality control. However, it is also generally recognized that densification only occurs when the soil contains little to no fines (soil passing US No. 200 sieve). When the foundation soil contains more than 15 to 25 percent fines, the design often has to rely primarily on the increase in stiffness and sometimes increased capacity for dissipation of pore water pressure, but construction quality control is more difficult to verify in this case.

In the stone column test program presented in this paper, the densification effect of soils in zones containing various amounts of fines was closely examined using penetration test data collected from four test areas. The purpose of the test program is to evaluate the feasibility of stone columns for subgrade improvement at this site, select final design geometry for the column installation, and evaluate the use of penetration tests for quality control during construction.

PROJECT DESCRIPTION

The project involves the design and construction of a 17-million-cubic-yard earth embankment located near Seattle. The earth embankment includes sections of 2 horizontal to 1 vertical (2H:1V) side slope and several mechanically stabilized earth (MSE) retaining walls.

Ground conditions below most of the 2H:1V slope and MSE wall areas generally consist of 10 to 20 feet of alluvial deposits, colluvium, and recessional outwash underlain by glacially overridden soils. The alluvial deposits typically include interlayered sand, silt, clay, and occasional peat. The colluvium and recessional outwash generally consist of medium dense to dense, slightly silty to silty, slightly gravelly to gravelly sand. Variable depositional processes have produce a wide range in soil properties over short distances, as indicated by comparison of side-by-side explorations described below. The glacially overridden soils are usually present in the form of dense to very dense gravelly, silty sand and very stiff to hard silt. Groundwater in the area is generally within 5 to 10 feet below existing ground surface.

Due to the presence of potentially liquefiable sand and compressible silt/clay in the foundation soils, several ground improvement techniques were considered to provide a stable foundation subgrade for the embankment slopes and MSE walls. A stone column test program was performed to evaluate the feasibility of the preferred alternative - vibro-replacement stone columns.

STONE COLUMN DESIGN

The stone column design for this project uses the concept of area replacement ratio (A_r) and includes both static and seismic considerations. The A_r is defined as the ratio of the stone column area to the total area of the soil/stone matrix. Barksdale and Bachus (1982) provided guidelines for estimating reduction in subgrade settlements based on A_r and stress concentration factor (n). The stress concentration factor can also be defined as the ratio of stone column's shear modulus to the shear modulus of the soil ($K_G = G_{stone} / G_{soil}$). For improvement of soil liquefaction resistance, Baez and Martin (1993) developed a design approach for determining A_r based on SPT data.

The design parameters for this project include an earthquake magnitude of 7.5, a peak ground acceleration of 0.32 g, and n (or K_G) = 3. For soils containing 15 percent fines, a minimum A_r of 17 percent was necessary to achieve a calculated SPT resistance of 22 or a CPT resistance of 110 tons per square foot (tsf) to prevent liquefaction. For soils containing 35 percent fines, a minimum A_r of 17 percent produced a calculated SPT resistance of 18 or a CPT resistance of 75 tsf to prevent liquefaction. For very silty sand or cohesive soils with no densification anticipated, a minimum A_r of 35 percent is necessary to achieve the same results. Once the A_r is determined, diameter and spacing of the column can be selected based on Eq. 1:

Eq. 1 $$A_r = 0.907\ (d/s)^2$$

where d is column diameter and s is center-to-center spacing between columns. Typical 42-inch-diameter stone columns installed in a triangular pattern at 8-foot center-to-center spacing would have an A_r of 17 percent.

In addition to the densification and stress concentration effects, dissipation of excess pore water pressure was examined using the one-dimensional model proposed by Seed and Booker (1976). Analysis indicated that the maximum pore water pressure of the soil never exceeded 45 percent of the effective overburden stress during and after ground shaking for the design earthquake.

STONE COLUMN INSTALLATION

Four test areas were selected for the stone column test programs. Test Areas 1 and 2 are immediately adjacent to each other. Test Areas 3 and 4 are approximately 300 and 500 feet north of the Test Areas 1 and 2, respectively. Figure 1 shows the view looking south to the test areas.

Vibro-replacement stone columns were constructed using crushed rock with the gradation that 90 percent passed a 1-inch size sieve and 5 percent or less passed a

1/2-inch size sieve. The stone was installed using an electric vibrator (165 horsepower) with bottom-feed attachment, as shown on Figure 2. The vibrator has a delivery of 120 kW at 1,800 rpm, a 3-phase current of 380 volts, and a frequency of 60 cycles per second. Free-hanging amperage draw is approximately 120 amps, with maximum amperage draw at 280 amps. The Manitowoc 3900 crane (100-ton capacity) has 80 feet of boom.

FIG. 1—View of Project Test Area

FIG. 2—Stone Column Vibrator with Bottom-Feed Attachment

It is commonly accepted that higher amperage is related to denser ground conditions, and in some cases stone column contracts require that a minimum amperage be attained while building the column up from the base. In addition, amperage is used to define probe refusal. During typical stone column construction in soft ground, relatively low amperage (on the order of about 150 to 180 amps) is often observed at the start of column construction. Where the probe encounters dense ground refusal at its tip elevation, amperage could be greater than 200 amps at the base of the column, and increasing to about 250 amps or higher as stone is compacted in successive lifts during column construction. Amperage may increase to over 300 amps during subsequent column construction due to densification caused by the initial column installations.

Table 1 shows the geometry of the test sections. Note Test Areas 1A and 2A were created by split spacing smaller diameter columns within a portion of Test Areas 1 and 2, respectively.

TABLE 1. Summary of Geometry of Test Sections

(1)	(2)	(3)	(4)	(5)	(6)	(7)
Test Area	1	1A	2	2A	3	4
Number of Stone Columns in Test Area	32	9	32	9	41	41
Column Diameter in Inches	42	28	36	25	42	42
Column Spacing, Center to Center in Feet	8	4.6	7	4	5.8	8
Ar in Percent	17	35	17	35	35	17
Column Depth in Feet	17 to 19	15 to 18	18 to 19	18 to 19	9 to 12	13 to 16

Figure 3 shows a plan of the columns and penetration tests in each test area.

Construction Observations

During installation of stone columns in Test Areas 1 and 2, water was observed flowing on the ground surface from newly constructed columns. This led to softening of the upper soils, a problem which was exacerbated by rainy weather that soaked the ground around the site. Because the upper soils in the work area became very soft, the front-end loader used during the stone column installation caused deep rutting in the ground surface. The ground surface surrounding the stone columns heaved on the order of about 1 to 3 feet, and this soil was continuously removed during installation to facilitate work activities. In addition to heave at the ground surface, the probe was observed to have substantial cohesive soils sticking to it as it was withdrawn from the hole. This "dirty" probe could be attributed to either the upper soft soils, or disturbance of relatively stiff cohesive soils at depth.

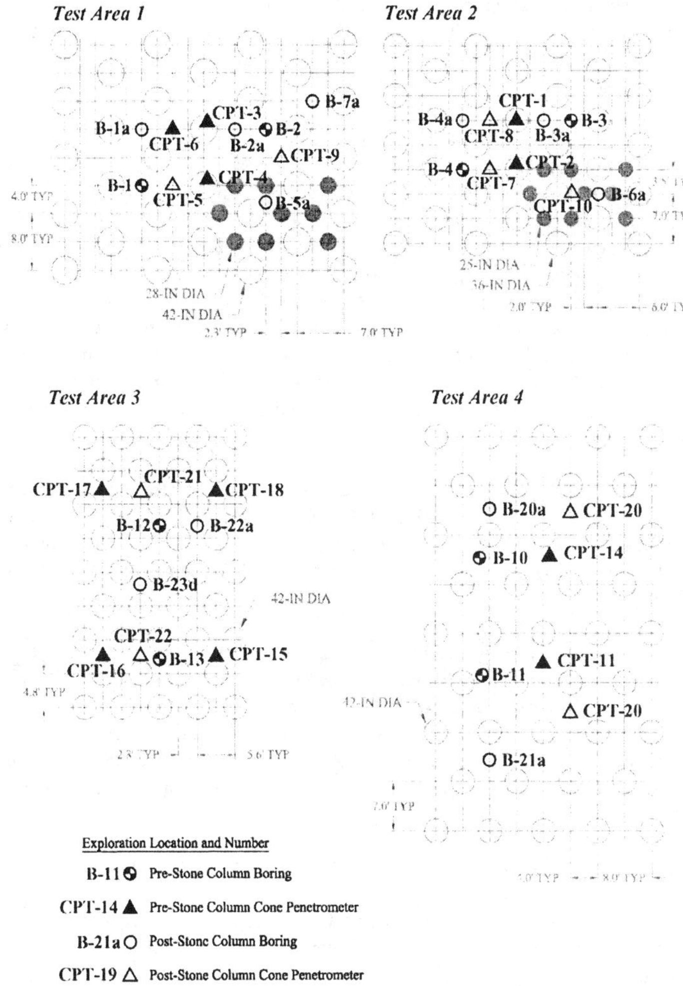

FIG. 3—Plans of Stone Columns and Penetration Tests in Each Test Area

During construction of the columns in Test Areas 1 and 2, the amperage of the probe was observed to remain relatively steady at about 140 to 220 amps as the column was being built to the ground surface. These amp readings are lower than one normally would expect to see as crushed rock in the column is compacted. The low amp readings are attributed to soft ground conditions encountered during construction, which may have resulted from low confinement and lateral spread as the crushed rock was placed and compacted.

During installation of stone columns in Test Area 3, the ground surface remained dry, and very little water was brought to the surface through existing columns. Special caution was taken to define refusal criteria that were intended to limit any disturbance of the stiff to hard cohesive soils that were present below depths of about 12 feet. These criteria were relatively subjective and required close observation of the behavior and sound of the vibrator probe. Although amperage was continuously monitored during the initial probe advance, the presence of the stiff to hard underlying soils was not definitively indicated by consistent higher amperage. Based on the cleanliness of the probe, as well as the lack of heave at the ground surface, we surmised there was not significant penetration of the vibrator probe into the stiff to hard underlying soils. However, this was refuted by comparison of pre- and post-installation penetration tests.

During construction of the columns in Test Area 3, the amperage of the probe was observed to range from about 250 to 300 amps as the column was being built up to the ground surface. These higher amperage readings indicate higher compaction within the stone column, which suggests that the surrounding soils were also being densified.

During installation of stone columns in Test Area 4, the ground surface remained dry, and very little water was brought to the surface through existing columns. At some column locations, a cone of depression formed around the probe during initial penetration, indication densification of the sands surrounding the probe. Other observations are similar to those in Test Area 3. The amperage of the probe was observed to range from about 220 to 300 amps as the column was being built up to the ground surface.

VERIFICATION TESTING

Within each test area, SPT borings and CPT probes were advanced prior to the installation of stone columns. After completing columns in the test area, about 1 foot of ground heave was bladed off the top of the test area to try to locate the top of the stone columns. The ground was allowed to dry for several days to facilitate access for the testing equipment. Table 2 summarizes the before and after SPT and CPT explorations in each of the four test areas. The locations of these explorations relative to the stone column layout are illustrated on Figure 3.

TABLE 2. Summary of Explorations in Test Areas

(1)	(2)	(3)	(4)	(5)
Test	Before		After	
Area	SPT	CPT	SPT	CPT
1	B-1	CPT-3	B-1a [6]	CPT-5 [7]
	B-2	CPT-4	B-2a [6]	CPT-6 [7]
			B-5a [6]	CPT-9 [15]
			B-7a [15]	
2	B-3	CPT-1	B-3a [5]	CPT-7 [6]
	B-4	CPT-2	B-4a [5]	CPT-8 [6]
			B-6a [14]	CPT-10 [14]
3	B-12	CPT-15	B-22a [6]	CPT-21 [6]
	B-13	CPT-16	B-23a [6]	CPT-22 [6]
		CPT-17		
		CPT-18		
4	B-10	CPT-11	B-20a [2]	CPT-19 [2]
	B-11	CPT-12	B-21a [2]	CPT-20 [2]
		CPT-13		
		CPT-14		
Note: Number in [ ] indicates the time interval in days, between completion of columns and post-construction penetration tests.				

Results of the penetration tests are shown on Figures 4 and 5. The depth axis on Figures 4 and 5 was adjusted where the surface was graded to remove soft soils during construction, to facilitate comparison of penetration resistance before and after construction.

In Test Area 1, as shown on Figure 4, CPT data recorded after stone column installation indicate that some densification occurred from 0 to 6 feet below the ground surface where clean sand is present. The CPT data did not indicate significant change in density for the silty sands present from 6 to 10 feet below the ground surface. In contrast to the CPT data, the SPT data indicated little improvement in the 0- to 6-foot-depth range after 7 days, and some possible improvement in this area 15 days after column completion. Below 6 feet, SPT data indicated that density might have actually decreased during stone column installation. Prior to stone column installation, CPT refusal occurred at a depth of 14 feet. After stone column installation, CPT refusal occurred between depths of 16 to 18 feet. This suggests that the deeper soils may have been disturbed due to stone column installation. Also, results of boring B-5 suggest a greater degree of disturbance occurred where the stone columns were closer together.

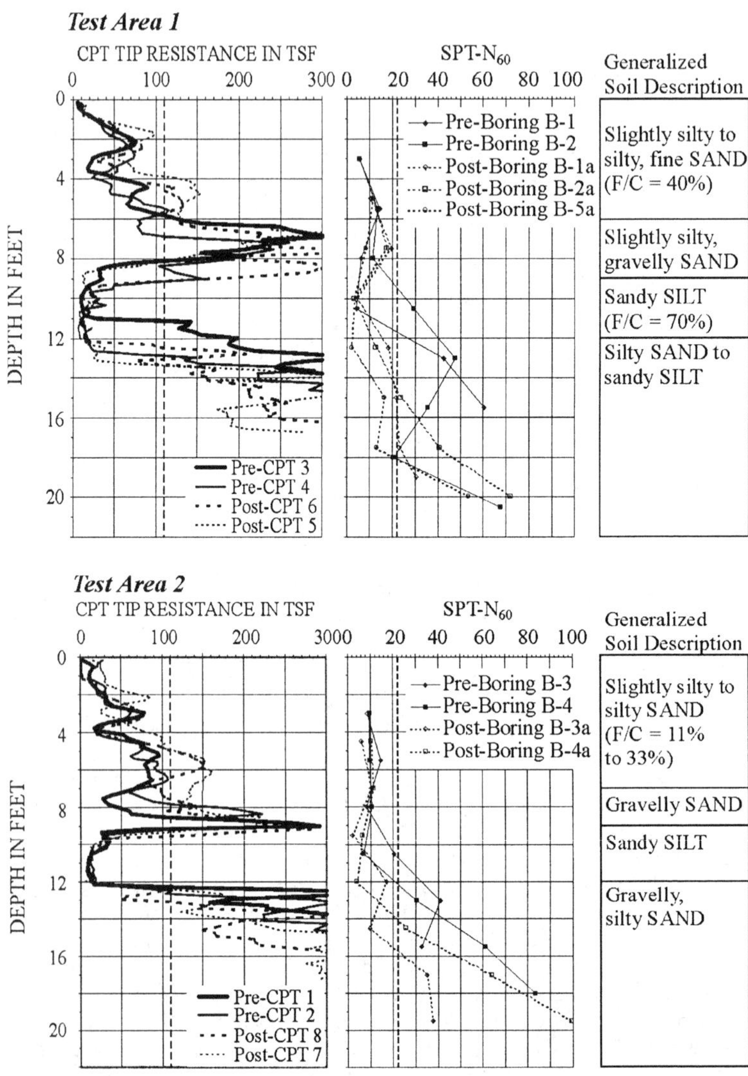

FIG. 4-Comparison of Before and After CPT and SPT in Test Areas 1 and 2

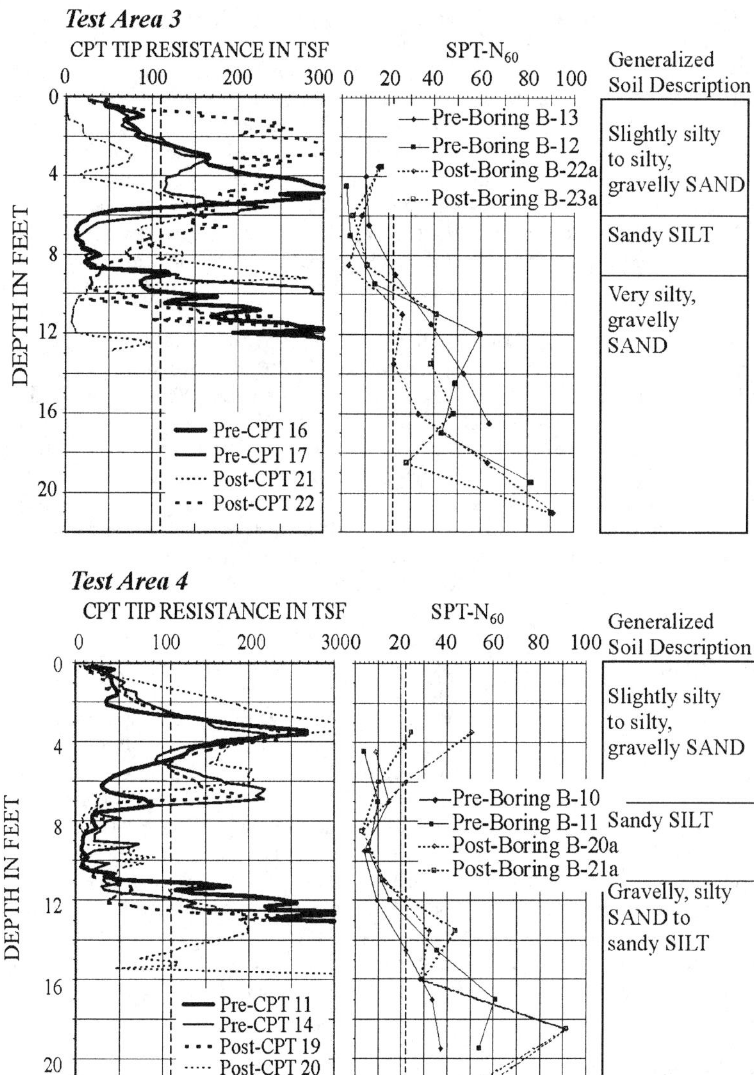

FIG. 5-Comparison of Before and After CPT and SPT in Test Areas 3 and 4

In Test Area 2, CPT tests [CPT-7 and CPT-8] show that the density of the upper soil, interpreted to be silty and clayey sand, was not much improved by the stone columns, as shown on Figure 4. Data from CPT-10 [not shown on Figure 4] showed apparent densification of sand from 0 to 5 feet below ground surface where clean sand is present. The CPT tests show good improvement from depths of 5 to 9 feet, no change in the sandy silt from 9 to 12 feet, and apparent reduction in penetration resistance below 12 feet in depth. Overall, the SPT blow counts appear very little changed down to about 12 feet, and somewhat reduced below 12 feet following stone column installation. Prior to stone column installation, CPT refusal occurred at a depth of 14 feet. After stone column installation, CPT refusal occurred at 17 feet. This suggests that the deeper soils may have been disturbed during stone column installation.

In Test Area 3, the four CPT tests performed prior to stone column work show inconsistent tip resistance in the depth interval from 2 to 6 feet in slightly silty to silty sand. Comparing CPT results after stone column installation, CPT-21 indicates the tip resistance decreased, while CPT-22 indicates large improvement in tip resistance in the same interval from depths of 2 to 6 feet. In the depth range of 6 to 12 feet, results from CPT-21 and CPT-22 also contrast one another, and alternatively indicate lower tip resistance when compared to the four CPT tests performed prior to stone column construction. [Only two of the CPT tests before construction are shown in Figure 5, for clarity, but all indicated relatively consistent conditions prior to stone column installation]. The SPT results indicate no improvement in density in the depth interval from 0 to 10 feet, while below a depth of 10 feet, blow counts decreased following stone column installation, indicating looser subgrade conditions in the very silty, gravelly sand.

In Test Area 4, the four CPT tests performed prior to stone column work show significant variability in tip resistance in the depth interval from 0 to 8 feet. Comparing CPT results after stone column installation, CPT-20 shows no improvement in tip resistance compared with CPT-11 down to a depth of 5 feet, with some improvement from depths of 5 to 7 feet as shown in Figure 5. CPT-19 indicates improvement in tip resistance from 0 to 6 feet when compared to CPT-14, but no change in tip resistance in the silts and clays down to depths of about 13 feet. After stone column installation, CPT-19 had deeper refusal compared to CPT-14 before installation, again indicating that there was apparently some disturbance of the stiff to hard cohesive soils at the base of the stone columns. The SPT results suggest some improvement in density of the surficial sand and silty sand following stone column installation.

CONCLUSIONS

1. Stone columns are an effective subgrade improvement technique in clean to slightly silty sands. However, for variably interbedded sands, silty sands, and

silts or clays, results are not consistent and this may not be an appropriate method of ground improvement where subgrade densification is required.

2. Where stone columns are used in very silty sand or cohesive soils, the engineer must be satisfied that the strength and stiffness of the stone columns is sufficient, even if shear strength of the cohesive matrix soils is reduced due to disturbance.
3. Penetration tests are a good means of quality control for stone column installation. Both CPT and SPT are recommended, as the two methods complement one another and neither method alone is as good as the combination.

ACKNOWLEDGMENTS

The authors wish to acknowledge their appreciation of the work by the ground improvement contractor on this project, GKN Hayward Baker. The authors also want to thank Phoebe Brandal, David Holmer, and Heidi LeVasseur of Hart Crowser for their assistance in preparation of this manuscript.

SYSTEM OF UNITS

Conversion from conventional US Units to International System of Units [SI] is as follows.

1.0 cubic yard	=	0.76 cubic meters
1.0 foot	=	0.30 meter
1.0 ton per square foot	=	0.10 mega Pascals
1.0 inch	=	2.5 centimeters
1.0 horsepower	=	750 watts
1.0 ton	=	910 kilograms

REFERENCES

Baez, J.I. and Martin, G. [1993]. "Advances in the Design of Vibro System for the Improvement of Liquefaction Resistance." *Proc. Symposium on Ground Improvement*, Vancouver, Canada.

Barksdale, R.D. and Bachus, R.C. [1983]. "Design and Construction of Stone Columns." Research Report, Federal Highway Administration, Contract No. DTFH61-80-00111.

Seed, H.B. and Booker, J.R. [1976]. "Stabilization of Potentially Liquefiable Sand Deposits Using Gravel Drain Systems." Report No. EERC 76-10, U.C. Berkeley.

DEVELOPMENT OF SLURRY WALL TECHNIQUE AND SLURRY WALL CONSTRUCTION EQUIPMENT

Wolfgang G. Brunner[1]

ABSTRACT: The slurry wall technique has undergone an evolution from its invention to its today`s status. In the early stages slurry walls were built with cable grab, later on with much stronger hydraulic grab and last but not least with the use of the cutter technique. The cutter allows thick walls with high verticality to extreme depths through hard soil and rock formations. With the rising demand of the construction of watertight diaphragm walls different joint systems have been developed. In addition to the rapid development of excavation systems it was necessary to improve slurry treatment and desanding plants, especially in conjunction with high performance trench cutters.

1. INTRODUCTION

The slurry wall or diaphragm wall technique was introduced in Europe in the 1950s for the construction of underground reinforced concrete structural wall systems. The technique initially evolved from the principal idea of stabilising open excavations with thixotropic fluids, a concept invented in the 1940s by Prof. Veder of Austria. During the past 20 years it has undergone a further dramatic evolution not only in terms of analysis and design methodologies, but also in terms of construction technology and construction equipment.

In the early stages of development, slurry walls were constructed by cable grab mounted on tripod rigs on rails (Figure 1). In subsequent years, specialist firms

[1] Dipl. ing, Bauer Maschinen GmbH, P.O. Box 1260, D- 86522 Schrobenhausen, Phone: +49-8252-971244, Fax: +49-8252-972625, Germany, Wolfgang.Brunner@bauer.de

developed alternative excavation equipment such as rope grabs suspended from crawler cranes (Figure 3).

FIG. 1 FIG. 2

The limitations of rope operated grabs, mainly due to their lack of power and speed when excavating stiff clayey soils and rock formations, were overcome by the development of hydraulically operated grabs capable of generating high closing forces (Figure 4).

In the 1960s, a milestone in the development of slurry wall excavation equipment was reached with the introduction, in Japan, of the reverse circulation technique in the field of slurry wall construction. Whilst one company developed a rig with blade cutters rotating around a vertical axis, another one designed a cutter with two cutting wheels rotating around a horizontal axis. Both systems used either a suction pump located outside the trench or an airlift system to remove the excavated material (Figure 2).

FIG. 3 FIG. 4

The arrival of the reverse circulation cutter technology in Europe resulted in intense research and development by several European geotechnical and foundation contractors.

2. DEVELOPMENT OF THE CUTTER

In the early 1980s, the first BC 30 trench cutter was developed and built for the construction of a cut-off wall in fractured sandstone below the main dam of the Brombach storage reservoir in Germany (Figure 5, 6,7, 8).

FIG. 5

FIG. 6

FIG. 7

FIG. 8

The introduction of the newly developed cutter system on the world market opened up new horizons for the uses and applications of the slurry wall technology:

- Increase in depth:
  Whilst the 1950s the average depth of slurry walls was in the range of 20 m in the 1950s, it is now possible to construct walls to depths of 100 m and more.

- Increase in wall thickness:
  When the slurry wall technique was first introduced, the equipment capacity was limited to widths of between 500 and 800 mm. Today, projects have been successfully executed with a wall thickness of 2500 mm.

- Wall construction in hard soil formations and rock:
  Trench excavation in rock has been made possible by the introduction of trench cutters capable of producing high torque (hydraulically power is often in excess of 600 kW) and also by the development of cutter wheels with excellent rock crushing capabilities (such as the roller bit).

- Wall construction in limited headroom conditions:
  Despite the fact that the capacity of trench cutters is still increasing, special cutters have also been developed which do not require more than 6 m headroom. These 'mini' cutters were specifically requested by the Japanese, Korean, Taiwanese and recently Singaporean markets for the construction of diaphragm walls in connection with mass rapid transit systems in the big cities (Figure 9, 10, 11, 12).

FIG. 9 CBC / MBC in Singapore
FIG. 10 CBC in Japan
FIG. 11 MBC in USA
FIG. 12 MBC in Korea

FIG. 9

FIG. 10

FIG. 11

FIG. 12

- Environmentally friendly construction process:
  Trench cutters have the added qualities of noise and vibration-free excavation, which is an important environmental factor when working in densely populated areas.

- High degree of verticality:
  One of the major advantages of trench cutters over other conventional cutter equipment is their ability to construct trenches to extremely high vertical tolerances. The verticality of cutters is controlled by inclinometers, which are integrated into the cutter system. For stringent verticality requirements a specially developed hydraulically operated cutter steering system is deployed. Adjustments of the cutter position in vertical direction of the trench are made via long hydraulically operated steering plates, whilst adjustments in longitudinal direction are made by pivoting the cutter head through up to +/- 3 degrees. Throughout the correction process, the rig operator is continuously guided by corrective measures displayed on his on- board monitor.

- Automatic quality assurance system:
  The entire slurry trench construction process is monitored and controlled by a PC-based electronic system. On completion of each diaphragm wall panel a full cutter report can be printed out inside the operator's cab or transmitted by remote data transfer direct to the site office, the company's head office or the consultant engineer's office for further analysis and evaluation (Figure 13).

In conclusion, the market required slurry trench or diaphragm wall construction equipment capable of excavating deeper and thicker trenches at greater speed through harder soil formations.

In the early days, the cutter frame as well as the hydraulic hoses and slurry hoses were simply suspended from a rope on a long crane jib and the depth of the trench was limited to twice the length of the jib of the base carrier. As a result of the increasing demands on slurry trenches as referred to above, the base machines for cutters also increased until gigantic cranes had to be used. But more and more frequently diaphragm walls had to be constructed on down- town city centre areas amidst bustling traffic and on very restricted sites. This led to the development of smaller, so called 'compact' cutters, with all hoses on hose drums, capable of constructing diaphragm and cut-off walls to greater depths on extremely restricted sites.

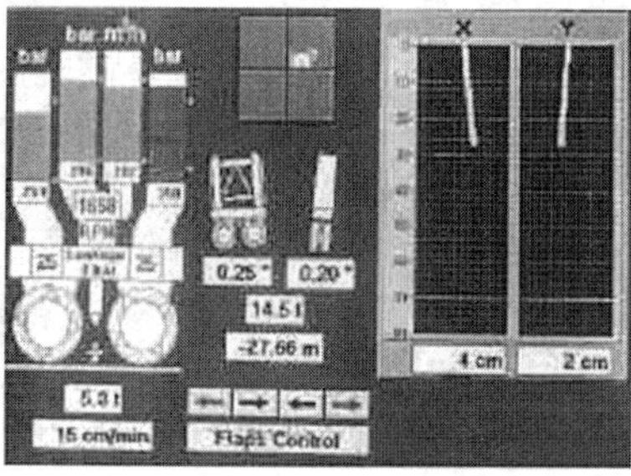

FIG. 13

FIG. 14

In addition to the rapid development of trench cutter excavation systems, it was also necessary to improve slurry treatment, desanding plants as well as decanters, especially in conjunction with high performance trench cutters (Figure 14).

3. WATER TIGHTNESS AND DIFFERENT JOINT SYSTEMS

Water tightness is particularly important for permanent diaphragm walls and is not always easy to achieve. The degree of water tightness that can be achieved depends primarily on the type of joint between individual panels, the joint details and adequacy of their construction.

3.1. Standard joints constructed with stop- end casings

This is the traditional and conventional system, using rope grabs for the excavation for a slurry wall depth of up to 25 m and for water pressures of 4 – 6 m. The stop-ends are pulled out with casing extractors. The guide wall has to be designed for the reaction forces resulting from the stop- end casing extractor.

3.2. Joints constructed with prefabricated concrete elements

This system is generally used for a slurry wall depth of up to 25 m or a lifting weight of the pre- cast concrete element of up to 25 tons. For greater depth the pre- cast element can be installed in several sections which will be connected during the installation. The pre- cast element guides the grab during the excavation of the secondary panel. The water tightness can be improved by placing additional rubber water stops in the concrete panel (Figure 15, 16).

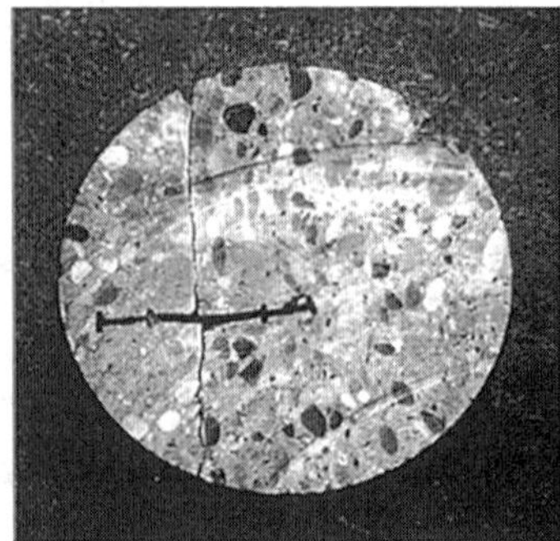

FIG. 15

FIG. 16

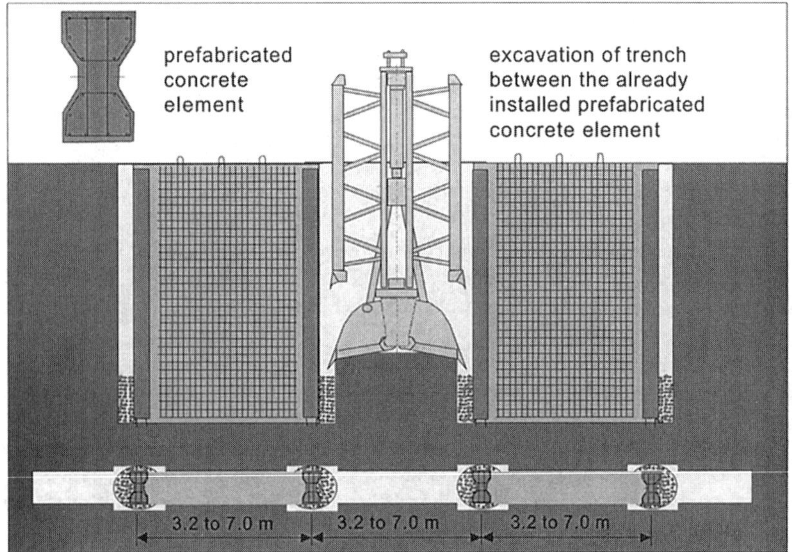

FIG. 17

3.3. Joints predrilled with prefabricated concrete elements

This system is used for a slurry wall depth of up to 25 m or a lifting weight of the precast element of up to 25 tons. For greater depth the precast element can be installed in several sections which will be connected together during the installation. The precast element guides the grab during the excavation of the secondary panel.

The water tightness can be improved by placing additional rubber water stops in the concrete panel (Figure 17).

3.4. Joints constructed with endplate reinforcement cage

This system is used for a slurry wall depth of up to 80 m. The endplates guide the grab during the excavation of the secondary panel. The water tightness can be improved by placing rubber water bars at the endplate (Figure 18).

FIG. 18

3.5. Joints constructed with CWS (continuous water stop)

This system is used for a slurry wall depth of up to 20 m. The water tightness is improved by the rubber water stops of the CWS system. 2 CWS forms should be placed in each primary panel. They can only be removed during the excavation of the secondary panels (Figure 19, 20).

FIG. 19

FIG. 20

3.6. Joints constructed by over cutting the primary panels

This system is used for a slurry wall depth of up to 120 m. The water tightness is improved by cleaning the overcut joint with a brush before installation of the reinforcement cage (Figure 21, 22, 23).

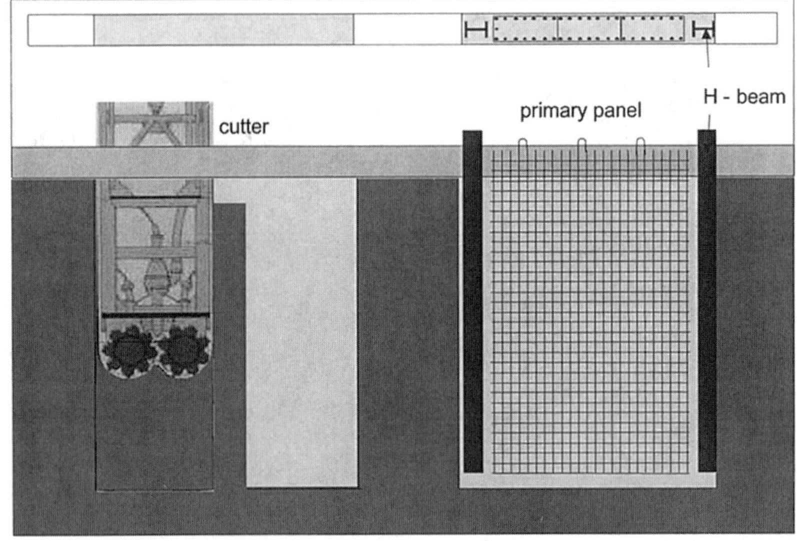

FIG. 21

Due to the cutter technique, the construction of watertight joints is now much easier and much more effective than with the conventional rope grab method. Instead of using stop- end- tubes, the cutter encroaches on the adjacent primary panels as it descends during excavation and thereby cuts back a fillet from each end of the previously concreted primary panels. The cut- back or overlap between primary and secondary panels ranges between 200 and 300 mm. Joints produced in this manner are watertight, because they consist of a serrated surface resulting from the formation of grooves cut into the concrete of the primary panels by the cutter wheels. This not only increases the seepage path through the joint, but also provides a clean and roughened surface with which the concrete of the secondary panel is then able to bond extremely well without the risk of the formation of a filter cake.

FIG. 22 FIG. 23

Limitations of the grab technology are:

The excavation output reduces with the increasing trench depth as a result of the intermittent nature of the process of excavation and in the case of the excavation of hard cemented soil formations. There are difficulties with the construction of joints for very deep slurry walls or cut- off walls.

4. Conclusion

The described advantages of the cutter technique led to the use of cutters in many projects around the world in particular because of high performance, greater depth, greater wall thickness, penetration through and into hard rock, noise and vibration-free excavation and high degree of verticality.

CONSTRUCTION OF THE DEEP CUT-OFF AT THE WALTER F. GEORGE DAM

Arturo Ressi di Cervia[1]

ABSTRACT: The Walter F. George dam was built in 1962 on the Chattahoochee River by the Corps of Engineers and it soon developed seepage problems.

Two partial cut-offs were built in the past to attempt to remedy the situation, but the problems continued, jeopardizing the safety of the structure as well as reducing its power generating potential.

In 2001 the Mobile district of the Corps advertised a design-build contract for the construction of a 200 feet deep cut-off in front of the concrete structure of the dam starting from 90 feet of water, cutting through a concrete lock structure, a submerged retaining wall and extending on both sides on the earthen embankments.

The winning proposal of the joint venture of Treviicos and Rodio anticipated the use of a secant pile wall in the portion to be built in front of the structure, working from barges with a reverse circulation drilling rig, and a conventional slurry wall built with a hydromill rig for the portion out of the water.

FIG. 1 Aerial View of Cut-Off Wall Construction

[1] Director, Treviicos Corporation.

1. INTRODUCTION

The Walter F. George lock and dam was constructed on the Chattahoochee River between 1955 and 1963 with the double function of improving navigation and generating power. Its 82 feet by 450 feet lock can accommodate large barge traffic and its four generating units with a plant capacity of 150 MW have an average energy output of 453,000 MWH. As a bonus to the local economy the 80 miles long lake offers various recreation opportunities as well as it harbors an 11,160 acres wildlife refuge, home to several bird species and a population of approximately 1,000 alligators.

The dam consists of a 1496 feet long concrete structure housing the spillway, a non overflow portion, the power units and two earthen wing dams, extending to the Georgia and Alabama sides, respectively 5810 feet and 6130 feet long.

As soon as the reservoir was impounded it became apparent that excessive seepage was occurring under the structure.

In October 1961, during stripping operations on the final stage of construction of Alabama dike, two sinkholes occurred.

In 1962 boils developed along the downstream toe of both Alabama and Georgia.

In order to control the under seepage, relief wells were installed at the toe of the embankments and the foundation beneath the dikes was drilled and grouted.

Subsequently, after completion of the project, a large spring developed near the downstream lock guide wall and sinkhole activity increased on the Georgia side.

In 1968 sinkhole activity increased and a new phase of remedial works began.

Initially, the remedial work consisted of grouting a small area of the foundation under the Georgia dike, filling the sinkholes with sand and grouting around their periphery.

In February 1969 a sinkhole was discovered in the reservoir near the Georgia dike and it was filled and grouted in a similar manner.

A surface investigation downstream of the Georgia dike was carried out, a number of piezometers were installed and information concerning solution channels was obtained. After the evaluation of the new data, a sand filled filter trench was constructed downstream of the lock guide wall in order to intercept the water bearing solutions channels which were discharging through a spring. The objective was to reduce the water velocity in order to prevent the movement of solid material.

In 1970 a line of grout holes was performed downstream of the Georgia dike.

During 1971, several small sinkholes were discovered at the Alabama dike toe and simultaneously pin boils were found in the relief well collector ditch in the same area.

In 1978 a supplemental design recommended the installation of a positive cut-off wall under the earth embankment to upgrade the dam to current standards.

This recommendation was adopted, and a 24 inches thick concrete cut-off wall was installed to the bottom of the shell layer within the Earthy Limestone. In limited areas, in accordance with a previous soil investigation, the cut-off was installed deeper, generally to the bottom of the Shell Limestone.

Phase I of the cut-off was installed under a portion of the Georgia dike in 1981, and phase II was installed beneath an additional portion of the Georgia dike and the Alabama dike between 1983 and 1985.

In 1982, between the two phases of construction, a boil was observed on the water surface immediately downstream of the Power House. The entrance point, called the "Hungry Hole", was plugged with a tremie pipe, and it and the eroded channel under the Power House were filled with 175 c.y. of tremied concrete. Additional grouting was performed along the upstream face of the power house and part of the spillway.

Those efforts to stop the seepage were only partially successful and resulted in a concentration of the seepage below the concrete structure. The seepage problems at the W. F. George dam continued to worsen.

Discharges from piezometer SP-5 and from the power House drainage system increased. As the Shell Limestone material was piped from under the concrete structure, the passage ways through which the flow was occurring were being enlarged resulting in greater erosion.

This situation compromised the generating potential of the dam as well as potentially jeopardized its stability.

Finally, the decision was taken to construct an additional, deeper cut-off to remedy the problem.

2. PROCUREMENT

On the 5th of July 2001 best and final offers were solicited by the Corps of Engineers for the construction of a deep cut-off to be built in front of the concrete portion of the dam from the bottom of the lake and tied into it. The cut-off had to cross an existing lock structure and an underwater retaining wall, as well as the remnants of a steel coffer cell left in place from the initial construction.

The specifications required that the wall be built with a minimum thickness of 24" and that it extended from the bottom of the lake to elevation -5 into the Providence formation, an impervious layer.

While the land portions of the wall were on average 208 feet deep, the marine portion started from up to 90 feet of water and continued through different rock formations for another 100 feet.

From a geological standpoint the major challenges consisted of the existence of a very hard stratum (up to 14000 PSI) overlaying a loose sand layer, the presence of voids, fractures and pathways in the rock mass and the existence of sediments and various debris at the lake bottom.

The specifications imposed limitations on the time utilized to cross the lock, as well as when work could be performed in front of the generating units and specifically prohibited the use of a grout curtain to construct the cut-off.

Otherwise contractors were invited to propose their own method of constructing the cut-off and offers would be evaluated using several criteria, technical, economical, organizational and experience.

Three proposals were received and on August 14, 2001 an award was made to the JV of Treviicos and Rodio.

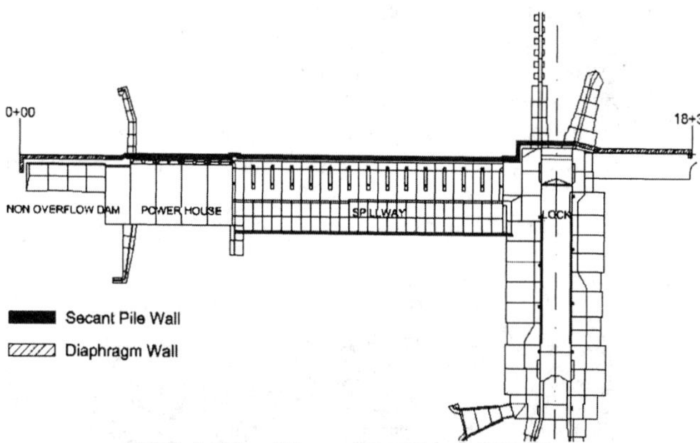

FIG. 2 Plan View of the Cut-Off Wall

3. JV's APPROACH TO THE WORK

The JV had been formed well ahead of the bid and a task force was assembled to study the best approach to this difficult procurement.

As a first guideline it was agreed that the method employed would make minimum use of diving in order to avoid both the high cost inherent in performing underwater work at depths of almost 100 feet and the risks involved in such operation.

It was then agreed that the most reliable and economical method to do the marine portion of the work would be secant piles, while the land portion would be a slurry wall done by hydromill, using bentonite as stabilizing fluid.

It was further decided, after long discussions, that the secant pile work, both primary and secondary elements, would be done by reverse circulation rigs mounted on top of the casing, utilizing lake water to lift the cuttings.

Lastly it was considered prudent to perform a preliminary drilling and grouting campaign in order to verify ground conditions and to fill any large voids which could disrupt the orderly progress of the work.

4. CONSTRUCTION METHOD

The work was performed in several principal phases. Due to the intricacy and complexity of some of the operations, and consequently this paper will just give a brief description of the salient points of interest, equipment used and results achieved in each of the phases.

4.1. Drilling and Grouting Campaign

In order to reduce the uncertainties over the extent of the Karstic phenomenon, a series of exploratory holes was done along the axis of the cut-off wall. They consisted of:

- Localized borings to ascertain:
  - Depth of sediments
  - Presence of fill materials placed upstream of the dam during construction
  - Thickness of hard sandy limestone layer
  - Characteristic of sand layer (apparently none cemented) within the sandy limestone
- Grout holes to fill major cavities, possibly with high water flows

FIG. 3 Drilling and Grouting

4.2. Cleaning the Lake Bottom and Constructing a Working Apron

The purpose of the apron installation was to create a zone, upstream of the dam monolith, free of obstruction and consisting of solid, homogeneous material forming the working surface used to install the cut-off wall and, ultimately, the cap beam connecting the cut-off to the dam.

To accomplish this, an under water trench was excavated, extending a minimum of five feet from the concrete monolith, of a minimum depth of six feet, to be filled with an engineered material.

4.3. Installing Casing Templates Tied to the Concrete Dam

Whenever possible, guide beams and templates were installed and secured to the concrete dam structures. The template is composed of one main beam, connected to the fender supports by pins and latches, and of one shutting beam, which closes the template.

The main beam was mounted on the props fixed to the fender of the dam. Its position and alignment was checked using reference points previously installed on the dam during the initial survey.

When it was impossible to attach the template to the dam, temporary piles were driven to support the original or a modified template. The templates were used to install and drive the temporary casing into the apron.

FIG. 4 Templates

4.4. Drilling and Concreting Piles

The 50 inches secant piles, spaced 33 inches center to center, formed the cut-off wall. The piles were installed in a sequence of primaries and secondaries, in which the secondary piles were drilled between the two adjacent primary piles thereby forming a cold joint.

The piles were installed to form a continuous wall, of the minimum thickness of 24 inches, from the lakebed to El – 5 MSL, into the Providence formation.

Typically, the execution of the piles entailed the following steps:

- Preparation of a working platform underwater (Apron)
- Installation of the temporary casing supported and guided for the steel template
- Setting the reverse circulation rig on top of the temporary casing
- Drilling to design depth
- Removal of the drilling string
- Setting tremie pipes and pouring concrete
- Withdrawal of casing and setting in a new pile position

Water pumped from the lake was used as drilling fluid.

The main equipment for the installation of the secant piles was the following:

- Pontoons and spoil barges
- Manitowoc 4000 crawler services cranes positioned on the pontoons
- Templates for the installations of casings
- Wirth PBA 612 reverse circulation rig
- ICE 66-80 vibro hammer to install the casing
- ICE 160 hydraulic pile hammers to install the casing trough hard strata
- IR 960 high pressure compressors
- Verticality control device

The secant pile cut-off wall was drilled to full depth using the Wirth PBA 612 reverse circulation rig mounted on top of the 50" casing. It has a rotary drill table mounted on the top of a frame, and a working platform. The rig is fixed on the top of the casing by means of clamps. A jib positioned on the top pile rig platform facilitates handling of rods.

On the top of rotary table there is a two way 8"swivel, discharging the cuttings into a 10" hose. The compressed air for drilling is supplied by means of 4"pipes and hoses, connected to a swivel below the rotary table.

A special stabilizer sets was used in the first rods in order to maintain the verticality of the drilling string during drilling operations.

Flanged double wall drillings rods, 10 feet long with 6 " inner cuttings conduit and air conduit and the drilling bit completed the drilling equipment

All the cuttings excavated by the reverse circulation rigs passed through a 10" discharge pipe and were delivered to the hopper barge. From this barge the cuttings were deposited to the bottom of the lake. Silt curtains, installed around the hopper barge, prevented the contamination of the lake water by the spoil material.

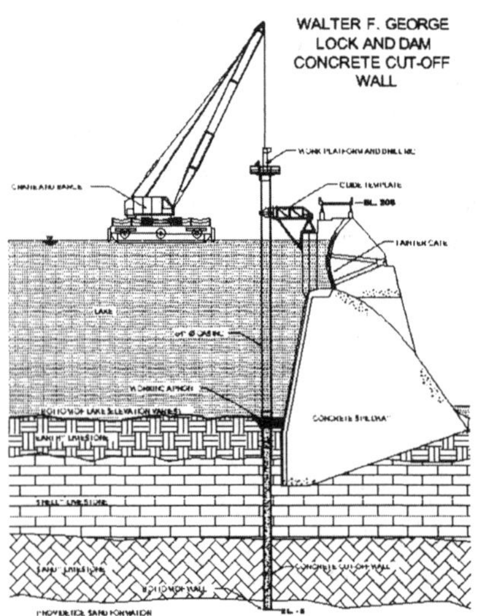

FIG. 5 Section of the Secant Pile Wall

FIG. 6 Reverse Circulation Rig

4.5. Constructing the Slurry Wall on the Georgia Side

The execution of the slurry wall entailed the following steps:

- Construction of a guide wall
- Installation of the temporary steel guide frame on top of guide wall
- Set up of the reverse circulation Hydromill
- Drilling to design depth using stabilizing slurry

- Testing stabilizing slurry before concreting
- Setting concrete pipes and pouring concrete
- Recovering slurry and sending it to tanks to be regenerated

Bentonite slurry was used to stabilize the panels during the excavation.

The diaphragm wall on the Georgia side was installed with the equipment operating from the existing parking lot.

4.6. Constructing the Slurry Wall on the Alabama Side

The cut-off in Alabama side was installed with the Hydromill placed on the crest of the dam. Consequently the working platform was only 35 feet wide by 240 feet long and the elevation of the working platform was El + 195.

The working platform was constructed as follows:

- Removal of the existing Rip Rap. This material was stock piled and then reinstalled at the completion of the work
- Sheet pile installation
- Placing flowable fill under water and continuing up to elevation 195
- Construction of guide walls in the flowable fill on the proposed alignment of the wall

In both cases, the following main equipment was available at the site to perform the trench excavation:

- Hydromill cutter to excavate the trench
- Clamshell to performed the pre-trench excavation
- Slurry mixing plant
- Desanding Plant
- Koden verticality testing device
- Manitowoc 4000 service crane

The slurry wall was excavated to the full depth using a Hydromill model HF 12000. The equipment is a continuous reverse circulation soil–rock cutter mounted on a Manitowoc 4100 crawler crane.

The slurry wall was built by primary and secondary panels.

The primary panel was formed by single or multiple bites with a total excavation length of 8 to 24 feet. The secondary panels overlapping the primaries were a single bite of 8 feet.

This excavating machine cuts a vertical panel of the proposed width of the trench in one full-depth pass while disposing of the excavated materials via a reverse mud circulation system.

At the base of a vertical steel frame, having the same width as the trench, two wheels fitted with tungsten carbide cutters rotate in opposite directions, excavating the soil or rock. Just above the wheels, a suction pump extracts the drilling fluids with the cuttings, and delivers them to the desanding plant through circulations lines.

The desanding plant was a Sotres-450 with a capacity of 500 cu. yd. per hour, consisting of a series of vibrating screens and cyclones capable of screening all the cutting sizes.

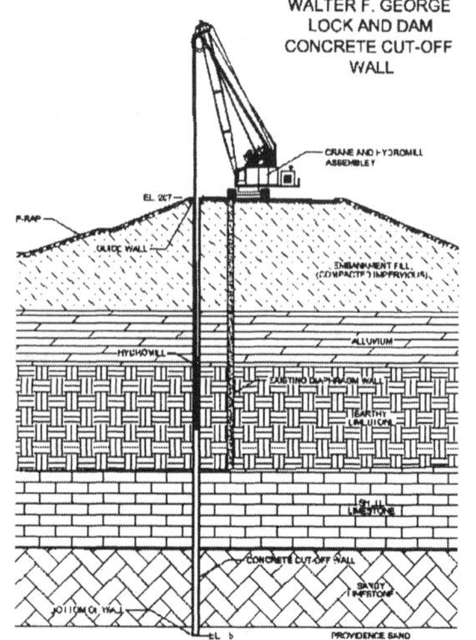

FIG. 7 Section of the Slurry Wall Cut-Off

FIG. 8 Hydromill

The automatic slurry mixing plant comprises:

- Dry platforms to store bentonite bags and additives
- Holding tank
- Unit composed by a hopper and a Venturi pipe
- Electrical control board

FIG. 9 Slurry Desanding Plant

Mobile steel thanks with a total capacity of 100,000 gals were installed to store the slurry. The circulation of the fresh slurry from the mixing plant to the hydration and storage tanks and from them to the holding tanks was done by pumps and 4" steel pipes. Six inches feeding and return slurry pipes were installed from the plant to the excavated panels and along the guide walls.

4.7. Equipment for Concrete Placement

The concrete used in pouring the diaphragm .wall and the secant piles was prepared on site by a batch plant. The concrete was delivery to the work area by truck mixers.

A complete set of the following equipment was available on site for the concreting operation:

- 10" diameter steel tremie pipes, equipped with a hopper.
- Service crane to raise and lower the tremie pipes
- Concrete pump to deliver the concrete from the trucks to piles and panels
- A pump to deliver the slurry mud to the desanding plant during concreting operation

4.8. Equipment for Verticality Control

During the Hydromill excavation, the panel verticality at each depth was constantly monitored by means of a precision encoder and two precision inclinometers fixed on the Hydromill frame.

While the encoder detects the actual depth, one inclinometer gives the verticality on the transversal axis and the other on the longitudinal one.

The readings of the instrumentation are displayed in real time inside the cabin, thus, as soon as the first sign of deviation is transmitted, the operator can take action to correct it.

After the completion of drilling by the Wirth rig, the verticality was checked in both axes from the bottom to the top of the pile at two feet intervals using an inclinometer installed inside of the drilling string.

FIG. 10 Verticality Measurement

4.9. Constructing the Cut-Off Through the Lock Structure

The crossing of the lock structure was accomplished by using the Hydromill on the Georgia side to cross the eastern retaining wall, while secant piles did the rest of the crossing

With the Hydromill equipment, vertical excavation within required tolerance is assured when both cutter wheels mill similar material in terms of hardness and consistency.

On the Georgia side, the sloping contact between the soil embankment and the monolithic wall did not allow for the wall to be built in the correct sequence of primary and secondary panels, due to the fact that no panel could be excavated through the inclined concrete surface without a vertical guide.

The solution was to re-create in advance a uniform condition where:

- The inclined surface had been removed by a preliminary reaming of the wall
- The adjacent soil had been replaced by concrete

Under this modified conditions, the standard sequence of overlapped primary and secondary panels was correctly carried out.

In details the following steps were adopted:

- Construction of the slurry wall up to the last panel touching the monolithic wall foundation
- Milling the first adjacent bite passing through the wall for a certain depth
- Note that when the Hydromill reached the inclined hard surface, it reacted against and was guided by the previously concreted panel acting as a shoulder. This did not allow any lateral deviation.
- After entering in the entire concrete mass for several feet, the excavation was stopped and the panel was concreted.
- The same procedure was repeated in single bites in an uphill progression until all of the inclined surface was entirely reamed and the soil replaced by concrete. The last panel in this sequence was excavated and left open.
- After this operation, the slurry wall was build at full depth, re-excavating the newly placed concrete as well as the remaining portions of the lock wall with the proper sequence of primary and secondary panels.

4.10. Connection to Secant Pile Wall

The connection between slurry wall and secant piles was located in the footprint of the eastern monolith. The following steps were carried out:

- The soil between the walls of the Lock was excavated under water from El 108 to the bottom of the walls at El 81, for a width of 15 feet. The lakebed westwards of the western monolith was dredged as well to the bottom of the wall.
- The excavation as filled with 3000 psi concrete to elevation 108
- The Hydromill then continued to remove the concrete from the center of the Eastern wall, progressing westward until the concrete has been removed to El 108. Cutter wheels on the Hydromill for this operation were 63 inches in width in order to accommodate the casings for the secant piles.
- The progression was aided by installing a steel casing to act as a vertical guide
- Construction of the wall by the secant pile method was advanced eastward using the guiding template across the lock, until it encountered the lock wall.

- A secondary pile overlapping the last panel and the first primary pile was the joint element between secant pile wall and slurry wall.
- At the completion of the secant pile portion, the cut of the retaining wall was repaired.

4.11. Crossing the Remaining Portion of the Lock

The cut-off in this section was installed in accordance with the followed phases:

- Removal a section of concrete approx. 12 feet high by 8 feet longs from the top of the wall.
- Fastening of a guide frame for the Hydromill to allow excavation of the first panel in the center of the lock
- Positioning the Hydromill on the top of the lock
- Placing of a casing in the open panel in order to aid in the excavation of the adjacent panel. The speed of the cutter wheels was adjusted to force the Hydromill to follow the guide and not the sloping lock wall. This was continued until the wall was removed to elevation 108.
- The progress was aided by installing additional casings to act as guides
- This work created a level concrete working surface at el. 108 from which the secant pile wall was constructed.

4.12. Constructing the Permanent Concrete Cap Tying the Cut-Off to the Dam

The construction of the cap beam was done upon completion of the cut-off installation. Depending on the cut-off wall's position and the apron elevation, a portion, or all, of the engineered backfilling between the cut-off wall and the dam was removed in order to install the cap beam.

As specified, the cap beam was poured two feet high above the cut-off wall, and at least six feet high between the cut-off and the dam's monoliths.

Before the pouring the 3000 psi. concrete of the cap beam, the surfaces of the monoliths and of the cut-off wall were cleaned to insure a proper joint.

5. CHANGES

It is inevitable that in a job of such complexity situations will arise requiring changes in the original scope of work. In some cases they are caused by changed conditions, in others by value engineering proposals, while in some other cases they involve additional work.

This contract had examples of all three cases, and the challenge was to integrate those changes in the schedule in order to minimize their impact in terms of both cost and time.

This can be best accomplished if the need for those changes is recognized, evaluated and contractually resolved in a timely fashion; in this respect an effective "partnering" atmosphere is critical in moving the process along.

We can proudly say that both the Corps and the JV worked hard to make partnering a reality on this project, and that, as a result, all necessary changes have been dealt expeditiously and in a mutually satisfactory manner.

As previously mentioned, a preliminary drilling and grouting campaign was carried out as initial phase of the project to verify the data supplied with the contract documents and to confirm the actual characteristic of the soil throughout the cut-off alignment. This campaign was the determining factor in some of the major changes.

5.1. Variation of the Soil Characteristics

The campaign evidenced considerable difference in the quality of the embankment fill and of the top portion of the earthy limestone formation at the two extremes of the underwater cut-off wall.

Consequently, in order to proceed with the cut-off wall installation by secant piles as planned on the majority of the underwater cut-off (2/3 of the alignment); two different techniques were adapted to overcome the unforeseen conditions:

- East area: (sta. 11+50 through sta. 14+00): an additional excavation by clamshell bucket was done to reach a sound formation on which the working apron was placed.
- West area: (sta. 2+30 through sta. 3+00): To perform additional excavation at the opposite end could not be done, The mass excavation needed in this area would have created a risk of piping due to the fact that the elevation of the foundation of the dam structure was higher then the bottom of the required excavation. Furthermore the required diving would add additional risk to the operation. As an alternate solution, consolidation by jet grouting was utilized to improve the characteristic of the natural soil formation.

5.2. Concrete Mass

During the preliminary drilling campaign a concrete mass approximately ten feet deep was encountered in the alignment of the cut-off wall.

To better determine the extent and the geometry of this mass additional investigation holes were completed in the area allowing a rough determination of the actual shape of the mass encountered.

These data were jointly reviewed by the COE and the JV to determine the steps needed to overcome this obstruction. Several options, from re-routing the cut-off wall by more than 100 feet to installing the secant piles through the newly discovered concrete, were evaluated. Finally, to avoid additional 'surprises', a Contract Modification was executed for the complete removal of the concrete mass.

Approximately 60 shifts of a combined effort of chisel and clamshell bucked were spent for the removal of the obstruction. The area was then backfilled with flowable fill and the apron was installed as in the other areas.

5.3. Obstructions

While these changes were discovered in advance and finalized before the affected work areas were needed, throughout the execution of the cut-off wall several steel obstructions were encountered.

These obstructions had a negative impact in the execution of both the secant piles and the slurry wall cut-off.

Delays in the execution of the work, additional work for the removal of the obstruction, excessive wear and tear and major break down of the equipment are the principal impacts caused by these obstructions. In one instance, at the junction between the power house and the spillway structure, the cut-off wall had to be re-aligned to overcome a steel obstruction extending almost to the final elevation of the cut-off.

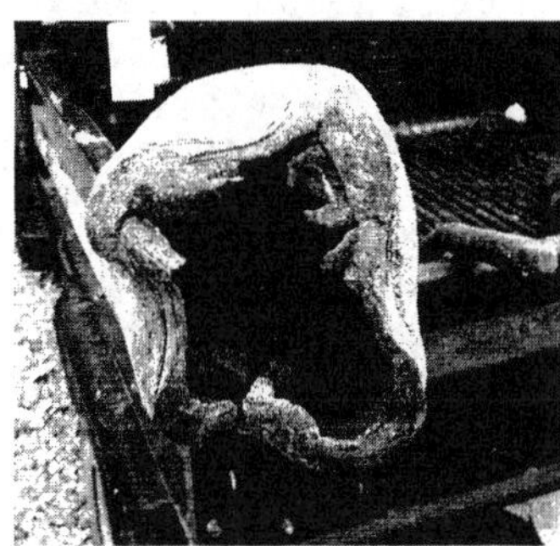

FIG. 11 Steel Obstruction

5.4. Value Engineering Proposal

The partnering and cooperation effort was used not only to overcome the unexpected but also when an alternate solution was developed with associated saving for all parties involved. This was the case in the Value Engineering Proposal for the open water disposal. The VECP proposed the modification of the contract for the disposal of the 50,000 c. y. of combined dredge and drill spoil. The open water disposal was selected in lieu of an up-land facility, resulting in substantial saving shared between the JV and the COE.

The VECP offered the following advantages:

- Minimized impact to the shoreline
- Centralized project supervision and marine traffic basically confined to the actual construction area in front of the dam
- Minimized impact to the powerhouse operations
- Increased safety by minimizing diving time

5.5. Additional Work

During the course of the work the JV analyzed the flow net around the Alabama side of the wall and discussed with the Corps the possibility of extending for some distance the cut-off in order to lengthen the flow path.

The Corps concurred with the findings of the JV and issued a contract modification to increase the length of the cut-off wall on the Alabama side.

6. CONCLUSIONS

The successful completion of this very challenging project has reinforced the validity of certain procedures which, although generally well understood, are worth repeating.

We will discuss them in the order in which they came to play:

- **Pre Bid Preparation**: by assembling a large team of experts the JV was able to explore different technical solutions, compare them by cost and reliability and consequently hone into the best possible approach. This exhaustive preliminary work had the additional benefit of comforting the JV that it had the right price when it became known that our competitor's proposals were considerably higher. Since the job was performed in line with the JV's expectations, it proved that in design-build contracts wide swings in prices reflect different solutions to the problem, not mistakes or greed.
- **Preliminary Investigation**: by conducting the exploratory campaign previously described, the JV gained vital information which allowed it to plan the work ahead, even when confronted with situations at variance from what the contract documents showed. The value of this approach was recognized by the Corps, who modified the contract to compensate the JV for such work, which was not a pay item. It reinforces the truism that advance information is relatively cheap to acquire and it pays for itself many times over in avoiding or mitigating delays and extra costs during the performance of the work.
- **Partnering**: much has been said about it which does not need repeating; it must be stressed, though, that partnering is both an attitude and a commitment and that it needs to be continuously reinforced. On this project both the Corps and the contractor performed admirably on both counts. It directly facilitated the resolution of the various issues caused by changed conditions, contract modifications and value engineering proposals by staging negotiations in an atmosphere of mutual trust and respect.
- **Project monitoring**: by the creation of a management committee drawn from the partners and by appointing an executive in charge, the JV was able to follow closely the progress of the job and to marshal quickly additional resources and expertise when required. This allowed for timely evaluation, study and implementation of procedures at variance from what was anticipated in the original proposal.

DEEP UNDERGROUND BASEMENTS FOR MAJOR URBAN BUILDING CONSTRUCTION

Seth L. Pearlman, P.E.[1], Member, ASCE, Michael P. Walker, P.E.[2], Member, ASCE, and Marco D. Boscardin, Ph.D. P.E.[3], Member, Geo-Institute

ABSTRACT: Deep underground basements that are integrated into urban development projects early in the overall project design offer many inherent improvements to the overall quality and value of the project and its surrounding community. Diaphragm walls combine into a single foundation unit the functions of temporary shoring, permanent basement walls, hydraulic cutoff, and vertical support elements/shear walls, and, because of this combination, have proven to be an economical alternative in many circumstances. This paper examines the evolution of deep underground basement construction, identifies considerations associated with diaphragm wall construction, and provides several case history examples to illustrate the resolution of key issues.

INTRODUCTION

Today, deep underground basements are an important component of new urban building construction. This is often because parking in most large cities is generally inadequate and often serviced by aging, outdated, and deteriorated above-grade parking structures that do not fit the surrounding architecture and occupy valuable aboveground space. The integration of underground parking into major, new building projects in urban environments can enhance the aesthetic and economic values of the overall development.

Conventional basement construction practices have evolved from drivers that include:

---

[1] Vice President, Nicholson Construction Company, 12 McClane St., Cuddy, PA 15031

[2] Project Manager, GEI Consultants, Inc., 1021 Main Street, Winchester, MA

[3] Principal, Boscardin Consulting Engineers, Inc., 53 Rolling Ridge Road, Amherst, MA 01002

- Economy
- Local expertise
- Local tradition
- Local geology
- Groundwater
- Site right-of-way

Evolution of Excavation Support Systems

Open Cuts

When space is available, the most economical way to build a permanent basement is often to construct the foundation in an open cut excavation. This requires sufficient right of way to construct safe slopes and access to excavation, dewatering, and backfill construction expertise. Limitations associated with an open cut solution are typically a construction site with restricted laydown areas far away from the building footprint, restricted crane access, and a need for dewatering. The latter is often associated with detrimental settlements of surrounding properties. In addition, the costs associated with handling, managing, and replacing the large volumes of earth located outside the building footprint increase rapidly as the excavations become deeper.

Soil Nailing

Soil nailing as temporary shoring is a relatively recent technique, introduced in the mid 1980s, which works very well under a specific set of conditions. These include stiff soils, adequate lateral right-of-way, no utilities at the perimeter below a depth of about 6 feet, and no running soils or water inflow issues. These are conditions that are somewhat ideal and not common in most major urban areas. As a consequence, soil nailing is not a widely used urban temporary shoring technique except in specific locales such as Seattle. Examples of successful soil-nailed temporary walls include those completed with a final shotcrete or cast-in-place concrete facing, and with the permanent lateral support provided by the basement floor system.

Soldier Piles and Lagging

Driven or drilled soldier piles, with wood or shotcrete lagging, supported by either drilled tiebacks (soil/rock anchors) or internal bracing supporting the piles are a simple temporary shoring type with a long history of usage in urban foundation construction. In many conditions with stiff soils, it is economical and safe. However, many urban centers in the United States are underlain by permeable soils with a high local groundwater table, particularly when compared to the depth of excavation required to provide needed parking. Soldier piles and lagging is a permeable wall system, and often accompanied by a lowering of the surrounding groundwater table. In addition, ground losses and raveling during lagging installation are inherent risks when using this system. Altering the water table around a deep excavation can cause elastic displacements of granular soils and consolidation settlements of clay or silt soils. Ground losses and raveling are also a source of settlement. If the surrounding buildings and their occupants cannot tolerate or are sensitive to the resulting settlements, then a shoring system that permits better water and ground loss control is warranted.

A second issue with soldier pile and lagging systems is that they are typically designed and constructed separate and independent of the permanent structure. They can serve as a back form for the permanent concrete basement walls, but the soldier piles are not easily integrated into the permanent structure. Most commonly, soldier pile and lagging walls are constructed 4 to 5 feet outside of the basement line to accommodate the double-sided concrete wall forms and the placement of an exterior waterproofing membrane. As a consequence, the impacts of construction at or outside of the property line must be considered. Impacts may include utility interference and additional excavation/backfill. Regardless, in cities with predominantly granular soils and a high groundwater table, soldier pile and lagging walls are not practical and are seldom used.

Tangent Auger Cast Piles or Tangent/Secant Drilled Shafts

Where running soils preclude the use of spaced soldier piles, tangent or secant walls made of auger cast in place piles (ACIPP) or drilled shafts can be used. Although ACIPP walls and tangent drilled shaft walls are not fully impermeable, they can be quickly excavated without major ground losses, except in the most severe conditons. A well constructed secant wall made of drilled shafts can result in a relatively water tight wall. To do this, softer concrete lagging shafts are used, and then cut into and bonded together with structural soldier piles that are designed like soldier piles.

Interlocking Steel Sheet Piles

Interlocking sheet piles have been used historically to shore excavations in granular soils below the water table, particularly when they can be toed into a relatively impermeable layer. However, a braced or anchored sheet pile wall excavation support system also requires a separate concrete basement structure. This structure needs to be designed to resist the uplift pressures from the surrounding groundwater table. If an underdrain system is used, long-term pumping can be very costly, and there is likely a permanent lowering of the surrounding groundwater table. The latter may result in settlements and possible deterioration of foundation elements such as timber piles. Many urban sites are not amenable to driven sheets due to obstructions, natural (boulders) and anthropogenic (utilities, miscellaneous fills, and remnants of prior construction), or vibration sensitive surroundings where building damage or settlements due to driving are an issue.

Soil Mixing Walls

Many of the advantages of a sheet pile wall system can be gained by constructing a soil mixed wall reinforced with closely spaced steel soldier piles (Pearlman and Himick, 1993). This is accompanied by the added benefits of a stiffer wall system and substantial reductions in vibrations during the temporary excavation support wall installation. However, the issues associated with the design and the construction of a permanent wall system remain. Installation of a soil mixed wall temporary excavation support system requires large equipment, and substantial spoils, that must be managed and removed for disposal, are generated during construction

Concrete Diaphragm (Slurry) Walls

Concrete diaphragm slurry walls were first introduced in the United States in the 1960s, and have found a niche in certain urban centers (such as Boston, MA). They are often attractive in granular soils with a high groundwater level and low permeability soils that underlie the granular soils that can serve as a cutoff; in areas with very dense and historic urban infrastructure; and where there exists a healthy demand for underground parking under buildings. Diaphragm walls are excavated through bentonite slurry using specially fabricated rectangular clamshell buckets to construct concrete panels of planned dimensions. After a panel excavation is completed, a carefully fabricated three-dimensional reinforcing cage is inserted into the panel excavation. Concrete is then placed around the reinforcing using tremie methods to form each concrete panel.

Permanent concrete diaphragm (slurry) walls are an ideal solution for structures requiring deep basements, particularly where a high groundwater table is present. Concrete diaphragm walls provide the following advantages for urban construction:

- Temporary and permanent groundwater cutoff
- Zero lot line construction
- Stiff structural capacity and superior resistance to movements
- Easily adapted to both anchors and internal structural bracing systems
- Expedited construction, because only interior columns and slabs need to be built.

Slurry wall installation requires sufficient work area to set up the slurry plant (de-sanders, etc.) as well as additional space to assemble the reinforcing cages prior to placement in the wall. Hence, this work may be difficult on congested sites.

Compared to wall types described above, diaphragm walls are considered to be very stiff with respect to ground movement control (Clough and O'Rourke, 1990).

Diaphragm walls are very amenable to a wide variety of ground conditions, and can be reinforced to allow incorporation of many structural configurations.

Demonstration of the Art

This paper discusses aspects of three successful projects with deep basements constructed in congested urban environments and the design approach that the authors developed and applied to these projects. The specific projects include 10 St. James in Boston's Back Bay, the new Harvard Medical School Research Facility in Boston, and the U.S. Capitol Visitor Center in Washington D.C.

All three projects are constructed adjacent to and below foundations supporting significant structures. In addition to designing for the earth and water pressure loadings, the appropriate consideration and modeling of surcharges and their effects are paramount for successful movement control by the excavation support system.

The 10 St. James project in Boston's Back Bay included excavation of narrow diaphragm wall panels directly adjacent to and below belled caissons supported with end bearing loads of 5 ksf. At the U.S. Capitol Visitor Center, a diaphragm wall excavation is constructed within 2 feet of the footings for the east front of the Capitol Building, with excavation to 50 feet below the existing grade. The 10 St. James project used internal cross-lot and corner braces, whereas the Harvard Medical School project used two types of support: an internal bracing system in a deep, constrained area at the end of the site; and tiebacks elsewhere. The U.S. Capitol project utilizes high capacity soil anchors to support the slurry wall.

The three project case histories are discussed, the design methodologies for the walls and wall support systems are described, and the movement performance is reviewed.

DESIGN METHODOLOGIES

The design analyses for excavation support systems can range from relatively simple empirical analyses to more complex computer analyses, where typically all stages of the excavation sequence are evaluated. The design considerations should include not only the stresses and loads on the support system, but also the affect of construction movements on the response of adjacent structures. The level of effort for the evaluation often depends on the stage of the project, proximity of structures, contractor's methods of construction, and known local practice. Our discussion of design methodologies will consider both structure loading and system movements.

Empirical Methods

Stress Analysis

Traditionally, apparent pressure envelope methods have been used successfully to design flexible wall systems such as soldier pile and lagging and steel sheetpile systems. The approach was developed based on data from flexible wall systems, and typically assumes that the wall acts as a simple beam spanning between the brace levels (Terzaghi et al., 1996). For the more rigid slurry wall system, the pattern of wall displacement that develops during the actual excavation and bracing sequence can have a major effect on the bending moments in the wall and the distribution of load to the bracing/anchors. Hence, use of apparent pressure envelopes for design of stiffer systems can be misleading. In general, apparent pressure envelope loadings are most appropriate as upper bounds for cases that match the bases of the empirical data, which include cases with relatively flexible walls and a stable subgrade.

The pressure envelope design approach is for a temporary support system and does not necessarily provide the long-term loading corresponding to the permanent condition after the end of excavation. When the temporary support system, such as a slurry wall system, is incorporated into the permanent building foundation, a staged analysis that includes loading at each stage is required to evaluate the built-in stresses and strains that are locked into the final structure at the end of construction.

Movement Analysis

The use of empirical data for the evaluations of movements is a useful tool in evaluating potential effects of a proposed excavation on adjacent buildings. Empirical data also allow the designer to validate the general magnitudes and patterns of the results of more sophisticated analyses. The empirical data can be used to estimate the zone of influence of the excavation as well as typical magnitudes of ground movements for various wall stiffness and subgrade stability conditions (e.g., Clough and O'Rourke, 1990).

Staged Excavation Analysis

Staged excavation analyses use numerical approaches to model the actual sequence of excavation and brace installation by considering each stage of the excavation as it is constructed, and the excavation support is installed and then removed. The soil and water pressures applied to the wall are representative of the actual pressures (not apparent pressure envelopes) expected in the system at each stage, and calculated loads are representative of the actual loads (not upper bound loads). The models can incorporate interaction of the soil and the structure as the earth pressures vary with displacement. The overall reliability of the structural requirements and displacement performance estimates determined from a staged excavation analysis is directly related to and very sensitive to the quality of the input parameters, particularly soil stiffness parameters.

Three general methods have been used for staged construction analyses:

- Equivalent Beam Method
- Beam on Elastic Foundation Method

- Finite Element Method

The equivalent beam method is outdated and rarely used in current practice. Our discussion will focus on the beam on elastic foundation and finite element methods. Both approaches can be used to predict stresses, loads, and system movements.

Beam on Elastic Foundation Method (BEF)

The earth pressures are modeled with a series of independent spring supports (Winkler elastic foundation model). At the start of the model, the springs are compressed to create an initial load equal to represent a state of at-rest pressure. At each stage of excavation or support system, the spring loads change as soil, water, and support system loads are applied or removed and lateral wall displacement occurs. The soil springs load-displacement relationship (modulus of subgrade reaction) is determined by the input soil stiffness and governs the spring displacement until the limiting value of active or passive pressure is reached.

The Winkler elastic foundation model approximates the wall-soil interaction with a one-dimensional model instead of a two-dimensional model that includes the soil mass, and hence does not include the effects of arching within the soil mass.

Typically, the required soil parameters include: unit weight; at-rest, active, and passive earth pressure coefficients; and values for the modulus of subgrade reaction for the various soils that may affect the system. The modulus of subgrade reaction is not a true soil property, but rather depends on both the soil conditions and the geometry of the excavation being modeled. To be representative, the modulus of subgrade reaction needs to be adjusted based on the effective influence zone, which varies with the size of the loaded area.

Typically, the predicted wall displacements are much more sensitive to the values of subgrade modulus used in the analysis than the predicted brace loads and wall bending moments. Hence, conservative selection of the modulus of subgrade values should provide conservative estimates of ground movements, without significantly increasing the structural demand of the wall and bracing system.

The BEF method does not directly estimate ground movements behind the wall. Ground movements behind the wall are evaluated using the calculated wall displacement from the model. An empirical relationship between wall movement and ground movements must then be used.

There are several computer programs that automate the analysis. Some use Young's modulus as input for the soil stiffness. The program then automatically converts the Young's modulus values for the various soils to adjusted values of subgrade reaction modulus using closed-form elastic solutions.

The BEF analytical model can provide useful insights into the behavior of the wall and the wall-soil boundary, and the automated computer programs make it easy to perform multiple analyses for optimizing the design and evaluating sensitivity to input parameters.

Finite Element Methods (FE)

Finite element models are typically two-dimensional models that include the soil mass surrounding the excavation. The stress-strain response of the soil is represented by a mathematical soil model that can vary from a simple linear-elastic model to a complex nonlinear elasto-plastic model. The stress-strain response can be defined in terms of effective stresses or total stresses. The required input parameters depend on the soil model used.

Generally, it is desirable to use a soil model that can model failure (plastic yield) when the soil strength is exceeded. In some problems, the ability to model volumetric changes in the soil (consolidation or dilation) may be important. A linear elastic, fully plastic Mohr-Coulomb soil model is often used. In this soil model, the soil acts linearly elastic until it reaches failure, defined by the Mohr-Coulomb criterion, where upon it becomes perfectly plastic.

In contrast to the BEF analysis, the FE analysis can provide direct information on the ground movements outside of and inside the excavation. It can also be used to model the soil-structure interaction response of nearby structures to the excavation-induced ground movements. In the past, performing FE analyses have been complex and time consuming to perform, but new, user-friendly programs (e.g., PLAXIS) are making their use more common. Another difference between the FE and BEF methods is that variations in the soil stiffness (modulus) can have a greater effect on predicted loadings and movements due to the inclusion of soil arching in the FE model.

FE models can be used to perform parametric studies to understand the relative effects of changes of parameters such as soil stiffness and excavation support stiffness and sequence on forces, stresses and displacements. They can also be used to estimate the absolute magnitudes and patterns of excavation support systems and ground movements which is much more difficult. A primary reason for the difficulty is the selection of reasonable stiffness values for the various materials that make up the soil mass. In general, values of stiffness based on laboratory and field tests tend to underestimate to a large degree the ground stiffness. This in turn can result in an overestimate of the magnitude of displacements, by two times or more, and the extent of the influence zone around an excavation. This tendency can be tempered to a great degree by using representative, local, field case history data during the selection of material parameters and to calibrate the numerical model to previous case histories.

CASE HISTORIES

We will discuss three case histories where slurry wall excavation support systems have been incorporated as permanent foundation elements in the final structure. In all three cases, the walls provide groundwater cutoff, and the excavation penetrated below the bearing level of the foundations of the adjacent buildings. The case studies will be used to illustrate design considerations for slurry wall systems as foundation elements. Nicholson Construction performed the wall construction and GEI Consultants provided design services on all three projects.

10 St. James Avenue

Project Description

10 St. James Avenue is a new 550,000-square-foot office complex constructed in Boston's Back Bay. The building includes a 19-story tower with a 280-foot-long by 170-foot-wide underground parking garage on 3-1/2 levels for 400 cars. The slurry wall excavation support system was included as the permanent basement wall, which was connected to a compensating mat foundation system, to create a watertight basement. Nicholson was engaged as a design-build package contractor with the responsibility for the site preparation, wall construction, mass excavation, lateral bracing, and waterproof concrete mat installation.

Overview of 10 St James foundation construction looking south from the Boylston Street side. Note excavation working to the north and the first mat placement has been installed in the center of the south wall

Design Considerations

The excavation was 50 feet deep and penetrated through a stiff crust into the underlying soft clay layer. Immediately adjacent to the excavation were two, approximately 15-story office buildings, which were supported by deep foundations bearing in the stiff crust. Existing concrete and timber pile foundation elements represented a challenge to the installation of the slurry wall along the property lines.

The analyses for the design of the support system were performed using a BEF model (WALLAP, 1997). During construction, additional modeling was performed to evaluate the affect of high bracing preloads on the wall stresses/movement performance, and to verify the initial design model using the actual construction sequence and the related instrumentation data.

Performance

The proximity of the adjacent structures with their foundation system support located above the subgrade level of the excavation prompted the construction team to re-examine the sequence and staging of the work, to balance costs with mitigation of potential impacts on the surrounding buildings. As the performance of the system was confirmed by the instrumentation data, the construction sequence was adjusted to expedite construction of the base mat, which in turn further limited lateral movements of the wall by acting as a subgrade level brace.

Numerous wood piles were present in the northeast portion of the excavation. At the start of their removal in preparation for slurry panel construction, inclinometer data showed horizontal movements near the building in the range of ¾ of an inch. This was approximately equal to the overall movements measured during the excavation construction. Revised procedures for the pile removal were then implemented to reduce overall movements of the buildings; however, it should be emphasized that site preparation and obstruction removal are activities that can create greater ground disturbance than the actual excavation sequence.

In general, the model movement predictions were conservative, especially for the cantilever excavation stage, where the predicted movements were two to three times larger than the measured movements. Assuming a 50 percent prestress in the bracing, the modeling predicted excavation wall movements on the order of 1.4 inches. During construction, a larger pre-load, to as much as 100 percent of the design load was employed, without adverse effects on the support system and the surrounding buildings and utilities. The result was excavation wall movements about half of those predicted (approx. 0.8 inch vs. 1.4 inches). When the models were rerun with the larger prestress, the calculated displacements more closely matched the measured wall displacements. It is also important to note that at the start of the construction, the design-build construction team performed additional field explorations and laboratory testing of the soil conditions at the site to determine input design parameters.

The authors' note that the literature suggests that there may be little value to using high values of prestress when using rigid bracing (O'Rourke, 1981). Using bracing loads closer to the total working load, as was done in this case, is uncommon and generally not necessary. If used, careful consideration needs to made for possible overload due to temperature effects and load shifting, and balanced against the benefits that an increased prestress may have on movement control. Certainly, when pre-loaded ground anchors are used, it is common practice to fully pre-load them to the fully anticipated design load, because their flexibility makes them less sensitive to load change with inward wall movement.

Harvard Medical School Research Facility

Project Description

One of the largest expansions to Harvard Medical School is a 430,000 SF biomedical research center shared by the medical school and affiliated hospital. Nicholson designed and constructed 123,000 SF (1,670 LF) of 2.5-foot-thick diaphragm wall composed of 84 panels with depths ranging from 65 ft to 93 ft; 178 temporary anchors (150-200 kips) and a temporary internal bracing system. A single row of tiebacks was installed in the shallow excavated area, while two levels of tiebacks and a third level of bracing were installed in the deepest areas.

The L-shaped structure consists of a 5-story building and a 10-story research tower. In addition, a 196,000 SF 2-level underground parking area to accommodate 500+ vehicles was built below the building. A deeper excavation under the research tower was required for mechanical equipment.

Design Considerations

The slurry walls for the shallow portion of the excavation were supported by a single row of high capacity tiebacks. Tiebacks were also used for the upper level support of the deep portion of the excavation. The foundations of one of the adjoining buildings consisted of steel piles to rock, which was located below the excavation subgrade. The tight spacing of these piles presented a challenge for the design and construction of the tiebacks for the support system. The second adjacent building was founded on belled caissons bearing approximately 25 feet below the ground surface and approximately 30 feet above the subgrade elevation of the deep portion of the building. Control of movement of these foundations was considered key to limiting damage to the building.

The analyses for the design of the support system were performed using a beam on elastic foundation model (WALLAP, 1997). During construction, additional modeling was performed to evaluate as-built conditions and to evaluate the design based on the results of instrumentation data.

Performance

The instrumentation data indicate that the wall performed better than was predicted by the model. In particular, the model over-estimated the cantilever movements of the wall by a factor of at least two. We note that for a variety of projects the cantilever case movement predictions from the model are higher than actual wall movements even when we get better agreement between the model predictions and actual movements for latter stages. Since the soil provides the sole support of the wall in the cantilever case, and conservative soil parameters are used in the model, it is not surprising that the cantilever case predictions tend to be higher than the actual movements of the wall. During an intermediate stage of construction for the deep excavation (the stage just before final excavation to subgrade), the inclinometer data indicated that a portion of the deep excavation was deflecting more than predicted by the model or observed at inclinometers in other portions of the deep excavation.

Disturbed subgrade (traffic and excess pore pressures in the silty subgrade) was identified as the potential reason for the larger deflections. We revised the model to include the effects of the disturbed subgrade on the next to last and final excavation stage, and concluded that additional bracing was not required to complete the excavation within the allowable movement criteria for the project. Table 1 lists modeled and measured deflections at several locations along the wall heights for the original and revised models and the inclinometer data.

High capacity tiebacks were installed with their bond zone in an inter-bedded deposit of fine sand and silty clay. The anchors were sized for a typical design load of 200 kips. Post grouting was used to achieve this capacity

Table 1. Modeled and Predicted Wall Movements

		Movement of the Wall			
Elevation (ft)	Model/Data	Cantilever (in)	Stage 1 (in)	Stage 2 (in)	Final Stage (in)
+10	Original	0.28	0.36	0.10	-0.08
	Revised	0.42	0.53	0.24	0.06
	Inclinometer	0.1	0.6	1.0	
-3	Original	0.11	0.46	0.46	0.28
	Revised	0.17	0.55	0.61	0.37
	Inclinometer	0	0.7	1.4	
-15	Original	.06	0.4	0.68	0.77
	Revised	0.08	0.41	0.83	0.79
	Inclinometer	0	0.5	1.1	
-33	Original	0.05	0.28	0.55	1.4
	Revised	0.07	0.25	0.55	1.34
	Inclinometer	0	0.3	0.7	

United States Capitol Visitor Center

Project Description

The U.S. Capitol Visitor Center (CVC) is the largest addition ever made the Capitol Building in Washington, DC. The CVC building, 50 feet underground, will front the entire East Front of the Capitol. The Capitol itself encompasses 775,000 SF, while the completed CVC will contain 580,000 SF on three levels. The CVC project footprint covers 193,000 SF, or approximately 5 acres, which is larger than the Capitol itself, whose footprint is 175,000 SF. Initial construction of the Capitol Building was in 1800.

The CVC will include space for exhibits, food service, two orientation theaters, an auditorium, gift shops, security, a service tunnel for truck loading and deliveries, mechanical facilities, storage, and much needed space for the House and Senate. When completed, the CVC will preserve and maximize public access to the Capitol, while greatly enhancing the experience for the millions who walk its historic corridors and experience its monumental spaces every year.

Excavation Support Wall Looking West
Courtesy of the Architect of the Capitol

Nicholson Construction Company was hired by Centex Construction Company, a general contractor working for the architect of the Capitol for the Phase I Structure package of the project. Nicholson's work includes design and construction of the 2400-foot-long permanent walls of the addition using 130,000 SF of 32-inch-thick concrete diaphragm walls, 500 drilled anchors (typically 200 to 300 kips design load), and jet grouting to complete the water cut-off in areas where the diaphragm wall abuts existing structures or its installation is restricted by other conditions. The entire support of the excavation system consists of the combination of the diaphragm wall, temporary tiebacks, cross bracing, and jet-grouted cutoffs.

The original project bid documents included a scheme for installing drilled shafts with permanent building columns from the same working grade as the diaphragm wall construction. Once the roof structure was installed over the columns, which was intended to act as an initial internal bracing level, then several rows of drilled anchors were to be installed along with the mass excavation. The mass excavation work was to occur under the deck while working around the pre-installed columns. Nicholson proposed to change the sequence of the work, allowing drilled shafts to be placed and columns erected from the final subgrade elevation. This eliminated the deck as an internal brace, but offered both economy and better site utilization. Extensive analyses, using both the BEF approach (WALLAP, 1997) and the FE approach (PLAXIS, 1998), were performed to verify that the alternative scheme would provide equivalent performance to the initial contract design.

Design Considerations

The primary design concern for this project is to control and minimize movements of the existing Capitol Building. The diaphragm wall foundation was designed with this goal and to act as a permanent water cut-off and structural wall for the three-level underground structure.

The foundations for the Capitol are within two feet of the wall in some locations. The foundation bearing level and loading varies significantly across the structure, presenting a wide variety of potential load cases and analysis profiles.

The analyses for the structural design of the support system were performed using both BEF and FE models. The BEF program (WALLAP, 1997) was easier and quicker to run than FE programs, so it was used for the structural design of the wall system. By using the BEF model, the design team could evaluate more design profiles. Two FE models were run to verify that the BEF model loadings and stresses were conservative, and to provide ground deformation predictions to compare to contract requirements. Soil properties were selected based on geotechnical laboratory testing (Weidlinger Associates, Inc. 2001), as well as published values from test section case histories in the Washington D.C. area (O'Rourke, 1975).

Performance

Figure 1 presents the predicted deflected shapes of the slurry wall for the BEF and FE model analyses, as well as inclinometer data for the most heavily loaded design sections. The FE model included the tieback anchors modeled within the soil mass. The difference between the movements predicted by the BEF model and the larger movements predicted by the FE model is essentially the free field movement behind the anchor zones of the tiebacks. In other words, the BEF and

FE model had good agreement in predicting the local movement of the wall. The actual wall movement is less than the values predicted by both models. This behavior is likely the result of the combination of conservative modulus values for the soils, and conservative estimates of building surcharges used in our models.

The authors note that overall the wall movements for the entire site are less than predicted, even in sections where there are no building surcharges.

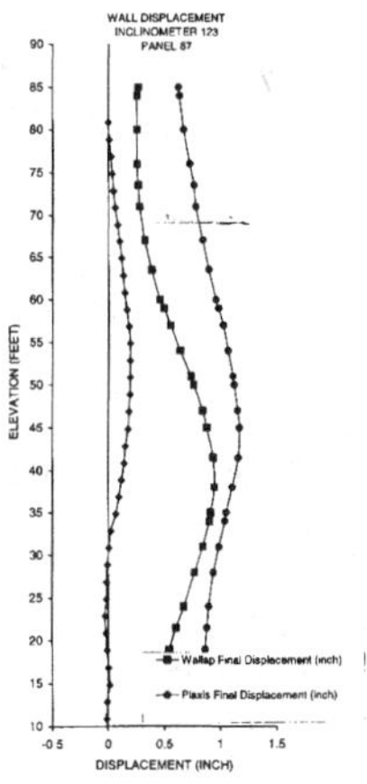

Figure 1: Modeled and Measured Wall Displacement Data

SUMMARY

Major building foundations in urban areas with high groundwater tables can be economically constructed using permanent concrete diaphragm walls. Temporary excavation support can be accomplished using drilled and grouted anchors when the lateral right of way is available, and with internal cross lots and corner struts when anchors are not practical.

Experienced design-build teams consisting of a specialty contractor, a design firm, and, where appropriate, other subcontractors (such as an excavator) can quickly deliver foundation packages in a cooperative and productive team environment.

Design techniques that involve sophisticated soil structure interaction models combined with local data and experience give a high level of confidence for predicting wall performance on projects surrounded by other structures, where control of building movement and damage are paramount to a successful project delivery. These models need to be calibrated to empirical predictions, and other case histories of successful excavation support projects in similar ground conditions.

The use of instrumentation and the reporting of the results are important to benefit the overall knowledge base of each region where deep basements are a popular choice for building owners and developers.

REFERENCES

Clough, W.G. and O'Rourke, T.D., 1990. "Construction induced movements of in-situ walls." *Design and Performance of Earth Retaining Structures*, ASCE GSP No.25, 439 - 470.

Pearlman, S.L. and Himick, D.E. 1993. "Anchored excavation support using SMW (Soil Mixed Wall)," Presented at the Deep Foundations Institute18th Annual Member's Conference, Oct. 18-20, 1993, Pittsburgh, PA.

O'Rourke, T. D., 1981. "Ground movements caused by braced excavations," *J. Geotech. Engrg Div.,* ASCE, 107 (GT9), 1159-1178.

O'Rourke, T. D. 1975. "A study of two braced excavations in sands and interbedded stiff clays," Ph.D. dissertation, Univ. of Illinois, 255 pp.

PLAXIS, 1998. *Finite Element Code for Soil and Rock Analyses.* Brokgreve and Vermeer, et al., (ed.), Balkema. Rotterdam, Brookfield, Version 7, A.A.

Terzaghi, K., Peck, R. B., and Mesri, G., 1996. *Soil Mechanics in Engineering Practice*, Third Edition, John Wiley & Sons, New York, NY, 349-360.

WALLAP, 1997. *Anchored and cantilevered retaining wall analysis program*, D.L. Borin, MA, Ph.D., CEng., MICE. Geosolve, Users Manual, Version 4.

Weidlinger Associates, Inc. 2001. "Geotechnical Engineering Study, United States Capitol Visitor Center", Prepared for RTKL Associates Inc. and The Architect of the Capitol.

Effectiveness of Toe-Grouting for Deep-Seated Bored Piles in Bangkok Subsoil

Narong Thasnanipan[1], Zaw Zaw Aye[2], Chanchai Submaneewong[3]

ABSTRACT: A comprehensive study on the effectiveness of two different toe-grouting methods, known as tube-â-manchette and drill-and-grout, commonly applied for deep-seated bored piles in Bangkok is presented in this paper. Mobilized unit end bearing of bored piles constructed by two toe-grouting methods are compared. The problems encountered in practical application of different methods for large diameter deep seated bored piles are highlighted. Despite considerable extra cost, various construction problems encountered and risks involved in successful execution of grouting, drill-and-grout method does not offer particular advantage over tube-â-manchette method as far as pile capacity improvement is concerned.

INTRODUCTION

In Bangkok, due to the prevailing subsoil and groundwater conditions, all the deep-seated bored piles (depth ranging from 25 to over 60 m) are constructed by wet-processed or slurry displacement method. Although the wet-processed method is most suitable and its application has been well established in Bangkok subsoil, the method itself cannot avoid undesired effects especially loosening and upheaval of soil at pile base as well as accumulated suspension of loose materials or sediments from drilling operation. These effects naturally may create a loose or soft pile toe causing negative impact on the pile capacity or load vs. pile head movement behavior. One of the special measures to improve the soft pile toe is to apply the post-grouting at pile toe. Two different methods, known as tube-â-manchette and drill-and-grout are commonly applied in Bangkok. A comprehensive study on the effectiveness of two toe-grouting techniques applied for deep-seated bored piles in Bangkok is presented in this paper.

---

[1]Managing Director, [2]Project Manager, [3]Geotechnical Engineer
SEAFCO Company Limited, Bangkok, Thailand, seafco@seafco.co.th

SUBSOIL PROFILE AND EXISTING PIEZOMETRIC PROFILE

Subsoil profile is relatively consistent at different localities in Bangkok. A typical subsoil profile is characterized by the alternating layers of clay and sand deposits as shown in Figure 1. Pore water pressure profile illustrating the current piezometric drawdown condition of Bangkok is also shown in Figure 1.

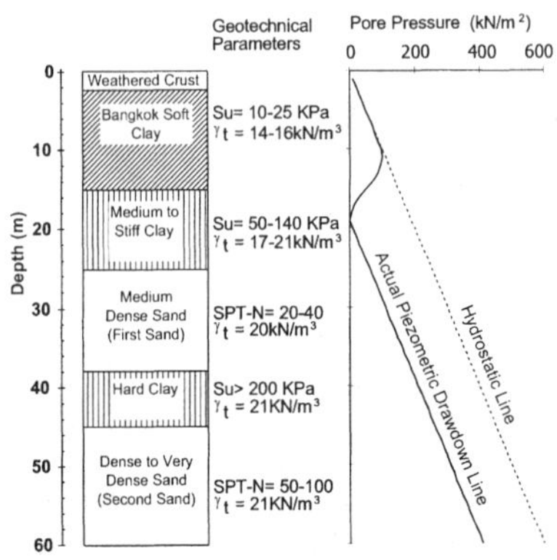

Figure 1. Typical soil profile of Bangkok with piezometric drawdown condition (Thasnanipan, 2002)

WET-PROCESS BORED PILE CONSTRUCTION METHOD

In Bangkok, medium to large diameter (0.6 m to 1.8 m diameter) bored piles with depth ranging from 24 m to over 60 m are constructed by wet process due to the prevailing subsoil and ground water condition. Temporary casing of 14-18 m in length is typically used as a support in soft clay layer. Drilling is commenced by dry process with auger applying rotary drilling action in the soft clay layer and stiff clay layer. Before reaching to the first sand layer, supporting fluid or slurry is fed to the borehole and drilling is continued with the bucket to the final depth. Depending on the type of slurry used, either a cleaning bucket or airlift method is commonly applied to clean the bottom of the borehole prior to reinforcement cage installation. Concreting is done by tremie method for wet-processed bored piles.

TOE GROUTING METHODS IN BANGKOK

Two grouting methods known as tube-â-manchette and drill-and-grout method used in some major projects where the authors were involved are presented in this paper. In Bangkok, both toe-grouting methods are mainly applied for the piles with the tip embedded in the sand layer though there were few cases where it was applied in clay layer.

Tube-â-manchette Toe-grouting Method (TAM Method)

This grouting method has been widely used in other parts of the world according to the available published papers. A review of past and present toe-grouting methods used throughout the world has been reported by Mullins et. al. (2000). In Bangkok, the first toe-grouting for large diameter bored piles diameter 2.0m seated at depth 35m in the first sand layer was carried out by tube-a-manchette method in 1985 for the construction of Rama IX cable stayed bridge (Morrison, 1987). Since then, tube-a-manchette method has been commonly used for toe-grouting of wet-processed bored piles in Bangkok. It is estimated that over 5,000 bored piles have been constructed using this toe-grouting method in Bangkok.

Grout pipes configuration for tube-â-manchette method. In normal practice of toe-grouting by this method in Bangkok, a system of grouting circuit consists of two U-shape loops formed by two pairs of PE pipes with manchette placed at 5 to 10 cm above pile tip. The manchette is formed by the perforated steel pipe wrapped with rubber sleeves as shown in Figure 2. The manchette allows the grouting material to flow out of the pipe from the perforations but prevents return of grouting material when the grout injection is stopped. PE tubes are fixed along the rebar cages to the top. In some projects where the specification required cross-hole sonic logging tests, same access tubes for the sonic test apparatus are used in lieu of PE pipes.

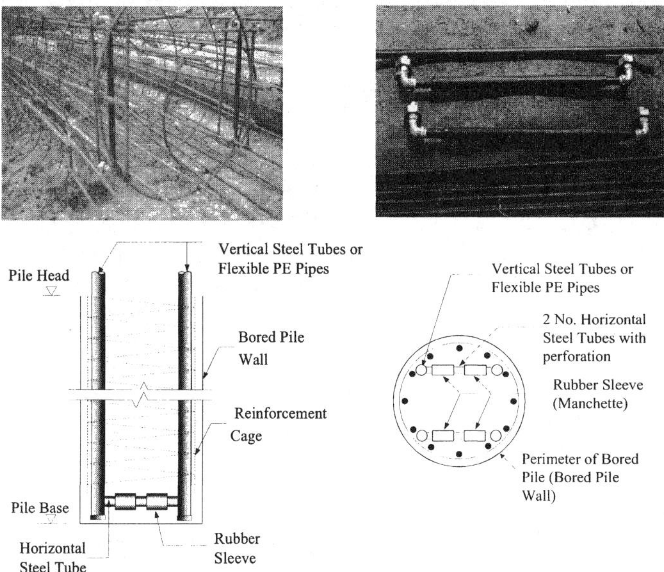

Figure 2. Position and layout configuration of grouting circuit in Tube-â-manchette method

Grout Mixture Composition and Mixing Process. The typical cement grout consists of cement and water with the ratio of 0.5-0.6 by weight. Bentonite or other reagent can also be added but trial mix should be made to find out the characteristics of the grout. Cement and water are added at right proportion in a high turbulence mixer and mixed about 2 to 4 minutes. Cement mixture is then conveyed to the agitator. The grout must be agitated continuously or at regular interval during the period between the time of grout mixing and the time it enters the pump. Usually, first grout injection should be started within 10 to 15 minutes after mixing the first batch to minimize the stiffening of the cement grout during grouting operation. It is essential that grouting team understands the required properties such as viscosity, density, setting time of grout are met to effectively grout the toe of the pile. Grouting operation should be completed within the initial setting time of the grout. Mixing rate, storage capacity and grouting rate should be well planned to complete the grouting work within the time limit.

Grouting procedure. The process involved in grouting by tube-a-manchettes method is illustrated in Figure 3. First, the grout pipes are flushed with water to ensure that the grouting loops are clear from any blockage. The pile toe is then cracked with relatively high pressure water flow, 12 to 24 hours after concreting, to open the manchettes and to make way for forth coming grout. The commencement of cement grouting depending on the preset pressure criteria. If specified maximum pressure is low (less than 20 bar / 2000 kN/m^2), grouting is commenced immediately after the pile toe cracking. Grouting is normally commenced 7 to 10 days after concreting if higher maximum control pressure is specified to avoid damage to pile by high pressure grouting. Grouting is stopped when the specified volume or pressure is achieved. In current practice, pressures ranging from 20 to 60 bar (2000 to 6000 kN/m^2) and grout volumes 500 to 1000 liters are normally applied.

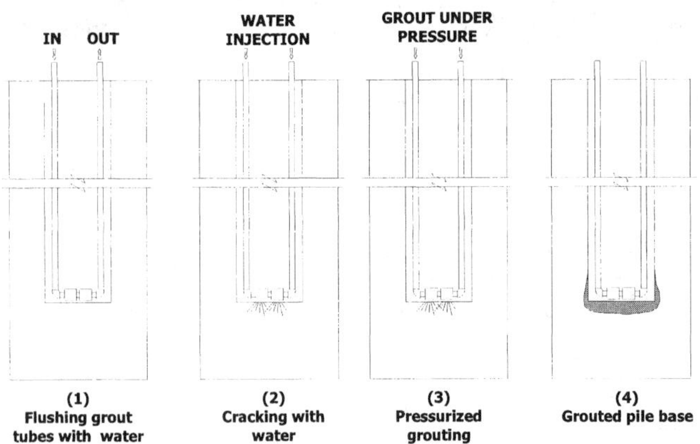

Figure 3. Grouting process in Tube-â-manchette method

Drill-and-grout Method (DAG Method)

Toe-grouting by this method is in fact not frequently used in Bangkok. It is to the author's knowledge, not commonly used in the current practice of toe-grouting of bored piles world wide. To date, less than 500 bored piles have been completed in Bangkok by this method.

Grout pipes configuration for drill-and-grout method. Depending on the size of bored piles, three to four black steel pipes used for sonic test access tubes are mostly utilized as grouting pipes in this method. Steel pipes are equally spaced and fixed around inside perimeter of reinforcement cage as shown in Figure 4. The bottom of steel pipes are securely closed with steel cap to prevent intrusion of foreign material during installation and concreting.

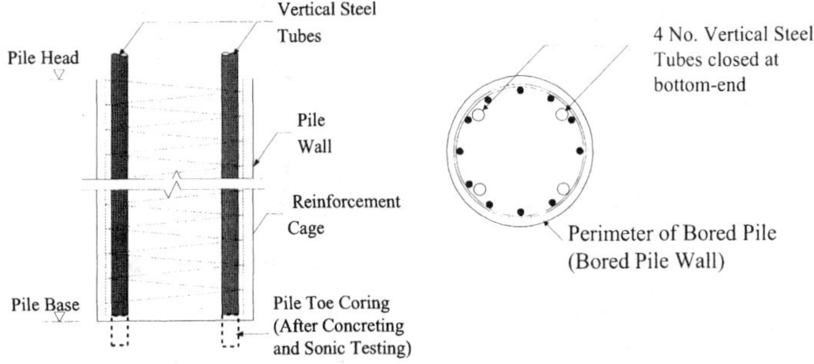

Figure 4. Position and layout configuration of grouting circuit in Drill-and-grout method

Grout Mixture Composition and Mixing Process. Cement to water ratio of 0.6-0.7 by weight is normally used in drill-and-grout method. Cement and water are filled at right proportion and mixed about 2 to 4 minutes in a high turbulence mixer. Bentonite is then added at 0.5 % of cement by weight and mixed for another 2 minutes and the readily mixed grout is conveyed to the storage tank equipped with agitator. The control factors such as viscosity, density and setting time are also equally important as mentioned in the tube-â-manchette method.

Grouting procedure. The procedure for drill-and-grout method presented in this paper is for diameter 1.50m bored pile with 4 sonic access tubes. The sonic logging test is performed minimum 7 days after casting of the pile. Coring is then carried out through the pile base to the depth 20 cm below the pile toe. First, drilling rod attached with core-barrel is lowered into the water-filled sonic access tubes (steel pipes) and drilled through the bottom steel cap and the concrete at pile base. Drill rod is then withdrawn to change core-barrel with the soil sampler and lower again to drill the soil beneath the pile base about 20 cm. It is important that sonic tube is completely filled with water during the drilling process to avoid excessive blow-in of soil into the tubes. After completion of toe coring and sampling, the grout pipes are flushed in a sequence to ensure that flow path exists between the tubes. To do this, after initial cleaning, two pipes (inlet pipes) are connected to the pressure pump and low pressure up to 10 bar (10 kN/^{2}) is applied using clean water. The other two pipes (outlet pipes) are filled with water and are left open. As the pressure from the pump is gradually increased, water will start to flow from one or more of the other two open pipes. Once the flow is observed from the first outlet pipe, it is closed off and the water flushing continued until water flow is observed from the second pipe. The process may be repeated in like sequence from each of the grout pipes until satisfactory flow is observed from each pipe. Once the flushing of the preinstalled grout pipes are completed, water-supply hoses are disconnected from 2 inlet grout pipes and connected with the grout-supply hoses. Two outlet pipes are then opened and cement grout is filled from two inlet pipes with low pressure until the grout discharge from other 2 outlet pipes.

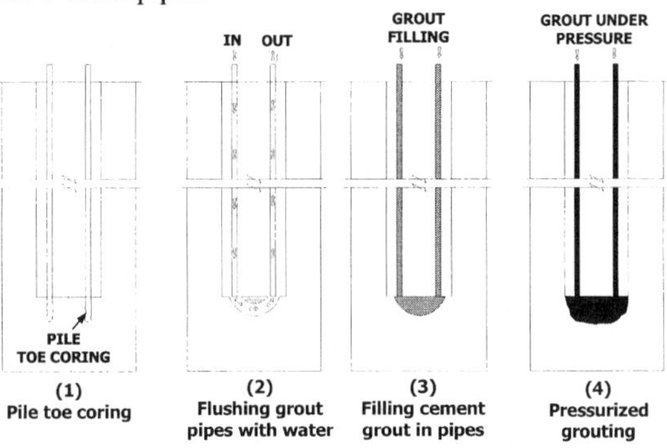

Figure 5. Grouting process in Drill-and-grout method

Initial discharged cement grout is normally contaminated with the in-situ sand so that it is required to check the purity of out coming grout. Once clean cement grout is flowed out, 2 outlet pipes are immediately connected with the grout-supply hoses and pressurized grouting commences. Grouting is started with the grout injection rate (IR) between 25 to 30 liter per minute. Grout pressure development is closely monitored from the pressure gauges installed at the top of each pipe. If the grouting pressure does not reach to 20 bar (2000 kN/m^2) after injecting the 300 to 500 liter of grout, the injection rate is reduced to 12-15 liter per minute. Otherwise grouting is continued until the control criteria is achieved. Figure 5 illustrates the working procedure of pile toe grouting by drill-and-grout method used in Bangkok subsoil for 3 major projects where the authors involved.

Grout Spreading Mechanism

Tube-â-manchette method. Teparaksa et al. (1999) investigated the penetration of grout into the sand layer from the toe-grouting of bored piles in Bangkok by tube-â-manchette method. Toe-grouting was performed 24 hours after concreting with maximum pressure 40 bar (4000 kN/m2) for grouting the base of large diameter bored piles constructed by wet-processed method under bentonite slurry seated in sand layer. From the series of SPT and soil sampling performed below the pile toe and immediate vicinity of the shaft, the authors concluded that grout did not permeate into the sand layer at pile base. Upward migration of grout along the pile shaft is most likely in tube-â-manchette method, as grout injection was applied under relatively high pressure within 24 hours after concreting of the piles, plane of weakness was formed at pile-soil interface which is most likely to be contributed by the following conditions.

- Stress relaxation from the drilling process was not yet fully restored along the shaft in relatively short time gap between pile drilling and grouting
- Presence of bentonite filter-cake along the shaft

Drill-and-grout method. Grout spreading mechanism of drill-and-grout method are discussed in relation to the steps involved in grouting operation (illustrated in Figure 5) as summarized below.

- Step 1 : drilling 4 boreholes of diameter 48 mm up to 20 cm below pile toe (using water-filled sonic tubes as access pipes to the bottom of the pile)
- Step 2 : water flushing the sonic tubes to remove the sand inside the tubes and to form flow-paths between the tubes
- Step 3 : filling the sonic tubes with the grout
- Step 4 : pressurized grouting until pressure reaches 6,000 kN/m2 and maintain for 5 minutes

Step 1 creates 4 boreholes in the soil underneath the pile base. These boreholes become larger as pocket of sand between them is washed out during Step 2 and created a cavity or relatively large opening beneath the pile base prior to the commencement of Step 3. This cavity is filled by grout in Step 3 which eventually formed a mass grout bulb underneath the pile base as illustrated in Figure 5. Pressurized grout injection in Step 4 thus compacted the grouted mass which in turn densified the grout-soil interface. Basically grout did not enter soil pores but remained in a compacted homogeneous mass. Upward migration of grout along the pile shaft is considered unlikely since toe-grouting was carried out minimum 14 days after pile concreting so that the time gap between concreting and grouting would have allowed some positive improvement of pile shaft condition (influence of stress relaxation and bentonite filter cake).

PERFORMANCE OF TOE-GROUTED BORED PILES

Typical Load Transfer Mechanism of Toe-Grouted Bored Piles in Bangkok

Table 1 shows the measured mobilized shaft resistance of bored piles with different toe grouting method at specific load in comparison with design working load and calculated ultimate values using empirical formula. The details of the empirical formula commonly used in Bangkok subsoil has been presented by

Thasnanipan et. al. (2002). As can be seen in the table under design working load, shaft friction carried the large portion of the load (over 90% of applied load). Only the small percentage of the total load was transferred to the pile tip. Therefore, even with the toe-grouting application, the end bearing component of bored piles is still under-utilized and thus remaining unmobilized end bearing capacity serves only as a reserve that forms the additional factor of safety.

Table 1. Summary of mobilized shaft resistance at specific load in comparison with design working load and calculated ultimate value

Test Pile No.	Pile Dia. x L (m)	Grouting method	Design Working Load, DL (kN)	Total shaft resistance, Q (KN)			Q_{fm} / DL	Ratio of shaft resistance (Q_{fm2} / Q_{fu})
				Calculated Ultimate Value, Q_{fu}	Measured at DL, Q_{fm}	Measured at 2 x DL, Q_{fm2}		
TT-1	1.2 x 44.6	TAM	5000	6900	4480	8560	0.90	1.24
TT-2	1.2 x 53.8		6500	9500	6120	10830	0.94	1.14
TT-3	1.2 x 46.0		5000	7900	4730	9250	0.95	1.17
TD-1	1.50 x 50.0	DAG	13000	14600	12200	22160	0.94	1.52
TD-2	1.50 x 56.5		13000	16500	12590	24270	0.97	1.47
TD-3	1.50 x 52.0		13000	16400	12570	24410	0.97	1.49

Mobilized Unit End Bearing

Figure 6 shows the mobilized unit end bearing vs. pile head movement of TAM and DAG grouted bored piles (same test piles as Table 1) . In general, as can be seen in the figure, mobilized unit end bearing at specific pile head movement of piles constructed with tube-â-manchette grouting method (TAM) is higher than those of piles constructed with drill-and-grout method (DAG). It is also evident from Figure 6 that large pile head movement is required to mobilize the end bearing despite toe-grouting applied for all piles. It should however to be noted that DAG grouted bored piles of Figure 6 (TD1, TD2 and TD3) were not tested to failure.

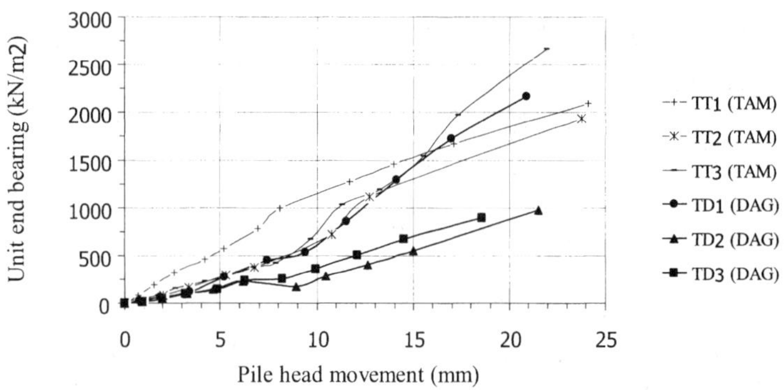

Figure 6. Mobilized unit end bearing of TAM and DAG grouted bored piles

Though it is not advisable to simply draw the conclusion on particular toe-grouting method effects on performance of piles solely by mobilized unit end bearing criteria from the available limited test results, the data provided in this research study however could be taken as preliminary guidance to assess the effectiveness of the different toe-grouting methods.

According to the recent full-scale instrumented static pile load tests carried out in late 2002 and early 2003 on 13 drill-and-grout bored piles of diameter 1.50m seated at depth over 50m (a complete set of data is not available while preparing this paper), developed end bearing values of these piles were found to be scattered or inconsistent despite the fact that grouting pressure of all piles were the same - reached 60 bars (6,000 KN/m^2). Therefore, in authors' opinion a careful justification is to be made to design the deep-seated bored piles in Bangkok subsoil with high end bearing values regardless of high pressure achievement in toe-grouting application.

Load vs. Pile Head Movement Comparison

Figure 7 shows the Load vs. Pile Head Movement of ungrouted, grouted piles of tube-â-manchette and drill-and-grout methods. Table 2 summarized and compared the load test results of the test piles of Figure 7. (test piles A1, A2, B1 and B2 are of additional test program different from that of Table 1; test pile C1 and D1 are TT2 and TT3 of Table 1 respectively).

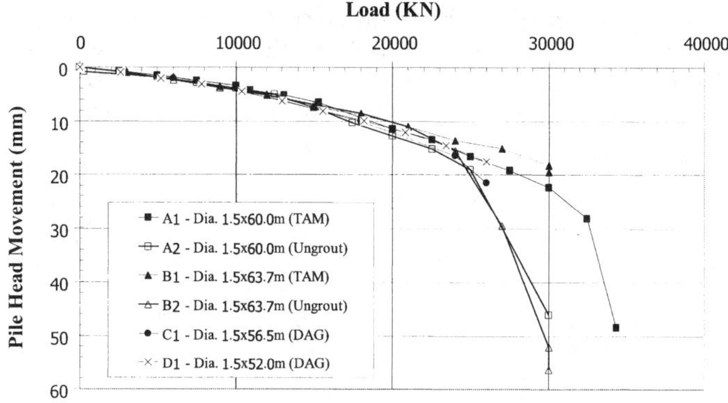

Figure 7. Load vs. pile head movement of non-grouted and toe-grouted piles

Following conclusions can be drawn from Figure 7 and Table 2.

- Limit loads at 20mm pile head movement are improved (14% and 21% for Project A and B respectively) by tube-a-manchette toe grouting method (TAM) in comparison with ungrouted piles.
- No significant difference of load-settlement behavior is found for non-grouted piles in comparison with TAM grouted and DAG grouted piles at design load and 2 times of design load.

Table 2. Summarized comparison of load test results

		Project A		Project B		Project C	Project D
		Pile No. A1 (TAM Grout)	Pile No. A2 (non-grout)	Pile No. B1 (TAM Grout)	Pile No. B2 (non-grout)	Pile No. C1 (DAG Grout)	Pile No. D1 (DAG Grout)
Diameter (m)		1500	1500	1500	1500	1500	1500
Length (m)		60.00	60.50	63.70	63.70	56.50	52.00
Q_d (KN)		10,000	10,000	10,000	10,000	13,000	13,000
Q_L (at 20mm total settlement)		26,500	23,200	34,280	21,750	25,500	26,000*
% increase in Q_L		14.20		21.57		N.A	N.A
Total pile head movement	at Q_d	3.50	3.00	4.20	3.75	6.10	5.80
	1.5 Q_d	6.50	6.50	7.20	6.00	11.90	11.00
	2 Q_d	10.50	12.00	10.00	9.70	21.50	17.50

Note : * At 17.50mm pile head movement

Therefore, despite the various construction problems encountered and risks involved in successful execution of grouting, drill-and-grout method does not offer particular advantages over tube-â-manchette method.

EFFECTIVENESS OF THE DIFFERENT METHOD IN PRACTICAL APPLICATION

Problems Encountered in Practical Application of Different Method

Table 3 summarizes the most critical problems encountered in practical application of different toe-grouting methods used in Bangkok. In summary, it is obvious that higher risks are involved in the drill-and-grout method. The most critical risk of this method for the long piles is to maintain a good verticality of the grout access tubes.

Table 3. Summaries of the problems encountered in practical application of different methods

Toe Grouting Method	Problems Encountered and Risks Involved to Complete Toe-grouting		
	Unsuccessful or ineffective cracking of manchette	Verticality of sonic / grout tube	Unsuccessful coring of pile toe
Tube-â-manchette	Possible but rarely occur	Not required	Not required
Drill-and-grout	Not required to crack manchette	High verticality is extremely important. If access tube is inclined drilling rod can not be lowered to the bottom of the pipe for toe coring	High risk. If coring is unsuccessful, toe grouting can't be performed.

Impact Caused by Different Toe-grouting Method on Bored Pile Construction

The procedure involved in preparation and execution of two methods of toe-grouting poses different impact on bored pile construction. Since verticality is the main governing criteria in the successful application of grouting by drill-and-grout method as explained in above section the major impact caused in this method is due to the need of significant extra time in the following activities to produce the good verticality of the pile itself and the grout access tubes.

- Drilling
- Extensive borehole monitoring for verticality by sonic caliper method
- Grout pipe installation

Due to the excessive time consumption in the above mentioned operations to further carry out the toe-grouting by drill-and-grout method, to produce 1 bored pile of diameter 1.50 m seated over 50m below ground, could take 2 to 3 days excluding time consumed for grouting. For normal (non-grouted) and TAM grouted bored pile of same size and length, it normally takes less than 24 hours to produce 1 pile. Therefore considerable additional cost would incur for applying drill-and-grout method due to the requirement of significant extra construction time.

Assessment on the Effective Toe-grouting Method in Practical Construction

Assessment on Constructability. Constructability is one of the most important issues for most construction projects. If the construction method is impractical or if it involved various problems and high risk in actual execution, it should not be employed in the first place. Careful justification on assessment of the constructability is essential in selection of the toe-grouting method and in establishing the technical criteria. The party who is responsible for drawing the specification such as project consultant or designer should take the following points into consideration.

- The potential risks involve in the practical execution and consequential impact in minimizing the risk imposed by the method itself (e.g. delay in construction, extra cost)
- Additional cost involved for toe-grouting against saving from higher bearing capacity achivable by toe-grouting
- Clear and practical acceptance and rejection criteria in case the problems are encountered in actual execution

Assessment on the Performance of Pile. As presented in the earlier section, deep-seated large diameter bored piles in Bangkok are mainly supported by shaft friction and very small percentage of the load is transferred to pile toe under working load and double of working load. Therefore, even with the toe-grouting application, the end bearing component of bored piles is still under-utilized. Thus remaining unmobilized end bearing capacity serves only as a reserve that forms the additional factor of safety.

CONCLUSION

Based on the field application and measured data of instrumented pile load tests, the following conclusions can be drawn.

(1) Under design working load for large diameter deep-seated bored piles with toe grouting in Bangkok subsoil, the large portion of load (over 90% of applied load) is carried by the shaft friction. Only the small percentage of the total load is transferred to the pile tip. No significant difference of load-settlement behavior is found for non-grouted piles as compared to TAM grouted and DAG grouted piles at design load and 2 times of design load. Therefore, for deep-seated bored piles in Bangkok subsoil, even with the toe grouting application, the end bearing component of bored piles could be still under-utilized. Thus the remaining unmobilized end bearing capacity serves only as a reserve which forms the additional factor of safety.

(2) The drill-and-grout method has higher risks than the tube-â-manchette method in successfully constructing the toe grouted bored pile. Considerable additional cost would incur in applying drill-and-grout method due to the requirement of significant extra construction time to complete a qualified pile.

(3) Despite the considerable extra cost, various construction problems encountered and risks involved in successful execution of grouting, drill-and-grout method does not offer particular advantage over tube-â-manchette method as far as pile capacity improvement is concerned.

(4) The potential risks involve in the practical execution and consequential impact in minimizing the risk imposed by the method itself should be taken into account in selecting the effective toe-grouting method.

REFERENCES

Morrison, I.M. (1987) "Bored piled foundation for Chao Phya river crossing at Wat Sai, Bangkok", *Proceedings of the 9th South East Asian Geotechnical Society Conference,* Bangkok, pp. 6-207 to 6-218.

Mullins A. G., Dapp S. D. and Lai P. (2000) "Pressure-Grouting drilled shaft tips in sand" New Technological and Design Developments in Deep Foundations, *Proceedings of the Sessions of Geo-Denver 2000, Geotechnical Special Pub. No. 100*, ASCE, August 5-8, 2000, Denver, Colorado, USA, p.p. 1-17

Teparaksa W., Thasnanipan N., and Anwar M. A. (1999). "Base grouting of wet process bored piles in Bangkok subsoils" *11th ARC. on soil mechanics and geotechnical Engineering.*, Seoul, Korea, pp. 269-272.

Thasnanipan, N., Zaw Zaw Aye, Chanchai Submaneewong and Wanchai Teparaksa (2002) "Performance of wet-process bored piles constructed with polymer-based slurry in Bangkok subsoil", *Proceedings of the International Deep Foundations Congress 2002, Geotechnical Special Publication No. 116, Volume One*, ASCE, February 14-16 2002, Orlando, Florida, USA, p.p. 143-157

FULL SCALE FIELD PERFORMANCE OF DRILLED SHAFTS CONSTRUCTED UTILIZING BENTONITE AND POLYMER SLURRIES

Rudolph P. Frizzi, P.E.[1], Matthew E. Meyer, P.E.[2], Members, GeoInstitute, and Lijian Zhou[3], Assoc. Member, ASCE

ABSTRACT: Three 6-ft (1.83-m) diameter, 120-ft (36.6-m) long test drilled shafts were constructed and load tested to evaluate the effects of using bentonite and polymer slurry on full-scale drilled shaft load carrying capacity. Ground conditions at the Miami, Florida site consisted of alternating layers of soft sedimentary rock and sand. Load tests were performed using embedded multi-level Osterberg-type loading jacks (O-cells) where maximum equivalent top loads of as much as 12,200 kips (54.3 kN) were applied. This paper presents the test drilled shaft construction procedures, and the load test procedures and results. Shaft drilling procedures, equipment, and bottom cleaning methods; slurry mixing and testing procedures; and construction time are also presented. Conclusions are drawn regarding the effects of construction methods and drilling slurries on observed side shear and end bearing load transfer.

INTRODUCTION

The tallest building currently in the State of Florida is a 776-foot-tall (237 m) office / hotel / residential tower constructed on a confined urban site in downtown Miami. In planning the building's design and construction, structural analysis dictated design average compressive area loads of 12.4 kips/ft^2 (595 kPa), and concentrated area loads approaching 16.2 kips/ft^2 (775 kPa) being transferred to the foundation level.

[1] Senior Associate, Langan Engineering and Environmental Services, River Drive Center 1, Elmwood Park, New Jersey 07407, (201) 794-6900, rfrizzi@langan.com

[2] Project Engineer, Langan Engineering and Environmental Services, 7900 Miami Lakes Drive West, Suite 102, Miami Lakes, Florida 33016, (305) 362-1166

[3] Senior Staff Engineer, Langan Engineering and Environmental Services, 7900 Miami Lakes Drive West, Suite 102, Miami Lakes, Florida 33016, (305) 362-1166

From the author's and their Firm's previous local geotechnical design and construction experience, the use of shallow foundations to support such loads would have resulted in unacceptable building settlements. Ground improvement alternatives to allow the use of a shallow foundation system were evaluated, analyzed, and cost-estimated and were deemed uneconomical and impractical. Therefore, a deep foundation support system was chosen.

Drilled shafts were considered to be ideally suited to support the high concentrated loads since fewer, highly loaded elements would minimize the size of overlying foundation caps thereby reducing overall foundation costs. However, previous South Florida drilled shaft construction experience raised several concerns including: load tests on smaller diameter drilled shafts were often successful yet some production shaft supported buildings settled 12 inches (300 mm) or more, and both water / casing and bentonite-stabilized shaft construction experienced varying degrees of success, concrete take, and construction delays. Polymer drilling slurry admixtures offered an economic advantage; however, slurries made with polymer had not been previously utilized for similar drilled shaft construction in this area. Advances in sampling, inspection, and load testing technology allowed full-scale production shaft inspection and testing. Since full-size test shafts could be constructed, potential scale-effect and constructability issues could be identified and addressed as part of a test program.

GEOLOGIC SETTING AND SITE SUBSURFACE CONDITIONS

The sedimentary geologic formations that underlie the Atlantic Coastal areas of south Florida are among the youngest in the United States. In the Miami-Dade County area, the uppermost geologic strata that most closely resemble commonly accepted rock-like material include the Fort Thompson and Miami geologic Formations. These Formations were contemporaneously deposited during the Pleistocene epoch which began about 2,000,000 years ago. A generalized geologic profile is shown in Figure 1. The silica sands, cemented sand and shell, and limestones of the Fort Thompson formation, the older of the two, is generally composed of relatively finer grained materials. Sea level and possibly other environmental fluctuations likely contributed to the varied composition of this Formation. As sea level rose during the most recent post-glacial epoch, low-lying mangrove swamps and tidal bays formed above the limestone. At oceanfront areas, Holocene sands of the Pamlico Formation were subsequently deposited above the organic soils.

The time since deposition of the above described geologic Formations has allowed some hardening of the parent materials to occur. This is expected to be a result of successive deposition, partial exposure and cementation, and subsequent inundation and sedimentation. However, the complete metamorphosis into a relatively uniform rock strata has not occurred in the geologically short time period from initial deposition to the present. In this geologic setting these varying interbedded materials can be classified by their appearance as ranging from soil to rock, and by their relative consistency as loose or soft, to very dense or very hard. Sand-filled vuggs are common within the rock-like zones. It is in these above described porous and vuggy

Formations that drilled shafts in South Florida derive their primary load carrying capacity.

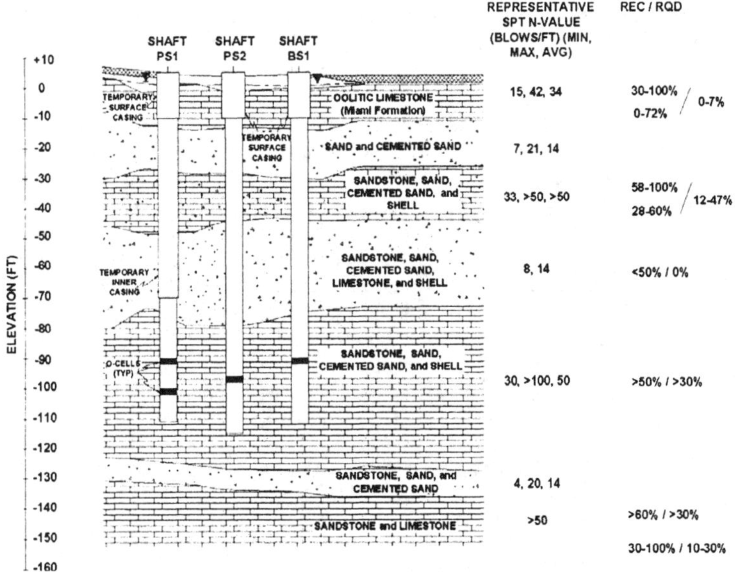

FIG. 1. Generalized Geologic Profile

Borings were performed to as deep as 220 feet (67 m) to: identify the subsurface materials and ground water conditions at the project site, differentiate the in-situ characteristics of the soil and soft rock layers, and obtain samples for testing. The boreholes were advanced using mud-rotary drilling techniques with casing installed as necessary to maintain a fluid pressure where more porous zones were encountered. The Standard Penetration Test (SPT) was done using a safety hammer to advance the split spoon at 5-ft (1.52-m) intervals in accordance with ASTM D1586. Careful observation of drill rig action and wash return were performed to assess variations in the subsurface conditions and delineate softer zones within the rock mass between sampling intervals. Continuous wire-line coring and 4-inch (97.6 mm) diameter sampling was done to obtain a continuous record of subsurface conditions and to obtain samples for laboratory testing. Representative SPT N-values, rock core recoveries (REC), and rock quality designations (RQD) are indicated on Figure 1

PRE-CONSTRUCTION SIDE SHEAR EVALUATION

Extensive sampling and unconfined compression testing of Fort Thompson Formation rock core samples were performed for the Dade County Metrorail project in the late 1970's and early 1980's. Laboratory unconfined compressive strength data reported by Preito-Portar (1982) for 110 samples taken and tested for the Metrorail Line Section nearest the project site averaged approx 510 kips/ft^2 (24.4 MPa); 90% of tested samples exceeded a strength of 90 kips/ft^2 (4.3 MPa). Laboratory unconfined (UC) strength tests were performed on soft rock samples obtained in site borings from depths of 80 to 120 ft (24.4 to 36.6 m) within the Lower Fort Thompson Formation. The UC test results averaged 222 kips/ft^2 and 460 kips/ft^2 (10.5 and 22 MPa) in the upper and lower portions, respectively. Correlations proposed by several authors between soft rock unconfined compressive strength and ultimate side shear are shown on Figure 2. The correlations presented in Reynolds and Kaderabek (1980) and Gupton and Logan (1984) were developed from full scale South Florida drilled shaft experience; however, the basis was top-load tests performed on shafts less than half the diameter and tested to less than ⅓ of the maximum loads presented in this paper. The drilled shaft construction methods associated with the observed correlations are not detailed in the references. Comparing the available average local rock strength data and the side shear correlations in the literature, lower-bound ultimate values of side shear in the range of about 10 to 20 kips/ft^2 (about 480 to 960 kPa) could have been considered for drilled shafts socketed into the Fort Thompson Formation limestone; the correlations imply that upper-bound values determined from local experience in the literature could be calculated to be about 4 to 8 times higher.

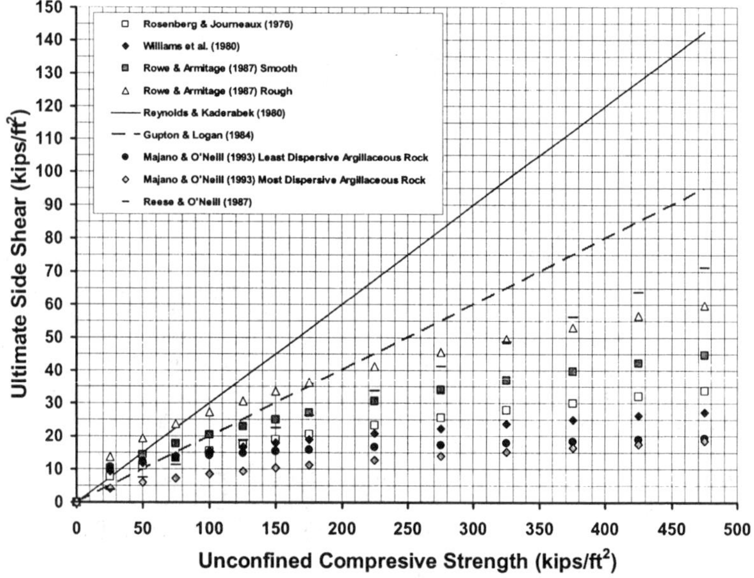

FIG. 2. Strength Data and Side Shear Correlations for Soft Rock

It should be noted that the UC strength approach is based on intact rock strengths. The Fort Thompson Formation is characterized by alternating layers of rock-like materials that range from soft to very hard with soil layers ranging from loose to very dense. Given the difficulty in obtaining representative rock core and laboratory test data in the Fort Thompson Formation, careful compatibility of strain evaluations and judicious use of judgment are necessary to obtain representative design side shear values following the aforementioned approach. Frizzi and Meyer (2000) discuss the difficulty obtaining core samples of south Florida rock suitable for laboratory unconfined compressive strength testing. Since only better quality samples tend to be recoverable, any test results would be biased towards these samples. Since the variability is significant for the Fort Thompson Formation, SPT boring data was considered due to the ability of the SPT, under careful engineering inspection, to identify variations in the relative hardness of the rock mass. Correlations proposed by several authors between SPT N-Value and ultimate side shear for drilled shafts in soft rock are shown in Figure 3. The drilled shaft construction methods associated with the observed correlations are not detailed in the references. A correlation developed from an auger cast-in-place (ACIP) pile database that included over sixty full-scale compressive and tension field load tests from Frizzi and Meyer (2000) is also shown in Figure 3. From the author's experience with ACIP piles, and understanding that drilled shaft construction would result in the shaft sockets being exposed to drilling slurry, it was anticipated that ultimate side shear values for drilled shafts could be lower than those observed in full-scale ACIP pile load tests. Considering a low and average SPT N-value in the Fort Thompson Formation drilled shaft socket of 30 and 50 blows/ft, respectively, lower-bound and average side shear could be less than 10.5 and 16 kips/ft^2 (about 505 and 765 kPa), respectively. The porous rock combined with multiple drilling tool passes associated with drilled shaft construction were expected to impact side shear – theory and practice indicated these factors could yield potential beneficial as well as detrimental results.

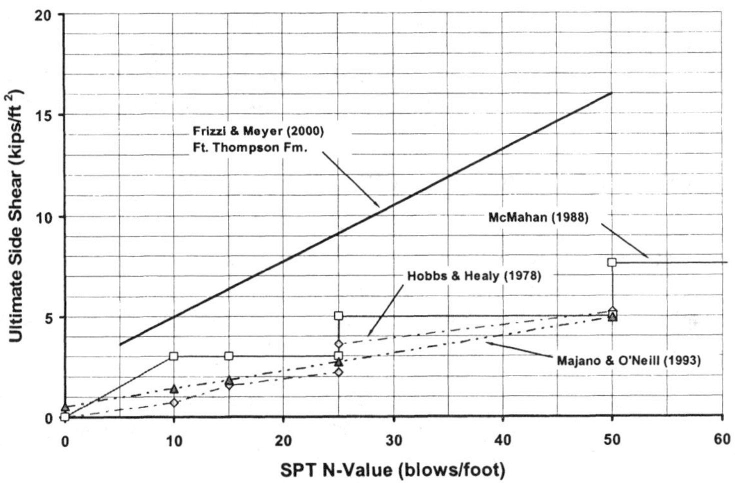

FIG. 3. SPT N-Value and Side Shear Correlations for Soft Rock

During the test drilled shaft bidding process, both cased and drilling fluid stabilized drilled shaft techniques were proposed and evaluated. Local experience was that bentonite stabilized drilled shafts tended to provide the required design load carrying capacity desired by the Engineer, while excessive concrete overtake typically observed in water-stabilized shafts was minimized and controlled which is desired by the Contractor. However, it was recognized that the bentonite-stabilized drilled shaft work would quickly cover the confined site with a mud-and-cutting mixture requiring a de-sanding operation and subsequent mud disposal that would require environmental permitting. A significant savings in both project cost and schedule would be realized if a polymer drilling admixture could be utilized in the work.

Although recent advances in polymer drilling fluid technology were recognized, these admixtures had not been previously used in similar South Florida drilled shaft construction. Little information was available in the technical literature in the late 1990's when the test program was initiated. Majano and O'Neill (1993) summarized the results of laboratory comparison tests on model shafts constructed using bentonite and polymer stabilizing slurries. From this reference, several applicable conclusions were considered. First, European experience with shafts in sand and gravel indicated an about 28% reduction in side shear between shafts constructed dry and those constructed with bentonite slurry, and a reduction in side shear was reported in shafts exposed to bentonite slurry over time (about 43% and 56% side shear reduction when exposed to slurry for 8 and 97 hours, respectively). In comparing bentonite and polymer stabilized model shafts, peak side shear was generally constant with varying dosage of bentonite, while about 10% lower peak side shear was observed for low dosage (0.0035 ppg) polymer stabilized shafts; about 40% greater side shear was observed for mid to high dosage (0.0080 and 0.0180 ppg) polymer stabilized shafts. Higher side shear was observed in polymer stabilized shafts, except in low dosage and long (24 hour) exposure conditions – in these cases higher side shear was observed in bentonite stabilized shafts. Lastly, slurry viscosities of 110 to 130 seconds/quart (116 to 137 seconds/liter) resulted in maximized interface friction in polymer stabilized shafts; a decrease in interface friction of about 5% was reported between shafts exposed to the polymer slurry for 4 and 24 hours.

TEST DRILLED SHAFT CONSTRUCTION

The drilled shaft test program consisted of installing 6-ft (1.83 m) diameter shafts: one un-reinforced method shaft, and three (3) fully instrumented and reinforced shafts. The method shaft and two of the instrumented and reinforced shafts were constructed using polymer slurry; one instrumented and reinforced shaft was constructed using bentonite slurry. As shown in Figure 1, the shafts were embedded into the sandstone of the Fort Thompson Formation. The shafts were constructed using an 80-inch (2 m) inside diameter, 15-ft (4.6 m) long surface casing. For the remainder of the shaft excavation, hole stabilization was provided by either polymer or bentonite drilling slurry as identified in Table 1. A 74-inch (1.88 m) inside diameter, 80 ft long inner casing was also utilized in shaft PS1.

The polymer slurry consisted of a proprietary polymer in combination with multiple proprietary additives. Actual field polymer dosing quantities were difficult to estimate as both pre-mixed slurry was injected at the top of the shaft excavation and multiple additional bags of slurry mix and other proprietary additives were placed within the slurry column during drilling operations. The bentonite slurry was prepared utilizing high yield bentonite, soda ash for pH adjustment, and a slurry additive. The bentonite was prepared in slurry tanks and allowed to hydrate prior to excavating the shaft.

Table 1. Drilling Slurry Properties.

Shaft	Viscosity (seconds)	pH	Sand Content	Density (lbs/ft^3)
Method Polymer Shaft (MS)	50 to 70	10 to 12	1.5%	63
Polymer Shaft 1 (PS1)	50 to 70	10 to 12	1.0%	63
Polymer Shaft 2 (PS2)	50 to 70	11 to 12	1.8%	63
Bentonite Shaft (BS1)	40	10	2.0%	67

Shaft excavation was performed using a fixed mast drilling rig with telescoping Kelly bar. Excavation tools included traditional drilling buckets, modified slurry pass-through drilling buckets with specially adapted bottom cutting teeth, rock augers, core barrels, and flat-bottomed clean out buckets. Drilling time in the socket, as well as time to place reinforcing and concrete the shafts are summarized in Table 2. The drill tools utilized are shown in Figure 4.

Table 2. Drilling Data.

Shaft	Time In Socket Drill / Reinforcing /Concreting			Bottom Sediment	Concrete Factor
Method Polymer Shaft (MS)	51 hrs	n/a	5 hrs	3 to 9 cm	1.21
Polymer Shaft 1 (PS1)	97 hrs	9.25 hrs	2.25 hrs	1 to 6 cm (avg 4 cm)	1.19
Polymer Shaft 2 (PS2)	21 hrs	2.42 hrs	2.17 hrs	7 cm	1.09
Bentonite Shaft (BS1)	31.5 hrs	1.25 hrs	2.75 hrs	1.5 to 2 cm (avg 1.75 cm)	1.07

The polymer slurry viscosity, unit weight, and sand content criteria were established based on information submitted by the foundation Contractor, discussion and submittals from the polymer slurry technical representative, and the state of practice for control of polymer slurries available at the time. The requirements established by the Florida Department of Transportation (1996) were utilized for the bentonite slurry quality control. Upon completion of drilling, the bentonite slurry was circulated and de-sanded to reduce its density and sand content to within specification limits. The proprietary additives were added to the polymer slurry stabilized shaft to facilitate rapid sediment settlement while still maintaining sidewall stability. Bottom materials were removed using the flat-bottomed clean-out bucket. There was a clear difference in the excavated materials: the polymer stabilized materials could be easily handled by excavation equipment right out of the drilling equipment, while the bentonite stabilized materials required de-sanding to facilitate subsequent handling. The slurry sand content and unit weight at the shaft bottom were tested immediately prior to reinforcement placement.

Core Barrel

Rock Auger

Digging Bucket

Flat-Bottom Clean-Out Bucket

FIG. 4. Drilling Tools

Shaft plumbness, sidewall characteristics, and bottom cleanliness were evaluated using sonar logging, sidewall sampling, and video inspection tools, respectively. A continuous shaft sidewall profile was obtained using an ultrasonic logging device that produced a continuous record of shaft diameter, verticality, and sidewall undulation versus depth. The logging results indicated shaft diameters consistent with the drilling tools utilized, and shaft verticality deviations typically less than 0.5%. Inferred sidewall undulations on the order of 1 to 2 inches (2.5 to 5 cm) in PS1 and BS1, and 4 to 16 inches (10 to 40 cm) in PS2 could be discerned within the Fort Thompson Formation socket. To assess filtercake formation on the bentonite-stabilized shaft sidewall, sidewall sampling was performed at selected intervals using a proprietary down-hole "sidewall sampler". Both samples of soft rock and cemented soil were obtained. The sampling results in the upper rock strata revealed an 8 mm soft layer of bentonite, which could be easily removed with little effort, and a 1 mm thick filtercake which had penetrated the surface of the soft rock; in the lower rock strata no well defined mud accumulation or filtercake could be discerned. Shaft bottom inspection was performed using a "Mini-SID" proprietary bottom camera / caliper. The Mini-SID allowed visual observation and direct measurement of shaft bottom materials and sediment thicknesses. Measured bottom sediment at each of the shafts is summarized in Table 2. Sidewall sampler and Mini-SID photographs are shown in Figure 5.

Sidewall Sampler Mini-SID

FIG. 5. Shaft Sampling and Inspection Tools

O-Cells and Instrumentation Reinforcing Cage With O-Cells

FIG. 6. Drilled Shaft O-Cells, Instrumentation, and Reinforcement

The load tested shafts were reinforced using cages consisting of 42, #9 grade 60 bars, and #5 bar ties spaced 10 inches (25 cm) on-center. Due to the shaft length, it was necessary to fabricate, pick, and install the reinforcing cages in two, 60-ft

(18.3 m) spliced sections. The reinforcing steel in BS1 is shown in Figure 6. Concrete was placed using tremie procedures; a 10-inch (25 cm) diameter tremie pipe was utilized. The concrete design compressive strength was 8000 lbs/in^2 (55 MPa) at 28 days, with slump in the range of 7.5 to 9 inches (19 to 23 cm) maintained for each 9 yd^3 (7 m^3) batch placed. The concrete depth was measured after each batch to monitor the concrete factor (the ratio of actual concrete volume placed to the shaft's theoretical drilled volume). Concrete factors are summarized in Table 2. The higher concrete volume in the MS is attributed to partial shaft collapse which occurred 38 hours after the polymer stabilized shaft was left undisturbed by drilling operations to evaluate hole stand-up time. The higher concrete volume in PS1 is attributed to volume change in the sandy soil zones above the socket resulting from casing vibration during installation and subsequent extraction.

Several excavated material handling and use observations were made. Materials excavated from the bentonite stabilized shaft required de-sanding to facilitate subsequent removal using conventional earth moving equipment. Off-site disposal of the residual bentonite mud was required. Materials excavated from the polymer stabilized shafts could be handled by conventional earthmoving equipment practically right after dumping from the drilled shaft excavation drilling / cleaning buckets. This allowed the subsequent re-use of the excavated soils for on-site compacted fill.

TEST SHAFT LOADING AND RESULTS

All three test shafts were load tested using groups of embedded Osterberg-type jacks (O-Cells) positioned at the approximate levels indicated on Figure 1. Each shaft was instrumented with vibrating wire "sister bar" strain gauges, telltales, and LVDT extensometers to determine the test load distribution along the shaft's socket. The O-Cell and instrumentation arrangement installed in BS1 is shown in Figure 6. The load tests were performed 6 to 9 days after construction. Laboratory results indicated concrete compressive strengths typically ranging from 5930 to 6737 lbs/in^2 (41 to 46 MPa) at the time of shaft testing. The test loads were applied following the quick loading procedures outlined in ASTM D1143.

The load test results are summarized in the plots of mobilized side shear and mobilized end bearing versus deflection given in Figures 7 and 8, respectively. Calculated equivalent top-load curves combining both end bearing and side shear along the entire shaft are presented in Figure 9.

The load test results indicate variation in mobilized side shear within the lower Fort Thompson Formation rock socket between about 5 and 20 kips/ft^2 (240 to 960 kPa) for the polymer-stabilized shafts, and about 4 and 23 kips/ft^2 (190 to 1100 kPa) for the bentonite-stabilized shaft. There appears to be a trend towards increasing mobilized side shear with depth in both the bentonite stabilized shaft and the polymer stabilized shafts. The average side shear observed in the bentonite stabilized shaft was about 25% to 50% lower as compared to the polymer stabilized shafts in the upper ½ of the socket. On average, the observed side shear was approximately similar for both the bentonite and polymer stabilized shafts in the lower ½ of the socket. The side shear values are considered to be ultimate values since they were measured at shaft movements of greater than 3 inches (75 mm).

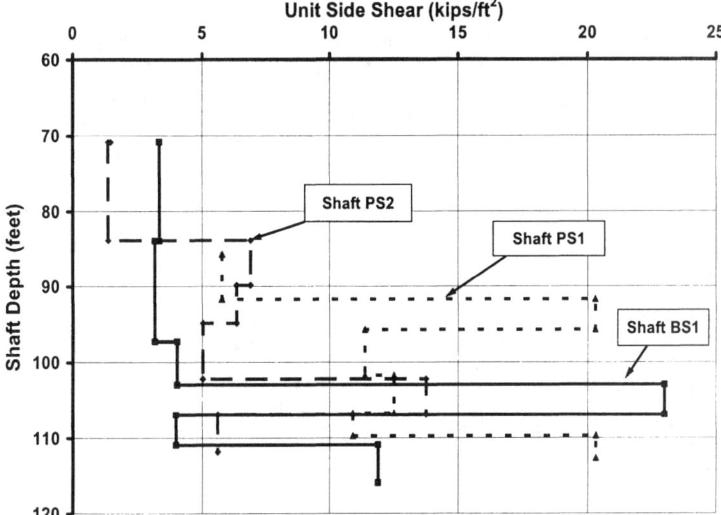

FIG. 7. Mobilized Side Shear In Lower Fort Thompson Formation Rock Socket

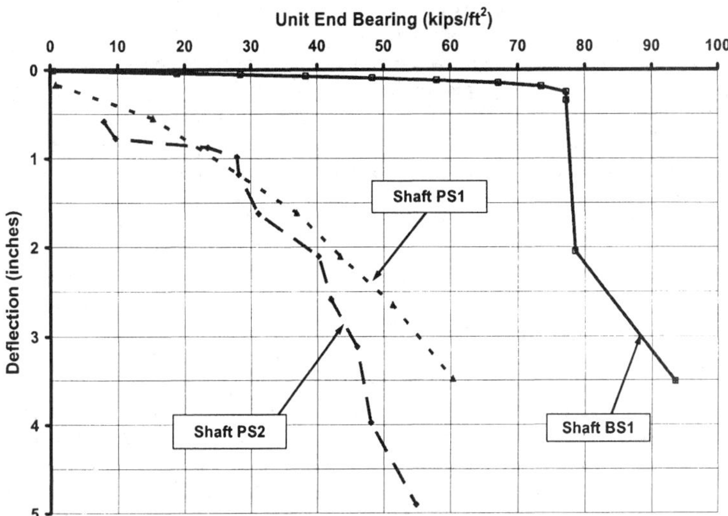

FIG. 8. Mobilized End Bearing In Lower Fort Thompson Formation Rock Socket

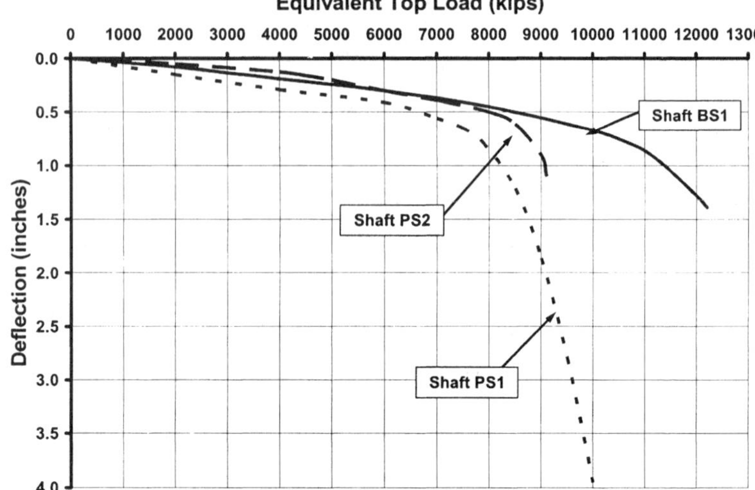

FIG. 9. Equivalent Top Load – Deflection Performance

Both polymer stabilized shafts exhibited similar end-bearing load-deflection characteristics, with calculated unit end bearing on the order of 25 to 28 kips/ft^2 (1200 to 1340 kPa) mobilized at a downward deflection of approx 1 inch (25 mm). At the same deflection, the bentonite-stabilized shaft exhibited a significantly higher mobilized unit end bearing on the order of 78 kips/ft^2 (3735 kPa). This significantly higher end bearing is reflected in the higher equivalent top load carrying characteristics shown in Figure 9 for the bentonite-stabilized shaft.

CONCLUSIONS

The following conclusions are drawn from the full-scale field construction and load testing of bentonite and polymer stabilized drilled shafts:

1. The trend of increasing unconfined rock compressive strength with depth in the Fort Thompson Socket was mirrored by the trend in increasing mobilized side shear with depth observed in both bentonite and polymer stabilized shafts. Utilizing average rock unconfined compressive strength data, an approximate correlation of ultimate shaft side shear to unconfined compressive strength of $f_{s,ult}$ = 0.03(q_u) was observed for the polymer stabilized shafts at this site. A reduction of 25% to 50% in side shear was observed in the upper ½ of the bentonite stabilized shaft. The use of other correlations in the literature based on lighter loaded, smaller diameter shafts (i.e. smaller scale tests) yield un-conservative side shear predictions. Model

tests indicate that perhaps a higher mobilized side shear could have been obtained if higher viscosity polymer slurry was utilized.

2. In comparing the observed low and average SPT N-values in the Fort Thompson Formation socket and the mobilized side shear with depth in both bentonite and polymer stabilized shafts, an approximate correlation of ultimate shaft side shear to SPT N-value of $f_{s,ult} = 0.25(N)$ was observed for the polymer stabilized shafts at this site. This presumes the lower bound N-values are utilized to calculate side shear in the upper ½ of the socket, and average N-values are utilized for the remainder of the socket. A reduction of 25% to 50% in side shear should be considered in the upper portion of bentonite stabilized shafts. The trend in SPT N-value versus mobilized side shear is similar to that observed for ACIP piles, with higher side shear mobilized in ACIP piles. This is believed to be due to the ACIP pile construction being such that the pile grout is cast practically immediately after drilling when compared to the drilled shaft. The trend in lower drilled shaft side shear is consistent with the European experience (Majano and O'Neill 1993) in dry versus slurry drilled shafts.
3. Significantly higher end bearing was observed in the bentonite stabilized shaft, as compared to the polymer stabilized shafts. This resulted in the bentonite stabilized shaft exhibiting a higher ultimate equivalent top load on the order of 25% to 40% higher than the polymer stabilized shafts.

REFERENCES

Florida Department of Transportation (1996). "Standard Specifications for Road and Bridge Construction".

Frizzi, R.P. and Meyer, M.E. (2000). "Augercast Piles – South Florida Experience." *Geotechnical Special Publication on Deep Foundations*, Geotechnical Special Publication No. 100, The Geo-Institute of ASCE, Reston, Virginia, pp 382-396.

Gupton, C. and Logan, T. (1984). "Design Guidelines for Drilled Shafts in Weak Rocks of South Florida." *Proc. South Florida Annual ASCE Meeting*, ASCE.

Hobbs, N.B., and Healy, P.R. (1979). "Piling in Chalk." Construction Industry Research and Information Association, London, United Kingdom.

Hoffmeister, J.E. (1974). *Land from the Sea - The Geologic Story of South Florida*, University of Miami Press, Coral Gables, Florida.

Majano, R.E. and O'Neill, M.W. (1993). "Effect of Mineral and Polymer Slurries on Perimeter Load Transfer in Drilled Shafts." *Research Report UHCE 93-1*, University of Houston, Houston, Texas.

McMahan, B. (1988). *Drilled Shaft Design and Construction in Florida.* Department of Civil Engineering, University of Florida, Gainesville, Florida.

McVay, M.C., Townsend, F.C., and Williams, R.C. (1992). "Design of Socketed Drilled Shafts in Limestone." *J. Geotech. Eng.*, ASCE, 118(10), 1626-1637.

Preito-Portar, L.A. (1982). "Elastic and Strength Properties of Calcareous Rocks of Dade County, Florida." *Geotechnical Properties, Behavior, and Performance of Calcareous Soils, ASTM STP 777*, ASTM, 359-381.

Reese, L.C. and O'Neill, M.W. (1988). "Drilled Shafts: Construction Procedures and Design Methods." *Publication No. FHWA-HI-88-042*, Federal Highway Administration, Washington, D.C.

Reynolds, R.T. and Kaderabek, T.J. (1980). "Miami Limestone Foundation Design and Construction." *Proc. ASCE Annual Meeting*, ASCE, New York, New York.

Rosenberg, P. and Journeaux, N.L. (1976). "Friction and End Bearing Tests on Bedrock for High Capacity Socket Design." *Canadian Geotechnical Journal*, 13:3, pp 324-333.

Rowe, R.K. and Armitage, H.H. (1987). "A Design Method for Drilled Piers in Soft Rock." *Canadian Geotechnical Journal*, 24:1, PP 126-142.

Williams, A.F., Johnston, I.W., and Donald, I.B. (1980). "The Design of Socketed Piles in Weak Rock". *Structural Foundations on Rock, Proceedings of the International Conference on Structural Foundations on Rock*, Sydney, Vol 1, pp 327-347.

APPENDIX I. Conversions

1 inch (in) x 25.4 = millimeter (mm)

1 foot (ft) x 0.3048 = meter (m)

1 pound mass (lbm) x 0.4536 = kilogram (kg)

1 pound force (lbf) x 4.448 = 1 newton (N)

1 pound force per square inch (lb/in^2) x 6.895 = kilopascal (kPa)

1 kip force per square foot (kip/ft^2) x 47.88 = kilopascal (kPa)

1 pound force per cubic foot (lb/ft^3) x 0.016 = gram per cubic centimeter (g/cm^3)

OSTERBERG LOAD CELL TEST RESULTS ON BASE GROUTED BORED PILES IN BANGLADESH

Raymond J. Castelli[1], P.E., Member, ASCE and Ed Wilkins[2], MIEAust, CPEng

ABSTRACT

The paper presents the results of an Osterberg load test program performed for the Paksey Bridge over the Padma (Ganges) River in western Bangladesh. The test program included the installation and testing of two 1.5-m diameter bored piles to depths of 65 m and 91 m. The paper discusses the design of the high capacity bored piles for a site susceptible to deep scour. The paper also discusses the installation of the test piles, and the application of base-grouting to improve end bearing resistance. Finally, the paper documents and evaluates the results of Osterberg load tests at the two test piles, including a comparison of pile capacity prior to and following base grouting.

INTRODUCTION

When completed, the Paksey Bridge will be the first roadway crossing of the Padma (Ganges) River in Bangladesh (Figure 1). As illustrated in Figure 2, the 1,786-m long structure includes 15 typical spans each 109.5 m, and two shorter end spans of 71.75 m each. The superstructure consists of a segmental concrete trapezoidal box, 18.0 m wide, and will carry two lanes of traffic, each 7.5 m wide, and two 1.0-m wide walkways.

Each bridge pier is founded on four 3.0-m diameter bored piles with a design capacity of 34 MN and length of 91 m. Abutments are each founded on two 2.5-m diameter bored piles with design capacity of 15 MN ana length of 68.6 m. All piles are founded in very dense, micaceous, medium to fine sand. To achieve the high design capacities, the bored piles were installed using base grouting, at grout pressures up to 100 bars.

Construction commenced in August 2000, and is scheduled to be completed in April 2004. The test pile program was conducted from April to October of 2001.

[1] Vice President, Parsons Brinckerhoff, One Penn Plaza, New York, NY 10119
e-mail: castelli@pbworld.com

[2] Geotechnical Consultant, Ed Wilkins Pty. Ltd., Melbourne, Australia

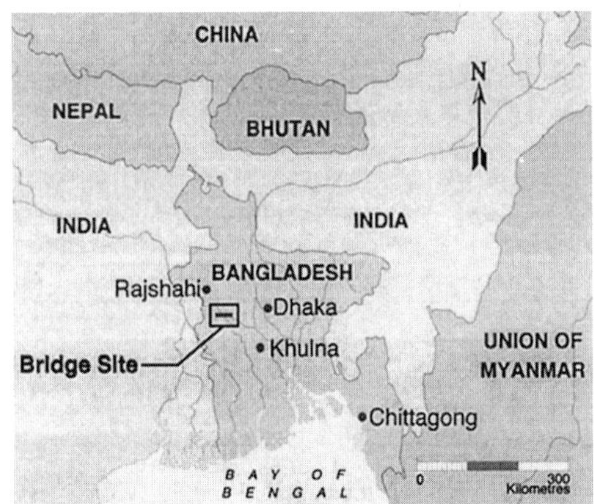

FIG. 1. Paksey Bridge Location Map

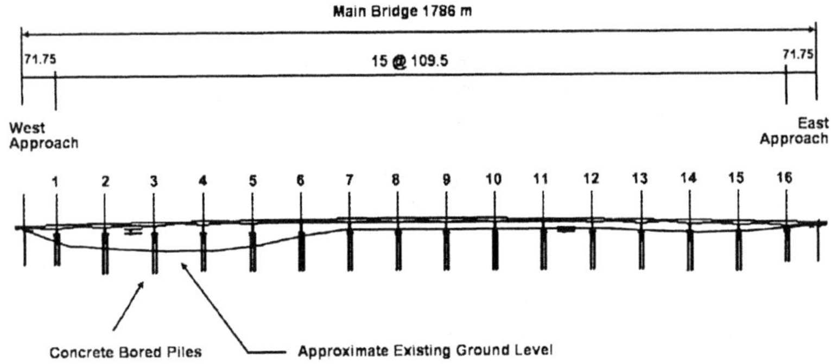

FIG. 2. Bridge Elevation

SUBSURFACE CONDITIONS

Bangladesh is located within the deltaic flood plain formed by three major rivers, including the Ganges, Brahmaputra and Meghna Rivers. The project site, located in western Bangladesh, lies within the active flood plain of the Ganges-Brahmaputra systems. These meandering rivers form a broad, low lying plain, underlain by more than 1000 m of alluvial sediments.

At the project site the Padma River is confined to a flood plain approximately 1,790-m wide between the river training works on both banks of the river. During the annual monsoon season the flooded river extends the full width of this flood plain. During the dry season, however, the river narrows to a shallow, meandering channel that at present is located on the western side of the crossing (Figure 2).

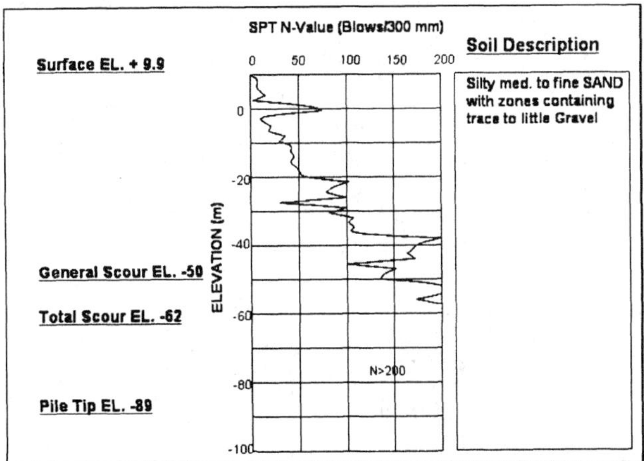

FIG. 3. Typical Boring Profile at Crossing Site

The existing soils generally consist of very dense micaceous, silty, medium to fine sand with occasional zones containing trace amounts of fine gravel. SPT N-values increase from less than 10 blows per 300 mm at the ground surface to about 50 blows at El. –20 m (Figure 3). Below about El. –30 m, SPT N-values are generally greater than 100 blows per 300 mm. Although these soils are very dense, a conservative friction value of 36° was used for design to reflect a mica content as high as 26 percent.

FOUNDATION DESIGN

The contract documents included both driven steel pipe pile and concrete bored pile foundation alternatives. The bored pile alternative was selected by the winning contractor, and is described in the remainder of this paper.

Loads

The bored pile foundations consisted of four vertical piles at each pier (Figure 4) and two vertical piles at each abutment. In addition to the structure dead and live loads, the foundations are subject to deep scour and stream flow forces, as discussed below.

Design scour depths were determined by both numerical analyses and physical model tests. The project design criteria included three cases (Table 1) with the foundation design governed by the 100-year scour case (Figure 5). During the useful life of the bridge, the main channel is predicted to migrate from its present location along the west bank eastward until it reaches the opposite bank. The channel will likely further migrate between the river training works, as it has at least twice since 1915. Consequently, all pier foundations were designed for the maximum anticipated scour depths.

For the design flood condition, the average river velocity was estimated to be 5 m/sec. The resulting stream flow load acting on the bored piles is illustrated in Figure 5(b).

TABLE 1. Design Scour Criteria

Design Case	Flood Level (m)	General Scour Elev. (m)	Total Scour Elev. (m)	Comments
Base Case	+15.6	-40.0	-40.0	No local scour
100-Year	+15.6	-50.0	-62.0	
500-Year	+16.0	-54.0	-67.0	Check safety factor > 1.0

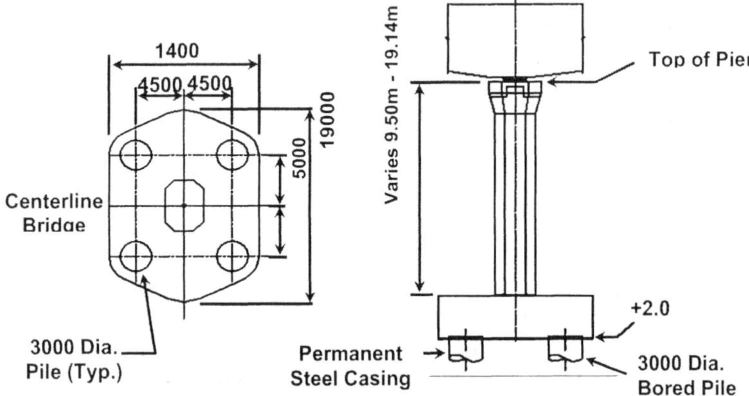

FIG. 4. Plan and Elevation of Typical Pier Foundation

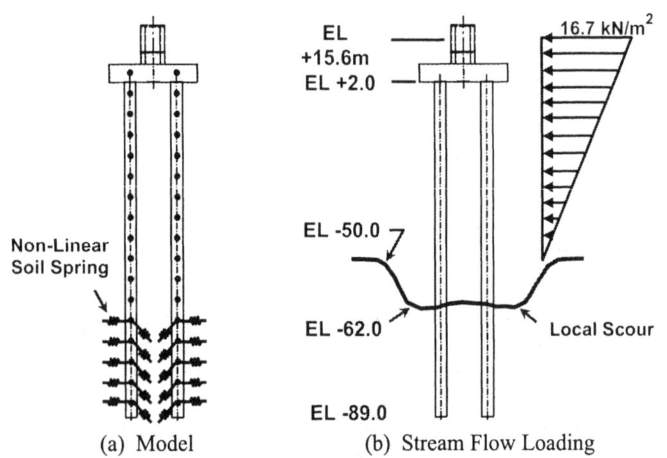

FIG. 5. Governing 100-Year Scour Condition and Design Stream Flow Loading

Pile Design

The axial pile loads and pile bending moment distribution were estimated using the computer program GROUP (Ensoft Inc., 1988). The GROUP program models the lateral soil resistance as non-linear soil springs (p-y curves) and axial stiffness as base springs (t-z curves) (Figure 5a). For this project the lateral soil springs were internally generated by the program assuming cohesionless soil with an angle of internal friction of 36° and a total unit weight of 19.8 kN/m^3. The axial springs were based on a subgrade modulus of 1,100 MN/m at the base of the pile.

Pile friction resistance, determined using procedures recommended by the U.S. Federal Highway Administration (FHWA, 1988), was estimated to be 67 to 83 kPa. The reliability of end bearing resistance was a concern since the anticipated installation time for these unusually long piles could result in accumulation of substantial sediments at the base prior to concrete placement. However, it was not economical to design the piles only for friction resistance. To assure reliable end bearing resistance, the design included provisions for post-installation pressure grouting at the base of the piles, a technique that had been successfully applied in similar soils at the recently completed Jamuna Bridge over the Brahmaputra River in Bangladesh. Using the FHWA design procedures, the ultimate end bearing resistance was estimated to be about 4,300 kPa. This value, however, was considered conservative since the FHWA procedures do not consider the contribution from base grouting. Using information from available references (Bruce, 1986a; Bruce, 1986b; Sliwinski and Fleming, 1984; Stocker, 1983; Teparaksa, 1992), an ultimate end bearing resistance of 7,700 kPa was selected for design. The allowable pile capacity was determined using factors of safety of 2.0 for AASHTO Group I loading, and 1.75 for AASHTO Groups II through VI loadings.

At the pier and abutment foundations, the settlement at the top of the pile was estimated to be less than 25 mm and 5 mm, respectively, for the design load.

DESCRIPTION OF O-CELL TEST PROGRAM

Prior to installing production piles, a test pile program was conducted to assess the proposed pile installation procedures and to confirm pile capacity. This test program included the following elements:

- Installation of two full-size trial piles to design tip elevation, including one at the west abutment and one near Pier No. 15 where soil penetration was greatest.
- Construction of two test piles extending to design depth with external diameter half that of the full size piles. Test pile B-1 was located on the west river bank to simulate an abutment pile, and test pile B-2 was located in the river near Pier No. 3. Pile B-2 was provided with an isolation casing to simulate the design scour depth.
- Load was applied to the test piles using Osterberg Cells (O-Cells®). Test Pile B-1 had two levels of O-Cells and Test Pile B-2 had one level (Table 2).
- Both test piles were base grouted.
- Load testing of pile B-1 was performed both before and after base grouting. Load testing on pile B-2 was performed only after base grouting.
- Integrity testing was performed by crosshole sonic logging (CSL).

TABLE 2. General Information for Load Test Piles

Pile No.	Location	Pile Diam. (mm)	Pile Length (m)	No. of O-Cell Levels	O-Cell Base Elev. (m)	Comment
B-1	West Abutment	1500	65.0	2	-38.3 -53.8	Test before and after base grouting.
B-2	River (Pier 3)	1500	91.0	1	-87.5	Test after base grouting. Isolation casing to El. -62

INSTALLATION OF THE TEST PILES

Schematic sections illustrating the test pile installations are presented in Figures 6 and 7, and the sequence of test pile construction is summarized in Table 3.

Construction Time

The test piles were constructed with reverse circulation drilling equipment using bentonite slurry as the circulating fluid and de-sanders to clean the bentonite during drilling. A flat cutter head was used for drilling and cleaning the base.

A key objective for construction was to minimize the time the pile walls were exposed to bentonite slurry since excessive exposure could result in softening of the sidewalls and build-up of filtercake, both potentially reducing the capacity of the piles. The specified maximum time from start of excavation below the toe of the permanent casing to the completion of concrete at that level was 36 hours. In addition, the bottom 1.5 m of the pile had to be excavated and concreted within 12 hours. However, during installation of two full scale trial piles, the contractor was unable to achieve these times. Therefore, the construction times of the two load test piles were deliberately lengthened to simulate the construction time of the corresponding production piles. The actual construction times for the test piles are summarized in Table 4.

Concrete Placement

Concrete was mixed at on-site batch plants, transported by agitator trucks and then pumped into a hopper above the pile. Concrete was placed using a 200-mm diameter tremie pipe immediately after installation of the pile reinforcement. During concrete placement the tremie tip was maintained at least 5 m below the concrete level to prevent segregation and ensure continuity. The specifications required a concrete mix with a minimum 28 days compressive strength of 30 MPa, maximum size aggregate of 20 mm, minimum cement content of 396 kg/m^3, and maximum water/cement ratio of 0.42.

At the start of placement the slump ranged from 210 to 230 mm at ambient temperatures between 25° and 32° C, and reduced to about 105 mm after a period of 11 hours. Maintaining the concrete slump was particularly important due to the long placement times required for the deep bored piles. A low slump could limit the ability of the concrete to flow through the reinforcement cage, potentially resulting in significant pile defects.

Sonic integrity testing between steel tubes attached to the outside of the reinforcement cage showed no major defects in the completed test piles.

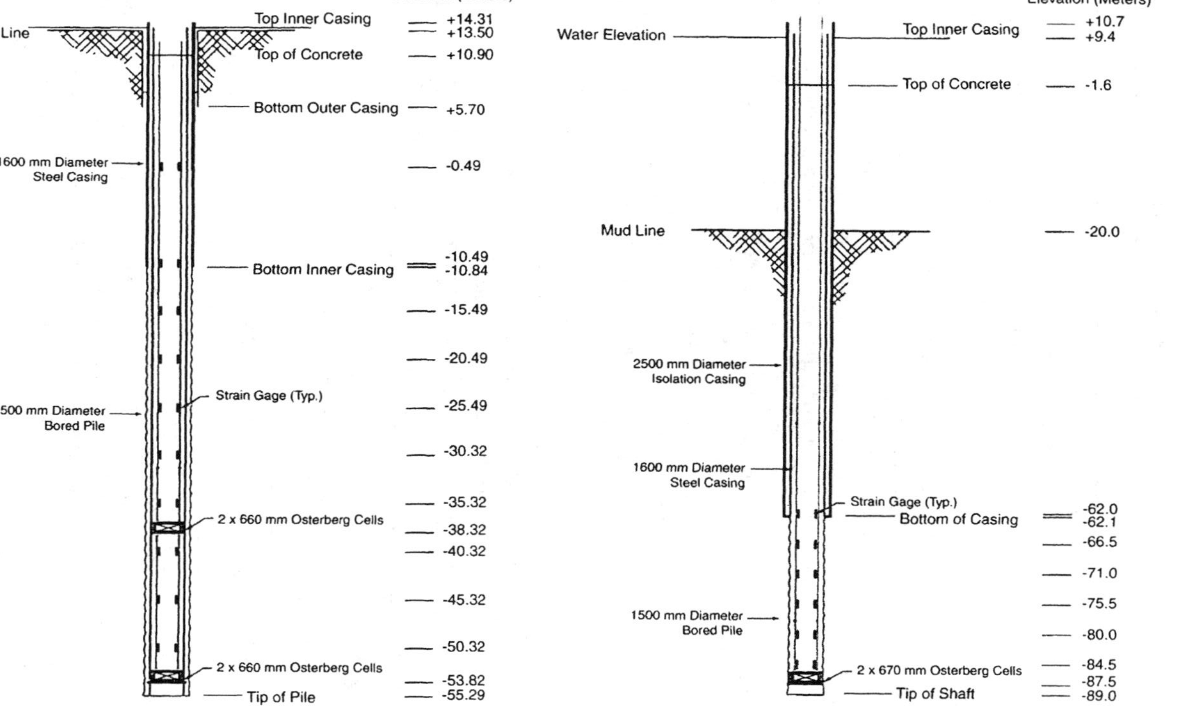

Figure 6. Schematic of Test Pile B-1

Figure 7. Schematic of Test Pile B-2

TABLE 3. Sequence of Test Pile Construction

Pile No.	Construction Sequence
B-1 (near west abutment)	• Install permanent casing by vibro-hammer to El. –10.67 m. • Drill to pile toe at El. –55 m. • Check depth, base sediment thickness, and slurry cleanliness. • Install reinforcement cage in three sections. • Place concrete by tremie pipe (125 m^3) • Crack concrete at base (24 hrs after concrete placement). • Integrity testing (7 days after concrete placement). • First Stage Load Test (9 days after concrete placement). • Base grouting (8 days after concrete placement). • Second Stage Load Test (18 days after concrete placement).
B-2 (in river near Pier No. 3, with bed level at El. –19 m)	• Install isolation casing to El. –62 m by vibro-hammer, airlifting inside casing when driving became difficult. Airlifted to –55m. • Install permanent casing to El. –62 m by vibro-hammer • Drill to toe of pile at El. –89 m. • Check depth, base sediment thickness, and slurry cleanliness. • Install reinforcement cage in four sections. • Place concrete by tremie pipe (173 m^3). • Crack concrete at base (24 hrs after concrete placement) • Integrity testing (7 days after concrete placement). • Base grouting (7.5 days after finishing concrete placement. • Remove material between permanent and isolation casings • Load tested pile 21 days after finishing concrete placement.

TABLE 4. Summary of Construction Times for Load Test Piles

Test Pile No.	Time to construct pile below permanent casing	Time to construct bottom 1.5 m of pile
B-1 (near west abutment)	54 hours	41 hours
B-2 (near Pier 3)	73 hours 30 min	55 hours

Base Grouting

Base grouting was performed through tube-a-manchettes in three pairs of grout circuits. Grout was injected in doses through each circuit in successive rounds of grouting, with individual doses limited to 50 liters per each circuit per round. All circuits were injected in turn with the specified dose or until a pressure of 100 bars was held for 10 minutes. Rounds commenced no less than six hours after completing the previous round. Rounds continued until a total of 600 liters had been injected to the base of the test pile, or until all lines sustained the required pressure of 100 bars, or until shaft uplift exceeded 15 mm. The grout consisted of a non-shrink bentonite/cement mix with a minimum compressive strength of 5 MPa at 7 days.

At test pile B-1, higher grout pressures were required to pump grout through the smaller diameter telescoping grout tubes at the O-cells, resulting in considerably more grout being injected into the base of the shaft than specified. The actual volume of grout injected was estimated to be almost 1,200 liters, twice the specified maximum volume of 600 liters. At pile B-2, grouting was stopped when the estimated total grout take reached 600 liters.

LOAD TEST PROCEDURES

Test loads were applied to the piles using Osterberg Load Cells after the concrete in the piles had attained a compressive strength greater than 30 MPa (Table 3). Loads were applied by pairs of 18 MN load cells at each level. The use of two cells at each level provided the necessary space to extend the concrete tremie pipe past the O-Cells.

Test pile instrumentation included pairs of strain gages at intervals along the length of the piles to determine the pile loads at these levels. Instrumentation also included telltales to determine top of load cell displacements, and LVDT gages to determine load cell extension. All instrumentation data were electronically monitored and recorded.

Test Pile B-2

The load cell level in pile B-2 was located 1.5 m above the base of the pile (Figure 7). Load was applied in increments until failure occurred in end bearing below the O-Cells. The applied load was then released incrementally, followed by a reload cycle which continued until the maximum stroke of the O-Cell (150 mm) was approached. The load was then progressively released and the test concluded.

Test Pile B-1

Pile B-1 had two levels of load cells, with one level 1.5 m above the pile base, and the other 15.5 m higher (Figure 6). Test pile B-1 was tested both before and following base grouting to:

- assess the influence of base grouting on side friction, and
- assess the influence of base grouting on the end bearing capacity of the shaft.

LOAD TEST RESULTS

Test Shaft B-2

For pile B-2 the combined maximum applied end bearing and side shear below the O-Cell was 24.5 MN at a total displacement of 142 mm (Figure 8). This corresponds to a maximum unit end bearing pressure of 10.8 MPa. At a base settlement equal to 5 percent of the pile diameter (used to define ultimate end bearing capacity) the applied load was 20 MN. This value exceeds the predicted ultimate end bearing value of 16.7 MN for test pile B-2 by about 20 percent.

The net side friction values estimated from the strain gage data are summarized in Table 5. However, since the pile failed in end bearing, the side shear values obtained from the test generally do not represent ultimate friction resistance. Only the lowest pile segment, directly above the Osterberg load cell, reached a peak resistance, computed to be 727 kPa.

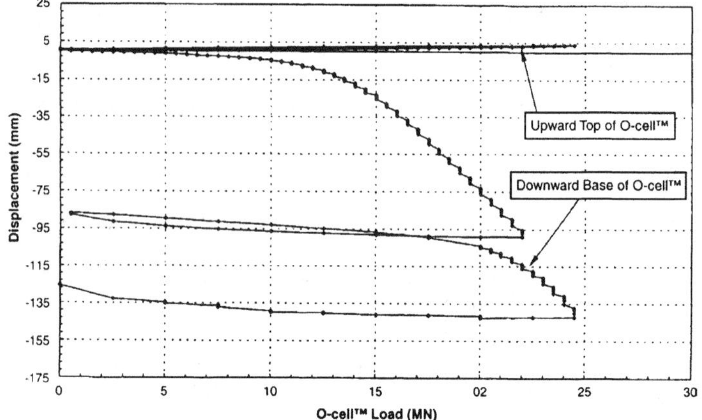

FIG. 8. Test Pile B-2, Load-Displacement Curves (LaodTest, Inc., 2001)

TABLE 5. Net Unit Side Shear, Test Pile B2

Load Transfer Zone	Net Unit Side Shear [1]
Top of shaft to stain gage level 6 [2]	0 kPa
Strain gage level 6 to strain gage level 5 [2]	57 kPa
Strain gage level 5 to strain gage level 4 [2]	66 kPa
Strain gage level 4 to strain gage level 3 [2]	130 kPa
Strain gage level 3 to strain gage level 2 [2]	145 kPa
Strain gage level 2 to strain gage level 1 [2]	311 kPa
Strain gage level 1 to O-Cell [3]	706 kPa

[1] The buoyant weight of pile was subtracted in this estimate
[2] Shear in these zones was still increasing at the maximum load
[3] Shear in this zone peaked at 727 kPa.

The average friction resistance of 208 kPa at pile B-2 was nearly 70 percent greater than the average ultimate friction resistance predicted by static analyses. However, this high friction resistance was primarily due to increased friction in the 7.5-m high zone directly above the O-Cells. These data suggest that base grouting significantly improved the friction resistance near the bottom of the pile. Above this zone, the average friction value of 100 kPa from the test was about 12 percent less than the average predicted ultimate friction resistance. However, the actual ultimate friction resistance is likely much higher since the peak resistance was not reached along most of the pile length.

The observed unit shear values for the two pile segments directly above the O-cells are unusually high for the existing soils. These results may have been due to penetration of grout around the lower portion of the pile shaft during base grouting operations.

Test Shaft B-1

The loading sequence at test pile B-1, and the maximum test load for each stage of the test are summarized in Table 6.

Figure 9 illustrates end bearing resistance during load Stage 1 before base grouting, and Stage 3 after base grouting. Before base grouting the maximum load below the lower O-Cells was 5.0 MN at a displacement of 27 mm, corresponding to a unit end bearing pressure of 2,521 kPa. In comparison, the maximum load after grouting (Stage 3a) was 12.0 MN at a net displacement of 17.5 mm (54.4 – 36.8 mm), corresponding to unit end bearing pressure is 6,137 kPa. For the reload Stage 3b, the maximum load was 13.0 MN, corresponding to unit end bearing pressure of 6,681 kPa at a net displacement of 44.6 mm. These results indicate a substantial improvement in end bearing resistance and base stiffness as a result of base grouting. These results also compared favorably with the ultimate end bearing resistance of 13.5 MN predicted for test pile B-1.

Figure 10 shows pile displacement below the upper level of O-Cells obtained from loading the upper O-Cells both before and following base grouting. The pre-grouting downward applied load (Stage 2) resulted in the pile base also being displaced downwards, indicating load transfer through the lower O-Cell level. The magnitude of the load transfer was estimated from strain gage data and the computed load transfer along the middle section of the pile was corrected accordingly. The maximum applied load to the middle section between the two levels of O-Cells was estimated to be 4.5 MN at an absolute downward movement of 20.4 mm. In comparison, the maximum post-grouting downward load (Stage 5) to the middle shear section was 14 MN at an absolute downward movement of only 7.5 mm. These results indicate a significant increase in side shear capacity occurred in the middle section after base grouting.

TABLE 6. Summary of Load Stages, Test Pile B1

Stage	Upper O-Cell		Lower O-Cell		Objective
	Max. Load (MN)	O-Cell System	Max. Load (MN)	O-Cell System	
1	0	Closed	5.0	Pressurized	End bearing before base grouting
2	10.0	Pressurized	0	Draining	Side shear between O-Cell levels before base grouting
3a	0	Closed	12.0	Pressurized	End bearing after base grouting
3b	0	Closed	13.0	Pressurized	Reload base
4	14.0	Pressurized	0	Draining	Side shear above & below upper O-Cell after grouting
5	0	Draining	12.0	Pressurized	Side shear of middle section after base grouting
6	11.4	Pressurized	11.4	Pressurized	Confined compression test

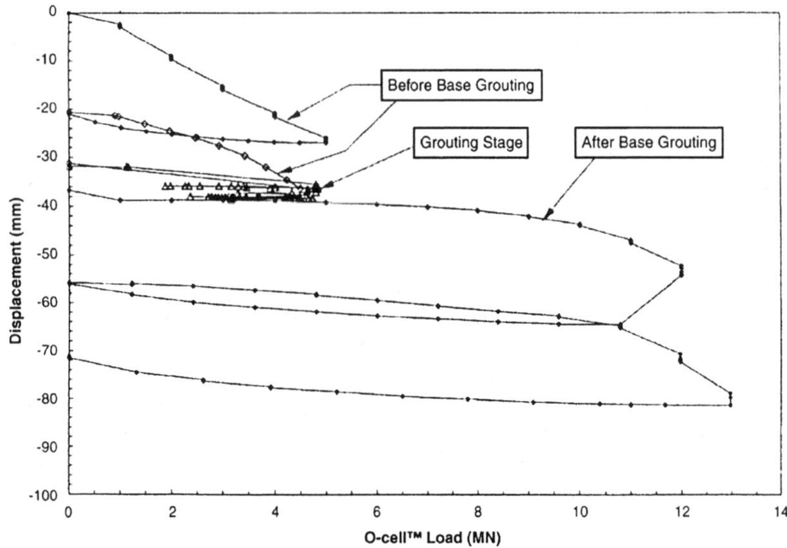

FIG. 9. Test Pile B-1, End Bearing Resistance Before and After Base Grouting (LoadTest, 2001a)

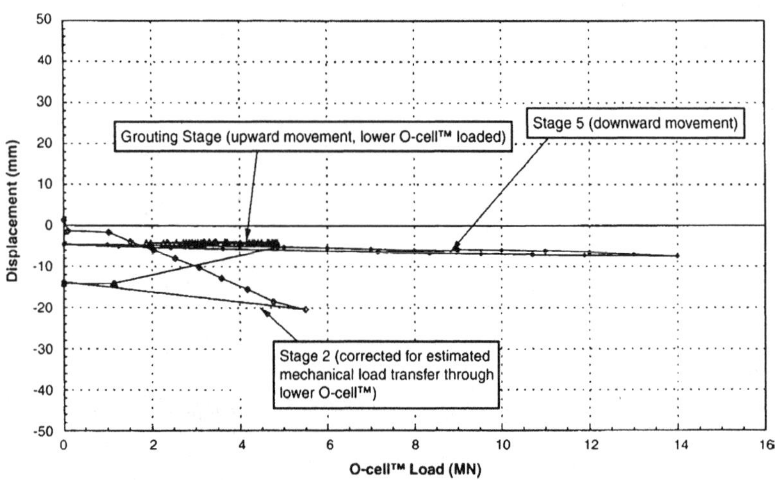

FIG. 10. Test Pile B1, Load-Displacement Below Upper O-Cell Before and After Base Grouting (LoadTest, 2001c)

Figure 11 shows pile displacement above the upper O-Cell level obtained from loading the upper O-Cells both before and after base grouting. The maximum pre-grouting upward applied net load (gross load minus buoyant shaft weight) during Stage 2 was 8.4 MN. At this loading the upward movement of the O-Cell was 17 mm. The maximum post-grouting upward applied net load was 12.4 MN at a corresponding O-Cell movement of 36.8 mm. The load-displacement data do not show any increase in friction due to base grouting.

Table 7 presents a summary of the estimated shear resistance along the length of pile B-1 as determined from the strain gage data both before and following base grouting. Before grouting, the average friction resistance was about one-half the average value of 103 kPa predicted by static analyses. As illustrated in Table 7, the post-grouting friction values varied more widely along the length of pile B-1, but the average friction resistance of 104 kPa was approximately the same as the predicted value. Further, the actual ultimate friction resistance is likely somewhat higher than the maximum test values shown in Table 7 since the peak friction resistance had not been reached for several of the pile segments.

INTERPRETATION OF RESULTS

Figure 12 presents plots of predicted load versus settlement for the production piles at the pier and abutment foundations based on the load test data. In developing these plots, the unit friction values obtained from the load tests were applied directly to the larger diameter production piles. Thus the total friction resistance for any length of the pile is equal to the unit friction resistance obtained from the load tests times the perimeter area of the selected pile length.

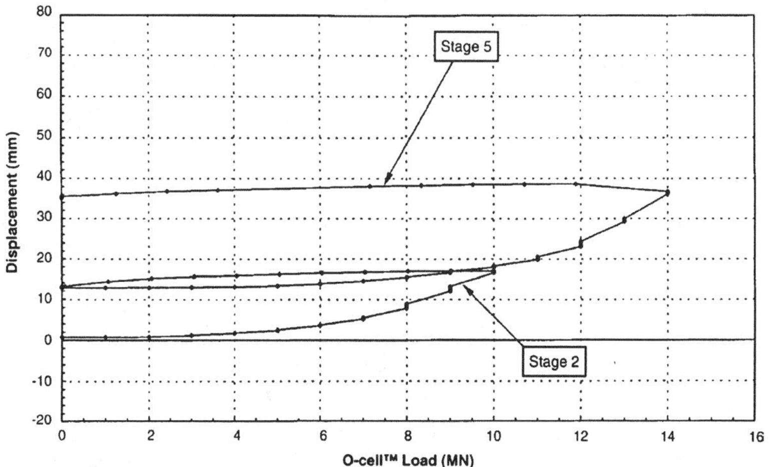

FIG. 11. Test Pile B1, Upward Load-Displacement above Upper O-Cell (LoadTest, 2001c)

TABLE 7. Net Unit Side Shear, Test Pile B1

Load Transfer Zone	Pre-Grouting Net Unit Side Shear[1]	Post Grouting Net Unit Side Shear[1]
Zero shear level to strain gage level 10	6.3 kPa	21.2 kPa
Strain gage level 10 to strain gage level 9	33.0 kPa	
Strain gage level 9 to strain gage level 8	36.7 kPa	22.6 kPa
Strain gage level 8 to strain gage level 7	36.8 kPa	23.9 kPa
Strain gage level 7 to strain gage level 6	24.0 kPa	47.6 kPa
Strain gage level 6 to strain gage level 5	12.9 kPa	42.7 kPa
Strain gage level 5 to strain gage level 4	66.9 kPa	102 kPa
Strain gage level 4 to Upper O-Cell	159 kPa	398 kPa
Upper O-Cell to strain gage level 3	182 kPa	476 kPa
Strain gage level 3 to strain gage level 2	40.6 kPa	135 kPa
Strain gage level 2 to strain gage level 1	69.7 kPa	145 kPa
Strain gage level 1 to Lower O-Cell		172 kPa

[1] The buoyant weight of pile was subtracted in this estimate

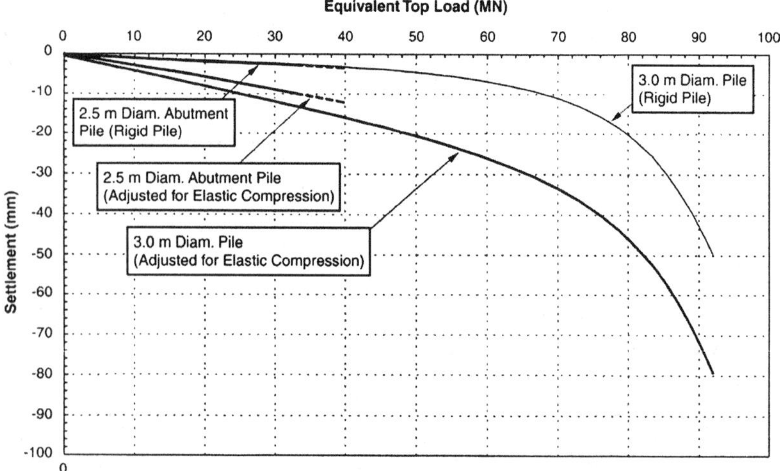

FIG. 12. Estimated Equivalent Top of Pile Settlement for the Production Piles (after LoadTest, 2001b,d)

The end bearing resistance for the production piles was determined by adjusting the load test data using Boussinesq solutions to address stress distribution effects beneath the pile base. Thus, the base area of the 3.0-m diameter pile is four times that of the 1.5-m diameter test pile, but the total end bearing resistance is only twice that of the smaller test pile for approximately the same magnitude of base settlement.

As illustrated in Figure 12, an abutment pile has a predicted top of pile settlement of only about 5 mm at the design load of 17.3 MN (15.0 MN service load plus 2.3 MN net pile weight). A pile at a pier foundation has a predicted settlement of approximately 17 mm at the design load of 43.0 MN (34.0 MN service load plus 9.0 MN net pile weight). These results are consistent with the pile settlements calculated during design.

CONCLUSIONS

The Paksey Bridge project in Bangladesh illustrates the successful construction of unusually deep, high capacity bored piles and the successful application of base grouting to improve end bearing resistance for such piles. The primary conclusions from the test pile program are as follows.

- The load tests confirmed the required pile capacity and design pile embedment, and verified the need for base grouting for these high capacity piles.
- Base grouting was successful despite the high overburden pressure at the pile tip.
- Osterberg load tests showed a significant increase in end bearing resistance and base stiffness after base grouting. End bearing resistance increased from 5 MN to more than 12 MN, while base settlement decreased from 27 mm to 18 mm.
- Base grouting also effectively increased friction resistance along the lower section of the test piles. The improved zone extended about 7.5 m above the base of one pile, and up to 20 m at the other pile.

ACKNOWLEDGEMENT

The Paksey Bridge Project was sponsored by the Bangladesh Roads and Highways Department (RHD), with funding from the Japan Bank for International Cooperation, or JBIC (formerly the Japanese Overseas Economic Cooperation Fund, or OECF). The bridge was originally designed by Mott MacDonald, Ltd. As part of its Consultancy Services for construction, Parsons Brinckerhoff International, Inc. served as prime consultant in association with Worley International LTD. (New Zealand), Kuljian Corporation (USA), Sarm Associates LTD. (Bangladesh), and KS Consultants LTD. (Bangladesh), and was responsible for design review, redesign of the bridge foundations and substructures, and construction engineering and inspection. Major Bridge Engineering Bureau (MBB) of the People's Republic of China was the construction contractor. The Osterberg load cell tests were conducted by LoadTest International under a subcontract to MBB and under the direction of Parsons Brinckerhoff.

REFERENCES

Bruce, D. A. (1986a). "Enhancing the Performance of Large Diameter Piles by Grouting – 1," Ground Engineering, Vol. 5, Pp 9 – 15.

Bruce, D. A. (1986b). "Enhancing the Performance of Large Diameter Piles by Grouting – 2," Ground Engineering, Vol. 6, Pp 11 – 18.

Ensoft, Inc. (1994). "GROUP Computer Program Version 3.0, Analyses of a Group of Piles Subjected to Axial and Lateral Loading," Austin, Texas.

Federal Highway Administration (1988). Drilled Shafts: Student Workbook, Publication No. FHWA HI-88-042, Washington, D.C..

LoadTest International Inc. (2001a). "Report on Drilled Pile Load Testing (Osterberg Method), Test Pile PB02 – Paksey Bridge, Bangladesh (LT-2101-2)," for Major Bridge Engineering Bureau.

LoadTest International Inc. (2001b). "Data Report: Test Pile PB02 – Paksey Bridge, Addendum A," for Major Bridge Engineering Bureau.

LoadTest International Inc. (2001c). "Report on Drilled Pile Load Testing (Osterberg Method), Test Pile B1 – Paksey Bridge, Bangladesh (LT-2101-1)," for Major Bridge Engineering Bureau.

LoadTest International Inc. (2001d). "Data Report: Test Pile B1 – Paksey Bridge, Addendum A," for Major Bridge Engineering Bureau.

Sliwinski, Z. J. and Fleming, W. G.K. (1984). "The Integrity and Performance of Bored Piles," Proceedings of the International Conference on Advances in Piling and Ground Treatment for Foundations, Institution of Civil Engineers, London, Pp 211 – 223.

Stocker, M. (1983). "The Influence of Post-Grouting on the Load-Bearing Capacity of Bored Piles," 8th European Conference on Soil Mechanics and Foundation Engineering, Helsinki, Pp 167 – 170.

Teparaksa, W. (1992). "Behaviour of Base-Grouted Bored Piles in Bangkok Subsoils," Proceedings of Conference on Piling: European Practice and Worldwide Trends, Institution of Civil Engineers, London, Pp. 296 – 301.

LOAD-SETTLEMENT CHARACTERISTICS OF DRILLED SHAFTS REINFORCED BY ROCKBOLTS

Sang-seom Jeong[1], Byung-chul Kim[2]
Dae-soo Lee[3], Dae-hak Kim[4]

ABSTRACT: This paper describes the load distribution and settlement of reinforced drilled shafts subjected to axial loads. The emphasis was on quantifying the reinforcing effects of rockbolts placed from the shafts to surrounding weathered rock based on a numerical analysis and 1/8 scale load tests performed on instrumented piles. The piles with and without rockbolts were instrumented with strain gages, load cell and tell tale, and load tests were performed on four piles: tension and compression. The results of the four load tests are presented in terms of top load-top movement curves. In addition numerical analyses are performed for the major influencing parameters on the behavior of reinforced drilled shafts such as the number of rockbolts, the positions on the shaft, reinforcement level, and the inclination angle at which the rockbolts are placed. As a result, the parametric study is highlighted. It was found that as the number of reinforcing levels increases, the incremental effect of reinforcement tends to increase, whereas the reinforcing effect on relative positions is negligible. In addition there is a reinforcing effect as the inclination angle increases up to 30 degrees.

INTRODUCTION

[1] Associate Professor, Yonsei University, Department of Civil Engineering, Seoul, 120-749, Korea
[2] Ph. D., Candidate, Yonsei University, Department of Civil Engineering, Seoul, 120-749, Korea
[3] Project Manager, Korea Electric Power Research Institute, Taejon, 305-380, Korea
[4] Vice-president, Backyoung Geotechnical & Construction Co. Ltd., Seoul, 156-095, Korea

In South Korea, a number of huge construction projects such as land reclamation projects for an international airport, high-speed railways and many harbor constructions are in progress in urban and coastal areas. Drilled shafts are frequently used in those areas as a viable replacement for driven piles for two applications: Deepwater offshore foundations, and foundations in urban areas where the noise and vibration are associated with pile driving.

Two general considerations are important in the construction and design of drilled shafts in rocks: The selection and implementation of construction details and consideration of load transfer in skin friction and in endbearing in making a design. A comprehensive study of drilled shafts has been reported by Reese and O'Neill (1988). They report that the magnitude and distribution of the side resistance transferred down to the tip are highly influenced by the compressive strength of rock and the settlement between the drilled shafts and the rock. This is based on the concept that both side resistance and endbearing will not develop simultaneously, and thus, detailed studies, including field tests, are needed in many instances to confirm a design.

In this study, fully instrumented drilled shafts reinforced by rockbolts were tested. The test results are presented in terms of top load-top movement curves to identify the incremental effects of reinforcement based on the shaft resistance before and after rockbolt installations.

FIELD LOAD TESTS

Field tests were performed on drilled shafts installed in weathered rocks of granite-gneiss. The subsurface investigation was performed on four boring holes and NX-size core samples were recovered up to 50% from the uppermost bedrock. Visual inspection of the bed rock samples indicated that the sample was difficult to core without fracture and was classified into highly weathered rocks.

To investigate the load-settlement curves of drilled shafts, small-scale (1/8 scale of prototype) load tests with and without rockbolts were performed for different loading conditions (compression and tension loads), as shown in Fig. 1. The 450 mm diameter and 1350 mm depth holes were drilled by wash and rotary techniques. Also, the 20 mm diameter and 250 mm depth holes were drilled along the inclination angle at which the rockbolts are placed. The steel and reinforcing bars were then inserted into the main and anchoring holes and grouted with cement milk. A hydraulic jack was used to apply loads to the piles. The applied load and displacement were measured using a load cell and LVDT installed at the pile top. All tests were conducted 28 days after grouting of the piles when the grout had developed a compressive strength of 2.8×10^4 kN/m^2. For load tests, a constant load was applied to the pile and sustained for 25 min at each load level. This procedure was repeated until the anticipated design load was reached. Material properties used in this study and geometries are described in Table 1.

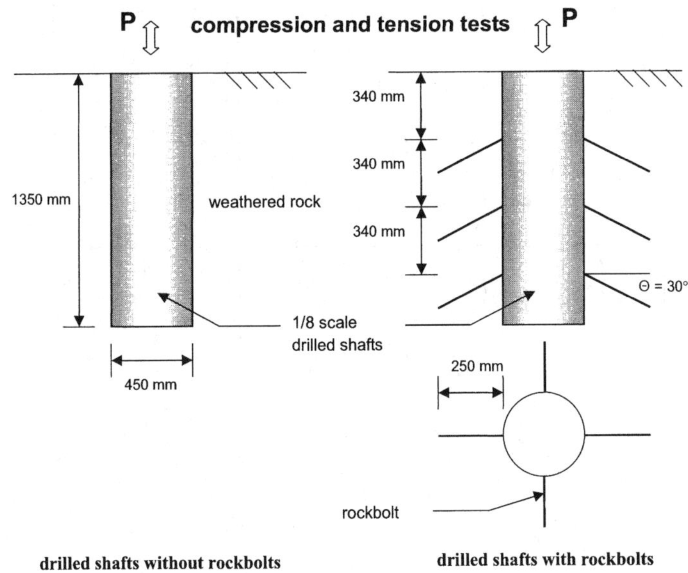

FIG. 1. Instrumented drilled shafts subjected to axial loading test

TEST RESULT AND DISCUSSION

Fig. 2 shows the load-settlement curves for tension piles with and without reinforcing bars. There is little difference in ultimate load level (about 300 kN) between the two tests. However, drilled shafts with reinforcing bars has a considerably smaller settlement to reach the ultimate level when compared with the test result by unreinforced drilled shafts.

On the other hand, the corresponding curve for compression piles show that there is a definite reinforcing effect, as shown in Fig. 3. For this test, a reinforcing effect of about 200 % was measured in compression. Note in the Figures 2 and 3 that the end bearing resistance is larger in the reinforced piles than in the unreinforced piles.

TABLE 1. Material Properties and Geometries

Wethered rock	unit weight(kN/m^3)	22
	cohesion(kPa)	49
	friction angle(°)	35
	poisson's ratio	0.3
	elastic modulus(kPa)	98×10^3
Drilled shaft	unit weight(kN/m^3)	25
	elastic modulus(kPa)	25×10^6
	diameter(mm)	450
	length(mm)	1350
Rock-bolt	elastic modulus(kPa)	5×10^5
	diameter(mm)	10
	length(mm)	250

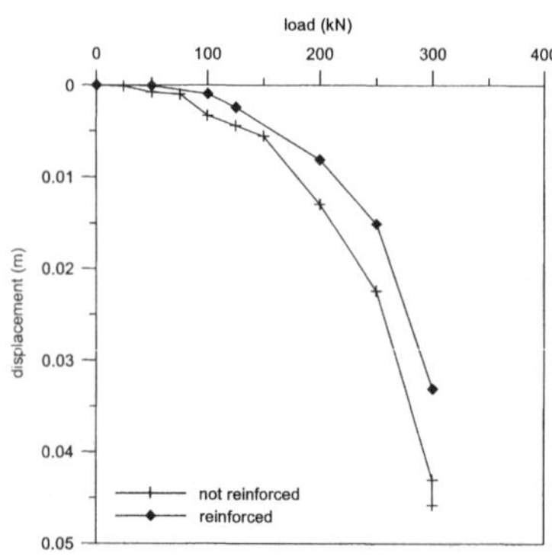

FIG. 2. Load-displacement curve (tensile load test)

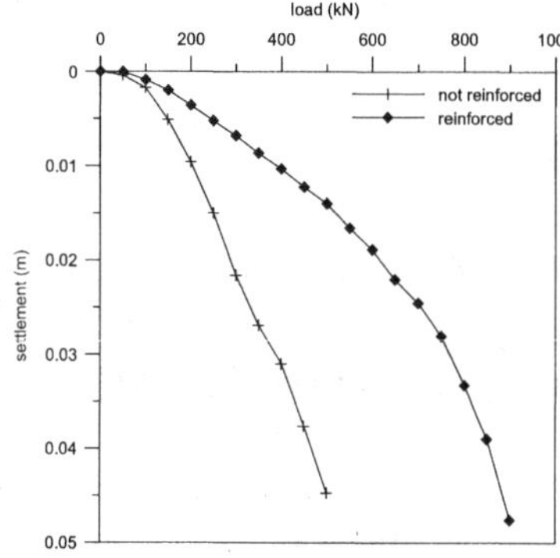

FIG. 3. Load-settlement curve (compressive load test)

NUMERICAL ANALYSIS

The load tests of this study were performed on small-scale models, and thus, are in conflict with the known behavior of full-scale drilled shafts. Major influencing parameters on reinforced drilled shafts include number of rock-bolts, the positions on the shaft, reinforcement level and the inclination angle at which the rock-bolts are placed. To obtain detailed information on the behavior of reinforced drilled shafts, a series of numerical analysis were performed for different loading conditions, level and inclination angle of reinforcing bars.

Finite Difference Model

The response of drilled shafts is analyzed by using a three-dimensional explicit-finite difference approach, FLAC 3D. The mesh of finite difference model for a typical case is shown in Fig. 4. The mesh is assumed to be resting on a rigid layer, and the vertical boundaries at the left- and right-hand side are assumed to be on rollers to allow movement of the soil layer. The interface elements between a pile and soil were two-dimensional elements comprising two surfaces compatible with the adjacent solid elements; the two surfaces coincide initially. The interface elements can transfer only shear forces across their surfaces when a compressive normal pressure acts on them.

The pile elements are assumed to remain elastic at all times, while the surrounding soil is idealized as an elastoplastic material. For certain group configurations, the use of symmetry reduced the size of the mesh. The actual size of the mesh is related to the pile length; the lower rigid boundary has been placed at a depth equal to 2.0 pile lengths and the side boundary is extended laterally to $r_m=2.5L(1-\nu)$ (Randolph and Wroth, 1978). The pile is 3 m in diameter, 12 m in length and surrounding soil has the size of 30m length ×15m width ×24m depth. Table 2 shows the material properties and geometries used in this study.

TABLE 2. Material Properties and Geometries

Wethered rock	unit weight(kN/m^3)	22
	cohesion(kPa)	49
	friction angle(°)	35
	poisson's ratio	0.3
	shear modulus(kPa)	37.7×10^3
	bulk modulus(kPa)	81.7×10^3
Drilled shaft	unit weight(kN/m^3)	25
	poisson's ratio	0.2
	shear modulus(kPa)	10.4×10^6
	bulk modulus(kPa)	13.9×10^6
Rock-bolt	unit weight(kN/m^3)	23
	poisson's ratio	0.2
	elastic modulus(kPa)	21×10^6

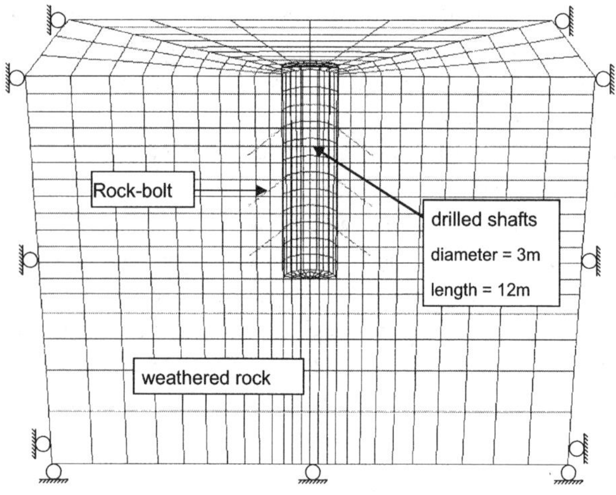

FIG. 4. Finite difference mesh for a typical case.

Parametric Study

The behavior of drilled shafts with reinforcing bars is influenced by various parameters on the reinforcement level, the positions on the shaft and the inclination angle at which the rock-bolts are placed. The numerical analyses are performed on the cases summarized in Table 3.

TABLE 3. Cases of numerical analyses

		Cases		
Load		Tensile load, Compressive load		
Rockbolt	Level of reinforcement	Level 1	Level 2	Level 3
	Relative position	upper	middle	lower
	Inclination angle	0°, 20°, 30°, 40°		

Figures 5 and 6 show the effect of reinforcing level on tensile piles (Fig. 5) and compressive piles (Fig. 6). As the number of level increases, the reinforcing effect also tends to increase. Next, figures 7 and 8 show the effect of relative positions of reinforcing bars on the shaft. There is little difference in load transfer and thus, the incremental effect of reinforcement on relative positions is negligible. Figures 9 and 10 show the effect of inclination angle. From this figures, there is a reinforcing effect as the inclination angle increases up to 30 degree. However for larger than 30 degree, the load does not increase any further and the incremental effect of reinforcement is almost constant.

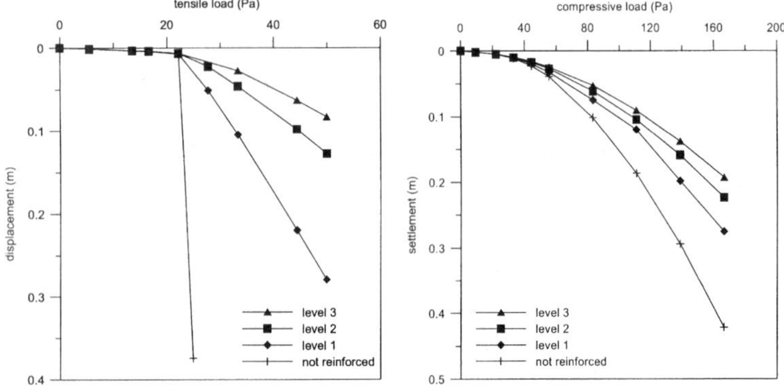

FIG. 5. Effect of level of reinforcement (tensile load test)

FIG. 6. Effect of level of reinforcement (compressive load test)

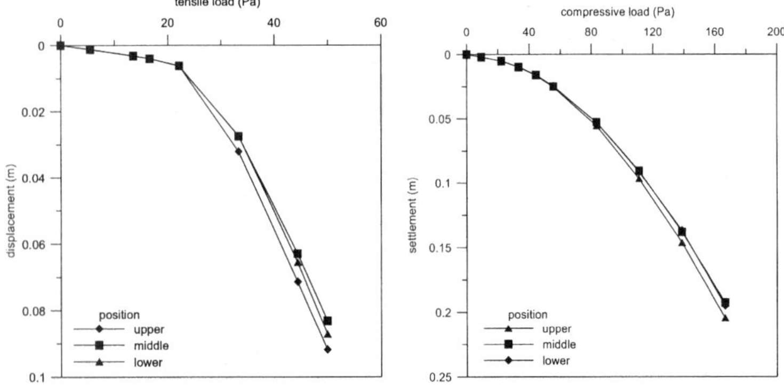

FIG. 7. Effect of different position (tensile load test)

FIG. 8. Effect of different position (compressive load test)

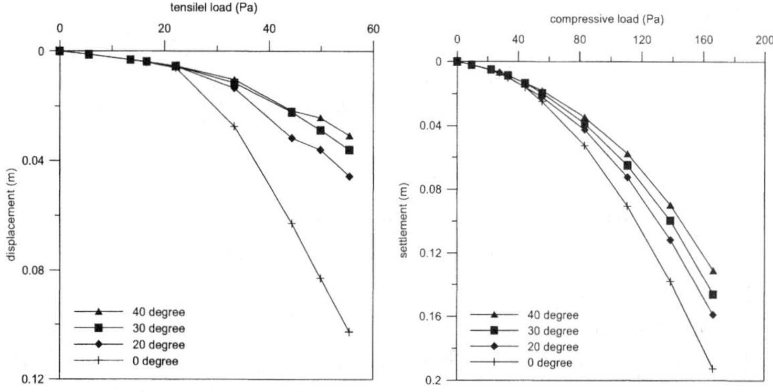

FIG. 9. Effect of installed angel (tensile load test)

FIG. 10. Effect of installed angel (compressive load test)

CONCLUSIONS

The main objective of this study was to investigate the load-settlement characteristics of drilled shafts reinforced by rockbolts based on field load tests and a numerical analysis. When applied to drilled shafts installed in the weathered rocks, the parametric studies have clearly demonstrated the reinforcing effect of drilled shafts. In this study, the reinforcing effect was found to about 15-25%. However, a cost/benefit analysis is needed before this technique can be applied in practice. From the findings of this study, the following conclusions are drawn:

1) Through the tensile and compressive load tests, it was identified that there is a definite reinforcing effect.
2) Based on a numerical analysis, it was found that as the number of reinforcing levels increases, the incremental effect of reinforcement tends to increase, whereas the reinforcing effect on relative positions is negligible. In addition there is a reinforcing effect as the inclination angle increases up to 30 degree. However for larger than 30 degrees, the load does not increase any further and the incremental effect of reinforcement is almost constant.

REFERENCES

Baquelin, F. (1982). "Rules for the structural design of foundations based on the selfboring pressuremeter test." *Symp. On the Pressuremeter and Its Marine Application*, IFP, Paris, 347

Coyle, H. M., Reese, L. C. (1966). "Load transfer for axially loaded piles in clay." *Journal of Soil Mech. And Found. Div.*, ASCE, 92(2), 1-26.

Kim, S. I., Jeong, S. S., Cho, S. H., Park, I. J. (1999). "Shear load transfer characteristics of drilled shafts in weathered rocks." *Journal of Geotechnical and Geoenvironmental Engineering*, ASCE, 125(11), 999-1010.

O'Neill, M. W., Hassan, K. M. (1994). "Drilled shafts: Effects of construction on performance and design criteria." *Proc., Int. Conf. on Design and Construction of Deep Foundation*, 1, Federal Highway Administration, Orlando, 137-187.

O'Neill, M. W., Reese, L. C. (1970). "Behavior of axially loaded drilled shafts in Beatmont clay." *Res. Rep. 89-8*, Ctr. for Hwy. Res., University of Texas, Austin, Tex.

Randolph, M. F., Wroth, C. P. (1978). "Analysis of deformation of vertically loaded piles." *Journal of Geotechnical Engineering*, ASCE, 104(12), 1465-1488.

Reese, L. C., O'Neill, M. W. (1988). "Drilled shafts: Construction procedures and design method." *Publ. No. FHWA-HI-88-042*, Federal Highway Administration, Washington, D.C.

Vijayvergiya, V. N. (1977). "Load-movement characteristics of piles." *Proc. 4th Annu. Symp. Of he Wtrwy., Port, Coast. And Oc. Div. of ASCE*, ASCE, Long Beach, Calif., 2, 269-284.

Response and Analysis of a Large Diameter Drilled Shaft Subjected to Lateral Loading

Tong Qiu[1], Patrick J. Fox[2], Kerop Janoyan[3], Jonathan P. Stewart[4], and John W. Wallace[5]

ABSTRACT

Numerical analyses are presented for the response of a 2 m diameter drilled shaft with flagpole configuration subjected to lateral loading. The analyses were conducted using three software programs: LPILE (subgrade reaction model), SWM (strain wedge model), and ABAQUS (3-D finite element model). The analyses consider nonlinear behavior of the reinforced concrete shaft, contact-slip-gap at the soil-shaft interface, and elastic-plastic properties of the soil. Numerical results are compared with measurements from a full-scale field test for the shaft and include lateral displacement at the top of the shaft, subgrade reaction (*p-y*) curves at three depths, and distributions of bending moment and shear force along the shaft.

INTRODUCTION

Observations of damage to shaft-supported bridges during recent earthquakes have highlighted the need for a better understanding of the behavior of large diameter drilled shafts under seismic loading conditions. During the 1989 Loma Prieta and the 1994 Northridge Earthquakes, for example, some bridge shafts experienced displacements that left gaps between the shaft and surrounding soil. Several aspects of the behavior of drilled shafts subjected to lateral loads are still not well understood,

[1]Grad. Res. Asst., Dept. of Civil & Envir. Engrg., Univ. of California, Los Angeles, CA, 90095 qiutong@ucla.edu

[2]Assoc. Prof., Dept. of Civil & Envir. Engrg. & Geodetic Sci., Ohio State University, Columbus, OH, 43210 fox.407@osu.edu

[3]Asst. Prof., Dept. of Civil & Envir. Engrg., Clarkson Univ., Potsdam, NY, 13699 kerop@clarkson.edu

[4]Assoc. Prof., Dept. of Civil & Envir. Engrg., Univ. of California, Los Angeles, CA, 90095 jstewart@seas.ucla.edu

[5]Assoc. Prof., Dept. of Civil & Envir. Engrg., Univ. of California, Los Angeles, CA, 90095 wallacej@ucla.edu

including the ultimate capacity of a shaft at large displacements and the nature of soil-shaft interaction. Numerical models of soil-shaft interaction often utilize simplifying assumptions including the specification of linear elastic materials, perfect bonding at the shaft-soil interface, and 2-D geometry for the shaft. These assumptions may, however, potentially lead to erroneous conclusions regarding the behavior of laterally-loaded drilled shafts (Trochanis *et al.* 1988).

Current methods for the analysis of lateral loads on drilled shafts can be divided into three categories: subgrade reaction models, the strain wedge model, and finite element models. Subgrade reaction models (also called Winkler spring models) are based on prescribed *p-y* curves representing soil lateral resistance per length of shaft *p* as a function of lateral shaft displacement *y* (McClelland and Focht 1958). Soil reaction along the shaft is characterized using a series of independent springs. Models for spring properties have been developed from the results of full-scale field tests. These models are relatively easy to use and can incorporate nonlinear shaft-soil interaction. However, as discussed by Reese (1983), empirically derived *p-y* curves do not account for the effects of soil layering, shaft stiffness, and shaft head conditions on the soil response. The subgrade reaction model program LPILE (Plus v4.0) was used in the current study. In LPILE, soil *p-y* curves are specified as a function of soil type and shaft properties. The shaft is modeled as an elastic column consisting of multiple sections, each of which can be assigned individual values of diameter and elastic modulus. Manual adjustment of section modulus values by iteration is required to simulate nonlinear shaft behavior in LPILE.

The strain wedge model (Norris 1986, Ashour *et al.* 1998, 2002, Ashour and Norris 2000) defines a 3-D passive soil wedge that resists lateral displacement of the shaft. The method can accommodate layered soil profiles, nonlinear soil behavior, variable shaft diameter and reinforcement distribution within the shaft, and the effect of soil slip at the soil-shaft interface. Required soil parameters are based on basic soil properties that are typically available to the designer. The program SWM has been written for the strain wedge model (v5.0) and was used in the current study.

The finite element method is capable of providing a more realistic model of soil behavior and soil-shaft interaction for a shaft under lateral loading conditions. The method has been used to model elastic piles at small lateral displacements (Trochanis *et al.* 1988, Koojiman 1989, and Brown *et al.* 1989) and the ultimate capacity of bored pile groups (Ng *et al.* 2001). Numerical difficulties are often encountered for highly nonlinear problems, including those involving large displacements. The ABAQUS STANDARD finite element analysis program (v5.8) was used in the current study. ABAQUS STANDARD offers the ability to account for full 3-D geometry, various soil constitutive models, contact-slip-gap at the soil-shaft interface, and large displacements. Procedures such as these are not widely used by design engineers.

In this paper, the response of a 2 m diameter laterally-loaded drilled shaft is analyzed using LPILE, SWM, and ABAQUS. The test site and field tests are first described, followed by the soil and shaft properties. Field measurements are then compared with simulated results for displacement at the top of the column, soil *p-y* curves at three depths, and distributions of lateral displacement and bending moment along the shaft.

TEST SITE

A field test of the response of a 2.0 m diameter flagpole shaft-column to lateral load was conducted during winter 2000 near the intersection of Interstate Highways 105 and 405 in Hawthorne, California (Janoyan 2001, Janoyan *et al.* 2001). Field and laboratory tests were used to characterize soil conditions at the site and included standard penetration tests (SPT), seismic cone penetration tests, down-hole suspension logging for shear wave velocities, pressuremeter tests, consolidation tests, and unconsolidated-undrained (U-U) triaxial tests. Fig. 1 shows the soil profile from the test site, in which $(N_1)_{60}$ is the SPT blow count corrected for energy and overburden stress, S_u is the undrained shear strength, and σ'_v and σ'_p are effective vertical stress and preconsolidation stress, respectively. Note that the water table occurs near the shaft toe.

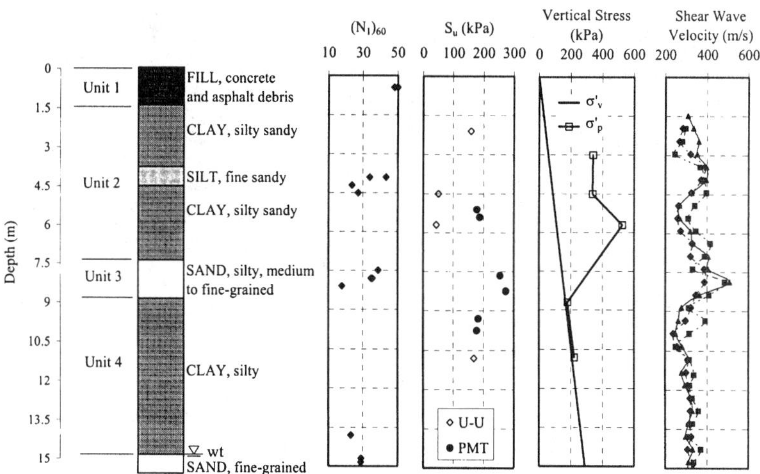

FIG. 1. Test site soil profile and results from laboratory and field tests

FIELD TEST

The geometry of the drilled shaft-column is shown in Fig. 2. The reinforced concrete shaft had a diameter D of 1.8 m above the ground surface and 2.0 m below the ground surface. The top of the shaft was 12.2 m above the ground surface and the tip of the shaft was at a depth of 14.6 m. The shaft was tested to failure over an 8-week period by applying successively increasing cyclic static loads to achieve target displacement levels ranging from 0.05 to 2.75 m. The loads were applied to the top of the shaft at an angle of 26.5° from the horizontal (2H:1V). The shaft was instrumented with embedded fiber optic sensors and extensometers, strain gages on the reinforcing bars, and down-shaft inclinometers to give redundant measurements of shaft section curvature with position along the shaft. Details of shaft design and installation, field test methods, field measurements, and data reduction techniques are presented by Janoyan (2001) and Janoyan *et al.* (2001).

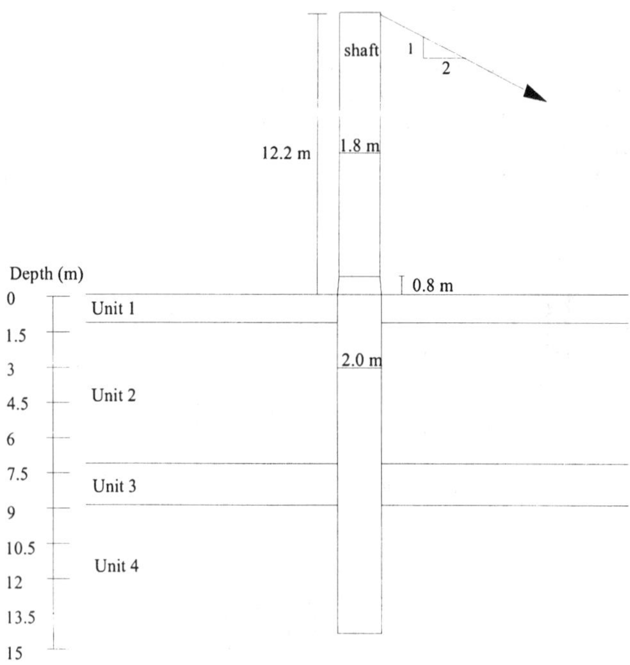

FIG. 2. Geometry of drilled shaft and idealized soil profile

NUMERICAL SIMULATIONS

Idealized Soil Profile

The soil profile was divided into four units, shown in Fig. 1, to simplify the numerical simulations. The fill layer was designated as Unit 1 and assumed to be drained with a cohesion value c = 0 and friction angle $\phi = 38^{\circ}$. Units 2 and 4 are predominantly clay layers and were modeled as undrained with $c = S_u$ and $\phi = 0$. U-U test data indicate no dependence of soil shear strength on confining pressure, which suggests that these layers are saturated. The S_u values for Units 2 and 4 were taken as 188 and 225 kPa, respectively. The sand layer (Unit 3) was assumed to be drained with $c = 0$ and $\phi = 38^{\circ}$ (estimated from $(N_1)_{60}$). All four units were assumed to have an average unit weight γ of 18.9 kN/m^3.

Reinforced Concrete Properties

The unit weight and compression strength of the reinforced concrete were 23.6 kN/m^3 and 42 MPa, respectively, and the yield stress of the reinforcing steel was 490 MPa. The cross section of the shaft had 2% reinforcing steel by area. The moment M vs. curvature φ relationship for the shaft as derived from field data is shown by the heavy line in Fig. 3. The relationship is nearly linear up to a yield point at $M = 16000$ kN-m, $\varphi = 2.54 \times 10^{-6}$ 1/m and then continues to an ultimate capacity of approximately $M = 16600$ kN-m. Fig. 3 also shows the $M - \varphi$ relationships that were used for the above-ground section ($D = 1.8$ m) and the below-ground section ($D = 2$ m) of the shaft in numerical simulations. The SWM curves were generated internally within the program based on the reinforced concrete section properties. The curves for LPILE and ABAQUS were obtained independently from the program BIAX (Wallace 1992). BIAX generates a moment-curvature relationship for a section by considering equilibrium and strain compatibility between tension and compression zones and assumes plane sections remain plane. For both LPILE and ABAQUS, the shaft was modeled as an equivalent linear material with multiple sections. For each applied load at the top of the shaft, manual iteration was required to determine a distribution of Young's modulus values along the shaft that was consistent with resulting φ values and Fig. 3.

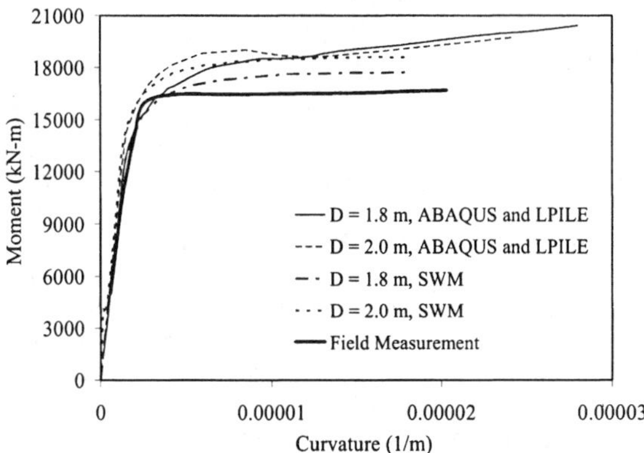

FIG. 3. Measured and calculated moment-curvature relationships for shaft

LPILE Simulations

LPILE simulations were performed for specified values of top-of-column displacement ranging from 0.15 to 1.22 m. For LPILE, the idealized soil profile was characterized as: silt (Unit 1), clay without free water (Unit 2), sand (Unit 3), clay without free water (Unit 4 above the water table), and clay with free water (Unit 4 below the water table). LPILE requires two additional soil parameters: k and ε_{50}. The value of k describes the increase of p-y curve secant modulus E_s with depth z in the linear relationship $E_s = kz$. Values of k for Units 1 – 4 were specified as 122 kN/m^3, 270 kN/m^3, 20 kN/m^3, and 405 kN/m^3, respectively, based on soil types shear strength parameters (LPILE 2000). The parameter ε_{50} represents the strain at 50 per cent stress level and, for all four Units, ε_{50} was estimated as 0.005 from the results of triaxial tests.

SWM Simulations

The SWM simulations were force-controlled and, as a result, manual iteration was required to obtain the desired specific displacements at the top of the shaft. The SWM simulations did not converge for top shaft displacements that exceeded 0.76 m in this study. The SWM simulations used the same units and soil parameters as were used in the LPILE simulations.

ABAQUS Simulations

The mesh for the ABAQUS simulations was composed of 6336 elements (9613 nodes) in which 688 second-order 15-node wedge elements were used for the shaft and 4720 first-order 8-node brick elements were used for the soil. The outer boundary of the soil was located at a radial distance of 5.5D (11 m) from the center of the shaft and was represented using 928 first-order 8-node infinite boundary elements. The bottom boundary for the ABAQUS simulations was located at a depth of 29.3 m.

Displacement-controlled ABAQUS simulations were performed using a Mohr-Coulomb elastic-plastic constitutive model for the soil. Soil parameters that were required in addition to those given above include Poisson's ratio ν, Young's modulus E, and dilation angle ψ. Difficulties in finite element simulations were encountered, and some trial-and-error was required to identify simulation types for which convergence was possible. Specifically, ABAQUS had difficulty converging when layers with zero cohesion or zero friction angle were included in the analysis. Furthermore, the specification of partial soil drainage along with a contact-slip-gap surface around the shaft precluded any solutions to be obtained. Thus, Units 1 and 3 were considered to be fully drained and assigned cohesion intercepts of 12 kPa and 0.24 kPa, respectively, and Units 2 and 4 were considered to be fully undrained and assigned $\phi = \psi = 1^\circ$. All soil units were modeled with $\psi = \phi$. Values of ν were chosen as 0.3 for the drained layers and 0.46 for the undrained layers. Young's modulus (E) values were evaluated from shear wave velocities and Poisson's ratio. Based on the shear wave velocity profile measured prior to shaft construction (Fig. 1), E was estimated to be 95 MPa for Unit 1 and 280 MPa for Units 2, 3, and 4. The effects of soil disturbance, stress relief, and water adsorption by soil near the shaft were expected to reduce the values of S_u and E near the shaft (Reese 1978, Kalinski and Stoke 1998, O' Neil 2001). Accordingly, S_u was reduced by 50 per cent and E was reduced by 40 per cent for all soil elements within 2 m (1*D*) of the shaft for the ABAQUS simulations. The soil between the tip of the shaft (depth = 14.6 m) and the base of the analysis domain (depth = 29.3 m) was assumed to have the same properties as Unit 4.

COMPARISON OF FIELD MEASUREMENTS AND SIMULATION RESULTS

Top of Column Displacement

Measured and computed relationships between top-of-column lateral load and lateral displacement are shown in Fig. 4. The field measurements indicate an initial response that is gradually softening and then yields at approximately 1310 kN. The ultimate capacity of the shaft was 1400 kPa at 1.2 m of displacement. The curves obtained from the numerical simulations are all in good agreement with this general trend. The stiffness of the initial response and the load at yield increases for LPILE, SWM, and ABAQUS, respectively. At large displacements, the ultimate shaft capacity for the ABAQUS simulations is slightly higher than the field measurements while the ultimate capacity given by LPILE is in very good agreement. Ultimate shaft capacity could not be obtained from SWM as previously discussed. It should be noted that top-of-column load-deflection relationship is relatively insensitive to details of soil-shaft interaction – thus the apparent good agreement among all methods in Figure 4.

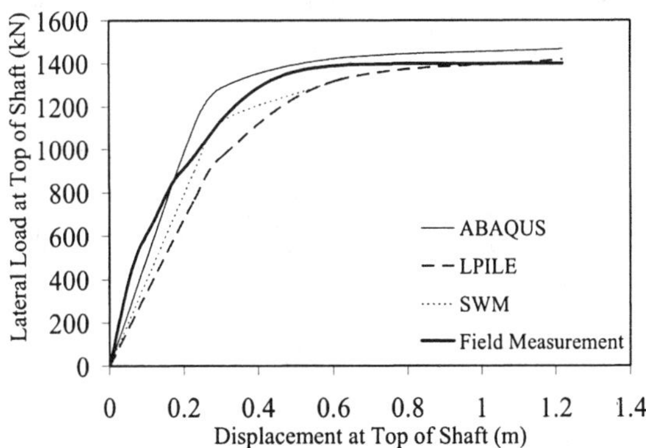

FIG. 4. Lateral load vs. lateral displacement at top of shaft

p-y Curves

Fig. 5 shows, for depths below ground surface of 0.9, 2.3, and 5.3 m, *p-y* curves estimated from field measurements (Janoyan, 2001), estimated from pressuremeter test (PMT) results, and calculated from the numerical models. The field p-y curves were estimated by back-calculation from the measured shaft curvature profiles. At 0.9 m, all the simulated curves yield well before the field measurements. This may be due to inaccuracy in the field data or deficiencies in the modeling effort; however the upper section of the fill was a tightly compacted soil with concrete and asphalt debris (Janoyan 2001) and may have had a substantially greater stiffness than that used for the numerical analyses. In general, there is significant variability among the curves at all depths. The ABAQUS results show the stiffest response and most closely match the field measurements at 0.9, while the LPILE *p-y* curves show the softest response and most closely matches the field measurements at 5.3 m. The PMT and field measurement curves are not in good agreement.

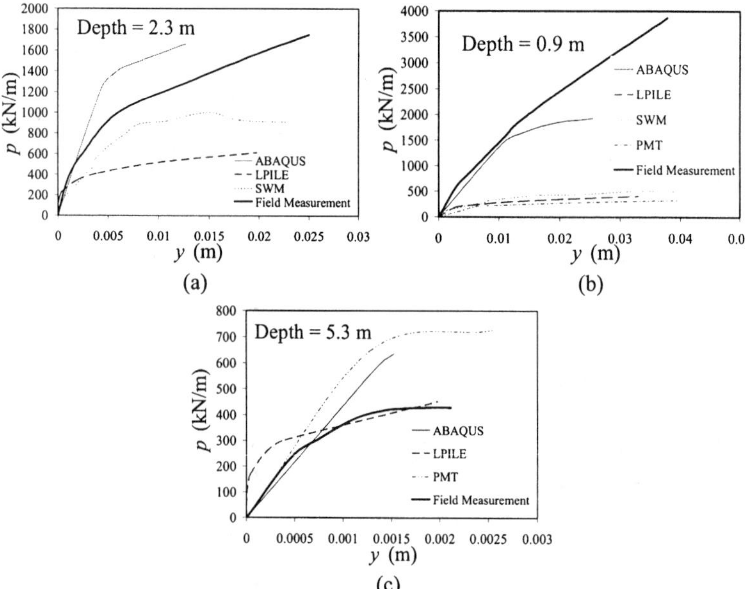

FIG. 5. *p-y* curves at depths of: (a) 0.9 m, (b) 2.3 m, and (c) 5.3 m

Displacement Profile

Lateral displacement profiles of the shaft are shown in Fig. 6 for top-of-column displacements of 0.30 m and 1.22 m. The numerical simulation results appear to be in good overall agreement with the field measurements. However, the match is not so close in detailed profiles drawn only below ground line, especially for ABAQUS and LPILE (Qiu 2002). Consistent with Fig. 5, shaft displacement profiles obtained from the ABAQUS simulations suggest a stiffer soil lateral resistance than those for LPILE or SWM.

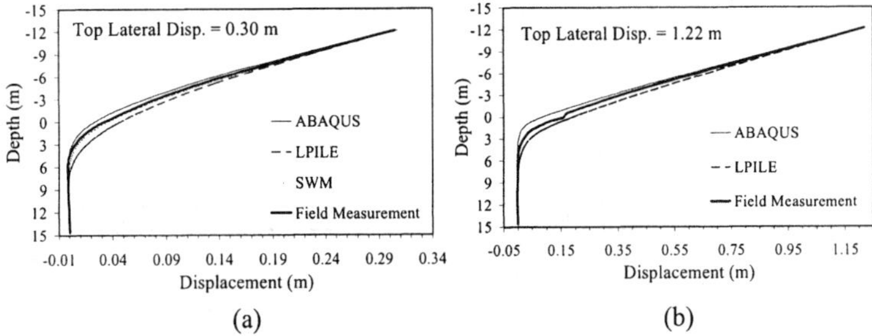

FIG. 6. Shaft displacement vs. depth for top shaft displacements of: (a) 0.3 m and (b) 1.22 m

Moment in Shaft

Bending moment profiles for top-of-column lateral displacements of 0.23 m, 0.30 m, and 0.61 m. are shown in Fig. 7. The field measurements indicate that the depth of maximum moment was 0.9 m below ground line for each displacement level. Negative moments, suggesting reverse curvature, occurred between depths of 10.0 and 13.4 m for top displacements of 0.23 and 0.30 m. Moments were positive at all depths for a top displacement of 0.61 m. The ABAQUS simulations provide the best estimate of maximum moment location, whereas LPILE simulations provide the least accurate estimate.

DISCUSSION AND CONCLUSIONS

The following conclusions are based on comparisons between numerical simulations and field measurements for a full-scale reinforced concrete drilled shaft subjected to lateral loading:

1. It is possible to account for nonlinear properties of the reinforced concrete shaft if values of concrete Young's modulus are varied along the shaft according to a specified moment-curvature relationship. By this method, the shaft is treated as an equivalent linear material and the appropriate modulus value for each shaft element is estimated by manual iteration in ABAQUS and LPILE. SWM incorporates this capability directly without the need for manual iteration.

2. Lateral resistance of the soil was simulated using nonlinear *p-y* curves in LPILE, a nonlinear soil stress-strain relationship in SWM, and the elastic-plastic Mohr-Coulomb constitutive model in ABAQUS. The numerical models indicated that in order to accurately model the lateral response of a drilled shaft, the strength and deformation properties of the soil near the ground surface must be accurately known.

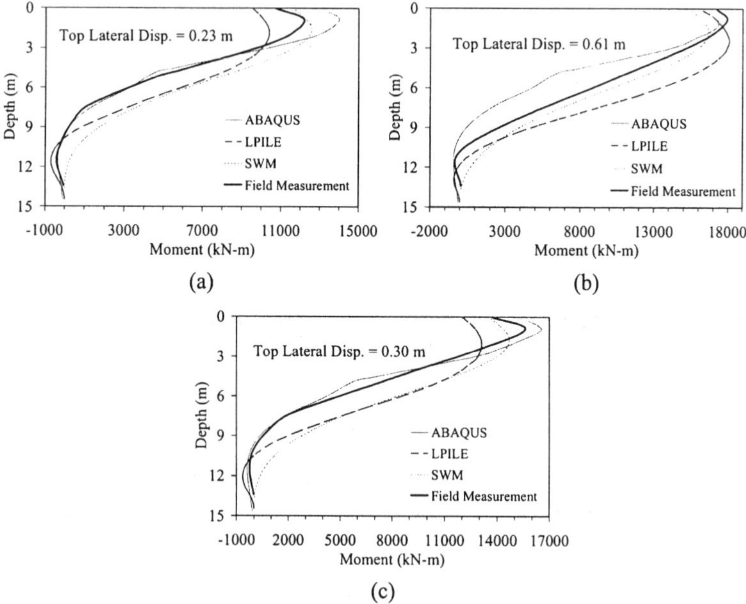

FIG. 7. Shaft moment vs. depth for top shaft displacements of: (a) 0.23 m, (b) 0.30 m, and (c) 0.61 m

3. All three models produced calculated values for lateral load vs. lateral displacement at the top of the shaft that were in close agreement with the field measurements. This indicates that each program was capable of reproducing the overall behavior of the shaft. The SWM program did not converge for top shaft displacements exceeding 0.76 m in this study.

4. The ABAQUS results generally displayed a stiffer response than LPILE or SWM for the input parameters provided in this study. LPILE produced the softest response with SWM in between. This resulted in stiffer p-y curves, shallower maximum bending moment points, and larger maximum moments and shear forces for the ABAQUS simulations. As a result, the ABAQUS simulations were in better overall

agreement with the field measurements that the LPILE or SWM simulations. The LPILE *p-y* curves were too soft for the shaft-column in this study. This suggests that there is a need to update these curves based on new field load tests with high quality curvature measurements.

ACKNOWLEDGMENT

Support for this research was provided by the California Department of Transportation under Research Contract No. 59A0183. This support is gratefully acknowledged.

REFERENCES

ABAQUS (1998). Hibbitt, Karlsson & Sorensen, Inc., Version 5.8.

Ashour, M., Norris, G., and Pilling, P. (1998). "Lateral Loading of a Pile in Layered Soil Using the Strain Wedge Model." *J. Geotech. and Geoenvir. Engrg.*, 124(4), 303-315.

Ashour, M., and Norris, G. (2000). "Modeling Lateral Soil-Pile Response Based on Soil-Pile Interaction." *J. Geotech. and Geoenvir. Engrg.*, ASCE, 126(5), 420-428.

Ashour, M., Norris, G., and Pilling, P. (2002). "Strain Wedge Model Capability of Analyzing Behavior of Laterally Loaded Isolated Piles, Drilled Shafts, and Pile Groups." *J. Geotech. and Geoenvir. Engrg.*, 128(4), 245-254.

Brown, D., Shie, C., and Kumar, M. (1989). "P-Y Curves for Laterally Loaded Piles Derived from Three Dimensional Finite Element Model." *Proc., 3rd International. Symposium on Numerical Models in Geomechanics*, Niagara Falls, 683-690.

Janoyan, K. (2001). "Interaction between Soil and Full-Scale Drilled Shaft under Cyclic Lateral Load." Ph.D. Dissertation, Dept. of Civil & Envir. Engrg., University of California, Los Angeles, CA.

Janoyan, K., Stewart, J. P., and Wallace, J. W. (2001). "Analysis of p-y Curves from Lateral Load Test of Large Diameter Drilled Shaft in Stiff Clay," *Proc., 6th Caltrans Workshop on Seismic Research*, Sacramento, CA, paper 5-105.

Kalinski, M.E., and Stokoe, K.H., II. (1998). ''Stress wave measurements in open holes in soil.'' *Final Rep., Dept. of Civ. Engrg.*, University of Texas at Austin, Austin, TX.

Koojiman, A. (1989). "Comparison of an Elastoplastic Quasi Three-Dimensional Model for Laterally Loaded Piles with Field Tests." *Proc., 3rd International Symposium on Numerical Models in Geomechanics*, Niagara Falls, 675-682.

McClelland, B. and Focht, J. (1958). "Soil Modulus for Laterally Loaded Piles," *Transactions of the ASCE*, Vol. 123, Paper 2954, 1049-1086.

LPILE Plus (2000). Ensoft, Inc., Version 4.0.

Ng, W.W.C., Zhang, L., and Nip, C.N. (2001). "Response of Laterally Loaded Large-Diameter Bored Pile Groups." *J. Geotech. and Geoenvir. Engrg.*, 127(8), 658-669.

Norris, G.M. (1986). "Theoretically based BEF laterally loaded pile analysis." *Proc., 3rd Int. Conf. on Num. Meth. in Offshore Piling*, Editions Technip, Paris, 361-386.

O'Neill, M.W. (2001). "Side Resistance in Piles and Drilled Shafts." *J. Geotech. and Geoenvir. Engrg.*, ASCE, 127(1), 3-16.

Qiu, T. (2002). "Finite Element Modeling of Lateral Loads on Large Diameter Drilled Shafts." M.S. Thesis, Dept. of Civil & Envir. Engrg., University of California, Los Angeles, CA.

Reese, L.C. (1978). "Design and construction of drilled shafts." *J. Geotech. Engrg. Div.*, 104(1), 91-116.

Reese, L.C. (1983). "Behavior of piles and pile groups under lateral load." *Rep. Prepared for U.S. Dept. of Transportation*, Federal Highway Administration, Office of Research, Development, and Technology, Washington, D.C.

SWM (2000). Dept. of Civil Engrg., University of Nevada, Version 5.0.

Trochianis, A., Bielak, J., and Christiano, P. (1988). "A Three-Dimensional Nonlinear Study of Piles Leading to the Development of a Simplified Model." *Rep. R-88-176*, Department of Civil Engineering, Carnegie Institute of Technology, Carnegie Mellon University, Pittsburgh, PA.

Wallace, J. (1992). BIAX, Revision 1: A Computer Program for the Analysis of Reinforced Concrete Masonry Sections, *Rep. No. CU/CEE-92/4*, Structural Engineering, Mechanics and Materials, Clarkson University, Potsdam, NY.

LATERAL OR TORSIONAL FAILURE MODES IN VERTICALLY LOADED DEFECTIVE PILE GROUPS

Linggang Kong [1], and Liming Zhang [2], Member ASCE

ABSTRACT: This paper focuses on investigating potential bending or torsional failure modes in axially loaded pile groups caused by defective piles and on investigating the effect of short piles on the response of vertically loaded bored pile groups socketed into rocks. A hypothetical group of six bored piles, 2.0 m in diameter and 25.5 m in length, was analyzed and six cases with different numbers and configurations of short piles in the pile group were considered using a three-dimensional pile foundation analysis program FB-Pier. It is found that the number and configuration of short piles have a significant effect on behavior of the pile group. The presence of short piles forces "intact" piles to carry larger loads. An important aspect of the group response is that the presence of short piles can result in induced lateral deflections of the pile group and additional bending moments in piles even under "concentric" axial loads at the pile cap. The lateral deflections and bending moments can cause lateral failures of the pile group in terms of both ultimate limits and serviceability limits. For the axially loaded cases analyzed, the torques developed in the piles are much smaller than the bending moments; a torsional failure mechanism is therefore not likely to occur.

INTRODUCTION

When large-diameter bored piles have been constructed, some quality assurance tests such as integrity tests and load tests are often conducted to check construction technique and workmanship and to confirm the performance of the piles. Load testing of piles is expensive and time consuming, so only a few piles at a site can be tested. Integrity tests, on the other hand, can be carried out on all working piles. Existing

[1] PhD Candidate, Hong Kong University of Science and Technology, Department of Civil Engineering, Clearwater Bay, Kowloon, Hong Kong, kongs@ust.hk
[2] Assistant Professor, Hong Kong University of Science and Technology, Department of Civil Engineering, Clearwater Bay, Kowloon, Hong Kong, cezhangl@ust.hk

integrity test techniques, however, are not able to detect every defect that may exist in a pile (Fleming, 1992; G-I Deep Foundations Committee, 2000; Hassan and O' Neill, 1998). So a risk of missing one or a few defective piles still exists.

Pile defects can affect structural safety. Tabsh and O'Neill (2001) investigated effects of minor defects on the structural safety of single bored piles. In practice, pile groups are commonly employed as foundations. Thus, it is necessary to analyze the behavior of pile groups with defects. Some researchers, such as Poulos (1997), Xu and Poulos (2001), and Abdrabbo and Abouseeda (2002) have paid attention to the problem of interactions among defective piles.

Poulos (1997) investigated the response of a six-pile group with geotechnical defects and structural defects using numerical methods. According to the study, the general characteristics of behavior of a group with piles containing geotechnical defects are similar to those with structural defects. The ability of the group to redistribute loads from defective piles to "intact" piles results in a less severe reduction in axial stiffness than is the case for a single defective pile. However, the presence of defective piles will generally lead to the development of lateral deflection and rotation of the group, and induces additional bending moments in the piles. Xu and Polous (2001) employed elastic continuum-based approaches to analyze the behavior of a pile group containing defective piles and introduced definitions of interaction factors. They described the axial interaction of two axially loaded piles with necking and showed the effect of the location of a neck along the pile depth.

A defect may have different effects on pile groups in different soil profiles. For a vertically loaded bored pile socketed into rocks, the toe resistance plays an important role if the pile is not very long. If the pile toe is not founded on the design rock level or there is a serious strength reduction in the pile shaft, the effect of this defective pile on the pile group behaviour will be significant. Abdrabbo and Abouseeda (2002) reported a foundation with bored piles of 0.5 m in diameter and 8 m in length. Because of insufficient sub-surface soil investigation, only part of the piles was founded on sand stone and the remaining piles were shorter than required. The building superstructure tilted noticeably during construction. Finally, the total differential settlement of the building was found to be 416 mm and the building tilted at an inclination angle of 0.014 radians with the vertical (1:70).

This paper focuses on investigating potential bending or torsional failure modes in axial loaded pile groups caused by defective piles and on investigating the effect of short piles on the response of vertically loaded bored pile groups socketed into rocks.

METHOD OF ANALYSIS

The finite element method was used to study the behavior of a 2 by 3 bored pile group containing different numbers and configurations of "short piles." The short piles are not founded on the bedrock and are therefore shorter than "normal piles" nearby that are socked into the bedrock. A normal pile group with all piles properly socketed into the bedrock was also analyzed as a reference case. The analysis was performed using a computer program FB-Pier, which was developed by the Florida Department of Transportation and the Federal Highway Administration (Hoit et al. 2001). The FB-Pier program is a nonlinear finite element analysis program. It couples nonlinear structural analysis with nonlinear soil models for the analysis of axial, lateral and torsional responses of superstructure-foundation systems. The program has the ability to model various pile configurations and soil layers at varying depths. Soil-pile interactions are modeled by nonlinear soil springs whose axial, lateral and torsional stiffness are defined by the *p-y*, *t-z* and *t-theta* curves, where *p, y, t, z* and *theta* are soil resistance per unit length, local horizontal displacement, side shear stress on pile shaft, local vertical displacement, and local torsional angle, respectively. The pile cap is modeled by special shell elements, which add normal rotational stiffness terms to account for torsional force transfer from piles.

SOIL AND ROCK PARAMETERS

To determine rational soil parameters representative of Hong Kong conditions, site investigation data at a test pile site (Interpretative Report 1998) were employed and corresponding load test results were used to verify the soil parameters. The test pile YCS1, which was 1.5 m in diameter and 50.6 m in length, was located in Yen Chow Street in Hong Kong. A summary of the materials encountered at the site is shown in Table 1. The unit weight of rocks was taken from laboratory test results with core samples. Assumed unit weight values were used for soils, as shown in Table 2.

TABLE 1. Summary of Materials Encountered at Test Site

Stratum (1)	Thickness (m) (2)	Level of Top of Stratum (mPD) (3)	SPT N (4)
Fill	18.0 – 18.5	+5.0	10~20
Marine Deposits	1.0 to 2.5	-11.0	10~20
Alluvium	11.0	-13.0	20~40
Completely decomposed granite (Grade V)	19.0 to 21.0	-22.5	<50
Highly decomposed granite (Grade IV)	1.4 to 6.0	-42.5	50 ~ >200
Moderately decomposed granite (Grade III)	1.8 to 2.3	-45.0	
Moderately to slightly decomposed granite (Grade III/II)	- to end of borehole	- 47.3	

Nonlinear *p-y* curves, *t-z* curves and *t-theta* curves are employed, respectively, to simulate the mobilized lateral soil resistance, shaft resistance, and torsional resistance of each of the soil layers in Table 1. The lateral soil-pile interaction is featured by *p-y* curves for sands suggested by Reese et al.(1974). The *p-y* curves require the input model parameters of coefficient of subgrade modulus (n_h), shear modulus (G) and soil unit weight (γ). The first two parameters can be evaluated from SPT N values based on empirical relations suggested in FB-Pier User's Manual (Hoit et al. 2001). The torsional soil-pile interaction, modeled by *t-theta* springs, is featured by hyperbolic curves proposed by Randolph (1981). Required input parameters include shear modulus (G), torsional shear stress (τ_0), friction angle (ϕ') and soil unit weight (γ). For the axial soil-pile interaction, the *t-z* curves suggested by O'Neill and Reese

(1999) are used to simulate the shaft resistance of the piles, which require the input parameters of soil unit weight (γ) and friction angle (ϕ'). These parameters are summarized in Table 2

TABLE 2. Soil Parameters Used for Analysis

Elevations (m) (1)	Unit Weight (kN/m^3) (2)	Friction Angle (degree) (3)	Coefficient of Subgrade Modulus (kN/m^3) (4)	Shear Modulus (kN/m^2) (5)	Torsional Shear Stress (kPa) (6)
+0.0 ~ -16.0	19.0	34.3	16300	9000	34.5
-16.0 ~ -18.0	18.0	34.3	16300	9000	55.2
-18.0 ~ -27.5	19.0	37	30000	18000	55.2
-27.5 ~ -49.4	19.5	40.5	41000	54000	55.2
-49.4 ~	25.3	-	-	480000	55.2

Load test results are used to develop a toe resistance model for the test pile, which was socketed into the bedrock. The ultimate unit toe resistance has to be estimated because it was not achieved in the load test. Littlechild et.al. (2000) proposed the following equation to evaluate the ultimate unit toe resistance of piles on rock (Q_b).

$$Q_b = 1.0 \text{ x UCS} \leq 25 \text{ MPa} \quad (1)$$

where UCS denotes the unconfined compressive strength of rock. The mean UCS of the rock near the toe is 22 MPa in the test. It is assumed that the stress-strain behavior of the rock is elastic, perfectly plastic. After yielding, the rock would deform like a perfectly plastic material. Fig.1 shows the toe model defined in terms of resistance force. As the length of the rock socket (0.5 m) is short, the shaft friction along the rock socket is neglected in the analysis.

If a pile is "short" and is founded on soils, then the *q-z* curve suggested by O'Neill and Reese (1999) will be used to simulate the toe resistance of the short pile. The model requires uncorrected SPT N as an input parameter. The toe resistance-displacement curve for the pile toe developed using a SPT N value of 90 is also plotted in Fig.1. Comparing the two curves in Fig.1, the toe resistance of the intact pile on rock is much larger than that of the short pile at the same displacement.

The reinforcement of the test pile from top to bottom comprised of 20T20 and 20T25. The concrete compressive strength and the concrete modulus are 40 MPa and 30 GPa, respectively. The steel yield stress and elastic modulus are 460 MPa and 210 GPa, respectively.

Fig.2 shows the measured and predicted load-settlement curves of the test pile. The predicted curve is close to the measured curve. Therefore, the soil and rock parameters utilized in the analysis are reasonable. These parameters will be used in subsequent analyses of pile groups.

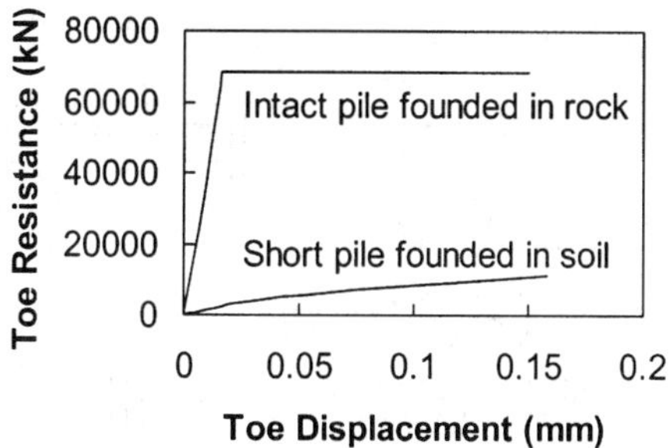

Fig. 1. Employed Pile Toe Models

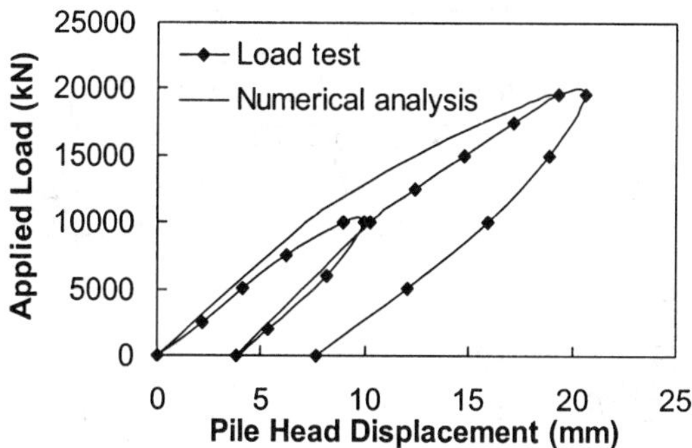

Fig. 2. Measured vs. Predicted Load – Displacement Curves

BEHAVIOR OF PILE GROUPS CONTAINING SHORT PILES

Configurations of Short Piles in Pile Groups

A hypothetical pile group illustrated in Fig.3 is analyzed. The group of six bored piles, 2.0 m in diameter and 25.5 m in length, is assumed. The pile length is approximately one half of the test pile described earlier. The soil and rock layers and material parameters for the pile group analysis are the same as those shown in Table 1, except that the thickness of each layer is only one half of that shown in Table 1. The pile group is subjected to vertical loads only. Six equal loads are applied at the pile cap at each of the centers of the piles (see Fig.3). The pile cap is 2.5 m in thickness and the bottom of the pile cap is 0.5 m off the ground.

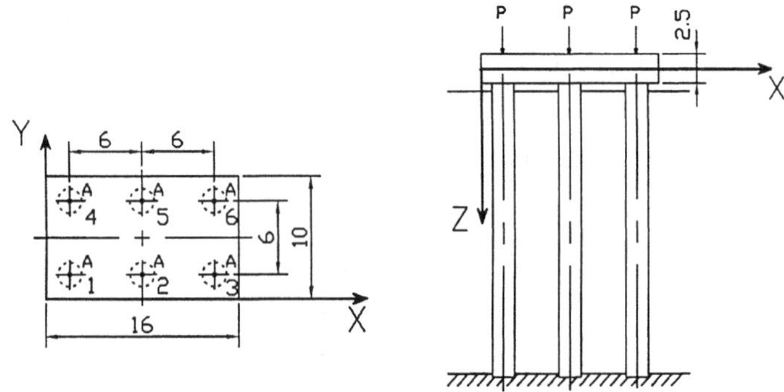

Fig. 3. Pile Group Analyzed

TABLE 3. Configurations of Short Piles in the Pile Group

Name (1)	Number of short piles (2)	Location (see Fig.3) (3)
Case 1	1	1
Case 2	2	1,2
Case 3		1,3
Case 4		1,4
Case 5	3	1,2,3
Case 6		1,2,4

Six group configurations listed in Table 3 are analyzed. The basic case of a pile group without any short piles is also analyzed as a reference case. The short piles in the pile group are 4 m shorter than the intact piles and are founded in the completely decomposed granite layer.

The responses of all the six cases under two loads of 90000 kN and 144000 kN were analyzed in particular. The two loads are approximately design workloads of the pile group in two different rock quality designations in Hong Kong. They are calculated using presumed bearing pressures of 5000 kPa and 7500 kPa for foundations on horizontal ground (GEO 1996).

Vertical Behavior

Changes in load distributions among the group piles will occur when some short piles are present in the pile group. Fig.4 shows the load distribution among the group piles in Case 1. The dashed line in the figure represents the average load carried by a pile in the reference case. It can be found that the load carried by the short pile decreases, while the normal piles are forced to carry larger loads, especially the normal piles next to the short pile, i.e., Pile 2 and Pile 4 in Fig. 4. Pile 6, located at the opposite side of the short pile carries a smaller load than other normal piles.

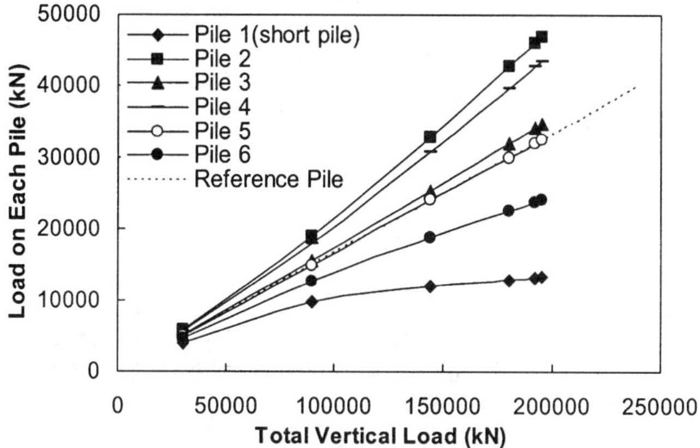

Fig. 4. Load Carried by Each Pile in Pile Group Case 1

Tilting of Pile Group

Fig. 5 shows the vertical displacements of the pile cap in Case 1. Because of the difference in the settlements of the short pile and the normal piles, the pile cap tilts towards the short pile side. Table 4 shows the maximum inclinations of the pile cap in the six cases under the two loads of 90000 kN and 144000 kN.

Coduto (2001) summarized allowable angular distortions of buildings. The range of allowable angular distortions for various types of buildings is from 1/2500 to 1/25. From Table 4, it can be seen that even the presence of one short pile in the pile group may cause serious serviceability problem for some structures.

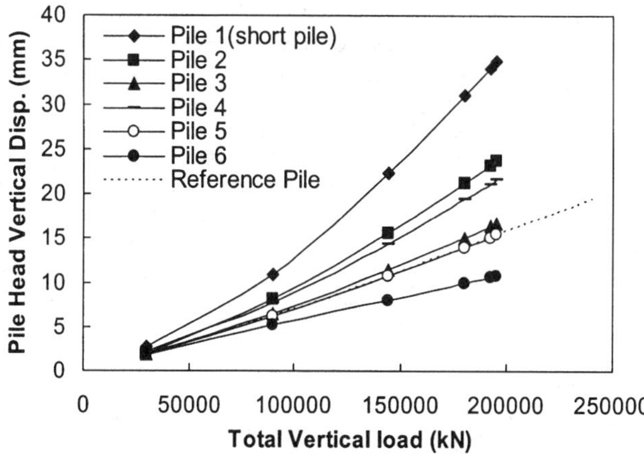

Fig. 5. Vertical Displacement Distributions at Pile Head in Pile Group Case 1

TABLE 4. Inclination of Pile Cap

Load (kN) (1)	Case 1 (2)	Case 2 (3)	Case 3 (4)	Case 4 (5)	Case 5 (6)	Case 6 (7)
90000	1:2350	1:1090	1:1430	1:1030	1:276	1:530
1440000	1:940	1:380	1:550	1:349	1:55 [a]	1:158

[a] Structural components failed.

Horizontal Behavior

Large bending moments are developed in all six cases of defective pile group cases. Figs. 6-8 show the development of bending moment with the applied load for some pile group cases. For example, the maximum bending moments about x-axis at the two loads of 90000 kN and 144000 kN are 7000 kN-m and 18500 kN-m, respectively in Case 5 as shown later in Fig. 8.

If configurations of short piles in the group are symmetric about an axis, like in Cases 3, 4 and 5, the critical direction of bending is the direction perpendicular to the axis. Take Case 5 as an example (Fig. 7), the short piles are symmetric about the x-axis, the critical direction of maximum bending moment is then the y-axis. In fact, for Case 3, Case 4 and Case 5, bending moment only develops along one direction. For the cases of short pile located at a corner of the group, bending moment develops along two horizontal directions, which is shown in Figs. 6 and 8.

To study lateral deflections of the pile group caused by defective piles, the displacements at the center of the pile cap in x-and y-axes are examined as shown in Fig.9. In Case 3, Case 4 and Case 5, large displacements only occur in one direction, because the short piles are located at one side of the pile group. For short piles located at one corner of the pile group, as in Case 1, Case 2 and Case 6, displacements develop in both horizontal directions. This can be analogous to the corresponding maximum bending moment distributions.

For all the six cases, large bending moments develop in both short piles and normal piles. Even in Case 1, in which only one short pile is present, the maximum bending moment at the load of 144000 kN is more than 5000 kN-m. The corresponding maximum horizontal displacement is 3.8 mm. For the extreme case (Case 5) where three piles at one side are short, the maximum bending moment is as much as 18500 kN-m and structural components of the group have failed. The corresponding maximum horizontal displacement is as large as 67.5 mm. Therefore, the presence of short piles can result in large lateral deflections and large bending moments in piles even under "concentric" axial loads at the pile cap. The induced lateral deflections and bending moments can cause lateral failure of the pile group in terms of both ultimate limits and serviceability limit.

Torsional Behavior

Torques also develop at some particular configurations of short piles in the pile group, including Case 1, Case 2 and Case 6. Figs 10 and 11 show the development of torques and rotational angles at pile heads with the vertical load in the three cases. The torques on the short piles and the normal piles are similar at high load levels. Therefore, only a representative curve for each case is plotted in Figs. 10 and 11. As the torques and rotational angles at the pile heads are small for the cases analyzed, a torsional failure mechanism is not likely to occur.

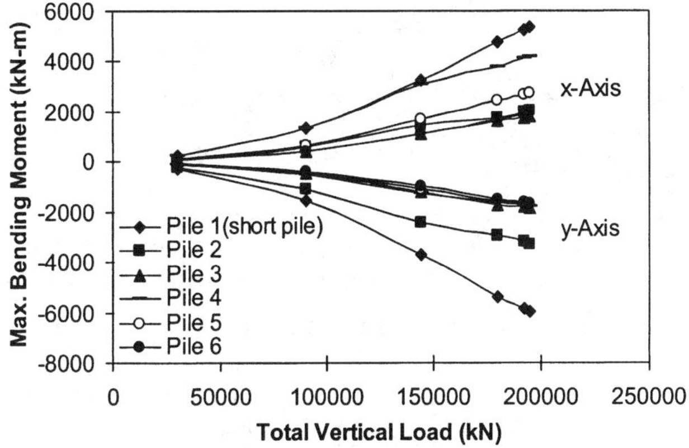

Fig. 6. Maximum Bending Moments about x- and y-Axes in Pile Group Case 1

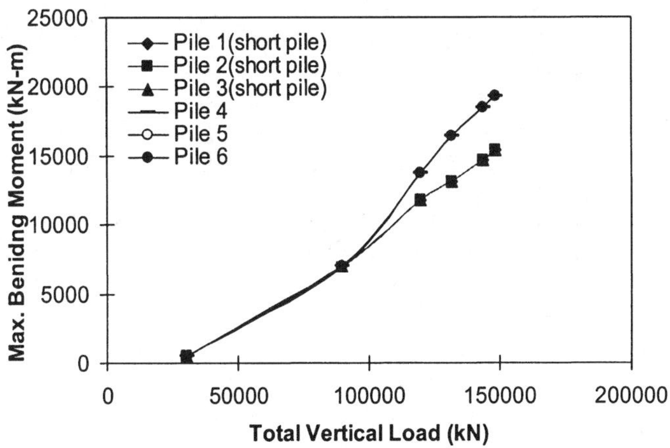

Fig. 7. Maximum Bending Moments about x-Axis in Pile Group Case 5

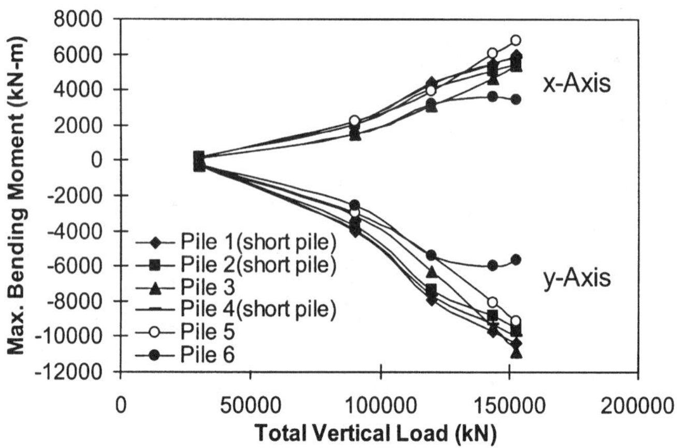

Fig. 8. Maximum Bending Moments about x- and y-Axes in Pile Group Case 6

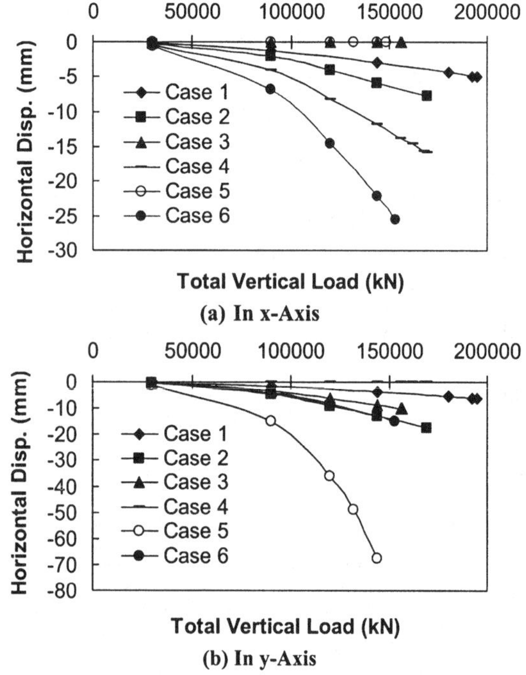

Fig. 9. Horizontal Displacements at Center of Pile Cap

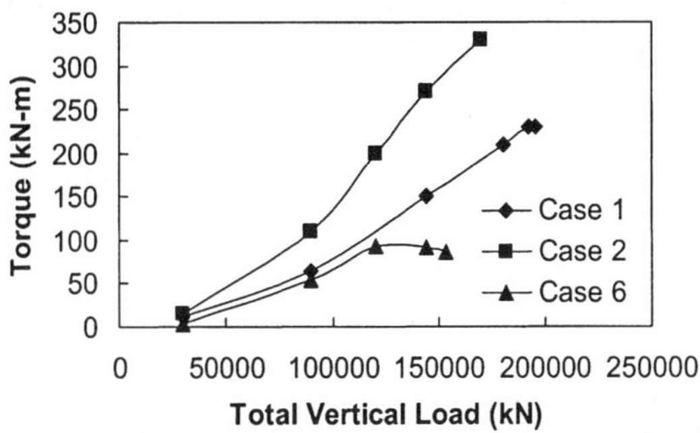

Fig. 10. Mobilized Torque at Pile Head

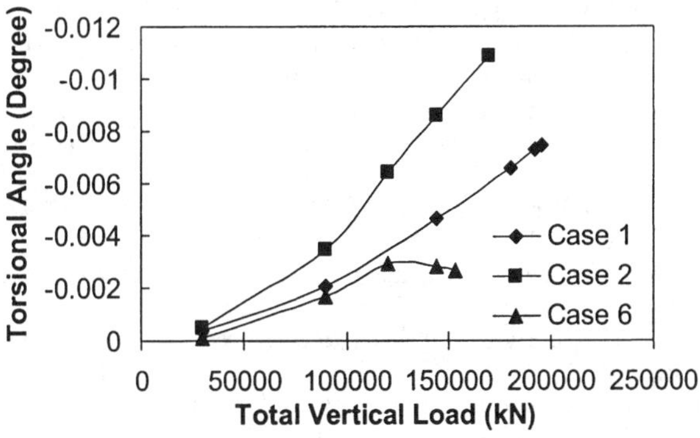

Fig. 11. Torsional Angle of Pile Cap

CONCLUSIONS

Analyses are conducted to examine possible failure modes in axially loaded pile groups containing short piles. A group of six piles is analyzed and six cases of short pile combinations are considered in this paper. The following characteristics are noted:

1. Load redistributions among the group piles cause normal piles to carry larger loads. Most of the additional loads transferred from defective piles are carried by normal piles adjacent to the defective piles.
2. An important aspect of the group response is that the presence of short piles can result in induced lateral deflections of the group, and additional bending moments in piles even under "concentric" axial loads at the pile cap. The induced lateral deflections and bending moments can cause lateral failures of the pile group in terms of both ultimate limits and serviceability limits.

3. The increase in bending moments in piles and tilting of the pile group become more marked as the number of defective piles, and /or the applied load level increase.
4. For the axially loaded pile group cases analyzed in this paper, the torques developed in the piles are much smaller than the bending moments. Therefore, a torsional failure mechanism is not likely to occur.

REFERENCES

Abdrabbo, F. M., and Abouseeda, H. (2002). "Effect of construction procedures on the performance of bored Piles." *Deep Foundation 2002,* ASCE, 1493-1454.

Coduto, D.P. (2001). *Foundation Design: Principle and Practices, Second Edition*, Prentice-Hall, Inc, New Jersey.

Fleming, W. G. K., Weltman, A. J., Randolph, M. F., and Elson, W. K. (1992). *Piling Engineering*, Blackie:Glasgow.

Geotechnical Control Office (GCO). (1996). "Pile design and construction." GEO Publ. No. 1/96, Geotechnical Engineering Office, Hong Kong.

Hassan, K. H., and O'Neill, M. W. (1998). "Structural resistance factors for drilled shafts with minor defects." *Final Rep., Phase I*, ADSC/FHWA, Dallas.

Hoit, M. I., McVay, M. C., Hays, C. O., and Williams, M. (2001). *FB-Pier Users Guide and Manual*, University of Florida, Gainesville.

Littlechild, B. D., Plumbridge, G. D., Hill, S. J., and Soon Chop L. (2000). "A rational design approach for foundations on rock." *Proceeding New Technological and Design Developments in Deep Foundations,* Denver.

O'Neil M. W. and Reese L.C. (1999). "Drilled shafts: Construction procedures and design methods." Publ. No. FHWA-IF-99-025, FHWA Office of Infrastructure/Office of Bridge Technology, McLean, Va.

Poulos, H. G. (1997). "Analysis of pile group with defect piles." *Proc. 14th ICSMFE,* Hamburg, Germany, A.A. Balkema, Rotterdam, 871-876.

Randolph, M.F. (1981). "Piles subjected to torsion." *Journal of the Geotechnical Division,* ASCE, 107 (8), 1095-1111

Report of a Task Sponsored by the G-I Deep Foundations Committee (2000). "Nondestructive evaluation of drilled shafts." *J. of Geotech. and Geoenver. Engrg.*, 126(1), 92~95.

Reese, L. C., Cox, W. R. and Koop, F. D. (1974). "Analysis of laterally loaded piles in sand." *Paper No. OTC 2080, Proceedings, Fifth Annual Offshore Technology Conference,* Houston, Texas (GESA Report No. D-75-9).

Tabsh, S. W., and O'Neill, N. W. (2001). "Structural safety of drilled shaft with minor defect." *Proc. ICCOSSAR'2001 Int. Asso. Structural Safety and Reliability,* Newport Beach, A. A. Balkema: Rotterdam.

Test Pile YCS1- Interpretative Report (Ref. TS500-REP-G-0095-1) (1998). *TS-500 Full Scale Deep Foundations Load Testing Programme*, Ove Arup & Partners Hong Kong Ltd.

Xu, K. J. and Poulos, H. G. (2001). "Interaction analysis of intact and defective piles." *Comp. Meth. and Advan. in Geomech.*, A. A. Balkema, Rotterdam, 1507-1514.

PERFORMANCE EVALUATION OF CFA VS. BENTONITE SLURRY DRILLED SHAFTS UTILIZING DROP WEIGHT TESTING

By Samuel G. Paikowsky[1], P.E. (Israel), Member, Geo-Institute, Israel Klar[2], P.E. (Israel), and Les R. Chernauskas[3], P.E., Member Geo-Institute

ABSTRACT: The performance of drilled shafts is known to be controlled by the type and quality of the construction methods. A comparison between different methods or different contractors is typically not available under the prevailing bidding and testing procedures.

Six drilled shafts 0.70 m in diameter and 25.0 m long were installed and tested to structural failure at a site in Haifa, Israel. Three of the shafts were constructed using the Continuous Flight Auger (CFA) construction method and three of the drilled shafts were constructed using bentonite slurry.

Site and shafts details are provided. The drop weight testing system is reviewed. Test results comparing the load carrying capacity of the shafts, load distribution and structural outlines are presented and discussed. While the bentonite slurry construction resulted in a more uniform constructed foundation capable of carrying higher load, the explanation to the variation at the given site conditions may be a result of the contractor's quality rather than advantages of one technology over another.

BACKGROUND

Overview

Drilled shafts are the prevailing deep foundation solutions in some parts of the world. Subsurface conditions consisting of hard clay, cemented sands, karstic limestone, low

---

[1] Professor, Geotechnical Engineering Research Laboratory. Dept. of Civil and Environmental Engineering, University of Massachusetts-Lowell, 1 University Avenue, Lowell, MA 01854, USA, Samuel_Paikowsky@uml.edu

[2] Geotechnical Consultant, Klar Foundation Consulting & Engineering Services, 23 Basri Street, Kiryat-Ata, 28706, Israel

[3] Project Manager, Geosciences Testing and Research Inc., GTR, 55 Middlesex St., Suite 225, N. Chelmsford, MA 01863, USA.

groundwater table, combined with high price of steel and fuel (transportation) in contrast with low price of concrete, make drilled shafts the preferable solution of deep foundations in Israel. As a result, various construction and quality control technologies are developed there and explored. Bentonite slurry drilled shaft (wet) construction is most commonly used while Continuous Flight Auger (CFA) is a relatively newcomer.

The bentonite method refers to the "wet" construction method (slurry-displaced method) in which bentonite slurry is used to keep the borehole stable during excavation. The construction process consists of the following stages:

(i) Drilling equipment is used to drill to the groundwater surface.
(ii) Bentonite slurry is introduced into the hole and the drilling is continued to the full depth of the hole. The slurry elevation is kept continuously above the groundwater surface elevation during construction and its quality is assured at the end of the construction.
(iii) A reinforcement steel cage is placed in the slurry.
(iv) Concrete is placed in the excavation using tremie with the bottom of the tremie remaining below the surface of the concrete. While the column of the concrete rises in the excavation, the slurry is displaced and pumped back into storage.

The Continuous Flight Auger (CFA) construction sequence is comprised of five stages:

(i) The digging head of the auger is fitted with an expendable cap.
(ii) The auger is screwed into the ground to the required depth.
(iii) Concrete is pumped through the hollow stem, blowing off the expendable cap under pressure.
(iv) Maintaining positive concrete pressure, the auger is withdrawn all the way to the surface.
(v) Reinforcement is placed into the pile up to the required depth.

The prices of both construction technologies are comparable (in Israel) with advocates to each of the techniques. Bentonite slurry construction is assumed to allow for better quality control (concrete volume per length, tubes for CSL etc.) while suspected of having potentially reduced interaction and strength at the interface with the subsurface soils, in particular at the tip. CFA method leaves more questions as to the quality of the construction (in spite of instrumentation monitoring auger and concrete pressure, and concrete volume), but is assumed to guarantee better interaction with the soil, i.e. higher friction and end bearing free of a possible slurry "cake" at the concrete-soil interface.

A large multi-story expansion of a pharmaceutical manufacturing complex called for the use of deep foundations. With similar cost estimation of bentonite and CFA construction and the need for performance verification, the owner had decided to conduct pre-bid testing of six deep foundation elements, constructed at the same location, three using bentonite slurry and three using CFA technique. Impact tests utilizing a drop weight system were used to examine the shafts.

Site and Subsurface Conditions

The site is located in Haifa Bay, west of Kiryat Ata intersection. The subsurface profile at the site is described in Figure 1. The upper 0.25 m consisted of an existing concrete slab. The subsurface itself consists of a yellow to gray uniform fine sand to a depth of 20 m; from 0 to 4 m the sand is medium dense and it becomes dense to very dense from

4 to 20 m. The sand layer is underlain by a gray to black, stiff, high plasticity clay from 20 to 22.7 m. A very dense yellow sand and calcareous sandstone underlies the clay layer to a depth of 30 m (which corresponds to the end of the borings). Groundwater was encountered during drilling at a depth of 5.8 m, which may not accurately represent stable groundwater level. Six separate test piles designated as B1 through B3 and C1 through C3 (see Figure 2 – Site Layout) were tested. The piles were located within the footprint of the proposed structure but were not planned to become a part of the permanent foundations.

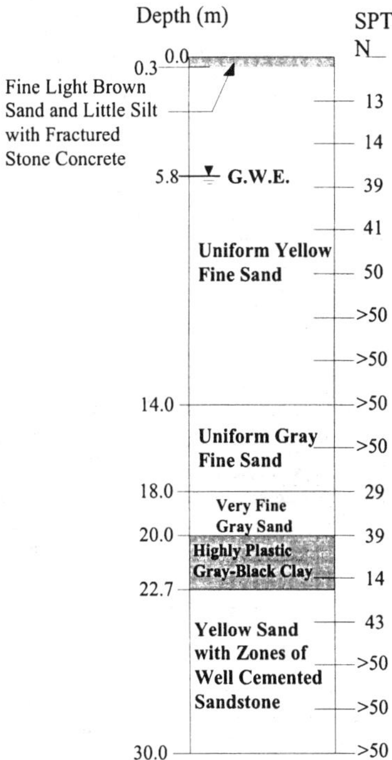

FIG. 1. Subsurface Conditions at the Test Site.

Pile Details

Details about the piles construction are provided in Table1. Three of the piles were constructed using the Continuous Flight Auger (CFA) technique between August 8 to 11, 2002 and built with type B-30 concrete (nominal strength of 30MPa). The CFA pile extensions above ground level (see details in Table 1) were built on August 21,

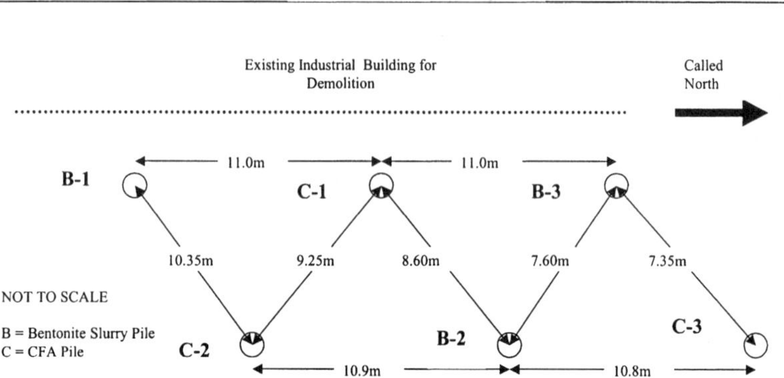

FIG. 2. Site Layout

TABLE 1. Summary of Drilled Shaft Construction Details

Shaft No.	Shaft Type	Date of Construction	Date of Extension Construction	Height of Extension (m)	Comments
C-1	CFA	8/11 to 8/13	8/21	1.40	Concrete mushroom around pile top was broken prior to extension construction, all steel sleeves removed prior to testing.
C-2	CFA	8/11 to 8/13	8/21	1.45	
C-3	CFA	8/11 to 8/13	8/21	1.34	
B-1	Bentonite Slurry	8/18	8/18	1.38 (1.12) [3]	No separation between pile and slab, poor quality concrete on top and circumference
B-2	Bentonite Slurry	8/19	8/19	1.50	Non-round extension
B-3	Bentonite Slurry	8/19	8/19	1.15	Non-round extension

Notes

1. All nominal pile sizes are 70 cm in diameter and 25 meters in length
2. All steel sleeve extensions of the bentonite piles are 1.7 meters long.
3. Number in parentheses is the height of the extension after the top was cut off to enable placement of the guide system

2002 using B-50 concrete. The remaining three piles were constructed using an auger bucket under wet drilling technique of bentonite slurry on August 18 and 19, 2002. The bentonite slurry pile extensions were built on the same day along with the pile construction. All piles were designed to have a nominal size of 70 cm in diameter and 25 m in length. The nominal cross-sectional area is 3847 square centimeters. The maximum allowable compressive stress limit is around 26 MPa, based on 85% of the 28-day compressive strength (0.85 f'c). The maximum allowable tensile impact stress is 1.47 MPa, based on three times the square root of the 28-day compressive strength ($3\ f'c^{1/2}$ in units of pound per square inch). The 28-day concrete compressive strength (fc) was reported to be around 30 MPa. The concrete extensions cast on top of the piles were approximately 1.5 meters in length.

IMPACT TESTING OF DRILLED SHAFTS USING DROP WEIGHT SYSTEMS

Background and Use

Increasingly, drop weight systems are being used to dynamically test cast-in-place deep foundations. Conventional pile driving hammers are often inadequate to test these deep foundation types since (i) they typically cannot deliver enough energy to mobilize the ultimate bearing capacity, and (ii) the size and location of the foundation member can present problems in adequately delivering the energy from the ram to the pile. Simple drop weight systems have therefore been developed to overcome the limitations of the conventional hammers and allow for dynamic testing of in place constructed deep foundations.

An in depth review of various available drop weight systems, and evaluation of the method is presented by Paikowsky et al. (2003). A typical drop weight system consists of four components: a frame or guide for the drop weight (ram), ram, a trip mechanism to release the weight, and a striker plate/cushion. Strain gages and accelerometers are placed at the pile top to obtain stress wave measurements utilizing available PDA's (Pile Driving Analyzers). Figure 3 shows the setup of an Israeli Drop Weight Impact Device, developed and used by GeoDynamica and GTR to test drilled shafts in Israel. This device is similar in principle to other drop weight systems presently in use with the distinction of modularity in ram weight as well as uniqueness in trip mechanism. The Israeli Drop Weight Impact Device uses modular weights that can be arranged into ram weights of 2, 4, 5, 7, or 9 tons and has an adjustable drop height of up to 4 meters thereby allowing for potential energy of up to 36 t·m (260 kip·ft.). Typically, a pile cushion is used to even pile stresses occurring during impact.

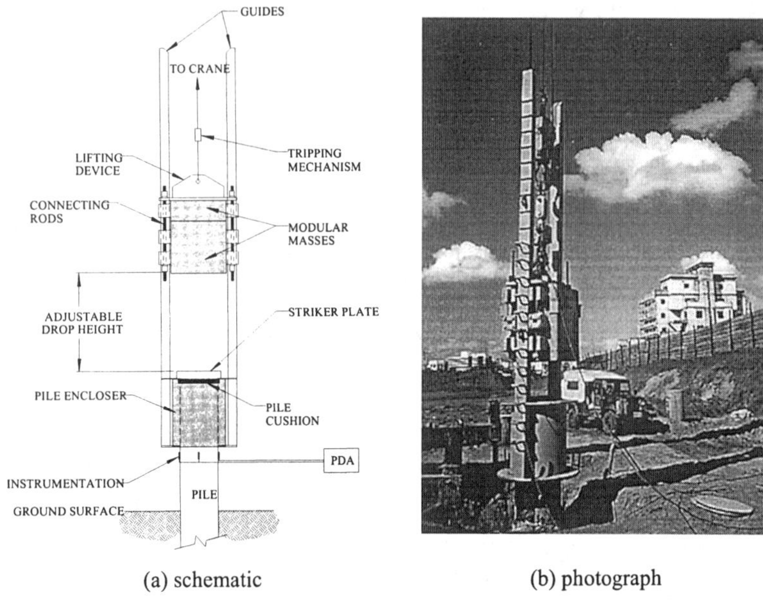

(a) schematic (b) photograph

FIG. 3. Israeli Drop Weight System (after GTR, 1997).

Advantages

Drop weight systems have several advantages when compared to traditional static load testing methods. Most of these advantages are similar to those presented by standard dynamic pile testing, namely:

- Rapid testing time allowing to carry out tests on several shafts in a single day.
- The ability to deliver high force and energy to mobilize the capacity, and hence test large deep foundations.
- The ability to use available transducers and data acquisition systems (such as the Pile Driving Analyzer of Pile Dynamics and the TNO Foundation Pile Diagnostic System).
- The test provides a means to conduct high strain integrity testing concurrent with capacity determination.
- Low test cost relative to standard static load test.
- Current analysis techniques include field methods (such as the Case Method and the Energy Approach) and signal matching (e.g. CAPWAP).

Disadvantages

There are several disadvantages and limitations relating to the use drop weight systems:

- The selected mass and drop height must be of sufficient magnitude to mobilize the resistance of the deep foundation shaft in order to obtain adequate capacity measurements.
- The installation process of in-situ deep foundations (drilled shafts, cast-in-place piles, etc) can cause irregularities in pile shape and homogeneity that can affect current analysis methods.
- For increased quality of the obtained measurements the gauges need to be away from the impact and as high as possible above the ground. This results in the need to create "extensions" of cast in place shafts and the use of multiple gauge systems.
- Although several studies have already been conducted comparing dynamic and static measurements of cast-in-place deep foundations (Rausche and Seidel, 1984, Jianren and Shihong, 1992, Townsend et al., 1991), a comprehensive comparison study has only recently been completed and presented by Paikowsky et al. (2003). Only a few studies are known to compare static and dynamic testing of CFA piles, one of which was presented by Cannon (2000).

FIELD TESTING

Impact Device

The aforementioned drop weight device was used to provide high force and energy impacts necessary to mobilize the soil resistance acting along the side (friction) and tip (end bearing) of the shafts. The device is typically adjusted to test shaft heads ranging from 60 to 80 cm in diameter. Due to non-round, non-uniform pile head extensions, on site modifications had to be performed specifically for the presented testing.

Ram weights of 7 tons (74 kN including attachment elements) were used during testing for this project. The strokes were typically varied between 0.25 and 2 meters, resulting in rated energies between 18.5 and 148.0 kN-m. Plywood sheets varying in total thickness between 40 to 100 millimeters in thickness were used for the pile cushion along with a 25 mm thick steel plate (for details see Table 2).

Instrumentation

The instrumentation consists of four strain gage and four accelerometer transducers attached approximately 1m below the top of the pile extension. A strain gage and accelerometer pair were bolted 90 degrees apart on the circumference of the pile to minimize the effects of uneven impact and pile bending. This instrumentation provides information about driving stresses (compressive and tensile), driving system performance (alignment of ram and transferred energy), and pile capacity. To further enhance the ability to monitor data quality, one accelerometer was attached to the ram itself, allowing measurement of the ram acceleration, and hence, the force developed at the top of the extension. This enabled independent measurement of the impact forces evaluated via the strain gages.

TABLE 2. Summary of Dynamic Load Test Results Taro Corporation, Haifa

Shaft No.	Depth (m)	Diam. (cm)	Blow No.	Ram Weight (kN)	Cushion Thick [1] (mm)	Stroke[2] (m)	Max Transferred Energy[3] (kN*m)	Max Displace-ment[3] (mm)	Pile Set[4] (mm)	Max Compress-ive Stress[3] (MPa)	Max Tensile Stress[5] (MPa)	Max Pile Top Force[3] (kN)	Case Method Capacity[6] (kN)	Energy Approach Capacity[7] (kN)	CAPWAP Capacity[8] (kN)	Predicted Ultimate Capacity[9] (kN)
B-1	25	70	2	74	3P+S+2P	1.01	38.2	8	2/3.0	15.8	2.5	6015	4005	6950	5600	5500
B-2	25	70	4	74	2P+S+2P	1.51	67.5	7	2/2.0	36.2	0.7	13749	5358	15000	5792	6000
B-3	25	70	6	74	2P+S+2P	2.01	114.0	15	3/5.0	30.9	5.5	11753	5778	11400	5695	6000
C-1	25	70	3	74	3P	0.95	31.9	6	2/3.0	24.8	0	9432	4220	7090	5450	5500
C-2	25	70	4	74	3P	1.33	40.9	6	1/1.0	22.6	0.8	8578	5325	11690	6145	6000
C-3	25	70	9	74	3P+S	2.00	77.0	13	7/na	29.3	2.6	11150	6553	7700	5050	5000

Notes

1. The striking plate for all tests was 46.5 cm diameter and 10 cm in thickness. The cushion system consisted of a combination of plywood and steel plates of the following thickness and diameter: Plywood (P) = diam = 70 cm and thick = 2 cm, Steel (S) = diam = 65 cm and thick = 2.5 cm
2. The stroke was typically increased from 0.25 m to 2.5 m during testing of each pile.
3. The maximum transferred energy; displacement, compressive stress, and force are determined by the PDA at the gage locations.
4. The pile set is presented as two values (1/1.0). The value on the left was determined from the PDA by integrating the acceleration measurements twice and the value on the right was measured independently using a level (accurate to 0.1 mm).
5. The maximum tensile stresses were calculated by the PDA and can be located anywhere along the pile shaft.
6. The Case Method was determined using the RMX method and a damping coefficient of 0.5 (RX5).
7. The Energy Approach Procedure was developed by GTR personnel and was proven to provide high accuracy long-term driven pile capacity at end of driving.
8. The CAPWAP capacity was determined using a computer program, which is capable of providing an estimate of the soil distribution.
9. The predicted pile capacity was determined based on the three methods described above.

The PDA is a computer fitted with a data acquisition and a signal conditioning system. The transducers are connected to the PDA via cables. During impact, the strain and acceleration signals are recorded and processed for each blow. The strain signal is converted to a force record and the acceleration signal is integrated to a velocity record. The PDA saves selected blows containing this information to disk and determines the compressive stresses, displacement, and energy at the point of measurement (pile top). In addition, the tensile stresses can be calculated and the pile bearing capacity determined using a procedure known as the Case Method. This information can be viewed on the computer screen during driving. Selected blows can be further processed to predict the static pile capacity using the Energy Approach method and CAPWAP analyses.

Testing Procedure

Shafts C1 to C3 were tested on August 28, 2002. Shafts B1 to B3 were tested the next day on August 29, 2002. A 7-ton ram (74 kN including attachments) was used to apply the impact force for all test piles. Between 5 and 13 blows were applied to each pile. The stroke was gradually increased from 0.25 meters to 2.5 meters in order to ensure that: (1) the impact stresses were as evenly distributed as possible, (2) the pile cushion was properly compressed prior to the last few blows, (3) allowable stress limits could be closely monitored, and (4) full mobilization of capacity could be observed prior to damaging the piles.

TESTING RESULTS

General

The results of the dynamic testing program are summarized in Table 2. Table 2 includes the pile depth below ground, shaft diameter, stroke, maximum transferred energy, maximum displacement, pile set, maximum compressive stress, maximum tensile stress, and maximum force for one selected blow on each pile. The maximum transferred energy, displacement, pile set, compressive stress, and force are determined by the PDA at the gage locations and are representative for the blow indicated. The ram stroke was measured in the field. The maximum tensile stress was estimated by the PDA and can occur at any location along the shaft. Also included in Table 2, are the pile bearing capacities as predicted by the Case and Energy Approach methods in the field and CAPWAP analyses in the office. Table 3 summarizes the CAPWAP results in more details separating the friction and end-bearing contributions as well as the contribution of the upper section of the shafts.

Field Observations and Driving System Performance

The pile set (permanent displacement) varied between 0 and 5 mm under each blow. The set was determined based on two procedures (1) double integration of the acceleration record (from the PDA) and (2) independent measurement using a level (accurate to 0.1 mm). The total set for all piles was relatively low, due to the high frictional resistance along the pile. For the 74-kN ram and a stroke height between approximately 1 and 2 m, the transferred energy ranged from 32 to 114 kN-m for the various analyzed blows. The overall driving system efficiency varied, therefore, between 42 and 77%, which is higher than that typically observed in drop weight systems (Paikowsky et al. 2003). The overall high driving system efficiency was a result of changes made to the system during this project.

TABLE 3. Summary of CAPWAP Results Taro Corporation, Haifa

Shaft No.	Depth (m)	Diam (cm)	Blow No.	Capacity (kN)				Skin Friction Total %	Quake (mm)		Damping (sec/m)	
				Upper 5m Friction	Lower 20m Friction	Tip	Total		Side	Tip	Side	Tip
B-1	25	70	2	1100	3000	1500	5600	73	3.0	6.0	0.460	0.126
B-2	25	70	4	1797	2996	999	5792	83	3.0	2.0	0.630	0.472
B-3	25	70	6	1000	3696	999	5695	82	3.0	13.0	0.442	0.151
C-1	25	70	3	2305	2545	600	5450	89	1.0	6.0	0.545	0.314
C-2	25	70	4	2198	2448	1499	6145	76	1.3	5.0	0.528	0.755
C-3	25	70	9	2200	2250	600	5050	88	1.0	10.0	0.593	0.629

Pile Integrity and Stresses

The maximum compressive stresses ranged between 16 and 25 MPa for strokes between approximately 1 and 1.5 meters and between 28 and 36 MPa using the higher strokes (approximately 1.5 to 2 m). The maximum tensile stresses ranged between approximately 1 and 8 MPa. All shafts were tested until signs of damage (crumbling) of the pile extensions were detected. Typical shafts' top conditions following the tests are shown in Figure 4.

Pile Construction

Evaluation of the shafts' construction can be obtained via the assessment of the variations (profile) of the impedance along the shaft used in the signal matching analysis (CAPWAP). The pile impedance is a measure of the concrete quality (through modulus and density) and cross sectional area in the following way:

$$I = EA/c \tag{1}$$

$$c = \sqrt{E/\rho} \tag{2}$$

for which: I = pile impedance
E = modulus of elasticity
A = cross-sectional area
c = speed of one-dimensional wave propagation

ρ = unit density

(a)

(b)

FIG. 4. Shaft's Top Extension Condition Following Drop Weight Tests (a) Shaft B3 Bentonite Slurry Cast in Place Shaft with a Steel Sleeve, (b) Shaft C3, CFA Constructed Shaft with Steel Sleeve Removed from the Above Ground Extension.

An expected impedance for a 70 cm diameter shaft ($A = 3848\ cm^2$) of concrete with $c = 4000$ m/s and $\rho = 2400\ kg/m^3$ is 3765 kN/m/s. Figure 5 describes the variation of the impedance along the shaft used in CAPWAP analyses for each of the shafts tested. Three distinct observations can be made in relation to the profiles presented in Figure 5; (i) the shafts constructed with bentonite slurry seem to be relatively uniform and present overall impedance equal or better than the one expected (excluding the lower part of B2), (ii) all shafts seem to have some type of increase in the impedance at the upper section, and (iii) all the CFA constructed shafts show a distinct decrease in the impedance, to a level lower than the expected value, below the depth of about 20 m. As the impedance profile is based on a combination of cross-sectional area and quality of concrete, a reduction below the expected value from any of the reasons is of concern. More so, the shaft model in the CAPWAP analyses was based on its anticipated (constructed) length. When the applied impact results with a high stress short duration stress wave, the stress reflection from the tip is clear and the shaft's length can be determined. Often in drop weight testing the produced stress wave is either not sharp enough (as in the presented tests), or the energy is not high enough to mobilize the shaft's tip and hence its clear detection. As such, the analysis utilized the designed length of 25 m. The above observation regarding the CFA shaft can in essence be interpreted that either the quality of the shaft in the lower 5 m was compromised, or the shaft was not constructed to the planned depth of 25 m.

To elucidate this situation, further analyses have been carried out on shafts B1, C1, C2, and C3. The dashed lines related to shafts C1, C2, and C3 in Figure 5 are the

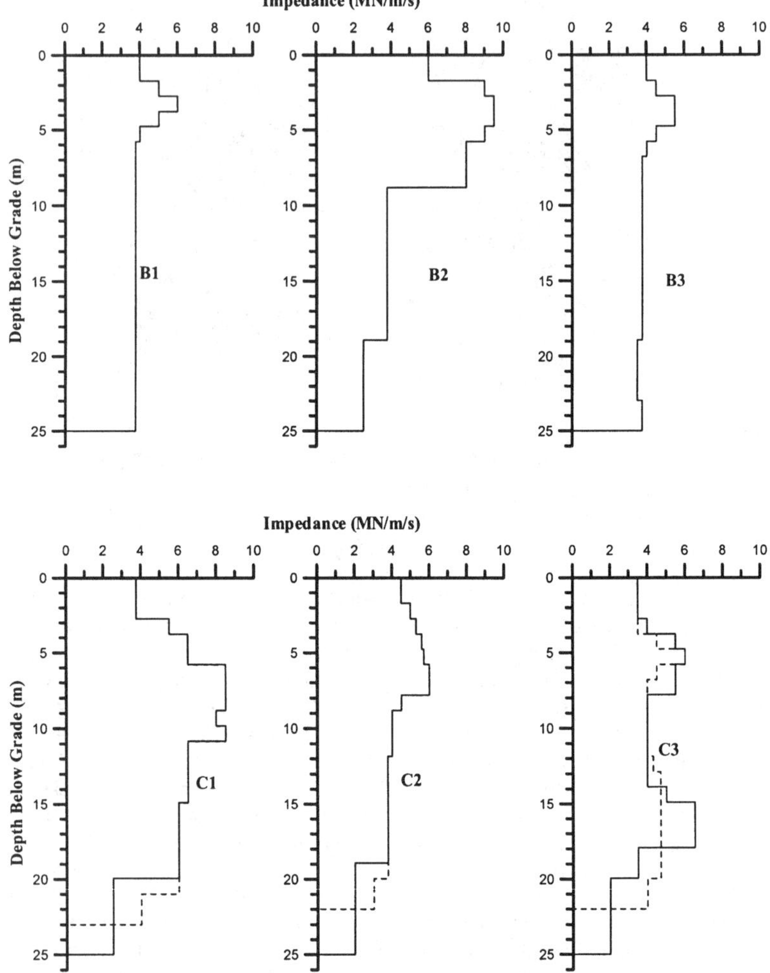

FIG. 5. Variation of Shaft Impedance with Depth for the Bentonite (B1 – B3) and CFA (C1 – C3) Constructed Shafts.

impedance used when assuming a shaft length below ground surface of only 22.0 m to 23.0 m. The revised analyses required small change in the capacity distribution and the use of the same soil parameters. The total resistance remained about the same (5,450, 6,045 and 5,030 kN vs. 5,450, 6,145, and 5,050 kN for C1, C2 and C3 in the new analysis vs. the original analysis, respectively), while the quality of the match obtained in the analyses increased. It was, therefore, concluded that the CFA shafts were constructed 22 – 23 m long, shorter than the planned length. To further examine the validity of this analysis, shaft B1 was reanalyzed assuming a shaft length of 22.0 m. In spite of many trials, the analysis resulted with a match of much lower quality affirming the inability to arbitrarily "shorten" a good quality shaft.

Pile Bearing Capacity

Pile bearing capacity was determined using the Case Method, Energy Approach, and CAPWAP procedures. Table 2 presents the capacities for the tested piles. The Case Method capacities (using the RMX procedure and a damping factor of 0.5), ranged between 4,005 kN and 5,563 kN. The use of the damping factor of Jc=0.5 for the predominantly granular subsurface provides value slightly conservative. The CAPWAP capacities varied between 5,050 and 6,145 kN. Table 3 present the results of the CAPWAP analyses in more detail. The total capacity, frictional capacity in the upper 5 meters, frictional capacity below a depth of 5 meters, end bearing (tip) capacity, and percentage of skin friction are included. The percent skin friction was consistently between 75 and 90%. Comparison of the dynamic prediction values of impacted drilled shafts to actual static load test results was presented by Paikowsky et al. (2003), and for driven piles under restrike conditions by Paikowsky and Stenerson (2000) and Paikowsky (2002). Both studies suggested high accuracy of this analysis for both cast in place piles (bias = 1.05, COV = 0.12, n = 39) and driven piles under restrike conditions (bias=1.16, COV = 0.34, n = 162). The bias in both cases represents the static capacity over the dynamic predicted value. Cannon (2000) presented a comparison between dynamic and static load tests on CFA shafts. His results suggested very good correlation between the two, including high accuracy of the modeled pile in the dynamic analysis, which matched well the recorded volume increase of 205% of the nominal design.

Figure 6 presents the distribution of the friction and accumulated load along the shaft including the tip resistance. The dashed lines in the distribution of shafts C1, C2, and C3 represent the revised length analysis discussed in the previous section. It can be concluded that within the accuracy of the differentiation (between the bearing components), a similar soil-pile interaction was observed for all piles regardless of the construction method. Inspecting the friction distribution along the shaft and the build up of the resistance (load) along it suggests, however, that the piles constructed with bentonite exhibit overall a much better distribution all along it, while the CFA constructed shafts exhibit high friction in the upper 12 to 14 m but very low frictional resistance below this depth. For one, this lack of friction and lower end bearing in the lower part affirms the aforementioned conclusions that the CFA shafts were constructed shorter than designed. Referring to Figure 1, it is possible that the shorter CFA shafts either ended in the clay layer or just about penetrated through it.

The results also suggests high soil mobilization around the areas of larger impedance, which can be associated with "bulging" out zones and/or some interaction between the shaft and the slab (see comments in Table 1).

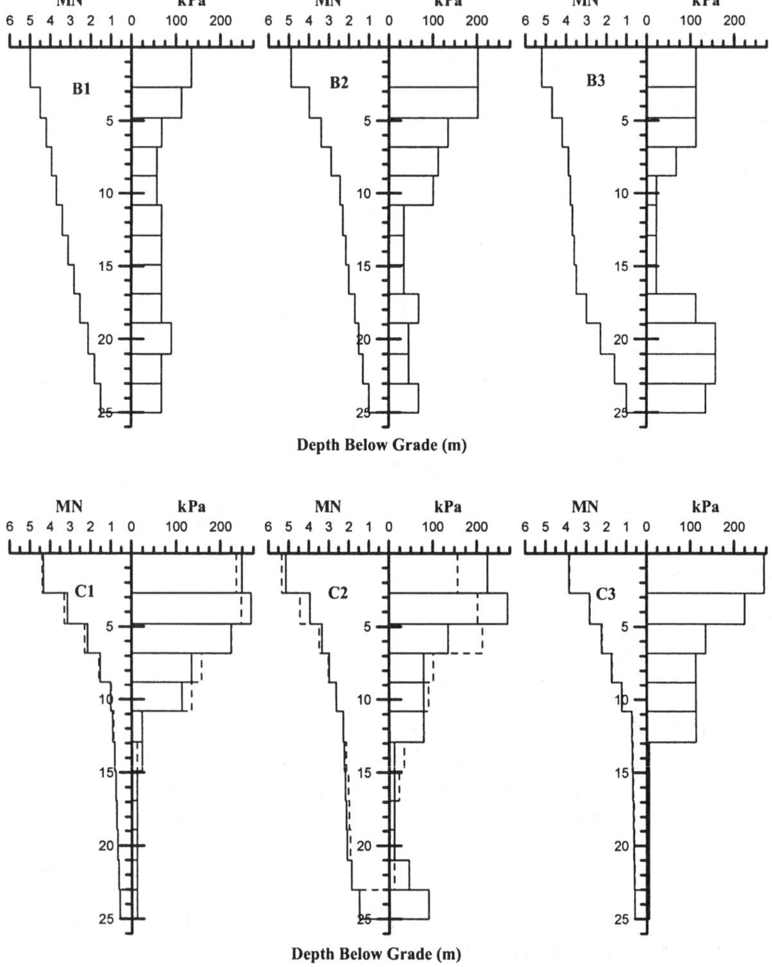

FIG. 6. Distribution of Friction and Accumulated Load Along the Shaft (Including Tip Resistance) for the Bentonite (B1 – B2) and CFA (C1 – C3) Constructed Shafts.

The Energy Approach method was developed by GTR personnel (Paikowsky, 1982, Paikowsky and Chernauskas, 1992, and Paikowsky and Stenerson, 2000) and was proven to provide high accuracy long-term pile capacity when applied to driven piles at the end of driving. This method relies on the relationship between the transferred energy and the work done by the pile during penetration. Measured values are used to determine the capacity (transferred energy, maximum pile displacement, and set). The method though not ideal for cast in place piles provides an indication for the upper end of the pile capacity in such cases. The Energy Approach capacities ranged between approximately 7,000 and 15,000 kN. In general, they were consistently higher than the Case method and CAPWAP capacities.

CONCLUSIONS

The presented data from the drop weight testing and the analyses leads to the following conclusions:

i. The predicted shaft capacities range between 5,000 and 6,000 kN (500 and 600 tons), thereby did not appear to be significantly different for either shaft type.
ii. The CFA shafts apparently developed more frictional resistance in the upper portions of the pile, while the bentonite slurry piles developed more evenly distributed friction. The percentage of tip resistance to the total capacity was lower for the CFA piles, but within a similar range for either pile type.
iii. Based on the CAWAP analyses, the unit skin friction over the lower 20 meters of the bentonite slurry shafts is around 75 kN/m^2 as compared to 55 kN/m^2 for the CFA piles. The unit end bearing resistance averaged around 2,500 kN/m^2.
iv. The shaft profile (based on CAPWAP) was more uniform for the bentonite slurry piles than for the CFA piles. The profile of the shaft impedance is based on material stiffness hence comprised of the combination of cross sectional area and quality of concrete. The CFA shafts were most likely constructed to a depth of 22 to 23 m only. In general, differences between the two pile types are expected, due to the different construction and concrete placement procedures.
v. This study appears to show that the tested bentonite slurry shaft construction resulted with a higher quality deep foundation element than that constructed by the CFA method. Considering, however, that the CFA shafts were shorter than designed, one may conclude that the importance of the quality of the construction and its monitoring are more significant than the specific construction technology. In other words, the lower performance of the CFA shafts may not be a testimony for a lower quality product, but rather to a lower quality craftsmanship.

ACKNOWLEDGMENT

The authors greatly appreciate the cooperation and support provided by Engineer Rafi Geisenberg, the project manager and Engineer Shaul Rosenfeld the site engineer. The collaboration of both contractors is acknowledged. The dynamic measurements were performed by the Technion Laboratory and the analyses were carried out by Geosciences Testing and Research of North Chelmsford, MA for Geodynamica of 27 Habanim Street, Herzelia, Israel.

REFERENCES

Cannon, J.G. (2000). "The Application of High Strain Dynamic Pile Testing to Screwed Steel Piles." *Sixth International Conference on the Application of Stress-Wave Theory to Piles*, S. Niyama and J. Beim eds., September 11-13, São Paulo, Brazil, pp. 393-398.

GTR. (1997). "Dynamic Pile Testing Report - Israel Electric Company, Haifa, Israel." Project 97.118, Geosciences Testing and Research, Inc., North Chelmsford, Massachusetts.

Jianren, D., and Shihong, Z. (1992). "The appraisal of results from PDA high strain dynamic tests on large and long drilled piles." *Fourth International Conference on the Application of Stress-Wave Theory to Piles*, F.B.J. Barends ed., September 21-24, 1992, The Hague, The Netherlands, pp. 271-278.

Paikowsky, S.G. (1982). "Use of Dynamic Measurements to Predict Pile Capacity Under Local Conditions." M.Sc. Thesis, Department of Civil Engineering, Technion-Israel Institute of Technology.

Paikowsky, S.G. (2002). "Load and Resistance Factor Design (LRFD) for Deep Foundations", Keynote lecture in the *Proceedings of Foundation Design Codes and Soil Investigation in View of International Harmonization and Performance*, Honjo et al. eds., IWS Kamakura 2002, Tokyo Japan, April 10-12, 2002, pp. 59-94.

Paikowsky, S.G. with contributions by Brown, D.A., Shi, L., Operstein, L., Mullins, G., Li, Z., Gorczyca, J.L., and Chernauskas, L. (2003). "Innovative Load Testin-g Systems". Final Research Report submitted to the National Cooperative Highway Research Program on Project NCHRP 24-08.

Paikowsky, S.G., and Chernauskas, L.R. (1992). "Energy Approach for Capacity Evaluation of Driven Piles." *Fourth International Conference on the Application of Stress-Wave Theory to Piles*, F.B.J. Barends ed., September 21-24, 1992, The Hague, The Netherlands, pp. 595-601.

Paikowsky, S.G., and Stenersen, K.L. (2000). "The Performance of the Dynamic Methods, their Controlling Parameters and Deep Foundation Specifications." Keynote lecture in the *Proceeding of the Sixth International Conference on the Application of Stress-Wave Theory to Piles*, S. Niyama and J. Beim eds., September 11- 13, 2000, São Paulo, BRAZIL, pp.281-304.

Rausche, F. and Seidel, J. (1984). "Design and performance of dynamic tests of large diameter drilled shafts." *Proceedings, Second International Conference on the Application of Stress-Wave Theory to Piles*, G. Holm, H. Bredenberg, and C-J Gravare eds., May 27-30, Stockholm, Sweden, pp. 9-16.

Townsend, F., Theos, J.F., and Sheilds, M.D. (1991). "Dynamic Load Testing of Drilled Shaft - Final Report." University of Florida, Department of Civil Engineering, Gainesville, Florida.

CASE HISTORY: MICROPILE USE FOR TEMPORARY EXCAVATION SUPPORT

Paul R. Macklin[1], PE, Member, ASCE, Donald Berger[2], PE, Member, ASCE, William Zietlow[3], PE, Wayne Herring[4], and Joshua Cullen[5]

ABSTRACT: Highway construction projects in the mountains typically present access challenges and slope stability risks during construction. Long narrow working corridors restrict activities complicating construction staging and lengthening schedules. On a recent Colorado Department of Transportation (CDOT) project near Aspen, Colorado, a micropile shoring system was built in-lieu of temporary tieback walls to improve on-site workability. The micropile shoring system consisted of approximately 7,400 square meters of retained face along 1,600 linear meters of wall with a maximum-shored height of 7.5 meters and an average height of 4.5 meters. Micropile shoring design methodologies are presented and compared to field measurements of wall performance collected from inclinometers and strain gauges.

INTRODUCTION

In the spring of 2000, the Colorado Department of Transportation awarded a $93.4 million dollar construction project to improve approximately 24 km of State Highway 82. Much of the project occurs along the Roaring Fork River on the western wall of Snowmass Canyon and ends approximately 1.5 km north of Aspen, Colorado, in Pitkin County. Temporary shoring was required for the majority of the up valley (southbound) roadway alignment. Excavations extending 1.5 m to 10.6 m were required to construct the CDOT designed roadway. The soils generally consisted of loose to very dense silty to gravelly sand with cobbles and boulders deposited as debris flows, sheet wash and colluvium over dense alluvial sandy gravel with cobbles

[1] Geotechnical Eng., Yenter Companies, Inc., 20300 West Highway 72, Arvada, CO 80007
[2] Geotechnical Eng., Yenter Companies, Inc., 1512 Grand Ave., Suite #104, Glenwood Springs, CO 81601
[3] Geotechnical Eng., Yeh & Associates, 2910 Tejon Street, Englewood, CO 80110
[4] Geotechnical Eng., ARM Group Inc., 1129 West Governor Road, Hershey, PA 17033
[5] Engineer, Colorado Department of Transportation, 25011 SH82, Snowmass, CO 81654

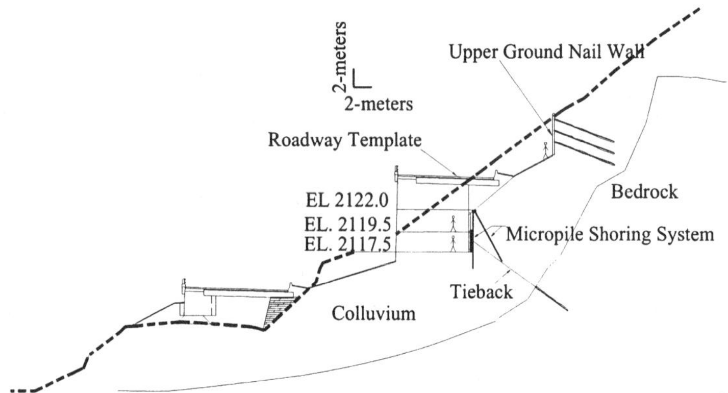

Fig. 1. Cross-Section Showing Steep Canyon Slope, Upper Ground Nail Wall and Temporary Micropile Shoring.

(Cogorno & Yeh 2000). To prevent distress of the finished roadway due to potential differential settlement within the debris flows, sheet wash and colluvium, the contract required removing all of the colluvial soils underlying the proposed roadway and recompacting these soils to 95% of the Modified Proctor maximum dry density. Default, temporary shoring, consisting of combinations of soil nail and tieback designs were included in the construction bid documents intended for use by the contractor. However, in an effort to improve the construction schedule and phasing, an alternate temporary shoring system, substituting micropiles for the default shoring systems, was eventually constructed.

The micropile shoring implemented on this project was essentially a hybrid between a soldier pile with lagging system and a soil nail stabilization system. Micropiles were drilled vertically down from the ground surface created by the pioneered roadway on a 455 mm center-to-center spacing along the desired temporary excavation alignment. Micropiles battered at 30 degrees from vertical (angled back into the hillside) were installed behind the front row on a 910 mm center-to-center spacing. The micropiles consisted of a centralized 32 mm diameter Grade 500 steel threadbar in a 133 mm diameter tremie grouted hole. A grade beam cast along the entire wall length connected the micropiles. The vertical micropiles provide the excavation limits and restraint of the soil mass between the micropiles while the battered piles resist the bending and overturning moments imposed by the soil mass behind the micropile shoring system. The grade beam rigidly connects all of the micropiles together.

Phasing of the project required constructing a number of bridges in a specific sequence. The micropile shoring system allowed the general contractor to transition access road grades where convenient, depending upon the excavation and access requirements for any phase of the work. An additional benefit was detection and subsequent profiling of the debris flow/colluvium thickness in the closely spaced vertical boreholes prior to actual production excavation.

In a few specific areas, CDOT directed the installation of a single row of temporary tiebacks, spaced at 2.5 m intervals, in front of the micropile shoring. The tiebacks were added in those areas where the colluvium extended deeper than anticipated requiring an extended excavation. These tiebacks improved the deep-seated global stability concern associated with a critical seismic event estimated to have a 5% probability of exceeding a 0.16 g peak horizontal ground acceleration in 50 years (Cogorno & Yeh 2000). Although the majority of the micropile shoring did not require tiebacks, the instrumented section that provided the performance data for this paper did have a tieback installed and tensioned approximately 3.7 m from the top of the micropile grade beam. However, prior to tensioning the tiebacks the micropiles maintained the 4.9 m excavation for approximately 6-weeks.

DESIGN APPROACHES

Opinions on the design of micropile shoring systems vary (eg Ueblacker 1996; Pearlman and Withiam 1992; Armour 1997; and Bruce and Juran 1997). No standardized design guideline was found that recommends a step-by-step analytical procedure. The lack of consensus in the geotechnical design community on how to handle micropile slope stabilization systems is evidenced by the fact that the first edition of the Federal Highway Administration micropile design manual has an incomplete section on the design of micropiles for slope stabilization and earth retention (Armour 2000).

Prior to this project, the shoring contractor had designed and constructed many micropile shoring walls in the mountains of Colorado using a system that consisted of one row of vertical micropiles at approximately 450 mm spacing with one or several rows of piles battered at angles ranging from 15 degrees to 30 degrees from vertical on a 900 mm center-to-center spacing. The micropiles generally consisted of a centralized 32 mm diameter, Grade 500, steel threadbar in a 90 to 130 mm diameter tremie grouted hole. The walls typically ranged in height from 3 to 6 meters. Most of these systems were designed assuming that the portion of the soil above the excavation line and between the vertical and battered micropiles behaves as a gravity structure. Those portions of the micropiles that extend into stable soil (below excavation depth for vertical micropile and beyond angle of repose for battered micropile) provide additional resistance to overturning and sliding.

For the State Highway 82 temporary micropile shoring a limit equilibrium analysis using a combination of gravity wall calculations, free earth support methods, lateral pile and slope stability analyses were used to evaluate the internal and external stability of the micropile shoring system. Also, finite element and finite difference models were used to predict deformation behavior, stresses in the micropiles and to analyze the potential for soil flow-around failure of the micropile shoring system. The design approach for this project is described in the following six steps.

1. Determine the structural capacity of typical, familiar micropile sections used on previous projects. Several combinations of centralized reinforcing bars and/or Schedule 40 or Schedule 80 steel pipe were analyzed for their ultimate bending capacity and flexural rigidity (EI). The flexural rigidity (EI) values were later used in the finite element and finite difference analyses. Micropiles

that consisted of a centralized reinforcing bar in a drilled and grouted hole were analyzed using LPILE's Ultimate Bending Analysis module, as suggested by Armour (1997). LPILE was used to determine the ultimate bending capacity of the cracked section. Micropiles that included a pipe section were analyzed for moment capacity using the strength of the steel, only. The compressive strength of the grout was ignored in this case. The tension capacity of each section was calculated using the factored tensile strength of the reinforcing bar, only. The compression capacity of each micropile was calculated as the sum of the factored compressive strength of the grout, bar and pipe, where applicable.

2. Determine the appropriate micropile sections, spacing and batter using a simplified method that treats the micropile wall as a gravity retaining wall. Earth pressures were calculated using classical earth pressure theories, assuming that the wall deformed sufficiently to allow the soil to reach an active state. Check horizontal sliding to make certain that the front micropiles provide a sufficient shear capacity and embedment depth. Sum moments about toe of micropile shoring wall and compute the required sectional capacity and anchorage length of the rear row of battered micropiles to safely resist the overturning moment with respect to yield and pullout failure. (The designer/contractor referred to this simplified method as the Y-Micro design method.)
3. Use a modified form of the free earth support (FES) method of anchored bulkhead design to check the adequacy of the design spacing and sections to resist the earth pressures applied. For design of a micropile system using the FES method, the front row of closely spaced micropiles are analogous to the sheet pile and the battered rows of micropiles are analogous to the anchors. Use of the FES method, in its classical form, for micropile shoring systems would assume that the battered piles assist only by providing a lateral restraining force at the top of the vertical pile similar to a "deadman" anchor. All bending moment would therefore be applied to the front, vertical micropile. This FES assumption is overly-conservative for the micropile system as the battered piles are closely spaced and it is likely that they provide soil reinforcing in a manner similar to soil nails The modified FES analysis assumed that one half of the calculated active earth load was applied to the vertical micropile row as a triangular pressure distribution and one half of the calculated active earth load was applied to the rear, battered micropile row. A lateral pile analysis was then completed for both the vertical and battered micropiles, using the LPILE software package.
4. After determining that the micropile shoring system is capable of safely resisting the applied earth loading, the global stability is checked. The resultant active earth pressure force was modeled in the slope stability analysis as an external line load acting horizontally at a location that is 1/3 of the height from the base of the excavation. The global stability of the section, with this added external force (which represents the calculated capacity of the micropile system) is then checked. The external line load is applied only to

those failure surfaces exiting at the base of the excavation and extending some distance behind the shoring.

5. Numerical analysis using simplified soil models was conducted to predict deformations in the structure. To better understand the complex interaction between the micropiles, soil, bedrock, soil nails and tiebacks proposed for the wall systems, a finite element model was developed using Plaxis software. Because of its simplicity, the Mohr-Coulomb elastic, perfectly plastic constitutive model was used to describe the behavior of the soil. Although more sophisticated constitutive models such as the hardening model would more accurately predict the soil response to excavation loading, they require more information about the elastic, plastic, and yield properties of the soil than were available. The models were developed to analyze the worst-case geometries and construction stages of the temporary excavation support for the walls.
6. Numerical analysis using simplified soil models was conducted to evaluate a potential soil flow around failure. A coupled analysis was used to assess the plastic flow between micropiles. In accordance with Ito and Matsui (1975) the pile diameter to spacing ratio was checked at various stress fields. The analyses indicated that the soil would arch between the micropiles. The cohesion and spacing were adjusted in the model to assess the sensitivity of the design to variability in the soil and construction methods. The results indicated, as the cohesion is reduced and the spacing is increased the potential for plastic flow between piles increases; however, complete loss of the soil arch did not occur at the stress fields observed within the analyses.

MICROPILE PERFORMANCE & PREDICTIONS

The data collected from the instrumentation program consisted of strain gauge data converted to bending and total stresses and inclinometer data reduced to resultant deflections. The data collected from pairs of strain gauges mounted on the vertical, micropile (32 mm threadbar) of the shoring system is compared to Plaxis and LPILE stress predictions at corresponding loading events. The strain gauge pairs were positioned at 0.8, 1.8, 3.1, 4.3 and 5.8 meters from the top of the micropile. Fig. 2 shows the relationship between the strain gauge locations, bending stress sign convention, micropile rotation and inclinometer deflection. Four loading events are described in Table 1 along with indications of the types of data, both measured and predicted, that are available for comparison. (The Plaxis and LPILE predictions presented in this paper were prepared and presented to the owner prior to construction of the micropiles.) Fig. 3 compares the stress predictions prepared for loading events 2 & 4 with the strain gauge measurements for all 4 loading events. Fig. 4 compares the deflections calculated from inclinometer data with the Plaxis and LPILE predictions.

The agreement between the measured and predicted bending stress distributions is very good at loading event 2 (4.9 m excavation) as indicated in Figs. 3a & 3b. The predicted values reach a maximum of about 240 MPa at the top of the micropile where a rotationally fixed boundary condition is assumed. The measured bending

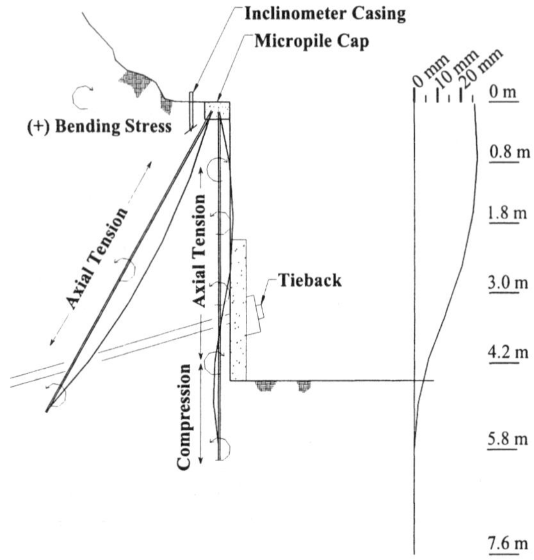

Fig. 2. Depiction of Micropiles in Bending based upon strain gauges and inclinometer data at 4.5 m excavation.

stresses reach a maximum of approximately 100 MPa at 0.8 m from the top of the micropile. The total stress maximum values (absolute values) are approximately 260 MPa and 210 MPa for predicted and measured data respectively.

The comparison is not as good between the measured and predicted data at loading event 4 (6.7 m excavation) as shown in Figs. 3c & 3d and Fig. 4. The effect of the tieback tensioning drastically unloaded the predicted stress distribution. Because the

Table 1. Description of the loading events and corresponding performance in the vertical micropiles, both measured and predicted plotted in Figs. 3 & 4.

Loading Event (1)	Description (2)	Measured Data (3)	Prediction (4)
1	Excavate to 2.4 meters	Strain Gauge	none
2	Extend excavation to 4.9 meters	Strain Gauge & Inclinometer	LPILE, PLAXIS
3	Tension tiebacks at bottom of the 4.9 meter excavation	Strain Gauge	none
4	After tensioning tiebacks, extend the excavation to 6.7 meters	Strain Gauge & Inclinometer	PLAXIS

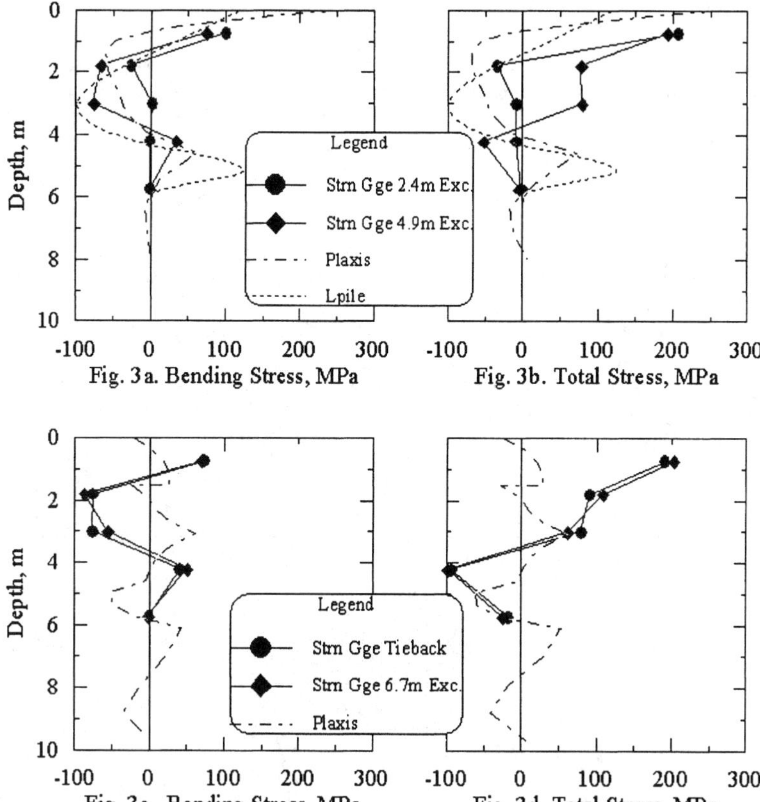

Fig. 3. Compares the measured bending and total stress distributions for the vertical (front row) micropiles with the Plaxis and LPILE predictions. Figures 3a & 3b compare data measured at excavation depths of 2.4 and 4.9 meters to the LPILE and Plaxis predictions. Figures 3c & 3d compare the measured data after post-tensioning the row of temporary tiebacks and then after excavating to 6.7 meters to the Plaxis predictions. Note: The "tieback" plot in Figures 3c & 3d is the distribution of stresses on the micropile (not the tieback) soon after the tieback was tensioned.

simplified elastic-perfectly plastic soil model has the same modulus for loading and unloading, applying the tieback load significantly unloads the micropiles and reduces the deformations. It should also be noted that because the predictions for axial stress on the micropile did not vary much along the micropile the shape and extent of the bending and total stress distribution does not change. However, the measured data does suggest that the axial stresses do vary along the micropile.

CONCLUSION

The measured data confirms that the methodology used for design was reasonable for this particular project. The LPILE analyses estimated the micropile performance reasonably well, supporting the method of simplifying and distributing loads between rows of micropiles as described in the paper. The FES method is more conservative than the soil-structure interaction analyses and as a result, the analyses presented here indicate that the designed micropile system behaves within the anticipated stress and deformation limits.

The use of micropile shoring expedited the construction process by eliminating the staged excavation and construction processes that is inherent in soil nail and/or tieback shoring systems.

The 1600 linear meters of micropile shoring performed well and provided a safe excavation.

Closely spaced tremi-grouted micropiles with centralized 32 mm reinforcing bars provided an effective shoring system for excavations that had maximum-shored heights of up to 4.9 meters without support from the added temporary tiebacks.

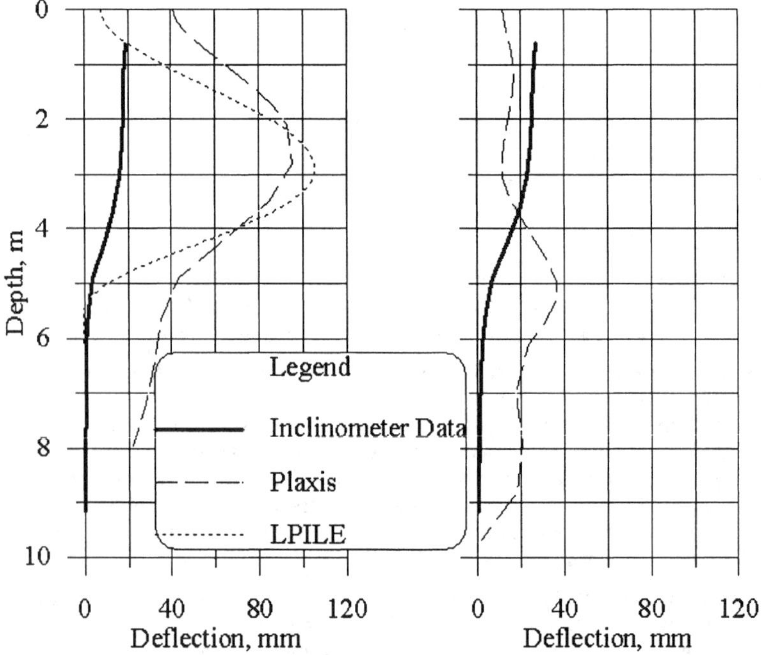

Fig. 4. Comparison between measured (inclinometer) and predicted (LPILE and Plaxis) deflections. The left graph is at the 4.9 m excavation stage and the right graph is after the end of construction (tieback tensioning and 6.7 m excavation.

REFERENCES

Allen, R.L, Duncan, J.M. and Sancio, R.T. (1987). "An Engineering Manual for Sheetpile Walls." Virginia Polytechnic Institute and State University.

Armour, T.A., (1997). "Design Methodology: Micropiles for Slope Stabilization and Earth Retention". *International Workshop on Micropiles,* Seattle, Washington, 329-411.

Armour, T.A., Groneck, P., Keeley, J. and Sharma, S., (2000). "Micropile – Design and Construction Guidelines". *Publication No. FHWA-SA-97-070*, Federal Highways Administration, US Dept. of Trans.

Brinkgreve, R.B.J, and Vermeer, P.A., (1998). "Plaxis ver. 7," *Users Manual*, Balkema, Rotterdam, Netherlands.

Bruce, D., and Juran, I., (1997). "Drilled and Grouted Micropiles, State-of-Practice Review." *Publication No. FHWA-RD-96-016*, Federal Highway Administration, US Dept. of Trans.

Cogorno, G. and Yeh, S.T., (2000). "Final Geotechnical Report for Earth Retaining Structures, SH 82 through Snowmass Canyon, Pitkin County, Colorado." *Consultant Report*, Yeh and Associates Inc.

"Foundations and Earth Structures." (1982). *Naval Facilities Engineering Command*, US Navy, Design Manual 7.02

Gabr, J.J., and Wang, M.Z., (1997). "Buckling of Piles with General Power Distribution of Lateral Subgrade Reaction." *J. of Geotech. and Geoenviron. Engrg.*, ASCE, 123(2).

Hassiotis, S., Chameau, J.L., and Gunaratne, M., (1997). "Design for Stabilization of Slopes with Piles." *J. of Geotech. and Geoenviron. Engrg.*, ASCE, 123(4).

Ito, T., and Matsui, T., (1977). "The Effects of Piles in a Row on the Slope Stability," Proceedings of the 9th International Conference on Soil Mechanics and Foundation Engineering, Tokyo, Japan.

Pearlman, S.L., Campbell, B.D., and Withiam, J.L., (1992). "Slope Stabilization Using In Situ Earth Reinforcements". *ASCE Specialty Conference On Stability and Performance of Slopes and Embankments - II*, June 29-July 1, Berkeley, CA, 16 p.

Porterfield, J., Cotton, D.M., and Byrne, R.J., (1994). "Soil Nail Inspectors Manual – Soil Nail Walls, Demonstration Project 103." *Publication No. FHWA-SA-93-068*, Federal Highways Administration, US Dept. of Trans.

Porterfield, J., Cotton, D.M., Byrne, R.J., Wolschlag, C. and Ueblacker, G., (1998). "Manual for Design and Construction Monitoring of Soil Nail Walls." *Publication No. FHWA-SA-96-069R,* Federal Highways Administration, US Dept. of Trans.

Reese, L., Wang, S.T., Arrellaga, J.A. and Hendrix, J., (1998). "Computer Program LPILE Plus Ver. 3." *User's Manual,* Ensoft Inc.

Ueblacker, G., (1996). "Portland Westside Lightrail Corridor Project Micropile Retaining Wall," *Foundation Drilling*, Association of Drilled Shaft Contractors, Dallas, November, pp. 8-12.

HIGH CAPACITY MICRO PILES FOR UTILITY RETROFIT: A CASE HISTORY AT D.E. KARN POWER PLANT IN BAY CITY, MICHIGAN

Timothy H. Bedenis[1], P.E., Member, ASCE,
Michael J. Thelen[2], PE, Member, ASCE, and Steve Maranowski[3]

ABSTRACT: The case history involves a major retrofit of the D. E. Karn power plant at the Consumers Energy's Karn/Weadock Generating Facility near Bay City, Michigan to meet current environmental requirements. The retrofit required the installation of heavily loaded columns above and between existing equipment, overhead piping, underground utilities and various structures without disruption to the operation of the plant. High capacity micropiles were proposed as the most practical alternative to meet the project budget and schedule requirements. Micropiles with a single high-strength steel bar and permanent upper steel casing (Type A) were installed in groups and extended into competent sandstone. An instrumented load test of a micropile was performed to determine the stress distribution in the rock socket and verify the design requirements. A total of 112 micropiles were installed in time for the steel erection to keep the project on schedule.

INTRODUCTION

Consumers Energy committed to a significant retrofit program for their older coal-burning power plants. The retrofit requires the installation of new equipment to improve the emissions from the plant. The required equipment is typically located within the existing plant complex at elevated levels. The retrofit at the D.E. Karn power plant required heavily loaded columns to be installed within relatively confined areas having limited headroom conditions.

[1] Principal, Soil and Materials Engineers, Inc. 43980 Plymouth Oaks Blvd., Plymouth, Michigan 48170, USA., bedenis@SME-USA.com
[2] Associate, , Soil and Materials Engineers, Inc., 2663 Eaton Rapids Road, Lansing, Michigan, USA, 48911, thelen@sme-usa.com
[3] President, Spartan Specialties, Ltd., Sterling Heights, Michigan , USA

The existing plant is supported on driven concrete filled pipe piles bearing on rock. Pipe piles were also originally planned to support retrofit structures. However, the limited space and headroom precluded driven piles with conventional equipment. Concrete piers were also considered, but the piers would also require special low headroom drilling equipment or would have to be hand excavated. The general contractor determined driven piles and concrete piers would be too costly to construct and would take too long to install. The owner was on tight schedule, and foundations had to be completed when the steel from the fabricator was due to arrive at the site.

Soil and Materials Engineers, Inc. (SME) and Spartan Specialties Ltd. proposed using high capacity micropiles as an alternative to the 610 mm driven pipe piles identified on the project documents. A high capacity single bar micropile was designed to reduce the total number of piles and to use the same pile cap design. An added benefit to using smaller diameter micropiles (nominal diameter of 152 mm) was greater flexibility to locate piles around existing abandoned piles and numerous known but uncharted utilities.

Based on the cost and schedule advantages, micropiles were selected as the most viable foundation for the project.

GEOLOGIC CONDITONS

The power plant is located in a glacial lacustrine deposit that has been previously inundated by water from Saginaw Bay of Lake Huron. Previous borings performed for the original construction of the power plant indicate the soils at the site consist of 1.8 meters to 2.7 meters of sand fill over layers of organic soils (silts and peats) that are underlain by a thick deposit of silty clays. The upper 15 meters of the clay deposit is very stiff to hard in consistency with the clays below this level becoming medium in consistency. A clastic sandstone formation exists about 23 to 24 meters below the ground surface. The previous soil borings indicate the sandstone formation is slightly weathered to fresh with infrequent joints.

SME performed an additional soil boring to determine the hardness of the sandstone, which was needed for design. The boring extended through the overburden soils and encountered sandstone at 24.7 meters below site grade level. A 3-meter core achieved 100 percent recovery of sandstone with a Rock Quality Designation (RQD) of 83 percent. Uniaxial compressive strength tests were performed on select sandstone core samples and results ranged from 25.5 MPa to 31.0 MPa. Groundwater was perched above the clay deposit and was encountered at a depth of about 1.5 meter below the ground surface.

MICRO-PILE DESIGN

The micropiles were spaced to accommodate the pile caps designed for the original drive pipe pile alternative. About half of the pile caps were rectangular to avoid adjacent structures and foundations. Square pile caps were used where there was sufficient space. Battered (angled) micropiles were designed to resist a majority of the anticipated lateral loads. However, conflicts with adjacent structures limited the use of battered piles beneath some pile caps. At these locations a series of grade beams was used to transfer anticipated lateral loads to pile caps where battered piles could be installed without interference.

Design Loads

Dead loads, live loads, wind loads and seismic loads were considered. The governing load combination resulted in a design column base vertical (compressive) loads ranging from 3550 kN to 9150 kN and design column base lateral loads ranging from 220 kN to 440 kN at some of the perimeter column locations.

Design Criteria for Micro Piles

The design of the micropiles was based on the criteria presented in FHWA's Micropile Design and Construction Guidelines (2000). More specifically, the design criteria used is summarized as follows:

- Type A Micro-Piles with Single Corrosion Protection.
- Service Load Design Method
- Use of a single high strength steel bar.
- No strength contribution from the grout.
- All load supported by the side friction in the rock socket.

Type A micropiles (gravity placed grout) were selected since the design accounted for bond strength in the rock. Full length DYWIDAG, ASTM A722 threaded bars with couplers were used to maximize the load capacity, facilitate installation, and reduce the number of micropiles required. The maximum strain of cement grout is not compatible with high yield capacity of ASTM A772 steel. Therefore, the strength of the grout was not used in the design. However, the grout provides lateral support and corrosion protection for the bars. Further corrosion protection was not needed based on the ground condition's low potential for soil corrosion. An upper permanent casing was used to provide protection and lateral resistance to the top of the pile.

The friction in the upper clay could be affected by the installation procedure. Therefore, only the side friction in the rock was used for the design. End bearing in the rock was also neglected due to the relatively small bearing area. Any additional resistance from the clay or end bearing would provide an additional factor of safety.

Bond Strength in Rock

There have been several empirical studies to determine the bond strength (friction) between rock and concrete based on the uniaxial compression strength of rock. Three of the more prominent studies were published by Horvath & Kenny (1979), Carter & Kalhawy (1988), and Rowe & Armitage (1987).

$$f_s = 0.5\sqrt{Q_u} \qquad Q_u \text{ (psi)} \qquad \text{(Horvath \& Kenny)} \qquad (1)$$

$$f_s/p_a = (0.63 \text{ to } 0.95)\sqrt{\frac{Q_u}{p_a}} \qquad \text{(Carter \& Kalhawy)} \qquad (2)$$

$$f_s = 0.45\sqrt{Q_u} \qquad Q_u \text{ (MPa)} \qquad \text{(Rowe \& Armitage)} \qquad (3)$$

where: Q_u = Uniaxial Compressive Strength, and
Pa = Atmospheric Pressure

The above studies were based on drilled shafts (caissons) socketed into soft rock. For drilled shaft construction, the lateral pressure on the sides of the rock sockets would be equal to the hydrostatic fluid weight of the concrete just after placement, which should also be similar for Type A micropiles. The uniaxial compression strength of the rock cores tested ranged from about 25.5 MPa to 31.0 MPa, which is considered a soft rock. Therefore, these relationships were judged to be valid and were used to determine the design side friction for the micropiles. Substituting the uniaxial compressive strength test results into Eqs. 1 through 3 results in bond strength values between about 1000 to 2300 kPa (refer to Fig. 1).

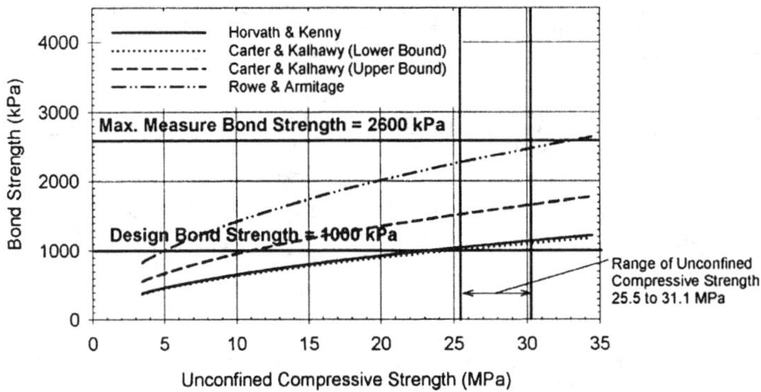

FIG 1. Bond Strength vs. Unconfined Compresssive Strength

It should be noted the above relationships are based on empirical tests on shafts with variable roughness on the sidewalls of the rock sockets. The side friction is very dependent on the roughness of the socket. For rough walls in relatively deep shafts, the side friction is much higher and can approach the shear strength of the rock or roughly one-half of the unconfined compression strength, Hassan and O'Niell (1997). However, much higher strains are required to mobilize the higher ultimate frictions for rough sockets than for smooth wall sockets. Since the roughness of the socket for the micropile would be difficult to determine, a lower bound side friction of 1000 kPa was chosen to design the micropiles, which is within the typical range published by FHWA of nominal bond strengths for Type A piles in sandstone.

Rock Socket Design

To simplify the design, the vertical compression loads were placed in 3 groups; heavily, moderately and lightly loaded. The maximum pile load in each group type was calculated based on the number of vertical piles used in each group. Battered piles were added to specific pile caps to provide resistance to the lateral loads. The number of piles per group was limited by the predesigned pile cap sizes. A minimum edge-to-edge spacing of 1 meter was also used to position the piles within the pile caps.

The length of the bonded zone for each loading group was calculated using the service load design (SLD) method provided by FHWA. A minimum factor-of-safety equal to 2.0 was applied to the design bond strength based on SME's previous load test results on various micropiles installed in soft sedimentary rocks. The calculated socket length was increased by 20 percent to account for any potential anomalies in the rock. Table 1 shows the design results.

TABLE No. 1 Pile Loads, Socket Lengths and Bar Sizes

Pile Group Type (1)	Max. Comp. Load (kN) (2)	No. of Piles (3)	Max Pile Load (kN) (4)	Length of Rock Socket (m) (5)	Bar Size (mm) (6)
A	3550	5	710	4.6	44
B	4730	4	1180	6.7	57
C	9150	6	1530	8.5	63
Battered	-	-	400	3.0	32

Structural Capacity

The required size (SLD Method) of the threaded steel bar is based on the following equation from the FHWA Manual.

$$P_{c\text{-}allowable} = 0.40f'_{c\text{-}grout}*Area_{grout} + 0.47F_{y\text{-}bar}*Area_{bar} \quad (4)$$

Since a casing was not used for structural support and since the strength of the grout was neglected, the above equation can be reduced to the following:

$$P_{c\text{-}allowable} = 0.47F_{y\text{-}bar}*Area_{bar} \quad (5)$$

As shown in equation (5), the required area of the bar is only dependent on the design load and the yield stress of the bar. The high strength DYWIDAG bars are produced from a standard smooth stock bar by cold froming the threads. Samples of standard smooth stock were tested for tensile capacity. A sample of the threaded bar delivered to the site was also tensile tested to determine the effective yeild stress. The results of the tensile test are shown in Table 2.

TABLE No.2 Tensile Test Results

SampleType (1)	Test Diameter (mm) (2)	Test Area (mm^2) (3)	Yield Load (kN) (4)	Failure Load (kN) (5)	Design Area (mm^2) (6)	Effective Yield Stress (MPa) (7)
Stock Bar A	47	1740	1700	1920	1450	1170
Stock Bar B	47	1740	1660	1890	1450	1150
Threaded Bar	44	1520	1490	1780	1450	1030

The diameter of the stock bars is larger than the final threaded bar due to the formation of the threads. However, using the nominal diameter size of 44 mm (as referenced by DYWIDAG), a minimum effective yield stress of 1030 kPa was determined for the threaded bar sample. Using the effective yield stress of 1030 MPa for the steel bars, the bar sizes shown in Table 1 resulted for each pile group type.

The grout coverage for the largest bar size is 44 mm (edge to edge) assuming the bar is centered in the borehole. This is well above the minimum coverage of 20 to 30 mm recommended by the FHWA manual.

Settlement and Lateral Deflection

The capacity of the socket is expected to occur at relatively low strains. Therefore, very little movement is expected in rock socket and the vertical deflection (settlement) is primarily based on the compressive stiffness (AE/L) of the vertical piles. Using this stiffness with an effective length (L) equal to the distance from the bottom of the cap to the top of the rock, the compression of the micropile under design loads range from 21 to 33 mm. These deflections are considered to be upper bound values since the actual effective length is probably shorter due to the side friction in the upper clays. Nevertheless, since the adjacent columns are similarly loaded, the differential settlement between columns is expected to be less than 6 to 10 mm.

Pile to Cap Connection

The pile to cap connection was made by placing two thick plates at the top of the threaded bar with nuts on the bottom and top of the plates. The plates were 57 to 63 mm thick. An upper square plate had a width of 300 mm and the lower square plate had a width of 230 mm. The stacked plate configuration was used to resist the high bending moments rather than a single very thick plate. The embedment and concrete cover over and adjacent to the plates where checked for shear capacity. Refer to Fig. 2 for a summary for the micro-pile design.

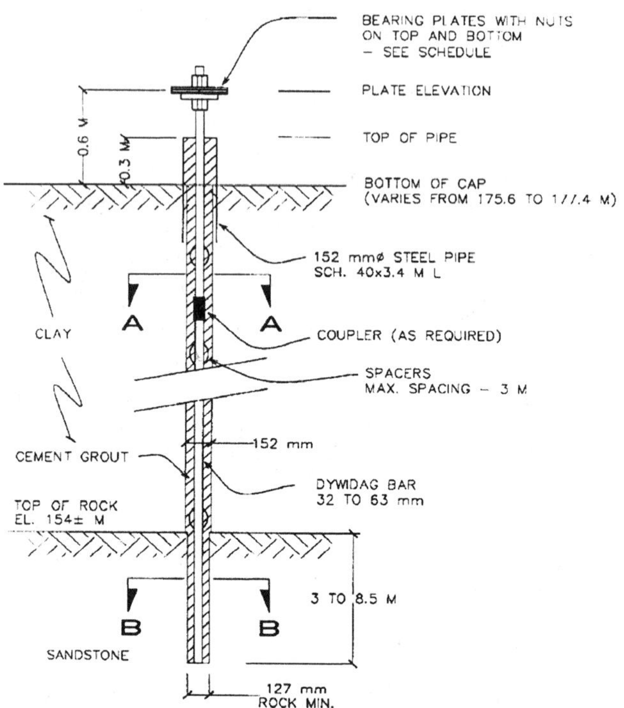

FIG 2. Micro-Pile Design

MICRO-PILE INSTALLATION

Specialty geotechnical contractor, Spartan Specialties, Ltd. of Sterling Heights, Michigan, used a compact high torque drill rig to install the micropiles within the physical constraints of low overhead and limited access, (see Fig 3). The pile caps were constructed below the existing grade. At most pile cap locations, the drill rig was positioned over the pile cap excavation on rails and/or steel plates. The drill rig was equipped with a duplex drilling system that allows a drill bit on a drill string and a casing to be advanced either separately or concurrently. In most cases, casing was not required beyond a depth of 4.6 to 6.1 m. The drilling could be performed vertically or on an angle (see Fig 4). Water was flushed continuously through the drill head and casing, which in turn washed the cuttings out away from the tips.

FIG 3: Compact Drill Rig

FIG 4. Drilling on a Batter

Once the borehole was completed, water was used to flush and clean the rock socket until the water was observed to be relatively free of soil. The bars were then inserted into the borehole with PVC spacers used to center the steel bar in the borehole. Various lengths of bars were used depending on the specific overhead constraint at each pile location. Couplers specifically made for the DYWIDAG bar where used to join the bars together.

FIG. 5. Grouting of Micro-Pile

FIG. 6. Completed Micro-Pile Group

After the full length of the steel bars was positioned, a tremie tube was placed in the annular space of the bore hole (see Fig. 5). Grout was introduced through the tremie pipe under gravity head displacing the water in the borehole from the bottom up. A neat cement grout mix (cement and water with no sand or admixtures) with a water to cement ratio of 0.5 was used. The grout strength required by the design was 34 MPa. Compression testing on the grout indicated unconfined strengths of over 40 to 50 MPa at 7 to 14 days.

Grout was placed through the tremie pipe until a clean return was observed. After grouting, the temporary steel-drilling casing was extracted. Then, a 3.4 m long permanent steel casing was installed with a stick-up of 0.3 meters above the bottom of the cap. The grout was placed to the top of casing with the bar extending above the casing. After the grout had set for a minimum of 24 hours, the bar was cut-off at the design elevation and nuts and plates installed to complete the installation (see Fig. 6).

VERIFICATION OF THE BOND STRENGTH

To verify the design bond strength in the rock, a load test was performed on one of the production piles. Since the piles are designed for friction only, an uplift test was performed to eliminate the end-bearing component. The tensile capacity of piles is typically less than in compression due to the "Poisson effect", i.e., the tendency for the cross sectional area to decrease under an axial tensile stress. However, the Poisson effect is much more pronounced for large cross sections (such as for a drilled shaft), and is also conservative for this application.

The test pile had a 3.1 m long, 127 mm diameter rock socket. A No. 14 (44 mm in diameter) threaded bar was installed in the test pile. Seven (7) spot weldable, vibrating wire strain gages were attached to the test pile steel bar at roughly equal distances along the socket. The test pile also included a bond braker within the overburden soils. The bond breaker consisted of a 22 m long, 152 mm diameter PVC pipe coated with grease. The purpose of the bond breaker was to reduce the side friction in the overburden soils, allowing a majority of the load to reach the socket.

An adjacent production pile and one sacrificial pile were used as reaction piles. A reaction beam connected the reaction pile and test pile and supported the hydraulic jack used to apply the test load. An electronic load cell was used to directly measure the applied load to the test pile. Electronic gauges were used to measure the deflection at the top of the test pile and reaction piles. All the instrumentation was connected to a Digitmate® dataloger which was connected to portable laptop computer. The dataloger processed and stored the information from the instruments. The computer provided quick checks of the data to determine if any significant errors occurred during loading. A schematic diagram of the load test set-up is shown in Fig. 7.

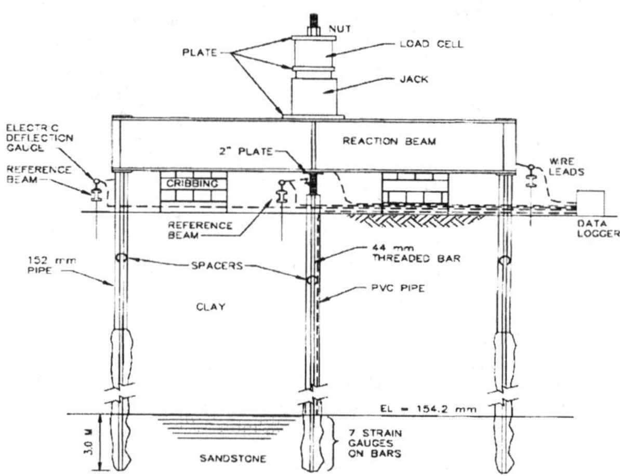

FIG 7. Load Test Set-Up

The load test was performed in general accordance with ASTM D 3689 using the "Quick Load Test" procedure. The test pile was loaded to 1330 kN, while strain measurements in the socket were obtained. The results of the load test are shown on Figs. 8 through 10. Fig. 8 shows the applied load versus measured movements at the test pile and the reaction piles, along with the theoretical elastic deflection line for the No. 14 bar. For comparison purposes, the elastic line for the composite section of steel bar and grout is also shown. The line for the test pile corresponded reasonably well with the elastic deformation (expansion) of the threaded bar while the line for composite section compares favorably with the deformation (compression) of the reaction piles. This is consistent with the expected behavior, where grout would have little effect in tension and a much greater affect on the deformation in compression.

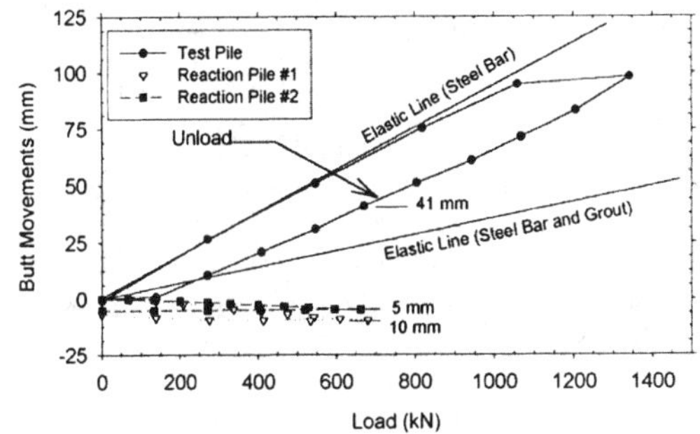

FIG. 8. Test Load vs. Movement

The loads at various elevations along the socket were calculated based on the measured strains (from the strain gauges), the estimated cross sectional areas, and the modulus of elasticity for the pile section. Fig. 9 presents the load distribution within the bar from the jack to the pile tip for each load and unload increment. Fig. 10 presents the average developed bond strength between strain gauge locations in the socket at various intervals based on the load distribution shown in Fig. 9.

As shown in Fig. 9, the maximum load transferred directly to the top of the rock socket is significantly less than the test load at the jack level. Therefore, it appears that even with the bond breaker there is significant side friction in the upper clays. Also, it is apparent from the load distribution within the socket that the side friction is mobilized only within the upper portion of the socket. At the maximum test load, the load is completely resisted by the side friction within the upper 0.5 to 1.0 m of the 3.1 m long socket. The maximum average developed bond strength was determined to be as much as about 2600 kPa. This value is 2.6 times higher than the nominal bond strength selected for design and exceeds the nominal values published by FHWA for Type A piles in sandstone. As can be seen in Fig. 1, the measured bond strength compares more favorable to the empirical correlations by Rowe & Armitage than from the lower values by Horvath & Kenny or Carter and Kalhawy.

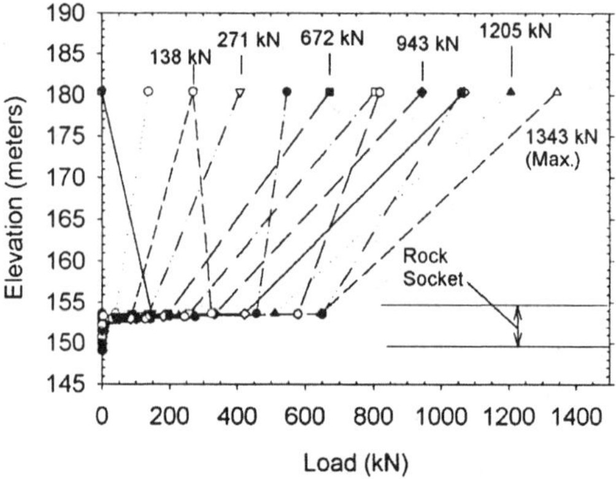

FIG. 9. Load Distribution vs. Elevation

Based on the results from the strain gauges, the bond strength in the rock socket was not fully mobilized. Therefore, the average nominal bond strength for the entire length of the socket could be even higher than the value measured by the load test. In any event, the load test was judged to be a success and the design bond strength within the sandstone formation was validated.

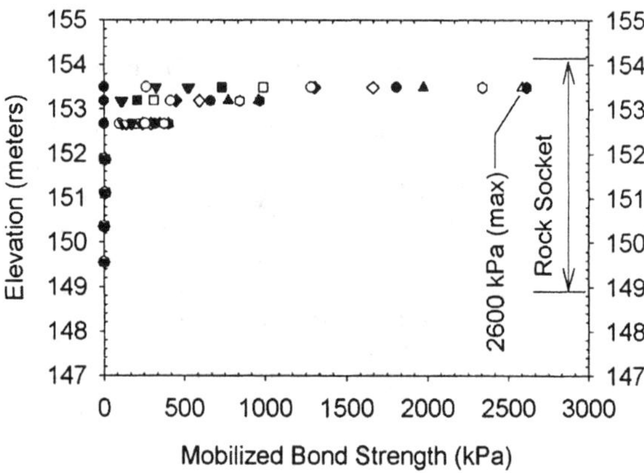

Figure 10: Bond Strength in Socket

CONCLUSIONS

The following conclusions can be made based on the information developed during the design, installation and verification testing of micropiles for the utility retrofit.

These conclusions may be helpful in the design of micropiles for similar applications and geologic conditions.

- Use of the high capacity single bar micropiles can be a cost efficient foundation for difficult site conditions.
- The small cross secitonal areas of micropiles allow horizontal adjustments around conflicts due to the previous foundations and existing utilities.
- To achieve the highest capacities, competent rock materials are necessary within reasonable depths.
- Bond strengths in excess of those reported by FHWA were achieved within the clastic sandstone formation tested at the D.E. Karn site with similar results achieved on other projects in Michigan Sedimentary Rock Formations.
- The Rowe & Armitage (1987) relationship more accurately estimates the nominal bond strength in Michigan Sedimentary Rock Formations than other reviewed methods.
- Verification testing of the pile design is still recommended for all high capacity applications.

ACKNOWLEDGMENTS

The authors would like to thank the Barton Malow Construction Co. and Consumers Power Company for their help and assistance in this challenging project.

REFERENCES

Carter, J.P. and Kalhawy, F.H., (1988). "Analysis and Design of Drilled Shaft Foundations Socketed into Rock", *Final Report, EPRI EL-5916, Proj. 1493-4*, Electric Power Res. Institute, Palo Alto, Calf.

Federal Highway Administration (2000). *Micropile Design and Construction Guidelines Implementation Manual. FHWA-SA-97-070*, Coauthored by T. Armor P.,Groneck, J. Keeley, S. Sharma

Hassan, K.M. and O'Niell, M. W. (1997). "Side Load-Transfer Mechanism Drilled Shafts in Soft Argillaceous Rock", *Journal of Geotechnical and Geoenvironmental Engineering*, ASCE, 123(2), 145-152.

Horvath, R.G. and Kenney T.C., (1979). "Shaft Resistance of Rock-Socketed Drilled Piers"., *Proceedings, Symposium on Deep Foundations*, ASCE, Atlanta, Georgia, 40-42.

Rowe, R.K., and Armitage, H.H. (1987). "Theoretical Solutions for Axial Deformation of Drilled Piers in Rock", *Can. Geotech. J.*, Ottawa, Canada, 24, 114-125.

MICROPILES IN KARSTIC DOLOMITE SIMILARITIES AND DIFFERENCES OF TWO CASE HISTORIES

Daniel D. Uranowski, P.E.[1], M. ASCE, Scott Dodds[2], M. ASCE, Scott Stonecheck, P.E.[3], M. ASCE

ABSTRACT:
Micropiles are being used today for a variety of applications. As the size and complexity of today's superstructures increase, difficult subsurface conditions in karstic dolomite preclude the use of traditional deep foundation systems. A report is presented on the redesign of the proposed deep foundations and installation of the micropile systems for two different projects. The similarities and differences of the two projects with regard to geology, design, load transfer, and installation are summarized.

INTRODUCTION (PROJECT NO. 1):

The first project involved a highway project in the Allentown, Pennsylvania (PA) area to extend Route 33 which will provide a connection between the town of Wilson and Interstate 78. As part of this project, a contract was awarded by PennDOT to Dick Corporation of Large, PA to construct a $55M, 6-span bridge crossing over the Lehigh River, Lehigh Canal, and Norfolk-Southern RR near Easton, PA.

[1] Vice President, Brayman Construction Corporation, 1000 John Roebling Way, Saxonburg, PA 16056, 724.443.1533, www.braymanconstruction.com
[2] Manager, Drilling & Grouting Group, Brayman Construction Corporation
[3] Project Engineer, Foundation Division, Brayman Construction Corporation

The project site is located in a geologically diverse area complicated by the presence of Karst deposits. Ground surface features, namely the Lehigh River, Lehigh Canal and Norfolk-Southern RR, provide challenges for accessing and constructing the new bridge. After performing an extensive geotechnical investigation, the section designer, URS Corporation of King of Prussia, PA selected a foundation system of drilled caissons and driven piling. Driven piling was used at Pier 3, which is located between the Lehigh River and the Lehigh Canal, and caissons were called for at the remaining pier and abutment locations.

Dick Corporation began work in the Spring of 2000. Initially work proceeded at the Abutments and Piers 1, 3 and 5. The original design of the foundation for both Piers 2 and 4 called for 54-inch diameter permanently steel cased "shaft" sections, and 48-inch diameter sockets drilled into competent rock. Forty caissons were required at each pier. Because of the variable nature of the ground conditions, the original contract required the Contractor to perform exploratory drilling at each caisson location to verify design assumptions and caisson tip elevations.

GEOLOGY (PROJECT NO. 1):

At Pier 4, results of the exploratory drilling not only confirmed the variable nature of the rock as anticipated, but also indicated that most caisson tip elevations would need to be lower than planned. Pinnacles, voids and clay-filled solution cavities were evident at variable depths, with a particularly troublesome area located in the southeast corner of the footing. The "as-bid" caisson lengths varied from 74-feet to 84-feet with an aggregate of 3,160-feet. After incorporating the new test drilling information new caisson lengths varied from 37-feet to 225-feet with an aggregate of 3,280-feet.

Realizing the schedule and cost impact of this new information, along with knowledge gained from the installation of caissons in the hard dolomite at Abutment No. 1 and Pier No. 1, Dick Corporation contacted Brayman Construction Corporation of Saxonburg, PA for assistance. Jointly, the two companies evaluated various methods of installing the foundations at Pier 4 and decided to propose a design change, utilizing a micropile substitution for the original caisson design.

VALUE ENGINEERED DESIGN (PROJECT NO. 1):

Brayman employed D'Appolonia Consulting Engineers of Monroeville, PA to design the micropile alternate, and together with PennDOT and URS, established design and installation parameters. The final micropile design at Pier 4 called for 160 piles of 200-kip design capacity. The piles consisted of 9 5/8-inch diameter pipes with a minimum wall thickness of 0.518-inches grouted into a minimum 12-inch diameter drill hole. Pile lengths varied from 46-feet to 191-feet and were designed to develop their full design capacity in end bearing resistance with only minimal side frictional resistance. The design is shown in Figure 1.

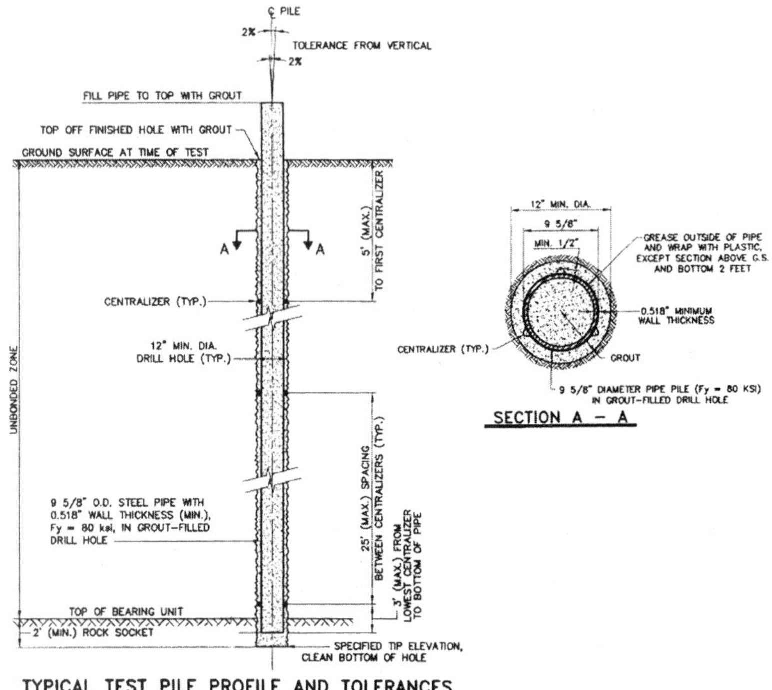

FIG. 1 – Typical Design of 9 5/8-inch diameter micropile

In order to verify design assumptions and installation procedures, a pre-production test program was performed. After successful completion of the test program, which included a full-scale pile load test to 600-kips, production pile installation began. During production pile installation, two additional pile load tests were performed. Recognizing the advantages and success of the micropile alternate at Pier 4, Dick Corporation and Brayman proposed a similar alternate for the Pier 2 caissons. Although Karst conditions were not as prevalent as Pier 4, the exploratory drilling performed at Pier 2 for each caisson confirmed the presence of a soft, water bearing shale layer "sandwiched" within the hard dolomite at a depth of about 40-feet below the top of caisson elevation. Therefore, the "as-designed" caissons would be required to extend an additional 10-feet below this soft layer for a total depth of 55-feet. Since most of the caisson length would be through dolomite, drilling the 54-inch shaft and 48-inch sockets would be slow and expensive.

Once again, after discussions to establish design parameters, Brayman and D'Appolonia prepared the alternate micropile design, utilizing 160 piles of 260-kip design capacity to substitute for the 40 caissons. Micropile lengths varied from 41-feet to 74-feet, and again consisted of 9 5/8-inch diameter pipes grouted in a minimum 12-inch diameter drill hole. A drilling and grouting test program was used to verify grout mixes and placement techniques used for the micropile construction. The test program demonstrated that grouting and placement techniques, in addition to water infiltrating through the soft shale zone, was not detrimental to micropile installation. Upon successful load test completion, the production pile installation for Pier 2 began. Numerous load test results were used to verify the redesign of the foundation system.

The different geologic conditions at Piers 2 and 4 required much different drilling and installation methods to construct the micropiles. With only a limited amount of Karst conditions evident across the Pier 2 footprint, the drilling and installation methods differed dramatically from the procedures, equipment, and tooling necessary for advancing through the Karst conditions at Pier 4.

As indicated by the exploratory drilling, a significant portion of the drilling at Pier 2 would be through varying layers of hard and soft bedrock. Unlike the conditions at Pier 4, the boreholes at Pier 2 would most likely remain "open" throughout the entire drilling and installation process, thereby permitting the use of open hole drilling techniques.

The drilling at Pier 2 was completed with a Drilltech D40K drill rig. Originally designed as a truck-mounted water well drill rig, the Drilltech D40K is equipped with a 1300/350 cfm on-board air compressor in addition to a drill rod carousel modified to handle up to 400-feet of 7-inch diameter drill rods. The 92-foot by 56-foot dimension for the concrete pier foundation provided ample space for the truck rig to maneuver and access each pile location. Set-up with an Ingersol-Rand QL120 down-the-hole hammer and the 7-inch diameter drill rods, the drill rig advanced each borehole to the final tip elevation using standard rotary percussive drilling methods without the need for temporary casing. Through a 3-inch diameter plastic pipe, a Schwing BP-500 concrete plant pumped the 4000-psi sand-cement mix from the bottom of the borehole to the top using tremie placement methods. The strength of the grout was verified from 2-inch by 2-inch grout cubes cast during each placement.

With the support of a 40-ton rough terrain crane, the steel pipe piles could be set through the freshly placed grout to the bottom of the hole. If the steel pile could not overcome the frictional forces between grout and steel or if the pile simply got hung up on the side of the borehole, a small D5 setting hammer placed on top of the pile tapped the piles to the final elevation.

The micropiles in Pier 2 ranged from 41-feet to 74-feet in length with an overall average of 55 lineal feet per pile and at Pier 4 from 46-feet to 191-feet with an average of 73-feet per pile. Delivered in lengths ranging from 38-feet to 45-feet, many of the steel pipe piles would require splices. Designed as "end bearing" piles, each splice was required to transfer 100% of the design load. Therefore a full penetration weld at each pile splice fulfilled this design criterion.

Unique to the circumstances at Pier 2, Pier 4 presented a completely different set of geologic conditions that directly impacted the equipment, productivity, drill tooling and drilling methods. A careful review of the exploratory borings performed at each of the original 48-inch diameter caisson locations revealed the variable nature of the rock and the presence of pinnacles and multiple soil lenses at varying depths within the rock strata. Such variable ground conditions made it impractical to drill and install the as-designed caissons at Pier 4 and presented some challenges with the installation of the micropiles.

DRILLING TECHNIQUES (PROJECT NO. 1):

Faced with the task of providing an open 12-inch minimum diameter borehole along the full pile length, Brayman utilized two types of drilling methods to overcome the difficult subsurface conditions present in Pier 4.

The Tubex XL-280® drilling system drilled a hole larger than the diameter of the surface casing, thus allowing the casing to advance simultaneously with the drilling of the hole. Equipped with a pilot and reamer bit set to drill a 15-inch diameter drill hole and 13-3/8-inch diameter temporary casing to follow, nearly 70% of the piles drilled at Pier 4 utilized the Tubex XL-280® drilling system. Though most effective for piles over 100 feet in length, pile locations encountering multiple soil/rock lenses also employed the Tubex drilling system.

Extending an SD-10 down-the-hole hammer with a 12-inch diameter bit on 7-inch drill rods slightly ahead of the temporary 13-3/8-inch surface casing provided another method of sealing off the soil seams and still provide the prescribed hole diameter. This method of drilling is commonly referred to as duplex drilling. Since the hammer drills a smaller diameter hole than the inside diameter of the surface casing, the casing must be drilled or "screwed" through the rock. Unfortunately, this procedure could only be applied in areas of slightly weathered to weathered rock which consisted of approximately 30% of the piles in Pier 4.

Both the Casagrande® C8 and MAIT® 130 hydraulic drill rigs were used for the Pier 4 drilling. During the early stages of the drilling, one rig would drill and install the surface casing while the other rig would retract the surface casing after the grouting and installation of the pipe pile. With piles extending 3-ft above the workbench elevation to tie into the 8-ft thick concrete pier footing, access for the two drill rigs quickly diminished in the excavation, limiting the remainder of the work to be performed using one drill rig operating two shifts.

The grouting and installation of the piles at Pier 4 mirrored the procedures used at Pier 2 with the exception of extracting the surface casing. All the piles were placed in one piece and the longer piles at Pier 4 were installed using a Manitowoc® 4100 crane provided by Dick Corporation.

SUMMARY (PROJECT NO. 1):

In summary the micropile alternate design provided an efficient design, particularly given the adverse geology and site conditions at the Route 33 bridge site. Cost savings on the caisson design were approximately $200,000 with 3 months saved on the project schedule.

INTRODUCTION (PROJECT NO. 2):

The second project involved the redesign of a deep foundation system for a new $58.5M building constructed by the Pennsylvania State University in State College,

PA. Both spread footings and micropiles were initially proposed but subsequent redesign of the micropile portion of the project led to significant cost savings to the University. Spread footings were installed for less critical applications.

The micropile foundations were used to support the higher compressive and tensile load areas of the Information Sciences & Technology Building, in addition to an elevated portion of the structure which spans an active segment of a 4-lane, State Business Route 322. The elevated portion of the building is comprised of classrooms and hallways. Battered and vertical micropiles were used to found the abutment walls of the elevated structure.

GEOLOGY (PROJECT NO. 2):

The bedrock for the site is comprised of crystalline dolomite with alternating beds of sandy, cherty dolomite. Compressive strength test results for 2-inch diameter rock cores obtained during the subsurface investigation were approximately 23,000 psi.

The interface between the bedrock and the soil mantle is characterized by pinnacle formations with joint channel development common. As expected, solution openings in the karstic region were encountered. Clay seams and voids were encountered throughout the bedrock to depths exceeding 100-feet. Brayman traversed these soil and bedrock conditions until the required rock socket length was penetrated in sound rock. Unlike Project No. 2, groundwater and artesian conditions were not encountered in the majority of the holes.

VALUE ENGINEERED DESIGN (PROJECT NO. 2):

Again, Brayman and D'Appolonia worked together to finalize a value-engineered design, that incorporated both 9 5/8-inch and 7-inch diameter casing. Finite element design and analysis incorporating Florida Pier® was also used for the project with field results verified through multiple full-scale load tests. The larger 9 5/8-inch casing was used primarily for lateral support through the soil and weathered rock seams in the upper strata at the site. Installation of the 9 5/8-inch diameter casing involved the use of percussive drilling methods with down hole hammers. Upon extraction of the down hole hammer tooling, the inner 7-inch diameter casing was advanced through the karstic dolomite, clay-filled seams, and voids by the overburden tubex drilling system described above. Unlike the previous design, the side friction of the grout-rock bond zone was used in the design of the socket length and end-bearing resistance was neglected. The lengths of the micropiles ranged from 13-ft to 110-ft.

Steel casing was advanced until the proper rock socket length was achieved and then extracted to 1-ft below the soil-rock interface. Next, tremie grouting methods were used to place a 4,000 psi sand cement mix in the rock socket prior to placing a 2.5-inch Grade 75 steel bar. Next, the steel casing was tremie filled with concrete from the top of the rock socket to the cut-off elevation. Finally, 90° reinforcing bars were installed to the proper elevations to transfer the design compressive and tensile loads to the micropiles. A typical micropile design for the side frictional resistance piles is shown in Figure 2.

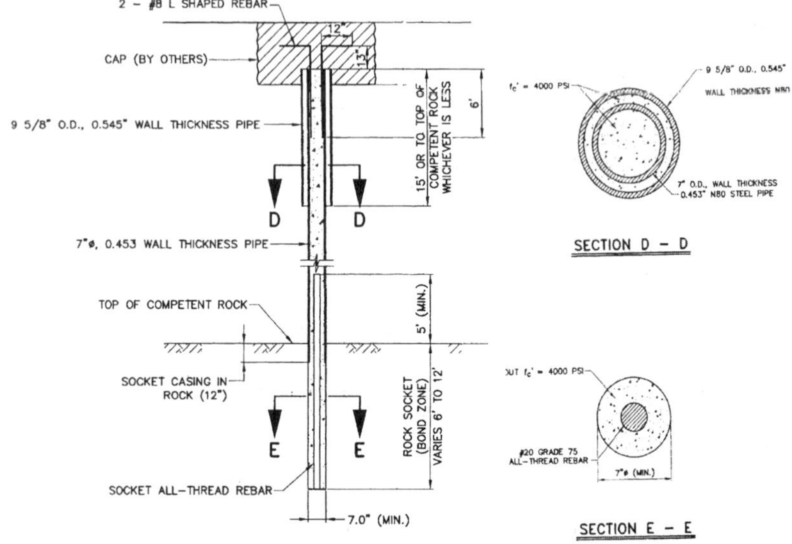

FIG. 2 – Typical Micropile Design for Side Friction Piles

SUMMARY (PROJECT NO. 2):

Nearly 9,000 lf of micropile was installed for the structure. The value engineered micropile system incorporated at the University provided an approximate $150,000 cost saving to the Owner. With these difficult geologic conditions, the deep foundation alternative selected for the site was a success.

SIMILARITIES AND DIFFERENCES:

Both of these unique projects involved the use of micropile technologies in karstic dolomite geologies. Soil and clay-filled seams, pinnacled rock formations, and voids left from the dissolving action of water in these carbonate bedrock settings precluded the use of conventional drilled shaft and spread footing foundation systems. While artesian conditions and high groundwater conditions were encountered at the SR 33 project, minimal groundwater was noticed for the Pennsylvania State University project discussed herein.

Although the SR 33 Easton project incorporated end bearing resistance of the micropile, load transfer of the steel casing to the grout and bedrock can also be accomplished through adhesion of the grout and bedrock interface along the socket. Either way, both projects met the design criteria set for the project and were verified through the performance of multiple pile load tests.

Tubex overburden drilling methodologies were used successfully to install both 13 3/8-inch and 7-inch diameter steel casing with overall lengths ranging from 13-ft to 190-ft through difficult subsurface conditions. And finally, both installations involved time and cost savings to the Pennsylvania Department of Transportation and the Pennsylvania State University.

With regard to the high vertical and lateral loads that today's superstructures transmit to the underlying bedrock and soil, conventional deep foundation systems (i.e. caissons, driven piling, etc.) may not be the best alternative for sites with karstic dolomitic geologies. Technologies exist for advancing deep foundation systems consisting of smaller diameter, high-strength steel elements through extremely hard bedrock and difficult subsurface conditions. Value engineering and alternative foundation designs for these unique projects should be encouraged by the foundation industry and all parties involved.

SPECIAL USE OF MICROPILES AND PERMANENT ANCHORS

Helmut Schwarz[1], Klaus Dietz[2], Horst Köster[3], Thomas Groß[4],
Stump Spezialtiefbau GmbH

ABSTRACT: Micropiles and permanent anchors are important elements in geotechnical engineering. To use them successfully under difficult site and soil conditions it is necessary to improve the normal systems in due compliance with project specific specialist measures. Necessary improvements for difficult site and soil conditions are explained by three case histories. The first shows the production of micropiles und permanent anchors in tidal-zone and the improvement of the load bearing zone to guarantee the load bearing capacity through a wharf foundation. In the second example it is described how to protect permanent anchors against lime-attacking carbon dioxide in the groundwater in rock. The concept for the post-foundation under organic soil conditions of the Berlin State Library "Unter den Linden" is the third case history.

1.0 INTRODUCTION

Micropiles and permanent anchors are important elements to provide increased safety in specialist underground works. Based on over 30 years of experience the scope of application is continuously extended. With the following three examples we show how permanent anchors and micropiles can be manufactured successfully even in an adverse subsoil environment, in due compliance with project specific specialist measures and a consistent quality assurance system.

[1]Dr.-Ing., CEO, Stump Spezialtiefbau GmbH, Fränkische Straße 11, 30455 Hannover/Germany, Dr.Helmut.Schwarz@stump.de
[2]Dipl.-Ing., technical manager, Stump Spezialtiefbau GmbH, Friedrich-Krupp-Straße 18, 40764 Langenfeld/Germany, Klaus.Dietz@Stump.de
[3]Dipl.-Ing., LU manager, Stump Spezialtiefbau GmbH, ZN Hannover , Fränkische Straße 11, 30455 Hannover/Germany, Horst.Koester@Stump.de
[4]Dipl.-Ing., LU manager, Stump Spezialtiefbau GmbH, ZN Langenfeld, Friedrich-Krupp-Straße 18, 40764 Langenfeld/Germany, Thomas.Gross@Stump.de

2.0 PRESTRESSED TENSION PILES IN HARBOR CONSTRUCTION

2.1 Project

The former AG Weser shipyard premises at Bremen are converted into a harbor and industrial park. Therefore, the existing quay-side installations must be strengthened and stabilized to accomodate deeper excavations. In one section, the so-called Pier II (FIG. 2), the existing sheet piling can still be used and only requires additional support by providing a permanent anchor in the tidal zone. In the area of Pier I, a completely new sheet piling is foreseen, with a micropile keeping the beam tie in place (FIG. 1).

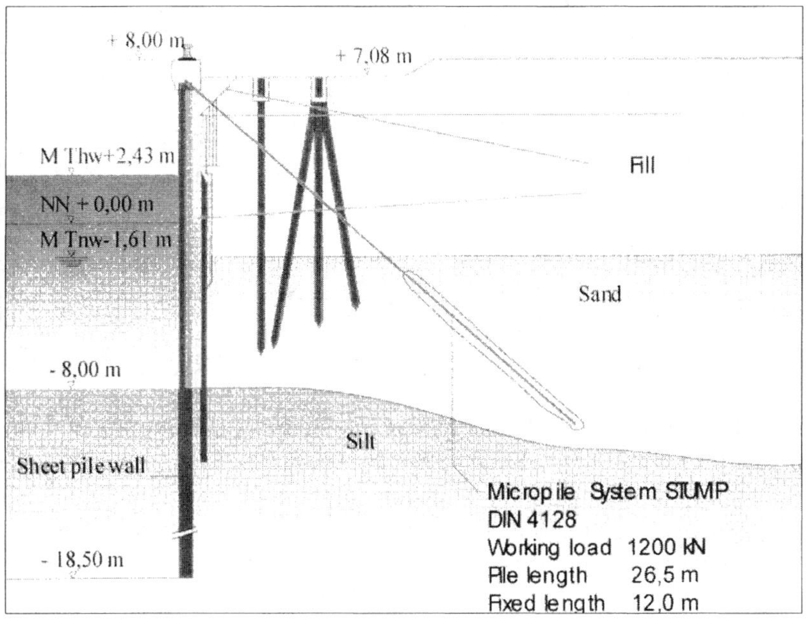

FIG. 1: Rehabilitation of AG Weser premises, Pier I

2.2 Foundation soil and boundary conditions

Owing to numerous structural changes on the premises the subsoil in the project area is covered entirely by fill material. This fill area with its brick and concrete remnants reaches down to elevation approx. – 7 m. The other sand-gravel fill layers further below had been placed to be used as foundation area. These layers are followed by a loose gravel-sand mix layer with clay and silt deposits between elevation – 8 to - 10 m. The loose gravel-sand mix layer necessitated additional measures, in the form of a jetted foot, for the 900 kN pile working load to be safely lowered into the ground. Owing to the fact that the anchor heads at Pier II were located in the tidal range area, anchor setting and tensioning work could only be carried out at low tide.

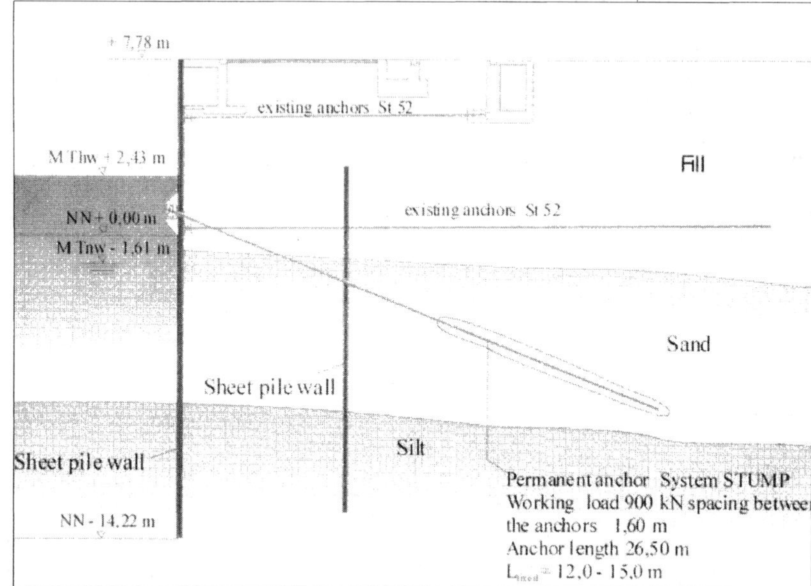

FIG. 2: Rehabilitation of AG Weser premises, Pier II

2.3 Special measures and quality assurance

The load bearing capacity of both piles and anchors was guaranteed through a jet-grouted tip. The twin-rotary-drive drilling method (overburden drilling) was used to drill the 160 mm diameter borehole. Upon reaaching the final depth of ca 27 m the last 3 m of the fixed length were enlarged by using the jet-grouted tip method. Following this, the piping was carried to the final depth again and the St 52 Ø 90 mm steel load bearing element inserted, which in turn was followed by the step-by-step grouting of the 12 m fixed length. All anchors were submitted to an acceptance test and tensioned to 1200 kN. All tension bars were found in compliance with the admissible deformation values.

In addition to ensuring the load bearing capacity of the piles, drilling proved equally difficult. In the area of Pier I, where the pile ended in the beam tie, a traveling platform was used which could be moved on the quay wall (FIG. 3).

FIG. 3: Drilling operations at the top end, from traveling platform (Pier I).

FIG. 4: Drilling operations in tidal zone, from pontoon (Pier II)

At Pier II, no additional load was admissible for the sheeting during construction work, so that the traveling cradle could not be used in this instance. For this reason, operations were carried out from the water side via a pontoon (FIG. 4). Owing to the tidal effects (ca. 3 m range of tide) the drill pipes had to be adjusted continuously during drilling. Owing to the fact that the drilling team carried out their work with utmost care, drill pipe twist-offs only occurred in exceptional cases.

3.0 PERMANENT ANCHORS IN AGGRESSIVE FOUNDATION ROCK

3.1 Project

DB AG (German Federal Railways) is constructing a new railway line between Cologne and Frankfurt for high-speed trains. Several special structures must be erected as part of this project some of which have to be secured by means of anchors. In one particular case a retaining wall had to be provided to secure a hillside cut. A buttress-type bored cut-off wall with permanent rock anchor (FIG. 5) was selected instead of the sloped retaining wall initially foreseen. A total of 431 permanent anchors with working loads of up to 1,130 kN and up to 45 m long had to be manufactured. In the case of approximately 290 permanent anchors the anchoring length reached below groundwater level. Special measures were required since the free carbon dioxide contained in the groundwater was aggressive to concrete.

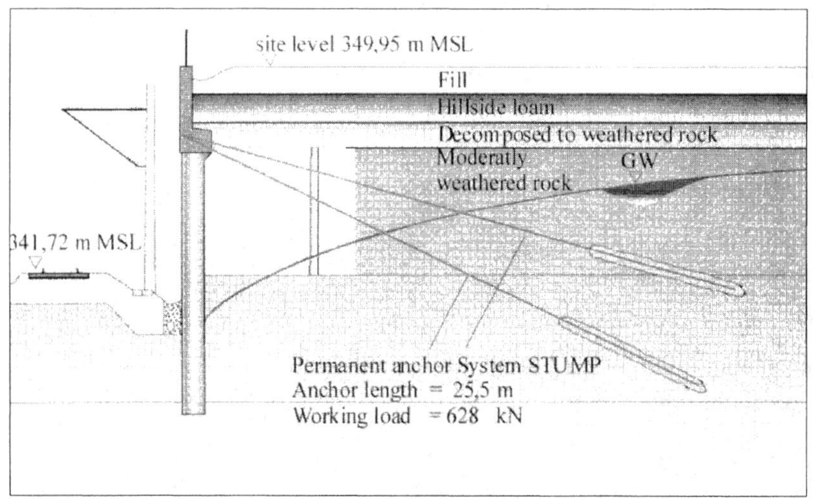

FIG. 5: Cross-section of Hombach retaining wall, new Cologne – Rhine/Main railway line

3.2 Foundation soil and boundary conditions

An approximately 1.5 m of hillside loam covered by approximately 1 m of fill. Below this, we find a layer of decomposed to weathered rock and, finally, moderately weathered Devonian rock consisting mainly of clay, silt and sandstone as well as sand and slate. Depending on the more or less open seams and fault zones this rock material of the Rhenish Slate Mountain Range normally constitutes a moderately permeable water-bearing horizon. Based on the foundation soil report the mean permeability was assumed to be 10-7 m/s. Owing to the lime-attacking carbon dioxide contained in the groundwater at an approximately 20 to 100 mg/l ratio, the groundwater must be classified as being slightly to strongly aggressive to concrete, according to DIN 4030.

3.3 SPECIAL MEASURES AND QUALITY ASSURANCE

Pursuant to DIN 4125, 5.1.3 permanent anchors may be installed in rock of a highly aggressive nature only in those cases where the load bearing capacity of the anchors is permanently protected through implementation of adequate special measures, e.g. bore hole improvement through grouting to DIN 4093, so that the grout body is protected against aggressive water attack. The de-stabilization of the hardened cement paste surface is only possible where the water flowing in the open seams and fault zones gets into contact with the grout body. A de-stabilization of the hardened cement paste surface is not to be expected in such areas of the grout body that are free of any open seams or faults. Initially, suitability tests were performed to find out whether and to which degree it was possible to successfully grout the solid rock.

For this purpose water permeability tests with a pressure of maximum 1.5 bar were carried out in the fixed anchor length after the bore holes had been drilled (FIG. 6). The quality criterion determined for installation of a permanent anchor was a flow of less than 3 l/min, under a pressure of 1 bar, throughout the entire 6 m bond length. The water permeability tests were recorded automatically and evaluated by computer. When the water permeability criterion was exceeded, pre-injection to the load application zone was effected using cement slurry. Depending on the injection material applied the water-cement ratio ranged between 1.0 and 0.45. CEM 42.5R was used, with an admixture of 20 % fly ash. The average grouting quantity used on the test anchors was in excess of 500 kg. The following day the grouted section was drilled open and another water permeability test carried out. Where the result was positive, the anchor could be installed, otherwise pre-grouting had to be repeated. Each work phase was recorded in a quality assurance plan and countersigned by the construction supervisor.

FIG. 6: Drilling operations using magazine drill

During execution it became apparent that the permeability of the weathered solid rock significantly exceeded the values given in the foundation soil report. As a result, a total of some 840 water permeability tests had to be carried out, in some cases up to seven times in one and the same bore. This led to major disturbances in the works progress. Even during construction execution each individual working step was recorded and countersigned, from permeability tests via pre-improvement and installation all the way up to the tensioning operations (FIG. 7). Drilling and installation of the permanent anchors involved a period of 8 months.

FIG. 7: Acceptance test to DIN 4125

4.0 MICROPILES IN SOILS WITH ORGANIC ADMIXTURES

4.1 Project

The "Unter den Linden" Berlin State Library is the main building of Germany's largest universal library of science. The building was erected between 1903 and 1904 and represents the most spacious historic building complex in Berlin. Due to structural retrofit requirements and a new intended use of the building, extensive post-foundation works had to be carried out on the building which is listed for preservation as a historical site. In addition to 8,000 m^3 of subsoil stabilization using the jet-grouting method and 2,000 m^3 of soil stabilization by means of ultra-fine cement injections a total of 50,000 lineal meters of micropiles were manufactured, with working loads of 860 kN and maximum lengths of 32 m.

4.2 Foundation soil and boundary conditions

The building complex with a floor area of 106 m x 170 m is founded in the Berlin 'Urstromtal' in the area of the river Spree lowlands. The foundation soil investigations have revealed a soft organic soil (Mudde) channel running in north-west direction underneath the building complex up to the well-yard. This channel is characterized by alternating layers of sand and peat. The sediments include digested sludge and non-cohesive soil material with considerable organic admixtures and are therefore to be considered as non-load bearing. The medium-dense layers of sand found further down are used as load bearing layers for building load transfer purposes (FIG. 8). Owing to the extremely varying foundation soil conditions three different foundation alternatives have been applied in the erection of the State Library:

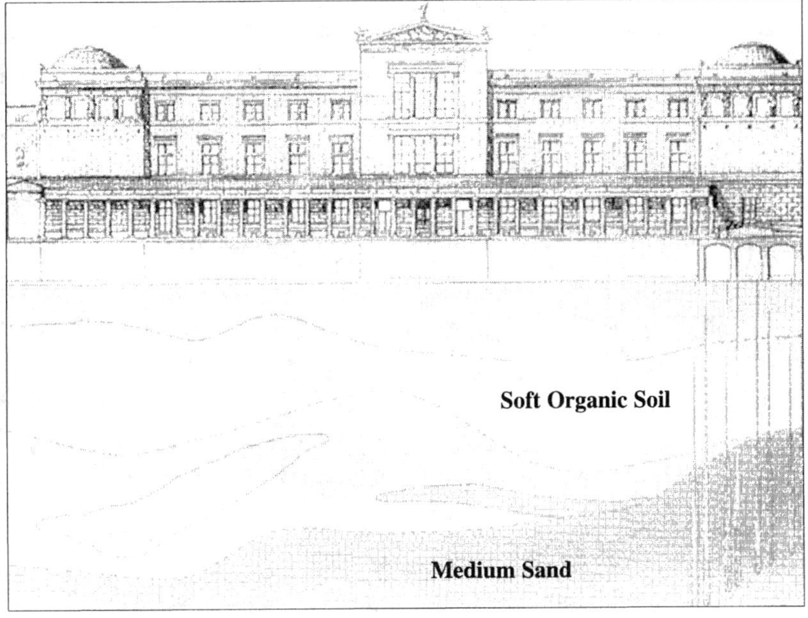

FIG. 8: Foundation soil situation at Berlin State Library, New Museum

1. Shallow foundation by means of strip footing
2. Deep foundation with ca 30 cm diameter pinewood piles
3. Deep foundation on timber caissons with rubble and concrete fill

In the course of construction works carried out nearby in the seventies and eighties the groundwater level which in this area reaches approximately 31 m above MSL was periodically lowered. As a result, the pile caps of the wooden pile foundation were exposed to constant changes from dampness to drying out, with the supply of oxygen. The resultant rotting had meanwhile reached a stage where the original cross-section was found to be damaged to 70 % on the average, in some areas however, to up to 95 %. It was determined that under these circumstances the foundation needed complete rehabilitation. The rehabilitation concept was based on relocation of the vertical loads from the existing defective wooden piles to new micropiles.

To protect the existing foundation situation against negative effects, one pile row each was arranged on both sides of the old beam ties (FIG. 9). Based on a special proposal submitted by Stump Spezialtiefbau GmbH "Stump" micropiles with GEWI steel load bearing elements 63.5 mm and tube piles system Stump 80 mm were used. The unfavorable foundation soil conditions were accounted for by a wider pile shaft diameter of 240 mm. A secure bond between foundation soil and a dense pile concrete was achieved by an intensive, phased grouting of the pile shaft. The test loads applied confirmed the required 800 kN load transfer with a safety factor of two.

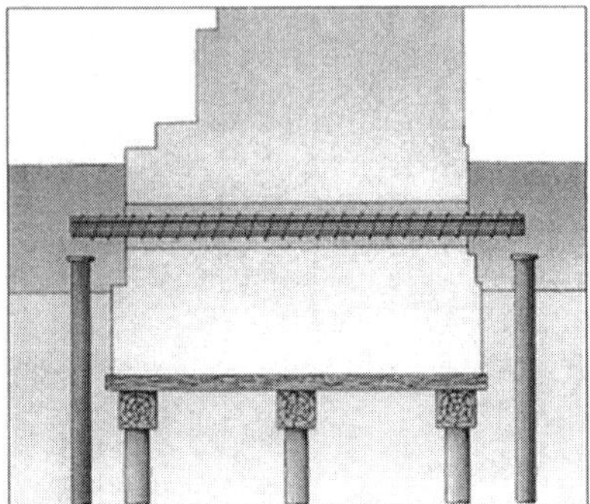

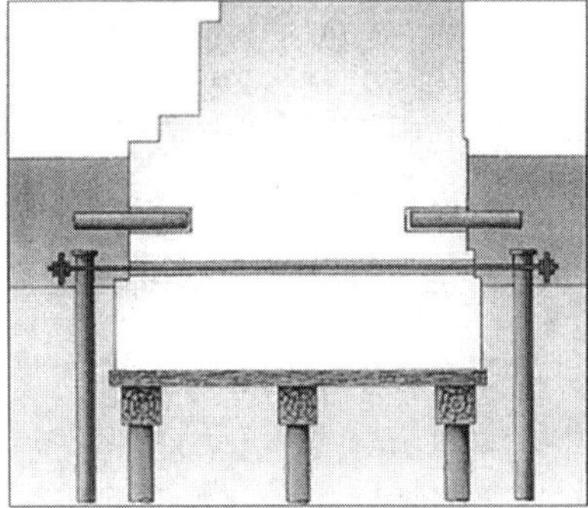

FIG. 9: Rehabilitation measures – Through-girders and insertion girders in connection with pipe piles or "Stump" GEWI-micropiles

The drillholes for the piles required inside the building had to be carried out using special drilling equipment, as the room height was only 2.30 m in most cases (FIG. 10).

FIG. 10: Drilling operations for pile manufacture

To check the foundation soil situation and identify drilling obstacles, all relevant drilling data such as feeding and feed pressure were continuously recorded during drilling. In view of the fact that a carefully executed injection was equally decisive for the durability of the foundation work, each injection operation was recorded in every detail via a pressure-quantity recording device.

The drillings for insertion and/or through-girders in the existing foundations were carried out as core drillings with particular care. Installation of the girders was followed by special mortar grouting to ensure an optimal frictional connection (FIG. 11).

FIG. 11: Tube piles System Stump and insertion girders prior to floor slab concreting

5.0 CONCLUSION

The examples given above have shown that a reliable manufacturing of anchors and micropiles is possible even under adverse environmental and subsoil conditions. This however, requires that project specific special measures are monitored based on a target-orientated quality assurance system.

BEARING CAPACITY IMPROVEMENT USING MICROPILES A CASE STUDY

G.L. Sivakumar Babu[1], B. Srinivasa Murthy[2], D.S. N. Murthy[3], M.S. Nataraj[4]

ABSTRACT

Micropiles have been used effectively in many applications of ground improvement to increase the bearing capacity and reduce the settlements particularly in strengthening the existing foundations. Frictional resistance between the surface of the pile and soil and the associated group/network effects of micropiles are considered as the possible mechanism for improvement. This paper deals with a case study in which micropiles of 100 mm diameter and 4 m long have been used to improve the bearing capacity of foundation soil and in the rehabilitation of the total building foundation system. The micropiles were inserted around the individual footings at inclination of 70^0 with the horizontal. The actual design for retrofitting was based on the assumption that the vertical component of the frictional force between the soil and the micropile resists the additional load coming from the structure over and above the bearing capacity. The technique was successful and the structure did not show any signs of distress later. Detailed finite element analysis conducted validated the suggested treatment. The paper describes the case study, the method of treatment adopted in the field and the results of numerical analysis.

---

[1] Assistant Professor, Department of Civil Engineering, Indian Institute of Science, Bangalore - 560 012, India, gls@civil.iisc.ernet.in

[2] Professor, Department of Civil Engineering, Indian Institute of Science, Bangalore - 560012, India.

[3] Research Scholar, Department of Civil Engineering, Indian Institute of Science, Bangalore - 560 012, India.

[4] Professor, Department of Civil and Environmental Engineering, University of New Orleans, New Orleans, LA 70148-2212.

INTRODUCTION

Micropiles are often used to improve the bearing capacity of the foundation against applied loading. In many cases, steel pipes of 50 to 200 mm diameters are used as micropiles. The strengthened ground acts as coherent mass and behaves remarkably well, capable of sustaining very high compressive loads at defined settlement or alternatively defined loads at reduced movement. Lizzi (1982) and Plumelle (1984) showed that micropiles create an in situ coherent composite reinforced soil system and the engineering behaviour of micropile-reinforced soil is highly dependent on the group and network effects that influence the overall resistance and shear strength of composite soil micropile system. Juran et al. (1999) presented an excellent state of art review, covering all the studies and contributions, on the state of practice using micropiles. Considerable information on single micropile design, evaluation of load bearing capacity, movement estimation models as well as effect of group and network effect have been covered in considerable detail. The authors also reviewed geotechnical design guidelines in different countries for axial, lateral load capacities and approach for movement estimation.

In India, in some circumstances steel pipes, coated wooden piles are used as cost-effective options in improving the bearing capacity of foundation or restrict displacements to tolerable levels and similar uses in stabilization of slopes, strengthening of foundations are common. Sridharan and Murthy (1993) described a case study in which a ten-storeyed building, originally in a precarious condition due to differential settlement, was restored to safety using micropiles. Galvanized steel pipes of 100 mm diameter and 10 m long with bottom end closed with shoe, driven at an angle of 60° with the horizontal were used and the friction between the pile and the soil was used as the design basis in evolving the remedial measures. A similar attempt was made in the present case study in which the bearing capacity of the existing foundation system of a building was restored to safety using micropiles.

DETAILS OF THE CASE STUDY

A two storeyed building rectangular in plan was constructed on a (loose sandy soil) filled up soil in a metropolitan city in India. The investigations revealed that the foundation soil is in relatively loose state (SPT values in the range of 6 to 8) and the results obtained from the laboratory tests indicated that the soil properties (effective stress parameters) viz., cohesion and friction angle can be taken as $c' = 0$ and $\phi = 25°$ respectively and the bulk density (γ_b) is in the range of 17 kN/m^3. The foundations were designed to carry expected column load of 600 kN, considering that the safe bearing pressure of the soil is 120 kN/m^2. Accordingly individual column footings of size 2.5 m × 2.0 m were proposed. Above the foundation level, compacted soil (bulk density in the range of 20 kN/m^3) was placed upto a height of 6.5 m to make up for the difference in level, which resulted in a further loading of 130 kN/m^2 on the foundation. This aspect was not considered in the original design and as a result, plinth beam (at the foundation level) and tie beams at the middle level showed considerable distress in the form of cracks when the filling was nearing completion. Foundations supporting columns tilted out of line and further construction was difficult. Later on soil investigation showed that the safe bearing pressure considered originally in the design is higher and its actual value is around 70 kN/m^2. It became necessary to restore the foundations and columns and hence micropiling has been chosen to strengthen the soil beneath the foundation. The treatment brought back the total foundation system to original requirements and was considered satisfactory. Fig. 1(a) shows the support system for retaining the soil outside the plinth beam and Fig. 1(b) shows the removal of the filled up soil before undertaking micropiling. Fig. 1(c) shows the micropiling in progress. The micropiling is done using a simple hammering system in which a mass of 2.6 kN falls through a guide over a height of 1.5 m.

FIG. 1. (a) Stage of construction at the time of micropiling;
(b) Excavation of soil before implementing micropiling technique;
(c) Micropiling at the site

ANALYTICAL CONSIDERATIONS

To design the micropile configuration to strengthen the foundation system the load on the footing from columns and filling are calculated and the required frictional resistance of the micropile system is evaluated. The calculations for a footing (2.5 m × 2.0 m) are as follows.

The total load coming on the system = column load +fill load
= 600 kN + 20 kN/m^3 × 6.5 m × 2.5 m × 2.0 m =1250 kN.
Required safe bearing pressure = 1250/2.5 × 2 = 250 kPa
Safe load on the foundation system = 70 kPa ×2.5 m ×2 m = 350 kN
Additional load for which the micropile system is designed = 1250 - 350
= 900 kN

The additional required frictional resistance of 900 kN from the micropile system is derived based on considering an element of soil at the micropile-soil interface and integrating the element resistance over the entire length of micropile. The vertical component of the frictional resistance at the interface opposes the applied load on the footing. The frictional resistance offered by each micropile (of 4 m length and 100 mm diameter) is obtained from the component of earth pressure parallel to the axis of the pile. The additional resistance required to be carried by micropiles is calculated and the number of micropiles are arrived at based on the contribution of each micropile. The calculations indicate that 100 mm diameter micropiles spaced 200 mm c/c provide the frictional resistance required for desired level of bearing capacity improvement. Foundation treatment consisted of driving the micropiles at an angle of 70° with the horizontal and close to the foundation as shown in Fig. 1(c). Plan and section of the structure and a schematic diagram of micropile system are shown in Figure 2. The remedial measures were implemented and the foundation has been retrofitted to the original requirements. While the solution that was suggested above was satisfactory in terms of immediate action of retrofitting, the case study also provided a good opportunity to examine the performance in terms of numerical analysis. Detailed finite element analysis has been conducted using PLAXIS to examine the above case study in terms of its overall performance.

GEOMETRY MODEL AND NUMERICAL ANALYSIS

PLAXIS is a 2-D finite element program, specially developed keeping in view geotechnical considerations. In the analysis, the footing is taken as beam element and assumed to behave as flexible and full footing of width 2.5 m is analysed as plane strain problem. Boundary effects are avoided, by keeping the ratio of mesh size to dimensions of footing at 12. Standard boundary conditions (viz., imposing horizontal as well as vertical fixity to all nodes at bottom of mesh, and arresting horizontal movement of all nodes at both sides of mesh) are applied. The foundation soil was modeled as Mohr-Coulomb material and the micropile is modeled as elastic material. The properties used for the materials are given in Table 1. Micropiles (mild steel pipes with closed ends) of 100 mm diameter and 6 mm thick, spaced uniformly

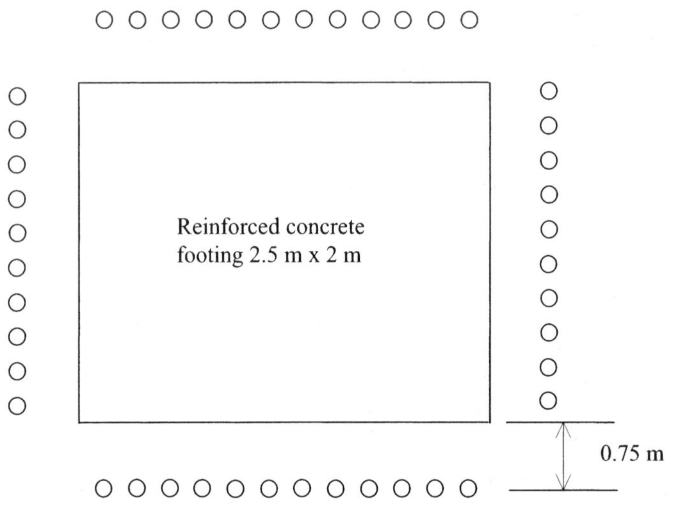

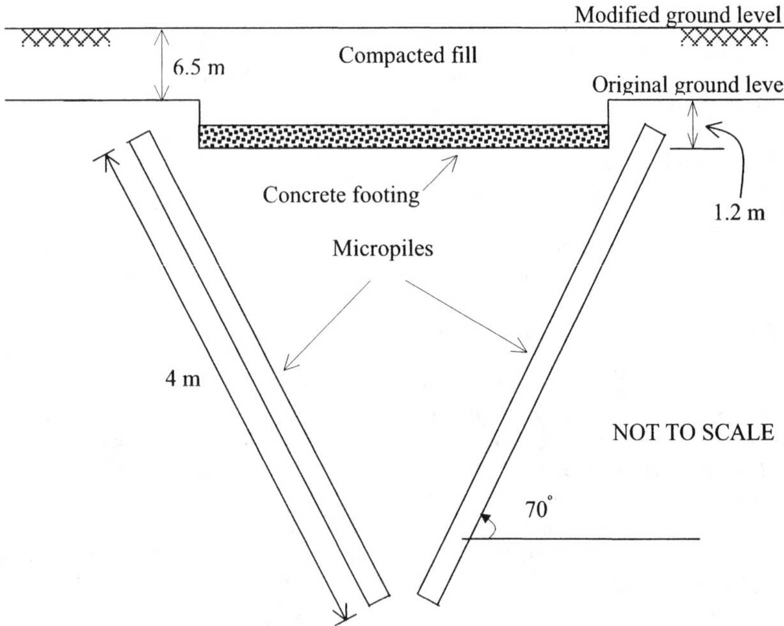

FIG. 2. Plan and cross sectional view of footing strengthened by micropile system

in the third direction are considered for the analysis. Reducing 3D problems with regularly spaced piles to 2D problems involves averaging the effect in 3D over the distance between the elements. Donovan et al. (1984) suggests linear scaling of material properties as a simple and convenient way of distributing the discrete effect of elements over the distance between elements in a regularly spaced pattern. Similar approach was used by Tan et al. (2000) to examine nail-soil interaction behaviour. This approach was used in the present study. Since the micropiles were closely spaced at a distance of twice the diameter of the pile, it was assumed that densification of the soil surrounding the piles and the corresponding group effect is significant. Fig. 3 shows the finite element mesh along with the boundary conditions.

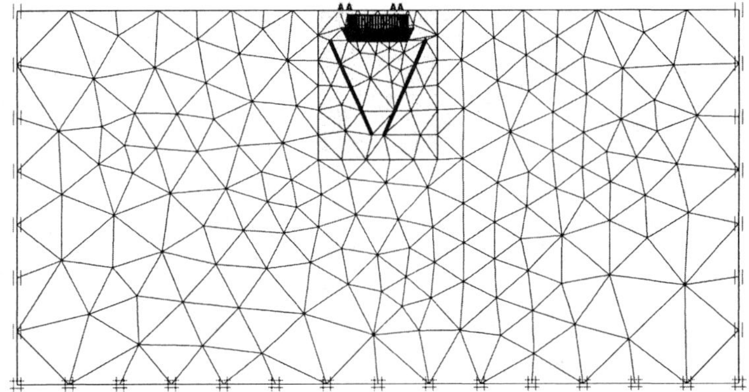

FIG. 3. Finite element mesh with boundary conditions

TABLE 1. Models and material properties used in the analysis

Material (1)	Model (2)	Cohesion (c') kPa (3)	φ' (°) (4)	E (MPa) (5)	ν (6)	γ_b (kN/m^3) (7)
Loose sand	Mohr-Coloumb	1	25	13	0.30	18
Dense sand	Mohr-Coloumb	1	35	30	0.30	20
Micropile	Elastic	-	-	2.1.x10^5	0.25	78.5

RESULTS AND DISCUSSION

Fig. 4 shows the load displacement response obtained from numerical simulations. Curve 1 is the load displacement curve obtained for the in-situ soil. The allowable bearing pressure corresponding to 25 mm is 66.8 kPa and is close to the allowable bearing pressure of 70 kPa considered in the revised design. Corresponding to 50 mm, the allowable bearing pressure is 106.3 kPa. Curve 2 is the load displacement curve obtained corresponding to micropiles. The curve includes the densification effect, with properties of the sand in dense state indicated in Table 1 for the analysis. Curve 2 shows the overall improvement in bearing capacity obtained. A value of 260 kPa is obtained. Hence the numerical results obtained in the present study validate the ground improvement adopted in the field using micropiles.

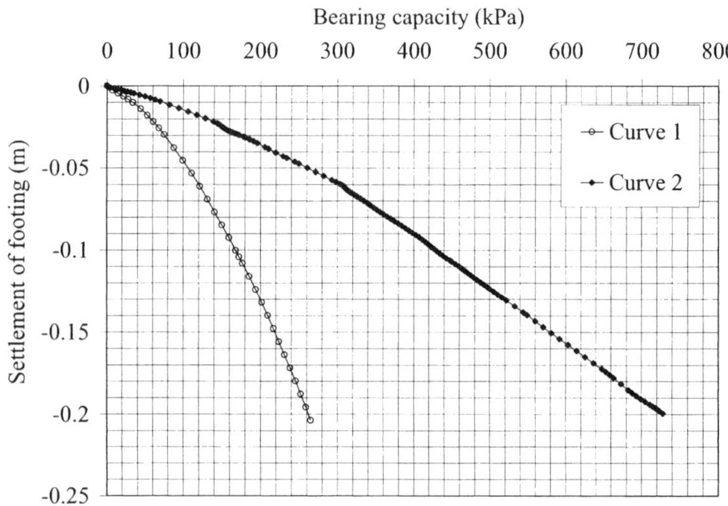

FIG. 4. Load-settlement curves with and without micropiling

CONCLUDING REMARKS

The paper describes a case study in which the bearing capacity of the foundation soil is improved using micropiles. Non-linear finite element analysis is carried out to examine the applicability and level of improvement obtained in the field. Densification of soil surrounding the micropiles and the frictional resistance between the micropiles and the foundation were given due consideration in the analysis. The results confirm that the methodology used was effective in obtaining the desired level of improvement.

REFERENCES

Donovan, K., Pariseau, W.G., and Cepak, M. (1984). "Finite element approach to cable bolting in steeply digging VCR slopes". *Geomechanics application in underground hardrock mining*, 65-90.

Juran, I., Bruce, D.A., Dimillio, A., and Benslimane, A. (1999). "Micropiles: the state of practice. Part II: design of single micropiles and groups and networks of micropiles". *Ground Improvement*, 3, 89-110.

Lizzi, F. (1982). "The pali radice (root piles)". *Symposium on soil and rock improvement techniques including geotextiles, reinforced earth and modern piling methods*, Bangkok, D-3.

Plumelle, C. (1984). "Improvement of the bearing capacity of the soil by inserts of group and reticulated micropiles". *International symposium on in-situ reinforcement of soils and rocks*, Paris, 83-89.

Sridharan, A., and Murthy, B.R.S. (1993). "Remedial measures to building settlement problem". *Proceedings of Third International conference on case histories in Geotechnical Engineering*, St. Louis, Missouri, 221-224.

Tan, S.A., Luo, S.Q., and Yong, K.Y. (2000). "Simplified models for soil-nail lateral interaction". *Ground Improvement*, Vol. 4, No. 4, 141-152.

FOUNDATION UNDERPINNING WITH MINI-PILES "A FIRST" IN GUYANA, SOUTH AMERICA

Richard P. Stulgis[1], Member ASCE, Brock E. Barry[2], Member ASCE, Francis S. Harvey, Jr.[3], Member, ASCE

ABSTRACT: In 1994, USAID provided funding to the Sisters of Mercy of the Americas for construction of a three-story expansion to the St. Joseph's Mercy Hospital in Georgetown, Guyana, South America. Subsurface conditions consist of approximately 16.5 m of very soft to soft marine clay overlying stiff to very stiff clays. The foundation system employed was a soil-supported, monolithically-cast, reinforced concrete mat foundation with down-turned beams arranged in a grid pattern below the mat. Damaging building settlement occurred within the first few years after completion of construction. In 1997, investigations were conducted to identify the cause of the on-going building settlement, and a remedial foundation design was subsequently developed. Foundation underpinning was performed in 2000, and consisted of installation of 102 grouted mini-piles (approximately 15 cm in diameter; 21 to 29 m long) drilled through the foundation mat and upper soft clay into the underlying very stiff/hard clay stratum. This is the first use of drilled, grouted mini-pile foundations in Guyana. Load transfer to the mini-pile foundations was achieved by constructing 28 reinforced concrete pile caps at existing column locations. The paper will describe: a) the nature of the building settlement problem, b) the basis for and details of the remedial foundation design, c) the unique aspects of mini-pile installation adopted by the Contractor, d) results of pile load tests, e) and post-construction settlement monitoring results.

[1] Director of Consulting Services; Geocomp Corporation, 1145 Massachusetts Ave, Boxborough, MA 01719

[2] Staff Engineer; Haley & Aldrich of New York, 200 Town Centre Dr., Rochester, NY 14623

[3] Principal; Harvey & Tracy Associates, Inc, 143 Dewey St., Worcester, MA 01610

PROJECT BACKGROUND

In 1994, a three-story hospital wing (1994 Addition) was constructed at the St. Joseph's Mercy Hospital in Georgetown, Guyana, South America. The 1994 Addition is located immediately south of and adjacent to the "old medical wing" (1945 Addition), see Figure 1. The 1994 Addition consists of three full floor levels, is of reinforced concrete construction, and occupies a footprint area of approximately 560 sq. m. Similarly, the 1945 Addition is also of reinforced concrete construction and consists of three building levels. There are no physical or structural connections between the two buildings. However, cantilever slabs project from the 1994 Addition and the 1945 Addition at the first and second floor levels to permit pedestrian traffic between both structures.

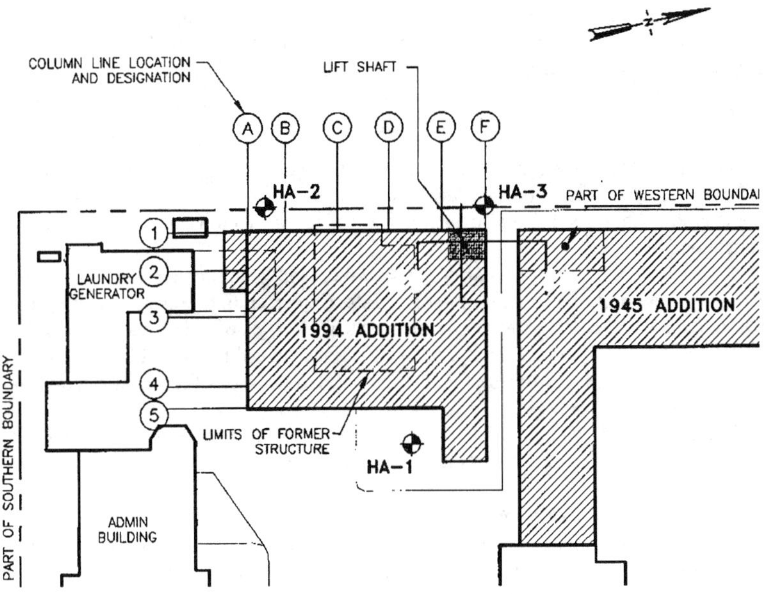

FIG. 1. Site Plan

Subsurface soil conditions in the upper zone of the coastal plain of Guyana consist of relatively deep deposits of soft to very soft marine clay, overlying extensive stiff to hard clay deposits. The 1994 Addition is supported on a reinforced concrete mat foundation (25 cm thick slab cast monolithically with 25 cm wide by 43 cm deep [projection below bottom of slab] down stand beams arranged on a grid pattern). The contact pressure at the base of the foundation mat assumed by the original project designer was 20 kPa. Subsequent independent analysis of the building loads concluded that the actual contact pressure was more likely 25 kPa. A lift shaft (elevator pit), approximately 2.4 m x 3 m in plan dimensions and 1.8 m deep, is located in the northwest corner of the building. The 1945 Addition is supported on timber pile foundations. Geotechnical studies were not undertaken as part of the 1994 Addition design.

Evidence of downward building movement was noticed in October 1995 (approximately one year after the completion of construction). However, in November 1994 (immediately after completion of construction) vertical movements had been noted at the roofline between the 1994 Addition and 1945 Addition.

Investigations were conducted in 1997 to evaluate building distress and recommend remedial measures. Building settlement was evident along the northern perimeter where the 1994 Addition abuts the 1945 Addition. The cantilever slabs of the 1994 Addition had dropped with respect to the corresponding floor levels of the 1945 Addition. The vertical movement varied from approximately 6.5 cm at the northeast building corner to about 9 cm at the northwest building corner.

GEOTECHNICAL EVALUATION

Geotechnical investigations were conducted in the Fall of 1997, and consisted of: a) test borings, b) in-situ vane shear tests and recovery of undisturbed tube samples of the clay foundation soils, and c) laboratory soil classification and one-dimensional consolidation tests.

As indicated by the test borings, subsurface soil conditions beneath the 1994 Addition consisted of a soft to very soft marine clay extending to depths on the order of 16.5m below existing ground surface. Stiff to very stiff clay was encountered below this depth. A relatively thin layer of surface fill (white sand) blankets the site and is the bedding layer for the foundation mat. Groundwater level is within several feet of ground surface.

The in-situ field vane and laboratory soil tests indicated the following relative to the engineering characteristics of the soft marine clay:

- It is highly plastic and extremely sensitive,
- It exhibits slight pre-consolidation in the upper 4.5 m (i.e. only about 9.5 kPa greater than the existing effective overburden pressure), while the lower portion of the deposit appears to exhibit slightly greater pre-consolidation (i.e. approximately 19 kPa greater than the existing effective overburden pressure),
- The shear strength within approximately 6 m of ground surface typically varies between about 7 and 9 kPa. The clays are sensitive and disturbance can reduce the shear strength to less than 2.5 kPa.

Based on site observations, review of construction records, and the results of settlement and bearing capacity analyses, it was concluded that: a) it was unlikely that consolidation settlement of the clay was responsible for the excessive building movements which have occurred along the north building line, b) settlement analyses indicated that the expected consolidation settlement at the building edges would be several centimeters, but would take many years to occur (local experience suggested that approximately one-half of the consolidation settlement would be expected to occur over about a 35-year time period after construction), and c) it was more likely that the excessive vertical movements experienced by the building were representative of a progressive local shear failure in the clay foundation soils beneath the northern portion of the building.

SETTLEMENT OBSERVATION PROGRAM

The geotechnical/structural engineering evaluation concluded that, although the structure was not considered to be in immediate peril, a detailed settlement survey should be undertaken to establish existing building distortion and the rate of any continuing building movement.

A settlement observation program was implemented in May 1998 and consisted of the following:

- A network of tilt plates was attached to vertical building surfaces at selected locations along the exterior building perimeter (northwest, north, northeast sides), to monitor any progressive horizontal building movements that might be continuing due to shear strains in the foundation soils.
- Both an optical level survey and water level survey system were established and calibrated. Once reproducible readings between the two systems were confirmed, the simpler, more economical, water level survey system was used for the duration of the monitoring program.
- Piezometers were installed in the soft clay deposit to determine if excess porewater pressure conditions exist that would be indicative of shear strains and/or on-going consolidation settlement.

Data from the instrumentation installed in May 1998 was obtained over about a six-month period, and identified troubling settlement trends:

- The entire structure was settling, but at different rates. The northern half of the structure was settling at rates that were in excess of twice that observed in the southern portion. The greater settlement rates observed in the northern portion of the structure were due to continuing shear deformations in the soft foundation soils, while the lesser rates occurring in other parts of the structure represented long-term consolidation of the underlying soft clay.
- Based on the settlement rates, it was estimated that an additional 2.5± to 7.5+ cm of settlement would occur during the next five-year period. This would result in a total building settlement, since the end of construction, of almost 18 cm at the NW building corner, about 13 cm at the NE building corner, and approximately 5 cm along the southern building line. While the total settlement was of concern, the magnitude and pattern of differential settlement (both to date and future estimated) was of greater concern. Such settlement could compromise the structural integrity of both the 1994 Addition and 1945

Addition toward which it was leaning, by inducing stresses that the structural members were not capable of resisting within allowable design parameters.

- The piezometer data confirmed that excess porewater pressures existed within the underlying clay. The data was indicative of the presence of both shear strains in the upper portion of the clay deposit and on-going long-term consolidation of the clay deposit as a whole.
- The tilt plate data for the north wall of the 1994 Addition and the south wall of the 1945 Addition did not indicate any discernible lateral movement trends. Tilt plate readings at the southwest building corner suggested rotation to the north, which was consistent with the pattern of differential settlement observed. Through a review of available drawings and on-site investigations it was concluded that the two buildings would, over time, likely come in structural contact and be subjected to forces beyond the capacity of their structural systems to resist.

The water level survey was continued through foundation remediation and for approximately one year after completion. As discussed later in the "Conclusions" to this paper, the readings confirmed that building movements had been arrested.

FOUNDATION REMEDIATION

Based on the magnitude and observed pattern of building movements, it was recommended that foundation underpinning be implemented immediately for the entire structure. Due to the need to install piles within the building proper, the underpinning approach selected consisted of small diameter grouted piles (mini-piles) drilled through the existing mat foundation into the stiff clay some 18 to 21m below the mat foundation. The remedial design also considered appropriate treatment for the existing utility lines within and below the existing mat foundation.

Several alternatives were evaluated to develop the required mini-pile capacities within the stiff/hard clay bearing stratum (individual structural pile loads varied from 11.5 to 34 metric tons). These included single stage tremie grouting, multiple stage post-grouting and mechanically over-reaming the embedded length within the bearing stratum. Based on our experience, increasing the ultimate side shear through the use of post-grouting was economically attractive, and the mini-pile design was based on this method of installation. An allowable side shear of 72 kPa was adopted, and the mini-pile embedment lengths were established on this basis.

The initial mini-pile design consisted of a NO. 11 DYWIDAG Reinforcing Steel Bar encapsulated in cement grout (34.5 mPa ultimate strength). A post grout tube was provided within the pile cross section. The upper length of the mini-pile was cased throughout the soft clay zone. Polyethylene-coated steel casing (15 cm O.D.) was selected, in order to reduce the potential for negative skin friction loading on the mini-pile. During the project bidding phase, the Contractor proposed modifications which were accepted and included: a) eliminating the DYWIDAG bar and using a fully cased composite mini-pile structural section with special post-grout injection ports in the casing within the embedment length, and b) increasing the embedment length to account for negative skin friction, thereby eliminating the need for polyethylene coating for the casing.

A number of methods were evaluated to transfer support of the building mass from the existing mat foundation to the proposed minipile system. The building loads consisted of the loads from framed levels above the ground floor which were transmitted through columns to the ground floor level foundation mat, of the loads at the ground floor level supported directly on the ground floor mat and of the foundation mat itself. Analyses indicated that the calculated center of the building mass was eccentric to the geometric center of the mat area in the direction of the greatest building settlement. Because the installation of the piles would result in load reversals in the foundation mat, it was necessary to determine the allowable spans of the mat in this mode. After it was determined that the mat structure could span between column locations, it was decided that pile groupings would be installed at column locations only.

The initial concept for load transfer consisted of a simple beam spanning between piles, with a concentrated load made up of its share of the column load from above, plus the end reaction load of the mat foundation haunch. The initial design for these conditions consisted of back-to-back steel channels, primarily for high shear capabilities available in standard sections. This also allowed for normal span variations, and bearing devices capable of loading and leveling at each column location. All of these elements were to be galvanized and encapsulated in non-structural concrete. It was also contemplated that excavation would be minimal and would only be that required to slip the back-to-back channels into place. Thus, quality control concerns would be kept to a minimum. In many instances the distance from column centerline to pile location was increased to minimize demolition of walls and partitions and relocation of fixed hospital equipment.

During bidding, the Contractor proposed a load transfer solution involving the use of reinforced concrete. The alternate was accepted, subject to design and installation approvals. The problems associated with handling the high shears required the use of 34.5 mPa concrete, an adequate configuration of the steel reinforcing, minimal excavation and minimal additional concrete.

FOUNDATION UNDERPINNING

Foundation underpinning operations were conducted during the period February through October 2000. The work included installing 102 minipiles (lengths varied from 21 to 29 m) and constructing 28 load transfer systems at existing column locations.

Underpinning operations encountered many impediments, some planned and some not planned:

- Phased conduct of the work and complete protection of the existing kitchen facilities (located on the ground floor), so as not to disrupt the Hospital's ability to provide food service to patients and staff. Temporary relocation and re-installation of the kitchen facilities were required.

- Selected demolition of portions of the existing structure to permit access for installation of mini-piles, and subsequent reconstruction.

- Underslab utilities and utilities embedded within the existing mat foundation.

Minipile Load Tests

Axial compression load tests were performed in general accordance with ASTM D 1143 prior to the start of production pile installation, to verify the assumed pile/soil design skin friction. Freely moving 1 cm diameter tell-tales were installed in casings mounted on the interior of the mini-pile. Tell-tales extended to the bottom of the mini-pile and to the top of the embedment length. A test boring was conducted adjacent to the test piles and undisturbed samples were obtained for laboratory testing. Mini-pile casing installation rates, grouting pressures, and grout volumes were closely monitored and documented for each test pile.

Figure 2 depicts the subsurface conditions at the test pile location, details of the test pile installation, and rate of advance of both the pilot hole and permanent casing installation.

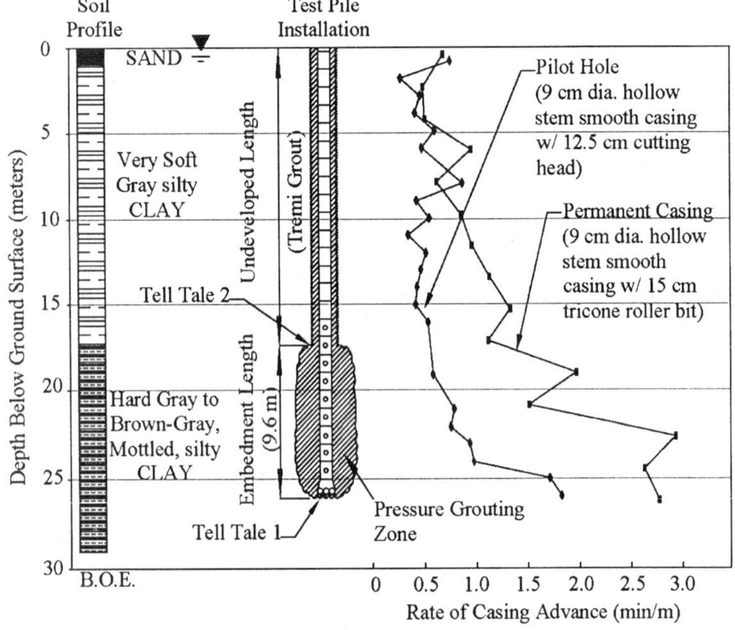

FIG. 2. Test Pile Details

Figure 3 presents the results for one of the completed load tests. The test pile was axially loaded to 68 metric tons; approximately twice design load capacity. At twice design load, significant downward movement was observed. The pile, in fact, began failing (plunging failure). Tell Tale 2 (bottom of mini-pile) indicates that load applied at the top of the pile was carried entirely in side shear within the embedment length until between 50 and 60 metric tons, at which time side shear was being overcome and load was being transferred to the pile tip. The back-calculated allowable side shear from the pile load test was approximately 76 kPa.

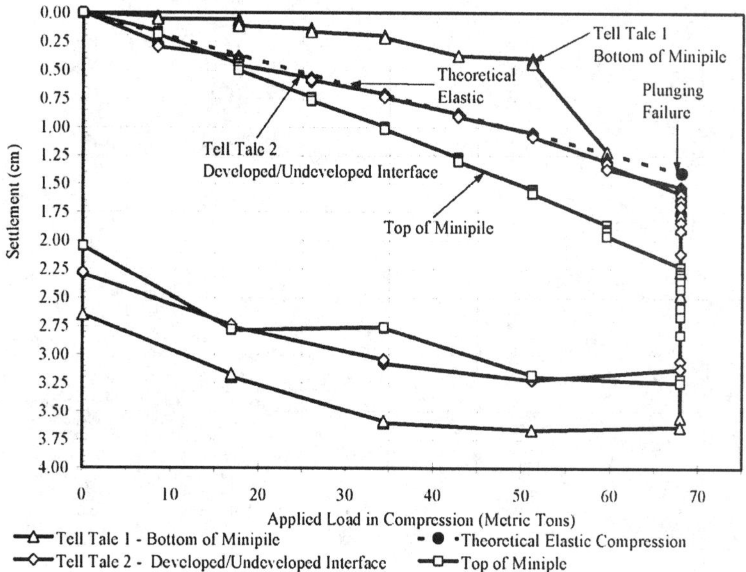

FIG. 3. Pile Load Test Results

Underpinning Sequence

Figure 4 depicts the sequence of underpinning operations at a typical building column. Details of the operations are discussed hereinafter.

Step 1 - Preparation for Underpinning

15 cm diameter openings were cored through the existing concrete floor slab at mini-pile locations. This maintained the structural integrity of the foundation mat and provided an even work surface for the mini-pile rigs. The cored holes also provided lateral restraint at the foundation mat level during pile installation. Photograph 1 of Figure 4 shows completed concrete coring at four pile locations adjacent to an interior column.

1. Preparing Cafeteria for Underpinning Construction

2. Installing Minipile Foundations

3. Removing Floor Slab to Construct Reinforced Concrete Pile Cap

4. Excavating for Pile Cap and Installing Reinforcing Steel

5. Assembled Load Transfer Cage

6. Completed Column Underpinning

FIG. 4. Underpinning Operation Sequence

Step 2 – Installing Mini-Piles

Mini-piles were installed by two hydraulically operated rigs designed specifically for mini-pile installation.

- Equipment - The smaller of the two rigs was a Contractor-fabricated unit aptly named the Minifor. The Minifor rig was a compact machine designed for work in confined areas. The narrow profile and independently controlled rubber tracks allowed the Minifor to easily maneuver through standard doorways and into small spaces. The larger mini-pile rig was a Puntel PX 310S. This rig had a longer stroke length and faster installation speed than the Minifor, but it was limited in accessing much of the building interior due to its overall size. Power to operate the hydraulics on the mini-pile rigs, as well as the grout batch plant and all other site equipment, was delivered by a temporary Caterpillar D343 diesel powered generator. In a location where city-wide power outages are commonplace, the construction generator allowed the Contractor to continue work independent of power outages.

- Pilot Hole - At each mini-pile location, a pilot hole was advanced the full depth of the proposed minipile design length. The pilot hole was created with a 12.5 cm cutting head attached to 9 cm diameter hollow stem smooth casing. The threaded casing sections had a length of approximately 1 m to facilitate installation in the low headroom conditions. A re-circulated bentonite slurry mixture was utilized to advance the pilot hole, preventing the pilot hole from collapsing, and carrying drill cuttings to the surface. The pilot hole was used to determine the top of the pile embedment length.

- Permanent Casing - Permanent casing was advanced to design bottom of pile depth with a sacrificial 15.5 cm tricone roller bit. Permanent casing sections were fabricated of Grade 2 steel, with an outside diameter of 9 cm and an inside diameter of 7 cm. The 1 m long sections of permanent casing were threaded together at the surface as each section was advanced. Casing sections within the embedment length of the minipile incorporated three small diameter rubberized injection ports at the mid-point of the long axis, spaced at 120-degrees around the circumference of the casing.

- Grouting - After the installation of permanent casing to the design depth, grout was tremied through the casing interior until returning at the surface in the roughly 6.7 cm annulus around the casing. Primary grouting volumes were on the order of 750 liters per mini-pile. The primary grout was allowed to partially set prior to flushing the grout from the interior of the casing, while maintaining the primary grout in the annulus around the casing.

  A double-packer system was then utilized to isolate 1 m sections of the casing in the embedment length and perform post-grouting through the casing injection ports. Post-grouting was performed for each 1 m casing section of the embedment length. Post-grouting was terminated at an injection resistance pressure of approximately 3 mPa or a total injection volume of approximately 0.5 m^3. The casing interior was then filled with secondary grout to complete the minipile installation process.

The primary and secondary grout consisted of a Type I cement mixture with a 0.5 water-cement ratio. Grout was mixed and delivered to the mini-piles, under pressure, by an on-site batch plant and pump. Regular density and viscosity measurements were made by an independent testing agency to ensure quality of the grout mixture. Grout cubes were prepared, cured, and broken to ensure the required 34,500 kPa 28-day design strength. A total of approximately 124,600 liters of grout was prepared on site and pumped during the underpinning operation.

Step 3 - Load Transfer (Pile Cap) Construction

After a pile group (typically consisting of four mini-piles) had been successfully installed at a column location, the installation of the load transfer system commenced. The load transfer installation at each column location required a portion of the foundation mat slab to be removed at each pile location. The predetermined extent of the slab removal varied and was kept to the minimum required for installation of the reinforcing and the required amount of concrete cover of the reinforcing for each of the beams comprising the load transfer system at each column location. The procedure consisted of: a) laying out the slab removal areas, b) saw cutting 13 mm deep into the foundation mat and then chiseling out of the remaining depth of concrete so as to maintain the integrity of the existing reinforcing at the perimeter of

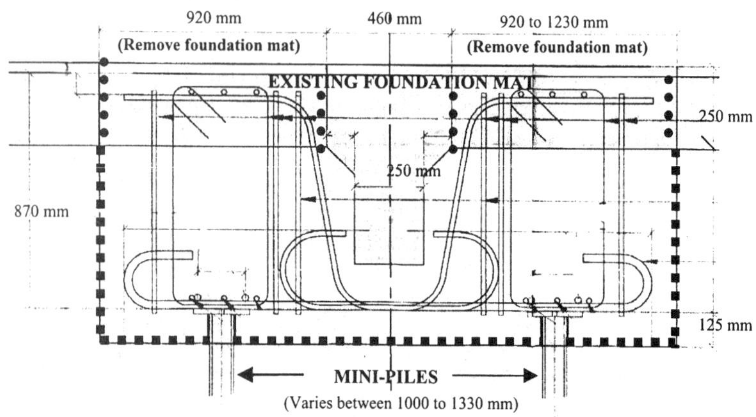

Fig. 5. Load Transfer System Details

the slab removal areas, c) carefully removing only the amount of earth necessary to construct the reinforcing assemblies (an above ground mockup of a reinforcing assembly was constructed to develop installation procedures), d) stabilizing the sides of the excavation below groundwater level, e) installing the reinforcing assemblies through constricted openings in the existing foundation mat, and f) placing concrete.

Two significant in-country quality control issues developed relative to the use of reinforced concrete beam assemblies for load transfer systems: a) fabricating bars within allowable bend tolerances, and b) batching, mixing and placing concrete to attain the required strength. The concrete quality issue was directly related to known difficulties in maintaining quality control for structural concrete in a tropical climate and the limitation of resources available to address these difficulties (i.e. ice or chilled water).

CONCLUSIONS

Innovative foundation remediation technology was implemented in Guyana, South America to stop potentially damaging building settlement at the St. Joseph's Mercy Hospital. As can be seen in Figure 6, the first time use of small diameter grouted piles (mini-piles) in Guyana was successful in averting a probable local bearing capacity failure and eliminating the effects of long term consolidation settlement. Building rededication ceremonies for the 1994 Addition were held in Guyana in February 2001.

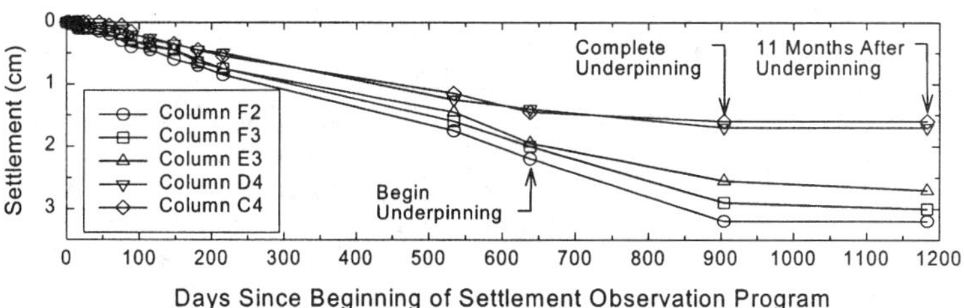

Fig. 6. Settlement vs. Time

ACKNOWLEDGMENTS

The authors would like to acknowledge Sister Susan Nowalis (Project Coordinator, Sisters of Mercy of the Americas) for her guidance throughout the project; Marvin Oosterbaan and Glen Zoladz (Haley & Aldrich, Inc.) for their technical contributions during design and construction; and Charles Ceres (Ground Structures Engineering Consultants LTD., Guyana) and his staff for local engineering support. The foundation underpinning was performed by Soletanche, Inc.

FOUR FALLS CORPORATE CENTER – A CASE STUDY

Iliana Alvarado [1], P.E., Member Geo-Institute; Allen Cadden [2], P.E., Member Geo-Institute; Jesús Gómez [3], Ph.D., P.E., Member Geo-Institute

ABSTRACT: Urban development in West Conshohocken, near Philadelphia, has been occurring at a fast pace over the past 10 years. The 300 Four Falls office complex was to be constructed in one of the few remaining undeveloped sites, near the original 100 Four Falls office building from the early 1980s. The site conditions were a major issue in the development. With an odd-shaped narrow lot, a creek across the site, and existing grade changes of nearly 18.3 m (60 ft), fitting a 26,942 m^2 (290,000 sf) building represented a major challenge. Exploration revealed further difficulties, including cuts of over 12.2 m (40 ft) into hard rock on one side of the site, and very soft soil materials to depths of over 24.4 m (80 ft) on the other.

Solutions included a combined foundation system consisting of spread footings on rock, micropiles, and 890 kN (100 ton) driven H-piles. To accommodate the deep excavations around three sides of the site and extensive filling on the fourth side, various earth support structures were built. These consisted of soldier piles and tiebacks, soil nails, rock bolts, and Mechanically Stabilized Earth (MSE) walls. This project was truly a study in geotechnical engineering challenges and solutions that met the needs of a difficult site. This paper describes the most relevant geotechnical aspects of the project, and the application of different geo-construction techniques to meet the demands of the site.

SITE AND PROJECT DESCRIPTION

The 18,580 m^2 (200,000 sf) site is located in the Borough of West Conshohocken in southeastern Pennsylvania. The project site is a narrow lot bounded by Conshohocken State Road (S.R. 23), Bliss Street, Woodmont Road and bridge, and

[1] Senior Engineer, Schnabel Engineering, Inc., 510 East Gay Street, West Chester, PA 19380; ialvarado@schnabel-eng.com

[2] Principal, Schnabel Engineering, Inc., 510 East Gay Street, West Chester, PA 19380; acadden@schnabel-eng.com

[3] Associate, Schnabel Engineering, Inc., 510 East Gay Street, West Chester, PA 19380: jgomez@schnabel-eng.com

railroad tracks, as shown on Figure 1. In addition, the site is bisected by Arrowmink Creek, which discharges to the Schuylkill River that is north of the railroad tracks.

The original ground surface slopes at grades steeper than 1H:1V from a high of about EL 37.8 (EL 124 ft) at the southeast corner of the site, and EL 30.5 (EL 100) at the southwest side of the site, to EL 15.3 (EL 50 ft) at the north central portion of the site, at Arrowmink Creek.

Even though the lot was narrow and had limited access, the site was attractive to local developers due to the strong real estate market in this area. The project consisted of constructing a seven-story 26,942 m^2 (290,000 sf), steel frame office building with a six level concrete parking garage beneath the office building. The parking garage lowest floor grade is at EL 22.6 (EL 74 ft), with a footprint of about 5,342 m^2 (57,500 sf).

GEOLOGIC CONDITIONS

The site is located within the Piedmont Physiographic Province. Geologic mapping of this area shows that the site is situated along the contact of three geologic units: Conestoga Limestone, Wissahickon Schist and Felsic Gneiss, and lies along the Cream Valley Fault crush zone. The schist and gneiss are of Precambrian Geologic Age, while the Conestoga Formation is of the Cambrian-Ordovician Age. Folding in this area often results in older rock overlying younger rock with mixed and interbedded geologic features. At the geologic contacts, the rocks are typically subjected to more folding and fracturing, often displaying more weathering, particularly along the crush zone of fault lines. Similarly, the soil profiles vary widely across the site.

Originally, the site was covered by grass and brush, with the exception of some remains of previous structures. Due to the difficult terrain, various exploration methods were used. The subsurface exploration program included soil borings, air percussion probes, test pits, and two-dimensional resistivity.

The exploration program showed that the subsurface conditions were different between the east and west portions of the site. The west side generally consisted of fill soils placed during adjacent site development and demolition. The fill contained brick, concrete, asphalt, rock fragments, and organic material. The bottom of the fill was encountered to EL 22.9 (EL 75 ft). Alluvial deposits and residual soils were present below the fill. The alluvial soils are associated with the meandering and flooding of the Schuylkill River and Arrowmink Creek.

The east side of the site generally consisted of fill less than 3.0 m (10 ft) thick, overlying residual soil. Figure 2 presents a generalized subsurface profile indicating the different strata along the southern side of the building.

SPECIAL CONSIDERATIONS

Design and construction considerations for site development included: a culvert under the building for Arrowmink Creek; relocating existing electric, sanitary, and water utilities to run either around or under the building; site construction with only one access road; and excavation near the century old Woodmont Road bridge.

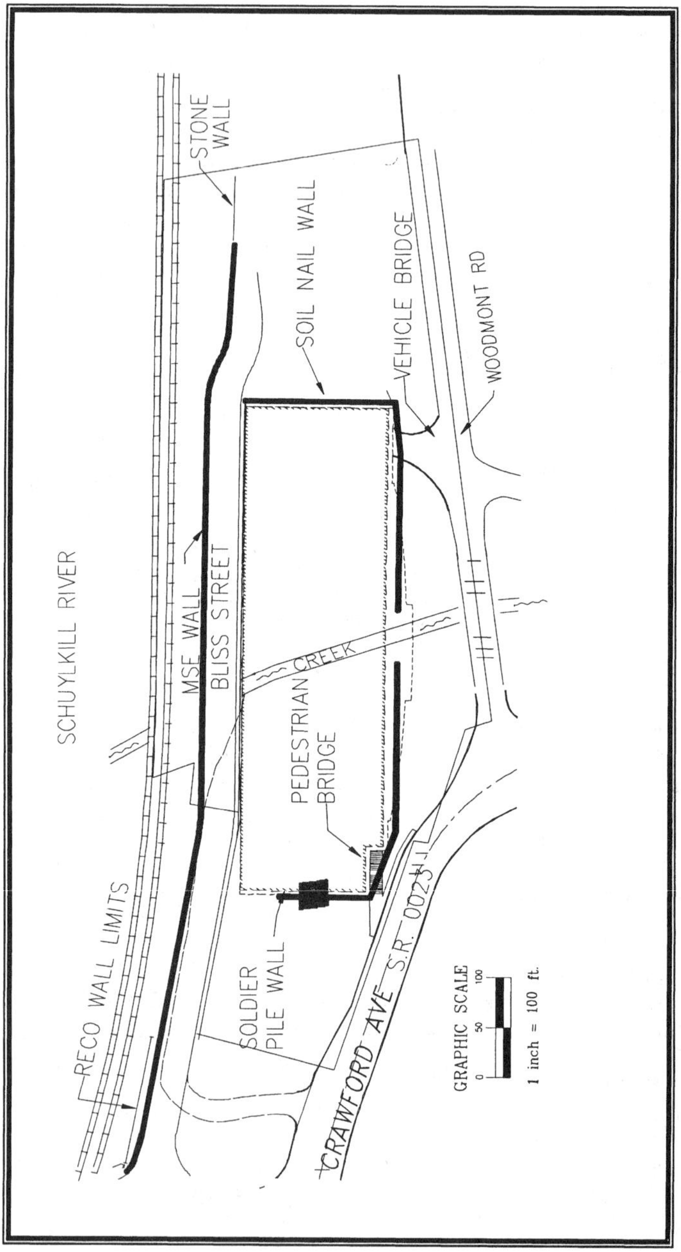

FIG. 1. Site Plan

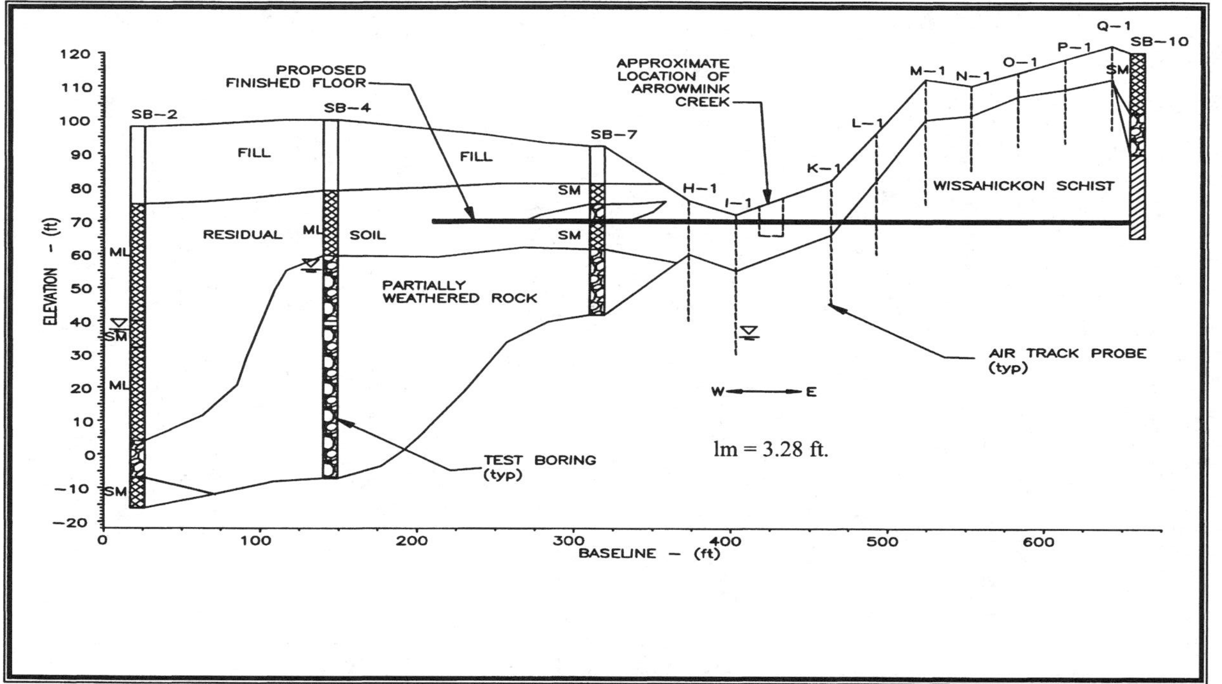

FIG. 2. Subsurface Profile

In addition, earth retention structures were necessary along all four sides of the building. Along the north side, fill walls up to 9.14 m (30 ft) high were needed. At the south side, along S.R. 0023 and Woodmont Road, and along most of the east and west sides of the proposed construction, retaining walls for cuts up to 14.94 m (49 ft) were necessary.

The owner selected to have the retaining walls not be in contact with the building wall to save costs related to garage ventilation and lateral load resistance of the building structure. A minimum 0.9 m (3 ft) gap was left between the lower garage level of the building and the retaining walls to provide natural ventilation. Two access bridges were necessary to span this gap: a vehicle bridge into the garage at the southeast corner, and a pedestrian bridge to the plaza at the southwest corner. These bridges would require special foundation systems, due to their close proximity to the retaining walls.

The subsurface conditions, the building layout, bridges, and structural loads resulted in four types of retaining walls and three types of foundation systems.

RETAINING WALL SOLUTIONS

MSE, soldier pile and lagging, soil nailed, gravity (gabions, concrete, and crib walls), and concrete cantilever walls were all evaluated for use at this site. Factors in evaluating feasible wall alternatives included: utility locations, right-of-ways (ROWs), wall heights, available area for open cuts, and subgrade materials. After meetings between the owner, geotechnical engineer, structural engineer, general contractor, and wall contractors, the wall systems selected included a 78 m (255 ft) long soldier pile and tieback wall, a 262 m (860 ft) long MSE wall, and a 76 m (250 ft) long soil nail wall over a bolted rock face. In addition, an existing RECO™ wall was reinforced with tiebacks to support additional loading due to grade changes. Figure 1 illustrates the location of the walls.

Soldier Pile and Tiebacks

West of the Arrowmink Creek along S.R. 0023, a soldier pile and tieback wall was constructed. The maximum wall height was 10.4 m (34 ft), with up to three levels of tiebacks. It was constructed with either two W10x33 welded soldier piles or one W10x49 soldier pile, with two or three levels of 169 to 623 kN (38 to 140 kip) capacity tiebacks. The tieback design included 1.52 cm (0.6 inch) diameter strands in groups of 2 to 4, with PTI Type I corrosion protection. Figure 3 contains photos of the soldier pile wall construction.

The soldier piles were installed by pre-drilling the holes and setting the piles in a low strength grout. The lagging was placed between the flanges of the piles. The tiebacks were installed between the two welded piles. Once the wall was in place, the permanent face was constructed by welding studs to the piles, then placing the permanent wall reinforcement. With the reinforcement complete, 250 mm (10 inches) of shotcrete were placed.

FIG. 3. Soldier Pile and Tieback Wall

Construction issues with this wall included encountering highly fractured rock during tieback installation in some anchors. This caused significant quantities of grout takes.

MSE Wall

The construction of Bliss Street along most of the north side required an MSE wall with a maximum height of 9.1 m (30 ft). Complications for construction of the MSE wall included installing a water line through the wall that met the water company specifications for access, placement of H-Piles through the geogrid reinforcement for the tower crane, relocation of power poles at the edge of the wall, and the installation of storm and sewer lines and manholes. In addition, the wall construction was a critical path element in order to construct Bliss Street and provide access to the east side of the site along the north side. The MSE wall tied into an old stone wall on the east end, and into a RECO™ wall on the west side of the site. Figure 4 shows the MSE and reinforced RECO wall.

RECO Wall Reinforced with Tiebacks

The RECO wall is located at the northwest end of the site. The existing wall needed to be reinforced because the site grades were raised in this area for Bliss Street access. The reinforcement consisted of tiebacks with capacities of 178 to 294 kN (40 to 66 kips). Large steel plates were placed on the face of the wall to limit the potential for damaging the RECO wall facing during stressing of the tiebacks. Figure 4 presents the reinforced RECO wall and the MSE wall tie in.

Soil Nail Wall over Bolted Rock Face

The eastern corner wall along S.R. 0023 is a soil nail wall over a bolted rock face (see Figure 5). The soil nail wall has a maximum height of 7.0 m (23 ft) over the rock face, with a maximum height of 7.9 m (26 ft). This wall combination was the highest cut on the site, at 14.9 m (49 ft). The soil nails were epoxy coated No. 8, Grade 75 bars, which ranged in length from 5.8 to 10.4 m (19 to 34 ft). The soil nail wall was constructed by first placing temporary shotcrete during excavation of the site, then a permanent facing once the soil nail wall was full height, but before the excavation for the lower level.

Upon completion of the soil nail wall, the rock wall was excavated. During excavation, the rock wall was reinforced with No. 8, Grade 75 bolts. In some areas, reinforced shotcrete was placed if the rock was highly weathered or fractured. A total of 210 rock bolts was installed in lengths between 3.0 to 6.1 m (10 to 20 ft).

FIG. 4. MSE Wall and Reinforced RECO™ Wall

FIG. 5. Soil Nail Wall over Rock Bolt Wall, Under Construction

FOUNDATION SOLUTIONS

As mentioned earlier, the site conditions vary dramatically from one end of the structure to the other. Consequently, the use of a single foundation system was not possible to accommodate the high foundation loads while limiting differential movements. For this reason, the site was divided into two general areas.

At the southeast corner of the site, where rock and partially weathered rock are present near or above the proposed foundation grades, the use of spread footings was the most practical foundation system. In the remainder of the structure, the suitable foundation materials are well below the floor grade; therefore, deep foundation systems were used for the remainder of the office building.

In addition to the subsurface conditions requiring two foundation systems, the two bridges on the site needed a rather unique foundation system. Both the vehicle and the pedestrian bridge required the use of micropiles to transmit the bridge loads to suitable soil and rock. The design of the micropiles considered transferring these loads below the soil and rock retained by the excavation support system.

Shallow Spread Footings

Spread footings of up to 3.8 x 3.8 m (12.5 x 12.5 ft) were used for the southeast corner of the site. The footings have an allowable bearing capacity of 957.6 kPa (10 tsf) on partially weathered rock, or 1915 kPa (20 tsf) where on rock.

Coordination of the bolted rock face and footing locations was critical during the design phase, since the rock wall was so close to the building in some areas. In some areas the rock wall was within 0.9 m (3 ft) of the building wall, which did not give the structural engineering very much space for the heavy building loads.

The structural design included a control joint through the middle of the building. This joint addressed potential differential settlements between the shallow and deep foundation system. In addition, this joint allowed construction of the west side of the building to start before the east side.

Pile Foundations

The deep foundation system selected for the project consisted of driven steel H-piles. The piles were HP12x53 with a yield stress of 345 MPa (50 ksi). An allowable capacity of 890 kN (100 tons) when driven to rock was selected for the project. The pile lengths ranged from 4.6 to 36 m (15 to 118 ft). Up to 13 piles per cap were required at some column locations.

The testing program for the piles included two static loads tests, as well as pile dynamic analysis (PDA) on six additional piles. This testing program was used to determine the driving criteria.

Micropiles

The micropiles for both bridge abutments were necessary to minimize the bridge loads on the top of the retaining walls. The micropiles needed to be supported on the residual soil material below the bottom of the excavation; however, the abutment was located behind the excavation support system.

A similar scenario occurred on the east end, at the vehicle bridge. However, this bridge had significantly higher loads. The rock present at the base of the excavation allowed much larger micropile capacities for the vehicle bridge.

The micropile design for each bridge varied significantly, due to the different loading and subsurface conditions. The pedestrian bridge micropiles had a 175 kN (20 ton) allowable pile capacity, and consisted of a No. 10, Grade 75 bar in a 14.0 cm (5.5 inches) drillhole and were bonded in 6.1 m (20 ft) of residual soil material. The vehicle bridge micropiles had a 756 kN (85 ton) allowable capacity, and consisted of a No. 20, Grade 75 bar in a 14.0 cm (5.5 inches) drillhole and were bonded in 4.6 m (15 ft) of rock. All bars were epoxy coated for additional corrosion protection.

Construction sequencing was an important factor for pile installation. At the pedestrian bridge, piles were installed after the soldier pile and tieback wall. The vehicle bridge micropiles were installed before the soil nail system due to the numerous nails that would need to be avoided. The contractor felt that it would be simpler to locate the micropiles during soil nail installation, than to try to avoid all of the soil nails with the pile drilling.

SUMMARY

At first glance, development of the site did not appear to be cost effective due to the site constraints, topography, existing utilities, and subsurface conditions. However, 300 Four Falls Corporate Center proves that, with a cooperative team, working together from design through construction, creative geotechnical solutions can be developed and adjusted to meet a difficult set of challenges. In the case presented herein, geotechnical difficulties and site constraints were addressed using a combination of systems that provided the necessary flexibility to the project.

Located on a lot visible from Interstate 76, S.R. 0023, Interstate 476, and from the offices and hotels across the Schuylkill River, 300 Four Falls will become a conspicuous benchmark of progress for West Conshohocken, and an example of effective use of foundation and earth support technologies.

ACKNOWLEDGMENTS

We would like to thank the owner of this project, Berwind Property Group and Acorn Development, for working closely with us through the design and construction phases. Additional thanks must also go to the design and construction team, which included Architectural Concepts LLP, Christakis VanOcker Morrison Engineers, Barclay White Skanska USA, and Schnabel Foundation Company.

SEISMIC RESPONSE AND EXTENDED LIFE ANALYSIS OF THE DEEPEST TOP-DOWN SOIL NAIL WALL IN THE U.S.

David M. Cotton[1], P.E., Principal, Golder Associates, Redmond, WA, USA
Richard D. Luark[2], P.E., Associate, Golder Associates, Redmond, WA, USA

ABSTRACT: Typically, temporary shoring support systems are not required to provide for design level earthquake occurrences consistent with the building or structure being constructed inside the excavation. This paper evaluates the seismic response of the deepest soil nailed walled constructed in the U.S. at the time of construction, for a Mw 6.8 earthquake in the Puget Sound Region in Washington State. In addition, an extended life analysis was required when the project was stopped in the middle of an economic downturn. The construction was stopped at a depth of 86.5 feet with the basement slab in place and no lateral support from the eight levels of planned reinforced concrete parking slabs. Project start up is not expected for 2 to 4 years; therefore, the life of the temporary soil nails had to be evaluated for their ability to support the cut for the interim time period. The construction design life was initially set for one year.

1. INTRODUCTION

The top-down soil nail wall for the Bellevue Technology tower included the following unique features:

- The use of strut nails for face stabilization and deflection control;
- The use of strut nails to carry vertical loads;

[1]David M. Cotton, P.E., Principal, Golder Associates, 18300 NE Union Hill, Suite 200, Redmond, WA 98052-3333, www.golder.com
[2]Richard D. Luark[2], P.E., Associate, 18300 NE Union Hill, Suite 200, Redmond, WA 98052-3333, www.golder.com

- The use of splayed nails for a re-entrant corner;
- A composite soil nail system with vertical elements to carry vertical loads and for face; stabilization; and
- The use of top-down construction to build the permanent basement wall with temporary nails to the deepest recorded depth in the U.S. to date at 86.5 feet.

However, since these features have been discussed in detail in other papers, which are referenced, the detailed discussion and presentation in this paper is centered around the seismic response of the soil nail wall during an Mw 6.8 earthquake event, and on evaluation of the potential extended life of the structure beyond the actual design life. The seismic evaluation required displacement analysis where the basic factor of safety criteria for seismic loadings could not be attained. The extended life analysis required an evaluation of the corrosion resistance of the bare steel bars. Both analysis required the use of Recommendations Clouterre 1991, from the French National Research Project Clouterre (1993), referenced below, to evaluate the extended life condition of the soil nail wall when an economic down turn stopped the project prior to completing the basement slab construction.

2. PROJECT BACKGROUND

The proposed Bellevue Technology Tower excavation encompassed an area slightly smaller than a city block, 194 by 167 feet in downtown Bellevue, Washington. A site plan is shown on Figure 1, which illustrates that the project site is bounded on two sides by city streets and existing buildings on the other two sides. The shape of the excavation also includes a reentrant corner. The Key Bank Tower located on the west boundary of the excavation has three levels of below grade parking garage. An 8 story office tower on top of the garage is set back from the excavation face approximately 20 feet. The planned development was to include a twenty story building with eight stories of below grade parking. The below grade floor slabs at 8 levels were intended to provide for the final lateral load support. However, due to an economic downturn in the local market the project construction temporarily ceased following the excavation and foundation construction. Therefore, an evaluation of the soil nail system for an extended period beyond the 12-month planned design life needed to be completed after construction ceased.

The base of the excavation was constructed to a depth of 86.5 feet, below 108th Avenue N.E. It is believed that this is the deepest soil nail wall, built to date in the US, with a top-down construction procedure. The top-down building procedure includes building the permanent 12-inch to 18-inch thick shotcrete basement wall, with a double curtain of steel reinforcing, as the excavation progresses, installing the temporary nail system to a maximum depth of 86.5 feet. A seven foot thick mat foundation was excavated approximately 20 feet inside the excavation line extending the total depth of the excavation to approximately 93.5 feet below street grade. The excavation required approximately 54,500 square feet of permanent shotcrete wall construction. During the excavation and shotcrete placement of the last level

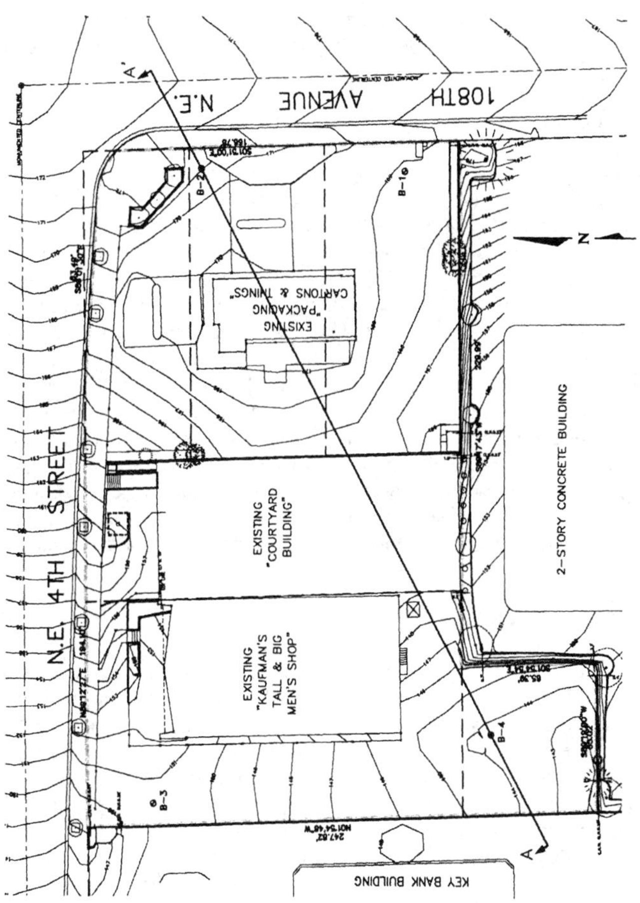

FIG. 1. Site Plan, Bellevue Technology Tower, Bellevue, Washington

of nails, at an average depth of 74.5 feet, the 2001 Nisqually Earthquake occurred with an Mw 6.8 event. No ground loss was experienced, and only nominal permanent lateral displacements were measured after the earthquake. DBM Contractors and Golder Associates worked together during construction to assure a successful project, economically, and also in terms of wall performance.

3. SUBSURFACE CONDITIONS

In general, the site is underlain by variable, minor amounts of fill, overlying dense till, dense advance outwash sands and gravels, and glaciolacustrine deposits of silt. The fill varied in thickness from less than one foot to 14 feet in the northwest corner of the property, and consisted of loose to dense gravel and sand on the west side of the site, and soft to hard silts with sand and gravel on the east side. The native glacial till is generally a dense sandy silt, with little gravel, and ranged from 30 feet thick on the east side of the site to zero feet on the west. The advance outwash gravel was generally a very dense gravel, with some fine to coarse sand and trace silt. Below the advance outwash gravels were the advance outwash sands, consisting of a very dense fine to coarse sand, with trace silt and some gravel. The advance outwash sand and gravels extended to an average depth of 72 feet where the glaciolacustrine deposits were encountered. The glaciolacustrine silts were generally hard and very dense silt and clayey silt with a trace of very fine sand and subrounded gravel. Groundwater was encountered at or near the base of the excavation, at an average depth of 72 feet below street level. Groundwater appears to be perched on top of the glaciolacustrine deposits.

4. SOIL NAIL WALL DESIGN

The soil nail wall design consisted of up to thirteen rows of nails on a 6-by 6-foot nail pattern. The upper row of nails included a strut nail system used to control deflection and ground loss at the face in the upper fill materials, as shown in Figure 2. Strut nails were also used to support the weight of a conveyor about midway down the north wall, which was being used to remove soil from the base of the excavation. The platform load was approximately 150 kips, supported by 5 strut nails, and is shown schematically in Figure 3. The maximum wall height on the north wall, 86.5 feet, included nail lengths that ranged from a maximum for 66 feet in the upper eight rows to decreasing lengths of 54, 50, 45 and 30 feet. The bar sizes ranged from 1-1/4-inch 150 grade steel to 1-inch 150 grade steel. The bars were typically inclined at 15 degrees, except for the upper row which was inclined at 20 degrees to avoid utilities. Strut nails were inclined at 45 degrees. A typical section is shown in Figure 2. The splayed layout for the southwest reentrant corner is shown in Figure 4. The splayed nail design is done simply by taking the 2D analysis performed using the program GoldNail, and then increasing the length of the nail according to the geometric orientation of each nail splayed. During excavation and construction of the excavation support wall, optical survey monitoring was performed that recorded maximum horizontal displacements on the order of 0.001H to 0.002H, where H is the maximum wall height. This resulted in permanent displacements at the top of the wall ranging from 0.25 inches (west wall) to 2 inches (north wall). These displacements during excavation and construction of the wall are normal and within the range for the soil type at the site.

The stability analyses were performed using the program GoldNail, Version 3.11 on five critical sections for each of the five walls surrounding the project. The soil design parameters used in the analysis were based on the field tests completed during the site investigation, laboratory testing, and Golders' experience in the Bellevue area and Puget Sound Region. Specifically the soil nail design parameters used in the final analysis were:

SOIL UNIT	UNIT WEIGHT (PCF)	FRICTION ANGLE (Degrees)	COHESION (PSF)	ULTIMATE OUT RESISTANCE (Kips/Ft)
Fill	120	34	100	6.3
Glacial Till	135	40	300	12
Advance Outwash	125	40	100	8.4
Lacustrine Silts and Clays	125	34	300	4.2

A factor of safety of 2 was utilized for nail pull out and a factor of safety of 1.35 on the soil strength parameters.

Another one of the unique features on this project was the use of vertical elements to stabilize a potential ground loss problem at the face of the excavation on the west side. The excavation section is shown in Figure 5, which illustrates the Key Bank Tower parking garage and tower footings. Anticipated lateral stresses at the face of the excavation combined with excavating 7 foot vertical faces in the relatively clean advance outwash sands and gravels, led to the decision to include the use of vertical face stabilization elements. These are shown in Figure 5 and consisted of 3-inch O.D. schedule 80 pipe in a 6-inch diameter drill hole filled with grout and spaced 36-inches on center. Vertical elements were extended 18 feet below the basement footing for the bank tower, which included three rows of nails. It was felt that adequate confinement would be provided at this level so that no loss of ground support would be experienced below the footing. Also shown in Figure 5 is the use of strut nails which were included to control deflection, carry imposed vertical loads from the adjacent building, and improve face stabilization during excavation.

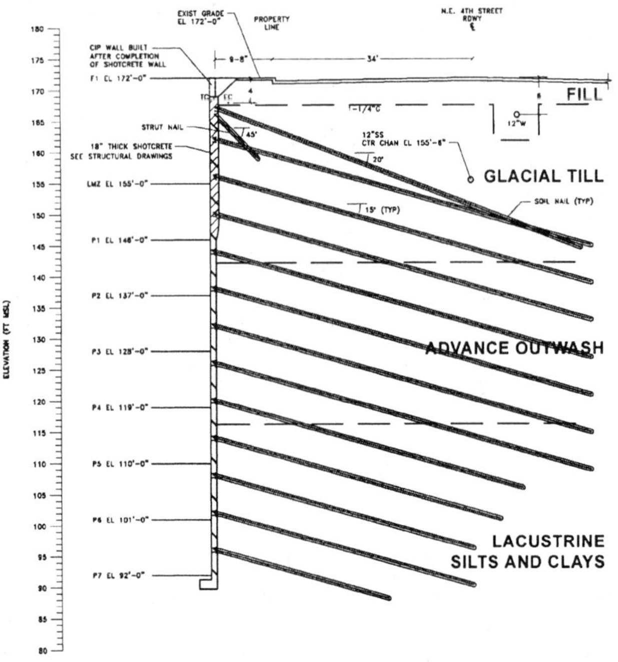

FIG. 2. Typical Wall Cross Section, Bellevue Technology Tower, Bellevue, Washington

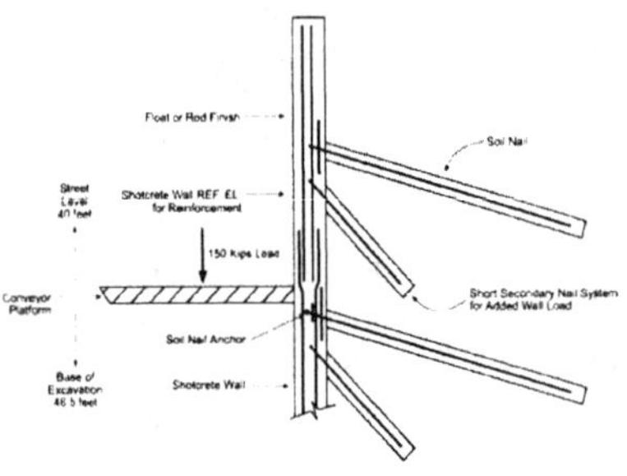

FIG. 3. Permanent Top-Down Method with Strut Nails to Support Conveyor Platform at 40 feet Below Street Level

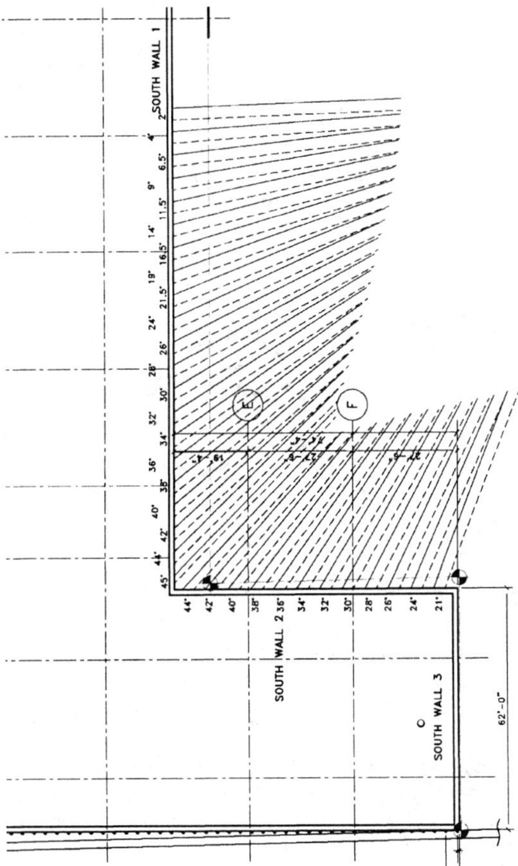

FIG. 4. Splayed Nail Layout Plan, Bellevue Technology Tower, Bellevue, Washington

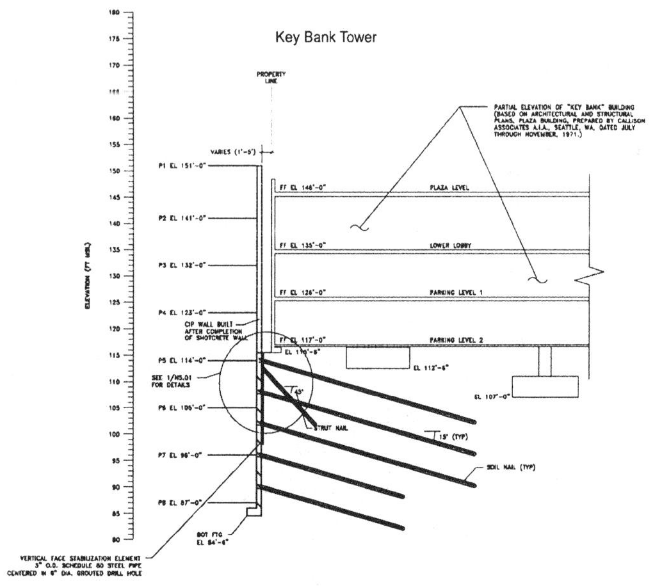

FIG. 5: Typical Cross Section West Wall, Bellevue Technology Tower, Bellevue, Washington

5. PERFORMANCE AND EXTENDED LIFE ANALYSIS

The excavation began in October 2000 and was finished in March 2001. As noted above, the soil nail system performed well during the Nisqually earthquake of February 28, 2001, which occurred during excavation of the deepest lift of the excavation with only nominal movements of the soil nail wall. Maximum movements of 0.25 inches were noted on the reentrant corner of the excavation, typically regarded as the weakest point of the structure. This movement was well within the design criteria for the wall. Additionally, no structural damage to neighboring buildings was discovered. The peak ground accelerations produced by the Nisqually Earthquake in Bellevue were on the order of .11g, about one-third of the design earthquake for permanent structures in the Seattle area.

Finally, an extended life study for the system had to be completed since the project was put on hold prior to completing the floor slabs for the garage. A temporary soil nail system is only designed for the life of the excavation which in this case was approximately one year. Therefore the owner requested an evaluation of the soil nail system for an estimated life of reasonable performance. A temporary wall is not designed for earthquake resistance, although the static factor of safety does provide for some inherent resistance. This was certainly the case for the resistance provided during the Nisqually earthquake. In addition, the soil nails are not designed with any corrosion protection for extended life performance. Based on the results of an extensive seismic analysis of the as built condition it was determined that the existing temporary excavation support system is adequate to resist the UBC design earthquake over the next 4 to 8 years. The following is our detailed analysis and approach toward solving this challenging problem.

A. Seismic Analysis

Overview

Our seismic evaluation consisted of internal and external stability analysis using limit equilibrium methods and the commercially available computer program GoldNail and developing a spreadsheet to calculate displacements, which could occur as a result of the design earthquake event. Both the stability and displacement calculations followed 1996 Federal Highway Administration (FHWA) Manual for Design and Construction of Soil Nail Walls guidelines. In addition, horizontal and vertical displacements, in response to the design earthquake, were estimated for the soil nail shoring wall and the reinforced soil mass.

In general, the walls that face the excavation consist of a nominal 12-inch thick layer of shotcrete with a double reinforcing mat. As the shoring facing consists of the permanent building walls, the structural capacity is well in excess of what is required for temporary face support. Temporary facing requirements for soil nail system typically consist of a single reinforcing mat enveloped in a four-inch thick shotcrete layer.

The shoring was completed in March of 2001, therefore, the design temporary construction life liability ended in March of 2002. A 12-month design life was chosen because the system was designed as a temporary system to maintain stability of the cut until the floor slab system for the garage was completed. The construction of the internal floor system was estimated to be completed in less than one year, at the time of our design. When construction ceased, a portion of three decks had been constructed, but the enclosure sections have not been finished. Because these decks are not connected to the surrounding basement wall, no bracing is provided to the basement walls by the floor system. The shoring system must now perform beyond the temporary period.

The differences between a temporary soil nail wall and a permanent soil nail wall are that corrosion protection is provided for a permanent soil nail and that permanent walls are designed to withstand earthquakes. Neither of these conditions were included in the original design.

Seismic Site Exposure

The Bellevue Technology Tower was designed in accordance with the 1997 Uniform Building Code (UBC 97). The peak ground acceleration (PGA) zoned by the UBC are probabilistically based, and relate to a risk level defined by a 10 percent probability of exceedance in a 50-year period. This is equivalent to a 475-year return period PGA for a specific site under consideration. The seismic hazard mapping data source for the 1997 UBC seismic zoning is Algermissen et al (1990). More recent seismic hazard mapping, completed by the U.S. Geological Survey (1996), was reviewed for comparison with the selected site PGA. The U.S. Geological Survey (1996) also indicates a 475-year return period PGA for the site area of about 0.33 g.

The open excavation at the site experienced the ground shaking effects from the February 28, 2001 Nisqually earthquake without significant distress (Tufenkjian, 2002). The 2001 Nisqually earthquake had a moment magnitude of M_W 6.8 and was located about 65 km (40 mi) from the site (U.S. Geological Survey, 2002; Tufenkjian, 2002). Based on Tufenkjian (2002) the PGA at and near the site from the Nisqually earthquake was 0.11 g.

As no structural damage was recorded during the Nisqually earthquake, it is of interest what the return period of a site PGA of 0.11 g is, and also what the annual probability of exceedance of such a level of shaking (i.e., 0.11 g) may be. An approximation of what the return period may be for a PGA of 0.11 g was developed from a qualitative evaluation of the published PGAs for the 475-, 1,000- and 2,475-year return periods (U.S. Geological Survey, 1996).

When these data are plotted and the hazard curve defined by the data is projected to a 0-year return period, the return period associated with a PGA of 0.11 g is about 70 years. That means that every 70 years, the site should experience one earthquake event of 0.11 g

GoldNail Seismic Analysis

For seismic conditions, as shown in the 1996 FHWA Manual for Design and Construction of Soil Nail Walls (FHWA 96), analyses were performed to evaluate the stabilized (nailed) block of soil with respect to both internal and external seismic forces. GoldNail was run in Factor of Safety mode to determine if the nail lengths specified in the original design would result in a factor of safety of 1.0 under seismic loading. It should be noted that "as built" nail lengths, bar sizes, and soil stratigraphy were used in calculating the factors of safety. The same soil strength parameters were used. However the soil units along the east wall were found to be flat lying instead of slightly dipping as the site was excavated. Therefore, our analysis included the revised stratigraphy identified during construction.

Two seismic checks are required based on 1996 FHWA design guidelines:

- Internal, which covers failure surfaces that primarily intersect the nail reinforcement. The seismic coefficient (A) for this analysis is calculated in accordance with section 5.8.10 of the AASHTO 1992 Standard Specifications for Highways Bridges, 15th Edition recommendations for MSE walls, and
- External, which covers potential failure surfaces that do not intersect the nail reinforcement or intersect them to a limited extent. The seismic coefficient (A_{pk}) for external seismic stability is calculated based on Section 6.3.2 Division 1A-Seismic Design, AASHTO 1992 (Figure 37, Seismic Design Commentary section, 15th Edition. The AASHTO code was used as a result

of the industry standard of care established by the 1996 FHWA design guidelines referenced below.

Internal Seismic Forces

For potential failure surfaces that are primarily "internal" in nature (i.e. intersect the nail reinforcement), the FHWA procedure defines the seismic coefficient A as:

A = (1.45-peak ground acceleration) (peak ground acceleration) = (1.45-0.33)(0.33) = 0.37
Where peak ground acceleration is taken from:

USGS report titled "Probabilistic Earthquake Acceleration and Velocity Maps for the United States and Puerto Rico," by Algermissen S. T., Perkins D. M., Thenhaus P.C., Hanson S. L., and Bender B.L., 1990.

A peak horizontal ground acceleration of 0.33g (33% of gravity) was used to evaluate the seismic stability of the soil nail wall, which is consistent for design of a permanent soil nail wall. This acceleration corresponds to an earthquake with a reoccurrence interval of 10% probability of being exceeded in 50 years, which is considered to be the current design event for office/residential/retail buildings, as based on the Uniform Building Code (UBC 1997).

The factors of safety for the internal seismic analysis as recommended in the 1996 FHWA Manual for Design and Construction of Soil Nail Walls is as follows:

Resistance Component	Factor of Safety
Soil Friction	1.01
Soil Cohesion	1.01
Soil-Grout Adhesion	1.50
Nail Tendon	1.37
Nail Head	1.12

Using the above factors of safety, ten wall sections were analyzed. Table 1 shows the results of the internal stability analyses. Note, when using the above recommended factors of safety on the various structural/soil components that comprise the soil nail wall, a factor of safety of one or greater satisfies the design criteria for permanent soil nail walls, as based on FHWA design guidelines.

External Seismic Forces

For potential failure surfaces that are primarily "external" in nature (i.e. do not intersect the nail reinforcement or only to a limited extent), the FHWA procedure defines the seismic coefficient Apk as:

A_{pk} = (1/2)(peak ground acceleration) = (1/2)(0.33) = 0.17

TABLE 1. Internal Stability Analysis (F.S. ≥ 1.0 Meets FHWA Criteria)

Wall Section	**Internal F.S.**
Northwr	1.01
Northmr	1.00
Norther	1.00
East	1.01
South1e	1.00
South1m	1.01
South1w	1.18
South2r	1.00
South3r	1.16
Westr	1.70

The factors of safety for the external seismic analysis as recommended in the 1996 FHWA Manual for Design and Construction of Soil Nail Walls is as follows:

Resistance Component	**Factor of Safety**
Soil Friction	1.01
Soil Cohesion	1.01
Soil-Grout Adhesion	2.00
Nail Tendon	1.82
Nail Head	1.67

Using the above-recommended factors of safety for external seismic stability, ten wall sections were analyzed. Table 2 shows the results of the external stability analyses. Note, when using the above recommended factors of safety on the various structural/soil components that comprise the soil nail wall, a factor of safety of one or greater satisfies the design criteria for permanent soil nail walls, as based on FHWA design guidelines.

TABLE 2. External Stability Analysis (F.S. ≥ 1.0 Meets FHWA Criteria)

Wall Section	External F.S.
Northwr	1.19
Northmr	0.99
Norther	1.06
East	0.98
South1e	1.02
South1m	1.21
South1w	1.18
South2r	1.04
South3r	1.13
Westr	1.26

Two sections, Northmr and East, have factors of safety less than 1, which is less than recommended by the FHWA design guidelines. This does not mean these sections will fail when subjected to the design earthquake, only that the deflections at the top of the wall will be greater than allowed under standard FHWA design criteria. Typically, a factor of safety of 1.0 will limit the permanent seismic deflections at the top of the soil nail wall to one-inch or less. Therefore, a permanent displacement analysis was performed to estimate the maximum permanent displacement of the walls in response to the design earthquake.

B. Displacement Analysis

Wall Maximum Displacement

Recommendations from the 1996 FHWA Manual for Design and Construction of Soil Nail Walls were followed in calculating displacements. The industry standard of care requires an evaluation of displacements if the factor of safety is less than 1.0 under the seismic design event. This leads to a broader understanding of the ultimate performance of the wall during the design seismic event. The FHWA manual recommended following AASHTO 1992 Division 1-A- Seismic Design AASHTO 15th Edition, which recommends that the permanent seismic displacement of a soil nail wall can be estimated using the following equation:

$$d = 0.087(V^2)/(Ag)(N/A)^{-4} \qquad (1)$$

where:

d = permanent displacement,
V = maximum (peak) ground velocity,
A = maximum (peak) ground acceleration,
N = the yield acceleration, and
g = gravity.

V and A came from the USGS Probabilistic Map for the 10% probability of being exceeded in 50-years (V equals 11.02 inches/sec and A = 0.33). N was calculated for each wall with GoldNail, where N is the seismic coefficient required for a factor of safety of 1 for external stability. The results of the displacement calculations are shown in Table 3.

TABLE 3. Estimated Permanent Displacements At The Top Of The Soil Wall Due To The Design Earthquake Event

Wall Section	N	d (inches)
Northwr	0.27	0.18
Northmr	0.165	1.33
Norther	0.2	0.61
East	0.15	1.94
South1e	0.18	0.94
South1m	0.28	0.16
South1w	0.265	0.20
South2r	0.19	0.75
South3r	0.24	0.30
Westr	0.5	0.02

The above horizontal displacements are calculated at the top of the wall where the greatest movement will occur. Vertical displacements are estimated to be equal to the horizontal. Movement of the wall is assumed to be zero at the bottom of the excavation.

Displacements of Adjacent Structures/Utilities

In addition we calculated the potential permanent lateral and vertical displacements which could occur at critical locations behind the wall in response to the design earthquake. Specifically, we examined the potential movement on streets, utilities, and structures adjacent to the wall, in the event of the design earthquake. We performed our analysis based on the guidelines presented in the Recommendations CLOUTERRE (1991), which presents guidelines for performing displacement analyses for soil nailed wall systems. Our analyses are summarized below.

North Wall Utilities (4th Street NE)

Five utilities are located beneath 4th street in the vicinity of the North wall. They are located at various depths, and consist of a 1.25-inch diameter gas line, an 8-inch diameter storm sewer, a 12-inch diameter sanitary sewer, a 12-inch water main, and an 18-inch diameter storm sewer on the opposite side of the street. The estimated displacements of these utilities at the wall midpoint, and the predicted radius of curvature that could develop in the event of displacement is summarized in Table 4.

TABLE 4. Estimated Displacement at Midpoint of North Wall (inches)

Utility	Estimated Displacement at Midpoint of North Wall (inches)	Predicted Resultant Radius of Curvature Induced in the Utility (feet)
1-1/4-inch Gas	0.80	1.5×10^5
8-inch Storm Sewer	0.90	1.35×10^5
12-inch Sanitary Sewer	0.20	6.2×10^5
12-inch Water Main	0.15	8.4×10^5
18-inch Storm Sewer	0	0

Note: This analysis assumes that the displacement of the utilities at the east and west ends of the North Wall are zero.

East Wall Utilities (108th Avenue NE)

Three utilities and a fire hydrant are located along 108th Avenue in the vicinity of the East Wall. They are located at various depths, and consist of a 6-inch diameter gas line, a 12-inch diameter water line, and an 8-inch diameter storm sewer. The fire hydrant is located at the surface and is situated approximately 8 feet from the top of the wall; the fire hydrant is connected to the water main in the middle of the street. The estimated displacements of these utilities at the wall midpoint, and the predicted radius of curvature which could develop in the event of displacement is summarized in Table 5.

In addition, we found that the fire hydrant may move approximately 0.7 inches more than the water main, in the event of the design earthquake.

TABLE 5. Estimated Displacement at Midpoint of North Wall (inches)

Utility	Estimated Displacement at Midpoint of North Wall (inches)	Predicted Resultant Radius of Curvature Induced in the Utility (feet)
6-inch Gas	1.2	2.9×10^4
8-inch Storm Sewer	0.5	1.2×10^5
12-inch Water main	1.0	2.2×10^5
Fire Hydrant	1.7	N/A

Note: This analysis assumes that the displacement of the utilities at the east and west ends of the North Wall are zero.

South Wall Displacement

While the south wall has no major utilities at risk in the event of the design seismic event, we examined the potential impact of movement in the south wall on the structure occupying the adjacent property. We found that in the event of the design earthquake, the footings closest to the south wall (approximately 32 feet south of the wall) may experience approximately 0.5 inch of settlement. Negligible settlement is predicted elsewhere in the building; as such, the building may experience differential settlement on the order of 0.5 inch if the south wall moves during the design seismic event.

West Wall Displacement

Negligible permanent displacements were calculated. This is primarily the result of the vertical stabilization elements and conservative design nail lengths.

C. Long-Term Corrosion Performance

Besides evaluating each walls ability to withstand earthquake loading, the other difference between a temporary and a permanent soil nail wall is that corrosion protection is provided for a permanent soil nail. Following FHWA design guidelines, corrosion protection includes epoxy-coated bars or a double corrosion PVC sheath system for soil nails. In this design, non-coated, non-covered, bare deformed reinforcing bars were used to construct the soil nails. Consequently, there will be corrosion of the bars dependent on the soil and groundwater conditions at the site.

At the Bellevue Technology Tower site, soil and groundwater conditions are not considered to be an aggressive corrosive environment. Corrosion generally requires the presence of water containing oxygen. The potential for corrosion of the nails is believed to be minimal for two reasons. First, groundwater at the site is below the bottom of the excavation and not expected to fluctuate greatly. Therefore, the nails are unlikely to be exposed to groundwater wetting. Second, the surrounding area is largely cover by concrete or asphalt and is immediately underlain by a dense, low permeability till. Therefore, surface waters will not have an opportunity to infiltrate and subject the nails to transient wetting conditions. Corrosion resistance testing and analysis of the soil was not done specifically at this site. However, the glacially deposited soils at this site, when tested locally at other sites, were determined to be non-aggressive. The ground is considered to be aggressive in regards to corrosion if low pH, low resistivity, or elevated sulfates or chlorides are present in the soil. Even though the site is considered a non-aggressive environment with respect to the corrosion of the bare bars, some thickness of steel will be lost over time.

In order to quantify the thickness of soil nail reinforcing which will be lost over time, a brief literature search was conducted, resulting in a procedure documented in the European design literature (Clouterre, 1991) to estimate the required thickness of sacrificial steel. The estimate is dependent on soil and groundwater conditions, proximity of critical structures, and the design life of the wall. Based on the conditions described above, considering the proximity of adjacent buildings, height of the walls, and a new design life of greater than 1.5 years, the procedure yields an estimate of four millimeters or approximately 0.16 inches of sacrificial steel for a design life up to 30-years. This thickness was then used to reduce the strength of the steel in the ground and analyze the factor of safety under static loading conditions.

At the Bellevue Technology Tower site, one inch and inch and a quarter, 150ksi, Dwidag bars were used as soil nails. Loss of 0.16 inches of steel, due to corrosion, reduces the ultimate strength of the one inch bar from 127.5 kips to 82.5 kips and the inch and a quarter from 187.5 kips to 139.5 kips. These reduced values were used in GoldNail to check the factor of safety under static loading for the East wall.

The East wall was chosen because it has the greatest height and lowest factor of safety. The factor of safety calculated for the East wall, under static loading with the full thirty-year design life reduction for corrosion of the nails, is 1.2 verses the industry standard of 1.35.

D. Monitoring

A monthly optical survey-monitoring program, is currently in place. Golder is performing a monthly evaluation of the results and quarterly survey which includes an onsite, visual evaluation of the performance of the wall.

During excavation and construction of the shoring wall, optical survey monitoring was performed that recorded maximum horizontal displacement on the order of 0.001H to 0.002H, where H is the maximum shoring wall height. This resulted in permanent displacements at the top of the wall ranging from 0.25 inches (west wall) to 2 inches (north wall). These displacements during excavation and construction of the shoring wall are normal and within the range for the soil type at the site. Once the excavation was completed, no further movement was recorded, as expected. The wall was not monitored between April 2001 and July 2002. However, the survey-monitoring program has been re-established, with no interpreted movement of the wall recorded to date.

E. Conclusions

Based on the results of our evaluation, the existing shoring system was determined to be adequate to resist the UBC design seismic event for 4 to 8 years past the 1 year temporary design life. After this period of time, corrosion of the soil nail steel will begin to lower the factor of safety. For the current conditions, our analysis indicated that the factor of safety for the section along the East wall and the middle section of the North wall required a displacement analysis. The displacement-based analysis indicates that the permanent displacement at the top of the wall in response to the UBC design earthquake is on the order of two-inches. Typically, the range of estimated permanent displacements were determine to be small enough not to cause widespread damage to surrounding streets, utilities, or buildings

It is anticipated that corrosion of the steel will occur in a non-linear fashion. As the steel corrodes, layers of oxidation will build-up, encapsulating the steel. Encapsulation of the steel in "rust" will limit the availability of oxygen to feed the chemical reaction and will slow the rate of the corrosion and loss of strength. Over time as corrosion progresses, factors of safety for both static and seismic loading will decrease, while wall displacements during the design seismic event increase. The reduction in strength of the soil nails will not only reduce the static factor of safety, but also decrease the earthquake yield acceleration and therefore increase both the temporary and permanent wall displacements during the design event. At this time, we do not consider that loss of steel due to corrosion has affected the wall performance. In addition, at four to eight years after the wall completion we estimate that the steel loss due to corrosion will only result in a negligible reduction in the factor of safety.

6. REFERENCES

Algermissen, Perkins, Thenhaus, Hanson, and Bender. USGS 1990 Probabilistic Earthquake Acceleration and Velocity Maps for the United States and Puerto Rico.

AASHTO Division 1A, Seismic Design Commentary, 15th Edition, 1992.

Armour, T.A., Cotton, D.M., "Recent Advances in Soil Nailed Retention". Presented at Earth Retention Systems 2003: A Joint Conference ASCE, DFI and ADSC, May 6 and 7, 2003, New York City.

Byrne, R. J. Cotton, D.M., Porterfield, J.A., Ueblacker, G., Wolschlag, C.J., Manual for the Design ad Construction Monitoring of Soil Nail Walls. Technical Publication for Office of Technology Publication for Office Technology Applications, Federal Highway Administration, May 1997.

Cotton, D.M., Byrne, R.J., Wolschlag, C.J., "Soil Nailing: The Recent Development of Design Innovations and Cost Saving Ideas Using the Top-Down Method of Permanent Wall Construction". Presented at the University of Wisconsin – Milwaukee short course on Specialty Geotechnical Construction in Urban Environments, San Francisco, California, February 1998.

Cotton, D.M., Soil Nailing: The development of the Top-Down Method of Permanent Wall Construction and Local Stability Problems and Resolutions in Fill Materials, Glacial Till, Out Wash and Lucustrine Deposits. Presented at 1992 ASCE Seattle Section Geotechnical Seminar, University of Washington, March, 1992.

French National Research Project CLOUTERRE (1993). "Recommendations CLOUTERRE 1991 – Soil Nailing Recommendations 1991 "*Presses de l'Ecole Nationale des Ponts et Claussees*" English Translation, July.

Tufenkjain, M., 2002. Performance of Soil Nailed Retaining Structure During the 2001 Nisqually Earthquake: Proceedings of the 7th U.S. National Conference on Earthquake Engineering, Boston, MS July 2002.

DESIGN, CONSTRUCTION, AND PERFORMANCE OF AN 18-METER SOIL NAIL WALL IN TUCSON, AZ

Edward Nowatzki, P.E., Fellow, ASCE[1] and Naresh Samtani, P.E., Member, ASCE.[2]

ABSTRACT: This paper describes the design, construction, and performance of a soil nail wall system used by the Pima County Department of Transportation (PCDOT) to stabilize and retain two steeply sloping bluffs up to 18 meters high as part of a major road widening and realignment project along River Road, a major arterial roadway in Tucson, Arizona. The use of a soil nail wall system precluded the need to relocate major components of the Altamira Apartments, a luxury apartment complex situated atop the bluff west of an access road (Campaña Drive) leading from River Road to the apartment complex. The impacted components included a large swimming pool, a clubhouse, and a two-story apartment building, all located close to the crest of the slope. Because of the high property values in the area, replacement costs of these components were prohibitive. In addition, the artificial sculpted-rock architectural facing appealed to a citizen's advisory group because of its aesthetics.

INTRODUCTION

General Geologic Setting

The city of Tucson is located in the Sonoran Desert in a broad basin surrounded by four mountain ranges. The general geologic setting of the Tucson Basin is typical of the Basin and Range Province that constitutes the southwestern half of the state of Arizona. The basin is filled with thick sequences of relatively young sediments, with depth to bedrock increasing laterally from the hard rock ranges toward the center of the basin where sediments are up to 500 m thick. The site is located on the edge of an alluvial fan at the base of the foothills of the Santa Catalina Mountains just north of the city of Tucson. Fig. 1 shows an overview of the site prior to construction including two of the impacted structures at the Altamira Apartment complex.

---

[1] Principal Engineer, NCS Consultants, 640 W. Paseo Rio Grande, Tucson, AZ 85737 USA; eanowatzki@msn.com;
[2] President, NCS Consultants, 640 W. Paseo Rio Grande, Tucson, AZ 85737 USA

Subsurface Conditions at the Site

As part of its scope of work under an "on-call" contract with PCDOT, Envirotech Southwest, LLC, (ESW) of Tucson, AZ researched the local literature to obtain information on the geotechnical properties of subsurface soils on or near the site. The quality of much of the existing data was not adequate for the proposed project. Therefore, ESW contracted with AGRA Earth and Environmental, Inc. (AGRA) to conduct a more site- and project-specific geotechnical field investigation. Unfortunately, most of the Altamira site was inaccessible to drilling equipment. Therefore, only one boring (B-1) was advanced at that site. Another boring (B-2) was drilled on the adjacent bluff approximately 100-m to the east. A summary of the results of geotechnical investigations relevant to the site is presented in Table 1. The values listed in Table 1 vary because of variations in soil types and because of differences in the methods used to sample and test the soils. Based on the results of the field and laboratory tests reported in Table 1, the geologic profile at the site appeared to be stratified with layers of varying thickness consisting of slightly- to heavily-cemented silty sands, gravels and river-run rounded cobbles, with the probability of seams of loose sand, gravel and cobbles. Standard penetration blow counts were typically in excess of 20 blows per foot for the full length of Boring B-1.

Fig.1 Overview of site prior to construction with apartment complex on bluff and River Road in the foreground.

DESIGN CONSIDERATIONS

Public Participation

The soil nail wall system was designed to blend in with the natural surroundings of the site in accordance with guidelines established by the Tucson Citizens Advisory Committee (TCAC). In the final design the architectural facing incorporated artificial naturalistic rockwork and sculpted shotcrete that mimicked eroded banks along nearby portions of the River Road corridor. The look of the natural slopes is shown in Fig. 2.

Encroachment

Fig. 3 shows a plan view of the site with the final configuration of the soil nail wall superimposed upon it. As indicated in the figure, only a small portion of the deck area around the swimming pool was lost because of the soil nail wall system. With the conventional reinforced concrete (RC) retaining wall system originally proposed, the swimming pool, recreation center, and the apartment building east of them would have had to been totally replaced. Except for Boring B-1 advanced by AGRA, the other borings listed in Table 1 are not shown in Fig. 3 because they are not within the work site. They are included only to show the range of values that can be expected for the variably-cemented soils typically encountered in the Tucson Basin.

Table 1. Summary of Soil Properties

Boring No.	Location (Depth in meters)	USCS Symbol	c′ (kPa)	ϕ′ (deg.)	γ′ (kN/m³)	ω (%)
Desert Earth Engineering (1987)						
No lab test data available - all values Based on SPT blow counts						
B-2	East of Campaña Dr.					
	0-6	GP-GM	0	32	17.27	Dry
	6-13.7	GP-GM	0	34	18.06	Dry
B-3	West of Campaña Dr.					
	0-1.5	SM	0	32	18.06	Dry
	1.5-7.6	GP-GM	0	34	18.06	Dry
	7.6-16.8	GP-GM	0	34	18.06	Dry
Engineers International, Inc. (1991)						
Values based on field and lab test data						
B-23	West of Campaña Dr. in Altamira parking lot.					
	0-1.2	N/A	207	39	17.27	Dry
	1.2-3.4	N/A	207	17	19.47	Dry
AGRA Earth and Environmental, Inc. (1998)						
Values based on field and lab test data						
B-1	West of Campaña Dr. in Altamira parking lot (Fig. 3)					
	0-3	SM	24	35	17.90	2
	3-10.7	SM	36	31	16.33	2-3
	10.7-15.3	SM	72	33	17.43	6-8
B-2	East of Campaña Dr.					
	0-9.75	SM	57	30	17.90	2
	9.75-22.9	SM	48	36	17.11	1

Fig. 2 Natural slope along River Road West of site

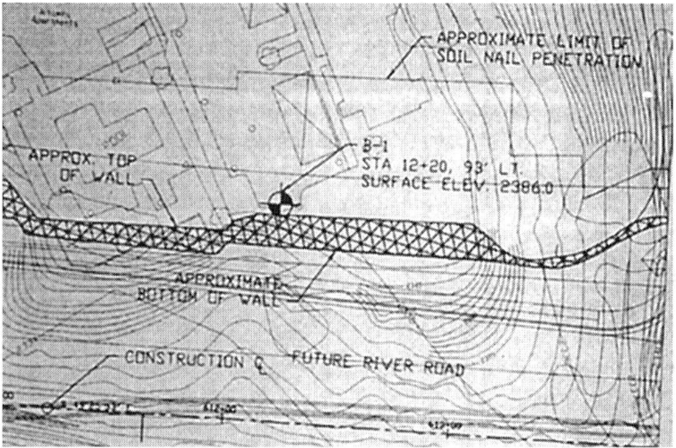

Fig. 3 Plan view showing encroachment onto Altamira Apartment complex (Boring B-1 was advanced by AGRA Earth and Environmental (Refer to Table 1))

Accessibility and Other Constraints

Accessibility impacted the design and construction in at least two ways. As indicated previously, site accessibility prevented the undertaking of a more robust geotechnical field investigation that would have resulted in less conservatism in the design. Accessibility also posed a problem during construction, especially in view of PCDOT's requirement to keep River Road open during the entire construction period and as free as possible of construction-caused delays. Fig. 4 shows the construction access road to the top level of the west wall at the start of construction. River Road is to the left and Campaña Drive is in the foreground. Because of the close proximity of the project to the apartment buildings work hours were limited from 7 AM to 3 PM so as not to disturb the residents. This imposed a severe constraint on the contractor since much of the construction work in the Tucson area during the summer months typically starts at 5 AM to avoid the hottest part of the day.

Fig. 4 Access road to top level of West wall – Campaña Drive in foreground.

DESIGN

Project Scope

The design and construction of the soil nail wall systems on the two bluffs near 1st Avenue were part of a widening and realignment project along River Road, a major arterial roadway in Tucson, Arizona. In addition to numerous conventional RC cantilever retaining walls, the project included two near-vertical soil nail walls ranging in height from 2 to 3-m. meters at their ends to more than 17-m. at their highest point. The west wall is approximately 130-m. long and the east wall is about 156-m. in length. The finished area of the walls is approximately 3,530 m^2.

Design Approach

Earth retention systems considered by PCDOT in the early planning stages for this project included crib walls, mechanically stabilized earth (MSE) walls, and a tiered system of standard RC cantilever walls. The costs of replacing impacted structures at the Altamira Apartments, including land acquisition costs, were prohibitive for all of these systems. A soil nail wall system was found to be most cost-effective. In addition, its architectural facing appealed to a citizen's advisory group. The soil nail system was designed in accordance with procedures described in the ***Manual for Design & Construction Monitoring of Soil Nail Walls*** (FHWA, 1998). The guidelines contained in FHWA's design manual are considered the standard of practice for soil nail walls constructed in the United States. A number of commercially available computer programs are currently used in practice for such designs, including **SNAIL**, developed by the California Department of Transportation (CALTRANS) and **GoldNail**, developed by Golder Associates (GA), Redmond, WA.

Soil Properties

The following soil parameters were used in the design of the soil nail wall system. These values were assigned based on the field and laboratory test results presented in Table 1.

Effective cohesion, $c' = 24$ kPa
Effective friction angle, $\varphi' = 33°$
Total unit weight, $\gamma' = 18$ kN/m^3

Bar Size, Length, and Spacing

The lead author, while a principal at ESW, performed the initial design based on an estimated pullout resistance of 14.6 kN/m. ESW's design called for nine (9) rows of 18.3-m. long cased nails (# 8 bars centered in 101.6-mm. diameter grout holes) installed at an inclination of 15° on a 1.524-m. x 1.524-m. grid. For the purpose of design, surcharges based on loads outlined in the ***Uniform Building Code*** (ICBO, 1994) were added to the crest of the wall at the locations of the swimming pool deck and apartment building. The top row of nails in the vicinity of the swimming pool was inclined at 30° in order to avoid hitting the pool substructure.

The ESW design was put out to bid by PCDOT. Only pre-qualified drilling contractors were invited to respond to the bid. Malcolm Drilling Co., Inc. (MCI) of Vista, CA, one of the specialty contractors to submit a bid, proposed a value engineered (VE) design developed by their sub-consultant Ground Support, PLLC (GS) of Redmond, WA. GS's design was based on an estimated pullout resistance of 29.2 kN/m that was subsequently confirmed by verification pullout tests conducted to twice the design load. The value-engineered design modified the original design by calling for nine (9) rows of variable diameter, self-drilling/grouting nails (IBO/Titan 40/20, 30/11) of variable length (4.877-m. to 15.24-m.) installed at an inclination of 15° on a 1.829-m. x 1.829-m. grid. The dual numbering of the hollow IBO/Titan nails refers to the outside/inside diameters in mm. The 40/20 rod is equivalent to a # 9 bar in terms of nominal cross sectional area. The 30/11 rod has a cross sectional area that is approximately the average of a # 7 and # 8 bar. The rods are threaded and have "purple marine" epoxy coating for corrosion protection. In both designs the top row of nails was assumed to be within 1-m. of the top of the slope, and in the area of the pool they were horizontally offset from the basic nail pattern and inclined at 30-degrees to avoid hitting the pool substructure. In both designs the toe of each soil nail wall extends a minimum of 0.61-m. below final grade. The elevation view of the mid section of the west wall in the area of the swimming pool is shown on Fig. 5.

IBO/Titan nails have a hollow core so as to allow injection of grout through the center during installation. A sacrificial drill bit larger in diameter than the nail advances the hole. Grout is pumped under low pressure through the hollow core of the nail. It flows through the nail bit and returns back to the nail head location at the wall face. Thus the nail itself serves as a casing and the larger nail hole created by the larger sacrificial drill bit is grouted at the same time during drilling. A schematic of the IBO/Titan nail and details of the installation procedure can be found on the following website: http://www.contechsystems.com/cts-cd/TITAN/TSoNa.pdf. The dynamic rotary pressure grouting penetrates into loose material around the drill hole thereby increasing the surface friction compared to traditionally installed "drill and grout" nails. This was reflected in greater pullout resistance during the verification tests and resulted in overall cost-savings. An attractive feature of the IBO/Titan system is that the installation is in essence a cased method. It was preferable to a wet rotary method on this project because wet drilling may have caused destabilization of the nail hole and necessitated the use of casing. The value-engineered design was accepted by PCDOT and resulted in savings of over $190,000.

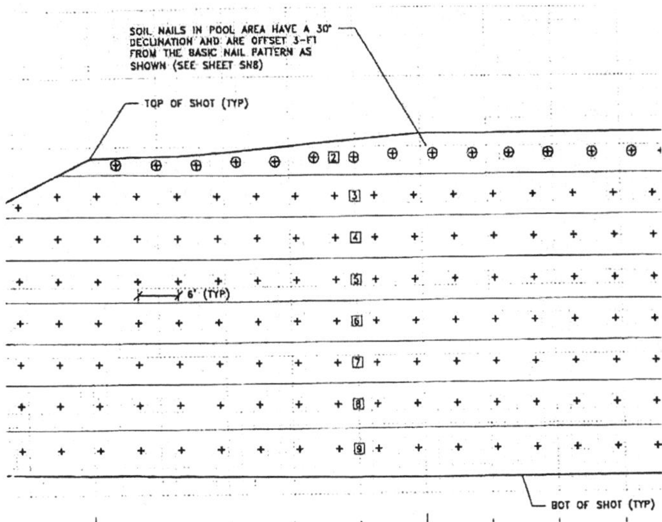

Fig. 5 Elevation view of middle-section of West wall

Corrosion Protection

The soils in the vicinity of the proposed soil nail slopes were found to have low to no corrosion potential. In accordance with FHWA guidelines, in the original design the soil nail tendons were specified as epoxy-coated with a minimum of 25.4-mm grout cover all around, but not double encapsulated. The "purple marine" epoxy-coated IBO/Titan nails used in the VE design satisfied this requirement. Additional corrosion protection was obtained from the minimum 102-mm diameter grout column specified by PCDOT to assure adequate pullout resistance.

Drainage

Subsurface drainage was designed based on FHWA requirements regarding geotextile face drains, shallow PVC drain pipes and weep holes, surface interceptor collector ditches, and surface waterproofing. Vertical geo-composite drain strips (0.3-m. wide) were installed between every column of soil nails to prevent hydrostatic pressure from building up behind the wall facing. The drain strips were connected to footing drains and weep holes that drain to the wall face. A concrete-lined gutter or V-ditch was installed along the base of the wall to collect drainage from weep holes and runoff from the face of the wall and direct them away from the site. Additional subsurface drainage was provided under the swimming pool by standard horizontal drains that are typically placed above impermeable clay layers to relieve hydrostatic pressure that may result from perched water. A 152-mm high PVC water-stop was installed continuously on the top of the shotcrete construction facing to prevent water from seeping between it and the architectural facing.

Shotcrete Facing

The design of the shotcrete construction facing was checked for punching shear and flexure in accordance with FHWA (1998) requirements. The facing design utilized 27.58-MPa (minimum 28-day strength) shotcrete with a minimum shotcrete thickness of 102-mm. The shotcrete was reinforced with two continuous # 4 bars in both the horizontal and vertical direction at each nail head and 6 x 6 - W1.4 x W1.4 welded wire mesh. Bearing plates connecting the nails to the facing were specified at 22.86-cm. x 22.86-cm. x 1.9-cm. at a minimum. All steel was Grade 60.

CONSTRUCTION

Prior to the start of construction, MDI developed an efficient construction sequence based on the time required for shotcrete to cure. The high mobility of the construction equipment enabled MDI to use the same equipment on both walls by alternating drilling and shotcreting between the two walls. Therefore construction of both walls proceeded concurrently, which explains the relatively short period of construction. A typical installation of nails, drainage strips, and reinforcing steel prior to shotcreting is shown in Fig. 6. The installation shown in this photo is for the final bench on the west wall. The vertical extension of the horizontal drains under the swimming pool are visible at the far end of the wall. The pilasters at the top of the soil nail wall in the background are part of the replaced section of the wall surrounding the swimming pool.

While installing nails on the 4th bench of the west wall, the contractor encountered a seam of dry loose sands, gravels and cobbles at the east end of the wall. Vibration of the drilling equipment caused the loose materials to ravel from behind the previously installed wall sections. Figs. 7 and 8 show the extent of the raveling and the size of the void created behind the wall. At some sections the voids were more than two feet deep and had propagated upward ("chimney void") towards the top. It was clear that continued construction activity would exacerbate the problem and possibly lead to loss of the entire wall.

Fig. 6 Typical installation of nails, drainage strips, and reinforcing steel

Fig.7 Raveling sand and gravel on 4th bench of West wall

Fig. 8 Void behind soil nail wall face from raveling of sand and gravel

Construction was halted while remedial measures were considered. A number of solutions were proposed including the installation of a vertical micro-pile wall to get the construction past the seam. A shallow test section of micro-pile wall was built and it performed satisfactorily, however the depth of the seam was unknown, so that method was abandoned. In the end, it was decided to build a berm against the exposed section of the wall and to water-soak the berm until enough apparent cohesion could be developed to complete the installation of the nails and excavate the berm and native material to the next bench level. This procedure worked very well and construction proceeded with its use until more competent materials were encountered.

After the wall had advanced to the point where the shotcrete facing on Bench 4 had reached sufficient strength, a lean 1:1:1 cement:sand:fly ash grout (12-13 kN/m^3) was introduced through the wall to fill the voids that had been created by the raveling. Injection of this flowable grout in these sections was staged to prevent "blowouts" at lower levels of the wall from fluid pressures caused by the grout. Other than for this problem, construction proceeded smoothly. During construction proof load tests were performed according to FHWA (1998) requirements to a maximum load of 1.5 times the design load on at least 5% of the nails in any given row. Only bonded lengths (3-m minimum) were tested. Unbonded lengths were obtained by using a PVC collar at the head of the nail as a bond breaker. Nails were sized to withstand test loads with a minimum movement of 80% of the theoretical elastic elongation of the unbonded length as an acceptance criterion. Each test also included a 10-minute creep test under maximum load with a failure criterion based on maximum movement of 1-mm. The results of the tests equaled or exceeded FHWA (1998) acceptability criteria.

WALL MOVEMENT AND MONITORING

Wall Movement

Because of the proximity of major structures to the edge of the bluff, potential movements during and after construction were major concerns. A comprehensive finite element analysis was performed as part of the original design to assess the stability of various slope configurations and to estimate the magnitude of potential horizontal and vertical ground movements at the crest of the wall and at strategic locations within the existing apartment complex (DeNatale, 1998a, 1998b). The analyses were performed by utilizing a two-dimensional finite element computer code and a basic isotropic, linear-elastic constitutive model for the soil. A minimum value of elastic modulus = 21 Mpa was used based on interpretation of data presented in Table 1. The results of the finite element analyses indicated that vertical ground movements in the vicinity of the swimming pool and apartment buildings because of construction of the soil nail wall should be less than 25-mm. Horizontal and vertical movements at the crest of the walls were calculated to be approximately 25-mm and 38-mm, respectively. These values are within the range of values expected at the crest of an 18-m high wall in dense granular soils (FHWA, 1998). The results of the analyses were used to develop an instrumentation program to monitor actual movements before, during, and after construction.

Monitoring Plan

A performance monitoring plan was developed to help avoid potential problems by detecting ground and/or structural movements before they became excessive so that remedial measures could be implemented to halt such movements. The following program to monitor movements was implemented for the west wall:

- Two 20-m long inclinometers were installed, one approximately 1-m. behind the location of the finished face of the wall just east of the swimming pool and the other approximately 3-m behind the location of the finished face of the wall between the wall and the impacted apartment building. Baseline measurements were taken before the start of construction and both inclinometers were read regularly during and after construction. Inclinometer measurements indicated that the ground surface at the location of the poolside inclinometer moved 4.5-mm horizontally in a direction perpendicular to the plane of the wall approximately 3-months after the start of construction, a value well below the calculated value. The difference is largely due to the conservative value of soil modulus used in the finite element analyses and the tighter soil nail grid spacing of the ESW design.

- Survey targets mounted on potentially impacted structures were regularly monitored before, during, and after construction.
- Nail heads on the wall itself were monitored during construction.
- All potentially impacted structures were inspected inside and outside prior to construction and all observed cracks were photo-documented. The ground surface behind the top of each wall was inspected at the start of construction and observed periodically during and after construction for cracks and other signs of disturbance. Such inspections included dated photo-documentation.

ARCHITECTURAL FACING

Fig. 9 shows the wall as it appears today. The artificial rockwork and sculpted shotcrete facing shown in Fig. 9 compare favorably with the natural slopes shown in Fig. 2. Construction of both walls started in June 2001. The shotcrete construction facing for both walls was done concurrently and the walls were completed in October 2001. The architectural facing was done by the Larson Company (LC) of Tucson, AZ one wall at a time and was not completed until June 2002.

Fig. 9 Completed soil nail wall

CONCLUSIONS

The design guidelines and procedures contained in the ***Manual for Design & Construction Monitoring of Soil Nail Walls*** (FHWA, 1998) provide the geotechnical engineer with the tools needed to design soil nail walls. Commercially available computer codes enhance the engineer's ability to perform such designs. However, field conditions can be encountered that may seriously impact the construction of such walls. Unless the engineer has a good grasp of geotechnical engineering principles and is able to apply them with confidence, those conditions may result in costly and unnecessary remediation procedures. One example of this is the solution that was ultimately used to correct the raveling problem that endangered the very wall itself. Another involves the correct interpretation of standard penetration test data to reconcile the fact that raveling occurred in a seam of dry, lightly- to non-cemented sands, gravels and cobbles that displayed standard penetration blow counts (N) in excess of 30 blows/foot. Such high values of N have meaning for foundation analysis and design where that type of soil remains confined under load just as it was during the SPT. However, such high values are meaningless when cuts are made into the same type of soil as is done during the installation of a soil nail wall system. Confinement is destroyed under such conditions and dry, lightly- to non-cemented granular soils will run with even the slightest disturbance as they did in this case. In hindsight, the use of water-soaked berms throughout construction would have been one way to circumvent potential raveling problems on this project.

ACKNOWLEDGMENTS

The authors wish to acknowledge the contributions of Mr. Robert Johnson, P.E., of PCDOT and Dr. Jay S. DeNatale, P.E., of the California Polytechnic State University in San Luis Obispo.

REFERENCES

AGRA Earth and Environmental (1998), ***Geotechnical Engineering Report - River Road Widening and Realignment Project, AEE Job No. 8-127-000-016,*** submitted to Pima County Dept. of Transportation, Tucson, AZ.

DeNatale, J.S. (1998a), ***Prediction of Slope Deformation: The River Road Realignment Project***, Report prepared for Envirotech Southwest, Tucson, AZ.

DeNatale, J.S. (1998b) ***Analysis of Slope Stability: The Altamira Section of River Road***, Report prepared for Envirotech Southwest, Tucson, AZ.

Desert Earth Engineering (1987), ***Geotechnical Engineering Slope Stability Analysis on 3 Subject Slopes Along the North Shoulder of River Road Between 1st Avenue and Via Entrada, Tucson, Arizona,*** a report submitted to Pima County Dept. of Transportation, Tucson, AZ.

Engineers International, Inc. (1991), ***River Road: First to Campbell, Slope Treatment Letter Report, Pima County Project TR-87-049, County W.0.4BRVCS,*** a report submitted to Rick Engineering, Inc., Tucson, AZ.

FHWA (1998), ***Manual for Design & Construction Monitoring of Soil Nail Walls,*** Federal Highway Administration Report FHWA-SA-96-096R, U.S. Department of Transportation, Washington, D.C.

ICBO (1994), ***Uniform Building Code***, International Conference of Building Officials, Whittier, CA.

SOIL NAILING – WALLS OF MANY FACADES

Daryl W. Wurster, P.E.[1], MASCE, ADSC

Abstract: Soil nailing can be used with a wide variety of facades to meet architectural and structural needs of the design team and owner. Soil nailing generally consists of small diameter, reinforced steel elements, grouted in approximately 4 to 6 inch diameter sub horizontal boreholes. A reinforced shotcrete façade is generally applied for surficial stability. A rough gun finish is usually satisfactory for temporary support. For permanent walls, it may be desirable to smooth or sculpt the shotcrete façade, form a concrete wall directly against the shotcrete, or build a separate wall structurally connected to the soil nail wall. Exploring the wide variety of facades that can be used with soil nail walls can create many job opportunities for the design team and contractor.

Introduction

The purpose of this paper is to illustrate how soil nailing can be used with a wide variety of facades to meet architectural and structural needs. When reviewing the use of soil nails for shoring, two requirements are often asked, 1) can soil nails provide the necessary support, and 2) how will they fit in with the overall project?

Following are four projects that illustrate a variety of ways that soil nailing can be used to stabilize the ground used in combination with a variety of facades. A partial list of possible soil nail facades that we have used include segmental block walls, smooth finished shotcrete, formed, patterned concrete walls cast directly against the soil nail wall and below grade building walls. One of the projects combined soil nail support with and another reinforcement method.

---

[1] President, Principal Engineer, Wurster Engineering & Construction, Inc., P.O. Box 25426, Greenville, SC 29616-0426, dwurster@wursterinc.com

US Hwy 501 – Myrtle Beach, SC

US Hwy 501 crosses the intercoastal waterway in Myrtle Beach, SC (reference Fig. 1). Approximately 50 feet of sandy fill soils had been placed about 20 years ago to create an approach to the bridge over the intercoastal waterway. A two-lane road loops around the toe of the fill embankment to provide access from the westbound lane of US Hwy 501 to the Waccamaw shopping area on the south side of US Hwy 501. Plans were to widen the road around the toe of the fill embankment by excavating the bottom portion of the approximately 1.75H:1V (horizontal to vertical) fill embankment and constructing a retaining wall. The retaining wall area would be about 14,000 ft2, about 830 ft long and range in height from approximately 15 to 30 ft.

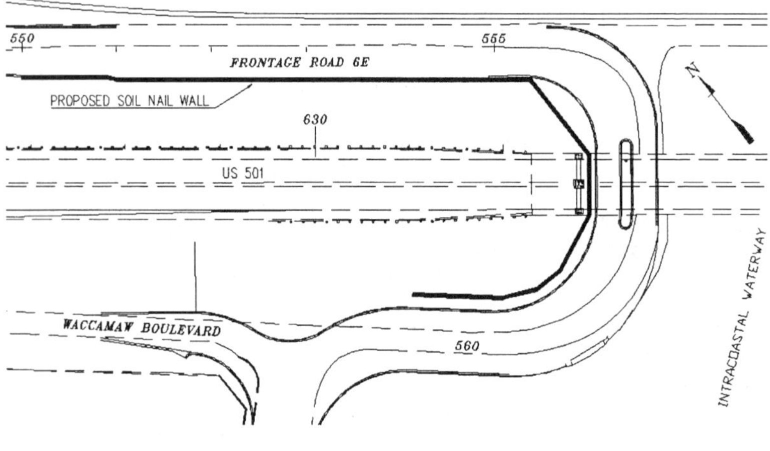

FIG. 1

Either soldier piles with tieback anchors and lagging or soil nailing had been recommended in the geotechnical report and bid documents. The project specifications indicated, "The final exposed anchored wall facings, color and texture, shall match the MSE walls on this project". We proposed to stabilize the proposed hillside cut using the soil nailing method and attachment of a segmental block wall façade of the same brand, color and texture as used in other portions of the project.

Fill consisted of brown and gray, medium dense, fine sand and coarse to fine sand, with little silt, some shells, and a trace of fine gravel. The fill had been in-place about 20 years and were modeled with the following parameters: ϕ = 34 degrees, c = 0, γ = 120 pcf.

The soil nails were constructed of #6, #7, and #8, 75 ksi, epoxy coated thread bar and a cement grout. Six-inch diameter bore holes were drilled using air to flush cuttings. A pullout value of 2,262 lb/ft (10 psi) was utilized for design. A 6-inch (min.) thick shotcrete façade was reinforced with 4x4 W4x W4 WWF augmented with 2 - #3 horizontal reinforcing bars and 2 - #3 vertical reinforcing bars at each row and column of nails. End hardware consisted of 8" x 8" x ¾" plates.

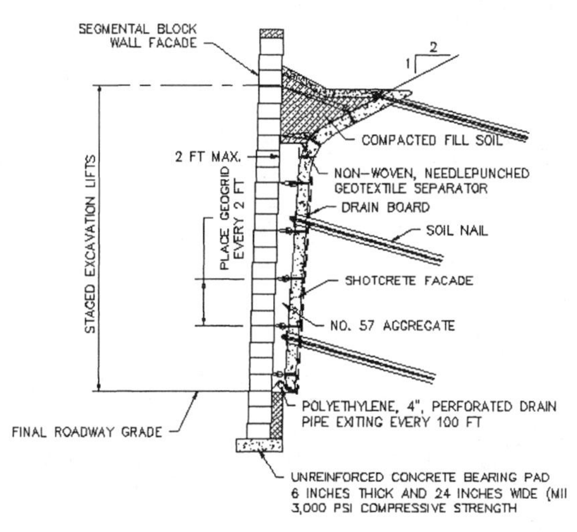

FIG. 2

Boreholes were drilled with a Klemm KR802 drill rig. Grout was mixed with a Colcrete, High Shear grout plant. Shotcrete was pumped with a Schwing BP-450 concrete pump.

Segmental blocks were attached to the soil nail shotcrete façade on a ¼" batter per 8 inch high block (reference Fig. 3). Three inch diameter, galvanized steel pipes were connected to the shotcrete façade with 5/8" diameter galvanized, carriage bolts, grade 36, set in ¾" diameter drilled holes and bonded with fast cure epoxy. Segmental blocks were installed to each level of galvanized steel pipe. Tensar BX1200 biaxial geogrid was connected to the segmental block; wrapped around the galvanized steel pipe and brought back to the segmental block. Aggregate meeting the gradation requirements of ASTM C-33, no. 57 was placed between the face of the soil nail wall and the back of the segmental blocks. Wall connections were made every three blocks (two ft) in height. [i]Lateral pressure on the segmental blocks and thus wall connection were calculated based on the assumption that arching occurred between the aggregate and the soil nail wall/segmental block wall.

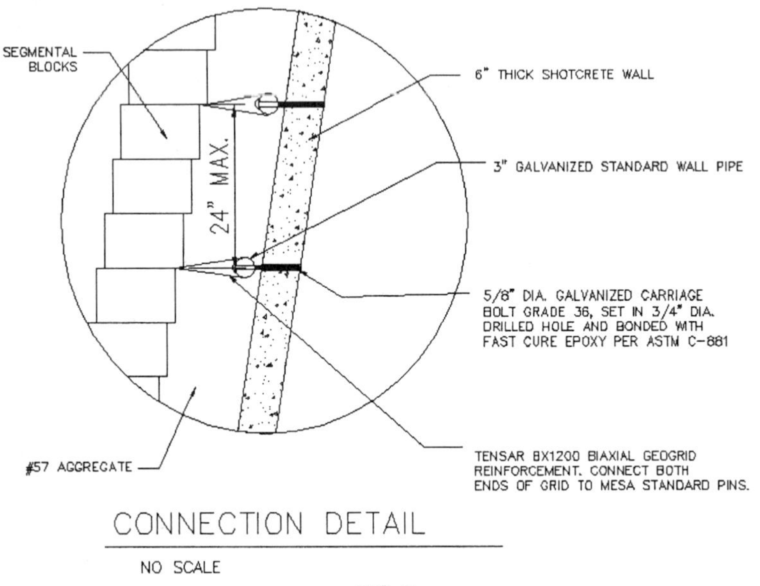

FIG. 3

The adjacent bridge abutment and approach roadway were monitored by survey as excavation proceeded. Although bridge abutment and roadway settlement was not detected during survey, it was observed that tension cracks appeared several feet behind the face of the soil nail wall each time additional excavations were made.

Boreholes in the nearly cohesionless sand fill stayed open long enough to grout without the need for casing. However, vertical cuts in these fill soils began to slough after several hours to a day. The resulting wall met the project requirements and resulted in an aesthetically pleasing and structurally sound wall.

Food Lion SkatePark – Asheville, NC

The Food Lion SkatePark is located in downtown Asheville, North Carolina at the corner of Flint and Cherry Streets, across I-240 from the Asheville Civic Center (reference Fig. 4). The Food Lion SkatePark is the premier facility of its kind throughout the region. Featuring 17,000 square feet of outdoor skateable surface housing three distinct areas; a shallow Warm-up area, the Street Course, and a Large Bowl. Excavation approximately 9 to 18 ft deep was required for the Street Course. Photos of the park can be seen at http://www.skateboardparks.com/northcarolina/foodlion/pics/index.html.

The public skate park was constructed below the surrounding ground surface and was constructed of reinforced concrete. The sides of the skate park are vertical. A sculpted skating surface including numerous ramps was formed between the vertical side walls.

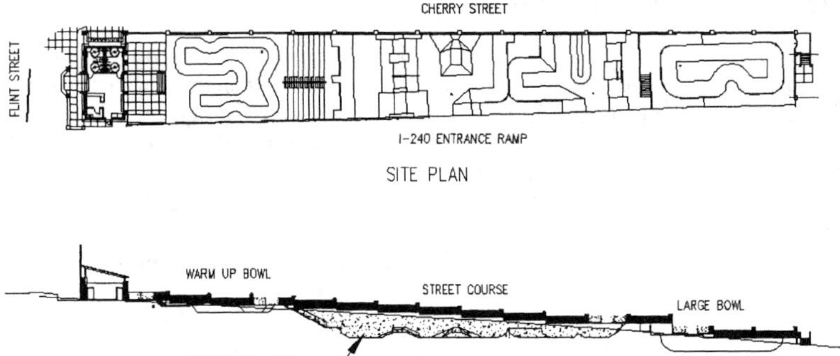

FIG. 4

Initial plans were to provide temporary shoring along the I-240 entrance ramp and along Cherry Street to allow construction of permanent, reinforced concrete retaining walls. Bids exceeded available funds. As an alternate, soil nailing was selected for both temporary construction shoring and for support of the completed wall face. Eliminating one set of retaining walls met the available budget. A total of approximately 3,130 ft^2 of soil nail wall was required.

Four-inch diameter holes were drilled using a TEI RDS528H Rotary Top Drive Drill mounted on a New Holland, LS 180 skid steer. A 375 cfm air compressor was used to remove cuttings. Shotcrete was pumped with a Schwing BP450 concrete pump.

Silty sands residual soils were exposed during excavation. The following soil parameters were assumed for design: ϕ = 28 degrees, c = 50 psf and γ = 120 pcf. Soil nails were constructed of #6, grade 75, epoxy coated threadbar and cement grout. Four-inch diameter bore holes were drilled using air to flush cuttings. An ultimate pullout value of 1,810 lb/ft was utilized for design. An initial four-inch thick layer of shotcrete was applied to stabilize the exposed soils. The initial shotcrete layer was reinforced with 4x4_W4xW4 WWF with additional 24" x 24" WWF at each nail head.

Piano wire was strung in front of the cured, initial shotcrete layer to mark the location of the desired wall face. A second layer of shotcrete was applied until the piano wire was covered, then cut back to the piano wire with a 4 ft gunite knife. The second layer of shotcrete was then hand troweled to create the desired smooth finish similar to a cast-in-place concrete wall.

Slight color variations resulted from one shotcrete lift to the next. If we were to provide a finished, shotcrete face on another project, it may be desirable to apply the entire final coat in one continuous operation to obtain a more uniform color. The various design team members and the City of Asheville were pleased with the final shotcrete walls. The alternate resulted in a single set of walls with a significant cost savings.

Woodberry Apartments – Asheville, NC

Woodberry Apartments in Asheville, North Carolina were constructed along a relatively steep hillside. Excavation into the hillside was required for construction. The hillside to the rear of one of the apartment buildings subsequently experienced slope failure. A soil scarp formed along the upper property line. The elevation of the failed slope ranged from approximately 1,000 ft directly behind the apartment buildings to approximately 1,075 ft along the upper property line. Portions of the failed slope surface were as steep as 1.5H:1V (horizontal to vertical). Residual soils with rock outcroppings were exposed along the hillside. A local area of the slope was actively sliding and contained loose soil and rock.

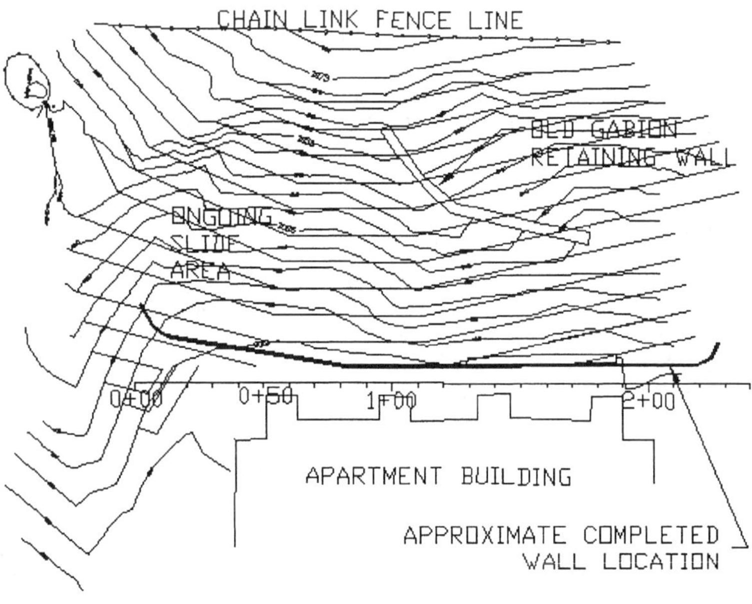

FIG. 5

The geotechnical engineer of record recommended that the slope be flattened to 2H:1V by fill placement from near the apartment building to the upper property line. A retaining wall ranging in height from approximately 8 to 20 ft, and 10 ft behind the apartment building was proposed to retain the new fill soils.

A segmental block wall was originally considered based on the assumption that it would cost less than other stabilization methods, access would not be a problem, and it would have a desirable appearance. However, excavation required for geogrid placement could result in further slope instability where rock was not exposed. Hard rock removal would be required in other areas.

Construction of a soldier pile, lagging and tieback wall was eventually specified to retain the new fill soils. Installation of soldier piles and tieback anchors would be difficult as a result of limited site access. Further, the wood lagging would degrade over time. Considering that a segmental block wall could be constructed with relatively small construction equipment, WEC submitted and was awarded an alternate slope stabilization method combining segmental block and soil nail construction methods. The segmental block wall had an area of about 4,050 ft^2. The base of the wall was constructed at about elevation 1,000 and the highest portion of the wall was constructed at elevation 1,020 ft. Soil nails with a reinforced shotcrete façade were installed at locations along the hillside where sufficient geogrid reinforcing could not be installed. The geogrid reinforcing was connected to the shotcrete façade.

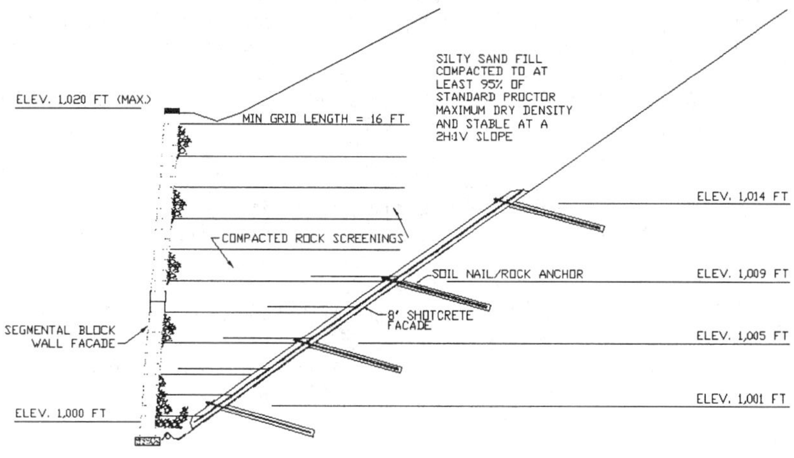

FIG. 6

A segmental block wall was designed as though sufficient room was available for geogrid placement at all locations. Calculations indicated that grid lengths would have to be 16 ft. The geogrid placement was superimposed on the face of the slope. Portions of the slope were marked for soil nail installation where 16 ft geogrid lengths could not be achieved.

Soil overburden depth ranged from approximately 10 ft to non-existent (rock outcroppings). Soil nails were installed through the soil overburden (where it existed) and at least 5 ft into the underlying rock. Soil nails were installed through the soil overburden and into the underlying rock. Soil nails consisted of #7 and #8, 75 ksi, epoxy coated thread bar and cement grout. Four-inch diameter boreholes were drilled using air to flush cuttings.

A reinforced shotcrete façade was constructed over the sloping hillside. 4x4, W4xW4 WWF and #4 rebar reinforcing was placed over the nailed slope. The proposed geogrid elevations were projected onto the slope with a laser level and marked with paint. Layers of horizontal geogrid reinforcement (Tensar UX1500SB) were tied to the steel reinforcement prior to shotcrete application. A #3 reinforcing bar was threaded through the geogrid reinforcing and tied to the WWF. #4 vertical reinforcing was tied over the horizontal #3 bars. Shotcrete was placed leaving the geogrid reinforcing to project out from the reinforced shotcrete façade.

Additional layers of geogrid reinforcing extending from the segmental block wall façade were overlapped with layers of geogrid reinforcing extending from the shotcrete façade providing the desired structural connection with the underlying rock. The overlap length was determined that would develop full geogrid strength as if they were one continuous layer. [ii]A minimum overlap length of at least 2.67 ft was calculated based on a maximum geogrid tensile strength of 1,460 lb/ft for Tensar UX1500SB using Equation 1. The design overlap was increased to 4 ft for added conservatism. A 4" min. to an 8" max. compacted rock screening lift was required between overlapping geogrid layers to fully transfer load from one geogrid layer to the other. Geogrid reinforcing from the nailed slope overlapped geogrid reinforcing from the segmental block wall to preclude a slope failure occurring between the layers of geogrid reinforcing. The following formula was used to calculate the required length of geogrid overlap.

$$L := \frac{T \cdot FS}{Ci \cdot \gamma \cdot z \cdot \tan(\phi)} \quad (1)$$

Where:
L = required geogrid overlap
T = maximum tensile strength of geogrid reinforcement
FS = safety factor
Ci = efficiency factor
γ = unit weight of backfill material
z = backfill height
ϕ = angle of internal friction of backfill material

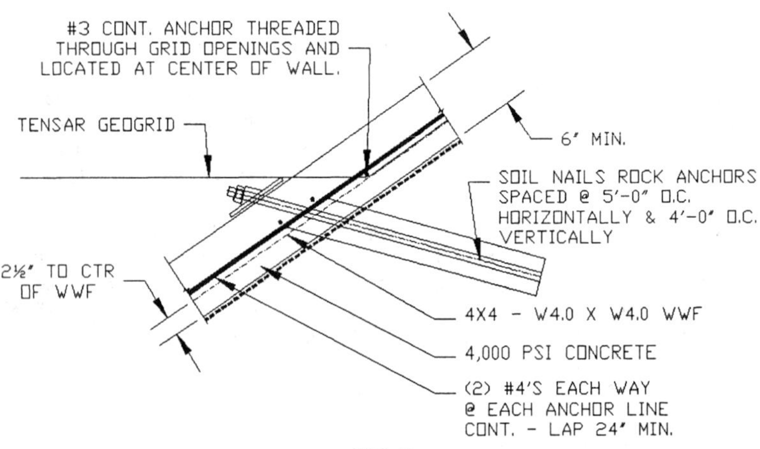

FIG. 7

In lower portions of the wall where less than 4 ft existed to overlap the geogrid reinforcing, one geogrid layer was extended from the soil nailed slope to the segmental block facing and concrete fill was placed in-lieu of rock screenings. No soil anchors were required in the upper portions of the wall where at least 16 ft existed between the segmental block wall and the hillside for full-length geogrid reinforcement.

Segmental blocks were embedded at least 12 inches. No. 57 stone was placed in the bottom of the wall for drainage and the drain board installed during soil nail wall construction was directed to the stone backfill. Compacted rock screenings were used to backfill the segmental block wall. Compacted silty sand fill was placed from the top of the segmental block wall to the upper property line at 2H:1V (horizontal to vertical) slope.

The slope, completed about July 2002 has performed satisfactorily and resulted in an aesthetically pleasing, cost effective slope stabilization repair.

First Presbyterian Church – Asheville, NC

A new parking deck was proposed for First Presbyterian Church in Asheville, North Carolina. The new parking deck would be constructed in the adjacent paved parking lot at a level 18 to 20 ft lower than the bottom floor of the adjacent, multistory church (reference Fig. 8). The grade difference was originally supported by a variety of retaining walls including soldier piles with lagging, stone, brick and concrete. Some of the walls had buttresses for lateral support. The layout of the new parking garage required removal of the existing retaining walls and construction of new retaining walls closer to the church. Soil nailing was selected as an economical and practical way to permanently support the proposed excavation.

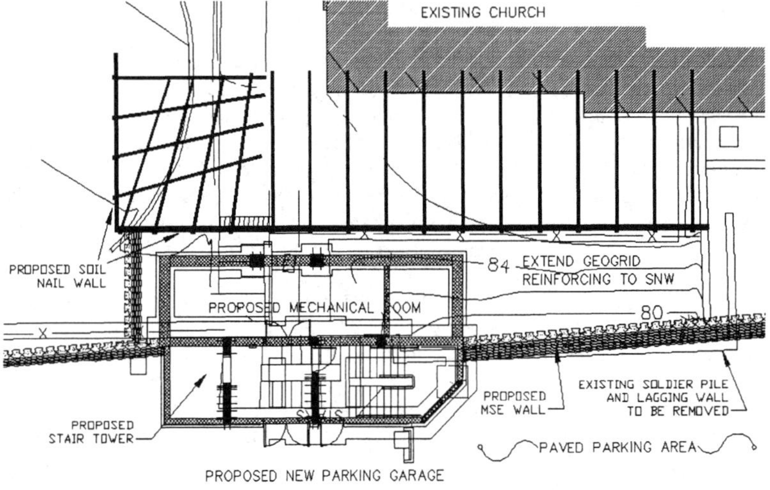

FIG. 8

New construction in the area of soil nailing included a new segmental block wall and a new stair tower/mechanical room. The soil nail wall was constructed about 14 to 16 ft from the existing multistory church. The existing soldier pile and lagging wall and stone, brick and concrete walls were removed in stages as the soil nail wall was installed from the top down. The mechanical room/stair tower was constructed about 20 ft from the church. Site grades were raised by fill placement in the left portion of the site prior to soil nail wall construction as shown in Figure 9.

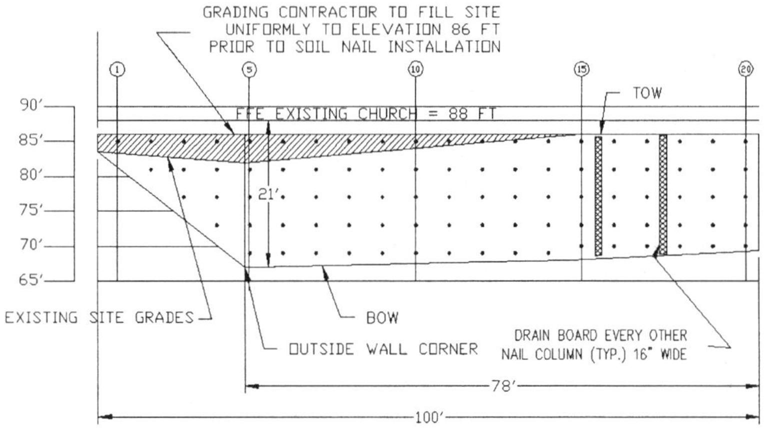

FIG. 9

Residual silty sand soils were exposed in the hillside excavation. The following soil parameters were assumed for design: ϕ = 30 degrees, c = 100 psf and γ = 120 pcf. The soil nails were constructed of #7, grade 75, epoxy coated threadbar and cement grout. Four-inch diameter bore holes were drilled using air to flush cuttings. Soil nail lengths ranged from 13 to 21 ft. An ultimate pullout value of 2,260 lb/ft (15 psi) was utilized for design. A six-inch thick layer of shotcrete was applied to stabilize the exposed soils. The shotcrete façade was reinforced with 4x4_W4xW4 WWF and 2-#4 vertical and horizontal reinforcing bars at each row and column of soil nails.

Boreholes were drilled with a TEI RDS528H Rotary Top Drive Drill mounted on a New Holland, LS 180 skid steer because of limited site access. A 375 cfm air compressor was used to remove cuttings. Shotcrete was pumped with a Schwing BP450 concrete pump.

The soil nails were left exposed in the location of the proposed stair tower. The stair tower walls were reinforced and grouted CMU. The soil nail bars were extended with couplers to structurally tie into the CMU wall as shown in Figure 10. Membrane waterproofing was applied to the CMU wall. The approximately 4 ft space between the shotcrete façade and the back of the CMU wall was backfilled with no. 57 stone.

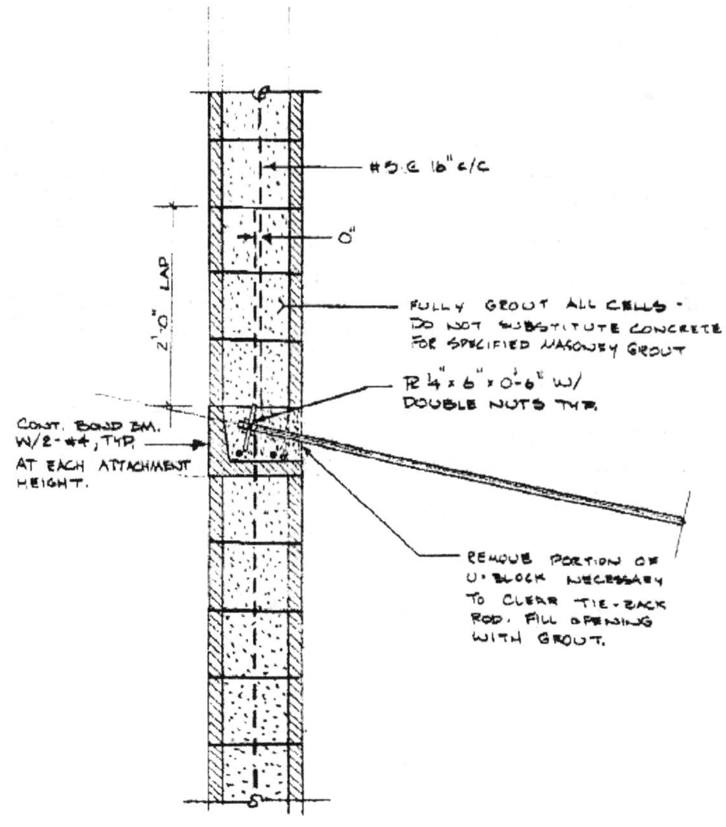

FIG. 10

Approximately 10 ft long geogrid for the adjacent segmental block wall were butted up to the shotcrete façade.

The soil nail wall has performed satisfactorily. Difficulties included limited site access and drilling through a drain adjacent to the church. It was later discovered that we grouted the kitchen drain solid. Fortunately, the drain was scheduled to be replaced.

Conclusions

Soil nailing is a versatile method of slope stabilization, which lends itself well to a variety of facades. Soil nailing can be used in combination with other forms of slope stabilization to create a hybrid slope stabilization method.

Soil nail facades may consist of additional layers of shotcrete that are hand troweled to form a finish, close to that of a formed and poured concrete wall. One-sided forms can be used to create a smooth concrete wall or a wall with patterns that might resemble a stone wall. An additional and separate façade can be constructed in front of the shotcrete façade. A frequently requested facade is segmental blocks constructed directly in front of and structurally attached to the soil nail wall. Below grade building walls can be constructed in front of and structurally connected to soil nail walls with backfill placed between the face of the soil nail wall and the back of the below grade building wall. Finally, soil nail reinforcing can be used in combination with other forms of reinforcing such as geogrid to form hybrid walls. Applying these and other facades opens up a wide range of soil nailing possibilities.

---

[i] Handy, Richard L. (1985). "The Arch in Soil Arching" *J. Geotech. Engrg.*, ASCE, Vol. 111, No. 3, 302-319.

[ii] Koerner, Robert M., (1999). "Designing with Geosynthetics", Fourth Edition, 221

SIDE RESISTANCE OF DRILLED SHAFT SOCKETED INTO WISSAHICKON MICA SCHIST

Michael Zhiqiang Yang[1], Ph. D., P.E., Member, Geo-Institute
M. Zia Islam[2], P.E., Member, Geo-Institute
Eric C. Drumm[3], Ph. D., P. E., Member, Geo-Institute,
Gang Zuo[4], Student Member, Geo-Institute

ABSTRACT: Drilled shafts socketed into bedrock are a common foundation type for support of large and/or important structures in downtown Philadelphia. The bedrock beneath the city is the Wissahickon Formation, a mica schist bedrock encountered with various degrees of weathering. The interface between soil and rock is known to be transitional, varying significantly with depth, and is usually difficult to identify. Based on an extensive geotechnical investigation program, the properties of mica schist rock in the city of Philadelphia area are summarized. The side resistance values in the rock socket measured from O-cell test results for a recent project in the city of Philadelphia are presented. The side resistance values are then compared with measured rock properties, such as, the unconfined compressive strength, RQD, and elastic modulus.

---

[1] Senior Geotechnical Engineer, Michael Baker Jr., Inc., 555 Business Center Dr. Suite 100, Horsham, PA 19044, E-mail: myang@mbakercorp.com
[2] Geotechnical Manager, Michael Baker Jr., Inc., 555 Business Center Dr. Suite 100, Horsham, PA 19044, E-mail: mzislam@mbakercorp.com
[3] Armour T. Granger Professor, Department of Civil & Environmental Engineering, The University of Tennessee, 223 Perkins Hall, Knoxville, TN 37996, E-mail: edrumm@utk.edu
[4] Graduate Research Assistant, Department of Civil & Environmental Engineering, The University of Tennessee, 223 Perkins Hall, Knoxville, TN 37996, E-mail: gzuo@engr.utk.edu

INTRODUCTION

Southeastern Pennsylvania Transportation Authority's (SEPTA) Market Street Elevated (MSE) reconstruction project is located in the west of downtown Philadelphia. The project involves the replacement of a 100-year-old existing steel framed, elevated light rail structure along Market Street, between 45th and 63rd streets. Due to the concerns related to seismic loads, a group of small diameter drilled shafts underneath a bent column were selected to provide the adequate lateral resistance. Approximately, 1,300 drilled shafts were designed for the entire project. The majority of the drilled shafts have a 610 mm shaft diameter in soil with a rock socket diameter of 560 mm. In each shaft, the construction specification requires that the shaft should socket into the competent bedrock with a minimum length of 3 meters.

As a part of the construction plan, Osterberg Cell (O-cell) tests on selected production shafts were required to verify the design load. Based on the O-cell test results with boring information close to the test shaft and an extensive geotechnical investigation program, this paper summarizes the mechanical properties of the mica schist bedrock in the city of Philadelphia. The side resistance of the drilled shafts socketed into mica schist rocks, with various degrees of weathering and strengths, were measured in O-cell tests. Previous studies on the performance of rock sockets in the city of Philadelphia area are also discussed.

SUBSURFACE CONDITIONS

The city of Philadelphia lies on the Coastal Plain and Piedmont physiographic provinces. The upper part of subsurface soil consists of heterogeneous alluvium. This alluvial material is inter-bedded with fine to medium sand, silty sand, and occasional silt layers. A very dense sandy gravel layer is usually encountered at the bottom of this layer. The soil-like underlying residual material is composed of completely decomposed to highly weathered mica schist with SPT blow counts ranging from 20 to more than 50 blows over 120mm. The thickness of the entire overburden soil varies from 3 to 10 meters at the project site.

The bedrock formation is the Wissahickon Formation of Cambrian and Ordovician Age. This formation consists of a thick sequence of metamorphosed pelitic (formerly clay-rich) and arenaceous mica schist rocks. The top of rock surface slopes eastward at about 0.6 degree (Glynn and Fergusson, 1991). The bedrock is dipping generally from the project site to the downtown Philadelphia.

Due to the different degrees of weathering of the mica schist bedrock, the interface between soil and rock is known to be transitional, varying significantly with depth, and usually difficult to identify. For pay item purposes, "rock" is usually defined by the split spoon refusal or auger refusal during the exploration, and by the earth-auger refusal during construction. As a result of these different "rock" definitions, the same material may be classified as "rock" in the design stage and become "soil" during the construction.

PROPERTIES OF WISSAHICKON MICA SCHIST

Wissahickon Formation mica schist is characterized by extremely variable physical properties dependent upon the orientation of the steeply dipping rock beds that are crosscut by closely spaced, steeply dipping and open joints. The measured relative foliation of the rock cores was very steep, between 30° and 90° to the horizontal. Fergusson and Glynn (1988) reported that the quartz content in the Wissahickon mica schist contributed to the higher strength. Thus, the greater mica contents, the lower the strength. Partos et al. (1989) studied the intact mica schist rock strength underneath downtown Philadelphia area. They found that the lowest strength samples had planes of foliation angle between 50°-59° to the horizontal.

The pre-construction exploratory borings indicated that the rock cores consist of mica-rich schists with occasional quartzite veins. Within the proposed rock socket influence zone, the borings indicated that the bedrock is relatively massive, and the RQD ranges from 0% to 100% with an average of 60%. The degree of weathering of bedrock varied significantly with boring locations as well as with depth within the same boring.

Figure 1 summarizes the unconfined compressive strength and the elastic modulus of intact mica schist rock samples from the project site. The rock samples tested have strength, q_u, varying from 3.5 to 80 MPa. The average modulus to strength ratio is 130. According to a rock classification criterion suggested by the International Society of Rock Mechanics (ISRM), samples tested could be classified from very low strength ($1 \leq q_u \leq 5$ MPa) to medium strength ($50 \leq q_u \leq 100$ MPa) rock. The strength of the majority of the intact rock samples lies between low ($5 \leq q_u \leq 25$MPa) and moderate

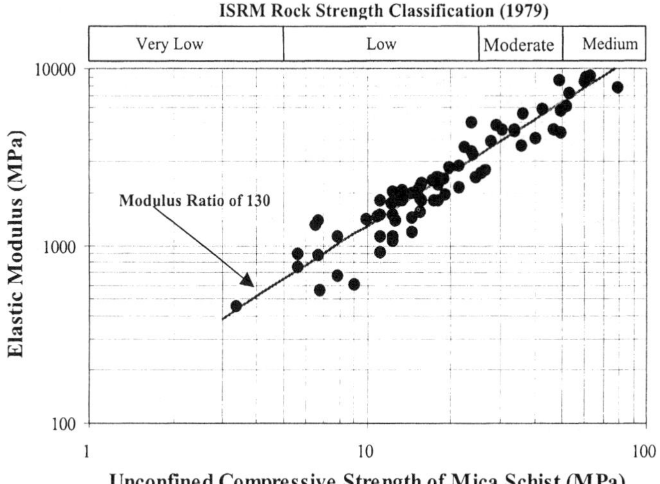

Figure 1 Elastic Modulus Versus Unconfined Compressive Strength of Intact Mica Schist

($25 \leq q_u \leq 50$ MPa). The relationship of strength versus sample depth is shown in Figure 2. The rock sample depths were measured from the top of rock surface where the auger refusal was encountered during the test boring exploration. The test results suggest that the rock strength varies significantly within the same sample depth, particularly at shallower depth. The anticipated increase in strength with depth was observed for only a few borings. The test results also indicate that the low (<20%) rock RQD is generally associated with lower strength. However, when the RQD is greater than 20%, the intact rock strength does not have a direct relationship with the rock RQD magnitude.

Typical consolidated drained (CD) triaxial shear tests of moderately weathered mica schist samples are presented in Figure 3. The rock sample usually demonstrated a brittle behavior with a sudden drop in deviator stress after the peak stress was reached. Usually, the strain at peak stress ranged from 0.7 to 1.5%. The strength parameters for the mica schist using different test methods are shown in Table 1.

It is noted that the triaxial or direct shear tested rock samples from the current project site were visually identified as highly to moderately weathered mica schist. Much higher cohesion values were obtained from triaxial tests from fresh or slightly weathered mica schist samples from downtown Philadelphia area (Partos, 1989).

DRILLED SHAFT SOCKETED INTO MICA SCHIST IN PHILADELPHIA

The use of drilled shafts with rock socket to carry high loads is a common construction practice in the city of Philadelphia. Many high-rise buildings along Market Street in the downtown area are supported by this foundation type. Due to the

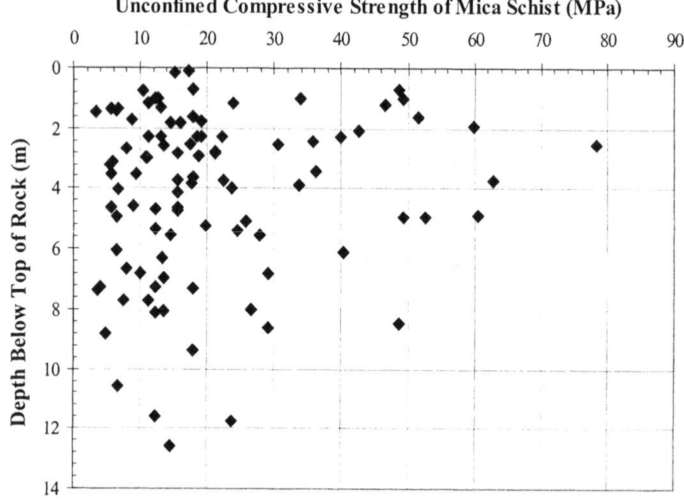

Figure 2 Unconfined Compressive Strength of Mica Schist Versus Sample Depth Measured From Top of Rock

Table 1. Rock Properties Obtained by Different Test Methods

	Triaxial (CD) Test		Direct Shear Test		Unconfined Compressive Test	
Number of Tests	23		19		64	
Strength Parameter	Cohesion c (MPa)	Internal Friction Angle φ (°)	Cohesion c (MPa)	Internal Friction Angle φ (°)	Unconfined Compressive Strength q_u (MPa)	Elastic Modulus of Intact Rock (MPa)
Minimum	0.07	6.0	0.00	12.0	3.36	456
Maximum	3.74	29.0	0.27	41.0	78.36	9,408
Average	0.64	22.2	0.12	31.1	23.93	3,078

transitional nature between the soil and rock, rock property characterization is a constant challenge to geotechnical engineers. Fergusson and Glynn (1988) reported a foundation failure in the downtown area associated with the erroneous estimation of rock mass strength as well as the poor drilled shaft construction.

The James A. Byrne Federal Court House is located at the intersection of Market and 7th streets. It is supported by drilled shafts socketed into mica schist (Koutsoftas, 1981), two of these were instrumented. The percussion drill as used to drill the rock socket in sound rock. Temporary casing was utilized to maintain the stability of overburden soil and weathered rock during the rock socket installation. The instrumented drilled shafts had a 9.45 m socket length with diameters of 610 and 762

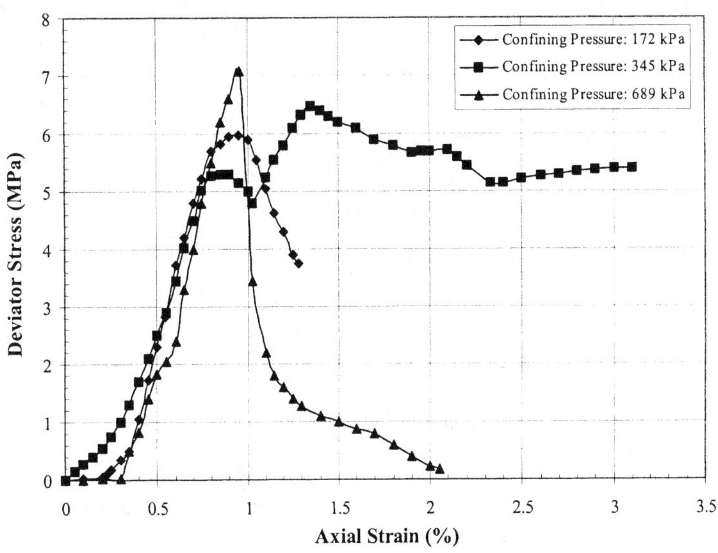

Figure 3 Typical Consolidated Drained Triaxial Test Results of Moderately Weathered Mica Schist

mm, respectively. The measured average side resistance over the entire socket area under the building load was 0.35 MPa, with highest measured side resistance in upper part of rock socket of 1.524 MPa. The average intact rock strength was 30 MPa, which is close to the strength of concrete shaft. The instruments in the socket indicated that the entire building load was carried by the side resistance only.

Another major building in the downtown area supported by rock sockets, One Liberty Place, is located at the intersection of Market and 17th streets. Drilled shafts with a diameter of 2,600 mm and rock socket length of 4.0 m were instrumented and the results were reported by Partos et al. (1989). Under a 27.8 MN building load, the measured average side resistance over entire socket length was 0.55 MPa with a measured rock socket compression of 1.7 mm. The rock socket tip provided an approximate base resistance of 10.0 MN, or about 36% of the total applied load. Mica schist rock samples had an average cohesion of 26.5 MPa with an internal friction angle of 25°. The strength parameters were measured by triaxial tests with a maximum confining pressure of 0.83 MPa. Average elastic modulus of intact rock obtained from unconfined compressive tests was 55,200 MPa. Unfortunately, the instrumentation was installed in a production shaft, the applied loads that strain gages responded were limited to building loads and results cannot be used to verify the appropriate design method.

When the bedrock is relatively massive, methods to predict the ultimate side resistance, f_s, are usually related to the unconfined compressive strength of intact rock or strength of concrete, q_u, which ever is weaker. The formula of these empirical correlations can be generalized as:

$$f_s = \alpha q_u^{\beta} \quad (1)$$

Where α and β are empirical constants determined from load tests. The f_s and q_u are in terms of MPa. Table 2 summarizes the various rock socket ultimate side resistance prediction methods.

Table 2 Summary of Design Methods of Ultimate Resistance Prediction

Design Method	α	β	Remarks
Rosenberg and Journeaux, 1976	0.375	0.515	Based on small diameter (200-610 mm) drilled shafts in shale
Horvath, Kenney and Kozicki, 1983	0.200	0.500	Developed from 710 mm diameter drilled shafts into mudstone, smooth shaft interface.
	0.300	0.500	Developed from 710 mm diameter drilled shaft into mudstone, rough shaft interface.
Rowe and Armitage, 1987	0.450	0.500	Drilled shaft into weak rock, smooth shaft interface
	0.600	0.500	Drilled shaft into weak rock, smooth shaft interface
AASHTO, 1998 LRFD	0.150	1.000	Drilled shaft into weak rock with $q_u \leq$ 1.9 MPa
	0.210	0.500	Drilled shaft into rock with $q_u \geq$ 1.9 MPa

As it can be seen from Table 2, the ultimate side resistances predicted by different design methods are significantly different. The magnitude of side resistance depends not only on the strength of rock mass/concrete, but also on the features of concrete and rock interface, which is the result of construction method. Artificially roughen the interface by cutting groves into the socket wall can significantly increase the side resistance. Design method proposed by Horvath et al (1983) has different α values to reflect effect of the shaft roughness on the side resistance.

Among various design methods, Horvath et al. (1983) method for smooth shaft interface represents the lower limit, whereas, Rowe and Armitage (1987) method for rough shaft interface is the upper limit. The current AASHTO (1998, LRFD version) suggested design method is close to the lower limit of ultimate side resistance prediction.

Glynn and Fergusson (1991) reported that design side resistances used in the drilled shafts socketed into sound mica schist in the city of Philadelphia area varied from 0.20 to 0.90 MPa for a customary practice in the area, which a smooth shaft wall was generally assumed.

TEST SHAFT CONSTRUCTION AND LOAD TEST SET-UP

Four O-cell test results on small diameter drilled shafts are available from the SEPTA MSE project. The shafts were started with a 610 mm O.D. steel casing. An earth auger was used to drill the alluvial soil and completely decomposed rock. According to general construction practice, the top of rock was defined at a depth where an earth auger lost its efficiency or reached refusal. The shaft was cleaned with a cleanout bucket and the casing was seated 150 mm into rock. The rock sockets had a nominal design diameter of 560 mm. The test shaft at bent 6304 had a 610 mm socket diameter due to some difficulties encountered in the shaft excavation. The shaft was excavated below the existing elevated light rail using a low-head-room equipment (Figure 4).

The rock socket at bent 6004 was the first shaft excavated. It was excavated with a rock core barrel. The penetration rate in the socket was between 240 and 2,700 seconds per 300 mm. The rest of the three shafts were excavated by a down-hole rock hammer with penetration rates ranging from 30 to 210 seconds per 300 mm. The rock hammer proved to be efficient in the rock socket installation, and the rest production rock sockets were excavated by the rock hammer. The temporary casing was used to maintain the stability of the overburden soil and weathered above the rock socket. No slurry was introduced.

After cleaning the socket by the air lifting method, the reinforcing cage with attached O-cell assembly was inserted into the base. Before placing concrete into the shaft, high strength grout was pumped through a PVC pipe down to the bottom of the shaft until the grout reached above the O-cell level. The grout served as a seating layer for the O-cell at the bottom of the shaft. Concrete was then placed by tremie method. The temporary steel casing was removed during the concrete placement.

Figure 5 illustrates the O-cell test set-up. To measure the load transfer within the socket, three levels of vibrating wire rebar strain gages (SG) were also installed above the O-cell. Typically, SG 1 was approximately one meter above the upper plate of O-

Figure 4 The Drilled Shaft Excavation Using Low Head Room Equipment

cell, SG 2 was placed one meter below top of rock and SG 3 located about one meter above the bedrock surface.

TEST RESULTS AND DISCUSSIONS

The construction specification requires the applied load on the O-cell should be a minimum of 2.2 MN, which is two times the 1.10 MN design load, and should not exceed 3.30 MN. The O-cell test results and measured properties related to the side resistance are summarized in Table 3. At each test shaft location, the rock strength at elevations within the socket was obtained from two pre-construction exploratory borings within 3-m of the test shaft. The maximum applied average side resistance was calculated using the maximum applied load divided by the side area of socket, assuming that the alluvium and decomposed soil did not carry the load. This assumption was confirmed by very small strains registered by SG 3 for all four shafts. For a comparison, the ultimate side resistances predicted by lower limit method (Eq. (1), Horvath et al., 1983) is also shown in Table 3. It is found that under the current maximum applied loads, the lower limit ultimate side resistance was not reached.

Table 3 Summary of O-cell Test Results

Bent Location	Rock Socket Length (m)	Socket Length Above O-cell (m)	Ave. Rock Strength within Socket (MPa)	Ave. RQD within Shaft (%)	Concrete Strength at Time of O-cell Test (MPa)	Max. Upward Load (MN)	Max. Measured Ave. Side Resist. (MPa)	Lower Limit Estimates of Ultimate Side Resist. (MPa)	Upward Disp. at Max. Load (mm)
4508	5.35	4.51	12.3	22	32.8	2.87	0.362	0.701	0.92
5604	4.34	3.42	50.3	71	32.1	2.88	0.441	1.133	0.27
6004	6.42	4.99	21.6	83	33.1	2.87	0.328	0.930	3.14
6304	5.58	4.05	8.3	48	31.0	3.35	0.456	0.576	0.69

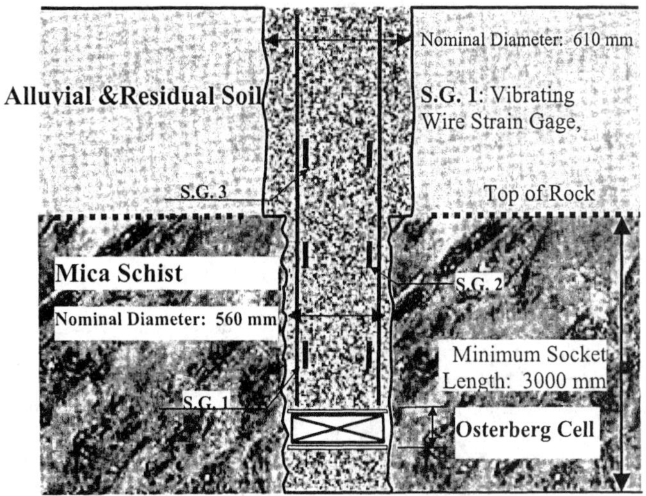

Figure 5 Illustration of O-cell Load Test Set-up

Figure 6 summarizes side resistance from four O-cell test results (solid symbols). All tests were terminated before the sockets mobilized the ultimate side resistance. The side resistance in this figure was obtained by the applied load over the entire rock

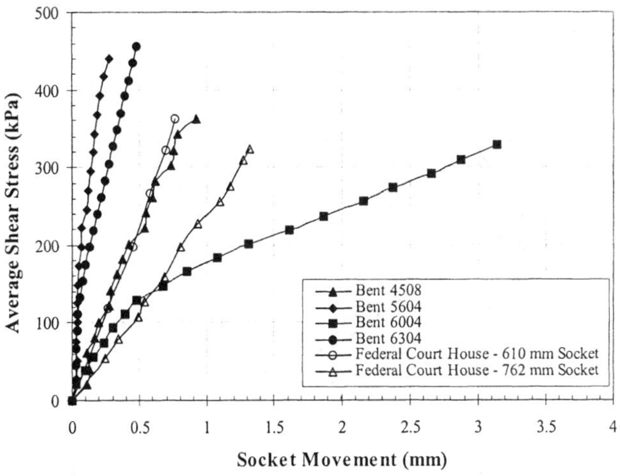

Figure 6 Average Side Resistances Versus Socket Movement

socket surface area. The measured average side resistance data from instrumented sockets under the Federal Court House (hollow symbols, Koutsoftas, 1981) are also presented in Figure 6 for the purpose of comparison.

Except for test shaft at bent 6004, all other shaft demonstrated a linear relationship between side resistance and socket displacements. It is of interest to note that the measured rock socket performance from the O-cell test on a shaft at bent 4508 in SEPTA MSE project is similar to the result at the Federal Court House.

At same average mobilized shear stress, the test shaft at bent 6004 demonstrated larger socket movement, although it has a longest socket length above the O-cell and relatively higher rock strength. This shaft was excavated by using a rock core barrel, which took a much longer time than other shafts excavated by a rock hammer. As a result, test shaft at bent 6004 might have smoother socket interface than the other test shafts. The smoother socket would result in a less side resistance. However, since the roughness of the test shaft was not measured, it is unknown whether the larger socket movement is caused by the different construction method.

Similar to the instrumentation results from rock socket underneath Federal Court House (Koutsoftas, 1981), the distribution of side resistance along the rock socket is not uniform. In Figure 7, the O-cell test results indicate that the distribution of side resistance within the rock socket decrease rapidly. The load transfer can be better interpreted when the distance along the shaft is normalized by the rock socket length (Figure 8).

It can be seen from Figures 7 and 8 that most of the applied load was resisted by the lower portion of rock socket close to the O-cell load device. Much lower side resistances were measured between strain gages at SG1 and SG2 than those between

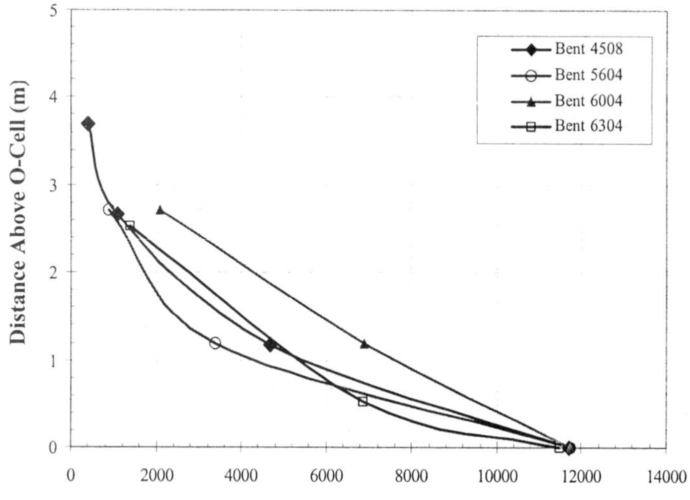

Figure 7 Vertical Stress Distribution in the Rock Socket Measured by Strain Gages

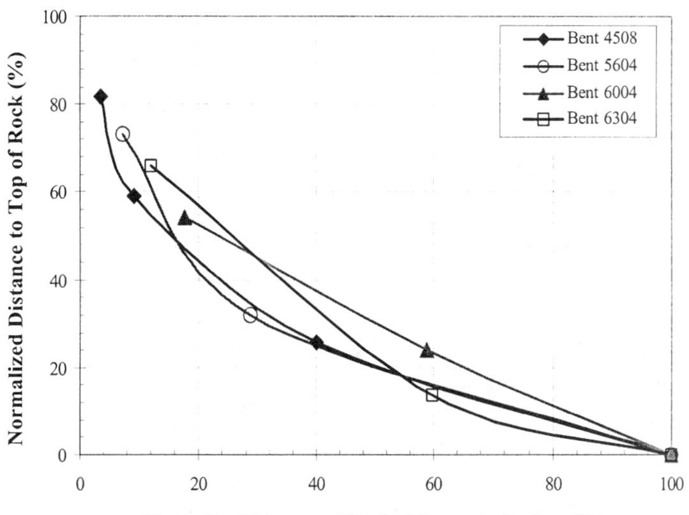

Figure 8 Normalized Vertical Stress Distribution in the Rock Socket above O-cell

the O-cell and SG 1. About 50% percent of the applied load was carried by the lower one third of the socket and the lower two third of socket resisted about 90% of the external load. In test shaft at bent 6004, the portion of the socket away from the O-cell carries more load than in other shafts.

The average side resistance in socket between the O-cell and SG1 (Figure 9) is much higher than the average value over the entire rock socket side area (Figure 6). In the shaft at bent 6004, the average side resistance at the socket segment between O-cell and SG 1 was 0.57 MPa when the O-cell was loaded to about 2.6 times of the design load. A high side resistance of 1.51 MPa between two layers of strain gages (Koutsoftas, 1981) was obtained in an instrumented rock socket in Federal Court House (hollow circles in Figure 9).

In a numerical analysis of rock socket at bent 6004 (Zuo et al., 2004), the material properties were first calibrated from the O-cell test results and then the load was applied at the top of drilled shaft. At a load magnitude of 2.87 MN acting on the top of the shaft, only 3% of total applied load was transferred to the base of rock socket. This result is consistent with the O-cell test results, in which a very small strain was recorded at SG3. At a computed failure load of 13.90 MN, the ultimate average side resistance over the entire socket surface area was 0.95 MPa with a side resistance of 1.410 MPa within the side area of first one meter from the top of rock socket. The corresponding displacement at the top of rock socket was about 25 mm.

Both the O-cell tests and numerical analysis results of the shaft at bent 6004 suggest that four rock sockets tested can carry an applied load greater than design load.

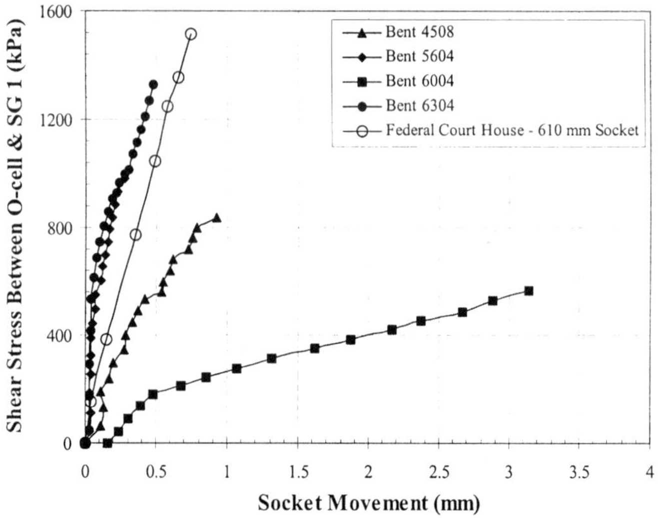

Figure 9 Maximum Average Side Resistance between Two Instrument Levels

CONCLUSIONS

Four O-cell load tests to measure the side resistance of drilled shafts socketed into mica schist bedrock are reported in this paper. Coupled with an extensive geotechnical investigation, the following conclusions are suggested from the study:

- The subsurface condition underneath the city of Philadelphia is complex, and the mechanical properties of the mica schist bedrock vary significantly.
- Under the specified design load, the displacements in the rock socket above the O-cell are small. Although the loading directions in the O-cell test and in service shaft are different, the O-cell test results are comparable to the instrumented rock socket under actual building loads in the downtown Philadelphia.
- Both the O-cell test and the previous instrumentation on the production socket indicate that the mobilized side resistance decreases rapidly with depth within the socket.
- The O-cell tests indicated that the drilled shaft socketed into rock could carry an applied load greater than design load.

ACKNOWLEDGEMENT

The authors are grateful to Southeastern Pennsylvania Transportation Authority (SEPTA) and its representative, the Market Street Elevated (MSE) reconstruction

Program Management team, for permission to release test data. The MSE Program Management service is provided by Jacobs Civil, Inc. in association with Michael Baker, Jr. Inc., Edwards & Kelcey, Inc., and Synterra, Ltd.. The general design consultant is DMJM+Harris. Gannett Fleming provides the geotechnical engineering service. Driscoll Construction Co., Inc. and PKF Mark III, Inc. are the drilled shaft contractors.

The first and second authors wish to express their gratitude to their employer, Michael Baker, Jr., Inc. for its support in preparing this manuscript.

REFERENCES

AASHTO (1998), *Standard Specifications for Highway Bridges*, LRFD Version, Washington, DC.

Fergusson, W. B. and Glynn, E. F. (1988). "A foundation failure at Philadelphia." Proc. 2nd International Conference on Case Histories in Geotechnical Engineering, St. Louise, MO, 1293-1296.

Glynn, E. F., and Fergusson, W. B., (1991). "Where does rock begin beneath Philadelphia?" *Detection of and Construction at the Soil/Rock Interface*, ASCE, GSP No. 28, 31-43

Horvath, R. G., Kenney, T. C., Kozicki, P. (1983) "Methods of improving the performance of drilled piers in weak rock." *Canadian Geotechnical Journal*, Vol. 20, 758-772.

Koutsoftas, D. C., (1981). "Caisson socketed in sound mica schist." *Journal of Geotechnical Engineering*, ASCE, Vol. 107, No. GT6, 743-757.

Partos, A. J., Sander, E. J., and Hungspruke, U. (1989). "Performance of a large diameter drilled pier." *Proc. International Conference on Piling and Deep Foundations*. London, 309-316.

Rosenberg, P. and Journeaux, N. L. (1976). "Friction and end bearing tests on bedrock for high capacity socket design." *Canadian Geotechnical Journal,* Vol. 13, 324-333.

Rowe, R. K., and Armitage, H. H., (1987). "A design method for piers in soft rock." *Canadian Geotechnical Journal,* Vol. 24, 126-142.

Zuo, G., Drumm, E. C., Islam, M. Z., and Yang, M. Z, (2004). "Numerical analysis of O-cell testing of drilled shafts in mica schist." *Geo-Support 2004: International Drilled Foundation Support Specialty Conference*, ASCE/ADSC, Orlando, Florida

NUMERICAL ANALYSIS OF DRILLED SHAFT O-CELL TESTING IN MICA SCHIST

Gang Zuo[1], Student Member, Geo-Institute,
Eric C. Drumm[2], Ph.D., P.E., Member, Geo-Institute,
M. Zia Islam[3], P.E., Member, Geo-Institute,
and Michael Zhiqiang Yang[4], Ph.D., P.E., Member, Geo-Institute

ABSTRACT: Osterberg-cell or O-cell tests are an effective method for load testing drilled shaft foundations, and can identify the relative contributions of side resistance and end bearing. A series of O-cell tests was conducted for several production shafts socketed into mica schist in downtown Philadelphia. The bearing material is the Wissahickon Formation, which is a mica schist bedrock typically encountered with various degrees of weathering. This paper describes a finite element model developed to simulate the drilled shaft rock socket subject to axial loadings. The material parameters used in the analyses were calibrated based on the O-cell test data during the "bottom up" loading and supported by results from shaft instrumentation. The calibrated model is used to investigate the load distribution in the weathered rock socket and the contribution of end bearing, both when loaded from the bottom during the load test and under the service loading from the top.

[1] Graduate Research Assistant, Department of Civil & Environmental Engineering, The University of Tennessee, 223 Perkins Hall, Knoxville, TN 37996, E-mail: gzuo@engr.utk.edu

[2] Armour T. Granger Professor, Department of Civil & Environmental Engineering, The University of Tennessee, 223 Perkins Hall, Knoxville, TN 37996, E-mail: edrumm@utk.edu

[3] Geotechnical Manager, Michael Baker Jr., Inc., 555 Business Center Dr. Suite 100, Horsham, PA 19044, E-mail: mzislam@mbakercorp.com

[4] Senior Geotechnical Engineer, Michael Baker Jr., Inc., 555 Business Center Dr. Suite 100, Horsham, PA 19044, E-mail: myang@mbakercorp.com

INTRODUCTION

The Osterberg method is an economical and effective technique, for load testing drilled shafts (Fellenius et al. 1999, Gunnink and Kiehne, 2002). In the Osterberg method, the Osterberg Cell (O-cell), a sacrificial hydraulic jack-like device is placed at the shaft tip. When the O-cell is pressurized, load is applied to the shaft and the base simultaneously. In comparison to the conventional top load test, the Osterberg method provides contributions of end bearing and side resistance separately. A series of O-cell tests was conducted on several production shafts socketed into mica schist in downtown Philadelphia (Yang et al., 2004). The bearing material is the Wissahickon Formation, which is a mica schist bedrock typically encountered with various degrees of weathering. The material properties of the rock obtained from laboratory testing are quite variable. Numerical analysis was used to obtain the overall material parameters of the rock by calibrating the numerical model to match the behavior obtained from O-cell testing. The calibrated model was then used to predict drilled shaft response under conventional top loading.

NUMERICAL MODEL

An axisymmetric model was adopted in the analysis using ABAQUS (2001), with the mesh extending 12 m laterally from the axis of symmetry and 9 m vertically from the bottom of the shaft. As shown in Figure 1, this width is about 20 times the socket diameter and the depth is 1.5 times the length of the rock socket. The soil and the part of the shaft above the rock socket were not modeled, since the side resistance on the soil-shaft interface is negligible with respect to that of the rock. However, the weight of the soil and the weight of the part of the shaft above the socket were applied as pressure to the model to simulate geostatic stress.

The concrete shaft was assumed to be linear elastic, with a Young's modulus of 9.74 GPa, a density of 2400 kg/m^3 and a Poisson's ratio of 0.15. It should be noted that the modulus of the shaft is somewhat lower than expected, consistent with the test report (LOADTEST, 2002) based on the tangent modulus analysis. The rock was assumed to be elastic-plastic and was represented by the Drucker-Prager model (ABAQUS, 2001), a simple constitutive model for geomaterials which requires relatively few parameters. The dilation angle (ψ) in the Drucker-Prager model was assumed to be half of the angle of friction to avoid unrealistically high dilation that takes placed in the classical Drucker-Prager model due to the normality rule. As discussed subsequently, a small sinusoidal asperity surface was introduced in the shaft socket (Hassan and O'Neill, 1997). Poisson's ratio of the rock was assumed to be 0.3 and the density of the rock was 2700 kg/m^3. The other material properties were varied within the range of laboratory values in a series of trial-and-error numerical analyses using ABAQUS (2001) to match the load-displacement relationship from the load testing of the relevant shaft.

As shown in Figure 1, a fine mesh was constructed for the shaft and the rock that was adjacent to the shaft, in which six elements were constructed for each wavelength (300 mm) of the sinusoidal socket asperity.

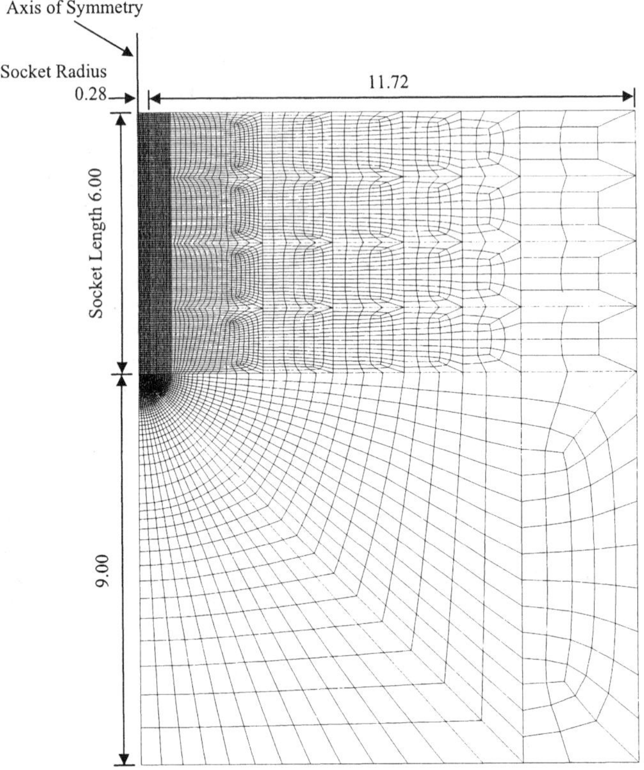

FIG. 1 Mesh of the Model (Unit: m)

In numerical analyses that include contacts, such as the shaft-rock interaction analysis here, the following two assumptions are usually made: (1) the side resistance is usually achieved by Coulomb friction, where the shear resistance is assumed to be proportional to the normal stress at the interface; (2) the contact interface is usually assumed to be planar with the surface roughness accounted for in the choice of friction coefficient. For shafts under top loading, stability of the numerical solution is almost guaranteed, as long as proper material model and properties are selected. However, the O-cell test induces a complication in the numerical modeling, since the upward load applied to the shaft is completely mobilized by side resistance. For a smooth interface with Coulomb friction, in order to provide sufficient resistance, either the friction coefficient or the normal stress which is induced by the initial geostress must be unreasonably high. To obtain realistic response, the shaft resistance must include a cohesion term.

Cohesion was incorporated into the contact model via a user material in ABAQUS (2001) to reduce the requirement for unreasonably high friction coefficient and/or geostress. However, a gap was opened close to the top of the contact interface under the friction, since the elastic modulus of the concrete shaft was much higher than that of the rock. The solution ended prematurely due to a phenomenon which is known as chattering, i.e., the solution failed to converge due to the difficulty in locating the initial point of the gap.

Hassan and O'Neill (1997) proposed a way to simulate the rough interface between a shaft and rock socket with sinusoidal profile. In their model, the rough interface was idealized by sequential curvilinear segments that formed a sinusoidal profile with a long wavelength. According to Hassan and O'Neill (1997), a sinusoidal curve with a wavelength of 0.305 m was considered the longest wavelength normally observed. The double amplitude of the sinusoidal wave used in Hassan and O'Neill (1997) was 25.4 mm. In this paper, a wavelength of 0.3 m was used, and the double amplitude was varied to achieve the best match between measured and computed displacements. The profile of the interface between the shaft and the rock socket is a series of concave and convex segments. Before the shaft is loaded, there is no relative movement and the shaft is in close contact with the rock with no gap in between. When the shaft is loaded, the front of a convex part on the shaft is pushed towards the relevant concave part in the rock, and it leaves a gap behind. Therefore, it would be anticipated that the stress distribution along the sinusoidal interface is not as gradual as along a straight interface. Part of the deformed mesh of the shaft-rock contact is shown in Figure 2. It should be noted that the deformation is exaggerated in the figure.

In reality, the interface is always rough to some extent, although the wavelength and the amplitude are neither constant nor usually known. A sinusoidal interface with constant wavelength and amplitude can be regarded as an average roughness over the entire contact interface.

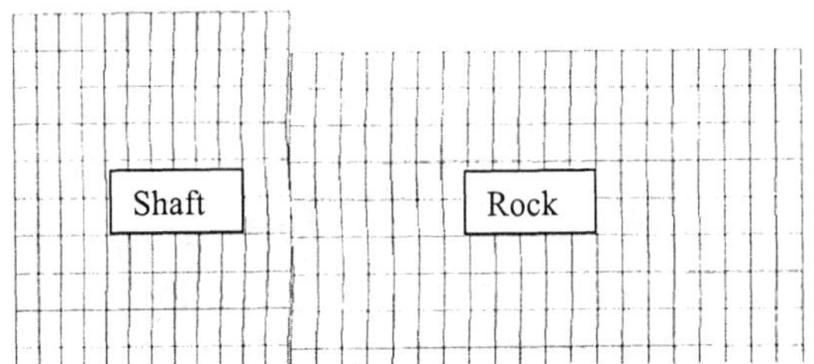

FIG. 2 Part of Deformed Mesh Showing the Rough Contact Interface

CALIBRATION OF THE MODEL WITH LOAD-DISPLACEMENT CURVES

The load-displacement curves of the O-cell test were used to calibrate the model and the following information was to be obtained from the calibration: (a) Young's modulus of the rock; (b) Drucker-Prager parameters (related to cohesion c and friction angle ϕ); (c) Double amplitude of the shaft-rock interface profile and (d) Coulomb friction coefficient of the interface.

The rock modulus was first obtained by matching the linear part of the load-displacement curve of the O-cell bottom plate. The nonlinear part of the same curve was used to determine Drucker-Prager parameters. Finally, a close match between calculated and measured load-displacement curve of the O-cell top plate was obtained by varying the double amplitude of the interface profile and the Coulomb friction coefficient.

RESULTS OF O-CELL TEST SIMULATION

A comparison of measured and computed O-cell displacements is shown in Figure 3. The maximum reported O-cell load was 2870 kN, but in the numerical analysis, the load was increased until the solution failed implying that the limit load was reached. The computed "ultimate" O-cell load in Figure 3 is 4355 kN. The parameters obtained from the numerical analysis which resulted in a close match of the load-displacement relationship as shown in Figure 3 are listed in Table 1. The material properties from laboratory testing are also listed in Table 1 for comparison. Rock properties obtained from the model calibration are all within the range of the laboratory testing. What should be noted in Table 1 is the low value of double amplitude of the sinusoidal interface profile. A double amplitude of 4.0 mm over a wavelength of 300 mm is barely discernible by eye. However this "negligible" roughness is able to provide side resistance which is impossible for a smooth interface with reasonable cohesion, friction coefficient and level of geostress.

An illustration of the development of yield zones under O-cell load is given in Figure 4. The maximum stress occurred at the lower corner of the interface. The sinusoidal interface is not apparent in Figure 4, due to the scale and low amplitude. However, the wavelength of the asperities can still be inferred from the yield zones developed along the interface. The yield zone started at the corner, and it increased as the simulated O-cell load increased under the bottom of the shaft. No yield zone was developed along the shaft interface until the load was high. The yield zone along the interface also started from the lower corner. As the load increased, the separate yield zones along the interface expanded, and at the same time, additional separate plastic zones appeared at the locations up the shaft far from the O-cell. The numerical solution failed (Figure 4(f)) when the entire area beneath the bottom plate of O-cell yielded, while the yield zone along the interface was still far from all the way to the top. This suggests that end bearing is more critical than side resistance for this case.

The comparison of the computed load and measured load at the strain gage locations is shown in Figure 5. The computed loads were obtained by multiplying the average vertical stress at the relevant location by the cross-sectional area of the shaft. Strain gage level 1 is 1.2 m above the shaft tip and level 2 is 2.7 m above the shaft tip. The match is not as good as that of the shaft displacement, but similar trends can be discerned from the comparison. Both the computed and measured load at different depths in the shaft increased slowly initially, but when the applied load reached a certain break point, the vertical stress increased more rapidly. Due to the side resistance, the gage farther away from the load has a higher break point load.

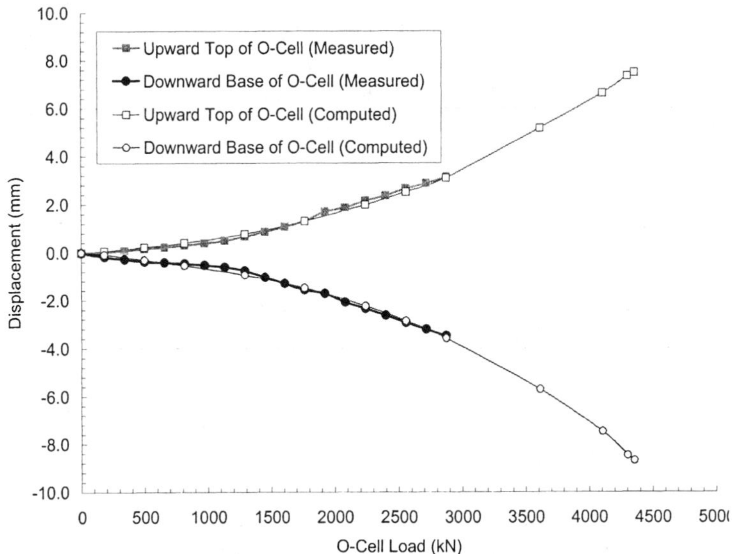

FIG. 3 Comparison of Measured and Calculated O-Cell Displacement

Table 1 Model Parameters Calibrated from Numerical Analysis

		Lab. Test			Calibrated from Numerical Analysis
		Minimum	Maximum	Average	
Young's modulus of the rock (GPa)		0.456	9.408	3.078	2.0
Drucker-Prager parameters	c (MPa)	0.07	3.74	0.64	1.4
	ϕ (°)	6.0	29.0	22.2	21.5
Double amplitude of the shaft-rock interface profile (mm)		-	-	-	4.0
Coulomb friction coefficient of the interface		-	-	-	0.6

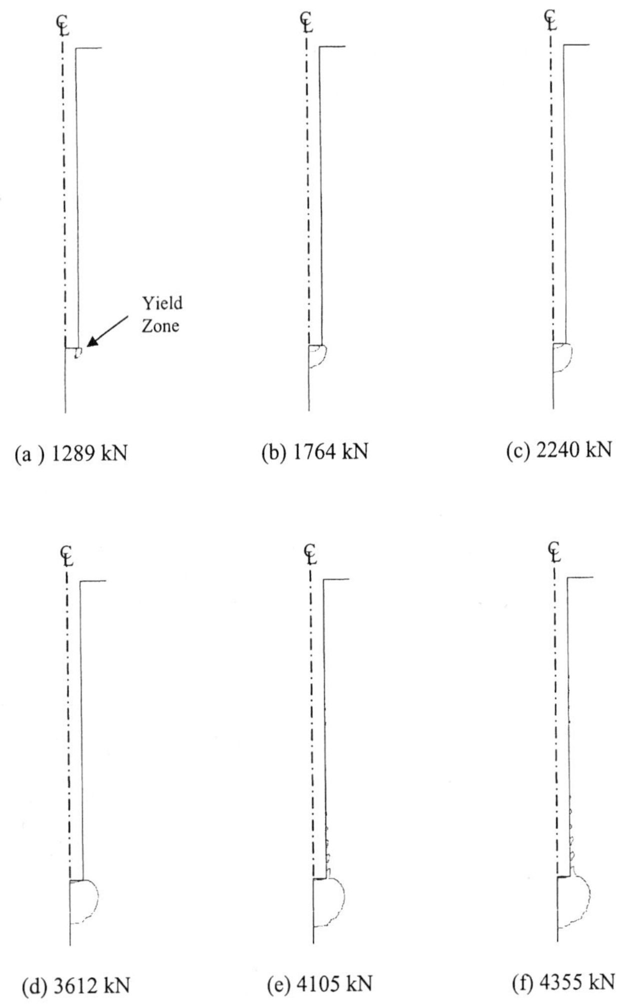

FIG. 4 Yield Zone Development in the Rock under Different Simulated O-cell Load Levels

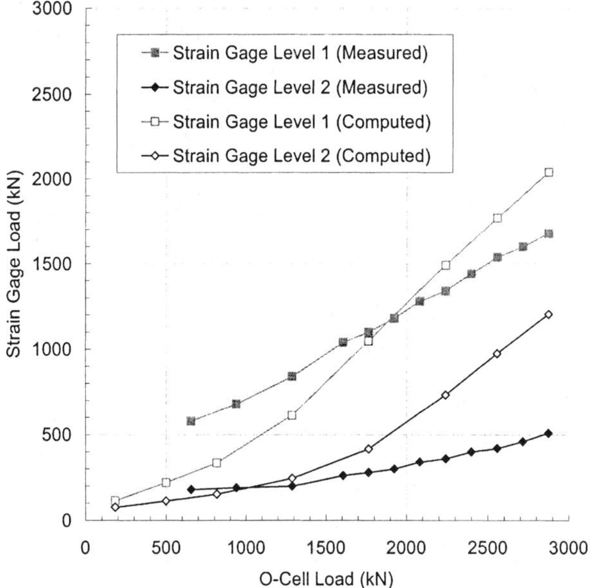

FIG. 5 Comparison of Vertical Load at the Strain Gage Locations: Gage 1 located 1.2 m above shaft tip, Gage 2 located 2.7 m above shaft tip.

RESULTS OF TOP LOAD SIMULATION

Since the O-cell test is different from a conventional load test, it is beneficial to obtain the load-displacement response for the same shaft under the top loading case using the calibrated model. The same model as used in the O-cell test simulation was used to study the behavior of the shaft under top loading. The only difference is that the load was applied to the top of the shaft and a hard contact was added to the shaft-rock interface at the bottom. A smooth contact interface was assumed for this contact and the friction coefficient was assumed to be 0.6, same as that of the sinusoidal interface.

For top loading, the maximum stress still occurred at the lower corner of the interface as under the O-cell load. However, the stress distribution along the interface is different from that of O-cell loading, since the strain in the shaft and interface slip is greater near the top of the socket. An illustration of the development of yield zones under the top loading is given in Figure 6. Yield zones started at two locations simultaneously: one at the lower corner as a result of end bearing, and the other at top of the socket as a result of side resistance. As the top load increased, both yield zones expanded. The yield zone under the shaft tip expanded continuously. Along the

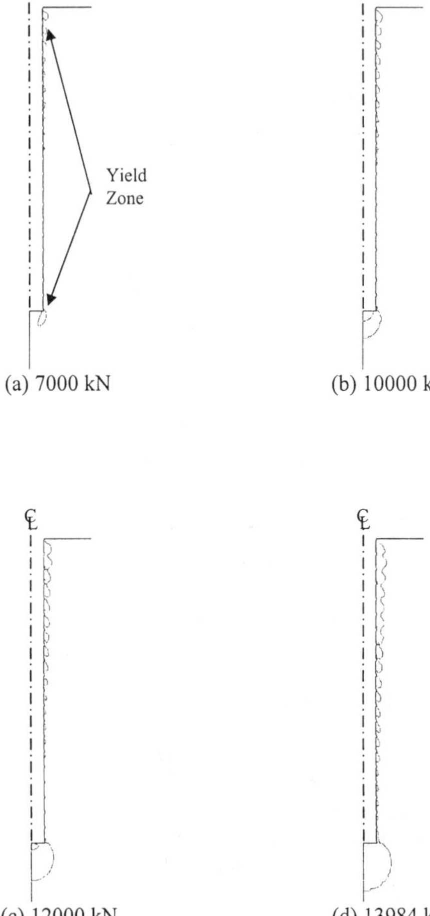

FIG. 6 Yield Zone in the Rock under Different Simulated Top Load Level

interface, more separate yield zones emerged while the existing yield zones expanded, and finally these separate yield zones connected and merged into a continuous yield zone. Similar to the O-cell load, the numerical analysis of the shaft under top load failed to converge (Figure 6(d)) when the entire area beneath shaft tip became plastic, while the plastic zone along the interface was not completely connected. Hence, the same shaft under top load still failed in end bearing.

The computed displacements of the shaft tip and socket top in the numerical model are shown in Figure 7. As mentioned priorly, the numerical analysis only modeled the part of the shaft that was in the rock socket. Therefore, the elastic shortening calculated for the portion of the shaft in the soil above the socket was added to the total vertical displacement to obtain the displacement of the shaft butt (Figure 7). The equivalent top loading load-displacement curve (LOADTEST, 2002) which was constructed from the O-cell test data using an empirical method is also shown in Figure 7. The butt settlement computed from the numerical analysis is lower than that of the empirical method. This is consistent with the fact that the empirical method is conservative. Figure 7 suggests that the settlement at the shaft tip or the compression of the socket is small. Most of the butt settlement can be attributed to the elastic compression of the shaft.

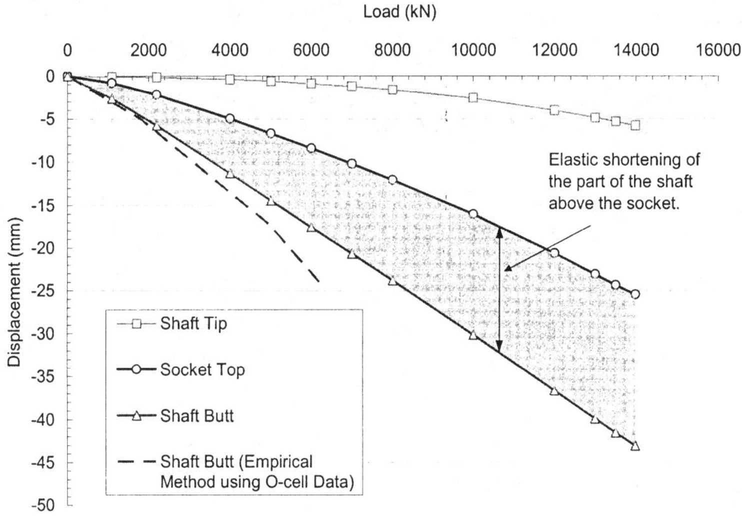

FIG. 7 Displacement of the Shaft Under Top Load

The computed vertical load in the shaft as a function of depth under top loading is shown in Figure 8. The percentage of load carried by end bearing is shown in Figure 9. Under the design load (1100 kN), only 3.5% of the load is carried by end bearing. Even under twice the design load (2200 kN), only 5.0% of the load is carried by end bearing. Side resistance takes up most of the load. With the increase of top load, a greater proportion of load is carried by end bearing. However, the percentage of the load carried by end bearing does not increase linearly with increasing load. It seems that for the rock socket discussed in this paper, the maximum percentage of load carried by end bearing is approximately 30%.

CONCLUSIONS

Numerical analysis was applied to back-calculate the response of a drilled shaft in weathered mica schist during an O-cell test. A trial-and-error approach was taken to obtain a close match between computed and measured O-cell top and bottom plate displacements at all the load steps. The rough contact interface between the shaft and rock, simulated by continuous sinusoidal profile, was shown to be an effective way to overcome the numerical difficulties in modeling contact problems in which high shear stress was required to be transferred across the contact interface. This analysis suggests that a small asperity along the contact interface (a sinusoidal wave curve with a double amplitude of 4 mm and a wavelength of 300 mm) can provide enough side resistance to assure the convergence of the numerical solution.

Numerical analysis is a promising approach to predict the shaft behavior under top load using the information obtained from O-cell load simulation. It indicated that the tip settlement of the shaft under top load was small, and most of the butt settlement was due to the elastic compression of the shaft. The load is mainly carried by side resistance. Under the design load, only 3.5% of the load was carried by end bearing. It remains to use the numerical method developed here to predict the response of shafts not used for the development of the material parameters.

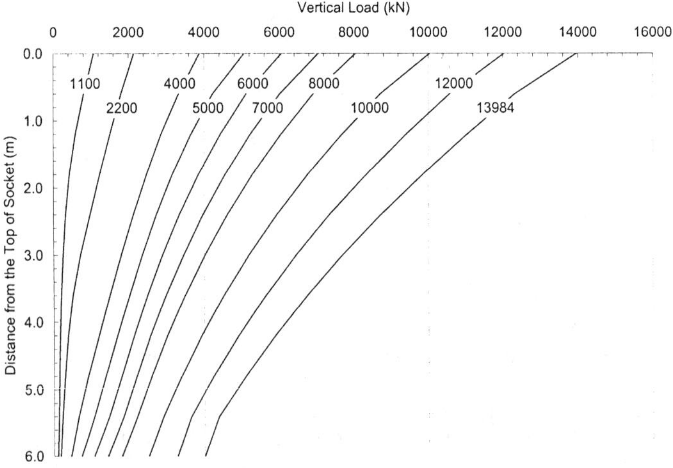

FIG. 8 Vertical Load vs. Distance from the Top of Socket (Numbers in on the curves indicating the magnitude of top load in the unit of kN)

ACKNOWLEDGMENTS

The authors are grateful to Southeastern Pennsylvania Transportation Authority (SEPTA) and its representative, the Market Street Elevated (MSE) reconstruction Program Management team, for the permission to release the load test data. The MSE Program Management service is provided by Jacobs Civil, Inc. in association with,

Michael Baker, Jr. Inc., Edwards & Kelcey, Inc., and Synterra, Ltd. The general design consultant is DMJM+Harris. Gannett Fleming provided geotechnical engineering services. Driscoll Construction Co., Inc. and PKF Mark III, Inc. are the drilled shaft contractors.

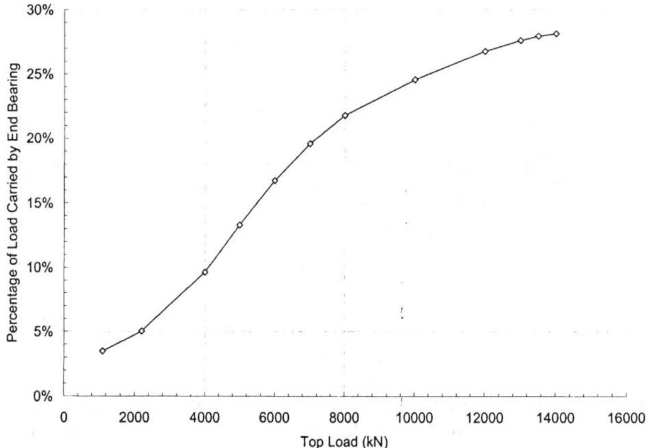

FIG. 9 Percentage of Load Carried by End Bearing

REFERENCES

ABAQUS 6.2-1, (2001). Habbitt, Karlsson & Sorensen, Inc., Rhode Island.

Fellenius, B. H., Altaee, A., Kulesza and R., Hayes, J., (1999). O-Cell Testing and FE Analysis of 28-m-Deep Barrette in Manila, Philippines. *Journal of Geotechnical and Geoenvironmental Engineering* Vol. 125, No. 7, 566-575.

Gunnink B. and Kiehne C., (2002). Capacity of Drilled Shafts in Burlington Limeston. *Journal of Geotechnical and Geoenvironmental Engineering* Vol. 128, No. 7, 539-545.

Hassan, K. and O'Neill, M. (1997). Side Load-Transfer Mechanisms in Drilled Shafts in Soft Argillaceous Rock. *Journal of Geotechnical and Geoenvironmental Engineering* Vol. 123, No. 2, 145-152.

LOADTEST Inc., (2002). Report on Drilled Shaft Load Testing (Osterberg Method) Test Shaft 60-04-2 – Market Street Guideway, Philadelphia, PA (LT 8813-1). Project Number: LT 8813-1, October 25, 2002.

Yang, M. Z., Islam, M. Z., Drumm, E. C., and Zuo, G., (2004). Side Resistance of Drilled Shaft Socketed into Wissahickon Mica Schist, *Geo-Support 2004: International Drilled Foundation Support Specialty Conference*, ASCE/ADSC, Orlando, Florida.

STRUCTURAL DAMPING CONCEPT FOR INTERPRETATION OF STATNAMIC PILE LOAD TEST RESULTS

San-Shyan Lin[1], J. L. Hong[2], Wei F. Lee[3], and Y. H. Chang[4]

ABSTRACT: A procedure that uses the structural damping concept for estimating the capacity of a pile based on the Statnamic pile load test results is proposed in this study. Although the interpreting procedure is similar to that of the commonly used UPM, however, displacement related soil damping is used instead of the velocity dependent damping. Hence, soil resistance is determined directly from the measured load-displacement curve. Three case studies are presented in this paper to study the validity and applicability of the present method. Comparison of predicted results with available test or analytical data shows that the proposed method accurately models the tested pile behavior.

INTRODUCTION

Since its inception in 1988, the Statnamic (STN) pile load test has often been used by geotechnical engineers around the world. Compared to other conventional static load test methods, the STN method is indeed a quicker and more efficient pile load test method.

Among the available methods for interpreting STN test results, especially for the purpose of determining an equivalent static capacity, the Unloading Point Method (UPM) is the most frequently used method (Brown, 1994, 1999; Chow et al 1998; Lin et al., 2001a, 2001b; Mullins et al. 2002). The key issue of UPM is to determine the unloading point that is based on the measured load-displacement curve. A

[1and 2]Porf. And Graduate Student, Dept. of Harbor & River Eng., Taiwan Ocean University, Keelung, Taiwan 20224; sslin@mail.ntou.edu.tw
[3]Associate Researcher, Taiwan Construction Research Institute, Taipei County, Taiwan.
[4]Vice President, Diagnostic Engineering Consultants, Ltd., Taipei, Taiwan.

constant static soil resistance under maximum displacement or maximum loading is assumed in the interpretation method. However, studies by Ealy and Justason (2000) have found that the overall shape of the derived static curves, based on UPM, does not follow the test results exactly. This suggests that the correlation could be improved by using a slightly more sophisticated damping constant than the one estimated by UPM.

In an effort to improve the accuracy and applications of the STN test, an interpretation method based on structural damping concept is proposed in this paper. In the proposed method, a displacement related soil damping is used to replace the velocity dependent damping of UPM. Thus, soil resistance can be determined directly from the measured load-displacement curve. Three case studies are also presented. The proposed method is used to verify its applicability and the results show an improved prediction of the STN tests.

BRIEF REVIEW OF THE UNLOADING POINT METHOD

The procedure for evaluating a STN load-displacement curve is usually based on a simplified pile and soil model that is subjected to rigid body translation during the loading and unloading stages. The pile is idealized as a rigid mass, while the soil is treated as a spring and a dashpot system, as shown in Fig. 1. The spring represents the static soil response, which includes the elastic deformation of the pile. The dashpot denotes the dynamic resistance, which depends on the rate of pile penetration. The time history measurements of the load and the vertical displacement on the pile head were taken using the load cell and the laser displacement meter, respectively, to obtain a load-displacement curve. The unloading point method is generally used for interpretation of the load-displacement curve.

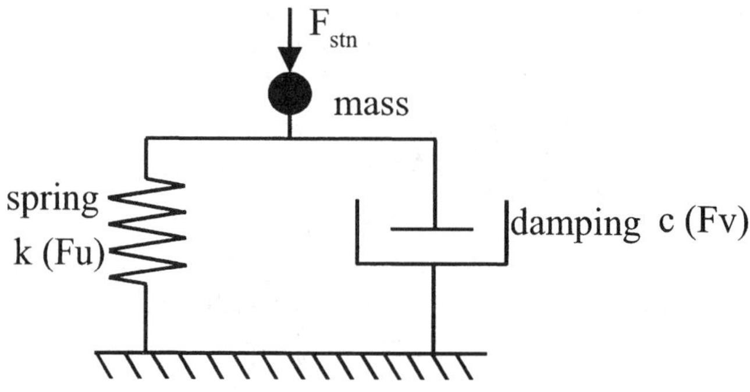

Fig 1 UPM Soil-Pile Model (Middendorp et al., 1992)

From equilibrium of the pile soil system idealized in Fig. 1, we can derive the following equation (Middendorp et al, 1992)

$$F_{stn}(t) = F_{soil}(t) + F_a(t) \tag{1}$$

where $F_{soil}(t) = F_u(t) + F_v(t)$; $F_{stn}(t)$ is the measured STN load; F_{soil} is the soil resistance; $F_u(t) = k \cdot u(t)$ is the static soil resistance; $F_v(t) = c \cdot a(t)$ is the damping force from soil; and $F_a(t) = m \cdot a(t)$ is the pile inertia force. Furthermore, k is soil spring stiffness; c is soil damping factor; m is pile mass; $u(t)$ is the measured displacement; $v(t)$ is the velocity; and $a(t)$ represents the acceleration.

The equation (1) can also be expressed as

$$F_{stn}(t) = F_u(t) + F_v(t) + F_a(t) \tag{2}$$

or

$$F_u(t) = F_{stn}(t) - c \cdot v(t) - m \cdot a(t) \tag{3}$$

At unloading point, where the velocity is zero, the damping force is also zero, and $F_u(t)$ is a maximum, then

$$F_{u-\max}(t) = F_{stn} \quad \text{(at unloading point)} \tag{4}$$

In addition, at maximum loading point, we have

$$F_u(t) = F_{u-\max} = F_{\max} - c \cdot v(t_{F-\max}) - m \cdot a(t_{F-\max}) \tag{5}$$

where $F_{\max}$ is the maximum applied STN load; $v(t_{F-\max})$ and $a(t_{F-\max})$ are the velocity and the acceleration at maximum applied STN load, respectively. Thus, from (2) and (3) the damping value can be obtained from

$$c = \frac{F\max - F_{u-\max} - m \cdot a(t_{F-\max})}{v(t_{F-\max})} \tag{6}$$

Once the damping value is determined, the static soil resistance, $F_u(t)$, can be obtained, and hence, from $F_u(t)$ and $u(t)$, a load-displacement curve can be drawn and yields a static soil resistance estimate as a function of displacement.

It is proposed that the soil damping force in Eq. (2) is replaced by the structural damping restoring force, $f_{SD}(t)$, thus, Eq. (2) can be rewritten as

$$F_u(t) = F_{stn}(t) - f_{SD}(t) \tag{7}$$

Detailed derivation of $f_{SD}(t)$ is given in the following section.

STRUCTURAL DAMPING CONCEPT

Conventionally, the form of structural damping restoring force has been postulated as (Meirovitch 1986)

$$F_{SD}(\omega) = i \cdot C_s \cdot x(\omega) \tag{8}$$

where x is the displacement and C_s is the structural damping constant. Both F_{SD} and x are complex-valued function of frequency, ω. Multiplication of the displacement by the imaginary, i, in Eq. (8) produces a force that rotates counter-clockwise by $\frac{\pi}{2}$ radians in the complex plane, as shown in Fig. 2(a) (Bronowicki 1981). This postulation allows the structural damping force applied at the time of maximum particle speed to have a magnitude in proportion to displacement. For positive or negative values of frequency, the structural damping force has either a 90 degree lead or lag over displacement, respectively. This condition leads to the presence of an unstable root in the simple spring-mass system equation of motion, which can be given as

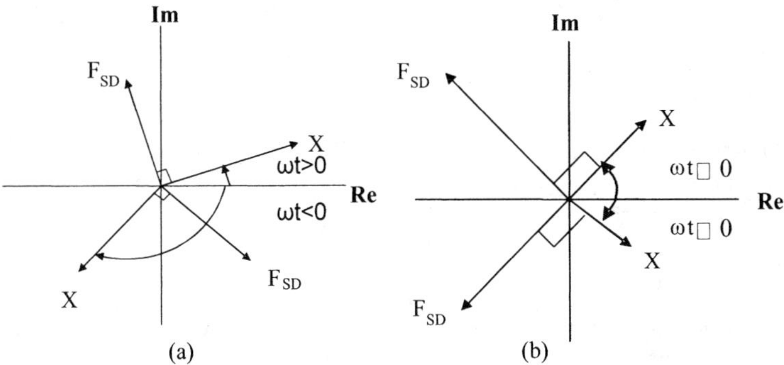

Fig. 2(a)Previous structural damping model (b)Structural damping model proposed by Bronowicki(1981)

$$m\ddot{x} + (i \cdot C_s + k) \cdot x(t) = 0 \tag{9}$$

The roots of the above equation are calculated using the definition of Eq. 8. We found that the roots lead to negative and positive values of damping (Soroka 1949).

To overcome the drawbacks of structural damping concept, a re-definition of structural damping was proposed by Bronowicki (1981), which can also be used in the time domain application. His definition produces positive damped motions in both frequency directions. As shown in Fig. 2(b), the structural damping restoring force can be expressed as (Bronowicki 1981)

$$F_{SD}(\omega) = -R(\omega) \cdot C_S \cdot X(\omega) \tag{10}$$

where C_S denotes structural damping; $X(\omega)$ is the displacement under frequency domain; and $R(\omega)$ is a differential filter as proposed by Papoulis (1961), which can be expressed as

$$R(\omega) = \begin{cases} i & \omega < 0 \\ 0 \quad if & \omega = 0 \\ -i & \omega > 0 \end{cases} \tag{11}$$

This filter provides a damping force that leads displacement by 90 degrees at any given non-zero frequency, as shown in Fig. 2.

To view Eq. 10 in the time domain, we can compute its inverse Fourier transformation to become

$$F_{SD}(t) = \frac{C_s}{2\pi} \int_{-\infty}^{\infty} -R(\omega) \cdot X(\omega) \cdot e^{i\omega t} d\omega \tag{12}$$

In addition, the impulse response of the filter, $R(\omega)$, after taking the inverse Fourier transformation, is (Papoulis 1961)

$$r(t) = \frac{1}{\pi \cdot t} \tag{13}$$

Hence, to compute the transient forces we can perform a convolution of the known function of time, i.e.,

$$f_{SD}(t) = \frac{-C_S}{2\pi} \sum_{\tau=-\infty}^{\infty} \frac{1}{\tau} X(t-\tau)\Delta\tau \tag{14}$$

The summation, Σ, of Eq. 14 considers the whole loading duration, based on the measured displacement versus time relationship. In addition, the only one unknown in Eq. 14 is the structural damping constant, C_s. Bronowicki (1981) defined the amount of energy dissipated by a single degree of freedom oscillator in one cycle as (Fig. 3a)

$$E_d = \pi \cdot C_s \cdot A_x^2 \tag{15}$$

where A_x is the amplitude of motion (Fig. 3a). Hence, the structural damping constant can be expressed as

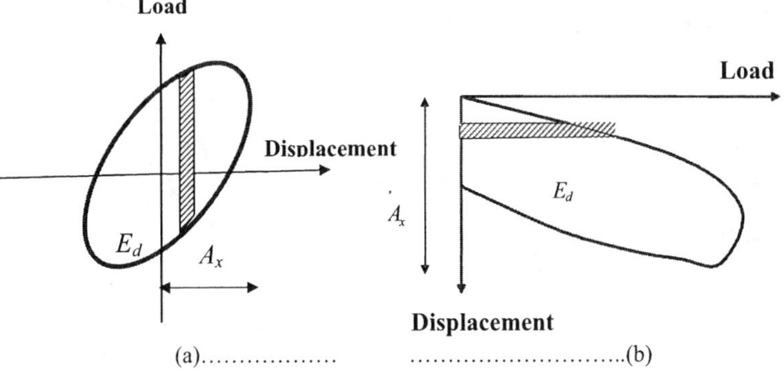

Fig.3.(a)A typical hysteresis loop.(b) Typical STN tested load-

$$C_s = \frac{E_d}{\pi \cdot A_x^2} \tag{16}$$

Since the load-displacement curve obtained from STN pile load test is not a closed curve, the energy dissipation is calculated by integrating the area enclosed by the obtained load-displacement curve and the vertical axis as shown in Fig. 3b. The amplitude of motion is assumed as the maximum displacement, $A_x^{'}$, also given in Fig. 3b.

CASE STUDIES

To verify the proposed method, four cases of STN pile load tests were analyzed and compared to the results obtained from the UPM. Results of the case studies are presented in the following sections.

Case 1- A Building Project, Taipei County, Taiwan 2002

A STN load test was conducted on an all casing method installed drilled shaft in Taipei county, 2002. The shaft is 1.2 m in diameter and 12.4 m in length, and is socketed 5 m into the rock. The tested load and displacement versus time relationships are shown in Fig. 4. Loading duration of the test is about 0.17 second. In addition to the derived result based on UPM, Fig. 5 also shows the sensibility of the derived static load–displacement curve to the damping constant. Changes in damping constant can have a dramatic effect on the shape of the load-displacement curve. An investigation on how the chosen damping constant on the derived results of UPM was conducted by Mullins et al (2002). Apparently, the predicted results using UPM appear to depend strongly on how the damping value is chosen. Fig. 6 gives the load-displacement curves obtained from measurement and from the predicted results using UPM and proposed method. Good agreement between the measured and the predicted, using the method described in this paper, is obtained. In addition, result obtained from the authors' method is unique.

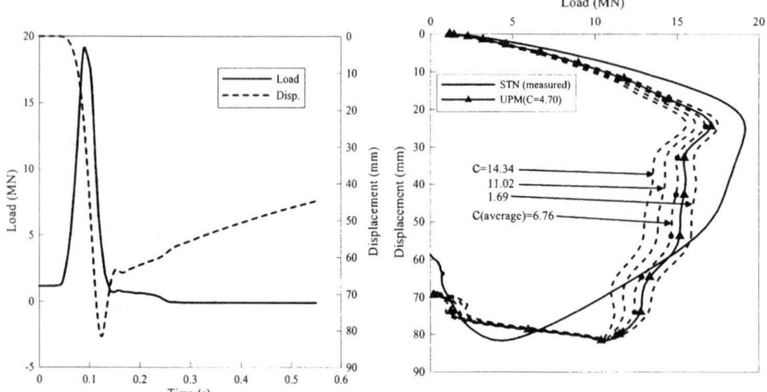

Fig. 4. Measured and Predicted results(Taipei County Case)

Fig. 5. Predicted v.s. Measured results(Taipei County Case)

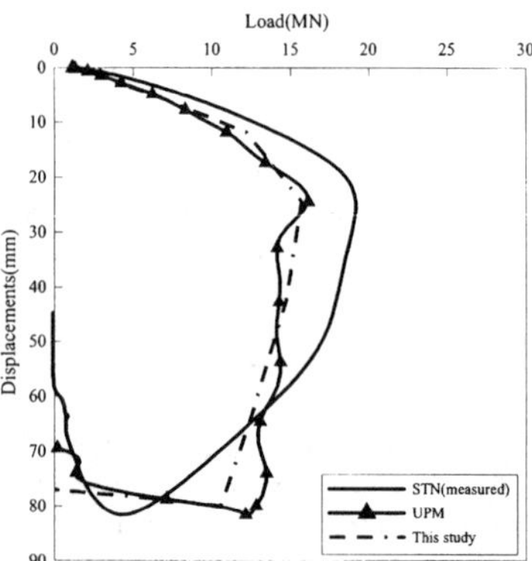

Fig. 6 Predicted v.s. Measured results(Taipei County Case)

Case 2- I-40 at Rio Grande, Albuquerque, New Mexico, USA (Brown 1994)

This case was firstly studied by Brown (1994) for a 0.76 m diameter, 18.3 m long drilled shaft installed in Rio Puerco in New Mexico, USA. The shaft was installed into a hard silty clay and/or clayey silt soil layer. The moisture content of the soil is 8 to 19%. The typical SPT-N value is in the range of 17 to 32 blow/ft. The STN test was conducted after a conventional static pile load test. The measured and the predicted results of this particular case are shown in Fig. 7. The capacity obtained from STN is higher than that of the static load test. The results interpreted using our method tend to be closer to the STN test results.

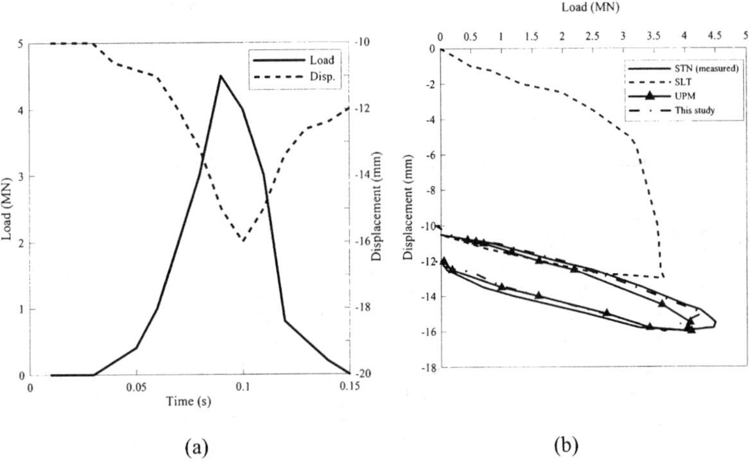

Fig. 7 I-40 New Mexico,USA(a)Load and Displacement v.s. Time (b)Predicted v.s. Measured results(Brown 1994)

Case 3 - Shonan test site in Japan (Sakimoto et al., 1995)

A series of pile load tests, including STN load tests, were carried out at the Shonan test site located in Shonan-Machi, 30 km east of Tokyo. This test site consists of humus, loam, and clay from the ground level to 6.6 m deep, and underlain by fine sand to 12.9 m depth. The upper stratum shows the SPT N-value ranging from 1 to 3. The fine sand stratum consists of a rather uniform particle size with SPT N-value ranging from 17 to 19. The test piles were cylindrical pre-stressed, precasted concrete piles with outside diameter of 300 mm. The embedded depth of the piles was 7 m. Fig. 8 gives the measured pile head load and the displacement versus time curves. The predicted load-displacement curves, based on UPM and the method proposed in this paper, are given in Fig. 8(b). Good agreement was observed between the measured and predicted results.

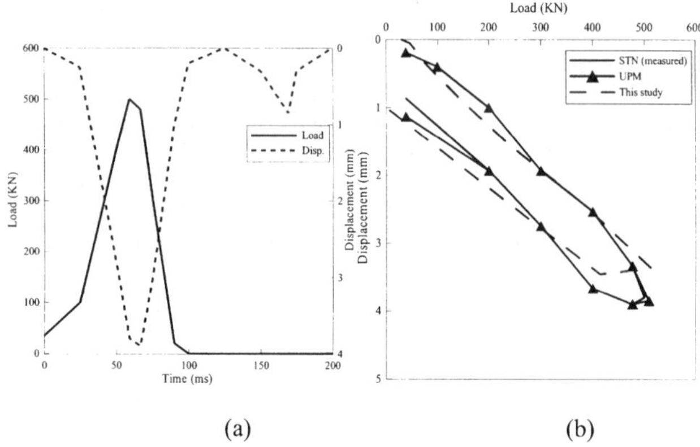

(a) (b)

Fig. 8 Shonan,Japan(a)Load and Displacement v.s. Time (b)Predicted v.s. Measured results

SUMMARY AND CONCLUSIONS

A simple and accurate approach for interpretation of pile capacity based on STN test results is outlined in the paper. According to the interpretation procedure of UPM, however, the soil resistance can be determined directly from the measured load-displacement curve, by replacing the velocity dependent damping by the displacement related structural damping. Comparison of the predicted results with available test data through several case studies shows reasonable agreement. The model used in this paper is capable of simulating the behaviors of the STN tested piles.

ACKNOWLEDGMENT

Support of this research was provided by the National Science Council of ROC under research contract NSC 88-2218-E-019-010, which is gratefully acknowledged.

REFERENCES

Bronowicki, A. J. (1981). "Structural Damping in The Time Domain," *Dynamic Response of Structures: Experimentation, Observation, Prediction and Control,* ASCE, 833-845.

Brown, D. A. (1994). "Evaluation of Static Capacity of Deep Foundations From Statnamic Testing" *Geotechnical Engineering Journal*, 17(4), 403-414.

Chow, Y. K., Chew, S. H., Chow, W. M., and Chuah, L. S. (1998). "Case Histories of Statnamic Pile Load Tests," *Proceedings of the 13th Southeast Asian Geotechnical Engineering Conference*, Taipei, Taiwan, 479-484.

Ealy, C. D., and Justason, M. D. (2000). “Statnamic and Static Load Testing of a Model Pile Group in Sand,” *Proceedings of the Second International Statnamic Seminar,* Tokyo, Japan, 169-177.

Lin, S. S., Yeh, S.C., Juang, C.H., and Chang, Y. H. (2001a). “Segmental Unloading Point Method for Interpretation of Some Statnamic Tested Shafts,” *Geotechnical Engineering Journal*, 32(1), 15-22

Lin, S. S., Yeh, S.C., Juang, C.H., and Chang, Y. H. (2001b). “Statnamic Tests for Six Long Rock Socketed Shafts,” *Foundations and Ground Improvement*, ASCE, GSP No. 113, 524-538.

Meirovitch, L. (1986) *Elements of vibration analysis*, McGraw Hill, 71-73.

Middendorp, P., Bermingham, P., and Kaiper, B. (1992) “Statnamic Load Testing of Foundation Piles,” *Proceedings of the 4th International Conference on Application of Stress Wave Theory to Piles*, Hague, Netherlands, 581-588.

Mullins, G., Lewis, C. L., and Justason, M. D. (2002). “Advancements in Statnamic Data Regression Techniques,” *Deep Foundations 2002*, ASCE, GSP No. 116, 915-919.

Papoulis, A. (1961) *The Fourier Integral and Its Applications*, McGraw Hill, NY, USA.

Sakimoto, Yamanaka, N, Omi, H., and Waki, T. (1995). A Comparative Study of the Statnamic Load Test on Driven and Bored Piles at the Shonan Test Site,” *Proceedings of the First International Statnamic Seminar*, Vancouver, Canada.

Soroka, W. W. (1949). “Note on The Relations Between Viscous and Structural Damping Coeffieients,” *Journal of the Aeronautical Sciences*, 16, 409-410.

LATERALLY LOADED ISOLATED PILES, DRILLED SHAFTS AND PILE GROUPS USING THE STRAIN WEDGE MODEL

Mohamed Ashour[1], Gary Norris[2], and Patrick Pilling[3]

ABSTRACT: This paper demonstrates the application of the strain wedge (SW) model to assess the response of laterally loaded isolated long piles, drilled shafts and pile groups in layered soil (sand and/or clay) and rock deposits. The basic goal of this paper is to illustrate the capabilities of the SW model versus other procedures and approaches. The SW model has been validated and verified through several comparison studies with model- and full-scale lateral load tests. Several factors and features related to the problem of a laterally loaded isolated pile and pile group are covered by the SW model. For example, the nonlinear behavior of both soil and pile material, soil-pile interaction (i.e. the assessment of the p-y curves rather than the adoption of empirical ones), the potential of soil to liquefy, the interaction among neighboring piles in a pile group and the pile cap contribution are considered in SW model analysis. The SW model analyzes the response of laterally loaded piles based on pile properties (pile stiffness, cross-sectional shape, pile-head conditions, etc.) as well as soil properties. The SW model has the capability of assessing the response of a laterally1 loaded pile group in layered soil based on more realistic assumptions of pile interaction as compared to techniques and procedures currently employed or proposed.

INTRODUCTION

The problem of a laterally loaded single pile and pile group has been under investigation and research for more than three decades. Many approaches, such as Broms= method (1964a and 1964b), the elastic method (Poulos 1971a and 1971b), and p-y curve approach (Matlock 1970, Reese 1977, Murchison and O=Neill 1984, etc.), have been employed in the analysis of

[1] Res. Assist. Professor, Dept. of Civil Engrg., Univ. Of Nevada, Reno, NV 89557, e-mail ashourm@unr.edu

[2] Prof., Dept. of Civil Engrg., Univ. Of Nevada, Reno, NV 89557

[3] Vice President, Black Eagle Consulting, Inc., 1345 Capital Blvd, Suite A, Reno, NV 89502

laterally loaded pile response. These methods consider some factors while neglecting others. Therefore, limitations exist with respect to all these approaches. As a result, designers have to switch from one method to another to best satisfy their needs.

The p-y method, which was developed by Matlock (1970) and Reese (1977), represents the most commonly used and convenient procedure for the analysis of laterally loaded piles. The confidence that designers have in this method derives from the fact that the p-y curves employed have been obtained (back calculated) from full scale field tests, albeit only a very few tests. However, the p-y method does not account for some significant factors such as various pile properties and soil continuity which could affect the predicted results tremendously. Researchers have attempted to improve the performance of the p-y method by evaluating the p-y curve based on the results of the pressuremeter test (Smith 1983) or dilatometer test (Robertson et al. 1989). The SW model analyzes the response of laterally loaded piles based on a representative soil-pile interaction that incorporates pile and soil properties (Ashour et al. 1998a). The SW model allows the designer to predict the associated p-y curve at any point along the deflected part of the loaded pile. The effect of pile properties and the surrounding soil profile on the nature of the p-y curve has been presented by (Ashour and Norris 2000a).

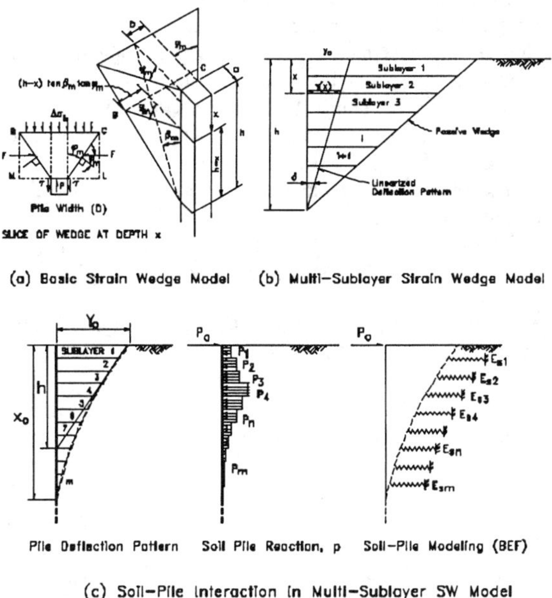

FIG. 1 Configuration of Strain Wedge (SW) Model

P-multipliers for piles in different rows have been suggested by Brown et al. (1988) for analysis of pile interaction effects. Recent tests (McVay et al. 1995 and 1998, and Rollins et al. 1998) indicate such multipliers are a function of pile stiffness, pile spacing, load or deflection level and soil type. Therefore, the p-y method is presently insufficient to the task of accurately evaluating pile group response. In contrast, the SW model utilizes the mobilized geometry of the passive wedge of soil in front of the pile (horizontally and vertically) to assess the interference of overlapping shear zones among the piles in the group. Consequently, the continually changing effect of pile group interaction on the associated p-y curves and (modulus of subgrade reaction, E_s) along every pile in the group can be evaluated.

The SW model has the capability of representing the undrained resistance of liquefiable soil (saturated sands) and the expected developing liquefaction on the behavior of laterally loaded piles as illustrated by Ashour and Norris (1998, 1999 and 2001) and Norris et al. (1997). The profession still lacks a realistic procedure for the design of pile foundations in liquefying or liquefied soil. The most common practice employed is that presented by Wang and Reese (1998) in which a traditional p-y curve shape is used but based on the undrained residual strength (S_r) of the sand. S_r can be related to the standard penetration test (SPT) corrected blowcount, $(N_1)_{60}$ (Seed and Harder 1990). However, a very large difference between values at the upper and lower limits at a particular $(N_1)_{60}$ value affects the assessment of S_r tremendously. Even if an accurate value of S_r is available, S_r occurs at a large value of soil strain and higher peak undrained resistance is ignored in such clay-type modeling. This is extremely conservative. Furthermore, the p-y curve reflects soil-pile-interaction, not just soil behavior. Therefore, the effect of soil liquefaction (i.e. degradation in soil resistance) does not reflect a one-to-one change in soil-pile or p-y curve response. Instead, the soil's undrained stress-strain relationship should be used in a true soil-pile interaction model to assess the corresponding p-y curve behavior. Because the traditional p-y curve is based on field data, a very large number of field tests for different pile types in liquefying sand would be required to develop a realistic, empirically based, p-y characterization.

Moreover, the nonlinear behavior of pile material (steel and/or concrete) based on soil-pile interaction is also analyzed using the SW model (Ashour et al. 2001a).

OVERVIEW OF LATERALLY LOADED PILE AND DRILLED SHAFT BEHAVIOR BASED ON THE SW MODEL

The SW model is an approach that has been developed to predict the response of a free-head flexible pile in uniform soil under lateral loading (Norris 1986). The main concept associated with the SW model is that traditional one-dimensional Beam on Elastic Foundation (BEF) pile response parameters can be characterized in terms of three-dimensional soil-pile interaction behavior (Fig. 1). In the last several years, the SW model has been improved and modified through additional series of research to accommodate a laterally loaded pile with different head conditions that is embedded in multiple soil layers (sand and clay) and rock (Ashour et al. 1998a and Ashour et al. 2001c). The main objective behind the development of the SW model is to evaluate the modulus of subgrade reaction (i.e. secant slope of p-y curve), E_s, in order to solve the BEF problem of a laterally loaded pile and pile group based on the envisioned soil-pile interaction and its dependence on both soil and pile properties in addition to the effect of interaction among the piles in a group.

Based on the basic properties from the soil profile (effective unit weight of the soil, γ, effective angle of internal friction, • , axial strain at 50% of stress level, ε_{50}, and undrained shear strength of clay, S_u), the SW model basic procedures presented by Ashour et al. (1998a) can be summarized as follows:

1. For a particular value of lateral strain (ε) in the developing passive wedge of soil in front of the pile, the increase in horizontal stress ($\Delta\sigma_h$), the stress level (SL) and the associated Young's modulus ($E = \Delta\sigma_h / \varepsilon$) are determined based on the stress-strain relationship of soil (Ashour et al. 1998) as assessed from the conventional triaxial testing.
2. The associated geometry of the passive wedge of soil (mobilized fan angle, φ_m, base angle, $\hat{a}_m$, and width of the wedge face, $\overline{BC}$) is assessed according to an assumed initial value (h) of the passive wedge depth (Fig. 1) which is related to the depth (X_o) of the zero deflection point (y = 0). The soil layers with the depth h are divided into thin sublayers, and steps 1 and 2 are applied to each sublayer (Fig. 1b).
3. The current variation of soil-pile line load (p) along depth h (Fig. 1c) is obtained as a function of soil and pile parameters ($\Delta\sigma_h$, $\overline{BC}$, D and τ) and the pile cross-section shape. D is the pile width and ô is the mobilized shear resistance along the pile sides (Fig. 1a).
4. Pile deflection (y) along the depth of the passive wedge is determined as a function of ε, Poisson's ratio, SL, and the size of the passive wedge. As a result, the associated profile of $E_s = p/y$ can be predicted (Fig. 1c).
5. Based on the current profile of E_s, the laterally loaded pile is analyzed as a BEF under an arbitrary pile-head lateral load (P_o). The values of pile-head deflection (Y_o) and X_o (i.e. h) assessed using BEF analysis are compared to those of the SW model analysis.
6. Through several iterative processes for the same value of soil strain å, converged values of h and Y_o are obtained. In addition, P_o is modified as a function of the values of Y_o from both the BEF and SW model analyses $[(P_o)_{modified} = (Y_o)_{SW\ Model}\ (P_o / Y_o)_{BEF}]$
8. For the next step of loading, a larger value for the horizontal soil strain (å) is used, and steps 1 through 7 are repeated.

It should be noted that the deflection pattern (Fig. 1c) is governed by pile-head fixity and pile bending stiffness (EI). Consequently, the shape and size of the passive wedge of soil and the associated E_s are affected by these pile properties. Different criteria for flow around failure, the stress-strain relationship of the soil, shear resistance along the pile sides, and the ultimate value of soil-pile reaction (p_{ult}) are employed in the SW model analysis (Ashour et al. 1998a).

Soil response in the SW model is computed over the full stress-strain (σ-ε) range of the soil (down to 10^{-4} % strain) instead of being projected from known empirical data (often obtained at larger strains) as in the COM624 and LPILE computer programs (Reese 1977 and 1985, and Reese and Sullivan 1980). This response exhibits very good agreement with the well known Seed and Idriss (1970) shear modulus reduction curve (Ashour and Norris 1999). The SW model approach has been employed in different projects and field tests such as the retrofit of the San Francisco Bay Bridge (by California Department of Transportation, Caltrans), Cypress overpass (Norris et al. 1993), and Oakland Outer Harbor Wharf (Norris et al. 1996).

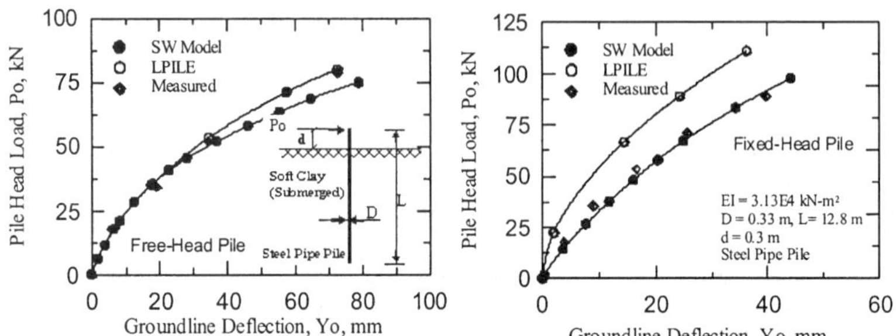

FIG. 2 A Comparison of Results of the SW Model, LPILE, and the Field Data for a Free- and a Fixed-Head Pile at Sabine River Test (Matlock 1970)

The SW model computer program has the efficiency to handle the response of free- and fixed-head piles as seen in Fig. 2 (after Matlock 1970). Also, it analyzes the behavior of laterally loaded piles in soft, medium stiff, and stiff clay (Matlock 1970 and Reuss et al. 1992) as shown by Figs. 2 and 3, respectively. The SW model program is also capable of analyzing the behavior of long drilled shafts although vertical side shear resistance due to lateral deflection is not considered in the current version of the program.

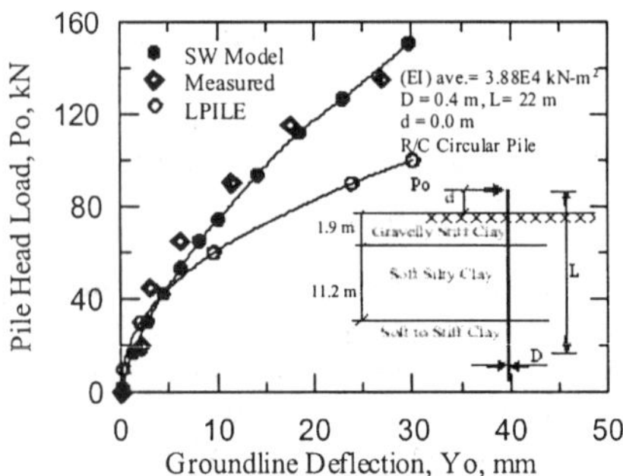

FIG. 3 Measured Pile-Head Response vs. Data Obtained from the SW Model and LPILE for the Pyramid Building, Memphis, Test (Reuss et al. 1992)

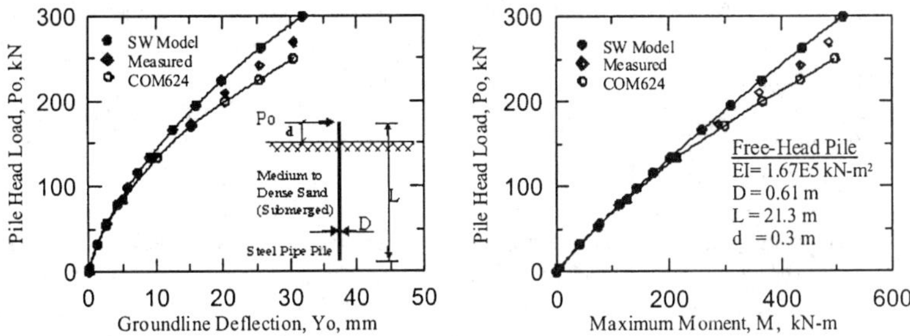

FIG. 4 Measured and Predicted Response of a Laterally Loaded Pile in Sand at the Mustang Island Test (Reese et al. 1974)

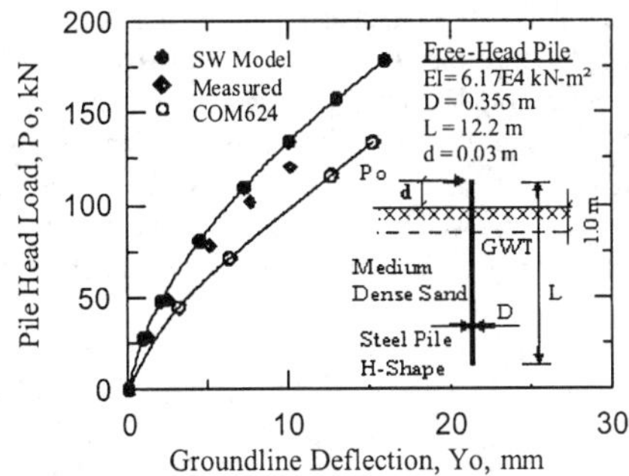

FIG. 5 A Comparison of Measured and the Predicted Response of Pile 6 of the Arkansas River Test (Alizadah and Davison 1970)

In sand soils, the SW model approach assesses the behavior of laterally loaded piles as seen with the Mustang Island test (Fig.4, Reese et al. 1974) and the Arkansas River test (Fig. 5, Alizadeh and Davisson 1970). Figure 5 features the capability of the SW model to consider pile cross section shape (H-shape and square piles as opposed to round). At the site of the Arkansas River test, the water table was approximately at 0.9 m below ground surface and the capillary tension influence above the water table, without doubt, must be considered in the analysis as seen in Fig. 5.

SOIL-PILE MODELING (P-Y CURVE)

Matlock-Reese p-y curves, which are employed in the computer programs COM624/GROUP (Reese 1977 and Reese and Sullivan 1980 / Reese and Wang 1996), LPILE1 (Reese 1987), PAR (PMB 1988), and FLPIER (University of Florida 1996), are a function of soil properties and only pile width. The problem of a laterally loaded pile is often solved as a BEF involving nonlinear modeling of the soil-pile interaction response (p-y curve). Currently employed p-y curve models were established/verified based on the results of field tests in uniform soils such as the Mustang Island (Reese et al. 1974), Sabine River (Matlock 1970) and Houston (Reese and Welch 1975) tests, and adjusted mathematically using empirical parameters to extrapolate beyond the soil=s specific field test conditions.

The traditional p-y curve models developed by Matlock (1970) and Reese et al. (1974) are semi-empirical models in which soil response is characterized as independent nonlinear springs (Winkler springs) at discrete locations. Therefore, the effect of a change in soil type of one layer on the response (p-y curve) of another is not specifically considered. In addition, the formulations for these p-y curve models do not account for a change in pile properties such as pile bending stiffness, pile cross-sectional shape, pile-head fixity and pile-head embedment below the ground surface. Soil-pile interaction or p-y curve behavior is not unique but a function of both soil and pile properties (Ashour and Norris 2000a). It would be prohibitively expensive to systematically evaluate all such effects through additional field tests; hence it behooves us to consider such influences based on available theoretical means (SW model formulation) that allows transformation of envisioned three-dimensional soil-pile interaction response to one-dimensional BEF parameters. As Terzaghi (1955) and Vesic (1961) stated, the subgrade modulus, E_s (and , therefore, the p-y curve), is not just a soil but, rather, a soil-pile interaction (and, therefore, a pile property dependent) response.

Unlike the Matlock-Reese p-y curves, a soil-pile interaction p-y curve should be affected by pile bending stiffness (EI) as seen in Fig. 6. Here, the stiffer the pile yields the stiffer associated p-y curve. The two piles employed by using the SW model in Fig. 6 are a steel pipe pile (stiff pile) and a timber pile (flexible pile). Both piles have the same shape, size, and head conditions (free-head pile).

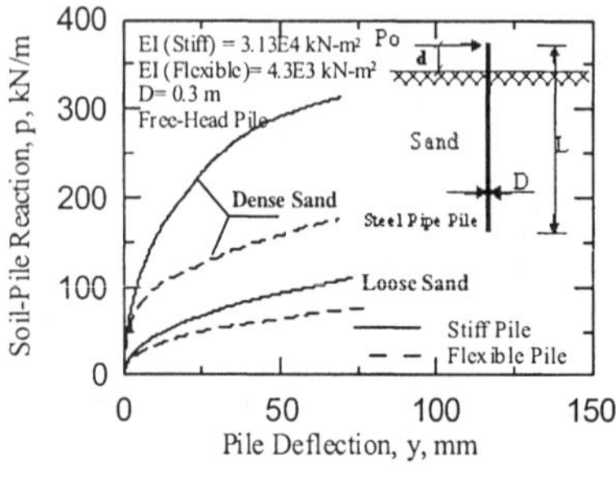

FIG. 6 Effect of Pile Stiffness on p-y Curve in Sand at a 1.22-m Depth

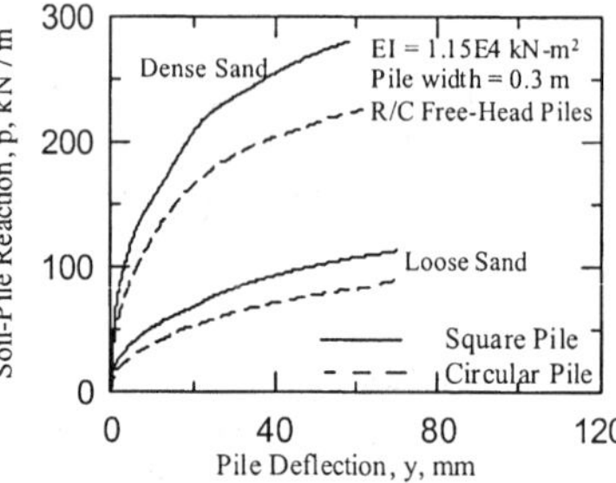

FIG. 7 Effect of Pile Cross-Section Shape on p-y Curve at a 1.22-m Depth

The significant influence of pile cross sectional shape on the nature of the p-y curve is seen in Fig. 7. The two piles employed (square pile and circular pile) are assumed to have the same width, bending stiffness and pile-head conditions, and are driven in the same soil. More details on the non-uniqueness of the p-y curve are presented by Ashour and Norris (2000a).

UNDRAINED RESPONSE OF PILES/SHAFTS IN LIQUEFIABLE SOIL (SATURATED SANDS)

Due to the shaking from an earthquake and the associated lateral load from the superstructure, excess porewater pressure in the free- and near-field develops and reduces the strength of loose to medium sand around a pile. The degradation in soil resistance and the induced excess porewater pressure in the free-field ($u_{xs,ff}$) is based on the procedures proposed by Seed et al. 1983. This is followed by the assessment of the excess porewater pressure ($u_{xs,nf}$) in the near-field soil region (adjacent to the pile, Fig. 8) induced by the lateral load from the superstructure. The variation in soil resistance (undrained stress-strain relationship) around the pile in the near-field zone is evaluated based on the undrained formulation for saturated sand presented by Ashour and Norris (1999).

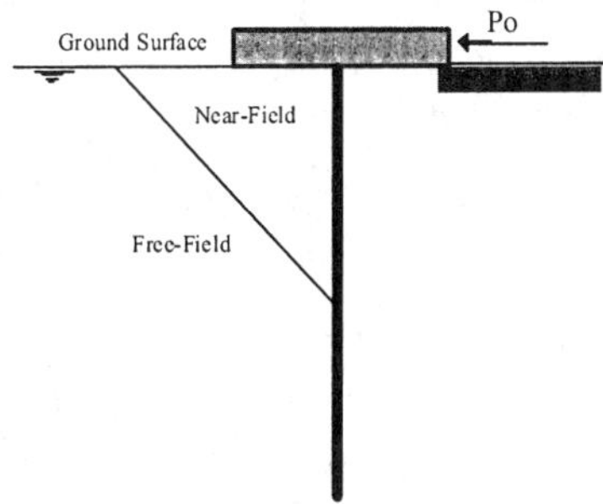

FIG. 8 Schematic Figure of the Excess Porewater Pressure Zones Around a Laterally Loaded Pile

The assessed value of the free-field excess porewater pressure ratio, r_u, induced by the earthquake is obtained using Seed's method (Seed et al. 1983). $u_{xs,ff}$ is calculated conservatively at the end of earthquake shaking corresponding to the number of equivalent uniform cycles produced over the full duration of the earthquake. Thereafter, the lateral load (from the superstructure) is applied at the pile head that generates additional porewater pressure ($u_{xs,nf}$) in the soil immediately around the pile, given the degradation in soil strength already caused by $u_{xs,ff}$. Thereafter, the undrained behavior due to an inertial induced lateral load is assessed using undrained stress-strain formulation in the SW model (Ashour and Norris 2001). Thus, the procedure accounts for both of $u_{xs,ff}$ and $u_{xs,nf}$. It should be noted that these procedures incorporate the whole undrained stress-strain curve (at any level of loading) not only the residual strength of the sand. The SW model analysis characterizes the reduction in E_s (i.e. the p-y curve) and pile response due to a drop in sand strength and Young's modulus as a result of developing liquefaction in the sand.

The SW model analysis of laterally loaded piles in liquefiable soil utilizes the basic properties of the sand such as its relative density (D_r), particle shape (roundness), and percentage of fines in addition to the characteristic of the earthquake (such as maximum ground acceleration, a_{max} and earthquake magnitude). The undrained stress-strain relationship, the effective stress path, and the potential of sand to liquefy or dilate can be predicted.

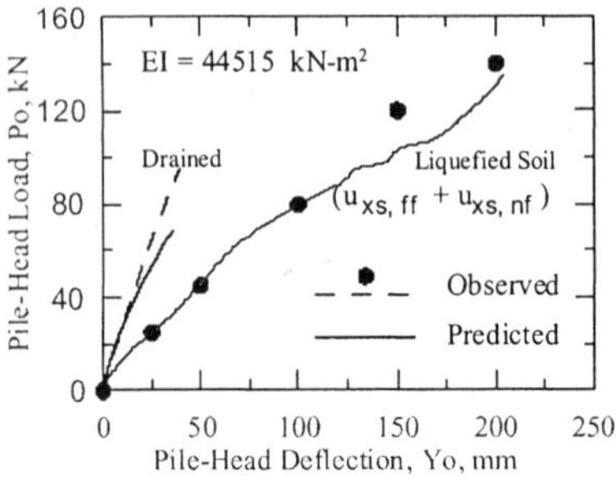

FIG. 9 CISS, 0.324 m-Diameter Pile at Treasure Island Test in Liquefied Soil

FIG. 10 H-Shape, 0.310 m Pile at Treasure Island Test in Liquefied Soil

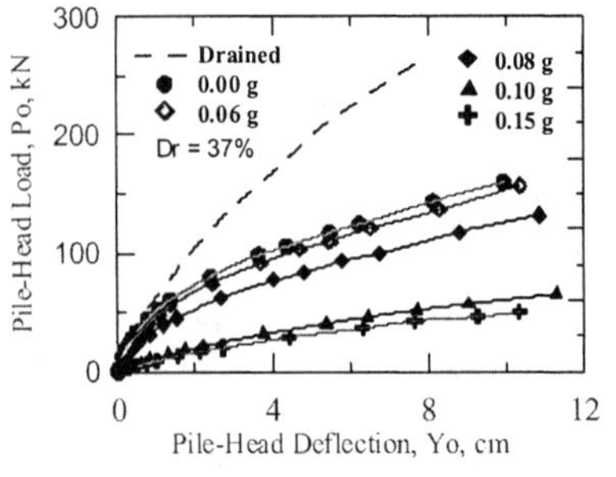

FIG. 11 Pile-Head Response in Banding Sand Under Developing Liquefaction

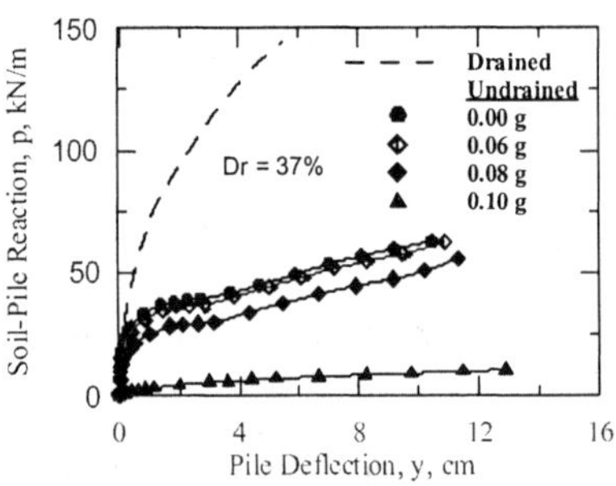

FIG. 12 p-y Curve in Liquefied Banding Sand

The full-scale load tests on the post-liquefaction lateral response of piles that were performed at Treasure Island (Ashford and Rollins 1999) are the most significant related tests. Figures 9 and 10 show a comparison between the Treasure Island test results (liquified soil) and the data predicted using the SW model analysis. This methodology also covers the behavior of piles in a group. More details are presented by Ashour and Norris (1998, 2000b and 2001). Figures 11 and 12 show the effect of different values of maximum ground surface acceleration, a_{max}, on the pile-head response and the associated p-y curves of a laterally loaded pile in saturated medium loose Banding sand of Dr = 37% (Ashour and Norris 1999 and 2001). Varying values of ($u_{xs,ff} + u_{xs,nf}$) develop at different levels of a_{max}.

SW MODEL ANALYSIS OF PILE GROUP

At present there is no particular technique that assesses the response of piles in a group using completely reasonable assumptions. The most common design procedure is to reduce the stiffness of the traditional (Matlock-Reese) p-y curve by using a multiplier ($f_m < 1$), as seen in Fig. 13. The value of the p-y curve multiplier (f_m) should be assumed and is based on the data collected from full-scale field tests on pile groups which are few (Brown et al. 1988). Consequently, a full-scale field test (which is costly) is strongly recommended in order to determine the value of f_m of the soil profile at the site under consideration. Moreover, the suggested value of the multiplier (f_m) is taken to be constant for each soil layer at all levels of loading (Fig. 13).

In reality, the p-multiplier should vary in accord with the pile and soil properties, the level of load or deflection (and thus, the current size or depth of the wedges), and the location of the pile in the pile group. The value of this multiplier is very difficult to evaluate from experiment and much more research is clearly needed to establish a methodology for empirically predicting the variation of f_m. The interaction among the piles in a group decreases with depth (Fig. 14) thus generating lower values of f_m near ground surface (or pile head) and greater ones at deeper points. Consequently, the values of f_m increase with depth below pile head in the same soil and it will be more complex in layered soils. The p-multipliers should be developed based on site specific field tests.

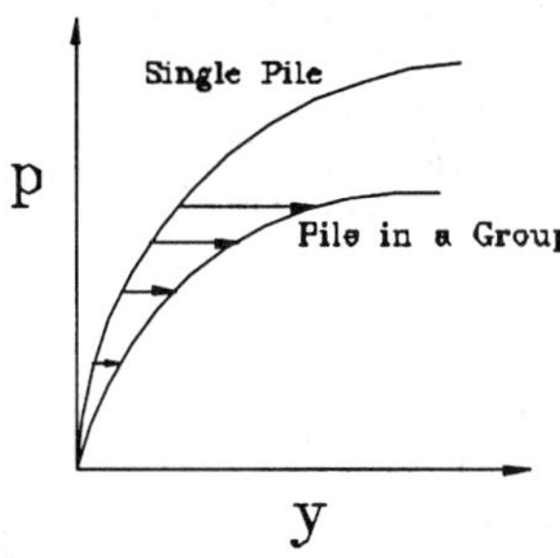

FIG. 13 Modified p-y Curve of an Individual Pile in a Pile Group

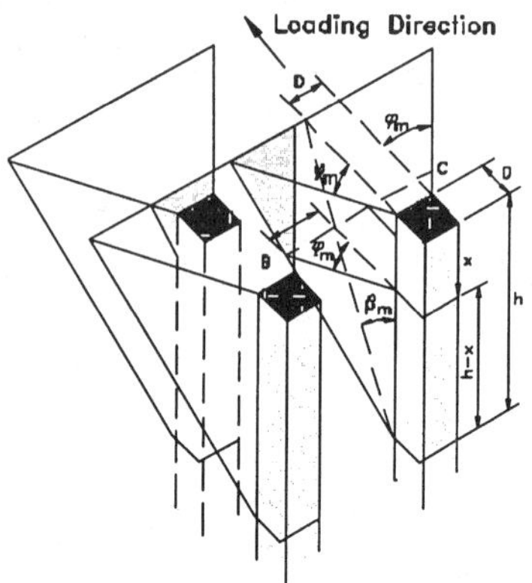

FIG. 14 Interaction Among Piles in a Group Based on the SW Model

As presented by Ashour et al. (2001b), the SW model approach calculates the ever changing geometry of the mobilized passive wedge of soil in front of the pile. The shape of the passive wedge of soil (Fig. 1) varies according to the associated level of loading and becomes deeper and wider (larger fanning angle) with increasing lateral load and deflection. This characterization allows the SW model to evaluate the interaction among neighboring piles in the pile group at different depths and increasing pile head deflection.

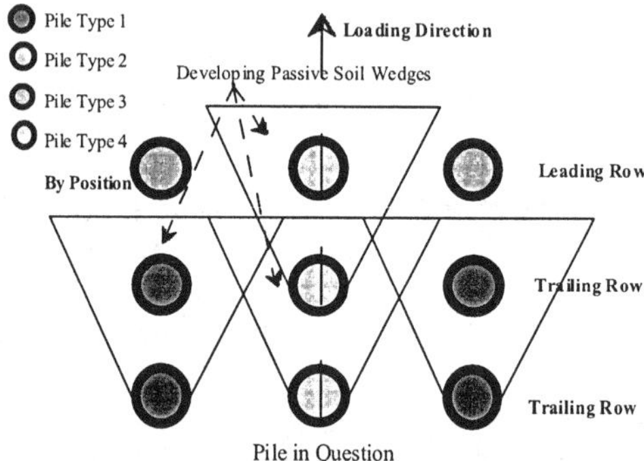

FIG. 15 Interaction Among Piles in a Group at a Given Depth

The changing value of, E_s, as determined by the SW model analysis, will account for the additional strains (i.e. stresses) in the adjacent soil due to passive wedge interference within the group (Figs. 14 and 15). Thus the E_s (i.e. the secant slope of the p-y curve) of an individual pile in a group will be reduced in a mobilized fashion according to pile and soil properties, pile spacing and position, the level of loading, and depth, x (Fig. 16). No single reduction factor (f_m) for the p-y curve (commonly, assumed to be a constant value with depth and level of loading) is needed or advised.

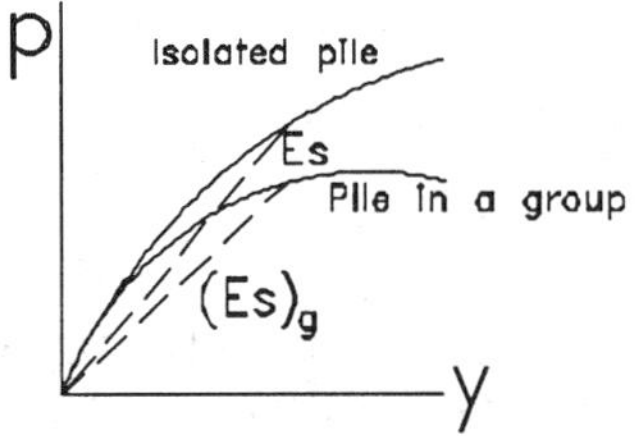

FIG. 16 E_s Variations for a Pile in a Group and an Isolated Pile in the SW Model Analysis

As seen in Fig. 15, the soil around the piles in the group interacts horizontally with that of adjacent piles by an amount that varies with depth. Therefore, the varying overlap of the wedges of neighboring piles in different sublayers over the depth of the interference and the associated increase in soil stress/strain can be determined as a function of the amount of overlap (Figs. 14). As pile lateral load increases, the wedges grow deeper and fan out horizontally thus causing a further change in overlap and group interference, all of which varies with a change in soil and pile properties. Such analysis is incorporated in the SWM3.2 and SWM6.0 computer programs (Ashour et al. 1998b). Figures 17 and 18 show the capability of the technique in comparison with observed data (Brown and Reese 1985 and McVay et al. 1995) for a pile group in layered clay soil and uniform sand, respectively.

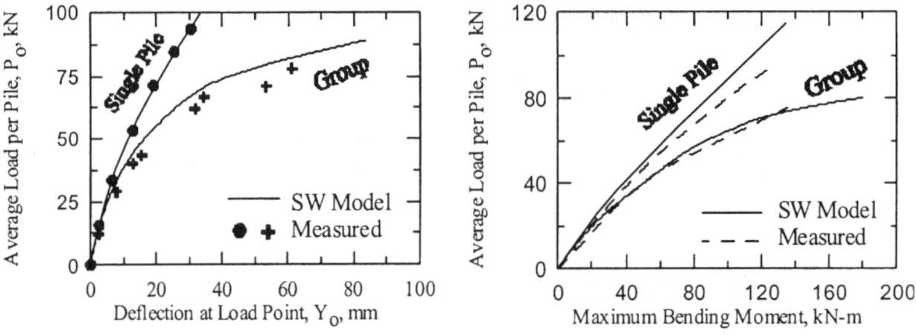

FIG. 17 Lateral Response of a 3 x 3 Free-Head Pile Group in Layered Soil (Sand and Caly) (Brown and Reese, 1985)

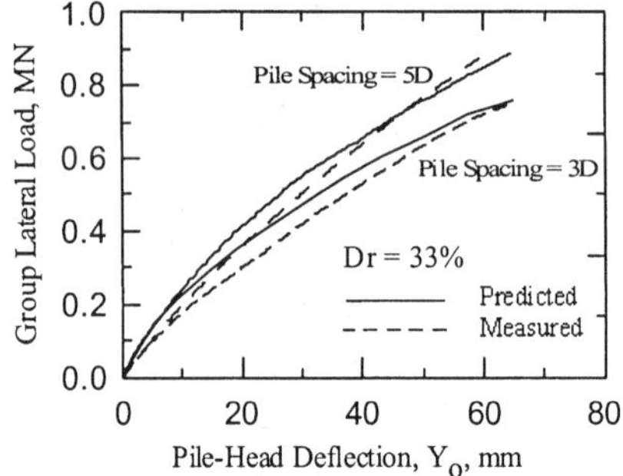

FIG. 18 A 3 x 3 Free-Head Pile Group Model (3D and 5D spacing) in Medium Loose Sand (McVay et al. 1995)

CONCLUSIONS

The SW model provides a realistic demonstration of soil-pile interaction by which the effect of such pile properties as pile size, pile cross sectional shape, bending stiffness, pile head conditions and pile interaction effects are taken into account in the development of the p-y curves. Such p-y curves are a product of, not the input to, the SW model. The SW model has been contrasted with the BEF analysis using the Matlock-Reese type p-y curves. The SW model approach provides a three-dimensional analysis of a laterally loaded pile group linked to the BEF analysis based on both soil and pile properties (soil-pile interaction) and pile spacing. Therefore, unlike the current procedures, no reduction factors or multipliers are required to reflect the interaction among the piles in the group. The SW model assesses the response of laterally loaded piles in liquefiable soils (post-liquefaction) with the incorporation of the induced excess porewater pressure in the free- and near field due to the shake of the earthquake and imposition of the superstructure load. Also, the nonlinear behavior of pile/shaft material can be accounted in the SW model to predict the ultimate capacity and the reduction in lateral resistance of piles/shafts and the associated p-y curves.

REFERENCES

Alizadeh, M. and Davisson, M. T. (1970). "Lateral Load Test on Piles-Arkansas River Project." J. of Soil Mechanics and Foundations Division, Proceedings of ASCE, Vol. 96, No. SM5, pp. 1583-1604.

Ashford, S. A., and Rollins, K. (1999). "Full-Scale Behavior of Laterally Loaded Deep Foundations in Liquefied Sands." Report No. TR - 99/03, Structural Engineering Dept., University of California, San Diego.

Ashour, M., Norris, G. M., and Shamsabadi, A. (2001a). "Effect of the Non-Linear Behavior of Pile Material on the Response of Laterally Loaded Piles." Fourth International Conference on Recent Advances in Geotechnical Earthquake Engineering and Soil Dynamics, San Diego, California, March 26-3, Paper 6.10.

Ashour, M., Pilling, P., and Norris, G. M. (2001b) "Assessment of Pile Group Response Under Lateral Load." Fourth International Conference on Recent Advances in Geotechnical Earthquake Engineering and Soil Dynamics, San Diego, California, March 26-31, Paper 6.11.

Ashour, M., Norris, G. M., Bowman, S., Beeston, H, Billing, P. and Shamsabadi, A. (2001c) "Modeling Pile Lateral Response in Weathered Rock." Proceeding 36st *Engineering Geology and Geotechnical Engineering Symposium,* March 28-30, Las Vegas, Nevada.

Ashour, M., and Norris, G. M. (2003). "Lateral Load Pile Response in Liquefied Soil." *J. of Geotechnical and Geoenvironmental Engineering*, ASCE, Vol. 129, No. 5.

Ashour, M., and Norris, G. M. (2000a). "Modeling Lateral Soil-Pile Response Based on Soil-Pile Interaction." *J. of Geotechnical and Geoenvironmental Engineering*, ASCE, Vol. 126, No. 5, pp. 420-428.

Ashour, M. and Norris, G. M. (2000b). "Undrained Lateral Pile and Pile Group Response in Saturated Sand." *Report No. CCEER-00-01*, Civil Engg Dept. University of Nevada, Reno, Fedral Study No. F98SD01, Submitted (Caltrans)

Ashour, M. and Norris, G. M. (1999). "Liquefaction and Undrained Response Evaluation of Sands From Drained Formulation." *J. of Geotechnical and Geoenvironmental Engineering*, ASCE, Vol. 125, No. 8 pp. 649-658.

Ashour, M., and Norris, G. (1998). "Undrained Laterally Loaded Pile Response in Sand" Special Publication, ASCE, *Geotechnical earthquake Engineering and Soil Dynamics Conference, Seattle*, Washington, August, Paper No. 63.

Ashour, M., Norris, G., and Pilling, P. (1998a). "Lateral Loading of a Pile in Layered Soil Using the Strain Wedge Model." *J. of Geotechnical and Geoenvironmental Engineering*, ASCE, Vol. 124, No. 4, pp. 303-315.

Ashour, M., Pilling, P., and Norris, G. (1998b). "Updated Documentation of the Strain Wedge Model Program for analyzing Laterally Loaded Piles and Pile Groups." *Proceeding 33rd Engineering Geology and Geotechnical Engineering Symposium*, March 25-27, Reno, Nevada, pp. 177-178.

Broms, Bengt B. (1964a). "Lateral Resistance of Piles in Cohesive Soils." *Proc.*, ASCE, Vol. 90, No. SM2, pp. 27-63.

Broms, Bengt B. (1964b). "Lateral Resistance of Piles in Cohesionless Soils." *Proc.*, ASCE, Vol. 90, No. SM3, pp. 123-156.

Brown, D. A. and Reese, L. C. (1985). "Behavior of a Large-Scale Pile Group Subjected to Cyclic Lateral Loading." *Report to Minerals Management Service*, U. S. Department of Interior, Reston, Virginia; Department of Research, Federal Highway Administration, Washington, D. C.; U. S. Army Engineer, Waterways Experiment Station, Vicksburg, Mississippi.

Brown, D. A., Morrison, C., and Reese, L. C. (1988). "Lateral Load Behavior of Pile Load in Sand." *J. of Geotechnical Engineering*, ASCE, Vol. 114, No. 11, 1261-1276.

Matlock, H. (1970). "Correlations for Design of Laterally Loaded Piles in Soft Clay." *The Second Annual Offshore Technology Conference*, Houston, Texas, April 22-24, OTC 1204, pp. 577-607.

McVay, M., Zhang, L., Molnit, T., and Lai, P. (1998). "Lateral Response of Three-Row Groups in Loose to Dense Sands at 3D and 5D Pile Spacing." *J. of Geotechnical*

Engineering, ASCE, Vol. 121, No. 5, pp. 436-441.

McVay, M., Casper, R., and Shang, T. (1995). "Lateral Response of Three-Row Groups in Loose to Dense Sands at 3D and 5D Pile Spacing." *J. of Geotechnical Engineering*, ASCE, Vol. 121, No. 5, pp. 436-441.

Morrison, C. and Reese, L. C. (1986). "Lateral-Load Test of a Full-Scale Pile Group in Sand." *Report to the Minerals Management Service*, U. S. Department of Interior, Reston, Virginia; Department of Research, Federal Highway Administration, Washington, D. C.; U. S. Army Engineer, Waterways Experiment Station, Vicksburg, Mississippi.

Murchison, J. M., and O=Neil, M. W. (1984). "Evaluation of p-y Relationships in Cohesionless Soils." *Analysis and Design of Pile Foundations*, Ed. By J. R. Meyer, ASCE, pp. 174-191.

Norris, G.M., Ashour, M., and Pilling, P. (1995). "The Non-Uniqueness of p-y Curves for Laterally Loaded Pile Analysis." *Proceeding 31st Engineering Geology and Geotechnical Engineering Symposium*, March, pp. 40-53.

Norris, G.M., Siddharthan, R., Zafir, Z., Abdel Ghaffar, S., and Gowda, P. (1996). "Soil-Foundation-Structure Behavior at the Oakland Outer Harbor Wharf." *Transportation Research Record*, TRB, 1546, National Academy Press, Washington, D.C., pp. 100-111.

Norris, G.M., Siddharthan, R., Zafir, Z., and Gowda, P. (1993). "Soil and Foundation Conditions and Ground Motions at Cypress." *Transportation Research Record*, TRB, No. 1411, pp. 61-69.

Norris, G., Siddharthan, R., Zafir, Z., and Madhu, R. (1997). "A Liquefaction and Residual Strength of Sands from Drained Triaxial Tests." *J. of Geotechnical Engineering*, ASCE, Vol. 123, No. 3, pp. 220-228.

Norris, G. M. (1986). "Theoretically Based BEF Laterally Loaded Pile Analysis." *Proceedings, Third International Conference on Numerical Methods in Offshore Piling*, Nantes, France, May, pp. 361-386.

PMB Engineering Inc. (1988). "Pile Analysis Routine (PAR)." San Francisco, California

Poulos, H. G. (1971a) ABehavior of Laterally Loaded Piles: I-Single Piles.@ *Proc., ASCE*, Vol. 97, No. SM5, pp. 711-731.

Poulos, H. G. (1971b). "Behavior of Laterally Loaded Piles: II-Pile Groups." *Proc., ASCE*, Vol. 97, No. SM5, pp. 733-751.

Reese, L. C. (1983). "Behavior of Piles and Pile Groups Under Lateral Load." Report to the U.S. Department of Transportation, Federal Highway Administration, Office of Research, Development, and Technology, Washington, D.C.

Reese, L. C. (1977). "Laterally Loaded Piles: Program Documentation." *J. of Geotechnical Engineering Division*, ASCE, Vol. 103, GT. 4, pp. 287-305.

Reese, L. C., and Sullivan, W. R. (1980). "Documentation of Computer Program COM624, Parts I and II, Analysis of Stresses and Deflections for Laterally Loaded Piles Including Generation of p-y Curves." *Geotechnical Engineering Software GS80-1*, Geotechnical Engineering Center, Bureau of Engineering Research, University of Texas at Austin.

Reese, L. C., Cox, W. R., and Koop, F. D. (1974). "Analysis of Laterally Loaded Piles in Sand." *The Six Annual Offshore Technology Conference*, Houston, Texas, May

6-8, OTC 2080, pp. 473-483.

Reese, L., and Wang, S. T. (1996). "Documentation of The Computer Program GROUP." *Ensoft Inc.*, Post Office Box 180348, Texas 78718.

Reese, L. C. (1987). "Documentation of The Computer Program LPILE1." *Ensoft Inc.*, Post Office Box 180348, Texas 78718.

Reese, L.C., and Welch, R. C. (1975). "Lateral Loading of Deep Foundations in Stiff Clay." *J. of Geotechnical Engineering Division*, ASCE, Vol. 101, GT. 7, pp. 633-649.

Reuss, R., Wang, S. T., Reese, L. C. (1992). "Tests of Piles Under Lateral Loading at the Pyramid Building, Memphis, Tennessee." *Geotechnical News*, December, pp. 44-46.

Robretson, P. K., Davies, M. P., and Campanella, R. G. (1989). ADesign of Laterally Loaded Driven Piles Using the Flat Dilatometer.@ *Geotechnical Testing J.,* ASTM, Vol. 12, No. 1, pp. 30-39.

Rollins, K. M., Peterson, K. T., and Weaver, T. J. (1998). "Lateral Load Behavior of Full-Scale
Pile Group in Clay." *J. of Geotechnical and Geoenvironmental Engg*, ASCE, Vol. 124, No. 6, pp. 468-478.

Seed, R. B. and Harder, L. F. (1990). "SPT-Based Analysis of Cyclic Pore Pressure Generation
and Undrained Residual Strength." *H. Bolton Seed Memorial Symposium Proceedings*, Vol. 2, BiTech Publishers Ltd, Vancouver, B. C., Canada.

Seed, H.B., Idriss, I.M., and Arango, I. (1983). "Evaluation of Liquefaction Potential Using Field Performance Data." J. of Geotechnical Division, ASCE, Vol. 109, No. 3, pp. 458-482.

Seed, H. B., and Idriss, I. M. (1970). "Soil Moduli and Damping Factors for Dynamic Response Analyses." *Report* No. EERC 70-10, College of Engineering, University of California, Berkeley, California.

Smith, T. D. (1983). "Pressuremeter Design Method for Single Piles Subjected to Static Lateral Load." *Ph.D. Dissertation*, Texas A&M University.

Terzaghi, K. (1955). "Evaluation of Coefficients of Subgrade Reaction." *Geotechnique*, Vol. 5, No.
4, pp. 297-326.

University of Florida (1996). "User Manual for FLORIDA-PIER Program." Dept. Of Civil Eng., University of Florida, Gainesville, Florida.

Vesic, A. (1961). "Bending of Beams Resting on Isotropic Elastic Solid." *J. ASCE, Engineering Mechanics Div.*, Vol. 87, SM 2, pp. 35-53.

Wang, S. T. and Reese, L. C. (1998). "Design of Pile Foundations in Liquefied Soils." Publication No. 75, ASCE, *Geotechnical Earthquake Engineering and Soil Dynamics Conference*, Seattle, Washington, Vol. II, August, pp. 1331-1343.

INTERFACE STRESSES BETWEEN SOIL AND LARGE DIAMETER DRILLED SHAFT UNDER LATERAL LOADING

Kerop D. Janoyan[1], P.E., Member, Geo-Institute, and Matthew J. Whelan[2]

ABSTRACT: Results are presented of a field testing program for a full-scale, large diameter cast-in-drilled-hole (CIDH) shaft/column under large displacement cyclic lateral loading. The test shaft was extensively instrumented to enable high-precision section curvature measurements in addition to measurements of contact pressure at the soil-shaft interface around the shaft perimeter. Among the principal objectives of the testing was to characterize the soil-shaft interaction across a wide displacement range in order to gain insight into the adequacy of existing design guidelines (which are based principally on the testing of small diameter piles) for the large diameter shafts commonly used to support highway bridges. The component stresses of resistance and the effects of nonlinear soil resistance to relative displacements between the soil and shaft are presented.

INTRODUCTION

Pile-supported bridges provide an economical option for highway construction, particularly in seismic regions. The inelastic deformations for a pile/column under lateral loading occur below grade; therefore, the overall lateral load behavior of the system is influenced by the interaction between the shaft and the surrounding soil, commonly modeled using *p-y* curves. The prediction of soil-pile-structure system behavior under lateral loading is among the most complex topics in geotechnical

---

[1]Assistant Professor, Clarkson University, Department of Civil and Environmental Engineering, Potsdam, NY 13699
kerop@clarkson.edu
[2]Undergraduate Research Assistant, Clarkson University, Department of Civil and Environmental Engineering, Potsdam, NY 13699
whelanmj@clarkson.edu

engineering. Current models for *p-y* curves are calibrated primarily from lateral load testing of relatively small diameter piles. Since the soil resistance is such a critical consideration, there is a need for better understanding of its component stresses, particularly for large-diameter piles and drilled shafts.

The beam on nonlinear Winkler foundation (BNWF) model that gives rise to the use of *p-y* curves in engineering design is based upon the concept that soil-shaft interaction can be represented by a compressive force per unit length of shaft (McClelland and Focht 1958). As seen in Fig. 1, this force is the resultant of several stresses, including:

- Normal compressive stress on the leading face of the shaft, which in general will be some percentage of the passive soil capacity.
- Shear stress along the sides of the shaft.
- Active normal soil stresses on the trailing face of the shaft.

Current sensor technology only allows measurements of normal stress in soil, so interface shear stresses can only be indirectly inferred.

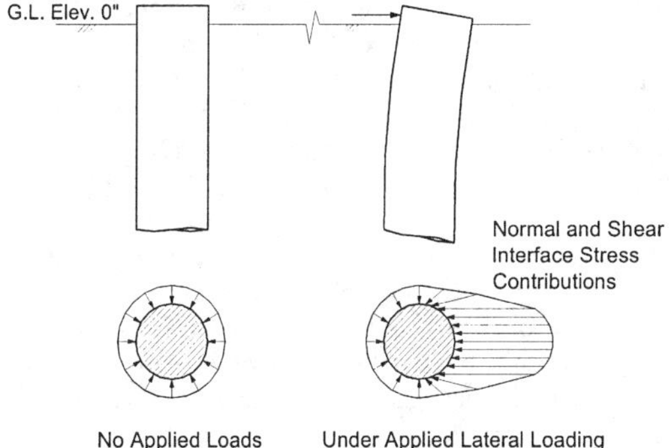

FIG. 1. Distribution of unit stresses against a pile before and after lateral displacement

This experimental study evaluated total soil reaction *p* from measurements of the internal bending deformation of the shaft through use of beam theory. The normal stresses at the soil-shaft interface, which were measured during field testing with non-displacement soil pressure cells (SPC), are compared to the total soil reaction and the interface shear stresses are calculated.

PREVIOUS STUDIES OF INTERFACE STRESSES

Few experimental studies of soil-shaft interface stresses have been performed because of difficulties associated with the use of SPCs. Such devices are obviously impractical for driven piles, but can be used in drilled shafts. The SPCs can be installed on the shaft wall after excavation of the shaft hole and placement of the reinforcing cage, but prior to concrete placement. Such installations are only practical when the dry method of construction is used, i.e. no slurry. Among the previous studies of laterally loaded drilled shafts which have utilized soil pressure cells (Kasch et al. 1977 and Briaud et al. 1983, 1985), the work by Bierschwale et al. (1981) is of particular interest because the pressure cell data are well documented, and data for several azimuthal angles relative to the direction of loading are available.

The study consisted of lateral load testing of three separate reinforced concrete drilled shafts of varying diameters of roughly 0.9 m (2.5 ft) installed at a stiff clay site and laterally loaded at ground line. The SPCs were generally installed in the line of loading, and the results were plotted as a function of depth for various levels of head load. The authors did not evaluate distributions of p for comparison to pressure profiles. A simple equilibrium check was performed by assuming that the measured pressures were uniformly applied over the projected diameter of the shafts, and then summing the moments about the point of zero lateral stress. The results indicated a lack of equilibrium, and that a substantial increase of soil pressure near the top of the shaft would be required to achieve equilibrium. Lateral pressures off the line of loading were also measured and the results were presented. The result shows a significant reduction of normal stress 45 degrees from the line of loading at small load levels, but more nearly uniform stresses at higher load levels.

TEST SITE CONIDITIONS, SPECIMENT SETUP AND INSTRUMENTATION OF PRESENT STUDY

Site Conditions

The present study was conducted at the interchange of Interstate Highways 105 and 405 in Hawthorne, California, just southeast of the Los Angeles International Airport (Janoyan 2001). The mapped local geology is Quaternary alluvium. Field in-situ and laboratory testing was performed to characterize the soil conditions at the site. The field testing included seismic cone penetration testing (SCPT), rotary-wash borings with standard penetration testing (SPT), down-hole suspension logging of shear wave velocities, pressuremeter testing (PMT), and test pit excavation mapping. Samples for laboratory testing were retrieved from the borings using thin-walled Pitcher tubes, and were hand-carved from the walls of the test pit. Laboratory testing was performed to evaluate particle size distribution, Atterberg limits, shear strength, and consolidation characteristics.

A generalized soil profile for the site is presented in Fig. 2 and consists of:

- 0.6 to 1.5 m (0 – 2 to 5 ft): Fill consisting of asphalt and concrete debris.
- Formation 1 (base of fill – approximately 5.5 m to 7.3 m (18 to 24 ft)): Silty clay. Classification testing indicates moderate plasticity (PI ~ 15), approximately 60% fines. A silty sand interbed with a thickness of about 0.6 m (2 ft) occurs at a depth of approximately 3 m (10 ft) within this layer.
- Formation 2 (base of Formation 1, about 0.6 m to 1.2 m (2 to 4 ft) in thickness): Medium- to fine-grained silty sand/sandy silt. Classification testing indicates 30% fines and PI ~ 12. SPT blow counts are approximately 20 to 45.

- Formation 3 (base of Formation 2 – approximately 14.6 m (48 ft)): Silty clay. Classification testing indicates 40% fines and PI ~ 13 to 14.
- A medium sand underlies Formation 3, and is water bearing (i.e., the groundwater table is located within this layer).

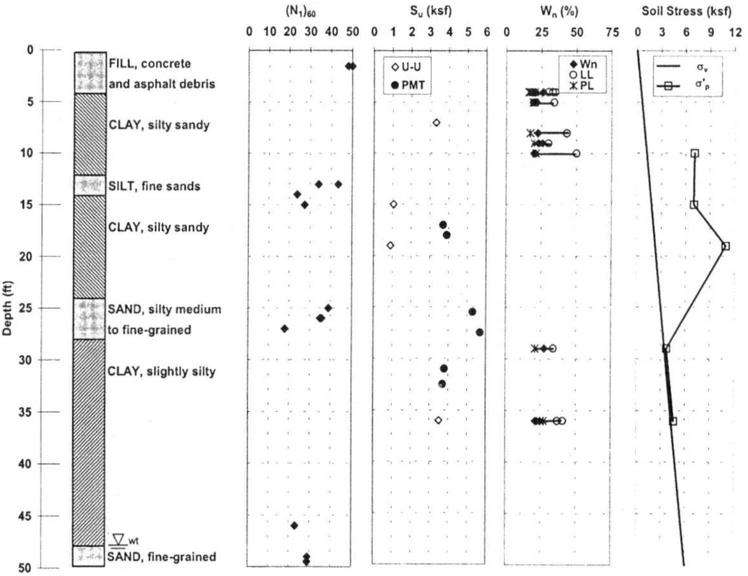

FIG. 2. Generalized test site soil profile along with some results from field and laboratory investigations

Specimen Setup

The CIDH shaft-column was designed according to standard Caltrans' Bridge Design Specifications (1995) using the Seismic Design Criteria (1999). The test specimen was comprised of a 1.8 m (6 ft) diameter column extending 12.2 m (40 ft) and a 2 m (6.5 ft) diameter shaft extending 14.6 m (48 ft) below ground line, as shown in Fig. 3. Steel reinforcement was comprised of 36-#14 longitudinal bars with average yield and ultimate tensile strengths of 490MPa (71 ksi) and 703 MPa (102 ksi), respectively; and shop-welded #8 hoops at 152 mm (6 inch) spacing with average yield and ultimate tensile strength of 496 MPa (72 ksi) and 738 MPa (107 ksi), respectively. The longitudinal and transverse steel ratios are roughly 2 per cent and 0.7 per cent, respectively. Normal-weight ready-mix concrete was used in the specimen, with an average tested cylinder compressive strength of 42 MPa (6.1 ksi).

Lateral loads were imposed at the top of the column using tension cables reacting against soil anchors at an angle of 26.6 degrees from the horizontal (2H:1V). Testing of the shaft/column was displacement controlled. Cyclic lateral loading was applied across an amplitude range of 50 mm to 2.75 m (2 to 108 inches). Two loading cycles were performed at most displacement levels. Cycles at displacement amplitudes greater than 152 mm (6 inches) included 16 stops to enable measurements of the shape of hysteretic response curves.

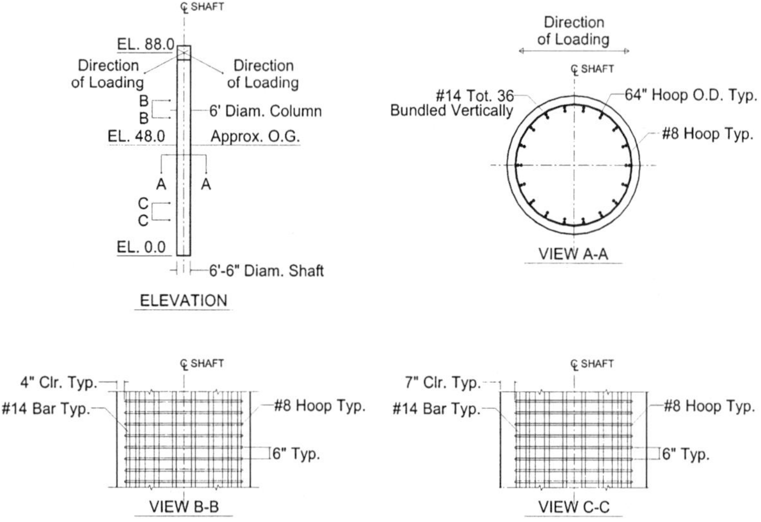

FIG. 3. Shaft/column section reinforcement details

Instrumentation

Extensive instrumentation, totaling over 200 channels of data, was used to monitor the response of the soil/shaft system and the local deformation of the column. Shaft bending deformations were measured with a redundant system of extensometers, inclinometers and high-precision fiber-optic sensors. The lateral displacement of the column (above ground) was measured using survey/total stations while the tilt at and below ground level shaft was measured using the inclinometers.

Profiles of soil reaction p and shaft displacement y were derived from the measured shaft curvature and displacement profiles, which were then used to generate p-y curves at various depths below ground. Janoyan (2001) and Janoyan et al. (2001) present details of the field measurements and data reduction techniques.

Other instrumentation included non-displacement soil pressure cells (i.e. cells with the approximate stiffness of concrete) which were installed at the interface between the concrete and soil. A total of 26 cells were installed at seven depths along the shaft both in the line of loading, and at several depths, at 30 and 60 degrees from the line of loading (Fig. 4). Each cell was installed by hand on the excavation wall after the cage had been placed and immediately prior to concrete placement.

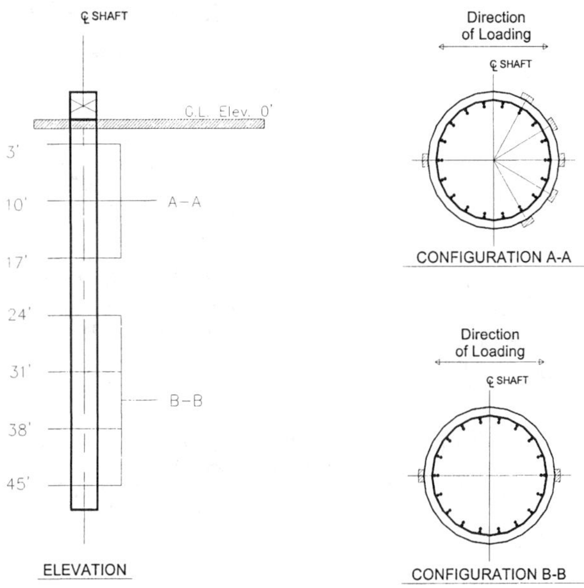

FIG. 4. Locations of soil pressure cells on test shaft

SOIL PRESSURE DISTRIBUTIONS FROM PRESENT TEST

Measured lateral soil pressures at the 22.9 cm (9 in) displacement level are plotted in Fig. 5 as a function of depth along with the lateral soil reaction profiles evaluated (from the analysis of *p-y* curves). The shapes of the profiles are comparable, and appear to have similar cross-over depths of about 3.7 m (12 ft). Similar plots were also generated at the other displacement levels.

Horizontal distributions of soil pressure around the shaft perimeter were measured at depths of 0.9 m (3 ft), 3.1 m (10 ft), and 5.2 m (17 ft). The horizontal distribution of pressures at the 0.9 m (3 ft) depth is shown in Fig. 6 for displacement levels up to 45.7 cm (18 in). As had been found in previous work (e.g., Bierschwale et al., 1981), normal soil stresses drop off significantly for azimuths off the line of loading. As seen in Fig. 6, the soil stresses along azimuths off the line of loading are not symmetric with respect to the loading direction. The lack of symmetry could have been due to some twisting of the reinforcing cage during placement; however the areas under the profiles are deemed to be reliable. Thus these readings were integrated around the shaft semi-perimeter to estimate the resultant normal stress acting against the shaft in the line of loading (p_σ),

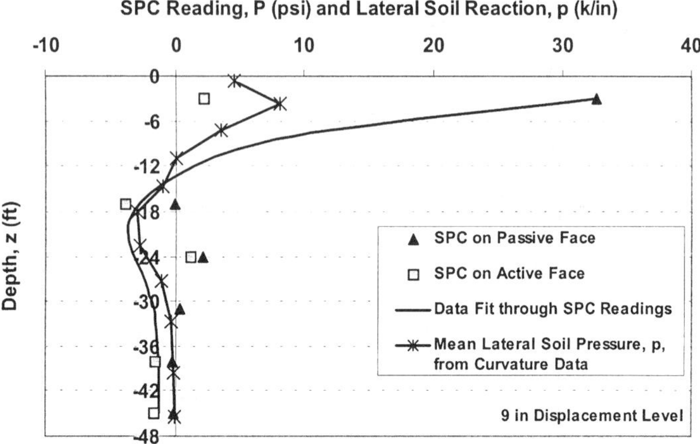

FIG. 5. Soil pressure cell reading, *P*, and soil reaction, *p*, distribution profiles for pull to west at 22.9 cm (9 inch) displacement level

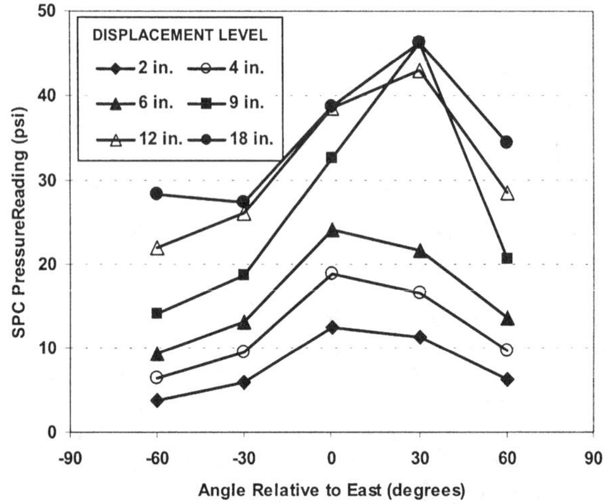

FIG. 6. Horizontal distribution of normal soil pressures against leading face of shaft at 0.9 m (3 ft) depth

$$p_\sigma = \int_{\theta=-\pi/2}^{\pi/2} \sigma(\theta)\cos\theta \cdot r \cdot d\theta \qquad (1)$$

where θ = azimuthal angle relative to line of loading. As noted earlier, the soil reaction against the shaft has contributions from passive soil resistance (i.e., the p_σ term), side shear resistance, and active pressures (which counteracts the resultant). As shown in Fig. 5, the active pressures are negligibly small. Hence, a good estimate of the side shear contribution (p_τ) can be obtained as the difference between p and p_σ.

To illustrate the effect of normal and side shear stresses on p at 0.9 m (3 ft) depth, the ratio of p_σ/p as a function of displacement level is plotted in Fig. 7. The results indicate that only about 20% of the soil reaction force p can be attributed to normal soil stress, and that this ratio does not vary with load level. Similar plots at the 3.1 m (10 ft) and 5.2 m (17 ft) depth levels (Fig. 7) reveal a different result, which is an increase of p_σ/p from about 15% to 25% at low displacement levels (i.e., displacements less than 22.9 cm or 9 in) to as high as 48% at the 45.7 cm (18 in) displacement level.

At all three depths, the modest contribution to p from p_σ is consistent with the previous results of Bierschwale et al. (1981) which found that substantial soil resistance beyond the normal stress was required to place the shaft in equilibrium. The trends at the 3.1 m (10 ft) and 5.2 m (17 ft) depths suggest that the side friction is mobilized early in the test and provides the majority of the resistance at low displacement levels. However, at larger displacements, the normal stresses account for an increasing fraction of p, suggesting a disproportionate build-up of p_σ with increasing shaft deflection.

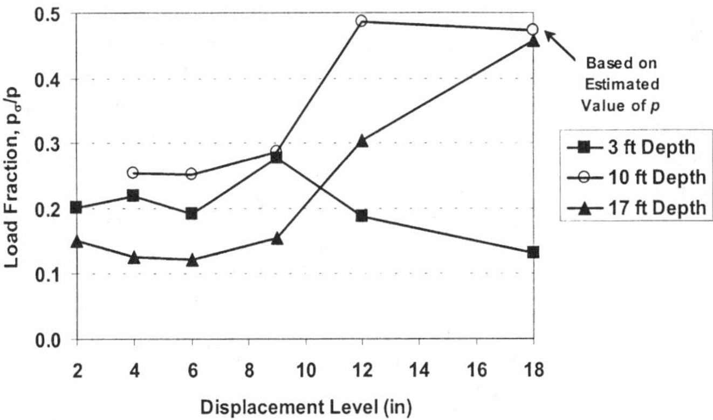

FIG. 7. Fraction of soil reaction force p that can be attributed to normal stresses at the soil-shaft interface as a function of displacement level at 0.9 m (3 ft) depth, 3.1 m (10 ft) depth, and 5.2 m (17 ft) depth

This observation is consistent with test results reported by Smith and Slyh (1986), who performed lateral load testing of a circular rigid concrete shaft, and found that side shear could account for 84% of the soil reaction at low displacement levels. Smith and Slyh (1986) provided a schematic illustration of this data trend, which is shown in Fig. 8.

The same trend is not apparent at the 0.9 m (3 ft) depth (Fig. 7). The soil reaction values, p, back-calculated in the p-y derivations increase nearly linearly with y. This increase essentially matches, on a percentage basis, the increase in cell pressures, which accounts for the uniform p_σ/p values shown at the 0.9m (3 ft) depth plot in Fig. 7. This result may not be reliable, however, because the p values at this depth were found to be unusually high (i.e., higher than the available soil resistance).

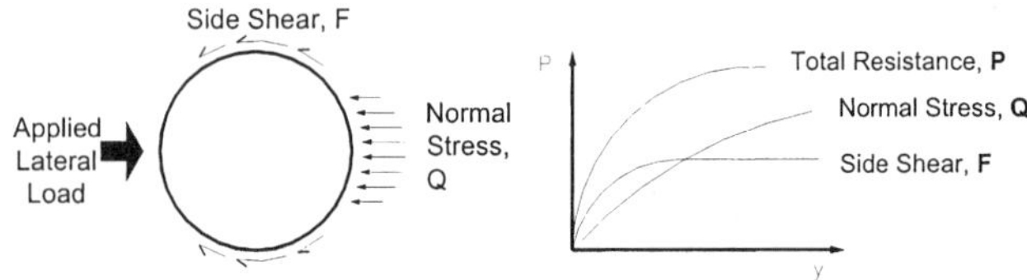

FIG. 8. The two components of soil resistance, p (after Smith and Slyh 1986)

CONCLUSIONS

In this study, p-y curves for a large diameter drilled shaft were directly inferred from field testing measurements through a carefully conceived data reduction program. Soil pressure cell data was then used to infer distributions of normal stress around the perimeter of the shaft at appropriate displacement levels. These pressure distributions were integrated around the shaft perimeter to evaluate the resultant soil reaction force per unit length at various depths. The results were used to evaluate the contribution of p_σ to p-values at each displacement level.

Insights into the mechanics of p-y response were obtained from the soil pressure cell data, which enabled the relative contributions of soil normal and shear stresses on p-values to be inferred. The relative contributions of normal stresses were seen to increase with shaft deflection, as passive pressures were mobilized on the face of the shaft. This study suggests that the side friction mobilized early in the test and provided the majority of the resistance at low displacement levels. However, at larger displacements, the normal stresses accounted for an increasing fraction of p, suggesting a disproportionate build-up of p_σ with increasing shaft deflection. Such behavior suggests that large diameter shafts could have added capacity from side friction, which the current design models do no incorporate.

ACKNOWLEDGMENT

Support for this research was in part provided by the California Department of Transportation (Caltrans) under Research Contract No. 59A0183 with contract monitors Craig Whitten and Anoosh Shamsabadi at Caltrans and principal investigators John Wallace and Jonathan Stewart at the University of California, Los Angeles. Their support is gratefully acknowledged.

REFERENCES

Bierschwale, M., Coyle, H., and Bartoskewitz, R. (1981). "Lateral Load Tests on Drilled Shafts Founded in Clay," *Drilled Piers and Caissons*, ASCE, 98-113.

Briaud, J., Smith, T., and Meyer, B. (1983). "Pressuremeter Gives Elementary Model For Laterally Loaded Piles," *Intl. Symp. on In Situ Testing of Soil and Rock*, Paris.

Briaud, J., Smith, T., and Tucker, L. (1985). "A Pressuremeter Method for Laterally Loaded Piles," *Intl. Conf. Soil Mechanics and Foundation Engineering*, San Francisco.

California Department of Transportation (Caltrans). (1995). "Bridge Design Specifications Manual".

California Department of Transportation (Caltrans). (1999). "Caltrans Seismic Design Criteria".

Janoyan, K. (2001). "Interaction between Soil and Full-Scale Drilled Shaft under Cyclic Lateral Load." Ph.D. Dissertation, Dept. of Civil & Envir. Engrg., University of California, Los Angeles, CA.

Janoyan, K., Stewart, J. P., and Wallace, J. W. (2001). "Analysis of p-y Curves from Lateral Load Test of Large Diameter Drilled Shaft in Stiff Clay," *Proc., 6th Caltrans Workshop on Seismic Research*, Sacramento, CA, paper 5-105.

Kasch, V., Coyle, H., Bartoskewitz, R., and Sarver, W. (1977). "Lateral Load Test of a Drilled Shaft in Clay," *Research Report No. 211-1*, Texas Transportation Institute, Texas A&M University, College Station.

McClelland, B. and Focht, J. (1958). "Soil Modulus for Laterally Loaded Piles," *Transactions,* ASCE, Vol. 123, Paper 2954, 1049-1086.

Smith, T. and Slyh, R. (1986). "Side Friction Mobilization Rates for Laterally Loaded Piles from the Pressuremeter," *The Pressuremeter and Its Marine Applications: Second Intl. Symposium*, ASTM STP 950, 478-491.

APPENDIX I. CONVERSION TO SI UNITS

Inches (in) × 0.0254 = meter (m)
Feet (ft) × 0.3048 = meter (m)
Kips per square foot (ksf) × 47.88 = kilopascals (kPa)
Kips per square inch (ksi) × 6.894 = Megapascals (MPa)
Pounds per square inch (psi) × 6.894 = kilopascals (kPa)
Kips per inch (k/in) × 175.1 = kinonewton per meter (kN/m)

INTERNATIONAL PERSPECTIVES ON QUALITY ASSESSMENT OF DEEP MIXING

Anand J. Puppala[1] and Ali Porbaha[2]

ABSTRACT

A wide range of in situ techniques has been developed in the last two decades in Europe and Asia to assess the quality of deep mixed soil columns. These methods use either conventional or specially designed tools to penetrate into the hard columns (such as pushing, pulling, rotary, driven, or displacement methods); or use geophysical concepts (such as seismic, resistivity, and echo-pulse methods). The main objective of this paper is to present the results of an international survey conducted for the quality assessment of deep mixed columns. The objectives were (a) to identify the most commonly accepted field methods of quality assessment practiced in different parts of the world, and (b) to understand the operational problems occurred when the method was applied for deep mixing projects. The respondents of the survey had diverse backgrounds, which included engineers, academicians, researchers, and contractors. The summary of the international survey provided valuable insights for the selection of in situ techniques for the quality assessment of deep mixing projects.

INTRODUCTION AND BACKGROUND

Deep mixing (DM) techniques, developed during 1960's, were first reported in the literature in the early 1970's (Broms and Boman, 1979). Deep mixing technology uses specially designed equipment with paddles and/or augers to mix in situ soil with

[1] Associate Professor, Department of Civil and Environmental Engineering, Box: 19308, The University of Texas at Arlington, Arlington, Texas 76019, Tel: 817-272-5821, Fax: 817-272-2630, anand@uta.edu;

[2] Assistant Professor, Department of Civil Engineering, California State University, 6000 J Street, Sacramento, California 95819-6029; porbaha@csus.edu

cementing materials such as cement, lime, or other types of binders. The choice of the binder to be used in the field depends on the requirements of the project. For example, if the strength of the soil was the main consideration as in the case of structures built on loose sandy soils, reclaimed soils, peats or soft clays, the use of deep cement mixing is normally preferred. In projects where soil compressibility properties need to be enhanced to reduce undesirable settlements, either lime or combinations of lime with cement or other additives are typically used.

In environmental related construction projects, where soil moisture barriers or soil solidification works are needed to reduce the hydraulic conductivity of the soil medium, cement stabilization is often recommended in DM treatments since cement treatment is known to reduce hydraulic conductivity of the soils by occupying the voids in the soils (Kamon, 1996; Okumara, 1996). Usually, the chemical stabilizer dosages used in DM projects are reported in the ranges of 150 to 200 kg/m^3, which usually represent 10 to 12% by dry weight of soil.

The deep mixing stabilizing process typically takes place by mechanical dry mixing, wet mixing or, grouting (Rathmayer, 1996; Holm, 1999; Porbaha, 1998; Bruce and Bruce, 2003). Dry mixing is usually preferred in project sites where water tables are high and close to the ground surface. Wet mixing is recommended for dry and arid environments or sites with deep water table locations. Grouting with or without jets has been used for ground strengthening, excavation support and ground water control in construction projects (Kamon, 1996; Porbaha, 1998).

Typical steps followed in a construction project involving deep mixing methods are: (1) Soil sampling and soil characterization at a field site, (2) Laboratory trial mix design, (3) Design of cement or a chemical binder of choice based on targeted soil properties of stabilized soil, (4) Selection of equipment and mixing processes for field installation, (5) Deep mixing in the field and cement or treated soil column installation, (6) Monitoring of grout and cement stabilizer magnitudes as well as installation-related equipment variables during installation as a part of Quality Control (QC) of soil treatment in the field, and (7) Quality Assessments (QA) of cement or stabilizer treated soil columns in the field.

In general, the strengths of treated soils in the field are equal to 20 to 80% of the corresponding strengths determined in laboratory settings (Holm et al. 1981; Rathmayer, 1996; Porbaha, 1998). The variations in strength are attributed to differences in soil mixing procedures as well as soil compositional and environmental conditions in both laboratory and field settings. These variations signify the need for quality assurance (QA) methods (as noted in Step 7) to assess deep mixing operations in field conditions. The QA methods are often used to address the quality of mixing methods to achieve uniform treatments at all depths. The QA step will also ascertain the role of the equipment or device used to achieve thorough field mixing.

Quality assurance methods typically employ either core sampling of cement columns followed by laboratory testing on the core samples, or field-tests utilizing intrusive and/or non-intrusive in situ methods for direct and indirect interpretation of soil improvements. Laboratory tests are influenced by sample disturbance and simplified boundary conditions. Additionally, laboratory tests are time consuming and provide verification only at discrete points, and may not detect localized weak zones along the treated soil column. Accordingly, several types of old and new in situ techniques have been used in different parts of the world to investigate the in situ quality of the treated soil column (Halkola, 1999; Holm, 1999). The effectiveness of a particular soil improvement technique can be evaluated by measuring in situ soil properties before and after soil improvements. Differences in these properties can be used to address QA aspects of DM treatment.

Several in situ methods have been used in various DM projects and there is an extensive literature available, which describes QA aspects of in situ methods and their procedural steps. A majority of the literature information is published abroad with two major international conferences on deep mixing held in Japan and Sweden in 1996 and 1999, respectively. One of the major limitations of this literature information is the lack of consistency or agreement between studies conducted in Europe, Asia and the United States. Additionally, a few novel methods developed for QA studies in one country are not well-known outside that country (Porbaha and Roblee, 2001). Hence, there is an important research need to compile the available information on in situ methods and disseminate them in a research report for future usage by practitioners and researchers.

A research study was hence conducted to accomplish this objective by first conducting surveys with practitioners from both abroad and within the U.S. to learn various in situ methods' practices for QA studies of DM columns. This paper provides an overview of these surveys' responses with insights learned regarding European and Asian perspectives on QA tasks, sampling methods used, percents of columns evaluated in a project site, types of in situ methods used and limitations noted in these in situ methods. Current US practices are also surveyed and presented for comparison purposes.

SURVEYS

A survey questionnaire was hence prepared to address the above-mentioned objectives to learn current and novel in situ methods and sampling practices for QA studies in Europe and Asia. Figure 1 presents the survey questionnaire and its details. Figure 2 presents various in situ penetration (intrusive type) and geophysical methods that were queried as part of question 4 in the survey. Acronyms used to identify these test methods are also listed in the same figure.

International Surveys

A total of 120 surveys were sent out to practitioners, contractors, academics, and researchers from several countries in Europe (Finland, Poland, Sweden, Italy, Netherlands, and the United Kingdom) and Asia (Thailand, Taiwan, and Japan). Out of 120 mailed, 28 of them responded either electronically or through postal mail. These responses, which represented approximately 25% of the survey mails sent, were compiled and analyzed. The distributions of survey respondents included engineers and practitioners (64%), academicians and researchers (32%), and contractors (25%). Some of the respondents were involved in more than one activity, i.e. research and engineering practice. Hence, the total percents do not add up to 100%.

Users were specifically asked about in situ penetration and geophysical test methods. In situ penetration tests noted in the surveys and their acronyms are already listed in Figure 2. The geophysical methods queried in the surveys were: Seismic

Reflection Method - PS Logging Test (PSL); Shear Wave Tomography Method (STM); DM Column Integrity Test (CIT); Electrical Resistivity Logging Test (ELT); and Ground Penetrating Radar Method (GPR).

1. What type(s) of QA studies were performed on DM columns?

☐ Field Sampling and Laboratory Testing ☐ Post Construction Instrumentation and Monitoring

☐ In Situ Testing on DM Columns

2. Please list "sampling" methods used in the field sampling of DM columns?

1) 2)

3. a) What type of DM method and b) binders were used in the project sites?

a) ☐ Wet Mixing ☐ Dry Mixing ☐ Both

b) ☐ Cement ☐ Lime ☐ Others

4. Please list various in-situ test methods used in the quality assessments of the DM sites? Please refer to the attached help sheet for the in situ methods.

1) 2) 3)

4) 5) 6)

5. How many DM columns (in percent) have to be checked for QA at a given project site? 1 in ?? ________

6. What was the approximate time lag used between DM installation and in situ testing for QA task?

☐ < One-Week ☐ One to Two Weeks ☐ 28 days ☐ Time Depends on Project

7. Which soil property was the major focus of your field investigations?

☐ Strength ☐ Compressibility ☐ Permeability

8. Were there any major (a) operational and (b) other problems experienced in performing in situ tests? Please note the in situ method(s) against each item.

(a) ☐ Difficult to Push Through High Strength Columns ________

☐ Buckling of Push Rods ________

☐ Damage to In Situ Device ________

(b) ☐ Expensive ________

☐ Time Consuming ________

☐ Poor Interpretations of Quality Assessments ________

9. Based on your QA experience, which in situ methods would you recommend using in DM projects?

1) 2) 3)

10. Please list any other in situ methods (other than those listed in the help sheet) used for QA tasks?

1) 2)

11. What was your assessment on the above (10th question) methods' performance?

☐ Good:________ ☐ Satisfactory:________ ☐ Poor:________

FIG 1. Survey Questionnaire and Details

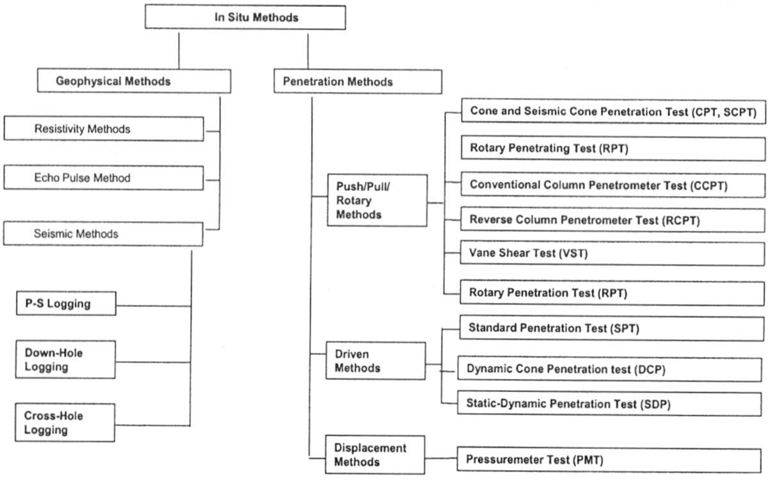

FIG 2. In Situ Methods Asked in the Present Survey

The first question in the survey queried the type of QA studies performed on DM treated soil columns. The choices were post construction instrumentation and monitoring, in situ tests, and laboratory tests on field cores. A few of them responded that all three choices presented were followed in their construction projects. Out of 28 responses received, 50% of them noted post construction instrumentation and monitoring methods, 79% noted field sampling and laboratory testing methods, and 89% reported using in situ testing methods. From these results, it can be concluded that the in situ testing method is the most popular procedure used in quality assessments of the DM columns.

The next question queried about the sampling methods used in collecting treated soil samples from the DM columns for laboratory testing. The majority of the respondents noted the utilization of core sampling, tube sampling, and excavation sampling to acquire field core samples from the DM columns. Table 1 provides a list of various sampling methods used to retrieve DM core samples throughout the world.

The following question queried about the type of mixing method used for DM columns and the type(s) of binder(s) used in the studies. Figure 3 presents both queries' results. A total of 54% of the respondents noted using the dry mixing method, whereas 21% of them noted using the wet mixing method. Approximately, 25% of the respondents noted using both methods. The majority of 86% of the respondents used cement and 46% used lime as the binder. Approximately 43% of them used other chemical admixtures including fly ash and granulated slag.

Table 1. Methods Used for Field Sampling of DM Columns

Sampling Method	# of Respondents
Core Sampling (Including Tube Sampling)	8 (2)
Double Tube Sampling	5
Triple Tube Sampling	3
Fresh Material Sampling	2
Excavation Sampling	3
Block Sampling	3
Wet Sampling	1

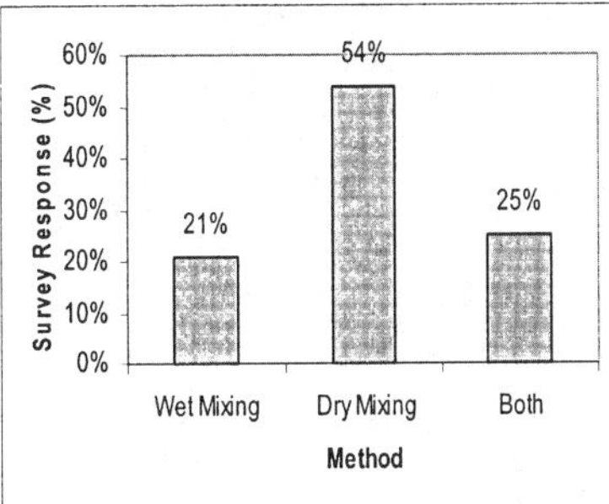

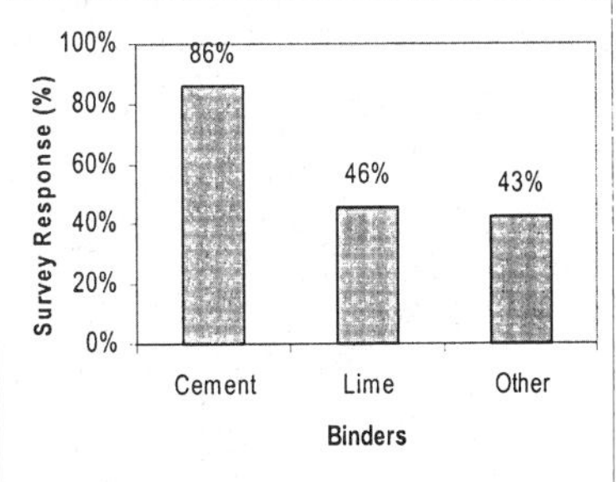

FIG 3. Mixing Methods and Binder Types Used in DM Project Sites

When queried about different types of in situ test methods used in their DM projects, respondents listed various in situ test methods (Table 2). The in situ test methods used to assess the quality of the DM sites vary a greater deal in Europe than in Asia, as seen in Table 2. Most of the respondents favored penetration tests over non-destructive tests. Also, a majority of the respondents listed cone penetration or CPT method as their first choice of QA method. Approximately 29% and 18% of them noted that they used two novel Swedish in situ methods, Conventional Column Penetrometer Test (CCPT) and Reverse Column Penetration Tests (RCPT) for QA studies on DM columns. These methods are not widely known outside European countries where they were developed. The SPT method has been primarily used for QA studies in Asian countries, whereas Column Vane Penetration (CVP) tests have been primarily used on low strength DM columns in Europe.

Approximately 22% of respondents also noted that they used a few geophysical test methods for QA studies. The seismic logging method was the most frequently used geophysical method, followed by the seismic integrity test method. Though respondents did not list other non-destructive test methods, several sources in the literature identified a few other geophysical methods used for QA studies. These were cross-hole and down-hole seismic logging methods and electrical resistivity methods.

Another important question is the number of columns to be tested for QA evaluations. According to the percentage of DM columns to be assessed for these studies in a given project site, 46% of the respondents check marked about 0.5 to 1 percent of the DM columns, 21% checked less than 0.5 percent, 11% checked 2 to 5 percent, and 4% checked 6 to 10 percent. These results are also shown in Figure 4. Approximately 18% of the respondents mentioned that the number of DM columns to undergo quality assessment studies depend on the project site. Overall, the majority of the respondents recommended the use of 1 out of every 100 columns for QA assessments.

The time lag that was used between installing the DM columns and in situ testing for quality assessment tasks was surveyed in another question. Approximately 64% of the respondents mentioned a time lag period of 28 days for performing in situ testing of DM columns. Approximately, 21% of them responded a time period of 1 to 2 days, 11% of them noted a time period of less than a week, and 32% of them noted that the time lag depended on the project objective. Figure 5 illustrates these findings.

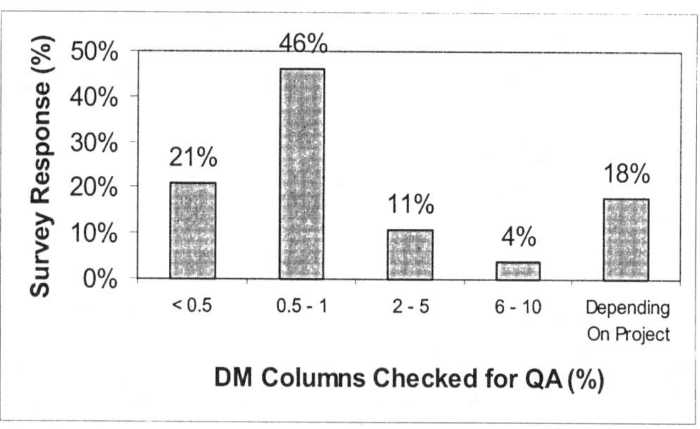

FIG. 4. Percent DM Columns in a Project Evaluated for QA Studies

Strength was the major focus of soil improvement in several DM project site investigations. An overwhelming majority of respondents (93% of them) noted that strength was the main focus of their projects. Compressibility was chosen by 18% of the respondents and permeability was mentioned by only 4% of the respondents.

Another important question in the survey was to learn various problems experienced during in situ testing of treated soil columns. Several problems were noted by the respondents, which were presented in Figure 6. A significant majority of respondents (68%) stated that it was difficult to push through the high strength columns. Poor interpretations of the test results and the high cost of performing these tests were also major concerns to 32% of the respondents. Approximately 25% of them noted that the in situ tests were time consuming, 14% of them noted that the push rods were buckled, and 7% of them responded that the tests were damaging to the in situ devices.

Table 2. In Situ Methods Used for Quality Assessments of DM Sites

	SURVEY RESPONSE		
In Situ Tests	**Europe**	**Asia**	**Total (%)**
Penetration Tests			
Standard Penetration Test (SPT)	0	2	7%
Cone Penetration Test (CPT) / Piezocone Test (PCPT or CPTU)	9	2	39%
Seismic Cone Penetration Test (SCPT)	3	0	11%
Pressuremeter Test (PMT)	1	0	4%
Dynamic Cone Penetration Test (DCP)	1	0	4%
Rotary Penetrating Test (RPT)	1	1	7%
Conventional Column Penetrometer Test (CCPT)	8	0	29%
Reverse Column Penetrometer Test (RCPT)	5	0	18%
Static-Dynamic Penetration Test (SDPT)	2	0	7%
Column Vane Penetrometer Test (CVP)	7	0	25%
Dilatometer Test (DMT)	1	0	4%
Geophysical Tests			
Seismic PS Logging Test (PSL)	3	2	18%
Shear Wave Tomography Method (STM)	0	0	0%
DM Column Integrity Test (CIT)	0	1	4%
Electrical Resistivity Logging Test (ELT)	0	0	0%
Ground Penetrating Radar Method (GPR)	0	0	0%
Others	11	4	54%

Note: Totals do not add up to 100% since respondents can list more than one in situ method in their surveys

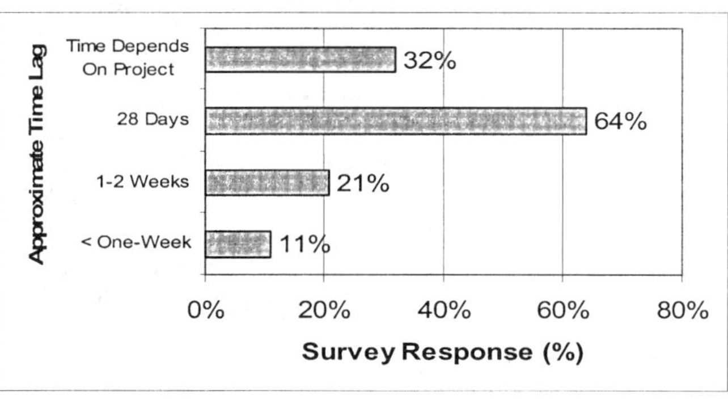

FIG 5. Approximate Time Lag Between DM Installation and In Situ Testing For QA Tasks

When asked to recommend in situ methods the respondents felt worked best, nine of the respondents chose CPT. Other recommended in situ penetration tests were

CCPT and RCPT. Several other respondents also recommended SPT, CVP, RCPT, PSL and CIT methods for QA studies.

US Surveys

The same questionnaire (Figure 1) was also mailed to practitioners, contractors and researchers in the United States, who were involved with deep mixing projects. The results from these surveys provided insights into QA techniques used within the United States. Out of 24 surveys mailed, a total of five replies were received. The first question of the survey queried the type of QA studies performed on the DM treated soil columns. Of the five respondents, four noted using field sampling and laboratory testing, three listed in situ testing methods, and one respondent used post construction instrumentation for QA studies. Survey results indicate that field sampling and laboratory testing was the most popular method used in the U.S. to perform QA studies on the DM columns. The following question queried about the sampling methods used in the field. The majority of the respondents noted that they utilized core sampling and a few of them noted that they utilized wet sampling methods for collecting samples from DM materials.

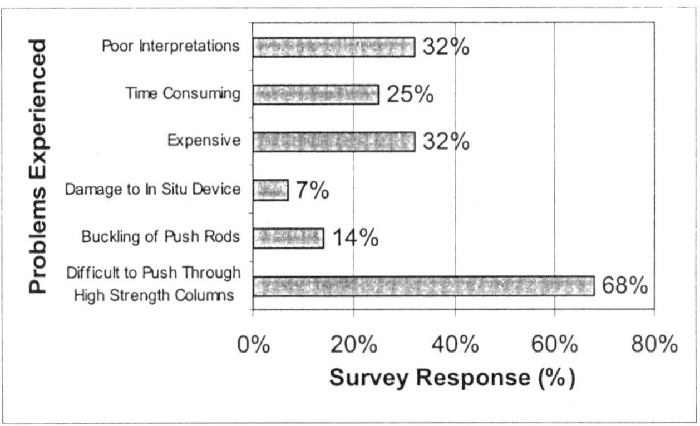

FIG. 6. Operational and Other Problems Experienced In Performing In Situ Tests

Based on survey responses, the wet mixing method was preferred in the U.S. over the dry mixing method for DM treatments and the binder frequently used was cement. When queried about different in situ test methods used in their QA studies, respondents again listed a few in situ methods including CPT and PMT methods, which were mentioned by three respondents each. The pressuremeter test is regarded as a better in situ test tool in the U.S. for quality assessments of high strength columns. Also, this method can provide both strength and stiffness property interpretations from the same set of test data at each location (Martin et al. 1999).

According to the percentage of DM columns to be assessed for QA studies, two of the respondents noted at least 2% of the total columns, one mentioned 5% of total columns, and another respondent noted 20% of the columns at the site. One respondent mentioned that the number of DM columns to undergo quality assessment studies depended on the type of construction project and the respondent did not mention any further details. Overall, it can be mentioned that the majority of respondents from the U.S. recommended 2 to 5% of DM columns at a site to perform QA assessments.

The approximate time lag between the installation of DM columns and in situ testing on them for quality assessments was queried. Two of the five respondents mentioned that the time lag depended on the objective of the construction project. Other respondents preferred one to two days, less than a week, and 28 days, respectively. Strength was listed as the major focus of soil improvement in the DM projects. Remaining respondents listed compressibility, permeability, and other soil properties.

The final important question is to learn various problems experienced during in situ testing of DM treated columns. The U.S. respondents listed several problems with the majority of the respondents stating that it was difficult to push through the high strength columns. Other comments included were expensive in situ tests and poor interpretations of the quality assessments. When asked to recommend in situ methods, the popular choices among U.S. respondents were load tests on columns and pressuremeter tests.

SUMMARY AND CONCLUSIONS

This paper provided a summary of various international surveys, including Asian and European perspectives on DM quality assessments practices. Surveys from the U.S. are also presented for comparison purposes. Several researchers, academics, contractors, and practitioners from different parts of the world provided input to these surveys. Respondents from abroad mentioned the use of several in situ destructive test methods for QA studies. Limited response of non-destructive methods was observed, which was consistent with the lack of sufficient literature information on the non-destructive methods for quality assessments of deep mixing.

The CPT method, followed by CCPT, CVP and RCPT methods have been frequently used in Europe for routine QA studies. The SPT method has been used in Asian countries with some success. Also, the majority of international respondents preferred testing of approximately 1% of the total columns in a project for QA studies and the time lag for these tests was preferred to be around 28 days after soil mixing. Surveys from the U.S. showed the use of in situ methods, in particular, pressuremeter and cone penetrometer tests for QA studies. In the U.S., 2% of the columns were preferred for QA testing and the time lag for testing depends on the objectives of the construction project.

Regarding limitations experienced the respondents from both the US and international surveys noted the buckling of the devices and the difficulty in pushing through hard cement columns. One of the common replies from the majority of the survey responses received was the need and the importance of in situ methods for better and reliable QA evaluations in DM projects.

ACKNOWLEDGMENTS

This project was sponsored by the National Deep Mixing (NDM) research program (www.deepmixing.org). The NDM is a research consortium of ten states, Federal Highway Administration (FHWA), and private sector. The authors are grateful to Al Dimillio with FHWA Office of Research in McLean, Virginia. The authors would like to appreciate several experts who kindly participated in both international and US surveys and the follow up interviews. Their names and their effort will be acknowledged in the final NDM report.

REFERENCES

Broms, B. and Boman, P. (1979). “Lime columns – a new foundation method.” *ASCE, Journal of Geotechnical Engineering*, Vol. 105, GT4, pp.539-556, 1979.

Bruce, D. A. and Bruce, M. E. C. “The Practitioner’s guide to deep mixing.” ASCE, Geotechnical Special Publication No.120. *Proceedings of the 3rd International Conference on Grouting and Ground Treatment,* Vol. 1, pp. 474-488. Feb 10-12, 2003.

Esrig, M., Mac Kenna P. and Forte, E. (2003). “Ground stabilization in the United States by the Scandinavian Lime Cement Dry Mix Process.” ASCE, Geotechnical Special Publication No.120. *Proceedings of the 3rd International Conference on Grouting and Ground Treatment,* Vol. 1, pp. 501-514. Feb 10-12, 2003.

Halkola, H. (1999). “Keynote lecture: Quality control for dry mix methods.” *Proceedings of the International Conference on Dry Mix Methods for Deep Soil Stabilization, 13-15 October 1999, Stockholm*, pp. 285-294. Balkema. 1999.

Holm, G., Bredenberg, H., and Broms, B. (1981). “Lime columns as foundations for light structures.” *Proceedings, 10th ICSMFE*, Stockholm, Sweden, pp. 687-693, 1981.

Holm, G. (1999). “Keynote lecture: Applications of dry mix methods for deep soil stabilization.” *Proceedings, International Conference on Dry Mix Methods for Deep Soil Stabilization, 13-15 October 1999, Stockholm*, pp. 3-14. Balkema, 1999.

Kamon, M. (1996). “Effects of grouting and DMM on big construction projects in Japan and the 1995 Hyogoken - Nambu earthquake.” *Proceedings of IS-Tokyo '96, The 2nd International Conference on Ground Improvement Geosystems, 14-17 May 1996, Tokyo*, pp. 807-824. Balkema. 1996.

Liu, S. and Hryciw, R.D. (2003) “Evaluation and quality control of dry jet mixed clay soil-cement columns by SPT.” *Transportation Research Board,* 2003 Annual Meeting, CD ROM, Washington, DC, 2003.

Martin, G. R., Arulmoli, K., Yan, L., Esrig, M. I. and Capelli, R.P. (1999). "Dry mix soil-cement walls: An application for mitigation of earthquake ground deformations in soft or liquefiable soils." *Proceedings, International Conference on Dry Mix Methods for Deep Soil Stabilization, 13-15 October 1999, Stockholm*, pp. 37-43. Balkema Publishers, 1999.

Okumara, T. (1996). "Deep mixing method of Japan." *Proceedings of IS-Tokyo '96, The 2nd International Conference on Ground Improvement Geosystems, 14-17 May 1996, Tokyo*, pp. 879-888. Balkema Publishers, 1996.

Porbaha, A. (1998). "State of the art in deep mixing technology, Part I: Basic concepts and overview of technology." *Ground Improvement*, 2, No. 2, pp.81-92, 1998.

Porbaha, A. and Roblee, C. (2001). "Challenges for implementation of deep mixing in the USA." *Proceedings, International Workshop on Deep Mixing Technology*, Oakland, CA., National Deep Mixing Program, 2 volumes, 2001.

Porbaha, A. and Puppala, A. J. (2003) "In situ techniques for quality assurance of deep treated soil columns." ASCE, Geotechnical Special Publication No.120. *Proceedings, 3rd International Conference on Grouting and Ground Treatment,* Vol. 1, February 10-12, 2003.

Rathmayer, H. (1996). "Deep mixing methods for soft subsoil improvement in the Nordic countries." *Proceedings of IS-Tokyo '96, The 2nd International Conference on Ground Improvement Geosystems, 14-17 May 1996, Tokyo*, pp. 869-878. Balkema, 1996.

NON-DESTRUCTIVE EVALUATION OF CEMENT-MIXED SOIL

David A. Staab[1], Member, ASCE, Tuncer B. Edil[2], Member , ASCE , David L. Alumbaugh[3]

ABSTRACT: This paper describes how electrical resistivity (ER) and time-domain reflectometry (TDR) were used to determine the electromagnetic (EM) properties, electrical resistivity (ρ) and apparent dielectric constant (K_a), of soil-cement mixtures. Laboratory tests simulating cement-mixed soil were performed on large-size specimens (286-mm diameter and 305-mm height) of different soil, cement, and water proportions at regular time intervals up to 56-days. Soil, cement, and water have measurably different EM properties, and as cement consumes water during hydration and bond formation continues during hardening the EM properties change over time. Water content and unconfined compressive strength (UCS) were also measured at 7-, 14-, 28-, and 56-days. For soil-cement samples K_a decreased and ρ increased with time. ρ was found to be directly related to strength gain. Increasing cement content increased strength but decreased electrical resistivity.

INTRODUCTION

The deep-mixing method (DMM) involves mixing a cemetitous material (e.g., lime, cement, lime/cement mixture) with soil *in situ* to create a stronger and less compressible soil-cement column. One challenge to greater acceptance of deep-mixing is quality control and quality assurance (QC/QA). The uniformity of mixing, degree of curing, and rate of strength gain of *in situ* columns is difficult to evaluate, which creates uncertainty regarding how well the *in situ* column matches the original design. Currently, QC/QA methods for deep-mixed systems are not standardized, and many QC/QA methods rely on destructive methods to evaluate the *in situ* soil-cement column. The most common QC/QA method is coring and testing unconfined compressive strength samples. Discrete core samples may not be representative of actual mixing and curing conditions *in situ* if variations in soil conditions or cement content exist in the unit, and therefore may not be the best way to evaluate deep-mixed columns.

[1] Grad. Res. Asst., Dept. of Civil & Environ. Eng., Univ. of Wisconsin, Madison, WI, 53706, dastaab@students.wis.edu

[2] Prof., Dept. of Civil & Environ. Eng., Univ. of Wisconsin, Madison, WI, 53706, edil@engr.wisc.edu

[3] Assnt. Prof., Dept. of Civil & Environ. Eng., Univ. of Wisconsin, Madison, WI, 53706, alumbaugh@engr.wisc.edu

This laboratory study investigated how the electromagnetic (EM) properties (dielectric constant (K) and electrical resistivity (ρ)) of deep-mixed systems varied with different cement content and water contents for two different soils. EM properties of soil-cement mixtures depend on cement content, water content, soil type, and time. EM methods might be useful for monitoring the uniformity and perhaps strength gain of *in situ* columns. Tamura et al. (2002) found that resistivity measurements might be useful for evaluating the degree of mixing and quality of soil-cement columns *in situ*.

Geophysical field applications could map the EM properties of columns to determine how well the column is mixed over its entire length. Non-destructive testing is preferable to destructive testing since the column is not altered, and changes spatially and with time could be monitored.

BACKGROUND

Electrical Resistivity

The resistivity of low clay-content formations depends on the porosity, degree of saturation (i.e. the amount of fluid filling pore spaces), pore water salinity, and temperature (Bassiouni 1994). Higher temperatures increase the ion mobility and decrease resistivity (Sharma 1997). Most mineral grains (except mineral ores and clay minerals) are insulators, and pore water is usually electrolytic (i.e., contains considerable ions), so electrical conduction is primarily related to the amount of pore water present. Temperature also affects soil conductivity.

For fine-grain soils, especially clay, other factors affect resistivity. In coarse-grain soils, the matrix is considered non-conductive. However, in clay, the matrix contributes to the soil conductivity due to clay chemistry. Clays generally have a net negative charge, which attracts pore water and the positive ions within the water. A bound water layer forms around clay particles, and electrical double-layer conductance occurs when a second, diffuse layer of loosely held cations can move in the presence of an electric field (Bassiouni 1994).

Dielectric Constant

The dielectric constant measures how well an electric field can polarize a material (Bottcher 1952). Water has a relatively large dielectric constant of approximately 80, whereas air has a relatively low dielectric constant of 1, and soil minerals have dielectric constants between 3-10. Bulk soil dielectric constant is dominated by water content, and Topp et al. (1980) developed empirical relationships between apparent dielectric constant and volumetric water content.

Water content changes over time were monitored with time-domain reflectometry (TDR). TDR and ground penetrating radar (GPR) are both sensitive to apparent dielectric constant contrasts. TDR is more suitable for laboratory-scale studies, whereas GPR is more suitable for field-scale studies. If TDR can identify apparent dielectric constant contrasts at the laboratory scale, GPR might be useful for identifying these contrasts at field scale.

METHODS AND MATERIALS

Sample Preparation

Although applicable to many soil types, deep mixing is generally used on weak, compressible soils. The soils selected for this study consisted of low plasticity clay and low plasticity silt (CL and ML in the Unified Soil Classification System) from Kenosha and Marshfield, Wisconsin, respectively. The stabilizing agent was Type I Portland cement.

Air-dry soil and cement were mixed until uniform, and water was then added. The resistivity probe, as described below, was inserted into a cylindrical container so the bottom electrode was 50-mm above the base of the container. The container was then filled around the resistivity probe with the mixed soil. The soil was placed in 50-mm lifts and compacted to achieve a uniform density (approximately 17-18 kN/m^3). When filled, the soil specimen had the final dimensions of 286-mm diameter and 305-mm height. Then, the TDR probes were inserted into the soil. Specimens were stored in a "wet room" at constant high humidity (~110%) and temperature (~23° F) in the containers. Plastic bags covered the specimen surface to prevent water infiltration and evaporation.

Multiple unconfined compressive strength (UCS) specimens were prepared from each batch of soil or soil-cement mixture by compacting into PVC molds. All UCS specimens for a given mixture had approximately the same density. Specimen dimensions were 100-mm in length and 50-mm in diameter. UCS molds were split along the length to minimize sample disturbance during extraction. UCS samples were removed from the molds after 24-hours, wrapped in plastic, and stored in the wet room.

Mix Proportions

Cement content was based on percent dry cement and soil solids. Cement contents investigated varied from 0% to 14% based on mass of dry cement per mass of dry soil (corresponding to approximately an equivalent cement dosage of 0-177 kg/m^3 based on mass of dry cement per volume of moist soil). Initial water contents varied from 18% to 31%.

For Kenosha soil (CL), cement contents were 0% and 10%, and initial water content varied from approximately 18% to 31%. For Marshfield soil (ML), water content was held constant at approximately 25%, while cement contents were 0%, 5%, 10%, and 14% for four samples. Additionally, a sample at 18% water content and 10% cement was prepared with Marshfield soil. A cement sample was prepared at a water-to-cement-ratio (w/c) of 0.35 (or water content of 35%) as a reference material.

Electrical Resistivity

Electrical resistivity methods are widely used for surface and down-hole geophysical studies. Current is applied through one pair of metallic electrodes, and the potential is measured between two additional metallic electrodes. The bulk resistivity of the media surrounding the resistivity tool can be calculated based on Ohm's law.

The resistivity probes used in this study were constructed with 6.35-mm diameter fiberglass rods. The electrodes were made of copper tubing with inner diameter of 6.35-mm, outer diameter of 9.3-mm, and thickness of 1.5-mm. The electrode height was 5-mm (see Fig. 1).

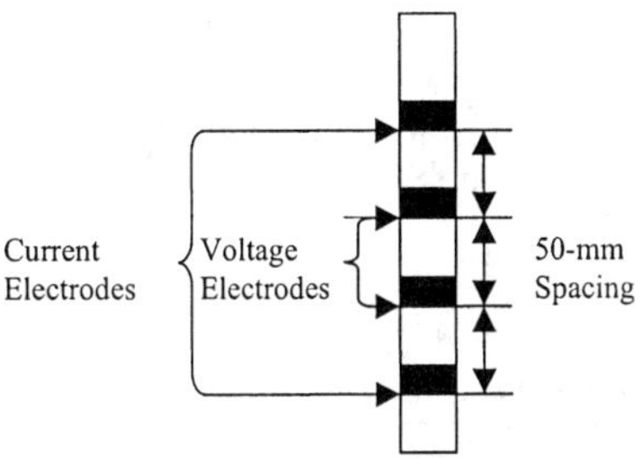

Fig. 1. Diagram of Wenner array electrode configuration.

A Tektronix CFG253 3-MHz Function Generator produced a 45-Hertz (Hz) sine wave. A Hewlett Packard (HP) 6824A DC Power Supply Amplifier increased the signal to 10-Volts. Fluke 87 True RMS Multimeters measured current (I) and voltage (V), which were recorded manually.

The resistivity tool used was a four-electrode Wenner array design with constant 50-mm electrode separation (Bassiouni 1994). Current was introduced through the outer two electrodes; the voltage was measured between the middle electrodes. Apparent electrical resistivity for the Wenner array is calculated with

$$\rho_a = \frac{V}{I} 4\pi a \tag{1}$$

where V and I are the measured voltage and current, respectively, and a is the electrode spacing. For heterogeneous structures, apparent resistivity is the resistivity of a homogenous medium that would produce the same V and I that were measured. The samples were assumed homogenous within the depth of investigation of the resistivity probe for this study, so ρ_a was assumed to equal ρ. The depth of investigation of resistivity tools is directly proportional to the electrode spacing. As discussed in Bassiouni (1994), the depth of investigation of a Wenner array is equal to twice the electrode spacing or 100-mm for this study.

Time Domain Reflectometry

The TDR technique is an electromagnetic technique originally developed to locate defects in buried cables, but was adapted to geo-engineering for water content determination (Topp et. al. 1980, Benson and Bosscher 1999). Metallic probes are inserted in the material of interest and connected to a waveform generator via a coaxial cable. The waveform generator sends an EM pulse through the coaxial cable and probes, and a reflected waveform is recorded. From the reflected waveform, the apparent dielectric constant, K_a, is determined as

$$K_a = \left(\frac{L_a}{L_p} \right)^2 \tag{2}$$

where L_p is the probe length and L_a is the horizontal distance between the maximum and minimum points of the waveform (Figure 2).

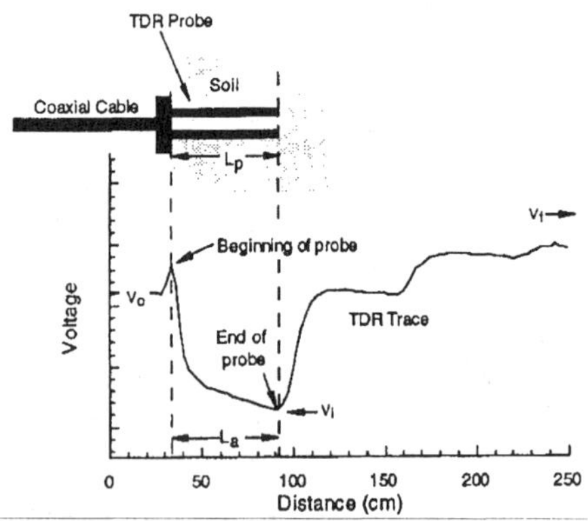

Fig. 2. TDR probe and waveform (from Benson and Bosscher 1999).

K_a is specific to TDR measurements and is based on certain assumptions. Suwansawat (1997) states that the complex dielectric constant is given as

$$K^* = K' + i(K'' + \frac{\sigma_{dc}}{\varepsilon_o \omega}) \tag{3}$$

where K' is the real part of the dielectric constant and K'' is the imaginary part of the dielectric constant (the dielectric loss), σ_{dc} is the zero frequency electrical conductivity, ε_o is the dielectric permittivity of free space, ω is the angular frequency, and i is the imaginary number ($\sqrt{-1}$). For high frequencies and low-loss medium, $K' >> K''$. However, some electrical losses do occur, so Topp, et. al. (1980) named the TDR-measured dielectric constant the "apparent dielectric constant," K_a, where

$$K_a \approx K' \tag{4}$$

Two, stainless-steel TDR probes measuring 286 mm long, 6.35 mm in diameter were used for testing. Acrylic spacers, 12-mm in height, were used to maintain a 50-mm spacing between the probes. The length of the probe in the soil was 274-mm. The TDR rod spacing and length were in the ranges used in previous studies conducted by Gibson (1999) and Suwansawat (1997).

To reduce noise in the TDR measurements, an impedance matching transformer (balun) was attached to the end of the coaxial cable. Gibson (1999) explains that the balun converts the electrical field from unbalanced to balanced before the signal enters the rods, which reduces errors introduced by a gradual transition.

To reduce or eliminate the conduction losses, the rod surface was coated with a non-conductive, epoxy coating (Champion Sprayon Interior/Exterior Epoxy). The coating will not drastically affect the dielectric constant provided the coating is thin, but will nearly eliminate attenuation (Suwansawat 1997).

A Tektronix 1502B Metallic Cable Tester produced waveforms, which were collected by a Campbell Scientific CR-10 datalogger and PC. The TDR waveform function of the PC208 software (P100) recorded the voltage at 256 equally spaced positions along a section of the waveform; the data was imported and graphed in a spreadsheet program (i.e., Microsoft Excel™).

Since the program recorded the relative distance between the data points collected and not the actual distance (L_a), which is need for moisture content calibration, manual TDR readings were also recorded, which were used to compute K_a from Eq. 2.

Unconfined Compressive Strength

UCS specimens were tested in duplicate at 7-, 14-, 28-, and 56-days. Sample height was 100-mm and diameter was 50-mm. Specimens were loaded at a displacement rate of 1mm/min until failure occurred, and the peak load was recorded.

Water Content Measurements

Water content determination was complicated due to the presence of cement. Cement binds water firmly; therefore, high energy is required to liberate the water. TDR is sensitive to free and bound water, so procedures were needed to measure both. There is no standardized test for determining the moisture content of cement.

Samples recovered from the containers or UCS specimens were dried at 105° C until mass was constant; the specimens were then placed in a desiccation chamber until cool before recording the mass. Several days were needed to reach constant mass for most samples containing cement.

RESULTS AND DISCUSSION

Electrical Resistivity

Figures 3 and 4 show resistivity values at the beginning and end of the testing. Resistivity appears inversely proportional to water content. Resistivity of all soil-cement samples increased over time. Specimens without cement had nearly constant resistivity though their resistivity as well as water content slightly decreased over time. It was expected that the resistivity of the 0% cement samples would have constant resistivity or have a slight resistivity increase since no hydration reactions were occurring and the water content decreased slightly. The observed opposite changes in resistivity might have resulted from the equilibration of water throughout the sample after placement. Following the initial measurements (e.g., after 7 days), the resistivity readings of 0% cement samples became quite stable. Two Kenosha samples with no-cement were prepared, one was placed in the container immediately after mixing and one hydrated in a sealed bag for 7-days prior to placement; the soil that was hydrated for 7-days showed less initial variation than the soil placed immediately after mixing.

Cement had the lowest initial resistivity and largest percent change in resistivity; the resistivity of cement increased 2300% over 56-days, whereas soil-cement soils increased by 13% to 113%. The largest changes in resistivity occurred in the lowest water content samples. This may be due to less free pore water as a percentage available after hydration.

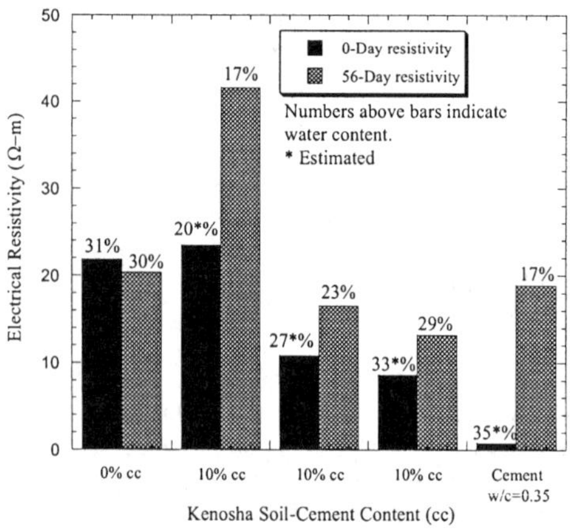

Fig. 3. Electrical resistivity at 0- and 56-days for Kenosha soil.

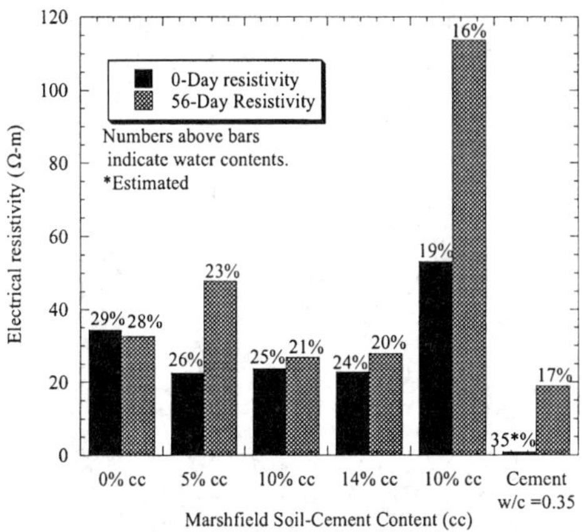

Fig. 4. Electrical resistivity at 0- and 56-days for Marshfield soil.

Time-Domain Reflectometry

Figures 5 and 6, show the measured K_a at 0- and 56-days for Kenosha and Marshfield soil, respectively. Apparent dielectric constant decreased with time for all soil-cement mixtures, but the dielectric constant was steady for non-cement specimens (i.e., soil). The pure cement specimen had the largest change in apparent dielectric constant of any specimen.

Cement content (10%) was kept constant for Kenosha specimens, and water content was varied (approximately 20% to 33%); the data indicated that larger initial water contents resulted in larger changes in apparent dielectric constant over the testing period. It may be argued that the lower water content samples might have lacked sufficient water, which limited the amount of cement hydration, whereas the higher initial water content samples may have had sufficient water for additional hydration and hardening reactions to occur. However, the unconfined compressive strength (UCS) tests (discussed below) indicated that the lower water content mixes had greater UCS than the higher water content samples.

Water content was held nearly constant (i.e., 24-29%) for four Marshfield samples, and cement content was varied (0%, 5%, 10%, and 14%). K_a increased when cement content was increased from 5% to 10%, but decreases when cement content went form 10% to 14%. An additional Marshfield sample was prepared at a lower water content of 19% and 10% cement content, which had lower apparent dielectric constant as expected but showed minor changes over time.

Water content changes affected K_a more than cement content, although greater cement content produced larger changes in K_a over time. The largest changes in K_a occurred in the highest water content samples, unlike electrical resistivity where the largest changes occurred in the lowest water content samples. Changes in K_a

mirrored changes in resistivity, but did not correlate well to water content. The explanation for this phenomenon is currently being investigated. Changes in K_a over time might be useful for qualitatively estimating cement content in deep-mixed soil columns if the initial water content is approximately uniform.

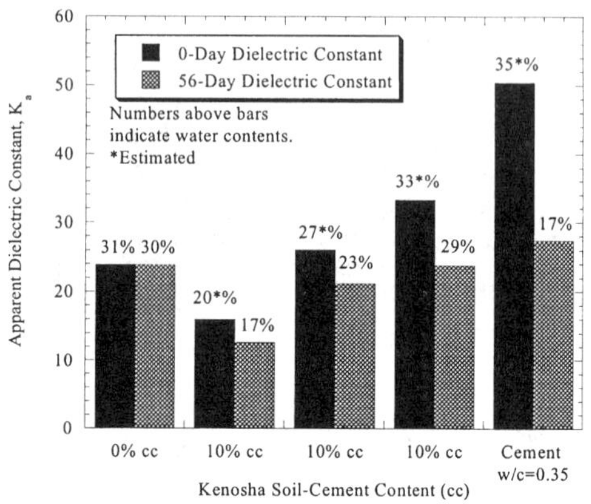

Fig. 5. Apparent dielectric constant at 0- and 56-days for Kenosha soil.

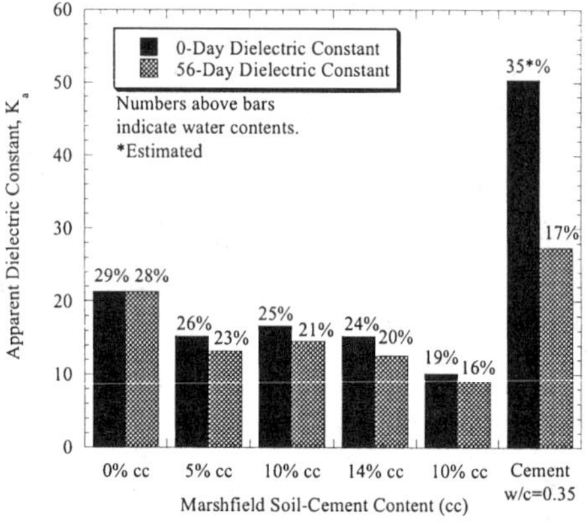

Fig. 6. Apparent dielectric constant at 0- and 56-days for Marshfield soil.

Unconfined Compressive Strength

Figures 7 and 8 show the relationship between electrical resistivity and unconfined compressive strength (UCS). UCS increased with increasing cement content in all soil-cement mixtures compared to the no-cement samples.

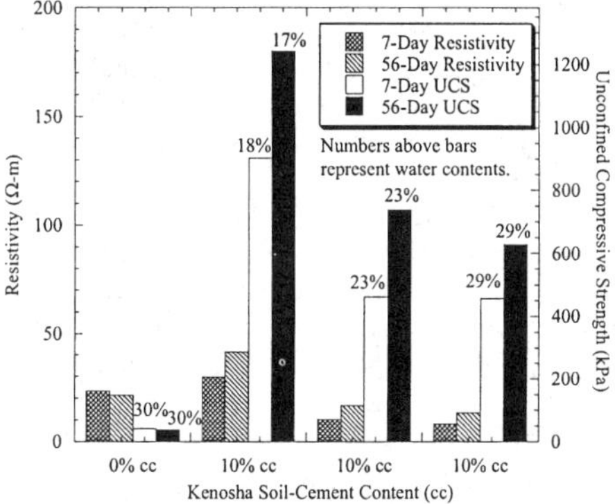

Fig. 7. Resistivity and unconfined compressive strength at 7-and 56-days for Kenosha soil.

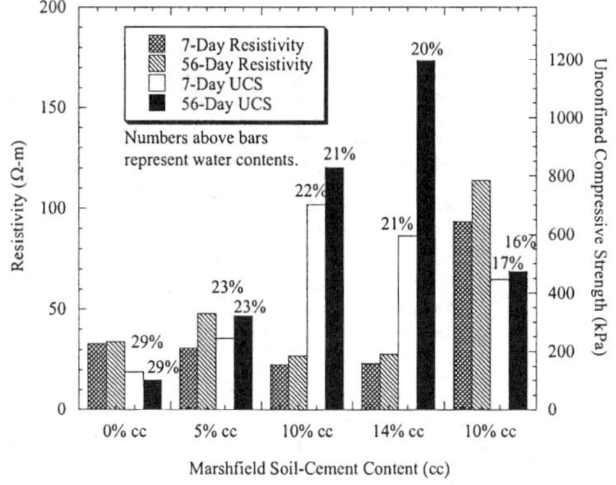

Fig. 8. Resistivity and unconfined compressive strength at 7-and 56-days for Marshfield soil.

Kenosha soils had constant cement content, and there was a direct relationship between electrical resistivity and UCS. In addition, UCS increased as initial water content decreased. Cement content varied in Marshfield specimens, and strength increased with increasing cement content. There was a larger difference between 5% cement and 10% cement than between 10% cement and 14% samples. The 5% cement content showed little strength gain over time. As cement content increased, strength increased and resistivity decreased. Decreased resistivity with increasing cement content is probably due to the higher ionic strength of cement compared to soil. For each 5% increase in cement content for Marshfield soil, final water content decreased by approximately 1-2%. The change in measured resistivity was more going from 5% cement content to 10% than going from 10% to 14%. Also, the change in strength from 5% to 10% cement content was greater than from 10% to 14%, implying that additional cemetitous reactions became limited as cement content increased.

Figures 9 and 10 show the relationship between K_a and UCS. For cement-mixed samples, K_a generally decreased in time as UCS increased, although there does not appear to be a strong correlation between K_a and UCS. K_a is a function of soil type, water content, cement content, and time (chemical reactions) and is not fully understood. Consequently, K_a may not be a reliable strength indicator.

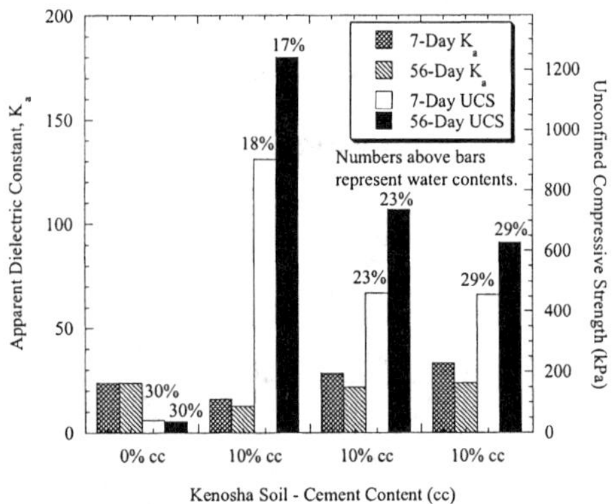

Fig. 9. K_a and unconfined compressive strength at 7-and 56-days for Kenosha soil.

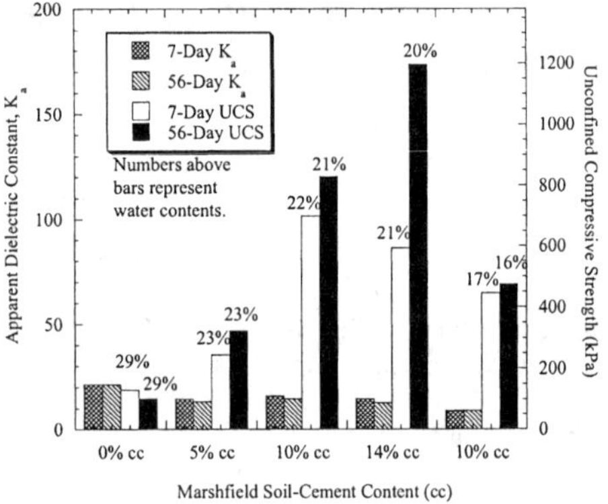

Fig.10. K_a and unconfined compressive strength at 7-and 56-days for Marshfield soil.

SUMMARY

This laboratory study investigated the electromagnetic (EM) properties of cement-mixed soils. Apparent dielectric constant and electrical resistivity were measured in two different soils mixed at different cement content and initial water content over a 56-day period to investigate how EM properties are affected by cement content, water content, and curing/hardening time. The EM properties depended on initial water content and cement content and changed as a result of hydration and hardening of cement and the subsequent reduction in water content and bond formation. Electrical resistivity increased in time, and dielectric constant decreased in time. Unconfined compressive strength (UCS) generally increased in time and appeared directly related to electrical resistivity. Increasing cement content increased strength but reduced resistivity since cement's resistivity was lower than the soils' resistivity. UCS generally increases as K_a decreases, although there does not appear to be a strong correlation between K_a and UCS. K_a appeared more suitable for measuring water content changes.

Acknowledgments

David Staab would like to thank the International Association of Foundation Drilling (ADSC) for a graduate research scholarship that aided in this project.

REFERENCES

Bassiouni, Z. Theory (1994), *Measurement, and Interpretation of Well Logs*, pp.92-128.

Bottcher, C.J.F. (1952), *Theory of Electric Polarisation.* Elsevier Publ. Co., Inc.

Benson, C.H., and Bosscher, P.J., "Time Domain Reflectometry (TDR) in Geotechnics: A Review" , *Nondestructive and Automated Testing for Soil and Rock Properties, ASTM STP 1350.* pp. 113-136.

Gibson, S., (1999), "Assessing Exploratory Borehole Seals With Electrical Geophysical Techniques," MS Thesis, University of Wisconsin-Madison.

Hime, W.G., (1999), "Chemical Methods of Analysis of Concrete", Handbook of Analytical Techniques in Concrete Science and Technology, pp.116-117.

Sharma, P.V. (1997) *Environmental and engineering geophysics*, pp.207-264.

Suwansawat, S., (1997), "Using Time Domain Reflectometry For Measuring Water Content in Compacted Clays," MS Thesis, University of Wisconsin- Madison.

Topp, G., Davis, J., and Annan, A. (1980), " Electromagnetic determination of soil water content," Water Resources Research, 16(3), 574-582.

Tamura, M., Hibino, S., Fujil M., Arai A, Sakai, Y., Kawamura, M., (2002), "Applicability of resistivity method for the quality evaluation of mechanical deep-mixing soil stabilizing method", 4th International conference on Ground Improvement techniques, 707-713.

ENGINEERING TOOLS FOR DESIGN OF EMBANKMENTS ON DEEP MIXED FOUNDATION SYSTEMS

Ali Porbaha [1] and Al Dimillio [2]

ABSTRACT: Construction of embankments on soft ground poses a challenge to the design engineer to address settlement and stability concerns. For this reason a wide range of technologies has been developed for support of embankment foundations on soft ground, including deep soil mixing. One barrier for implementation of these technologies is lack of engineering guides for feasibility study, preliminary evaluation, simplified analysis and design tools, quality assessment and monitoring of the constructed facility. This paper aims to presents several initiatives for developing engineering guides to address these deficiencies when deep mixing and geosynthetics is used for support of embankments. The initiatives include, synthesizing the-state-of-the practice in the US and overseas, developing design charts for geosynthetically reinforced embankments on deep mixed columns, evaluating in situ methods for quality assessment, and developing a simplified tool to handle uncertainties associated with interpretation of strength and stiffness of soil cement using reliability theory. This effort is expected to provide practitioners with engineering tools, advance the state of current practice, and enhance the user's confidence in applying deep mixing and geosynthetics system for embankment support.

Introduction

Embankment construction is an integral part of highway infrastructure system. Design engineers dealing with construction of embankments on soft ground need to take measures to address settlement and stability concerns, in addition to shear deformations caused by dynamic/seismic loads. One remedial measure is to transfer

[1] **Ali Porbaha** (corresponding author)
Department of Civil Engineering, California State University, Sacramento, 6000 J Street, Sacramento, California 95819-6029, Phone: (916)-278-6120, E-mail:porbaha@ecs.csus.edu

[2] **Al Dimillio**
Office of Infrastructure Research, Federal Highway Administration, 6300 Georgetown Pike, McLean, VA 22101, Phone: (202) 493-3035, E-mail: al.dimillio@fhwa.dot.gov

the embankment load to the bearing layers using vertical column elements (see Figure 1). These elements could be constructed with a variety of techniques based on the cemetation, pore pressure reduction and densification concepts. Deep mixing is one technique of ground modification in which soil is blended in situ with a cementitious material to produce vertical columns of treated columns using specially designed equipment with paddles and augers.

The National Deep Mixing (NDM) research program is a consortium of ten states, federal government and private sector with the mission to facilitate advancement and implementation of deep mixing through partnered research and technology transfer. An international workshop was organized in Oakland (Porbaha and Roblee, 2001) in which experts from industry, government and academics examined the barriers for implementation of deep mixing geo-support system. One area of concern was lack of design guides and design tools for practitioners.

The objective of this paper is to present several initiatives developed by the NDM program related to support of embankments on deep mixed foundation systems. The iniitatives include development of design guidelines, design manual, guide specification, review of international practice on in situ techniques to assess quality of treated soil, review of literature on current state of the practice for quality assessment, development of protocols for in situ techniques; develop criteria for non-compliance based on the reliability theory. The incentives for developing each initiative along with the objectives for each study are presented here.

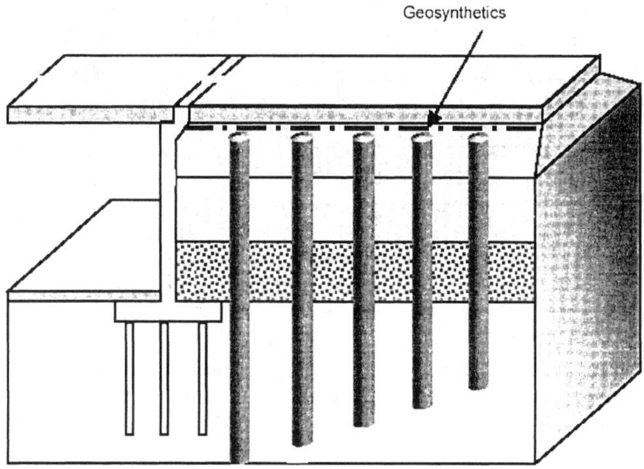

Figure 1: Application of deep mixing for bridge approach embankment

Synthesizing the state-of-the practice

Deep mixing technology has been used for over two decades to install stabilized soil elements to support embankments placed soft ground. Much of the early soft ground stabilization work for embankments was in Scandinavia, where dry quicklime was initially the major binder of choice. However, the current trend in Europe is to use cement and or a mixture of cement and lime, or just dry cement to achieve higher strength than provided by lime alone. In Asia, most notably in Japan, the deep mixing practice use mainly cement (both as slurry and dry) to stabilize soft soil. Both methods, wet and dry, however, apply the similar design principles, and the stiffness and strength of stabilized soil element by either method are becoming comparable.

In the early years, slope stability design methods were based on use of a composite averaged strength. However, ongoing studies in Scandinavian countries are demonstrating that individual deep mixed elements subjected to horizontal shear forces develop only small amounts of their potential full cross-sectional shear resistance. Current research and design application is applying overlapped elements in the form of shear wall buttresses to support side slopes of embankment fills. Individual deep mixed elements are used under the main body of the embankment to control settlement, which also relieves what would otherwise be considerable in-situ stress increase. The reduced stress increase therefore reduces horizontal forces that must be carried by rows of overlapped elements in buttress panels or shear walls. The horizontal loads to be resisted by the deep mix elements have also been subject to recent discussion. The classic sliding block mode with active and passive soil loadings has frequently been applied, but may not be appropriate when it is critical to limit horizontal displacements.

When the body of the embankment behind the slope is supported on treated columns to control settlement, then the driving forces can be substantially reduced by considering that only a fraction of the embankment load actually goes into the soft soils (mostly carried by the columns). Thus the buttress receives a much reduced lateral loading, and this must be considered in the analysis of horizontal stability. Several modes of instability must be assessed in the design of treated buttresses for stabilizing side slopes. These include global circular (and irregular) surfaces through the buttress, and the sliding block model with realistic application of expected lateral soil driving pressures and resisting forces.

Also important to design of embankments on deep mixed stabilized ground is the potential settlement of the embankment and effects of treated column spacing. Compressibility parameters appropriate to settlement analysis is also an important element to be incorporated into the embankment design synthesis.

A systematic synthesis of evaluation of the stability issues that incorporates the current state-of-the-practice and research is needed by the geotechnical community. Such a design synthesis is particularly important for the highway/transportation engineers who are being faced with less than desirable sites and compressed construction schedules that necessitate use of soil stabilization techniques such as deep mixing that can be installed and ready for use in short time frames. The analytical stage of design must be carried into construction via construction drawings and specifications. The degree of sophistication of the construction documents varies among projects, however, there are minimum requirements that must always be incorporated into plans and specifications. Then there are other details and requirements that may only be needed on certain types of projects or for especially challenging site conditions. Along with the design, there must be performance confirmation, which will necessitate instrumentation installation and monitoring. In some instances, initial test embankments may be warranted to give insight as to actual performance, thus verifying the design and demonstrating that it is buildable.

Here are several philosophical and practical issues related to design of deep mixing for embankment support to be addressed:

- What alternate technologies or foundation systems (shallow/deep) are available to support embankments? And what are their advantages and limitations compared to deep mixing?
- How is the design of deep mixing for embankment support over soft ground different from design using these other support methods?
- In what ground conditions and site situations is the deep mixing support solution viable?
- In what situations should deep mixing not be used for embankment support (based both on constructibility and in-situ performance)?
- In what situations should buttresses be included in the embankment foundation as opposed to simply using individual column elements?
- What are the design considerations for soil-cement wall stiffness/flexibility, and how is it specified (if needed in the deisgn)? Also, how is the wall brittleness (post-yield behavior) addressed in design considerations?
- How is stability affected when soil-cement columns are not installed down to the bearing stratum?
- How are the different soil-cement failure modes considered and addressed in design?

Accordingly, the main objective is to provide highway/transportation design engineers, and others in geotechnical practice with a design guide manual for using deep mixing to stabilize soft soils to support side slopes of embankments, and greatly reduce settlement of the embankment. As part of this design synthesis, the steps of initial feasibility assessment with preliminary cost estimate will be provided in detail outline to enable early go- or no-go decision making regarding applicability of deep mixing for embankment support. The current procedures in design/analysis methods (including recommended practice for determining appropriate loadings) slope and sliding block stability, and settlement calculation procedures will be presented. Also critical to the design synthesis is guidance on selection of appropriate treated column properties and parameters. Both published literature on these topics, and interviews and other correspondence with practitioners will serve as the basis for developing the state-of-the-practice design guideline.

Developing design charts

Geosynthetics (such as geogrids and geotextiles, made of polymer materials) as reinforcement have been adopted to stabilize soft foundations, slopes and embankments. Geosyntehtics have a high tensile capacity that soils do not have. The benefits associated with the use of geosynthetics for embankments over soft soils are to increase bearing capacity, reduce differential settlement, and prevent global instability.

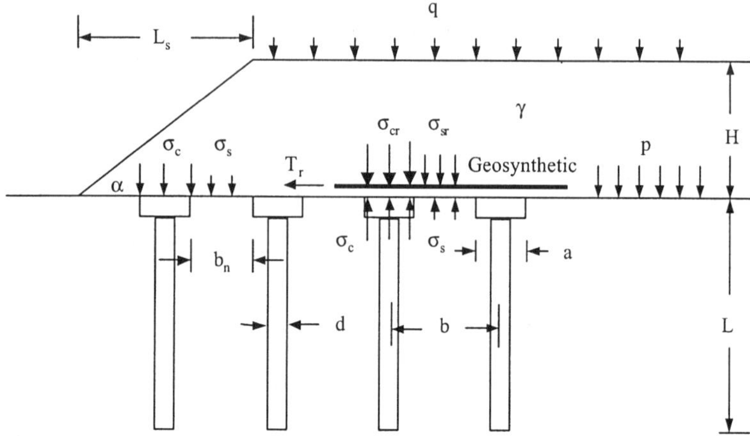

Figure 2.:Several design parameters used for analysis of column-supported embankment (Han et al 2003)

The bearing capacity of deep mixed columns can be significantly reduced when the columns are subjected to lateral forces from the embankments. The deflection between the deep mixed columns may be reflected to the ground surface, if the columns are spaced at a larger distance. Many studies have shown that geosynthetics can be used for reducing differential settlements, however, they have limited contributions in reducing total settlements of embankments over soft soils (e.g., Han and Gabr, 2002).

In a column supported embankment the stiffened earth platforms can span weak soils and prevent deflection between deep mixed columns from being reflected to the ground surface. Since deep mixed columns have large diameters and necessary strength and stiffness, when combined with geosynthetics, they are ideal for constituting a composite foundation system for embankment support. The inclusion of geosynthetic layers can reduce the lateral forces applied onto the columns, transfer more vertical loads onto the columns and reduce the vertical loads applied onto the soft soil, and minimize the deflection to be reflected to the ground surface. The deep mixed geosynthetic system can also increase the spacing between treated columns. These benefits, along with reduced construction time, can provide a cost-effective solution for support of embankment on compressible ground.

The deep mixing and geosynthetic system has been successfully used in Japan to support a relatively low (1.5m) embankment on a soft subgrade. Only 11% coverage of soil-cement columns were used in conjunction with one layer geogrid. This is in comparison to the 50 to 70% coverage of pile caps that would be required in accordance to Rathmayer (1975). Deep mixing columns and geosynthetics were also used in China to support embankments approaching bridges (Lin and Wong 1999). Numerical analyses are available to investigate load transfer mechanisms between the columns and the soil (Han and Gabr 2002). Today, however, no design guideline exists to assist highway and geotechnical engineers to apply deep mixing and geosynthetics to embankment construction over soft soils. Therefore, there is a need for developing rational design guides and design charts for this emerging composite foundation system.

Accordingly, the prime objectives are to develop charts and guidelines for design of deep mixing and geosynthetics for embankment support to reduce settlement and to maintain stability; and to provide step-by-step examples to use charts for design of deep mixing and geosynthetics composite system. The design approach/charts will include considerations required for design of geosynthetic and treated columns to address settlement of the composite system and stability of the side slopes.

The results of this study is expected to advance the knowledge and user's confidence in applying deep mixing and geosynthetic system for embankment support by providing highway and geotechnical engineers with design methods/charts and guidelines necessary to apply this composite system. This research is also anticipated to be helpful to deep mixing and geosynthetic industries to seek more business opportunities (Han et al, 2003).

Quality assessment

Quality assurance (QA) of deep mixing methods used in the construction projects must be addressed to evaluate the effectiveness of treatment methods adapted in the field. Hence, selection of QA is essential in the preliminary stage of the project. Typically, QA methods employ either core sampling of DM cement columns followed by laboratory testing on the core samples, and/or field-tests utilizing load tests, intrusive and non-intrusive in situ test methods for direct and indirect assessments of soil improvement effects.

Laboratory tests are influenced by the sample disturbance and simplified boundary conditions. Cracks or microcracks occur in the cores during sampling due to a variety of reasons, such as bend in the bore hole, rigidity of the sampler, locking of the sampler, and rotation of the sampling core with the sampler. Additionally, laboratory tests are time consuming and provide verification only at discrete points, and may not incorporate localized weak zones along the treated soil column. Because of these reasons, several types of old and new in situ techniques have been developed and used to investigate the in situ quality of the treated ground. There is an extensive literature available on various in situ methods and their methodologies for QA studies on DM columns. One of the limitations of the information collected from literature is the lack of agreements on conclusions of various in situ tests used in QA evaluation studies conducted across the world. In addition, several novel in situ methods have been exclusively developed in Europe and Japan for quality assessment studies and these methods are not well known to practitioners and researchers from other parts of the world.

Figure 3: Reverse column penetrometer used in Europe
(Courtesy of Hakan Erickson, Hercules Grundläggning, Sweden)

Several non-destructive and destructive field test methods are utilized to conduct rapid quality assessment studies. In US, these in-situ practices are still new and are not frequently used in the DM based construction projects. Other reasons for not using are due to lack of knowledge base on the field tests that have been successfully used elsewhere in the world. Hence, it is necessary to develop a state-of-the-art report on in situ technologies and their applications in quality assessments of DM techniques.

The present research is an attempt to address and accomplish this research need by conducting extensive literature reviews, surveying DM practitioners and users in both U.S. and abroad, conducting interviews to learn various in situ methods used for QA studies, and studying advantages and limitations of the in situ methods. The main intent is to provide knowledge base on various in situ methods (Standard Penetration Test, Cone Penetration Test, Dynamic Cone Penetrometer, Pressuremeter, Vane Shear Test, Rotary Penetration Test, Static Dynamic Penetration Test, Conventional Column Penetration Test, Reverse Column Penetration Test, as well as three geophysical tests) to practitioners and provide directions for the selection and execution of in situ methods for QA studies in DM projects (Puppala et al. 2003).

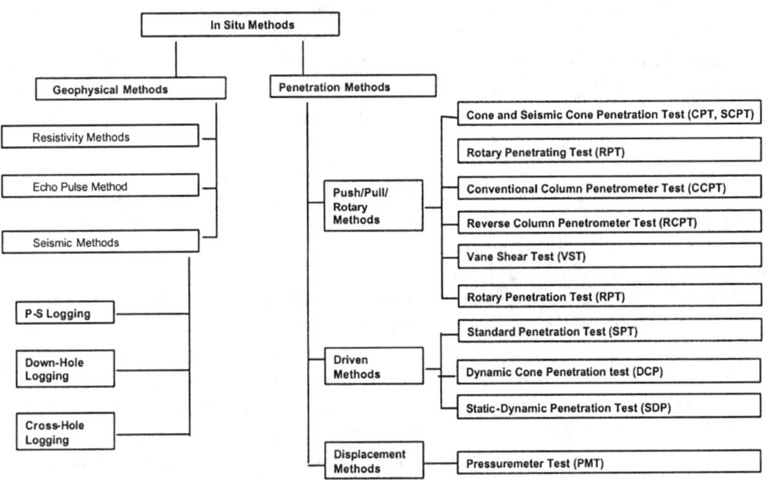

Figure 4: In situ methods for quality assessment of deep mixed columns

Handling uncertainty through reliability theory

The most common laboratory test for quality assessment of deep mixing is unconfined compression test. This is an inexpensive type of test, but the lack of confinement produces a bias towards laboratory strengths that are lower than field strengths. Furthermore, in the case of core samples, the strength determination process may be biased by at least two other conflicting factors: (1) strengths from core samples may underrepresent in situ strengths because of potential damage to specimens during coring, and (2) only intact core pieces that are sufficiently long can be tested, which may bias the results on the high side.

The measured strength of soil-cement mixtures depends not only on the sampling and testing method, but also on such factors as the water-cement ratio of the slurry, the cement factor (i.e., the dose rate of cement left in the ground), the characteristics of the soil, the type of mixing, the mixing energy, the curing time, the curing temperature, the consolidation pressure during curing, and the rate of field loading. Relevant soil characteristics include water content, particle-size distribution, mineralogy of fines, and amount and type of organics and other constituents.

A consequence of variability in these factors is that the measured strength and stiffness of soil-cement mixtures exhibit substantial variability that must be quantified to permit reliability-based design of foundation systems incorporating soil-cement columns. Values of the coefficients of variation of strength and stiffness are needed to permit calculating the reliability against collapse and against excessive deformation.

Furthermore, construction specifications must be established to provide assurance that the intent of the design will be achieved during construction. This is a critical point because of its impact on performance and construction cost. If specifications are written such that all strengths measured during QA/QC must meet or exceed the design strength, the average strength will be very high and construction costs will also be high. If specifications are written such that the average strength meets or exceeds the design strength, but without specification controls on variability, low strength values can result in a higher probability of failure than intended in design.

As one example, soil-cement column foundations were designed for several embankments at the I-95/Route 1 interchange project in Alexandria, Virginia in which the probability of embankment side slope failure was calculated to be less than one percent using the procedures in Duncan (2000). These calculations were performed using an average soil-cement strength of 160 psi (1100 kPa) and a standard deviation of 37 psi (255 kPa), which corresponds to a coefficient of variation equal to 23 percent. The construction specifications for the project are based on unconfined compression tests performed on core samples. For a given column, the specifications require that the average unconfined compression strength must not be less than 160 psi (1100 kPa) and that the minimum strength not be less than 100 psi (690 kPa). For the purpose of computing the average strength, the measured strength values are capped at 220 psi (1520 kPa). This specification has the advantage of simplicity, and it allows individual columns to be accepted or rejected. However, this specification does not have a direct relationship to the probability-of-failure calculations performed in design. For example, a set of five core test results with measured strengths of 120, 120, 120, 220, and 220 psi would satisfy the specification while having a standard deviation of 55 psi when calculated as a representative sample of strength tests. Such a set of strength test results demonstrates that this specification does not necessarily achieve the project's design intent. This example demonstrates that practical procedures are needed for reliability-based design and for writing construction specifications that correspond to the design intent.

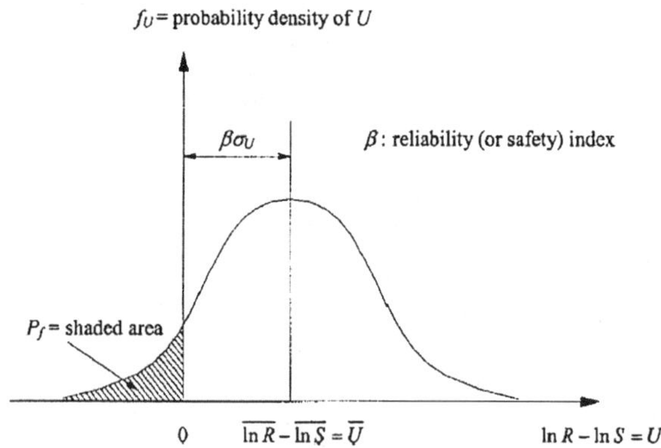

Figure 5: Load effects, resistance, and reliability (Scott et al., 2003)

It is worthwhile to provide a small amount of background and definition of terms for reliability assessment. The probability of failure, P_f, is the probability that unsatisfactory performance, such as instability or excessive total or differential settlement, will occur. The reliability, R, is the probability that unsatisfactory performance will not occur. Thus, $P_f + R = 1$. The reliability index, β, is another normalized measure of reliability. As an example, if the application concerns slope stability as evaluated by the factor of safety, F, then β?is defined according to $\beta = (F_m - 1)/F_\sigma$, where F_m = the mean value of F, F_σ= the standard deviation of F, and the value of F corresponding to unsatisfactory performance is one. As β?increases, P_f decreases and R increases. The exact relationship between β?and P_f depends on the type of statistical distribution. As discussed by Scott et al. (2003) and others, a lognormal distribution is appropriate for many of the quantities calculated in geotechnical engineering, such as settlement and factor of safety against instability. For a lognormal distribution, β?is related to P_f and R as shown in the following table, which is adapted from the U.S. Army Corps of Engineers (1998):

Table 1: Performance and reliability

Performance Level	β	P_f	R
High	5	0.0000003	0.9999997
Good	4	0.00003	0.99997
Above Average	3	0.001	0.999
Below Average	2.5	0.006	0.994
Poor	2	0.023	0.977
Unsatisfactory	1.5	0.07	0.93
Hazardous	1	0.16	0.84

The overall goal is to provide simple and practical means for design engineers to incorporate reliability concepts in design and construction of deep mixing projects. The goal is not to replace deterministic design approaches, e.g., requiring a factor of safety for slope stability greater than, say, 1.3 for an embankment supported on deep-mixing-method columns. Instead, the reliability-based design will supplement deterministic design. In summary, the principal objectives are to (1) compile databases of strength and stiffness measurements; (2) determine the best fitting type of distribution and the mean, standard deviation, and coefficient of variation for each data set; (3) attempt to determine correlations among values of strength and stiffness obtained from laboratory mixed samples, grab samples, core samples, and in situ tests; (4) develop reliability-based design examples for settlement and stability of embankments using the method in Duncan (2000); (5) develop recommendations for construction specifications that meet the design intent without incurring excessive construction costs. One of the deliverables of this project is to provide the owner with a simplified tool to use reliability analyses during construction to determine the impact on probability of failure of QA/QC test results that do not meet specifications (Filtz et al, 2003).

Concluding remarks

Implementation of new technologies requires users confidence in the quality of the final product and the availability of engineering tools for analysis and design, among others. Deep mixing technology, developed primarily overseas, is gaining acceptance in the US as evidenced by the increased number of deep mixing projects executed in the last five years,. Challenges to construct facilities on soft ground and timely delivery of the projects are the main reasons to use deep mixing. The initiatives of the National Deep Mixing (*www.deepmixing.org*) research program are aimed to assist owners, contractors and design engineers when deep mixing is used as a geo-support system. The initiatives presented here are expected to provide the basic tools for geotechnical and highway engineers when deep mixing geosynthetic system is used for support of embankment construction projects.

Acknowledgements

The authors are grateful to (alphabetically) George Filz, Jie Han, Jim Lambrechts, and Anand Puppala, among others, for their contribution to the NDM initiatives presented here. Appreciation is extended to the steering committee members, the project advisory panel members, independent reviewers, and to the administrative support. Photo is the courtesy of Hakan Erickson, Hercules Grundläggning, Sweden.

References

Amano, H., Morita, T., Tsukada, H. and Takashi, M. (1986) Design of deep mixing method for high road embankment, Hayashima I.C.-: Proc. of the 21th JSSMFE: 1999-2002 (in Japanese).

Duncan J. M. (2000) Factors of safety and reliability in geotechnical engineering, Journal of Geotechnical and Geoenvironmental Eng., Vol. 126, No.4, April 307-316.

Filtz, G. et al. (2003) Simplified reliability-based method for interpretation of strength and stiffness of soil cement, NDM research in progress.

Han, J. and Gabr, M.A. (2002) A numerical study of load transfer mechanisms in geosynthetic reinforced and pile supported embankments over soft soil. Journal of Geotechnical and Geoenvironmental Engineering, ASCE, 128(1), 44-53.

Han, J. et al. (2003) Development of design chats for geosynthetically reinforced embankments on deep mixed columns, NDM research in progress.

Lambrechts J., et al (2003) Manual for design of embankments on deep mixed columns, NDM research in progress.

Lin Q. K., and Wong H., I. (1999) Use of deep cement mixing to reduce settlement at Bridge Approaches." *Journal of Geotechnical and Environmental Engineering*, American Society of Civil Engineers, April, pp. 309-320.

Porbaha, A. and Roblee, C. (2001) Barriers for implementation of deep mixing technology in the US, Proceedings of international workshop for deep mixing, published by NDM program, Oakland, California, July, 2 Volumes.

Puppala, A. et al. (2003) In situ methods for quality assessment of deep mixed columns, NDM research in progress.

Rathmayer, H. (1975) Piled embankment supported by single pile caps. Proceedings, Istanbul C onference on Soil Mechanics and Foundation Engineering, 283-290.

Scott, B., Kim, B.J., and Salgado, R. (2003) Assessment of current load factors for use of geotechnical load and resistance factor design, Journal of Geotechnical and Geoenvironmental Engineering, Vol. 129, No. 4, 287-295.

US Army Corps of Engineers (1998) Risk-based analysis in geotechnical engineering for support of planning studies, Eng. Circular No. 1110-2-5554, Department of the Army, Washington, D.C.

SIMPLIFIED PROCEDURE FOR SEISMIC ANALYSIS OF SITES WITH DEEP SOIL MIXING

Raj V. Siddharthan[1], Member, Geo-Institute, Magdy El-Desouky[2], and Ali Porbaha[3], Member, ASCE

ABSTRACT: The paper reports the results of a parametric study undertaken to investigate the influence of deep mixing (DM) of soil columns on the porewater pressure response in the vicinity of the treatment. The study utilized a two-dimensional effective stress program, TARA-2M. The predictive capability of TARA-2M was initially verified by using a well-documented centrifuge database generated as a part of the NSF funded study VELACS (Arulanandan and Scott 1993). The porewater pressure responses have been provided for two levels of excitation (a_{max}= 0.2g and 0.4g) and for two DM column spacing distances. Computed results show the extent of the influence of the treated columns in reducing the excess porewater pressures. The results are presented in normalized form such that the estimates of porewater pressures at interior locations can be made, if porewater pressure response is known in the free-field.

INTRODUCTION

The remedial solutions that meet design requirements on poor quality ground are accomplished by either improving the compressible soil found near the surface or by installing deep foundations. In many cases, where issues such as installation noise and bearing strata found at much deeper location, foundation ground improvement methods often become more attractive. One such method of ground improvement is deep mixing (DM), which is categorized as a solidification technique. Deep mixing

---

[1] Professor, Department of Civil Engineering, University of Nevada Reno, Reno, NV 89557, siddhart@unr.edu

[2] Graduate Student, Department of Civil Engineering, University of Nevada Reno, Reno, NV 89557

[3] Assistant Professor, Department of Civil Engineering, California State University, 6000 J St., Sacramento, CA 95819, porbahaa@ecs.csus.edu

is also known as deep soil mixing but the later designation has been avoided in the recent times because of copyright issues (Porbaha 1999). The DM methodology has been evolving over the last three decades and extensive research has been undertaken to gain insight into different aspects of DM. Many important design issues such as appropriate construction methods and their extent of applicability (e.g. soft saturated ground), laboratory and field material characterization, and full-scale field demonstration projects have been undertaken. Many details on this technique, including its historical developments, applicability, and design have been well documented by Porbaha (1998), Porbaha *et al.* (1998), Porbaha *et al.* (1999) and O'Rourke *et al.* (1997), among others.

The most common reason for poor performance of foundations under seismic loading has been the loss of strength and stiffness of saturated foundation soil caused by liquefaction. It is a phenomenon that is associated with the behavior of saturated loose to medium dense cohesionless soils subjected to repeated loading. Such soils under earthquake loading give rise to excess porewater pressures u_{ex} (in excess of static) and in level ground when u_{ex} becomes equal to the initial vertical effective stress, the soil losses all its strength (i.e. liquefaction). The DM treated soils provide higher resistance relative to two important seismic soil performance/response parameters, namely liquefaction and settlement. These performance parameters have to be evaluated for improved and unimproved (original) soil masses to ascertain the effectiveness of the improvement.

The DM treatment generally involves a rectangular grid (or lattice) pattern and the design dimensions such as cell width b, and thickness, d of treatment and depth of treatment (D) need to be specified to achieve a desired level of improvement (Fig. 1). These design dimensions are often controlled by many site-specific issues that include design level of excitation, existing untreated soil layering and properties, equipment to be used with DM, thickness of liquefiable layers, areal extent of treatment etc. A verified analytical procedure that is flexible enough to accommodate these variables is necessary to investigate many options before arriving at a set of optimum design dimensions (e.g. spacing, s; and thickness, d etc.) for the configuration of the DM treatment.

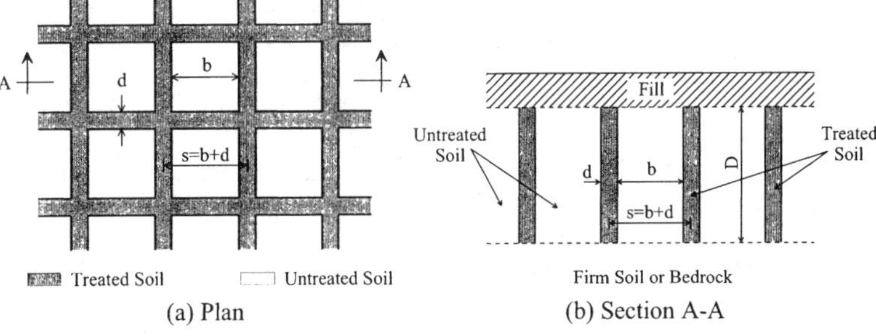

Fig. 1. Typical configuration of Deep Mixing

The paper reports the results of a parametric study that was undertaken as a part of the development of a simplified design procedure to aid the designers in the selection of optimum DM treatment plan (design dimensions). The study utilized a two-dimensional effective stress program (TARA-2M) to study the behavior of soils adjacent to the DM treated soil columns. The predictive capability of TARA-2M was initially verified by using a well-documented centrifuge database generated as a part of a NSF funded study - *V*erification of *L*iquefaction *A*nalysis by *C*entrifuge *S*tudies (VELACS) (Arulanandan and Scott 1993). Subsequently, a parametric study that focused on the investigation of the development of excess porewater pressure in the presence of DM soil columns is presented.

BRIEF DESCRIPTION OF ANALYTICAL MODEL

The methodology and application of TARA family of computer programs have been reported in the literature (Finn 1988; Siddharthan and Norris 1988, 1990). In these publications, the applicability of the TARA to a variety of problems under liquefying soil conditions has been documented. The problems reported include buried heavy structures simulating nuclear power plants, dams (e.g. the Sardis Dam in Mississippi), and rigid surface foundations. Therefore, only a brief description of the methodology is presented below.

This method is based on the finite element method, and solutions to the dynamic equilibrium equations are obtained in the time domain. It is basically an extension of the method of nonlinear dynamic effective stress analysis (DESRA2) developed by Finn *et al.* (1977) for level ground conditions. The soil response is modeled by combining the effects of shear and normal stresses. While extending the one-dimensional method to two dimensions (plane strain), an additional material parameter is necessary. The tangent bulk modulus or Poisson's ratio can be selected for this purpose. Soil behavior in relation to changes in effective mean normal stresses has been taken to be nonlinear and effective stress dependent, but essentially elastic, compared to shear response.

The original version of the program (TARA-2) has undergone a number of modifications at the University of British Columbia (UBC) and at the University of Nevada, Reno. Modifications to the original program were such that some of the recently available nonlinear soil behavior data can readily be incorporated. The main changes to the original TARA-2 include new characterization for stress-strain and volumetric-strain relationships, since better characterizations have been proposed (EPRI 1993; Byrne 1991) and allowance for dissipation of porewater pressure. The modifications have been chosen such that only a minimum number of soil parameters are used in the model, while retaining many of the convenient features of the original model.

The new shear stress-strain equation for initial loading in TARA-2M is defined as,

$$G = \frac{\tau}{\gamma} = \frac{G_{max}}{1 + \left| \frac{\gamma}{\gamma_y} \right|^n} \qquad (1)$$

in which τ and γ are shear stress and strain; G_{max} is the shear modulus at very low strain level; γ_y is the reference strain, defined as τ_{max}/G_{max}; and n is a constant. It may be noted that in the original model, n was set to equal unity. Studies by Nakagawa and Soga (1995) and Ni *et al.* (1997) also proposed this type of variation based on a large database of laboratory soil behavior. The subsequent unloading and reloading are given by Masing stress-strain curves (Masing 1926). For cohesionless soils, the G_{max} depends only on the relative density and it is given by,

$$G_{max} = 218.8(K_2)_{max}(\sigma'_m)^{1/2} \qquad (2)$$

in which σ'_m is the effective mean normal stress; and $(K_2)_{max}$ is a constant that depends on the relative density (Seed and Idriss 1970). Here G_{max} and σ'_m are given in kPa.

During the strong shaking, the effective stresses decrease because of excess porewater pressure generation. The increment in excess (or residual) porewater pressure, Δu_{ex}, is evaluated using the porewater pressure model of Martin *et al.* (1975), given by,

$$\Delta u_{ex} = E_r \Delta \varepsilon_{vd} \qquad (3)$$

in which $\Delta\varepsilon_{vd}$ is the increment in volumetric compaction strain (given in %), and E_r is the one-dimensional rebound modulus. The $\Delta\varepsilon_{vd}$ is a function of accumulated volumetric strain ε_{vd} and shear strain γ. Martin *et al.* (1975) used four coefficients to evaluate $\Delta\varepsilon_{vd}$. Byrne (1991) provided a new, much simpler relation,

$$\frac{\Delta\varepsilon_{vd}}{\gamma} = c_1 \exp\left(-c_2 \frac{\varepsilon_{vd}}{\gamma}\right) \qquad (4)$$

in which c_1, c_2 are constants that depend on the relative density of the sand. The rebound modulus E_r is a function of effective stress level σ'_v, and the relation given by Byrne (1991) can be written as,

$$E_r = \frac{(\sigma'_v)^{1/2}}{K_r} \qquad (5)$$

in which K_r is an experimental constants for a given sand.

These equations are solved numerically in conjunction with the dynamic equations of motion using a step-by-step integration procedure in the time domain. More details on the procedure may be found in Siddharthan and Norris (1998, 1990) and Finn (1988).

SELECTION OF MATERIAL PROPERTIES FOR TARA-2M

Before the computation of the dynamic response, a static analysis is performed with TARA-2M to estimate the in-situ static stresses. The procedures adopted are similar to those outlined by Duncan and Chang (1970), in which a hyperbolic relationship is used. Layer by layer construction can also be simulated. The procedures to obtain the material parameters required to do this evaluation can be easily estimated from static monotonic drained triaxial test results (Duncan *et al.* 1980).

The material properties that are required for the computation of the dynamic response are evaluated using the procedures outlined below. Traditionally, the nonlinear behavior of soil is defined by two strain-dependent soil parameters: normalized shear modulus ratio, G/G_{max}, and damping ratio, ζ (Seed and Idriss 1970). There is a large database for these parameters for many types of soil. However, a vast majority of these laboratory tests used to estimate these parameters were obtained from tests conducted at a confining pressure range of 100 to 200 kPa. Therefore, the applicability of these parameters is limited to shallow deposits of depth up to 10 to 20 m or so.

The EPRI has recommended the use of depth (or stress-level) -dependent soil properties (G/G_{max} and ζ) in the evaluation of deep soil response (EPRI 1993). This recommendation was arrived at based on many laboratory tests carried out under low and high confining pressures. Fig. 2 shows the EPRI recommendations for sand at different depths as a function of shear strain (EPRI 1993). It can be seen that as the depth of soil increases, the values of G/G_{max} and ζ for a given strain level increases and decreases, respectively. In other words, the soil elements under large confining pressure have lower damping and do not exhibit a strong nonlinear behavior.

The soil parameter γ_y and n in Eq. 1 have been evaluated such that it provides a best fit to the G/G_{max} and ζ variations given in Fig. 2. The Masing criteria that define the unloading and reloading gives the area enclosed by a strain cycle, which is proportional to ζ. An optimization technique has been utilized to estimate different sets of γ_y and n values for two depths (or stress level). The predicted G/G_{max} and ζ values along with optimum values of γ_y and n are also provided in Fig. 2. It may be noted from Fig. 2 that there is a certain damping ζ_o at the low (10^{-4}%) strain level. This damping amount is considered not to be due to hysteretic soil behavior and therefore was subtracted from EPRI data in the optimization. This damping, which is around 0.6% to 1.5%, is quite small and is incorporated through the damping matrix in the dynamic equilibrium equations (Finn *et al.*, 1977). The figure reveals that the predicted values provide a good fit to EPRI data. These two sets of stress-strain relationships were used to characterize the dynamic shear stress-strain relationship.

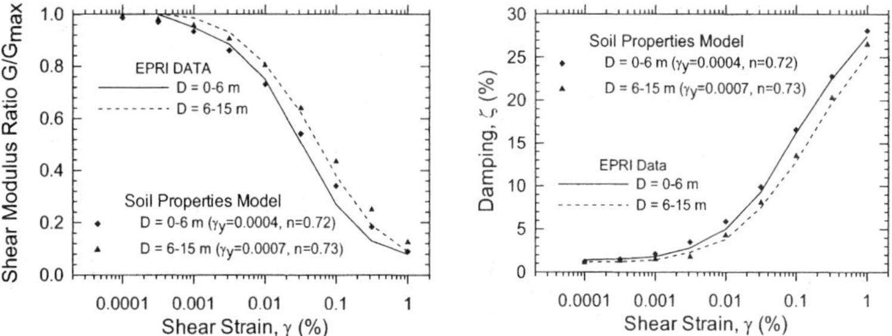

Fig. 2. EPRI recommended material properties G/G_{max} and ζ for cohesionless soils

Other important model parameters are those that define the volume change and porewater pressure generation behavior. The volume change behavior (constants c_1 and c_2) has been extensively studied by Byrne (1991) and he recommended values as a function of the relative density of soil. A convenient way of obtaining porewater pressure model parameters is to match a specified liquefaction potential curve with the one predicted by the porewater pressure generation model used in the approach (Finn *et al.* 1982). In using the TARA-2M model, it is customary to select model constant K_r (equn. 5) such that there is a close match between the predicted and specified liquefaction potential curves.

VALIDATION OF APPLICABILITY OF TARA-2M MODEL

As pointed out earlier, the first part of the paper is devoted to validating the applicability of TARA-2M model to predict saturated soil response. Results from centrifuge tests, which were made available to VELACS project participants, were selected for this purpose. Nevada sand was used in the tests and an extensive database on this sand under static and dynamic testing conditions are available (Arulmoli *et al.* 1992). As a part of the VELACS project, centrifuge models (Model 1) with a prototype thickness of 10 m instrumented to measure porewater pressure, accelerations, and horizontal and vertical displacements were tested at many institutions around the country. The test selected here was carried out at RPI (Model 1) under a centrifugal acceleration, N = 50g (Fig. 3). The centrifuge container is a laminar box, and the test model was subjected mainly to horizontal base shaking of as many as 20 cycles with a maximum acceleration of $a_{max} = 0.23$g (Fig. 4). The sand used was water-saturated Nevada sand with a relative density of $D_r = 45\%$. As a first approximation, the response of the soil deposit may be assumed to be a one-dimensional response (lateral) caused essentially by vertically propagating shear waves. Under this assumption, no vertical accelerations are assumed to occur. However, a large settlement in sand which is the result of plastic compaction (volume change) in the sand may be expected. More details on the sample preparation, test procedure, and test results and their interpretation are presented in Taboada and Dobry (1993).

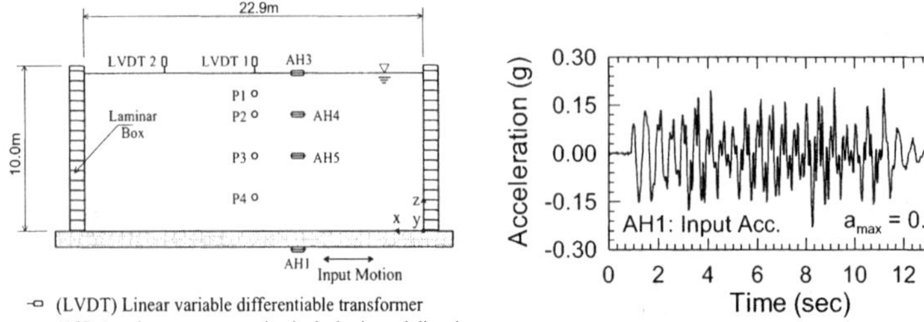

Fig. 3. Test setup and instrumentation in RPI test - Model 1

Fig. 4. Input acceleration history used in RPI test - Model 1

The constant $(K_2)_{max}$ interpreted from resonance column tests on Nevada sand is 35.8 (Arulmoli *et al.* 1992). Recommended values for c_1 and c_2 are 0.56 and 0.71, respectively, for a sand deposited at a relative density $D_r = 45\%$ (Byrne 1991). The rebound modulus constant K_r was obtained by matching the liquefaction potential curve as shown in Fig. 5. The liquefaction potential curve for $D_r = 45\%$ was interpolated from the test data provided by Arulmoli *et al.* (1992) for $D_r = 40$ and 60%.

Though many computed responses such as histories of acceleration, settlement, porewater pressure can be used in the validation, only the porewater pressure responses computed by TARA-2M are presented here. This is because the focus of this study is limited to porewater pressure prediction in soil with and without soil improvement. The computed porewater pressure histories at four locations (see Fig. 3) in Model 1 are presented in Figs. 6a through 6d along with those measured in the test. The

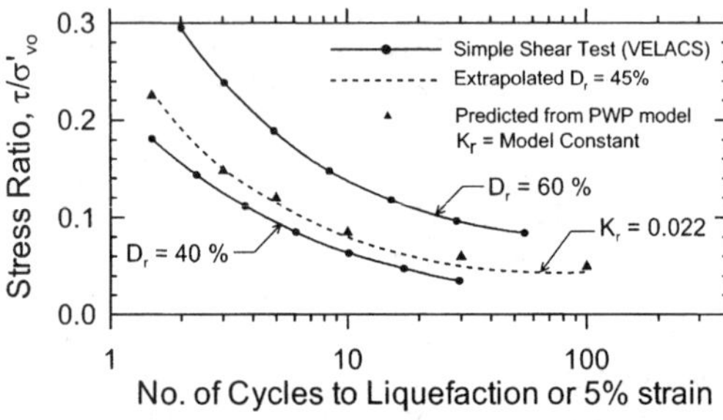

Fig. 5. Matching of liquefaction potential curve for Nevada sand ($D_r = 45\%$)

predictive capability of TARA-2M is very good. The plots clearly show that the liquefaction ($u_{ex}/\sigma'_{vo} = 1$) in the deposit is initiated quite early in the excitation and subsequently, total liquefaction of the deposit occurred.

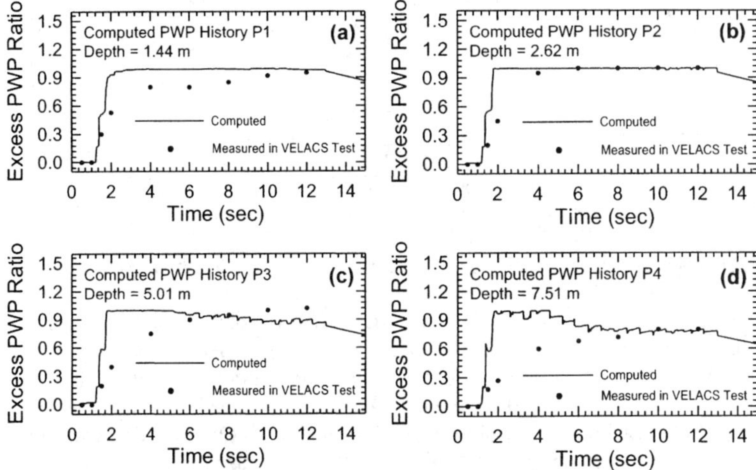

Figs. 6a-6d. Computed and measured excess porewater pressure histories

A closer look at the computed and measured porewater pressure responses reveals that though both indicate liquefaction, the computed responses predict occurrence of liquefaction slightly sooner. This may be attributed to a complication that occurs when liquefaction is initiated. Scott (1986) pointed out that the liquefied layers undergo a combined process of solidification and consolidation, which makes the characterization of the liquefied soil very difficult. The VELACS laboratory data provided by Arulmoli *et al.* (1992) does not provide any information to characterize the liquefied soil or soil at low effective stresses. Though comparison of acceleration histories are not reported here, these results show a substantial reduction in acceleration at the surface after the initiation of liquefaction.

APPLICATION TO DEEP MIX FIELD CASES

The study reported above lends credibility to the applicability of the TARA-2M model to undertake porewater pressure response of field cases that contain DM soil. Fig. 7 shows a typical DM field configuration in which the depth D to the firm soil is 10 m, the untreated soil has a $D_r = 40\%$, the thickness of treatment d = 0.9m, and the water table is at the top of the DM treated soil. A fill of height 1 m is provided at the top of the existing soil. Two field cases in which cell width b = 9d (Case 1) and b = 3d (Case 2) have been reported.

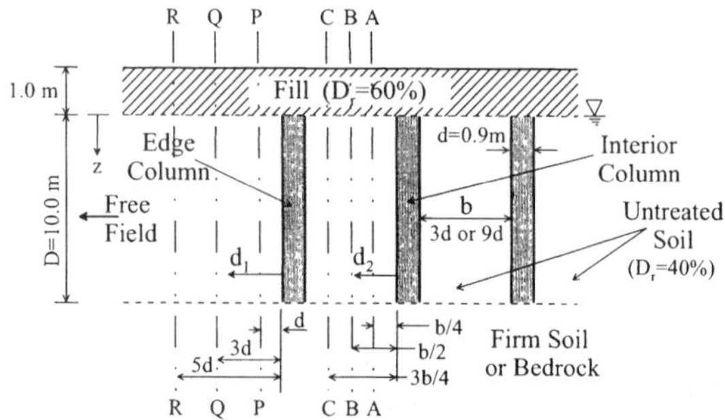

Fig. 7. DM configuration considered in the field case study

The material properties chosen for use with TARA-2M were obtained by utilizing the similar procedure outlined in the validation study above. The liquefaction potential data for D_r = 40% (N_1 = 7.1) was deduced from the widely-used field liquefaction database provided by Youd *et al.* (2001) in their state-of-practice of liquefaction evaluation for level ground. The properties of DM soil columns are much stiffer (in excess of 10 times the soil) and were obtained from measured data provided by Shibuya *et al.* (1992) and Probaha *et al.* (2000). The procedure outlined above for cohesionless soils (see Fig. 2) was used to obtain the model constants γ_y and n for DM treated soil and the corresponding values are $16x10^{-4}$ and 0.62, respectively. Table 1 gives all material constants used with TARA-2M.

The base motion used in the study is a motion recorded at Rio Dell, California in the Petrolia earthquake (M = 7.0) of 1992. This motion was scaled to yield 0.2g and 0.4g, respectively and the scaled motions were applied at the base of the DM columns.

The maximum computed porewater pressures for both cases (Case 1 and Case 2) along three vertical sections located between the free-field and the treated edge

Table 1. Soil parameters used in the analysis of field cases

Parameter	Untreated Soil (D_r = 40%)	Treated Soil	Fill Material (D_r = 60%)
Unit Weight of Soil,γ (kN/m^3)	18.0	18.5	18.5
Skeleton Curve Constants (γ_y, n)	$4x10^{-4}$, 0.72	$16x10^{-4}$, 0.62	$4x10^{-4}$, 0.72
$(K_2)_{max}$	40.0	454.0	52.0
Volume Change Constants (c_1 and c_2)	0.75, 0.53	0.0, 0.0	0.24, 1.66
Rebound Modulus Constant, K_r	0.03	----*	----*

* No excess porewater pressure develops

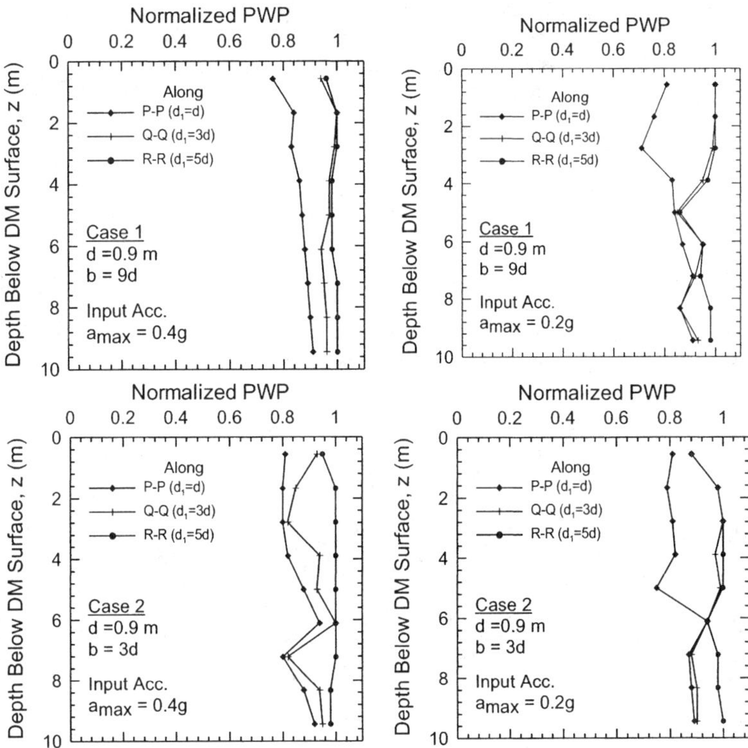

Figs. 8a-8d. Porewater pressure ratios with depth between edge column and free-field

column are presented in Figs. 8a through 8d. These vertical sections are placed (see Fig. 7) at $d_1 = d$ (Section P-P), $d_1 = 3d$ (Section Q-Q) and $d_1 = 5d$ (Section R-R). The computed porewater pressure responses shown in the figures have been normalized by dividing by the corresponding computed porewater pressure at the same horizontal level in the free-field. This way of presentation was adopted because these normalized ratios can be subsequently used to obtain porewater pressure response at any interior location, if the free-field response is known. When the maximum base acceleration $a_{max} = 0.4g$, a substantial liquefaction was observed in the free-field in both cases. On the other hand, when $a_{max} = 0.2g$, liquefaction was limited only to the layers located near the surface.

A closer examination of Figs. 8a through 8d reveals that the vertical section closest to the edge column (Section P-P) experienced the lowest amount of porewater pressure response, while the Section R-R located the farthest showed the highest. This indicates the effectiveness of the treated columns in reducing the porewater pressure response at closer locations. However, the indication of similar porewater pressure response between Section R-R and the free-field (i.e. Ratio = 1),

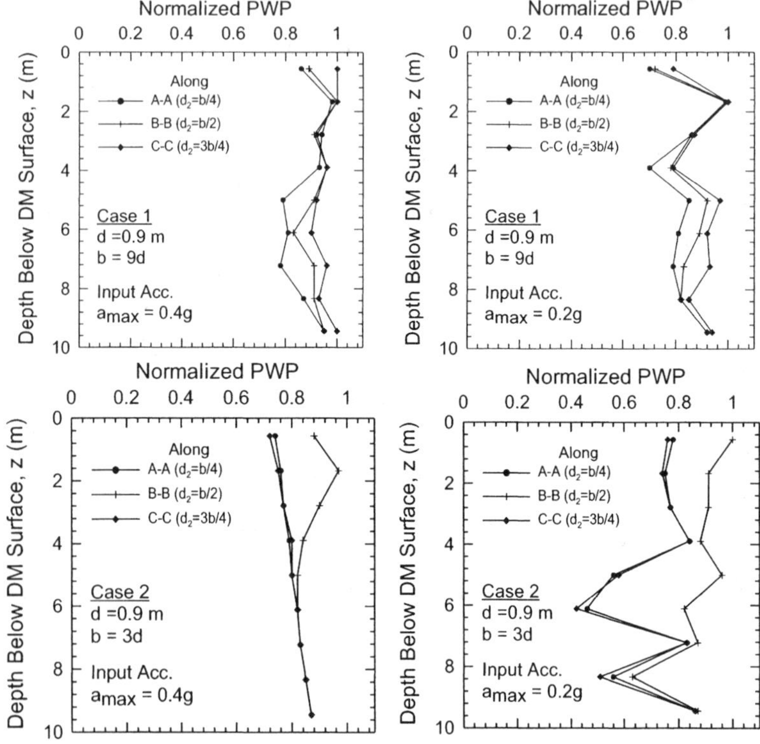

Figs. 9a-9d. Porewater pressure ratios with depth between treated columns

reveals that the effectiveness of treatment is limited to within about 5d from the edge column.

On the other hand, the observations relative to porewater pressure response between the treated columns are much more complex (Figs. 9a through 9d). The vertical sections here are equally spaced at b/4, as shown in Fig. 7. In general, the following observations can be made: (1) the porewater pressure within the treated zone is smaller than those computed in the free-field, (2) the porewater pressure response is consistently lower (Case 1 and Case 2) along Section A-A, which is located closest to the interior column, and (3) when the treated columns are closer (Case 2), the porewater pressure responses are similar between the Sections A-A and C-C. The reason for the second observation above can be attributed to the fact that unlike the soil elements near the edge column, the elements near the interior column are influenced by other treated columns.

CONCLUSIONS

The study utilized a two-dimensional effective stress program, TARA-2M to investigate the influence of DM treated soil columns on seismic soil response. The predictive capability of TARA-2M was initially verified by using a well-documented centrifuge database generated as a part of the NSF funded study VELACS (Arulanandan and Scott 1993). The porewater pressure responses have been provided for two levels of excitation (a_{max}= 0.2g and 0.4g) and for two DM column spacing distances. Computed results clearly show the extent of the influence of the treated columns in reducing the excess porewater pressures.

The results presented above are for a representative field case. Investigations focusing on the relative influence of other additional parameters such as different base excitations, levels of excitations, depth of treated columns, location of the specified base excitations, stiffness values of treated columns etc. are underway.

REFERENCES

Arulanandan, K., and Scott, R.F. (1993). *Verification of numerical procedures for the analysis of soil liquefaction problems*, Editors, A.A. Balkema Publishing.

Arulmoli, K., Muraleetharan, K.K., Hossain, M.M., and Fruth, L.S. (1992). "VELACS: Verification of liquefaction analyses by centrifuge studies - laboratory testing program soil data report," The Earth Technology Corporation, Irvine, CA.

Byrne, P.M. (1991). "A cyclic shear-volume coupling and pore pressure model for sand," *Proc., 2nd Int. Conf. on Recent advances in Geotech. Earth. Engrg. and Soil Dyn.*, St. Louis, Missouri, March 47-55.

Duncan, J.M. and Chang, C.Y. (1970). "Nonlinear analysis of stress and strain in soils," *J. of the Soil Mech. and Found. Engrg.* Vol. 96(5), ASCE, 1629-1653.

EPRI (1993). "Guidelines for determining design basic ground motions. Volume 1: Method of guidelines for estimating earthquake ground motions in Eastern North America," Report EPRI TR-102293, November.

Finn, W.D.L, Lee, K.W., and Martin, G.R. (1977). "An effective stress model for liquefaction," *J. of Geotech. Engrg.*, ASCE, Vol. 103(6), 517-533.

Finn, W.D.L. (1988). "Dynamic analysis in geotechnical engineering," State-of-the-Art Report, *Earthquake Engrg. and Soil Dynamics II Conf.*, Geotechnical Special Pub. No. 20, ASCE, 523-591.

Martin, G.R., Finn, W.D.L, and Seed, H.B. (1975). "Fundamentals of liquefaction under cyclic loading," *J. Geotech. Engrg.*, Vol. 101(5), ASCE, 423-438.

Masing, G. (1926). "Eigenspannungen and Verfestigung beim Messing," *Proc., 2nd Int. Congress of Applied Mech.*, 323-335.

Nakagama, K., and Soga, K. (1995) "Nonlinear cyclic stress-strain relation," *Proc., 3rd Int. Conf. on Recent Ad. in Geotech. Engrg. and Soil Dyn.*, Vol. I, 57-60.

Ni, S.D., Siddharthan, R.V., and Anderson, J.G. (1997). "Characteristics of nonlinear response of deep saturated soil deposits," *Bull. of the Seismological Soc. of America* Vol. 87(2), 342-355.

O'Rourke, T.D., and Goh, S.H. (1997). "Reduction of liquefaction hazard by deep soil mixing," *Proc. Earthquake Engrg. Frontiers in Transportation Facilities*, Eds: Lee, G.C., and Friedland, I.M., NCEER Technical Rep. 97-0005, 87-105.

Porbaha, A. (1998). "State of the art in deep mixing technology. part I: Basic concepts and overview," *Ground Improvement*, Vol. 2, 81-92.

Porbaha, A., Tanaka, H., and Kobayashi, M. (1998). "State of the art in deep mixing technology. part II: Applications," *Ground Improvement*, Vol. 2, 125-139.

Porbaha, A., Zen K., and Kobayashi, M. (1999). "Deep mixing technology for liquefaction mitigation," *J. of Infrastructure Systems*, Vol. 5(1), ASCE, 21-33.

Porbaha, A., Shibuya, S., and Kishida., T. (2000). "State of the art in deep mixing technology. part III: geomaterials characterization," *Ground Imp.*, (3), 91-110.

Scott, R.F. (1986). "Solidification and consolidation of a liquefied sand column," *Soils and Foundations,* Vol. 26(14), 23-31.

Seed, H.B., and Idriss, I.M. (1970). "Soil moduli and damping of soils, design equations, and curves," Report EERC 70-10, Earthquake Engineering Research Center, University of California, Berkeley.

Shibuya, S., Tatsuoka, F., Teachavorasinskun S., King, X.J., Abe, F., Kim, Y.S., and Park, C.S. (1992). "Elastic deformation properties of geomaterials," *Soils and Foundations*, Vol. 32(3), 26-46.

Siddharthan, R. (1984). "Two-dimensional nonlinear static and dynamic response analysis of soil structures," *Ph.D. Thesis*. Univ. of British Columbia, May.

Siddharthan, R. and Norris, G.M. (1988). "Performance of foundations resting on saturated sands," *Proc. Earthquake Engrg. and Soil Dynamics II Conf.,* Geotechnical Special Pub. No. 20, ASCE, 508-522.

Siddharthan, R.V., and Norris, G.M. (1990). "Residual porewater pressure and structural response," *Int. J. of Soil Dyn. and Earth. Engrg.*, Vol. 9(5), 265-271.

Siddharthan, R.V., and El-Gamal, M. (1993). "Numerical predictions for Model No. 1, 2, 3, and 4A," *Int. Conf. on Verification of Numerical Procedures for the Analysis of Soil Liquefaction Problems*, Vol. I, A.A. Balkema Publishing, 221-246; 395-413; 561-582; and 651-664.

Taboada, V.M., and Dobry, R. (1993). "Experimental results of Model No. 1 at RPI," *Int. Conference on Verification of Numerical Procedures for the Analysis of Soil Liquefaction Problems*, Vol. I, A.A. Balkema Publishing, 3-24.

Youd et al. (2001). "Liquefaction resistance of soils: summary report from the 1996 NCEER and 1998 NCER/NSF Workshops on evaluation of liquefaction resistance of soils," *J. Geotech. and Geoenvironmental Engrg.*, Vol. 127(10), ASCE, 817-833.

JET GROUTING SYSTEMS: ADVANTAGES AND DISADVANTAGES

George K. Burke, P.E.[1], Member, ASCE

ABSTRACT: The process of jet grouting has become a worldwide system of ground treatment. Specialized equipment vendors abound, each focusing on a niche of the jet grouting market. Specialty contractors build their experience using certain equipment and procedures so that they may be better apt to understand the risks that they assume.

This paper will compare the jet grouting systems available today, and bracket the parameters most often utilized. A discussion on the advantages and disadvantages of each system will be presented as well as how they may influence performance for a variety of applications. This evaluation should offer a guide for a better understanding of this often misunderstood technology.

INTRODUCTION

Jet grouting technologies continue to prove their effectiveness in the most difficult challenges. Enhancements of many kinds have been implemented to improve soilcrete quality, reduce cost, and advance production speed. But with so many systems of jet grouting, and specialist contractors all applying different parameters in their pursuit to achieve the design requirement at the lowest cost, how can one select what might work best for their project, and, how can a designer be assured his design will or can be met? What system and what parameters work best

---

[1]Vice President Engineering, Hayward Baker Inc., 1130 Annapolis Road, Odenton, MD 21113, gkburke@haywardbaker.com

for each application and soil type? There are many parameters and systems available, so how is the layman to choose? These questions are not easily answered, and in fact there is likely to be more than one system and one set of parameters that can achieve the design requirements.

This paper will break down the systems and parameters, compare the nuances of different soil types, and offer some guidance for the application of jet grouting.

JET GROUTING SYSTEMS

There are three primary systems of jet grouting, as illustrated in Figure 1.

Single Fluid System: the injection of cementitious grout slurry at high velocity to erode and mix with the soil.

Double Fluid System: the injection of cementitious grout slurry at high velocity, sheathed in a cone of air at an equally high velocity, to erode and mix with the soil.

Triple Fluid System: the injection of water at high velocity, sheathed in a cone of air at an equally high velocity, to erode the soil while simultaneously tremie injecting a cementitious grout slurry from beneath the erosion jets.

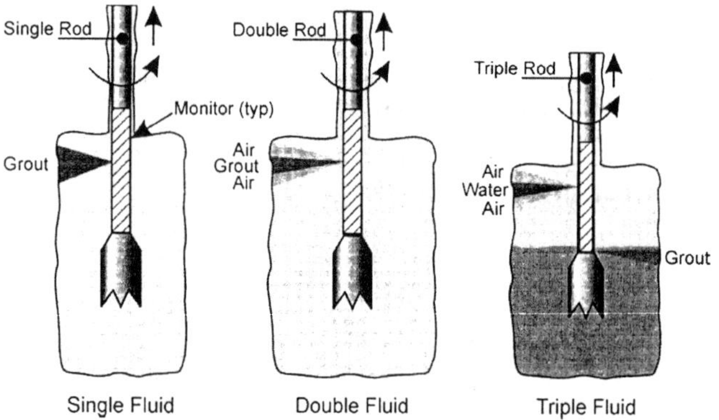

FIG 1: The three most common Jet Grouting systems

There are more variations of these systems than there are systems themselves, but in most cases they are a "bottom-up" process. That is to say, they use hydraulic rotary drilling to reach the design depth, and at that point initiate jet grouting parameters and procedures to create a cementitious soil matrix commonly called soilcrete, as shown in Figure 2.

During grouting, the borehole annulus must be large enough to permit unimpeded up-hole spoil return. This allows for control of the in situ stress environment. A lack of this spoil return will result in hydrofracturing the ground and loss of control. Loss of this control can lead to extreme inconsistencies in the soilcrete quality and geometry.

Other emerging jet grouting systems include SuperJet and X-Jet grouting. SuperJet System: a double fluid system reliant on specialized tooling and high injection energy for enhanced erosion capability (up to 5 m diameter) (Figure 3) (Ref 1, 5). X-Jet System: a triple fluid system using a pair of colliding erosion jets to create a more uniform and controlled diameter of treatment (Figure 4) (Ref, 3).

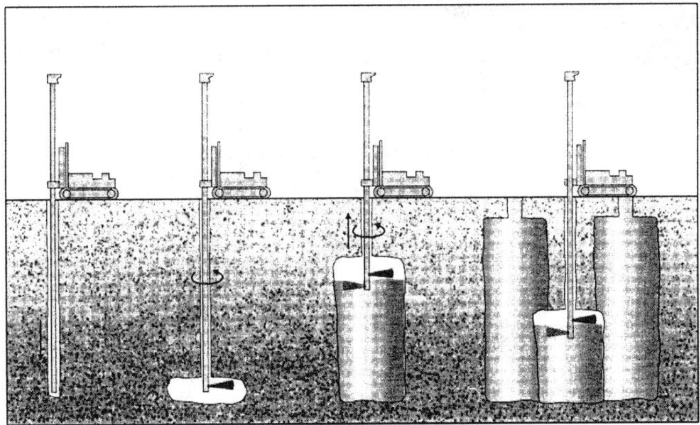

FIG 2: The Jet Grouting Process

CONSIDERING THE GROUND

Soil type and stratigraphy may be influencing factors in which systems of jet grouting are appropriate. From a practical standpoint, no other grouting system can be as effective across the range of soils as jet grouting. However, there are many competing systems that can effectively solve the problem that should be considered relative to cost and schedule.

When considering the soil, focus on the "erodibility" of the soil… how easily can it be broken down with the hydraulic energy of jet grouting? As shown in Figure 5, cohesionless soils are most easily eroded (Ref 4). In the local region of fluid injection, the turbulence created alone is enough to break down cohesionless soil types. As plasticity and stiffness increase, erodibility decreases to a point where jet grouting will not be effective in most stiff clays.

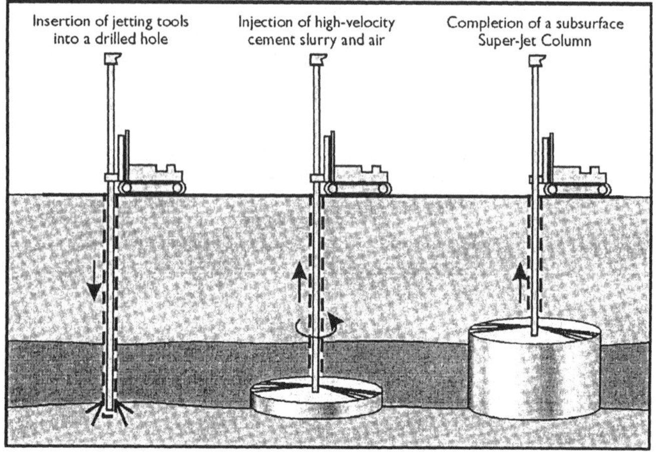

FIG 3: The SuperJet System

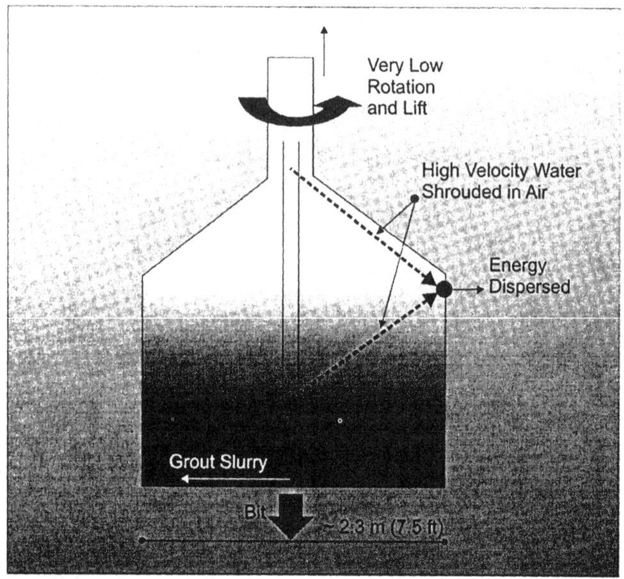

FIG 4: X-Jetting

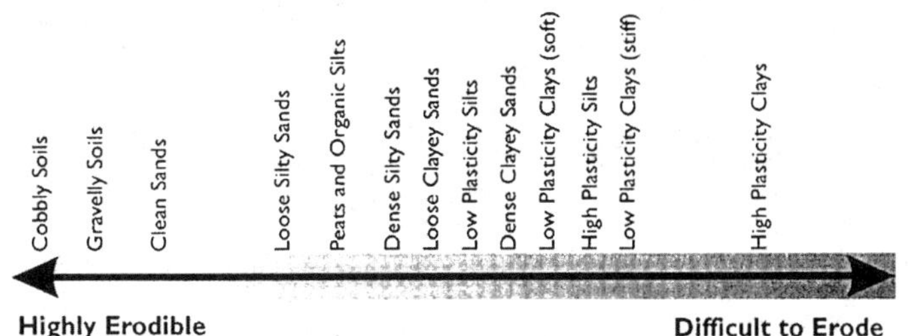

FIG 5: Jet Grouting Soil Erodibility Scale

Soil stratification is also a consideration, as variable soil conditions lead to variable soilcrete quality and parameters. Also, the jet grouting parameters may also need to change versus depth to create a uniform geometry, or one must simply accept variable geometry and install more soilcrete.

Gravels, cobbles and boulders, although considered cohesionless, may range from highly erodible to very difficult to erode, depending on in situ density, soil matrix and other conditions. Boulders will block the jet stream and a "shadow" of untreated soil will exist beyond. Buried trees and cemented soil can also be cause for poor erosion.

EQUIPMENT IMPACTS

The in situ erosion of soil is all about energy. The kinetic energy of the high velocity fluid impacting the soil causes a breakdown in the soil structure. This energy is dependent on (Ref. 9):

- The dynamic pressure of the fluid jet. The tooling design affects this value. SuperJet technology offers the highest dynamic pressures in situ.

- Air containment. The sheath of air surrounding the fluid jet can offer a 5 times increase in cutting energy, simply due to the protection from friction effects versus distance from the exit point.

- The mass of the fluid. The volume and density of the eroding fluid, per unit time and volume, will affect the geometry of soilcrete created.

- Rotation and lift speed. Most equipment is designed to rotate at a desired speed, and the rods are lifted in steps at time intervals. Slower rotation means more energy per unit volume of erosion. In general, rotation is set for 1 revolution per step, unless multiple nozzles are employed. Step height is adjustable, and is usually small for cohesive soils (2 – 3 cm per nozzle per rotation) and larger for cohesionless soils (3 – 5 cm per nozzle per rotation).

- The viscosity of the fluid. Higher viscosity fluids usually disperse more than lower viscosity fluids, and since this reduces fluid focus, it affects erosion energy.

- The velocity of the fluid. Higher velocity means higher energy, but if not focused will not yield commensurate geometry increases.

The equipment supporting jet grouting must consider all of these issues, from preparation of the fluid to the pumping pressures and volumes, to the drill controls and tooling. There are a number of parameters that require a degree of quality control, and instrumentation is now available to sense and record these versus time:

TABLE 1: Recordable Jet Grouting Parameters.

Aspect	**Parameter**
Water	Pressure, Volume
Air	Pressure, Volume
Cementitious Slurry	Pressure, Volume, Density
Rotation	Rate
Lift	Step, Height, Step Time, Rate
Depth	Depth

JET GROUTING PROCEDURES

Jet grouting is usually employed as a "bottom-up" procedure. However, there are many variations. Precutting during drilling can be used to open a larger borehole annulus. Double cutting the design zone can be utilized to remove a greater percentage of cohesive soil and/or to enhance the quality of soilcrete.

Double cutting is when the erosive action of the jet grouting process is repeated immediately after the first cut. This second cut can cover the entire design zone, and, like the first cut, is done with calculated lift and rotation rates in order to ensure thorough mixing of the grout slurry and in situ soil, and the removal of cohesive soil. Double cutting is most commonly used in cohesive soils, where large chunks of highly plastic clays can remain in the design zone after the first cut.

Care must be taken when double cutting depending on the cutting fluid used. For example, too much water can yield lower strength, while too much grout can simply be a waste of materials. Also, procedures can easily be adjusted to accommodate changes in soil conditions, but this requires detailed knowledge of the subsurface conditions, and concentration by the jet grout practitioner to enact any changes.

Spoil return and its control may also be cause to alter procedures. This is the most important aspect, and many things can be adjusted to assist with its return and control, such as (Ref. 1):

- changing grout viscosity
- changing air pressure and flow rate
- use of casing to reduce up-hole friction
- precutting measures
- auxiliary air lift system
- changing borehole size
- manual reaming of the hole
- reduced jetting energy

Many things can be responsible for a loss of spoil return, such as:

- a borehole restriction… too small a hole through a footing; soft squeezing clays; gravels that will not arch
- loss of air return… open porous gravelly zones; very soft clays; fibrous peats
- cohesive soil erosion… if plastic, cuts up in chunks and can block the annulus; if the step height is too high, it will cut into larger chunks and block up; if the spoil is so thick that the air stops returning up the borehole and takes a path of lesser resistance

JET GROUTING PARAMETERS

The jet grouting parameters employed are often somewhat controlled by the equipment. Every part of the process must be in good working order, and even so, has its limitations. It should be noted that, in order to create good quality soilcrete, several systems may yield excellent results using the appropriate procedures. Due to the equipment limitations and the experience with systems and procedures, only specialized contractors should make the selection of parameters. With this knowledge, performance specifications and pre-qualifying specialist contractors will be the preferred method of assuring the work. In general, jet grouting parameters fall into the ranges given in Table 2.

The determination of soilcrete quality is usually based on experience with parameters and soil types. Most specialist contractors evaluate the cement content of the soilcrete as part of their assessment. To do this, a set of parameters that will erode a target geometry in the soil must be selected. From these parameters and geometry, the following calculation can be performed:

TABLE 2. Ranges of Jet Grouting Parameters

		Single Fluid	Double Fluid	Triple Fluid
Water	Pressure (bar)	na	na	300 – 400
	Volume (l/min)	na	na	80 – 200
	No. Nozzles	na	na	1 – 2
	Nozzle Sizes (mm)	na	Na	1.5 – 3.0
Air	Pressure (bar)	na	7 – 15	7 – 15
	Volume (m^3/min)	na	8 – 30	4 – 15
Grout Slurry	Pressure (bar)	400 – 700	300 – 700	7 – 100
	Volume (l/min)	100 – 300	100 – 600	120 – 200
	Density (S.G)	1.25 – 1.6	1.25 – 1.8	1.5 – 2.0
	No. Nozzles	1 – 6	1 – 2	1 – 3
	Nozzle Sizes (mm)	1.0 – 4	2 – 7	5 – 10
Lift	Step Height (cm)	0.5 – 60	2.5 – 40	2 – 5
	Step Time (sec)	4 – 30	4 – 30	4 – 20
Rotation	Speed (rpm)	7 – 20	2 – 20	7 – 15

(Ref. 2, 6, 7, 8, 9, 10)

Cement Content = (Single Fluid, Double Fluid, and SuperJet systems)
wt. dry cement/(vol. soilcrete + vol. grout)

Cement Content = (Triple Fluid, X-Jet)
wt. dry cement/(vol. soilcrete + vol. grout + vol. jetting water

The geometry expected can be assessed from Table 3. There is a lot of overlap between Double Fluid and Triple Fluid systems. In general, Double Fluid will create larger column diameters, very much dependent on the parameters used and the soils being eroded. The opposite can be said as well, but not as often. It should be noted that partial column and panels can also be created in lieu of full diameter columns. X-Jet creates a single column size due to the diffusion of the colliding jets. In general, the strength of the soilcrete that can be expected is identified in Figure 6.

TABLE 3. Soilcrete Column Diameter Expectations

System	Soft Clays	Silts	Sands
Single Fluid	1.5 – 3.0 ft 0.4 – 0.9 m	2.0 – 3.5 ft 0.6 – 1.1 m	2.5-2.4 ft 0.8 – 1.2 m
Double Fluid	3.0 – 6.0 ft 0.9 – 1.8 m	3.0 – 6.0 ft 0.9 – 1.8 m	4.0 – 7.0 ft 1.2 – 2.1 m
SuperJet	10.0 – 14.0 ft 3.0 – 4.3 m	11.0 – 15.0 ft 3.3 – 4.6 m	11.5 – 16 ft 3.5 – 5.0 m
Triple Fluid	3.0 – 4.0 ft 0.9 – 1.2 m	3.0 – 4.5 ft 0.9 – 1.4 m	3.0 – 6.0 ft 0.9 – 2.5 m
X-Jet	7.5 ft 2.3 m	7.5 ft 2.3 m	7.5 ft 2.3 m

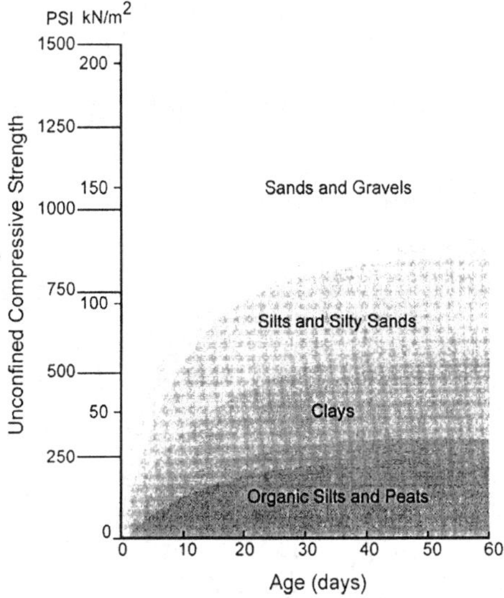

FIG 6: Typical Soilcrete Strengths

TABLE 4. Typical Cement Contents for Soilcrete

Soil Type	Cement Content (kg/m^3) of Soilcrete
Sands	150 – 250
Silts and Silty Sands	200 – 275
Clays	250 – 350
Organic Silts and Peats	300 – 400

ADVANTAGES AND DISADVANTAGES FOR SYSTEM SELECTION

There is no single answer to the question of which system is the best for an application. Depending on experience, system, parameters, and procedures used there may be many choices that will furnish the product desired. The key is to prepare a performance specification that includes approximate verification methods and monitoring to maximize the degree of success.

To offer some guidance, the advantages and disadvantages of each system are addressed in Table 5.

TABLE 5. Jet Grouting System Advantages and Disadvantages

System	Advantages	Disadvantages
Single Fluid	-Simplest system and equipment -Good to seal vertical joints -Good in cohesionless soil	-Smallest geometry created -Hardest to control heave -Difficult to control quality in cohesive soils
Double Fluid	-Most utilized system -Availability of equipment and tooling -High energy, good geometry achieved -Most experience -Often most economical	-Very difficult to control heave in cohesive soils -Spoil handling can be difficult -Not usually considered for underpinning
Triple Fluid	-Most controllable system -Highest quality in difficult soils -Best underpinning system -Easiest to control spoil and heave	-Complex system and equipment -Requires significant experience
SuperJet	-Lowest cost per volume treated -Best mixing achieved	-Requires special equipment and tooling -Difficult to control heave in cohesive soils -Spoil handling difficult -Cannot work near surface without support -Highest logistical problems
X-Jet	-Confidence of geometry -Controllable materials cost -Best for soft cohesive soils	-Very specialized equipment that requires daily calibration -Limited experience available

SUMMARY

Jet grouting as a soil treatment system is unsurpassed in the diverse range of applications and soil types that it is applicable to. Of all the ground treatment systems, it is the most complex, requiring experience in order to predict product quality, as well as to determine the method to achieve it. Soil type and stratigraphy evaluation are critical to this determination. Parameters, equipment, and procedure alterations also play roles in the selection of means and methods.

Since most of this experience is at the disposal of specialist contractors, specifications developed for this work should be performance-based, with a method of considering experience as well as cost in the selection process. Special conditions will exist for every project, and these must be assessed also.

APPENDIX. CONVERSION TO SI UNITS

Feet (ft) x 0.305 = meter (m)
bar x .01 = kNm^2
kips x 4.448222 = kN

REFERENCES

1. Brill, G. T., Burke, G. K., Ringen, A.R. (2003). "A Ten Year Perspective on Jet Grouting: Advancements in Applications and Technology." *3rd International Conference on Grouting and Ground Treatment*, ASCE/DFI, New Orleans.

2. Bruce, D. A., Hague, S. T., Hitt, R. (2001). "The Treatment by Jet Grouting of a Bridge Foundation on Karstic Limestone." *Foundations and Ground Improvement*, GSP No. 113, ASCE, Blacksburg, Virginia.

3. Burke, G. K., Welsh, J. P. (2000). "Advances in Grouting Technology." *GeoEng 2000*, DFI, Melbourne, Australia.

4. Burke, G. K. (2002). "The State of the Art of Jet Grouting in the United States." *9th International Conference on Piling and Deep Foundations,* DFI, Nice, France.

5. Burke, G. K., Toth, P., Weber, A. (2000). "Soilcrete-DS: Enhanced Technology for Surgical Soil Improvement." *4th International Conference on Ground Improvement,* Helsinki, FI.

6. Croce, P., Flora, A., Modoni, G. (2001). "Experimental Investigations of Jet Grouting." *Foundations and Ground Improvement*, GSP No. 113, ASCE, Blacksburg, Virginia.

7. Marolanda, A., Marolanda, An., Amaya, F., Millan, M. (2001). "Jet Grouting at el "Tambor"." *Foundations and Ground Improvement*, GSP No. 113, ASCE, Blacksburg, Virginia.

8. Morioka, B.T., Kwong, J. K. P., Kalani, J.K. (2001). "Jet Grouting for Microtunneling and Shaft Construction." *Foundations and Ground Improvement*, GSP No. 113, ASCE, Blacksburg, Virginia.

9. Shibazaki, M. (2003). "State of Practice of Jet Grouting." 3rd *International Conference on Grouting and Ground Treatment*, ASCE/DFI, New Orleans.

10. Personal Experience from over 200 Jet Grouting projects carried out for HBI.

FEATURES AND RESULTS OF A JET-GROUTING TRIAL FIELD IN VERY SOFT PEATY SOILS

Tiziano Collotta[1], Andrea Frediani[2], Vittorio Manassero[3]

ABSTRACT: The paper deals with the main features and the results of a field trial related to a jet grouting treatment in very soft peaty soil; for this soil the technical literature offers few reference examples in terms of operative parameters and treatment effectiveness: actual average diameter, compressive strength and deformability parameters of the treated soil. The trial field was instrumented to monitor the induced vertical and horizontal displacement, the excess pore pressure values and the dissipation velocities. The actual column diameters were evaluated by means of direct measurements of the top of the columns and by means of a cross holes technique. The test was carried out by Rodio S.p.A., one of the largest foundation contractor in Italy.

INTRODUCTION

The Rome-Naples motorway between km 722 and km 724, crosses two hollows; the foundation soil consists of very soft clayey-peaty deposits, down to about 17 m below the ground level.

These lithological and stratigraphical features, and the presence of embankments about 3.5 m height, implied high settlements and several repeated pavement rehabilitation works, to maintain the motorway serviceability.

Despite the motorway was built from 1956 to 1958, and opened to traffic in 1959, the consolidation phenomena are still in progress; settlements of centimeters per year are observed where thicker are the peaty layers.

[1] Chief of Geotechnical Office, SPEA Ingegneria Europea SpA, via Vida No.11, 20127, Milan, Italy, tiziano.collotta@spea.autostrade.it

[2] Technical Management Department, Autostrade SpA, via A.Bergamini, No. 50, 00154, Rome, Italy,

[3] Engineering Department Manager, RODIO S.p.A., via XXV Aprile 2, 20097 San Donato Milanese, Milan, Italy

Among the different rehabilitation technologies, to control settlements the design attention was focused to the following:

- partial replacement of the motorway embankment with lightweight fill (lightweight expanded clay aggregate);
- soil improvement by means of jet grouting columns, to decrease the foundation soil deformability.

Whilst several experiences relevant to expanded clay are reported in the technical literature, the operative parameters and the results of jet grouting treatments for peats are not well documented; the undesirable effects of jet grouting to highway facilities or structures close to the intervention areas are not well documented too.

Considering the large extension of the area under study and the relevant economic implications related to the jet grouting treatment, a trial field was designed; aim of the test was to evaluate and define:

- treatment effectiveness in terms of operative parameters: actual column diameters and strength/deformability characteristics of the treated soil;
- ranges for the operative parameters to be included in the technical specifications;
- specific operative aspects, i.e. management of spoil;
- stress and deformation level induced to the highway structures.

Referring to these aspects, the trial field was instrumented to monitor the induced vertical and horizontal displacements, the instantaneous pore pressure values and the dissipation velocities.

The geometrical and operative features of the trial field and the obtained results, in particular relevant to the geotechnical monitoring, are shown. To allow comparative analyses for similar applications a schematic geotechnical soil profile is reported.

GEOTECHNICAL CHARACTERIZATION OF NATURAL SOIL

The geotechnical parameters were defined according to in situ and laboratory tests:

- No. 3 geotechnical boreholes, 35.0 m deep, carried out at the highway embankment toe;
- No. 14 stratigraphic boreholes, 8.0 m deep and No. 1 stratigraphic borehole, 20 m deep, along the highway axis;
- No. 36 static electrical cone penetrometric tests (CPTE).

Laboratory tests, performed by the Autostrade S.p.A. Testing Materials Laboratory, consisted of:

- classification tests: grain size distribution, natural unit weight, water content, Atterberg limits, carbonate content;
- oedometric tests, to evaluate deformability and consolidation parameters;
- unconsolidated undrained triaxial tests;
- direct shear tests and torsional shear tests to evaluate peak and residual strength.

A typical CPTE profile is reported on FIG. 1. The softer (0.2-0.6 MPa point resistance registered in CPTE) thin layers are constituted by a succession of peats, silts and organic clays with peaty layers.

The bed-rock is constituted by pozzolanic (gravelly-silty sands) medium dense soils.

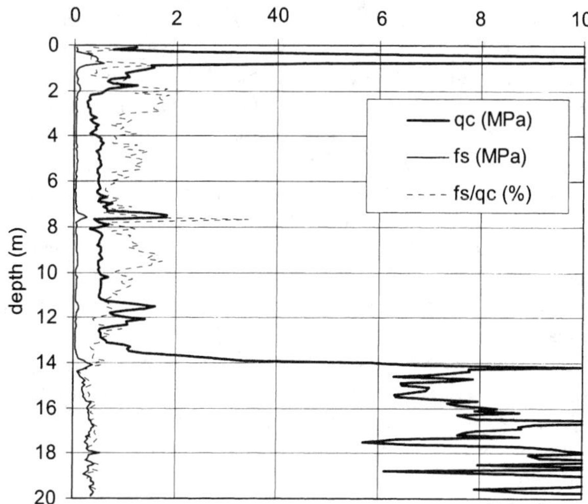

FIG. 1. Typical CPTE profile

The average geotechnical parameters of the peaty soil are:

- specific weight of solids = 18÷20 kN/m^3
- natural unit weight = 9÷12 kN/m^3.
- clay content = 45÷60%
- organic content = 20÷80%
- liquid limit = 60÷120%
- water content = 110÷490%
- undrained cohesion: c_u = 10÷20 kPa.

The primary and secondary consolidation coefficients can be respectively assumed 1x10^{-4}÷5x10^{-4} cm^2/s and 0.02÷0.05, depending on the stress level.

JET GROUTING

Soil improvement is achieved by means of high pressure fluids used as a cutting agent and the simultaneous placement and mixing of a cement grout. The obtained treated soil columns present a diameter that is a function of the adopted injection system and operative parameters.

Four jet-grouting systems may be adopted:

- *single-fluid system*: one type of fluid only is utilized to cut and stabilize the soil; the "fluid" is represented by the cement grout. Injection pressure is usually in the range 30 to 50 MPa.
- *double-fluid system (stabilizing grout and compressed air)*: soil breaking is enhanced enveloping the grout jet with a coaxial compressed air jet. As per single-fluid system, the grout injection pressure ranges between 30÷50 MPa; the compressed air pressure ranges between 0.5÷1.7 MPa.

- *double-fluid system (stabilizing grout and water)*: the water jet cuts the soil and allows a partial removal of fines content too; the cement grout, simultaneously injected by a nozzle installed underneath, has a stabilizing role, mixing the disaggregated and partially washed out soil. Water pressure is usually in the range of 30 to 50 MPa; the grout pressure ranges between 2 and 10 MPa.
- *triple-fluid system*: the soil cutting due to the water jet is enhanced enveloping it with a coaxial compressed air jet; the coaxial water and compressed air jets imply a partial removal of the soil fine content too. The grout, simultaneously injected by a nozzle installed underneath, acts in this case as a stabilizing element, mixing with the disaggregated and partially washed out soil. Water and grout pressures are within the ranges 30 to 50 MPa and 2 to 10 MPa respectively; compressed air pressure is usually in the range 0.5 to 1.7 MPa.

The jet grouting effect may be enhanced by a preliminary treatment, called "pre-cutting"; it has the aim to increase soil disaggregation and partial fines removal. The pre-cutting treatment can be carried out upward, at the end of drilling, or downward, during the drilling phase at controlled speed. The adopted water pressure ranges between 20 to 40 MPa.

The double-fluid system (stabilizing grout and compressed air) was adopted in the trial field; some columns were carried out by using pre-cutting.

JET GROUTING TRIAL FIELD

Geometrical Features

The geometrical features of the trial field are reported in Fig. 2.

Seven columns have been installed on the motorway platform (zone A) and five columns have been installed at the motorway embankment toe (zone B). In zone A the columns are 9 to 15 m deep (the top of columns is located 5.0 m below ground level); in zone B the columns are 4 m deep (the top of columns is located 1.0 m below ground level). In Fig. 2 the geotechnical monitoring and topographic reference points are also reported.

Geotechnical monitoring consisted in the installation of one inclinometric tube, 26 m deep, and two piezometric verticals 12.0 m (Pz1) and 20.5 m (Pz2) deep. Two electric piezometers were installed at depths of 8 and 12 m from ground level at the location Pz1; at location Pz2 the piezometers were installed at depths of 8 and 20 m from ground level.

In the trial field 5 further inclinometric tubes were installed for tomographic cross-holes tests, to evaluate the grouted columns diameters.

A vertical continuous coring was carried out for each column, in order to recover grouted soil samples to be tested in laboratory for the evaluation of the physical and mechanical soil features.

Equipment

The main equipment utilized to install the jet grouting columns are the grout batching and storage plant, the high pressure pump, the air compressor and the drilling equipment.

An automatic drilling and injection parameter recording system was operative.

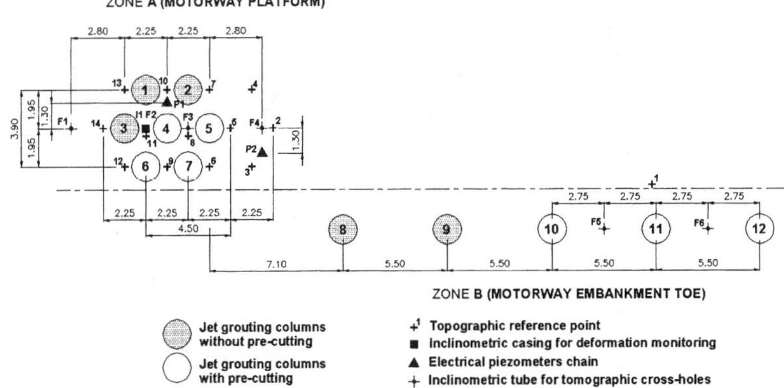

FIG. 2. Geometrical features of the trial field and monitoring layout

The following parameters were continuously recorded during drilling operations: time, drilling tool depth, penetration rate, rotational speed, thrust, torque, drilling fluid pressure and the drilling fluid flow.

An automatic control system of the verticality during the various drilling operations steps was utilized.

During the high pressure injection the following parameters have been registered: time, nozzles depth, withdrawal speed, rotational speed, cement grout pressure, grout flow rate, compressed air flow rate, compressed air pressure and the total injected grout volume

Operative Parameters

A picture of the operative parameters is shown in Table 1;

According to different combinations, the influence of the following parameters was analyzed:

- presence or not of pre-cutting and, if present, the pre-cutting specific energy (fluid pressure x unitary fluid volume);
- type of stabilizing grout mix;
- jetting specific energy and unitary grout consumption per meter of column;
- nozzles combinations and diameters and the corresponding stabilizing fluid flow rate;
- total specific energy (pre-cutting and jetting specific energies).

All the specific energies disregard the compressed air contribution.

Results Evaluation

The controls of the consolidated columns were carried out to evaluate:

- columns diameter;
- columns continuity (absence of interruptions);
- physical and mechanical characteristics of the treated soil.

Jet grouting type:	Double fluids system (stabilizing grout + compressed air)
High pressure pump type:	Geoastra 5T302
Drilling rig type:	Casagrande C11
Parametres recording system:	Lutz CL88

col. n°	phase	depth		date	time		C/W grout		nozzles		high pressure pump			drilling rig								unitary fluid volume	unitary cement consumption	specific energy	
		from	to		start	end	nominal	actual	n°	φ	fluid pressure	rotational speed	calculated fluid flow (η=0.9)	fluid pressure manom.	fluid pressure CL88	air pressure manom.	air pressure CL88	fluid flow CL88	air flow CL88	time per step	average rotational speed			phase	total
		(m)	(m)							(mm)	(bar)	(pist/min)	(l/min)	(bar)	(bar)	(bar)	(bar)	(l/min)	(hl/min)	(s/4cm)	(rpm)	(l/m)	(kg/m)	(MJ/m)	(MJ/m)
1	pre-cutting	14,0	5,0	-	-	-	-	-	-	-	-	-	-	-	-	-	-	-	-	-	-	-	-	0,00	34,47
	jetting			11/30/99	10.30	11.15	1,2	1,182	2	3,0	400	66	184	400	395	8,0	7,5	183	76	11,3	15	862	737	34,47	
2	pre-cutting	14,0	5,0	-	-	-	-	-	-	-	-	-	-	-	-	-	-	-	-	-	-	-	-	0,00	42,44
	jetting			11/29/99	15.52	16.40	1,2	1,197	1	4,5	400	74	206	400	405	8,0	7,5	207	73	12,3	15	1061	916	42,44	
3	pre-cutting	20,0	5,0	-	-	-	-	-	-	-	-	-	-	-	-	-	-	-	-	-	-	-	-	0,00	43,11
	jetting			11/26/99	-	18.33	1,4	1,446	2	3,0	400	62	172	400	420	7,8	7,5	-	67	15,0	14	1078	1063	43,11	
4	pre-cutting	14,0	5,0	12/1/99	10.51	11.15	-	-	1	4,5	310	77	214	300	305	8,0	7,5	218	76	6,2	14	563	-	16,90	48,67
	jetting			1/12/99	11.28	12.05	1,2	1,197	1	4,5	400	73	203	400	405	8,0	7,8	205	73	9,3	15	794	686	31,78	
5	pre-cutting	20,0	5,0	11/27/99	11.25	12.08	-	-	2	3,0	320	70	195	300	330	8,0	7,8	200	70	6,5	14	542	-	16,25	45,42
	jetting			11/27/99	12.20	15.40	1,4	1,583	2	3,0	400	60	167	400	405	8,0	8,0	175	64	10,0	15	729	764	29,17	
6	pre-cutting	14,0	5,0	11/29/99	11.05	11.40	-	-	1	4,5	400	90	250	400	410	8,0	7,5	250	73	8,5	14	885	-	35,42	67,19
	jetting			11/29/99	11.50	12.36	1,2	1,136	1	4,5	400	72	200	400	405	8,0	7,5	205	73	9,3	15	794	660	31,78	
7	pre-cutting	14,0	5,0	11/30/99	14.27	15.06	-	-	2	3,0	400	82	228	400	405	8,0	7,5	230	76	9,8	14	939	-	37,57	66,97
	jetting			11/30/99	15.18	16.49	1,4	1,390	2	3,0	400	63	175	400	405	8,0	7,5	180	74	9,8	16	735	705	29,40	
8	pre-cutting	5,0	1,0	-	-	-	-	-	-	-	-	-	-	-	-	-	-	-	-	-	-	-	-	0,00	35,53
	jetting			12/2/99	12.37	12.57	1,2	1,182	2	3,0	400	66	184	400	405	8,0	7,5	187	76	11,4	15	888	760	35,53	
9	pre-cutting	5,0	1,0	-	-	-	-	-	-	-	-	-	-	-	-	-	-	-	-	-	-	-	-	0,00	42,85
	jetting			12/1/99	17.36	17.58	1,2	1,182	1	4,5	400	74	206	400	405	8,0	7,5	209	77	12,3	15	1071	917	42,85	
10	pre-cutting	5,0	1,0	12/2/99	15.44	15.55	-	-	2	3,0	300	72	200	300	310	8,0	7,5	204	76	6,4	15	544	-	16,32	46,82
	jetting			12/2/99	16.04	16.21	1,4	1,420	2	3,0	405	66	184	400	405	8,0	7,5	183	76	10,0	15	763	743	30,50	
11	pre-cutting	5,0	1,0	12/2/99	9.11	9.26	-	-	1	4,5	400	92	256	400	405	8,0	7,5	254	76	8,4	14	889	-	35,56	68,11
	jetting			12/2/99	11.22	11.38	1,2	1,182	1	4,5	400	74	206	400	405	8,0	7,5	210	76	9,3	15	814	696	32,55	
12	pre-cutting	5,0	1,0	12/2/99	17.02	17.18	-	-	2	3,0	405	82	228	400	405	8,0	7,5	225	76	10,0	16	938	-	37,50	66,63
	jetting			12/2/99	17.26	18.03	1,4	1,390	2	3,0	400	65	181	400	405	8,0	7,5	184	73	9,5	15	728	699	29,13	

TAB. 1. Trial field operative parameters

The control of the columns dimension was carried out by means of two different methods:

• direct method: columns excavation and visual check
• indirect method: cross-hole seismic tomography

The continuous coring boreholes carried out through the columns allowed to verify the treatment continuity and to recover consolidated soil samples for the laboratory analyses. An accurate stratigraphic log and photographs of the recovered materials were carried out for each borehole.

Columns Diameter

Chronologically, the cross-hole seismic tomography was the first method adopted to check the columns diameter: Tomography was carried out for columns number 3, 4, 5 (zone A) and 11 (zone B); the accuracy was estimated in the order of 10-15%. Then, excavations around the columns were executed; all the columns in zone B were interested (numbers 8 to 12).

The columns diameter is usually function of the total cutting specific energy; the presence or not of pre-cutting and the actual fluid flow rate may also affect the results. The actual diameter obtained is reported on fig. 3 versus the total cutting specific energy (for each diameter the unitary cement consumption values, referred to both the linear meter of column and the cubic meter of column, are also specified).

Fig. 4 shows the actual column diameters versus jetting energies.

The actual diameters versus the unitary cement consumption, referred to linear cubic meter of column, are reported on Figs. 5 and 6.

The columns installed at the center of zone A and B are also pointed out; for these columns the actual diameter appears to be smaller if compared to all the other columns.

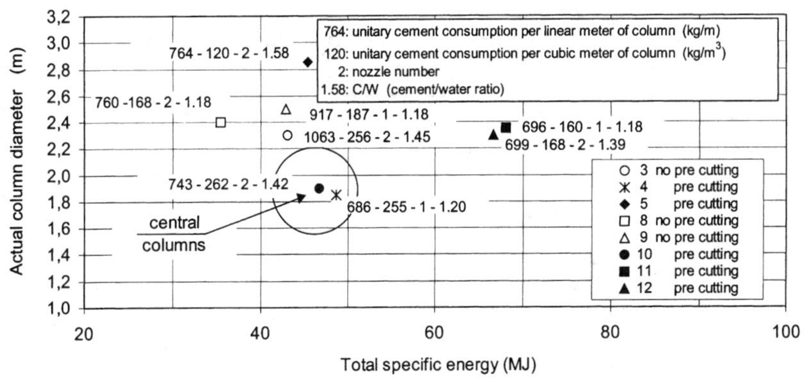

Fig. 3. Actual columns diameter versus total specific energy

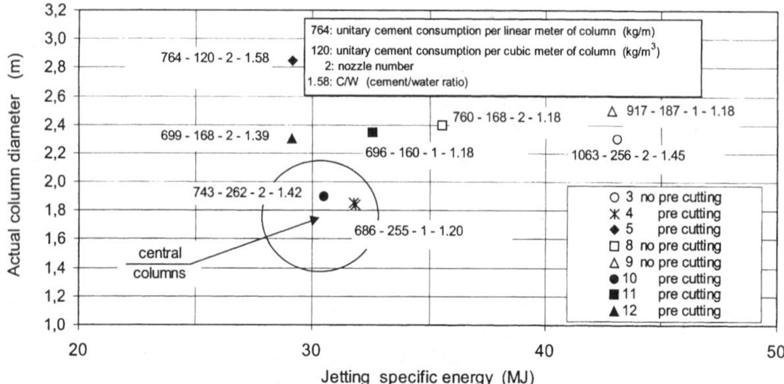

Fig. 4. Actual columns diameter versus jetting specific energy

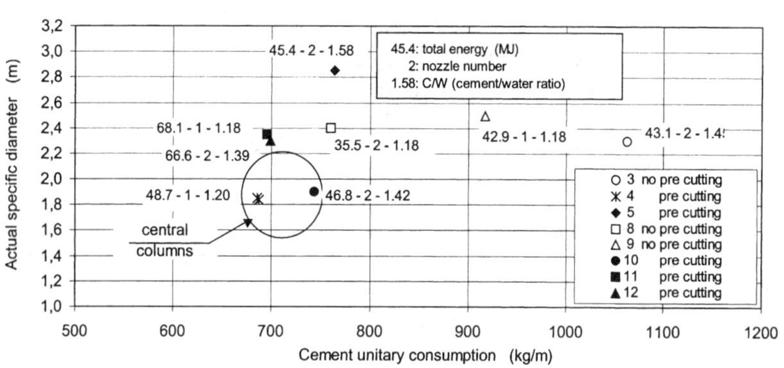

Fig. 5. Actual columns diameter vs cement unitary consumption (kg/m)

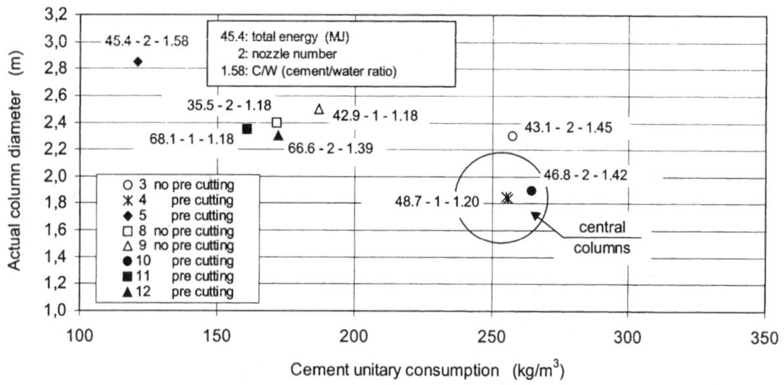

Fig. 6. Actual columns diameter versus cement unitary consumption (kg/m^3)

A few remarks referred to the above mentioned results are listed below:

- The actual diameter does not appear to be much influenced by the total energy or the jetting energy, at least for the adopted energies range.
- Column No. 4, installed in the center of zone A, is characterized by a considerable reduction of the actual diameter as regard to the surrounding columns; column 4 was the last executed and is likely to be affected by the soil compaction/improvement due to the already installed columns.
- The confining effect results to be more clear if columns 4 and 5 (Fig. 3), carried out with pre-cutting and same energies, are compared. Column n.10 (the last executed column in zone B), despite the higher interaxes (5.0 m) with the surrounding columns, appears also to be affected by the confining effect.
- Fig. 5 shows that columns having equal or similar actual diameter, carried out with pre-cutting, have had a unitary cement consumption, referred to linear meter of column, lower as regard to the columns without pre-cutting; this trend is not clear from the comparison of the unitary consumption referred to the cubic meter of column.
- The actual diameter of column 5, carried out with pre-cutting, is 30% higher than the observed average value for the remaining columns. As shown on Fig. 6, the specific cement consumption referred to column 5 is lower than all the other columns carried out without pre-cutting.
- The results of columns 11 and 12, carried out with similar pre-cutting and jetting energies, but with different nozzles combinations, do not appear to show effect of the nozzles number and diameter to the effective diameter.

Strength

Thirty-five treated soil specimens to be tested in laboratory were recovered: bulk density was determined for all the specimens- and ranges between 10 and 13 kN/m^3. Twenty-four specimens were tested for unconfined compressive strength with stress-strain relationship; due to the treated soil weakness, it was not possible to test the remaining eleven specimens.

The core cylindrical compressive strength was obtained from samples characterized by H/D ratios ranging between 2 and 0.6. Since the sample strength is affected by H/D ratio, to allow a results comparison, the core cylindrical compressive strength was converted in cubic compressive strength according to British Standard 1881, Part 120 (1983), where a relationship developed for concrete specimens recovered by coring is given:

$$S_{cubic} = S_{core}\, D/(1.5 + 1/\lambda)$$

where:

D = 2.5 (for cores drilled vertically or parallel to the height of casting);

λ = height/diameter ratio of the specimen after preparation.

The cubic compressive strength is plotted versus bulk density in Fig. 7.

In general, from the above mentioned diagram we can observe:

- proportionality between cubic compressive strength and bulk density;
- under the same bulk density value, higher strength values for the columns carried out with pre-cutting compared with columns without pre-cutting.

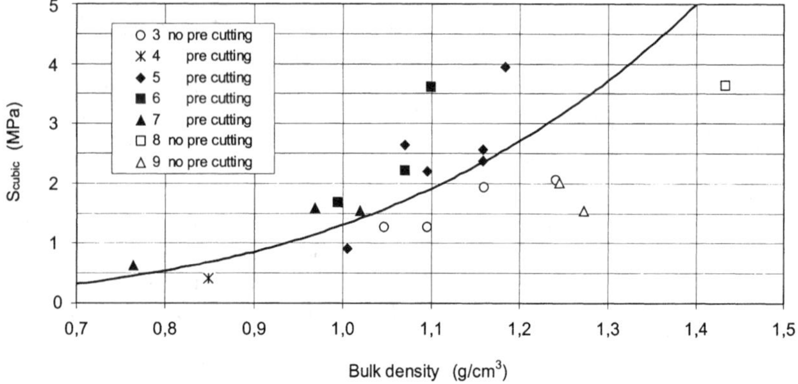

Fig. 7. Cubic compressive strength versus bulk density

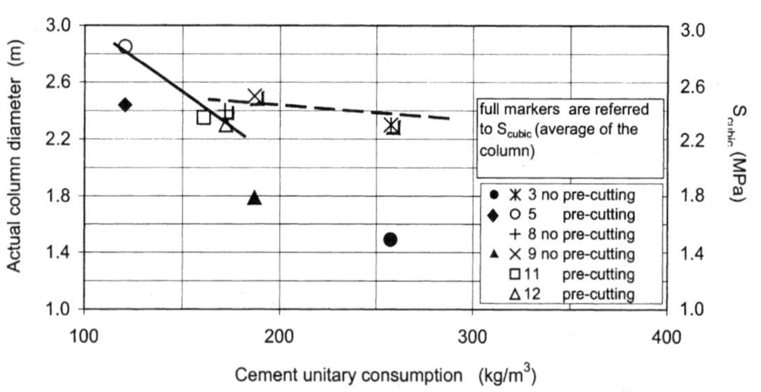

Fig. 8. Actual columns diameter and cubic compressive strength versus cement unitary consumption (kg/m^3 of column)

In detail, the values referred to column 5 (average 2438 kPa, minimum 918 kPa, maximum 3615 kPa) and column 6 (average 2505 kPa, minimum 1683 kPa, maximum 3615 kPa) are higher than the values relevant to the other columns.

We can highlight that for columns 4, 8 e 9 the low number of recovered or tested samples represent an indirect index of lower quality of the coring product.

The cubic compressive strength and the actual diameter values versus unitary cement consumption (kg/m^3 of column) are plotted on Fig. 8; the figure shows the behavior of column 5, carried out with pre-cutting, to be better than the ones without pre-cutting, despite this column is characterized by the lowest unitary cement consumption. The cutting action of the high pressure water jet during pre-cutting improves the high pressure grout jet cutting effect in the subsequent jetting phase, thus reducing grout consumption.

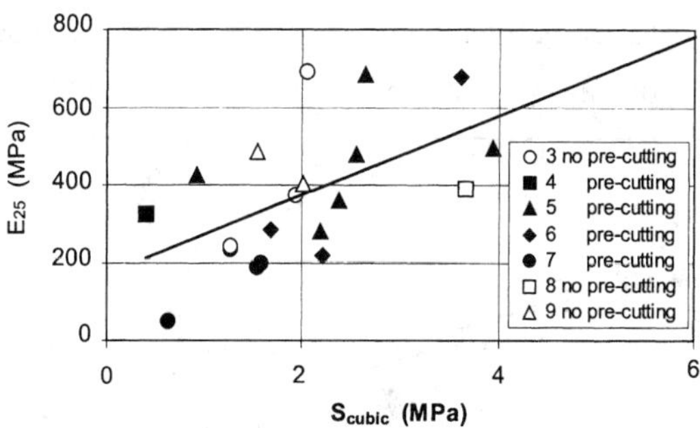

Fig. 9a. Elastic modulus E_{25} versus cubic compressive strength

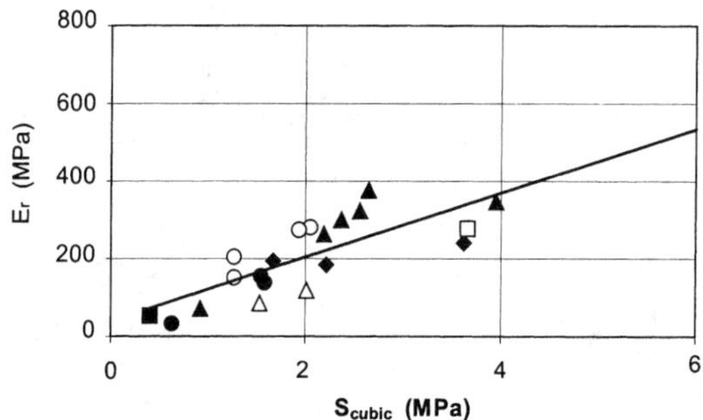

Fig. 9b. Elastic modulus E_r versus cubic compressive strength

Elastic Modulus

Elastic modulus values referred to 25% (E_{25}) and 100% (E_r) of the rupture condition stress versus cubic compressive strength, are reported in Figs. 9a and 9b respectively.

The modulus in correspondence of the rupture conditions is about 100 times the cubic compressive strength, while E_{25} / cubic compressive strength ratio is about 200.

Geotechnical Monitoring and Topographic Survey Results

Horizontal Displacements

Horizontal displacements induced by the jet grouting treatment were monitored by means of inclinometric survey.

As shown in Fig. 10, maximum horizontal displacements of the order of 15 cm were recorded (distance between column axis and inclinometer tube = 1.1 m).

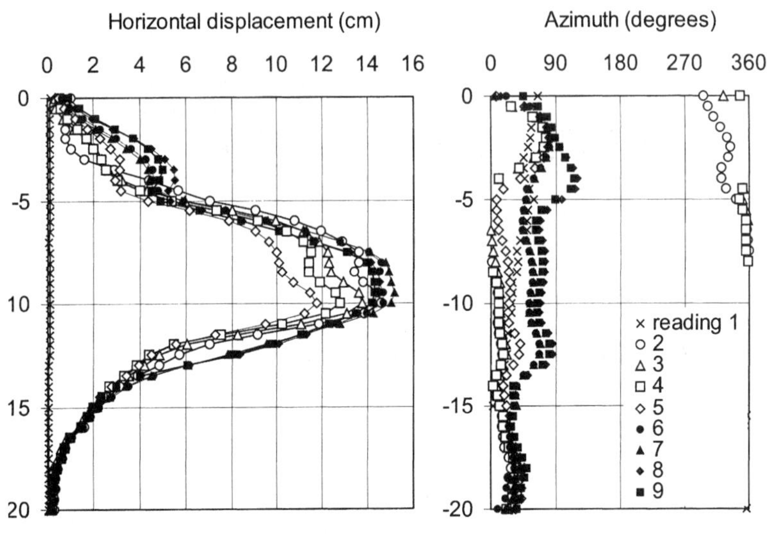

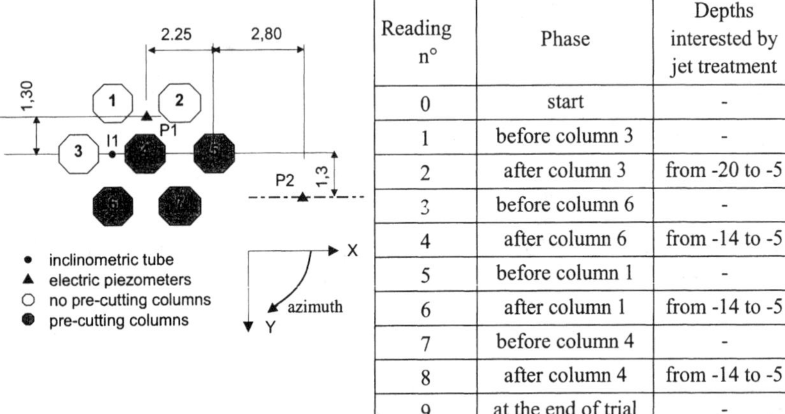

Reading n°	Phase	Depths interested by jet treatment
0	start	-
1	before column 3	-
2	after column 3	from -20 to -5
3	before column 6	-
4	after column 6	from -14 to -5
5	before column 1	-
6	after column 1	from -14 to -5
7	before column 4	-
8	after column 4	from -14 to -5
9	at the end of trial	-

Fig. 10. Horizontal displacements versus depth and versus time

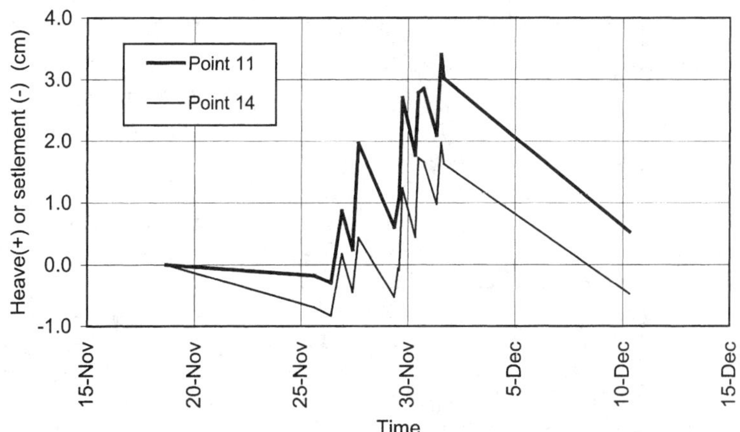

Fig.11. Vertical deformations versus time

Uplift / Settlements

High accuracy topographic survey were carried out during the trial field by checking 13 topographic reference points (Fig. 2). Two benchmarks, one each side of the field, were located 20 m far from the last reference point. Readings were carried out before drilling and at the end of injection phase of each column.

The analyses of the results show that the jetting undesirable effects on vertical movements can be regarded as negligible. This remark is emphasized by Fig. 11 where the recorded deformations versus time are reported; the data are referred to point 14, inside the group of columns, and point 11, outside the group.

Induced Pore Pressures

Piezometric readings referred to vertical P1, located inside area A, show maximum induced pore pressures values of the order of 150-200 kPa (Fig. 12); the two piezometers of vertical P2, outside area A, 3.0 m apart from the center of the nearest column, show maximum induced pore pressures values of the order of 20-30 kPa (Fig. 13).

The complete dissipation of the induced pore pressures was registered at 4-5 days after the end of the injection operations.

CONCLUSIONS

The trial field considered several different combinations of:

- presence or not of pre-cutting and, if present, pre-cutting specific energy (specific energy = fluid pressure x unitary fluid volume);
- specific jetting energy and unitary grout consumption;
- specific total energy (specific pre-cutting and jetting energies);
- stabilizing grout mix composition;
- unitary cement consumption.

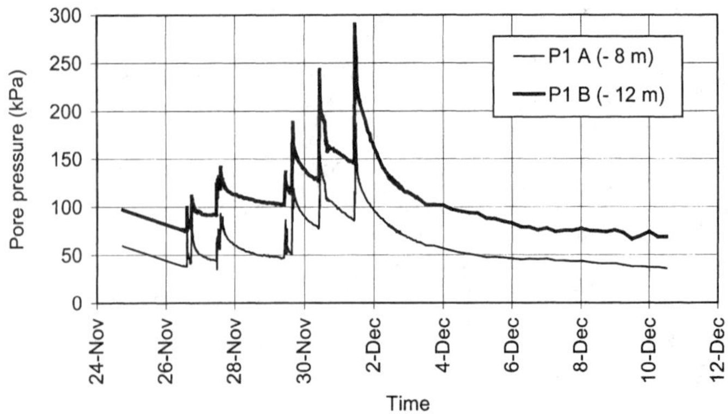

Fig. 12. Pore pressure versus time - Location Pz1

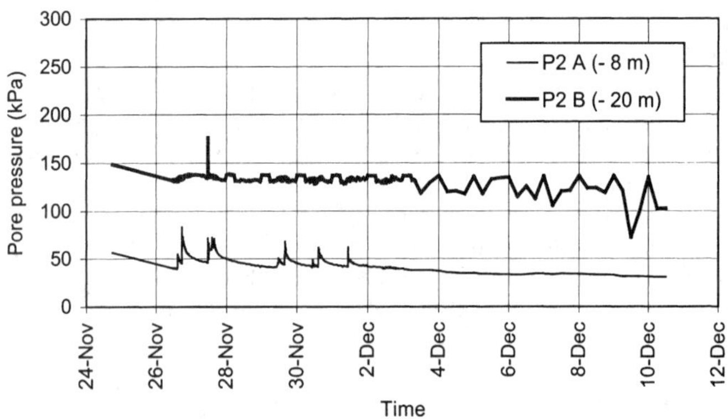

Fig. 13. Pore pressure versus time - Location Pz2

The average actual diameters are of the order of 2.2 ÷ 2.3 m, with maximum values of 2.8 m for columns 1 e 5. Central columns of areas A and B are characterized by 1.8 ÷ 1.9 m diameters.

As shown in Fig. 6 pre-cutting appears to allow the achievement of similar diameters as obtained without pre-cutting, despite a lower unitary cement contents, referred to linear meter of column.

According to Fig. 7, the columns installed by using pre-cutting show the highest strength values; conversely, this behaviour is not clear as far as it concerns the elastic moduli values.

As far as it concerns geometrical and mechanical characteristics (in terms of cubic compressive strength and elastic modulus), column 5 appears to present the best results.

According to the above mentioned remarks, the following guidelines were established:

- installation of columns by using pre-cutting;
- the provided pre-cutting has to be carried out upward, immediately before jetting;
- choose of nozzles combination has to be carried out in order to limit the fluid flow rate and the jet cutting power, to reduce the radius of action;
- pre-cutting and jetting energies have to be chosen sufficiently low to avoid large dimension columns;
- adoption of a stabilizing grout with high C/W ratio.

The operative parameters to be included in the design specifications should fit the requirements illustrated in Table 2.

It is interesting to point out another important remark coming from the trial field: the jet grouting implies to treat and carry out to a controlled waste a large volume of slurry, in the order of 0.4 m^3/m of column.

Table 2. Suggested operative parameters

Type of grout		$C / W \geq 1.6$ ($W / C \leq 0.63$)	
Grout nozzles (No. and diameter)		n° 2	$2.6 \leq D \leq 3.0$
Compressed air nozzles (No. and type)		n° 2	Annular and coaxial
Pre-cutting	Water pressure (MPa)	$p_w = 30$	
	Water flow rate (l/min)	$150 \leq q_w \leq 220$	
	Air pressure (MPa)	$0.8 \leq p_a \leq 1.2$	
	Air flow rate (l/min)	$6000 \leq q_a \leq 10000$	
	Unitary water consumption (l/m)	$V_{uw} = 500$	
	Withdrawal speed (m/min)	$v_e = q_w / V_{uw}$	
	Rotational speed (rpm)	$10 \leq v_r \leq 15$	
Jetting	Grout pressure (MPa)	$p_m = 40$	
	Grout flow rate (l/min)	$140 \leq q_m \leq 190$	
	Air pressure (MPa)	$0.8 \leq p_a \leq 1.2$	
	Air flow rate (l/min)	$6000 \leq q_a \leq 10000$	
	Unitary grout consumption (l/m)	$V_{um} \geq 750$	
	Unitary cement consumption (kg/m)	$Q_c \geq 790$	
	Withdrawal speed (m/min)	$v_e = q_m / V_{um}$	
	Rotational speed (rpm)	$10 \leq v_r \leq 15$	
Total cutting specific energy (MJ/m)		$45 \leq E_s \leq 60$	

Grouting Repair of Seawall and Revetment, Dana Point Harbor, California

Jeffrey Geraci[1], and Frank Nonamaker[2]

ABSTRACT: Located at the westerly end of Dana Point Harbor, the Ocean Institute annually educates over 78,000 children through "immersion" expeditions into the wonders of the marine environment. A long awaited expansion of the facilities was begun in early 2001, to create the Ocean Education Center. This new waterfront campus is a 3,066 m^2 working marine research station, intended to maximize student participation in the learning process. Remedial grading for construction of the campus expansion involved removal and recompaction of old fill soils, immediately inland of the adjacent seawall and revetment. Excavation procedures included a shoring wall for protection of the seawall during soil removals. A sudden breach in the shoring system during a high-tide event flooded the construction excavation with over 10,000 m^3 of seawater. Local erosion during this event resulted in subsidence of the revetment and a voided condition in the vicinity of the breach, leaving an approximately 8-m section of the seawall foundation unsupported. Consequently, repair of the breach was essential for construction to proceed. The repair process involved application of four discrete grouting techniques. Macroscopic voids were filled with a bentonite/cement slurry grout in the immediate vicinity of the shoring wall. Limited-mobility displacement grouting was used to compensate for local soil loss and to supplement inbound support of the revetment. Gravel was then packed beneath the seawall behind a bulkhead, and permeated with a Type-III cement grout. Smaller voids beneath the seawall were grouted with a fly ash/cement mix. The water was then successfully pumped from the excavation, and construction of the campus was able to resume. The Institute's Ocean Education Center opened in October 2002.

----

[1]Member, ASCE, Project Manager, CEG, Moore & Taber, 1290 North Hancock Street, Suite 202, Anaheim, CA 92807; phone 714-779-0681; j.geraci@mooreandtaber.com
[2]Affiliate Member, ASCE, Operations Manager, Moore & Taber, f.nonamaker@mooreandtaber.com

INTRODUCTION

Dana Point was named for the famous seafaring novelist Richard Henry Dana, Jr., whose classic work "Two Years Before the Mast" (Dana, 1841) provides a detailed account of his 1834 – 1836 round-trip journey from Boston, Massachusetts to the California coast. A full-scale replica of Dana's trading brig "Pilgrim" is anchored in the westerly end of the harbor, and is today the Orange County Ocean Institute's largest classroom, providing a national award-winning living history program to thousands of students each year (Ocean Institute, 2002, Dana Point Harbor Association, 2003).

Construction for the Ocean Institute's facility expansion began in early 2001. Project specifications called for remedial grading, involving removal and recompaction of older fill soils to the bedrock surface, approximately 5 m below sea level. Project planning called for a vertical excavation, immediately west of the existing seawall and revetment in the west end of the harbor. A shoring system, consisting of timber lagging between 0.46-m cast-in-drilled-hole (CIDH) piles, spaced at 2.4 m on-center was implemented to provide support for the sea wall during the excavation and recompaction phase (Figure 1). This shoring system was breached during a high tide event in the spring of 2001, flooding the nearly completed excavation with approximately 10,000 m^3 of seawater (Figure 2).

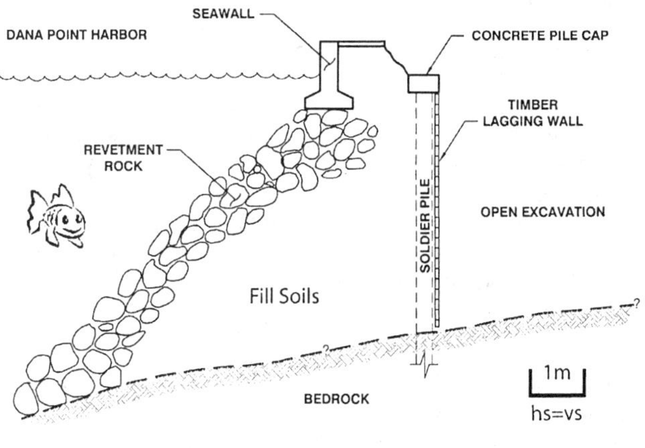

FIG 1. Generalized cross-section, depicting pre-failure condition at the shoring wall.

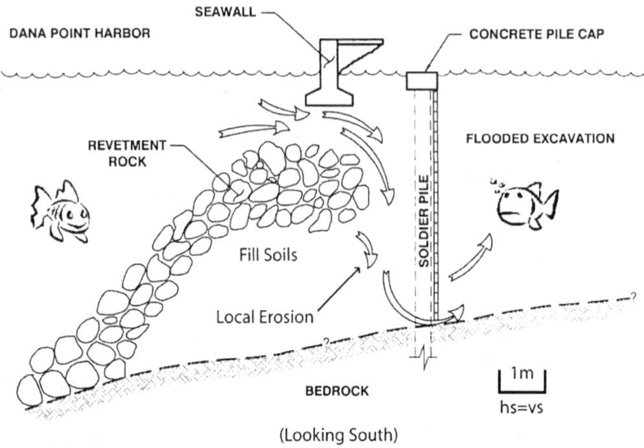

FIG 2. Generalized cross-section, illustrating breached condition.

Local erosion resulting from the rapid inflow of water resulted in a breached zone of approximately 8 m to 10 m wide. Revetment rock above the eroded fill soils subsequently receded into the void, leaving an approximately 8-m section of the seawall footing unsupported. Consequently, emergency repair measures were required to seal the breach.

REMEDIAL DESIGN AND CONSTRUCTION

The first consideration in developing a remedial approach was to provide access for construction crews and equipment to the landward side of the shoring wall. Gravel was consequently imported to the site and placed along the excavation side of the lagging wall. Steel plates were then driven alongside the timber lagging to the bedrock surface, within the area of the breach. Gravel was also placed within the eroded area between the pile cap and the seawall. The intent of installing steel plates was not to mitigate the flow of water, but to provide a means of physically confining the lateral migration of slurry grout into the adjacent gravel. Grout casing was then installed between the lagging wall and the steel plates to the bedrock surface. Slurry grout, consisting of 10 to 12 kg bentonite, 80 to 90 kg Portland cement, and sufficient water to achieve a 14- to 16-second flow, as measured in accordance with ASTM C 939-97 (ASTM, 1997) was then injected through the casing to fill macroscopic voids between the steel plates, lagging wall and revetment fill (Figure 3).

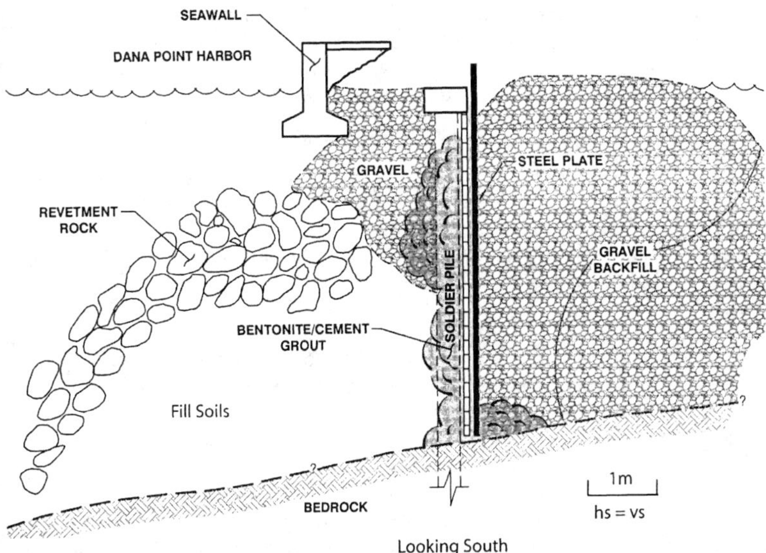

FIG 3. Generalized cross-section, showing placement of gravel, installation of steel plates, and injection of slurry grout at the breached zone.

Grout was injected at relatively low pressures (typically less than 200 kpa) until surface return was observed west of the pile cap. A total of 15,886 liters (4,197 gallons) of slurry grout were injected into this area.

The next phase of grouting involved pressure injection of a limited-mobility displacement (LMD) grout, consisting of silty sand, 8 to 10 percent Portland cement and sufficient water to achieve a relatively low slump (approximately 5 cm to 8 cm).

The intent of LMD grouting is to inject grout that stays where it is placed, while displacing a portion of the material into which it is injected (Byle, 1997). Casings were installed through holes drilled at the top of the pile cap to bedrock, spaced at approximately 0.8 m, and positioned between the CIDH piles throughout the 100-m length of the shoring wall. The intent here was to displace fill soils beneath the subsided revetment, not only in the area of the breach, but in order to identify and address any additional areas of instability along the excavation's margin.

The ground surface was monitored locally for uplift with manometers, while the railing at the top of the seawall was monitored for both vertical and lateral movement with optical levels. Ground movements on the order of 2 mm per grout injection stage were allowed, whereas a zero-movement tolerance was established for the seawall. These tolerances were established to restore positive contact between the base of the seawall footing and the subgrade, while preventing damage to the concrete seawall sections.

Typical injection pressures ranged from 600 to 1400 kpa. Grouting was performed in a "bottom-up" sequence, that is, completing grout injection stages at the base of the treatment profile prior to completing subsequent overlying intervals. Figure 4 shows the positioning of LMD grout injections in the context of the previous remedial phases. Several soft soil zones were identified and addressed throughout the length of the excavation's margin during LMD grouting operations.

A total of 76 m^3 of LMD grout was injected in the target treatment zone, including 12 m^3 at the breached section. Comparison of tide level fluctuations in the harbor with water levels in the flooded excavation began to show a marked disparity at the conclusion of the LMD grouting phase, indicating a significant reduction in hydraulic communication.

A permeation grouting approach was selected to address support of the seawall footing in the area of the breach. This remedial phase was particularly challenging, as access to the work zone could only be obtained at low tide.

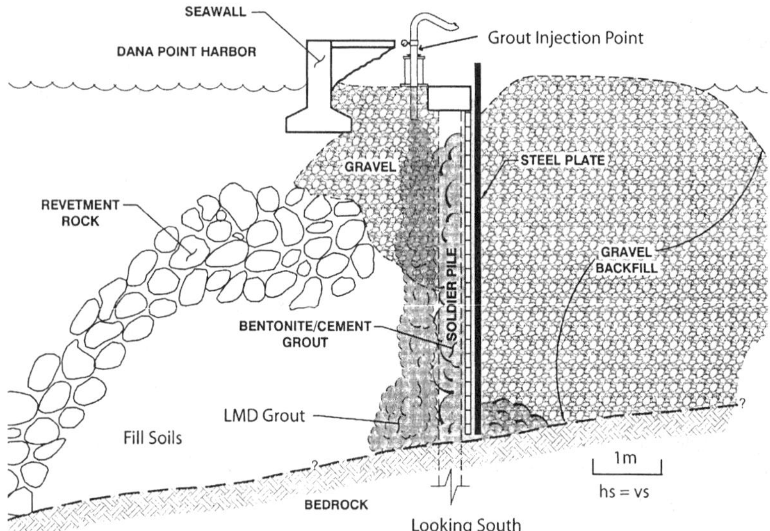

FIG 4. Generalized cross-section, showing placement of LMD grout relative to previous remedial construction phases.

Approximately 20 mm diameter open-graded gravel was first packed beneath the footing, behind a temporary bulkhead. The bulkhead was fitted with grout injection ports at the outer face. The plan was to then inject permeation grout through the bulkhead to treat pore space within the gravel packing as shown in Figure 5.

An initial attempt at grouting was made with a mix consisting of 9 kg bentonite, 125 to 130 kg Type-III cement and sufficient water to produce a 12 to 16 second flow (ASTM, 1997). Bentonite was initially selected as a grout additive to control bleed water. Upon removal of the bulkhead, it was discovered that the initial mix design either lacked the ability to permeate the gravel below the level of the rising tide, or was being physically washed out. Consequently, the gravel packing and bulkhead were replaced, and the mix design revised under the assumption that tide action had contributed to grout washout.

The revised grout mix design consisted of 125 to 130 kg Type-III cement, 15 kg undensified silica fume and sufficient water to achieve similar flow rates as with the initial attempt. Silica fume was selected as a grout additive in the revised mix because of its favorable cohesiveness and tendency to prevent washout of the cement during underwater placement (ACI Committee 234, 1996). Removal of the bulkhead 24 hours after treatment with this mix revealed thorough permeation of the gravel beneath the seawall footing, and a solid grouted mass (Figure 6). A total of 15,744 liters of permeation grout were injected.

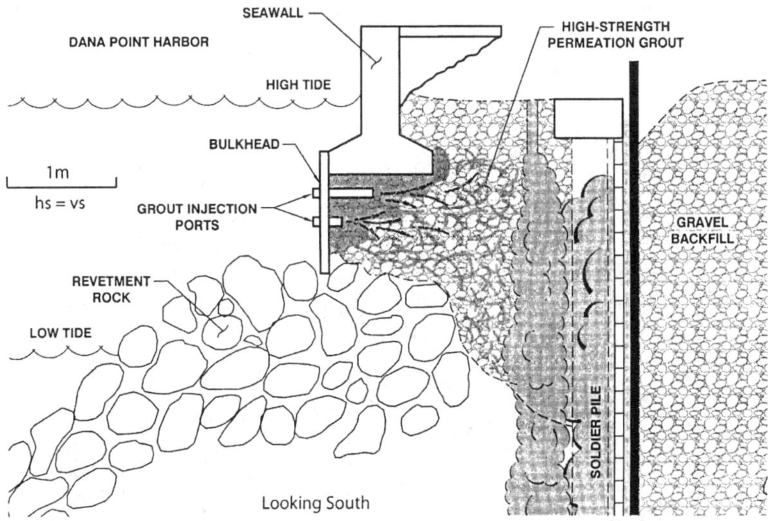

FIG 5. Generalized cross-section, showing permeation grouting beneath the seawall in the vicinity of the former breach.

Smaller voids (on the order of 1 to 3 cm) were identified along the face of the seawall footing. These anomalies were addressed with a contact grouting approach, using a mixture of 100 kg Type-F fly ash, 45 kg Portland cement and sufficient water to achieve a 10 to 16 second flow (ASTM, 1997). The specific mix design used is provided in Caltrans Standard Specifications (1999). Bulkheading of the seaward section was accomplished with sand bags, and grout was pumped through sacrificial injection tubes.

Floodwater was pumped from the excavation at the conclusion of all remedial grouting operations, and remained dry for the remainder of the construction phase. Replacing the boulders that were lost at the initial breach subsequently repaired the revetment.

CONCLUSIONS

The presence of soft soil zones, identified during LMD grouting procedures, along the excavation's margin indicated that less-than-favorable bearing conditions had developed beneath the seawall since its original construction in 1971. Consequently, the engineer's decision to address the entire length of the excavation with LMD grout treatment proved to be quite beneficial to this project's successful outcome.

Substitution of silica fume for bentonite in the permeation grout mix design provided sufficient cohesiveness to prevent washout by tidal action during the set, while maintaining its ability to control bleed.

FIG 6. Permeation grouted gravel beneath the seawall footing.

Frequent site meetings were conducted, involving the owner's representative, the architect, general contractor, engineer, and grouting subcontractor. Good jobsite communication practices and cooperation among all parties contributed to the overall success of this remedial repair project.

ACKNOWLEDGEMENTS

We would first like to acknowledge Iraj Poormand, GE, of Leighton & Associates, who provided the engineering and design parameters for this remedial repair. We also acknowledge Jim DePoorter, who drafted the figures for this manuscript, and Karen Geraci, PE, for her thoughtful review comments. Finally, we would like to thank Mr. Dan Gee, President of the Ocean Institute, for authorizing the presentation of this case study.

REFERENCES

ACI Committee 234 (1996). "Guide for Use of Silica Fume in Concrete" *American Concrete Institute Committee 234*, ACI 234R-96, pp 1-53

ASTM (1997). "Standard Test Method for Flow for Preplaced Aggregate Concrete (Flow Cone Method)", Annual Book of ASTM Standards, Volume 4.02, 'Concrete and Aggregates', American Society for Testing and Materials, 1998, pp 469-471

Byle, M. J. (1997). "Limited Mobility Displacement Grouting: When 'Compaction Grout' is Not 'Compaction Grout'", Proceedings of sessions sponsored by the Grouting Committee of the Geo-Institute of the American Society of Civil Engineers in conjunction with the Geo-Logan '97 Conference, Grouting, Compaction, Remediation and Testing, Geotechnical Special Publication No. 66, pp 32-42

Caltrans (1999). "Section 41: Pavement Subsealing and Jacking", Standard Specifications, State of California Business, Transportation and Housing Agency, Department of Transportation, pp 311-315

Dana, Jr., R. H. (1841). "Two Years Before the Mast", P.F. Collier & Son, 1909-14, New York, Volume 23 of 51

Dana Point Harbor Association (2003). "Welcome to Dana Point Harbor", Dana Point Harbor, Dana Point, CA, http://www.danapointharbor.com>(March 28, 2003)

Ocean Institute (2002). "Dana Point Tallships", Ocean Institute Visitor Info, http://www.ocean-institute.org/html/tallships.html>(March 29, 2003)

JET-GROUTING PERFORMANCE IN TUNNELLING

Paolo Croce[1], Giuseppe Modoni[2], Giacomo Russo[3]

ABSTRACT: The paper focuses on the role of jet grouting on tunnel support, although the influence of supplementary techniques is also considered. For this purpose, experience gained from two well documented case histories is presented. Reported observations and experimental data have been selected in order to help in the identification of proper design limit states. Jet grouting has proven to be quite effective, allowing tunnelling in very difficult conditions, but some failures have also been recorded. Failure mechanisms can be classified according to three different modes: excavation face failure, canopy tip failure, provisional lining failure. With regard to the control of settlements, jet grouting has proven to be very effective but some ground uplift can be temporarily induced by soil treatment.

INTRODUCTION

The construction of shallow tunnels through loose soils is still one of the most challenging problems of civil engineering and different techniques have thus been proposed, in several countries, each of them having its own advantages and shortcomings (Mair and Taylor, 1997). In Italy, the morphological and geological features of the ground are extremely variable and therefore flexible construction methods are usually preferred (A.G.I., 1985).

In fact, tunnel design in poor soil conditions is typically achieved by combining different soil improvement and soil support techniques: grouting, jet grouting, micropiles, etc. In principle, each technique is best suited for particular soil types and

[1] Associate Professor, ASCE Member, University of Cassino, Via Di Biasio 43, 03043 Cassino (Fr), Italy, phone +390776299644; e-mail croce@unicas.it

[2] Researcher, University of Cassino, Via Di Biasio 43, 03043 Cassino (Fr), Italy, phone +390776299738; e-mail modoni@unicas.it

[3] Researcher, University of Cassino, Via Di Biasio 43, 03043 Cassino (Fr), Italy, phone +390776299385; e-mail giarusso@unicas.it

seepage conditions. The construction procedure can be thus customized, for each particular case, allowing also for rapid changes on site in case unforeseen conditions were encountered. The paper focuses on the role of jet grouting for tunnel support, although the influence of supplementary techniques (reinforcing steel or fibreglass elements, shotcrete, steel ribs, etc.) is also considered where appropriate.

It is recalled that, in a limited number of cases (Lunardi et al., 1986; Cresta and Serra, 1991), jet grouting has been performed from the ground level, before excavating the tunnel. In general, however, jet grouting treatments are carried out underground. In this case, typical design procedures can be subdivided in two main categories (Fig. 1), according to the construction sequence: divided face (e.g. Ceppi et al. 1989) and full face (e.g. Bertoli et al.1991). In both cases, jet grouting is performed in sub-horizontal direction arranging the columns in order to obtain a sort of canopy, covering the soil to be excavated. By this means, temporary support of the underground opening can be provided in advance and excavation can proceed at consecutive spans usually ranging from 6 to 10 m. Supplementary horizontal treatments may also be added, in order to create a block of rigid material behind the tunnel face (Cresta and Serra, 1991). The latter method is rather expensive and time consuming but can provide effective support and minimize surface settlements as well.

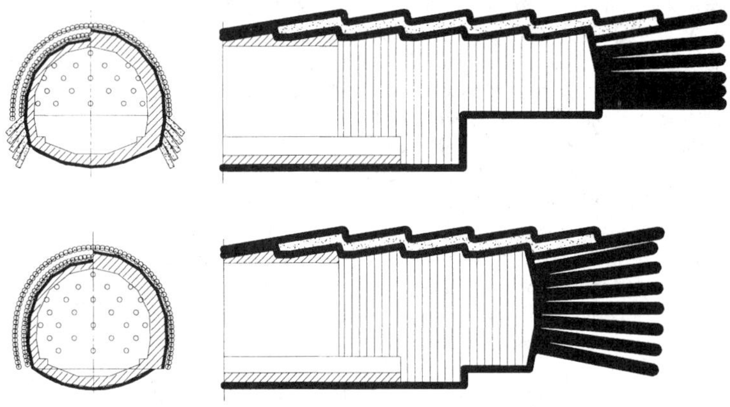

FIG. 1 Typical construction sequences of jet-grouted tunnels.

Jet grouting is usually performed by the single-fluid method, which does not require very heavy equipment. After soil treatment, the jet columns forming the canopy are often reinforced by steel tubes or bars, while the jet columns of the frontal block may be reinforced by fibreglass tubes or bars. These latter can then be easily removed during subsequent excavation.

As it usually happens with new construction techniques, knowledge in the use of jet grouting for tunnel support has been progressively gained by means of a trial and error procedure. A rather small part of this experience, however, has been diffused throughout the engineering community. It is known, in particular, that some failures have occurred in the past, but such failures have not been reported in the literature.

In the following, some relevant experience gained from two well documented case histories is presented. Reported observations and experimental data, taken from both cases, have been selected in order to help in the identification of proper design limit states.

LES CRETES TUNNEL

The Les Cretes case history regards two parallel highway tunnels, built some years ago in the north-western Italian Alps, between the city of Aosta and Mont Blanc. The tunnels, spanned at a distance of 24 m from each other, follow a slightly curved and gently inclined outline, for a total length of about 1500 m. The cross section of each tunnel is about 13 m wide and 9 m high, while the soil cover ranges from a minimum of 3 m to a maximum of 50 m approximately.

On the Aosta side, the tunnels were built by cut and cover, for a stretch of about 200 m, where the soil cover was less then 10 m (Fig.2.a). The remaining 1300 m were excavated through a soil deposit of glacial origin (moraine), mainly composed of dense sandy gravels and silty sands (Fig.3) with erratic rock boulders. Excavation was accomplished by divided face sequence (Fig.2.b), using the canopy technique.

In particular, the top section was first excavated along the entire tunnel length and the bottom was then dug in a subsequent phase. Most of the soil strata were quite pervious and the water level was highly variable, due to sharp seasonal climatic changes. However, the water level was generally located above the tunnel crown and therefore most canopies were made by jet columns, aiming to provide protection from water inflow as well as mechanical support of the underground openings. A few canopies were made only by steel tubes, when finer and less pervious soil strata were occasionally encountered along the tunnel axis.

The jet grouting treatments were performed by the single fluid system, as usual in tunnelling, and some of the jet columns were also reinforced by steel tubes. Provisional lining was provided by steel ribs (IPN 180) placed at one meter span, plus a 20 cm thick fibre-reinforced shotcrete layer. The ribs were inserted during excavation, at a maximum distance of two meters from the tunnel face.

The most critical stage was the excavation of the top section, when some unexpected failures occurred at different locations. All failures developed abruptly and caused a large amount of soil to flow into the tunnel. In particular, for each failure, a sinkhole was progressively generated, starting from the tunnel and rapidly developing towards the ground surface.

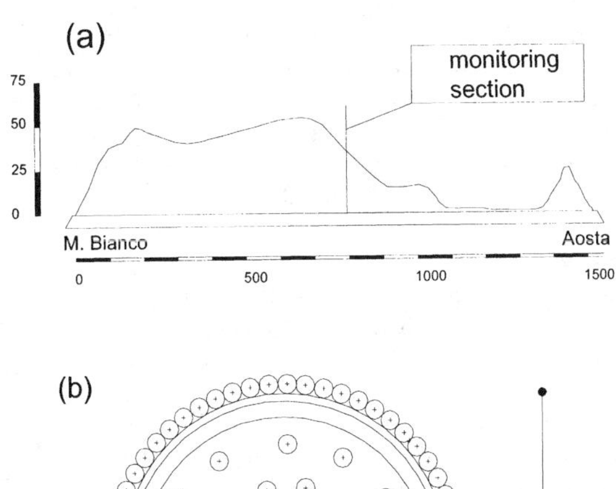

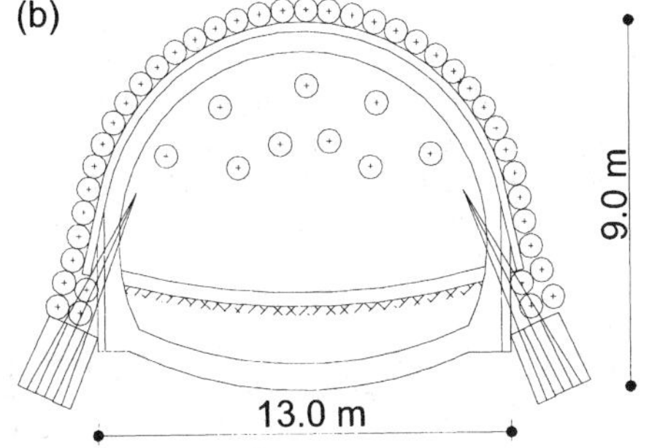

FIG. 2 Longitudinal profile (a) and cross section (b) of Les Cretes tunnel.

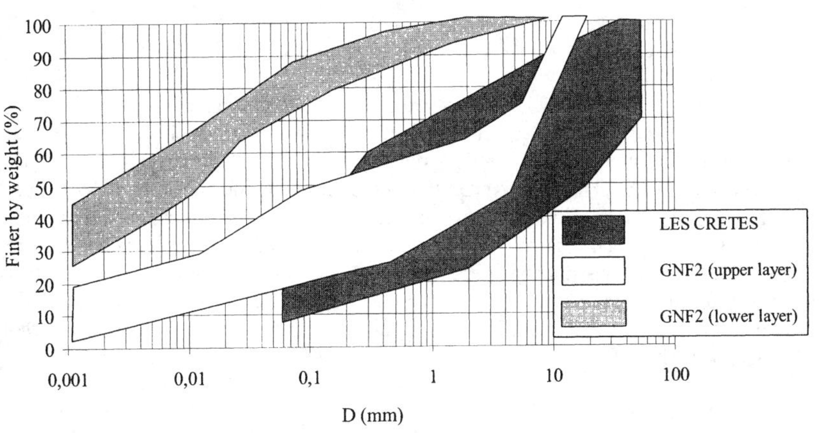

FIG. 3 Grain size distributions of Les Cretes and GNF2 soils.

The observed failure mechanisms can be classified according to three different modes. The first mode (Fig.4.a) consisted in the soil collapse at the excavation face and did not involve the canopy nor the provisional lining. This failure mode can be attributed to very weak layers of soil and is enhanced by piping induced by water seepage. It is also worth reporting that a reduction of pore water pressures was attempted by inserting drain tubes all around the tunnel contour, but this expedient was not effective due to soil heterogeneity and to pipe clogging induced by previously injected grout.

The second failure mechanism (Fig.4.b) consisted in the collapse of the canopy tip, just behind the tunnel face. Such failures, which involve also the excavation face, are mainly due to local defects of the jet canopy. In fact, it has already been noted that both diameter and mechanical properties of jet columns are highly variable due to natural soil heterogeneity (Croce et al., 2001). As a consequence, the overall resistance of the canopy may be reduced by relevant discontinuities.

The third failure mechanism (Fig.4.c) was observed on the tunnel crown and can be attributed to local defects of the provisional lining, including the jet canopy, the steel ribs and the shotcrete layer. This type of failure occurred only once in this case history.

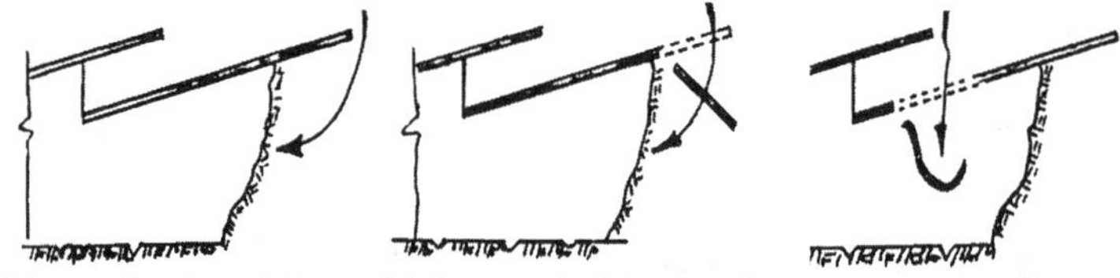

a) Excavation face failure b) Canopy tip failure c) Provisional lining failure

FIG. 4 Typical failure mechanisms observed on Les Cretes tunnel.

In the Les Cretes project, the movements induced by tunnelling were carefully recorded in a cross section where the soil cover was 29 m thick (see Fig.2). The position of the monitoring section, with respect to the construction spans, is also indicated in Fig.5. The soil movements were recorded by topographical observations of the ground surface, plus inclinometric and assestimetric measurements taken along five PVC tubes. The piezometric head was measured by one Casagrande type piezometer. These instruments were obviously placed before excavation. Three load cells, three strain gauges and three convergence bolts were also placed on a steel rib, immediately after excavation. Monitoring details and results were reported in detail by Croce et al. (1993).

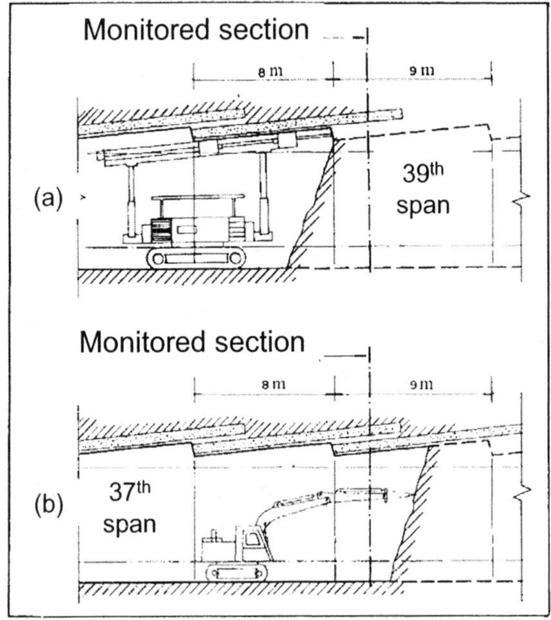

FIG. 5 Construction stages near the monitored section.

The settlements on the tunnel axis, taken at the ground level and at an intermediate level between tunnel and surface, as well as the piezometric head are plotted in Fig.6, as a function of the distance from the excavation face. The two settlement profiles show an almost rigid downward movement of the soil located above the tunnel. Both profiles could be subdivided in four subsequent stages. The first stage shows small initial settlements, induced during the previous excavation span, starting at the distance of about ten meters from the tunnel face. The second stage is characterised by a sudden but relatively small lift, occurred during jet grouting. The third stage, which shows the highest settlement rate, corresponds to the excavation span crossing the monitoring section. Finally a fourth long lasting increment of downward movements is observed during further excavation.

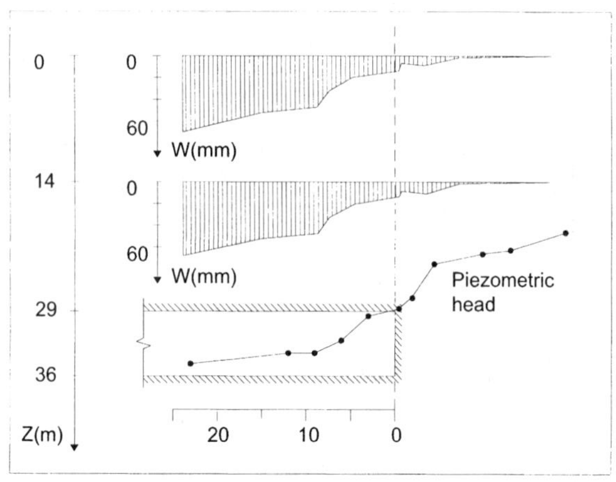

FIG. 6 Settlements profiles and water head at the monitoring section of Les Cretes tunnel.

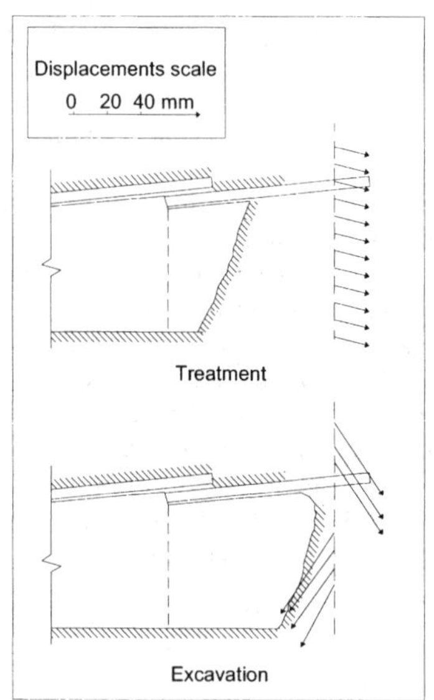

FIG. 7 Vertical and horizontal displacements measured at the monitoring section of Les Cretes tunnel.

The soil movements induced just behind the face are plotted in Fig.7 in two subsequent construction stages, namely immediately after the jet-grouting treatments and just before excavation. The plots show a clear correlation between soil displacements and construction stages. In fact, after jet grouting treatment, an almost horizontal forward displacement of about 20 mm was observed in all the monitored points. Such response can be regarded as an effect of the jet grouting and excavations performed before the monitored section. During the subsequent excavation step, the displacements of the soil elements respectively located above and below the jet canopy show different trends and a global rotation of the soil mass around the canopy tip can be recognised. This deformation trend, which confirms the bearing action of the canopy, is in substantial agreement with the face failure mechanism previously suggested.

GNF2 TUNNEL

The GNF2 tunnel belongs to the new "High Velocity" railway line, crossing the city of Florence. The tunnel is about 50 m long and connects two cut and cover sections of the same project. It was excavated, under very low soil cover (5÷7 m), in order to underpass a main urban street and its neighbouring buildings (Fig.8). The tunnel, which has a multi-centred cross section about 13 m wide and 10 m high (Fig.9), was recently completed with the aid of jet grouting treatments.

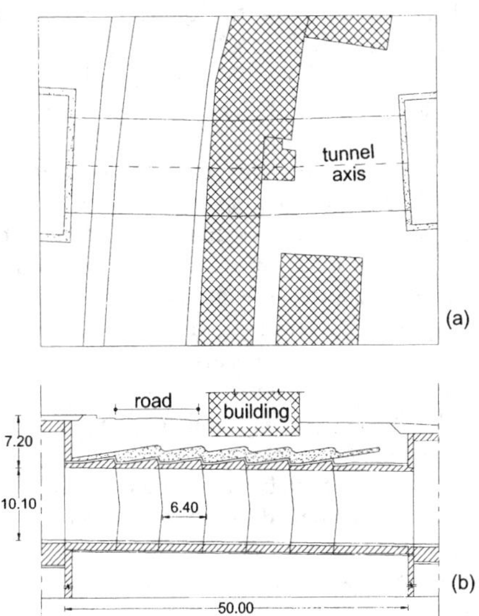

FIG. 8 Plan view (a) and longitudinal section (b) of the GNF2 tunnel.

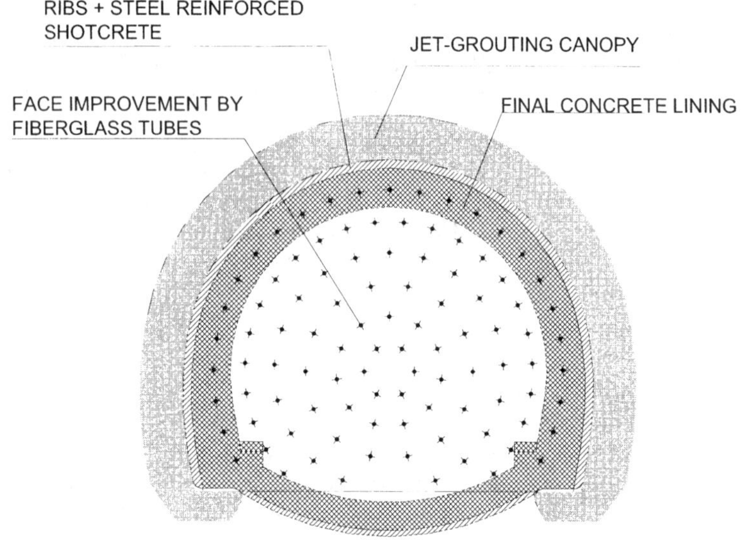

FIG. 9 Typical cross section of the GNF2 tunnel.

Design investigations conducted in the tunnel area detected a sequence of heterogeneous alluvial layers, which can be roughly grouped in two main strata (Fig.3). The upper stratum is about eight meters thick and is composed by heterogeneous angular gravel immersed in a silty matrix. The lower stratum is made by clayey and sandy silt.

The construction procedure was designed in order to ensure tunnel stability and to reduce surface settlements to allowable limits, with respect to the urban environment. In particular, the whole tunnel length was divided into eight excavation spans, each of them being about six meters long. For each span, a supporting canopy was made by 71 overlapping jet columns, each of them reinforced by steel tubes. Each canopy was about twice as long as the excavation span and thus a double supporting lining was provided during excavation. Soil improvement behind the tunnel face was also achieved, by means of 99 fibreglass tubes, injected with grout. The fibreglass tubes were placed in a single shot for the entire tunnel length. Excavation support was completed by a provisional lining consisting of 2 IPN200 steel ribs, 0.75 meter spanned, plus a 30 cm thick shotcrete layer reinforced by steel net. Definitive lining was made by cast in place reinforced concrete, conforming to the shape of the canopies, with thickness ranging from 55 to 110 cm.

Particular attention was paid to the technological details. For this purpose, preliminary field trials were made next to the construction site by performing vertical jet grouting treatments from the ground level. Minimum recorded values of column diameter and soilcrete uniaxial compressive strength were 0.55 m and 20 MPa respectively.

Ground movements were monitored during the field trials, in order to detect possible jet grouting effects at the surface level. In fact, upward vertical displacements were observed, ranging from few millimetres to some centimetres, and suggested some significant adjustments of the grouting procedures. In particular, additional holes were drilled next to the grouting holes to allow for quicker spoil discharge. Moreover, preliminary soil cutting by water jet treatment was performed, in order to reduce the grout pressure in the subsoil. Treatment parameters are reported in Table I.

Table I. Jet-grouting parameters adopted in GNF2 tunnel.

Water/ Cement ratio (1)	Grout pressure (MPa) (2)	Water pressure (MPa) (3)	Number of nozzles (4)	Diameters of nozzles (mm) (5)	Rotation speed (r.p.m.) (6)	Withdrawal speed (cm/min) (7)
0.95:1	35	30	2	2	10	48

Ground settlements continuously monitored, during tunnel construction, by means of topographical and assestimetric measurements, suggested, after the first excavation span to tune again the treatment procedures. The grout pressure was in fact reduced to 32 MPa on the side piers and to 25 MPa on the tunnel crown. Moreover, two grouting cycles were performed in each borehole, one with water and one with grout, by inserting and withdrawing the monitor twice. During grouting the monitor forwarding speed was increased to 80 cm/min.

Upward and downward movements along the longitudinal tunnel section (a) and in a monitored cross section, located at the beginning of the sixth span (b), are plotted in Figs.10. In particular, Fig. 10a shows the results of assestimetric measurements taken in the monitored section at a depth of two meters from ground surface. The vertical movements are reported as a function of the distance between the measuring point and the excavation face. The settlements profile shows that soil improvement by fibreglass bars and by jet grouting produced a significant lift which reached the maximum value of about 25 mm. As expected, the subsequent excavation produced a downward settlement. The final vertical displacements above the tunnel axis were downward oriented, in the range of 10 mm.

A more detailed picture of ground improvement and excavation effects can be observed in Fig.10.b, showing the vertical settlements at different points of the ground surface. These data were obtained by means of topographical measurements. In particular, four curves are reported, each of them related to different construction stages. The first curve relates to the introduction of the grout injected fibreglass tube behind the tunnel face. The corresponding displacements are upward oriented, with a maximum value of about 15 mm occurring above the tunnel axis (curve a in Fig. 10.b). Additional upward movements, shaped as the previous ones, were then caused by jet grouting (curve b in Fig.10.b). The maximum lift increment was about 10 mm, and determined a maximum global upward movement at the end of this phase equal to 25 mm. It is noted that these movements were not perfectly symmetrical. The third curve (curve c in Fig.10.b) shows the downward movements induced by excavation, just above the tunnel face. Compared to the previous curves a downward translation occurred in all the monitored points. However, downward movements continued as excavation progressed further away from the monitoring section as shown by curve d (Fig.10.b).

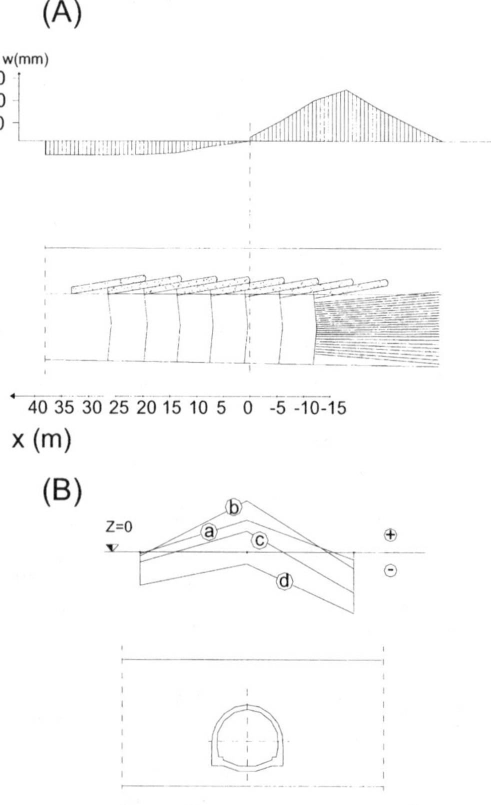

FIG. 10 Ground settlements longitudinal (a) and cross sectional (b) profiles of the GNF2 tunnel.

As a final comment about this case history, it can be concluded that the adopted ground improvement techniques allowed tunnelling in very difficult conditions, since a large diameter tunnel was successfully excavated in loose soils at a very shallow depth. However, as a secondary effect, some ground uplift was temporarily induced by face reinforcements and jet grouting.

CONCLUSIONS

The experience gained from the two reported tunnelling projects has pointed out that jet grouting is a very flexible method, which can be used also in conjunction with other ground improvement techniques. However different failure mechanisms may occur, as was observed in the Les Cretes case.

The observed failure mechanisms can be classified according to three different modes. The first mode consists in the soil collapse at the excavation face and does not involve the canopy nor the provisional lining. This failure mode can be attributed to very weak layers of soil and/or piping induced by water seepage. The most logical countermeasure for preventing this type of failure should consist in front reinforcement capable of providing tensile strength to the soil mass behind the excavation face. However, a suitable stabilising effect can also be obtained by increasing the distance from the tunnel face to the canopy tip, i.e. by making longer columns or by reducing the excavation span.

The second failure mechanism consists in the collapse of the canopy tip, just behind the tunnel face. Such failures, which involve also the excavation face, are mainly due to local defects of the jet canopy. In fact, it has already been noted that both diameter and mechanical properties of jet columns are highly variable due to natural soil heterogeneity (Xanthakos et al., 1994; Croce et al., 2001). As a consequence, the overall resistance of the canopy may be reduced by relevant discontinuities. This kind of failure could be avoided by careful construction control, canopy over-sizing and systematic introduction of reinforcing bars or tubes, being all these measures not mutually exclusive.

The third failure mechanism, observed on the tunnel crown, can be attributed to local defects of the provisional lining, including the jet canopy, the steel ribs and the shotcrete layer. This type of failure, which was observed only once, could be easily avoided by careful construction control.

At present, the risk of collapse is dealt with by oversizing the elements forming the canopy and the face reinforcement. This approach has been very successful in the GNF2 case, as well as in a large number of projects (i.e. Ceppi et al., 1989; Lunardi et al., 1986, Cresta and Serra, 1991). However, design reliability should be further investigated on the basis of probabilistic analysis (Katzenback et al., 2001, Croce and Modoni, 2002).

With regard to settlement control, jet grouting has proven to be quite effective allowing tunnelling in very difficult conditions, i.e. large multi-centred cross sections, loose soils and very shallow covers. However, as a secondary effect, some ground uplift can be temporarily induced by soil treatment. It is hoped that future research on jet grouting mechanisms will clarify the causes of local defects and soil bulging.

REFERENCES

A.G.I. Italian Geotechnical Society (1985), "Geotechnical Engineering in Italy", *Published on the occasion of ISSMFE Golden Jubilee*, pp.111-138.

Bertoli, A., Leone, A., Lombardi, V. (1991) "Aspetti tecnici e realizzativi di un sottopassaggio ferroviario ad un binario, attraverso l'esistente rilevato della linea Tirrenica a La Spezia realizzato con limitato spessore di copertura e senza interruzione del traffico ferroviario", *Proc. Conference Il Consolidamento del Suolo e delle Rocce nelle Realizzazioni in Sotterraneo, Milano, Società Italiana Gallerie,* vol. II

Ceppi, G., Maggioni, F., De Paoli, B., Stella, C., Lotti, A., Pedemonte, S. (1989) "Horizontal jet grouting as a temporary support for the Monte Olimpino 2 tunnel", *Proc. of Int. Conference on Progress and Innovation in Tunnelling*, Toronto

Cresta, L., Serra, A., (1991) "L'uso del jet-grouting nell'esecuzione di gallerie in terreni sciolti", *Proc. of the Conference Il consolidamento del suolo e delle rocce nelle realizzazioni in sotterraneo*, Società Italiana Gallerie, Milano

Croce, P., Francia, S., Giambanco, G., Pasquali, M., Speciale, G., (1993), "Analisi sperimentale di una galleria autostradale eseguita con l'ausilio del jet grouting", *Proc. of the XIII National Conference of Geotechnical Engineering (AGI - Italian Geotechnical Association)*, May 11-13th, Rimini (Italy).

Croce, P., Flora, A., Modoni, G. (2001) "Experimental investigations of jet-grouting", *Proc. of Conference on Foundations and Ground Improvement*, American Society of Civil Engineers (ASCE), Virginia Tech.

Croce, P., Modoni, G. (2002), "Numerical modelling of jet-grouted foundations", *Proc. of the V European Conference on Numerical Methods in Geotechnical Engineering*, Paris.

Katzenback R., Weidle A., Hoffmann, H. (2001), "Jet-grouting – Chance of risk assessment based on probabilistic methods", *Proc. of the XV ICSMFE*, pp.1763-1766

Lunardi, P., Mongilardi, E., Tornaghi, R. (1986) "Il preconsolidamento mediante jet-grouting nella realizzazione di opere in sotterraneo", *Proc. of the Conference Grandi opere sotterranee*, Florence, 2, pp. 601-612.

Mair, R.J. and Taylor, R.N. (1997), "Bored tunnelling in the urban environment", *State-of-the-art Report and Theme Lecture, Proceedings of 14th International Conference on Soil Mechanics and Foundation Engineering*, Hamburg, Balkema, 4., 2353-2385.

Xanthakos, P. P., Abramson, W. A., Bruce, D. A. (1994) "Ground control and improvement", John Wiley & Sons Inc.

GROUND IMPROVEMENT OPERATIONS FOR THE CONSTRUCTION OF THE "VIA GRAMSCI RAILWAY UNDERPASS"

Giovanna Cassani[1], Project manager, Andrea Bellocchio[2], Team manager, Luigi Nardacci[3]

ABSTRACT: The project involves the construction of a natural tunnel under Via Gramsci to connect the new high capacity railway line between Bologna and Florence with the Florence Railway Station located in the city centre. The ground involved belongs to the geological Formation of the *Bacino di Firenze*, consisting of silty sands and sandy silts with a varying content of clay and with lithic elements varying in size from the millimetric to centimetric. It is below the water table.

Located between two stretches of artificial tunnel and diaphragms consisting of columns of ground reinforced by jet-grouting, the tunnel is approximately 50 m. in length and has an overburden of just 5 m. Excavation is performed according to "ADECO-RS" principles which consists of:

- reinforcement of the face-core using fibreglass structural elements injected with expansive mortars;
- improvement of the ground around the tunnel using horizontal jet-grouting technology.

A field test was carried out to fine tune the operating parameters of the ground improvement works. The results obtained are shown in the paper.

INTRODUCTION

To connect the Milan-Naples High Capacity railway line with the Florence railway station it must pass under a road (Via Gramsci) that carries heavy traffic running between the centre of the city and towns in the North of the province. To achieve this, the design consisted of a natural tunnel, 50 m. long with artificial tunnels at each end and excavated between diaphragm walls. The tunnel is located in a heavily built up area (public and private buildings, underground utilities, etc.) and in an extremely difficult geological and geotechnical environment.

[1] Dott. Ing., ROCKSOIL S.p.A., Piazza San Marco 1, I-20121 Milano, Italy
[2] Dott. Ing., ROCKSOIL S.p.A., Via Nazionale 160, I-40065 Pianoro (BO), Italy
[3] SGF-INC S.p.A., Viale Italia 1, I-20099 Sesto S. Giovanni (MI), Italy

It was the poor quality of the ground which made resort to ground improvement ahead of the face necessary to guarantee the short and long term stability of the tunnel.
This paper illustrates how the type of ground improvement to be employed was chosen, how this was calibrated and the technology used to implement it.

GEOLOGICAL AND GEOTECHNICAL CONDITIONS

In the part that interests us, the route of the railway runs in a N/NE-S/SW direction, hydrographically on the left bank of the River Zambra, across the almost flat area of the built up Sesto Fiorentino area. The tunnel is located on the strip that joins the plane of Florence with the neighbouring hills and consists of numerous fluvial terraces on different levels. Geologically, the underpass passes through the recent fluvial-lacustrian deposits of the *Bacino di Firenze* (BF - Florence Basin), consisting of sandy silts and silty sands with a varying clay content. Lithic elements are present varying in size from the millimetric to the centimetric, sub-angular and sub-round in shape. The geological survey located the contact between the gravelly facies and the clayey silt facies right where the natural tunnel runs at a depth of 8.50 m. The upper ground is in contact with the gravels, mainly angular and sub-angular (∅ 10-60 mm) moderately weathered. The lower ground consists of locally sandy clayey silts with consistencey varying with depth (SPT increasing from 25 to 65), incorporating prevalently angular calcareous marly gravel with greatly reduced permeability than the ground of the upper facies (10^{-4} m./sec on average). On site and laboratory tests on core samples found that this ground possessed the following deformation and strength properties:

TABLE 1. Bacino di Firenze Formation: geotechnical parameters

Unit weight (kN/m^3)	True peak cohesion (Mpa)	True residual cohesion (Mpa)	True peak frictional angle (°)	True residual frictional angle (°)	Elasticity modulus (Gpa)	Poisson's ratio
21-22	0.01-0.02	0.0	29-32	18-20	0.06-0.08	0.25

DESIGN FORECASTS

The design and construction of the tunnel was performed according to the principles of the ADECO-RS approach which involves the design and construction of a tunnel in the phases outlined below.
The design stage consists of:

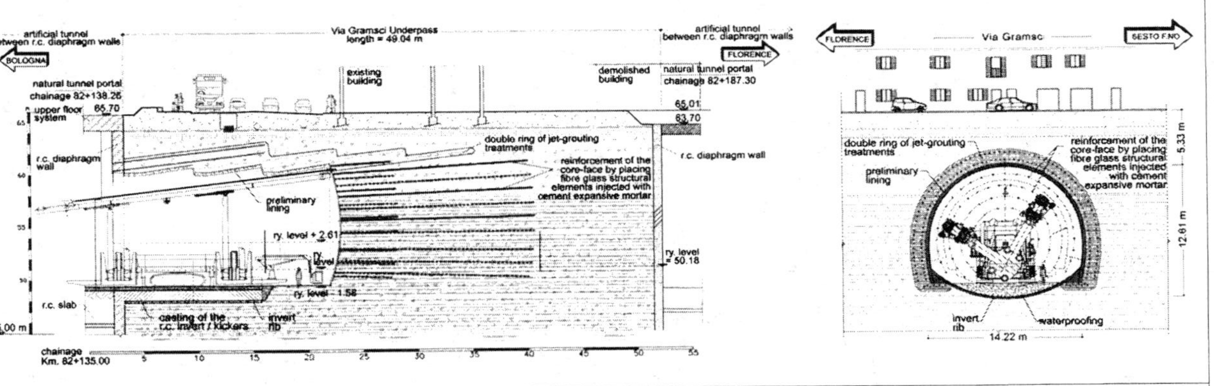

FIG.1. Scheme of the full section advance after having improved the ground ahead of the face and around the excavation.

The tunnel section type designed consisted of:

- improvement of the ground ahead of the face by placing 99 fibre glass structural elements set firmly into the ground by injecting them with low pressure expansive mortar;
- improvement of the ground around and ahead of the tunnel with a double ring of jet-grouting treatments or, alternatively, by means of fibre glass structural elements injected with mortar at low pressures;
- the preliminary lining consisting of 2IPN200 steel ribs at intervals of 0.75 m. and a 30 cm. layer of shotcrete reinforced either with steel fibre or mesh;
- water proofing consisting of non woven textiles and a layer of PVC sheets;
- final lining of reinforced concrete, 90 cm thick in the roof and 100 cm. thick for the tunnel invert.

- a survey phase to gain knowledge of the geology, geomechanics, geotechnics and hydrogeology of the medium to be excavated;
- a diagnostic phase using theory to predict the behaviour of the medium in terms of the deformation response of the face-core in the absence of stabilisation measures;
- a therapy phase to first define how to excavate and stabilise the medium in order to regulate the deformation response and to assess, using theoretical tools, the effectiveness of the stabilisation measures chosen;

The construction stage consists of:

- an operational phase during which the stabilisation instruments are put into operation in accordance with the design forecasts;
- a monitoring phase using empirical instruments to check the actual behaviour of the tunnel in terms of the deformation response by experimental means for the final calibration of the excavation and stabilisation systems adopted.

In the case of the tunnel in question, the stability analysis of the face-core in the absence of stabilisation performed in the diagnosis phase predicted unstable behaviour of the face (behaviour category C). On the basis of this prediction, full face tunnel advance was decided in the therapy stage after first improving the ground ahead of the face and around the excavation.

The tunnel section type designed consisted of (figure 1):

- improvement of the ground ahead of the face by placing 99 fibre glass structural elements set firmly into the ground by injecting them with low pressure expansive mortar;
- improvement of the ground around and ahead of the tunnel with a double ring of jet-grouting treatments or, alternatively, by means of fibre glass structural elements injected with mortar at low pressures;
- the preliminary lining consisted of 2IPN200 steel ribs at intervals of 0.75 m. and a 30 cm. layer of shotcrete reinforced either with steel fibre or mesh;
- water proofing consisting of non woven textiles and a layer of PVC sheets;
- final lining of reinforced concrete, 90 cm thick in the roof and 100 cm. thick for the tunnel invert.

TEST FIELD FOR THE GROUND IMPROVEMENT TECHNOLOGIES

The requirement to guarantee effective implementation of the ground improvement operations without causing damage to the underground utilities (gas and power lines, fibre optics, etc.) located just a few metres above the ground to be improved and without raising the street level and interrupting traffic flows suggested the idea of a "test field" in which to perform extensive experimentation of the ground improvement operations around the tunnel. The purpose was to define the technology and the most appropriate operating parameters for the conditions to be tackled.

Location and method

The "test field" was set up right next to the location of the actual underpass by excavating a trench between the diaphragms of the portal of the natural tunnel. The walls of the trench were protected by a 10 cm layer of shotcrete, reinforced with steel mesh.

Three zones were identified on this wall where the ground was to be reinforced in three different ways (figures 2 and 3):

- Zone A: with mortar injections from valved tubes positioned in a grid pattern of 1.20 x 1.20 m.
- Zone B: with injections form valved tubes as above but positioned in a grid pattern of 0.80 x 0.80 m. as above
- Zone C: with jet-grouting.

The volume of ground reinforced was measured for each type of treatment as was the effect in terms of strength and deformation properties by comparing the treated ground with natural undisturbed ground.

Test field for the jet-grouting technology

The experimentation was performed in two stages. In the first stage the technology to be employed and the operating parameters most appropriate to the ground improvement were identified (diameter and strength of the columns) which would not cause excessive disturbance to the surrounding ground. In the second stage a portion of roof treatment was performed to check the result.

The ground treatment was performed using a normal drilling rig equipped with devices to regulate the speed of extraction of the rods during the very high pressure injections. Special pressure meters measured the continuity of the pressure of the grout as it was injected and recorded any falls in pressure along the circuit of the cement in suspension.

During the tests, the delivery pressure was maintained within the 25-40 Mpa range.

The grout mix employed consisted of water and cement in a ratio of between 1.0 and 1.1.

In the first phase, two isolated columns 10 m. in length with a minimum diameter of 127 mm. were created . The first part of each was created using standard jet-grouting, while the part near the end of the hole was created using the hydraulic precutting technique where the actual ground improvement treatment itself is preceded by a high pressure injection of water which breaks up and carries away the finer particles of the ground.

While the standard technique resulted in greater raising of the ground at street level because the hole drilled tended to block causing excess pressure with uncontrollable effects, the hydraulic precutting technique, on the other hand, tended to blow material away from the delivery outlet and create columns of ground with a diameter of 40-80 cm. without any appreciable raising of the ground at street level.

The flow rate of the mix was measured volumetrically and the injection pressure was also measured constantly during the operational phases.

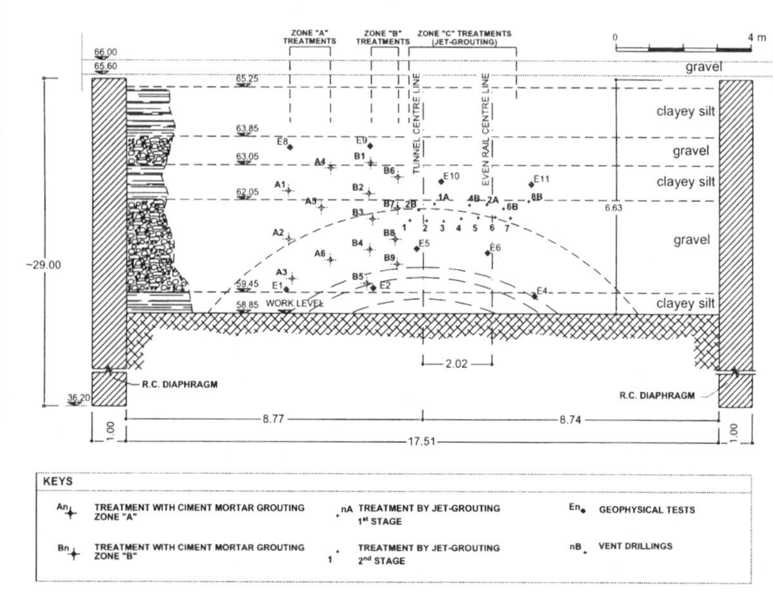

FIG. 2. Scheme of the test field

FIG. 3. Photo of the test field

The operating parameters for the test are given in the table below.

TABLE 2. Jet-grouting technology test field: geometries and operating parameters

COL. No.	Injection length (m.)	Nozzles No.	Nozzles Diam (mm.)	Suspension pressure (Mpa)	Extraction speed (cm/min)	$V_{ROT.}$ (rev/min)	Suspension Density (kN/m³)	Suspension Doses (C/W)
1	10.0	2	2.2	40	32	10	14.81	~1
2	10.0	2	2.0	40	26	8	14.81	~1
3_P	7.0	2	2.2	25	36	10	9.81	~1
3	10.0	2	2.2	40	32	10	14.81	~1
4_P	7.0	2	2.4	25	44	10	9.81	~1
4	10.0	2	2.4	40	40	10	14.81	~1
5_P	10.0	2	2.4	30	32	10	14.81	~1
6_P	10.0	2	2.0	35	36	10	9.81	~1
7_P	10.0	2	2.0	30	26	8	14.81	~1

Consumption

COL. No.	Flow rate (l/min)	Volumes (l/m)	Cement (kg/m)
1	95	300	~230
2	80	300	~230
3_P	95	250	~230
3	95	300	~230
4_P	115	250	~230
4	115	290	~230
5_P	95	300	~230
6_P	80	250	~230
7_P	95	290	~230

("$_p$" indicates the use of hydraulic pre-cutting technique)

Field test for the "mortar injections from fibre glass structural elements fitted with valved tubes"

As previously stated, two different types of geometry were employed in the experiment:

- the first geometry consisted of 6 fibre glass structural elements consisting of 3 plates 40x6 mm, 10 m in length (A1 - A6) set in a grid pattern (1.20 x 1.20 m.) injected with a cement mortar through PVC tubes fitted with three valves "a manchette" each metre;
- the second geometry consisted of 9 fibre glass structural elements similar to those above (B1 - B9) but set in a grid pattern of 0.80 x 0.80 m. injected through PVC tubes fitted with only two valves each metre;

The injection holes were drilled using a machine capable of performing the rotary drilling or rotary percussion drilling and of placing a temporary lining for the circulation of fluid, water or air (according to the conditions of the ground and the stability of the walls of the hole drilled).

The holes were drilled in a simple alternating sequence of 1-3-5-7-9 and then 2-4-6-8.

A fairly fluid mortar mix was used for the injections, given the shallow overburden, the presence of underground utilities and the rather "closed" nature of the ground with low absorption properties.

Once the sheath of cement inside the hole drilled (casing grout) had set sufficiently, the grout was injected into the ground in stages, proceeding valve by valve, starting with the valve at the end of the hole by means of a double packer connected to the injection circuit. A pressure metre between the injector and the PVC tube was used to monitor the injection pressure. The volume absorbed, the pressures and the water to cement ratios of the injections are given in the table below.

TABLE 3: Mortar injections test field: geometries and operating parameters

Description		**1st Geometry (Zone "A")**	**2nd Geometry (Zone "B")**
Number of holes drilled		6	9
Length		10.00 m	10.00 m
Diameter		>114 mm	>114 mm
Angle		~0%	~0%
Distance between centres		1.20 m	0.80 m
Distance between rows		1.20 m	0.80 m
Material injected		Cement suspension	Cement suspension
Number of valves		3 valves/m	2 valves/m
SHEATH	Water/Cement ratio	0.9<W/C<1.1	0.9<W/C<1.1
	Bentonite/Cement ratio	0.01<B/C<0.03	0.01<B/C<0.03
INJECTION	Water/Cement ratio	0.7<W/C<0.8	0.7<W/C<0.8
	Bentonite/Cement ratio	0	0
	Volumes	65<V<110 l/m	30<V<50 l/m
	Pressures	< 1 Mpa	< 1 Mpa
	Flow rate	< 25 l/min	< 25 l/min

Final measurements

Once the last field test had been completed, the next phase involved checking the results obtained with the different technologies and types of action employed in the experiment in terms of geometry, mechanics and the volumes of improved ground produced. Ample use of on site and laboratory tests were employed for this purpose.

The first included

- *3D electrical surveys (horizontal cross-hole and tomographic)* to locate and determine the shape and extension of the treatments performed in the field tests;
- *Down hole seismic surveys*, to measure the before-after treatment change in the speed of seismic waves, both horizontally and vertically, and to measure the increase in the dynamic elastic moduli obtained with the different types of ground improvement (figure 4).

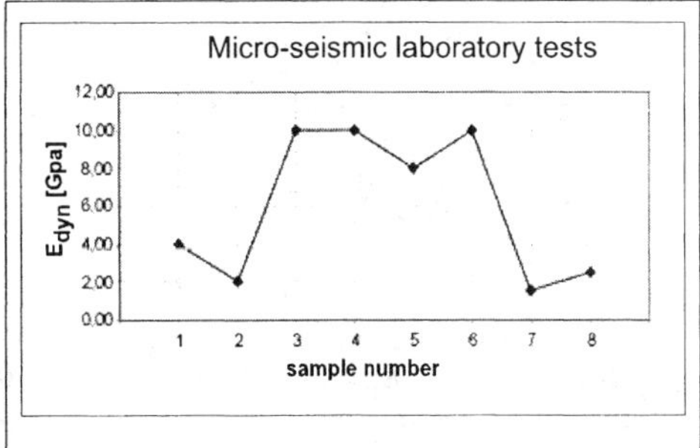

FIG. 4. Dynamic elastic modulus (column No. 7)

In the zones treated with mortar injections (zones A and B), there was an average increase in the dynamic elastic modulus (E_{dyn}) equal to 50-60%, while in the jet- grouted zone this was 30-40%. If, however, we consider the different thicknesses of the ground treated, it is clear that the contribution to the increase in E_{dyn} per unit of thickness using jet-grouting was much greater than that per unit of thickness with the mortar injections. This, however, is intrinsic to the nature of the two types of ground improvement: "diffuse" through manchettes and "concentrated" using jet- grouting.

The laboratory tests were performed on samples consisting of blocks of ground improved by jet-grouting. The following were performed:

- monoaxial compression tests;
- ultrasonic wave velocity tests.

The monoaxial compression tests were performed on 8 core samples treated using jet-grouting with a diameter of 97.4 mm. They gave average failure values of approximately 6.5 N/mm^2. The average modulus of elasticity of the treatment was 5 Gpa.

The results obtained from the field test showed that both types of technology investigated were suitable with the following differences:

- the injection technology requires the treatment of greater volumes of ground because the injection of the mortar that binds with the ground through tubes with *manchettes* is not very precise and controllable;
- the jet-grouting system projects the grout into the ground at extremely high pressure by rotating and at the same time extracting the injection tool. It therefore gives maximum control over the injection energy and makes it possible to improve smaller volumes of ground (50-80 cm in thickness), but with greater strength.

Given the location of the underpass to be constructed, with shallow overburdens and a concentration of important underground utilities, it was decided that for the reasons cited above, ground improvement using jet-grouting would be most appropriate.

The results of the field test not only demonstrated the importance of drilling excess pressure holes adjacent and parallel to the hole used for the jet-grouting treatment to prevent excess pressure from raising the ground at street level, but also showed that the following parameters would be the most appropriate for the jet- grouting treatment:

TABLE 4. Operating parameters chosen for jet-grouting.

INITIAL PRESSURE	C/W VOLUME RATIO	QUANTITY OF MIX INJECTED (l/m)
≥ 30	1.05 ÷1.15	≥ 250

CONSTRUCTING THE UNDERPASS AND THE RESULTS OF MONITORING

Once the results of the field test were acquired, the natural tunnel to pass under Via Gramsci was driven according to the design specifications. The specified tunnel section type was employed after first treating the ground ahead of the face and around the tunnel using jet-grouting (figures 5 and 6). Intense monitoring was performed during tunnel advance for the purpose of controlling the stress-strain behaviour of the ground and the stress states of the stabilisation structures and of the preliminary and final lining.

Monitoring consisted of the following:

- measurement of extrusion at the face-core;
- systematic observation of the geology at the tunnel faces;
- systematic measurement of convergence of the tunnel;
- measurement of the pressure on the linings using load cells and bar extensometers
- a surface monitoring network with topographical, extensometer, inclinometer and piezometer measurements and installation of the necessary instrumentation in buildings in the proximity of the area concerned.

The data acquired made it possible to ascertain the appropriateness of the intensity of the stabilisation operations and of the tunnel section type employed.

FIG. 5. Works for the core-face reinforcement

Extrusion of the core-face measurements

Maximum differences measured were 2 mm. (over a distance of 1 m.) at 20 m. ahead of the first tube installed at chainage 82+138. Maximum cumulative values measured with the same sliding micrometer were approximately 2 cm.

FIG. 6. View of the core-face reinforced

Convergence of the cavity measurements

Three convergence measuring stations were installed at chainages 82+140, 82+152 and 82+168.

- The station at chainage 82+140 monitored the excavation of the first field (see Table 5). Maximum convergence of the tunnel at this station reached nearly one centimetre and average diametrical convergence was 0.73 cm. The maximum lowering of the target placed in the roof was 0.35 cm.
- The station located at chainage 82+152 was monitored during excavation of the third to the sixth field and of the ground improvement by jet-grouting of the fourth to the seventh field. Maximum convergence of the tunnel at this station was a little greater than one centimetre and average diametrical convergence was 0.34 cm. The maximum lowering of the target placed in the roof was 0.42 cm.
- The station located at chainage 82+168 was monitored during the ground improvement by jet-grouting of the sixth and seventh field and excavation of the sixth field. Maximum convergence of the tunnel at this station was around half a centimetre and average diametrical convergence was 0.26 cm. The maximum lowering of the target placed in the roof was 0.14 cm.

The convergence measured corresponded to surface subsidence at street level of a little greater than one centimetre.

TABLE 5. Advance field location

Advance Field No.	Initial chainage [m]	Final chainage [m]	Advance field length [m]
1	82,138.26	82,144.51	6.25
2	82,144.51	82,150.76	6.25
3	82,150.76	82,157.01	6.25
4	82,157.01	82,163.26	6.25
5	82,163.26	82,169.51	6.25
6	82,169.51	82,175.76	6.25
7	82,175.76	82,182.01	6.25
8	82,182.01	82,187.30	5.29

CONCLUSIONS

The natural tunnel that passes under Via Gramsci in Florence as part of the new Milan to Naples High Capacity railway line was a very difficult and delicate task to complete because of the highly built up area in which it is set and the extremely poor quality of the ground. The tunnel was driven according to the principles of the ADECO-RS approach with full face advance after ground improvement of the face-core using fibre glass structural elements and horizontal jet-grouting around the tunnel ahead of the face. A field test was used to calibrate the techniques employed in the design and to acquire a detailed knowledge of the performance of each technique so that the best technique and the most efficient operating parameters could be selected for the context of the ground to be tunnelled.

The authors give their thanks to the CAVET Consortium, SGF-INC S.p.A., Dott. G. Romano (Rocksoil S.p.A.) and to Geom. G. Melani (Rocksoil S.p.A.).

REFERENCES

Lunardi, P. (1997). "Ground improvement by means of jet-grouting." *Ground Improvement,* 1(2), 65-85.

Lunardi, P. (2000). "Design and constructing tunnels - ADECO-RS approach." *Tunnels&Tunnelling International special supplement*, May 2000.

Lunardi, P. (2001). "The ADECO-RS approach in the design and construction of the underground works of Rome to Naples High Speed Railway Line: a comparison between final design specifications, construction design and "as built." *Proc. AITES-ITA 2001 World Tunnel Congress on Progress in tunnelling after 2000*, Pàtron Editore, Bologna, Italy, vol. 3, 329-340.

Cassani, G., Morelli S., Bindi R., Nicola A. (2003). "Analysis of the controlled deformation in rock and soil (ADECO-RS) tunneling approach - Performance of an alternate tunneling approach in the construction of 100 kilometers of tunnel for the Bologna to Florence high speed rail line." *Proc. Rapid Excavation and Tunneling Conference*, New Orleans, 2003, 245-261.

PILE-BOTTOM GROUTING TECHNOLOGY FOR BORED CAST-IN-SITU PILE FOUNDATION

Zhiguo Zhang[1], Member, ASCE, Tiejun Lu[1], Yucheng Zhao[2], Jingchun Wang[1], Member, CRMEI, Youdao Li[3], P.E.

ABSTRACT: Grouting at the bottom of a pile may considerably increase the bearing capacity, reduce settlement, and improve the property of pile-soil working together, hence its wide application in recent years in China. This paper, based on a specific case of construction, explains the mechanism of how grouting improves the bearing capacity, the technical parameters of grouting, quality guarantee measures and common problems in grouting. It is proved through test piles that grouting technology is not only effective, but also economical and simple, therefore it is a worthwhile technology to consider.

INTRODUCTION

Pile foundations are presently the most important type of deep foundation in civil engineering construction in China and abroad. Especially within the recent one or two decades, pile foundation technology has been developing very fast with the development of collective, large or heavy-type structures in civil construction and tall and super-tall buildings in city modification projects. Pile foundations have been more and more popular thanks to high single-pile capacity, good adaptability to

[1]Associate Professor and [2] Senior Civil Engineer, Shijiazhuang Railway Inst., Dept. of Civil Engrg., Post Code 050043, Shijiazhuang City, China.

[3] Civil Engineer, 11th China Railway Construction Bureau, Railway Construction Administration Center, Post Code 441003, Xiangfan City, China

various ground conditions and simplicity of construction machinery, without serious squeezed soil effect and vibration or noise pollution. According to incomplete statistics (W.X. Fu and Q.Y. Wang 1996b), more than 800 thousand piles are constructed annually in China, among which 70% are bored cast-in-situ piles. However, one problem that exists with the cast-in-situ piles is that it is difficult to clear away the slimes at the bottom of the pile. Since the pile hole is usually bored below the underground water level, mud is used to protect the hole-wall. In the course of boring, local collapse of the hole-wall due to different soil layers plus deposit of the mud often lead to slimes as thick as 0.2~0.6m at the bottom, causing the pile to stand on a soft soil base and producing a pile bearing capacity of only 150~250 kN/m^2, which is about 10~30% the capacity of pre-cast or sunk pipe cast piles (W.M. Gong and Z.T. Lu 1996a). This creates the potential for downward movement into the soft soil under working loads, causing a relatively large settlement of the pile. In order to meet the requirements of the bearing capacity and settlement, the level of the pile end is usually designed deep enough to enter the base rock, increasing costs and creating more difficulty in construction if the elevation of the base rock is low. It is more difficult to guarantee the quality of the pile as the pile length increases. Therefore it has been a topic of general interest for years among engineers how to eliminate the potential soft spot at the bottom of the pile, improve the single-pile capacity and reduce pile settlement, and making better use of the strength of the pile itself.

In order to overcome this disadvantage of the pile foundation, pressure grouting has been applied to the bottom and/or the lateral sides of the pile in many countries. In China, grouting technologies for piles have experienced a wide application and fast development in the past decade (B.H. Shen 2001a). In the following example of pile-bottom grouting construction, the mechanism of capacity increase, construction technology, grouting parameters, quality guarantee measures, common problems and their solutions, and the effect of the application, etc, are explained and analyzed.

CAPACITY-INCREASE MECHANISM AND ADAPTABILITY OF PILE-BOTTOM GROUTING FOR BORED CAST-IN-SITU PILES

Mechanism

The concept of pile-bottom-grouting is as follows: during concrete placement, a grouting pipe is placed at the bottom end of the pile. After the concrete reaches required strength, a specially mixed grout of cement (or other chemicals like silicate) is injected by means of a high-pressure pump through the grouting tube to the bottom of the pile, so as to increase the resistance against the pile tip and the pile capacity considerably.

The factors related to the grouting effect mainly include the type of soil at the bottom of the pile, the pressure of grouting, the quantity of grouting, the diameter and length of the pile, etc. Due to the various factors that are interrelated and the limited number of tests conducted, no method of calculation of single-pile capacity of the bottom-grouted pile is given in the present specifications.

The function and capacity-increase mechanism of pile-bottom grouting of bored cast-in-situ piles can be summarized in the following four aspects:

(1) The grout under high pressure tends to split, penetrate, fill, compact and solidify the slimes and the surrounding soil at the bottom of the pile so as to form a new type soil mixture with a much higher strength and resistance to loads transmitted from the pile tip.

(2) The soil around the pile end is considerably compressed and deformed by the high-pressure grout to form an expanded pile end and hence the effective area of the pile end is increased.

(3) With the high-pressure grout, the strength of the pile section of mixed slime and concrete is improved and its compression deformation is completed earlier, hence the vertical deformation of the pile under design load is reduced and full use is made of the bearing capacity of the pile-end.

(4) Part of the high-pressure grout seeps upward along the pile-soil interface to form an integrated body of the lower section of the pile with the mud peel and adjacent soil so that lateral resistance of the soil near the pile tip is improved.

Applicable Ground Conditions

Pile-bottom grouting technology is applicable to soil layers with good permeability, such as gravel, cobble, crushed stone soil, various sands and light clay, etc., but not to silty soils. With soil layers of larger sized grains (such as cobble, gravel, sand, etc.), the increase of bearing capacity of the pile is relatively larger due to a higher rate of grout seepage. By contrast, with finer-grained soils (such as clay or silty soil), the increase of the bearing capacity is smaller. Pile-bottom grouting technology can also be applied to piles in rock stratum to eliminate the negative influence of the slimes and the crushed rocks around the pile bottom so as to make the most of the resistance against the pile end.

CASE STUDY

Background

The project is a high-rise residential building located in Xiangfan City of Hubei province in China. It has 22 floors above the ground level and a basement under the ground level, with a construction area of 23569m^2, and the skeleton works adopts the shear wall structure. In order to increase the single-pile capacity and reduce the lengths and settlement of the piles, pile-bottom grouting is adopted in the design of the bored pile cast-in-situ foundation. The foundation works consists of 146 piles, including 2 test piles (dia = 800mm, L = 30.5m), 8 anchor piles (dia = 1000mm, L = 30.5m), and 136 engineering piles (dia = 800mm, L = 29.5m). The designed pile concrete is C30 and the single-pile capacity should be no less than 7500 kN.

Geological conditions: This building is located at the middle edge of a Grade I terrace deposit of the Hanjiang River, which consists of two distinct geological layers in the investigated area, with the upper layers being silty clay, silts and muddy soil, and the lower layers being well-rounded gravel. Hydrogeological conditions are: perched water in the upper clayey soil layers and artesian water in the lower gravel layers. The gravel layer is the target bearing stratum for the pile end.

Procedure of Pile-bottom Grouting Technology for Bored Cast-in-situ Piles

The procedure consists of the following steps: survey and locating, installation of the casing, boring the hole (meanwhile preparing the reinforcement cage, grouting pipes, and components), clearing the hole first time, installing reinforcement and grouting system, clearing the hole second time, and casting concrete. The concrete is cured for 3 days, the grouting system is connected and the pipes are flushed through with pressurized water. Finally, the grout is mixed and placed, and the temporary pipes are removed and cleaned.

Arrangement and Installation of the Grouting Pipes

Three grouting pipes, which are made of 25-mm diameter × 3-mm seamless steel pipes, are evenly arranged along the outside of the reinforcement cage, with 20 cm extending below the cage at the bottom and 20 cm above the natural ground level. The grouting device at the bottom is 40 cm long and is inserted directly to the bottom of the slime. Four 8-mm diameter grouting holes are evenly drilled on the grouting device vertically every 5 m from the upper edge of the device, totaling 20 grouting holes for each grouting device. The grouting holes are wrapped tightly with rubber and binding wire in order to prevent them from being blocked. The lowest 10 cm part of the grouting device assumes a reversed conic shape, closed and welded at the bottom with 25-mm steel tube as a connecting joint. The grouting pipes are welded to the stirrups and fixed to the main bars with #10 iron wire at every 2~4 m, and installed together with the reinforcement cage.

Grouting Parameters

Mix Design

Ordinary #425 portland silicate cement is adopted with water-cement ratios of 1:1 ~ 0.5 :1. The mixed grout must be filtered before use. Grout mix parameters are shown in Table 1.

TABLE 1. Mixing of Grout

W/c Ratio (1)	Cement Quantity(kg) (2)	Water(L) (3)	Grout Mixed(L) (4)
1:1	200	200	264.52
0.8:1	250	200	280.65
0.5:1	300	150	276.77

Time of Grouting

After 3 days of pile curing, water is circulated through the grouting pipes to clear the slimes at the hole bottom. Grouting is done around the 7th day of curing. Grouting too early might damage the integration of the pile bottom due to insufficient strength of pile concrete; but if grouting is too late, the grouting device would be solidified in concrete and grout would be blocked, or too great grouting pressure would cause an accident.

Grouting Pressure

In the round gravel stratum with relatively large pores, the grout cannot both seep and split into the gravel unless the grouting pressure reaches a certain value. The grouting pressure can be calculated according to Coulomb-More damage theory (as shown in Figure 1).

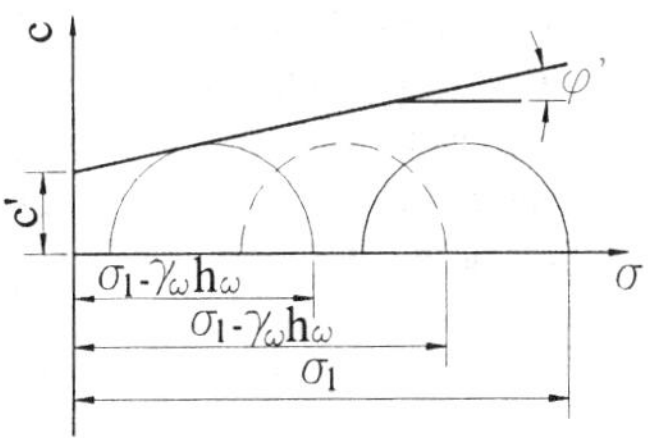

FIG. 1. Grouting Pressure Calculation Model

The grouting pressure causes the effective stress to decrease. The geological layer shall be damaged when the grouting pressure P_0 reaches the standard as decided by the following formula:

$$P_0 = 0.5\omega(1+\eta) - 0.5\omega\csc\varphi'(1-\eta) + c'ctg\varphi' \tag{1}$$

Where: P_0--grouting pressure (MPa), η--main stress ratio, c'--effective cohesion (kPa), φ' --effective internal friction angle(°), ω --pressure difference(kPa), $\omega = \gamma h - \gamma_\omega h_\omega$, γ_ω--unit weight of water(kN/m^3), h_ω--water level depth(m).

It can be seen from the above that the grouting pressure is an important index to control in pile-bottom grouting for bored cast-in-situ piles. In this project, based on the geological and hydrological conditions and the test pile grouting, the grouting pressure at the beginning is 2-3 MPa, and when the grouting passages are through, the pressure reading on the gauge drops sharply. Then the grouting pressure should be controlled manually at 1-2 MPa.

Quantity of Grout

Under normal conditions, when the pump pressure is no more than 2 MPa, the quantity of grout for each pile is controlled at 2.5 t (cement). If no more grout can be pumped in due to special conditions after 80% of that quantity is grouted, the grouting for the concerned pile can also be regarded as acceptable.

Quality Control Measures

(1) A seamless steel pipe is required with an inner diameter of 25 mm and a thickness of 3 mm as the grouting pipe. Each grouting pipe must be pressure tested and approved before use. Use a threaded connection at the joints of pipes and the net thickness at the threads should be no less than 2/3 of the original thickness of the pipe so as to avoid damage to the threaded part.

(2) The grouting pipes should be point-welded firm to the stirrups of the reinforcement cage. Before putting it down into the hole, check carefully the sealing of the grouting nozzle to prevent blockage of the nozzle due to leakage of sand or sand.

(3) The grouting pipes should be lowered down together with the reinforcement skeleton, and sealing material should be applied to the threads at the threaded connection of the pipes. After each segment of cage is lowered into the hole, water is injected into the grouting pipes to check and ensure their watertightness. After the

pipes are filled with water, the water level should be stable and not descend, which shows the pipes are reliably watertight, otherwise, the cage should be lifted and the cause of the leakage must be determined and corrected. Then lower the cage and check again until it is satisfactory. Only after that can the next segment be installed. The cage must be lowered into the hole slowly and lightly to avoid damage to the grouting devices and pipes. After the installation is completed, the mouths of the pipes should be covered to prevent anything from entering them and causing blockade of the pipes. The grouting pipe should extend above the ground by 20cm. Too much surplus length is not only unnecessary but may be liable to be damaged in the construction of other piles.

(4) In order to guarantee the through passage of grout, inject pressure water into the pipes to make the passages through after 3 days of curing of the pile.

(5) BW-150 or other types of grout pump that can provide a grouting pressure over 10MPa should be used. Free fall type JZ-350 mixer may be adopted to mix the grout. The steel wire woven rubber tubes to deliver the grout should have a pressure resistance over 10MPa.

(6) Grouting is usually done after 7 days of curing of the pile. First break through the grouting holes with clear water. Before starting the pump, open the return valve. When water injection is begun, close the return valve slowly and observe the changes of the gauge reading and record the pressure to force through the grouting path. In case of difficulty in breaking through, then press the water in intermittently to create pulse pressure to break through the holes. In case of failure to break through even with a pressure as high as 10MPa, then investigations must be made to make clear the cause. After the grouting path is forced through, first inject thinner grout with a w/c ratio of 1:1, then gradually transcend to thicker grout till a w/c ratio of 0.5:1 in the end. A specific person should be responsible for the observation and recording of the changes of readings of the gauge in the course of grouting.

(7) The mixing of the grout must be in accordance with the mix design. A specific person should be assigned to add water, with marks made on the water barrel to control the water quantity. The mixed grout should go through a wire net before entering the grout storage barrel in order to filter out the bigger sized grains to avoid blockade of the nozzle of the grouting nozzle. In order to prevent the ready-made grout from depositing, an agitator should be installed in the storage barrel to stir the grout from time to time.

(8) Control of the grouting pressure: A specific person should be assigned to observe the readings of the gauge, which should be normally controlled at 1MPa~2MPa. If the pressure is smaller than 0.4MPa, the grout can not be pressed to the pile bottom. Then check and see whether the grout comes back to the ground, and if it does, immediately stop grouting and wait until after about 7 days to do the make-up grouting till the designed quantity. On the other hand, too great pressure that

is over 3MPa may destroy the soil structure.

(9) Completion: The completion of grouting is mainly controlled by the cement quantity that is grouted, with the grouting pressure as an auxiliary means of control. Grouting is finished when the designed quantity of grout is injected. Or if the grouting pressure is over 3 MPa after 80% of designed quantity is grouted and it becomes very difficult to inject more grout, then grouting can also be regarded as completed.

(10) After grouting is finished, wash clean the pump and high-pressure rubber tubes with water. Make proper records and submit for approval.

Common Problems and Relevant Measures in Pile-Bottom Grouting

Inter-Pile Grouting

Sometimes while grouting is being applieed for one pile, the grout suddenly shoots out from another pile, and this is called inter-pile grouting. The causes of inter-pile grouting may include too high grouting pressure, the existence of cavities in the soil layers that are disturbed by boring, too small distance between two piles, too tick loose soil layers like sand or sandy cobble at the pile bottom that are easy to collapse, etc.

Once inter-pile grouting is observed, stop the grouting immediately. Firmly block the pipe mouths of "visited" pile where grout is coming out. Then resume grouting until finished. After that, proceed with the grouting for the "visited" pile immediately and before the initial setting of the "visiting" grout inside the grouting pipes of the visited pile. Inter-pile grouting may produce a group-piles effect and affect the bearing capacity and deformation of the piles, therefore sufficient attention should be paid to it.

Dead Pile

If the grouting pipe refuses to take in grout just after the grouting is begun and at the same time the pressure is increased sharply to a very high value, then this is a dead pile phenomenon. The causes of a dead pile may include: (1) The pipe or the joint between pipes are blocked in the course of the construction of the bored cast-in-situ pile; and (2) The lower mouth or lower segment of the grouting pipe has been blocked by cement grout from the grouting of other piles, forming a state of inter-pile grouting or semi-inter-pile grouting.

In case of the dead pile, immediately stop grouting, insert a φ6 reinforcement bar into the pipe till the blocked part, hammer the bar to break through the blockade, then resume grouting.

Sudden Increase or Drop of Pump Pressure

Normally the grouting pressure remains at 1-2MPa. In case of sudden pressure change, stop grouting immediately and analyze the cause before making the decision in order to prevent the cracks in the earth layer are excessively expanded or blocked, which would cause loss or waste of cement grout.

Effect of Pile-bottom Grouting

In the example of the high–rise building, after grouting was applied to the bored cast-in-situ piles, according to the results of test pile vertical static load tests and high strain dynamic tests, the single-pile bearing capacities all exceeded 8000 kN, higher than the designed value of 7500 kN. The detailed test results are shown in Table 2 and Table 3.

TABLE 2. Vertical Static Load Test Results of the Test Piles

Test Pile No. (1)	Pile Diameter (mm) (2)	Pile Length (m) (3)	Maximum Load (kN) (4)	Maximum Deformation (mm) (5)	Residual Deformation (mm) (6)	Ultimate Capacity (kN) (7)
S4	800	31.7	8000	3.01	0.19	≥8000
S5	800	30.4	8000	7.97	2.03	≥8000

TABLE 3. High Strain Dynamic Test Results

No. (1)	Pile No. (2)	Pile Length Tested (m) (3)	Pile Diameter (mm) (4)	Hammering Energy WE (kN-m) (5)	Deformation (mm) (6)	Ultimate Capacity (kN) (7)
1	4	29.5	800	62.8	6.1	13512.2
2	8	29.5	800	59.2	4.9	14395.4
3	118	29.5	800	61.9	6.4	13748.4
4	116	29.5	800	92.9	9.7	13559.8
5	S4	29.5	800	32.2	6.6	11735.5

It can be seen from the test results in Table 2 and 3 that the single-pile vertical ultimate bearing capacities all exceed 11735 kN, much greater than the design standard of 7500 kN, showing that pile-bottom grouting can considerably increase the single-pile ultimate bearing capacity of bored cast-in-situ piles. This new technology can be used in engineering design and construction to replace longer piles with shorter ones, and reduce the quantity of work and construction period. And if the pile length remains unchanged, the adoption of this technology may increase the single-pile capacity, improve the seismic and pullout resistances of the pile, and reduce cost.

CONCLUSIONS

(1) Engineering practice has proved that pile-bottom grouting for bored cast-in-situ piles is an effective means to increase the capacity, rigidity, and stability of the pile, reduce pile settlement, improve construction quality and lower the cost of foundation construction, hence it is worthy of consideration by contractors and engineers.

(2) The theories and discussions about pile-bottom grouting technology for bored cast-in-situ piles are mainly qualitative rather than quantitative. Therefore further study is needed on the following issues: (i) relationship between grout quantity and the bearing capacity, (ii) the optimum ratio of grouting pressure and grout quantity in different geological conditions, and (iii) the magnitude of bearing capacity increase.

(3) There is not yet a general specification in China on pile-bottom grouting technology, which presents difficulty to foundation designers, who have to guarantee the safety of the structures on one hand and reduce as much as possible the diameter and length and quantity of the piles on the other hand. Hence it is suggested here that concerned experts be organized to make related studies and analyses and develop general national specifications for pile-grouting technology for bored cast-in-situ piles to be referred to in engineering construction, management and design.

REFERENCES

Weiming Gong, Zhitao Lu (1996a). "Innovation of cast-in-place pile--grouting at bottom." *Industrial Construction,* 26(3), 33-37.

Wenxun Fu, Qingyou Wang (1996b). "Review on pressure grouting pile." *Beijing hydraulics*, 12(2), 35-40

Baohan Shen (2001a). "Post grouting pile technique—the invention and development of post grouting piles." *Industrial Construction,* 31(5), 64-66.

Ping Yang, Zhenbin Peng (2001b). "Mechanism and calculation model of post-grouting technique." *Geotechnical Engineering World*, 5(3),54-56.

Yaofeng Xie, Yunqiu Wang (2002a). "Research on rising the effect of post-grouting bored pile." *Port and Waterway Engrg.,* 340(5), 47-50.

Shuting Jia(2002b). "Discussion on the construction technique of post-grouting bored piles." *Geotechnical Engineering World*, 5(4),59-61.

EVALUATION OF DEEP-SEATED SLOPE STABILITY OF EMBANKMENTS OVER DEEP MIXED FOUNDATIONS

Jie Han[1], Jin-Chun Chai[2], Dov Leshchinsky[3], and Shui-Long Shen[4]

ABSTRACT: When embankments are constructed over soft foundations, deep-seated slope stability often becomes one of the controlling factors in design. Deep mixing methods have been commonly used as an alternative to solve the deep-seated slope stability problem. Bishop's modified method is a commonly adopted approach for analyzing the slope stability of embankments on deep mixed foundations. Bishop's modified method assumed slopes fail along a circular slip surface and the soils along this slip surface provide shear resistance. However, experimental studies have showed that deep mixed columns under a combination of vertical and horizontal forces could fail due to shearing or bending. The possible failure modes depend on the combination of the forces, the strengths of soft soils and deep mixed columns, dimensions and arrangements of deep mixed columns. A numerical method was used in this study to evaluate the factors of safety varying with the strengths of deep mixed columns and their arrangements with three rows of columns having two different thickness. The numerical analysis indicated that the critical slip surface of the deep-seated slope failure was not circular when the deep mixed columns were used. The factors of safety obtained using the numerical method were compared with those using Bishop's modified method and Spencer's three-part wedge method. The comparisons indicated that Bishop's modified method yielded significantly higher factors of safety than the numerical method, especially when the deep mixed columns had higher strengths. The Spencer's three-part wedge method yielded lower factors of safety than the numerical method.

---

[1]Ph.D., PE, Assistant Professor, Dept. of Civil Engineering, Widener University, One University Place, PA 19013, USA, e-mail: jxh0305@mail.widener.edu

[2]Ph.D., Associate Professor, Institute of Lowland Technology, Saga University, 1 Honjo, Saga 840-8502, JAPAN, e-mail: chai@cc.saga-u.ac.jp

[3]Professor, Dept. of Civil and Environmental Engineering, University of Delaware, Newark, DE 19716, USA

[4]Ph.D., Associate Professor, School of Civil Engineering and Mechanics, Shanghai Jiao Tong University, Shanghai 200030, CHINA, e-mail: slshen@sjtu.edu.cn

INTRODUCTION

When designing embankments over soft foundations, geotechnical engineers may face a number of challenges, which include bearing capacity failure, excessive total and differential settlements, and slope instability, etc. The slope instability of embankments may develop locally, near the facing, inside the embankment, or through the foundation soil as local failure, surficial failure, general slope failure, or deep-seated failure as shown in Figure 1. The deep-seated slope failure is also referred as a global slope failure, mainly induced by a soft foundation existing under the embankment. A number of techniques have been successfully adopted to prevent deep-seated slope failure, such as ground improvement techniques and use of geosynthetics or piles. As one of ground improvement techniques, deep mixed (DM) columns have been commonly used as an alternative to solve deep-seated slope stability problems. Terashi (2002) indicated that "nearly 60% of on-land application in Japan and perhaps roughly 85% of Nordic applications are for the settlement reduction and improvement of stability of embankment by means of group of treated soil columns". This paper focuses on the evaluation of deep-seated slope stability of embankments over deep mixed foundations using limit equilibrium methods and a numerical solution.

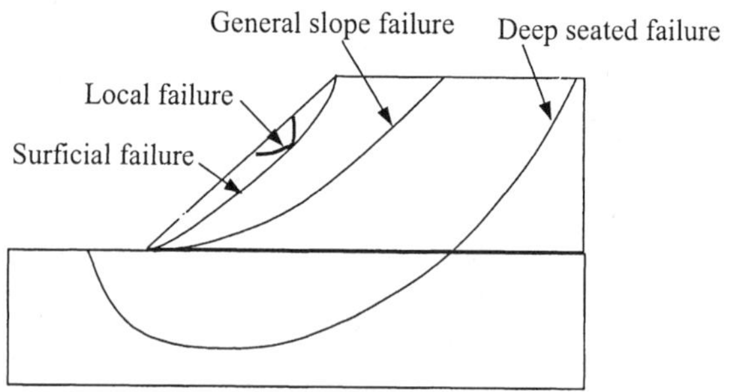

FIG. 1. Potential Slope Stability Failures

Limit equilibrium methods have been commonly adopted for analyzing the deep-seated slope stability of embankments over deep mixed foundations. Bishop's modified method with a circular slip surface is probably the most commonly used limit equilibrium method. In the analysis of DM foundations, the DM columns and the soil are either treated as individual components or as a composite ground to resist the shear stresses. In this study, the DM columns are treated as individual DM walls in 2-D analyses. Limit equilibrium methods assume that the shear strengths of the columns are always fully mobilized if a slip surface cuts through any part of the columns. In reality, the resistance of the columns depends on the intersected location by the slip surface. As shown in Figure 2, the columns may only provide very limited resistance at the locations A and C because the soil around the columns may fail prior to the failure of the columns or the slip surfaces may go around the columns to create noncircular slip surface or the columns may behave as piles punching into the upper or lower soil layer as described by Broms and Wong (1985). In addition, the centrifuge model tests done by Kitazume and Terashi (1991) indicated that the DM columns failed by bending under a combination of vertical and horizontal forces. Broms (1999) also indicated that horizontal forces would reduce the bearing capacity of DM columns. Since the bending strengths of the DM columns are much lower than their shear strengths, Kitazume et al. (1997) was concerned about the possibility of overestimation using Bishop's slip circle analysis in the current design.

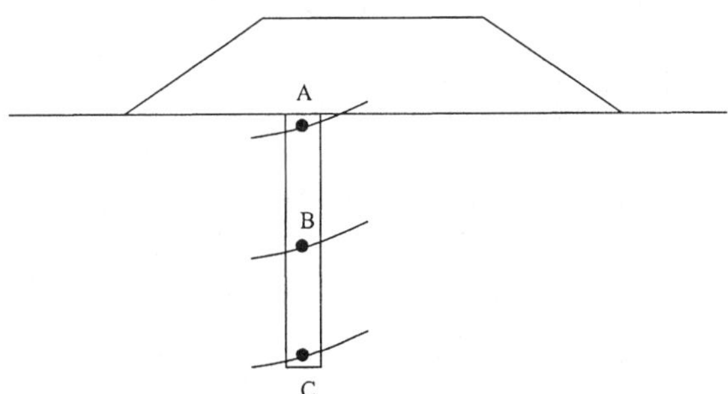

FIG. 2. Potential Slip Locations through DM Columns

In recent years, numerical methods have been increasingly used for analyzing slope stability including the computation of its factor of safety. San et al. (1994) indicated that finite element and limit equilibrium methods could consistently determine the locations of critical slip surfaces and required tensile strength of reinforcement in geosynthetic-reinforced slopes. Dawson et al. (1999) concluded that the factors of safety of unreinforced slopes obtained using a finite difference method (FLAC - Fast Lagrangian Analysis of Continua) were in good agreement with those using the limit equilibrium method with a log-spiral slip surface. Han et al. (2002) used the same finite difference software (FLAC) to obtain the identical corresponding factors of safety of unreinforced and geosynthetic-reinforced slopes as the Bishop's modified method (limit equilibrium method). The technique used for computing the factor of safety of slope stability in the numerical method is discussed in the following section. As compared with limit equilibrium methods, numerical methods have the following advantages in solving the factor of safety of slope stability (Cundall, 2002): (1) no pre-defined slip surface is needed; (2) the slip surface can be any shapes; (3) no assumptions are needed for functions of inter-slice force angles; (4) multiple failure surfaces are possible; (5) structures (such as footings, tunnels, etc.) and/or structural elements (such as beams, cables, etc.) and interfaces can be included; and (6) kinematics is satisfied. However, numerical methods generally require the user to have more knowledge and experience in order to properly use it. A complicated and large size problem may require significant computation time. The inclusion of structural elements and interfaces may create numerical instability. It is difficult to perform a specific search when it is sometimes needed (for example, surficial slope stability needs to be prevented in order to study the deep-seated slope stability). The comparisons of numerical solutions and limit equilibrium methods are presented in Table 1. Considering the complexity of the failure mechanisms, the finite difference method (FLAC) was adopted in this study to evaluate the deep-seated slope stability of embankments over DM foundations. The computed factors of safety by the numerical method were compared with those obtained using Bishop's circular slip surface method and the Spencer's three-part wedge method.

TABLE 1. Comparisons of Numerical Solutions and Limit Equilibrium Methods (Cundall, 2002)

	Numerical solution	Limit equilibrium
Equilibrium	Satisfied everywhere	Satisfied only for specific objects (slices)
Stresses	Computed everywhere using field equations	Computed approximately on certain surfaces
Deformation	Part of the solution	Not considered
Failure	Yield condition satisfied everywhere; failure surfaces develop "automatically" as condition dictate	Failure allowed only on certain pre-defined surfaces; no check on yield condition elsewhere
Kinematics	The "mechanisms" that develop satisfy kinematic constraints	Kinematics are not considered – mechanisms may not be feasible

NUMERICAL METHOD

The finite difference program (FLAC 2D Version 4.0) developed by the Itasca Consulting Group, Inc. was adopted in this study for numerical analyses of slope stability of embankments over DM foundations. A shear strength reduction technique was adopted in this program to solve for a factor of safety of slope stability. Dawson et al. (1999) exhibited the use of the shear strength reduction technique in this finite difference program and verified numerical results with limit equilibrium results for simple slopes. In this technique, a series of trial factors of safety are used to adjust the cohesion, c and the friction angle, ϕ, of soil as follows:

$$c_{trial} = \frac{1}{FS_{trial}} c \tag{1}$$

$$\phi_{trial} = \arctan\left(\frac{1}{FS_{trial}} \tan\phi\right) \tag{2}$$

Adjusted cohesion and friction angle of soil layers are re-inputted in the model for equilibrium analysis. The factor of safety is sought when the specific adjusted cohesion and friction angle make the slope become instability from a verge stable condition or verge stable from an unstable condition.

MODELING

The geometry and material properties of the models used in this study are shown in Figure 3. Deep mixed columns were modeled as continuous walls. Mohr-Coulomb failure criteria were used for embankment fill, soft soil, firm soil, and deep mixed walls. The properties of embankment fill, soft soil, and firm soil were kept constant. The unit weight of all the soil layers including DM walls was 18kN/m^3. In this study, DM walls are installed at the toe, the shoulder, and the mid-point between the toe and the shoulder with a thickness of 1.0m or 2.0m although they can be installed under the whole embankment in practice. For a parametric study, the DM walls have an undrained shear strength varying from 100kPa to 1000kPa and friction angle equal to zero. There was no interface assumed between the bottom of columns and the soil. The slope stability analysis would take the lower value of their shear strengths. The same models were used in numerical and limit equilibrium analyses. Since the numerical analysis was conducted to determine the factors of safety of embankments against the global failure, no deformation properties were needed.

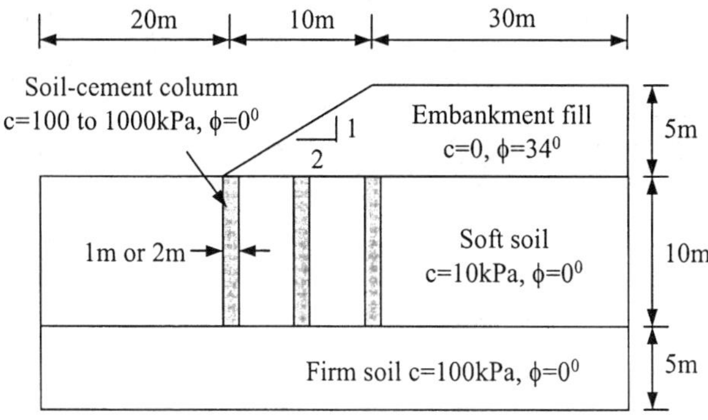

FIG. 3. Numerical Analysis Model

ANALYSIS OF RESULTS

Numerical Analyses

An embankment over untreated ground was selected as a baseline case. The shear-strain rate contours of this case were shown in Figure 4. It is shown that high shear strain rates developed in the embankment and the soft soil, which created a critical slip zone. The shape of this critical slip zone is circular, which is consistent with the slip surface assumed in Bishop's method. It is apparent that the critical slip zone was bottomed out by the firm soil underneath the soft soil. Figure 5 presents shear-strain rate contours for an embankment over a DM foundation. There is no continuous shear-strain rate zone. It is shown that the shear-strain rate contours intercepted by soil-cement walls. The high shear-strain rate zones developed in the embankment and soft soil in front and behind the DM walls. In front of the column at the toe, the high shear-strain rate zone was caused by the rotation of the column towards the soft soil.

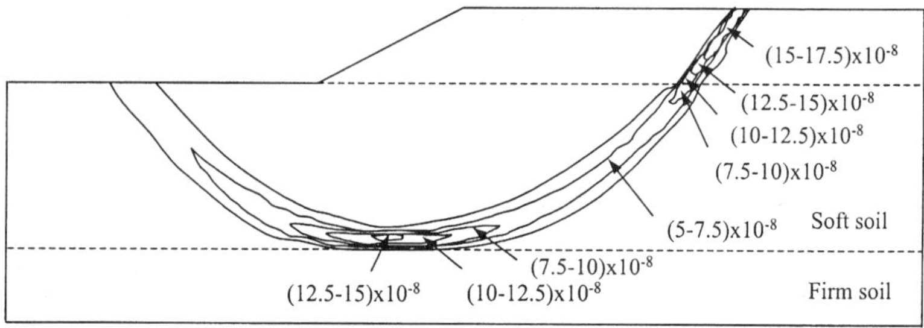

FIG. 4. Shear Strain Rates Developed in the Embankment and Untreated Ground

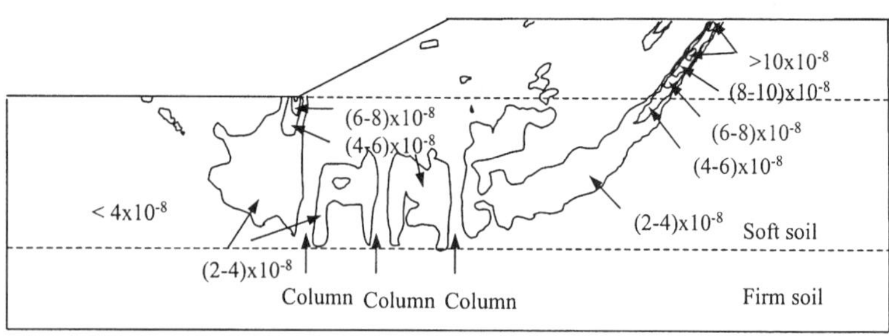

FIG. 5. Shear Strain Rates Developed in the Embankment and the DM Foundation (wall thickness = 1.0m; undrained shear strength of DM column, c_u = 100kPa)

Since the critical slip surface was not very obvious based on the shear strain rate contours in the embankment and the DM foundation, the velocity vectors obtained by the numerical method were plotted in Figure 6 to assist in evaluating the critical slip surface. The velocity vectors shown in Figure 6 represent the movement of the slope and the soft soil with DM columns. The purpose of this velocity vector plot was to show two distinguished velocity zones and the potential slip surface between these two zones, therefore, the details of the velocity vectors were not critical for the analysis purpose. The shape of the critical slip surface can be approximately considered as a three-part wedge type. This critical slip surface was simulated in the slope stability analysis using Spencer's three-part wedge method in the next section.

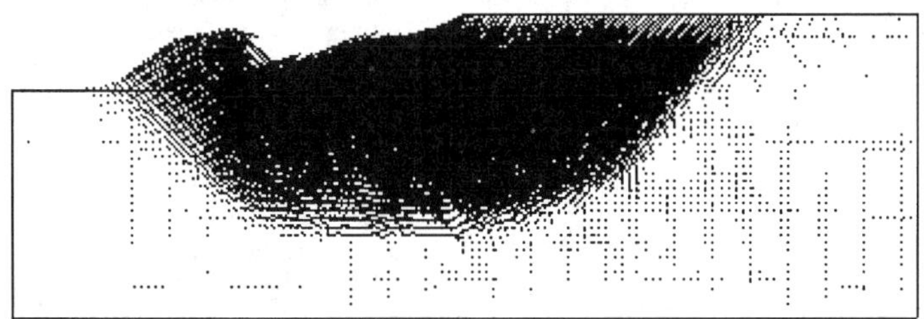

FIG. 6. Velocity Vectors Developed in the Embankment and the DM Foundation (wall thickness = 1.0m; undrained shear strength of DM column, c_u = 100kPa)

Limit Equilibrium Analyses

The limit equilibrium program, ReSSA2.0, developed by ADAMA Engineering, Inc., was used in this study to conduct slope stability analyses using Bishop's modified method and Spencer's three-part wedge method. The tangent to critical circles at their points of exit for Bishop's method is limited in ReSSA to a maximum of 50 degrees. This limitation is to avoid numerical errors leading to misleading values of factor of safety when the circles emerge too steeply (Whitman and Bailey, 1967). The critical slip surface based on Bishop's modified method for the embankment over DM columns is shown in Figure 7. Since the slip surface cut through three rows of DM walls, all DM walls mobilized their full strengths. The critical slip surface without DM columns is slightly different from that with DM columns.

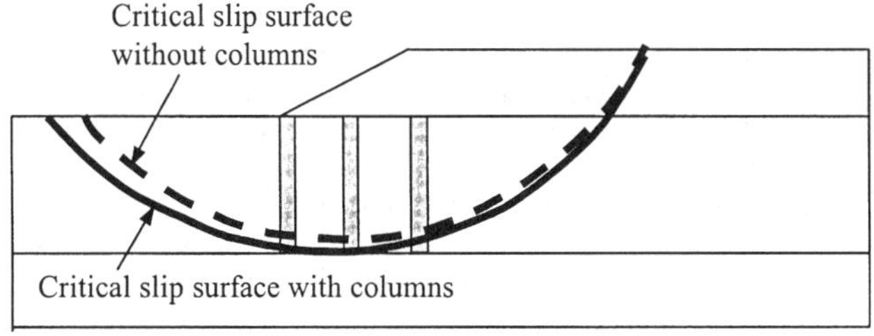

FIG. 7. Bishop's Circular Slip Surface Analysis (wall thickness = 1.0m; undrained shear strength of DM column, c_u = 100kPa)

The critical slip surface based on Spencer's three-part wedge method for the embankment over DM columns is shown in Figure 8. As compared with the shape of critical slip surface based on the velocity vectors, the three-part wedge slip surface has a good representation of the critical slip surface obtained by the numerical method. The critical slip surface without DM columns is almost identical to that with DM columns.

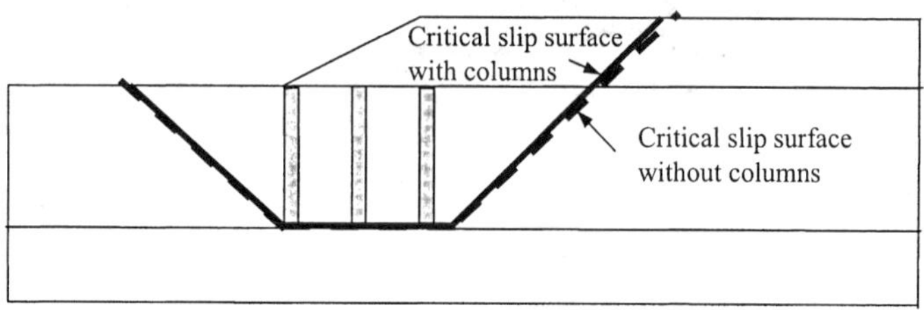

FIG. 8. Critical Slip Surface in Three-Part Wedge Analysis (undrained shear strength of DM column, c_u = 100kPa)

Factor of safety

The computed factors of safety against deep-seated slope failure using the FLAC numerical method, Bishop's modified method, and Spencer's three-part wedge method are plotted in Figure 9. When the undrained shear strength of DM walls is equal to 10kPa, it represents a case with an untreated ground. It is shown that all three methods computed almost identical factors of safety. With an increase of the undrained shear strength of DM walls, the factors of safety computed by these three methods become different. Bishop's modified method yielded significantly higher factors of safety than those obtained by the numerical method when the undrained shear strength of the DM walls is higher. In the Bishop analysis, the lowest point of the potential failure surface was limited below the toe of the slope to ensure a global failure. The dash lines in Figure 9 represent the cases with a surficial slope failure, which govern the failure of the embankments when the factor of safety against the global failure is higher than that against the surficial slope failure. The differences in FoS values from Bishop's method and the numerical method may result from the misrepresentation of the circular slip surface and fully mobilized strength of DM walls assumed in Bishop's method for this specific problem. However, the three-part wedge method yielded lower or conservative factors of safety as compared with the numerical method. Spencer's method assumed that the interslice force angles from the left and right vertical sides of each slice are equal. The inclusion of vertical strong elements (DM walls) may change functions of interslice force angles between the left and right vertical sides of slice.

CONCLUSIONS

Numerical analyses indicated that the critical slip surface of the embankment over a deep mixed foundation for this specific problem was not circular. Bishop's modified method computed significantly higher factors of safety than the numerical method when the undrained shear strength of the deep mixed walls is high. Spencer's three-part wedge method yielded lower factors of safety than the numerical method. These conclusions are applicable to the modes of failure that were considered for this study.

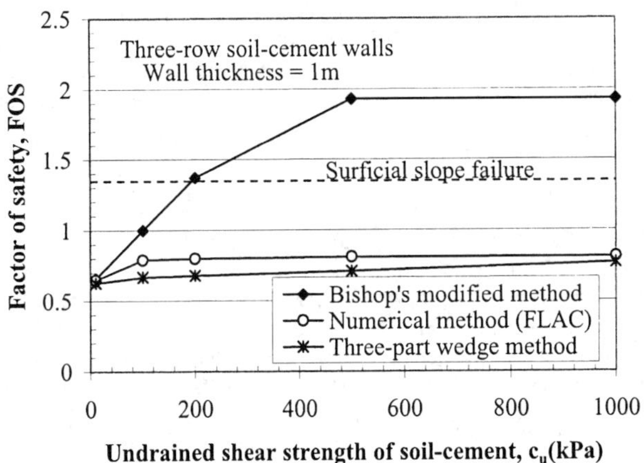

(a)

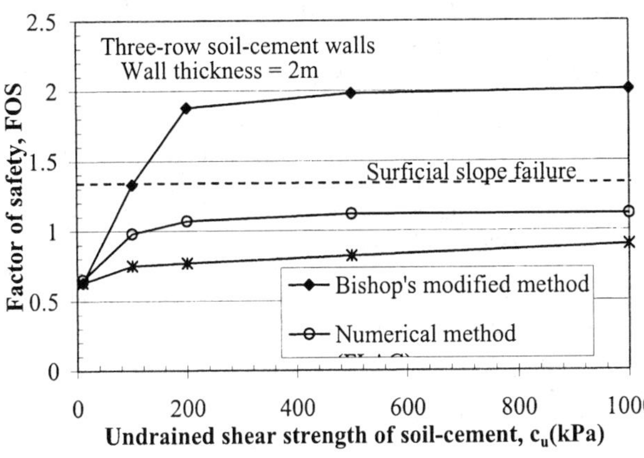

(b)

FIG. 9. Computed Factors of Safety using Limit Equilibrium and Numerical Methods (wall thickness = 2.0m)

ACKNOWLEDGEMENT

This research work was conducted mainly under the research fellowship provided by the Japan Society for Promotion of Science (JSPS) for the first author as a visiting associate professor at Saga University, Japan. It is also partially supported by the National Deep Mixing Cooperative Research Program managed by Dr. Ali Porbaha. These supports are greatly appreciated. The authors are thankful for Prof. N. Miura at Saga University for valuable discussions on this specific research topic and Dr. Roger Hart at Itasca Consulting Group, Inc. for his help in the use of FLAC program.

REFERENCES

Bishop, A.W. (1955). "The use of the slip circle in the stability analysis of slopes." *Geotechnique*, 5, 7-17.

Broms, B.B. and Wong, I.H. (1985). "Embankment piles." Third International Geotechnical Seminar – Soil Improvement Methods, Singapore, 27-29 November.

Broms, B.B. (1999). "Can lime/cement columns be used in Singapore and Southeast Asia?" 3rd GRC Lecture, Nov. 19, Nanyang Technological University and NTU-PWD Geotechnical research Centre, 214p.

Cundall, P.A. (2002). "The replacement of limit equilibrium methods in design with numerical solutions for factor of safety." Powerpoint presentation, Itasca Consulting Group, Inc.

Dawson, E.M., Roth, W.H., and Drescher, A. (1999). "Slope stability analysis by strength reduction." *Geotechnique* 49(6), 835-840.

Han, J., Leshchinsky, D., and Shao, Y. (2002). "Influence of tensile stiffness of geosynthetic reinforcements on performance of reinforced slopes." *Proceedings of Geosynthetics – 7th ICG*, Delmas, Gourc & Girard (eds), Swets & Zeitlinger, Lisse, 197-200.

Itasca Consulting Group, Inc. (2002). *FLAC/Slope User's Guide*, 1st Edition, 82p.

Kitazume, M. and Terashi, M. (1991). "Effect of local soil improvement on the behavior of revetment." *Proc. Geo-Coast'91*, 1, 341-346.

Kitazume, M., Omine, K., Miyake, M., and Fujisawa, H. (1997). "JGS TC Report: Japanese design procedures and recent activities of DMM." *Grouting and Deep Mixing*, Yonekura, Terashi, and Shibazaki (eds), Balkema, Rotterdam, 925-930.

San, K.C., Leshchinsky, D., and Matsui, T. (1994). "Geosynthetic reinforced slopes: limit equilibrium and finite element analyses." *Soils and Foundations*, 34(2), 79-85.

Terashi, M. (2002). "The state of practice in deep mixing methods." *ASCE Geotechnical Special Publication* No. 120, L.F. Johnsen, D.A. Bruce, and M. J. Byle (eds.), Vol. 1, 25-49.

Whitman and Bailey (1967). "Use of computers for slope stability analysis." Journal of Soil Mechanics and Foundation Engineering, ASCE, 93(SM4), 475-498.

DEEP SOIL MIXING FOR FOUNDATION SUPPORT OF A PARKING GARAGE

Joseph Cavey, P.E.[1], Member, ASCE, Lawrence F. Johnsen[2], Member, ASCE, Jeffrey DiStasi[3], Member, ASCE

ABSTRACT: A design-build foundation system consisting of 6 and 8 ft. diameter soilcrete columns was constructed to support spread footings for a five level parking garage in Allentown, Pennsylvania. The soil mixing method was selected over stone columns/aggregate piers because it provided a more rigid foundation system, and would not create vertical drainage paths in the potentially karst terrain. The load on the supported footings varied between 200 and 2,200 kips. The design was subject to settlement criteria of 1.25 in. total movement and 0.75 in. differential movement. The soilcrete columns were installed through medium and stiff clay, bearing on dense sand or weathered limestone rock. A total of 120 columns were constructed below the parking structure foundations. Three soilcrete columns were core drilled using a 2 in. diameter barrel, with each achieving over 95% recovery. Twenty-eight day unconfined compressive strengths of the cores and wet-grab samples averaged 900 psi.

INTRODUCTION

The Pennsylvania Power and Light (PP & L) office building and parking garage were constructed in downtown Allentown, PA between 2002 and 2003. The garage foundation would be founded in a relatively inactive sinkhole area, and therefore

---

[1]Senior Project Manager, Hayward Baker Inc., 1875 Mayfield Road, Odenton, MD 21113, 410-551-1980, jkcavey@haywardbaker.com
[2]Principal, Heller and Johnsen, Foot of Broad Street, Stratford, CT 06514, 203-380-8188, HJ30115@aol.com
[3]Project Engineer, Hayward Baker Inc., 17-17 Route 208, Fair Lawn, NJ 07410, 201-797-1985, jmdistasi@haywardbaker.com

stone columns/aggregate piers were originally considered to increase bearing capacity and limit settlement. After further analysis, stone columns/aggregate piers were not installed for the following reasons:

1. Differential settlement was expected to exceed .75" as the depth to competent soil ranged from 9 to 43'. The compression of a deep stone column would be much greater than a shallow column. Similarly, the deformation of a large footing would be greater than a small footing.
2. The cost of the soil mixing alternative discussed herein was less than the other alternatives.
3. Introducing a column of stone in this karst region could initiate a sinkhole due to the greatly enhanced vertical drainage.

Driven and drilled piles were not considered due to cost and installation concerns over the pinnacled limestone

The design was based upon actual column loadings provided by the design- build pre-cast concrete panel contractor. The layout of the soilcrete columns beneath the garage footings, including number, size, and orientation of the columns, was determined using a design strength of 100 psi. This layout of soilcrete columns is shown in Figure 1. This design criteria was established using the contractor's past soil mixing experience along with soil information obtained by subsurface investigation of the site. It should be noted that the soilcrete columns were designed for compression loads only. No reinforcing steel was placed in the soilcrete. Typical soilcrete column loads varied from 400 to 700 kips, corresponding to an allowable bearing capacity of 14 ksf in the end-bearing columns. The layout of columns below shallow column footings is shown in Figure 2. The design was based on limiting settlements to 1.25 in. total and 0.75 in. differential.

SUBSURFACE EXPLORATION

Cone Penetration Tests (CPTs) were performed at each column location and strategically along the strip footings to determine depth to competent bearing material and to identify potential obstructions prior to production work. The results of these tests served two purposes: (a) comparison with SPT results from previous subsurface investigations and (b) determination of soil parameters to assist in calculation of the end bearing of the soil cement columns. Test columns were soil-mixed prior to production work to verify that established grouting parameters could be maintained and design strengths achieved.

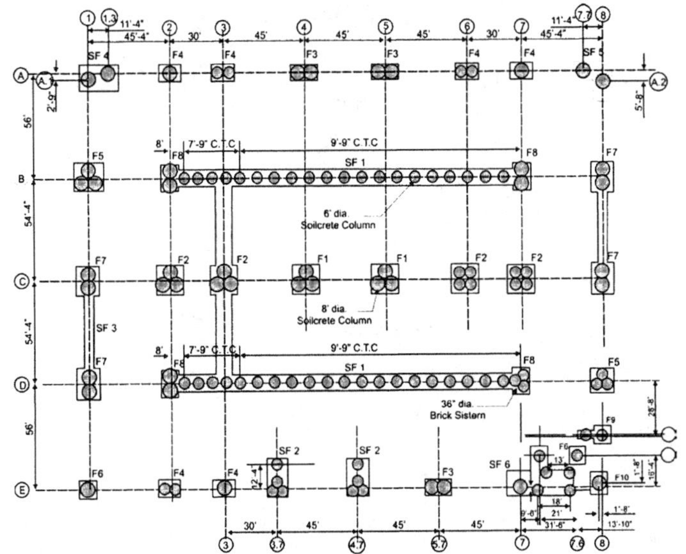

FIG 1: Plan View.

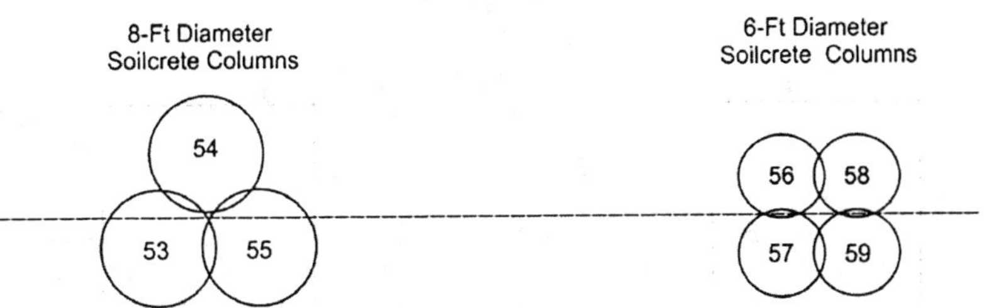

FIG 2: Layout of Columns below Shallow Column Footings.

Site Geology

Soil borings from previous geotechnical explorations showed the site consisted predominantly of residual clayey silt and silty clay overlying limestone rock of average quality. The depth to the limestone rock varied greatly across the site. A typical soil profile is shown in Figure 3. Groundwater was not encountered in any of the borings.

Cone Penetrometer soundings taken at the site for the contractor's investigation gave results consistent with those shown in the borings, and verified the widely

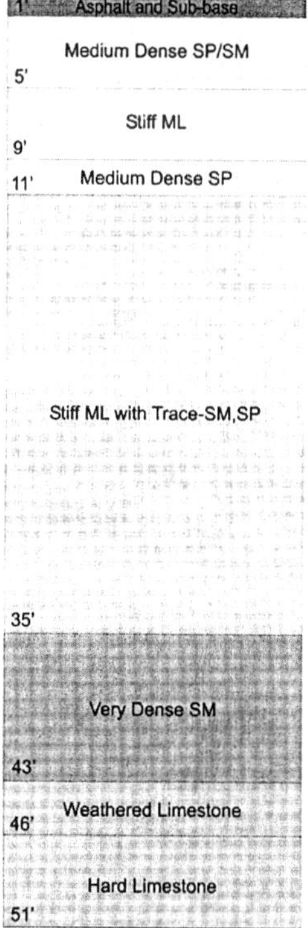

FIG 3: Typical Soil Profile.

varying depths to rock across the site, which were anywhere between 5 and 48 feet. Locations where CPTs terminated quite shallow or were inconsistent with the borings were either probed again or excavated for visual examination. In some instances, slabs of concrete were found 10 feet deep. Concrete beneath the footprint of any columns was removed prior to pre-augering.

PRODUCTION WORK

Pre-augering

Column locations were pre-augered to identify depths to competent bearing material and to identify and remove potential obstructions to the mixing operations. The site had seen extensive construction in the past. Consequently, several underground storage tanks and a few cisterns were discovered during production.

After pre-augering each hole and determining the column length, three-quarters of the column was immediately filled with the augered soil, after culling any boulders. This proportion was later increased to 80% of the column as higher than expected strengths of soilcrete cores permitted a reduction in column cement content, and, accordingly, grout slurry volume. The remaining volume would serve as a repository for the injected neat cement grout during mixing. Pre-augered columns ranged in depth from 5 ft to 48 ft, with an average of 18 ft. In addition to cement content, the placement and distribution of the grout in the soilcrete column is vital in achieving the design strength.

Soil Mixing

Based on the contractor's soil mixing experience and the laboratory testing conducted prior to production work, column cement content and mixing energy were identified as the two most important parameters during mixing of the soilcrete columns. The project began with target values of 260 kg/m^3 and a 40-rpm mixing tool rotation. This requires consistent placement of the grout throughout the column and the ability to adequately shear the soil as grout is being pumped to create a relatively uniform soilcrete product. The cement content of the column is controlled by the cement content of the grout and the volume of augered soil that is placed back into the pre-drilled bore-hole. If a low cement content grout is pumped, a high grout volume is necessary. Therefore, the volume of soil placed into the hole would have to be reduced (eg. from 80% soil to 70% of the volume to fill the pre-drilled column). Likewise, the volume of soil could be reduced if a higher cement content grout were utilized. In this case, it is important to introduce enough moisture to aid in mixing.

The driller is aided by onboard instrumentation that records and displays grout specific gravity, flow rate, and volume pumped. This real-time information is extremely valuable in that mixing of a column can be easily monitored and problems identified and resolved more quickly. A sensor and transmitter mounted to the rig displayed the real-time data, graphically such that the operator could easily read the actual and target grout volume along the length of the column.

The longest soilcrete columns were 48 feet. For very short columns, it was easier to simply pour the column with neat grout without mixing. The average soilcrete column length was 18 feet. In some instances, mixing rotation rate slowed significantly at great depths due to the increased weight on the mixing tool and greater stiffness of the soil. This was resolved by reaming the hole (moving up and down the column quickly with the mixing tool), which increased the fluid nature of the column as additional grout was pumped, and improved the tool's ability to shear the soil-cement mixture. The mixing tool and drill rig are shown in Figures 4 and 5, respectively.

FIG 4: Mixing Tool.

FIG 5: Drill Rig.

Data Analysis

The cement content of the columns was back-calculated based on the volume of the soilcrete column, rather than the combined volume of soil and grout. Although the contractor had a target percentage of soil to place in the hole after pre-augering, the volume of soil placed is approximate, due to difficulties in measuring the volume, and the effects of soil porosity and subsequent compaction. The porosity of the replaced soil, varies with large depths as the soil "compacts" itself after being dropped from any height. Because all holes were predrilled and then soil was replaced, deeper holes contained more compact soil. In the event the column is terminated below the desired elevation, neat grout can be used to top off the column. If the column was too high, it was scraped down. Either way, the volume of the column is set and the grout volume within the column is essentially set. Therefore, it is more accurate to back calculate the column's cement content using the column's final volume instead of the two independent calculations of soil volume and grout volume.

Cement contents of soilcrete columns typically ranged from 280 kg/m^3 to 400 kg/m^3. For the entire job, the weighted average, based upon column volume, was 320 kg/m^3. These values, while higher than the original design cement content, can be attributed to a number of factors. Shortly after the start of work, grout flow was slightly reduced when preliminary core breaks showed more than adequate strengths. Short columns tended to have much higher cement contents, as the mixing time for these columns was short and it was imperative the driller pumped a certain volume of grout, such that the column was properly seated on the bearing layer. Over the course of the project, cement contents generally decreased as the crew gained experience. Towards the end of the project, however, efforts were made to terminate the column at the proper elevation instead of scraping down the hardened column the following day. To accomplish this, several columns were "topped off" with neat cement grout in the event that the freshly-mixed column was below the proper elevation. This procedure slightly skewed the cement content values upward for some columns.

Sampling and Testing

Two-inch cores, taken in 5-foot runs, were sampled continuously over the column length of three soilcrete columns. The cores were taken at least one week after the column was mixed. Core recoveries averaged greater than 95% recovery. Core samples were then trimmed to a 2:1 height-to-diameter ratio and strength tested. Laboratory results of the cores' unconfined compressive strengths are shown in Figure 6. Extrapolating the data, 28-day strengths of all the core breaks were at least 400 psi, or a minimum of 4 times the design strength. Compressive strength results from the two-inch core samples were high enough to reduce the cement content of columns installed later in the project, which was accomplished by a slight reduction in the grout flow rate.

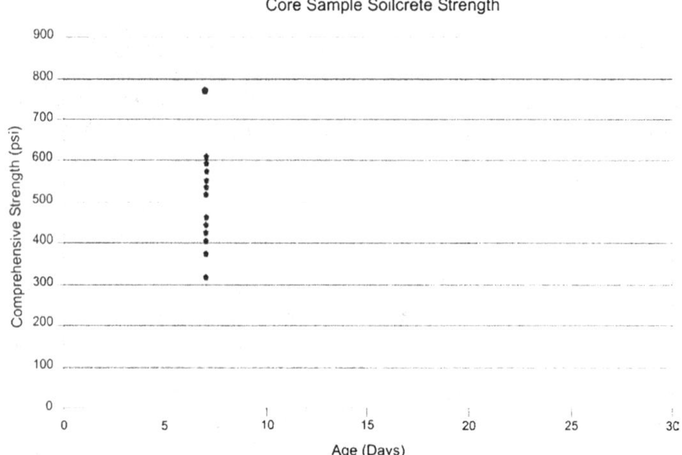

FIG 6: Core Sample Soilcrete Strength.

Wet grab samples were taken with an in-situ sampler. The in-situ sampler was air-actuated using a small air compressor. After pushing the sampler into the freshly-mixed column, its piston was retracted, creating a vacuum and sucking in some of the uncured soilcrete product through ports on the side of the sampler. The sampler was then pulled out of the column, the piston extended, and the sample retrieved and placed in 2" cubes. These samples were used to make 2" cubes and were placed in water to cure. Cube breaks were conducted at 3, 5, 7, 14, and 28 days. Figure 7 plots the strength of the soilcrete cubes versus age. Overall, the average 28-day compressive strength was 1,200 psi, greatly exceeding expectations. These strengths are not typically achieved in most soil mixing applications. The high strengths are attributed to increased cement content and excellent shearing of the loose pre-augered (and backfilled) soil material.

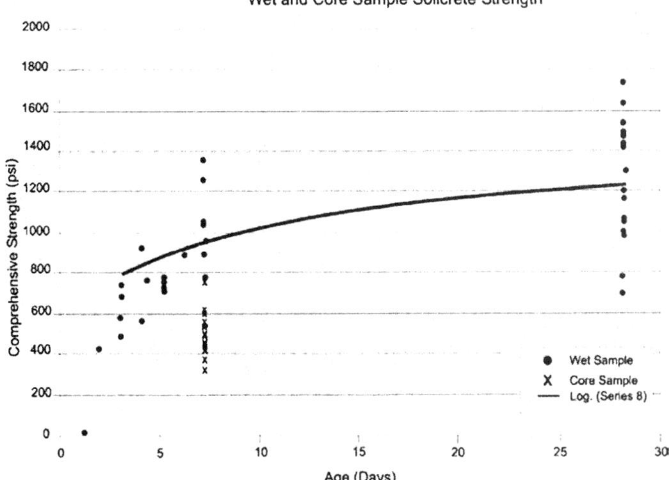

FIG 7: Wet and Core Sample Soilcrete Strength.

Conclusions

The following conclusions can be drawn from this soil mixing experience.

1. Soil mixing is a viable ground improvement alternative for supporting high column loads.

2. Deep soil mixing can be advantageous in comparison to drilled shafts in that it limits the amount of spoils trucked off-site. This attribute became even more important when previously undiscovered tanks were encountered during installation of soil cement columns, some of which had leached contaminants into the ground. A large percentage of the contaminated soil was mixed into the columns, minimizing the amount of contaminated soil that had to be removed from the site.

3. Soilcrete columns (via soil mixing) proved advantageous to stone columns/aggregate piers, in the karst terrain where vertical drainage paths could initiate the formation of sinkholes. Additionally, the increased stiffness of the soilcrete columns in comparison with stone columns/aggregate piers was essential in meeting the differential settlement criteria for cases where the depth to rock varied significantly between adjacent columns.

APPENDIX. CONVERSION TO SI
Feet (ft) x 0.305 = meter (m)
Inch (in) x 0.025400 = meter (m)
Psi x 6.894757 = kN/m^2
Kips x 4.448222 = kN
Ksf x 47.880260 = kNm^2

ACKNOWLEDGMENTS

The authors would like to acknowledge Dr. Al Sehn for testing and analysis of samples that were used to predict cement contents. This project was a success due to the commitment and support of Pennoni Associates and the L.F. Driscoll Company.

INSTALLATION, LOAD-TESTING AND DESIGN OF GEO-JET® SOIL-CEMENT PILES

Jonathan R. Craft[1], Member ASCE and Geo-Institute

ABSTRACT: Deep mixing techniques for mass ground improvement are now widely accepted and are used most commonly for bearing capacity enhancement, liquefaction mitigation, and for slope stabilization. The employment of secant soil-cement columns, alternately with and without a steel profile inserted, is also well-known in the industry, the Central Artery & Tunnel Project ("Big Dig") in Boston being the most notable example of such an application here in the United States. However, the use of individual soil-cement columns for load-bearing purposes while less common is no less adaptable, either with steel reinforcement or without, and this paper will describe the particular merits of the patented Geo-Jet® system of deep mixing for such an application including consideration of design issues, testing and installation. Case histories of vertically loaded soil-cement columns using the patented Geo-Jet® system, including projects such as San Francisco International Airport, and the Millbrae BART Station and other projects in the Bay Area, will also be described.

1. INTRODUCTION

The use of soil-cement columns, either singly or in groups beneath a pile cap, is a well-established technique in Japan (Taki and Bell, 1998; Taki, 2003) while here in the United States, only a few such applications exist, some of which will be described herein. However, with ever stiffer environmental controls and noise restrictions being imposed upon urban construction projects in particular, the substitution of driven piles and auger-cast piles by soil-cement columns, reinforced (for uplift and lateral loading) or unreinforced (for compression loading only) is predicted to increase at an exponential rate.

---

[1] Geotechnical Division Manager, Condon-Johnson & Associates, Inc., P.O. Box 12368, Oakland, CA 94604.

Deep mixing ("DMM") techniques that are most readily adaptable for installing load-bearing elements, are those based on single axis auger systems. In turn, this category may be sub-divided into two sub-categories, 1) that comprises mechanical mixing only, and, 2) that uses mechanical mixing augmented by the use of high pressure jetting of grout into the augered soil-mass, also known as "Geo-Jet®" (a system patented by V. Schellhorn). The advantages of the latter system - which has been defined (Bruce and Bruce, 2003; FHWA, 2000) as a "rotary + jet" DMM technique - over the former, are speed of installation and more efficient mixing in situ of the grout with the soil.

2. INSTALLATION

The Geo-Jet® soil processor shown in Figure 1, consists of a single auger (with teeth) having the diameter of the desired column (typically 600 mm to 1,500 mm) mounted either on a hydraulic CIDH unit as seen in Figure 2 or, for drilled depths in excess of 23 metres, on a set of leads fixed to a crawler-crane. Below the auger at the tip of the drilling rod is a drag bit that incorporates three downward pointing slurry jets to aid penetration, while mounted in the arms of the auger itself, there is an array of sub-vertical downward pointing jets (up to 8 No. depending on the soil conditions and column diameter) which break up the lumps of soil cut by the auger and aid the mixing of the soil. Above the auger there is a blending paddle incorporating two horizontal slurry jets which as well as enhancing the soil-cement mixture, ensure the safe retrieval of the mixing tool on completion of the column. The effect of the jets may be seen in Figure 3.

The cement-water slurry or grout is batched remotely and continuously using a nuclear densometer to monitor the slurry density. The slurry is then transferred at low pressure to a high pressure triplex pump, which may be located on or nearby the rig. These pumps generate pressures in the range of 150 ~ 300 bars and flow rates of the order of 500 ~ 1,000 litres per minute.

The triplex pump is started immediately before the tip of the monitor enters the soil and the column of soil-cement is thereby formed as the soil processor is advanced into the ground. Upon reaching tip, cement slurry continues to be pumped at high pressure to ensure complete mixing of the soil at the base and continues during the extraction phase at reduced pressure, only to ensure that the jets do not get plugged. Any structural insert is then placed inside the column i.e. before initial set of the soil-cement occurs.

The whole process is computer-controlled with pre-determined rates of processor advance and slurry flow keyed-in at the start of the shift by the operator in the cab. A print-out of salient parameters (e.g. torque, rpm, rate of advance, flow rate, pressure, slurry density, etc) as shown in Figure 4 may be obtained by downloading data produced by the on-board computer. Further quality control may be effected by sampling the soil-cement in the field and testing such samples in the laboratory to measure the unconfined compressive strength of this material.

FIG.1. Geo-Jet® Soil Processor

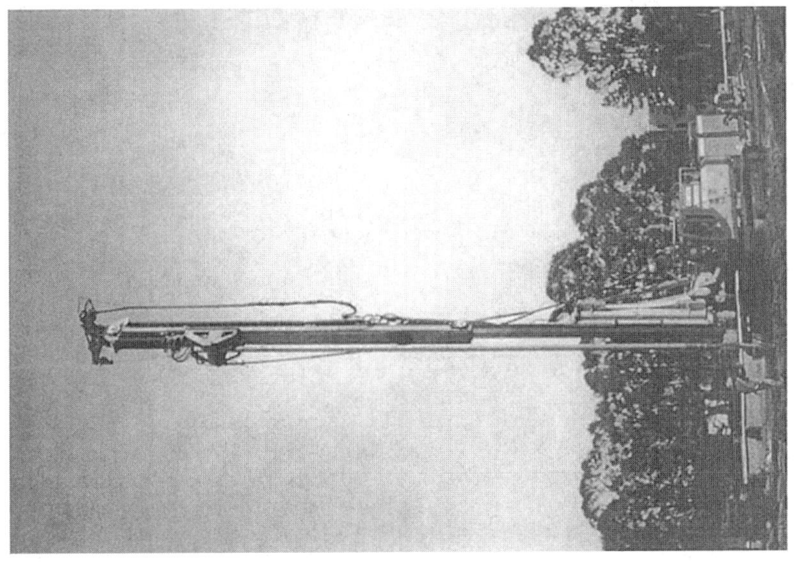

FIG.2. Hydraulic Self-erect Drilling Rig

FIG.3. Pre-drilling Check of Jetting Function

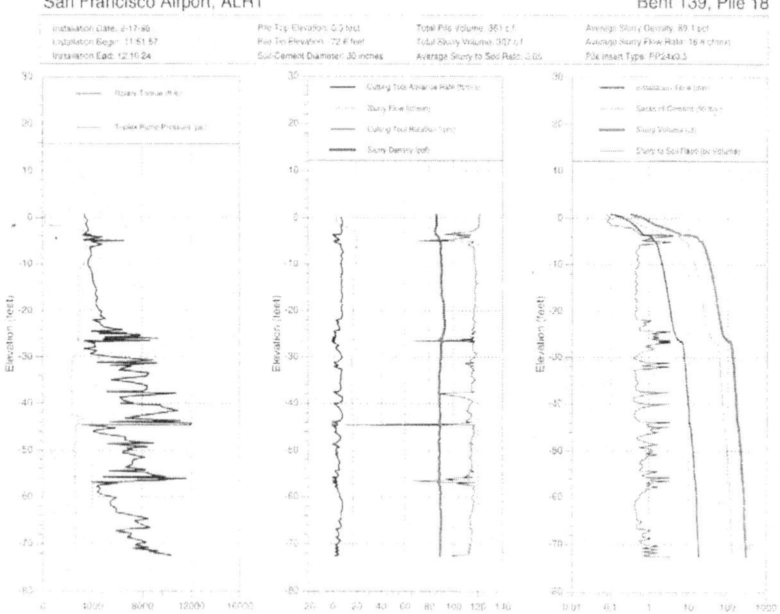

FIG.4. Typical Output from Real Time Geo-Jet® Data Logger

3. APPLICATIONS AND LOAD-TESTING

Millbrae Station for the BART Airport Extension, San Francisco

The proximity of other structures to the proposed station structure, led to the substitution of a driven pile foundation design by one based on Geo-Jet® soil-cement columns. The soil profile found at the Millbrae Station location varies from south to north across the site and comprises 1.5 to 2.4 m of Fill overlying 0.0 to 1.5m of Bay Mud, over 3.0 to 2.4 m of Alluvium, which in turn overlies the dense silty sands of the Colma Formation. The design loads for these Geo-Jet® soil-cement columns were

- 670 kN compression (i.e. DL + LL)
- 200 kN tension (earthquake loading)
- 270 kN lateral (earthquake loading)
-

and this required that the foundation elements penetrate the Colma Formation Sand. A typical soil profile for design purposes is shown in Figure 5. To allow for the local variations in the profile, the 750 mm diameter Geo-Jet® columns were designed to penetrate the Colma Formation an average of 2.7 m, giving a total depth of 9.1 m. A structural steel member in the form of a 400 mm O.D pipe of wall thickness 10 mm, was inserted into the freshly-formed soil-cement column to provide vertical load transfer and lateral stiffness.

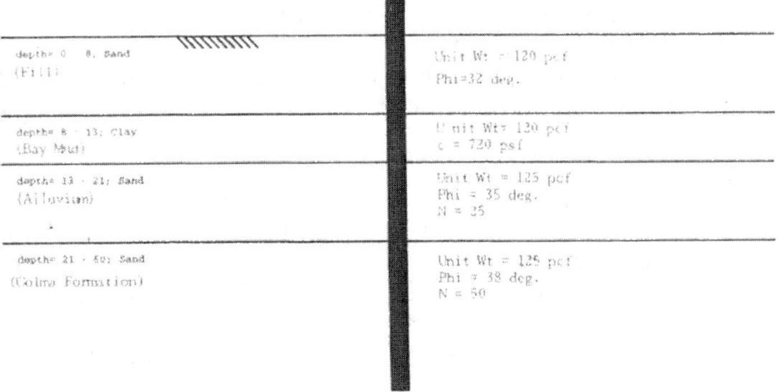

FIG.5. Soil Profile at North End of Millbrae BART Station

The Geo-Jet® soil-cement column is designed to fail "geotechnically", that is to say, failure at the soil / column interface should occur before that between the steel insert and the soil-cement. In this case, with a factor of safety of 2.0 on maximum vertical loading, the ultimate average bond stress between the steel pipe (inside as well as outside surfaces) and the soil-cement computes to be about 60 kPa. Laboratory tests done at the time indicated that the bond stress is typically of the order of 10% of the average unconfined compressive strength of the soil-cement mix. However, given the inherent variability of all DSM systems, combining as they do in situ soils with albeit known quantities of cement slurry, it is prudent to apply a safety factor of at least 2.0 to the bond stress to derive a target design strength for the soil cement mix. For this project therefore, the target unconfined compressive strength was 1.4 MPa.

A series of load tests were carried out to verify the design assumptions the results of which are summarized in Figure 6. The compression load test indicates that failure occurred at a test load of about 2,000 kN while the tension test indicates a failure load of about 1,200 kN. The difference could be said to represent the end bearing capacity of the column.

Analyzing these results, the average ultimate shaft friction between the soil-cement column and the ground is 56 kPa, while the ultimate end-bearing capacity is about 1,800 kPa.

Ignoring the contribution of the Fill (shallow) and Bay Mud (soft), the average ultimate shaft friction in the Alluvium and Colma Formation is about 100 kPa. Thus given that

$$q_s = K \, . \, \sigma_v \, . \, \tan\delta \qquad (1)$$

where,

q_s = average shaft friction

K = coefficient of lateral earth pressure

σ_v = effective overburden pressure

δ = angle of friction between soil-cement and ground

and, if it is assumed that $\delta = \Phi = 38^o$ (ref. Figure 5), $K \approx 1.4$ which is, on the face of it, a curious result, implying as it does, a load-transfer behavior more akin to that of a driven pile.

As to the end-bearing contribution, bearing capacity factor N_q would compute to a value of < 20, which would imply a value for Φ somewhat lower than that assumed for design purposes. Unfortunately, in the absence of any instrumentation (i.e. strain gauge data) in this case, it is difficult to be more rigorous. However, one possible explanation for the apparent reduction in end-bearing might be that after failing in shaft friction, the steel pipe debonded from the soil cement leading to a punching failure of the steel pipe itself.

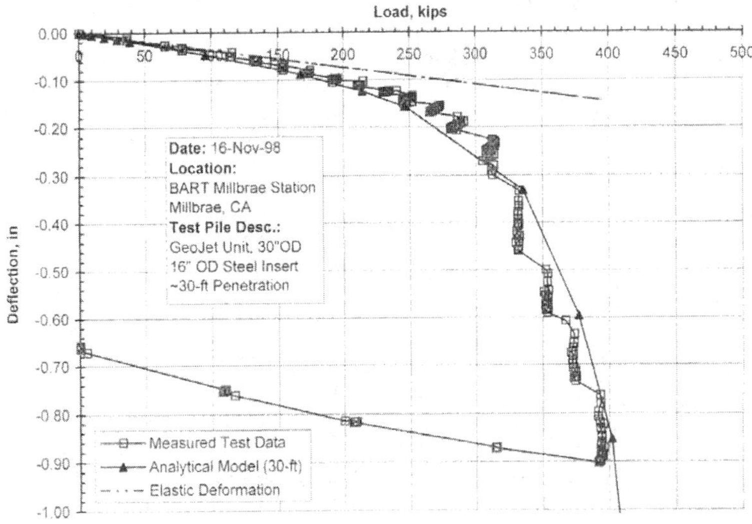

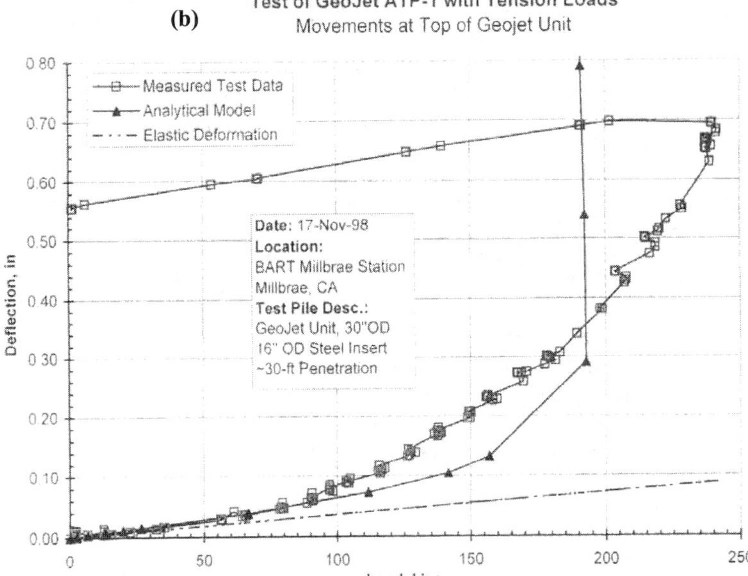

FIG. 6. Load Tests on Geo-Jet® Test Pile for Millbrae BART Station

ALRT, San Francisco International Airport

The foundations of the new automatic light rail transit system that links the BART Airport Extension with the passenger terminals, were designed to be built on driven piles. However, where the new structure alignment came close to existing buildings or near sensitive underground utilities, a vibration-less and noise-free system of pile installation was required and in such locations the pile-caps were built upon steel pipe reinforced Geo-Jet® soil-cement columns typically arranged in groups (6 x 6, 7 x 5, etc.). A total of 500 No. load-bearing Geo-Jet® columns were installed on this project.

Table 1. Soil Profile at ALRT Geo-Jet Test Site

Depth	Description	c_u (kPa) / N (blows/ft)
0.0 to 2.7 m	Fill (silty sand, etc)	38
2.7 to 9.4 m	Bay Mud	0 ~ 25
9.4 to 12.0 m	stiff clayey silt with some silty sand	8
12.0 to 19.0 m	dense / very dense silty / clayey sands	40 ~ 60

Prior to commencing the works, a preliminary test program was undertaken at this site comprising one test pile with four reaction piles. One compression and one tension load test was carried out on this test pile, both to failure. The test column was, like the production columns to follow, 900 mm in diameter and incorporated a 600 mm diameter by 12.5 mm thick steel pipe by way of reinforcement. The Geo-Jet® column was installed to a depth of approximately 19.0 m.

Load testing followed ASTM Quick Load Test Methods with load and displacement measurements logged by a data acquisition system connected to a laptop computer. The compression load test was carried out first and as can be seen from the load – displacement curve in Figure 7(a), plunging failure occurred at a load of about 4,400 kN at a maximum displacement of approximately 21 mm. Yield appeared to be reached at about half the ultimate load and at a displacement of approximately 5 mm. In the pull-out test, from the plot in Figure 7(b) a marked elasto-plastic behavior may be noted with yield occurring at a load of about 1,600 kN and an elongation of about 13 mm, i.e. at about two-thirds of the ultimate failure load of 2,400 kN. A further unloading and reloading cycle was undertaken following which it was recorded by the testing engineer that soil-cement had separated from one side of the steel pipe pile. Tests carried out on samples of the soil-cement taken from various depths in the column soon after its installation, revealed unconfined compressive strengths at 28 days in the range of 700 kPa to 5,500 kPa with an average just below 3,000 kPa, the range of results being indicative of the very varied soil profile here.

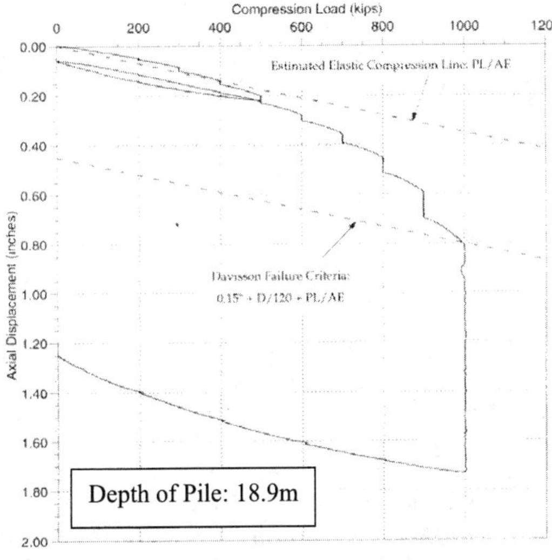

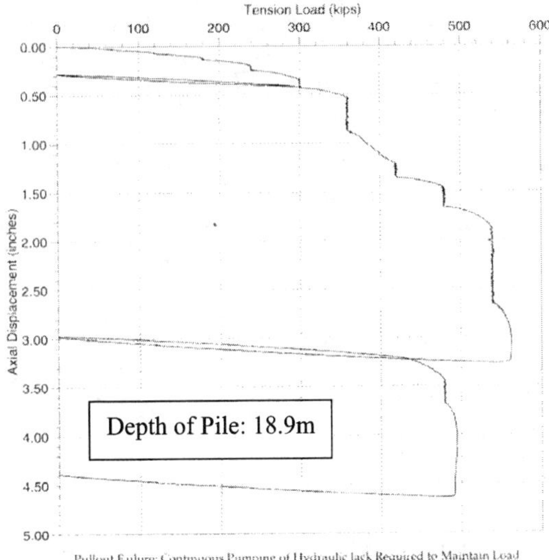

FIG.7. Load Test on Geo-Jet® Test Pile for ALRT, SFIA

In the pull-out test, and neglecting the Bay Mud, the contribution of cohesive soils to shaft friction, based on laboratory unconfined compressive test result, is of the order of 300 kN. Therefore, the contribution of the mostly dense granular soils is about 2,100 kN. In which case, the average value for shaft friction q_s, is approximately 150 kPa, and, from the relationship given in (1) above, where, $\delta = \Phi = 35^o$, and $\sigma_v = 200$ kPa, then $K \approx 1.1$.

Given that plunging seems to have occurred during the compression test, it is reasonable to assume that a) the column has failed in shaft friction and b) that end-bearing has been mobilized here. In which case, $Q_b =$ (4,400 – 2,400) kN and from

$$Q_b = N_q \cdot \sigma_v \cdot \pi D^2 / 4 \qquad (2)$$

$N_q = 2{,}000 / 250 \cdot \pi \cdot 0.375^2 = 18$, which is on the low side for what is described as a “very dense clayey SAND” yet is similar to the outcome of the test at Millbrae BART Station referred to above.

Subsequently, three load tests – two compression and one tension – were carried out on three production piles at Bent 4L (1 No. compression test on Pile 23) and Bent 139 (1 No. compression test on Pile 18 and 1 No. tension test on Pile 25) to verify performance. The soil conditions at the test sites may be summarized as follows:

at Bent 4L, Fill (0.0 to 3.3 m) over Bay Mud (3.3 to 13.4 m) over sandy clays / clayey sands (13.4 to 24.7 m) over very dense silty sands;

at Bent 139, Fill (0.0 to 2.1 m) over Bay Mud (2.1 to 9.7 m) over gravel and sand / silty sands / silty clay (9.7 to 24.0 m) over silty and clayey sand.

The piles are 750 mm diameter and 21.3 m long, with a 600 mm diameter steel pipe of wall thickness 12.5 mm inserted to the full depth of the Geo-Jet® columns. It should be noted that the tops of the piles were lowered to approximately 3.3 m (Bent 4L) and 2.3 m Bent 139) below existing grade to allow for the depth of the footings.

Print-outs from the on-board computer of the Geo-Jet® test column installations, one of which is shown in Figure 4, all show a significant increase in torque as the soil processor exits the Bay Mud and enters the underlying stiffer soils. This is a very valuable and useful characteristic of the Geo-Jet® system, offering as it does additional site investigation data to the designer that allows him or her to make adjustments and refinements as the work proceeds.

Based on the preliminary test program, the predicted ultimate capacity of these piles would have been:

Bent 4L / TP23: from boring B-15 and CPT C-14 (ref. App.A) of the 11.0 m of competent soils beneath the Bay Mud, 4.5 m are very stiff silty clays having a $c_u \geq$ 120 kPa; these layers should contribute approximately 1,300 kN in shaft friction. The very dense clayey sands in turn, assuming K.tan $\delta = 0.9$, should contribute approximately 1,700 kN, making a total shaft friction component of 3,000 kN.

Bent 139 / TP18 & TP25: from boring RAC-7 of the 13.5 m of competent soils beneath the Bay Mud, 6.0 m is very stiff silty clay having a c_u = 134 kPa (from laboratory test); this layer would contribute approximately 1,900 kN in shaft friction. The dense to very dense gravels, sands and silty sands assuming K . tan $\delta = 0.8$, should contribute approximately 2,100 kN making a total shaft friction component of 4,000 kN.

Load testing was carried out using Quick Load Test Methods (ASTM D1143-1 and D3689-90 refer) and two telltales were installed on each steel pipe inserted in he test columns. The load displacement plots for all three tests are shown in Figure 8.

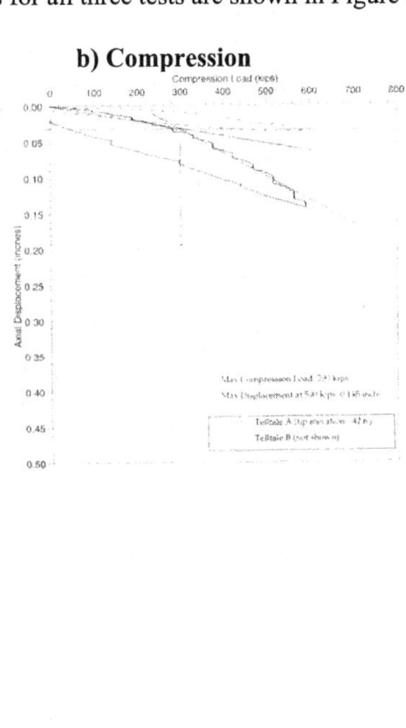

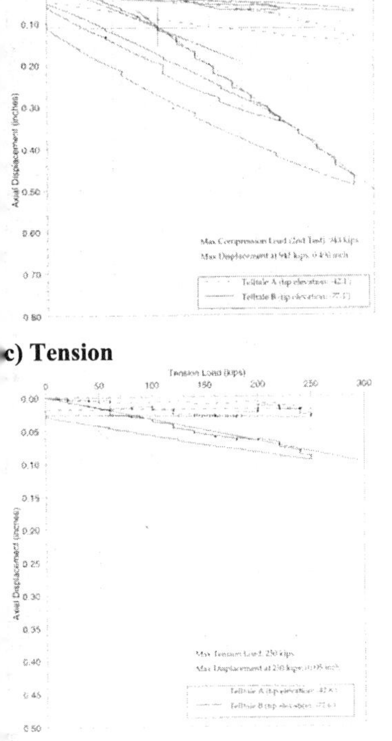

FIG.8. Load Tests on Geo-Jet® Production Piles for ALRT, SFIA

Test Pile 23 at Bent 4L was loaded to 4,200 kN (against an ultimate design load of 2,500 kN) and yet the telltale at the tip of the pile shows less than 2 mm of displacement, from which it may be deduced that most of the load is being carried by shaft friction. However, a yield point is apparent at a load of about 1,500 kN (60% of ultimate design load) and a corresponding total displacement of about 3 mm. The apparently large difference between the total measured axial displacement and that at Telltale A, equates exactly with the elastic shortening of the steel pipe through the Bay Mud. Some movement is evidenced at the tip (Telltale B) in which case it can be assumed that not only has shaft friction been mobilized for the full length of the pile but that there is a significant contribution from end-bearing.

This would explain the greater than predicted capacity of this pile. For example, assuming $c_u \approx 150$ kPa at the tip, $Q_b \approx 600$ kN, which thereby puts the predicted shaft friction total within 20% of that measured.

Test Pile 18 at Bent 139 was loaded to 2,600 kN (against an ultimate design load of 2,100 kN). In this case a yield point is apparent at a load of about 1,300 kN (which is also about 60% of ultimate design load) and a corresponding displacement of less than 1 mm. Unfortunately the telltale at the tip did not function in this case but it seems unlikely that any end-bearing was mobilized at such a small displacement. Given this premise, it is unlikely that even shaft friction has been fully-mobilized in this case, and so any comparison with predicted capacity is futile.

The tension load (1,100 kN) applied to Test Pile 25 at Bent 139 exceeded the ultimate design load of 900 kN by about 25% yet did not demonstrate any elasto-plastic post yield behavior. This is superficially at least, consistent with the compression test on nearby Test Pile 18 where yield did not begin to occur before a load of 1,300 kN was applied. However, the strains are not consistent, being larger in the tension test. If it is assumed that there is no measurable shaft friction between the Geo-Jet® column and the Bay Mud, the elastic strain of the pipe in this zone would compute to an extension of 1.8 mm at the maximum load applied which, coincidentally, is equivalent to the elastic rebound upon unloading. The net displacement taking place in the "anchored" part of the pile is less than 1 mm and not surprisingly therefore, the telltales reveal little as to the mechanism of the load transfer in the more competent soils beneath the Bay Mud.

West Field Detention Basin, San Francisco International Airport

Geo-Jet® columns were designed to support an underground flood retention facility at the airport in San Francisco and over 400 N°. such columns were eventually constructed here. But before construction could start, a test program was instigated. Three test columns of differing depths (12.2, 13.7 and 15.2 m) were installed. All were 750 mm in diameter and all had full-length 400 mm diameter by 9.5 mm structural steel pipes inserted in them while the soil-cement was still wet. The piles were instrumented with telltales at their tips and strain gauges along their lengths. Two were tested in compression and one in tension.

The objective was to identify the optimum length for production piles. Due to site constraints, only two piles could be tested, the 12.2 m and 13.7 m deep piles. Each pile was tested first in compression and then in tension, using ASTM "Quick Methods". The program is discussed in detail by Isenhower, Arrellaga, Wang and Johnson 1998.

Table 2. Subsurface Conditions at Test Site After Excavation

Depth	Description	c_u (kPa) / N (blows/ft)
0.0 to 2.1 m	soft Bay Mud	20 – 33
2.1 to 9.1 m	dense clayey / silty SAND	17 - 34
9.1 to 10.6 m	very stiff sandy CLAY	100
10.6 to 12.2 m	very dense silty SAND	46
12.2 to 24.0 m	stiff Old Bay Mud	77

The load – displacement curves for these test piles are shown in Figure 9, and they matched almost exactly the predictions Messrs. Isenhower et al had made, as can be seen in with both piles failing at about 2,000 kN. However, the authors comment that there was perhaps a 5% contribution of end-bearing to the shallower test-pile result, based on the strain gauge and telltale information.

(a) Test results for compression loads on 40 and 45 ft GeoJet® Piles

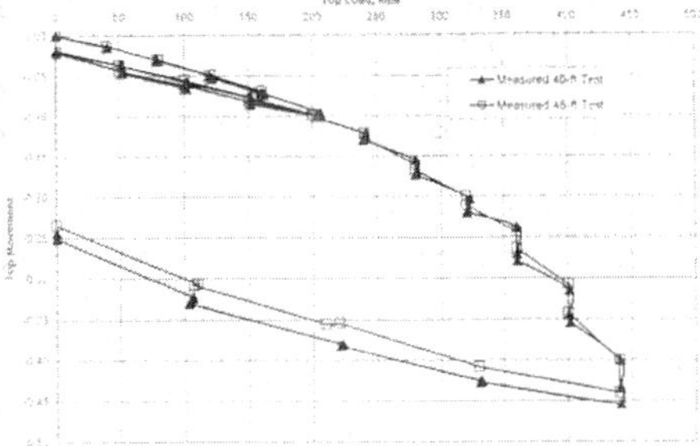

(b) Test results for tension loads on 40 and 45 ft GeoJet® Piles

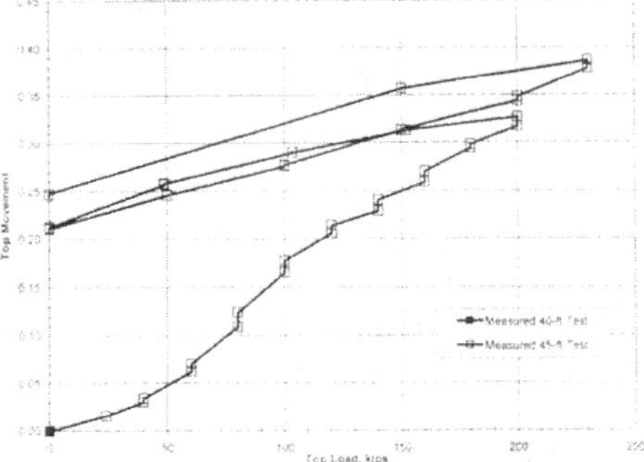

FIG.9. Load Tests on Geo-Jet® Test Piles for West Field Detention Basin. SFIA

If these results are analyzed as before, using the parameters included in Table 4, values for the coefficient of lateral earth pressure K, can again be deduced, which for the 12.2 m pile computes to be 1.4 while for the 13.7 m pile it is 1.2.

4. DESIGN

In all cases examined here, the piles have appeared to fail "geotechnically", that is to say, the failure has occurred at the soil / soil-cement interface, not at the steel / soil-cement interface, even though the range of unconfined compressive strengths of the soil-cement sampled in these test columns is 0.5 ~ 6.0 MPa, depending on the soil conditions. Design can therefore be based on the assumption of complete load transfer from structural element to the ground.

Furthermore, due to the unique process of simultaneous augering and jetting of the column, it is safe to assume in the case of cohesive soils, the full, unfactored undrained shear strength in computing ultimate shaft friction values, and, in the case of granular soils, $\delta = \Phi$, as well as a lateral effective earth pressure at least equivalent to the effective overburden pressure.

5. CONCLUSIONS

A number of major projects incorporating Geo-Jet® columns as load-bearing elements have been carried out by the author's company and the foregoing test program results demonstrate more than adequately the consistency and repeatability of the product in varying soil profiles in the San Francisco Bay Area.

It has been demonstrated here, somewhat counter-intuitively perhaps, that the Geo-Jet® process of pile installation is not at all comparable with a cast in situ drilled shaft, but rather results in a pile that behaves much more like a driven pile. In each case history considered in this paper, the computation of the coefficient of lateral earth pressure in granular soils, K, has produced a number in the range 1.0 ~ 1.5.

In other words, the soil is exhibiting mildly "passive" behavior rather than the "at rest" condition that is perhaps more characteristic of drilled shafts of similar diameters. This may be explained by the jetting process that is a) very rapid (typically, piles can be installed at the rate of 1.5 ~ 3.0 meters per minute), and b) does not allow the soil to relax due to the constant presence of the high pressure jets (typically, 150 ~ 200 bars) of grout that break up and mix the soil with effectively no reduction in density.

In conclusion therefore, the application of Geo-Jet® columns as load-bearing elements by way of straight substitution for driven piles, is a natural choice especially in sensitive environments where noise and vibration controls must be considered. They also represent a cost-effective and resource-sensitive solution for the design and realization of piled foundations in difficult and even contaminated ground.

ACKNOWLEDGMENTS

The author wishes to acknowledge the support of his colleagues at Condon-Johnson & Associates, Inc, notably Messrs. James Johnson and Alan Macnab, P.E. together with that of Mr. Lonnie Schellhorn who invented the system known as "Geo-Jet®" here (and "Turbo-Jet" elsewhere), a personal communication from Mr. Jose Arrellaga, P.E. of Lymon C. Reese & Associates concerning the Millbrae test results, and, the excellent test data produced by Mr. John Lemke, P.E. previously of Delta Geotechnical Services and now of GeoDAQ, Inc, in Sacramento.

REFERENCES

Bruce, D.A. and Bruce, M.E.C. (2003). "The Practitioner's Guide to Deep Mixing". *Proceedings of the 3rd International Conference on Grouting and Ground Treatment*, ASCE Geotechnical Special Publication No. 120, 474-488.

Delta Geotechnical Services. (1997). "Pile Load Test Report West Field Detension Basin, San Francisco Airport". Report for Kulchin-Condon & Associates, Inc.

Delta Geotechnical Services. (1998). "Pile Load Test Reports ALRT Bents 4L and 139, San Francisco Airport". Report for Condon-Johnson & Associates, Inc.

Federal Highway Administration. (2000). "An Introduction to the Deep Mixing Method as used in Geotechnical Applications". Prepared by Geosystems, L.P., Document No. FHWA-RD-99-138, March, 143 p.

Isenhower, W.M., Arrellaga, J.A., Wang, S.T. and Johnson, J.O. (1998). "Experiences with GeoJet Piles in Sandy Clays". *Proceedings of the Geo-Institute Conference on Soil Improvement for Big Digs,* ASCE Geotechnical Special Publication No. 81, 72-83.

Lymon C. Reese & Associates. (1998). "Design Aspects of GeoJet Piles for BART Airport Extensions / Millbrae Station". Report for Condon-Johnson & Associates, Inc.

Taki, O. (2003). "Strength and Properties of Soil Cement Produced by Deep Mixing". *Proceedings of the 3rd International Conference on Grouting and Ground Treatment*, ASCE Geotechnical Special Publication No. 120, 646-657.

Taki, O. and Bell, R.A. (1998). "Soil-Cement Pile / Column – A System of Deep Mixing". *Proceedings of the Geo-Institute Conference on Soil Improvement for Big Digs,* ASCE Geotechnical Special Publication No. 81, 59-71.

CMC FOUNDATION SYSTEM FOR EMBANKMENT SUPPORT -- A CASE HISTORY[1]

Cyril Plomteux, Ali Porbaha, and Charles Spaulding

ABSTRACT:A wide range of deep foundation systems has recently been developed for construction of embankments on soft soils. Controlled Modulus Columns (CMC) is one technique of ground modification, originally developed in France, for support of light structures such as highway and railway embankments. This paper presents the case history of a column supported embankment project in which 2193 CMC columns were installed up to a depth of 12.5 m to minimize settlement of an embankment. Main challenges were soft ground condition of the site, and the need for an accelerated construction technology for timely delivery of the project. Evaluation of various alternatives, design considerations, construction related issues and monitoring system are discussed in detail.

INTRODUCTION

Embankment construction is an essential element of any highway and railway construction. The problem arises when the embankment passes through soft ground conditions such as soft clay, bay mud, organic soil/peat, chalk or loose fine sand. In

---

[1] **Cyril Plomteux**
Drainage & Ground Improvement, Inc., 275 Millers Run Road, Bridgeville, PA 15017, Phone: 800 326-6015, Fax: 412 257-8455, E-mail: cyril@wickdrain.com

Ali Porbaha (Corresponding author)
Department of Civil Engineering, California State University, Sacramento, 6000 J Street, Sacramento, California 95819-6029, Phone: (916)-278-6120, E-mail:porbaha@ecs.csus.edu

Charles A. Spaulding
Menard Soiltreatment, Inc. - 1331 Airport Freeway - Suite 302 - Euless, TX 76020 - Phone: (817) 283-5503 fax:(817) 283-6931 E-mail: cspaulding@menardusa.com

that case the soil is subject to settlement or stability problem due to lack of bearing capacity; and/or in case of loose saturated fine sand subject to liquefaction due to ground shaking. To overcome these difficulties, a wide range of deep foundation systems has recently been developed for construction of embankments on soft soils (Porbaha et al, 2002a and b). These techniques take advantages of various ground modification concepts such as densification, reinforcement, solidification, etc. (see Figure 1).

The objective of this paper is to present a new deep foundation system, namely Controlled Modulus Columns (CMC), for support of embankments. The detail of design, construction, and monitoring of an embankment project supported by the CMC system is outlined here.

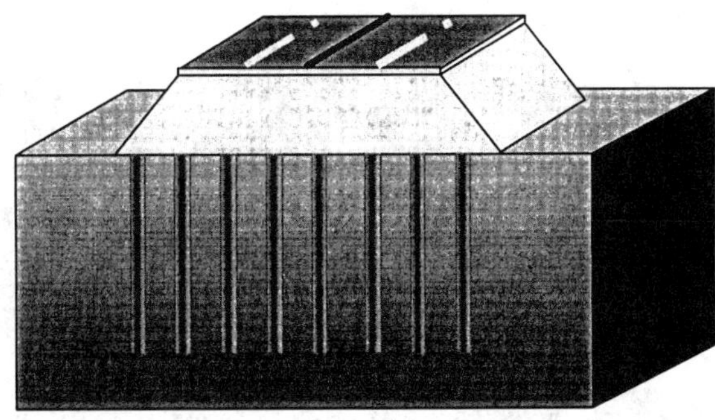

Figure 1: Concept of column-supported embankment

CONTROLLED MODULUS COLUMNS

The Controlled Modulus Columns (CMC) is a ground modification system that reinforces soil by screwing a hollow auger into the soft soil and installing a low-pressure cement-based grout column through the hollow auger (see Figure 2). The combined effect of densification and reinforcement improves characteristics of the soft ground due to composite action. The CMC system uses a displacement auger powered by equipment with very large torque capacity and very high downward thrust, which displace the soil laterally with virtually no spoil or vibration. The auger is screwed into the soil to the required depth and as such it increases the density of the surrounding soil and thus increases its load bearing capacity. When the required depth or a preset drilling criteria (usually rotational torque) is reached, a highly workable grout-cement mixture is pumped through the center of the hollow auger. The grout mixture then flows under low pressure out of the auger base as it is retracting to obtain a high capacity column that can be used in close vicinity of sensitive structures. The grout is injected under low pressure, typically less than 10 bars (145 psi) and no soil mixing takes place during the pressure grouting. To ensure that the soil above the auger remains compacted, the top of the auger is equipped with reverse direction flights. The result is a composite system with column reinforcements bonded to the surrounding soil. The main features of CMC technology are:

- Material is grouted in place with the use of a displacement auger without spoil.
- Deformation modulus is 100 to 3000 times that of soil.
- Soil properties are improved around the columns by compression resulting from the lateral displacement.
- Diameter is determined based on size of the auger (usually in range from 350 to 500 mm).
- Common installation practice is based upon square grids with center-to-center spacing in range from 1.2 to 3 m.
- Maximum length of treatment is 25 m.

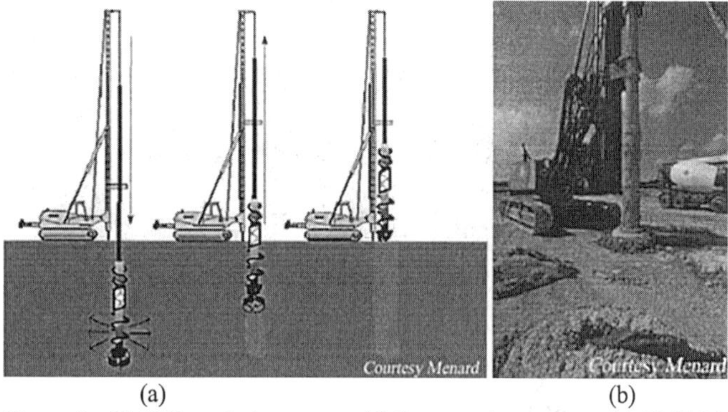

(a) (b)

Figure 2: CMC Foundation system (a) Construction process (b) CMC Rig

The case history of a project in which CMC foundation system was used for embankment support is discussed here.

PROJECT DESCRIPTION

The project consists of improving the ground under the access embankment of a road crossing over the Channel Tunnel Rail Link (CTRL) covering 17 kilometers of railways construction. The embankment height ranged up to 7.5 meters and live load of 20 kPa from the road traffic was taken into account. The main design requirement for the ground improvement was to limit the post-construction settlement of the embankment to 10 mm per decade. The works have been done in a design and built basis, using a Controlled Modulus Columns solution.

SUBSURFACE CONDITION

The geotechnical data for this project is derived from the Cone Penetration Tests (CPT's) performed along the axis of the anticipated embankment from station 0+420 to 0+600. The CPT results are presented in Figure 3.

A very soft alluvium layer, made of a succession of very soft clay (w_p=60%, w_L=150%, 60%<w<150%) and fibrous peat (w_L=480%, 250%<w<450%, C_u=33 kPa), was extending all over the site with a variable thickness ranging from 11 m at station 0+420 to 6 m at station 0+600. In this alluvium layer, CPT cone resistance values q_c were less than 0.5 MPa with typical values around 0.3 MPa. Under this very soft alluvium layer was a medium density structureless chalk layer with silt fragments between station 0+420 and 0+500, mixed with medium dense sand and silty sand layer between station 0+500 and 0+600. In those layers, the CPT cone resistances q_c were greater than 3 MPa. As a consequence, according to the anticipated loads and subsequent design requirements, those layers were able to be considered as suitably competent layers for the CMC. At certain locations, a 1 or 2 m thick stiff clay layer (with q_c>20 MPa) was present at the bottom of the soft alluvium layer. Despite being stiff, the clay layer was not considered as a suitable competent layer because of its limited thickness.

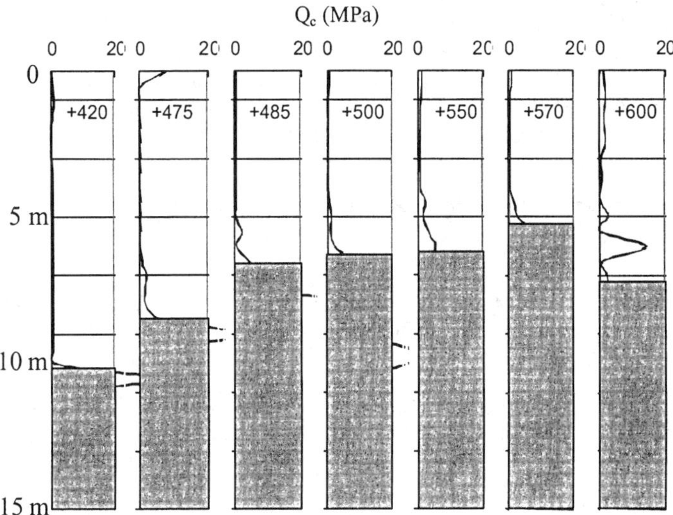

Figure 3: CPT's performed along the embankment axis at station 0+420 to 0+600

ALTERNATE FOUNDATION SYSTEMS

Three foundation systems considered for this project includes: conventional vertical (wick) drains, stone columns, and the CMC systems. "Time" was critical for this project, and thus a wick drain and surcharge solution was not acceptable to the project owners. Moreover, a CMC solution was preferred to a stone columns solution for the following reasons:

- Predicted settlement: The predicted settlement for stone columns was more than twice the predicted settlement with CMC. The settlement requirement (less than 10 mm per decade) was too strict to allow an economical solution using stone columns.

- Field performance: The alluvium layer was too soft to allow a safe implementation of stone columns to support the embankment. The presence of critically soft clay and peat, associated with high loading from the embankment (up to 155 kPa), raised the possibility of bulging problems. Indeed, the very soft soil did not provide enough lateral confinement to ensure the lateral stability of the stone columns.

DESIGN OF THE GROUND IMPROVEMENT

The ground improvement solution takes into account the different loading conditions; the settlement tolerance associated with those conditions; and the variability in different soil characteristics. The CMC ground improvement was thus adapted from this project. The design parameters needed to define the CMC reinforcement system were:

- The depth of the columns and their possible anchorage length in the competent layer
- The diameter of the columns
- The pattern & configuration of the columns grids
- The modulus of the grout used to make the CMC. This parameter is adapted for each new project, but for practical reason, is constant over a particular project.

The complete design of the CMC includes estimation of bearing capacity of single columns, checking pre-design parameters by a numerical procedure, stability analysis, and development of final design, as discussed in the following sections.

Bearing capacity of single CMC column

The bearing capacity of a single CMC column was evaluated according to the equation of the ultimate capacity of piles in chalk proposed by Sanglerat (1972):

$$Q=\left(kAq_c+k_l\frac{kq_c}{10}Z\pi\phi\right)/FOS \quad (1)$$

in which, $k = 0.5$; $k_l = 0.5$ (usually $0.5 < k_l < 0.9$), A is the section of the CMC (0.138 m^2 for diameter 420 mm, 0.101 m^2 for diameter 360 mm); Z is the anchorage-length in the competent layer; and ϕ is the diameter of the CMC. A factor of safety (FOS) of 2 was applied to evaluate the bearing capacity of the CMC's.

The project was divided in two different areas, corresponding to the two main soil profiles of the site:

- The first case was the most general case (from station 0+460 to 0+600). In this area, the thickness of the soft soil was less than 10 m, and anchorage-length of 1.5 m were able to be executed. The stiff layer reached q_c values greater than 4 MPa, with a diameter of 360 mm, each CMC from this area was able to support $Q_{allowable}$ = 185 kN/CMC. (taking into account a factor of safety of 2)
- The second case was related to CPT in station 0+420 which showed alluvium thickness of about 11 m followed by a stiff layer having a cone resistance q_c of 3 MPa. This case was limited to a small part of the job. However, due to the soil profile, the anchorage-length of the CMC had to be limited to only 1 m to avoid penetration into a less stiff layer. To compensate this limitation, CMC's with larger diameter (420 mm) were chosen. In those conditions, each of these CMC was able to support $Q_{allowable}$ = 153 kN/CMC with a factor of safety of 2.

Max thickness of soft soil (m)	CMC diameter (mm)	Maximum allowable load (kN)	Maximum height of embankment (m)			
			1.0	3.0	5.0	7.5
			adopted mesh (m)	adopted mesh (m)	adopted mesh (m)	adopted mesh (m)
11	420	153	-	1.7	1.3	1.0
10	360	185	-	1.7	1.4	1.2

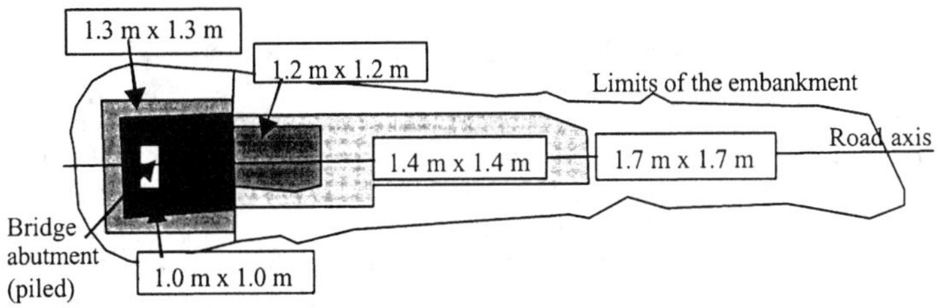

Figure 4: Layout of the CMC

The CMC pattern could thus be estimated from those bearing capacities; simply by dividing them by the applied load. The result is the maximum area of influence that can be associated to one single CMC. For example: for an embankment of 5 m high the applied load was 5 m x 18 kN/m^3 = 90 kN/m^2 . In the general case (i.e., less than 10 m of soft soil), the area of influence of a single CMC was 185 kN/CMC / 90 kN/m^2 = 2.06 m^2/CMC corresponding to a square grid of 1.40 m x 1.40 m = 1.96 m^2/CMC. The pattern of the ground improvement, depending on height of embankment and thickness of treated ground, is shown in figure 4.

Design Check

To check this preliminary design, axial-symmetrical finite difference method calculations have been implemented with the program PLAXIS. Under uniform loads, the problem can be studied by an axial-symmetrical calculation as shown in figure 5.

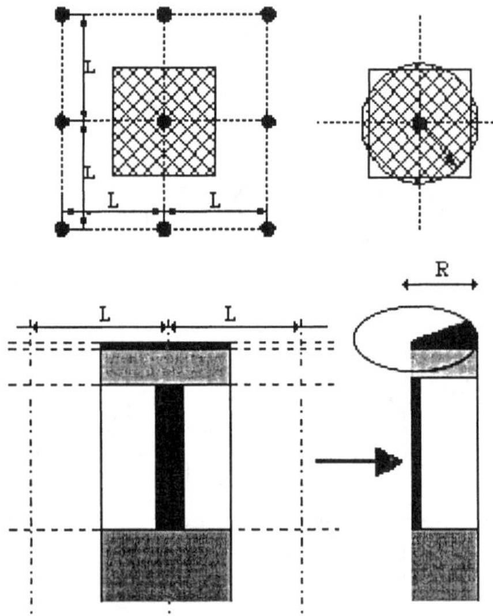

Figure 5 :Principle of the axial-symmetrical calculation

This approach to the problem allows taking into account, under vertical loadings, the reinforced soil as well as the embankment, the geotextile (with ultimate tensile strength 84 kN/m and modulus 630 kN/m) and the transition layer. Thus it gives the strain and stress distribution between the soil and the columns. These FEM calculations were made in short term and long term conditions alike, in order to assess the time related behavior of the improved ground. For FEM calculation purposes, the parameters used in the numerical calculation for long-term behavior in correlation with available CPT's cone resistance are shown in Table 1.

Table 1: Input parameters for FEM analysis

Material	Alluvium	Competent Layer	Transition layer	Embankment	CMC
Modulus,E (MPa)	1.5	15	35	80	11,000
Thickness (m)	7.0 to 11.0	3.0	0.5	0 to 7.5	7.5 to 12.5
Cohesion (kN/m²)	0	0	-	0	-
Friction Angle (°)	18	25	-	33	-

Note: a grout with unconfined compressive strength at 28 days f_{28} greater than 11 MPa was chosen, the Young's modulus of the CMC was then assumed to be at least 11,000 Mpa.

For short-term behavior, the geotechnical characteristics of the alluvium competent layers were assessed using the following equation:

$$\frac{E}{1+\nu}=\frac{E'}{1+\nu'} \quad (2)$$

in which the short-term elastic modulus of 1.7 and 17 Mpa was adopted for the alluvium and stiff layers, respectively. To address all possibilities, two different cases, corresponding to the worst case scenarios, were studied:

Case 1:

- 7 m of soft soil
- CMC diameter 360 mm
- CMC mesh 1.40 m x 1.40 m
- Embankment height (including transition layer): 5.00 m
- Road surcharge: 20 kPa

Case 2:

- 11 m of soft soil
- CMC diameter 420 mm
- CMC mesh 1.00 m x 1.00 m
- Embankment height (including transition layer): 8.00 m
- Road surcharge: 20 kPa

The main results of these axial-symmetrical FEM calculation are presented in Table 2.

The geotextile layer has been placed originally to increase the factor of safety against shear stress and was a requirement of the British Standard. However, the calculation showed that the tensile stresses in the geotextile were very limited.

The time related aspect of settlement was assessed by comparing the computed settlement for short and long term behavior. The difference between short and long-term settlement was of about 4 mm, and the typical duration for this settlement to occur was between 5 and 10 years to ensure that the settlement requirement of 10 mm per decade is respected.

Table 2 : Results of numerical analysis

Case #			Case 1		Case 2	
Studied case			Short term	Long term	Short term	Long term
CMC datas	CMC diameter (mm)		360 mm		420 mm	
	CMC mesh (m)		1.4 m		1.0 m	
	Alluvium thickness (m)		7.0 m		11.0 m	
FEM results	Embankment construction	Settlement	29 mm	32 mm	31 mm	34 mm
		Stress in geotextile	1.2 kN/m	1.2 kN/m	0.63 kN/m	0.63 kN/m
	Road loading	Settlement	+ 9 mm	+ 10 mm	+ 5.5 mm	+ 6 mm
		Stress in geotextile	1.4 kN/m	1.4 kN/m	0.72 kN/m	0.72 kN/m

Note: The other cases with similar soft soil conditions have been designed so that the load per CMC was comparable to the studied cases.

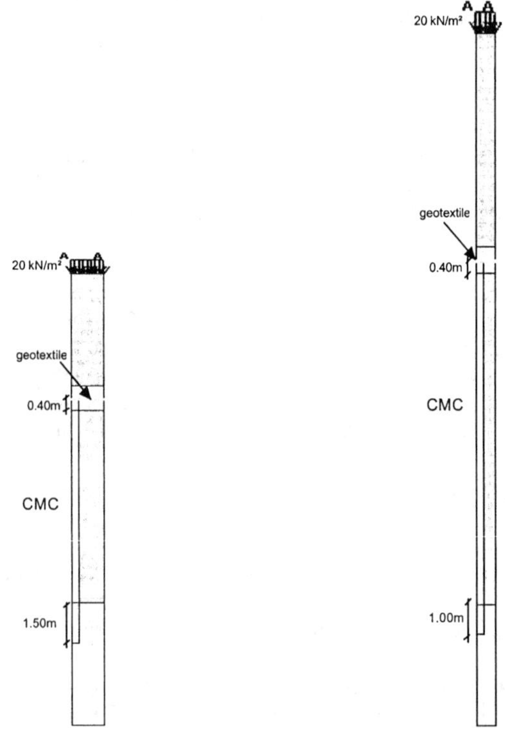

Figure 6: Axial-symmetrical finite difference models, case 1 and 2

Stability Calculations

Stability calculation was performed using TALREN program to check possible slope failure problem during embankment construction. The results of these calculations showed a factor of safety against slip circle failure of 1.39.

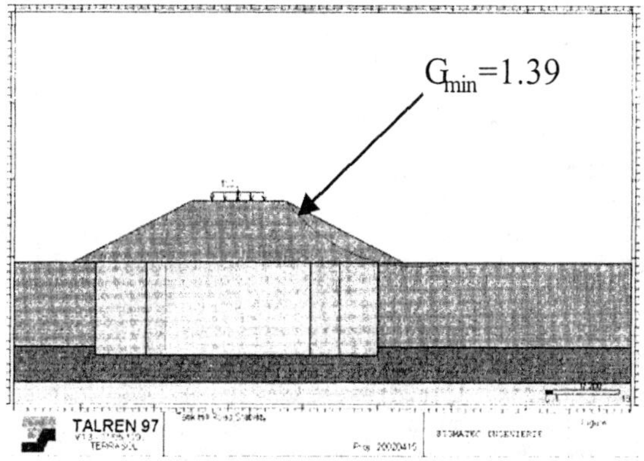

Figure 7: Stability calculation

Final Design

Due to very soft ground condition for this project (particularly the chalk layer), a very dense column spacing of 1.0 m square grid spacing was initially adopted to correspond with an area replacement ratio of 13.9 %. The area replacement ratios for CMC system are typically between 2 to 8 %, with CMC single columns designed to support loading of 150 to 350 kN. The installation process use a displacement auger, and thus the unusually high CMC density may have a risk of damaging the freshly grouted surrounding columns during the installation of a new CMC. Consequently, the conventional construction method was modified for the high density area. The CMC columns were installed in two different interleave passes, each with 1.4 m x 1.4 m grids as shown in figure 8; corresponding to an area replacement ratio of 6.9 %. The CMC columns were anchored in the chalk or sand layers, resulting in columns in length from 7.5 to 12.5 m. 2193 CMC columns were installed in two months for Tank Hill Road South Embankment project.

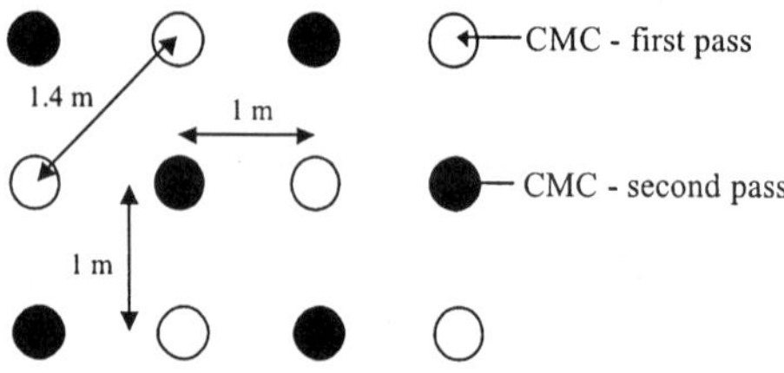

Figure 8: Modification of CMC layout

CONSTRUCTION CHALLENGES

The presence of existing facilities imposed construction changes of the soil reinforcement system in some areas. The main adaptation was related to the existing drainage culvert at the site. This drainage culvert with a total width of about 2 m was crossing the working area transversally. The requirement was to protect the existing culvert against differential settlement. The decision was made to install additional CMC to bridge the culvert and protect it. The strategy to reduce differential settlement was to adapt the grid in order to have an almost constant area of influence for each CMC even if the square pattern is modified (see figure 9). With those construction changes, the predicted differential settlement along the culvert was less than 3 mm, totally compatible with the rigidity of the culvert.

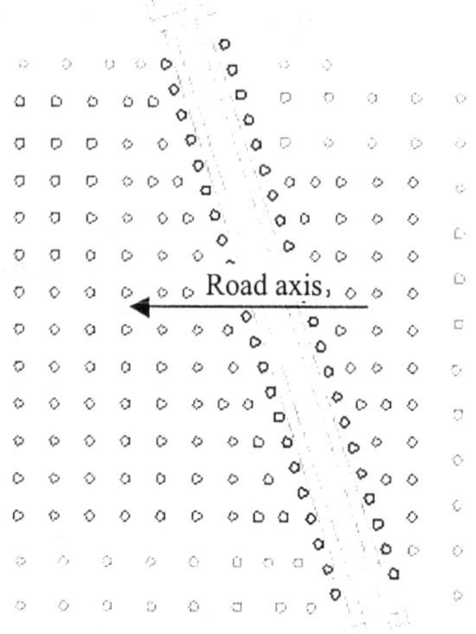

Figure 9: CMC grid modification along the culvert

Existing overhead electrical cables were also crossing the construction site, requiring special attention and safety measures. Equipment modification were also required in order to shorten the mast of the CMC drilling rig to leave a safety distance between the top of the rig and the live electrical cables, while still being able to go to the required depths (see figure 10).

QUALITY CONTROL

The quality of execution of each of those CMC columns was controlled by monitoring and recording the followings for each individual column:

- Speed of rotation and advancement of the auger.
- Torque, down-thrust and drilling energy applied during advancement.
- Pressure and volume of injected grout, from which the profile of the columns are determined.

The quality of the grout was controlled regularly by unconfined compressive strength cube tests at 7, 14 and 28 days, four cube samples for testing were taken every 100 m^3 of grout. The bearing capacity of the installed CMC columns was controlled by the execution of 11 vertical load tests on isolated columns (3 on CMC with diameter 420 mm and 8 on CMC with diameter 360 mm). Those test were carried out until one-and-a-half times the design load of the columns.

Figure 10: Construction challenges near the culvert and the electrical cables

A complete settlement monitoring system including settlement plates and settlement pegs were installed over the treated area in order to record the potential settlement during and after construction of the embankment. The average settlement measured under the design loads of 185 kN and 153 kN was around 10 mm, equivalent to one third of the calculated settlement. However, the load test on an isolated CMC, in which only the CMC column is loaded, does not mobilize negative skin friction. Therefore, additional downward forces are induced by the differential settlement between the surrounding soil and the CMC column. As a difference, the uniform loading of the ground improvement system by the embankment does mobilize negative skin friction. These additional downward forces may result in additional settlement of the CMC columns.

CONCLUDING REMARKS

The case history of the Tank Hill Road South Embankment project demonstrates that CMC foundation system was an effective solution in timely project delivery and meeting the serviceability requirements of the project. The main challenges were associated with construction time limitation and very soft ground condition of the site. In evaluating the alternatives stone column was not feasible; and the conventional vertical drain solution was not practical due to time constraints.

REFERENCES

Plomteux, C. and Spaulding, C. (2003) Reinforcement of Soft Soils by Means of Controlled Modulus Columns- 12th Pan-American Conference on Soil Mechanics and Geotechnical Engineering – Cambridge, USA - June 22-26.

Porbaha, A., Brown, D., Macnab, A., Short, R. (2002a) Innovative European technologies to accelerate construction of embankment foundations- part I: GEC, AuGeo, and CFA. Proceedings of Time Factor in Design and Construction of Deep Foundations, Deep Foundation Institute, San Diego, CA, October, 3-14.

Porbaha, A., Brown, D., Macnab, A., Short, R. (2002b) Innovative European technologies to accelerate construction of embankment foundations- part II: DM, FMI, Mass Stabilization, and CSV. Proceedings of Time Factor in Design and Construction of Deep Foundations, Deep Foundation Institute, San Diego, CA, October, 15-28.

Rogbeck, Y. & al. (1998) *Two and three dimensional numerical analysis of the performance of piled embankment*, 6th International Conference on Geosynthetics, Atlanta

Rogbeck, Y. & al. (1998) *Reinforced piled embankments in Sweden – Design Aspects*, 6th International Conference on Geosynthetics, Atlanta

Sanglerat, G. (1972) The Penetrometer and soil exploration Interpretation of penetration diagrams - Theory and Practice, Part 3 – Page 285.

SLAB CONSTRUCTION-30 YEARS LATE

W. Tom Witherspoon[1] and Jason Taylor[2]

ABSTRACT: A residential foundation repair project is described. Poor construction of the original foundation, consisting of slab on grade and perimeter grade beams, resulted in damage in the form of settlement, cracking, misalignment of doors, and plumbing leaks. The initial remediation plan involved construction of underpinning piers. Pier installation beneath perimeter grade beams was accomplished, however it was discovered that interior slabs that were designed to be 6-inches thick were in fact only 2-inches thick and not properly reinforced. A 2-stage repair was used consisting of grout injection to increase the effective slab thickness followed by mudjacking to establish the original slab elevations. This process was highly successful and suggests an approach for addressing similar residential foundation repairs.

BACKGROUND

The house owners paid a house contractor to construct their dream house in 1965 in Wills Point, Texas. At approximate 2,700 s.f., this house sits on a large flat corner lot. Surface drainage on the property is marginal for slope away from the house but appears to be within accepted tolerances. There are also large pecan trees that appear to be a safe distance away from the house perimeter.

The owners remember watching construction of the foundation, which was poured in August and on a particularly hot day. They also remember that the contractor laid down a thick cushion of crushed stone for a base/moisture barrier but differ on whether polyethylene sheeting was placed upon the base course.

The owners remember deep grade beams with what they described to the writers as #4 reinforcing steel bars in the top and bottom of the beam with #3 stirrups tying the bars together. They also remember that wire mesh was used for the slab reinforcement and recall being told by the contractor that it was as good as

[1]President and [2] Project Engineer, S&W Foundation Contractors, 1030 East Belt Line Rd., Richardson, TX 75081-3703 tomw5@ix.netcom.com

conventional rebar. The terms of their contract with the builder specified a 6" thick slab with polyethylene sheeting over the crushed stone. Despite such a large foundation pour, the concrete contractor's crew consisted of a small group of workers who had to use water to complete the finish on the slab because of the extreme temperature and the large finished area.

In the summer of 2000 the owners contacted their insurance company because the foundation was moving, with damaging effects, and they thought they had a plumbing leak. Cracking of the interior and exterior began to appear and the doors failed to function because of misalignment. At this time 3 leaks were discovered and the insurance company sent an engineer to evaluate the problem and the effect of the leaks. The engineer hired by the insurance company performed an evaluation of the damage and cited a probable cause. The engineer's foundation layout and interior floor elevations indicated areas of distress and differential movement that caused door function problems and excessive cosmetic finish problems. It was the engineer's opinion that the leaks caused at least some of the damage and that foundation remediation was necessary to restore function. Original elevations provided by the engineer are shown in Figure 1.

It was the decision of the insurance company that they would pay for 100% of the expected cost of repairing the foundation, consisting of the installation of 44 piers at the perimeter and 2 at the front bathroom. The owners selected the authors' firm for this repair and the work was commenced in November of 2000.

INITIAL REMEDIATION EFFORTS

Underpinning at the perimeter went as expected with the discovery that the grade beam was an approximate 30" deep and appeared to be well constructed. Pier installation for the interior garage perimeter was to be under an interior grade beam, which would provide ample strength for lifting. Breakout, however, revealed that the slab was only 2" thick with wire mesh at the bottom of this section and what appeared to be 3"- 4" of concrete aggregate below this mesh (Figure 2). A later probe at the front bathroom showed an even more disturbing section of 1" to 1-1/2" thickness of concrete with aggregate above and below the wire mesh. Also, there was no polyethylene barrier use observed at any breakout.

Such a thickness of concrete guaranteed that the interior piers would be of little value at this bathroom. The floor was very low at this section and door function was unusable. Figures 3 and 4 show how badly the bathroom doors were out of square and why they would not function.

Visual observations indicated that the original contractor had intended to provide a 6" thickness of slab but without a polyethylene barrier on the crushed stone and in the heat of summer, the lower 3" to 5" of cement and water permeated into the stone and left only the concrete aggregate. This was further evident by obvious cement stained gravel in the intended slab section and on top of the crushed stone.

It also appeared that the only reason this slab performed as well as it did was because the aggregate and crushed stone were weakly cemented together. A viewing of the slab also revealed the presence of hydration cracking, which would confirm the use of water as a finishing aid and increase the potential for defects in the concrete surface.

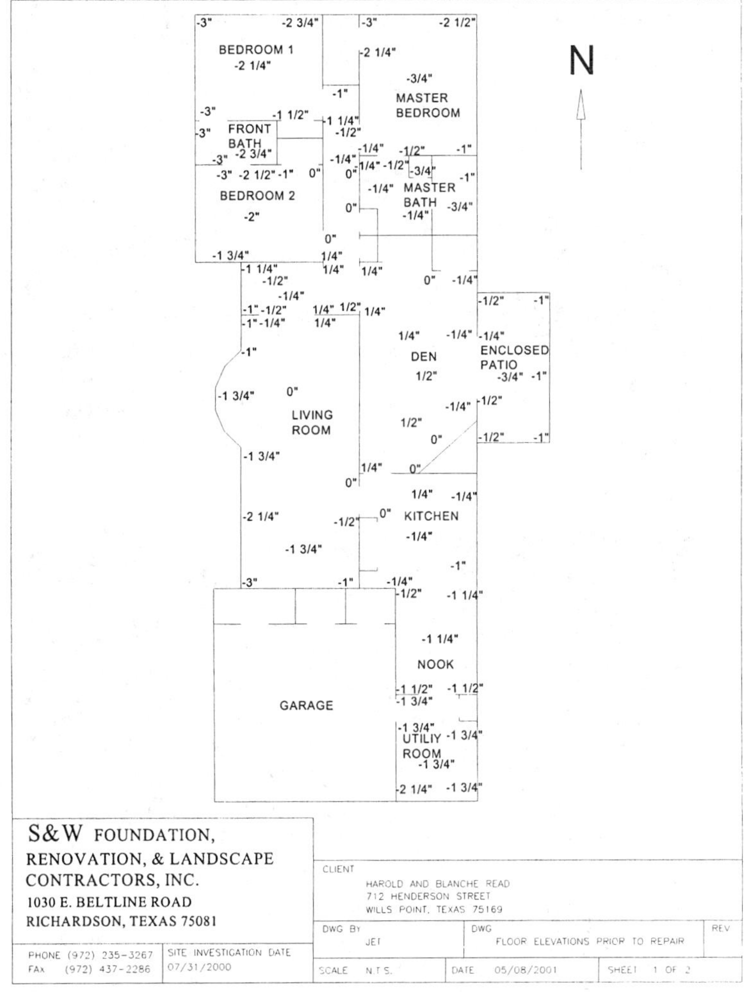

FIG. 1. Original elevation information, September 6, 2000.

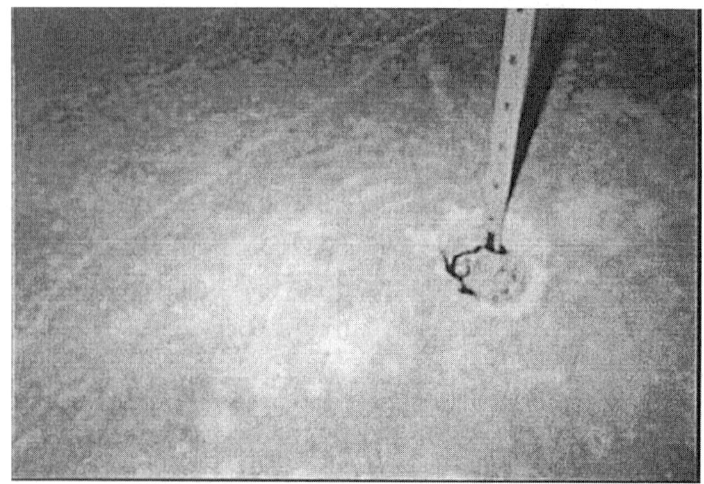

FIG. 2. Slab thickness at south side of the front bathroom.

FIG. 3. View from bedroom 1 of north front bathroom door.

FIG. 4. View from front bathroom of South front bathroom door.

THE PROBLEM

As previously stated, conventional foundation repair methods were out of the question with such a thin, unreinforced slab. These repair methods would normally consist of exterior and interior piers to stabilize the foundation, and pressure injection of a soil/cement grout to lift low areas and fill voids created by lifting the slab on the remedial piers. This process of injecting a soil/cement grout under pressure is also known as mudjacking. Interior piers would be very ineffective due to the lack of resistance to punching shear that a thin, unreinforced slab would provide. Even with steel "I" beams under the floor to distribute the load, lifting of the foundation would surely break the slab and worsen the problem. Pressure injecting the slab to lift low areas would impose bending moments on the slab. A calculation of bending moments imposed by grout injection was computed modeling the slab as a square flat plate that experiences two-way bending using the following assumptions.

1. Slab on grade bearing soils are incompressible.
2. No pressure losses due to friction.
3. Slab behavior consistent with two-way bending of a square flat plate that has pinned supports all around the perimeter. (due to extensive cracking of the slab).
4. Spread of grout is uniform from point of introduction, thus uniform loading.
5. Nonreinforced concrete.
6. Compressive strength of existing concrete = 3000psi.

7. Loading imposed by pressure injection is approximately 30 pounds per square inch.

The moment was calculated for this slab section using the following formula: $M = 0.036wl^2$. Where w is the uniform load and l is the length of the slab section. The area of pressure reaction is determined by resistance of the grout to flow away from the injection point. Experience in this operation indicates that an area approximately 2 ft. in diameter will initially be affected but, as resistance builds up by the filling of the subslab void, this pressure gradient will expand with the affecting areas increasing to greater than a 5 ft. radius. Because the rupture radius is near the injection port, early in the injection process, a small radius is the actual design parameter that will determine failure of the slab. Therefore, using l equal to 2 ft. yields a moment of 622 inch*pounds.

The modulus of rupture for concrete is established in the "Building code Requirements for Structural Concrete" (ACI 318-95) as 7.5 times the square root of the compressive strength. (equation 9-9) This value corresponds to 0.41 kips per square inch (ksi). For a 1-1/2" thick slab, the loading would create stresses equal to 1.66 ksi. This value is over 4 times greater than the modulus of rupture, and shows that this injection process would have severely cracked the floor. Clearly this would not be a functional slab.

Further evaluation revealed that a minimum 3" thickness of slab was required to reduce the stresses to equal the modulus of rupture. For a 3" thick slab the loading would create stresses equal to 0.415 ksi. This value is slightly higher than the modulus of rupture (0.1%). One factor that increases slab strength in this situation is the fact that a base course of rock exists that, if the dispersion depth is satisfactorily below the base course, will act with some support strength during the mudjacking operation. Because of this stratum of material, the slab will have additional strength to resist vertical pressure.

What was needed was to repour the slab so that the wire mesh, in conjunction with a 3" slab thickness would be sufficient to lift and restore the door and slab function. We could not remove the entire slab without severe damage to the interior walls, plumbing and extensive cabinet work and finishes. We also had a problem with extensive plumbing in the front bathroom that would be very susceptible to vertical and lateral pressures.

THE SOLUTION

Mudjacking has been in use to lift slab foundations for over 40 years and is very effective in lifting slabs upon their bearing soils, improving floor slope and function. This process involves injection of a grout solution, normally Portland cement and sandy loam soil (soil/cement) at low pressure of approximately 30 p.s.i. The problem with this foundation, however, was even that with a uniform injection the slab was too thin to lift and future performance was shaky at best. Therefore, the process must first involve restoration of the slab thickness by injection.

To accomplish this, the injection was with type I Portland cement through 2" holes in the bedroom floor at a spacing of 3' o.c. and in a triangular spacing from bathroom perimeter outward for some 6'. The injection was at a maximum of 30 p.s.i. just below the thin slab section and in the original concrete aggregate. The intention was to replace the missing cement into the original concrete aggregate, which would rebuild the slab. Because of the close proximity of the base stone an even thicker slab section was possible. Because of the injection hole concentration, saturating possibilities under the bathroom were increased and the floor would at least be stronger than prior to this operation.

Injection was done on December 8, 2000 and lasted almost 6 hours with a total injected cement of 57 c.f. The mudjack supervisor also noted that cement was heard flowing some 20' from the injection points with some reaching the perimeter. Because sink and tub water was kept flowing during this process, the sanitary lines could be monitored by noting changes in water pressure or blockages of water drainage to ensure that no leaks/intrusion of material entered the lines. No problems were recorded. An inspection of the injection points, some 10 days later revealed that the slab thickness had been increased to approximately 3" to 4" as evidenced by the darkened homogeneous section below the original slab and wire mesh.

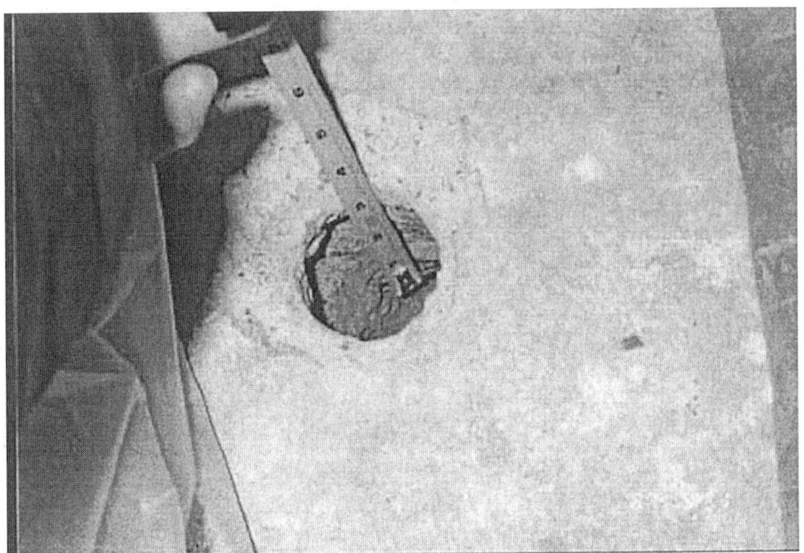

FIG. 5. Slab section after pressure injection of cement

The second phase of this operation consisted of an injection of 12% cement to sandy loam below the cemented section to provide uniform lift to the bottom of the slab and correct the slab's angular distortion. Figure 6 shows the final elevations recorded after this process.

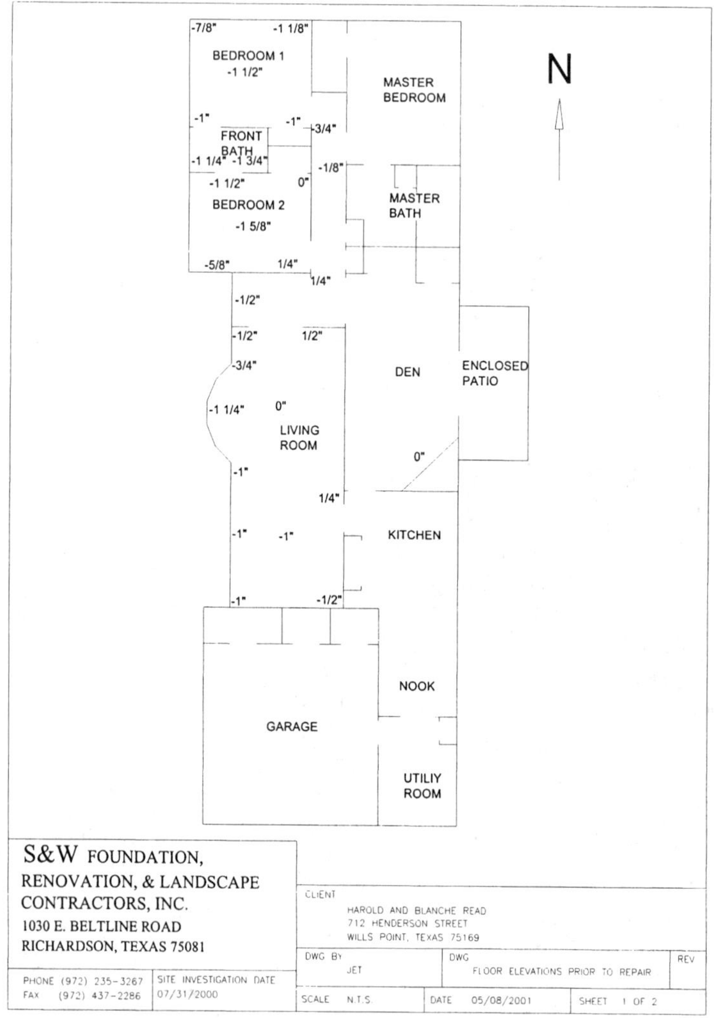

FIG. 6. Final elevations recorded January 4, 2001

The following photographs show the difference in door function after the injection process.

FIG. 7. View from bedroom 2 of South front bathroom door.

FIG. 8. View from bedroom 1 of North front bathroom door.

The only negative action was a rupture of the slab some 12 ft away from the bathroom, which was later repaired by chipping out the loose concrete and repouring this section. This defect occurred in an area where no direct cement injection was performed and where the slab was much weaker than the point of injection. This was to be expected and indicates what might have happened if mudjacking or other lifting had been performed without rebuilding the slab thickness.

CONCLUSIONS

Although the authors are not aware of this two-step process being performed previously, it now appears that its application can be successful in the right situation. The reason this operation was successful on this house was an existing concrete aggregate section that only needed Portland cement to restore its original strength and function. Therefore, other situations will require individual evaluation to determine if this process might be successful within the boundaries.

BACK ANALYZED PARAMETERS FROM PILED FOUNDATIONS FOUNDED ON TROPICAL POROUS CLAY

R. P. Cunha[1], Member Brazilian Geotechnical Society, J. E. Bezerra[2], H. H. Zhang[3], and J. C. Small[4], Member ASCE

ABSTRACT: This paper presents the results of numerical analyses carried out with a program exclusively developed for simulating piled raft foundations. It was developed by Zhang (2000) and it extends the use of the finite layer method to the analysis of a general loading system applied to piled raft foundations. It allows the establishment of a coupled relationship between displacements, rotations, and external loads and it furnishes displacements, moments, shears and forces within the analyzed system. The paper explores the applicability of this program by using it to simulate the behavior of experimental field loading tests carried out at the University of Brasília (UnB) experimental site. At this site, Cunha and Sales (1998) carried out several vertical loading tests on a small scale isolated raft and piles, group of piles, and piled rafts. Zhang's program was adopted to back analyze this series of experimental results, furnishing preliminary geotechnical parameters for use in future designs with this or similar soils. The analyses also highlighted the fact that it is possible to numerically simulate the complex behavior of such mixed foundation systems founded on "non classical" materials such as the Brasília porous clay.

INTRODUCTION

In the past decade, several papers have been published with emphasis on what are now called "piled-rafts", i.e., pile groups in which the raft connecting the pile heads positively contributes to the overall foundation behavior (for example see Mandolini and Viggiani 1997, Poulos 1998 or Cunha et al. 2001). The International Society for Soil Mechanics and Foundation Engineering (ISSMFE) also focused the activities of one of its Technical Committees on the study of piled raft foundations.

---

[1] Adjunct Professor, Civil Engineer, Ph.D., University of Brasília, Dep. of Civil and Environmental Engineering, Brasília-DF, Brazil, rpcunha@unb.br
[2] Civil Engineer, M.Sc., Geotechnical Consultant, Brasília-DF, Brazil
[3] Civil Engineer, Ph.D., SMEC Australia Pty Ltd., Sydney, Australia
[4] Associate Professor, Ph.D., Dept. of Civil Engineering, University of Sydney, Sydney, Australia

Poulos (1998) has examined a number of situations and suggested ideal and adverse conditions for the use of piled rafts. He considered that soil profiles containing soft clays or loose sands near the surface, and those containing compressible soils, or collapsible or swelling layers, are not suitable for the use of piled rafts. This author has also pointed out that, in some cases, the concept of "creep piling" could be used in a more extreme version, in which some of the piles would operate at full load capacity, although the entire system would still have an acceptable factor of safety.

Other recent papers have expanded upon these ideas, such as those by Cunha and Sales (1998), Cunha et al. (2000a and b), and Cunha et al. (2001). Cunha and Sales (1998) presented a paper describing and discussing field loading tests carried out on small scale footings supported by a reduced number of piles. These tests were performed at the University of Brasília research site, and have confirmed that this design methodology has a large potential (although with some restrictions) to be adopted for use with the collapsible porous clay of the Federal District. Cunha et al. (2000a) have focused on the influence of the input variables on the numerical analysis of piled rafts, whereas Cunha et al. (2000b) have advocated an "optimized" parametric procedure for the preliminary design of both piled rafts and standard deep foundations. Cunha et al. (2001) demonstrated the use of such an "optimized" procedure, by testing it against a real (and instrumented) case history in Tokyo. The optimized procedure has also been shown to lead to savings in the final foundation design.

Hence, this paper continues on this line of research, exploring the numerical evaluation of piled rafts. In the present paper the influence of the geometric distribution of the piles underneath the raft, on the resulting geotechnical parameters (for use in design) is evaluated. The paper calibrates the adopted numerical tool with an existing case history of small scale field loaded piled rafts, showing how to use this program to derive design parameters for a future stage. Although the parameters have to be considered as preliminary values, they can be used in practice. Besides, they can also provide an idea of the effect of the piled raft system (in particular the effect of considering raft/soil interaction), and the soil's non-linear behavior, upon the derived back analyzed design parameters.

NUMERICAL TOOL

The method developed by Hain and Lee (1978) considered the interactions of the piles, raft and soil, but the rotations and horizontal movements of a pile head induced by a vertical load applied to an adjacent pile or the soil surface were ignored. Ta and Small (1997) developed a method for analysis of a piled raft (with the raft on or off the ground) and, as for Hain and Lee's method, the solutions were only for vertical loads. Zhang and Small (2000a) subsequently developed a method for analysis of piled raft foundations subjected to both vertical and horizontal loadings. In this method, the interactions between raft and piles, raft and soil, piles and piles, piles and soil, and soil and soil were fully considered. However, the method could only deal with piled raft foundations clear of the ground.

Zhang and Small (2000b) have introduced an extension of the method presented by Zhang and Small (2000a), where the raft can be in contact with the ground surface. The approach uses a combination of the finite layer method for modeling the soil and the finite element method for simulating the raft and piles. The piled raft foundations can be subjected to horizontal and vertical loads as well as moments, and the movements of the piled raft in three directions (x, y, z) and rotations in two directions (x, y) may be computed by the program APRAF (analysis of piled raft foundations). Comparisons of these new solutions with those of the finite element method have been successfully made by these authors, and the effects of parameters (adopted for the soil and raft) on the behavior of piled rafts have been examined.

As shown in Figure 1, reproduced from Zhang and Small (2000b), the problem of the piled raft foundation can be solved by assuming that the forces between the piles and layered soil can be treated as a series of ring loads applied to 'nodes' along the pile shaft. These loads are both horizontal and vertical. The contact stresses that act between the raft and the soil can be considered to be made up of uniform rectangular blocks of pressure that approximate the actual stress distribution. These can be considered uniform vertical blocks of pressure or uniform shear stresses. The displacement of the layered soil can then easily be computed, as the solution for a layered soil subjected to ring loads at the layer interfaces may be obtained from finite layer theory.

Firstly the response of the piles and soil (with no raft) is computed by applying unit surface loads to the rectangular regions on the ground surface or unit ring loads to the soil along the pile shaft, or a unit uniform circular load at the base of the pile. The deflections so computed can be used to form the influence matrix for the soil. For the piles, a stiffness relationship may be written based on shaft loads and on applied load at the pile heads. Three noded linear bending elements are used to model the piles.

Deflections of the soil or of the piles can be obtained for loads applied to the pile heads from the final stiffness relationship for the pile-soil continuum. This method is not as efficient computationally as computing the interaction between two piles only (i.e. the interaction method). However it is much more accurate, especially for piles at close spacing because all the piles are considered at once.

Because the deflection of the piles can be computed when one is loaded at the head, or when the ground surface is loaded, this can be used to determine the behavior of the raft. Thus, by applying unit loads to the raft, its influence matrix may be obtained. By applying unit pressures to the ground, or unit pressures and moments to the pile heads, an influence matrix for the soil-piles may be obtained. Finally, by considering equilibrium of applied forces and moments acting on the piles and raft, and compatibility of displacements of the soil and raft (and of displacement and rotation of the pile head and raft) enough equations may be assembled to obtain the solution under general loading.

Hence, similar solutions to those presented by Zhang and Small (2000b), and incorporated within the APRAF software, are adopted herein to back analyze and simulate the experimental data obtained from field loading tests in Brazil.

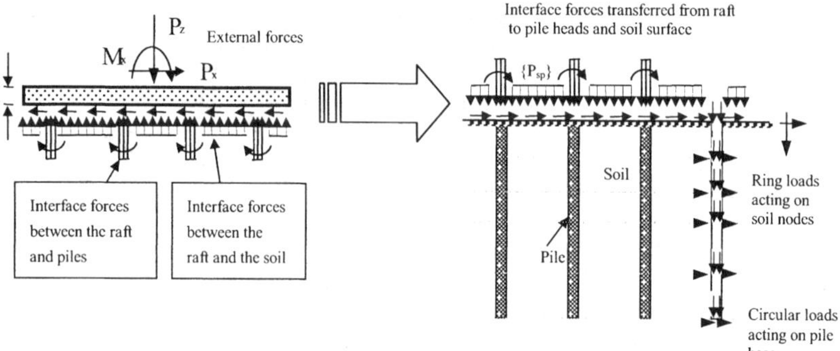

FIG. 1. Piled raft under external loading conditions

EXPERIMENTAL RESEARCH SITE

The behavior of deep foundations founded in the local unsaturated, tropical and collapsible clay of the Brazilian capital, Brasília, is one of the most important research topics in the Geotechnical Post-Graduate Program of the University of Brasília. Therefore, in 1995 this University started a major research project in the foundation area, in order to enhance knowledge of the behavior of the distinct foundation types that were founded in the predominant subsoil of the Federal District. It was decided to carry out horizontal and vertical field loading tests on the different kinds of locally used deep foundations, as well as new foundation techniques (such as piled rafts). These foundations were constructed within the University of Brasília campus, at the experimental research site of the Geotechnical Post-Graduate Program.

A large effort was also undertaken in association with local engineering contractors to characterize this experimental site, by performing advanced in situ testing techniques (Marchetti Dilatometer, Ménard Pressuremeter and Standard Penetration tests) and standard/advanced laboratory tests (Soil Characterization, Consolidation K0 and CK0D Triaxial tests). This joint research effort allowed a better knowledge of some of the locally used foundation and in situ testing technologies, and resulted in a number of congress papers and some of the University of Brasília M.Sc. and D.Sc. Theses. An effort was also undertaken to associate with other universities in Brazil and abroad (for instance with University of Sydney) in order to exchange experience and technology, besides sending and receiving scholars and students.

As shown in Figure 2, Brasília and its neighboring areas (Federal District) are localized in the Central Plateau of Brazil. This district has a total area of 5814 km^2 and is limited in the north by the 15°30' parallel and in the south by the 16°03' parallel. The regional geology and geomorphology of this area have already been described in detail by Araki (1997) and others. The area is predominantly flat and it is covered by latosols and lateritic soils from the "Paranoah" formation of the upper Precambrian.

The city of Brasília has an airplane like shape, and it contains two main (North and South) wings. The University of Brasília research site is located at the university campus, between the North wing and the Paranoah lake. The geotechnical profile of this site is composed of the typical "porous", unsaturated and tropical clay of Brasília, found in large areas of the Federal District. This clay is collapsible under saturation, and it is underlain by a folded slate, also typical in the region.

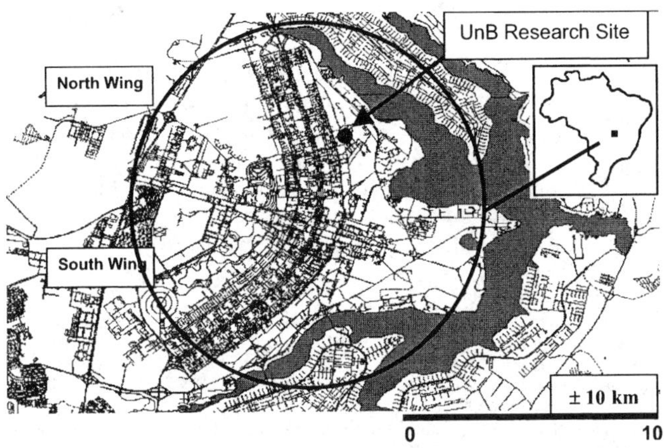

FIG. 2. Location map of the Brazilian capital Brasília and the research site

The porous clay is a weathered latosol with a variable thickness that ranges from 10 to 30 m. This clay displays a reduced Standard Penetration Test blow count (from 2 to around 15 in the deeper layers) and it is susceptible to a collapsible behavior under saturation (or decrease of suction). According to Cunha et al. (1999), the weak particle cement bonding and collapsibility of this soil are caused by the combined lixiviation and laterization processes that have taken place during the geological eras of its formation. These processes are intrinsically associated with well defined "wet" and "dry" seasons that normally occur in Central Brazil.

In geotechnical terms, the studied latosol may be classified as a "collapsible" sandy clay with traces of silt, with a high void ratio and coefficient of collapse. Its coefficient of permeability is also high for a typical clay, being close to those found for fine to silty sands. The major geotechnical parameters of this soil are displayed in Table 1, modified after Cunha et al. (1999).

These parameters were obtained by a comprehensive laboratory testing program which included characterization tests, double oedometer tests, permeability tests, triaxial K0 and triaxial CK0D tests (the triaxial tests used soil samples at natural moisture content – unsaturated conditions). Some of the laboratory tests were carried out with undisturbed block samples taken from an inspection pit dug at the research site. These samples were collected at 3, 6 and 9 m below the soil surface.

Table 1. Geotechnical parameters of the porous clay (after Cunha et al. 1999)

Parameter (1)	Unity (2)	Range of Values (3)	Parameter (4)	Unity (5)	Range of Values (6)
Sand percentage	%	12-27	Effective cohesion	kPa	9-20
Silt percentage	%	8-36	Friction angle	degrees	10-34
Clay percentage	%	80-37	Drained cohesion[a]	kPa	10-34
Dry unit weight	kN/m^3	10-17	Drained Friction angle[a]	degrees	26-34
Moisture content	%	20-33	Young's Modulus[b]	MPa	1-8
Degree of saturation	%	50-86	Coefficient of Collapse	%	0-12
Void ratio	--	0.9-2.0	Coefficient of earth pressure at rest[c]	--	0.5-0.6
Liquid limit	%	25-78			
Plasticity limit	%	21-35	Coefficient of permeability	cm/s	10^{-6}-10^{-3}

a. Range from triaxial CK_0D tests-soil at both natural moisture and saturated conditions;
b. Range from triaxial CK_0D tests-soil at natural moisture& 50% of failure dev. stress;
c. Range from triaxial K_0 tests with the soil at natural moisture content conditions.

FIELD LOADING TESTS

All the field load tests used herein were fully detailed in Sales (2000), whereas some of them were also presented in Cunha and Sales (1998). In summary, the tested foundations consisted of 15cm thick footings (rafts) 1m x 1m in plan, overlying bored floating piles which were manually augered to a circular cross section of 15cm in diameter and 5m in length.

The field loadings were carried out with the recommendations of the Brazilian testing standard NBR12131 (ABNT 1996), and were of the "slow maintained load" type tests with loading intervals of 20% of the calculated working load. The applied force on top of the raft was monitored by a 1kN precision load cell, and the displacements at the corners and center were measured by five 0.01mm precision dial gauges. A 10m steel frame, anchored at both extremities by two reaction piles, was used to react against the loads imposed by the hydraulic jack.

Most of the tested piles had load cells on their top and their tip to monitor the load transfer along the shaft, as well as to monitor the load absorbed by the raft. The raft load was simply obtained by the subtraction of the loads on top of the piles from the total applied load. The following foundation systems were tested:

- One isolated raft: Herein called Test I;
- One free standing pile: Test II;
- One piled raft with one pile underneath the raft: Test III;

- One standard pile group-four piles underneath the raft (off the ground): Test IV;
- The same as above, field loaded again after the piles had already failed: Test V;
- One piled raft with four piles underneath the raft: Test VI.

The aforementioned piled rafts had a perfect contact between the ground and the raft, whereas in the standard pile group there was a gap between the base of the raft and the ground surface. This testing set up allowed the applied load to be exclusively absorbed by the freestanding pile group, thus enhancing the interpretation of results from the other foundation systems.

Although Sales (2000) has carried out tests with the soil both at its natural water content and inundated conditions, only the former tests are analyzed herein. The effect of the soil collapse on the behavior of the foundation systems, and its numerical simulation, is not one of the objectives of the present paper. Nevertheless, this subject has been extensively discussed by this author given the collapsible nature of the Brasília porous clay.

Figure 3 presents the experimental load versus displacement curve of both Test IV and Test VI systems, modified after Cunha and Sales (1998), whereas Figure 4 presents the distribution of total applied load between the pile and the raft for Test VI. In this particular case (Test VI), it is noticed that the distribution is approximately constant for all the loading stages of the test, i.e. the piles absorbed ≈ 70 % of the applied load (or around 17.5 % for each pile), and the raft absorbed the remaining 30 %. Moreover it can be seen that the piled raft system had a higher failure load than the standard group of piles, although this former system has displaced more in the early stages of the loading test. It seems, then, that one gains in bearing capacity by changing the design system from a standard pile group to a piled raft. However by allowing the raft to absorb load, the average displacement of the system increases, given the fact that a pressure bulb is developed underneath the raft - contributing to the overall settlement of the soil mass under the foundation.

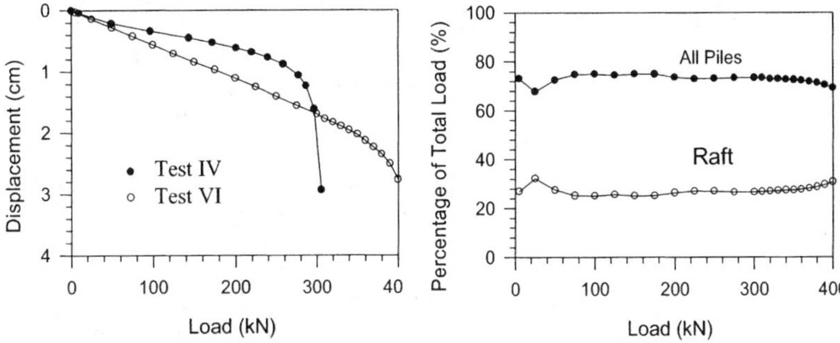

FIG. 3. Experimental Curves

FIG. 4. Load distribution-Test VI

BACK ANALYSIS OF EXPERIMENTAL DATA

All the field loading tests carried out in the Experimental Research Site were herein back analyzed with the APRAF software. These tests were carried out until a geotechnical failure was initiated in the soil surrounding the piles and raft. This can be observed by the non-linearity of the final part of the load-displacement curves of Tests IV and VI, presented in Figure 3.

However, since the APRAF software assumes the behavior of the soil as linearly elastic, with a constant Young's modulus (E_s) during the loading process, it can not be directly used to simulate the whole spectrum of the experimental load-displacement curve. At least not with a constant Young's modulus.

Therefore, in order to back analyze the experimental data, the Young's modulus of the soil was assumed to be variable in the non linear (plastic) region according to the following power law:

$$E_s = \left[E_0 - C_1 \cdot \left(\frac{P_a}{P_u} \right)^{C_2} \right] \tag{1}$$

Where E_0 is the initial modulus; P_a and P_u are, respectively, the applied and failure loads of the foundation system; C_1 and C_2 are regression coefficients.

The values of C_1 and C_2, as well as the value of the initial modulus E_0, were obtained by trial and error until a reasonably good match between the experimental and the theoretical load-deflection curve was obtained. Moreover, a Poisson´s coefficient of 0.2 was assumed for both raft and piles, and a value of 0.35 assumed for the soil.

Figure 5 presents the back analyzed moduli obtained for all the experimental tests, as a function of the applied load on top of the foundation system. These moduli were obtained after matching both predicted and measured load-deflection curves with the assumed E_s (non linear) variation. Hence, it is possible to notice that:

- All the tests furnished a constant value of the modulus E_s in the initial loading stages. Nevertheless, this modulus has considerably decreased after some threshold loading point (variable for each test), which is indicative of the non-linear behavior of the load-displacement curve. The threshold loading value is also variable for each of the tests, which is related to the distinct stiffness each foundation system and the surrounding soil can withstand before the geotechnical failure (or initial yield) of the soil. It should be pointed out that, with one exception to be mentioned below, all the foundations underwent geotechnical rather than structural failure;
- All the tests, with the exception of Test VI, furnished close values of the initial modulus E_0 (in the range of 4 to 9 MPa), indicating reasonable homogeneity of the soil around the Experimental Site. The higher result for the modulus obtained for Test VI was not expected, and it may be indicative of differences between the tests for either the foundation system or the subsoil. Sales (2000) has also experienced spurious results in his analysis, which have been explained by the structural cracking of the raft during one of the field loading tests (and its subsequent reinforcement for

the following tests). This hypothesis was not numerically tested herein, given all the difficulties involved, but, perhaps, could explain part of the observed differences.

Figure 6 and 7 present the distribution of vertical load along a typical pile, respectively for Tests IV and VI. The distribution is shown as a function of the pile depth, and for each loading stage of the test (25 to 100% of the pile failure load, P_u).

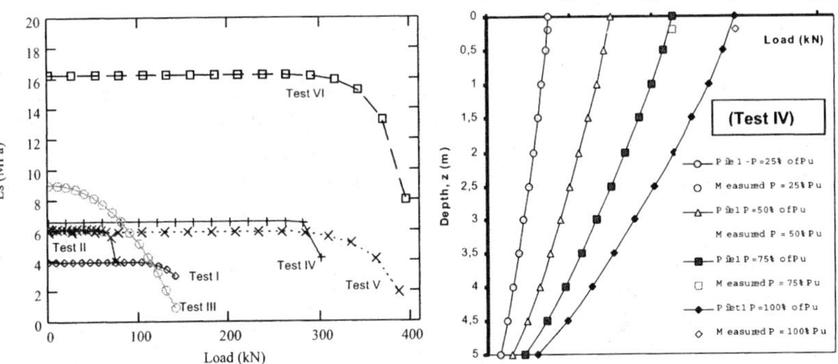

FIG. 5. Back analyzed E_s moduli

FIG. 6. Distribution of vertical load

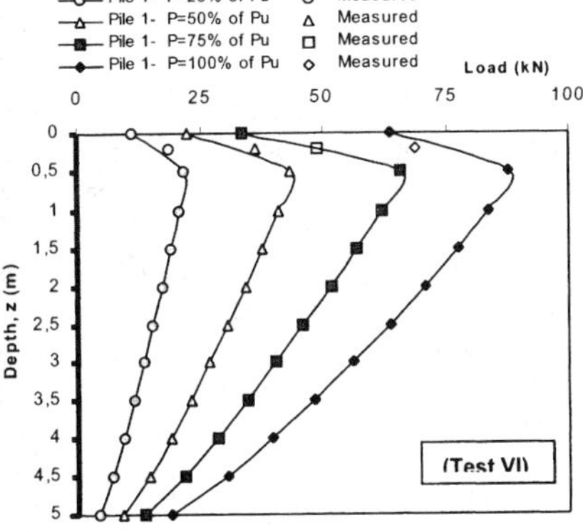

FIG. 7. Distribution of vertical load

With regard to Figures 6 and 7, it can be observed that:

- Both analyses yielded a smooth decrease of the structural load of the pile along its length, indicating that this load was continuously absorbed by friction in the pile's shaft/soil interface;

•Both analyses yielded a good agreement between the measured and the predicted loads on top of the piles, indicating that the software was able to well define the load share between the piles and raft (in the case of Test IV the load in the raft is nil, given the fact it was off the ground);

•The analysis of Test VI indicates an increase of vertical load within the pile at a certain depth below its top. This result clearly indicates the effect of the contact between the soil and the raft, since it allows for an increase of the level of mean and lateral stresses of the soil around the upper portion of the pile – given the raft's generated pressure bulb. This increase generates a higher "confinement" of the pile within this region, increasing its internal load. It also has led to a higher average unit lateral friction for the pile within the piled raft system when compared to a similar pile within the standard pile group. The average lateral friction for Test VI was 38 kPa, whereas this same variable was 25 kPa for Test IV, hence, an increase of 52%;

•The standard pile group had also a lower unit bearing pressure at the bottom of the pile when compared to the value measured for the piled raft system. This is consistent with the fact that the piled raft system showed a higher overall bearing capacity when compared to the standard pile group. The average unit bearing pressure for the pile of Test VI was 1000 kPa whereas this same variable was 850 kPa for Test IV, hence, an increase of 17%.

It shall be finally emphasized that the back analysis procedure presented in this paper can be used in actual design for pile foundations in this, or other, regions. For instance, it is possible to back analyze the results of a field loading test on a full scale single pile in order to obtain the magnitude of the main (input) variables of the problem. These variables are then further used to simulate a pile group in the same geotechnical material.

Besides, with the aforementioned preliminary values of Young's modulus, unit lateral friction and bearing pressure it is also possible to design real foundations in soils similar to the one presented herein. In Brasília, for instance, it is common to use small bored piles placed in a pile group or in a piled raft to transmit, to the subsoil, the loads that come from one or two pillars of (residential) units. It shall be also pointed out that, although the piled raft system is widely used in the city as a low cost foundation solution, especially for low cost or standard residential units (also used in swimming pools, given the collapsible nature of the superficial soil), consideration of the raft to absorb pillar loads is still not employed. It is designed as a standard pile group, assuming no contact between the soil and the raft. This is so given some conservatism of local foundation designers, lack of knowledge about the behavior of this particular type of foundation system, and because of the collapsible nature of the superficial Brasília soil. Nevertheless, recent studies of the University of Brasília (Sales 2000) have also demonstrated that such a design consideration is viable even with an inundated superficial soil in which part of it collapses under applied load. In this case the raft continues to absorb and transmit loads to the underlying soil and bored piles, although these loads are less in comparison to those carried by the soil at its natural moisture content.

CONCLUSIONS

The results of this paper highlight the fact that it is possible to numerically simulate the behaviour of a foundation system (piled raft or standard group of piles) founded on "non classical" materials, such as the tropical porous clay of Brasília. This can be achieved by a back analysis procedure, in order to obtain the input variables of the problem and their (non linear) variation – which is the case for the Young's modulus of the soil. This modulus can also be employed by using it with a rather simple, but powerful, numerical technique incorporated in the (APRAF) software that was developed with a sound theoretical basis. It should then be pointed out that this conclusion was established based on the limited conditions of the cases in this study. Although more research is necessary in this area, it is believed that such a conclusion could be generalized to other foundation cases similar to those presented herein. Ongoing M.Sc. and D.Sc. research theses of the University of Brasília aim to verify this and other aspects involving such "mixed" foundation systems.

The results also demonstrate that the piled raft with small diameter floating piles has a better mechanical behavior in comparison to small diameter pile groups. A large increase in unit lateral skin friction and base resistance is observed for the piles within the piled raft system, when compared to similar piles of the piled group system. The use of this foundation solution should be encouraged in tropical soils as a means to found low cost or standard residential units with floating piles. This is so, however, provided that a proper research effort is taken at a previous stage to obtain some design guidelines, and define typical behavior under existing subsoil conditions. The use of a back analysis procedure associated with field experimentation is therefore suggested and advocated herein in practical applications as a means of obtaining preliminary geotechnical parameters (as well as to calibrate existing numerical tools) to be further used in design. These loading tests also serve to provide actual data on the possible comparative behavior, and indicate economic disadvantages and advantages of distinct design solutions – for instance, piled rafts with piles underneath the raft at distinct configurations (see Cunha et al. 2001).

Finally, it is clear from the results of this paper that, in the case of the piled raft system, a beneficial load share and interaction may occur between raft, piles and soil, leading to a gain in terms of bearing capacity and the foundation's overall behavior.

ACKNOWLEDGMENTS

The authors would like to thank Prof. Mauricio Sales from Federal University of Goiás for providing technical assistance, and both the mechanical and electronic laboratories of the University of Brasília for their maintenance support. The scholarship provided by CNPq and CAPES to the Geotechnical Post Graduation Program of the University of Brasilia, which allowed the establishment of the experimental research described herein and in other related papers is also gratefully acknowledged. Finally, the authors acknowledge the extensive aid provided by colleagues from the University of Sydney, in terms of fruitful discussions.

REFERENCES

ABNT. (1996). "NBR12131:Piles – Static load test". *Brazilian Association of Standards,* Rio de Janeiro, 12p. (In Portuguese).

Araki, M.S. (1997). "Aspects related to the proprieties of the collapsible soils of the Federal District". *M.Sc. Thesis*. University of Brasilia. Department of Civil and Environmental Engineering. (In Portuguese).

Cunha, R.P. and Sales, M.M. (1998). "Field load tests of piled footings founded on a tropical porous clay". *Proceedings of the 3rd Intern. Geotechnical Seminar on Deep Foundation on Bored and Auger Piles*, Ghent, Vol. 1, 433-438.

Cunha, R.P., Jardim, N.A. and Pereira, J.H.F. (1999). "In situ characterization of a tropical porous clay via dilatometer tests". *Geo-Congress 99 on Behavioural Characteristics of Residual Soils*, ASCE Geotechnical Special Publication 92, Charlotte, 113-122.

Cunha, R.P., Poulos, H.G. and Small, J.C. (2001). "Investigation of design alternatives for a piled raft case history". *ASCE Journal of Geotechnical and Geoenvironmental Engineering*, August, 127(8), 635-641.

Cunha, R.P., Small, J.C., and Poulos, H.G. (2000a). "Class C analysis of a piled raft case history in Gothenburg, Sweden". *Year 2000 Geotechnics – Geotechnical Engineering Conference*, Bangkok, Asian Institute of Technology, 1, 271-280.

Cunha, R.P., Small, J.C., and Poulos, H.G. (2000b). "Parametric analysis of a piled raft case history in Uppsala, Sweden". *IV Seminar of Special Foundations and Geotechnics*, São Paulo, Brazilian Soil Mechanics Society, 2, 381-390.

Hain, S.J. and Lee, I.K. (1978). "The analysis of flexible raft-pile systems". *Géotechnique*, Vol. 28, 1, 65-83.

Mandolini, A., and Viggiani, C. (1997). "Settlement of piled foundations". *Géotechnique*, Vol. 47(4), 791-816.

Poulos, H.G. (1998). "The pile-enhanced raft - an economical foundation system". Keynote Lecture, *Proceedings of the XI Brazilian Congress on Soil Mechanics and Geotechnical Engineering*, Brasilia, Brazilian Soil Mechanics Society, 4, 27-43.

Sales, M.M. (2000). "Analysis of the behavior of piled rafts". *Ph.D. Thesis*, University of Brasília, Department of Civil and Environmental Engineering, Publication G.TD – 002A/00, 229 p. (In Portuguese).

Ta, L.D. and Small, J.C. (1997). "An approximation for analysis of raft and piled raft foundations". *Computers and Geotechnics*, Vol. 20(2), 105-123.

Zhang, H.H. (2000) "Finite Layer Method for Analysis of Piled Raft Foundations", PhD Thesis, University of Sydney.

Zhang, H.H. and Small, J.C. (2000a). "Analysis of capped pile groups subjected to horizontal and vertical loads". *Computers and Geotechnics*, Vol.26(1),1-21.

Zhang, H.H. and Small, J.C. (2000b). "Piled Raft Foundations Subjected to General Loadings", *John Booker Memorial Symposium, 2000*, Eds. D.W. Smith and J.P. Carter, A.A.Balkema, 431-444.

BEHAVIOR OF SQUARE FOOTINGS ON SINGLE REINFORCED SOIL

Fathi M. Abdrabbo[1], Member, ASCE, Khaled E. Gaaver[2], and Amr Z. Elwakil[3]

ABSTRACT: Increasing the bearing capacity of soils has been an important issue thousands of years ago till the time being. In 1960's the French Road Research Laboratory conducted an extensive study to evaluate the benefits of using reinforced soil as a construction technique. Nowadays, it is still an aim to increase the soil bearing capacity either through adding inclusions or through reinforcing sheets. The study aims to demonstrate the effects of embedment ratio and the length ratio of the reinforcing element on the bearing capacity of sand. About fifty plate-loading tests were conducted on steel square model footing of side dimension (B) equal to 100 mm. The effect of relative density of sand, the embedment ratio, and the length ratio of the reinforcing material were studied. It was revealed that the soil reinforcement might successfully increase the bearing capacity. Also it was found that the improvement in the bearing capacity of sand of lower relative density is more pronounced compared with higher relative density sand. The bearing capacity ratio (BCR) was found to be inversely proportional to the relative density of the sand. The most efficient length ratio (L/B) of the soil reinforcement was found to be 3.00, and the optimum embedment ratio (d/B) was found to be 0.30. Finally, the reinforced sand exhibited a failure mode that resembles typical punching-shear failure

INTRODUCTION

Many attempts occurred thousands of years ago aiming to increase the soil bearing capacity. Pharaohs were the pioneers to perform those attempts by mixing ashes with clay as a reinforcing material to increase the bearing capacity of the clay and they succeeded in that attempt. Now the use of geosynthetics in ground improvement is increasing annually. The geosynthetics is a name given to a family of man-made sheet or net-like products derived from plastics, petroleum, or fiberglass compounds.

[1] Professor, Head of Structural Engineering Department, Faculty of Engineering, Alexandria University, Alexandria, Egypt, P.O Box 21544, E-mail: F.M.Abdrabbo@excite.com.

[2] Lecturer, Faculty of Engineering, Alexandria University, Alexandria, Egypt, P.O Box 21544, E-mail: Khaledg2000@hotmail.com.

[3] Assistant Lecturer, Faculty of Engineering, Alexandria University, Alexandria, Egypt, P.O Box 21544, E-mail: Elwakilamr@hotmail.com.

Belongings of this group are geotextiles, geogrids, geonets, and geomembranes. In soil reinforcement technique, one or more layers of a geosynthetic sheet and controlled fill material are placed beneath the footing to create a composite geotechnical material with improved performance characteristics. The gain in popularity of soil reinforcement technique is largely due to cost effectiveness, simplicity of construction process and large selection of materials available. It is attractive for contractors, designers and owners. Although the use of geosynthetics is entirely compatible with conventional construction procedures, some important precautions need to be taken into consideration (Pinto 2002).

Geotextiles may be woven, knitted, or nonwoven. In fact that geotextiles caught researchers attention in the late 50s when woven fabrics played an important role in coastal projects in the Netherlands and in the USA (Richardson et al. 1992). In the late 60s, Richardson mentioned that geotextiles found a role as separators that enhance the performance of roadbeds for highway and track support systems. Das (1999) quoted that; in the 1960's, the French Road Research Laboratory conducted an extensive study to evaluate the benefits of using reinforced soil as a construction technique. This research has been documented in detail by Schlosser and Vidal (1969), Darbin (1970), and Schlosser and Long (1974). Kurian et al. (1997) mentioned that the technique of reinforcing soil is one of the most and fast-growing techniques of soil improvement methods, and it is realized that reinforcing soil is so useful in solving many foundation problems. They also mentioned that Binquet and Lee (1975) were the first to perform significant study on the use of reinforcing elements in foundations, they proved that the bearing capacity of sand increases as much as three times when using reinforcement. In the last four decades, there has been continued development of high performance geotextile products that have proven to be more economical, easier to handle, stronger, and longer lasting than traditionally used construction materials (Fowler 1992).

Guido et al. (1986) conducted laboratory model tests on reinforced soil foundations and compared the performance of geotextiles and geogrids as soil reinforcement. Their study revealed that the geogrid reinforcement was more effective than the geotextile from the standpoint of improving the bearing capacity of footings on reinforced sand. Yetimoglu et al. (1994) stated that, Jewell et al. (1984) and Milligan and Palmeira (1987) pointed out that the mobilization mechanism of frictional resistance in geogrids was different from that of geotextiles. Wasti and Butun (1997) mentioned that Alrefeai (1991) performed triaxial tests on two sands reinforced with glass fiber and mesh elements. It was concluded that mesh elements were superior to fiber, especially in case of fine sand. Also Wasti and Butun (1997) reported that Uysal (1993) performed comparison triaxial tests on sand reinforced with meshes, and also with 5 mm long fibers cut from the meshes. The sand was compacted at the optimum water content. Failure strains of fiber-reinforced sand samples were generally lower than failure strains for mesh-reinforced samples. Adams and Collin (1997) conducted thirty-four large model load tests to evaluate the potential benefits of geosynthetic-reinforced spread footings. They concluded that the bearing capacity ratio (BCR) is more than 2.60 for three layers of grid. Wang and Richwien (2002) stated that the soil-reinforcement interface friction is one of the basic factors influencing the strength and the deformation of reinforced soils. Direct shear tests and pullout tests are the available methods to determine the value of soil-reinforcement interface friction. They concluded that the soil-reinforcement interface friction is dependent on the reinforcement roughness, the friction angle of sand, and the dilatancy.

The beneficial effects of soil reinforcement drive from the soil's increased tensile strength and the shear resistance developed from the friction at the soil-reinforcement interfaces. For a geosynthetic acting as reinforcement, the most important properties are: tensile strength, tensile modulus, and interface shear strength (Pinto 2002). Tensile strength and tensile modulus are important because the reinforcing element needs to resist the tensile stresses, transferred from the soil, under deformations compatible with those allowable for the soil. Interface shear strength is an important parameter, as it is responsible for the transfer of stresses from the soil to the geosynthetic. Here we investigate the influence of adding a reinforcing sheet to sandy soil. The study aims to illustrate the effects of the embedment ratio (d/B), the length ratio (L/B) of the reinforcing element, and the relative density on the bearing capacity of sand.

TESTING EQUIPMENT

To study the effects of adding single reinforcing sheet on the bearing capacity of sand about fifty laboratory experiments were conducted on steel square model footing of side dimension (B) equals to 100 mm and of thickness 25 mm. The model footing had smooth faces and a notch at the center of the top face for mounting a calibrated proving ring of 28 kN maximum capacity and 10 N accuracy via ball bearing. Two-dial gauges, of accuracy 0.01 mm, were used to measure the footing vertical displacement. The general layout of the equipment used in the present study is illustrated in figure (1). The soil bin is made out of three steel rings each of 300 mm height and 750 mm in diameter. These rings were connected together using steel bolts to form soil bin of total height 900 mm. A woven reinforcing material (Stabilenka 200/200) of thickness 0.12 mm was used. The ultimate longitudinal and transversal tensile strengths of the reinforcing sheets are 200 kN/m, and the longitudinal strain at nominal tensile strength is 10%. The sheets were cut into different dimensions giving different length ratios (L/B) of 1.20, 1.50, 3.00, and 5.00. The side dimensions of the reinforcing sheets were measured within an accuracy of 1 mm.

EXPERIMENTAL PROCEDURE

The sand used in this study was a medium-grained with a minimum dry unit weight of 15.60 kN/m^3, maximum dry unit weight of 18.20 kN/m^3 and uniformity coefficient of 2.95. The specific gravity of sand particles is 2.66. A designed weight of sand, with an accuracy of 010 kg, was formed into a certain volume of the soil bin by compaction to give the target relative density. The relative densities were 60%, 70%, and 88%. The reinforcing material was placed in the required depth according to a planned testing program shown in Table (1). Then the sand layer was completed by the same procedure up to the design level. The top surface was leveled and the model footing was placed on a predefined alignment such that the applied loads from the loading device would be transferred concentrically to the footing.

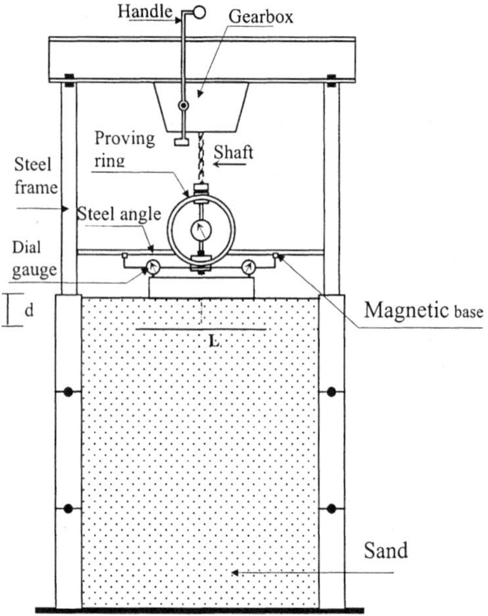

FIG. 1. Complete set-up of testing procedures

The applied loads were measured with an accuracy of 10 N using a calibrated proving ring of 28 kN maximum capacity. The load was applied incrementally. Each increment was kept constant till no significant change occurs in settlement, that's to say the difference between two successive readings is less than 0.01 mm per 5 minutes for three consequative readings. The corresponding footing settlement was measured within an accuracy of 0.01 mm using two-dial gauges. After each test was completed, the sand was carefully excavated and the geosynthetic reinforcement visually inspected. Reference test was carried out on unreinforced soil for each relative density of sand to explore the effect of reinforcing sheet on the bearing capacity of sand. It is important to note that, both diameter and depth of the used soil pin are greater than six times of the footing width, so the boundary effect on the test results was considered insignificant. Moreover, during the tests it was observed that the extent of soil bulging on the sides of the footing is less than two times of the footing width, indicating that the boundary effect on the test results was likely to be insignificant.

TABLE 1. Testing Program and Results

Test No. (1)	Dr % (2)	L/B (3)	d/B (4)	BCR (5)	Remarks (6)
1	60	--	--	1.00	No reinforcement
2	60	3.00	1.00	1.50	Single reinforcement
3	60	3.00	0.75	1.70	
4	60	3.00	0.50	2.10	
5	60	3.00	0.25	2.45	
6	60	1.50	1.00	1.30	
7	60	1.50	0.75	1.52	
8	60	1.50	0.50	1.92	
9	60	1.50	0.25	2.25	
10	60	1.20	1.00	1.20	
11	60	1.20	0.75	1.35	
12	60	1.20	0.50	1.55	
13	60	1.20	0.25	1.80	
14	60	5.00	1.00	1.50	
15	60	5.00	0.75	1.69	
16	60	5.00	0.50	2.10	
17	60	5.00	0.25	2.45	
18	70	--	--	1.00	No reinforcement
19	70	3.00	1.00	1.48	Single reinforcement
20	70	3.00	0.75	1.68	
21	70	3.00	0.50	2.00	
22	70	3.00	0.25	2.25	
23	70	1.50	1.00	1.28	
24	70	1.50	0.75	1.50	
25	70	1.50	0.50	1.88	
26	70	1.50	0.25	2.13	
27	70	1.20	1.00	1.15	
28	70	1.20	0.75	1.34	
29	70	1.20	0.50	1.52	
30	70	1.20	0.25	1.70	
31	70	5.00	1.00	1.48	
32	70	5.00	0.75	1.68	
33	70	5.00	0.50	2.05	
34	70	5.00	0.25	2.18	
35	88	--	--	1.00	No reinforcement
36	88	3.00	1.00	1.48	Single reinforcement
37	88	3.00	0.75	1.65	
38	88	3.00	0.50	1.94	
39	88	3.00	0.25	2.18	
40	88	1.50	1.00	1.25	
41	88	1.50	0.75	1.40	
42	88	1.50	0.50	1.80	
43	88	1.50	0.25	2.00	
44	88	1.20	1.00	1.05	
45	88	1.20	0.75	1.19	
46	88	1.20	0.50	1.50	
47	88	1.20	0.25	1.55	
48	88	5.00	1.00	1.49	
49	88	5.00	0.75	1.65	
50	88	5.00	0.50	1.95	
51	88	5.00	0.25	2.20	

TEST RESULTS AND DISCUSSION

1. Response of Square Footings on Single Reinforced Soil

To study the effect of a single reinforcing layer on the behavior of square footings, three references tests were conducted on unreinforced sand with relative densities 60%, 70%, and 88%. For the sake of comparisons, other tests were performed with the same relative densities but on reinforced sand. Different embedment ratios of the reinforcing sheet (d/B) of 0.25, 0.50, 0.75, and 1.00, as shown in Table (1), were considered. The reinforcing elements were of square dimensions and with different length ratios (L/B) of 1.20, 1.50, 3.00, and 5.00. The load was applied incrementally and the average corresponding settlements were recorded. The load-settlement relationships were plotted for all tests. Sample of the obtained load-settlement relationships is shown in figure (2). The ultimate load was defined as the load where settlement continued without further increase of loads. The settlement corresponding to the ultimate load was found to be about 8% of the footing width (S/B=8%). The ratio between the ultimate load attained from loading test on reinforced sand to that of the reference test is defined as the bearing capacity ratio (BCR). Table (1) shows the test results.

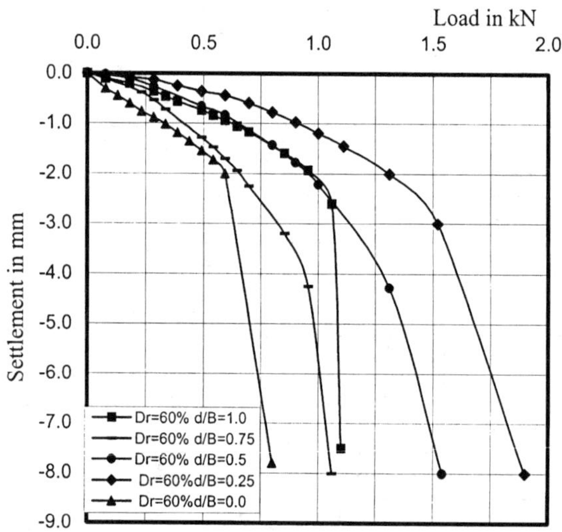

FIG. 2. Load-settlement relationships, L/B=1.5

It was found that single reinforcing layer increases the bearing capacity ratio (BCR) from 1.05 to 2.45, depending upon the depth ratio (d/B), the length ratio (L/B), and the relative density (Dr%) of sand as well. This improved performance can be attributed to an increase in shear strength in the reinforced soil mass from the inclusion of the geosynthetic reinforcement. The soil-geosynthtic system creates a composite material that inhibits development of the soil-failure wedge beneath shallow spread footings. The inclusion of reinforcement in sand changed the mechanism of failure from general shear failure assumed by Prandtl to punching failure. Definitely tension force induced in reinforcement alters stress and strain fields in soil underneath the footing, and consequently changed the location of failure planes. So, the improvement of bearing capacity depends upon the extent of reinforcing element beyond planes of failure. Clearly, when the reinforcement is deformed, their tensile stress is substantially mobilized, and thus they serve as reinforcement. So, the more geotextile is deformed, the greater the load carried, hence more bearing capacity of footing soil system is attained. Several early researches found that reinforcement in sand could increase the cohesion component of soil strength. That is to say reinforcement imparted cohesion to a mass of granular soil particles.

2. Effect of Relative Density on The Bearing Capacity Ratio

The load-settlement relationships of footing on reinforced sand at the same length ratio (L/B) of the reinforcement and with different relative densities (Dr%) were grouped. It can be noticed from the plotted relationships– shown in Figures (3-a, b, c, and d)– that the effect of reinforcing material is more feasible in soil with lower relative density. This can be attributed to the modulus of deformation of soil. So, the higher the relative density of soil, the higher the modulus of deformation is. For higher relative density soil its modulus of deformation is approaching that of the reinforcing material so the reinforcing may not be highly effective. On the other hand, for soil of low relative density and low modulus of deformation, it will be more affected by the reinforcing material. In a loose sand-reinforced footing, large deformations of the footing are required to mobilize the beneficial effects of the reinforcement. This observation confirms Adams and Colline (1997). But one should realize that in cohesionless soils, allowable settlement and not ultimate bearing capacity often governs the design.

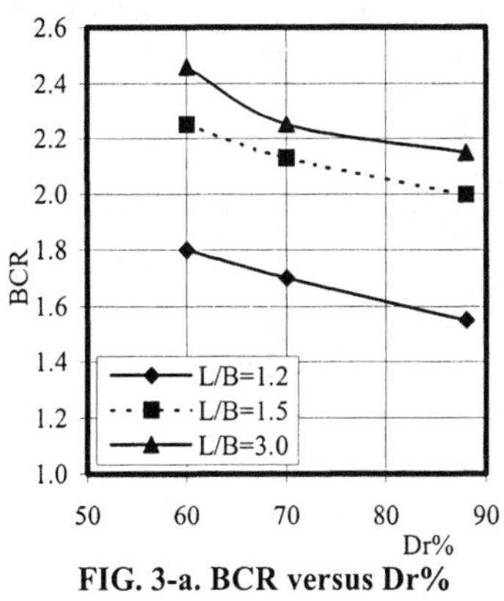

FIG. 3-a. BCR versus Dr% for d/B = 0.25

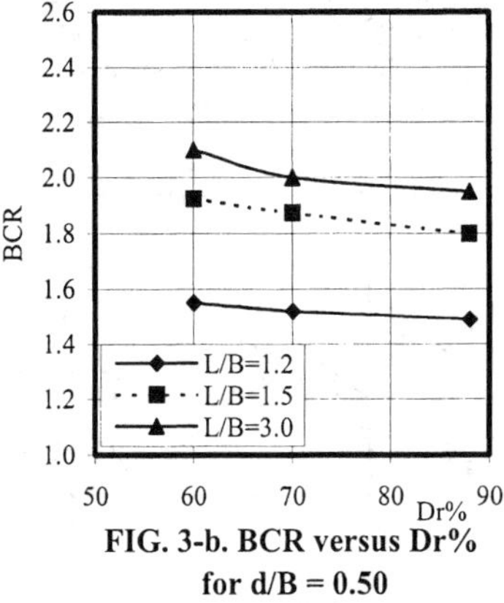

FIG. 3-b. BCR versus Dr% for d/B = 0.50

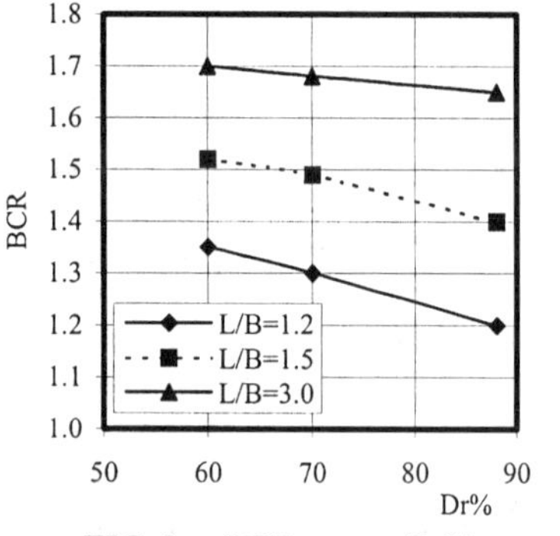

FIG. 3-c. BCR versus Dr% for d/B = 0.75

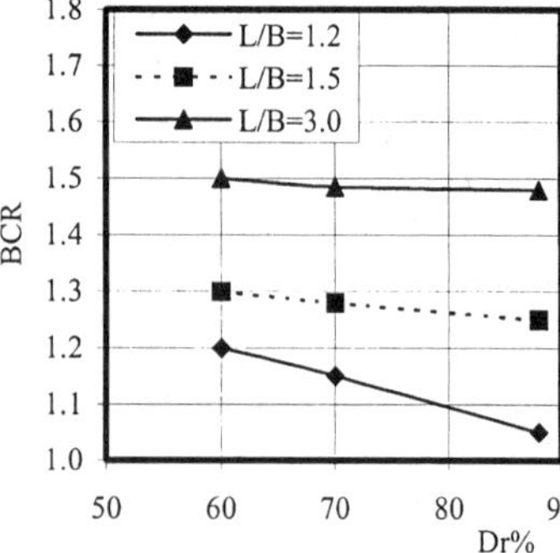

FIG. 3-d. BCR versus Dr% for d/B = 1.00

3. Effect of Embedment Ratio on The Bearing Capacity Ratio

In order to investigate the effect of the embedment ratio (d/B) on the bearing capacity ratio (BCR), Figures (4-a, b, and c) were plotted. The figures illustrate that (BCR) increases as (d/B) increases up to (d/B) = 0.30 where (BCR) reaches its peak value irrespective of (Dr) and (L/B). These relationships show that the maximum (BCR) reached at the most effective embedment ratio (d/B) of 0.30. These results conform Yetimoglu et al. (1994) in which they mentioned that for single reinforced sand, the optimum depth was approximately 0.30B. On the other hand, Das (1999) stated that the most effective depth is about 0.40 the footing width. Yetimoglu et al. (1994) quoted that Singh (1988) conducted an experimental study of square footings on sand reinforced by mild steel grids (welded mesh). He concluded that the optimum depth ratio was 0.25 for both single layer and multilayer reinforced sand. Huang and Menq (1997) stated that Binquet and Lee (1975) attributed the increase in the ultimate load of the reinforced soil to the tensile force developed in the vertically bending part of the reinforcement across the assumed shear band.

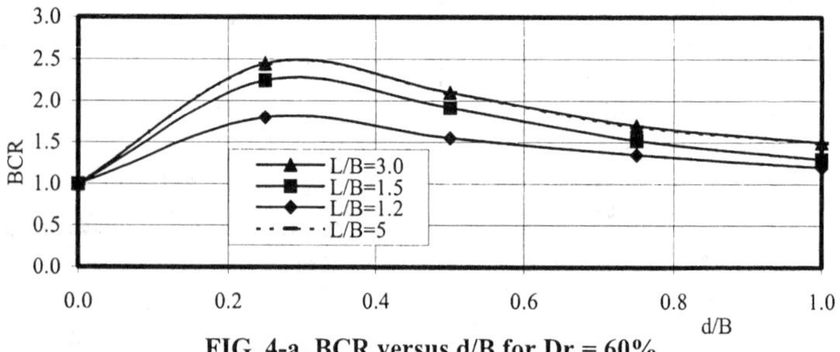

FIG. 4-a. BCR versus d/B for Dr = 60%

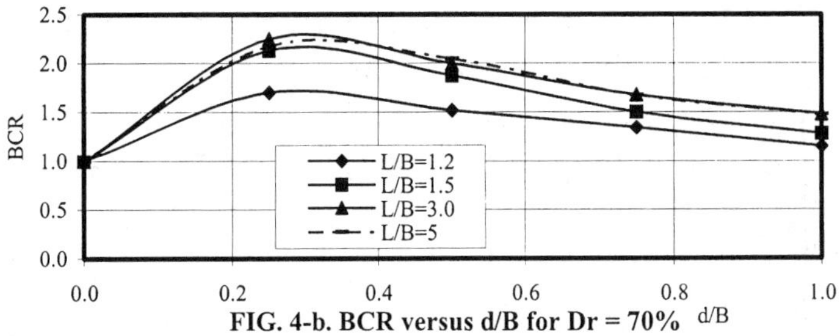

FIG. 4-b. BCR versus d/B for Dr = 70%

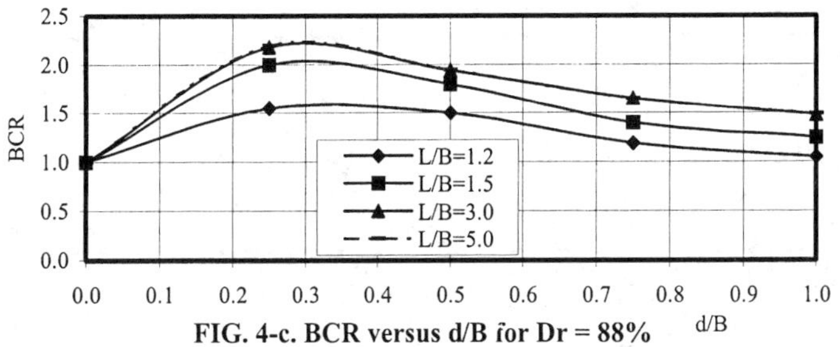

FIG. 4-c. BCR versus d/B for Dr = 88%

4. Effect of Length Ratio on The Bearing Capacity Ratio

As mentioned before (BCR) increases as the reinforcing material extends beyond the failure surface, so the length of the reinforcing material is a main factor affecting the (BCR). To study the effect of length ratio on (BCR) we used reinforcing material of different length ratios (L/B) of 1.20, 1.50, 3.00, and 5.00. The relationships between the (BCR) attained from the loading tests and (L/B) were plotted and shown in figures (5-a, b, c, and d). The figure illustrates that the bearing capacity ratio

(BCR) of shallow footings increased with length ratio (L/B) of the reinforcing element up to an effective value of (L/B) and remains practically constant thereafter. The most efficient length ratio (L/B) is 3.00, hence its more economical to use reinforcing material of L/B = 3.00. This result conforms to Guido et al. (1985) in which they concluded that the optimum size of reinforcement is about three times of the footing width that should allow sufficient tensile stresses development. On the other hand, Omar et al. (1993), based on geogrid-reinforced sand study, reported that the effective reinforcement length ratio was around 8 for strip foundations ad 4.50 for square foundations.

Moreover, the figures show that the most efficient length ratio (L/B) of reinforcement is independent of both the reinforcement depth ratio (d/B) and the relative density of sand (Dr). This finding confirmed that failure mechanism in reinforced-soil foundation is punching. Yetimoglu et al. (1994) quoted that

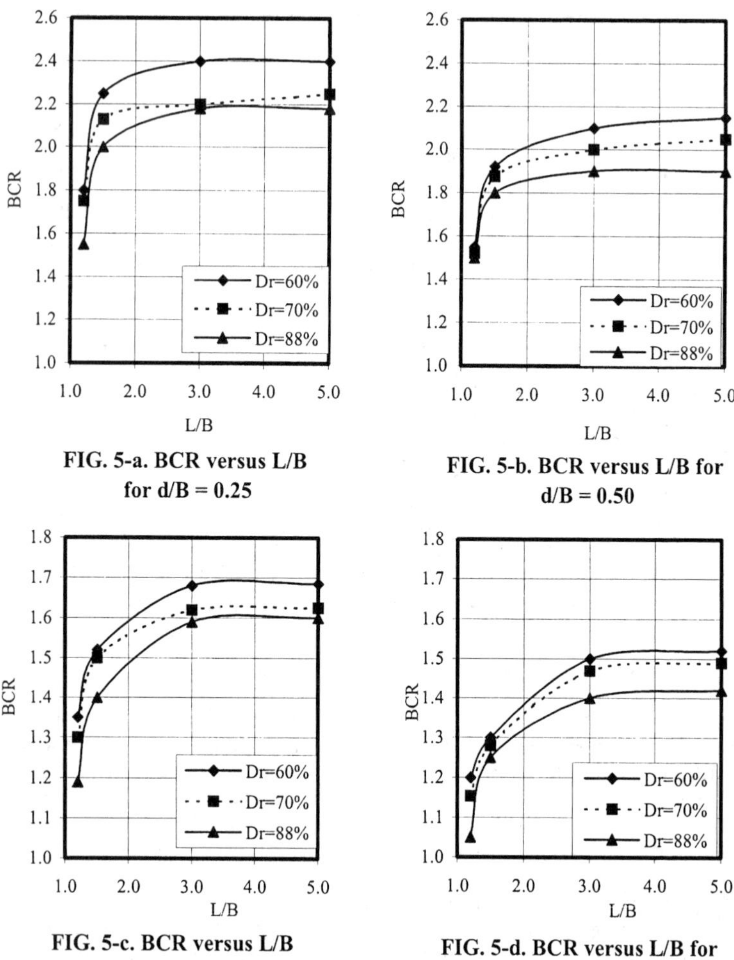

FIG. 5-a. BCR versus L/B for d/B = 0.25

FIG. 5-b. BCR versus L/B for d/B = 0.50

FIG. 5-c. BCR versus L/B for d/B = 0.75

FIG. 5-d. BCR versus L/B for d/B = 1.00

Akinmusuru and Akinbolade (1981) and Guido et al. (1987) reported that the geogrid-reinforced sand exhibited a failure mode that resembles typical punching-shear failure. It important to note that during the achieved tests no failure in reinforcement was noticed. The figures confirmed an improvement in the bearing capacity of a footing on a reinforced-soil foundation at L/B = 1.20, and the improvement decreases as the depth of reinforcing layer below the footing increases. This is because the anchorage length of the reinforcement decreases, and consequently the pullout resistance decreases. The coefficient of pullout friction is dependent on the roughness of reinforcement, the friction angle of the sand, and the dilatancy (Wang and Richwien, 2002).

CONCLUSIONS

From the accomplished laboratory tests, it can be proved that the use of geotextiles is effective in the improvement of the bearing capacity of sand. The bearing capacity ratio (BCR) varied from 1.05 to 2.45 for single reinforced soil. The most effective depth ratio (d/B) of the reinforcing material was found to be 0.30 irrespective from the relative density of sand, or the length ratio ($L/B \geq 1.20$). The most effective length ratio (L/B) was proved to be 3.00. It was found that the effect of reinforcement on the bearing capacity of sand is more pronounced in soils with lower relative density than soils with higher relative density. The reinforced sand exhibited a failure mode that resembles typical punching-shear failure

REFERENCES

Adams, M. T., and Collin, J. G. (1997). "Large footing load tests on geosynthetic reinforced soil foundations." *J. of Geotech. and Geoenv. Engrg.,* ASCE, 123 (1), 66-72.

Alrefeai, T. O. (1991). "Behavior of granular soils reinforced with discrete randomly oriented inclusions." *Geotextiles and Geomembranes* 10, 319-333.

Binquet, J., and Lee, K. (1975). "Bearing capacity analysis of reinforced earth slabs." *J. of Geotech. Engrg.*, ASCE, 110 (10), 1500-1511.

Das, B. M. (1999). "*Shallow fundations - bearing capacity and settlement.*" CRC press LLC.

Fowler, J. (1992). "Embankments over soft soils." *A Design primer: Geotextiles and Related Materials*. Industrial Fabric Association International.

Guido, V. A., Biesiadecki, G. L., and Sullivan, M. J. (1985). "Bearing capacity of a geotextile-reinforced foundation." *Proc. 11 th Int. Conference on Soil Mechanics and Foundation Engineering*, San Francisco, Vol. 3, 1777-1780.

Guido, V. A., Dong, K. G., and Sweeny, A. (1986). "Comparison of geogrid and geotextile reinforced earth slabs." *Canadian Geotechnical J.,* 23 (1), 435-440.

Huang, C. C., and Menq, F. Y. (1997). "Deep-footing and wide-slab effects in reinforced sandy ground." *J. of Geotech. Engrg.*, ASCE, 123 (1), 30-36.

Kurian, N. P., Beena, K. S. and Kumar, R. K. (1997). "Settlement of reinforced sand in foundations." *J. of Geotech. Engrg.*, ASCE, 101 (9), 818-827.

Omar, M. T., Das, B. M., Puri, V. K.and Yen, S. C. (1993). "Ultimate bearing capacity of shallow foundations on sand with geogrid reinforcement." *Canadian Geotechnical J.,* 30 (3), 545-549.

Pinto, M. I. (2002). "Applications of geosynthetics for soil reinforcement." *Proc. 4 th Int. Conference on Ground Improvement Techniques*, Malaysia, 147-162.

Richardson, G. N. (1992). "Introduction to geotextiles and related materials." *A Design primer: Geotextiles and Related Materials*. Industrial Fabric Association International.

Uysal, C. (1993). "Stress strain behavior of randomly distributed discrete fibers and mesh reinforced sand." *M.Sc. Thesis*, Middle East Technical University, Ankara, Turkey.

Wang, Z. and Richwien, W. (2002). "A study of soil-reinforcement interface friction." *J. of Geotech. and Geoenv. Engrg.*, ASCE, 128 (1), 92-94.

Yetimoglu, T., Jonathan, T. H. Wu, and Saglamer, A. (1994). "Bearing capacity of rectangular footings on geogrid-reinforced sand." *J. of Geotech. Engrg.*, ASCE, 120 (12), 2083-2099.

Yildiz, W., and Mustafa, D. (1996). "Behavior of model footings on sand reinforced with discrete inclusions." *Geotextiles and Geomembranes* 14, 575-584.

STUDENT VIEWS OF THE GEO-INDUSTRY

Andrea L. Welker[1], P.E., Associate Member, Geo-Institute and ASCE and Cynthia Finley[2], Associate Member, Geo-Institute and ASCE

ABSTRACT: A survey was performed at two universities to gather information on undergraduate's perceptions of the geo-industry. The surveys were administered to students in a required soil mechanics course and in geotechnical elective courses at The University of Texas at Austin and Villanova University. These two universities - a large, public institution and a small, private institution - provided an interesting range of responses. Example questions include: What do you think geotechnical engineers do?; What is appealing about a career in the geo-industry?; and What is the highest degree you intend to seek?. Demographic information, such as the gender and GPA of the respondent, was also collected to provide a context for the survey data. Academics can use the results of the survey to improve their geotechnical course materials, and practitioners will find the results useful when recruiting students for summer internships and permanent employment after graduation.

INTRODUCTION

A survey was administered at Villanova University (VU) in Villanova, PA and at The University of Texas at Austin (UT). VU is a private, Augustinian Catholic university located about 15 miles west of Philadelphia, PA. There are approximately 6,000 undergraduates at VU with about 900 in the College of Engineering. The Civil and Environmental Engineering (CEE) Department awards about 50 Bachelor of Civil Engineering degrees a year. The highest degree awarded by VU's CEE

[1] Assistant Professor, Villanova University, 800 Lancaster Ave., Villanova, PA, 19087, andrea.welker@villanova.edu
[2] Graduate Student, The University of Texas at Austin, ECJ 9.227, Austin, TX, 78712

Department is a Master of Civil Engineering. UT is a public, land-grant university located in Austin, TX. There are approximately 37,000 undergraduates with about 5,000 in the College of Engineering. The Civil Engineering Department, which includes Environmental and Architectural Engineering, awards about 100 Bachelor of Science degrees each year. The highest degree awarded by UT's Civil Engineering department is a Doctor of Philosophy.

Students in a required, introductory soils class and in elective classes were asked about their impressions of geotechnical engineering. The elective classes were Foundation Design at VU and Foundation Design and Retaining Structures at UT. Their responses are enlightening for both academics and practitioners. Most of the students in the introductory classes were juniors, whereas most of the students in the elective classes were seniors. The surveys were conducted at the end of the Fall 2002 semester and approximately half-way through the Spring 2003 semester.

The eleven questions asked of the students in the introductory classes were:

1. What do you think geotechnical engineers do?
2. Are you interested in learning more about geotechnical engineering? Why or why not?
3. Would you consider a career in geotechnical engineering? Why or why not?
4. What is appealing about a career as a geotechnical engineer?
5. What is unappealing about a career as a geotechnical engineer?
6. Do you have any experience working the geotechnical engineering field? If so, please describe what you do?
7. What is your major?
8. Have you decided on a specialty yet? If so, what is it?
9. What is the highest degree you intend to seek?
10. What is your GPA?
11. What is your gender?

The students in the elective geotechnical classes were asked the same eleven questions as the introductory classes, as well as these additional questions:

12. Why did you decide to take this elective?
13. What other technical electives are you currently taking, or plan to take in the future?
14. Do you wish there were more geotechnical engineering electives offered? If so, what topics should they cover?

The students wrote their own responses to all of the questions except 7, 9, and 10, where choices were given for students to circle. Students' answers to the free-response questions were categorized and tabulated by the authors.

RESULTS AND DISCUSSION OF RESULTS

Basic Statistics

A total of 159 civil engineering students responded to the survey: 61 from VU and 98 from UT. Of the students surveyed, 92 were enrolled in an introductory class, while 67 were taking a geotechnical elective. The results of questions 7 through 11, which asked students for basic information, are described below.

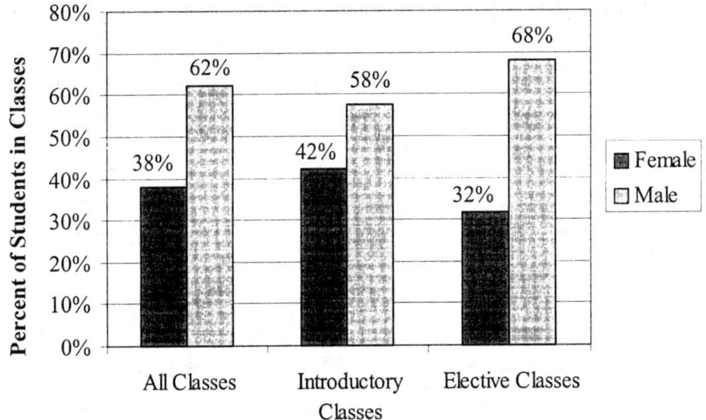

FIGURE 1. Composition of Surveyed Classes by Gender

Approximately 38% of all respondents were women (Figure 1), overall and at each institution. To place this statistic in context, in 1998 approximately 20% of all engineering students in the United States were women (National Science Foundation 2000). Unfortunately, the percentage of women in the elective classes was less than the percentage in the introductory classes.

Most of the students surveyed recognize the importance of continuing education. Ten percent of the VU students and 19% of the UT students stated that a bachelor's degree is the highest degree they intend to seek. Forty-five percent of the students in the introductory class intend to pursue a master's degree. This number rises to 62% for the students in the elective classes.

Even as seniors, many students have not decided on a specialty area of civil engineering. In the introductory classes, 36% had not decided; that number decreased to 30% in the elective classes. Only 3.3% of the students in the introductory classes indicated that geotechnical engineering was their preferred specialty area; however, 18% of those in the elective courses stated this preference. This dramatic change has an important implication. Most students do not know what geotechnical engineering is before taking an introductory course in the area, which is typically junior year. Consequently, the quality of that course is pivotal in generating interest in geotechnical engineering.

To place the statistics gathered from the students on their preferred specialty area into context, ASCE was contacted to provide information on specialty areas of their members. Of 96,000 ASCE members, 52,549 selected an area of primary interest on their membership profiles (ASCE 2003). The areas of primary interest for ASCE members are compared to those given by the students surveyed in Table 1.

TABLE 1. Comparison of Chosen Civil Engineering Specialty Areas

Civil Engineering Area (1)	Students in Introductory Courses[a] (2)	Students in Elective Courses[a] (3)	ASCE Members[b] (4)
Structural	18%	36%	34%
Environmental	13%	3%	24%
Geotechnical	3%	18%	16%
Water Resources	7%	3%	20%
Construction	14%	3%	--
Transportation/Urban Planning	9%	7%	6%

[a] Percent of students who had already chosen a specialty.
[b] Percent of ASCE members who indicated a specialty on their membership profiles.

A Geotechnical Education

Question 2 and the three additional questions asked of the students taking geotechnical electives are discussed in this section. The overwhelming majority of students responded positively when asked if they wanted to learn more about geotechnical engineering. Their responses are summarized for each university and for each type of course (introductory or elective) in Table 2. As expected, a very high percentage of those taking geotechnical electives (91%) wanted to learn more. However, 84% of those taking introductory courses also wanted to learn more. This result is an endorsement of the instruction the students are receiving at VU and UT, since students will generally want to learn more about a subject when the quality of instruction they receive is high.

A large percentage of all students cited the importance of geotechnical engineering to all civil engineering projects, with its connection to structural engineering cited especially by the VU students. A large percentage of all students also said they wanted to learn more because geotechnical engineering is interesting and it contains variety. These results show that the geotechnical engineering courses are successful in sparking students' interest in the field, particularly as it pertains to other areas of civil engineering.

The explanations given by the students who did not want to learn more about geotechnical engineering were that they had already decided on another specialty and that other topics were more interesting. These results, especially when viewed in light of the 3% of students in the introductory courses that selected geotechnical engineering as their specialty, indicate that students need more exposure to geotechnical engineering at an earlier stage in their education.

TABLE 2: Reasons Students Do or Do Not Want to Learn More about Geotechnical Engineering

Reasons Students Do or Do Not Want to Learn More (1)	UT (2)	VU (3)	Introductory Courses (4)	Elective Courses (5)
Interesting field and variety in work	28%	22%	20%	34%
Geotechnical engineering is important to all CE projects	34%	22%	36%	20%
Want to know other things geotechnical engineers do	10%	20%	14%	15%
Couples well with structures	5%	17%	8%	11%
Creative/innovative discipline	3%	7%	2%	8%
Enjoy empirically based calculations	4%	2%	3%	3%
Total students who want to learn more	**85%**	**90%**	**84%**	**91%**
Already decided on another specialty	10%	3%	7%	7%
Other topics are more interesting	6%	7%	10%	2%
Total students who do not want to learn more	**15%**	**10%**	**16%**	**9%**

The reasons that students gave for taking a geotechnical engineering elective are summarized in Table 3. The most common reason given for why students decided to take an elective course in geotechnical engineering was that they liked soil mechanics. This again illustrates the importance of the introductory geotechnical engineering class in generating students' interest in the field. Students were much more likely to take the elective courses because of the relationship between geotechnical engineering and structural engineering at UT (29%) than at VU (12%), while wanting to learn more about foundations was cited approximately equally at both institutions (13% at UT and 15% at VU). A greater percentage of students at UT (16%) than at VU (3%) took the elective because it fit into their schedule.

As shown in Table 3, the role of the professor in the students' decision to take an elective course was more important at VU than at UT. At VU, over one-quarter of students gave reasons relating to the professor for why they took the elective, with 15% saying that they liked the professor and 12% saying they did not like the other professors teaching electives. Only 7% of students at UT took the elective because they liked the professor, and none said they did not like the other professors teaching electives. There are several possible reasons for this difference in students' emphasis on the professors. UT is a much larger university than VU, with larger class sizes. Students at UT often only take one course from a professor, and often do not know the professor outside of class. At VU, the professors and students have more opportunities to interact both in and outside of class. In addition, it is our experience

TABLE 3. Reasons Students Enrolled in a Geotechnical Engineering Course

Reasons for Taking Elective Course (1)	UT (2)	VU (3)
Liked soil mechanics class	36%	39%
Importance to structural engineering	29%	12%
Wanted to learn more about foundations	13%	15%
Liked professor	7%	15%
Fit into schedule	16%	3%
Did not like other classes or professors	0%	12%

that students at VU place a high value on interpersonal relationships with professors. At UT, there are six geotechnical engineering professors, and undergraduate teaching assignments rotate. The professors teaching electives may not have taught an introductory course for several previous semesters. Therefore, students enrolling in electives are unlikely to know the professor from a previous course. At VU, there are two geotechnical engineering professors. In addition to the geotechnical courses, these professors also teach introductory engineering courses such as Statics and senior capstone design projects. It also worth noting that the students in the elective course at VU that participated in this survey had the same professor for the elective as they had for the introductory course.

The other questions asked of the students taking a geotechnical elective were what other electives they were taking and what other geotechnical electives they would like to see, if any. There was no clear relationship between the other electives students chose to take. Over half of the students (55%) stated their desire to have more geotechnical electives offered. Their wide range of suggestions included a geotechnical engineering capstone design course, a course in geosynthetics, and offshore geotechnical engineering.

A Geotechnical Career

Students in both the introductory and elective classes were asked several questions about their perceptions of a career in geotechnical engineering. They were asked what they think geotechnical engineers do, and their responses are summarized in Figure 2. Multiple responses were recorded for each student when appropriate, since many students listed several tasks in their answer to this question.

The overwhelming majority of student responses were that geotechnical engineers evaluate soils and design foundations or solve soil-structure problems. The percentage of responses for evaluating soil conditions was substantially higher for the introductory courses (48%) than for the elective courses (35%), while the percentage of responses for designing foundations or solving soil-structure problems was higher

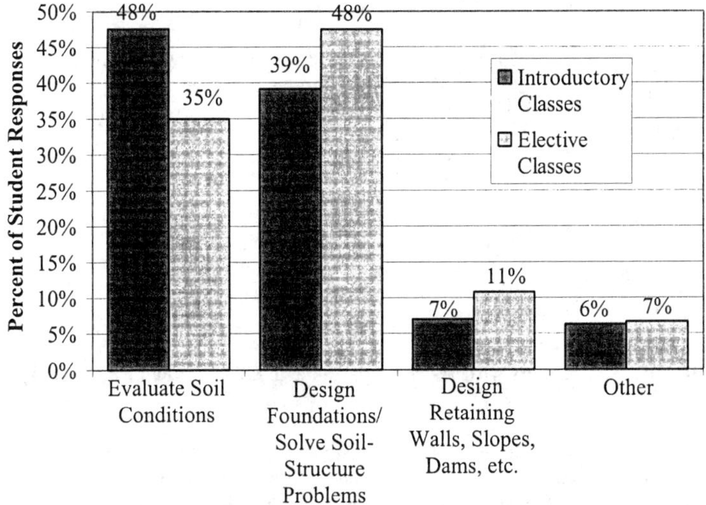

FIGURE 2. Student Responses for What Geotechnical Engineers Do

for the elective courses (48%) than for the introductory courses (39%). This reflects the emphasis on soil classification and soil mechanics in the introductory courses and the larger design component of elective courses. A very small percentage of students mentioned other specific aspects of geotechnical engineering, such as retaining wall design, slope stability, and landfill design. These results indicate that while students have a general idea of what geotechnical engineers do in practice, they need to be exposed to design problems other than foundations.

Less than half (49%) of UT students said they would consider a career in geotechnical engineering, while 69% of VU students said they would. As expected, a much greater percentage (70%) of students in the elective courses stated they would consider a geotechnical career as compared to those in the introductory courses (48%). Students at both institutions indicated that the variety of interesting work, including both field and office work, was the main reason (58%) for that choice. More students at VU (28%) than at UT (11%) stated that they would consider a career in geotechnical engineering because it is close to other fields they found interesting. The major reason given by those that would not consider a career in geotechnical engineering was that they (60%) had already chosen another field.

A very small number of students have work experience in geotechnical engineering, as shown in Figure 3. Only 3% of students in the introductory classes had work experience in geotechnical engineering, compared to 15% of students in the elective classes. The increase in the elective classes could be because students with positive work experiences wanted to take more classes in geotechnical engineering, or because students with an interest in geotechnical engineering sought out work

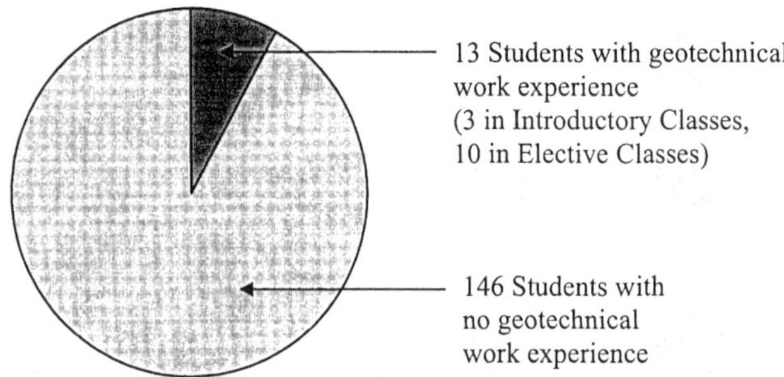

FIGURE 3. Students with Work Experience in Geotechnical Engineering

experience in this field. Because of the few students surveyed with work experience, it is difficult to draw conclusions about the effects of their experiences on their opinions about geotechnical engineering. One student with work experience as both a laboratory and field technician wants to pursue a career in geotechnical engineering. This student wrote that appealing aspects of the field were "working outside and variety of projects", but unappealing aspects were "usually do not get to be in charge of projects...[or] do a lot of design work." Another student with work experience only in lab testing also wants to pursue a career in geotechnical engineering "because there are many aspects involved: field work, writing, calculations, etc."

The questions about what is appealing and unappealing about a career in geotechnical engineering yielded similar results to those reported above, as shown in Figures 4 and 5. The combination of field and office work and the variety of projects that students anticipate working on are real draws to the field. In addition, seeing the impact that soils have on a project is also appealing to students. Students at VU (37%) described monotonous work/lack of variety as an unappealing characteristic of geotechnical engineering more often than students at UT (14%). As expected, this concern was not expressed by many of those taking a geotechnical elective (9%); 33% of students in the introductory course indicated that lack of variety was unappealing. These results again point to the need for academics to expose students to the many applications of geotechnical engineering principles in introductory courses. In addition, the need for interesting internships is also highlighted by these responses.

The uncertainty inherent in geotechnical engineering was viewed as both an appealing and unappealing aspect of geotechnical engineering. Of students taking a geotechnical elective, 34% described uncertainty as an unappealing aspect of geotechnical engineering. One elective student stated "...the design process isn't always as 'clear-cut' as it is in structural design." In contrast, 10% of those in the introductory course described uncertainty as unappealing. Typically, students are not exposed to the uncertainty and judgment involved in geotechnical engineering in

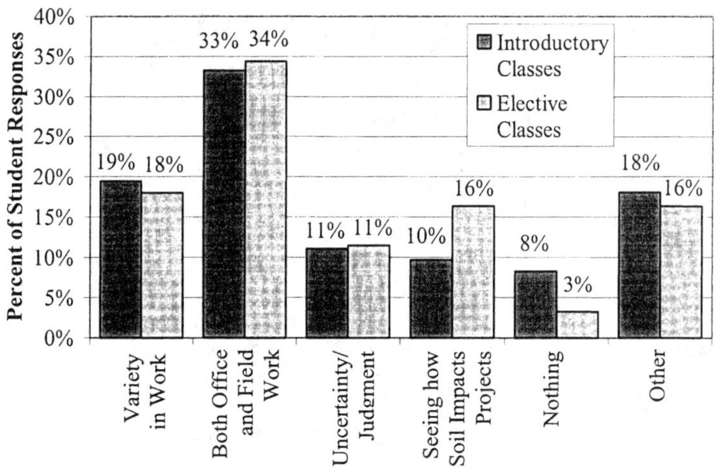

FIGURE 4. Student Responses for Appealing Aspects of Geotechnical Engineering

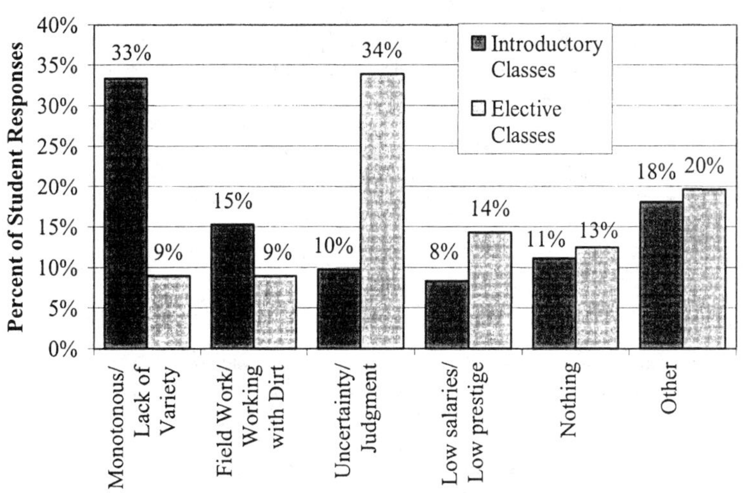

FIGURE 5. Student Responses for Unappealing Aspects of Geotechnical Engineering

introductory courses. These results indicate that academics need to do a better job in the elective courses of presenting uncertainty and how to manage it in a positive manner. Paradoxically, the judgment required in and the complexity of geotechnical engineering is also appealing to many students. Nineteen percent of students in electives, as opposed to 4% in introductory courses, described judgment/complexity as an appealing part of geotechnical engineering. An example of this paradox is that the same elective student quoted above also wrote that an appealing aspect of geotechnical engineering was "...you don't need a code book for everything as in structural design."

WHAT SHOULD WE DO (IF ANYTHING)?

Educators

The survey results described above can be used by educators to modify their courses to appeal to more students. After reviewing the surveys, we recommend the following

- Expose students to a variety of problems in geotechnical projects in introductory courses so that students can see the many different aspects of geotechnical engineering.
- Stress design in the introductory courses.
- Ensure that the quality of introductory courses is high. Many students stated that they continued taking geotechnical courses because they liked soil mechanics.
- Develop geotechnically-related examples and problems for use in early engineering courses (i.e. freshman design, statics, etc.) in order to expose students to the field earlier.
- Present uncertainty in a positive light and provide students with information on how to manage the uncertainty. This is especially important in the elective courses.

Employers

Likewise, employers can use the surveys to improve the view students have of geotechnical engineering. We recommend:

- More high-quality internship opportunities for students. While interning is intrinsically worthwhile, students need to be exposed to design, not just drilling inspection and laboratory work.
- Present your interesting, complex projects to students at a local university; ASCE student chapters are always in need of speakers.

CONCLUSIONS

A survey was conducted at two very different universities to ascertain student perceptions of geotechnical engineering. The respondents were students in introductory soils classes and in geotechnical electives. The key findings of the study are:

- Most students recognize the importance of continuing education.
- The introductory soils class is pivotal in generating interest in geotechnical engineering.
- Most students want to learn more about geotechnical engineering.
- The variety of work, the combination of field and office work, and the judgment required are very appealing to students.
- Paradoxically, students find the uncertainty inherent in geotechnical engineering unappealing. They are also concerned that geotechnical engineering would be monotonous or boring.
- Very few students had work experience in geotechnical engineering.

Recommendations to educators include exposing students to the breadth of geotechnical engineering problems in introductory courses and teaching students how to manage uncertainty. Recommendations to employers include providing meaningful internships and volunteering to speak at local universities.

ACKNOWLEDGEMENTS

The authors wish to thank the students that participated in this study and the professors that agreed to administer this survey in their classes: Edward Glynn, Rominder Suri, Ellen Rathje, Alan Rauch, Brent Rosenblad, Stephen Wright, and Kenneth Stokoe.

REFERENCES

ASCE (2003) ASCE membership office. Personal communication with author.

National Science Foundation (2000) National Science Board, Science and Engineering Indicators-2000, Arlington, VA, Appendix 4-33. Available on-line at http://www.swe.org/SWE/ProgDev/stat/statundergrad_graph.html.

Subject Index

Page number refers to the first page of paper

Author Index

Page number refers to the first page of paper